Periodic Table of the Elements

Legend:
- Metals (main-group)
- Metals (transition)
- Metals (inner transition)
- Metalloids
- Nonmetals

MAIN–GROUP ELEMENTS

TRANSITION ELEMENTS

INNER TRANSITION ELEMENTS

Period	IA (1)	IIA (2)	IIIB (3)	IVB (4)	VB (5)	VIB (6)	VIIB (7)	VIIIB (8)	VIIIB (9)	VIIIB (10)	IB (11)	IIB (12)	IIIA (13)	IVA (14)	VA (15)	VIA (16)	VIIA (17)	VIIIA (18)
1	1 H 1.008																	2 He 4.003
2	3 Li 6.941	4 Be 9.012											5 B 10.81	6 C 12.01	7 N 14.01	8 O 16.00	9 F 19.00	10 Ne 20.18
3	11 Na 22.99	12 Mg 24.31											13 Al 26.98	14 Si 28.09	15 P 30.97	16 S 32.07	17 Cl 35.45	18 Ar 39.95
4	19 K 39.10	20 Ca 40.08	21 Sc 44.96	22 Ti 47.88	23 V 50.94	24 Cr 52.00	25 Mn 54.94	26 Fe 55.85	27 Co 58.93	28 Ni 58.69	29 Cu 63.55	30 Zn 65.39	31 Ga 69.72	32 Ge 72.61	33 As 74.92	34 Se 78.96	35 Br 79.90	36 Kr 83.80
5	37 Rb 85.47	38 Sr 87.62	39 Y 88.91	40 Zr 91.22	41 Nb 92.91	42 Mo 95.94	43 Tc (98)	44 Ru 101.1	45 Rh 102.9	46 Pd 106.4	47 Ag 107.9	48 Cd 112.4	49 In 114.8	50 Sn 118.7	51 Sb 121.8	52 Te 127.6	53 I 126.9	54 Xe 131.3
6	55 Cs 132.9	56 Ba 137.3	57 La 138.9	72 Hf 178.5	73 Ta 180.9	74 W 183.9	75 Re 186.2	76 Os 190.2	77 Ir 192.2	78 Pt 195.1	79 Au 197.0	80 Hg 200.6	81 Tl 204.4	82 Pb 207.2	83 Bi 209.0	84 Po (209)	85 At (210)	86 Rn (222)
7	87 Fr (223)	88 Ra (226)	89 Ac (227)	104 Rf (261)	105 Db (262)	106 Sg (266)	107 Bh (262)	108 Hs (265)	109 Mt (266)	110 Uun (269)	111 Uuu (272)	112 Uub (277)	113	114 Uuq (285)	115	116 Uuh (289)	117	118 Uuo

Lanthanides (6):

58 Ce 140.1	59 Pr 140.9	60 Nd 144.2	61 Pm (145)	62 Sm 150.4	63 Eu 152.0	64 Gd 157.3	65 Tb 158.9	66 Dy 162.5	67 Ho 164.9	68 Er 167.3	69 Tm 168.9	70 Yb 173.0	71 Lu 175.0

Actinides (7):

90 Th 232.0	91 Pa (231)	92 U 238.0	93 Np (237)	94 Pu (242)	95 Am (243)	96 Cm (247)	97 Bk (247)	98 Cf (251)	99 Es (252)	100 Fm (257)	101 Md (258)	102 No (259)	103 Lr (260)

Source: Davis, M., and Davis, R., *Fundamentals of Chemical Reaction Engineering*, McGraw-Hill, 2003.

W9-BYN-990

Foundations of Materials Science and Engineering

McGraw-Hill Series in Materials Science

Budynas
Advanced Strength and Applied Stress Analysis

Courtney
Mechanical Behavior of Materials

DeHoff
Thermodynamics in Materials Science

Dieter
Mechanical Metallurgy

Henkel, Pense
Structure and Properties of Engineering Materials

Hyer
Stress Analysis of Fiber Reinforces Composite Materials

Kasap
Principles of Electrical Engineering Materials: Devices

Schaffer, Saxena, Antolovich, Sanders, Warner
The Science and Design of Engineering Materials

Smith, Hashemi
Foundations of Materials Science and Engineering

Smith
Structure and Properties of Engineering Alloys

Foundations of Materials Science and Engineering

Fourth Edition

William F. Smith
Professor Emeritus of Engineering
University of Central Florida

Javad Hashemi, Ph.D.
Professor of Mechanical Engineering
Texas Tech University

 **Higher Education**

Boston   Burr Ridge, IL   Dubuque, IA   Madison, WI   New York   San Francisco   St. Louis
Bangkok   Bogotá   Caracas   Kuala Lumpur   Lisbon   London   Madrid   Mexico City
Milan   Montreal   New Delhi   Santiago   Seoul   Singapore   Sydney   Taipei   Toronto

Higher Education

FOUNDATIONS OF MATERIALS SCIENCE AND ENGINEERING, FOURTH EDITION

Published by McGraw-Hill, a business unit of The McGraw-Hill Companies, Inc., 1221 Avenue of the Americas, New York, NY 10020. Copyright © 2006, 2004, 1993 by The McGraw-Hill Companies, Inc. All rights reserved. Portions of this text have been taken from *Principles of Materials Science and Engineering*. Copyright © 1990, 1986 by McGraw-Hill, Inc. All rights reserved. No part of this publication may be reproduced or distributed in any form or by any means, or stored in a database or retrieval system, without the prior written consent of The McGraw-Hill Companies, Inc., including, but not limited to, in any network or other electronic storage or transmission, or broadcast for distance learning.

Some ancillaries, including electronic and print components, may not be available to customers outside the United States.

This book is printed on acid-free paper.

2 3 4 5 6 7 8 9 0 DOC/DOC 0 9 8 7 6

ISBN–13: 978–0–07–295358–9

ISBN–10: 0–07–295358–6

Publisher: *Suzanne Jeans*
Senior Sponsoring Editor: *Bill Stenquist*
Developmental Editor: *Kathleen L. White*
Senior Marketing Manager: *Mary K. Kittell*
Senior Project Manager: *Sheila M. Frank*
Senior Production Supervisor: *Laura Fuller*
Media Technology Producer: *Eric A. Weber*
Senior Coordinator of Freelance Design: *Michelle D. Whitaker*
Cover Designer: *Kelly Fassbinder/Imagine Design Studio*
(USE) Cover Images: © *Bayer MaterialScience*
Senior Photo Research Coordinator: *Lori Hancock*
Photo Research: *Jerry Marshall*
Compositor: *The GTS Companies*
Typeface: *10.5/12 Times Roman*
Printer: *R. R. Donnelley Crawfordsville, IN*

The innovative design of the golf tee on the front cover allows the ball to rest at a point 8 mm from the center of the tee, towards the striking side, on an offset ball rest. The club thus hits the ball a fraction of a second before it hits the tee itself, which means that the tee, unlike conventional wooden tees, does not affect the trajectory. Makrolon® 2405 was selected as the material, since it has excellent flowability and offers sufficient impact strength so the tees do not break in most cases when hit by the club. Instead, they just fly out of the ground as illustrated in the cover image.

Bayer MaterialScience produces a wide range of proven and innovative materials including polycarbonates, polyurethanes, thermoplastic polyurethanes, coatings, adhesives and sealants. Bayer's materials and solutions are used to develop the next generation products in the automotive, construction, information technology, appliance, furniture as well as sports and leisure industries. For more information about Bayer MaterialScience, visit www.Bayermaterialsciencenafta.com or call 1-800-662-2927.

Library of Congress Cataloging-in-Publication Data

Smith, William Fortune, 1931–
    Foundations of materials science and engineering / William F. Smith, Javad Hashemi. — 4th ed.
        p. cm.—(McGraw-Hill series in materials science)
    Includes bibliographical references and index.
    ISBN 0–07–295358–6
    1. Materials science. 2. Materials. I. Hashemi, Javad. II. Title. III. McGraw-Hill series in materials science and engineering.

TA403.S5955   2006
620.1′1—dc22   2005043865
                                                        CIP

www.mhhe.com

ABOUT THE AUTHORS

William F. Smith is Professor Emeritus of engineering in the Mechanical and Aerospace Engineering Department of the University of Central Florida at Orlando, Florida. He was awarded an M.S. degree in metallurgical engineering from Purdue University and a Sc.D. degree in metallurgy from Massachusetts Institute of Technology. Dr. Smith, who is a registered professional engineer in the states of California and Florida, has been teaching undergraduate and graduate materials science and engineering courses and actively writing textbooks for many years. He is also the author of *Structure and Properties of Engineering Alloys,* Second Edition (McGraw-Hill, 1993).

Javad Hashemi is a Professor of Mechanical Engineering at Texas Tech University, where he has taught introduction to materials science since 1991. Javad received his Ph.D. in Mechanical Engineering from Drexel University in 1988. Dr. Hashemi has been teaching undergraduate and graduate materials and mechanics courses, as well as laboratories at Texas Tech University. He is also the principal developer of the virtual laboratory modules accompanying this textbook as part of a pilot project funded by National Science Foundation. Dr. Hashemi's current research focus is the areas of materials, biomechanics, and education for engineers.

TABLE OF CONTENTS

Preface   xvii

CHAPTER **1**

Introduction to Materials Science and Engineering   2

1.1   Materials and Engineering   3
1.2   Materials Science and Engineering   6
1.3   Types of Materials   8
   1.3.1 *Metallic materials*   8
   1.3.2 *Polymeric Materials*   10
   1.3.3 *Ceramic Materials*   11
   1.3.4 *Composite Materials*   13
   1.3.5 *Electronic Materials*   15
1.4   Competition Among Materials   16
1.5   Recent Advances in Materials Science and Technology and Future Trends   18
   1.5.1 *Smart Materials*   18
   1.5.2 *Nanomaterials*   19
1.6   Design and Selection   20
1.7   Summary   21
1.8   Definitions   21
1.9   Problems   22
1.10 Materials Selection and Design Problems   23

CHAPTER **2**

Atomic Structure and Bonding   24

2.1   The Structure of Atoms   25
2.2   Atomic Numbers and Atomic Masses   26
   2.2.1 *Atomic Numbers*   26
   2.2.2 *Atomic Masses*   26
2.3   The Electronic Structure of Atoms   29
   2.3.1 *The Hydrogen Atom*   29
   2.3.2 *Quantum Numbers of Electrons of Atoms*   33
   2.3.3 *Electronic Structure of Multielectron Atoms*   35
   2.3.4 *Electronic Structure and Chemical Reactivity*   39
2.4   Types of Atomic and Molecular Bonds   41
   2.4.1 *Primary Atomic Bonds*   42
   2.4.2 *Secondary Atomic and Molecular Bonds*   42
2.5   Ionic Bonding   42
   2.5.1 *Ionic Bonding in General*   42
   2.5.2 *Interionic Forces for an Ion Pair*   43
   2.5.3 *Interionic Energies for an Ion Pair*   46
   2.5.4 *Ion Arrangements in Ionic Solids*   47
   2.5.5 *Bonding Energies of Ionic Solids*   48
2.6   Covalent Bonding   49
   2.6.1 *Covalent Bonding in the Hydrogen Molecule*   49
   2.6.2 *Covalent Bonding in Other Diatomic Molecules*   50
   2.6.3 *Covalent Bonding by Carbon*   51
   2.6.4 *Covalent Bonding in Carbon-Containing Molecules*   53
   2.6.5 *Benzene*   53
2.7   Metallic Bonding   55
2.8   Secondary Bonding   59
   2.8.1 *Fluctuating Dipoles*   60
   2.8.2 *Permanent Dipoles*   61
2.9   Mixed Bonding   62
   2.9.1 *Ionic-Covalent Mixed Bonding*   62
   2.9.2 *Metallic-Covalent Mixed Bonding*   63
   2.9.3 *Metallic-Ionic Mixed Bonding*   64
2.10 Summary   64
2.11 Definitions   65
2.12 Problems   66
2.13 Materials Selection and Design Problems   70

CHAPTER **3**

Crystal and Amorphous Structure in Materials   72

3.1   The Space Lattice and Unit Cells   73

3.2   Crystal Systems and Bravais Lattices   74

3.3   Principal Metallic Crystal Structures   75

   3.3.1   *Body-Centered Cubic (BCC) Crystal Structure   77*

   3.3.2   *Face-Centered Cubic (FCC) Crystal Structure   80*

   3.3.3   *Hexagonal Close-Packed (HCP) Crystal Structure   81*

3.4   Atom Positions in Cubic Unit Cells   83

3.5   Directions in Cubic Unit Cells   84

3.6   Miller Indices for Crystallographic Planes in Cubic Unit Cells   88

3.7   Crystallographic Planes and Directions in Hexagonal Crystal Structure   93

   3.7.1   *Indices for Crystal Planes in HCP Unit Cells   93*

   3.7.2   *Direction Indices in HCP Unit Cells   94*

3.8   Comparison of FCC, HCP, and BCC Crystal Structures   96

   3.8.1   *FCC and HCP Crystal Structures   96*

   3.8.2   *BCC Crystal Structure   98*

3.9   Volume, Planar, and Linear Density Unit-Cell Calculations   98

   3.9.1   *Volume Density   98*

   3.9.2   *Planar Atomic Density   99*

   3.9.3   *Linear Atomic Density   101*

3.10   Polymorphism or Allotropy   102

3.11   Crystal Structure Analysis   103

   3.11.1   *X-Ray Sources   104*

   3.11.2   *X-Ray Diffraction   105*

   3.11.3   *X-Ray Diffraction Analysis of Crystal Structures   107*

3.12   Amorphous Materials   113

3.13   Summary   114

3.14   Definitions   115

3.15   Problems   116

3.16   Materials Selection and Design Problems   122

CHAPTER **4**

Solidification and Crystalline Imperfections   124

4.1   Solidification of Metals   125

   4.1.1   *The Formation of Stable Nuclei in Liquid Metals   127*

   4.1.2   *Growth of Crystals in Liquid Metal and Formation of a Grain Structure   132*

   4.1.3   *Grain Structure of Industrial Castings   133*

4.2   Solidification of Single Crystals   134

4.3   Metallic Solid Solutions   138

   4.3.1   *Substitutional Solid Solutions   139*

   4.3.2   *Interstitial Solid Solutions   141*

4.4   Crystalline Imperfections   143

   4.4.1   *Point Defects   143*

   4.4.2   *Line Defects (Dislocations)   144*

   4.4.3   *Planar Defects   147*

   4.4.4   *Volume Defects   150*

4.5   Experimental Techniques for Identification of Microstructure and Defects   151

   4.5.1   *Optical Metallography, ASTM Grain Size, and Grain Diameter Determination   151*

   4.5.2   *Scanning Electron Microscopy (SEM)   156*

   4.5.3   *Transmission Electron Microscopy (TEM)   158*

   4.5.4   *High-Resolution Transmission Electron Microscopy (HRTEM)   159*

   4.5.5   *Scanning Probe Microscopes and Atomic Resolution   161*

4.6   Summary   166

4.7   Definitions   166

4.8   Problems   168

4.9   Materials Selection and Design Problems   170

CHAPTER **5**

Thermally Activated Processes and Diffusion in Solids   172

5.1   Rate Processes in Solids   173
5.2   Atomic Diffusion in Solids   177
    5.2.1 *Diffusion in Solids in General   177*
    5.2.2 *Diffusion Mechanisms   177*
    5.2.3 *Steady-State Diffusion   180*
    5.2.4 *Non–Steady-State Diffusion   182*
5.3   Industrial Applications of Diffusion Processes   184
    5.3.1 *Case Hardening of Steel by Gas Carburizing   184*
    5.3.2 *Impurity Diffusion into Silicon Wafers for Integrated Circuits   188*
5.4   Effect of Temperature on Diffusion in Solids   191
5.5   Summary   195
5.6   Definitions   195
5.7   Problems   196
5.8   Materials Selection and Design Problems   198

CHAPTER **6**

Mechanical Properties of Metals I   200

6.1   The Processing of Metals and Alloys   201
    6.1.1 *The Casting of Metals and Alloys   201*
    6.1.2 *Hot and Cold Rolling of Metals and Alloys   203*
    6.1.3 *Extrusion of Metals and Alloys   208*
    6.1.4 *Forging   209*
    6.1.5 *Other Metal-Forming Processes   211*
6.2   Stress and Strain in Metals   212
    6.2.1 *Elastic and Plastic Deformation   213*
    6.2.2 *Engineering Stress and Engineering Strain   213*
    6.2.3 *Poisson's Ratio   216*
    6.2.4 *Shear Stress and Shear Strain   216*

6.3   The Tensile Test and the Engineering Stress-Strain Diagram   217
    6.3.1 *Mechanical Property Data Obtained from the Tensile Test and the Engineering Stress-Strain Diagram   220*
    6.3.2 *Comparison of Engineering Stress-Strain Curves for Selected Alloys   225*
    6.3.3 *True Stress and True Strain   225*
6.4   Hardness and Hardness Testing   227
6.5   Plastic Deformation of Metal Single Crystals   229
    6.5.1 *Slipbands and Slip Lines on the Surface of Metal Crystals   229*
    6.5.2 *Plastic Deformation in Metal Crystals by the Slip Mechanism   232*
    6.5.3 *Slip Systems   234*
    6.5.4 *Critical Resolved Shear Stress for Metal Single Crystals   235*
    6.5.5 *Schmid's Law   237*
    6.5.6 *Twinning   240*
6.6   Plastic Deformation of Polycrystalline Metals   242
    6.6.1 *Effect of Grain Boundaries on the Strength of Metals   242*
    6.6.2 *Effect of Plastic Deformation on Grain Shape and Dislocation Arrangements   244*
    6.6.3 *Effect of Cold Plastic Deformation on Increasing the Strength of Metals   246*
6.7   Solid-Solution Strengthening of Metals   247
6.8   Recovery and Recrystallization of Plastically Deformed Metals   249
    6.8.1 *Structure of a Heavily Cold-Worked Metal before Reheating   250*
    6.8.2 *Recovery   251*
    6.8.3 *Recrystallization   252*
6.9   Superplasticity in Metals   257
6.10   Nanocrystalline Metals   259
6.11   Summary   261
6.12   Definitions   262
6.13   Problems   263
6.14   Materials Selection and Design Problems   268

CHAPTER **7**

Mechanical Properties of Metals II   270

7.1   Fracture of Metals   271

   7.1.1 *Ductile Fracture*   272

   7.1.2 *Brittle Fracture*   273

   7.1.3 *Toughness and Impact Testing*   276

   7.1.4 *Ductile to Brittle Transition Temperature*   276

   7.1.5 *Fracture Toughness*   279

7.2   Fatigue of Metals   281

   7.2.1 *Cyclic Stresses*   285

   7.2.2 *Basic Structural Changes that Occur in a Ductile Metal in the Fatigue Process*   286

   7.2.3 *Some Major Factors that Affect the Fatigue Strength of a Metal*   287

7.3   Fatigue Crack Propagation Rate   288

   7.3.1 *Correlation of Fatigue Crack Propagation with Stress and Crack Length*   288

   7.3.2 *Fatigue Crack Growth Rate versus Stress-Intensity Factor Range Plots*   290

   7.3.3 *Fatigue Life Calculations*   292

7.4   Creep and Stress Rupture of Metals   294

   7.4.1 *Creep of Metals*   294

   7.4.2 *The Creep Test*   296

   7.4.3 *Creep-Rupture Test*   297

7.5   Graphical representation of Creep- and Stress-Rupture Time-Temperature Data Using the Larsen-Miller Parameter   298

7.6   A Case Study in Failure of Metallic Components   300

7.7   Recent Advances and Future Directions in Improving the Mechanical Performance of Metals   303

   7.7.1 *Improving Ductility and Strength Simultaneously*   303

   7.7.2 *Fatigue Behavior in Nanocrystalline Metals*   305

7.8   Summary   305

7.9   Definitions   306

7.10   Problems   307

7.11   Materials Selection and Design Problems   309

CHAPTER **8**

Phase Diagrams   310

8.1   Phase Diagrams of Pure Substances   311

8.2   Gibbs Phase Rule   313

8.3   Cooling Curves   314

8.4   Binary Isomorphous Alloy Systems   315

8.5   The Lever Rule   318

8.6   Nonequilibrium Solidification of Alloys   322

8.7   Binary Eutectic Alloy Systems   326

8.8   Binary Peritectic Alloy Systems   333

8.9   Binary Monotectic Systems   338

8.10   Invariant Reactions   339

8.11   Phase Diagrams with Intermediate Phases and Compounds   341

8.12   Ternary Phase Diagrams   345

8.13   Summary   348

8.14   Definitions   349

8.15   Problems   351

8.16   Materials Selection and Design Problems   355

CHAPTER **9**

Engineering Alloys   358

9.1   Production of Iron and Steel   360

   9.1.1 *Production of Pig Iron in a Blast Furnace*   360

   9.1.2 *Steelmaking and Processing of Major Steel Product Forms*   361

9.2   The Iron–Iron-Carbide System   363

   9.2.1 *The Iron–Iron-Carbide Phase Diagram*   363

   9.2.2 *Solid Phases in the $Fe\text{-}Fe_3C$ Phase Diagram*   363

   9.2.3 *Invariant Reactions in the $Fe\text{-}Fe_3C$ Phase Diagram*   364

   9.2.4 *Slow Cooling of Plain-Carbon Steels*   366

9.3   Heat Treatment of Plain-Carbon Steels   373

   9.3.1 *Martensite   373*

   9.3.2 *Isothermal Decomposition of Austenite   378*

   9.3.3 *Continuous-Cooling Transformation Diagram for a Eutectoid Plain-Carbon Steel   383*

   9.3.4 *Annealing and Normalizing of Plain-Carbon Steels   386*

   9.3.5 *Tempering of Plain-Carbon Steels   387*

   9.3.6 *Classification of Plain-Carbon Steels and Typical Mechanical Properties   391*

9.4   Low-Alloy Steels   392

   9.4.1 *Classification of Alloy Steels   392*

   9.4.2 *Distribution of Alloying Elements in Alloy Steels   394*

   9.4.3 *Effects of Alloying Elements on the Eutectoid Temperature of Steels   395*

   9.4.4 *Hardenability   396*

   9.4.5 *Typical Mechanical Properties and Applications for Low-Alloy Steels   401*

9.5   Aluminum Alloys   401

   9.5.1 *Precipitation Strengthening (Hardening)   403*

   9.5.2 *General Properties of Aluminum and Its Production   410*

   9.5.3 *Wrought Aluminum Alloys   411*

   9.5.4 *Aluminum Casting Alloys   416*

9.6   Copper Alloys   418

   9.6.1 *General Properties of Copper   418*

   9.6.2 *Production of Copper   419*

   9.6.3 *Classification of Copper Alloys   419*

   9.6.4 *Wrought Copper Alloys   422*

9.7   Stainless Steels   424

   9.7.1 *Ferritic Stainless Steels   424*

   9.7.2 *Martensitic Stainless Steels   425*

   9.7.3 *Austenitic Stainless Steels   427*

9.8   Cast Irons   429

   9.8.1 *General Properties   429*

   9.8.2 *Types of Cast Irons   429*

   9.8.3 *White Cast Iron   429*

   9.8.4 *Gray Cast Iron   431*

   9.8.5 *Ductile Cast Irons   432*

   9.8.6 *Malleable Cast Irons   435*

9.9   Magnesium, Titanium, and Nickel Alloys   436

   9.9.1 *Magnesium Alloys   436*

   9.9.2 *Titanium Alloys   438*

   9.9.3 *Nickel Alloys   440*

9.10   Special-Purpose Alloys and Applications   441

   9.10.1 *Intermetallics   441*

   9.10.2 *Shape-Memory Alloys   442*

   9.10.3 *Amorphous Metals   446*

9.11   Metals in Biomedical Applications— Biometals   448

   9.11.1 *Stainless Steels   449*

   9.11.2 *Cobalt-Based Alloys   449*

   9.11.3 *Titanium Alloys   451*

9.12   Some Issues in the Orthopedic Application of Metals   452

9.13   Summary   454

9.14   Definitions   455

9.15   Problems   457

9.16   Materials Selection and Design Problems   465

CHAPTER **10**

Polymeric Materials   468

10.1   Introduction   469

10.2   Polymerization Reactions   471

   10.2.1 *Covalent Bonding Structure of an Ethylene Molecule   471*

   10.2.2 *Covalent Bonding Structure of an Activated Ethylene Molecule   472*

   10.2.3 *General Reaction for the Polymerization of Polyethylene and the Degree of Polymerization   473*

   10.2.4 *Chain Polymerization Steps   473*

   10.2.5 *Average Molecular Weight for Thermoplastics   475*

10.2.6 *Functionality of a Monomer 476*

10.2.7 *Structure of Noncrystalline Linear Polymers 476*

10.2.8 *Vinyl and Vinylidene Polymers 478*

10.2.9 *Homopolymers and Copolymers 479*

10.2.10 *Other Methods of Polymerization 482*

10.3 Industrial Polymerization Methods 484

10.4 Crystallinity and Stereoisomerism in Some Thermoplastics 486

10.4.1 *Solidification of Noncrystalline Thermoplastics 486*

10.4.2 *Solidification of Partly Crystalline Thermoplastics 486*

10.4.3 *Structure of Partly Crystalline Thermoplastic Materials 488*

10.4.4 *Stereoisomerism in Thermoplastics 489*

10.4.5 *Ziegler and Natta Catalysts 490*

10.5 Processing of Plastic Materials 491

10.5.1 *Processes Used for Thermoplastic Materials 492*

10.5.2 *Processes Used for Thermosetting Materials 496*

10.6 General-Purpose Thermoplastics 498

10.6.1 *Polyethylene 500*

10.6.2 *Polyvinyl Chloride and Copolymers 503*

10.6.3 *Polypropylene 505*

10.6.4 *Polystyrene 505*

10.6.5 *Polyacrylonitrile 506*

10.6.6 *Styrene-Acrylonitrile (SAN) 507*

10.6.7 *ABS 507*

10.6.8 *Polymethyl Methacrylate (PMMA) 509*

10.6.9 *Fluoroplastics 510*

10.7 Engineering Thermoplastics 511

10.7.1 *Polyamides (Nylons) 512*

10.7.2 *Polycarbonate 515*

10.7.3 *Phenylene Oxide–Based Resins 516*

10.7.4 *Acetals 517*

10.7.5 *Thermoplastic Polyesters 518*

10.7.6 *Polyphenylene Sulfide 519*

10.7.7 *Polyetherimide 520*

10.7.8 *Polymer Alloys 521*

10.8 Thermosetting Plastics (Thermosets) 521

10.8.1 *Phenolics 523*

10.8.2 *Epoxy Resins 525*

10.8.3 *Unsaturated Polyesters 527*

10.8.4 *Amino Resins (Ureas and Melamines) 529*

10.9 Elastomers (Rubbers) 531

10.9.1 *Natural Rubber 531*

10.9.2 *Synthetic Rubbers 534*

10.9.3 *Properties of Polychloroprene Elastomers 536*

10.9.4 *Vulcanization of Polychloroprene Elastomers 536*

10.10 Deformation and Strengthening of Plastic Materials 539

10.10.1 *Deformation Mechanisms for Thermoplastics 539*

10.10.2 *Strengthening of Thermoplastics 541*

10.10.3 *Strengthening of Thermosetting Plastics 545*

10.10.4 *Effect of Temperature on the Strength of Plastic Materials 545*

10.11 Creep and Fracture of Polymeric Materials 546

10.11.1 *Creep of Polymeric Materials 546*

10.11.2 *Stress Relaxation of Polymeric Materials 547*

10.11.3 *Fracture of Polymeric Materials 550*

10.12 Polymers in Biomedical Applications— Biopolymers 552

10.12.1 *Cardiovascular Applications of Polymers 553*

10.12.2 *Ophthalmic Applications 554*

10.12.3 *Drug-Delivery Systems 555*

10.12.4 *Suture Materials 556*

10.12.5 *Orthopedic Applications 556*

10.13 Summary 557

10.14 Definitions 558

10.15 Problems 560

10.16 Materials Selection and Design Problems 570

CHAPTER **11**
Ceramics   572

11.1   Introduction   573
11.2   Simple Ceramic Crystal
             Structures   575
  11.2.1 *Ionic and Covalent Bonding in Simple
              Ceramic Compounds   575*
  11.2.2 *Simple Ionic Arrangements Found in
              Ionically Bonded Solids   576*
  11.2.3 *Cesium Chloride (CsCl) Crystal
              Structure   579*
  11.2.4 *Sodium Chloride (NaCl) Crystal
              Structure   580*
  11.2.5 *Interstitial Sites in FCC and HCP Crystal
              Lattices   584*
  11.2.6 *Zinc Blende (ZnS) Crystal
              Structure   586*
  11.2.7 *Calcium Fluoride (CaF$_2$) Crystal
              Structure   588*
  11.2.8 *Antifluorite Crystal Structure   590*
  11.2.9 *Corundum (Al$_2$O$_3$) Crystal
              Structure   590*
  11.2.10 *Spinel (MgAl$_2$O$_4$) Crystal
              Structure   590*
  11.2.11 *Perovskite (CaTiO$_3$) Crystal
              Structure   590*
  11.2.12 *Carbon and Its Allotropes   591*
11.3   Silicate Structures   595
  11.3.1 *Basic Structural Unit of the Silicate
              Structures   595*
  11.3.2 *Island, Chain, and Ring Structures of
              Silicates   595*
  11.3.3 *Sheet Structures of Silicates   595*
  11.3.4 *Silicate Networks   597*
11.4   Processing of Ceramics   598
  11.4.1 *Materials Preparation   599*
  11.4.2 *Forming   599*
  11.4.3 *Thermal Treatments   604*
11.5   Traditional and Engineering
             Ceramics   606
  11.5.1 *Traditional Ceramics   606*
  11.5.2 *Engineering Ceramics   609*

11.6   Mechanical Properties of Ceramics   611
  11.6.1 *General   611*
  11.6.2 *Mechanisms for the Deformation of
              Ceramic Materials   611*
  11.6.3 *Factors Affecting the Strength of Ceramic
              Materials   612*
  11.6.4 *Toughness of Ceramic Materials   613*
  11.6.5 *Transformation Toughening of Partially
              Stabilized Zirconia (PSZ)   615*
  11.6.6 *Fatigue Failure of Ceramics   615*
  11.6.7 *Ceramic Abrasive Materials   617*
11.7   Thermal Properties of Ceramics   618
  11.7.1 *Ceramic Refractory Materials   619*
  11.7.2 *Acidic Refractories   620*
  11.7.3 *Basic Refractories   620*
  11.7.4 *Ceramic Tile Insulation for the Space
              Shuttle Orbiter   620*
11.8   Glasses   620
  11.8.1 *Definition of a Glass   622*
  11.8.2 *Glass Transition Temperature   622*
  11.8.3 *Structure of Glasses   623*
  11.8.4 *Composition of Glasses   624*
  11.8.5 *Viscous Deformation of Glasses   626*
  11.8.6 *Forming Methods for Glasses   628*
  11.8.7 *Tempered Glass   630*
  11.8.8 *Chemically Strengthened Glass   630*
11.9   Ceramic Coatings and Surface
             Engineering   632
  11.9.1 *Silicate Glasses   632*
  11.9.2 *Oxides and Carbides   632*
11.10 Ceramics in Biomedical Applications   634
  11.10.1 *Alumina in Orthopedic
              Implants   634*
  11.10.2 *Alumina in Dental Implants   636*
  11.10.3 *Ceramic Implants and Tissue
              Connectivity   636*
11.11 Nanotechnology and Ceramics   637
11.12 Summary   639
11.13 Definitions   640
11.14 Problems   642
11.15 Materials Selection and Design
             Problems   646

CHAPTER **12**
Composite Materials 648

12.1 Introduction 649
12.2 Fibers for Reinforced-Plastic Composite Materials 651
    12.2.1 *Glass Fibers for Reinforcing Plastic Resins 651*
    12.2.2 *Carbon Fibers for Reinforced Plastics 653*
    12.2.3 *Aramid Fibers for Reinforcing Plastic Resins 654*
    12.2.4 *Comparison of Mechanical Properties of Carbon, Aramid, and Glass Fibers for Reinforced-Plastic Composite Materials 655*
12.3 Fiber-Reinforced-Plastic Composite Materials 657
    12.3.1 *Matrix Materials for Fiber-Reinforced Plastic Composite Materials 657*
    12.3.2 *Fiber-Reinforced-Plastic Composite Materials 658*
    12.3.3 *Equations for Elastic Modulus of a Lamellar Continuous-Fiber-Plastic Matrix Composite for Isostrain and Isostress Conditions 662*
12.4 Open-Mold Processes for Fiber-Reinforced-Plastic Composite Materials 667
    12.4.1 *Hand Lay-Up Process 667*
    12.4.2 *Spray-Up Process 667*
    12.4.3 *Vacuum Bag–Autoclave Process 668*
    12.4.4 *Filament-Winding Process 670*
12.5 Closed-Mold Processes for Fiber-Reinforced Plastic Composite Materials 672
    12.5.1 *Compression and Injection Molding 672*
    12.5.2 *The Sheet-Molding Compound (SMC) Process 672*
    12.5.3 *Continuous-Prorusion Process 674*
12.6 Concrete 674
    12.6.1 *Portland Cement 675*
    12.6.2 *Mixing Water for Concrete 678*
    12.6.3 *Aggregates for Concrete 679*
    12.6.4 *Air Entrainment 679*
    12.6.5 *Compressive Strength of Concrete 679*
    12.6.6 *Proportioning of Concrete Mixtures 679*
    12.6.7 *Reinforced and Prestressed Concrete 682*
    12.6.8 *Prestressed Concrete 683*
12.7 Asphalt and Asphalt Mixes 684
12.8 Wood 685
    12.8.1 *Macrostructure of Wood 685*
    12.8.2 *Microstructure of Softwoods 688*
    12.8.3 *Microstructure of Hardwoods 689*
    12.8.4 *Cell-Wall Ultrastructure 690*
    12.8.5 *Properties of Wood 692*
12.9 Sandwich Structures 695
    12.9.1 *Honeycomb Sandwich Structure 695*
    12.9.2 *Cladded Metal Structures 695*
12.10 Metal-Matrix and Ceramic-Matrix Composites 696
    12.10.1 *Metal-Matrix Composites (MMCs) 696*
    12.10.2 *Ceramic-Matrix Composites (CMCs) 700*
    12.10.3 *Ceramic Composites and Nanotechnology 703*
12.11 Bone: A Natural Composite Material 703
    12.11.1 *Composition 703*
    12.11.2 *Macrostructure 703*
    12.11.3 *Mechanical Properties 705*
    12.11.4 *Biomechanics of Bone Fracture 706*
    12.11.5 *Viscoelasticity of the Bone 707*
    12.11.6 *Bone Remodeling 707*
    12.11.7 *Nanotechnology and Bone Repair 708*
12.12 Summary 708
12.13 Definitions 709
12.14 Problems 712
12.15 Materials Selection and Design Problems 716

CHAPTER **13**
Corrosion 718

13.1 General 719
13.2 Electrochemical Corrosion of Metals 720
    13.2.1 *Oxidation-Reduction Reactions 720*
    13.2.2 *Standard Electrode Half-Cell Potentials for Metals 722*

13.3 Galvanic Cells   724

13.3.1 *Macroscopic Galvanic Cells with Electrolytes That Are One Molar   724*

13.3.2 *Galvanic Cells with Electrolytes That Are Not One Molar   726*

13.3.3 *Galvanic Cells with Acid or Alkaline Electrolytes with No Metal Ions Present   727*

13.3.4 *Microscopic Galvanic Cell Corrosion of Single Electrodes   729*

13.3.5 *Concentration Galvanic Cells   730*

13.3.6 *Galvanic Cells Created by Differences in Composition, Structure, and Stress   733*

13.4 Corrosion Rates (Kinetics)   735

13.4.1 *Rate of Uniform Corrosion or Electroplating of a Metal in an Aqueous Solution   736*

13.4.2 *Corrosion Reactions and Polarization   739*

13.4.3 *Passivation   742*

13.4.4 *The Galvanic Series   743*

13.5 Types of Corrosion   745

13.5.1 *Uniform or General Attack Corrosion   745*

13.5.2 *Galvanic or Two-Metal Corrosion   745*

13.5.3 *Pitting Corrosion   746*

13.5.4 *Crevice Corrosion   749*

13.5.5 *Intergranular Corrosion   751*

13.5.6 *Stress Corrosion   753*

13.5.7 *Erosion Corrosion   756*

13.5.8 *Cavitation Damage   756*

13.5.9 *Fretting Corrosion   757*

13.5.10 *Selective Leaching   757*

13.5.11 *Hydrogen Damage   758*

13.6 Oxidation of Metals   759

13.6.1 *Protective Oxide Films   759*

13.6.2 *Mechanisms of Oxidation   761*

13.6.3 *Oxidation Rates (Kinetics)   762*

13.7 Corrosion Control   764

13.7.1 *Materials Selection   764*

13.7.2 *Coatings   765*

13.7.3 *Design   766*

13.7.4 *Alteration of Environment   767*

13.7.5 *Cathodic and Anodic Protection   768*

13.8 Summary   770

13.9 Definitions   770

13.10 Problems   771

13.11 Materials Selection and Design Problems   776

CHAPTER **14**
Electrical Properties of Materials   778

14.1 Electrical Conduction in Metals   779

14.1.1 *The Classical Model for Electrical Conduction in Metals   779*

14.1.2 *Ohm's Law   781*

14.1.3 *Drift Velocity of Electrons in a Conducting Metal   785*

14.1.4 *Electrical Resistivity of Metals   786*

14.2 Energy-Band Model for Electrical Conduction   790

14.2.1 *Energy-Band Model for Metals   790*

14.2.2 *Energy-Band Model for Insulators   792*

14.3 Intrinsic Semiconductors   792

14.3.1 *The Mechanism of Electrical Conduction in Intrinsic Semiconductors   792*

14.3.2 *Electrical Charge Transport in the Crystal Lattice of Pure Silicon   793*

14.3.3 *Energy-Band Diagram for Intrinsic Elemental Semiconductors   794*

14.3.4 *Quantitative Relationships for Electrical Conduction in Elemental Intrinsic Semiconductors   795*

14.3.5 *Effect of Temperature on Intrinsic Semiconductivity   797*

14.4 Extrinsic Semiconductors   799

14.4.1 *n-Type (Negative-Type) Extrinsic Semiconductors   799*

14.4.2 *p-Type (Positive-Type) Extrinsic Semiconductors   801*

14.4.3 *Doping of Extrinsic Silicon Semiconductor Material   803*

14.4.4 *Effect of Doping on Carrier Concentrations in Extrinsic Semiconductors   803*

14.4.5 *Effect of Total Ionized Impurity Concentration on the Mobility of Charge Carriers in Silicon at Room Temperature 806*

14.4.6 *Effect of Temperature on the Electrical Conductivity of Extrinsic Semiconductors 807*

14.5 Semiconductor Devices 809

14.5.1 *The pn Junction 810*

14.5.2 *Some Application for pn Junction Diodes 813*

14.5.3 *The Bipolar Junction Transistor 815*

14.6 Microelectronics 816

14.6.1 *Microelectronic Planar Bipolar Transistors 818*

14.6.2 *Microelectronic Planar Field-Effect Transistors 819*

14.6.3 *Fabrication of Microelectronic Integrated Circuits 821*

14.7 Compound Semiconductors 828

14.8 Electrical Properties of Ceramics 831

14.8.1 *Basic Properties of Dielectrics 831*

14.8.2 *Ceramic Insulator Materials 834*

14.8.3 *Ceramic Materials for Capacitors 835*

14.8.4 *Ceramic Semiconductors 836*

14.8.5 *Ferroelectric Ceramics 838*

14.9 Nanoelectronics 841

14.10 Summary 842

14.11 Definitions 843

14.12 Problems 846

14.13 Materials Selection and Design Problems 850

CHAPTER **15**

Optical Properties and Superconductive Materials 852

15.1 Introduction 853

15.2 Light and the Electromagnetic Spectrum 853

15.3 Refraction of Light 856

15.3.1 *Index of refraction 856*

15.3.2 *Snell's Law of Light Refraction 857*

15.4 Absorption, Transmission, and Reflection of Light 859

15.4.1 *Metals 859*

15.4.2 *Silicate Glasses 859*

15.4.3 *Plastics 862*

15.4.4 *Semiconductors 862*

15.5 Luminescence 863

15.5.1 *Photoluminescence 864*

15.5.2 *Cathodoluminescence 864*

15.6 Stimulated Emission of Radiation and Lasers 866

15.6.1 *Types of Lasers 868*

15.7 Optical Fibers 870

15.7.1 *Light Loss in Optical Fibers 870*

15.7.2 *Single-Mode and Multimode Optical Fibers 871*

15.7.3 *Fabrication of Optical Fibers 872*

15.7.4 *Modern Optical-Fiber Communication Systems 874*

15.8 Superconducting Materials 875

15.8.1 *The Superconducting State 875*

15.8.2 *Magnetic Properties of Superconductors 876*

15.8.3 *Current Flow and Magnetic Fields in Superconductors 878*

15.8.4 *High-Current, High-Field Superconductors 879*

15.8.5 *High Critical Temperature (T_c) Superconducting Oxides 881*

15.9 Definitions 883

15.10 Problems 884

15.11 Materials Selection and Design Problems 886

CHAPTER **16**

Magnetic Properties 888

16.1 Introduction 889

16.2 Magnetic Fields and Quantities 889

16.2.1 *Magnetic Fields 889*

16.2.2 *Magnetic Induction 892*

16.2.3 *Magnetic Permeability 892*

16.2.4 *Magnetic Susceptibility 894*

16.3   Types of Magnetism   894
   16.3.1 *Diamagnetism   895*
   16.3.2 *Paramagnetism   895*
   16.3.3 *Ferromagnetism   895*
   16.3.4 *Magnetic Moment of a Single Unpaired Atomic Electron   897*
   16.3.5 *Antiferromagnetism   899*
   16.3.6 *Ferrimagnetism   899*
16.4   Effect of Temperature on Ferromagnetism   899
16.5   Ferromagnetic Domains   900
16.6   Types of Energies That Determine the Structure of Ferromagnetic Domains   902
   16.6.1 *Exchange Energy   902*
   16.6.2 *Magnetostatic Energy   903*
   16.6.3 *Magnetocrystalline Anisotropy Energy   903*
   16.6.4 *Domain Wall Energy   904*
   16.6.5 *Magnetostrictive Energy   905*
16.7   The Magnetization and Demagnetization of a Ferromagnetic Metal   907
16.8   Soft Magnetic Materials   908
   16.8.1 *Desirable Properties for Soft Magnetic Materials   909*
   16.8.2 *Energy Losses for Soft Magnetic Materials   909*
   16.8.3 *Iron-Silicon Alloys   910*
   16.8.4 *Metallic Glasses   911*
   16.8.5 *Nickel-Iron Alloys   912*
16.9   Hard Magnetic Materials   915
   16.9.1 *Properties of Hard Magnetic Materials   915*
   16.9.2 *Alnico Alloys   917*
   16.9.3 *Rare earth Alloys   919*
   16.9.4 *Neodymium-Iron-Boron Magnetic Alloys   921*
   16.9.5 *Iron-Chromium-Cobalt Magnetic Alloys   921*
16.10   Ferrites   923
   16.10.1 *Magnetically Soft Ferrites   923*
   16.10.2 *Magnetically Hard Ferrites   928*
16.11   Summary   928
16.12   Definitions   929
16.13   Problems   932
16.14   Materials Selection and Design Problems   936

APPENDIX **I**
Important Properties of Selected Engineering Materials   937

APPENDIX **II**
Some Properties of Selected Elements   992

APPENDIX **III**
Ionic Radii of the Elements   994

APPENDIX **IV**
Selected Physical Quantities and Their Units   997

References for Further Study by Chapter   999

Glossary   1002

Answers   1013

Index   1016

PREFACE

Foundations of Materials Science and Engineering, Fourth Edition is designed for a first course in materials science and engineering for engineering students. Understanding that this might be a student's first exposure to materials science, the book presents essential topics in a clear, concise manner, without extraneous details to overwhelm newcomers. Industrial examples and photographs used throughout the book give students a look at the many ways material science and engineering are applied in the real world.

NEW FEATURES OF THE FOURTH EDITION

In addition to its already renowned student friendly writing style and applications to industry, the fourth edition offers new features including a thorough coverage of modern materials science topics that prepare students for life outside the classroom. The new sections are:

- New reference to smart materials/devices, MEMs, and nanomaterials (1.1)
- New reference to superalloys and their biomedical applications (1.3)
- Added discussion of engineering plastics and applications in automobiles (1.3.2)
- Added discussion of engineering ceramics and applications (1.3.3)
- Added discussion of composite materials (1.3.4)
- New coverage of smart materials and nanomaterials (1.5)
- New section featuring a simplified case study in selection of materials for the frame and forks of a bicycle (1.6)
- New coverage of amorphous materials was added in chapter 3
- Added references to long and short range order (SRO also known as amorphous materials) (3.1)
- Chapter 4 has been split into chapters 4 and 5 for the fourth edition so diffusion can be covered in a stand-alone chapter
- Coverage of microscopes added to the end of chapter 4
- Added coverage of planar defects and twin boundaries (4.4.3)
- New section on volume defects (4.4.4)
- New section on experimental techniques for identification of microstructure and defects (4.5)
- Added coverage of fine-grained metals and the Hall-Petch equation in chapter 6
- New case study in failure and coverage of recent advances in improving mechanical performance in chapter 7

- Added coverage of failure and fracture of metals (7.1)
- New section on ductile to brittle transition temperature (7.1.4)
- New section on recent advances and future directions in improving the mechanical performance of metals (7.7)
- New coverage of cooling curves in chapter 8
- Added coverage of intermediate compounds (8.11.1)
- Three new sections devoted to advanced alloys and their application in biomedical engineering have been added to chapter 9
- New section devoted to biomedical applications of polymeric materials added to chapter 10
- New section with coverage of bucky balls and carbon nano-tubes (11.2.12)
- New section on ceramic coatings and surface engineering (11.9)
- New section on ceramics in biomedical applications (11.10)
- New section on nanotechnology and ceramics (11.11)
- New section on bone: a natural composite material (12.11)
- New section on hydrogen damage (13.5.11)
- New section on nanoelectronics (14.9)
- New appendix featuring extensive materials properties reference

Other New Features:

- Learning objectives have been added to every chapter
- Icons have been added to highlight the supplemental media resources
- Many new chapter openers and interior photos are included

Retained Features:

- Over 1200 end-of-chapter problems and over 180 materials selection and design problems are offered
- Over 140 example problems
- Modern applications of materials
- A concise, readable style is used throughout; readers are given understandable explanations without excessive detail

SUPPLEMENTS

Student CD-ROM

There is a new student CD-ROM to accompany the fourth edition of *Foundations of Materials Science and Engineering* featuring a wealth of visualization and study materials.

- Three Virtual Labs, with video, interactive quizzing, and step-by-step real-life processes (with possible pitfalls), are included to better prepare students for the laboratory or serve as a simulation of an actual laboratory experience for students. Lab topics: Measurement of Hardness Using Rockwell Hardness test, Metallography, Tensile Testing.

- An extensive lab manual in PDF and Word formats coordinates with the Virtual Labs.

- MatVis is 2.0 crystal visualization software allows students to create different crystal structures and molecules, as well as view already created structures.

- ICENine Phase Diagram software allows students to search a list of phase diagrams and find the composition at any given temperature.

- EES (Engineering Equation Solver) computational software provides students with capability to solve materials science problems.

- Animations cover bonds, bond forces, rotating crystals, atomic packing arrangements, different crystal planes and coordination members, among other topics. Likewise, tutorials with sound are provided.

- Extensive, searchable materials properties database.

- Bonus chapters on "Materials for MEMS and Microsystems," a chapter from Tai-Ran Hsu's *MEMS and Microsystems: Design and Manufacture.*

- Numerous web links to professional sites and reference materials

Instructor's Resource CD-ROM

- Lecture PowerPoint slides with added animations, videos, and images not included in the text
- Text images in Jpeg format
- Solutions manual in PDF and Word formats
- Teaching resources to help instructors incorporate the Virtual Labs and other media into their course
- Lab manual to accompany the Virtual Labs
- Sample syllabi
- Transition guides to help instructors transition from their current text to the fourth edition of Smith/Hashemi

COSMOS CD-ROM for Instructors   The detailed solutions for all text problems are delivered in our new electronic Complete Online Solutions Manual Organization System. COSMOS is a database management tool geared towards assembling homework assignments, tests and quizzes. No longer do instructors need to wade through thick solutions manuals or huge Word files. COSMOS helps them to quickly find solutions, and also keeps a record of problems assigned to avoid duplication in subsequent semesters. Instructors can contact their McGraw-Hill sales representative at www.mhhe.com/catalogs/rep/ to obtain a copy of the COSMOS solutions manual.

Online Learning Center   Web support is provided for the book at the website www.mhhe.com/smithmaterials. Visit this site for book and supplement information, errata, author information, and resources for further study or reference.

ACKNOWLEDGMENTS

The co-author, Javad Hashemi, would like to acknowledge the everlasting love, support, encouragement, and guidance of his mother, Sedigheh, throughout the course of his life and career. He dedicates this textbook to her, to the love of his life Eva, to the most precious gifts of his life, Evan and Jonathon, to his siblings, and last but not least to the memory of his father.

The authors would like to acknowledge with appreciation the numerous and valuable comments, suggestions, constructive criticisms, and praise from the following evaluators and reviewers:

Raul Bargiola, *University of Virginia*

Deepak Bhat, *University of Arkansas*

Nigel Browning, *University of California, Davis*

David Cann, *Iowa State University*

Nikhilesh Chawla, *Arizona State University*

Deborah D.L. Chung, *University at Buffalo, The State University of New York*

James H. Edgar, *Kansas State University*

Jeffrey W. Fergus, *Auburn University*

Raymond A. Fournelle, *Marquette University*

Randall M. German, *Penn State University*

Stacy Gleixner, *San Jose State University*

Brian P. Grady, *University of Oklahoma*

Richard B. Griffin, *Texas A&M University*

Masanori Hara, *Rutgers University*

Lee Hornberger, *Santa Clara University*

Osman T. Inal, *New Mexico Tech*

Stephen Kampe, *Virginia Tech*

Ibrahim Karaman, *Texas A&M University*

Steve Krause, *Arizona State University*

David Kunz, *University of Wisconsin, Platteville*

Gladius Lewis, *University of Memphis*

Robert W. Messler, Jr., *Renesselaer Polytechnic Institute*

John Mirth, *University of Wisconsin, Platteville*

Devesh Misra, *University of Louisiana at Lafayette*

Kay Mortensen, *Brigham Young University*

David Niebuhr, *California Polytechnic State University*

Kanji Ono, *University of California, Los Angeles*

Martin Pugh, *Concordia University, Canada*

Susil Putatunda, *Wayne State University*

Lew Rabenberg, *University of Texas at Austin*

Wayne Reitz, *North Dakota State University*

James Michael Rigsbee, *North Carolina State University*

Jay M. Samuel, *University of Wisconsin*

John Schlup, *Kansas State University*

Yu-Lin Shen, *University of New Mexico*

Scott Short, *Northern Illinois University*

Susan B. Sinnott, *University of Florida*

Thomas Staley, *Virginia Tech*

Christopher Steinbruchel, *Renesselaer Polytechnic Institute*

Alexey Sverdlin, *Bradley University*

David S. Wilkinson, *McMaster University, Canada*

Chris Wise, *New Mexico State University*

Henry Daniel Young, *Wright State University*

Jiaxon Zhao, *Indiana University—Purdue University Fort Wayne*

Naveen Chandrashekar—*Texas Tech University*

Javad Hashemi

GUIDED TOUR

Modern materials science topics have been added throughout the text including:

- Smart materials/devices, MEMs and nanomaterials
- Biomedical applications
- Chapter 4 has been split into two chapters to allow diffusion to be covered alone
- Three new sections devoted to advanced alloys and their application in biomedical engineering
- Ceramics coverage has been expanded to include nanotechnology and ceramics

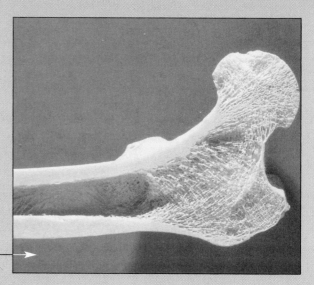

Materials selection and design problems follow the end-of-chapter problems. These problems challenge students to create engineering solutions through materials selection and process design.

198   CHAPTER 5   Thermally Activated Processes and Diffusion in Solids

5.8 MATERIALS SELECTION AND DESIGN PROBLEMS

1. In integrated circuits, dissimilar metal wires such as gold and aluminum are joined at the cross section to form a bonded interface. At elevated temperatures, the interface starts to move or shift in the direction of one of the metals, called the Kirkendall effect. (a) Can you explain this phenomenon? (b) Is the direction of the shift random? (c) What are the negative effects of this process?
2. (a) Design a process that would be used to make a solid steel component out of fine steel powder. The density of the formed solid should be very close to the density of the physical metal. (b) Explain how this process works at both macro and micro levels. (c) What are some of the obstacles that you may encounter in your process?
3. Explain what happens if carbon steel is exposed to an oxygen-rich atmosphere at elevated temperatures inside a furnace.
4. Based on the data in Table 5.2, is it easier for carbon atoms to diffuse in the structure of BCC iron or FCC iron? What is the physical reason for this?
5. Classify the mechanism of diffusion in all 12 solute/solvent pairs given in Table 5.2 (interstitial or substitutional). Compare the diffusivity values and draw a conclusion.
6. The activation energy for the diffusion of hydrogen in steel at room temperature is 3600 cal/mol, which is significantly lower than that of, for example, carbon at 20,900 cal/mol. Investigate the effect of diffused hydrogen on the mechanical behavior of steel.
7. Investigate the role of solid state diffusion in manufacturing of ceramic components using the powder metallurgy process.

Icons have been added to highlight the media supplement resources.

2.3 The Electronic Structure of Atoms   35

2.3.3 Electronic Structure of Multielectron Atoms

Maximum Number of Electrons for Each Principal Shell   Atoms consist of principal shells* of high electron densities as dictated by the laws of quantum mechanics. There are seven of these principal **electron shells** when the atomic number of the atom reaches 87 for the element francium (Fr). Each shell can only contain a maximum number of electrons, which is again dictated by the laws of quantum mechanics. The maximum number of electrons that can be contained in each shell in an atom is defined by different sets of the four quantum numbers (Pauli principle) and is $2n^2$, where n is the principal quantum number. Thus, there can be only a maximum of 2 electrons for the first principal shell, 8 for the second, 18 for the third, 32 for the fourth, etc., as indicated in Table 2.3.

Atomic Size   Each atom can be considered to a first approximation as a sphere with a definite radius. The radius of the atomic sphere is not a constant but depends to some extent on its environment. Figure 2.6 shows the relative atomic sizes of many of the elements along with their atomic radii. Many of the values of the atomic radii are still not fully agreed upon and vary to some extent depending on the reference source.

From Fig. 2.6, some trends in atomic size are evident. In general, as the principal quantum number increases, the size of the atom increases. There are, however, a few exceptions where the atomic size actually gets smaller. The alkali elements of group 1A of the periodic table (Fig. 2.1) are a good example of atoms whose size increases with increasing n. For example, lithium ($n = 2$) has an atomic radius of 0.157 nm, whereas cesium ($n = 6$) has an atomic radius of 0.270 nm. In progressing across the period table from an alkali group 1A element to a noble-gas element 8A, the atomic size

Learning Objectives have been added to each chapter to guide students' comprehension of the material.

LEARNING OBJECTIVES

By the end of this chapter, students will be able to . . .

1. Describe the subject of materials science and engineering as a scientific discipline.
2. Cite the primary classification of solid materials.
3. Give distinctive features of each group of materials.
4. Cite one material from each group. Give some applications of different types of materials.
5. Evaluate how much you know, and how much you do not know about materials.
6. Establish the importance of materials science and engineering in selection of materials for various applications.

INSTRUCTOR'S RESOURCE CD-ROM

Offering a wide range of class preparation and presentation resources, this CD helps instructors to easily transition to Smith/Hashemi's new edition.

- Lecture PowerPoint slides with added animations, videos, and images not included in the text
- Text images in Jpeg format
- Solutions manual in PDF and Word formats
- Teaching resources to help instructors incorporate the Virtual Labs and other media into their course
- Lab manual to accompany the Virtual Labs
- Sample syllabi
- Transition guides—to help instructors transition from their current text to the fourth edition of Smith/Hashemi

COSMOS CD-ROM

This CD, also available to instructors only, provides electronic solutions delivered via our database management tool. McGraw-Hill's COSMOS is a powerful solutions manual tool to help instructors streamline the creation of assignments, quizzes, and tests by using problems and solutions from the textbook—as well as their own custom material.

ENGINEERING EQUATION SOLVER (EES)

EES is a powerful equation solver with built-in functions and property tables for thermodynamic and transport properties as well as automatic unit checking capability. It requires less time than a calculator for data entry and allows more time for thinking critically about modeling and solving engineering problems. While every new copy of our text is packaged with a CD-ROM containing the Limited Academic Version of EES, the full academic version of the EES engine is available free to departments of educational institutions who adopt this text.

COURSE MANAGEMENT SYSTEMS—WEBCT AND BLACKBOARD

With help from our partners, WebCT and Blackboard, professors can take complete control over their course content. These course cartridges also provide online testing and powerful student tracking features.

LEARNING SUPPLEMENTS FOR THE STUDENT

There is a new student CD-ROM to accompany the fourth edition of *Foundations of Materials Science and Engineering*. This CD-ROM features a wealth of visualization and study materials.

VIRTUAL LABS

Three Virtual Labs, with video, interactive quizzing, and step-by-step real-life processes (with possible pitfalls), are included to better prepare students for the laboratory or serve as a simulation of an actual laboratory experience for students. Lab topics: Measurement of Hardness Using Rockwell Hardness test, Metallography, Tensile Testing. An extensive lab manual in PDF and Word formats coordinates with the Virtual Labs.

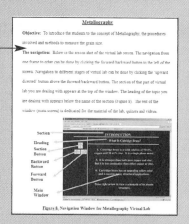

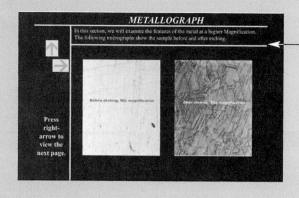

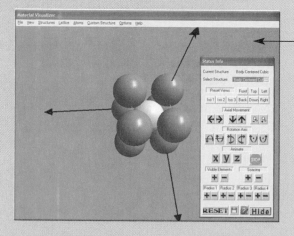

VISUALIZATION SOFTWARE

MatVis 2.0 crystal visualization software allows students to create different crystal structures and molecules, as well as view already created structures. ICENine Phase Diagram software allows students to search a list of phase diagrams and find the composition at any given temperature.

LEARNING SUPPLEMENTS FOR THE STUDENT

ANIMATIONS

Animations cover bonds, bond forces, rotating crystals, atomic packing arrangements, different crystal planes and coordination members, among other topics. Likewise, tutorials with sound are provided.

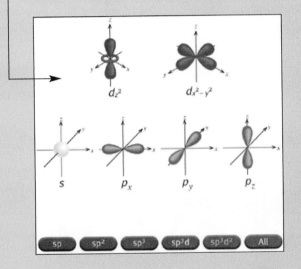

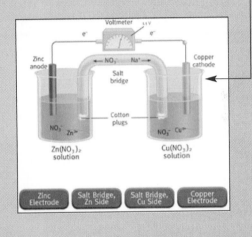

QUIZZING

Multiple choice quizzes on atomic numbers/masses, electronic structure, galvanic cell/corrosion, ionic bonding and Rutherford's Experiment along with FE exam review questions.

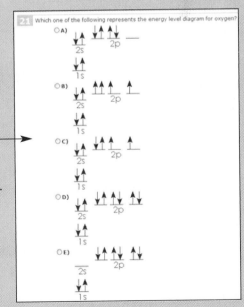

Introduction to Materials Science and Engineering

NASA (Source: *www.nasa.gov*)

Opportunity and *Spirit,* two "robot geologists" designed and constructed by NASA, have finally provided irrefutable evidence that some regions of Mars were once covered with water. The rovers are masterpieces of engineering achievement whose design and construction required expert scientists, designers, and engineers in a variety of fields. One such field is the field of materials science. Consider the materials engineering and selection issues in constructing the rovers, delivering them to Mars using launch vehicles, landing the rovers safely on the Mars surface, controlling the movement and activities of the rovers, performing scientific experiments, collecting and storing data, and, finally, communicating the gathered data back to earth. All major classes of materials including metals, polymers, ceramics, composites, and electronic are used in the structure of the rovers. The Mars missions push the boundaries of knowledge and creativity in the field of materials science and engineering. Through the study of materials science and engineering, we can understand and appreciate the challenges and complexities involved in such missions. ∎

By the end of this chapter, students will be able to . . .

1. Describe the subject of materials science and engineering as a scientific discipline.

2. Cite the primary classification of solid materials.

3. Give distinctive features of each group of materials.

4. Cite one material from each group. Give some applications of different types of materials.

5. Evaluate how much you know, and how much you do not know about materials.

6. Establish the importance of materials science and engineering in selection of materials for various applications.

1.1 MATERIALS AND ENGINEERING

Humankind, materials, and engineering have evolved over the passage of time and are continuing to do so. All of us live in a world of dynamic change, and materials are no exception. The advancement of civilization has historically depended on the improvement of materials to work with. Prehistoric humans were restricted to naturally accessible materials such as stone, wood, bones, and fur. Over time, they moved from the materials Stone Age into the newer Copper (Bronze) and Iron Ages. Note that this advance did not take place uniformly everywhere—we shall see that this is true in nature even down to the microscopic scale. Even today we are restricted to the materials we can obtain from earth's crust and atmosphere (Table 1.1). According to Webster's dictionary, materials may be defined as substances of which something is composed or made. Although this definition is broad, from an engineering application point of view, it covers almost all relevant situations.

The production and processing of materials into finished goods constitutes a large part of our present economy. Engineers design most manufactured products and the processing systems required for their production. Since products require materials, engineers should be knowledgeable about the internal structure and properties of materials so that they can choose the most suitable ones for each application and develop the best processing methods.

Research and development engineers create new materials or modify the properties of existing ones. Design engineers use existing, modified, or new materials to design and create new products and systems. Sometimes design engineers have a problem in their design that requires a new material to be created by research scientists and engineers. For example, engineers designing a *high-speed civil transport* (HSCT) (Fig. 1.1) will have to develop new high-temperature materials that will withstand temperatures as high as 1800°C (3250°F) so that airspeeds as high as Mach 12 to 25 can be attained.[1] Research is currently underway to develop new

---

[1]Mach 1 equals the speed of sound in air.

Table 1.1 The most common elements in planet earth's crust and atmosphere by weight percentage and volume

Element	Weight percentage of the earth's crust
Oxygen (O)	46.60
Silicon (Si)	27.72
Aluminum (Al)	8.13
Iron (Fe)	5.00
Calcium (Ca)	3.63
Sodium (Na)	2.83
Potassium (K)	2.70
Magnesium (Mg)	2.09
Total	98.70
Gas	**Percent of dry air by volume**
Nitrogen (N_2)	78.08
Oxygen (O_2)	20.95
Argon (Ar)	0.93
Carbon dioxide (CO_2)	0.03

Figure 1.1
High-speed civil transport image shows the Hyper-X at Mach 7 with the engines operating. Rings indicate surface flow speeds.
(© *The Boeing Company.*)

Figure 1.2
The International Space Station
(© *AFP/CORBIS*)

ceramic-matrix composites, refractory intermetallic compounds, and single-crystal superalloys for this and other similar applications.

One area that demands the most from materials scientists and engineers is space exploration. The design and construction of the *International Space Station* (ISS) and *Mars Exploration Rover* (MER) missions are examples of space research and exploration activities that require the absolute best from our materials scientists and engineers. The construction of ISS, a large research laboratory moving at a speed of 27,000 km/h through space, required selection of materials that would function in an environment far different than ours on earth (Fig. 1.2). The materials must be lightweight to minimize payload weight during liftoff. The outer shell must protect against the impact of tiny meteoroids and man-made debris. The internal air pressure of roughly 15 psi is constantly stressing the modules. Additionally, the modules must withstand the massive stresses at launch. Materials selection for MERs is also a challenge, especially considering that they must survive an environment in which night temperatures could be as low as −96°C. These and other constraints push the limits of material selection in the design of complex systems.

We must remember that materials usage and engineering designs are constantly changing. This change continues to accelerate. No one can predict the long-term future advances in material design and usage. In 1943 the prediction was made that successful people in the United States would own their own autogyros (auto-airplanes).

How wrong that prediction was! At the same time, the transistor, the integrated circuit, and television (color and high definition included) were neglected. Thirty years ago, many people would not have believed that some day computers would become a common household item similar to a telephone or a refrigerator. And today, we still find it hard to believe that some day space travel will be commercialized and we may even colonize Mars. Nevertheless, science and engineering push and transform our most unachievable dreams to reality.

The search for new materials goes on continuously. For example, mechanical engineers search for higher-temperature materials so that jet engines can operate more efficiently. Electrical engineers search for new materials so that electronic devices can operate faster and at higher temperatures. Aerospace engineers search for materials with higher strength-to-weight ratios for aerospace vehicles. Chemical and materials engineers look for more highly corrosion-resistant materials. Various industries look for smart materials and devices and microelectromechanical systems (MEMs) to be used as sensors and actuators in their respective applications. More recently, the field of nanomaterials has attracted a great deal of attention from scientists and engineers all over the world. Novel structural, chemical, and mechanical properties of nanomaterials have opened new and exciting possibilities in the application of these materials to a variety of engineering and medical problems. These are only a few examples of the search by engineers and scientists for new and improved materials and processes for a multitude of applications. In many cases what was impossible yesterday is a reality today!

Engineers in all disciplines should have some basic and applied knowledge of engineering materials so that they will be able to do their work more effectively when using them. The purpose of this book is to serve as an introduction to the internal structure, properties, processing, and applications of engineering materials. Because of the enormous amount of information available about engineering materials and due to the limitations of this book, the presentation has had to be selective.

1.2 MATERIALS SCIENCE AND ENGINEERING

Materials science is primarily concerned with the search for basic knowledge about the internal structure, properties, and processing of materials. **Materials engineering** is mainly concerned with the use of fundamental and applied knowledge of materials so that the materials can be converted into products needed or desired by society. The term *materials science and engineering* combines both materials science and materials engineering and is the subject of this book. Materials science is at the basic knowledge end of the materials knowledge spectrum and materials engineering is at the applied knowledge end, and there is no demarcation line between the two (Fig. 1.3).

Figure 1.4 shows a three-ringed diagram that indicates the relationship among the basic sciences (and mathematics), materials science and engineering, and the other engineering disciplines. The basic sciences are located within the inner ring or

Figure 1.3
Materials knowledge spectrum. Using the combined knowledge of materials from materials science and materials engineering enables engineers to convert materials into the products needed by society.

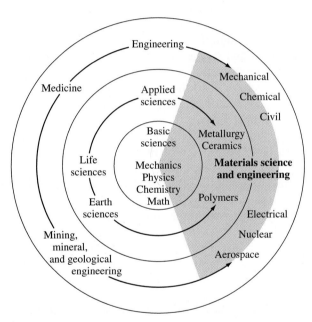

Figure 1.4
This diagram illustrates how materials science and engineering form a bridge of knowledge from the basic sciences to the engineering disciplines.
(*Courtesy of the National Academy of Science.*)

core of the diagram, while the various engineering disciplines (mechanical, electrical, civil, chemical, etc.) are located in the outermost third ring. The applied sciences, metallurgy, ceramics, and polymer science are located in the middle ring. Materials science and engineering is shown to form a bridge of materials knowledge from the basic sciences (and mathematics) to the engineering disciplines.

1.3 TYPES OF MATERIALS

For convenience most engineering materials are divided into *three* main or fundamental classes: **metallic materials, polymeric materials,** and **ceramic materials.** In this chapter we shall distinguish among them on the basis of some of their important mechanical, electrical, and physical properties. In subsequent chapters we shall study the internal structural differences among these types of materials. In addition to the three main classes of materials, we shall consider two processing or applicational classes, **composite materials** and **electronic materials,** because of their great engineering importance.

1.3.1 Metallic Materials

These materials are inorganic substances that are composed of one or more metallic elements and may also contain some nonmetallic elements. Examples of metallic elements are iron, copper, aluminum, nickel, and titanium. Nonmetallic elements such as carbon, nitrogen, and oxygen may also be contained in metallic materials. Metals have a crystalline structure in which the atoms are arranged in an orderly manner. Metals in general are good thermal and electrical conductors. Many metals are relatively strong and ductile at room temperature, and many maintain good strength even at high temperatures.

Metals and alloys[2] are commonly divided into two classes: **ferrous metals and alloys** that contain a large percentage of iron such as the steels and cast irons and **nonferrous metals and alloys** that do not contain iron or contain only a relatively small amount of iron. Examples of nonferrous metals are aluminum, copper, zinc, titanium, and nickel. The distinction between ferrous and nonferrous alloys is made because of the significantly higher usage and production of steels and cast irons when compared to other alloys.

Metals in their alloyed and pure forms are used in numerous industries including aerospace, biomedical, semiconductor, electronic, energy, civil structural, and transport. The U.S. production of basic metals such as aluminum, copper, zinc, and magnesium is expected to follow the U.S. economy fairly closely. However, the production of iron and steel has been less than expected because of global competition and the always-important economical reasons. Materials scientists and engineers are constantly attempting to improve the properties of existing alloys and to design and produce new alloys with improved strength, high temperature strength, creep (see Sec. 7.4), and fatigue (see Sec. 7.2) properties. The existing alloys may be improved by better chemistry, composition control, and processing techniques. For example, by 1961, new and improved nickel-base, iron-nickel-cobalt-base **superalloys** were available for use in high-pressure turbine airfoils in aircraft gas turbines. The term superalloy was used because of their improved performance at elevated temperatures of approximately 540°C (1000°F) and high stress levels. Figures 1.5 and 1.6 show a PW-400 gas turbine engine that is made primarily of metal alloys and superalloys. The metals used inside the engine must be

---

[2]A metal alloy is a combination of two or more metals or a metal (metals) and a nonmetal (nonmetals).

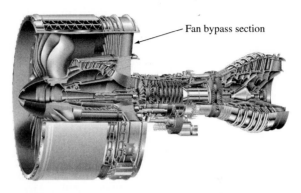

Fan bypass section

Figure 1.5
The aircraft turbine engine (PW 4000) shown
is made principally of metal alloys. The latest
high-temperature, heat resistant, high-strength
nickel-base alloys are used in this engine. This
engine has many advanced, service-proven
technologies to enhance operational performance
and durability. These include second-generation
single-crystal turbine blade materials, powder
metal disks, and an improved full authority digital
electronic control.
(*Courtesy of Pratt & Whitney Co.*)

Figure 1.6
Cutaway of the PW 4000 112-in. (284.48 cm) gas
turbine engine, showing fan bypass section.
(*Courtesy of Pratt & Whitney Co.*)

able to withstand high temperatures and pressures generated during its operation. By
1980, casting techniques were improved to produce directionally solidified columnar
grain (see Sec. 4.2) and single crystal casting nickel-base alloys (see Sec. 4.2). By the
1990s, single crystal directionally solidified cast alloys were standard in many aircraft
gas turbine applications. The better performance of superalloys at elevated operating
temperatures, significantly improved the efficiency of the aircraft engines.

Many metal alloys such as titanium alloys, stainless steel, and cobalt-base alloys
are also used in biomedical applications including orthopedic implants, heart valves,
fixations devices, and screws. These materials offer high strength, stiffness, and biocompatibility. Biocompatibility is important because the environment inside the
human body is extremely corrosive, and therefore materials used for such applications must be effectively impervious to this environment.

In addition to better chemistry and composition control, researchers and engineers also concentrate on improving new processing techniques of these materials.
Processes such as hot isostatic pressing (see Sec. 11.4) and isothermal forging have

led to improved fatigue life of many alloys. Also, powder metallurgy techniques (see Sec. 10.4) will continue to be important since improved properties can be obtained for some alloys with lower finished product cost.

1.3.2 Polymeric Materials

Most polymeric materials consist of long molecular chains or networks that are usually based on organics (carbon-containing precursors). Structurally, most polymeric materials are noncrystalline, but some consist of mixtures of crystalline and noncrystalline regions. The strength and ductility of polymeric materials vary greatly. Because of the nature of their internal structure, most polymeric materials are poor conductors of electricity. Some of these materials are good insulators and are used for electrical insulative applications. One of the more recent applications of polymeric materials has been in manufacture of *digital video disks* (DVDs) (Fig. 1.7). In general, polymeric materials have low densities and relatively low softening or decomposition temperatures.

Historically, plastic materials have been the fastest growing basic material in the United States, with a growth rate of 9 percent per year on a weight basis Fig. (1.14). However, the growth rate of plastics through 1995 dropped to below 5 percent, a significant decrease. This drop was expected because plastics have already substituted for metals, glass, and paper in most of the main volume markets, such as packaging and construction, for which plastics are suitable.

Engineering plastics such as nylon are expected to remain competitive with metals, according to some predictions. Figure 1.8 shows expected costs for engineering plastic resins versus some common metals. The polymer-supplying industries are increasingly focusing development of polymer-polymer mixtures, also known as *alloys* or **blends,** to fit specific applications for which no other single polymer is suitable. Because blends are produced based on existing polymers with well-known properties,

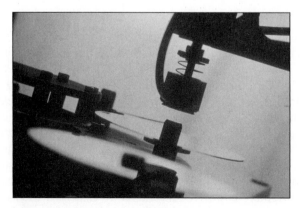

Figure 1.7
Plastic resin producers are developing ultrapure,
high-flow grades of polycarbonate plastic for DVDs.
(© *George B. Diebold/CORBIS.*)

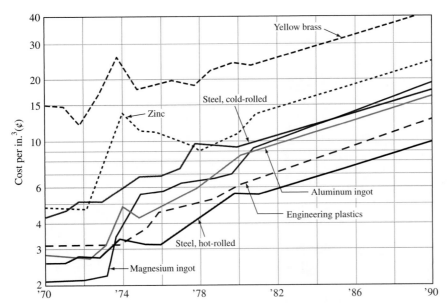

Figure 1.8

Historical and expected competitive costs of engineering plastic resins versus some common metals from 1970 to 1990. Engineering plastics are expected to remain competitive with cold-rolled steel and other metals.

(*After Modern Plastics, August 1982, p. 12, and new data, 1998.*)

their development is less costly and more reliable than synthesizing a new single polymer for a specific application. For example, elastomers (a highly deformable type of polymer) are often blended with other plastics to improve the impact strength of the material. Such blends have important usage in automotive bumpers, power tool housing, sporting goods, and the synthetic components used in many indoor track facilities, which are generally made of a combination of rubber and polyeurathane. Acrylic coatings blended with various fibers and fillers and with brilliant colors are used as coating material for tennis courts and playgrounds. Yet, other polymer-coating materials are being used for protection against corrosion, aggressive chemical environments, thermal shock, impact, wear, and abrasion. Search for new polymers and blends continues because of their low cost and suitable properties for many applications.

1.3.3 Ceramic Materials

Ceramic materials are inorganic materials that consist of metallic and nonmetallic elements chemically bonded together. Ceramic materials can be crystalline, noncrystalline, or mixtures of both. Most ceramic materials have high-hardness and high-temperature strength but tend to be brittle (little or no deformation prior to fracture). Advantages of ceramic materials for engineering applications include light weight, high strength and hardness, good heat and wear resistance, reduced friction, and insulative properties (Figs. 1.9 and 1.10). The insulative property along with high heat and wear resistance

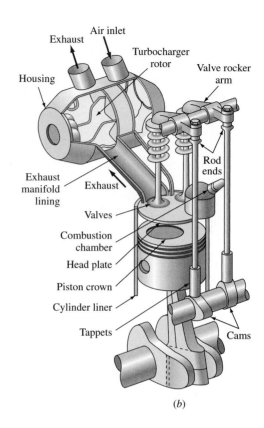

(a)   (b)

Figure 1.9
(a) Examples of a newly developed generation of engineered ceramic materials for advanced engine applications. The black items include engine valves, valve seat inserts, and piston pins made of silicon nitride. The white item is a port-manifold liner made of an alumina ceramic material.
(*Courtesy of Kyocera Industrial Ceramics Corp.*)
(b) Potential ceramic component applications in a turbocharged diesel engine.
(*After Metals and Materials, December 1988.*)

of many ceramics make them useful for furnace linings for heat treatment and melting of metals such as steel. The search for new plastics and alloys continues because of their lower cost and suitable properties for many applications.

The historic rate of growth of traditional ceramic materials such as clay, glass, and stone in the United States has been 3.6 percent (1966 to 1980). The expected growth rate of these materials from 1982 to 1995 followed the U.S. economy. In the past few decades an entirely new family of ceramics of oxides, nitrides, and carbides, with improved properties, have been produced. The new generation of ceramic materials called *engineering ceramics, structural ceramics,* or **advanced ceramics** has higher strength, better wear and corrosion resistance (even at higher temperatures), and enhanced thermal shock (due to sudden exposure to very high or very low temperatures) resistance. Among the established advanced ceramic materials are alumina (oxide), silicon nitride (nitride) and silicon carbide (carbide).

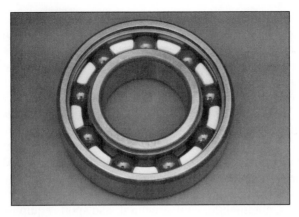

Figure 1.10
High-performance ceramic ball bearings and races
are made from titanium and carbon nitride
feedstocks through power metal technology.
(© *Tom Pantages photo/Courtesy Bearing Works.*)

An important aerospace application for advanced ceramics is the use of ceramic tiles for the space shuttle. The ceramic tiles are made of silicon carbide because of its ability to act as a heat shield and to quickly return to normal temperatures upon removal of the heat source. These ceramic materials thermally protect the aluminum internal sub-structure of the space shuttle during ascent and reentry into the earth's atmosphere (see Figs. 11.50 and 11.51). Another application for advanced ceramics that points to the versatility, importance, and future growth of this class of materials is its use as a cutting tool material. For instance, silicon nitride with a high thermal shock resistance and fracture toughness is an excellent cutting tool material.

The applications for ceramic materials are truly unlimited as they can be applied to aerospace, metal manufacturing, biomedical, automobile, and numerous other industries. The two main drawbacks for this class of materials are that they are (1) difficult to process into finished products and are therefore expensive and (2) brittle and have low fracture toughness compared to metals. If techniques for developing high toughness ceramics are developed further, these materials could show a tremendous upsurge for engineering applications.

1.3.4 Composite Materials

A composite material may be defined as two or more materials (phases or constituents) integrated to form a new one. The constituents keep their properties and the overall composite will have properties different than each of them. Most composite materials consist of a selected filler or reinforcing material and a compatible resin binder to obtain the specific characteristics and properties desired. Usually, the components do not dissolve in each other, and they can be physically identified by an interface between them. Composites can be of many types. Some of the predominant types are fibrous (composed of fibers in a matrix) and particulate (composed of particles in a matrix). Many different combinations of reinforcements and matrix materials are used to produce composite materials. For example, the matrix material may be a metal such as aluminum, a ceramic such as alumina, or a polymer such as epoxy. Depending on the type of matrix used, the composite may be classified as a *metal matrix composite* (MMC), *ceramic matrix composite* (CMC), or a *polymer matrix composite* (PMC). The fiber or particulate materials may also be selected from any of the three main classes of materials with examples such as carbon, glass, aramid, silicon carbide, and others. The combinations of materials utilized in the design of composites depend mainly on the type of application and the environment in which the material will be used.

Composite materials have replaced numerous metallic components especially in aerospace, avionics, automobile, civil structural, and sports equipment industries. An average annual gain of about 5 percent is predicted for the future usage of these materials. One reason is due to their high strength and stiffness-to-weight ratio. Some advanced composites have stiffness and strength similar to some metals but with significantly lower density, and therefore lower overall component weight. These characteristics make advanced composites extremely attractive in situations where component weight is critical. Generally speaking, similar to ceramic materials, the main disadvantage of most composite materials is their brittleness and low fracture toughness. Some of these shortcomings may be improved, in certain situations, by the proper selection of the matrix material.

Two outstanding types of *modern composite materials* used for engineering applications are fiberglass-reinforcing material in a polyester or epoxy matrix and carbon fibers in an epoxy matrix. Figure 1.11 shows schematically where carbon-fiber-epoxy

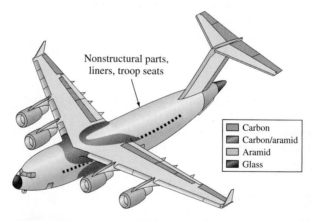

Figure 1.11
Overview of the wide variety of composite parts used in the Air Force's C-17 transport. This airplane has a wingspan of 165 ft and uses 15,000 lb of advanced composites.
(*After Advanced Composites, May/June 1988, p. 53.*)

composite material was used for the wings and engines of the C-17 transport plane. Since these airplanes have been constructed, new cost-saving procedures and modifications have been introduced (See *Aviation Week & Space Technology* for June 9, 1997, p. 30).

1.3.5 Electronic Materials

Electronic materials are not a major type of material by production volume but are an extremely important type of material for advanced engineering technology. The most important electronic material is pure silicon that is modified in various ways to change its electrical characteristics. A multitude of complex electronic circuits can be miniaturized on a silicon chip that is about 3/4 in. square (1.90 cm square) (Fig. 1.12). Microelectronic devices have made possible such new products as communication satellites, advanced computers, handheld calculators, digital watches, and robots (Fig. 1.13).

The use of silicon and other semiconductor materials in solid-state and microelectronics has shown a tremendous growth since 1970, and this growth pattern is expected to continue. The impact of computers and other industrial types of equipment using integrated circuits made from silicon chips has been spectacular. The full effect of computerized robots in modern manufacturing is yet to be determined. Electronic materials will undoubtedly play a vital role in the "factories of the future" in which almost all manufacturing may be done by robots assisted by computer-controlled machine tools.

Over the years integrated circuits have been made with a greater and greater density of transistors located on a single silicon chip with a corresponding decrease in transistor width. For example, in 1998 the point-to-point resolution for the smallest measurement on a silicon chip was 0.18 μm and the diameter of the silicon wafer used was 12 in. (300 mm). Another improvement may be the replacement of aluminum by copper for interconnects because of the higher conductivity of copper.

Figure 1.12
Modern microprocessors have a multitude of outlets, as indicated on this picture of Intel's Pentium II microprocessor.
(© *Don Mason/CORBIS.*)

Figure 1.13
Computerized robots weld a 1994 GM vehicle in Shreveport, LA.
(© *Charles O'Rear/CORBIS.*)

1.4 COMPETITION AMONG MATERIALS

Materials compete with each other for existing and new markets. Over a period of time many factors arise that make it possible for one material to replace another for certain applications. Certainly cost is a factor. If a breakthrough is made in the processing of a certain type of material so that its cost is decreased substantially, this material may replace another for some applications. Another factor that causes material replacement changes is the development of a new material with special properties for some applications. As a result, over a period of time, the usage of different materials changes.

Figure 1.14 shows graphically how the production of six materials in the United States on a weight basis varied over the past years. Aluminum and polymers show an outstanding increase in production since 1930. On a volume basis the production increases for aluminum and polymers are even more accentuated since these are light materials.

The competition among materials is evident in the composition of the U.S. auto. In 1978 the average U.S. auto weighed 4000 lb (1800 kg) and consisted of about 60 percent cast iron and steel, 10 to 20 percent plastics, and 3 to 5 percent aluminum. The 1985 U.S. auto by comparison weighed an average of 3100 lb (1400 kg) and consisted of 50 to 60 percent cast iron and steel, 10 to 20 percent plastics, and

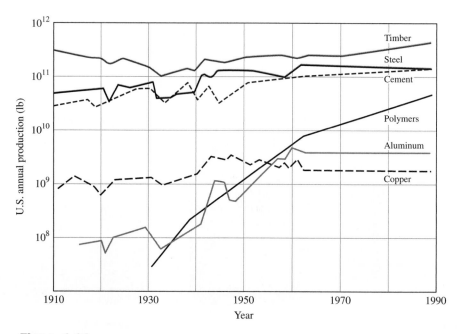

Figure 1.14
Competition of six major materials produced in the United States on a weight (pound) basis. The rapid rise in the production of aluminum and polymers (plastics) is evident.

5 to 10 percent aluminum. Thus in the period 1978–1985 the percentage of steel declined, that of polymers increased, and that of aluminum remained about the same. In 1997 the domestic U.S. auto weighed an average of 3248 lb (1476 kg), and plastics comprised about 7.4 percent of it (Fig. 1.15). The trend in the usage of materials appears to be more aluminum and steel in autos and less *cast* iron. The amount of plastics (by percentage) in autos appears to be about the same (Fig. 1.16).

For some applications only certain materials can meet the engineering requirements for a design, and these materials may be relatively expensive. For example, the modern jet engine (Fig. 1.5) requires high-temperature nickel-base superalloys to function. These materials are expensive, and no cheap substitute has been found to replace them. Thus, although cost is an important factor in engineering design, the materials used must also meet performance specifications. Replacement of one material by another will continue in the future since new materials are being discovered and new processes are being developed.

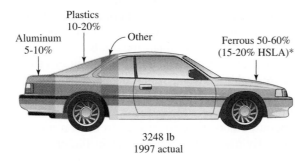

Figure 1.15
Breakdown of weight percentages of major materials used in the average 1985 U.S. automobile.
HSLA—High Strength Low Alloy Steel

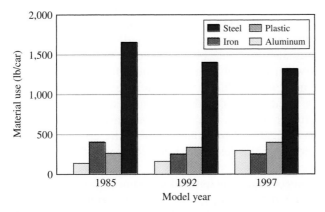

Figure 1.16
Predictions and use of materials in U.S. automobile.
(After J.G. Simon, Adv. Mat. & Proc., 133:63(1988) and new data.)

1.5 RECENT ADVANCES IN MATERIALS SCIENCE AND TECHNOLOGY AND FUTURE TRENDS

In recent decades, a number of exciting initiatives in materials science have been undertaken that could potentially revolutionize the future of the field. Smart materials and devices at micrometer size scale and nanomaterials are two classes of materials that will critically affect all major industries.

1.5.1 Smart Materials

Some smart materials have been around for years but are finding more applications. They have the ability to sense external environmental stimuli (temperature, stress, light, humidity, and electric and magnetic fields) and respond to them by changing their properties (mechanical, electrical, or appearance), structure, or functions. These materials are generically called **smart materials.** Smart materials or systems that use smart materials consist of sensors and actuators. The sensory component detects a change in the environment, and the actuator component performs a specific function or a response. For instance, some smart materials change or produce color when exposed to changes in temperature, light intensity, or an electric current.

Some of the more technologically important smart materials that can function as actuators are **shape-memory alloys** and **piezoelectric** ceramics. Shape-memory alloys are metal alloys that, once strained, revert back to their original shape upon an increase in temperature above a critical transformation temperature. The change in shape back to the original is due to a change in the crystal structure above the transformation temperature. One biomedical application of a shape-memory alloys is as a stent for supporting weakened artery walls or for expanding narrowed arteries (Fig. 1.17). The deformed stent is first delivered in the appropriate position in the artery using a probe Fig. 1.17a. The stent expands to its original shape and size after increasing its temperature to body temperature Fig. 1.17b. For comparison, the conventional method of expanding or supporting an artery is through the use of a stainless steel tube that is expanded using a balloon. Examples of shape-memory alloys are nickel-titanium and copper-zinc-aluminum alloys.

Actuators may also be made of piezoelectric materials. The materials produce an electric field when exposed to a mechanical force. Conversely, a change in an external electric field will produce a mechanical response in the same material. Such materials may be used to sense and reduce undesirable vibrations of a component through their actuator response. Once a vibration is detected, a current is applied to produce a mechanical response that counters the effect of the vibration.

Now let us consider the design and development of systems at micrometer size scale that use smart materials and devices to sense, communicate, and actuate: such is the world of **microelectromechanical systems** (MEMs). Originally, MEMs were devices that integrated technology, electronic materials, and smart materials on a semiconductor chip to produce what were commonly known as a **micromachines.** The original MEMs device had the microscopic mechanical elements fabricated on silicon chips using integrated circuits technology; MEMs were used as sensors or

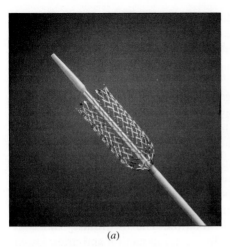
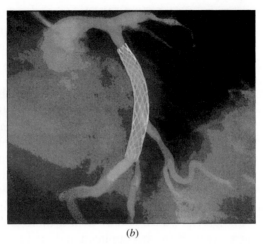

(a)   (b)

Figure 1.17
Shape-memory alloys used as a stent to expand narrowed arteries or support weakened
ones: (a) stent on a probe and (b) stent positioned in a damaged artery for support.
(Source: *http://www.designinsite.dk/htmsider/inspmat.htm.*)
Courtesy of Nitinol Devices & Components © Sovereign/Phototake NYC

actuators. But today the term "MEMs" is extended to any miniaturized device. The
applications of MEMs are also numerous, including but not limited to micropumps,
locking systems, motors, mirrors, and sensors. For instance, MEMs are used in automobile airbags to sense both the deceleration and the size of the person sitting in
the car and to deploy the airbag at a proper speed.

1.5.2 Nanomaterials

Nanomaterials are generally defined as those materials that have a characteristic length
scale (that is, particle diameter, grain size, layer thickness, etc.) smaller than 100 nm
($1 \text{ nm} = 10^{-9}$ m). Nanomaterials can be metallic, polymeric, ceramic, electronic, or
composite. In this respect, ceramic powder aggregates of less than 100 nm in size, bulk
metals with grain size less than 100 nm, thin polymeric films with thickness less than
100 nm, and electronic wires with diameter less than 100 nm are all considered nanomaterials or nanostructured materials. At the nanoscale, the properties of the material are
neither that of the molecular or atomic level nor that of the bulk material. Although
a tremendous amount of research and development activities has been devoted to this
topic in the past decade, early research on nanomaterials dates back to the 1960s
when chemical flame furnaces were used to produce particles smaller than one
micron ($1 \text{ micron} = 10^{-6} \text{ m} = 10^3$ nm) in size. The early applications of nanomaterials were as chemical catalysts and pigments. Metallurgists have always been
aware that by refining the grain structure of a metal to ultrafine (submicron) levels,
its strength and hardness increases significantly in comparison to the coarse-grained
(micron-size) bulk metal. For example, nanostructured pure copper has a yield
strength six times that of coarse-grained copper.

The reasons for the recent extraordinary attention to these materials may be due to the development of (1) new tools that make the observation and characterization of these materials possible and (2) new methods of processing and synthesizing nanostructured materials that enable researchers to produce these materials more easily and at a higher yield rate.

The future applications of nanomaterials are only limited to the imagination, and one of the major obstacle in fulfilling this potential is the ability to efficiently and inexpensively produce these materials. Consider the manufacturing of orthopaedic and dental implants from nanomaterials with better biocompatibility characteristics, better strength, and better wear characteristics than metals. One such material is nanocrystalline zirconia (zirconium oxide), a hard and wear resistant ceramic that is chemically stable and biocompatible. This material can be processed in a porous form, and when it is used as implant material, it allows for bone to grow into its pores, resulting in a more stable fixation. The metal alloys that are currently used for this application do not allow for such interaction and often loosen over time, requiring further surgery. Nanomaterials may also be used in producing paint or coating materials that are significantly more resistant to scratching and environmental damage. Also, electronic devices such as transistors, diodes and even lasers may be developed on a nanowire. Such materials science advancements will have both technological and economical impact on all areas of engineering and industries.

Welcome to the fascinating and exceedingly interesting world of materials science and engineering!

1.6 DESIGN AND SELECTION

Material engineers should be knowledgeable of various classes of materials, their properties, structure, manufacturing processes involved, environmental issues, economic issues, and more. As the complexity of the component under consideration increases, the complexity of the analysis and the factors involved in the materials selection process also increase. Consider the materials selection issues for the frame and forks of a bicycle. The selected material must be strong enough to support the load without yielding (permanent deformation) or fracture. The chosen material must be stiff to resist excessive elastic deformation and fatigue failure (due to repeated loading). The corrosion resistance of the material may be a consideration over the life of the bicycle. Also, the weight of frame is important if the bicycle is used for racing: It must be lightweight. What materials will satisfy all of the above requirements? A proper materials selection process must consider the issues of strength, stiffness, weight, and shape of the component (shape factor) and utilize materials selection charts in order to determine the most suitable material for the application. The detailed selection process is outside the scope of this textbook, but we use this example as an exercise in identifying various material candidates for this application. It turns out that a number of materials may satisfy the strength, stiffness, and weight considerations including some aluminum alloys, titanium alloys, magnesium alloys, steel, carbon fiber reinforced plastic (CFRP), and even wood. Wood has excellent properties for our application but it cannot be easily shaped to from a frame and the forks. Further analysis shows CFRP is the best choice; it offers a strong, stiff, and lightweight frame that is both fatigue and corrosion resistant. However, the fabrication process is costly. Therefore, if cost is an issue, this material may not be the most suitable choice. The remaining materials, all metal alloys, are all suitable and comparatively easy to manufacture into the desired shape. If cost is a major issue, steel emerges as the most suitable choice. On the other hand, if lower bicycle weight is important, the aluminum alloy emerge as the most suitable material. Titanium and magnesium alloys are more expensive than both aluminum and steel alloys and are lighter than steel; they, however, do not offer significant advantages over aluminum.

1.7 SUMMARY

Materials science and materials engineering (collectively, materials science and engineering) form a bridge of materials knowledge between the basic sciences (and mathematics) and the engineering disciplines. Materials science is concerned primarily with the search for basic knowledge about materials, whereas materials engineering is concerned mainly with using applied knowledge about materials.

The three main types of materials are metallic, polymeric, and ceramic materials. Two other types of materials that are very important for modern engineering technology are composite and electronic materials. All these types of materials will be dealt with in this book. Smart materials and devices in micrometer size scale and nanomaterials are presented as new classes of materials with novel and important applications in many industries.

Materials compete with each other for existing and new markets, and so the replacement of one material by another for some applications occurs. The availability of raw materials, cost of manufacturing, and the development of new materials and processes for products are major factors that cause changes in materials usage.

1.8 DEFINITIONS

Sec. 1.1

Materials: substances of which something is composed or made. The term *engineering materials* is sometimes used to refer specifically to materials used to produce technical products. However, there is no clear demarcation line between the two terms, and they are used interchangeably.

Sec. 1.2

Materials science: a scientific discipline that is primarily concerned with the search for basic knowledge about the internal structure, properties, and processing of materials.

Materials engineering: an engineering discipline that is primarily concerned with the use of fundamental and applied knowledge of materials so that they can be converted into products needed or desired by society.

Metallic materials (metals and metal alloys): inorganic materials that are characterized by high thermal and electrical conductivities. Examples are iron, steel, aluminum, and copper.

Sec. 1.3

Ferrous metals and alloys: metals and alloys that contain a large percentage of iron such as steels and cast irons.

Nonferrous metals and alloys: metals and alloys that do not contain iron, or if they do contain iron, it is only in a relatively small percentage. Examples of nonferrous metals are aluminum, copper, zinc, titanium, and nickel.

Superalloys: metal alloys with improved performance at elevated temperatures and high stress levels.

Ceramic materials: materials consisting of compounds of metals and nonmetals. Ceramic materials are usually hard and brittle. Examples are clay products, glass, and pure aluminum oxide that has been compacted and densified.

Polymeric materials: materials consisting of long molecular chains or networks of low-weight elements such as carbon, hydrogen, oxygen, and nitrogen. Most polymeric materials have low electrical conductivities. Examples are polyethylene and *polyvinyl chloride* (PVC).

Composite materials: materials that are mixtures of two or more materials. Examples are fiberglass-reinforcing material in a polyester or epoxy matrix.

Electronic materials: materials used in electronics, especially microelectronics. Examples are silicon and gallium arsenide.

Blends: mixture of two or more polymers, also called polymer alloys.

Advanced Ceramics: new generation of ceramics with improved strength, corrosion resistance, and thermal shock properties, also called engineering or structural ceramics.

Sec. 1.5

Smart materials: materials with the ability to sense and respond to external stimuli.

Shape-memory alloys: materials that can be deformed but return to their original shape upon an increase in temperature.

Piezoelectric ceramics: materials that produce an electric field when subjected to mechanical force (and vice versa).

Microelectromechanical systems (MEMs): any miniaturized device that performs sensing and/or actuating function.

Micromachine: MEMs that perform a specific function or task.

Nanomaterials: materials with a characteristic length scale smaller than 100 nm.

1.9 PROBLEMS

1.1   What are materials? List eight commonly encountered engineering materials.

1.2   Define materials science and materials engineering.

1.3   What are the main classes of engineering materials?

1.4   What are some of the important properties of each of these engineering materials?

1.5   Define a composite material. Give an example of a composite material.

1.6   List some materials usage changes that you have observed over a period of time in some manufactured products. What reasons can you give for the changes that have occurred?

1.7   What factors might cause materials usage predictions to be incorrect?

1.10 MATERIALS SELECTION AND DESIGN PROBLEMS

1. Consider the common household component in a lightbulb: (a) identify various critical components of this item, (b) determine the material selected for each critical component, and (c) design a process that would be used to assemble the lightbulb.

2. *Total hip arthroplasty* (THA) is the process of total replacement of a damaged hip with an artificial prosthesis. (a) Identify the replacement components used in THA. (b) Identify the material(s) used in the manufacture of each component and why they are used. (c) What are some of the factors that the materials engineer should consider in selecting these materials?

3. Transistors are considered to have caused a revolution in electronics and consequently in many other industries. (a) Identify the critical components of a junction transistor. (b) Identify the material used in the manufacture of each component.

4. (a) Name the important factors in selecting materials for the frame of a mountain bike. (b) Steel, aluminum, and titanium alloys have all been used as the primary metals in the structure of a bicycle; determine the major weaknesses and strengths of each. (c) The more modern bikes are made of advanced composites. Explain why and name a specific composite used in the structure of a bike.

5. (a) Name the important criteria for selecting materials to use in a protective sports helmet. (b) Identify materials that would satisfy these criteria. (c) Why would a solid metal helmet not be a good choice?

6. (a) Determine the needed properties of the material(s) used as the heat shield in the structure of a space shuttle. (b) Identify materials that would satisfy these requirements. (c) Why would titanium alloys not be a good choice for this application?

7. (a) What kind of material is OFHC copper? (b) What are the desirable properties of OFHC copper? (c) What are the applications of OFHC copper in the power industry?

8. (a) To which class of materials does PTFE belong? (b) What are its desirable properties? (c) What are its applications in cookware manufacturing industries?

9. (a) To which class of materials does *cubic boron nitride* (cBN) belong? (b) What are its desirable properties? (c) What are its applications in the metal machining industries?

10. (a) What are aramids? (b) What are their desirable properties? (c) What are their applications in sports equipment industries?

11. (a) In what class of materials does GaAs (gallium arsenide) belong? (b) What are its desirable properties? (c) What are its applications in electronic industries?

12. Nickel-base superalloys are used in the structure of aircraft turbine engines. What are the major properties of this metal that make it suitable for this application?

13. Identify various sports equipment that could benefit from smart materials or MEMs technology. Give specific reasons for the suitability of the application.

14. What are nanotubes? Give some examples of their application to structural materials such as composites.

Atomic Structure and Bonding

(© Tom Pantages adaptation/courtesy Prof. J. Spence)

Atomic orbitals represent the statistical likelihood that electrons will occupy various points in space. Except for the innermost electrons of the atoms, the shapes of the orbitals are nonspherical. Until recently, we have only been able to imagine the existence and shape of these orbitals because no experimental verifications were available. Recently, scientists have been able to create a three-dimensional image of these orbitals using a combination of x-ray diffraction and electron microscopy techniques. The chapter-opening image shows the orbital of d state electrons of the copper-oxygen bond in Cu_2O. Through an understanding of the bonding in copper oxides, using the techniques just described, researchers edge closer to explaining the nature of high-temperature superconductivity in copper oxides.[1] ■

---

[1]www.aip.org/physnews/graphics/html/orbital.html

By the end of this chapter, students will be able to . . .

1. Describe the nature and structure of an atom as well as its electronic structure.

2. Describe various types of primary bonds including ionic, covalent and metallic.

3. Describe covalent bonding by carbon.

4. Describe various types of secondary bonds and differentiate between these and primary bonds.

5. Describe the effect of bond type and strength on the mechanical and electrical performance of various classes of materials.

6. Describe mixed bonding in materials.

2.1 THE STRUCTURE OF ATOMS

Let us now review some of the fundamentals of atomic structure since **atoms** are the basic structural unit of all engineering materials.

Animation

Atoms consist primarily of three basic subatomic particles: *protons, neutrons,* and *electrons.* The current simple model of an atom envisions a very small nucleus of about 10^{-14} m in diameter surrounded by a relatively thinly dispersed electron cloud of varying density so that the diameter of the atom is of the order of 10^{-10} m. The nucleus accounts for almost all the mass of the atom and contains the protons and neutrons. A proton has a mass of 1.673×10^{-24} g and a unit charge of $+1.602 \times 10^{-19}$ coulombs (C). The neutron is slightly heavier than the proton and has a mass of 1.675×10^{-24} g but no charge. The electron has a relatively small mass of 9.109×10^{-28} g ($\frac{1}{1836}$ that of the proton) and a unit charge of -1.602×10^{-19} C (equal in charge but opposite in sign from the proton). Table 2.1 summarizes these properties of subatomic particles.

The electron charge cloud thus constitutes almost all the volume of the atom but accounts for only a very small part of its mass. The electrons, particularly the outer ones, determine most of the electrical, mechanical, chemical, and thermal properties of the atoms, and thus a basic knowledge of atomic structure is important in the study of engineering materials.

Table 2.1 The mass and charge of the proton, neutron, and electron

	Mass (g)	Charge (C)
Proton	1.673×10^{-24}	$+1.602 \times 10^{-19}$
Neutron	1.675×10^{-24}	0
Electron	9.109×10^{-28}	-1.602×10^{-19}

2.2 ATOMIC NUMBERS AND ATOMIC MASSES

2.2.1 Atomic Numbers

Quiz

The **atomic number** of an atom indicates the number of protons (positively charged particles) that are in its nucleus, and in a neutral atom the atomic number is also equal to the number of electrons in its charge cloud. Each element has its own characteristic atomic number, and thus the atomic number identifies the element. Atomic numbers of the elements from hydrogen, which has an atomic number of 1, to hahnium, which has an atomic number of 105, are located above the atomic symbols of the elements in the periodic table of Fig. 2.1.

2.2.2 Atomic Masses

The *relative atomic mass* of an element is the mass in grams of 6.023×10^{23} atoms (**Avogadro's number** N_A) of that element. The relative atomic masses of the elements from 1 to 105 are located below the atomic symbols in the periodic table of the elements (Fig. 2.1). The carbon atom with 6 protons and 6 neutrons is the carbon-12 atom and is the reference mass for atomic masses. One **atomic mass unit (u)** is defined as exactly one-twelfth of the mass of a carbon atom, which has a mass of 12 u. One molar relative atomic mass of carbon 12 has a mass of 12 g on this scale. *One gram-mole* or *mole* (abbreviated mol) of an element is defined as having the mass in grams of the relative molar atomic mass of that element. Thus, for example, 1 gram-mole of aluminum has a mass of 26.98 g and contains 6.023×10^{23} atoms.

EXAMPLE PROBLEM 2.1

a.   What is the mass in grams of one atom of copper?
b.   How many copper atoms are in 1 g of copper?

■ Solution

a.   The atomic mass of copper is 63.54 g/mol. Since in 63.54 g of copper there are 6.02×10^{23} atoms, the number of grams in one atom of copper is

$$\frac{63.54 \text{ g/mol Cu}}{6.02 \times 10^{23} \text{ atoms/mol}} = \frac{x \text{ g Cu}}{1 \text{ atom}}$$

$$x = \text{mass of 1 Cu atom}$$

$$= \frac{63.54 \text{ g/mol}}{6.02 \times 10^{23} \text{ atoms/mol}} \times 1 \text{ atom} = 1.05 \times 10^{-22} \text{ g} \blacktriangleleft$$

b.   The number of copper atoms in 1 g of copper is

$$\frac{6.02 \times 10^{23} \text{ atoms/mol}}{63.54 \text{ g/mol Cu}} = \frac{x \text{ atoms of Cu}}{1 \text{ g Cu}}$$

$$x = \text{no. of Cu atoms} = \frac{(6.02 \times 10^{23} \text{ atoms/mol})(1 \text{ g Cu})}{63.54 \text{ g/mol Cu}}$$

$$= 9.47 \times 10^{21} \text{ atoms!} \blacktriangleleft$$

Periodic Table of the Elements

(Source: *Davis, M., and Davis, R.,* Fundamentals of Chemical Reaction Engineering, *McGraw-Hill, 2003.*)

MAIN–GROUP ELEMENTS

Legend:
- Metals (main-group)
- Metals (transition)
- Metals (inner transition)
- Metalloids
- Nonmetals

Period	IA (1)	IIA (2)	IIIB (3)	IVB (4)	VB (5)	VIB (6)	VIIB (7)	VIIIB (8)	VIIIB (9)	VIIIB (10)	IB (11)	IIB (12)	IIIA (13)	IVA (14)	VA (15)	VIA (16)	VIIA (17)	VIIIA (18)
1	1 H 1.008																	2 He 4.003
2	3 Li 6.941	4 Be 9.012											5 B 10.81	6 C 12.01	7 N 14.01	8 O 16.00	9 F 19.00	10 Ne 20.18
3	11 Na 22.99	12 Mg 24.31											13 Al 26.98	14 Si 28.09	15 P 30.97	16 S 32.07	17 Cl 35.45	18 Ar 39.95
4	19 K 39.10	20 Ca 40.08	21 Sc 44.96	22 Ti 47.88	23 V 50.94	24 Cr 52.00	25 Mn 54.94	26 Fe 55.85	27 Co 58.93	28 Ni 58.69	29 Cu 63.55	30 Zn 65.39	31 Ga 69.72	32 Ge 72.61	33 As 74.92	34 Se 78.96	35 Br 79.90	36 Kr 83.80
5	37 Rb 85.47	38 Sr 87.62	39 Y 88.91	40 Zr 91.22	41 Nb 92.91	42 Mo 95.94	43 Tc (98)	44 Ru 101.1	45 Rh 102.9	46 Pd 106.4	47 Ag 107.9	48 Cd 112.4	49 In 114.8	50 Sn 118.7	51 Sb 121.8	52 Te 127.6	53 I 126.9	54 Xe 131.3
6	55 Cs 132.9	56 Ba 137.3	57 La 138.9	72 Hf 178.5	73 Ta 180.9	74 W 183.9	75 Re 186.2	76 Os 190.2	77 Ir 192.2	78 Pt 195.1	79 Au 197.0	80 Hg 200.6	81 Tl 204.4	82 Pb 207.2	83 Bi 209.0	84 Po (209)	85 At (210)	86 Rn (222)
7	87 Fr (223)	88 Ra (226)	89 Ac (227)	104 Rf (261)	105 Db (262)	106 Sg (266)	107 Bh (262)	108 Hs (265)	109 Mt (266)	110 Uun (269)	111 Uuu (272)	112 Uub (277)	113	114 Uuq (285)	115	116 Uuh (289)	117	118 Uuo

— TRANSITION ELEMENTS —

INNER TRANSITION ELEMENTS

Lanthanides (6)	58 Ce 140.1	59 Pr 140.9	60 Nd 144.2	61 Pm (145)	62 Sm 150.4	63 Eu 152.0	64 Gd 157.3	65 Tb 158.9	66 Dy 162.5	67 Ho 164.9	68 Er 167.3	69 Tm 168.9	70 Yb 173.0	71 Lu 175.0
Actinides (7)	90 Th 232.0	91 Pa (231)	92 U 238.0	93 Np (237)	94 Pu (242)	95 Am (243)	96 Cm (247)	97 Bk (247)	98 Cf (251)	99 Es (252)	100 Fm (257)	101 Md (258)	102 No (259)	103 Lr (260)

Figure 2.1

The periodic table of the elements.

EXAMPLE PROBLEM 2.2

The cladding (outside layers) of the U.S. quarter coin consists of an alloy[2] of 75 wt % copper and 25 wt % nickel. What are the atomic percent Cu and atomic percent Ni contents of this material?

■ **Solution**

Using a basis of 100 g of the 75 wt % Cu and 25 wt % Ni alloy, there are 75 g Cu and 25 g Ni. Thus, the number of gram-moles of copper and nickel is

$$\text{No. of gram-moles of Cu} = \frac{75 \text{ g}}{63.54 \text{ g/mol}} = 1.1803 \text{ mol}$$

$$\text{No. of gram-moles of Ni} = \frac{25 \text{ g}}{58.69 \text{ g/mol}} = \underline{0.4260 \text{ mol}}$$

$$\text{Total gram-moles} = 1.6063 \text{ mol}$$

Thus, the atomic percentages of Cu and Ni are

$$\text{Atomic \% Cu} = \left(\frac{1.1803 \text{ mol}}{1.6063 \text{ mol}}\right)(100\%) = 73.5 \text{ at \%} \blacktriangleleft$$

$$\text{Atomic \% Ni} = \left(\frac{0.4260 \text{ mol}}{1.6063 \text{ mol}}\right)(100\%) = 26.5 \text{ at \%} \blacktriangleleft$$

---
[2]An alloy is a combination of two or more metals or a metal (metals) and a nonmetal (nonmetals).

EXAMPLE PROBLEM 2.3

An intermetallic compound has the general chemical formula $Ni_x Al_y$, where x and y are simple integers, and consists of 42.04 wt % nickel and 57.96 wt % aluminum. What is the simplest formula of this nickel aluminide?

■ **Solution**

We first determine the gram-mole fractions of nickel and aluminum in this compound. Using a basis of 100 g of the compound, we have 42.04 g of Ni and 57.96 g of Al. Thus,

$$\text{No. of gram-moles of Ni} = \frac{42.04 \text{ g}}{58.71 \text{ g/mol}} = 0.7160 \text{ mol}$$

$$\text{No. of gram-moles of Al} = \frac{57.96 \text{ g}}{26.98 \text{ g/mol}} = \underline{2.1483 \text{ mol}}$$

$$\text{Total gram-moles} = 2.8643 \text{ mol}$$

Thus,

$$\text{Gram-mole fraction of Ni} = \frac{0.7160 \text{ mol}}{2.8643 \text{ mol}} = 0.25$$

$$\text{Gram-mole fraction of Al} = \frac{2.1483 \text{ mol}}{2.8643 \text{ mol}} = 0.75$$

Next, we replace the x and y in the $Ni_x Al_y$ compound with 0.25 and 0.75, respectively, to give $Ni_{0.25} Al_{0.75}$, which is the simplest chemical formula on a mole-fraction basis. The simplest chemical formula on an integral basis is obtained by multiplying both the 0.25 and 0.75 by 4 to give $NiAl_3$ for the simplest chemical formula of this nickel aluminide.

2.3 THE ELECTRONIC STRUCTURE OF ATOMS

2.3.1 The Hydrogen Atom

The hydrogen atom is the simplest atom and consists of one electron surrounding a nucleus of one proton. If we consider the motion of the hydrogen electron around its nucleus, only certain definite energy levels (orbitals) are allowed. It is important to note that the term atomic orbital does not represent the orbit of an electron around the nucleus; it simply represents the energy state of the electron. The reason for the restricted energy values is that electrons obey the laws of **quantum mechanics,** which only allow certain energy values and not any arbitrary ones. Thus, if the hydrogen electron is excited to a higher energy level, energy is absorbed by a discrete amount (Fig. 2.2a). Similarly, if the electron drops to a lower energy level, energy is emitted by a discrete amount (Fig. 2.2b).

Animation

During the transition to a lower energy, the hydrogen electron will emit a discrete amount (quantum) of energy in the form of electromagnetic radiation called a **photon.** The energy change ΔE associated with the transition of the electron from

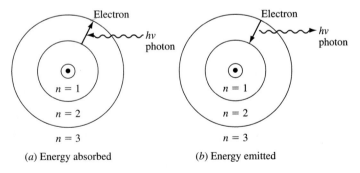

(a) Energy absorbed   (b) Energy emitted

Figure 2.2
(a) The hydrogen electron being excited into a higher orbit.
(b) An electron in a higher energy orbit dropping to a lower orbit, resulting in the emission of a photon of energy $h\nu$.
(note that the rings shown around the nucleus are not strictly indicative of electron position in space but rather energy level.)

one level to another is related to the frequency ν (nu) of the photon by Planck's[3] equation

$$\Delta E = h\nu \tag{2.1}$$

where h = Planck's constant = 6.63×10^{-34} joule-second (J · s). Since for electromagnetic radiation $c = \lambda\nu$, where c is the velocity of light equal to 3.00×10^8 meters/second (m/s) and λ (lambda) is its wavelength, the energy change ΔE associated with a photon can be expressed as

$$\Delta E = \frac{hc}{\lambda} \tag{2.2}$$

EXAMPLE PROBLEM 2.4

Calculate the energy in joules (J) and electron volts (eV) of the photon whose wavelength λ is 121.6 nanometers (nm). (1.00 eV = 1.60×10^{-19} J; $h = 6.63 \times 10^{-34}$ J · s; 1 nm = 10^{-9} m.)

■ **Solution**

$$\Delta E = \frac{hc}{\lambda} \tag{2.2}$$

$$\Delta E = \frac{(6.63 \times 10^{-34}\ \text{J} \cdot \text{s})(3.00 \times 10^8\ \text{m/s})}{(121.6\ \text{nm})(10^{-9}\ \text{m/nm})}$$

$$= 1.63 \times 10^{-18}\ \text{J} \blacktriangleleft$$

$$= 1.63 \times 10^{-18}\ \text{J} \times \frac{1\ \text{eV}}{1.60 \times 10^{-19}\ \text{J}} = 10.2\ \text{eV} \blacktriangleleft$$

Experimental verification of the energies associated with electrons being excited to discrete higher energy levels or losing energy and dropping to lower discrete levels is obtained mainly by the analyses of the wavelengths and intensities of spectral lines. Using hydrogen spectral data, Niels Bohr[4] in 1913 developed a model for the hydrogen atom that consisted of a single electron orbiting a proton at a fixed radius (Fig. 2.3). A good approximation of the energy of the hydrogen electron for allowed energy levels is the Bohr equation

$$E = -\frac{2\pi^2 m e^4}{n^2 h^2} = -\frac{13.6}{n^2}\ \text{eV} \qquad (n = 1, 2, 3, 4, 5, \ldots) \tag{2.3}$$

where e = electron charge

m = electron mass

n = an integer called the *principal quantum number*

---

[3]Max Ernst Planck (1858–1947). German physicist who received the Nobel Prize in Physics in 1918 for his quantum theory.

[4]Niels Henrik Bohr (1885–1962). Danish physicist who was one of the founders of modern physics. In 1922 he received the Nobel Prize in Physics for his theory of the hydrogen atom.

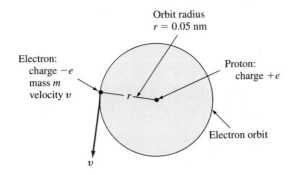

Orbit radius
r = 0.05 nm

Electron:
charge −e
mass m
velocity v

Proton:
charge +e

r

Electron orbit

v

Figure 2.3
The Bohr hydrogen atom. In the Bohr hydrogen
atom, an electron revolves in a circular orbit of
0.05 nm radius around a central proton.

Quiz

A hydrogen atom exists with its electron in the $n = 3$ state. The electron undergoes a transition to the $n = 2$ state. Calculate (a) the energy of the photon emitted, (b) its frequency, and (c) its wavelength.

**EXAMPLE
PROBLEM 2.5**

■ **Solution**
a.   Energy of the photon emitted is

$$E = \frac{-13.6 \text{ eV}}{n^2} \qquad (2.3)$$

$$\Delta E = E_3 - E_2$$

$$= \frac{-13.6}{3^2} - \frac{-13.6}{2^2} = 1.89 \text{ eV} \blacktriangleleft$$

$$= 1.89 \text{ eV} \times \frac{1.60 \times 10^{-19} \text{ J}}{\text{eV}} = 3.02 \times 10^{-19} \text{ J} \blacktriangleleft$$

b.   The frequency of the photon is

$$\Delta E = h\nu \qquad (2.1)$$

$$\nu = \frac{\Delta E}{h} = \frac{3.02 \times 10^{-19} \text{ J}}{6.63 \times 10^{-34} \text{ J} \cdot \text{s}}$$

$$= 4.55 \times 10^{14} \text{ s}^{-1} = 4.55 \times 10^{14} \text{ Hz} \blacktriangleleft$$

c.   The wavelength of the photon is

$$\Delta E = \frac{hc}{\lambda} \qquad (2.2)$$

or

$$\lambda = \frac{hc}{\Delta E} = \frac{(6.63 \times 10^{-34} \text{ J} \cdot \text{s})(3.00 \times 10^8 \text{ m/s})}{3.02 \times 10^{-19} \text{ J}}$$

$$= 6.59 \times 10^{-7} \text{ m}$$

$$= 6.59 \times 10^{-7} \text{ m} \times \frac{1 \text{ nm}}{10^{-9} \text{ m}} = 659 \text{ nm} \blacktriangleleft$$

d.   Energy is emitted by this transition.

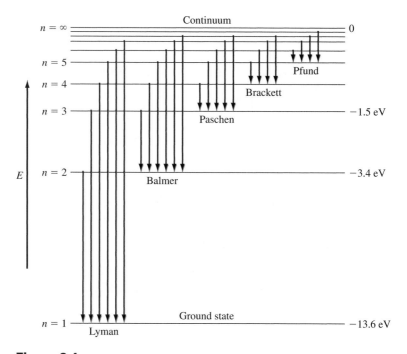

Figure 2.4

Energy-level diagram for the line spectrum of hydrogen.

(After F. M. Miller, "Chemistry: Structure and Dynamics," McGraw-Hill, 1984, p. 141.)

Quiz
Animation

In modern atomic theory, the n in Bohr's equation is designated the *principal quantum number* and represents the principal energy levels for electrons in atoms. From Bohr's equation (2.3) the energy level of the hydrogen electron in its **ground state** is -13.6 eV and corresponds to the line where $n = 1$ on the hydrogen energy-level diagram (Fig. 2.4). When the hydrogen electron is excited to higher energy levels, its energy is raised but its numerical value is less. For example, when the hydrogen electron is excited to the second principal quantum level, its energy is -3.4 eV, and if the hydrogen electron is excited to the free state where $n = \infty$, the electron has a zero value of energy. The energy required to remove the electron completely from the hydrogen atom is 13.6 eV, which is the **ionization energy** of the hydrogen electron.

The motion of electrons in atoms is more complicated than that presented by the simple Bohr atomic model. Electrons can have noncircular orbits around the nucleus, and according to **Heisenberg's**[5] **uncertainty principle** the position and momentum (mass $\times$ velocity) of a small particle such as an electron cannot be determined simultaneously. Thus, the exact position of the electron at any instant

---

[5] Werner Karl Heisenberg (1901–1976). German physicist who was one of the founders of the modern quantum theory. In 1932 he received the Nobel Prize in Physics.

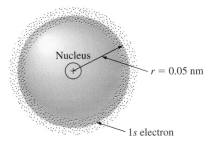

Figure 2.5
Electron charge cloud (schematic) surrounding the nucleus of a hydrogen atom in the ground state. The outer circle of $r = 0.05$ nm corresponds to the radius of the first Bohr orbit (i.e., for $n = 1$) and indicates the most probable region for finding the electron.

cannot be determined because the electron is so small a particle. Since the position of the hydrogen electron cannot be precisely determined, an electron charge cloud density distribution is sometimes used to represent the position of the hydrogen electron in its motion about its nucleus (Fig. 2.5). The highest electron density is at the nucleus. However, if the volume around the nucleus is divided into many concentric spherical layers, the possibility of finding electrons is highest at a radius of about .05 nm, which corresponds to the Bohr radius of the hydrogen atom.

2.3.2 Quantum Numbers of Electrons of Atoms

Modern atomic theory states that the motion of an electron about its nucleus and its energy is characterized by not just one **quantum number** but by four quantum numbers: the principal quantum number n, the subsidiary quantum number l, the magnetic quantum number m_l, and the electron spin quantum number m_s.

Quiz

The Principal Quantum Number n   The principal quantum number n corresponds to the n in the Bohr equation. It represents the main energy levels for the electron or shells. The higher the value of n, the higher the energy of the electron and the higher the possibility of the electron being farther away from the nucleus. The values for n are positive integers and range from 1 to 7.

The Subsidiary Quantum Number l   The second quantum number is the subsidiary quantum number l. This quantum number specifies subenergy levels within the main energy levels (subshell) where the probability of finding an electron is high if that energy level is occupied. The allowed values of l are $l = 0, 1, 2, 3, \ldots, n - 1$. The letters s, p, d, and f are used[6] to designate the l subenergy levels as follows:

$$\text{Number designation } l = 0 \quad 1 \quad 2 \quad 3$$
$$\text{Letter designation } l = s \quad p \quad d \quad f$$

The s, p, d, and f subenergy levels of an electron are termed **orbitals,** and so one speaks, for example, of an s or p subenergy level. The term *orbital* also refers to a subshell of an atom where the density of a particular electron or pair of electrons is high. Thus, one may speak of an s or p subshell of a particular atom.

The Magnetic Quantum Number m_l   The third quantum number, the magnetic quantum number m_l, specifies the spatial orientation of a single atomic orbital and has little effect on the energy of an electron. The number of different permissible orientations of an orbital depends on the l value of a particular orbital. The m_l quantum number has permissible values of $-l$ to $+l$, including zero. When $l = 0$, there is only one allowed value for m_l, which is zero. When $l = 1$, there are three permissible values for m_l, which are $-1, 0$, and $+1$. In general, there are $2l + 1$ allowed values for m_l. In terms of s, p, d, and f orbital notation, there is a maximum of one s orbital, three p orbitals, five d orbitals, and seven f orbitals for each allowed s, p, d, and f subenergy level.

Electron Spin Quantum Number m_s   The fourth quantum number, the electron spin quantum number m_s, specifies two allowed spin directions for an electron spinning on its own axis. The directions are clockwise and counterclockwise rotation, and their allowed values are $+\frac{1}{2}$ and $-\frac{1}{2}$. The spin quantum number has only a very minor effect on the energy of an electron. It should be pointed out that two electrons may occupy the same orbital, and if they do, they must have opposed spins.

Table 2.2 summarizes the allowed values for the four quantum numbers of electrons. According to the **Pauli**[7] **exclusion principle** of atomic theory, *no two electrons can have the same set of four quantum numbers.*

---

[6]The letters s, p, d, and f were adopted from the first letters of the old spectral-line intensity designation: *s*harp, *p*rincipal, *d*iffuse, and *f*undamental.

[7]Wolfgang Pauli (1900–1958). Austrian physicist who was one of the founders of the modern quantum theory. In 1945 he received the Nobel Prize in Physics.

Table 2.2 Allowed values for the quantum numbers of electrons

n	Principal quantum number	$n = 1, 2, 3, 4, \ldots$	All positive integers
l	Subsidiary quantum number	$l = 0, 1, 2, 3, \ldots, n - 1$	n allowed values of l
m_l	Magnetic quantum number	Integral values from $-l$ to $+l$, including zero	$2l + 1$
m_s	Spin quantum number	$+\frac{1}{2}, -\frac{1}{2}$	2

2.3.3 Electronic Structure of Multielectron Atoms

Maximum Number of Electrons for Each Principal Shell   Atoms consist of principal shells* of high electron densities as dictated by the laws of quantum mechanics. There are seven of these principal **electron shells** when the atomic number of the atom reaches 87 for the element francium (Fr). Each shell can only contain a maximum number of electrons, which is again dictated by the laws of quantum mechanics. The maximum number of electrons that can be contained in each shell in an atom is defined by different sets of the four quantum numbers (Pauli principle) and is $2n^2$, where n is the principal quantum number. Thus, there can be only a maximum of 2 electrons for the first principal shell, 8 for the second, 18 for the third, 32 for the fourth, etc., as indicated in Table 2.3.

Quiz

Atomic Size   Each atom can be considered to a first approximation as a sphere with a definite radius. The radius of the atomic sphere is not a constant but depends to some extent on its environment. Figure 2.6 shows the relative atomic sizes of many of the elements along with their atomic radii. Many of the values of the atomic radii are still not fully agreed upon and vary to some extent depending on the reference source.

Animation

From Fig. 2.6, some trends in atomic size are evident. In general, as the principal quantum number increases, the size of the atom increases. There are, however, a few exceptions where the atomic size actually gets smaller. The alkali elements of group 1A of the periodic table (Fig. 2.1) are a good example of atoms whose size increases with increasing n. For example, lithium ($n = 2$) has an atomic radius of 0.157 nm, whereas cesium ($n = 6$) has an atomic radius of 0.270 nm. In progressing across the period table from an alkali group 1A element to a noble gas of group 8A, the atomic size, in general, decreases. However, again there are some small exceptions. Atomic size will be important to us in the study of atomic diffusion in metal alloys.

Electron Configurations of the Elements   The **electron configuration** of an atom describes the way in which electrons are arranged in orbitals in an atom. Electron

---

*The term shell is not indicative of space but rather the energy level.

Table 2.3 Maximum number of electrons for each principal atomic shell

Shell number, n (principal quantum number)	Maximum number of electrons in each shell $(2n^2)$	Maximum number of electrons in orbitals
1	$2(1^2) = 2$	s^2
2	$2(2^2) = 8$	s^2p^6
3	$2(3^2) = 18$	$s^2p^6d^{10}$
4	$2(4^2) = 32$	$s^2p^6d^{10}f^{14}$
5	$2(5^2) = 50$	$s^2p^6d^{10}f^{14}\ldots$
6	$2(6^2) = 72$	$s^2p^6\ldots$
7	$2(7^2) = 98$	$s^2\ldots$

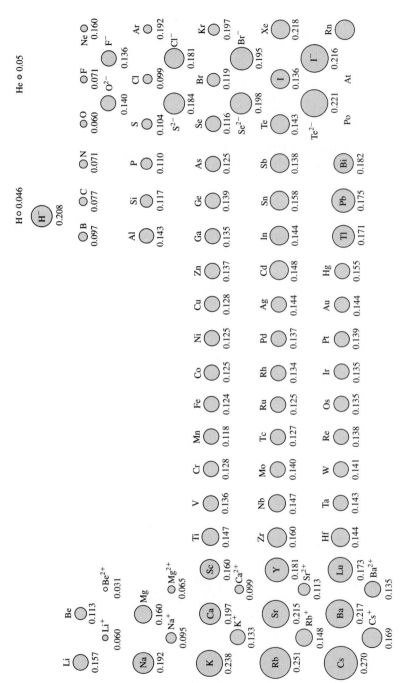

Figure 2.6

Relative sizes of some atoms and ions. Values are given in nanometers for the radii of the atoms and ions. Metallic radii are given for atoms where applicable.

(Adapted from F. M. Miller, "Chemistry: Structure and Dynamics," McGraw-Hill, 1984, p. 176.)

Animation

configurations are written in a conventional notation that lists the principal quantum number first, which is then followed by an orbital letter s, p, d, or f. A superscript above the orbital letter indicates the number of electrons it contains. The order by which the electrons fill up the orbitals is as follows:[8]

$$1s^2 2s^2 2p^6 3s^2 3p^6 4s^2 3d^{10} 4p^6 5s^2 4d^{10} 5p^6 6s^2 4f^{14} 5d^{10} 6p^6 7s^2 5f^{14} 6d^{10} 7p^6$$

The order for writing the orbitals for the electron configurations (for this book) will be by increasing principal quantum number, as

$$1s^2 2s^2 2p^6 3s^2 3p^6 3d^{10} 4s^2 4p^6 4d^{10} 4f^{14} 5s^2 5p^6 5d^{10} 5f^{14} 6s^2 6p^6 6d^{10} 7s^2$$

Write the electron configurations of the elements (*a*) iron, $Z = 26$ and (*b*) samarium, $Z = 62$.

EXAMPLE PROBLEM 2.6

■ **Solution**

Using the order of filling the orbitals listed previously, the following electron configurations for these elements are

Iron ($Z = 26$): $\qquad\qquad 1s^2 2s^2 2p^6 3s^2 3p^6 3d^6 4s^2$

Note the incomplete $3d$ orbitals. The electron arrangement of the $3d^6$ electrons in the iron atom will be important in the study of ferromagnetism.

Samarium ($Z = 62$): $\qquad 1s^2 2s^2 2p^6 3s^2 3p^6 3d^{10} 4s^2 4p^6 4d^{10} 4f^6 5s^2 5p^6 6s^2$

Note that for this rare-earth element the $4f$ orbitals are incomplete. The close similarity of chemical properties of the rare-earth elements is due to the filling in of the $4f$ orbitals, which are two shells below the outer shell containing two $6s$ electrons.

Table 2.4 lists the electron configurations of the elements as determined experimentally. It is noted that there are some irregularities inconsistent with the previously listed system. For example, copper ($Z = 29$) has the outer electron configuration of $3d^{10} 4s^1$. One would expect the outer configuration to be $3d^9 4s^2$. The reason for these irregularities is not precisely known.

Experimental evidence also shows that electrons with the same subsidiary quantum number have as many parallel spins as possible. Thus, if there are five electrons in d orbitals, there will be one electron in each of the d orbitals and the spin directions of all the electrons will be parallel, as shown in Fig. 2.7.

---

[8]A convenient memory device for this order is to list the orbitals as shown below and to use a series of arrows as drawn on the orbitals. By following the arrows from tail to head, the order is indicated.

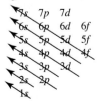

Table 2.4 Electron configurations of the elements

Z	Element	Electron configuration	Z	Element	Electron configuration	Z	Element	Electron configuration
1	H	$1s$	38	Sr	$[Kr]5s^2$	75	Re	$[Xe]4f^{14}5d^56s^2$
2	He	$1s^2$	39	Y	$[Kr]4d5s^2$	76	Os	$[Xe]4f^{14}5d^66s^2$
Quiz 3	Li	$[He]2s$	40	Zr	$[Kr]4d^25s^2$	77	Ir	$[Xe]4f^{14}5d^76s^2$
4	Be	$[He]2s^2$	41	Nb	$[Kr]4d^45s$	78	Pt	$[Xe]4f^{14}5d^96s$
5	B	$[He]2s^22p$	42	Mo	$[Kr]4d^55s$	79	Au	$[Xe]4f^{14}5d^{10}6s$
6	C	$[He]2s^22p^2$	43	Tc	$[Kr]4d^55s^2$	80	Hg	$[Xe]4f^{14}5d^{10}6s^2$
7	N	$[He]2s^22p^3$	44	Ru	$[Kr]4d^75s$	81	Tl	$[Xe]4f^{14}5d^{10}6s^26p$
8	O	$[He]2s^22p^4$	45	Rh	$[Kr]4d^85s$	82	Pb	$[Xe]4f^{14}5d^{10}6s^26p^2$
9	F	$[He]2s^22p^5$	46	Pd	$[Kr]4d^{10}$	83	Bi	$[Xe]4f^{14}5d^{10}6s^26p^3$
10	Ne	$[He]2s^22p^6$	47	Ag	$[Kr]4d^{10}5s$	84	Po	$[Xe]4f^{14}5d^{10}6s^26p^4$
11	Na	$[Ne]3s$	48	Cd	$[Kr]4d^{10}5s^2$	85	At	$[Xe]4f^{14}5d^{10}6s^26p^5$
12	Mg	$[Ne]3s^2$	49	In	$[Kr]4d^{10}5s^25p$	86	Rn	$[Xe]4f^{14}5d^{10}6s^26p^6$
13	Al	$[Ne]3s^23p$	50	Sn	$[Kr]4d^{10}5s^25p^2$	87	Fr	$[Rn]7s$
14	Si	$[Ne]3s^23p^2$	51	Sb	$[Kr]4d^{10}5s^25p^3$	88	Ra	$[Rn]7s^2$
15	P	$[Ne]3s^23p^3$	52	Te	$[Kr]4d^{10}5s^25p^4$	89	Ac	$[Rn]6d7s^2$
16	S	$[Ne]3s^23p^4$	53	I	$[Kr]4d^{10}5s^25p^5$	90	Th	$[Rn]6d^27s^2$
17	Cl	$[Ne]3s^23p^5$	54	Xe	$[Kr]4d^{10}5s^25p^6$	91	Pa	$[Rn]5f^26d7s^2$
18	Ar	$[Ne]3s^23p^6$	55	Cs	$[Xe]6s$	92	U	$[Rn]5f^36d7s^2$
19	K	$[Ar]4s$	56	Ba	$[Xe]6s^2$	93	Np	$[Rn]5f^46d7s^2$
20	Ca	$[Ar]4s^2$	57	La	$[Xe]5d6s^2$	94	Pu	$[Rn]5f^67s^2$
21	Sc	$[Ar]3d4s^2$	58	Ce	$[Xe]4f5d6s^2$	95	Am	$[Rn]5f^77s^2$
22	Ti	$[Ar]3d^24s^2$	59	Pr	$[Xe]4f^26s^2$	96	Cm	$[Rn]5f^76d7s^2$
23	V	$[Ar]3d^34s^2$	60	Nd	$[Xe]4f^46s^2$	97	Bk	$[Rn]5f^97s^2$
24	Cr	$[Ar]3d^54s$	61	Pm	$[Xe]4f^56s^2$	98	Cf	$[Rn]5f^{10}7s^2$
25	Mn	$[Ar]3d^54s^2$	62	Sm	$[Xe]4f^66s^2$	99	Es	$[Rn]5f^{11}7s^2$
26	Fe	$[Ar]3d^64s^2$	63	Eu	$[Xe]4f^76s^2$	100	Fm	$[Rn]5f^{12}7s^2$
27	Co	$[Ar]3d^74s^2$	64	Gd	$[Xe]4f^75d6s^2$	101	Md	$[Rn]5f^{13}7s^2$
28	Ni	$[Ar]3d^84s^2$	65	Tb	$[Xe]4f^96s^2$	102	No	$[Rn]5f^{14}7s^2$
29	Cu	$[Ar]3d^{10}4s$	66	Dy	$[Xe]4f^{10}6s^2$	103	Lr	$[Rn]5f^{14}6d7s^2$
30	Zn	$[Ar]3d^{10}4s^2$	67	Ho	$[Xe]4f^{11}6s^2$	104	Rf	$[Rn]5f^{14}6d^27s^2$
31	Ga	$[Ar]3d^{10}4s^24p$	68	Er	$[Xe]4f^{12}6s^2$	105	Db	$[Rn]5f^{14}6d^37s^2$
32	Ge	$[Ar]3d^{10}4s^24p^2$	69	Tm	$[Xe]4f^{13}6s^2$	106	Sg	$[Rn]5f^{14}6d^47s^2$
33	As	$[Ar]3d^{10}4s^24p^3$	70	Yb	$[Xe]4f^{14}6s^2$	107	Bh	$[Rn]5f^{14}6d^57s^2$
34	Se	$[Ar]3d^{10}4s^24p^4$	71	Lu	$[Xe]4f^{14}5d6s^2$	108	Hs	$[Rn]5f^{14}6d^67s^2$
35	Br	$[Ar]3d^{10}4s^24p^5$	72	Hf	$[Xe]4f^{14}5d^26s^2$	109	Mt	$[Rn]5f^{14}6d^77s^2$
36	Kr	$[Ar]3d^{10}4s^24p^6$	73	Ta	$[Xe]4f^{14}5d^36s^2$			
37	Rb	$[Kr]5s$	74	W	$[Xe]4f^{14}5d^46s^2$			

Handbook of Chemistry & Physics 76th ed. CRC Press 1995–1996.

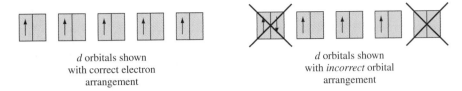

d orbitals shown
with correct electron
arrangement

d orbitals shown
with *incorrect* orbital
arrangement

Figure 2.7
Spin directions of unpaired electrons in *d* orbitals.

2.3.4 Electronic Structure and Chemical Reactivity

Noble Gases   The chemical properties of the atoms of the elements depend principally on the reactivity of their outermost electrons. The most stable and least reactive of all the elements are the noble gases. With the exception of helium, which has a $1s^2$ electron configuration, the outermost shell of all the other noble gases (Ne, Ar, Kr, Xe, and Rn) has an s^2p^6 electron configuration. The s^2p^6 configuration for the outermost shell has high chemical stability as indicated by the relative chemical inactivity of the noble gases.

Electropositive and Electronegative Elements   Electropositive elements are metallic in nature and give up electrons in chemical reactions to produce *positive ions,* or **cations.** The number of electrons given up by an electropositive atom of an element is indicated by a *positive oxidation number.* Figure 2.8 lists the oxidation numbers for the elements. Note that some elements have more than one oxidation number. The most electropositive elements are in groups 1A and 2A of the periodic table.

Electronegative elements are nonmetallic in nature and accept electrons in chemical reactions to produce *negative ions,* or **anions.** The number of electrons accepted by an electronegative atom of an element is indicated by a *negative oxidation number* (Fig. 2.8). The most electronegative elements are in groups 6A and 7A of the periodic table in Fig. 2.1.

Some elements in groups 4A to 7A of the periodic table can behave in either an electronegative or an electropositive manner. This dual behavior is shown by elements such as carbon, silicon, germanium, arsenic, antimony, and phosphorus. Thus, in some reactions they have positive oxidation numbers, where they show electropositive behavior, and in others they have negative oxidation numbers, where they show electronegative behavior.

Write the electron configuration for the Fe atom (Z = 26) and the Fe^{2+} and Fe^{3+} ions by using conventional *spdf* notation.

EXAMPLE PROBLEM 2.7

■ **Solution**

$$Fe: \quad 1s^2 2s^2 2p^6 3s^2 3p^6 3d^6 4s^2$$
$$Fe^{2+}: 1s^2 2s^2 2p^6 3s^2 3p^6 3d^6$$
$$Fe^{3+}: 1s^2 2s^2 2p^6 3s^2 3p^6 3d^5$$

Note that the outer two 4s electrons are lost first since they have the highest energy and are the easiest to remove.

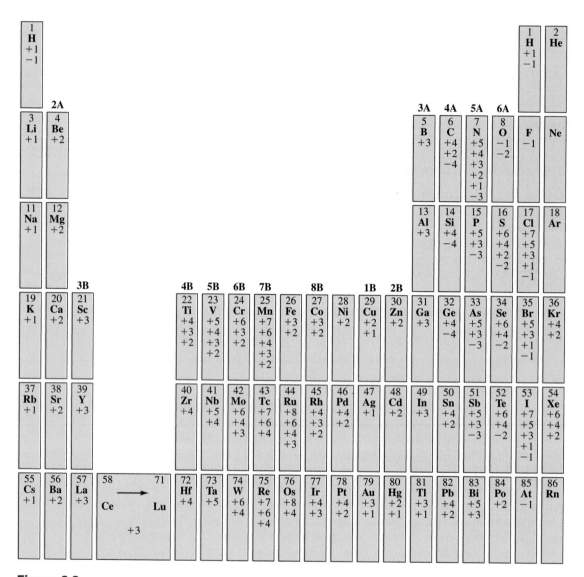

Figure 2.8
Oxidation numbers of the elements with respect to their positions in the periodic table.
(After R. E. Davis, K. D. Gailey, and K. W. Whitten, "Principles of Chemistry," CBS College Publishing, 1984, p. 299.)

Electronegativity   *Electronegativity* is defined as the degree to which an atom attracts electrons to itself. The comparative tendency of an atom to show electropositive or electronegative behavior can be made quantitative by assigning each element an electronegativity number. Electronegativity is measured on a scale from

															H 2.1		
Li 1.0	Be 1.5											B 2.0	C 2.5	N 3.1	O 3.5	F 4.1	
Na 1.0	Mg 1.3											Al 1.5	Si 1.8	P 2.1	S 2.4	Cl 2.9	
K 0.9	Ca 1.1	Sc 1.2	Ti 1.3	V 1.5	Cr 1.6	Mn 1.6	Fe 1.7	Co 1.7	Ni 1.8	Cu 1.8	Zn 1.7	Ga 1.8	Ge 2.0	As 2.2	Se 2.5	Br 2.8	
Rb 0.9	Sr 1.0	Y 1.1	Zr 1.2	Nb 1.3	Mo 1.3	Tc 1.4	Ru 1.4	Rh 1.5	Pd 1.4	Ag 1.4	Cd 1.5	In 1.5	Sn 1.7	Sb 1.8	Te 2.0	I 2.2	
Cs 0.9	Ba 0.9	La 1.1	Hf 1.2	Ta 1.4	W 1.4	Re 1.5	Os 1.5	Ir 1.6	Pt 1.5	Au 1.4	Hg 1.5	Tl 1.5	Pb 1.6	Bi 1.7	Po 1.8	At 2.0	
Fr 0.9	Ra 0.9	Ac 1.0	Lanthanides: 1.0–1.2														
			Actinides: 1.0–1.2														

Figure 2.9
The electronegativities of the elements.
(After F. M. Miller, "Chemistry: Structure and Dynamics," McGraw-Hill, 1984, p. 185.)

0 to 4.1, and each element is assigned a value on this scale, as shown in Fig. 2.9. The most electropositive elements are the alkali metals, which have electronegativities ranging from 0.9 for cesium, rubidium, and potassium to 1.0 for sodium and lithium. The most electronegative elements are fluorine, oxygen, and nitrogen, which have electronegativities of 4.1, 3.5, and 3.1, respectively. The electronegativity concept helps in understanding the bonding behavior of the elements.

Summary of some of the electronic structure–chemical property relationships for metals and nonmetals

Metals	Nonmetals
1. Have few electrons in outer shells, usually three or less	1. Have four or more electrons in outer shells
2. Form cations by losing electrons	2. Form anions by gaining electrons
3. Have low electronegativities	3. Have high electronegativities

2.4 TYPES OF ATOMIC AND MOLECULAR BONDS

Chemical bonding between atoms occurs since there is a net decrease in the potential energy of atoms in the bonded state. That is, atoms in the bonded state are in a more stable energy condition than when they are unbonded. In general, chemical bonds between atoms can be divided into two groups: primary (or strong bonds) and secondary (or weak bonds).

2.4.1 Primary Atomic Bonds

Primary atomic bonds in which relatively large interatomic forces develop can be subdivided into the following three classes:

1. *Ionic bonds.* Relatively large interatomic forces are set up in this type of bonding by an electron transfer from one atom to another to produce ions that are bonded together by coulombic forces (attraction of positively and negatively charged ions). The ionic bond is a relatively strong nondirectional bond.

2. *Covalent bonds.* Relatively large interatomic forces are created by the sharing of electrons to form a bond with a localized direction.

3. *Metallic bonds.* Relatively large interatomic forces are created by the sharing of electrons in a delocalized manner to form strong nondirectional bonding between atoms.

2.4.2 Secondary Atomic and Molecular Bonds

1. *Permanent dipole bonds.* Relatively weak intermolecular bonds are formed between molecules that possess permanent dipoles. A dipole in a molecule exists due to asymmetry in its electron density distribution.

2. *Fluctuating dipole bonds.* Very weak electric dipole bonding can take place among atoms due to the asymmetrical distribution of electron densities around their nuclei. This type of bonding is termed *fluctuating* since the electron density is continuously changing with time.

2.5 IONIC BONDING

2.5.1 Ionic Bonding in General

Animation

Quiz

Ionic bonds can form between highly electropositive (metallic) elements and highly electronegative (nonmetallic) elements. In the ionization process electrons are transferred from atoms of electropositive elements to atoms of electronegative elements, producing positively charged cations and negatively charged anions. The ionic bonding forces are due to the electrostatic or coulombic force attraction of oppositely charged ions. Ionic bonds form between oppositely charged ions because there is a net decrease in potential energy of the ions after bonding.

An example of a solid that has a high degree of ionic bonding is sodium chloride (NaCl). In the ionization process to form a Na^+Cl^- ion pair, a sodium atom gives up its outer $3s^1$ electron and transfers it to a half-filled $3p$ orbital of a chlorine atom, producing a pair of Na^+ and Cl^- ions (Fig. 2.10).

In the ionization process the sodium atom, which originally had a 0.192 nm radius, is reduced in size to a sodium cation with a radius of 0.095 nm, and the chlorine atom that originally had a radius of 0.099 nm is expanded into a chloride anion with a radius of 0.181 nm.

The sodium atom is reduced in size when the ion is formed because of the loss of its outer-shell $3s^1$ electron and because of the decrease in the electron-to-proton

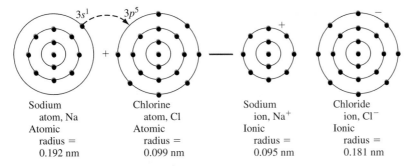

Figure 2.10
Formation of a sodium chloride ion pair from sodium and chlorine atoms. In the ionization process a $3s^1$ electron from the sodium atom is *transferred* to a 3p orbital of the chlorine atom. (Rings indicate energy level of the electron not position)

ratio. The higher positively charged nucleus of the sodium ion attracts the electron charge cloud closer to itself, causing the size of the atom to decrease during ionization. In contrast, during ionization the chlorine atom expands due to an increase in the electron-to-proton ratio. During ionization, atoms decrease in size when they form cations and increase in size when they form anions, as shown for some atoms in Fig. 2.6.

2.5.2 Interionic Forces for an Ion Pair

Consider a pair of oppositely charged ions (for example, a Na^+Cl^- ion pair) approaching each other from a large distance of separation a. As the ions come closer together, they will be attracted to each other by coulombic forces. That is, the nucleus of one ion will attract the electron charge cloud of the other, and vice versa. When the ions come still closer together, eventually their two electron charge clouds will interact and repulsive forces will arise. When the attractive forces equal the repulsive forces, there will be no net force between the ions and they will remain at an equilibrium separation distance, the interionic distance a_0. Figure 2.11 schematically shows force versus separation distance curves for an ion pair.

Animation

The net force between a pair of oppositely charged ions is equal to the sum of the attractive and repulsive forces. Thus,

$$F_{net} = F_{attractive} + F_{repulsive} \qquad (2.4)$$

The attractive force between the ion pair is the coulombic force that results when the ions are considered as point charges. Using Coulomb's[9] law with SI units, the

[9]Charles Augustin Coulomb (1736–1806). French physicist who demonstrated experimentally that the force between two charged bodies varied inversely as the square of the distance between them.

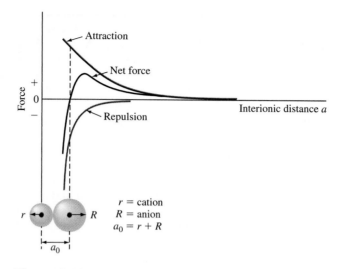

Animation

Figure 2.11
Force versus separation distance for a pair of oppositely
charged ions. The equilibrium interionic separation distance
a_0 is reached when the force between the ions is zero.

following equation can be written:

$$F_{\text{attractive}} = -\frac{(Z_1 e)(Z_2 e)}{4\pi\epsilon_0 a^2} = -\frac{Z_1 Z_2 e^2}{4\pi\epsilon_0 a^2} \qquad (2.5)$$

where Z_1, Z_2 = number of electrons removed or added from the atoms
   during the ion formation
   e = electron charge
   a = interionic separation distance
   ϵ_0 = permittivity of free space = 8.85×10^{-12} C^2/(N · m^2)

The repulsive force between an ion pair is found by experiment to be inversely
proportional to the interionic separation distance a and can be described by the
equation

$$F_{\text{repulsive}} = -\frac{nb}{a^{n+1}} \qquad (2.6)$$

where a is the interionic separation distance and b and n are constants; n usually
ranges from 7 to 9 and is 9 for NaCl.
   Substituting Eqs. 2.5 and 2.6 into 2.4 gives the net force between an ion pair:

$$F_{\text{net}} = -\frac{Z_1 Z_2 e^2}{4\pi\epsilon_0 a^2} - \frac{nb}{a^{n+1}} \qquad (2.7)$$

Calculate the coulombic attractive force ($\bullet\!\!\rightarrow \leftarrow\!\!\bullet$) between a pair of Na^+ and Cl^- ions that just touch each other. Assume the ionic radius of the Na^+ ion to be 0.095 nm and that of the Cl^- ion to be 0.181 nm.

EXAMPLE PROBLEM 2.8

■ **Solution**

The attractive force between the Na^+ and Cl^- ions can be calculated by substituting the appropriate values into the Coulomb's law equation (2.5).

$$Z_1 = +1 \text{ for } Na^+ \qquad Z_2 = -1 \text{ for } Cl^-$$

$$e = 1.60 \times 10^{-19}\,C \qquad \epsilon_0 = 8.85 \times 10^{-12}\,C^2/(N \cdot m^2)$$

$$a_0 = \text{sum of the radii of } Na^+ \text{ and } Cl^- \text{ ions}$$

$$= 0.095 \text{ nm} + 0.181 \text{ nm}$$

$$= 0.276 \text{ nm} \times 10^{-9}\,m/nm = 2.76 \times 10^{-10}\,m$$

$$F_{\text{attractive}} = -\frac{Z_1 Z_2 e^2}{4\pi\epsilon_0 a_0^2}$$

$$= -\frac{(+1)(-1)(1.60 \times 10^{-19}\,C)^2}{4\pi[8.85 \times 10^{-12}\,C^2/(N \cdot m^2)](2.76 \times 10^{-10}\,m)^2}$$

$$= +3.02 \times 10^{-9}\,N \blacktriangleleft$$

Thus, the attractive force ($\bullet\!\!\rightarrow \leftarrow\!\!\bullet$) between the ions is $+3.02 \times 10^{-9}\,N$. The repulsive force ($\leftarrow\!\!\bullet\,\bullet\!\!\rightarrow$) will be equal and opposite in sign and thus will be $-3.02 \times 10^{-9}\,N$.

If the attractive force between Mg^{2+} and S^{2-} ions is $1.49 \times 10^{-8}\,N$ and if the S^{2-} ion has a radius of 0.184 nm, calculate a value for the ionic radius of the Mg^{2+} ion in nanometers.

EXAMPLE PROBLEM 2.9

■ **Solution**

The value of a_0, the sum of the Mg^{2+} and S^{2-} ionic radii, can be calculated from a rearranged form of Coulomb's law equation (2.5):

$$a_0 = \sqrt{\frac{-Z_1 Z_2 e^2}{4\pi\epsilon_0 F_{\text{attractive}}}}$$

$$Z_1 = +2 \text{ for } Mg^{2+} \qquad Z_2 = -2 \text{ for } S^{2-}$$

$$|e| = 1.60 \times 10^{-19}\,C \qquad \epsilon_0 = 8.85 \times 10^{-12}\,C^2/(N \cdot m^2)$$

$$F_{\text{attractive}} = 1.49 \times 10^{-9}\,N$$

Thus,

$$a_0 = \sqrt{\frac{-(2)(-2)(1.60 \times 10^{-19}\,C)^2}{4\pi[8.85 \times 10^{-12}\,C^2/(N \cdot m^2)](1.49 \times 10^{-8}\,N)}}$$

$$= 2.49 \times 10^{-10}\,m = 0.249 \text{ nm}$$

$$a_0 = r_{Mg^{2+}} + r_{S^{2-}}$$

or

$$0.249 \text{ nm} = r_{Mg^{2+}} + 0.184 \text{ nm}$$

$$r_{Mg^{2+}} = 0.065 \text{ nm} \blacktriangleleft$$

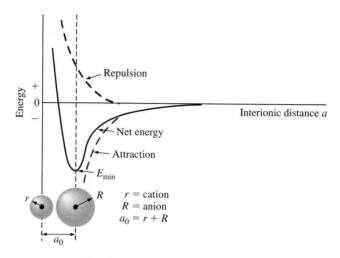

Figure 2.12
Energy versus separation distance for a pair of oppositely charged ions. The equilibrium interionic separation distance a_0 is reached when the net potential energy is a minimum.

2.5.3 Interionic Energies for an Ion Pair

Animation

The net potential energy E between a pair of oppositely charged ions, for example, Na^+Cl^-, which are brought closer together is equal to the sum of the energies associated with the attraction and repulsion of the ions and can be written in equation form as

$$E_{net} = +\frac{Z_1 Z_2 e^2}{4\pi\epsilon_0 a} + \frac{b}{a^n}$$ (2.8)

$$\underbrace{\phantom{+\frac{Z_1 Z_2 e^2}{4\pi\epsilon_0 a}}}_{\substack{\text{Attractive}\\\text{energy}}} \underbrace{\phantom{\frac{b}{a^n}}}_{\substack{\text{Repulsive}\\\text{energy}}}$$

The attractive-energy term of Eq. 2.8 represents the energy released when the ions come close together and is negative because the product of $(+Z_1)(-Z_2)$ is negative. The repulsive-energy term of Eq. 2.8 represents the energy absorbed as the ions come close together and is positive. The sum of the energies associated with the attraction and repulsion of the ions equals the net energy, which is a minimum when the ions are at their equilibrium separation distance a_0. Figure 2.12 shows the relationship among these three energies and indicates the minimum energy E_{min}. At the minimum energy, the force between the ions is zero.

EXAMPLE PROBLEM 2.10

Calculate the net potential energy of a simple Na^+Cl^- ion pair by using the equation

$$E_{net} = \frac{+Z_1 Z_2 e^2}{4\pi\epsilon_0 a} + \frac{b}{a^n}$$

and by using the b value obtained from the repulsive force calculated for the Na^+Cl^- ion pair in Example Problem 2.8. Assume $n = 9$ for NaCl.

■ Solution

a. To determine the b value for a NaCl ion pair,

$$F = -\frac{nb}{a^{n+1}} \tag{2.6}$$

The repulsive force between a Na^+Cl^- ion pair from Example Problem 2.8 is -3.02×10^{-9} N. Thus,

$$-3.02 \times 10^{-9}\,N = \frac{-9b}{(2.76 \times 10^{-10}\,m)^{10}}$$

$$b = 8.59 \times 10^{-106}\,N \cdot m^{10} \blacktriangleleft$$

b. To calculate the potential energy of the Na^+Cl^- ion pair,

$$E_{Na^+Cl^-} = \frac{+Z_1Z_2e^2}{4\pi\epsilon_0 a} + \frac{b}{a^n}$$

$$= \frac{(+1)(-1)(1.60 \times 10^{-19}\,C)^2}{4\pi[8.85 \times 10^{-12}\,C^2/(N \cdot m^2)](2.76 \times 10^{-10}\,m)} + \frac{8.59 \times 10^{-106}\,N \cdot m^{10}}{(2.76 \times 10^{-10}\,m)^9}$$

$$= -8.34 \times 10^{-19}\,J^* + 0.92 \times 10^{-19}\,J^*$$

$$= -7.42 \times 10^{-19}\,J \blacktriangleleft$$

$^*1\,J = 1\,N \cdot m.$

2.5.4 Ion Arrangements in Ionic Solids

Until now our discussion of ionic bonding has been for a pair of ions. We shall now briefly discuss ionic bonding in three-dimensional ionic solids. Since elemental ions have approximately spherical symmetrical charge distributions, they can be considered spherical with a characteristic radius. Ionic radii of some selected elemental cations and anions are listed in Table 2.5. As previously stated, elemental cations are smaller than their atoms and anions are larger. Also, as in the case of atoms, elemental ions, in general, increase in size as their principal quantum number increases (i.e., as the number of electron shells increases).

Animation

Table 2.5 Ionic radii of selected elements

Ionic	Ionic radius (nm)	Ion	Ionic radius (nm)
Li^+	0.060	F^-	0.136
Na^+	0.095	Cl^-	0.181
K^+	0.133	Br^-	0.195
Rb^+	0.148	I^-	0.216
Cs^+	0.169		

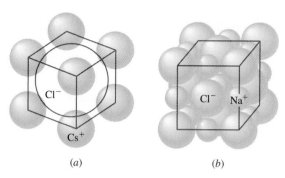

(a)   (b)

Figure 2.13

Ionic packing arrangements in (*a*) CsCl and (*b*) NaCl.
Eight Cl^- ions can pack around a Cs^+ ion, but only
six Cl^- ions can pack around a Na^+ ion.

(After C. R. Barrett, W. D. Nix, and A. S. Tetelman,
"The Principles of Engineering Materials," Prentice-Hall,
1973, p. 27.)

When ions pack together in a solid, they do so with no preferred orientation since electrostatic attraction of symmetrical charges is independent of the orientation of the charges. Thus the ionic bond is *nondirectional* in character. However, the packing of ions in an ionic solid is governed by the geometric arrangement of the ions that is possible and the necessity to maintain electrical neutrality of the solid.

Geometric Arrangement of Ions in an Ionic Solid   Ionic crystals may have extremely complex structures. At this point we shall only consider the *geometric arrangement* of the ions in two simple ionic structures, CsCl and NaCl. In the case of CsCl, *eight* Cl^- ions ($r = 0.181$ nm) pack around a central Cs^+ ion ($r = 0.169$ nm), as shown in Fig. 2.13*a*. However, in NaCl, only *six* Cl^- ions ($r = 0.181$ nm) can pack around a central Na^+ ion ($r = 0.095$ nm), as shown in Fig. 2.13*b*. For CsCl, the ratio of the cation-to-anion radius is $0.169/0.181 = 0.934$, whereas for NaCl the ratio is $0.095/0.181 = 0.525$. Thus, as the ratio of the cation-to-anion radius decreases, fewer anions can surround a central cation in this type of structure. Ionic crystal structures will be discussed in more detail in Chap. 10 on ceramic materials.

Electrical Neutrality of Ionic Solids   The ions in an ionic solid must be arranged in a structure so that local charge neutrality is maintained. Thus, in an ionic solid such as CaF_2, the ionic arrangement will be partly governed by the fact that there must be two fluoride ions for every calcium ion.

2.5.5 Bonding Energies of Ionic Solids

The lattice energies and melting points of ionically bonded solids are relatively high, as indicated in Table 2.6, which lists the lattice energies and melting points of

Table 2.6 Lattice energies and melting points of selected ionic solids

Ionic solid	Lattice energy*		Melting point (°C)
	kJ/mol	kcal/mol	
LiCl	829	198	613
NaCl	766	183	801
KCl	686	164	776
RbCl	670	160	715
CsCl	649	155	646
MgO	3932	940	2800
CaO	3583	846	2580
SrO	3311	791	2430
BaO	3127	747	1923

*All values are negative for bond formation (energy is released).

selected ionic solids. As the size of the ion increases in a group in the periodic table, the lattice energy decreases. For example, the lattice energy for LiCl is 829 kJ/mol (198 kcal/mol), whereas that of CsCl is only 649 kJ/mol (155 kcal/mol). The reason for this decrease in lattice energy is that the bonding electrons in the larger ions are farther away from the attractive influence of the positive nucleus. Also, multiple bonding electrons in an ionic solid increase the lattice energy, as exemplified by MgO, which has a lattice energy of 3932 kJ/mol (940 kcal/mol).

2.6 COVALENT BONDING

A second type of primary atomic bonding is **covalent bonding.** Whereas ionic bonding involves highly electropositive and electronegative atoms, covalent bonding takes place between atoms with small differences in electronegativity and which are close to each other in the periodic table. In covalent bonding the atoms most commonly share their outer s and p electrons with other atoms so that each atom attains the noble-gas electron configuration. In a single covalent bond, each of the two atoms contributes one electron to form an electron-pair bond, and the energies of the two atoms associated with the covalent bond are lowered (made more stable) because of the electron interaction. In covalent bonding, multiple electron-pair bonds can be formed by one atom with itself or other atoms.

Animation

2.6.1 Covalent Bonding in the Hydrogen Molecule

The simplest case of covalent bonding occurs in the hydrogen molecule in which two hydrogen atoms contribute their $1s^1$ electrons to form an electron-pair covalent

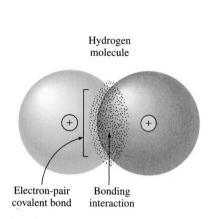

Hydrogen
molecule

Electron-pair
covalent bond

Bonding
interaction

Figure 2.14
Covalent bonding in the hydrogen
molecule. The highest electron
charge cloud density is in the region
of overlap between the hydrogen
atom nuclei.

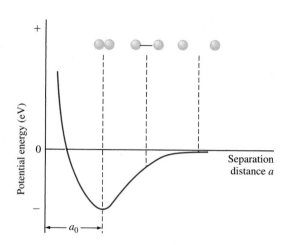

Figure 2.15
Potential energy versus separation distance
for two hydrogen atoms. The equilibrium
interatomic distance a_0 in the hydrogen
molecule occurs at the minimum potential
energy E_{min}.

bond, as indicated by the electron-dot notation reaction

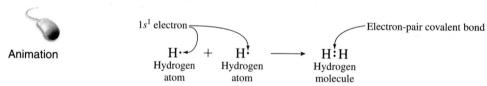

Animation

Although the electron-dot notation is useful to represent the covalent bond, it does
not consider the valence-electron density distribution. In Fig. 2.14 the two hydrogen
atoms are shown as forming an electron-pair covalent bond with a high electron
charge cloud density between the two nuclei of the hydrogen atoms. As the two
hydrogen atoms come together to form the hydrogen molecule, their electron charge
clouds interact and overlap, creating a high probability of finding the $1s^1$ electrons
of the atoms between the two nuclei in the molecule (Fig. 2.14). In the bonding
process of forming the hydrogen molecule, the potential energy of the hydrogen
atoms is lowered, as indicated in Fig. 2.15, and energy is released. If the hydrogen
atoms in the hydrogen molecule are separated, energy is required since the atoms
would then be in a higher-energy state.

2.6.2 Covalent Bonding in Other Diatomic Molecules

Electron-pair covalent bonds are also formed in other diatomic molecules such
as F_2, O_2, and N_2. In these cases p electrons are shared between the atoms. The

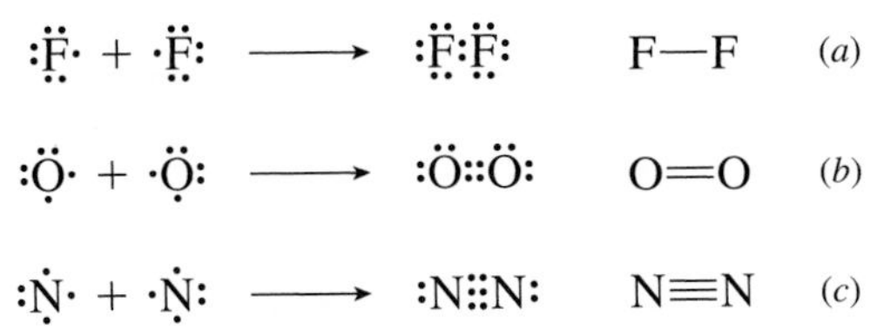

$$:\ddot{F}\cdot + \cdot\ddot{F}: \longrightarrow :\ddot{F}:\ddot{F}: \qquad F{-}F \qquad (a)$$

$$:\ddot{O}\cdot + \cdot\ddot{O}: \longrightarrow :\ddot{O}::\ddot{O}: \qquad O{=}O \qquad (b)$$

$$:\dot{N}\cdot + \cdot\dot{N}: \longrightarrow :N{:\!:\!:}N: \qquad N{\equiv}N \qquad (c)$$

Figure 2.16
Covalent bonding in molecules of (*a*) fluorine (single bond), (*b*) oxygen (double bond), and (*c*) nitrogen (triple bond). The electron-pair covalent bonds between the atoms are shown on the left by electron-dot notation and on the right by straight-line notation.

fluorine atom with its outer seven electrons $(2s^2 2p^5)$ can attain the neon noble-gas electron configuration by sharing one $2p$ electron with another fluorine atom, as shown in the electron-dot notation reaction in Fig. 2.16*a*. Similarly, the oxygen atom with its outer six electrons $(2s^2 2p^4)$ can attain the neon noble-gas electron configuration $(2s^2 2p^6)$ by sharing *two* $2p$ electrons with another oxygen atom to form the diatomic O_2 molecule (Fig. 2.16*b*). Nitrogen also with its five outer valence electrons $(2s^2 2p^3)$ can attain the neon noble-gas electron configuration of $(2s^2 2p^6)$ by sharing three $2p$ electrons with another nitrogen atom to form the diatomic nitrogen molecule (Fig. 2.16*c*).

Chemical reactions involving covalent bonds are sometimes written by using electron-dot notation for the covalent bonds, as shown in Fig. 2.16. However, more commonly the straight-line notation for covalent bonds (Fig. 2.16) is used. We shall encounter both types of these notations in Chap. 10 on Polymeric Materials.

Table 2.7 lists approximate bond energies and lengths of selected covalent bonds. Note that higher bond energies are associated with multiple bonds. For example, the C—C bond has an energy of 370 kJ/mol (88 kcal/mol), whereas the C=C bond has a much higher energy of 680 kJ/mol (162 kcal/mol).

2.6.3 Covalent Bonding by Carbon

In the study of engineering materials carbon is very important since it is the basic element in most polymeric materials. The carbon atom in the ground state has the electron configuration $1s^2 2s^2 2p^2$. This electron arrangement indicates that carbon should form *two covalent bonds* with its two half-filled $2p$ orbitals. However, in many cases carbon forms *four covalent bonds* of equal strength. The explanation for the four carbon covalent bonds is provided by the concept of *hybridization* whereby upon bonding, one of the $2s$ orbitals is promoted to a $2p$ orbital so that *four equivalent sp^3* **hybrid orbitals** are produced, as indicated in the orbital diagrams of Fig. 2.17. Even though energy is required to promote the $2s$ electron to

Table 2.7 Bond energies and lengths of selected covalent bonds

Bond	Bond energy*		Bond length (nm)
	kcal/mol	kJ/mol	
C—C	88	370	0.154
C=C	162	680	0.13
C≡C	213	890	0.12
C—H	104	435	0.11
C—N	73	305	0.15
C—O	86	360	0.14
C=O	128	535	0.12
C—F	108	450	0.14
C—Cl	81	340	0.18
O—H	119	500	0.10
O—O	52	220	0.15
O—Si	90	375	0.16
N—O	60	250	0.12
N—H	103	430	0.10
F—F	38	160	0.14
H—H	104	435	0.074

*Approximate values since environment changes energy. All values are negative for bond formation (energy is released).

Source: L. H. Van Vlack, "Elements of Materials Science," 4th ed., Addison-Wesley, 1980.

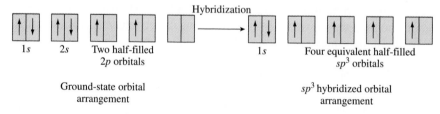

Figure 2.17

Hybridization of carbon orbitals for the formation of single covalent bonds.

Animation

the $2p$ state in the hybridization process, the energy necessary for the promotion is more than compensated for by the decrease in energy accompanying the bonding process.

Carbon in the form of diamond exhibits sp^3 tetrahedral covalent bonding. The four sp^3 hybrid orbitals are directed symmetrically toward the corners of a regular tetrahedron, as shown in Fig. 2.18. The structure of diamond consists of a massive molecule with sp^3 tetrahedral covalent bonding, as shown in Fig. 2.19. This structure accounts for the extremely high hardness of diamond and its high bond strength

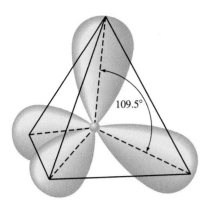

Figure 2.18
A carbon atom with four equivalent sp^3 orbitals directed symmetrically toward the corners of a tetrahedron. The angle between the orbitals is 109.5°.

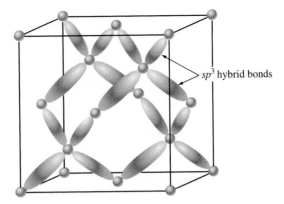

Figure 2.19
Tetrahedral sp^3 covalent bonds in the diamond structure. Each shaded region represents an electron-pair covalent bond.

Animation

and melting temperature. Diamond has a bond energy of 711 kJ/mol (170 kcal/mol) and a melting temperature of 3550°C.

2.6.4 Covalent Bonding in Carbon-Containing Molecules

Covalently bonded molecules containing only carbon and hydrogen are called *hydrocarbons*. The simplest hydrocarbon is methane in which carbon forms four sp^3 tetrahedral covalent bonds with hydrogen atoms, as shown in Fig. 2.20. The intramolecular bonding energy of methane is relatively high at 1650 kJ/mol (396 kcal/mol), but the intermolecular bond energy is very low at about 8 kJ/mol (2 kcal/mol). Thus, the methane molecules are very weakly bonded together, resulting in a low melting point of −183°C.

Figure 2.21 shows structural formulas for methane, ethane, and normal (*n*-) butane, which are single covalently bonded hydrocarbons. As the molecular mass of the molecule increases, so does its stability and melting point.

Carbon can also bond to itself to form double and triple bonds in molecules, as indicated in the structural formulas for ethylene and acetylene shown in Fig. 2.22. Double and triple carbon-carbon bonds are chemically more reactive than single carbon-carbon bonds. Multiple carbon-carbon bonds in carbon-containing molecules are referred to as *unsaturated bonds*.

2.6.5 Benzene

An important molecular structure for some polymeric materials is the benzene structure. The benzene molecule has the chemical composition C_6H_6, with the carbon

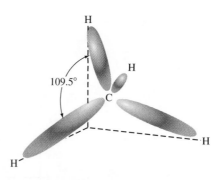

Figure 2.20
The methane molecule, CH$_4$, contains four tetrahedral sp^3 covalent bonds. Each shaded region represents an electron-pair covalent bond.

Animation

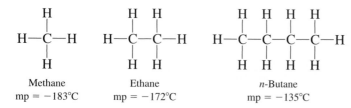

Figure 2.21
Structural formulas for simple covalently bonded hydrocarbons. Note that as the molecular mass increases, the melting point increases.

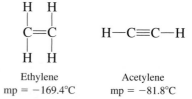

Figure 2.22
Structural formulas for two covalently bonded molecules that have multiple carbon-carbon covalent bonds.

atoms forming a hexagonal-shaped ring referred to sometimes as the *benzene ring* (Fig. 2.23). The six hydrogen atoms of benzene are covalently bonded with single bonds to the six carbon atoms of the ring. However, the bonding arrangement between the carbon atoms in the ring is complex. The simplest way of satisfying the requirement that each carbon atom have four covalent bonds is to assign alternating single and double bonds to the carbon atoms in the ring itself (Fig. 2.23*a*). This structure can be represented more simply by omitting the external hydrogen atoms (Fig. 2.23*b*). This structural formula for benzene will be used in this book since it more clearly indicates the bonding arrangement in benzene.

However, experimental evidence indicates that a normal reactive double carbon-carbon bond does not exist in benzene and that the bonding electrons

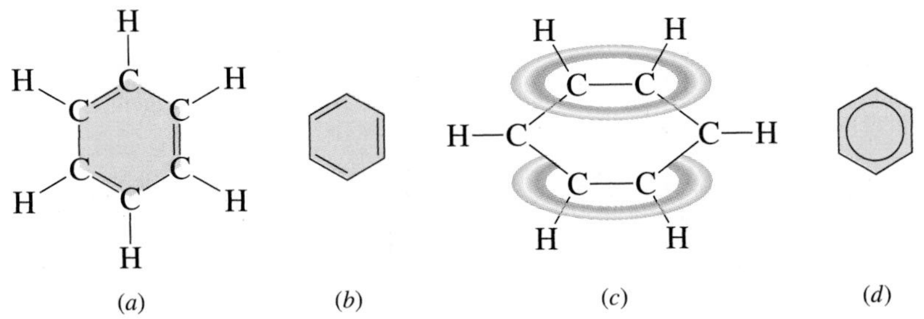

Figure 2.23
Structural formula representations for the covalently bonded molecule benzene. (*a*) Structural formula using straight-line bonding notation. (*b*) Simplified notation not showing the hydrogen atoms. (*c*) Bonding arrangement indicating the delocalization of the carbon-carbon bonding electrons within the benzene ring.
(After R. E. Davis, K. D. Gailey, and K. W. Whitten, "Principles of Chemistry," Saunders College Publishing, 1984, p. 830.)
(*d*) Simplified notation indicating delocalized bonding of electrons within the benzene ring.

within the benzene ring are delocalized, forming an overall bonding structure intermediate in chemical reactivity to that between single and double carbon-carbon bonds (Fig. 2.23*c*). Thus, most chemistry books are written using a circle inside a hexagon to represent the structure of benzene (Fig. 2.23*d*).

2.7 METALLIC BONDING

A third primary type of atomic bonding is **metallic bonding** that occurs in solid metals. In metals in the solid state, atoms are packed relatively close together in a systematic pattern or crystal structure. For example, the arrangement of copper atoms in crystalline copper is shown in Fig. 2.24*a*. In this structure the atoms are so close together that their outer valence electrons are attracted to the nuclei of their numerous neighbors. In the case of solid copper, each atom is surrounded by *12* nearest neighbors. The valence electrons are therefore not closely associated with any particular nucleus and are thus spread out among the atoms in the form of a low-density electron charge cloud, or "electron gas."

Solid metals are therefore visualized as consisting of **positive-ion cores** (atoms without their valence electrons) and of **valence electrons** dispersed in the form of an electron cloud that covers a large extent of space (Fig. 2.24*b*). The valence electrons are weakly bonded to the positive-ion cores and can readily move in the metal crystal; they are generally referred to as *free electrons.** The high thermal and electrical

---

*Free or delocalized electrons may also exist in some crystals that primarilly contain covalent or ionic bonds; for instance, graphite.

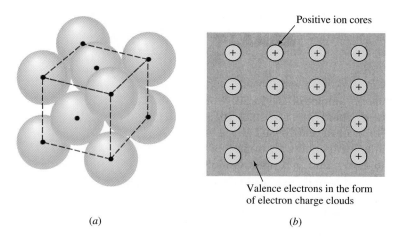

(a)   (b)

Figure 2.24

(a) Atomic arrangement in a metallic copper crystal. Each copper atom is coordinated with 12 other copper atoms, producing a crystal structure called the *face-centered-cubic structure*. The atoms are bonded together by an "electron gas" of delocalized valence electrons. (b) Two-dimensional schematic diagram of metallically bonded atoms. The circles with the inner positive signs represent positive-ion cores, and the charge clouds around the ion cores represent the dispersed valence electrons.

conductivities of metals support the theory that some electrons are free to move through the metal crystal lattice. Most metals can be deformed a considerable amount without fracturing because the metal atoms can slide past each other without completely disrupting the metallically bonded structure.

Atoms in a solid metal bond themselves together by metallic bonding to achieve a lower energy (or more stable) state. For metallic bonding there are *no electron-pair restrictions* as for the covalent bond or *charge neutrality restrictions* as for the ionic bond. In metallic bonding the outer valence electrons of the atoms are shared by many surrounding atoms, and so, in general, metallic bonding is *nondirectional*.

When metallic atoms bond together by valence-electron sharing to form a solid crystal, the overall energy of the individual atoms is lowered by the bonding process. As is the case for ionic and covalent bonding, a minimum in energy between a pair of atoms is attained when the equilibrium atomic separation distance a_0 is reached, as shown in Fig. 2.25. The magnitude of $E_0 - E_{min}$ in Fig. 2.25 is a measure of the bonding energy between the atoms of a particular metal.

The energy levels in multiatom metal crystals differ from those of single atoms. When metal atoms bond together to form a metal crystal, their energies are lowered but to slightly different levels. The valence electrons in a metal crystal therefore form an energy "band." The energy-band theory of metals will be discussed further in Sec. 6.2.

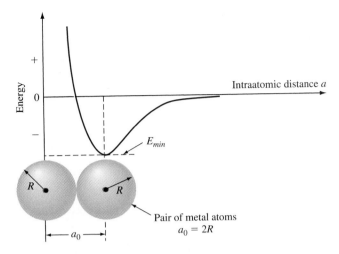

Figure 2.25
Energy versus separation distance for a pair of metal
atoms. The equilibrium atomic separation distance a_0 is
reached when the net potential energy is a minimum.

The bond energies and melting points of different metals vary greatly. In
general, the fewer valence electrons per atom involved in metallic bonding, the more
metallic the bonding is. That is, the valence electrons are freer to move. The highest degree of metallic bonding is found in the alkali metals, which have just one
bonding electron outside a noble-gas electron configuration. Thus, the bonding energies and melting points of the alkali metals are relatively low. For example, the
bonding energy for sodium is 108 kJ/mol (25.9 kcal/mol) and that for potassium is
89.6 kJ/mol (21.4 kcal/mol). The melting points of sodium (97.9°C) and potassium
(63.5°C) are also relatively low.

However, as the number of bonding electrons increases, the bonding energies
and melting points of the metals also increase, as indicated in Table 2.8 for the metals of the fourth period. Calcium with two valence electrons per atom has its bonding electrons more tightly bound than potassium, and as a result calcium's bonding
energy of 177 kJ/mol (42.2 kcal/mol) and melting point (851°C) are both considerably higher than those of potassium.

With the introduction of the 3d electrons in the transition metals of the fourth
period from scandium to nickel, the bonding energies and melting points of these
elements are raised even much higher. For example, titanium has a bonding energy
of 473 kJ/mol (113 kcal/mol) and a melting point of 1660°C. The increased
bonding energies and melting points of the transition metals are attributed to *dsp
hybridized bonding,* which involves a significant fraction of covalent bonding. An
attempt has been made by the author to quantify the effect of the covalent bonding on the metallic bonding of most metals by assigning *covalent-metallitivities*

Table 2.8 Bonding energies, melting points, and electron configurations of the fourth-period metals of the periodic table

Element	Electron configuration	Bonding energy		Melting point (°C)
		kJ/mol	kcal/mol	
K	$4s^1$	89.6	21.4	63.5
Ca	$4s^2$	177	42.2	851
Sc	$3d^14s^2$	342	82	1397
Ti	$3d^24s^2$	473	113	1660
V	$3d^34s^2$	515	123	1730
Cr	$3d^54s^1$	398	95	1903
Mn	$3d^54s^2$	279	66.7	1244
Fe	$3d^64s^2$	418	99.8	1535
Co	$3d^74s^2$	383	91.4	1490
Ni	$3d^84s^2$	423	101	1455
Cu	$3d^{10}4s^1$	339	81.1	1083
Zn	$4s^2$	131	31.2	419
Ga	$4s^24p^1$	272	65	29.8
Ge	$4s^24p^2$	377	90	960

Table 2.9 Covalent-metallitivities of some solid elements*

																	H
Li 0.21	Be 1.47												B 2.7	C 4.0	N	O	F
Na 0.11	Mg 0.74												Al 0.75	Si	P	S	Cl
K 0.07	Ca 0.97	Sc 1.60	Ti 1.91	V 1.98	Cr 2.17	Mn 1.42	Fe 1.75	Co 1.70	Ni 1.66	Cu 1.24	Zn 0.48		Ga	Ge	As	Se	Br
Rb 0.05	Sr 0.88	Y 1.69	Zr 2.12	Nb 2.23	Mo 2.98	Tc	Ru 2.86	Rh 2.24	Pd 1.98	Ag 1.10	Cd 0.37		In 0.18	Sn 0.26	Sb	Te	I
Cs 0.03	Ba 0.97	La	Hf 2.55	Ta 3.40	W 3.86	Re 3.60	Os 3.09	Ir 2.79	Pt 2.02	Au 1.21	Hg		Tl 0.35	Pb 0.37	Bi	Po	At
Fr	Ra	Ac															

*Carbon is chosen as completely covalently bonded with a melting point of 3500°C and a C-M value of 4.0.

(C-M) values to some of the solid elements based on their melting temperatures in °C as compared to carbon (diamond), with a melting temperature of 3500°C, which is assumed to be 100% covalently bonded and have a covalent-metallitivity of 4.0 (Table 2.9). For example, the percent metallic bonding in pure iron is 56.3 percent using the C-M value for iron listed in Table 2.9, as shown in Example Problem 2.11.

EXAMPLE PROBLEM 2.11

Calculate the percent metallic and covalent bonding in solid iron, which has a melting temperature of 1535°C and a C-M of 1.75.

$$\% \text{ covalent bonding} = \frac{1.75}{4.00} \times 100\% = 43.7\%$$

$$\% \text{ metallic bonding} = 100\% - 43.7\% = 56.3\%$$

In this calculation we are assuming that only metallic and covalent bonding are involved in the chemical bonding of iron atoms.

When the $3d$ and $4s$ orbitals are filled, the outer electrons become more loosely bound and the bonding energies and melting points of the metals decrease again. For example, zinc with a $3d^{10}4s^2$ electron configuration has a relatively low bonding energy of 131 kJ/mol (31.2 kcal/mol) and a melting temperature of 419°C.

2.8 SECONDARY BONDING

Until now we have considered only primary bonding between atoms and showed that it depends on the interaction of their valence electrons. The driving force for primary atomic bonding is the lowering of the energy of the bonding electrons. Secondary bonds are relatively weak in contrast to primary bonds and have energies of only about 4 to 42 kJ/mol (1 to 10 kcal/mol). The driving force for secondary bonding is the attraction of the electric dipoles contained in atoms or molecules.

An electric dipole moment is created when two equal and opposite charges are separated, as shown in Fig. 2.26a. Electric dipoles are created in atoms or molecules when positive and negative charge centers exist (Fig. 2.26b).

Dipoles in atoms or molecules create dipole moments. A *dipole moment* is defined as the charge value multiplied by the separation distance between positive

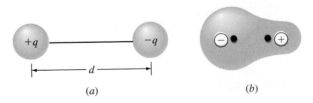

(a)   (b)

Figure 2.26
(a) An electric dipole. The dipole moment is *qd*.
(b) An electric dipole moment in a covalently bonded molecule. Note the separation of the positive and negative charge centers.

and negative charges, or

$$\mu = qd \tag{2.9}$$

where μ = dipole moment

q = magnitude of electric charge

d = separation distance between the charge centers

Dipole moments in atoms and molecules are measured in coulomb-meters (C · m) or in debye units, where 1 debye = 3.34×10^{-30} C · m.

Electric dipoles interact with each other by electrostatic (coulombic) forces, and thus atoms or molecules containing dipoles are attracted to each other by these forces. Even though the bonding energies of secondary bonds are weak, they become important when they are the only bonds available to bond atoms or molecules together.

In general, there are two main kinds of secondary bonds between atoms or molecules involving electric dipoles: fluctuating dipoles and permanent dipoles. Collectively, these secondary dipole bonds are sometimes called *van der Waals bonds (forces)*.

2.8.1 Fluctuating Dipoles

Very weak secondary bonding forces can develop between the atoms of noble-gas elements that have complete outer-valence-electron shells (s^2 for helium and s^2p^6 for Ne, Ar, Kr, Xe, and Rn). These bonding forces arise because the asymmetrical distribution of electron charges in these atoms creates electric dipoles. At any instant there is a high probability that there will be more electron charge on one side of an atom than on the other (Fig. 2.27). Thus, in a particular atom, the electron charge cloud will change with time, creating a "fluctuating dipole." Fluctuating dipoles of

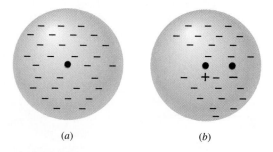

(a)   (b)

Figure 2.27
Electron charge cloud distribution in a noble-gas atom. (*a*) Idealized symmetrical electron charge cloud distribution. (*b*) Real case with asymmetrical electron charge cloud distribution that changes with time, creating a "fluctuating electric dipole."

Table 2.10 Melting and boiling points of noble gases at atmospheric pressure

Noble gas	Melting point (°C)	Boiling point (°C)
Helium	−272.2	−268.9
Neon	−248.7	−245.9
Argon	−189.2	−185.7
Krypton	−157.0	−152.9
Xenon	−112.0	−107.1
Radon	−71.0	−61.8

nearby atoms can attract each other, creating weak interatomic nondirectional bonds. The liquefaction and solidification of the noble gases at low temperatures and high pressures are attributed to fluctuating dipole bonds. The melting and boiling points of the noble gases at atmospheric pressure are listed in Table 2.10. Note that as the atomic size of the noble gases increases, the melting and boiling points also increase due to stronger bonding forces since the electrons have more freedom to create stronger dipole moments.

2.8.2 Permanent Dipoles

Weak bonding forces among covalently bonded molecules can be created if the molecules contain **permanent dipoles.** For example, the methane molecule, CH_4, with its four C—H bonds arranged in a tetrahedral structure (Fig. 2.20), has a zero dipole moment because of its symmetrical arrangement of four C—H bonds. That is, the vectorial addition of its four dipole moments is zero. The chloromethane molecule, CH_3Cl, in contrast, has an asymmetrical tetrahedral arrangement of three C—H bonds and one C—Cl bond, resulting in a net dipole moment of 2.0 debyes. The replacement of one hydrogen atom in methane with one chlorine atom raises the boiling point from −128°C for methane to −14°C for chloromethane. The much higher boiling point of chloromethane is due to the permanent dipole bonding forces among the chloromethane molecules. Table 2.11 lists some experimental dipole moments of some compounds.

The **hydrogen bond** is a special case of a permanent dipole-dipole interaction between polar molecules. Hydrogen bonding occurs when a polar bond containing the hydrogen atom, O—H or N—H, interacts with the electronegative atoms O, N, F, or Cl. For example, the water molecule, H_2O, has a permanent dipole moment of 1.84 debyes due to its asymmetrical structure with its two hydrogen atoms at an angle of 105° with respect to its oxygen atom (Fig. 2.28a).

The hydrogen atomic regions of the water molecule have positively charged centers, and the opposite end region of the oxygen atom has a negatively charged center (Fig. 2.28a). In hydrogen bonding between water molecules the negatively charged region of one molecule is attracted by coulombic forces to the positively charged region of another molecule (Fig. 2.28b).

In liquid and solid water, relatively strong intermolecular permanent dipole forces (hydrogen bonding) are formed among the water molecules. The energy

Table 2.11 Experimental dipole moments of some compounds (debyes)

H_2O	1.84	CH_3Cl	2.00
H_2	0.00	$CHCl_3$	1.10
CO_2	0.00	HCl	1.03
CCl_4	0.00	NH_3	1.46

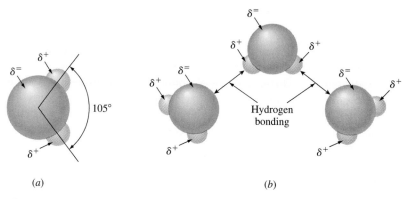

(a)   (b)

Figure 2.28
(*a*) Permanent dipole nature of the water molecule. (*b*) Hydrogen bonding among water molecules due to permanent dipole attraction.

associated with the hydrogen bond is about 29 kJ/mol (7 kcal/mol) as compared to about 2 to 8 kJ/mol (0.5 to 2 kcal/mol) for fluctuating dipole forces in the noble gases. The exceptionally high boiling point of water (100°C) for its molecular mass is attributed to the effect of hydrogen bonding. Hydrogen bonding is also very important for strengthening the bonding between molecular chains of some types of polymeric materials.

2.9 MIXED BONDING

The chemical bonding of atoms or ions can involve more than one type of primary bond and can also involve secondary dipole bonds. For primary bonding there can be the following combinations of mixed-bond types: (1) ionic-covalent, (2) metallic-covalent, (3) metallic-ionic, and (4) ionic-covalent-metallic.

2.9.1 Ionic-Covalent Mixed Bonding

Most covalent-bonded molecules have some ionic binding, and vice versa. The partial ionic character of covalent bonds can be interpreted in terms of the electronegativity scale of Fig. 2.9. The greater the difference in the electronegativities of the elements involved in a mixed ionic-covalent bond, the greater the degree of ionic character of the bond. Pauling[10] proposed the following equation to determine the percentage ionic character of bonding in a compound AB:

$$\% \text{ ionic character} = (1 - e^{(-1/4)(X_A - X_B)^2})(100\%) \tag{2.10}$$

where X_A and X_B are the electronegativities of the atoms A and B in the compound.

Many semiconducting compounds have mixed ionic-covalent bonding. For example, GaAs is a 3–5 compound (Ga is in group 3A and As is in group 5A of the periodic table) and ZnSe is a 2–6 compound. The degree of ionic character in the bonding of these compounds increases as the electronegativity between the atoms in the compounds increases. Thus, one would expect a 2–6 compound to have more ionic character than a 3–5 compound because of the greater electronegativity difference in the 2–6 compound. Example Problem 2.12 illustrates this.

Calculate the percentage ionic character of the semiconducting compounds GaAs (3–5) and ZnSe (2–6) by using Pauling's equation

$$\% \text{ ionic character} = (1 - e^{(-1/4)(X_A - X_B)^2})(100\%)$$

a. For GaAs, electronegativities from Fig. 2.9 are $X_{Ga} = 1.8$ and $X_{As} = 2.2$. Thus,

$$\% \text{ ionic character} = (1 - e^{(-1/4)(1.8-2.2)^2})(100\%)$$
$$= (1 - e^{(-1/4)(-0.4)^2})(100\%)$$
$$= (1 - 0.96)(100\%) = 4\%$$

b. For ZnSe, electronegativities from Fig. 2.9 are $X_{Zn} = 1.7$ and $X_{Se} = 2.5$. Thus,

$$\% \text{ ionic character} = (1 - e^{(-1/4)(1.7-2.5)^2})(100\%)$$
$$= (1 - e^{(-1/4)(-0.8)^2})(100\%)$$
$$= (1 - 0.85)(100\%) = 15\%$$

EXAMPLE PROBLEM 2.12

Note that as the electronegativities differ more for the 2–6 compound, the percentage ionic character increases.

2.9.2 Metallic-Covalent Mixed Bonding

Mixed metallic-covalent bonding occurs commonly. For example, the transition metals have mixed metallic-covalent bonding involving *dsp* bonding orbitals (Table 2.9). The high melting points of the transition metals are attributed to mixed metallic-covalent bonding. Also in group 4A of the periodic table there is a gradual transition from pure covalent bonding in carbon (diamond) to some metallic character in silicon and germanium. Tin and lead are primarily metallically bonded.

---

[10]Linus Carl Pauling (1901–1994). American chemist whose pioneer work led to a greater understanding of chemical bonding. He received the Nobel Prize in Chemistry in 1954.

2.9.3 Metallic-Ionic Mixed Bonding

If there is a significant difference in electronegativity in the elements that form an intermetallic compound, there may be a significant amount of electron transfer (ionic binding) in the compound. Thus, some intermetallic compounds are good examples for mixed metallic-ionic bonding. Electron transfer is especially important for intermetallic compounds such as $NaZn_{13}$ and less important for compounds Al_9Co_3 and Fe_5Zn_{21} since the electronegativity differences for the latter two compounds are much less.

2.10 SUMMARY

Atoms consist mainly of three basic subatomic particles: *protons, neutrons,* and *electrons.* The electrons are envisaged as forming a cloud of varying density around a denser atomic nucleus containing almost all the mass of the atom. The outer electrons (high energy electrons) are the valence electrons, and it is mainly their behavior that determines the chemical reactivity of each atom.

Electrons obey the laws of quantum mechanics, and as a result the energies of electrons are *quantized.* That is, an electron can have only certain allowed values of energies. If an electron changes its energy, it must change to a new allowed energy level. During an energy change, an electron emits or absorbs a photon of energy according to Planck's equation $\Delta E = h\nu$, where ν is the frequency of the radiation. Each electron is associated with four quantum numbers: the principal quantum number n, the subsidiary quantum number l, the magnetic quantum number m_l, and the spin quantum number m_s. According to Pauli's exclusion principle, *no two electrons in the same atom can have all four quantum numbers the same.* Electrons also obey Heisenberg's uncertainty principle, which states that it is impossible to determine the momentum and position of an electron simultaneously. Thus, the location of electrons in atoms must be considered in terms of electron density distributions.

There are two main types of atomic bonds: (1) *strong primary bonds* and (2) *weak secondary bonds.* Primary bonds can be subdivided into (1) *ionic,* (2) *covalent,* and (3) *metallic bonds,* and secondary bonds can be subdivided into (1) *fluctuating dipoles* and (2) *permanent dipoles.*

Ionic bonds are formed by the transfer of one or more electrons from an electropositive atom to an electronegative one. The ions are bonded together in a solid crystal by electrostatic (coulombic) forces and are *nondirectional.* The size of the ions (geometric factor) and electrical neutrality are the two main factors that determine the ion packing arrangement. *Covalent bonds* are formed by the sharing of electrons in pairs by half-filled orbitals. The more the bonding orbitals overlap, the stronger the bond. Covalent bonds are *directional. Metallic bonds* are formed by metal atoms by a mutual sharing of valence electrons in the form of delocalized electron charge clouds. In general the fewer the valence electrons, the more delocalized they are and the more metallic the bonding. *Metallic bonding* only occurs among an aggregate of atoms and is *nondirectional.*

Secondary bonds are formed by the electrostatic attraction of electric dipoles within atoms or molecules. *Fluctuating dipoles* bond atoms together due to an asymmetrical distribution of electron charge within atoms. These bonding forces are important for the liquefaction and solidification of noble gases. *Permanent dipole* bonds are important in the bonding of polar covalently bonded molecules such as water and hydrocarbons.

Mixed bonding commonly occurs between atoms and in molecules. For example, metals such as titanium and iron have mixed metallic-covalent bonds; covalently bonded compounds such as GaAs and ZnSe have a certain amount of ionic character; some inter-metallic compounds such as $NaZn_{13}$ have some ionic bonding mixed with metallic bonding. In general, bonding occurs between atoms or molecules because their energies are lowered by the bonding process.

2.11 DEFINITIONS

Sec. 2.1
Atom: the basic unit of an element that can undergo chemical change.

Sec. 2.2
Atomic number: the number of protons in the nucleus of an atom of an element.
Atomic mass unit (u): mass unit based on the mass of exactly 12 for $^{12}_{6}C$.
Avogadro's number: 6.023×10^{23} atoms/mol; the number of atoms in one relative gram-mole or mole of an element.

Sec. 2.3
Quantum mechanics: a branch of physics in which systems under investigation can have only discrete allowed energy values that are separated by forbidden regions.
Ground state: the quantum state with the lowest energy.
Photon: a particle of radiation with an associated wavelength and frequency. Also referred to as a *quantum* of radiation.
Ionization energy: the energy required to remove an electron from its ground state in an atom to infinity.
Heisenberg's uncertainty principle: the statement that it is impossible to determine accurately at the same time the position and momentum of a small particle such as an electron.
Quantum numbers: the set of four numbers necessary to characterize each electron in an atom. These are the principal quantum number n, the orbital quantum number l, the magnetic quantum number m_l, and the spin quantum number m_s.
Atomic orbital: the region in space about the nucleus of an atom in which an electron with a given set of quantum numbers is most likely to be found. An atomic orbital is also associated with a certain energy level.
Pauli exclusion principle: the statement that no two electrons can have the same four quantum numbers.
Electron shell: a group of electrons with the same principal quantum number n.
Electron configuration: the distribution of all the electrons in an atom according to their atomic orbitals.
Anion: an ion with a negative charge.
Cation: an ion with a positive charge.

Sec. 2.5
Ionic bond: a primary bond resulting from the electrostatic attraction of oppositely charged ions. It is a nondirectional bond. An example of an ionically bonded material is a NaCl crystal.

Sec. 2.6

Covalent bond: a primary bond resulting from the sharing of electrons. In most cases the covalent bond involves the overlapping of half-filled orbitals of two atoms. It is a directional bond. An example of a covalently bonded material is diamond.

Hybrid orbital: an atomic orbital obtained when two or more nonequivalent orbitals of an atom combine. The process of the rearrangement of the orbitals is called *hybridization.*

Sec. 2.7

Valence electrons: electrons in the outermost shells that are most often involved in bonding.

Positive-ion core: an atom without its valence electrons.

Metallic bond: a primary bond resulting from the sharing of delocalized outer electrons in the form of an electron charge cloud by an aggregate of metal atoms. It is a nondirectional bond. An example of a metallically bonded material is elemental sodium.

Sec. 2.8

Permanent dipole bond: a secondary bond created by the attraction of molecules that have permanent dipoles. That is, each molecule has positive and negative charge centers separated by a distance.

Hydrogen bond: a special type of intermolecular permanent dipole attraction that occurs between a hydrogen atom bonded to a highly electronegative element (F, O, N, or Cl) and another atom of a highly electronegative element.

2.12 PROBLEMS

Answers to problems marked with an asterisk are given at the end of the book.

2.1   What is the mass in grams of (*a*) a proton, (*b*) a neutron, and (*c*) an electron?

2.2   What is the electric charge in coulombs of (*a*) a proton, (*b*) a neutron, and (*c*) an electron?

2.3   Define (*a*) atomic number, (*b*) atomic mass unit, (*c*) Avogadro's number, and (*d*) relative gram atomic mass.

2.4   What is the mass in grams of one atom of gold?

2.5   How many atoms are there in 1 g of gold?

*2.6   A gold wire is 0.70 mm in diameter and 8.0 cm in length. How many atoms does it contain? The density of gold is 19.3 g/cm^3.

2.7   What is the mass in grams of one atom of molybdenum?

*2.8   How many atoms are there in 1 g of molybdenum?

*2.9   A solder contains 52 wt % tin and 48 wt % lead. What are the atomic percentages of Sn and Pb in the solder?

2.10   A monel alloy consists of 70 wt % Ni and 30 wt % Cu. What are the atomic percentages of Ni and Cu in this alloy?

2.11   A cupronickel alloy consists of 80 wt % Cu and 20 wt % Ni. What are the atomic percentages of Cu and Ni in the alloy?

*2.12   What is the chemical formula of an intermetallic compound that consists of 49.18 wt % Cu and 50.82 wt % Au?

2.13 What is the chemical formula of an intermetallic compound that consists of 15.68 wt % Mg and 84.32 wt % Al?

2.14 Define a photon.

*2.15 Calculate the energy in joules and electron volts of the photon whose wavelength is 303.4 nm.

2.16 Calculate the energy in joules and electron volts of the photon whose wavelength is 226.4 nm.

*2.17 A hydrogen atom exists with its electron in the $n = 4$ state. The electron undergoes a transition to the $n = 3$ state. Calculate (*a*) the energy of the photon emitted, (*b*) its frequency, and (*c*) its wavelength in nanometers (nm).

2.18 A hydrogen atom exists with its electron in the $n = 6$ state. The electron undergoes a transition to the $n = 2$ state. Calculate (*a*) the energy of the photon emitted, (*b*) its frequency, and (*c*) its wavelength in nanometers.

*2.19 In a commercial x-ray generator, a stable metal such as copper (Cu) or tungsten (W) is exposed to an intense beam of high-energy electrons. These electrons cause ionization events in the metal atoms. When the metal atoms regain their ground state they emit x-rays of characteristic energy and wavelength. For example, a "tungsten" atom struck by a high-energy electron may lose one of its K shell electrons. When this happens, another electron, probably from the tungsten L shell will "fall" into the vacant site in the K shell. If such a $2p \rightarrow 1s$ transition occurs in tungsten, a tungsten K_α x-ray is emitted. A tungsten K_α x-ray has a wavelength λ of 0.02138 nm. What is its energy? What is its frequency?

2.20 Most modern *scanning electron microscopes* (SEMs) are equipped with energy-dispersive x-ray detectors for the purpose of chemical analysis of the specimens. This x-ray analysis is a natural extension of the capability of the SEM because the electrons that are used to form the image are also capable of creating characteristic x-rays in the sample. When the electron beam hits the specimen, x-rays specific to the elements in the specimen are created. These can be detected and used to deduce the composition of the specimen from the well-known wavelengths of the characteristic x-rays of the elements. For example:

Element	Wavelength of K_α x-rays
Cr	0.2291 nm
Mn	0.2103 nm
Fe	0.1937 nm
Co	0.1790 nm
Ni	0.1659 nm
Cu	0.1542 nm
Zn	0.1436 nm

Suppose a metallic alloy is examined in an SEM and three different x-ray energies are detected. If the three energies are 7492, 5426, and 6417 eV, what elements are present in the sample? What would you call such an alloy? (Look ahead to Chap. 9 in the textbook.)

2.21 Describe the Bohr model of the hydrogen atom. What are the major shortcomings of the Bohr model?

2.22 What is the ionization energy in electron volts for a hydrogen electron in the ground state?

2.23 Describe the four quantum numbers of an electron and give their allowed values.

2.24 Write the electron configurations of the following elements by using *spdf* notion: (*a*) yttrium, (*b*) hafnium, (*c*) samarium, (*d*) rhenium.

2.25 What is the outer electron configuration for all the noble gases except helium?

2.26 Of the noble gases Ne, Ar, Kr, and Xe, which should be the most chemically reactive?

2.27 Define the term *electronegativity.*

2.28 Which five elements are the most electropositive according to the electronegativity scale?

2.29 Which five elements are the most electronegative according to the electronegativity scale?

2.30 Write the electron configuration of the following ions using the *spdf* notation: (*a*) Cr^{2+}, Cr^{3+}, Cr^{6+}; (*b*) *Mo^{3+}, Mo^{4+}, Mo^{6+}; (*c*) Se^{4+}, Se^{6+}, Se^{2-}.

2.31 Briefly describe the following types of primary bonding: (*a*) ionic, (*b*) covalent, and (*c*) metallic.

2.32 Briefly describe the following types of secondary bonding: (*a*) fluctuating dipole and (*b*) permanent dipole.

2.33 In general, why does bonding between atoms occur?

2.34 Describe the ionic bonding process between a pair of Na and Cl atoms. Which electrons are involved in the bonding process?

2.35 After ionization, why is the sodium ion smaller than the sodium atom?

2.36 After ionization, why is the chloride ion larger than the chlorine atom?

2.37 Calculate the attractive force (•→ ←•) between a pair of K^+ and Br^- ions that just touch each other. Assume the ionic radius of the K^+ ion to be 0.133 nm and that of the Br^- ion to be 0.196 nm.

*2.38 Calculate the attractive force (•→ ←•) between a pair of Ba^{2+} and S^{2-} ions that just touch each other. Assume the ionic radius of the Ba^{2+} ion to be 0.143 nm and that of the S^{2-} ion to be 0.174 nm.

*2.39 Calculate the net potential energy for a K^+Br^- pair by using the *b* constant calculated from Prob. 2.37. Assume $n = 9.5$.

2.40 Calculate the net potential energy for a $Ba^{2+}S^{2-}$ ion pair by using the *b* constant calculated from Prob. 2.38. Assume $n = 10.5$.

2.41 If the attractive force between a pair of Cs^+ and I^- ions is 2.83×10^{-9} N and the ionic radius of the Cs^+ ion is 0.165 nm, calculate the ionic radius of the I^- ion in nanometers.

*2.42 If the attractive force between a pair of Sr^{2+} and O^{2-} ions is 1.29×10^{-8} N and the ionic radius of the O^{2-} ion is 0.132 nm, calculate the ionic radius of the Sr^{2+} ion in nanometers.

2.43 Describe the two major factors that must be taken into account in the packing of ions in an ionic crystal.

2.44 Describe the covalent bonding process between a pair of hydrogen atoms. What is the driving energy for the formation of a diatomic molecule?

2.45 Describe the covalent bonding electron arrangement in the following diatomic molecules: (*a*) fluorine, (*b*) oxygen, (*c*) nitrogen.

2.46 Describe the hybridization process for the formation of four equivalent sp^3 hybrid orbitals in carbon during covalent bonding. Use orbital diagrams.

2.47 List the number of atoms bonded to a C atom that exhibits sp^3, sp^2, and sp hybridization. For each, give the geometrical arrangement of the atoms in the molecule.

2.48 Why is diamond such a hard material?

2.49 Describe the metallic bonding process among an aggregate of copper atoms.

2.50 How can the high electrical and thermal conductivities of metals be explained by the "electron gas" model of metallic bonding? Ductility?

2.51 The melting point of the metal potassium is 63.5°C, while that of titanium is 1660°C. What explanation can be given for this great difference in melting temperatures?

2.52 Is there a correlation between the electron configurations of the elements potassium ($Z = 19$) through copper ($Z = 29$) and their melting points? (See Tables 2.8 and 2.9.)

*2.53 Using the covalent-metallitivity values of Table 2.9, calculate values for the percent metallic and covalent bonding in the structural metal titanium.

2.54 Using the covalent-metallitivity values of Table 2.9, calculate values for the percent metallic and covalent bonding in the metal tungsten.

2.55 Define an electric dipole moment.

2.56 Describe fluctuating dipole bonding among the atoms of the noble gas neon. Of a choice between the noble gases krypton and xenon, which noble gas would be expected to have the strongest dipole bonding and why?

2.57 Describe permanent dipole bonding among polar covalent molecules.

2.58 Carbon tetrachloride (CCl_4) has a zero dipole moment. What does this tell us about the C—Cl bonding arrangement in this molecule?

2.59 Describe the hydrogen bond. Among what elements is this bond restricted?

2.60 Describe hydrogen bonding among water molecules.

2.61 Methane (CH_4) has a much lower boiling temperature than does water (H_2O). Explain why this is true in terms of the bonding between molecules in each of these two substances.

2.62 Sketch a tetrahedron and make a small model of one from cardboard and tape with a 2-in. side. Do the same for an octahedron.

2.63 What is Pauling's equation for determining the percentage ionic character in a mixed ionic-covalently bonded compound?

*2.64   Compare the percentage ionic character in the semiconducting compounds CdTe and InP.

2.65   Compare the percentage ionic character in the semiconducting compounds InSb and ZnTe.

2.66   For each of the following compounds, state whether the bonding is essentially metallic, covalent, ionic, van der Waals, or hydrogen: (*a*) Ni, (*b*) ZrO_2, (*c*) graphite, (*d*) solid Kr, (*e*) Si, (*f*) BN, (*g*) SiC, (*h*) Fe_2O_3, (*i*) MgO, (*j*) W, (*k*) H_2O within the molecules, (*l*) H_2O between the molecules.

   If ionic and covalent bonding are involved in the bonding of any of the compounds listed, calculate the percentage ionic character in the compound.

2.13 MATERIALS SELECTION AND DESIGN PROBLEMS

1.   Cartridge brass is an alloy of two metals: 70 wt % copper and 30 wt % zinc. Discuss the nature of the bonds between copper and zinc in this alloy.

2.   Pure aluminum is a ductile metal with low tensile strength and hardness. Its oxide Al_2O_3 (alumina) is extremely strong, hard, and brittle. Can you explain this difference from an atomic bonding point of view?

3.   Graphite and diamond are both made from carbon atoms. (*a*) List some of the physical characteristics of each. (*b*) Give one application for graphite and one for diamond. (*c*) If both materials are made of carbon, why does such a difference in properties exist?

4.   Silicon is extensively used in the manufacture of integrated circuit devices such as transistors and light-emitting diodes. It is often necessary to develop a thin oxide layer (SiO_2) on silicon wafers. (*a*) What are the differences in properties between the silicon substrate and the oxide layer? (*b*) Design a process that produces the oxide layer on a silicon wafer. (*c*) Design a process that forms the oxide layer only in certain desired areas.

5.   In the manufacturing of a light bulb, the bulb is evacuated of air and then filled with argon gas. What is the purpose of this?

6.   Stainless steel is a corrosion-resistant metal because it contains large amounts of chromium. How does chromium protect the metal from corrosion?

7.   Robots are used in auto industries to weld two components at specific locations. Clearly, the end position of the arm must be determined accurately in order to weld the components at the precise position. (*a*) In selecting the material for the arm of such robots, what factors must be considered? (*b*) Select a proper material for this application.

8.   (*a*) Explain why contact between a hard and soft metal is not suitable for applications where "abrasive wear" is of great concern. (*b*) How would you protect a steel cutting tool from premature wear? (*c*) Design a process for your solution.

9.   A certain application requires a material that must be very hard and corrosion resistant at room temperature and atmosphere. It would be beneficial, but not

necessary, if the material were also impact resistant. (*a*) If you only consider the major requirements, which classes of materials would you search for this selection? (*b*) If you consider both major and minor requirements, which classes would you search? (*c*) Suggest a material.

10. A certain application requires a material that is lightweight, an electrical insulator, and has some flexibility. (*a*) Which class of materials would you search for this selection? (*b*) Explain your answer from a bonding point of view.

11. A certain application requires a material that is electrically nonconductive (insulator), extremely stiff, and lightweight. Which classes of materials would you search for this selection? (*b*) Explain your answer from a bonding point of view.

3 Crystal and Amorphous Structure in Materials

(*a*)   (*b*)   (*c*)   (*d*)

(© *Paul Silverman/Fundamental Photographs*)

Solids may be categorized broadly into crystalline and amorphous solids. Crystalline solids, due to orderly structure of their atoms, molecules, or ions possess well-defined shapes. Metals are crystalline and are composed of well-defined crystals or grains. The grains are small and are not clearly observable due to the opaque nature of metals. In minerals, mostly translucent to transparent nature, the well-defined crystalline shapes are clearly observable. The following images show the crystalline nature of minerals such as (*a*) celestite ($SrSo_4$) with a sky blue or celestial color, (*b*) pyrite (FeS_2), also called the "fool's gold" due to its brassy yellow color, (*c*) amethyst (SiO_2), a purple variety of quartz, and (*d*) halite ($NaCl$), better known as rock salt. In contrast, amorphous solids have poor or no long-range order and do not solidify with the symmetry and regularity of crystalline solids. ∎

By the end of this chapter, students will be able to. . .

1. Describe what crystalline and noncrystalline (amorphous) materials are.

2. Learn how atoms and ions in solids are arranged in space and identify the basic building blocks of solids.

3. Describe the difference between atomic structure and crystal structure for solid material.

4. Distinguish between crystal structure and crystal system.

5. Explain why plastics cannot be 100% crystalline in structure.

6. Explain polymorphism or allotropy in materials.

7. Compute the densities for metals having body-centered, and face-centered cubic structures.

8. Describe how to use the x-ray diffraction method for material characterization.

9. Write the designation for atom position, direction indices, and Miller indices for cubic crystals. Specify what are the three densely packed structures for most metals. Determine Miller-Bravais indices for hexagonal-closed packed structure. Be able to draw directions and planes in cubic and hexagonal crystals.

3.1 THE SPACE LATTICE AND UNIT CELLS

The physical structure of solid materials of engineering importance depends mainly on the arrangements of the atoms, ions, or molecules that make up the solid and the bonding forces between them. If the atoms or ions of a solid are arranged in a pattern that repeats itself in three dimensions, they form a solid that has *long-range order* (LRO) and is referred to as a *crystalline solid* or *crystalline material*. Examples of crystalline materials are metals, alloys, and some ceramic materials. In contrast to crystalline materials, there are some materials whose atoms and ions are not arranged in a long-range, periodic, and repeatable manner and possess only *short-range order* (SRO). This means that order exists only in the immediate neighborhood of an atom or a molecule. As an example, liquid water has short-range order in its molecules in which one oxygen atom is covalently bonded to two hydrogen atoms. But this order disappears, as each molecule is bonded to other molecules through weak secondary bonds in a random manner. Materials with only short-range order are classified as *amorphous* (without form) or noncrystalline. A more detailed definition and some examples of amorphous materials are given in Sec. 3.12.

Atomic arrangements in crystalline solids can be described by referring the atoms to the points of intersection of a network of lines in three dimensions. Such

Animation
Tutorial

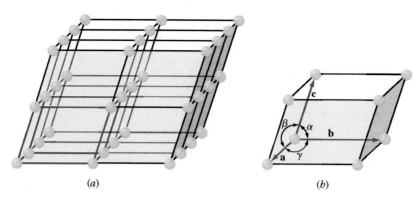

(a)   (b)

Figure 3.1
(a) Space lattice of ideal crystalline solid. (b) Unit cell showing lattice constants.

a network is called a **space lattice** (Fig. 3.1a), and it can be described as an infinite three-dimensional array of points. Each point in the space lattice has identical surroundings. In an ideal **crystal** the grouping of **lattice points** about any given point are identical with the grouping about any other lattice point in the crystal lattice. Each space lattice can thus be described by specifying the atom positions in a repeating **unit cell,** such as the one heavily outlined in Fig. 3.1a. A group of atoms organized in a certain arrangement relative to each other and associated with lattice points constitues the **motif** or basis. The crystal structure may then be defined as the collection of lattice and basis. It is important to note that atoms do not necessarily coincide with lattice points. The size and shape of the unit cell can be described by three lattice vectors **a**, **b**, and **c**, originating from one corner of the unit cell (Fig. 3.1b). The axial lengths a, b, and c and the interaxial angles α, β, and γ are the *lattice constants* of the unit cell.

3.2 CRYSTAL SYSTEMS AND BRAVAIS LATTICES

Tutorial

By assigning specific values for axial lengths and interaxial angles, unit cells of different types can be constructed. Crystallographers have shown that only seven different types of unit cells are necessary to create all space lattices. These crystal systems are listed in Table 3.1.

Many of the seven crystal systems have variations of the basic unit cell. A. J. Bravais[1] showed that 14 standard unit cells could describe all possible lattice networks. These Bravais lattices are illustrated in Fig. 3.2. There are four basic types of unit cells: (1) simple, (2) body-centered, (3) face-centered, and (4) basecentered.

---

[1]August Bravais (1811–1863). French crystallographer who derived the 14 possible arrangements of points in space.

Table 3.1 Classification of space lattices by crystal system

MatVis

Crystal system	Axial lengths and interaxial angles	Space lattice
Cubic	Three equal axes at right angles $a = b = c, \alpha = \beta = \gamma = 90°$	Simple cubic Body-centered cubic Face-centered cubic
Tetragonal	Three axes at right angles, two equal $a = b \neq c, \alpha = \beta = \gamma = 90°$	Simple tetragonal Body-centered tetragonal
Orthorhombic	Three unequal axes at right angles $a \neq b \neq c, \alpha = \beta = \gamma = 90°$	Simple orthorhombic Body-centered orthorhombic Base-centered orthorhombic Face-centered orthorhombic
Rhombohedral	Three equal axes, equally inclined $a = b = c, \alpha = \beta = \gamma \neq 90°$	Simple rhombohedral
Hexagonal	Two equal axes at 120°, third axis at right angles $a = b \neq c, \alpha = \beta = 90°, \gamma = 120°$	Simple hexagonal
Monoclinic	Three unequal axes, one pair not at right angles $a \neq b \neq c, \alpha = \gamma = 90° \neq \beta$	Simple monoclinic Base-centered monoclinic
Triclinic	Three unequal axes, unequally inclined and none at right angles $a \neq b \neq c, \alpha \neq \beta \neq \gamma \neq 90°$	Simple triclinic

In the cubic system there are three types of unit cells: simple cubic, bodycentered cubic, and face-centered cubic. In the orthorhombic system all four types are represented. In the tetragonal system there are only two: simple and body-centered. The face-centered tetragonal unit cell appears to be missing but can be constructed from four body-centered tetragonal unit cells. The monoclinic system has simple and base-centered unit cells, and the rhombohedral, hexagonal, and triclinic systems have only one simple type of unit cell.

3.3 PRINCIPAL METALLIC CRYSTAL STRUCTURES

In this chapter the principal crystal structures of elemental metals will be discussed in detail. In Chap. 10 the principal ionic and covalent crystal structures that occur in ceramic materials will be treated.

Most elemental metals (about 90 percent) crystallize upon solidification into three densely packed crystal structures: **body-centered cubic (BCC)** (Fig. 3.3a), **face-centered cubic (FCC)** (Fig. 3.3b) and **hexagonal close-packed (HCP)** (Fig. 3.3c). The HCP structure is a denser modification of the simple hexagonal crystal structure shown in Fig. 3.2. Most metals crystallize in these dense-packed structures because energy is released as the atoms come closer together and bond more tightly with each other. Thus, the densely packed structures are in lower and more stable energy arrangements.

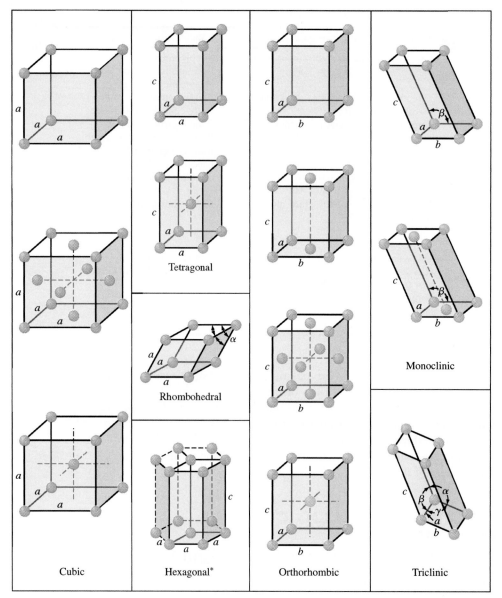

Figure 3.2

The 14 Bravais conventional unit cells grouped according to crystal system. The dots indicate lattice points that, when located on faces or at corners, are shared by other identical lattice unit cells.

Tutorial
Animation
MatVis

(After W. G. Moffatt, G. W. Pearsall, and J. Wulff, "The Structure and Properties of Materials," vol. I: "Structure," Wiley, 1964, p. 47.)

The unit cell is represented by solid lines.

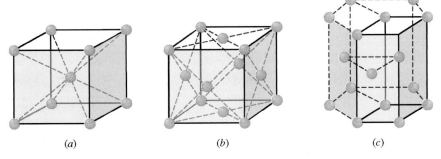

(a)   (b)   (c)

Figure 3.3
Principal metal crystal structure and unit cells: (a) body-centered cubic,
(b) face-centered cubic, (c) hexagonal close-packed crystal structure (the unit
cell is shown by solid lines).

The extremely small size of the unit cells of crystalline metals that are shown in Fig. 3.3 should be emphasized. The cube side of the unit cell of body-centered cubic iron, for example, at room temperature is equal to 0.287×10^{-9} m, or 0.287 nanometer (nm).[2] Therefore, if unit cells of pure iron are lined up side by side, in 1 mm there will be

$$1 \text{ mm} \times \frac{1 \text{ unit cell}}{0.287 \text{ nm} \times 10^{-6} \text{ mm/nm}} = 3.48 \times 10^6 \text{ unit cells!}$$

Let us now examine in detail the arrangement of the atoms in the three principal crystal structure unit cells. Although an approximation, we shall consider atoms in these crystal structures to be hard spheres. The distance between the atoms (interatomic distance) in crystal structures can be determined experimentally by x-ray diffraction analysis.[3] For example, the interatomic distance between two aluminum atoms in a piece of pure aluminum at 20°C is 0.2862 nm. The radius of the aluminum atom in the aluminum metal is assumed to be half the interatomic distance, or 0.143 nm. The atomic radii of selected metals are listed in Tables 3.2 to 3.4.

3.3.1 Body-Centered Cubic (BCC) Crystal Structure

First, consider the atomic-site unit cell for the BCC crystal structure shown in Fig. 3.4a. In this unit cell the solid spheres represent the centers where atoms are located and clearly indicate their relative positions. If we represent the atoms in this cell as hard spheres, then the unit cell appears as shown in Fig. 3.4b. In this unit cell we see that the central atom is surrounded by eight nearest neighbors and is said to have a coordination number of 8.

Animation
Tutorial

---

[2]1 nanometer $= 10^{-9}$ meter.

[3]Some of the principles of x-ray diffraction analysis will be studied in Sec. 3.11.

Table 3.2 Selected metals that have the BCC crystal structure at room temperature (20°C) and their lattice constants and atomic radii

Metal	Lattice constant a (nm)	Atomic radius $R*$ (nm)
Chromium	0.289	0.125
Iron	0.287	0.124
Molybdenum	0.315	0.136
Potassium	0.533	0.231
Sodium	0.429	0.186
Tantalum	0.330	0.143
Tungsten	0.316	0.137
Vanadium	0.304	0.132

*Calculated from lattice constants by using Eq. (3.1), $R = \sqrt{3}a/4$.

Tutorial

Table 3.3 Selected metals that have the FCC crystal structure at room temperature (20°C) and their lattice constants and atomic radii

Metal	Lattice constant a (nm)	Atomic radius $R*$ (nm)
Aluminum	0.405	0.143
Copper	0.3615	0.128
Gold	0.408	0.144
Lead	0.495	0.175
Nickel	0.352	0.125
Platinum	0.393	0.139
Silver	0.409	0.144

*Calculated from lattice constants by using Eq. (3.3), $R = \sqrt{2}a/4$.

Table 3.4 Selected metals that have the HCP crystal structure at room temperature (20°C) and their lattice constants, atomic radii, and c/a ratios

Metal	Lattice constants (nm)		Atomic radius R (nm)	c/a ratio	% deviation from ideality
	a	c			
Cadmium	0.2973	0.5618	0.149	1.890	+15.7
Zinc	0.2665	0.4947	0.133	1.856	+13.6
Ideal HCP				1.633	0
Magnesium	0.3209	0.5209	0.160	1.623	−0.66
Cobalt	0.2507	0.4069	0.125	1.623	−0.66
Zirconium	0.3231	0.5148	0.160	1.593	−2.45
Titanium	0.2950	0.4683	0.147	1.587	−2.81
Beryllium	0.2286	0.3584	0.113	1.568	−3.98

If we isolate a single hard-sphere unit cell, we obtain the model shown in Fig. 3.4c. Each of these cells has the equivalent of two atoms per unit cell. One complete atom is located at the center of the unit cell, and an eighth of a sphere is located at each corner of the cell, making the equivalent of another atom. Thus there is a total of 1 (at the center) + 8 × $\frac{1}{8}$ (at the corners) = 2 atoms per unit cell. The atoms in the BCC unit cell contact each other across the cube diagonal, as indicated in

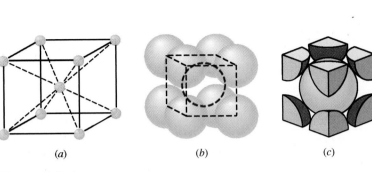

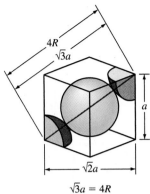

Figure 3.5

$\sqrt{3}a = 4R$

(a)   (b)   (c)

Figure 3.4
BCC unit cells: (*a*) atomic-site unit cell, (*b*) hard-sphere unit cell, and
(*c*) isolated unit cell.

Figure 3.5
BCC unit cell showing
relationship between the
lattice constant *a* and
the atomic radius *R*.

Tutorial

Animation

MatVis

Fig. 3.5, so that the relationship between the length of the cube side *a* and the atomic
radius *R* is

$$\sqrt{3}a = 4R \quad \text{or} \quad a = \frac{4R}{\sqrt{3}} \tag{3.1}$$

Iron at 20°C is BCC with atoms of atomic radius 0.124 nm. Calculate the lattice constant *a* for the cube edge of the iron unit cell.

**EXAMPLE
PROBLEM 3.1**

■ **Solution**
From Fig. 3.5 it is seen that the atoms in the BCC unit cell touch across the cube diagonals. Thus, if *a* is the length of the cube edge, then

$$\sqrt{3}a = 4R \tag{3.1}$$

where *R* is the radius of the iron atom. Therefore

$$a = \frac{4R}{\sqrt{3}} = \frac{4(0.124 \text{ nm})}{\sqrt{3}} = 0.2864 \text{ nm} \blacktriangleleft$$

If the atoms in the BCC unit cell are considered to be spherical, an **atomic packing
factor** (APF) can be calculated by using the equation

$$\text{APF} = \frac{\text{volume of atoms in unit cell}}{\text{volume of unit cell}} \tag{3.2}$$

Using this equation, the APF for the BCC unit cell (Fig. 3.3*a*) is calculated to be 68 percent (see Example Problem 3.2). That is, 68 percent of the volume of the BCC unit cell is occupied by atoms and the remaining 32 percent is empty space. The BCC crystal structure is *not* a close-packed structure since the atoms could be packed closer together. Many metals such as iron, chromium, tungsten, molybdenum, and vanadium have the BCC crystal structure at room temperature. Table 3.2 lists the lattice constants and atomic radii of selected BCC metals.

EXAMPLE PROBLEM 3.2

Tutorial

Calculate the atomic packing factor (APF) for the BCC unit cell, assuming the atoms to be hard spheres.

■ **Solution**

$$APF = \frac{\text{volume of atoms in BCC unit cell}}{\text{volume of BCC unit cell}} \tag{3.2}$$

Since there are two atoms per BCC unit cell, the volume of atoms in the unit cell of radius R is

$$V_{\text{atoms}} = (2)(\tfrac{4}{3}\pi R^3) = 8.373R^3$$

The volume of the BCC unit cell is

$$V_{\text{unit cell}} = a^3$$

where a is the lattice constant. The relationship between a and R is obtained from Fig. 3.5, which shows that the atoms in the BCC unit cell touch each other across the cubic diagonal. Thus

$$\sqrt{3}a = 4R \quad \text{or} \quad a = \frac{4R}{\sqrt{3}} \tag{3.1}$$

Thus

$$V_{\text{unit cell}} = a^3 = 12.32R^3$$

The atomic packing factor for the BCC unit cell is, therefore,

$$APF = \frac{V_{\text{atoms}}/\text{unit cell}}{V_{\text{unit cell}}} = \frac{8.373R^3}{12.32R^3} = 0.68 \blacktriangleleft$$

3.3.2 Face-Centered Cubic (FCC) Crystal Structure

Consider next the FCC lattice-point unit cell of Fig. 3.6*a*. In this unit cell there is one lattice point at each corner of the cube and one at the center of each cube face. The hard-sphere model of Fig. 3.6*b* indicates that the atoms in the FCC crystal structure are packed as close together as possible. The APF for this close-packed structure is 0.74 as compared to 0.68 for the BCC structure, which is not close-packed.

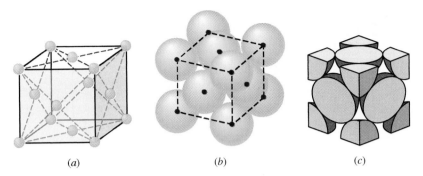

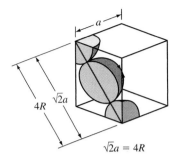

(a)   (b)   (c) $\sqrt{2}a = 4R$

Figure 3.6
FCC unit cells: (a) atomic-site unit cell, (b) hard-sphere unit cell, and (c) isolated unit cell.

Figure 3.7
FCC unit cell showing relationship between the lattice constant a and atomic radius R. Since the atoms touch across the face diagonals, $\sqrt{2}a = 4R$.

Tutorial

Animation

MatVis

The FCC unit cell as shown in Fig. 3.6c has the equivalent of four atoms per unit cell. The eight corner octants account for one atom $(8 \times \frac{1}{8} = 1)$, and the six half-atoms on the cube faces contribute another three atoms, making a total of four atoms per unit cell. The atoms in the FCC unit cell contact each other across the cubic face diagonal, as indicated in Fig. 3.7, so that the relationship between the length of the cube side a and the atomic radius R is

$$\sqrt{2}a = 4R \quad \text{or} \quad a = \frac{4R}{\sqrt{2}} \tag{3.3}$$

The APF for the FCC crystal structure is 0.74, which is greater than the 0.68 factor for the BCC structure. The APF of 0.74 is for the closest packing possible of "spherical atoms." Many metals such as aluminum, copper, lead, nickel, and iron at elevated temperatures (912 to 1394°C) crystallize with the FCC crystal structure. Table 3.3 lists the lattice constants and atomic radii for some selected FCC metals.

3.3.3 Hexagonal Close-Packed (HCP) Crystal Structure

The third common metallic crystal structure is the HCP structure shown in Fig. 3.8a & b. Metals do not crystallize into the simple hexagonal crystal structure shown in Fig. 3.2 because the APF is too low. The atoms can attain a lower energy and a more stable condition by forming the HCP structure of Fig. 3.8b. The APF of the HCP crystal structure is 0.74, the same as that for the FCC crystal structure since in both structures the atoms are packed as tightly as possible. In both the HCP and FCC crystal structures each atom is surrounded by 12 other atoms, and thus both structures have a coordination number of 12. The differences in the atomic packing in FCC and HCP crystal structures will be discussed in Sec. 3.8.

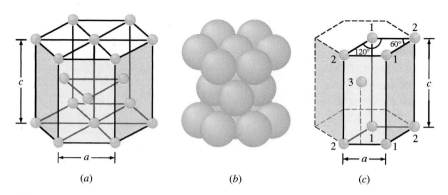

(a)   (b)   (c)

Tutorial

MatVis

Figure 3.8
HCP crystal structure: (*a*) schematic of the crystal structure, (*b*) hard-sphere
model, and (*c*) isolated unit cell schematic.
[(*b*) and (*c*) after F. M. Miller, "Chemistry: Structure and Dynamics," McGraw-Hill, 1984,
p. 296.]

The isolated HCP unit cell also called the primitive cell is shown in Fig. 3.8*c*.
The atoms at locations marked "1" on figure 3.8*c* contribute $\frac{1}{6}$ of an atom to the unit
cell. The atoms at locations marked "2" contribute $\frac{1}{12}$ of an atom to the unit cell.
Thus the atoms at the eight corners of the unit cell collectively contribute one atom
$(4(\frac{1}{6}) + 4(\frac{1}{12}) = 1)$. The atom at location "3" is centered inside the unit cell but
extends beyond the boundary of the cell. The total number of atoms inside an HCP
unit cell* is therefore 2 (1 at corners and 1 center).

The ratio of the height *c* of the hexagonal prism of the HCP crystal structure
to its basal side *a* is called the *c/a ratio* (Fig. 3.8*a*). The *c/a ratio* for an ideal
HCP crystal structure consisting of uniform spheres packed as tightly together as
possible is 1.633. Table 3.4 lists some important HCP metals and their *c/a* ratios.
Of the metals listed, cadmium and zinc have *c/a* ratios higher than ideality, which
indicates that the atoms in these structures are slightly elongated along the *c* axis
of the HCP unit cell. The metals magnesium, cobalt, zirconium, titanium, and
beryllium have *c/a* ratios less than the ideal ratio. Therefore, in these metals the
atoms are slightly compressed in the direction along the *c* axis. Thus, for the HCP
metals listed in Table 3.4 there is a certain amount of deviation from the ideal
hard-sphere model.

---

*In some text books the HCP unit cell is represented by figure 3.8*a* and is called the "larger cell". In such a
case one finds 6 atoms per unit cell. This is mostly for convenience and the true unit cell is presented in
figure 3.8*c* by the solid lines. When presenting the topics of crystal directions and planes we will also use
the larger cell for convenience.

**EXAMPLE
PROBLEM 3.3**   Calculate the volume of the zinc crystal structure unit cell by using the following data:
pure zinc has the HCP crystal structure with lattice constants $a = 0.2665$ nm and
$c = 0.4947$ nm.

■ **Solution**

The volume of the zinc HCP unit cell can be obtained by determining the area of the base of the unit cell and then multiplying this by its height (Fig. 3.9).

The area of the base of the unit cell is area *ABDEFG* of Fig. 3.9*a* and *b*. This total area consists of the areas of six equilateral triangles of area *ABC* of Fig. 3.9*b*. From Fig. 3.9*c*,

$$\text{Area of trianlge } ABC = \tfrac{1}{2}(\text{base})(\text{height})$$
$$= \tfrac{1}{2}(a)(a \sin 60°) = \tfrac{1}{2}a^2 \sin 60°$$

From Fig. 3.9*b*,

$$\text{Total area of HCP base} = (6)(\tfrac{1}{2}a^2 \sin 60°)$$
$$= 3a^2 \sin 60°$$

From Fig. 3.9*a*,

$$\text{Volume of zinc HCP unit cell} = (3a^2 \sin 60°)(c)$$
$$= (3)(0.2665 \text{ nm})^2(0.8660)(0.4947 \text{ nm})$$
$$= 0.0913 \text{ nm}^3 \blacktriangleleft$$

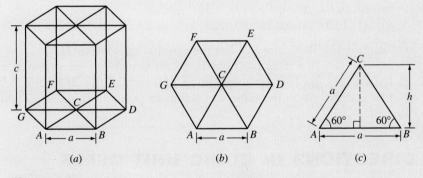

Figure 3.9
Diagrams for calculating the volume of an HCP unit cell. (*a*) HCP unit cell.
(*b*) Base of HCP unit cell. (*c*) Triangle *ABC* removed from base of unit cell.

3.4 ATOM POSITIONS IN CUBIC UNIT CELLS

To locate atom positions in cubic unit cells, we use rectangular x, y, and z axes. In crystallography the positive x axis is usually the direction coming out of the paper, the positive y axis is the direction to the right of the paper, and the positive z axis is the direction to the top (Fig. 3.10). Negative directions are opposite to those just described.

Atom positions in unit cells are located by using unit distances along the x, y, and z axes, as indicated in Fig. 3.10*a*. For example, the position coordinates for the

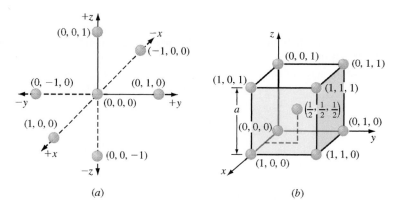

Tutorial

Figure 3.10
(*a*) Rectangular *x, y,* and *z* axes for locating atom positions in cubic unit cells. (*b*) Atom positions in a BCC unit cell.

atoms in the BCC unit cell are shown in Fig. 3.10*b*. The atom positions for the eight corner atoms of the BCC unit cell are

(0, 0, 0)   (1, 0, 0)   (0, 1, 0)   (0, 0, 1)
(1, 1, 1)   (1, 1, 0)   (1, 0, 1)   (0, 1, 1)

The center atom in the BCC unit cell has the position coordinates $(\frac{1}{2}, \frac{1}{2}, \frac{1}{2})$. For simplicity sometimes only two atom positions in the BCC unit cell are specified which are (0, 0, 0) and $(\frac{1}{2}, \frac{1}{2}, \frac{1}{2})$. The remaining atom positions of the BCC unit cell are assumed to be understood. In the same way the atom positions in the FCC unit cell can be located.

3.5 DIRECTIONS IN CUBIC UNIT CELLS

Often it is necessary to refer to specific directions in crystal lattices. This is especially important for metals and alloys with properties that vary with crystallographic orientation. *For cubic crystals the crystallographic* **direction indices** *are the vector components of the direction resolved along each of the coordinate axes and reduced to the smallest integers.*

To diagrammatically indicate a direction in a cubic unit cell, we draw a direction vector from an origin, which is usually a corner of the cubic cell, until it emerges from the cube surface (Fig. 3.11). The position coordinates of the unit cell where the direction vector emerges from the cube surface after being converted to integers are the direction indices. The direction indices are enclosed by square brackets with no separating commas.

For example, the position coordinates of the direction vector *OR* in Fig. 3.11*a* where it emerges from the cube surface are (1, 0, 0), and so the direction indices for the direction vector *OR* are [100]. The position coordinates of the direction vector

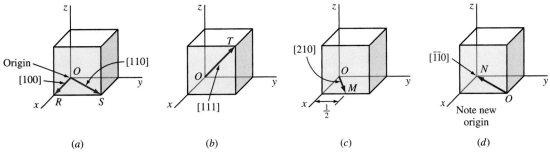

Figure 3.11
Some directions in cubic unit cells.

Tutorial

OS (Fig. 3.11*a*) are (1, 1, 0), and so the direction indices for *OS* are [110]. The position coordinates for the direction vector *OT* (Fig. 3.11*b*) are (1, 1, 1), and so the direction indices of *OT* are [111].

The position coordinates of the direction vector *OM* (Fig. 3.11*c*) are $(1, \frac{1}{2}, 0)$, and since the direction vectors must be integers, these position coordinates must be multiplied by 2 to obtain integers. Thus, the direction indices of *OM* become $2(1, \frac{1}{2}, 0) = [210]$. The position coordinates of the vector *ON* (Fig. 3.11*d*) are (−1, −1, 0). A negative direction index is written with a bar over the index. Thus, the direction indices for the vector *ON* are [$\bar{1}$10]. Note that to draw the direction *ON* inside the cube the origin of the direction vector had to be moved to the front lower-right corner of the unit cube (Fig. 3.11*d*). Further examples of cubic direction vectors are given in Example Problem 3.4.

The letters *u*, *v*, *w* are used in a general sense for the direction indices in the *x*, *y*, and *z* directions, respectively, and are written as [*uvw*]. It is also important to note that *all parallel direction vectors have the same direction indices.*

Directions are said to be *crystallographically equivalent* if the atom spacing along each direction is the same. For example, the following cubic edge directions are crystallographic equivalent directions:

$$[100], [010], [001], [0\bar{1}0], [00\bar{1}], [\bar{1}00] \equiv \langle 100 \rangle$$

Equivalent directions are called *indices of a family or form*. The notation $\langle 100 \rangle$ is used to indicate cubic edge directions collectively. Other directions of a form are the cubic body diagonals $\langle 111 \rangle$ and the cubic face diagonals $\langle 110 \rangle$.

Draw the following direction vectors in cubic unit cells:

a.   [100] and [110]
b.   [112]
c.   [$\bar{1}$10]
d.   [$\bar{3}2\bar{1}$]

**EXAMPLE
PROBLEM 3.4**

■ **Solution**

a. The position coordinates for the [100] direction are (1, 0, 0) (Fig. 3.12a). The position coordinates for the [110] direction are (1, 1, 0) (Fig. 3.12a).

b. The position coordinates for the [112] direction are obtained by dividing the direction indices by 2 so that they will lie within the unit cube. Thus they are $(\frac{1}{2}, \frac{1}{2}, 1)$ (Fig. 3.12b).

c. The position coordinates for the [$\bar{1}$10] direction are (-1, 1, 0) (Fig. 3.12c). Note that the origin for the direction vector must be moved to the lower-left front corner of the cube.

d. The position coordinates for the [$\bar{3}2\bar{1}$] direction are obtained by first dividing all the indices by 3, the largest index. This gives $-1, \frac{2}{3}, -\frac{1}{3}$ for the position coordinates of the exit point of the direction [$\bar{3}2\bar{1}$], which are shown in Fig. 3.12d.

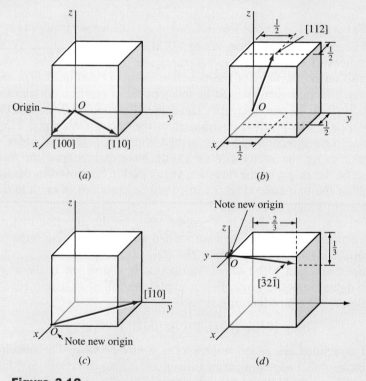

Figure 3.12
Direction vectors in cubic unit cells.

Tutorial

EXAMPLE PROBLEM 3.5

Determine the direction indices of the cubic direction shown in Fig. EP3.5a.

■ **Solution**

Parallel directions have the same direction indices, and so we move the direction vector in a parallel manner until its tail reaches the nearest corner of the cube, still keeping the vector within the cube. Thus, in this case, the upper-left front corner becomes the new origin for the direction vector (Fig. EP3.5*b*). We can now determine the position coordinates where the direction vector leaves the unit cube. These are $x = -1$, $y = +1$, and $z = -\frac{1}{6}$. The position coordinates of the direction where it leaves the unit cube are thus $(-1, +1, -\frac{1}{6})$. The direction indices for this direction are, after clearing the fraction $6x$, $(-1, +1, -\frac{1}{6})$, or $[\bar{6}6\bar{1}]$.

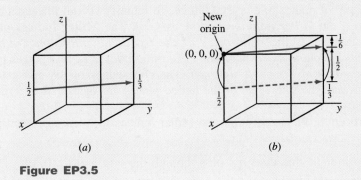

(a)   (b)

Figure EP3.5

Tutorial

Determine the direction indices of the cubic direction between the position coordinates $(\frac{3}{4}, 0, \frac{1}{4})$ and $(\frac{1}{4}, \frac{1}{2}, \frac{1}{2})$.

**EXAMPLE
PROBLEM 3.6**

■ **Solution**

First we locate the origin and termination points of the direction vector in a unit cube, as shown in Fig. EP3.6. The fraction vector components for this direction are

$$x = -(\tfrac{3}{4} - \tfrac{1}{4}) = -\tfrac{1}{2}$$
$$y = (\tfrac{1}{2} - 0) = \tfrac{1}{2}$$
$$z = (\tfrac{1}{2} - \tfrac{1}{4}) = \tfrac{1}{4}$$

Thus, the vector direction has fractional vector components of $-\frac{1}{2}, \frac{1}{2}, \frac{1}{4}$. The direction indices will be in the same ratio as their fractional components. By multiplying the fraction vector components by 4, we obtain $[\bar{2}21]$ for the direction indices of this vector direction.

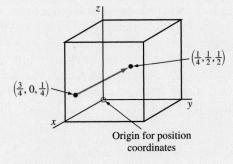

Origin for position
coordinates

Figure EP3.6

3.6 MILLER INDICES FOR CRYSTALLOGRAPHIC PLANES IN CUBIC UNIT CELLS

Sometimes it is necessary to refer to specific lattice planes of atoms within a crystal structure, or it may be of interest to know the crystallographic orientation of a plane or group of planes in a crystal lattice. To identify crystal planes in cubic crystal structures, the *Miller notation system*[4] is used. The **Miller indices of a crystal plane** are defined as the *reciprocals of the fractional intercepts (with fractions cleared) that the plane makes with the crystallographic x, y, and z axes of the three nonparallel edges of the cubic unit cell.* The cube edges of the unit cell represent unit lengths, and the intercepts of the lattice planes are measured in terms of these unit lengths.

The procedure for determining the Miller indices for a cubic crystal plane is as follows:

Tutorial

1. Choose a plane that does *not* pass through the origin at (0, 0, 0).

2. Determine the intercepts of the plane in terms of the crystallographic *x*, *y*, and *z* axes for a unit cube. These intercepts may be fractions.

3. Form the reciprocals of these intercepts.

4. Clear fractions and determine the *smallest* set of whole numbers that are in the same ratio as the intercepts. These whole numbers are the Miller indices of the crystallographic plane and are enclosed in parentheses without the use of commas. The notation (*hkl*) is used to indicate Miller indices in a general sense, where *h*, *k*, and *l* are the Miller indices of a cubic crystal plane for the *x*, *y*, and *z* axes, respectively.

Figure 3.13 shows three of the most important crystallographic planes of cubic crystal structures. Let us first consider the shaded crystal plane in Fig. 3.13*a*, which has the intercepts 1, ∞, ∞ for the *x*, *y*, and *z* axes, respectively. We take the reciprocals of these intercepts to obtain the Miller indices, which are therefore 1, 0, 0. Since these numbers do not involve fractions, the Miller indices for this plane are (100), which is read as the one-zero-zero plane. Next let us consider the second plane shown in Fig. 3.13*b*. The intercepts of this plane are 1, 1, ∞. Since the reciprocals of these numbers are 1, 1, 0, which do not involve fractions, the Miller indices of this plane are (110). Finally, the third plane (Fig. 3.13*c*) has the intercepts 1, 1, 1, which give the Miller indices (111) for this plane.

Consider now the cubic crystal plane shown in Fig. 3.14 which has the intercepts $\frac{1}{3}, \frac{2}{3}, 1$. The reciprocals of these intercepts are $3, \frac{3}{2}, 1$. Since fractional intercepts are not allowed, these fractional intercepts must be multiplied by 2 to clear the $\frac{3}{2}$ fraction. Thus, the reciprocal intercepts become 6, 3, 2 and the Miller

---

[4]William Hallowes Miller (1801–1880). English crystallographer who published a "Treatise on Crystallography" in 1839, using crystallographic reference axes that were parallel to the crystal edges and using reciprocal indices.

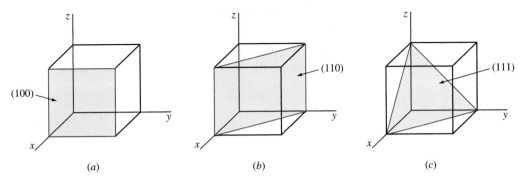

Figure 3.13
Miller indices of some important cubic crystal planes: (*a*) (100), (*b*) (110), and (*c*) (111).

Tutorial

MatVis

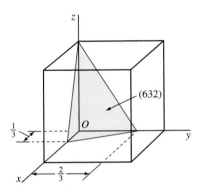

Figure 3.14
Cubic crystal plane (632), which
has fractional intercepts.

indices are (632). Further examples of cubic crystal planes are shown in Example Problem 3.7.

If the crystal plane being considered passes through the origin so that one or more intercepts are zero, the plane must be moved to an equivalent position in the same unit cell and the plane must remain parallel to the original plane. This is possible because all equispaced parallel planes are indicated by the same Miller indices.

If sets of equivalent lattice planes are related by the symmetry of the crystal system, they are called *planes of a family or form,* and the indices of one plane of the family are enclosed in braces as {*hkl*} to represent the indices of a family of symmetrical planes. For example, the Miller indices of the cubic surface planes (100), (010), and (001) are designated collectively as a family or form by the notation {100}.

EXAMPLE PROBLEM 3.7

Draw the following crystallographic planes in cubic unit cells:

a. (101)   b. (1$\bar{1}$0)   c. (221)

d. Draw a (110) plane in a BCC atomic-site unit cell, and list the position coordinates of the atoms whose centers are intersected by this plane.

■ **Solutions**

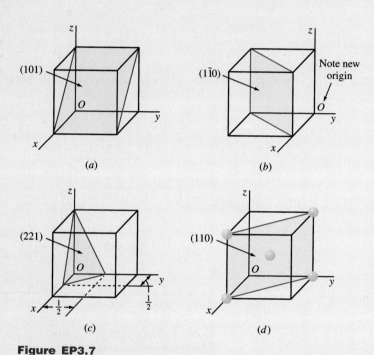

Figure EP3.7
Various important cubic crystal planes.

MatVis

a. First determine the reciprocals of the Miller indices of the (101) plane. These are 1, ∞, 1. The (101) plane must pass through a unit cube at intercepts $x = 1$ and $z = 1$ and be parallel to the y axis.

b. First determine the reciprocals of the Miller indices of the (1$\bar{1}$0) plane. These are 1, −1, ∞. The (1$\bar{1}$0) plane must pass through a unit cube at intercepts $x = 1$ and $y = -1$ and be parallel to the z axis. Note that the origin of axes must be moved to the lower-right back side of the cube.

c. First determine the reciprocals of the Miller indices of the (221) plane. These are $\frac{1}{2}, \frac{1}{2}, 1$. The (221) plane must pass through a unit cube at intercepts $x = \frac{1}{2}$, $y = \frac{1}{2}$, and $z = 1$.

d. Atom positions whose centers are intersected by the (110) plane are (1, 0, 0), (0, 1, 0), (1, 0, 1), (0, 1, 1), and ($\frac{1}{2}, \frac{1}{2}, \frac{1}{2}$). These positions are indicated by the solid circles.

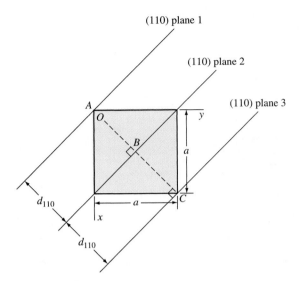

Figure 3.15

Top view of cubic unit cell showing the distance between (110) crystal planes, d_{110}.

An important relationship for the cubic system, and *only the cubic system,* is that the direction indices of a direction *perpendicular* to a crystal plane are the same as the Miller indices of that plane. For example, the [100] direction is perpendicular to the (100) crystal plane.

In cubic crystal structures the *interplanar spacing* between two closest parallel planes with the same Miller indices is designated d_{hkl}, where h, k, and l are the Miller indices of the planes. This spacing represents the distance from a selected origin containing one plane and another parallel plane with the same indices that is closest to it. For example, the distance between (110) planes 1 and 2, d_{110}, in Fig. 3.15 is *AB*. Also, the distance between (110) planes 2 and 3 is d_{110} and is length *BC* in Fig. 3.15. From simple geometry, it can be shown that for cubic crystal structures

$$d_{hkl} = \frac{a}{\sqrt{h^2 + k^2 + l^2}} \tag{3.4}$$

where d_{hkl} = interplanar spacing between parallel closest planes with Miller indices h, k, and l

a = lattice constant (edge of unit cube)

h, k, l = Miller indices of cubic planes being considered

Determine the Miller indices of the cubic crystallographic plane shown in Fig. EP3.8a.

EXAMPLE PROBLEM 3.8

■ **Solution**

First, transpose the plane parallel to the z axis $\frac{1}{4}$ unit to the right along the y axis as shown in Fig. EP3.8b so that the plane intersects the x axis at a unit distance from the new origin

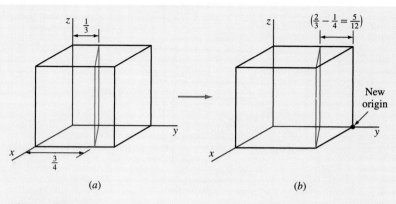

(a)   *(b)*

Figure EP3.8

located at the lower-right back corner of the cube. The new intercepts of the transposed plane with the coordinate axes are now $(+1, -\frac{5}{12}, \infty)$. Next, we take the reciprocals of these intercepts to give $(1, -\frac{12}{5}, 0)$. Finally, we clear the $\frac{12}{5}$ fraction to obtain $(5\bar{12}0)$ for the Miller indices of this plane.

EXAMPLE PROBLEM 3.9

Determine the Miller indices of the cubic crystal plane that intersects the position coordinates $(1, \frac{1}{4}, 0)$, $(1, 1, \frac{1}{2})$, $(\frac{3}{4}, 1, \frac{1}{4})$, and all coordinate axes.

■ **Solution**

First, we locate the three position coordinates as indicated in Fig. EP3.9 at A, B, and C. Next, we join A and B, extend AB to D, and then join A and C. Finally, we join A to C to complete plane ACD. The origin for this plane in the cube can be chosen at E, which gives axial intercepts for plane ACD at $x = -\frac{1}{2}$, $y = -\frac{3}{4}$, and $z = \frac{1}{2}$. The reciprocals of these axial intercepts are -2, $-\frac{4}{3}$, and 2. Multiplying these intercepts by 3 clears the fraction, giving Miller indices for the plane of $(\bar{6}\bar{4}6)$.

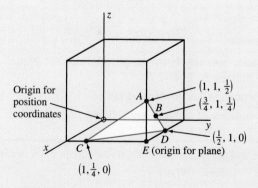

Figure EP3.9

Copper has an FCC crystal structure and a unit cell with a lattice constant of 0.361 nm. What is its interplanar spacing d_{220}?

■ **Solution**

$$d_{hkl} = \frac{a}{\sqrt{h^2 + k^2 + l^2}} = \frac{0.361 \text{ nm}}{\sqrt{(2)^2 + (2)^2 + (0)^2}} = 0.128 \text{ nm} \blacktriangleleft$$

3.7 CRYSTALLOGRAPHIC PLANES AND DIRECTIONS IN HEXAGONAL CRYSTAL STRUCTURE

3.7.1 Indices for Crystal Planes in HCP Unit Cells*

Crystal planes in HCP unit cells are commonly identified by using four indices instead of three. The HCP crystal plane indices, called *Miller-Bravais indices,* are denoted by the letters h, k, i, and l and are enclosed in parentheses as ($hkil$). These four-digit hexagonal indices are based on a coordinate system with four axes, as shown in Fig. 3.16 in an HCP unit cell. There are three basal axes, a_1, a_2, and a_3, which make 120° with each other. The fourth axis or c axis is the vertical axis located at the center of the unit cell. The a unit of measurement along the a_1, a_2, and a_3 axes is the distance between the atoms along these axes and is indicated in Fig. 3.16. The unit of measurement along the c axis is the height of the unit cell. The reciprocals of the intercepts that a crystal plane makes with the a_1, a_2, and a_3 axes give the h, k, and i indices, while the reciprocal of the intercept with the c axis gives the l index.

---

*In discussion of planes and directions we will use the "larger cell" for presentation of concepts.

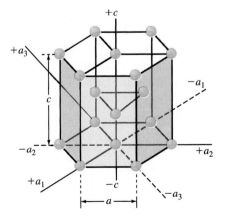

Figure 3.16
The four coordinate axes (a_1, a_2, a_3, and c) of the HCP crystal structure.

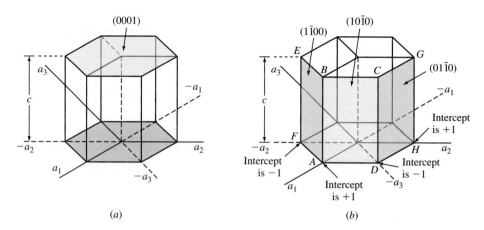

MatVis

Figure 3.17
Miller-Bravais indices of hexagonal crystal planes: (*a*) basal planes and (*b*) prism planes.

Basal Planes   The basal planes of the HCP unit cell are very important planes for this unit cell and are indicated in Fig. 3.17*a*. Since the basal plane on the top of the HCP unit cell in Fig. 3.17*a* is parallel to the a_1, a_2, and a_3 axes, the intercepts of this plane with these axes will all be infinite. Thus, $a_1 = \infty$, $a_2 = \infty$, and $a_3 = \infty$. The *c* axis, however, is unity since the top basal plane intersects the *c* axis at unit distance. Taking the reciprocals of these intercepts gives the Miller-Bravais indices for the HCP basal plane. Thus $h = 0$, $k = 0$, $i = 0$, and $l = 1$. The HCP basal plane is, therefore, a zero-zero-zero-one or (0001) plane.

Prism Planes   Using the same method, the intercepts of the front prism plane (*ABCD*) of Fig. 3.17*b* are $a_1 = +1$, $a_2 = \infty$, $a_3 = -1$, and $c = \infty$. Taking the reciprocals of these intercepts gives $h = 1$, $k = 0$, $i = -1$, and $l = 0$, or the ($10\bar{1}0$) plane. Similarly, the *ABEF* prism plane of Fig. 3.17*b* has the indices ($1\bar{1}00$) and the *DCGH* plane the indices ($01\bar{1}0$). All HCP prism planes can be identified collectively as the $\{10\bar{1}0\}$ family of planes.

   Sometimes HCP planes are identified only by three indices (*hkl*) since $h + k = -i$. However, the (*hkil*) indices are used more commonly because they reveal the hexagonal symmetry of the HCP unit cell.

3.7.2 Direction Indices in HCP Unit Cells[5]

Directions in HCP unit cells are also usually indicated by four indices *u*, *v*, *t*, and *w* enclosed by square brackets as [*uvtw*]. The *u*, *v*, and *t* indices are lattice vectors

---

[5]The topic of direction indices for hexagonal unit cells is not normally presented in an introductory course in materials but is included here for advanced students.

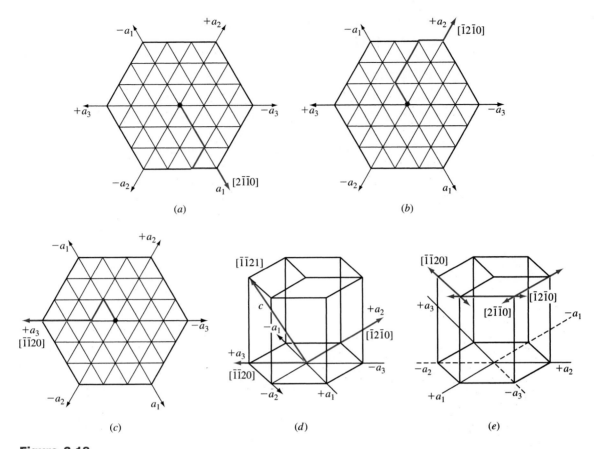

Figure 3.18
Miller-Bravais hexagonal crystal structure direction indices for principal directions: (*a*) $+a_1$ axis direction on basal plane, (*b*) $+a_2$ axis direction on basal plane, (*c*) $+a_3$ direction axis on basal plane, and (*d*) $+a_3$ direction axis incorporating *c* axis. (*e*) Positive and negative Miller-Bravais directions are indicated in simple hexagonal crystal structure on upper basal plane.

in the a_1, a_2, and a_3 directions, respectively (Fig. 3.16), and the w index is a lattice vector in the c direction. To maintain uniformity for both HCP indices for planes and directions, it has been agreed that $u + v = -t$ for directions.

Let us now determine the Miller-Bravais hexagonal indices for the directions a_1, a_2, and a_3, which are the positive basal axes of the hexagonal unit cell. The a_1 direction indices are given in Fig. 3.18*a*, the a_2 direction indices in Fig. 3.18*b*, and the a_3 direction indices in Fig. 3.18*c*. If we need to indicate a c direction also for the a_3 direction, this is shown in Fig. 3.18*d*. Fig. 3.18*e* summarizes the positive and negative directions on the upper basal plane of the simple hexagonal crystal structure.

3.8 COMPARISON OF FCC, HCP, AND BCC CRYSTAL STRUCTURES

3.8.1 FCC and HCP Crystal Structures

As previously pointed out, both the HCP and FCC crystal structures are close-packed structures. That is, their atoms, which are considered approximate "spheres," are packed together as closely as possible so that an atomic packing factor of 0.74 is attained.[6] The (111) planes of the FCC crystal structure shown in Fig. 3.19a have the identical packing arrangement as the (0001) planes of the HCP crystal structure shown in Fig. 3.19b. However, the three-dimensional FCC and HCP crystal structures are not identical because there is a difference in the stacking arrangement of their atomic planes, which can best be described by considering the stacking of hard spheres representing atoms. As a useful analogy, one can imagine the stacking of planes of equal-sized marbles on top of each other, minimizing the space between the marbles.

Consider first a plane of close-packed atoms designated the *A* plane, as shown in Fig. 3.20a. Note that there are two different types of empty spaces or voids

---

[6]As pointed out in Sec. 3.3, the atoms in the HCP structure deviate to varying degrees from ideality. In some HCP metals the atoms are elongated along the *c* axis, and in other cases they are compressed along the *c* axis (see Table 3.4).

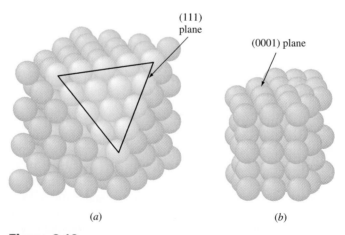

(a)   (b)

Figure 3.19

Comparison of the (*a*) FCC crystal structure showing a close-packed (111) plane and (*b*) a HCP crystal structure showing the close-packed (0001) plane.

(After W. G. Moffatt, G. W. Pearsall, and J. Wulff, "The Structure and Properties of Materials," vol. I: "Structure," Wiley, 1964, p. 51.)

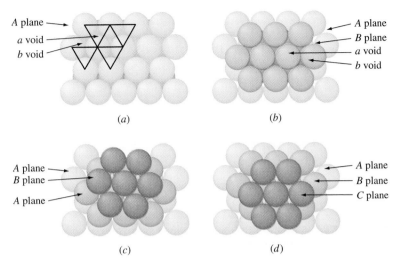

Animation

Figure 3.20
Formation of the HCP and FCC crystal structures by the stacking
of atomic planes. (*a*) *A* plane showing the *a* and *b* voids. (*b*) *B* plane
placed in *a* voids of plane *A*. (*c*) Third plane placed in *b* voids of
B plane, making another *A* plane and forming the HCP crystal
structure. (*d*) Third plane placed in the *a* voids of *B* plane, making
a new *C* plane and forming the FCC crystal structure.
(*Adapted from P. Ander and A. J. Sonnessa, "Principles of Chemistry," Macmillan,
1965, p. 661.*)

between the atoms. The voids pointing to the top of the page are designated *a* voids
and those pointing to the bottom of the page, *b* voids. A second plane of atoms can
be placed over the *a* or *b* voids and the same three-dimensional structure will be
produced. Let us place plane *B* over the *a* voids, as shown in Fig. 3.20*b*. Now if a
third plane of atoms is placed over plane *B* to form a closest-packed structure, it is
possible to form two different close-packed structures. One possibility is to place the
atoms of the third plane in the *b* voids of the *B* plane. Then the atoms of this third
plane will lie directly over those of the *A* plane and thus can be designated another
A plane (Fig. 3.20*c*). If subsequent planes of atoms are placed in this same alternating stacking arrangement, then the stacking sequence of the three-dimensional
structure produced can be denoted by *ABABAB*. . . . Such a stacking sequence leads
to the HCP crystal structure (Fig. 3.19*b*).

The second possibility for forming a simple close-packed structure is to
place the third plane in the *a* voids of plane *B* (Fig. 3.20*d*). This third plane is
designated the *C* plane since its atoms do not lie directly above those of the
B plane or the *A* plane. The stacking sequence in this close-packed structure
is thus designated *ABCABCABC*. . . and leads to the FCC structure shown in
Fig. 3.19*a*.

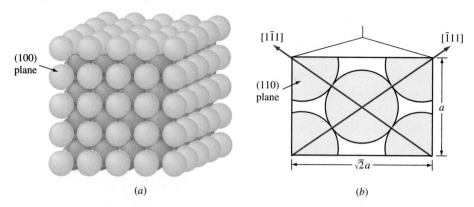

Figure 3.21
BCC crystal structure showing (*a*) the (100) plane and (*b*) a section of the (110) plane. Note that this is not a close-packed structure but that the diagonals have close-packed directions.
[(*a*) after W. G. Moffatt, G. W. Pearsall, and J. Wulff, "The Structure and Properties of Materials," vol. I: "Structure," Wiley, 1964, p. 51.]

3.8.2 BCC Crystal Structure

The BCC structure is not a close-packed structure and hence does not have close-packed planes like the {111} planes in the FCC structure and the {0001} planes in the HCP structure. The most densely packed planes in the BCC structure are the {110} family of planes of which the (110) plane is shown in Fig. 3.21*b*. However, the atoms in the BCC structure do have close-packed directions along the cube diagonals, which are the ⟨111⟩ directions.

3.9 VOLUME, PLANAR, AND LINEAR DENSITY UNIT-CELL CALCULATIONS

3.9.1 Volume Density

Using the hard-sphere atomic model for the crystal structure unit cell of a metal and a value for the atomic radius of the metal obtained from x-ray diffraction analysis, a value for the **volume density** of a metal can be obtained by using the equation

$$\text{Volume density of metal} = \rho_v = \frac{\text{mass/unit cell}}{\text{volume/unit cell}} \tag{3.5}$$

In Example Problem 3.11 a value of 8.98 Mg/m^3 (8.98 g/cm^3) is obtained for the density of copper. The handbook experimental value for the density of copper is 8.96 Mg/m^3 (8.96 g/cm^3). The slightly lower density of the experimental value could be attributed to the absence of atoms at some atomic sites (vacancies), line defects, and mismatch where grains meet (grain boundaries). These crystalline defects are discussed in Chap. 4. Another cause of the discrepancy could also be due to the atoms not being perfect spheres.

Copper has an FCC crystal structure and an atomic radius of 0.1278 nm. Assuming the atoms to be hard spheres that touch each other along the face diagonals of the FCC unit cell as shown in Fig. 3.7, calculate a theoretical value for the density of copper in megagrams per cubic meter. The atomic mass of copper is 63.54 g/mol.

EXAMPLE PROBLEM 3.11

■ Solution

For the FCC unit cell, $\sqrt{2}a = 4R$, where a is the lattice constant of the unit cell and R is the atomic radius of the copper atom. Thus,

$$a = \frac{4R}{\sqrt{2}} = \frac{(4)(0.1278 \text{ nm})}{\sqrt{2}} = 0.361 \text{ nm}$$

$$\text{Volume density of copper} = \rho_v = \frac{\text{mass/unit cell}}{\text{volume/unit cell}} \tag{3.5}$$

In the FCC unit cell there are four atoms/unit cell. Each copper atom has a mass of $(63.54 \text{ g/mol})/(6.02 \times 10^{23} \text{ atoms/mol})$. Thus, the mass m of Cu atoms in the FCC unit cell is

$$m = \frac{(4 \text{ atoms})(63.54 \text{ g/mol})}{6.02 \times 10^{23} \text{ atoms/mol}} \left(\frac{10^{-6} \text{ Mg}}{\text{g}} \right) = 4.22 \times 10^{-28} \text{ Mg}$$

The volume V of the Cu unit cell is

$$V = a^3 = \left(0.361 \text{ nm} \times \frac{10^{-9} \text{ m}}{\text{nm}} \right)^3 = 4.70 \times 10^{-29} \text{ m}^3$$

Thus, the density of copper is

$$\rho_v = \frac{m}{V} = \frac{4.22 \times 10^{-28} \text{ Mg}}{4.70 \times 10^{-29} \text{ m}^3} = 8.98 \text{ Mg/m}^3 \quad (8.98 \text{ g/cm}^3) \blacktriangleleft$$

3.9.2 Planar Atomic Density

Sometimes it is important to determine the atomic densities on various crystal planes. To do this a quantity called the **planar atomic density** is calculated by using the relationship

$$\text{Planar atomic density} = \rho_p = \frac{\text{equiv. no. of atoms whose centers are intersected by selected area}}{\text{selected area}} \tag{3.6}$$

For convenience the area of a plane that intersects a unit cell is usually used in these calculations, as shown, for example, in Fig. 3.22 for the (110) plane in a BCC unit cell.

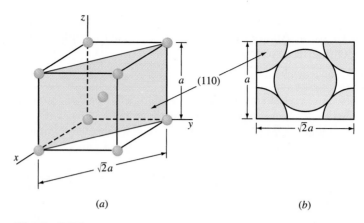

Figure 3.22
(a) A BCC atomic-site unit cell showing a shaded (110) plane.
(b) Areas of atoms in BCC unit cell cut by the (110) plane.

In order for an atom area to be counted in this calculation, the plane of interest must intersect the center of an atom. In Example Problem 3.12 the (110) plane intersects the centers of five atoms, but the equivalent of only two atoms is counted since only one-quarter of each of the four corner atoms is included in the area inside the unit cell.

EXAMPLE PROBLEM 3.12

Calculate the planar atomic density ρ_p on the (110) plane of the α iron BCC lattice in atoms per square millimeter. The lattice constant of α iron is 0.287 nm.

■ **Solution**

$$\rho_p = \frac{\text{equiv. no. of atoms whose centers are intersected by selected area}}{\text{selected area}} \tag{3.6}$$

The equivalent number of atoms intersected by the (110) plane in terms of the surface area inside the BCC unit cell is shown in Fig. 3.22 and is

$$1 \text{ atom at center} + 4 \times \tfrac{1}{4} \text{ atoms at four corners of plane} = 2 \text{ atoms}$$

The area intersected by the (110) plane inside the unit cell (selected area) is

$$(\sqrt{2}a)(a) = \sqrt{2}a^2$$

Thus, the planar atomic density is

$$\rho_p = \frac{2 \text{ atoms}}{\sqrt{2}(0.287 \text{ nm})^2} = \frac{17.2 \text{ atoms}}{\text{nm}^2}$$

$$= \frac{17.2 \text{ atoms}}{\text{nm}^2} \times \frac{10^{12} \text{ nm}^2}{\text{mm}^2}$$

$$= 1.72 \times 10^{13} \text{ atoms/mm}^2 \blacktriangleleft$$

3.9.3 Linear Atomic Density

Sometimes it is important to determine the atomic densities in various directions in crystal structures. To do this a quantity called the **linear atomic density** is calculated by using the relationship

$$\text{Linear atomic density} = \rho_l = \frac{\substack{\text{no. of atomic diam. intersected by selected}\\ \text{length of line in direction of interest}}}{\text{selected length of line}}$$

(3.7)

Example Problem 3.13 shows how the linear atomic density can be calculated in the [110] direction in a pure copper crystal lattice.

Calculate the linear atomic density ρ_l in the [110] direction in the copper crystal lattice in atoms per millimeter. Copper is FCC and has a lattice constant of 0.361 nm.

**EXAMPLE
PROBLEM 3.13**

■ **Solution**

The atoms whose centers the [110] direction intersects are shown in Fig. 3.23. We shall select the length of the line to be the length of the face diagonal of the FCC unit cell, which is $\sqrt{2}a$. The number of atomic diameters intersected by this length of line are $\frac{1}{2} + 1 + \frac{1}{2} = 2$ atoms. Thus using Eq. 3.7, the linear atomic density is

$$\rho_l = \frac{2 \text{ atoms}}{\sqrt{2}a} = \frac{2 \text{ atoms}}{\sqrt{2}(0.361 \text{ nm})} = \frac{3.92 \text{ atoms}}{\text{nm}}$$

$$= \frac{3.92 \text{ atoms}}{\text{nm}} \times \frac{10^6 \text{ nm}}{\text{mm}}$$

$$= 3.92 \times 10^6 \text{ atoms/mm} \blacktriangleleft$$

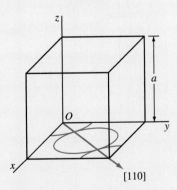

Figure 3.23
Diagram for calculating the atomic linear density in the [110] direction in an FCC unit cell.

3.10 POLYMORPHISM OR ALLOTROPY

Many elements and compounds exist in more than one crystalline form under different conditions of temperature and pressure. This phenomenon is termed **polymorphism,** or *allotropy*. Many industrially important metals such as iron, titanium, and cobalt undergo allotropic transformations at elevated temperatures at atmospheric pressure. Table 3.5 lists some selected metals that show allotropic transformations and the structure changes that occur.

Iron exists in both BCC and FCC crystal structures over the temperature range from room temperature to its melting point at 1539°C, as shown in Fig. 3.24. Alpha (α) iron exists from −273 to 912°C and has the BCC crystal structure. Gamma (γ)

Table 3.5 Allotropic crystalline forms of some metals

Metal	Crystal structure at room temperature	At other temperature
Ca	FCC	BCC (> 447°C)
Co	HCP	FCC (> 427°C)
Hf	HCP	BCC (> 1742°C)
Fe	BCC	FCC (912–1394°C)
		BCC (> 1394°C)
Li	BCC	HCP (< −193°C)
Na	BCC	HCP (< −233°C)
Tl	HCP	BCC (> 234°C)
Ti	HCP	BCC (> 883°C)
Y	HCP	BCC (> 1481°C)
Zr	HCP	BCC (> 872°C)

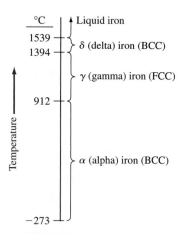

Figure 3.24
Allotropic crystalline forms of
iron over temperature ranges
at atmospheric pressure

iron exists from 912 to 1394°C and has the FCC crystal structure. Delta (δ) iron exists from 1394 to 1539°C, which is the melting point of iron. The crystal structure of δ iron is also BCC but with a larger lattice constant than α iron.

Calculate the theoretical volume change accompanying a polymorphic transformation in a pure metal from the FCC to BCC crystal structure. Assume the hard-sphere atomic model and that there is no change in atomic volume before and after the transformation.

EXAMPLE PROBLEM 3.14

■ **Solution**

In the FCC crystal structure unit cell, the atoms are in contact along the face diagonal of the unit cell, as shown in Fig. 3.7. Hence,

$$\sqrt{2}a = 4R \quad \text{or} \quad a = \frac{4R}{\sqrt{2}} \tag{3.3}$$

In the BCC crystal structure unit cell, the atoms are in contact along the body diagonal of the unit cell as shown in Fig. 3.5. Hence,

$$\sqrt{3}a = 4R \quad \text{or} \quad a = \frac{4R}{\sqrt{3}} \tag{3.1}$$

The volume per atom for the FCC crystal lattice, since it has four atoms per unit cell, is

$$V_{\text{FCC}} = \frac{a^3}{4} = \left(\frac{4R}{\sqrt{2}}\right)^3\left(\frac{1}{4}\right) = 5.66R^3$$

The volume per atom for the BCC crystal lattice, since it has two atoms per unit cell, is

$$V_{\text{BCC}} = \frac{a^3}{2} = \left(\frac{4R}{\sqrt{3}}\right)^3\left(\frac{1}{2}\right) = 6.16R^3$$

The change in volume associated with the transformation from the FCC to BCC crystal structure, assuming no change in atomic radius, is

$$\frac{\Delta V}{V_{\text{FCC}}} = \frac{V_{\text{BCC}} - V_{\text{FCC}}}{V_{\text{FCC}}}$$

$$= \left(\frac{6.16R^3 - 5.66R^3}{5.66R^3}\right)100\% = +8.8\% \blacktriangleleft$$

3.11 CRYSTAL STRUCTURE ANALYSIS

Our present knowledge of crystal structures has been obtained mainly by x-ray diffraction techniques that use x-rays whose wavelength are the same as the distance between crystal lattice planes. However, before discussing the manner in which

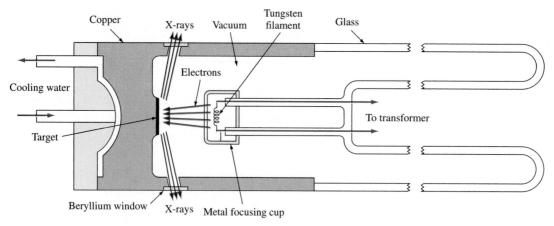

Figure 3.25
Schematic diagram of the cross section of a sealed-off filament x-ray tube.
(*After B. D. Cullity, "Elements of X-Ray Diffraction," 2d ed., Addison-Wesley, 1978, p. 23.*)

x-rays are diffracted in crystals, let us consider how x-rays are produced for experimental purposes.

3.11.1 X-Ray Sources

X-rays used for diffraction are electromagnetic waves with wavelengths in the range 0.05 to 0.25 nm (0.5 to 2.5 Å). By comparison, the wavelength of visible light is of the order of 600 nm (6000 Å). In order to produce x-rays for diffraction purposes, a voltage of about 35 kV is necessary and is applied between a cathode and an anode target metal, both of which are contained in a vacuum, as shown in Fig. 3.25. When the tungsten filament of the cathode is heated, electrons are released by thermionic emission and accelerated through the vacuum by the large voltage difference between the cathode and anode, thereby gaining kinetic energy. When the electrons strike the target metal (e.g., molybdenum), x-rays are given off. However, most of the kinetic energy (about 98 percent) is converted into heat, so the target metal must be cooled externally.

The x-ray spectrum emitted at 35 kV using a molybdenum target is shown in Fig. 3.26. The spectrum shows continuous x-ray radiation in the wavelength range from about 0.2 to 1.4 Å (0.02 to 0.14 nm) and two spikes of characteristic radiation that are designated the K_α and K_β lines. The wavelengths of the K_α and K_β lines are characteristic for an element. For molybdenum, the K_α line occurs at a wavelength of about 0.7 Å (0.07 nm). The origin of the characteristic radiation is explained as follows. First, K electrons (electrons in the $n = 1$ shell) are knocked out of the atom by highly energetic electrons bombarding the target, leaving excited atoms. Next, some electrons in higher shells (that is, $n = 2$ or 3) drop down to lower energy levels to replace the lost K electrons, emitting energy of a

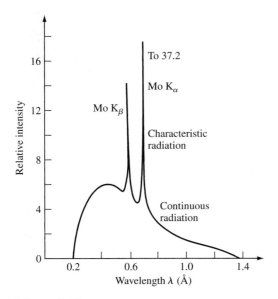

Figure 3.26
X-ray emission spectrum produced when
molybdenum metal is used as the target metal
in an x-ray tube operating at 35 kV.

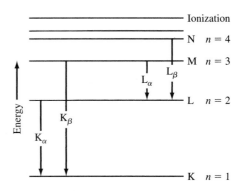

Figure 3.27
Energy levels of electrons in
molybdenum showing the origin of K_α
and K_β radiation.

characteristic wavelength. The transition of electrons from the L ($n = 2$) shell to
the K ($n = 1$) shell creates energy of the wavelength of the K_α line, as indicated
in Fig. 3.27.

3.11.2 X-Ray Diffraction

Since the wavelengths of some x-rays are about equal to the distance between planes
of atoms in crystalline solids, reinforced diffraction peaks of radiation of varying
intensities can be produced when a beam of x-rays strikes a crystalline solid. However, before considering the application of x-ray diffraction techniques to crystal
structure analysis, let us examine the geometric conditions necessary to produce diffracted or reinforced beams of reflected x-rays.

Consider a monochromatic (single-wavelength) beam of x-rays to be incident on
a crystal, as shown in Fig. 3.28. For simplification let us allow the crystal planes of
atomic scattering centers to be replaced by crystal planes that act as mirrors in
reflecting the incident x-ray beam. In Fig. 3.28 the horizontal lines represent a set
of parallel crystal planes with Miller indices (hkl). When an incident beam of monochromatic x-rays of wavelength λ strikes this set of planes at an angle such that the
wave patterns of the beam leaving the various planes are *not in phase, no reinforced
beam will be produced* (Fig. 3.28*a*). Thus, destructive interference occurs. If the
reflected wave patterns of the beam leaving the various planes are in phase, then
reinforcement of the beam or constructive interference occurs (Fig. 3.28*b*).

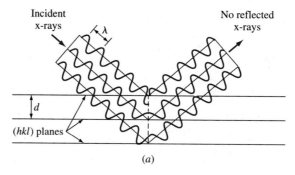

Incident x-rays · No reflected x-rays · λ · d · (hkl) planes

(a)

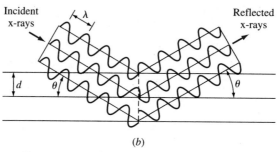

Incident x-rays · Reflected x-rays · λ · d · θ · θ

(b)

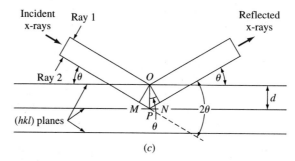

Incident x-rays · Ray 1 · Reflected x-rays · Ray 2 · θ · O · θ · d · (hkl) planes · M · N · 2θ · P · θ

(c)

Figure 3.28

The reflection of an x-ray beam by the (hkl) planes of a crystal. (a) No reflected beam is produced at an arbitrary angle of incidence. (b) At the Bragg angle θ, the reflected rays are in phase and reinforce one another. (c) Similar to (b) except that the wave representation has been omitted.

(After A. G. Guy and J. J. Hren, "Elements of Physical Metallurgy," 3d ed., Addison-Wesley, 1974, p. 201.)

Let us now consider incident x-rays 1 and 2 as indicated in Fig. 3.28c. For these rays to be in phase, the extra distance of travel of ray 2 is equal to $MP + PN$, which must be an integral number of wavelengths λ. Thus,

$$n\lambda = MP + PN \qquad (3.8)$$

where $n = 1, 2, 3, \ldots$ and is called the *order of the diffraction*. Since both *MP* and *PN* equal $d_{hkl} \sin \theta$, where d_{hkl} is the interplanar spacing of the crystal planes of indices (hkl), the condition for constructive interference (i.e., the production of a diffraction peak of intense radiation) must be

$$n\lambda = 2d_{hkl} \sin \theta \qquad (3.9)$$

This equation, known as *Bragg's law,*[7] gives the relationship among the angular positions of the reinforced diffracted beams in terms of the wavelength λ of the incoming x-ray radiation and of the interplanar spacings d_{hkl} of the crystal planes. In most cases, the first order of diffraction where $n = 1$ is used, and so for this case Bragg's law takes the form

$$\lambda = 2d_{hkl} \sin \theta \qquad (3.10)$$

A sample of BCC iron was placed in an x-ray diffractometer using incoming x-rays with a wavelength $\lambda = 0.1541$ nm. Diffraction from the $\{110\}$ planes was obtained at $2\theta = 44.704°$. Calculate a value for the lattice constant a of BCC iron. (Assume firstorder diffraction with $n = 1$.)	**EXAMPLE PROBLEM 3.15**

■ **Solution**

$$2\theta = 44.704° \qquad \theta = 22.35°$$

$$\lambda = 2d_{hkl} \sin \theta$$

$$d_{110} = \frac{\lambda}{2 \sin \theta} = \frac{0.1541 \text{ nm}}{2(\sin 22.35°)} \qquad (3.10)$$

$$= \frac{0.1541 \text{ nm}}{2(0.3803)} = 0.2026 \text{ nm}$$

Rearranging Eq. 3.4 gives

$$a = d_{hkl}\sqrt{h^2 + k^2 + l^2}$$

Thus,

$$a(\text{Fe}) = d_{110}\sqrt{1^2 + 1^2 + 0^2}$$
$$= (0.2026 \text{ nm})(1.414) = 0.287 \text{ nm} \blacktriangleleft$$

3.11.3 X-Ray Diffraction Analysis of Crystal Structures

The Powder Method of X-Ray Diffraction Analysis   The most commonly used x-ray diffraction technique is the *powder method*. In this technique a powdered specimen is utilized so that there will be a random orientation of many crystals to ensure that some of the particles will be oriented in the x-ray beam to satisfy the diffraction

---

[7]William Henry Bragg (1862–1942). English physicist who worked on x-ray crystallography.

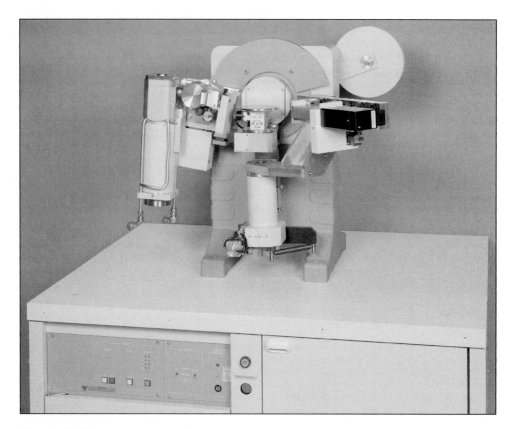

Figure 3.29
An x-ray diffractometer (with x-radiation shields removed).
(*Courtesy of Rigaku*)

conditions of Bragg's law. Modern x-ray crystal analysis uses an x-ray diffractometer that has a radiation counter to detect the angle and intensity of the diffracted beam (Fig. 3.29). A recorder automatically plots the intensity of the diffracted beam as the counter moves on a goniometer[8] circle (Fig. 3.30) that is in synchronization with the specimen over a range of 2θ values. Figure 3.31 shows an x-ray diffraction recorder chart for the intensity of the diffracted beam versus the diffraction angles 2θ for a powdered pure-metal specimen. In this way both the angles of the diffracted beams and their intensities can be recorded at one time. Sometimes a powder camera with an enclosed filmstrip is used instead of the diffractometer, but this method is much slower and in most cases less convenient.

Diffraction Conditions for Cubic Unit Cells   X-ray diffraction techniques enable the structures of crystalline solids to be determined. The interpretation of x-ray

---

[8]A goniometer is an instrument for measuring angles.

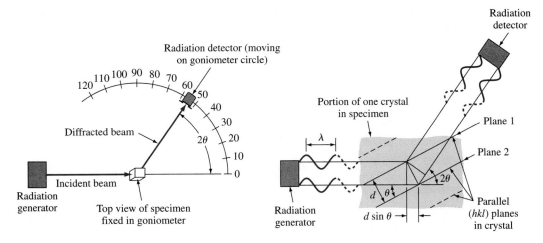

Figure 3.30
Schematic illustration of the diffractometer method of crystal analysis and of the conditions necessary for diffraction.
(*After A. G. Guy, "Essentials of Materials Science," McGraw-Hill, 1976.*)

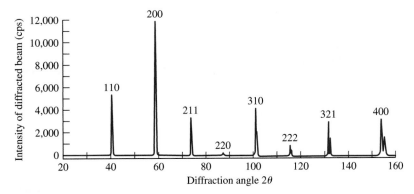

Figure 3.31
Record of the diffraction angles for a tungsten sample obtained by the use of a diffractometer with copper radiation.
(*After A. G. Guy and J. J. Hren, "Elements of Physical Metallurgy," 3d ed., Addison-Wesley, 1974, p. 208.*)

diffraction data for most crystalline substances is complex and beyond the scope of this book, and so only the simple case of diffraction in pure cubic metals will be considered. The analysis of x-ray diffraction data for cubic unit cells can be simplified by combining Eq. 3.4,

$$d_{hkl} = \frac{a}{\sqrt{h^2 + k^2 + l^2}}$$

with the Bragg equation $\lambda = 2d \sin \theta$, giving

$$\lambda = \frac{2a \sin \theta}{\sqrt{h^2 + k^2 + l^2}} \tag{3.11}$$

This equation can be used along with x-ray diffraction data to determine if a cubic crystal structure is body-centered or face-centered cubic. The rest of this subsection will describe how this is done.

To use Eq. 3.11 for diffraction analysis, we must know which crystal planes are the diffracting planes for each type of crystal structure. For the simple cubic lattice, reflections from all (hkl) planes are possible. However, for the BCC structure diffraction occurs only on planes whose Miller indices when added together $(h + k + l)$ total to an even number (Table 3.6). Thus, for the BCC crystal structure the principal diffracting planes are {110}, {200}, {211}, etc., which are listed in Table 3.7. In the case of the FCC crystal structure, the principal diffracting planes are those whose Miller indices are either all even or all odd (zero is considered even). Thus, for the FCC crystal structure the diffracting planes are {111}, {200}, {220}, etc., which are listed in Table 3.7.

Interpreting Experimental X-Ray Diffraction Data for Metals with Cubic Crystal Structures We can use x-ray diffractometer data to determine crystal structures. A simple case to illustrate how this analysis can be used is to distinguish between the

Table 3.6 Rules for determining the diffracting {hkl} planes in cubic crystals

Bravais lattice	Reflections present	Reflections absent
BCC	$(h + k + l) =$ even	$(h + k + l) =$ odd
FCC	(h, k, l) all odd or all even	(h, k, l) not all odd or all even

Table 3.7 Miller indices of the diffracting planes for BCC and FCC lattices

Cubic planes {hkl}	$h^2 + k^2 + l^2$	Sum $\Sigma[h^2 + k^2 + l^2]$	Cubic diffracting planes {hkl} FCC	BCC
{100}	$1^2 + 0^2 + 0^2$	1		
{110}	$1^2 + 1^2 + 0^2$	2	...	110
{111}	$1^2 + 1^2 + 1^2$	3	111	
{200}	$2^2 + 0^2 + 0^2$	4	200	200
{210}	$2^2 + 1^2 + 0^2$	5		
{211}	$2^2 + 1^2 + 1^2$	6	...	211
...		7		
{220}	$2^2 + 2^2 + 0^2$	8	220	220
{221}	$2^2 + 2^2 + 1^2$	9		
{310}	$3^2 + 1^2 + 0^2$	10	...	310

BCC and FCC crystal structures of a cubic metal. Let us assume that we have a metal with either a BCC or an FCC crystal structure and that we can identify the principal diffracting planes and their corresponding 2θ values, as indicated for the metal tungsten in Fig. 3.3.

By squaring both sides of Eq. 3.11 and solving for $\sin^2 \theta$, we obtain

$$\sin^2 \theta = \frac{\lambda^2 (h^2 + k^2 + l^2)}{4a^2} \tag{3.12}$$

From x-ray diffraction data we can obtain experimental values of 2θ for a series of principal diffracting $\{hkl\}$ planes. Since the wavelength of the incoming radiation and the lattice constant a are both constants, we can eliminate these quantities by forming the ratio of two $\sin^2 \theta$ values as

$$\frac{\sin^2 \theta_A}{\sin^2 \theta_B} = \frac{h_A^2 + k_A^2 + l_A^2}{h_B^2 + k_B^2 + l_B^2} \tag{3.13}$$

where θ_A and θ_B are two diffracting angles associated with the principal diffracting planes $\{h_A k_A l_A\}$ and $\{h_B k_B l_B\}$, respectively.

Using Eq. 3.13 and the Miller indices of the first two sets of principal diffracting planes listed in Table 3.7 for BCC and FCC crystal structures, we can determine values for the $\sin^2 \theta$ ratios for both BCC and FCC structures.

For the BCC crystal structure the first two sets of principal diffracting planes are the $\{110\}$ and $\{200\}$ planes (Table 3.7). Substitution of the Miller $\{hkl\}$ indices of these planes into Eq. 3.13 gives

$$\frac{\sin^2 \theta_A}{\sin^2 \theta_B} = \frac{1^2 + 1^2 + 0^2}{2^2 + 0^2 + 0^2} = 0.5 \tag{3.14}$$

Thus, if the crystal structure of the unknown cubic metal is BCC, the ratio of the $\sin^2 \theta$ values that correspond to the first two principal diffracting planes will be 0.5.

For the FCC crystal structure the first two sets of principal diffracting planes are the $\{111\}$ and $\{200\}$ planes (Table 3.7). Substitution of the Miller $\{hkl\}$ indices of these planes into Eq. 3.13 gives

$$\frac{\sin^2 \theta_A}{\sin^2 \theta_B} = \frac{1^2 + 1^2 + 1^2}{2^2 + 0^2 + 0^2} = 0.75 \tag{3.15}$$

Thus, if the crystal structure of the unknown cubic metal is FCC, the ratio of the $\sin^2 \theta$ values that correspond to the first two principal diffracting planes will be 0.75.

Example Problem 3.16 uses Eq. 3.13 and experimental x-ray diffraction data for the 2θ values for the principal diffracting planes to determine whether an unknown cubic metal is BCC or FCC. X-ray diffraction analysis is usually much more complicated than Example Problem 3.16, but the principles used are the same. Both experimental and theoretical x-ray diffraction analysis has been and continues to be used for the determination of the crystal structure of materials.

**EXAMPLE
PROBLEM 3.16**

An x-ray diffractometer recorder chart for an element that has either the BCC or the FCC crystal structure shows diffraction peaks at the following 2θ angles: 40, 58, 73, 86.8, 100.4, and 114.7. The wavelength of the incoming x-ray used was 0.154 nm.

a. Determine the cubic structure of the element.
b. Determine the lattice constant of the element.
c. Identify the element.

■ **Solution**

a. *Determination of the crystal structure of the element.* First, the $\sin^2\theta$ values are calculated from the 2θ diffraction angles.

2θ(deg)	θ(deg)	$\sin\theta$	$\sin^2\theta$
40	20	0.3420	0.1170
58	29	0.4848	0.2350
73	36.5	0.5948	0.3538
86.8	43.4	0.6871	0.4721
100.4	50.2	0.7683	0.5903
114.7	57.35	0.8420	0.7090

Next the ratio of the $\sin^2\theta$ values of the first and second angles is calculated:

$$\frac{\sin^2\theta}{\sin^2\theta} = \frac{0.117}{0.235} = 0.498 \approx 0.5$$

The crystal structure is BCC since this ratio is ≈ 0.5. If the ratio had been ≈ 0.75, the structure would have been FCC.

b. *Determination of the lattice constant.* Rearranging Eq. 3.12 and solving for a^2 gives

$$a^2 = \frac{\lambda^2}{4}\frac{h^2 + k^2 + l^2}{\sin^2\theta} \tag{3.16}$$

or

$$a = \frac{\lambda}{2}\sqrt{\frac{h^2 + k^2 + l^2}{\sin^2\theta}} \tag{3.17}$$

Substituting into Eq. 3.17 $h = 1, k = 1$, and $l = 0$ for the h, k, l Miller indices of the first set of principal diffracting planes for the BCC crystal structure, which are the {110} planes, the corresponding value for $\sin^2\theta$, which is 0.117, and 0.154 nm for λ, the incoming radiation, gives

$$a = \frac{0.154 \text{ nm}}{2}\sqrt{\frac{1^2 + 1^2 + 0^2}{0.117}} = 0.318 \text{ nm} \blacktriangleleft$$

c. *Identification of the element.* The element is tungsten since this element has a lattice constant of 0.316 nm and is BCC.

3.12 AMORPHOUS MATERIALS

As discussed previously, some materials are called amorphous or noncrystalline because they lack long-range order in their atomic structure. It should be noted that in general materials have a tendency to achieve a crystalline state because that is the most stable state and corresponds to the lowest energy level. However, atoms in amorphous materials are bonded in a disordered manner because of factors that inhibit the formation of a periodic arrangement. Atoms in amorphous materials, therefore, occupy random spatial positions as opposed to specific positions in crystalline solids. For clarity, various degrees of order (or disorder) are shown in Fig. 3.32.

Most polymers, glasses, and some metals are members of the amorphous class of materials. In polymers, the secondary bonds among molecules do not allow for the formation of parallel and tightly packed chains during solidification. As a result, polymers such as polyvinylchloride consist of long, twisted molecular chains that are entangled to form a solid with amorphous structure, similar to Fig. 3.32c. In some polymers such as polyethylene, the molecules are more efficiently and tightly packed in some regions of the material and produce a higher degree of regional long-range order. As a result, these polymers are often classified as *semicrystalline*. A more detailed discussion of semicrystalline polymers will be given in Chap. 10.

Inorganic glass based on glass-forming oxide, silica (SiO_2), is generally characterized as a ceramic material (ceramic glass) and is another example of a material with an amorphous structure. In this type of glass, the fundamental subunit in the molecules is the SiO_4^{4-} tetrahedron. The ideal crystalline structure of this glass is shown in Fig. 3.32a. The schematic shows the Si–O tetrahedrons joined corner to corner to form long-range order. In its viscous liquid state, the molecules have limited mobility, and, in general, crystallization occurs slowly. Therefore, a modest cooling rate suppresses the formation of the crystal structure and instead the tetrahedra join corner to corner to form a network lacking in long-range order (Fig. 3.32b).

In addition to polymers and glasses, some metals also have the ability to form amorphous structures (*metallic glass*) under strict and often difficult to achieve conditions. Unlike glasses, metals have very small and mobile building blocks under molten

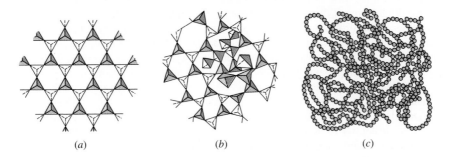

    (a)   (b)   (c)

Figure 3.32
A schematic showing various degrees of order in materials: (*a*) long-range order in crystalline silica, (*b*) silica glass without long-range order, and (*c*) amorphous structure in polymers.

conditions. As a result, it is difficult to prevent metals from crystallizing. However, alloys such as 78%Fe–9%Si–13%B that contain a high percentage of semimetals, Si and B, may form metallic glasses through rapid solidification at cooling rates in excess of 10^8°C/s. At such high cooling rates, the atoms simply do not have enough time to form a crystalline structure and instead form a metal with an amorphous structure, that is, they are highly disordered. In theory, any crystalline material can form a noncrystalline structure if solidified rapidly enough from a molten state.

Amorphous materials, because of their structure, possess properties that are superior. For instance, metallic glasses possess higher strength, better corrosion characteristics, and magnetic properties when compared to their crystalline counterparts. Finally, it is important to note that amorphous materials do not show sharp diffraction patterns when analyzed using x-ray diffraction techniques. This is due to a lack of order and periodicity in the atomic structure. In future chapters, the role of structure of the material on its properties will be explained in detail.

3.13 SUMMARY

Atomic arrangements in crystalline solids can be described by a network of lines called a *space lattice.* Each space lattice can be described by specifying the atom positions in a repeating *unit cell.* The crystal structure consists of space lattice and *motif* or *basis.* Crystalline materials possess long range atomic order such as most metals. But some materials such as many polymers and glasses possess only short range order. Such materials are called semicrystalline or amorphous. There are seven crystal systems based on the geometry of the axial lengths and interaxial angles of the unit cells. These seven systems have a total of 14 sublattices (unit cells) based on the internal arrangements of atomic sites within the unit cells.

In metals the most common crystal structure unit cells are: *body-centered cubic* (BCC), *face-centered cubic* (FCC), and *hexagonal close-packed* (HCP) (which is a dense variation of the simple hexagonal structure).

Crystal directions in cubic crystals are the vector components of the directions resolved along each of the component axes and reduced to smallest integers. They are indicated as [*uvw*]. Families of directions are indexed by the direction indices enclosed by pointed brackets as ⟨*uvw*⟩. *Crystal planes* in cubic crystals are indexed by the reciprocals of the axial intercepts of the plane (followed by the elimination of fractions) as (*hkl*). Cubic crystal planes of a form (family) are indexed with braces as {*hkl*}. Crystal planes in hexagonal crystals are commonly indexed by four indices *h, k, i,* and *l* enclosed in parentheses as (*hkil*). These indices are the reciprocals of the intercepts of the plane on the $a_1, a_2, a_3,$ and *c* axes of the hexagonal crystal structure unit cell. Crystal directions in hexagonal crystals are the vector components of the direction resolved along each of the four coordinate axes and reduced to smallest integers as [*uvtw*].

Using the hard-sphere model for atoms, calculations can be made for the volume, planar, and linear density of atoms in unit cells. Planes in which atoms are packed as tightly as possible are called *close-packed planes,* and directions in which atoms are in closest contact are called *close-packed directions.* Atomic packing factors for different crystal structures can also be determined by assuming the hard-sphere atomic model. Some metals have different crystal structures at different ranges of temperature and pressure, a phenomenon called *polymorphism.*

Crystal structures of crystalline solids can be determined by using x-ray diffraction analysis techniques. X-rays are diffracted in crystals when the *Bragg's law* ($n\lambda = 2d \sin \theta$) conditions are satisfied. By using the x-ray diffractometer and the *powder method*, the crystal structure of many crystalline solids can be determined.

3.14 DEFINITIONS

Sec. 3.1

Amorphous: lacking in long range atomic order

Crystal: a solid composed of atoms, ions, or molecules arranged in a pattern that is repeated in three dimensions.

Crystal structure: a regular three-dimensional pattern of atoms or ions in space.

Space lattice: a three-dimensional array of points each of which has identical surroundings.

Lattice point: one point in an array in which all the points have identical surroundings.

Unit cell: a convenient repeating unit of a space lattice. The axial lengths and axial angles are the lattice constants of the unit cell.

Motif: a group of atoms that or (basis) are organized relative to each other and are associated with corresponding lattice points.

Sec. 3.3

Body-centered cubic (BCC) unit cell: a unit cell with an atomic packing arrangement in which one atom is in contact with eight identical atoms located at the corners of an imaginary cube.

Face-centered cubic (FCC) unit cell: a unit cell with an atomic packing arrangement in which 12 atoms surround a central atom. The stacking sequence of layers of close-packed planes in the FCC crystal structure is *ABCABC.* . . .

Hexagonal close-packed (HCP) unit cell: a unit cell with an atomic packing arrangement in which 12 atoms surround a central identical atom. The stacking sequence of layers of close-packed planes in the HCP crystal structure is *ABABAB.* . . .

Atomic packing factor (APF): the volume of atoms in a selected unit cell divided by the volume of the unit cell.

Sec. 3.5

Indices of direction in a cubic crystal: a direction in a cubic unit cell is indicated by a vector drawn from the origin at one point in a unit cell through the surface of the unit cell; the position coordinates (x, y, and z) of the vector where it leaves the surface of the unit cell (with fractions cleared) are the indices of direction. These indices, designated u, v, and w are enclosed in brackets as [uvw]. Negative indices are indicated by a bar over the index.

Sec. 3.6

Indices for cubic crystal planes (Miller indices): the reciprocals of the intercepts (with fractions cleared) of a crystal plane with the x, y, and z axes of a unit cube are called the Miller indices of that plane. They are designated h, k, and l for the x, y, and z axes, respectively, and are enclosed in parentheses as (hkl). Note that the selected crystal plane must *not* pass through the origin of the x, y, and z axes.

Sec. 3.9

Volume density ρ_v: mass per unit volume; this quantity is usually expressed in Mg/m^3 or g/cm^3.

Planar density ρ_p: the equivalent number of atoms whose centers are intersected by a selected area divided by the selected area.

Linear density ρ_l: the number of atoms whose centers lie on a specific direction on a specific length of line in a unit cube.

Sec. 3.10

Polymorphism (as pertains to metals): the ability of a metal to exist in two or more crystal structures. For example, iron can have a BCC or an FCC crystal structure, depending on the temperature.

Sec. 3.12

Semicrystalline: materials with regions of crystalline structure dispersed in the surrounding, amorphous region; for instance some polymers.

Metallic Glass: metals with an amorphous atomic structure.

3.15 PROBLEMS

Answers to problems marked with an asterisk are given at the end of the book.

3.1   Define (*a*) crystalline solid and (*b*) amorphous solid.

3.2   Define a crystal structure. Give examples of materials that have crystal structures.

3.3   Define (*a*) space lattice and (*b*) motif.

3.4   Define a unit cell of a space lattice. What lattice constants define a unit cell?

3.5   What are the 14 Bravais unit cells?

3.6   What are the three most common metal crystal structures? List five metals that have each of these crystal structures.

3.7   How many atoms per unit cell are there in the BCC crystal structure?

3.8   What is the coordination number for the atoms in the BCC crystal structure?

3.9   What is the relationship between the length of the side *a* of the BCC unit cell and the radius of its atoms?

3.10 Molybdenum at 20°C is BCC and has an atomic radius of 0.140 nm. Calculate a value for its lattice constant *a* in nanometers.

*3.11 Niobium at 20°C is BCC and has an atomic radius of 0.143 nm. Calculate a value for its lattice constant *a* in nanometers.

3.12 Lithium at 20°C is BCC and has a lattice constant of 0.35092 nm. Calculate a value for the atomic radius of a lithium atom in nanometers.

*3.13 Sodium at 20°C is BCC and has a lattice constant of 0.42906 nm. Calculate a value for the atomic radius of a sodium atom in nanometers.

3.14 How many atoms per unit cell are there in the FCC crystal structure?

3.15 What is the coordination number for the atoms in the FCC crystal structure?

*3.16 Gold is FCC and has a lattice constant of 0.40788 nm. Calculate a value for the atomic radius of a gold atom in nanometers.

3.17 Platinum is FCC and has a lattice constant of 0.39239 nm. Calculate a value for the atomic radius of a platinum atom in nanometers.

*3.18 Palladium is FCC and has an atomic radius of 0.137 nm. Calculate a value for its lattice constant *a* in nanometers.

3.19 Strontium is FCC and has an atomic radius of 0.215 nm. Calculate a value for its lattice constant *a* in nanometers.

3.20 Calculate the atomic packing factor for the FCC structure.

3.21 How many atoms per unit cell are there in the HCP crystal structure? Give your answer for both the primitive cell and the "larger cell".

3.22 What is the coordination number for the atoms in the HCP crystal structure?

3.23 What is the ideal c/a ratio for HCP metals?

3.24 Of the following HCP metals, which have higher or lower c/a ratios than the ideal ratio: Zr, Ti, Zn, Mg, Co, Cd, and Be?

*3.25 Calculate the volume in cubic nanometers of the titanium crystal structure unit cell. Titanium is HCP at 20°C with $a = 0.29504$ nm and $c = 0.46833$ nm (use the larger cell).

*3.26 Rhenium at 20°C is HCP. The height c of its unit cell is 0.44583 nm and its c/a ratio is 1.633. Calculate a value for its lattice constant a in nanometers (use the larger cell).

3.27 Osmium at 20°C is HCP. Using a value of 0.135 nm for the atomic radius of osmium atoms, calculate a value for its unit-cell volume (use the larger cell). Assume a packing factor of 0.74.

3.28 How are atomic positions located in cubic unit cells?

3.29 List the atom positions for the eight corner and six face-centered atoms of the FCC unit cell.

3.30 How are the indices for a crystallographic direction in a cubic unit cell determined?   Tutorial

*3.31 Draw the following directions in a BCC unit cell and list the position coordinates of the atoms whose centers are intersected by the direction vector:

(a) [100]   (b) [110]   (c) [111]

3.32 Draw direction vectors in unit cubes for the following cubic directions:

(a) [1$\bar{1}\bar{1}$]   (b) [1$\bar{1}$0]   (c) [$\bar{1}$2$\bar{1}$]   (d) [$\bar{1}\bar{1}$3]

3.33 Draw direction vectors in unit cubes for the following cubic directions:

(a) [1$\bar{1}\bar{2}$]   (c) [$\bar{3}$31]   (e) [2$\bar{1}$2]   (g) [$\bar{1}$01]   (i) [321]   (k) [1$\bar{2}\bar{2}$]

(b) [1$\bar{2}$3]   (d) [0$\bar{2}$1]   (f) [2$\bar{3}$3]   (h) [12$\bar{1}$]   (j) [10$\bar{3}$]   (l) [$\bar{2}\bar{2}$3]

3.34 What are the indices of the directions shown in the unit cubes of Fig. P3.34?

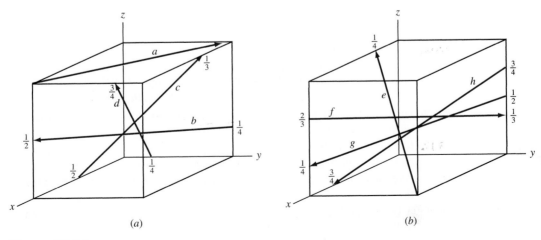

(a)   (b)

Figure P3.34

*3.35 A direction vector passes through a unit cube from the $(\frac{3}{4}, 0, \frac{1}{4})$ to the $(\frac{1}{2}, 1, 0)$ positions. What are its direction indices?

3.36 A direction vector passes through a unit cube from the $(1, 0, \frac{3}{4})$ to the $(\frac{1}{4}, 1, \frac{1}{4})$ positions. What are its direction indices?

3.37 What are the crystallographic directions of a family or form? What generalized notation is used to indicate them?

3.38 What are the directions of the $\langle 100 \rangle$ family or form for a unit cube?

3.39 What are the directions of the $\langle 111 \rangle$ family or form for a unit cube?

3.40 What $\langle 110 \rangle$-type directions lie on the (111) plane of a cubic unit cell?

3.41 What $\langle 111 \rangle$-type directions lie on the (110) plane of a cubic unit cell?

3.42 How are the Miller indices for a crystallographic plane in a cubic unit cell determined? What generalized notation is used to indicate them?

3.43 Draw in unit cubes the crystal planes that have the following Miller indices:
   (a) $(1\bar{1}\bar{1})$ (c) $(1\bar{2}\bar{1})$ (e) $(3\bar{2}1)$ (g) $(20\bar{1})$ (i) $(\bar{2}32)$ (k) $(3\bar{1}2)$
   (b) $(10\bar{2})$ (d) $(21\bar{3})$ (f) $(30\bar{2})$ (h) $(\bar{2}1\bar{2})$ (j) $(13\bar{3})$ (l) $(\bar{3}3\bar{1})$

3.44 What are the Miller indices of the cubic crystallographic planes shown in Fig. P3.44?

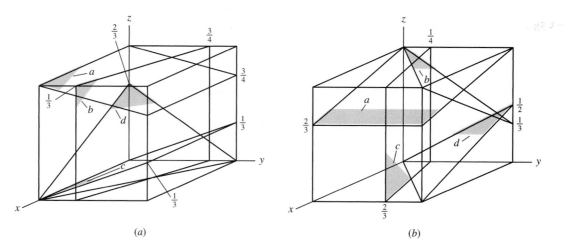

(a)   (b)

Figure P3.44

Tutorial
MatVis

3.45 What is the notation used to indicate a family or form of cubic crystallographic planes?

*3.46 What are the {100} family of planes of the cubic system?

3.47 Draw the following crystallographic planes in a BCC unit cell and list the position of the atoms whose centers are intersected by each of the planes:
   (a) (100)   (b) (110)   (c) (111)

*3.48 Draw the following crystallographic planes in an FCC unit cell and list the position coordinates of the atoms whose centers are intersected by each of the planes:
   (a) (100)   (b) (110)   (c) (111)

*3.49 A cubic plane has the following axial intercepts: $a = \frac{1}{3}$, $b = -\frac{2}{3}$, $c = \frac{1}{2}$. What are the Miller indices of this plane?

3.50 A cubic plane has the following axial intercepts: $a = -\frac{1}{2}, b = -\frac{1}{2}, c = \frac{2}{3}$. What are the Miller indices of this plane?

*3.51 A cubic plane has the following axial intercepts: $a = 1, b = \frac{2}{3}, c = -\frac{1}{2}$. What are the Miller indices of this plane?

*3.52 Determine the Miller indices of the cubic crystal plane that intersects the following position coordinates: $(1, 0, 0)$; $(1, \frac{1}{2}, \frac{1}{4})$; $(\frac{1}{2}, \frac{1}{2}, 0)$.

3.53 Determine the Miller indices of the cubic crystal plane that intersects the following position coordinates: $(\frac{1}{2}, 0, \frac{1}{2})$; $(0, 0, 1)$; $(1, 1, 1)$.

*3.54 Determine the Miller indices of the cubic crystal plane that intersects the following position coordinates: $(1, \frac{1}{2}, 1)$; $(\frac{1}{2}, 0, \frac{3}{4})$; $(1, 0, \frac{1}{2})$.

3.55 Determine the Miller indices of the cubic crystal plane that intersects the following position coordinates: $(0, 0, \frac{1}{2})$; $(1, 0, 0)$; $(\frac{1}{2}, \frac{1}{4}, 0)$.

3.56 Rhodium is FCC and has a lattice constant a of 0.38044 nm. Calculate the following interplanar spacings:

(a) d_{111} (b) d_{200} (c) d_{220}

*3.57 Tungsten is BCC and has a lattice constant a of 0.31648 nm. Calculate the following interplanar spacings:

(a) d_{110} (b) d_{220} (c) d_{310}

*3.58 The d_{310} interplanar spacing in a BCC element is 0.1587 nm. (a) What is its lattice constant a? (b) What is the atomic radius of the element? (c) What could this element be?

*3.59 The d_{422} interplanar spacing in an FCC metal is 0.083397 nm. (a) What is its lattice constant a? (b) What is the atomic radius of the metal? (c) What could this metal be?

3.60 How are crystallographic planes determined in HCP unit cells?

3.61 What notation is used to describe HCP crystal planes?

3.62 Draw the hexagonal crystal planes whose Miller-Bravais indices are:

(a) $(10\bar{1}1)$ (d) $(1\bar{2}12)$ (g) $(\bar{1}2\bar{1}2)$ (j) $(\bar{1}100)$

(b) $(01\bar{1}1)$ (e) $(2\bar{1}\bar{1}1)$ (h) $(2\bar{2}00)$ (k) $(\bar{2}111)$

(c) $(\bar{1}2\bar{1}0)$ (f) $(1\bar{1}01)$ (i) $(10\bar{1}2)$ (l) $(\bar{1}012)$

MatVis

*3.63 Determine the Miller-Bravais indices of the hexagonal crystal planes in Fig. P3.63.

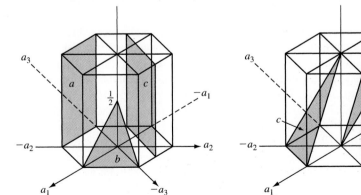

(a)   (b)

Figure P3.63

3.64 Determine the Miller-Bravais direction indices of the $-a_1$, $-a_2$, and $-a_3$ directions.

*3.65 Determine the Miller-Bravais direction indices of the vectors originating at the center of the lower basal plane and ending at the endpoints of the upper basal plane as indicated in Fig. 3.18d.

3.66 Determine the Miller-Bravais direction indices of the basal plane of the vectors originating at the center of the lower basal plane and exiting at the midpoints between the principal planar axes.

*3.67 Determine the Miller-Bravais direction indices of the directions indicated in Fig. P3.67.

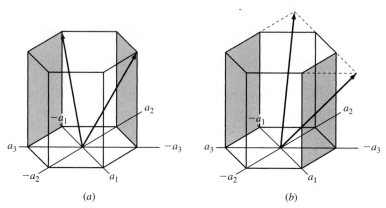

(a) (b)

Figure P3.67

3.68 What is the difference in the stacking arrangement of close-packed planes in (a) the HCP crystal structure and (b) the FCC crystal structure?

3.69 What are the most densely packed planes in (a) the FCC structure and (b) the HCP structure?

3.70 What are the closest-packed directions in (a) the FCC structure and (b) the HCP structure?

*3.71 The lattice constant for BCC tantalum at 20°C is 0.33026 nm and its density is 16.6 g/cm^3. Calculate a value for its atomic mass.

3.72 Calculate a value for the density of FCC platinum in grams per cubic centimeter from its lattice constant a of 0.39239 nm and its atomic mass of 195.09 g/mol.

3.73 Calculate the planar atomic density in atoms per square millimeter for the following crystal planes in BCC chromium, which has a lattice constant of 0.28846 nm: (a) (100), (b) (110), (c) (111).

*3.74 Calculate the planar atomic density in atoms per square millimeter for the following crystal planes in FCC gold, which has a lattice constant of 0.40788 nm: (a) (100), (b) (110), (c) (111).

3.75 Calculate the planar atomic density in atoms per square millimeter for the (0001) plane in HCP beryllium, which has a constant $a = 0.22856$ nm and a c constant of 0.35832 nm.

*3.76 Calculate the linear atomic density in atoms per millimeter for the following directions in BCC vanadium, which has a lattice constant of 0.3039 nm: (a) [100], (b) [110], (c) [111].

3.77 Calculate the linear atomic density in atoms per millimeter for the following directions in FCC iridium, which has a lattice constant of 0.38389 nm:
(*a*) [100],   (*b*) [110],   (*c*) [111].

3.78 What is polymorphism with respect to metals?

3.79 Titanium goes through a polymorphic change from BCC to HCP crystal structure upon cooling through 882°C. Calculate the percentage change in volume when the crystal structure changes from BCC to HCP. The lattice constant *a* of the BCC unit cell at 882°C is 0.332 nm, and the HCP unit cell has $a = 0.2950$ nm and $c = 0.4683$ nm.

*3.80 Pure iron goes through a polymorphic change from BCC to FCC upon heating through 912°C. Calculate the volume change associated with the change in crystal structure from BCC to FCC if at 912°C the BCC unit cell has a lattice constant $a = 0.293$ nm and the FCC unit cell $a = 0.363$ nm.

3.81 What are x-rays, and how are they produced?

3.82 Draw a schematic diagram of an x-ray tube used for x-ray diffraction, and indicate on it the path of the electrons and x-rays.

3.83 What is the characteristic x-ray radiation? What is its origin?

3.84 Distinguish between destructive interference and constructive interference of reflected x-ray beams through crystals.

3.85 Derive Bragg's law by using the simple case of incident x-ray beams being diffracted by parallel planes in a crystal.

*3.86 A sample of BCC metal was placed in an x-ray diffractometer using x-rays with a wavelength of $\lambda = 0.1541$ nm. Diffraction from the {221} planes was obtained at $2\theta = 88.838°$. Calculate a value for the lattice constant *a* for this BCC elemental metal. (Assume first-order diffraction, $n = 1$.)

3.87 X-rays of an unknown wavelength are diffracted by a gold sample. The 2θ angle was 64.582° for the {220} planes. What is the wavelength of the x-rays used? (The lattice constant of gold = 0.40788 nm; assume first-order diffraction, $n = 1$.)

3.88 An x-ray diffractometer recorder chart for an element that has either the BCC or the FCC crystal structure showed diffraction peaks at the following 2θ angles: 41.069°, 47.782°, 69.879°, and 84.396°. (The wavelength of the incoming radiation was 0.15405 nm.)[9]

(*a*) Determine the crystal structure of the element.

(*b*) Determine the lattice constant of the element.

(*c*) Identify the element.

*3.89 An x-ray diffractometer recorder chart for an element that has either the BCC or the FCC crystal structure showed diffraction peaks at the following 2θ angles: 38.60°, 55.71°, 69.70°, 82.55°, 95.00°, and 107.67°. (Wavelength λ of the incoming radiation was 0.15405 nm.)

(*a*) Determine the crystal structure of the element.

(*b*) Determine the lattice constant of the element.

(*c*) Identify the element.

---

[9]X-ray diffraction data courtesy of the International Centre for Diffraction Data.

3.90 An x-ray diffractometer recorder chart for an element that has either the BCC or the FCC crystal structure showed diffraction peaks at the following 2θ angles: 36.191°, 51.974°, 64.982°, and 76.663°. (The wavelength of the incoming radiation was 0.15405 nm.)

(a) Determine the crystal structure of the element.

(b) Determine the lattice constant of the element.

(c) Identify the element.

*3.91 An x-ray diffractometer recorder chart for an element that has either the BCC or the FCC crystal structure showed diffraction peaks at the following 2θ angles: 40.663°, 47.314°, 69.144°, and 83.448°. (The wavelength λ of the incoming radiation was 0.15405 nm.)

(a) Determine the crystal structure of the element.

(b) Determine the lattice constant of the element.

(c) Identify the element.

3.92 Explain, in general terms, why many polymers have an amorphous or semicrystalline structure.

3.93 Explain, in general terms, why ceramic glasses have an amorphous structure.

3.94 Explain the process by which some metals can achieve amorphous structure.

3.16 MATERIALS SELECTION AND DESIGN PROBLEMS

1. In the design of computer chips and microelectronic devices, single crystal silicon wafers are used as the building blocks of the system. (a) To which class of materials does silicon belong? (b) Discuss the bonding and crystal structure of the silicon crystal. (c) Propose a process by which single silicon crystals can be manufactured.

2. Steel is manufactured by adding smaller carbon atoms to the crystal structure of iron. It is possible to add more carbon to the structure when the structure of iron is FCC. However, the normal room-temperature structure of iron is BCC. Design a process that allows the introduction of more carbon to the structure of iron in a solid state.

3. You are given an unknown material and are asked to identify it to the best of your ability. What are some of the tests that you can perform to help identify the material?

4. Often, turbine blades operating at high temperature and high stress levels are manufactured in the form of a large single crystal. (a) Speculate on the advantages of a single-crystal turbine blade. (b) What properties should the selected material have? (c) What specific material would you select to make the single-crystal turbine blade?

5. Name as many carbon allotropes as you can and discuss their crystal structure.

6. Silicon wafers are sometimes coated with a thin layer of aluminum nitride at high temperatures (1000°C). The coefficient of thermal expansion of the silicon crystal is significantly different than that of aluminum nitride. Will this cause a problem? Explain.

4

Solidification and Crystalline Imperfections

(*Photo courtesy of Stan David and Lynn Boatner, Oak Ridge National Library*)

When molten alloys are cast, solidification starts at the walls of the mold as it is being cooled. The solidification of the alloy takes place not at a specific temperature but over a range of temperatures. While the alloy is in this range, it has a pasty form that consists of solid, tree-like structures called *dendrites* (meaning *tree-like*) and liquid metal. The size and shape of the dendrite depends on the cooling rate. The liquid metal existing among these three-dimensional dendritic structures eventually solidifies to form a completely solid structure that we refer to as the grain structure. The study of dendrites is important because they influence compositional variations, porosity, and segregation and therefore the properties of the cast metal. The figure on the previous page shows the three-dimensional structure of dendrites. The figure shows a "forest" of dendrites formed during the solidification of a nickel-based superalloy.[1] ■

---

[1]http://mgnews.msfc.nasa.gov/IDGE/IDGE.html

By the end of this chapter, students will be able to . . .

1. Describe the process of the solidification of metals distinguishing between homogeneous and heterogeneous nucleation.

2. Describe the two energies involved in the solidification process of a pure metal, and write the equation for the total free-energy change associated with the transformation of the liquid state to solid nucleus.

3. Distinguish between equiaxed and columnar grains and the advantage of the former over the latter.

4. Distinguish between single crystal and polycrystalline materials and explain why single crystal and polycrystalline forms of the material have different mechanical properties.

5. Describe various forms of metallic solid solutions and explain the differences between solid solution and mixture alloys.

6. Classify various types of crystalline imperfections and explain the role of defects on the mechanical and electrical properties of crystalline materials.

7. Determine the ASTM grain size number and average grain size diameter and describe the importance of grain size and grain boundary density on the behavior of crystalline materials.

8. Learn how and why optical microscopy, SEM, TEM, HRTEM, AFM, and STM techniques are used to understand more about the internal and surface structures of materials at various magnifications.

9. Explain, in general terms, why alloys are preferred materials over pure metals for structural applications.

4.1 SOLIDIFICATION OF METALS

The solidification of metals and alloys is an important industrial process since most metals are melted and then cast into a semifinished or finished shape. Figure 4.1 shows a large, semicontinuously[2] cast aluminum ingot that will be further fabricated into aluminum alloy flat products. It illustrates the large scale on which the casting process (solidification) of metals is sometimes carried out.

In general the solidification of a metal or alloy can be divided into the following steps:

1. The formation of stable **nuclei** in the melt (nucleation) (Fig. 4.2a)
2. The growth of nuclei into crystals (Fig. 4.2b) and the formation of a grain structure (Fig. 4.2c)

---

[2]A semicontinuously cast ingot is produced by solidifying molten metal (e.g., aluminum or copper alloys) in a mold that has a movable bottom block (see Fig. 4.8) that is slowly lowered as the metal is solidified. The prefix *semi-* is used since the maximum length of the ingot produced is determined by the depth of the pit into which the bottom block is lowered.

Figure 4.1
Large, semicontinuously cast aluminum alloy ingot being removed from casting pit. Ingots of this type are subsequently hot- and cold-rolled into plate or sheet.
(*Courtesy of Reynolds Metals Co.*)

Animation

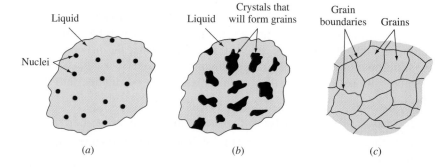

(a)   (b)   (c)

Figure 4.2
Schematic illustration showing the several stages in the solidification of metals: (*a*) formation of nuclei, (*b*) growth of nuclei into crystals, and (*c*) joining together of crystals to form grains and associated grain boundaries. Note that the grains are randomly oriented.

Virtual Lab

Figure 4.3
A grain grouping parted from an arc-cast titanium alloy ingot under the blows of a hammer. The grouping has preserved the true bonding facets of the individual grains of the original cast structure. (Magnification $\frac{1}{6}\times$.)
(*After W. Rostoker and J. R. Dvorak, "Interpretation of Metallographic Structures," Academic, 1965, p. 7.*)

The shapes of some real grains formed by the solidification of a titanium alloy are shown in Fig. 4.3. The shape that each grain acquires after solidification of the metal depends on many factors, of which thermal gradients are important. The grains shown in Fig. 4.3 are *equiaxed* since their growth is about equal in all directions.

4.1.1 The Formation of Stable Nuclei in Liquid Metals

The two main mechanisms by which the nucleation of solid particles in liquid metal occurs are homogeneous nucleation and heterogeneous nucleation.

Homogeneous Nucleation Homogeneous nucleation is considered first since it is the simplest case of nucleation. **Homogeneous nucleation** in a liquid melt occurs when the metal itself provides the atoms needed to form a nuclei. Let us consider the case of a pure metal solidifying. When a pure liquid metal is cooled below its equilibrium freezing temperature to a sufficient degree, many homogeneous nuclei are created by slow-moving atoms bonding together. Homogeneous nucleation usually requires a considerable amount of undercooling, which may be as much as several hundred degrees Celsius for some metals (see Table 4.1). For a nucleus to be stable so that it can grow into a crystal, it must reach a *critical size*. A cluster of atoms bonded together that is less than the critical size is called an **embryo,** and one that is larger than the critical size is called a *nucleus.* Because of their instability, embryos are continuously being formed and redissolved in the molten metal due to the agitation of the atoms.

Energies Involved in Homogeneous Nucleation In the homogeneous nucleation of a solidifying pure metal, two kinds of energy changes must be considered: (1) the *volume (or bulk) free energy* released by the liquid-to-solid transformation and (2) *the surface energy* required to form the new solid surfaces of the solidified particles.

Table 4.1 Values for the freezing temperature, heat of fusion, surface energy, and maximum undercooling for selected metals

Metal	Freezing temp. °C	K	Heat of fusion (J/cm³)	Surface energy (J/cm²)	Maximum undercooling, observed (ΔT[°C])
Pb	327	600	280	33.3×10^{-7}	80
Al	660	933	1066	93×10^{-7}	130
Ag	962	1235	1097	126×10^{-7}	227
Cu	1083	1356	1826	177×10^{-7}	236
Ni	1453	1726	2660	255×10^{-7}	319
Fe	1535	1808	2098	204×10^{-7}	295
Pt	1772	2045	2160	240×10^{-7}	332

Source: B. Chalmers, "Solidification of Metals," Wiley, 1964.

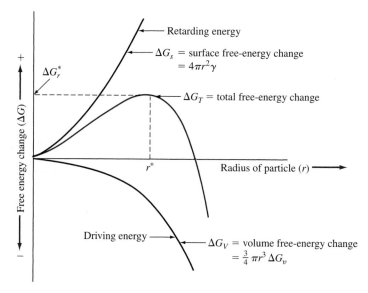

Figure 4.4
Free-energy change ΔG versus radius of embryo or nucleus created by the solidifying of a pure metal. If the radius of the particle is greater than r^*, a stable nucleus will continue to grow.

When a pure liquid metal such as lead is cooled below its equilibrium freezing temperature, the driving energy for the liquid-to-solid transformation is the difference in the volume (bulk) free energy ΔG_v of the liquid and that of the solid. If ΔG_v is the change in free energy between the liquid and solid per unit volume of metal, then the free-energy change for a *spherical nucleus* of radius r is $\frac{4}{3}\pi r^3 \Delta G_v$ since the volume of a sphere is $\frac{4}{3}\pi r^3$. The change in volume free energy versus radius of an embryo or nucleus is shown schematically in Fig. 4.4 as the lower curve and is a negative quantity since energy is released by the liquid-to-solid transformation.

However, there is an opposing energy to the formation of embryos and nuclei, the energy required to form the surface of these particles. The energy needed to create a surface for these spherical particles, ΔG_s, is equal to the specific surface free energy of the particle, γ, times the area of the surface of the sphere, or $4\pi r^2 \gamma$, where $4\pi r^2$ is the surface area of a sphere. This retarding energy ΔG_s for the formation of the solid particles is shown graphically in Fig. 4.4 by an upward curve in the positive upper half of the figure. The total free energy associated with the formation of an embryo or nucleus, which is the sum of the volume free-energy and surface free-energy changes, is shown in Fig. 4.4 as the middle curve. In equation form the total free-energy change for the formation of a spherical embryo or nucleus of radius r formed in a freezing pure metal is

$$\Delta G_T = \tfrac{4}{3}\pi r^3\,\Delta G_v + 4\pi r^2\gamma \qquad (4.1)$$

where ΔG_T = total free-energy change

r = radius of embryo or nucleus

ΔG_v = volume free energy

γ = specific surface free energy

In nature, a system can change spontaneously from a higher- to a lower-energy state. In the case of the freezing of a pure metal, if the solid particles formed upon freezing have radii less than the **critical radius** r^*, the energy of the system will be lowered if they redissolve. These small embryos can, therefore, redissolve in the liquid metal. However, if the solid particles have radii greater than r^*, the energy of the system will be lowered when these particles (nuclei) grow into larger particles or crystals (Fig. 4.2b). When r reaches the critical radius r^*, ΔG_T has its maximum value of ΔG^* (Fig. 4.4).

A relationship among the size of the critical nucleus, surface free energy, and volume free energy for the solidification of a pure metal can be obtained by differentiating Eq. 4.1. The differential of the total free energy ΔG_T with respect to r is zero when $r = r^*$ since the total free energy versus radius of the embryo or nucleus plot is then at a maximum and the slope $d(\Delta G_T)/dr = 0$. Thus,

$$\frac{d(\Delta G_T)}{dr} = \frac{d}{dr}\left(\frac{4}{3}\pi r^3 \Delta G_v + 4\pi r^2\gamma\right)$$

$$\frac{12}{3}\pi r^{*2}\,\Delta G_V + 8\pi r^*\gamma = 0 \qquad (4.1a)$$

$$r^* = -\frac{2\gamma}{\Delta G_v}.$$

Critical Radius versus Undercooling   The greater the degree of undercooling ΔT below the equilibrium melting temperature of the metal, the greater the change in volume free energy ΔG_V. However, the change in free energy due to the surface

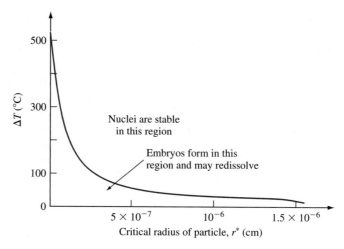

Figure 4.5
Critical radius of copper nuclei versus degree of
undercooling ΔT.
(*After B. Chalmers, "Principles of Solidification," Wiley, 1964.*)

energy ΔG_s does not change much with temperature. Thus, the critical nucleus size is determined mainly by ΔG_V. Near the freezing temperature, the critical nucleus size must be infinite since ΔT approaches zero. As the amount of undercooling increases, the critical nucleus size decreases. Figure 4.5 shows the variation in critical nucleus size for copper as a function of undercooling. The maximum amount of undercooling for homogeneous nucleation in the pure metals listed in Table 4.1 is from 327 to 1772°C. The critical-sized nucleus is related to the amount of undercooling by the relation

$$r^* = \frac{2\gamma T_m}{\Delta H_f \Delta T} \tag{4.2}$$

where r^* = critical radius of nucleus
γ = surface free energy
ΔH_f = latent heat of fusion
ΔT = amount of undercooling at which nucleus is formed

Example Problem 4.1 shows how a value for the number of atoms in a critical nucleus can be calculated from experimental data.

EXAMPLE PROBLEM 4.1

a. Calculate the critical radius (in centimeters) of a homogeneous nucleus that forms when pure liquid copper solidifies. Assume ΔT (undercooling) = $0.2T_m$. Use data from Table 4.1.

b. Calculate the number of atoms in the critical-sized nucleus at this undercooling.

■ **Solution**

a. Calculation of critical radius of nucleus:

$$r^* = \frac{2\gamma T_m}{\Delta H_f \Delta T} \tag{4.2}$$

$$\Delta T = 0.2 T_m = 0.2(1083°C + 273) = (0.2 \times 1356 \text{ K}) = 271 \text{ K}$$

$$\gamma = 177 \times 10^{-7} \text{ J/cm}^2 \quad \Delta H_f = 1826 \text{ J/cm}^3 \quad T_m = 1083°C = 1356 \text{ K}$$

$$r^* = \frac{2(177 \times 10^{-7} \text{ J/cm}^2)(1356 \text{ K})}{(1826 \text{ J/cm}^3)(271 \text{ K})} = 9.70 \times 10^{-8} \text{ cm} \blacktriangleleft$$

b. Calculation of number of atoms in critical-sized nucleus:

$$\text{Vol. of critical-sized nucleus} = \tfrac{4}{3}\pi r^{*3} = \tfrac{4}{3}\pi (9.70 \times 10^{-8} \text{ cm})^3$$
$$= 3.82 \times 10^{-21} \text{ cm}^3$$

$$\text{Vol. of unit cell of Cu}(a = 0.361 \text{ nm}) = a^3 = (3.61 \times 10^{-8} \text{ cm})^3$$
$$= 4.70 \times 10^{-23} \text{ cm}^3$$

Since there are four atoms per FCC unit cell,

$$\text{Volume/atom} = \frac{4.70 \times 10^{-23} \text{ cm}^3}{4} = 1.175 \times 10^{-23} \text{ cm}^3$$

Thus, the number of atoms per homogeneous critical nucleus is

$$\frac{\text{Volume of nucleus}}{\text{Volume/atom}} = \frac{3.82 \times 10^{-21} \text{ cm}^3}{1.175 \times 10^{-23} \text{ cm}^3} = 325 \text{ atoms} \blacktriangleleft$$

Heterogeneous Nucleation   **Heterogeneous nucleation** is nucleation that occurs in a liquid on the surfaces of its container, insoluble impurities, and other structural material that lower the critical free energy required to form a stable nucleus. Since large amounts of undercooling do not occur during industrial casting operations and usually range between 0.1 and 10°C, the nucleation must be heterogeneous and not homogeneous.

For heterogeneous nucleation to take place, the solid nucleating agent (impurity solid or container) must be wetted by the liquid metal. Also the liquid should solidify easily on the nucleating agent. Figure 4.6 shows a nucleating agent (substrate) that is wetted by the solidifying liquid, creating a low contact angle θ between the solid metal and the nucleating agent. Heterogeneous nucleation takes place on the nucleating agent because the surface energy to form a stable nucleus is lower on this material than in the pure liquid itself (homogeneous nucleation). Since the surface energy is lower for heterogeneous nucleation, the total free-energy change for the formation of a stable nucleus will be lower and the critical size of the nucleus will be smaller. Thus, a much smaller amount of undercooling is required to form a stable nucleus produced by heterogeneous nucleation.

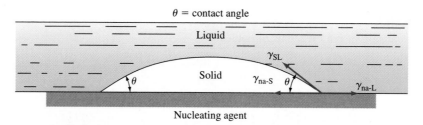

Figure 4.6
Heterogeneous nucleation of a solid on a nucleating agent. na = nucleating agent, SL = solid-liquid, S = solid, L = liquid; θ = contact angle.
(After J. H. Brophy, R. M. Rose, and John Wulff, "Structure and Properties of Materials," vol. II: "Thermodynamics of Structure," Wiley, 1964, p. 105.)

4.1.2 Growth of Crystals in Liquid Metal and Formation of a Grain Structure

After stable nuclei have been formed in a solidifying metal, these nuclei grow into crystals, as shown in Fig. 4.2*b*. In each solidifying crystal the atoms are arranged in an essentially regular pattern, but the orientation of each crystal varies (Fig. 4.2*b*). When solidification of the metal is finally completed, the crystals join together in different orientations and form crystal boundaries at which changes in orientation take place over a distance of a few atoms (Fig. 4.2*c*). Solidified metal containing many crystals is said to be *polycrystalline*. The crystals in the solidified metal are called **grains,** and the surfaces between them, *grain boundaries.*

The number of nucleation sites available to the freezing metal will affect the grain structure of the solid metal produced. If relatively few nucleation sites are available during solidification, a coarse, or large-grain, structure will be produced. If many nucleation sites are available during solidification, a fine-grain structure will result. Almost all engineering metals and alloys are cast with a fine-grain structure since this is the most desirable type for strength and uniformity of finished metal products.

When a relatively pure metal is cast into a stationary mold without the use of *grain refiners,*[3] two major types of grain structures are usually produced:

1. Equiaxed grains
2. Columnar grains

If the nucleation and growth conditions in the liquid metal during solidification are such that the crystals can grow about equally in all directions, **equiaxed grains** will be produced. Equiaxed grains are commonly found adjacent to a cold mold wall, as shown in Fig. 4.7. Large amounts of undercooling near the wall create a relatively high concentration of nuclei during solidification, a condition necessary to produce the equiaxed gain structure.

---

[3]A grain refiner is a material added to a molten metal to attain finer grains in the final grain structure.

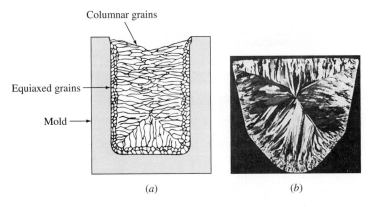

Columnar grains

Equiaxed grains

Mold

(a)   (b)

Figure 4.7
(a) Schematic drawing of a solidified metal grain structure
produced by using a cold mold. (b) Transverse section through
an ingot of aluminum alloy 1100 (99.0% Al) cast by the Properzi
method (a wheel and belt method). Note the consistency with
which columnar grains have grown perpendicular to each
mold face.
*(After "Metals Handbook," vol. 8, 8th ed., American Society for Metals,
1973, p. 164.)*

Columnar grains are long, thin, coarse grains created when a metal solidifies
rather slowly in the presence of a steep temperature gradient. Relatively few nuclei
are available when columnar grains are produced. Equiaxed and columnar grains
are shown in Fig. 4.7. Note that in Fig. 4.7b the columnar grains have grown perpendicular to the mold faces since large thermal gradients were present in those
directions.

4.1.3 Grain Structure of Industrial Castings

In industry, metals and alloys are cast into various shapes. If the metal is to be further fabricated after casting, large castings of simple shapes are produced first and
then fabricated further into semifinished products. For example, in the aluminum
industry, common shapes for further fabrication are sheet ingots (Fig. 4.1), which
have rectangular cross sections, and extrusion[4] ingots, which have circular cross sections. For some applications, the molten metal is cast into essentially its final shape
as, for example, an automobile piston (see Fig. 6.3).

The large aluminum alloy sheet ingot in Fig. 4.1 was cast by a direct-chill semicontinuous casting process. In this casting method the molten metal is cast into a

---

[4]*Extrusion* is the process of converting a metal ingot into lengths of uniform cross section by forcing solid
plastic metal through a die or orifice of the desired cross-sectional outline.

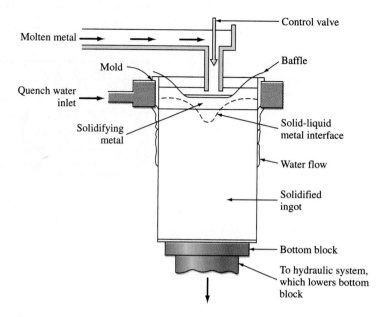

Figure 4.8
Schematic of an aluminum alloy ingot being cast in a direct-chill
semicontinuous casting unit.

mold with a movable bottom block that is slowly lowered after the mold is filled
(Fig. 4.8). The mold is water-cooled by a water box, and water is also sprayed down
the sides of the solidified surface of the ingot. In this way large ingots about 15 ft
long can be cast continuously, as shown in Fig. 4.1. In the steel industry about 60 percent of the metal is cast into stationary molds, with the remaining 40 percent being
continuously cast, as shown in Fig. 4.9.

To produce cast ingots with a fine grain size, grain refiners are usually added to
the liquid metal before casting. For aluminum alloys, small amounts of grainrefin-
ing elements such as titanium, boron, or zirconium are included in the liquid metal
just before the casting operation so that a fine dispersion of heterogeneous nuclei
will be available during solidification. Figure 4.10 shows the effect of using a grain
refiner while casting 6-in.-diameter aluminum extrusion ingots. The ingot section
cast without the grain refiner has large columnar grains (Fig. 4.10a), and the section
cast with the grain refiner has a fine, equiaxed grain structure (Fig. 4.10b).

4.2 SOLIDIFICATION OF SINGLE CRYSTALS

Almost all engineering crystalline materials are composed of many crystals and are
therefore **polycrystalline.** However, there are a few that consist of only one crystal
and are therefore *single crystals.* For example, high-temperature creepresistant gas

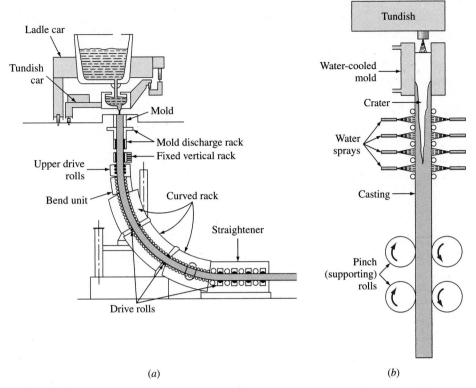

(a)   (b)

Figure 4.9
Continuous casting of steel ingots. (a) General setup and (b) close-up of the mold arrangement.
(*After "Making, Shaping, and Treating of Steel," 10th ed., Association of Iron and Steel Engineers, 1985.*)

turbine blades are sometimes made of single crystals, as shown in Fig. 4.11c. Single-crystal turbine blades are more creep resistant at high temperatures than the same blades made with an equiaxed grain structure (Fig. 4.11a) or a columnar grain structure (Fig. 4.11b) because at high temperatures above about half the absolute melting temperature of a metal the grain boundaries become weaker than the grain bodies.

In growing single crystals, solidification must take place around a single nucleus so that no other crystals are nucleated and grow. To accomplish this, the interface temperature between the solid and liquid must be slightly lower than the melting point of the solid, and the liquid temperature must increase beyond the interface. To achieve this temperature gradient, the latent heat of solidification[5] must be conducted through the solidifying solid crystal. The growth rate of the crystal must be slow so

---

[5]The latent heat of solidification is the thermal energy released when a metal solidifies.

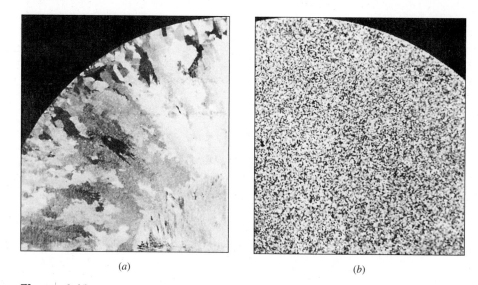

(a)   (b)

Figure 4.10

Parts of transverse sections through two 6-in-diameter ingots of alloy 6063
(Al–0.7% Mg–0.4% Si) that were direct-chill semicontinuous cast. (*a*) Ingot section
was cast without the addition of a grain refiner; note columnar grains and
colonies of featherlike crystals near the center of the section. (*b*) Ingot section
was cast with the addition of a grain refiner and shows a fine, equiaxed grain
structure. (Tucker's reagent; actual size.)

(After "Metals Handbook," vol. 8, 8th ed., American Society for Metals, 1973, p. 164.)

that the temperature at the liquid-solid interface is slightly below the melting point
of the solidifying solid. Figure 4.12*a* illustrates how single-crystal turbine blades can
be cast, and Fig. 4.12*b* and *c* show how competitive grain growth is reduced to a
single grain by using a "pigtail" selector.

Another example of an industrial use of single crystals is the silicon single
crystals that are sliced into wafers for solid-state electronic integrated circuit chips
(see Fig. 13.1). Single crystals are necessary for this application since grain
boundaries would disrupt the flow of electrons in devices made from semiconductor silicon. In industry, single crystals of silicon 8 to 12 in. (20 to 25 cm) in
diameter have been grown for semiconducting device applications. One of the
commonly used techniques to produce high-quality (minimization of defects)
silicon single crystals is the Czochralski method. In this process high-purity polycrystalline silicon is first melted in a nonreactive crucible and held at a temperature just above the melting point. A high-quality seed crystal of silicon of the
desired orientation is lowered into the melt while it is rotated. Part of the surface
of the seed crystal is melted in the liquid to remove the outer strained region and
to produce a surface for the liquid to solidify on. The seed crystal continues to
rotate and is slowly raised from the melt. As it is raised from the melt, silicon

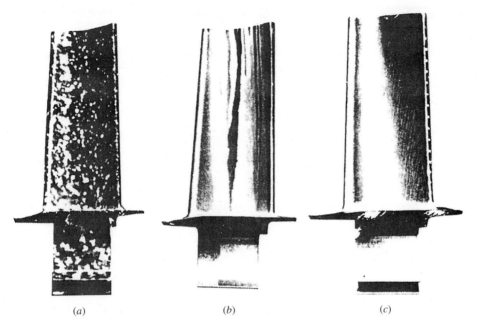

Figure 4.11

Different grain structures of gas turbine airfoil blades: (*a*) Polycrystal equiaxed, (*b*) polycrystal columnar, and (*c*) single crystal.
(*Courtesy of Pratt and Whitney Co.*)

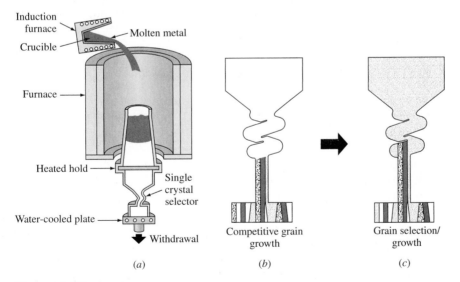

Figure 4.12

(*a*) A process schematic for producing a single-crystal gas turbine airfoil. (*b*) Starter section of casting for producing a single-crystal airfoil showing competitive growth during solidification below the single-crystal selector ("pigtail"). (*c*) Same as (*b*) but showing the survival of only one grain during solidification through the single-crystal selector.
(*After Pratt and Whitney Co.*)

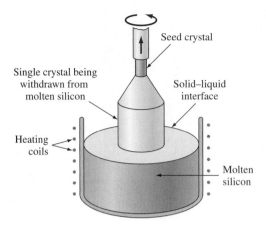

Figure 4.13
Formation of single crystal of silicon by the Czochralski process.

from the liquid in the crucible adheres and grows on the seed crystal, producing a much larger diameter single crystal of silicon (Fig. 4.13). With this process large single-crystal silicon ingots up to about 12 in. ($\cong$30 cm) in diameter can and have been made.

4.3 METALLIC SOLID SOLUTIONS

Although very few metals are used in the pure or nearly pure state, a few are used in the nearly pure form. For example, high-purity copper of 99.99 percent purity is used for electronic wires because of its very high electrical conductivity. High-purity aluminum (99.99% Al) (called *superpure aluminum*) is used for decorative purposes because it can be finished with a very bright metallic surface. However, most engineering metals are combined with other metals or nonmetals to provide increased strength, higher corrosion resistance, or other desired properties.

A *metal alloy,* or simply an **alloy,** is a mixture of two or more metals or a metal (metals) and a nonmetal (nonmetals). Alloys can have structures that are relatively simple, such as that of cartridge brass, which is essentially a binary alloy (two metals) of 70% Cu and 30% Zn. On the other hand, alloys can be extremely complex, such as the nickel-base superalloy Inconel 718 used for jet engine parts, which has about 10 elements in its nominal composition.

The simplest type of alloy is that of the solid solution. A **solid solution** is a *solid* that consists of two or more elements atomically dispersed in a single-phase structure. In general there are two types of solid solutions: *substitutional* and *interstitial.*

Figure 4.14
Substitutional solid
solution. The dark circles
represent one type of
atom and the light
another. The plane of
atoms is a (111) plane in
an FCC crystal lattice.

4.3.1 Substitutional Solid Solutions

In **substitutional solid solutions** formed by two elements, solute atoms can substitute for parent solvent atoms in a crystal lattice. Figure 4.14 shows a (111) plane in an FCC crystal lattice in which some solute atoms of one element have substituted for solvent atoms of the parent element. The crystal structure of the parent element or solvent is unchanged, but the lattice may be distorted by the presence of the solute atoms, particularly if there is a significant difference in atomic diameters of the solute and solvent atoms.

The fraction of atoms of one element that can dissolve in another can vary from a fraction of an atomic percent to 100 percent. The following conditions, known as Hume–Rothery rules, are favorable for extensive solid solubility of one element in another:

1. The diameters of the atoms of the elements must not differ by more than about 15 percent.
2. The crystal structures of the two elements must be the same.
3. There should be no appreciable difference in the electronegativities of the two elements so that compounds will not form.
4. The two elements should have the same valence.

If the atomic diameters of the two elements that form a solid solution differ, there will be a distortion of the crystal lattice. Since the atomic lattice can only sustain a limited amount of contraction or expansion, there is a limit in the difference in atomic diameters that atoms can have and still maintain a solid solution with the same kind of crystal structure. When the atomic diameters differ by more than about 15 percent, the "size factor" becomes unfavorable for extensive solid solubility.

Using the data in the following table, predict the relative degree of atomic solid solubility of the following elements in copper:

a. Zinc   d. Nickel
b. Lead   e. Aluminum
c. Silicon   f. Beryllium

Use the scale very high, 70–100%; high, 30–70%; moderate, 10–30%; low, 1–10%; and very low, <1%.

Element	Atom radius (nm)	Crystal structure	Electro-negativity	Valence
Copper	0.128	FCC	1.8	+2
Zinc	0.133	HCP	1.7	+2
Lead	0.175	FCC	1.6	+2, +4
Silicon	0.117	Diamond cubic	1.8	+4
Nickel	0.125	FCC	1.8	+2
Aluminum	0.143	FCC	1.5	+3
Beryllium	0.114	HCP	1.5	+2

■ **Solution**

A sample calculation for the atomic radius difference for the Cu–Zn system is

$$\text{Atomic radius difference} = \frac{\text{final radius} - \text{initial radius}}{\text{initial radius}}(100\%)$$

$$= \frac{R_{Zn} - R_{Cu}}{R_{Cu}}(100\%) \tag{4.3}$$

$$= \frac{0.133 - 0.128}{0.128}(100\%) = +3.9\%$$

System	Atomic radius difference (%)	Electronegativity difference	Predicted relative degree of solid solubility	Observed maximum solid solubility (at %)
Cu–Zn	+3.9	0.1	High	38.3
Cu–Pb	+36.7	0.2	Very low	0.1
Cu–Si	−8.6	0	Moderate	11.2
Cu–Ni	−2.3	0	Very high	100
Cu–Al	+11.7	0.3	Moderate	19.6
Cu–Be	−10.9	0.3	Moderate	16.4

The predictions can be made principally on the atomic radius difference. In the case of the Cu–Si system, the difference in the crystal structures is important. There is very little electronegativity difference for all these systems. The valences are all the same except for Al and Si. In the final analysis the experimental data must be referred to.

If the solute and solvent atoms have the same crystal structure, then extensive solid solubility is favorable. If the two elements are to show complete solid solubility in all proportions, then both elements must have the same crystal structure. Also, there cannot be too great a difference in the electronegativities of the two elements forming solid solutions, or else the highly electropositive element will lose electrons, the highly electronegative element will acquire electrons, and compound formation will result. Finally, if the two solid elements have the same valence, solid solubility will be favored. If there is a shortage of electrons between the atoms, the binding between them will be upset, resulting in conditions unfavorable for solid solubility.

4.3.2 Interstitial Solid Solutions

In interstitial solutions the solute atoms fit into the spaces between the solvent or parent atoms. These spaces or voids are called *interstices*. **Interstitial solid solutions** can form when one atom is much larger than another. Examples of atoms that can form interstitial solid solutions due to their small size are hydrogen, carbon, nitrogen, and oxygen.

An important example of an interstitial solid solution is that formed by carbon in FCC γ iron that is stable between 912 and 1394°C. The atomic radius of γ iron is 0.129 nm and that of carbon is 0.075 nm, and so there is an atomic radius difference of 42 percent. However, in spite of this difference, a maximum of 2.08 percent of the carbon can dissolve interstitially in iron at 1148°C. Figure 4.15a illustrates this schematically by showing distortion around the carbon atoms in the γ iron lattice.

The radius of the largest interstitial hole in FCC γ iron is 0.053 nm (see Example Problem 4.3), and since the atomic radius of the carbon atom is 0.075 nm, it is not surprising that the maximum solid solubility of carbon in γ iron is only 2.08 percent. The radius of the largest interstitial void in BCC α iron is only 0.036 nm, and as a result just below 723°C only 0.025 percent of the carbon can be dissolved interstitially.

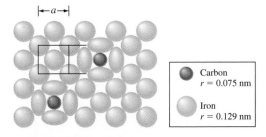

Figure 4.15a
Schematic illustration of an interstitial solid solution of carbon in FCC γ iron just above 912°C showing a (100) plane. Note the distortion of the iron atoms (0.129 nm radius) around the carbon atoms (0.075 nm radius), fitting into voids of 0.053 nm radius.
(*After L. H. Van Vlack, "Elements of Materials Science and Engineering," 4th ed., Addison-Wesley, 1980, p. 113.*)

Calculate the radius of the largest interstitial void in the FCC γ iron lattice. The atomic radius of the iron atom is 0.129 nm in the FCC lattice, and the largest interstitial voids occur at the $(\frac{1}{2}, 0, 0)$, $(0, \frac{1}{2}, 0)$, $(0, 0, \frac{1}{2})$, etc., type positions.

■ **Solution**

Figure 4.15b shows a (100) FCC lattice plane on the yz plane. Let the radius of an iron atom be R and that of the interstitial void at the position $(0, \frac{1}{2}, 0)$ be r. Then, from Fig. 4.15b,

$$2R + 2r = a \tag{4.4}$$

Also from Fig. 4.15b,

$$(2R)^2 = (\tfrac{1}{2}a)^2 + (\tfrac{1}{2}a)^2 = \tfrac{1}{2}a^2 \tag{4.5}$$

Solving for a gives

$$2R = \frac{1}{\sqrt{2}}a \quad \text{or} \quad a = 2\sqrt{2}R \tag{4.6}$$

Combining Eqs. 4.4 and 4.6 gives

$$2R + 2r = 2\sqrt{2}R$$
$$r = (\sqrt{2} - 1)R = 0.414R$$
$$= (0.414)(0.129 \text{ nm}) = 0.053 \text{ nm} \blacktriangleleft$$

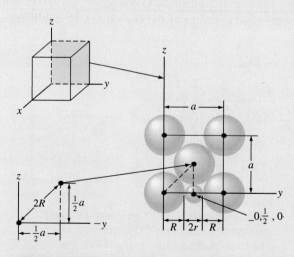

Figure 4.15b
(100) plane of the FCC lattice containing an interstitial atom at the $(0, \frac{1}{2}, 0)$ position coordinate.

4.4 CRYSTALLINE IMPERFECTIONS

In reality, crystals are never perfect and contain various types of imperfections and defects that affect many of their physical and mechanical properties, which in turn affect many important engineering properties of materials such as the cold formability of alloys, the electronic conductivity of semiconductors, the rate of migration of atoms in alloys, and the corrosion of metals.

Crystal lattice imperfections are classified according to their geometry and shape. The three main divisions are (1) zero-dimensional or point defects, (2) one-dimensional or line defects (dislocations), and (3) two-dimensional defects, that include external surfaces, grain boundaries, twins, low-angle boundaries, high-angle boundaries, twists, stacking faults, voids, and precipitates. Three-dimensional macroscopic or bulk defects could also be included. Examples of these defects are pores, cracks, and foreign inclusions.

4.4.1 Point Defects

The simplest point defect is the vacancy, an atom site from which an atom is missing (Fig. 4.16a). **Vacancies** may be produced during solidification as a result of local disturbances during the growth of crystals, or they may be created by atomic rearrangements in an existing crystal due to atomic mobility. In metals the equilibrium concentration of vacancies rarely exceeds about 1 in 10,000 atoms. Vacancies are equilibrium defects in metals, and their energy of formation is about 1 eV.

Additional vacancies in metals can be introduced by plastic deformation, rapid cooling from higher temperatures to lower ones to entrap the vacancies, and by bombardment with energetic particles such as neutrons. Nonequilibrium vacancies have a tendency to cluster, causing divacancies or trivacancies to form. Vacancies can move by exchanging positions with their neighbors. This process is important in the

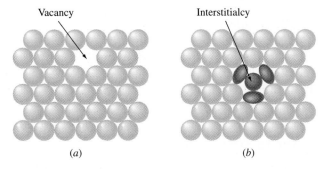

(a) *(b)*

Figure 4.16
(*a*) Vacancy point defect. (*b*) Self-interstitial, or interstitialcy, point defect in a close-packed solid-metal lattice.

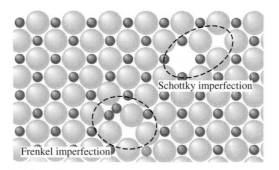

Figure 4.17
Two-dimensional representation of an ionic
crystal illustrating a Schottky defect and a
Frenkel defect.
*(After Wulff et al., "Structure and Properties of Materials,"
vol. 1: "Structure," Wiley, 1964, p. 78.)*

migration or diffusion of atoms in the solid state, particularly at elevated temperatures where atomic mobility is greater.

Sometimes an atom in a crystal can occupy an interstitial site between surrounding atoms in normal atom sites (Fig. 4.16*b*). This type of point defect is called
a **self-interstitial,** or **interstitialcy.** These defects do not generally occur naturally
because of the structural distortion they cause, but they can be introduced into a
structure by irradiation.

In ionic crystals point defects are more complex due to the necessity to maintain electrical neutrality. When two oppositely charged ions are missing from an ionic
crystal, a cation-anion divacancy is created that is known as a **Schottky imperfection**
(Fig. 4.17). If a positive cation moves into an interstitial site in an ionic crystal, a
cation vacancy is created in the normal ion site. This vacancy-interstitialcy pair is
called a **Frenkel[6] imperfection** (Fig. 4.17). The presence of these defects in ionic
crystals increases their electrical conductivity.

Impurity atoms of the substitutional or interstitial type are also point defects and
may be present in metallic or covalently bonded crystals. For example, very small
amounts of substitutional impurity atoms in pure silicon can greatly affect its electrical conductivity for use in electronic devices. Impurity ions are also point defects
in ionic crystals.

4.4.2 Line Defects (Dislocations)

Line imperfections, or **dislocations,** in crystalline solids are defects that cause lattice
distortion centered around a line. Dislocations are created during the solidification

---

[6]Yakov Ilyich Frenkel (1894–1954). Russian physicist who studied defects in crystals. His name is associated
with the vacancy-interstitialcy defect found in some ionic crystals.

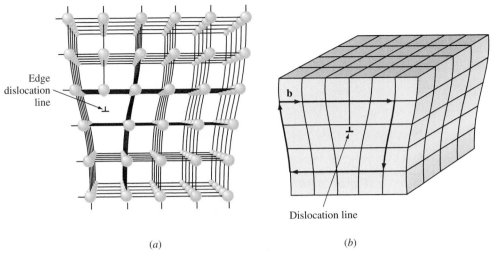

Edge
dislocation
line

b

Dislocation line

(a)   (b)

Figure 4.18

(a) Positive edge dislocation in a crystalline lattice. A linear defect occurs in the region just above the inverted "tee," ⊥, where an extra half plane of atoms has been wedged in.

(After A. G. Guy, "Essentials of Materials Science," McGraw-Hill, 1976, p. 153.)

(b) Edge dislocation that indicates the orientation of its Burgers or slip vector **b**.

(After M. Eisenstadt, "Introduction to Mechanical Properties of Materials," Macmillan, 1971, p. 117.)

Animation

of crystalline solids. They are also formed by the permanent or plastic deformation of crystalline solids and by vacancy condensation and by atomic mismatch in solid solutions.

The two main types of dislocations are the *edge* and *screw types*. A combination of the two gives *mixed dislocations,* which have edge and screw components. An edge dislocation is created in a crystal by the insertion of an extra half plane of atoms, as shown in Fig. 4.18a just above the symbol ⊥. The inverted "tee," ⊥, indicates a positive edge dislocation, whereas the upright "tee," ⊤, indicates a negative edge dislocation.

The displacement distance of the atoms around the dislocation is called the *slip* or *Burgers vector* **b** and is *perpendicular* to the edge-dislocation line (Fig. 4.18b). Dislocations are nonequilibrium defects, and they store energy in the distorted region of the crystal lattice around the dislocation. The edge dislocation has a region of compressive strain where the extra half plane is and a region of tensile strain below the extra half plane of atoms (Fig. 4.19a).

The screw dislocation can be formed in a perfect crystal by applying upward and downward shear stresses to regions of a perfect crystal that have been separated by a cutting plane, as shown in Fig. 4.20a. These shear stresses introduce a region of distorted crystal lattice in the form of a spiral ramp of distorted atoms or screw dislocation (Fig. 4.20b). The region of distorted crystal is not well

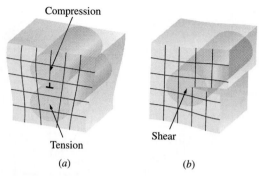

Figure 4.19
Strain fields surrounding (*a*) an edge
dislocation and (*b*) a screw dislocation.
*(After John Wulff et al., "The Structure and Properties
of Materials," vol. 3: "Mechanical Behavior," Wiley,
1965, p. 69.)*

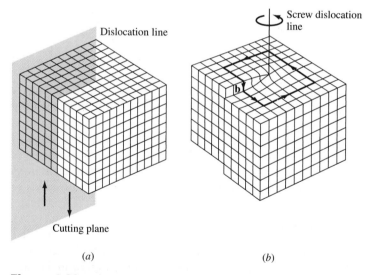

Figure 4.20
Formation of a screw dislocation. (*a*) A perfect crystal is sliced
by a cutting plane, and up and down shear stresses are applied
parallel to the cutting plane to form the screw dislocation in (*b*).
(*b*) A screw dislocation is shown with its slip or Burgers vector **b**
parallel to the dislocation line.
*(After M. Eisenstadt, "Mechanical Properties of Materials," Macmillan, 1971,
p. 118.)*

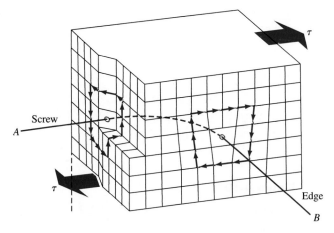

Figure 4.21
Mixed dislocation in a crystal. Dislocation line *AB* is pure
screw type where it enters the crystal on left and pure
edge type where it leaves the crystal on right.
(After John Wulff et al., "Structure and Properties of Materials,"
vol. 3: "Mechanical Properties," Wiley, 1965, p. 65.)

defined and is at least several atoms in diameter. A region of shear strain is created
around the screw dislocation in which energy is stored (Fig. 4.19*b*). The slip or
Burgers vector of the screw dislocation is *parallel* to the dislocation line, as shown
in Fig. 4.20*b*.

Most dislocations in crystals are of the mixed type, having edge and screw components. In the curved dislocation line *AB* in Fig. 4.21, the dislocation is of the pure
screw type at the left where it enters the crystal and of the pure edge type on the
right where it leaves the crystal. Within the crystal, the dislocation is of the mixed
type, with edge and screw components.

4.4.3 Planar Defects

Planar defects include external surfaces, **grain boundaries, twins, low-angle
boundaries, high-angle boundaries, twists,** and **stacking faults.** The free or external surface of any material is the most common type of planar defect. External
surfaces are considered defects because the atoms on the surface are bonded to
other atoms only on one side. Therefore, the surface atoms have a lower number
of neighbors. As a result these atoms have a higher state of energy when compared to the atoms positioned inside the crystal with an optimal number of
neighbors. The higher energy associated with the atoms on the surface of a material makes the surface susceptible to erosion and reaction with elements in the

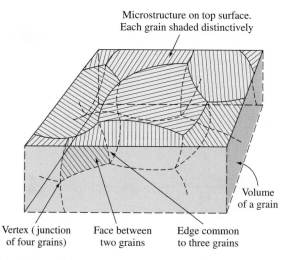

Microstructure on top surface.
Each grain shaded distinctively

Volume
of a grain

Vertex (junction
of four grains)

Face between
two grains

Edge common
to three grains

Figure 4.22
Sketch showing the relation of the two-dimensional
microstructure of a crystalline material to the
underlying three-dimensional network. Only
portions of the total volume and total face of any
one grain are shown.
*(After A. G. Guy, "Essentials of Materials Science,"
McGraw-Hill, 1976.)*

environment. This point further illustrates the importance of defects in the behavior of materials.

Grain boundaries are surface imperfections in polycrystalline materials that separate grains (crystals) of different orientations. In metals grain boundaries are created during solidification when crystals formed from different nuclei grow simultaneously and meet each other (Fig. 4.2). The shape of the grain boundaries is determined by the restrictions imposed by the growth of neighboring grains. Grain-boundary surfaces of an approximately equiaxed grain structure are shown schematically in Fig. 4.22 and of real grains in Fig. 4.3.

The grain boundary itself is a narrow region between two grains of about two to five atomic diameters in width and is a region of atomic mismatch between adjacent grains. The atomic packing in grain boundaries is lower than within the grains because of the atomic mismatch. Grain boundaries also have some atoms in strained positions that raise the energy of the grain-boundary region.

The higher energy of the grain boundaries and their more open structure make them a more favorable region for the nucleation and growth of precipitates (see Sec. 9.5). The lower atomic packing of the grain boundaries also allows for more rapid diffusion of atoms in the grain boundary region. At ordinary temperatures grain boundaries also restrict plastic flow by making it difficult for the movement of dislocations in the grain boundary region.

Figure 4.23
Twin boundaries in the grain structure of brass.

Virtual Lab

Twins or *twin boundaries* are another example of a two-dimensional defect. A twin is defined as a region in which a mirror image of the structure exists across a plane or a boundary. Twin boundaries form when a material is permanently or plastically deformed (*deformation twin*). They can also appear during the recrystallization process in which atoms reposition themselves in a deformed crystal (*annealing twin*), but this happens only in some FCC alloys. A number of annealing twins formed in the microstructure of brass are shown in Fig. 4.23. As the name indicates, twin boundaries form in pairs. Similar to dislocations, twin boundaries tend to strengthen a material. A more detailed explanation of twin boundaries is given in Sec. 6.5.

When an array of edge dislocations are oriented in a crystal in a manner that seems to misorient or tilt two regions of a crystal (Fig. 4.24*a*), a two-dimensional defect called a *small-angle tilt boundary* is formed. A similar phenomenon can occur when a network of screw dislocations create a small-angle twist boundary (Fig 4.24*b*). The misorientation angle θ for a small-angle boundary is generally less than 10 degrees. As the density of dislocations in small-angle boundaries (tilt or twist) increases, the misorientation angle θ becomes larger. If θ exceeds 20 degrees, the boundary is no longer characterized as a small-angle boundary but

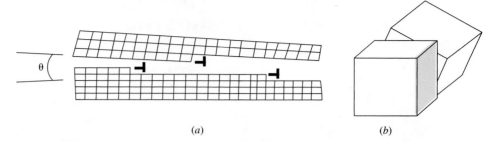

(a)   (b)

Figure 4.24
(a) Edge dislocations in an array forming a small-angle tilt boundary. (b) Schematic of a small-angle twist boundary.

is considered a general grain boundary. Similar to dislocations and twins, small-angle boundaries are regions of high energy due to local lattice distortions and tend to strengthen a metal.

In Sec. 3.8, we discussed formation of FCC and HCP crystal structures by the stacking of atomic planes. It was noted that the stacking sequence ABABAB . . . leads to the formation of an HCP crystal structure while the sequence ABCABCABC . . . leads to the FCC structure. Sometimes during the growth of a crystalline material, collapse of a vacancy cluster, or interaction of dislocations, one or more of the stacking planes may be missing, giving rise to another two-dimensional defect called a *stacking fault* or a *piling-up fault*. Stacking faults ABCABA**A**CBABC and ABA**A**BBAB are typical in FCC and HCP crystals, respectively. The bold-faced planes indicate the faults. Stacking faults also tend to strengthen the material.

It is important to note that, generally speaking, of the two-dimensional defects discussed here, grain boundaries are most effective in strengthening a metal; however stacking faults, twin boundaries, and small-angle boundaries often also serve a similar purpose. The reason why these defects tend to strengthen a metal will be discussed in more detail in Chap. 6.

4.4.4 Volume Defects

Volume or three-dimensional defects form when a cluster of point defects join to form a three-dimensional void or a pore. Conversely, a cluster of impurity atoms may join to form a three-dimensional precipitate. The size of a volume defect may range from a few nanometers to centimeters or sometimes larger. Such defects have a tremendous effect or influence on the behavior and performance of the material. Finally, the concept of a three-dimensional or volume defect may be extended to an amorphous region within a polycrystalline material. Such materials were briefly discussed in Chap. 3 and will be more extensively discussed in future chapters.

4.5 EXPERIMENTAL TECHNIQUES FOR IDENTIFICATION OF MICROSTRUCTURE AND DEFECTS

Material scientists and engineers use various instruments to study and understand the behavior of materials based on their microstructures, existing defects, microconstituents, and other features and characteristics specific to the internal structure. The instruments reveal information about the internal makeup and structure of the material at various length scales extending from the micro- to nano range. In this range, the structure of grains, grain boundaries, various microphases, line defects, surface defects and their effect on material behavior may be studied by various instruments. In the following sections, we will discuss the use of optical metallography, scanning electron microscopy, transmission electron microscopy, high-resolution transmission electron microscopy, and scanning probe microscopy techniques to learn about the internal and surface features of materials.

4.5.1 Optical Metallography, ASTM Grain Size, and Grain Diameter Determination

Optical metallography techniques are used to study the features and internal makeup of materials at the micrometer level (magnification level of around 2000X). Qualitative and quantitative information pertaining to grains size, grain boundary, existence of various phases, internal damage, and some defects may be extracted using optical metallography techniques. In this technique the surface of a small sample of a material such as a metal or a ceramic is first prepared through a detailed and rather lengthy procedure. The preparation process includes numerous surface grinding stages (usually four) that remove large scratches and thin plastically deformed layers from the surface of the specimen. The grinding stage is followed with a number of polishing stages (usually four) that remove fine scratches formed during the grinding stage. The quality of the surface is extremely important in the outcome of the process, and generally speaking, a smooth, mirror-like surface without scratches must be produced at the end of the polishing stage. These steps are necessary to minimize topographic contrast. The polished surface is then exposed to chemical etchants. The choice of the etchant and the etching time (the time interval in which the sample will remain in contact with the etchant) are two critical factors that depend on the specific material under study. The atoms at the grain boundary will be attacked at a much more rapid rate by the etchant than those atoms inside the grain. This is because the atoms at the grain boundary possess a higher state of energy because of the less efficient packing. As a result, the etchant produces tiny groves along the boundaries of the grains. The prepared sample is then examined using a metallurgical microscope (inverted microscope) based on visible incident light. A schematic representation of the metallurgical microscope is given in Fig. 4.25. When exposed to incident light in an optical microscope, these groves do not reflect the light as intensely as the remainder of the grain material

Virtual Lab

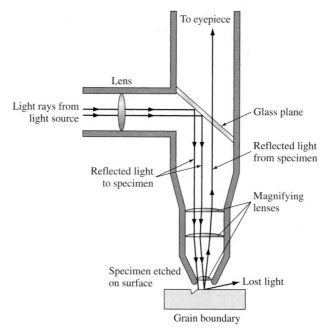

Figure 4.25

Schematic diagram illustrating how light is reflected from the surface of a polished and etched metal. The irregular surface of the etched-out grain boundary does not reflect light.

(After M. Eisenstadt, "Mechanical Properties of Materials," Macmillan, 1971, p. 126.)

Virtual Lab

(Fig. 4.26). Because of the reduced light reflection, the tiny grooves appear as dark lines to the observer, thus revealing the grain boundaries (Fig. 4.27). Additionally, impurities, other existing phases, and internal defects also react differently to the etchant and reveal themselves in photomicrographs taken from the sample surface. Overall, this technique provides a great deal of qualitative information about the material.

In addition to the qualitative information that is extracted from the photomicrographs, some limited quantitative information may also be extracted. Grain size and average grain diameter of the material may be determined using the photomicrographs obtained by this technique.

The grain size of polycrystalline metals is important since the amount of grain boundary surface has a significant effect on many properties of metals, especially strength. At lower temperatures (less than about one-half of their melting temperature) grain boundaries strengthen metals by restricting dislocation movement under stress. At elevated temperatures grain boundary sliding may occur, and grain boundaries can become regions of weakness in polycrystalline metals.

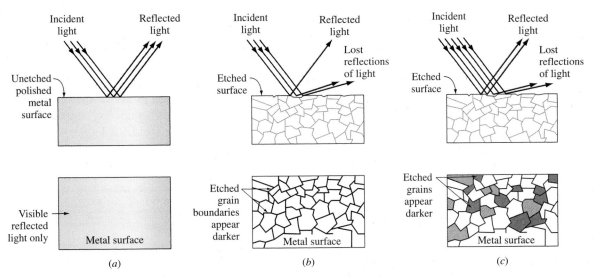

Virtual Lab

Figure 4.26
The effect of etching a polished surface of a steel metal sample on the microstructure observed in the optical microscope. (*a*) In the as-polished condition, no microstructural features are observed. (*b*) After etching a very low-carbon steel, only grain boundaries are chemically attacked severely, and so they appear as dark lines in the optical microstructure. (*c*) After etching a medium-carbon steel polished sample, dark (pearlite) and light (ferrite) regions are observed in the microstructure. The darker pearlite regions have been more severely attacked by the etchant and thus do not reflect much light.

Figure 4.27
Grain boundaries on the surface of polished and etched samples as revealed in the optical microscope. (*a*) Low-carbon steel (magnification 100×).
(*After "Metals Handbook," vol. 7, 8th ed., American Society for Metals, 1972, p. 4.*)
(*b*) Magnesium oxide (magnification 225×).
[*After R. E. Gardner and G. W. Robinson, J. Am. Ceram. Soc., 45:46 (1962).*]

One method of measuring grain size is the *American Society for Testing and Materials* (ASTM) method, in which the **grain-size number** *n* is defined by

$$N = 2^{n-1} \qquad (4.7)$$

Virtual Lab

Tutorial

where *N* is the number of grains per square inch on a polished and etched material surface at a magnification of $100\times$ and *n* is an integer referred to as the *ASTM grain-size number*. Grain-size numbers with the nominal number of grains per square inch at $100\times$ and grains per square millimeter at $1\times$ are listed in Table 4.2. Figure 4.28 shows some examples of nominal grain sizes for low-carbon sheet steel samples. Generally

Table 4.2 ASTM grain sizes

Grain-size no.	Nominal number of grains	
	Per sq mm at 1×	Per sq in. at 100×
1	15.5	1.0
2	31.0	2.0
3	62.0	4.0
4	124	8.0
5	248	16.0
6	496	32.0
7	992	64.0
8	1980	128
9	3970	256
10	7940	512

Source: "Metals Handbook," vol. 7, 8th ed., American Society for Metals, 1972, p. 4.

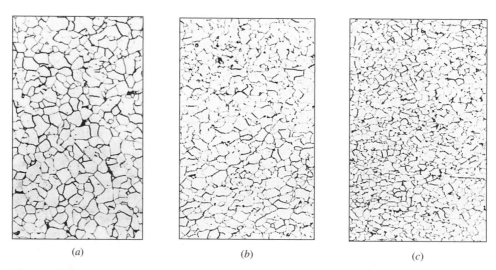

(a) (b) (c)

Figure 4.28
Several nominal ASTM grain sizes of low-carbon sheet steels: (*a*) no. 7, (*b*) no. 8, and (*c*) no. 9. (Etch: nital; magnification 100×.)
(*After "Metals Handbook," vol. 7, 8th ed., American Society for Metals, 1972, p. 4.*)

speaking, a material may be classified as coarse-grained when $n < 3$; medium-grained, $4 < n < 6$; fine-grained, $7 < n < 9$, and ultrafine-grained, $n > 10$.

A more direct approach of assessing the grain size of a material would be to determine the actual average grain diameter. This offers clear advantages to the ASTM grain-size number that in reality does not offer any direct information about the actual size of the grain. In this approach, once a photomicrograph is prepared at a specific magnification, a random line of known length is drawn on the photomicrograph. The number of grains intersected by this line is then determined, and the ratio of the number of grains to the actual length of the line is determined, n_L. The average grain diameter d is determined using the equation,

$$d = C/(n_L M) \tag{4.8}$$

where C is a constant ($C = 1.5$ for typical microstructures) and M is the maginification at which the photomicrograph is taken.

EXAMPLE PROBLEM 4.4

Tutorial

An ASTM grain size determination is being made from a photomicrograph of a metal at a magnification of 100×. What is the ASTM grain-size number of the metal if there are 64 grains per square inch?

■ **Solution**

$$N = 2^{n-1}$$

where N = no. of grains per square inch at 100×
$\quad n$ = ASTM grain-size number

Thus,

$$64 \text{ grains/in}^2 = 2^{n-1}$$
$$\log 64 = (n - 1)(\log 2)$$
$$1.806 = (n - 1)(0.301)$$
$$n = 7 \blacktriangleleft$$

EXAMPLE PROBLEM 4.5

If there are 60 grains per square inch on a photomicrograph of a metal at 200×, what is the ASTM grain-size number of the metal?

■ **Solution**

If there are 60 grains per square inch at 200×, then at 100× we will have

$$N = \left(\frac{200}{100}\right)^2 (60 \text{ grains/in}^2) = 240 = 2^{n-1}$$
$$\log 240 = (n - 1)(\log 2)$$
$$2.380 = (n - 1)(0.301)$$
$$n = 8.91 \blacktriangleleft$$

Note that the ratio of the magnification change must be squared since we are concerned with the number of grains per square inch.

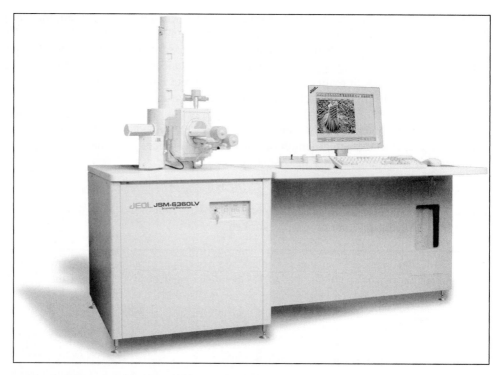

Figure 4.29
Photograph of a commercial scanning electron microscope. Note the column (which is evacuated during use) on the left and the viewing screen on the right.
(*Courtesy of JEOL USA, INC.*)

4.5.2 Scanning Electron Microscopy (SEM)

Scanning electron microscope is an important tool in materials science and engineering; it is used for microscopic feature measurement, fracture characterization, microstructure studies, thin coating evaluations, surface contamination examination, and failure analysis of materials. As opposed to optical microscopy where the sample's surface is exposed to incident visible light, the SEM impinges a beam of electrons in a pinpointed spot on the surface of a target specimen and collects and displays the electronic signals given off by the target material. Figure 4.29 is a photograph of a recent model of a scanning electron microscope, and Fig. 4.30 schematically illustrates its principles of operation. Basically, an electron gun produces an electron beam in an evacuated column that is focused and directed so that it impinges on a small spot on the target. Scanning coils allow the beam to scan a small area of the surface of the sample. Low-angle backscattered electrons interact with the protuberances of the surface and generate secondary[7] backscattered electrons to produce an electronic signal, which in turn produces an image having a depth of field of up to about 300 times that

---

[7]Secondary electrons are electrons that are ejected from the target metal atoms after being struck by primary electrons from the electron beam.

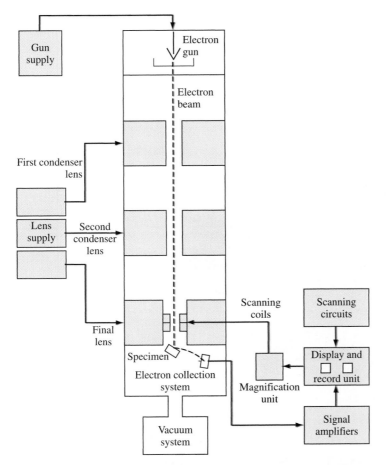

Figure 4.30
Schematic diagram of the basic design of a scanning electron microscope.
(*After V. A. Phillips, "Modern Metallographic Techniques and Their Applications," Wiley, 1971, p. 425.*)

Figure 4.31
Scanning electron fractograph of intergranular corrosion fracture near a circumferential weld in a thick-wall tube made of type 304 stainless steel. (Magnification 180×.)
(*After "Metals Handbook," vol. 9: "Fractography and Atlas of Fractographs," 8th ed., American Society for Metals, 1974, p. 77.*)

of the optical microscope (about 10 μm at 10,000 diameters magnification). The resolution of many SEM instruments is about 5 nm, with a wide range of magnification (about 15 to 100,000×).

The SEM is particularly useful in materials analysis for the examination of fractured surfaces of metals. Figure 4.31 shows an SEM fractograph of an intergranular corrosion fracture. Notice how clearly the metal grain surfaces are delineated and the depth of perception. SEM fractographs are used to determine whether a fractured surface is intergranular (along the grain boundary) transgranular (across the grain), or a mixture of both. The samples to be analyzed using standard SEM are often coated with gold or other heavy metals to achieve better resolution and signal quality. This is especially important if the sample is made of a nonconducting material.

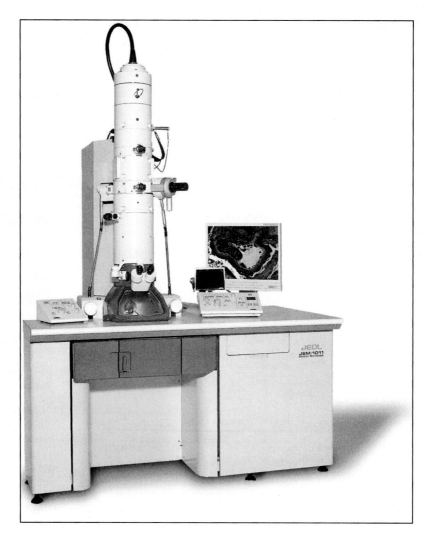

Figure 4.32
Photograph of a modern transmission electron microscope.
(*Courtesy of JEOL USA, INC.*)

Qualitative and quantitative information relating to the makeup of the sample may also be obtained when the SEM is equipped with an x-ray spectrometer.

4.5.3 Transmission Electron Microscopy (TEM)

Transmission electron microscopy (Fig. 4.32) is an important technique for studying defects and precipitates (secondary phases) in materials. Much of what is known about defects would be speculative theory and would have never been verified without the use of TEM, which resolves features in the nanometer range.

Defects such as dislocations can be observed on the image screen of a TEM. Unlike optical microscopy and SEM techniques where sample preparation is rather basic and easy to achieve, sample preparation for TEM analysis is complex and requires highly specialized instruments. Specimens to be analyzed using a TEM must have a thickness of several hundred nanometers or less depending on the operating voltage of the instrument. A properly prepared specimen is not only thin but also has flat parallel surfaces. To achieve this, a thin section (3 to 0.5 mm) is cut out of the bulk material using techniques such as electric-discharge machining (used for conducting samples) and a rotating wire saw, among others. The specimen is then reduced to 50 μm thickness while keeping the faces parallel using machine milling or lapping processes with fine abrasives. Other more advanced techniques such as electropolishing and ion-beam thinning are used to thin a sample to its final thickness.

In the TEM, an electron beam is produced by a heated tungsten filament at the top of an evacuated column and is accelerated down the column by high voltage (usually from 100 to 300 kV). Electromagnetic coils are used to condense the electron beam, which is then passed through the thin specimen placed on the specimen stage. As the electrons pass through the specimen, some are absorbed and some are scattered so that they change direction. It is now clear that the sample thickness is critical: A thick sample will not allow the passage of electrons due to excessive absorption and diffraction. Differences in crystal atomic arrangements will cause electron scattering. After the electron beam has passed through the specimen, it is focused with the objective coil (magnetic lens) and then enlarged and projected on a fluorescent screen (Fig. 4.33). An image can be formed either by collecting the direct electrons or the scattered electrons. The choice is made by inserting an aperture into the back focal plane of the objective lens. The aperture is maneuvered so that either the direct electrons or scattered electrons pass through it. If the direct beam is selected, the resultant image is called a *bright-field image,* and if the scattered electrons are selected, a *dark-field image* is produced.

In a bright field mode, a region in a metal specimen that tends to scatter electrons to a higher degree will appear dark on the viewing screen. Thus, dislocations that have an irregular linear atomic arrangement will appear as dark lines on the electron microscope screen. A TEM image of the dislocation structure in a thin foil of iron deformed 14 percent at $-195°$C is shown in Fig 4.34.

4.5.4 High-Resolution Transmission Electron Microscopy (HRTEM)

Another important tool in the analysis of defects and crystal structure is the high-resolution transmission electron microscopy. The instrument has a resolution of about 0.1 nm, which allows viewing of the crystal structure and defects at the atomic level. To grasp what this degree of resolution may reveal about a structure, consider that the lattice constant of the silicon unit cell at approximately 0.543 nm is five times larger than the resolution offered by HRTEM. The basic concepts behind this technique are similar to those of the TEM. However, the sample must be significantly thinner—on the order of 10 to 15 nm. In some situations, it is possible to view a two-dimensional

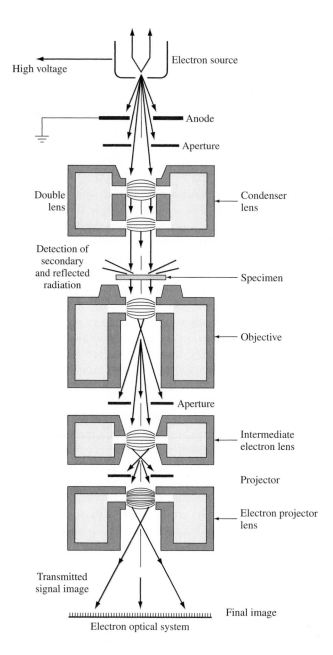

Figure 4.33

Schematic arrangement of electron lens system in a transmission electron microscope. All the lenses are enclosed in a column that is evacuated during operation. The path of the electron beam from the electron source to the final projected transmitted image is indicated by arrows. A specimen thin enough to allow an electron beam to be transmitted through it is placed between the condenser and objective lenses as indicated.

(*After L. E. Murr, "Electron and Ion Microscopy and Microanalysis," Marcel Decker, 1982, p. 105.*)

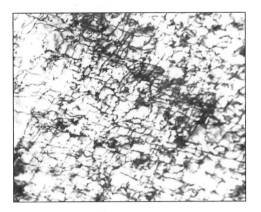

Figure 4.34
Dislocation structure in iron deformed 14 percent at −195°C. The dislocations appear as dark lines because electrons have been scattered along the linear irregular atomic arrangements of the dislocations. (Thin foil specimen; magnification 40,000×.)
(After "Metals Handbook," vol. 8, 8th ed., American Society for Metals, 1973, p. 219.)

projection of a crystal with the accompanying defects. To achieve this, the thin sample is tilted so that a low-index direction in the plane is perpendicular to the direction of electron beam (atoms are exactly on top of each other relative to the beam). The diffraction pattern is representative of the periodic potential for the electrons in two dimensions. The interference of all the diffracted beams and the primary beam, when brought together again using the objective lens, provides an enlarged picture of the periodic potential. Figure 4.35 shows the HRTEM image of several edge dislocations forming a small-angle boundary. In the figure, the top half is tilted relative to the bottom half. It should be noted that due to limitations in the objective lens in HRTEM, accurate quantitative analysis of images is not easy to achieve and must be done with care.

4.5.5 Scanning Probe Microscopes and Atomic Resolution

Scanning tunneling microscope (STM) and *atomic force microscope* (AFM) are two of many recently developed tools that allow scientists to analyze and image materials at the atomic level. These instruments and others with similar capabilities are collectively classified as *scanning probe microscopy* (SPM) techniques. The SPM systems have the ability to magnify the features of a surface to the subnanometer scale, producing an atomic-scale topographic map of the surface. These instruments have important applications in many areas of science including but not limited to surface sciences where the arrangement of atoms and their bonding are important; metrology where surface roughness of materials needs to be analyzed; and nanotechnology where the position of individual atoms or molecules may be manipulated and new nanoscale phenomena may be investigated. It is appropriate to discuss these systems, how they function, the nature of information they provide, and their applications.

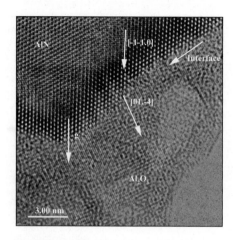

Figure 4.35
HRTEM image of a low-angle
boundary.
(*Work done at Lawrence Berkeley National
Laboratory, Berkeley, CA. J. Chaudhuri,
Department of Mechanical Engineering, Texas
Tech University, Lubbock, TX and J. Perrin,
Department of Mechanical Engineering,
Wichita State University, Wichita, KS*)

Scanning Tunneling Microscope   IBM researchers G. Binnig and H. Rohrer developed the STM technique in the early 1980s and later received the Nobel Prize in Physics in 1986 for their invention. In this technique, an extremely sharp tip (Fig. 4.36), traditionally made of metals such as tungsten, nickel, platinum-iridium, or gold, and more recently out of carbon nanotubes (see Sec. 11.11), is used to probe the surface of a sample.

The tip is first positioned a distance in the order of an atom diameter ($\cong$0.1 to 0.2 nm) from the surface of the sample. At such proximity, the electron clouds of the atoms in the tip of the probe interact with the electron clouds of the atoms on the surface of the sample. If at this point a small voltage is applied across the tip and the surface, the electrons will "tunnel" the gap and, therefore, produce a small current that may be detected and monitored. Generally, the sample is analyzed under ultrahigh vaccum to avoid contamination and oxidation of its surface.

The produced current is extremely sensitive to the gap size between the tip and the surface. Small changes in the gap size produce an exponential increase in the detected current. As a result, small changes (less than 0.1 nm) in the position of the tip relative to the surface may be detected. The magnitude of current is measured when the tip is positioned directly above an atom (its electron cloud). This current is maintained at the same level as the tip moves over the atoms and valleys between the atoms (*constant current mode*) (Fig. 4.37*a*). This is accomplished by adjusting the vertical position of the tip. The small movements required to adjust and maintain the current through the tip is then used to map the surface. The surface may also be mapped using

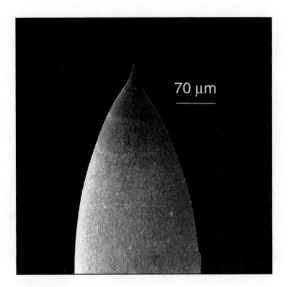

Figure 4.36
STM tip made of Pt-Ir alloy. The tip is
sharpened using chemical etching techniques.
(*Courtesy of Molecular Corp.*)

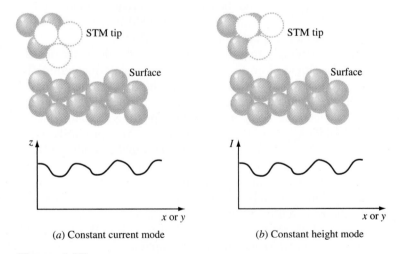

(*a*) Constant current mode   (*b*) Constant height mode

Figure 4.37
Schematics showing the STM modes of operation. (*a*) Adjust
the *z* coordinate of the tip to maintain constant current (record *z*
adjustments), (*b*) Adjust the current in the tip to maintain constant
height (record *I* adjustments).

Figure 4.38
High resolution bright field TEM image from the AIN/Al$_2$O$_3$ interface region of an oxidized AIN single crystal.
(*IBM Research, Almaden Research Center*)

a *constant height mode* in which the relative distance between the tip and the surface is maintained at a constant value and the changes in current are monitored (Fig. 4.37b). The quality of the surface topography achieved by STM is striking as observed in the STM images of the surface of platinum and silicon (Fig. 4.38).

Clearly, what is extremely important here is that the diameter of the tip should be of the order of a single atom to maintain atomic-scale resolution. The conventional metal tips can easily be worn and damaged during the scanning process, which results in poor image quality. More recently, carbon nanotubes of about one to tens of nanometer in diameter are being used as nanotips for STM and AFM applications because of their slender structure and strength. The STM is primarily used for topography purposes, and it offers no quantitative insight into the bonding nature and properties of the material. Because the apparatus's function is based on creating and monitoring small amounts of current, only those materials that can conduct electricity may be mapped, including metals and semiconductors. However, many materials of high interest to the research community, such as biological materials or polymers, are not conductive and, therefore, cannot be analyzed using this technique. For nonconducting materials, the AFM techniques are applied.

Atomic Force Microscope The AFM uses a similar approach as the STM in that it uses a tip to probe the surface. However, in this case, the tip is attached to a small cantilever beam. As the tip interacts with the surface of the sample, the forces (Van der Waals forces) acting on the tip deflect the cantilever beam. The interaction may be a short-range repulsive force (*contact mode AFM*) or a long-range attractive force (*non–contact mode AFM*). The deflection of the beam is monitored using a laser and

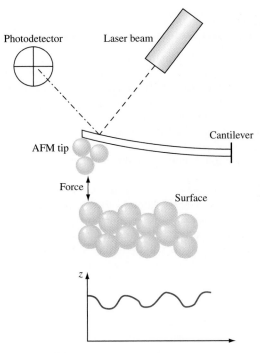

Figure 4.39
A schematic showing the basic AFM
technique.

a photodetector set up as shown in Fig. 4.39. The deflection is used to calculate the force acting on the tip. During scanning, the force will be maintained at a constant level (similar to the constant current mode in STM), and the displacement of the tip will be monitored. The surface topography is determined from these small displacements. Unlike STM, the AFM approach does not rely on a current tunneling through the tip and can therefore be applied to all materials even nonconductors. This is the main advantage of AFM over its predecessor, STM. There are currently numerous other AFM-based techniques available with various imaging modes, including magnetic and acoustic. AFM in various imaging modes is being used in areas such as DNA research, in situ monitoring of corrosion in material, in situ annealing of polymers, and polymer-coating technology. The fundamental understanding of important issues in the above areas has been significantly enhanced because of the application of such techniques.

Understanding the behavior of advanced materials at the atomic level drives the state-of-the-art in high-resolution electron microscopy, which in turn provides an opportunity for the development of new materials. Electron and scanning probe microscopy techniques are and will be particularly important in nanotechnology and nanostructured materials.

4.6 SUMMARY

Most metals and alloys are melted and cast into semifinished or finished shapes. During the solidification of a metal into a casting, nuclei are formed that grow into grains, creating solidified cast metal with a polycrystalline grain structure. For most industrial applications a very small grain size is desirable. The grain size may be indirectly determined by ASTM grain size number n or directly determined by finding the average grain diameter. Large single crystals are rarely made in industry. However, an exception is the large single crystals of silicon produced for the semiconductor industry. For this material special solidification conditions must be used and the silicon must be of very high purity.

Crystal imperfections are present in all real crystalline materials, even at the atomic or ionic size level. Vacancies or empty atomic sites in metals can be explained in terms of the thermal agitation of atoms and are considered equilibrium lattice defects. Dislocations (line defects) occur in metal crystals and are created in large numbers by the solidification process. Dislocations are not equilibrium defects and increase the internal energy of the metal. Images of dislocations can be observed in the transmission electron microscope. Grain boundaries are surface imperfections in metals created by crystals of different orientations meeting each other during solidification. Other important types of defects that affect the properties of materials are twins, low angle boundaries, high angle boundaries, twists, stacking faults, and precipitates.

Materials scientists and engineers use high-tech instruments to learn about the internal structure (including defect structure), behavior, and failure of materials. Instruments such as metallographs, SEM, TEM (HRTEM), and SPM allow analysis of materials from macro to nano range. Without such instruments understanding the behavior of materials would be impossible.

4.7 DEFINITIONS

Sec. 4.1

Nuclei: small particles of a new phase formed by a phase change (e.g., solidification) that can grow until the phase change is complete.

Homogeneous nucleation (as pertains to the solidification of metals): the formation of very small regions of a new solid phase (called *nuclei*) in a pure metal that can grow until solidification is complete. The pure homogeneous metal itself provides the atoms that make up the nuclei.

Embryos: small particles of a new phase formed by a phase change (e.g., solidification) that are not of critical size and that can redissolve.

Critical radius r^* of nucleus: the minimum radius that a particle of a new phase formed by nucleation must have to become a stable nucleus.

Heterogeneous nucleation (as pertains to the solidification of metals): the formation of very small regions (called *nuclei*) of a new solid phase at the interfaces of solid impurities. These impurities lower the critical size at a particular temperature of stable solid nuclei.

Grain: a single crystal in a polycrystalline aggregate.

Equiaxed grains: grains that are approximately equal in all directions and have random crystallographic orientations.

Columnar grains: long, thin grains in a solidified polycrystalline structure. These grains are formed in the interior of solidified metal ingots when heat flow is slow and uniaxial during solidification.

Sec. 4.2

Polycrystalline structure: a crystalline structure that contains many grains.

Sec. 4.3

Alloy: a mixture of two or more metals or a metal (metals) and a nonmetal (nonmetals).

Solid solution: an alloy of two or more metals or a metal(s) and a nonmetal(s) that is a single-phase atomic mixture.

Substitutional solid solution: a solid solution in which solute atoms of one element can replace those of solvent atoms of another element. For example, in a Cu–Ni solid solution the copper atoms can replace the nickel atoms in the solid-solution crystal lattice.

Interstitial solid solution: a solid solution formed in which the solute atoms can enter the interstices or holes in the solvent-atom lattice.

Sec. 4.4

Vacancy: a point imperfection in a crystal lattice where an atom is missing from an atomic site.

Interstitialcy (self-interstitial): a point imperfection in a crystal lattice where an atom of the same kind as those of the matrix lattice is positioned in an interstitial site between the matrix atoms.

Frenkel imperfection: a point imperfection in an ionic crystal in which a cation vacancy is associated with an interstitial cation.

Schottky imperfection: a point imperfection in an ionic crystal in which a cation vacancy is associated with an anion vacancy.

Dislocation: a crystalline imperfection in which a lattice distortion is centered around a line. The displacement distance of the atoms around the dislocation is called the *slip* or *Burgers vector* **b.** For an *edge dislocation* the slip vector is perpendicular to the dislocation line, while for a *screw dislocation* the slip vector is parallel to the dislocation line. A *mixed dislocation* has both edge and screw components.

Grain boundary: a surface imperfection that separates crystals (grains) of different orientations in a polycrystalline aggregate.

Grain-size number: a nominal (average) number of grains per unit area at a particular magnification.

Twin boundary: a mirror image misorientation of the crystal structure which is considered a surface defect.

Low angle boundary (Tilt): an array of dislocations forming angular mismatch inside a crystal.

Twist boundary: an array of screw dislocations creating mismatch inside a crystal.

Stacking fault: a surface defect formed due to improper (out of place) stacking of atomic planes.

Sec. 4.5

Scanning electron microscope (SEM): an instrument used to examine the surface of a material at very high magnifications by impinging electrons.

Transmission electron microscope (TEM): an instrument used to study, the internal defect structures based on passage of electron through a thin films of materials.

High resolution transmission electron microscope (HRTEM): a technique based on TEM but with significantly higher resolution by using significantly thinner samples.

Scanning probe microscopy (SPM): microscopy techniques such as STM and AFM that allow mapping of the surface of a material at the atomic level.

4.8 PROBLEMS

Answers to problems marked with an asterisk are given at the end of the book.

4.1 Describe and illustrate the solidification process of a pure metal in terms of the nucleation and growth of crystals.

4.2 Define the homogeneous nucleation process for the solidification of a pure metal.

4.3 In the solidification of a pure metal, what are the two energies involved in the transformation? Write the equation for the total free-energy change involved in the transformation of liquid to produce a strain-free solid nucleus by homogeneous nucleation. Also illustrate graphically the energy changes associated with the formation of a nucleus during solidification.

4.4 In the solidification of a metal, what is the difference between an embryo and a nucleus? What is the critical radius of a solidifying particle?

4.5 During solidification, how does the degree of undercooling affect the critical nucleus size? Assume homogeneous nucleation.

4.6 Distinguish between homogeneous and heterogeneous nucleation for the solidification of a pure metal.

*4.7 Calculate the size (radius) of the critically sized nucleus for pure platinum when homogeneous nucleation takes place.

4.8 Calculate the number of atoms in a critically sized nucleus for the homogeneous nucleation of pure platinum.

4.9 Calculate the size (radius) of the critical nucleus for pure iron when nucleation takes place homogeneously.

*4.10 Calculate the number of atoms in a critically sized nucleus for the homogeneous nucleation of pure iron.

4.11 Describe the grain structure of a metal ingot that was produced by slow-cooling the metal in a stationary open mold.

4.12 Distinguish between equiaxed and columnar grains in a solidified metal structure.

4.13 How can the grain size of a cast ingot be refined? How is grain refining accomplished industrially for aluminum alloy ingots?

4.14 What special techniques must be used to produce single crystals?

4.15 How are large silicon single crystals for the semiconductor industry produced?

4.16 What is a metal alloy? What is a solid solution?

4.17 Distinguish between a substitutional solid solution and an interstitial solid solution.

4.18 What are the conditions that are favorable for extensive solid solubility of one element in another?

4.19 Using the data in the following table, predict the relative degree of solid solubility of the following elements in aluminum:

    *a. copper   d. zinc

    b. manganese   *e. silicon

    *c. magnesium

Use the scale very high, 70–100%; high, 30–70%; moderate, 10–30%; low, 1–10%; and very low, <1%.

Element	Atom radius (nm)	Crystal structure	Electro-negativity	Valence
Aluminum	0.143	FCC	1.5	+3
Copper	0.128	FCC	1.8	+2
Manganese	0.112	Cubic	1.6	+2, +3, +6, +7
Magnesium	0.160	HCP	1.3	+2
Zinc	0.133	HCP	1.7	+2
Silicon	0.117	Diamond cubic	1.8	+4

4.20 Using the data in the following table, predict the relative degree of atomic solid solubility of the following elements in iron:

a. nickel   d. titanium

b. chromium   e. manganese

c. molybdenum

Use the scale very high, 70–100%; high, 30–70%; moderate, 10–30%; low, 1–10%; and very low, <1%.

Element	Atom radius (nm)	Crystal structure	Electro-negativity	Valence
Iron	0.124	BCC	1.7	+2, +3
Nickel	0.125	FCC	1.8	+2
Chromium	0.125	BCC	1.6	+2, +3, +6
Molybdenum	0.136	BCC	1.3	+3, +4, +6
Titanium	0.147	HCP	1.3	+2, +3, +4
Manganese	0.112	Cubic	1.6	+2, +3, +6, +7

*4.21 Calculate the radius of the largest interstitial void in the BCC α iron lattice. The atomic radius of the iron atom in this lattice is 0.124 nm, and the largest interstitial voids occur at the $(\frac{1}{4}, \frac{1}{2}, 0)$; $(\frac{1}{2}, \frac{3}{4}, 0)$; $(\frac{3}{4}, \frac{1}{2}, 0)$; and $(\frac{1}{2}, \frac{1}{4}, 0)$, etc., type positions.

4.22 Describe and illustrate the following types of point imperfections that can be present in metal lattices: (*a*) vacancy, (*b*) divacancy, and (*c*) interstitialcy.

4.23 Describe and illustrate the following imperfections that can exist in crystal lattices: (*a*) Frenkel imperfection and (*b*) Schottky imperfection.

4.24 Describe and illustrate the edge- and screw-type dislocations. What type of strain fields surround both types of dislocations?

4.25 Describe the structure of a grain boundary. Why are grain boundaries favorable sites for the nucleation and growth of precipitates?

4.26 Why are grain boundaries easily observed in the optical microscope?

4.27 How is the grain size of polycrystalline materials measured by the ASTM method?

*4.28 If there are 600 grains per square inch on a photomicrograph of a metal at 100×, what is its ASTM grain-size number?

4.29 If there are 400 grains per square inch on a photomicrograph of a ceramic material at 200×, what is the ASTM grain-size number of the material?

4.30 Determine, by counting, the ASTM grain-size number of the low-carbon sheet steel shown in Fig. P4.30. This micrograph is at 100×.

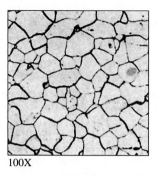

100X

Figure P4.30
(After "Metals Handbook," vol.
7, 8th ed., American Society for
Metals, 1972, p. 4.)

Tutorial

*4.31 Determine the ASTM grain-size number of the type 430 stainless steel micrograph shown in Fig. P4.31. This micrograph is at 200×.

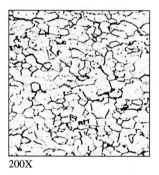

200X

Figure P4.31
(After "Metals Handbook," vol.
7, 8th ed., American Society for
Metals, 1972, p. 4.)

4.9 MATERIALS SELECTION AND DESIGN PROBLEMS

1. Refer to Fig. 4.11*b* in which a turbine blade is made from long columnar grains; propose a process that was used to accomplish the directional solidification.
2. Design a process by which high-purity copper wires can be produced continuously.
3. A low-carbon steel bar of circular cross section is cast such that its grain structure is equiaxed. Your application requires that the grain dimensions be longer along the longitudinal axis of the bar. How would you accomplish this?

4. You need to select the metal with larger grains between copper with ASTM grain size 7 and mild steel with ASTM grain size 4 for a given application. Which metal will you select and why?

5. In the welding process, two components are joined by filling space between the two with molten metal. (*a*) Explain how this process works from a solidification point of view. (*b*) What are some factors that are crucial in having a strong welded joint?

6. What is the significance or impact of Schottky and Frenkel imperfections on the behavior of ionic materials?

7. For a certain application, at room temperature, you must select a known copper alloy, but you have two choices: alloy of ASTM grain size 4 and alloy of ASTM grain size 8. If strength is important, which one would you choose?

8. Answer question 7 assuming that your application was at elevated temperatures (below melting temperature).

9. What differences in behavior would you expect to exist between a single crystal metal component and a polycrystalline one?

10. Pure silver is soft, expensive, chemically unstable, and not durable. Select an alloying element and the corresponding weight percent that would improve the characteristics of silver without compromising its structure and precious appeal.

Thermally Activated Processes and Diffusion in Solids

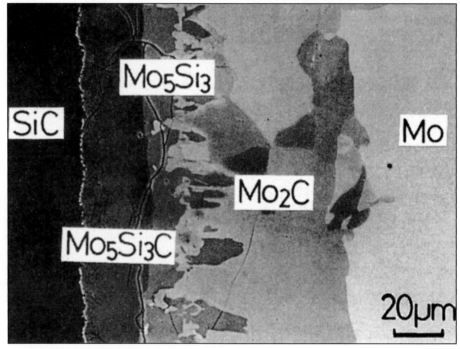

SiC

Mo₅Si₃

Mo

Mo₂C

Mo₅Si₃C

20μm

(After "Engineered Materials Handbook vol. 4: Ceramics and Glasses", American Society for Metals, p. 525. ISBN 0-87170-282-7. ASM International.)

Automobile engine components are often made from a combination of metals and ceramics. Metals offer high strength and ductility and ceramics offer high temperature strength, chemical stability, and low wear. In many situations it is necessary to join a metallic part to a thin layer of ceramic to make the best part for the application. The ceramic layer will serve to protect the inner metallic part from corrosive environments at high temperature. One method of joining metallic and ceramic components together is through solid-state bonding. The process works based on simultaneous application of pressure and high temperature. Externally applied pressure assures contact between the joining surfaces, and high temperature facilitates diffusion across the contact surface. The figure shows the interfacial microstructure when the metal molybdenum (Mo) is joined wih a thin layer of silicon carbide (SiC) at a bonding temperature of 1700°C and a pressure of 100 MPa for a period of one hour. Note that a transition region exists that contains a layer of mostly Mo_2C (carbide) and Mo_5Si_3 (silicide). These products form due to diffusion and form a strong bond. ■

By the end of this chapter, students will be able to . . .

1. Describe rate processes in solids involving movement of atoms in a solid state based on Boltzmann's relationship. Explain the concept of activation energy, E*, and determine the fraction of atoms or molecules having energy greater than E* at a given temperature.

2. Describe the effect of temperature on reaction rates based on Arrhenius rate equation.

3. Describe the two main mechanisms of diffusion.

4. Distinguish between steady and nonsteady diffusion and apply Fick's first and second law to the solution of related problems.

5. Describe the industrial applications of the diffusion process.

5.1 RATE PROCESSES IN SOLIDS

Many processes involved in the production and utilization of engineering materials are concerned with the rate at which atoms move in the solid state. In many of these processes reactions occur in the solid state that involve the spontaneous rearrangement of atoms into new and more stable atomic arrangements. In order for these reactions to proceed from the unreacted to the reacted state, the reacting atoms must have sufficient energy to overcome an activation energy barrier. The additional energy required above the average energy of the atoms is called the **activation energy** ΔE^*, which is usually measured in joules per mole or calories per mole. Figure 5.1 illustrates the activation energy for a thermally activated solid-state reaction. Atoms possessing an energy level E_r (energy of the reactants) $+\Delta E^*$ (activation energy) will have sufficient energy to react spontaneously to reach the reacted state E_p (energy of the products). The reaction shown in Fig. 5.1 is exothermic since energy is given off in the reaction.

At any temperature only a fraction of the molecules or atoms in a system will have sufficient energy to reach the activation energy level of E^*. As the temperature of the system is increased, more and more molecules or atoms will attain the activation energy level. Boltzmann studied the effect of temperature on increasing the energies of gas molecules. On the basis of statistical analysis, Boltzmann's results showed that the probability of finding a molecule or atom at an energy level E^* greater than the average energy E of all the molecules or atoms in a system at a particular temperature T in kelvins was

$$\text{Probability} \propto e^{-(E^* - E)/kT} \tag{5.1}$$

where k = Boltzmann's constant = 1.38×10^{-23} J/(atom $\cdot$ K).

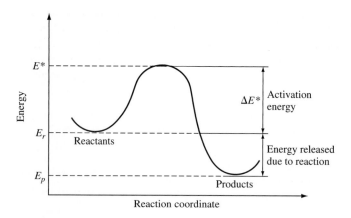

Figure 5.1
Energy of reacting species as it proceeds from the unreacted to the reacted state.

The fraction of atoms or molecules in a system having energies greater than E^* in a system where E^* is much greater than the average energy of any atom or molecule can be written as

$$\frac{n}{N_{\text{total}}} = Ce^{-E^*/kT} \tag{5.2}$$

where n = number of atoms or molecules with an energy greater than E^*

N_{total} = total number of atoms or molecules present in system

k = Boltzmann's constant = 8.62×10^{-5} eV/K

T = temperature, K

C = a constant

The number of vacancies at equilibrium at a particular temperature in a metallic crystal lattice can be expressed by the following relationship, which is similar to Eq. 5.2:

$$\frac{n_v}{N} = Ce^{-E_v/kT} \tag{5.3}$$

where n_v = number of vacancies per cubic meter of metal

N = total number of atom sites per cubic meter of metal

E_v = activation energy to form a vacancy (eV)

T = absolute temperature (K)

k = Boltzmann's constant = 8.62×10^{-5} eV/K

C = constant

In Example Problem 5.1 the equilibrium concentration of vacancies present in pure copper at 500°C is calculated by using Eq. 5.3 and assuming that $C = 1$. According to this calculation, there is only about one vacancy for every one million atoms!

Calculate (*a*) the equilibrium number of vacancies per cubic meter in pure copper at 500°C and (*b*) the vacancy fraction at 500°C in pure copper. Assume the energy of formation of a vacancy in pure copper is 0.90 eV. Use Eq. 5.3 with $C = 1$. (Boltzmann's constant $k = 8.62 \times 10^{-5}$ eV/K)

■ **Solution**

a.   The equilibrium number of vacancies per cubic meter in pure copper at 500°C is

$$n_v = Ne^{-E_v/kT} \quad \text{(assume } C = 1\text{)} \qquad (5.3a)$$

where n_v = no. of vacancies/m^3
  N = no. of atom sites/m^3
  E_v = energy of formation of a vacancy in pure copper at 500°C (eV)
  k = Boltzmann's constant
  T = temperature (K)

First, we determine a value for N by using the equation

$$N = \frac{N_0 \rho_{Cu}}{\text{at. mass Cu}} \qquad (5.4)$$

where N_0 = Avogadro's constant and ρ_{Cu} = density of Cu = 8.96 Mg/m^3. Thus,

$$N = \frac{6.02 \times 10^{23} \text{ atoms}}{\text{at. mass}} \times \frac{1}{63.54 \text{ g/at. mass}} \times \frac{8.96 \times 10^6 \text{ g}}{\text{m}^3}$$
$$= 8.49 \times 10^{28} \text{ atoms/m}^3$$

Substituting the values of N, E_v, k, and T into Eq. 5.3a gives

$$n_v = Ne^{-E_v/kT}$$
$$= (8.49 \times 10^{28}) \left\{ \exp\left[-\frac{0.90 \text{ eV}}{(8.62 \times 10^{-5} \text{ eV/K})(773 \text{ K})} \right] \right\}$$
$$= (8.49 \times 10^{28})(e^{-13.5}) = (8.49 \times 10^{28})(1.37 \times 10^{-6})$$
$$= 1.2 \times 10^{23} \text{ vacancies/m}^3 \blacktriangleleft$$

b.   The vacancy fraction in pure copper at 500°C is found from the ratio n_v/N from Eq. 5.3a:

$$\frac{n_v}{N} = \exp\left[-\frac{0.90 \text{ eV}}{(8.62 \times 10^{-5} \text{ eV/K})(773 \text{ K})} \right]$$
$$= e^{-13.5} = 1.4 \times 10^{-6} \blacktriangleleft$$

Thus, there is only one vacancy in every 10^6 atom sites!

A similar expression to the Boltzmann relationship for the energies of molecules in a gas was arrived at by Arrhenius[1] experimentally for the effect of temperature on chemical reaction rates. Arrhenius found that the rate of many chemical reactions

---

[1]Svante August Arrhenius (1859–1927). Swedish physical chemist who was one of the founders of modern physical chemistry and who studied reaction rates.

as a function of temperature could be expressed by the relationship

Arrhenius rate equation: Rate of reaction $= Ce^{-Q/RT}$ **(5.5)**

where Q = activation energy, J/mol or cal/mol

R = molar gas constant

$\quad = 8.314$ J/(mol $\cdot$ K) or 1.987 cal/(mol $\cdot$ K)

T = temperature (K)

C = rate constant, independent of temperature

In working with liquids and solids the activation energy is usually expressed in terms of a mole, or 6.02×10^{23} atoms or molecules. The activation energy is also usually given the symbol Q and expressed in joules per mole or calories per mole.

The Boltzmann equation (5.2) and the Arrhenius equation (5.5) both imply that the reaction rate among atoms or molecules in many cases depends on the number of reacting atoms or molecules that have activation energies of E^* or greater. The rates of many solid-state reactions of particular interest to materials scientists and engineers obey the Arrhenius rate law, and so the Arrhenius equation is often used to analyze experimental solid-state rate data.

The Arrhenius equation (5.5) is commonly rewritten in natural logarithmic form as

$$\ln \text{ rate} = \ln C - \frac{Q}{RT}$$ **(5.6)**

This equation is that of a straight line of the type

$$y = b + mx$$ **(5.7)**

where b is the y intercept and m is the slope of the line. The ln rate term of Eq. 5.6 is equivalent to the y term of Eq. 5.7, and the ln constant term of Eq. 5.6 is equivalent to the b term of Eq. 5.7. The $-Q/R$ quantity of Eq. 5.6 is equivalent to the slope m of Eq. 5.7. Thus a plot of ln rate versus $1/T$ produces a straight line of slope $-Q/R$.

The Arrhenius equation (5.5) can also be rewritten in common logarithmic form as

$$\log_{10} \text{ rate} = \log_{10} C - \frac{Q}{2.303 \, RT}$$ **(5.8)**

The 2.303 is the conversion factor from natural to common logarithms. This equation is also an equation of a straight line. A schematic plot of $\log_{10}$ rate versus $1/T$ is given in Fig. 5.2.

Thus, if a plot of experimental ln reaction rate versus $1/T$ data produces a straight line, an activation energy for the process involved can be calculated from the slope of the line. We shall use the Arrhenius equation to explore the effects of temperature on the diffusion of atoms and the electrical conductivity of pure elemental semiconductors later.

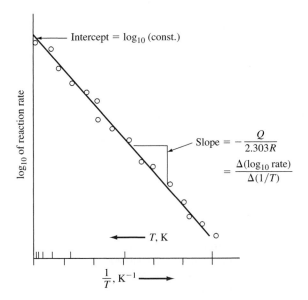

Figure 5.2
Typical Arrhenius plot of experimental rate data.
(*After J. Wulff et al., "Structure and Properties of Materials,"*
vol. II: "Thermodynamics of Structure," Wiley, 1966, p. 64.)

5.2 ATOMIC DIFFUSION IN SOLIDS

5.2.1 Diffusion in Solids in General

Diffusion can be defined as the mechanism by which matter is transported through matter. Atoms in gases, liquids, and solids are in constant motion and migrate over a period of time. In gases atomic movement is relatively rapid, as indicated by the rapid movement of cooking odors or smoke particles. Atomic movements in liquids are in general slower than in gases, as evidenced by the movement of colored dye in liquid water. In solids atomic movements are restricted due to bonding to equilibrium positions. However, thermal vibrations occurring in solids do allow some atoms to move. Diffusion of atoms in metals and alloys is particularly important since most solid-state reactions involve atomic movements. Examples of solid-state reactions are the precipitation of a second phase from solid solution Sec. 9.5.1 and the nucleation and growth of new grains in the recrystallization of a cold-worked metal Sec. 7.1.3.

5.2.2 Diffusion Mechanisms

There are two main mechanisms of diffusion of atoms in a crystalline lattice: (1) the *vacancy or substitutional mechanism* and (2) the *interstitial mechanism.*

Vacancy or Substitutional Diffusion Mechanism Atoms can move in crystal lattices from one atomic site to another if there is enough activation energy provided

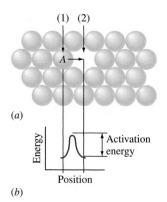

(a)

(b)

Animation

Figure 5.3
Activation energy associated
with the movement of atoms
in a metal. (a) Diffusion of
copper atom A at position 1
on the (111) plane of a
copper crystal lattice to
position 2 (a vacancy site) if
sufficient activation energy is
provided as indicated in (b).

by the thermal vibration of the atoms and if there are vacancies or other crystal defects in the lattice for atoms to move into. Vacancies in metals and alloys are equilibrium defects, and therefore some are always present to enable **substitutional diffusion** of atoms to take place. As the temperature of the metal increases, more vacancies are present and more thermal energy is available, and so the diffusion rate is higher at higher temperatures.

Consider the example of vacancy diffusion shown in Fig. 5.3 on a (111) plane of copper atoms in a copper crystal lattice. If an atom next to the vacancy has sufficient activation energy, it can move into the vacant site and thereby contribute to the **self-diffusion** of copper atoms in the lattice. The activation energy for self-diffusion is equal to the sum of the activation energy to form a vacancy and the activation energy to move the vacancy. Table 5.1 lists some activation energies for self-diffusion in pure metals. Note that in general as the melting point of the metal is increased, the activation energy is also. This relationship exists because the higher-melting-temperature metals tend to have stronger bonding energies between their atoms.

During self-diffusion or substitutional solid-state diffusion, atoms must break the original bonds among atoms and replace these with new bonds. This process is assisted by having vacancies present, and thus it can occur at lower activation energies (Fig. 5.3). In order for this process to occur in alloys, there must exist solid solubility of one type of atom in another. Thus, this process is dependent on the rules of solid solubility, which are listed in Sec. 4.3. Because of these differences in chemical bonding and solid solubility and other factors, substitutional diffusion data must

Table 5.1 Self-diffusion activation energies for some pure metals

Metal	Melting point (°C)	Crystal structure	Temperature range studied (°C)	Activation energy	
				kJ/mol	kcal/mol
Zinc	419	HCP	240–418	91.6	21.9
Aluminum	660	FCC	400–610	165	39.5
Copper	1083	FCC	700–990	196	46.9
Nickel	1452	FCC	900–1200	293	70.1
α iron	1530	BCC	808–884	240	57.5
Molybdenum	2600	BCC	2155–2540	460	110

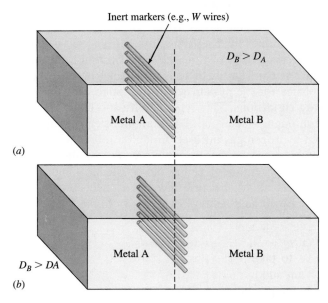

Figure 5.4
Experiment to illustrate the Kirkendall effect. (*a*) At start of diffusion experiment ($t = 0$). (*b*) After time *t*, markers move in the direction opposite the most rapidly diffusing species, *B*.

be obtained experimentally. With time these measurements are made more precisely, and hence these data may change with time.

One of the major breakthroughs in diffusion measurements occurred in the 1940s when the Kirkendall effect was discovered. This effect showed that the markers at the diffusion interface moved slightly in the opposite direction to the most rapidly moving (faster diffusing) species of a binary diffusion couple (Fig. 5.4*a*). After much discussion, it was concluded that the presence of vacancies allowed this phenomenon to occur.

Diffusion can also occur by the vacancy mechanism in solid solutions. Atomic size differences and bonding energy differences between the atoms are factors that affect the diffusion rate.

Interstitial Diffusion Mechanisms   The **interstitial diffusion** of atoms in crystal lattices takes place when atoms move from one interstitial site to another neighboring interstitial site without permanently displacing any of the atoms in the matrix crystal lattice (Fig. 5.5). For the interstitial mechanism to be operative the size of the diffusing atoms must be relatively small compared to the matrix atoms. Small atoms such as hydrogen, oxygen, nitrogen, and carbon can diffuse interstitially in some metallic crystal lattices. For example, carbon can diffuse interstitially in BCC α iron and FCC γ iron (see Fig. 4.15a). In the interstitial diffusion of carbon in iron, the carbon atoms must squeeze between the iron matrix atoms.

5.2.3 Steady-State Diffusion

Consider the diffusion of solute atoms in the x direction between two parallel atomic planes perpendicular to the paper separated by a distance x as shown in Fig. 5.6. We will assume that over a period of time the concentration of atoms at plane 1 is C_1 and that of plane 2 is C_2. That is, there is no change in the concentration of solute atoms at these planes for the system with time. Such diffusion conditions are said to be **steady-state conditions.** This type of diffusion takes place when a nonreacting gas diffuses through a metal foil. For example, steady-state diffusion conditions are attained when hydrogen gas diffuses through a foil of palladium if the hydrogen gas is at high pressure on one side and low pressure on the other.

If in the diffusion system shown in Fig. 5.6 no chemical interaction occurs between the solute and solvent atoms, because there is a concentration difference between planes 1 and 2, there will be a net flow of atoms from the higher concentration to the

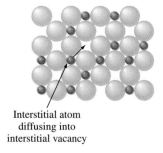

Interstitial atom
diffusing into
interstitial vacancy

Animation

Figure 5.5
A schematic diagram of an interstitial solid solution. The large circles represent atoms on a (100) plane of an FCC crystal lattice. The small, dark circles are interstitial atoms that occupy interstitial sites. The interstitial atoms can move into adjacent interstitial sites that are vacant. There is an activation energy associated with interstitial diffusion.

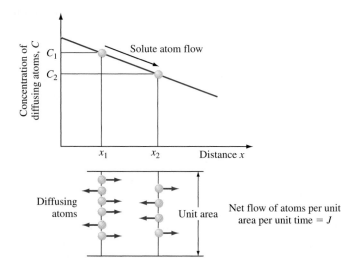

Figure 5.6
Steady-state diffusion of atoms in a concentration gradient. An example is hydrogen gas diffusing through a palladium metal foil.

lower concentration. The *flux* or flow of atoms in this type of system can be represented by the equation

$$J = -D\frac{dC}{dx} \tag{5.9}$$

where J = flux or net flow of atoms

D = proportionality constant called the **diffusivity** (atomic conductivity) or *diffusion coefficient*

$\dfrac{dC}{dx}$ = concentration gradient

A negative sign is used because the diffusion is from a higher to a lower concentration; i.e., there is a negative diffusion gradient.

This equation is called **Fick's[2] first law of diffusion** and states that for steady-state diffusion conditions (i.e., no change in system with time), the net flow of atoms by atomic diffusion is equal to the diffusivity D times the diffusion gradient dC/dx. The SI units for this equation are

$$J\left(\frac{\text{atoms}}{\text{m}^2 \cdot \text{s}}\right) = D\left(\frac{\text{m}^2}{\text{s}}\right)\frac{dC}{dx}\left(\frac{\text{atoms}}{\text{m}^3} \times \frac{1}{\text{m}}\right) \tag{5.10}$$

Table 5.2 lists some values of atomic diffusivities of selected interstitial and substitutional diffusion systems. The diffusivity values depend on many variables, of which the following are important:

1. *The type of diffusion mechanism.* Whether the diffusion is interstitial or substitutional will affect the diffusivity. Small atoms can diffuse interstitially in the crystal lattice of larger solvent atoms. For example, carbon diffuses interstitially in the BCC or FCC iron lattices. Copper atoms diffuse substitutionally in an aluminum solvent lattice since both the copper and the aluminum atoms are about the same size.

2. *The temperature at which the diffusion takes place* greatly affects the value of the diffusivity. As the temperature is increased, the diffusivity also increases, as shown in Table 5.2 for all the systems by comparing the 500°C values with those for 1000°C. The effect of temperature on diffusivity in diffusion systems will be discussed further in Sec. 5.4.

3. *The type of crystal structure of the solvent lattice* is important. For example, the diffusivity of carbon in BCC iron is 10^{-12} m²/s at 500°C, which is much *greater* than 5×10^{-15} m²/s, the value for the diffusivity of carbon in FCC iron at the same temperature. The reason for this difference is that the BCC crystal structure has a lower atomic packing factor of 0.68 as compared to that of the FCC crystal structure, which is 0.74. Also, the interatomic spaces between the iron atoms are wider in the BCC crystal structure than in the FCC one, and so the carbon atoms can diffuse between the iron atoms in the BCC structure more easily than in the FCC one.

4. *The type of crystal imperfections present* in the region of solid-state diffusion is also important. More open structures allow for more rapid diffusion of atoms. For

---

[2]Adolf Eugen Fick (1829–1901). German physiologist who first put diffusion on a quantitative basis by using mathematical equations. Some of his work was published in the *Annals of Physics (Leipzig)*, **170:**59 (1855).

Table 5.2 Diffusivities at 500°C and 1000°C for selected solute-solvent diffusion systems

Solute	Solvent (host structure)	Diffusivity (m²/s) 500°C (930°F)	1000°C (1830°F)
1. Carbon	FCC iron	(5×10^{-15})*	3×10^{-11}
2. Carbon	BCC iron	10^{-12}	(2×10^{-9})
3. Iron	FCC iron	(2×10^{-23})	2×10^{-16}
4. Iron	BCC iron	10^{-20}	(3×10^{-14})
5. Nickel	FCC iron	10^{-23}	2×10^{-16}
6. Manganese	FCC iron	(3×10^{-24})	10^{-16}
7. Zinc	Copper	4×10^{-18}	5×10^{-13}
8. Copper	Aluminum	4×10^{-14}	10^{-10} M[†]
9. Copper	Copper	10^{-18}	2×10^{-13}
10. Silver	Silver (crystal)	10^{-17}	10^{-12} M
11. Silver	Silver (grain boundary)	10^{-11}	
12. Carbon	HCP titanium	3×10^{-16}	(2×10^{-11})

*Parentheses indicate that the phase is metastable.
[†]M—Calculated, although temperature is above melting point.
Source: L. H. Van Vlack, "Elements of Materials Science and Engineering," 5th ed., Addison-Wesley, 1985.

example, diffusion takes place more rapidly along grain boundaries than in the grain matrix in metals and ceramics. Excess vacancies will increase diffusion rates in metals and alloys.

5. *The concentration of the diffusing species* is important in that higher concentrations of diffusing solute atoms will affect the diffusivity. This aspect of solid-state diffusion is very complex.

5.2.4 Non–Steady-State Diffusion

Steady-state diffusion in which conditions do not change with time is not commonly encountered with engineering materials. In most cases **non–steady-state diffusion** in which the concentration of solute atoms at any point in the material changes with time takes place. For example, if carbon is being diffused into the surface of a steel camshaft to harden its surface, the concentration of carbon under the surface at any point will change with time as the diffusion process progresses. For cases of non–steady-state diffusion in which the diffusivity is independent of time, **Fick's second law of diffusion** applies, which is

$$\frac{dC_x}{dt} = \frac{d}{dx}\left(D\frac{dC_x}{dx}\right) \tag{5.11}$$

This law states that the rate of compositional change is equal to the diffusivity times the rate of change of the concentration gradient. The derivation and solving of this differential equation is beyond the scope of this book. However, the particular solution to this equation in which a gas is diffusing into a solid is of great importance for some engineering diffusion processes and will be used to solve some practical industrial diffusion problems.

Let us consider the case of a gas A diffusing into a solid B, as illustrated in Fig. 5.7a. As the time of diffusion increases the concentration of solute atoms at any point in the x direction will also increase, as indicated for times t_1 and t_2 in Fig. 5.7b. If the diffusivity of gas A in solid B is independent of position, then the solution to Fick's second law (Eq. 5.11) is

$$\frac{C_s - C_x}{C_s - C_0} = \text{erf}\left(\frac{x}{2\sqrt{Dt}}\right) \tag{5.12}$$

where C_s = surface concentration of element in gas diffusing into the surface

C_0 = initial uniform concentration of element in solid

C_x = concentration of element at distance x from surface at time t

x = distance from surface

D = diffusivity of diffusing solute element

t = time

erf is a mathematical function called error function.

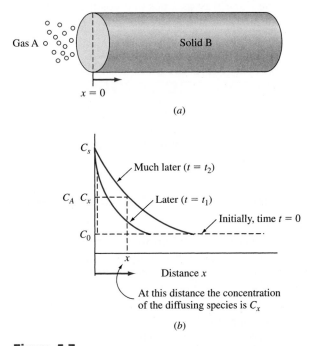

(a)

(b)

Figure 5.7
Diffusion of a gas into a solid. (*a*) Gas A diffuses into solid B at the surface where $x = 0$. The gas maintains a concentration of A atoms, called C_s, on this surface. (*b*) Concentration profiles of element A at various times along the solid in the x direction. The solid contains a uniform concentration of element A, called C_0, before diffusion starts.

Table 5.3 Table of the error function

z	erf z	z	erf z	z	erf z	z	erf z
0	0	0.40	0.4284	0.85	0.7707	1.6	0.9763
0.025	0.0282	0.45	0.4755	0.90	0.7970	1.7	0.9838
0.05	0.0564	0.50	0.5205	0.95	0.8209	1.8	0.9891
0.10	0.1125	0.55	0.5633	1.0	0.8427	1.9	0.9928
0.15	0.1680	0.60	0.6039	1.1	0.8802	2.0	0.9953
0.20	0.2227	0.65	0.6420	1.2	0.9103	2.2	0.9981
0.25	0.2763	0.70	0.6778	1.3	0.9340	2.4	0.9993
0.30	0.3286	0.75	0.7112	1.4	0.9523	2.6	0.9998
0.35	0.3794	0.80	0.7421	1.5	0.9661	2.8	0.9999

Source: R. A. Flinn and P. K. Trojan, "Engineering Materials and Their Applications," 2nd ed., Houghton Mifflin, 1981, p. 137.

The error function, erf, is a mathematical function existing by agreed definition and is used in some solutions of Fick's second law. The error function can be found in standard tables in the same way as sines and cosines. Table 5.3 is an abbreviated table of the error function.

5.3 INDUSTRIAL APPLICATIONS OF DIFFUSION PROCESSES

Many industrial manufacturing processes utilize solid-state diffusion. In this section we will consider the following two diffusion processes: (1) case hardening of steel by gas carburizing and (2) the impurity doping of silicon wafers for integrated electronic circuits.

5.3.1 Case Hardening of Steel by Gas Carburizing

Many rotating or sliding steel parts such as gears and shafts must have a hard outside case for wear resistance and a tough inner core for fracture resistance. In the manufacture of a carburized steel part, usually the part is machined first in the soft condition, and then, after machining, the outer layer is hardened by some casehardening treatment such as gas carburizing. Carburized steels are low-carbon steels that have about 0.10 to 0.25% C. However, the alloy content of the carburized steels can vary considerably depending on the application for which the steel will be used. Some typical gas-carburized parts are shown in Fig. 5.8.

In the first part of the gas-carburizing process, the steel parts are placed in a furnace in contact with gases containing methane (CH_4) or other hydrocarbon gases at about 927°C (1700°F). Figure 5.9 shows some gears about to be gascarburized in a furnace with a nitrogen-methanol mixture for an atmosphere. The carbon from the atmosphere diffuses into the surface of the gears so that after subsequent heat treatments the gears are left with high-carbon hard cases as indicated, for example, by the darkened surface areas of the macrosection of the gear shown in Fig. 5.10.

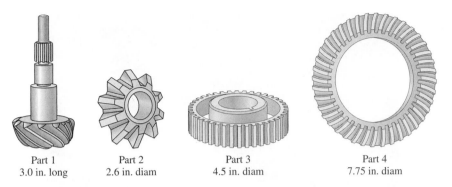

Part 1
3.0 in. long

Part 2
2.6 in. diam

Part 3
4.5 in. diam

Part 4
7.75 in. diam

Figure 5.8
Typical gas-carburized steel parts.
(*After "Metals Handbook," vol. 2: "Heat Treating," 8th ed., American Society for Metals, 1964, p. 108.*)

Figure 5.9
Parts to be carburized in a nitrogen-methanol carburizing atmosphere.
(*After B. J. Sheehy, Met. Prog., September 1981, p. 120.*)

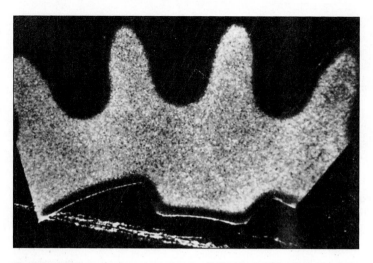

Animation

Figure 5.10
Macrosection of nitrogen-methanol-carburized SAE 8620
final-drive pinion gear.
(*After B. J. Sheehy, Met. Prog., September 1981, p. 120.*)

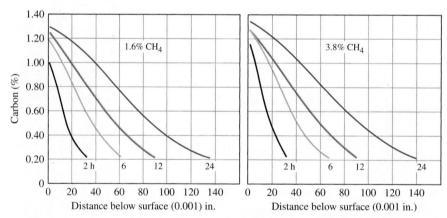

Figure 5.11
Carbon gradients in test bars of 1022 steel carburized at 918°C (1685°F) in a
20% CO–40% H_2 gas with 1.6 and 3.8% methane (CH_4) added.
(*After "Metals Handbook," vol. 2: "Heat Treating," 8th ed., American Society for Metals, 1964, p. 100.*)

Figure 5.11 shows some typical carbon gradients in test bars of AISI 1022
(0.22% C) plain-carbon steel carburized at 1685°F (918°C) by a carburizing atmosphere with 20% CO. Notice how the carburizing time greatly affects the carbon content versus distance-below-the-surface profile. Example Problems 5.2 and 5.3 illustrate how the diffusion equation (5.12) can be used to determine one unknown
variable, such as time of diffusion or carbon content at a particular depth below the
surface of the part being carburized.

Consider the gas carburizing of a gear of 1020 steel at 927°C (1700°F). Calculate the time in minutes necessary to increase the carbon content to 0.40% at 0.50 mm below the surface. Assume that the carbon content at the surface is 0.90% and that the steel has a nominal carbon content of 0.20%.

EXAMPLE PROBLEM 5.2

$$D_{927°C} = 1.28 \times 10^{-11} \text{ m}^2/\text{s}$$

■ **Solution**

$$\frac{C_s - C_x}{C_s - C_0} = \text{erf}\left(\frac{x}{2\sqrt{Dt}}\right) \qquad (5.12)$$

$C_s = 0.90\%$ $x = 0.5 \text{ mm} = 5.0 \times 10^{-4} \text{ m}$

$C_0 = 0.20\%$ $D_{927°C} = 1.28 \times 10^{-11} \text{ m}^2/\text{s}$

$C_x = 0.40\%$ $t = ? \text{ s}$

Substituting these values in Eq. 5.12 gives

$$\frac{0.90 - 0.40}{0.90 - 0.20} = \text{erf}\left[\frac{5.0 \times 10^{-4} \text{ m}}{2\sqrt{(1.28 \times 10^{-11} \text{ m}^2/\text{s})(t)}}\right]$$

$$\frac{0.50}{0.70} = \text{erf}\left(\frac{69.88}{\sqrt{t}}\right) = 0.7143$$

Let

$$Z = \frac{69.88}{\sqrt{t}} \quad \text{then erf } Z = 0.7143$$

We need a number for Z whose error function (erf) is 0.7143. From Table 5.3 we find this number by interpolation to be 0.755:

$$\frac{0.7143 - 0.7112}{0.7421 - 0.7112} = \frac{x - 0.75}{0.80 - 0.75}$$

$$x - 0.75 = (0.1003)(0.05)$$

$$x = 0.75 + 0.005 = 0.755$$

erf Z	Z
0.7112	0.75
0.7143	x
0.7421	0.80

Thus,

$$Z = \frac{69.88}{\sqrt{t}} = 0.755$$

$$\sqrt{t} = \frac{69.88}{0.755} = 92.6$$

$$t = 8567 \text{ s} = 143 \text{ min} \blacktriangleleft$$

Consider the gas carburizing of a gear of 1020 steel at 927°C (1700°F) as in Example Problem 5.2. Only in this problem calculate the *carbon content* at 0.50 mm beneath the surface of the gear after 5 h carburizing time. Assume that the carbon content of the surface of the gear is 0.90% and that the steel has a nominal carbon content of 0.20%.

EXAMPLE PROBLEM 5.3

■ **Solution**

$$D_{927°C} = 1.28 \times 10^{-11} \text{ m}^2/\text{s}$$

$$\frac{C_s - C_x}{C_s - C_0} = \text{erf}\left(\frac{x}{2\sqrt{Dt}}\right) \tag{5.12}$$

$$C_s = 0.90\% \qquad x = 0.50 \text{ mm} = 5.0 \times 10^{-4} \text{ m}$$
$$C_0 = 0.20\% \qquad D_{927°C} = 1.28 \times 10^{-11} \text{ m}^2/\text{s}$$
$$C_x = ?\% \qquad t = 5 \text{ h} = 5 \text{ h} \times 3600 \text{ s/h} = 1.8 \times 10^4 \text{ s}$$

$$\frac{0.90 - C_x}{0.90 - 0.20} = \text{erf}\left[\frac{5.0 \times 10^{-4} \text{ m}}{2\sqrt{(1.28 \times 10^{-11} \text{ m/s})(1.8 \times 10^4 \text{ s})}}\right]$$

$$\frac{0.90 - C_x}{0.70} = \text{erf } 0.521$$

Let $Z = 0.521$. We need to know what the corresponding error function for the Z value of 0.521 is. To determine this number from Table 5.3, we must interpolate the data as shown in the accompanying table.

Z	erf Z
0.500	0.5205
0.521	x
0.550	0.5633

$$\frac{0.521 - 0.500}{0.550 - 0.500} = \frac{x - 0.5205}{0.5633 - 0.5205}$$

$$0.42 = \frac{x - 0.5205}{0.0428}$$

$$x - 0.5205 = (0.42)(0.0428)$$

$$x = 0.0180 + 0.5205$$

$$= 0.538$$

Therefore,

$$\frac{0.90 - C_x}{0.70} = \text{erf } 0.521 = 0.538$$

$$C_x = 0.90 - (0.70)(0.538)$$

$$= 0.52\% \blacktriangleleft$$

Note that by increasing the carburizing time from about 2.4 to 5 h for the 1020 steel, the carbon content at 0.5 mm below the surface of the gear is increased from 0.4 to only 0.52%.

5.3.2 Impurity Diffusion into Silicon Wafers for Integrated Circuits

Impurity diffusion into silicon wafers to change their electrical conducting characteristics is an important phase in the production of modern integrated electronic circuits. In one method of impurity diffusion into silicon wafers the silicon surface is exposed to the vapor of an appropriate impurity at a temperature above about 1100°C in a quartz tube furnace, as shown schematically in Fig. 5.12. The part of

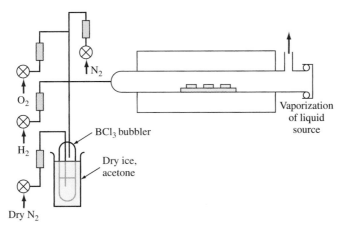

Figure 5.12
Diffusion method for diffusing boron into silicon wafers.
(*After W. R. Runyan, "Silicon Semiconductor Technology," McGraw-Hill, 1965.*)

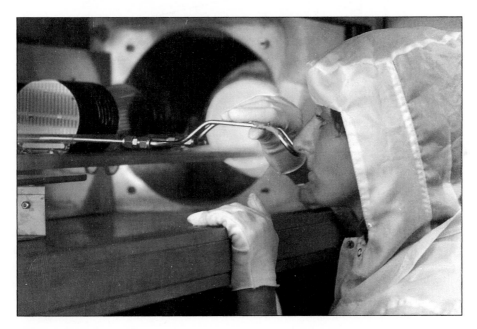

Figure 5.13
Loading a rack of silicon wafers into a tube furnace for impurity diffusion.
(*Courtesy of Harris Corporation.*)

the silicon surface not to be exposed to the impurity diffusion must be masked off so that the impurities diffuse into the parts selected by the design engineer for conductivity change. Figure 5.13 shows a technician loading a rack of silicon wafers into a tube furnace for impurity diffusion.

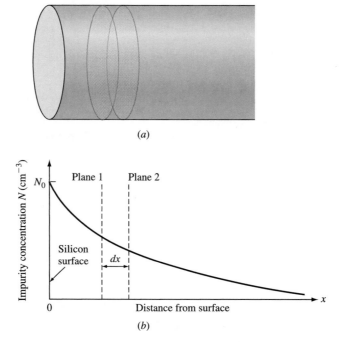

Figure 5.14
Impurity diffusion into a silicon wafer from one face.
(*a*) A silicon wafer with thickness greatly exaggerated
having an impurity concentration that diminishes from
the left face toward the interior. (*b*) Graphical
representation of the same impurity distribution.
(*After R. M. Warner, "Integrated Circuits," McGraw-Hill, 1965, p. 70.*)

As in the case of the gas carburizing of a steel surface, the concentration of
impurities diffused into the silicon surface decreases as the depth of penetration
increases, as shown in Fig. 5.14. Changing the time of diffusion will also change
the concentration of impurities versus depth-of-penetration profile, as shown qualitatively in Fig. 5.7. Example Problem 5.3 illustrates how Eq. 5.12 can be used quantitatively to determine one unknown variable, such as time of diffusion or depth of
penetration at a particular concentration level.

**EXAMPLE
PROBLEM 5.4**
Consider the impurity diffusion of gallium into a silicon wafer. If gallium is diffused into
a silicon wafer with no previous gallium in it at a temperature of 1100°C for 3 h, what is
the depth below the surface at which the concentration is 10^{22} atoms/m^3 if the surface concentration is 10^{24} atoms/m^3? For gallium diffusing into silicon at 1100°C, the solution is
as follows:

■ **Solution**

$$D_{1100°C} = 7.0 \times 10^{-17} \text{ m}^2/\text{s}$$

$$\frac{C_s - C_x}{C_s - C_0} = \text{erf}\left(\frac{x}{2\sqrt{Dt}}\right) \tag{5.12}$$

$C_s = 10^{24}$ atoms/m^3 $x = ?$ m   (depth at which $C_x = 10^{22}$ atoms/m^3)

$C_x = 10^{22}$ atoms/m^3 $D_{1100°C} = 7.0 \times 10^{-17}$ m^2/s

$C_0 = 0$ atoms/m^3 $t = 3$ h $= 3$ h $\times 3600$ s/h $= 1.08 \times 10^4$ s

Substituting these values into Eq. 5.12 gives

$$\frac{10^{24} - 10^{22}}{10^{24} - 0} = \text{erf}\left[\frac{x \text{ m}}{2\sqrt{(7.0 \times 10^{-17} \text{ m}^2/\text{s})(1.08 \times 10^4 \text{ s})}}\right]$$

$$1 - 0.01 = \text{erf}\left(\frac{x \text{ m}}{1.74 \times 10^{-6} \text{ m}}\right) = 0.99$$

Let

$$Z = \frac{x}{1.74 \times 10^{-6} \text{ m}}$$

Thus,

$$\text{erf } Z = 0.99 \quad \text{and} \quad Z = 1.82$$

(from Table 5.3 using interpolation). Therefore,

$$x = (Z)(1.74 \times 10^{-6} \text{ m}) = (1.82)(1.74 \times 10^{-6} \text{ m})$$
$$= 3.17 \times 10^{-6} \text{ m} \blacktriangleleft$$

Note: Typical diffusion depths in silicon wafers are of the order of a few micrometers (i.e., about 10^{-6} m), while the wafer is usually several hundred micrometers thick.

5.4 EFFECT OF TEMPERATURE ON DIFFUSION IN SOLIDS

Since atomic diffusion involves atomic movements, it is to be expected that increasing the temperature of a diffusion system will increase the diffusion rate. By experiment, it has been found that the temperature dependence of the diffusion rate of many diffusion systems can be expressed by the following Arrhenius-type equation:

$$D = D_0 e^{-Q/RT} \tag{5.13}$$

where D = diffusivity, m^2/s

D_0 = proportionality constant, m^2/s, independent of temperature in range for which equation is valid

Q = activation energy of diffusing species, J/mol or cal/mol

R = molar gas constant

    = 8.314 J/(mol · K) or 1.987 cal/(mol · K)

T = temperature, K

Example Problem 5.4 applies Eq. 5.13 to determine the diffusivity of carbon diffusing in γ iron at 927°C when given values for D_0 and the activation energy Q.

EXAMPLE PROBLEM 5.5

Calculate the value of the diffusivity D in meters squared per second for the diffusion of carbon in γ iron (FCC) at 927°C (1700°F). Use values of $D_0 = 2.0 \times 10^{-5}$ m²/s, $Q = 142$ kJ/mol, and $R = 8.314$ J/(mol·K).

■ **Solution**

$$D = D_0 e^{-Q/RT} \tag{5.13}$$

$$= (2.0 \times 10^{-5} \text{ m}^2/\text{s}) \left\{ \exp \frac{-142{,}000 \text{ J/mol}}{[8.314 \text{ J/(mol} \cdot \text{K)}](1200 \text{ K})} \right\}$$

$$= (2.0 \times 10^{-5} \text{ m}^2/\text{s})(e^{-14.23})$$

$$= (2.0 \times 10^{-5} \text{ m}^2/\text{s})(0.661 \times 10^{-6})$$

$$= 1.32 \times 10^{-11} \text{ m}^2/\text{s} \blacktriangleleft$$

The diffusion equation $D = D_0 e^{-Q/RT}$ (Eq. 5.13) can be written in logarithmic form as the equation of a straight line as was done in Eqs. 5.6 and 5.8 for the general Arrhenius rate law equation:

$$\ln D = \ln D_0 - \frac{Q}{RT} \tag{5.14}$$

or

$$\log_{10} D = \log_{10} D_0 - \frac{Q}{2.303RT} \tag{5.15}$$

Table 5.4 Diffusivity data for some metallic systems

Solute	Solvent	D_0(m²/s)	Q kJ/mol	kcal/mol
Carbon	FCC iron	2.0×10^{-5}	142	34.0
Carbon	BCC iron	22.0×10^{-5}	122	29.3
Iron	FCC iron	2.2×10^{-5}	268	64.0
Iron	BCC iron	20.0×10^{-5}	240	57.5
Nickel	FCC iron	7.7×10^{-5}	280	67.0
Manganese	FCC iron	3.5×10^{-5}	282	67.5
Zinc	Copper	3.4×10^{-5}	191	45.6
Copper	Aluminum	1.5×10^{-5}	126	30.2
Copper	Copper	2.0×10^{-5}	197	47.1
Silver	Silver	4.0×10^{-5}	184	44.1
Carbon	HCP titanium	51.0×10^{-5}	182	43.5

Source: Data from L. H. Van Vlack, "Elements of Materials Science and Engineering," 5th ed., Addison-Wesley, 1985.

If diffusivity values for a diffusion system are determined at two temperatures, values for Q and D_0 can be determined by solving two simultaneous equations of the Eq. 5.15 type. If these Q and D_0 values are substituted into Eq. 5.15, a general equation for $\log_{10} D$ versus $1/T$ over the temperature range investigated can be created. Example Problem 5.5 shows how the activation energy for a binary diffusion system can be calculated directly by using the relationship $D = D_0 e^{-Q/RT}$ (Eq. 5.13) when the diffusivities are known for two temperatures.

Table 5.4 lists D_0 and Q values for some metallic systems used to produce the Arrhenius diffusivity plots in Fig. 5.15. Figure 5.16 shows similar plots for the diffusion of impurity elements into silicon, which are useful for the fabrication of integrated circuits for the electronics industry.

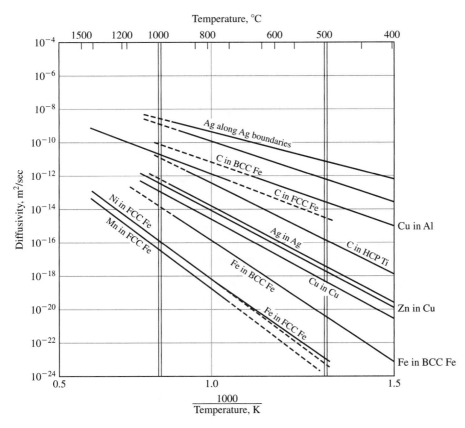

Figure 5.15
Arrhenius plots of diffusivity data for some metallic systems.
(*After L. H. Van Vlack, "Elements of Materials Science and Engineering," 5th ed., Addison-Wesley, 1985. p. 137.*)

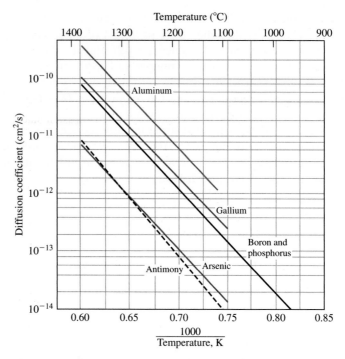

Figure 5.16
Diffusion coefficients as a function of temperature for some impurity elements in silicon.
[*After C. S. Fuller and J. A. Ditzenberger, J. Appl. Phys.,* **27:** *544(1956).*]

EXAMPLE PROBLEM 5.6

The diffusivity of silver atoms in solid silver metal is 1.0×10^{-17} m^2/s at 500°C and 7.0×10^{-13} m^2/s at 1000°C. Calculate the activation energy (joules per mole) for the diffusion of Ag in Ag in the temperature range 500 to 1000°C.

■ **Solution**

Using Eq. 5.13, $T_2 = 1000°C + 273 = 1273$ K, $T_1 = 500°C + 273 = 773$ K, and $R = 8.314$ J/(mol · K):

$$\frac{D_{1000°C}}{D_{500°C}} = \frac{\exp(-Q/RT_2)}{\exp(-Q/RT_1)} = \exp\left[-\frac{Q}{R}\left(\frac{1}{T_2} - \frac{1}{T_1}\right)\right]$$

$$\frac{7.0 \times 10^{-13}}{1.0 \times 10^{-17}} = \exp\left\{-\frac{Q}{R}\left[\left(\frac{1}{1273 \text{ K}} - \frac{1}{773 \text{ K}}\right)\right]\right\}$$

$$\ln(7.0 \times 10^4) = -\frac{Q}{R}(7.855 \times 10^{-4} - 12.94 \times 10^{-4}) = \frac{Q}{8.314}(5.08 \times 10^{-4})$$

$$11.16 = Q(6.11 \times 10^{-5})$$

$$Q = 183,000 \text{ J/mol} = 183 \text{ kJ/mol} \blacktriangleleft$$

5.5 SUMMARY

Atomic diffusion occurs in metallic solids mainly by (1) a vacancy or substitution mechanism and (2) an interstitial mechanism. In the vacancy mechanism atoms of about the same size jump from one position to another, using the vacant atomic sites. In the interstitial mechanism very small atoms move through the interstitial spaces between the larger atoms of the parent matrix. Fick's first law of diffusion states that diffusion takes place because of a difference in concentration of a diffusing species from one place to another and is applicable for steady-state conditions (i.e., conditions that do not change with time). Fick's second law of diffusion is applicable for non–steady-state conditions (i.e., conditions in which the concentrations of the diffusing species change with time). In this book the use of Fick's second law has been restricted to the case of a gas diffusing into a solid. The rate of diffusion depends greatly on temperature, and this dependence is expressed by the diffusivity, a measure of the diffusion rate: Diffusivity $D = D_0 e^{-Q/RT}$. Diffusion processes are used commonly in industry. In this chapter we have examined the diffusion-gas-carburizing process for surface-hardening steel and the diffusion of controlled amounts of impurities into silicon wafers for integrated circuits.

5.6 DEFINITIONS

Sec. 5.1

Activation energy: the additional energy required above the average energy for a thermally activated reaction to take place.

Arrhenius rate equation: an empirical equation that describes the rate of a reaction as a function of temperature and an activation energy barrier.

Sec. 5.2

Substitutional diffusion: the migration of solute atoms in a solvent lattice in which the solute and solvent atoms are approximately the same size. The presence of vacancies makes the diffusion possible.

Self-diffusion: the migration of atoms in a pure material.

Interstitial diffusion: the migration of interstitial atoms in a matrix lattice.

Fick's first law of diffusion in solids: the flux of a diffusing species is proportional to the concentration gradient at constant temperature.

Fick's second law of diffusion in solids: the rate of change of composition is equal to the diffusivity times the rate of change of the concentration gradient at constant temperature.

Diffusivity: a measure of the rate of diffusion in solids at a constant temperature. Diffusivity D can be expressed by the equation $D = D_0 e^{-Q/RT}$, where Q is the activation energy and T is the temperature in Kelvins. D_0 and R are constants.

Steady-state conditions: for a diffusing system there is no change in the concentration of the diffusing species with time at different places in the system.

Non–steady-state conditions: for a diffusing system the concentration of the diffusing species changes with time at different places in the system.

5.7 PROBLEMS

Answers to problems marked with an asterisk are given at the end of the book.

5.1 What is a thermally activated process? What is the activation energy for such a process?

5.2 Write an equation for the number of vacancies present in a metal at equilibrium at a particular temperature and define each of the terms. Give the units for each term and use electron volts for the activation energy.

*5.3 (*a*) Calculate the equilibrium concentration of vacancies per cubic meter in pure copper at 850°C. Assume that the energy of formation of a vacancy in pure copper is 1.00 eV. (*b*) What is the vacancy fraction at 800°C?

5.4 (*a*) Calculate the equilibrium concentration of vacancies per cubic meter in pure silver at 750°C. Assume the energy of formation of a vacancy in pure silver is 1.10 eV. (*b*) What is the vacancy fraction at 700°C?

5.5 Write the Arrhenius rate equation in the (*a*) exponential and (*b*) common logarithmic forms.

5.6 Draw a typical Arrhenius plot of $\log_{10}$ of the reaction rate versus reciprocal absolute temperature, and indicate the slope of the plot.

5.7 Describe the substitutional and interstitial diffusion mechanisms in solid metals.

5.8 Write the equation for Fick's first law of diffusion, and define each of the terms in SI units.

5.9 What factors affect the diffusion rate in solid metal crystals?

5.10 Write the equation for Fick's second law of diffusion in solids and define each of the terms.

5.11 Write the equation for the solution to Fick's second law for the diffusion of a gas into the surface of a solid metal crystal lattice.

5.12 Describe the gas-carburizing process for steel parts. Why is the carburization of steel parts carried out?

*5.13 Consider the gas carburizing of a gear of 1018 steel (0.18 wt %) at 927°C (1700°F). Calculate the time necessary to increase the carbon content to 0.35 wt % at 0.40 mm below the surface of the gear. Assume the carbon content at the surface to be 1.15 wt % and that the nominal carbon content of the steel gear before carburizing is 0.18 wt %. D (C in γ iron) at 927°C = 1.28×10^{-11} m^2/s.

5.14 The surface of a steel gear made of 1022 steel (0.22 wt % C) is to be gas-carburized at 927°C (1700°F). Calculate the time necessary to increase the carbon content to 0.30 wt % at 0.030 in. below the surface of the gear. Assume the carbon content of the surface to be 1.20 wt %. D (C in γ iron) at 927°C = 1.28×10^{-11} m^2/s.

5.15 A gear made of 1020 steel (0.20 wt % C) is to be gas-carburized at 927°C (1700°F). Calculate the carbon content at 0.90 mm below the surface of the gear after a 4.0-hour carburizing time. Assume the carbon content at the surface of the gear is 1.00 wt %. D (C in γ iron) at 927°C = 1.28×10^{-11} m^2/s.

*5.16 A gear made of 1020 steel (0.20 wt % C) is to be gas-carburized at 927°C (1700°F). Calculate the carbon content at 0.040 in. below the surface of the gear after a 7.0-hour carburizing time. Assume the carbon content at the surface of the gear is 1.15 wt %. D (C in γ iron) at 927°C = 1.28×10^{-11} m^2/s.

*5.17 The surface of a steel gear made of 1018 steel (0.18 wt % C) is to be gas-carburized at 927°C. Calculate the time necessary to increase the carbon content to 0.35 wt %

at 1.00 mm below the surface. Assume the carbon content of the surface of the gear is 1.20 wt %. D (C in γ iron) at 927°C $= 1.28 \times 10^{-11}$ m²/s.

5.18 A gear made of 1020 steel (0.20 wt % C) is to be gas-carburized at 927°C. Calculate the carbon content at 0.95 mm below the surface of the gear after an 8.0-hour carburizing time. Assume the carbon content at the surface of the gear is 1.25 wt %. D (C in γ iron) at 927°C $= 1.28 \times 10^{-11}$ m²/s.

*5.19 A gear made of 1018 steel (0.18 wt % C) is to be gas-carburized at 927°C. If the carburizing time is 7.5 h, at what depth in millimeters will the carbon content be 0.40 wt %? Assume the carbon content at the surface of the gear is 1.20 wt %. D (C in γ iron) at 927°C $= 1.28 \times 10^{-11}$ m²/s.

*5.20 If boron is diffused into a thick slice of silicon with no previous boron in it at a temperature of 1100°C for 5 h, what is the depth below the surface at which the concentration is 10^{17} atoms/cm³ if the surface concentration is 10^{18} atoms/cm³? $D = 4 \times 10^{-13}$ cm²/s for boron diffusing in silicon at 1100°C.

5.21 If aluminum is diffused into a thick slice of silicon with no previous aluminum in it at a temperature of 1100°C for 6 h, what is the depth below the surface at which the concentration is 10^{16} atoms/cm³ if the surface concentration is 10^{18} atoms/cm³? $D = 2 \times 10^{-12}$ cm²/s for aluminum diffusing in silicon at 1100°C.

5.22 Phosphorus is diffused into a thick slice of silicon with no previous phosphorus in it at a temperature of 1100°C. If the surface concentration of the phosphorus is 1×10^{18} atoms/cm³ and its concentration at 1 μm is 1×10^{15} atoms/cm³, how long must the diffusion time be? $D = 3.0 \times 10^{-13}$ cm²/s for P diffusing in Si at 1100°C.

*5.23 If the diffusivity in Prob. 5.22 had been 1.5×10^{-13} cm²/s, at what depth in micrometers would the phosphorus concentration be 1×10^{15} atoms/cm³?

5.24 Arsenic is diffused into a thick slice of silicon with no previous arsenic in it at 1100°C. If the surface concentration of the arsenic is 5.0×10^{18} atoms/cm³ and its concentration at 1.2 μm below the silicon surface is 1.5×10^{16} atoms/cm³, how long must the diffusion time be? ($D = 3.0 \times 10^{-14}$ cm²/s for As diffusing in Si at 1100°C.)

*5.25 Calculate the diffusivity D in square meters per second for the diffusion of nickel in FCC iron at 1100°C. Use values of $D_0 = 7.7 \times 10^{-5}$ m²/s; $Q = 280$ kJ/mol; $R = 8.314$ J/(mol · K).

*5.26 Calculate the diffusivity in m²/s of carbon in HCP titanium at 700°C. Use $D_0 = 5.10 \times 10^{-4}$ m²/s; $Q = 182$ kJ/mol; $R = 8.314$ J/(mol · K).

5.27 Calculate the diffusivity in m²/s for the diffusion of zinc in copper at 350°C. Use $D_0 = 3.4 \times 10^{-5}$ m²/s; $Q = 191$ kJ/mol.

5.28 The diffusivity of manganese atoms in the FCC iron lattice is 1.50×10^{-14} m²/s at 1300°C and 1.50×10^{-15} m²/s at 400°C. Calculate the activation energy in kJ/mol for this case in this temperature range. Data: $R = 8.314$ J/(mol · K).

*5.29 The diffusivity of copper atoms in the aluminum lattice is 7.50×10^{-13} m²/s at 600°C and 2.50×10^{-15} m²/s at 400°C. Calculate the activation energy for this case in this temperature range. [$R = 8.314$ J/(mol · K).]

5.30 The diffusivity of iron atoms in the BCC iron lattice is 4.5×10^{-23} m²/s at 400°C and 5.9×10^{-16} m²/s at 800°C. Calculate the activation energy in kJ/mol for this case in this temperature range. [$R = 8.314$ J/(mol · K).]

5.8 MATERIALS SELECTION AND DESIGN PROBLEMS

1. In integrated circuits, dissimilar metal wires such as gold and aluminum are joined at the cross section to form a bonded interface. At elevated temperatures, the interface starts to move or shift in the direction of one of the metals, called the Kirkendall effect. (*a*) Can you explain this phenomenon? (*b*) Is the direction of the shift random? (*c*) What are the negative effects of this process?

2. (*a*) Design a process that would be used to make a solid steel component out of fine steel powder. The density of the formed solid should be very close to the density of the physical metal. (*b*) Explain how this process works at both macro and micro levels. (*c*) What are some of the obstacles that you may encounter in your process?

3. Explain what happens if carbon steel is exposed to an oxygen-rich atmosphere at elevated temperatures inside a furnace.

4. Based on the data in Table 5.2, is it easier for carbon atoms to diffuse in the structure of BCC iron or FCC iron? What is the physical reason for this?

5. Classify the mechanism of diffusion in all 12 solute/solvent pairs given in Table 5.2 (interstitial or substitutional). Compare the diffusivity values and draw a conclusion.

6. The activation energy for the diffusion of hydrogen in steel at room temperature is 3600 cal/mol, which is significantly lower than that of, for example, carbon at 20,900 cal/mol. Investigate the effect of diffused hydrogen on the mechanical behavior of steel.

7. Investigate the role of solid state diffusion in manufacturing of ceramic components using the powder metallurgy process.

Mechanical Properties of Metals I

(© AP/Wide World Photos)

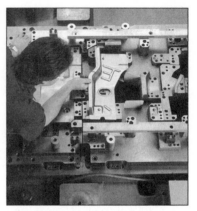

(© Roger Ball/Corbis)

Metals are formed into functional shapes using a wide variety of metal forming operations under both cold and hot conditions. Perhaps one of the most important examples, revealing the use of metal forming operations, is in manufacturing of automobile parts (both body and engine). The engine block is usually made of cast iron or aluminum alloys; the cylinder and other openings in the block are made by drilling, boring and tapping operations; the cylinder heads are also cast of aluminum alloys; connecting rods, crankshafts, and cams are forged (sometimes cast) and are then finish ground; the body panels including the roof, trunk lid, doors, and side panels are stamped from steel sheets and are then spot-welded together, (left figure). As the number of operations to produce a part increases, so does the cost of the part and therefore the overall product. To reduce the cost, manufacturers follow the "Near Net Shape" manufacturing concepts, in which the product is formed with the least number of operations and with the least amount of finish machining or grinding required. Automotive parts with complex and nonsymmetrical shapes such as bevel gears or universal joints are forged almost ready-to-install, (right figure). ■

By the end of this chapter, students will be able to . . .

1. Describe the forming operations that are used to shape metals into functional shapes. Differentiate between wrought alloy and cast products. Differentiate between hot and cold forming processes.

2. Describe the constitutive behavior of metals in terms of the engineering and true definition of stress and strain.

3. Explain the differences between elastic and plastic deformation at the atomic, micro, and macro scales.

4. Explain the differences between normal and sharing stresses and strains.

5. Explain what a tensile test is, what type of a machine is used to perform the tensile tests, and what information regarding the properties of a material can be extracted from such tests.

6. Define hardness and explain how is it measured. Describe various available hardness scales.

7. Describe the plastic deformation of a single crystal at the atomic level. Describe the concept of a slip, dislocations, twins, and their role in plastic deformation of a single crystal.

8. Define critical slip systems in BCC, FCC, and HCP single crystals.

9. Describe Schmid's law and its application in determination of the critical resolved shear stress.

10. Describe the effect of plastic deformation process on properties and grain structure of polycrystalline materials.

11. Explain the effect of grain size (Hall-Petch equation) and grain boundary on the plastic deformation and properties of a metal.

12. Describe various strengthening mechanisms used for metals.

13. Describe the annealing process and its impact on properties and microstructure of a cold worked metal.

14. Describe the superplastic behavior in metals.

15. Describe what a nanocrystalline metal is and what advantages it may offer.

This chapter first examines some of the basic methods of processing metals and alloys into useful shapes. Then stress and strain in metals are defined, and the tensile test used to determine these properties is described. This is followed by a treatment of hardness and hardness testing of metals. Next, the plastic deformation of metal single crystals and polycrystalline metals are examined. The concepts of superplasticity and Hall-Petch equation are introduced. Finally, the solid-solution strengthening and annealing of metals are considered.

6.1 THE PROCESSING OF METALS AND ALLOYS

6.1.1 The Casting of Metals and Alloys

Most metals are processed by first melting the metal in a furnace that functions as a reservoir for the molten metal. Alloying elements can be added to the molten metal to produce various alloy compositions. For example, solid magnesium metal may be added to molten aluminum and, after melting, can be mechanically mixed with the aluminum to produce a homogeneous melt of an aluminum-magnesium alloy. After oxide impurities and unwanted hydrogen gas are removed from the molten Al–Mg alloy, it is cast into a mold of a direct-chill semicontinuous casting unit, as shown in Fig. 4.8. Huge sheet ingots, such as those shown in Fig. 4.1, are produced in this way. Other types of ingots with different cross sections are cast in a similar way, for example, extrusion ingots are cast with circular cross sections.

Semifinished products are manufactured from the basic shaped ingots. Sheet[1] and plate[2] are produced by rolling sheet ingots to reduced thicknesses (Fig. 6.1).

---

[1]For this book *sheet* is defined as a rolled product rectangular in cross section and form of thickness 0.006 to 0.249 in. (0.015 to 0.063 cm).

[2]For this book *plate* is defined as a rolled product rectangular in cross section and form of thickness 0.250 in. (0.635 cm) or more.

Figure 6.1
Large aluminum four-high rolling mill is shown hot rolling a slab of an aluminum alloy that will eventually be reduced in thickness to make aluminum sheet.
(*Courtesy of Reynolds Metals Co.*)

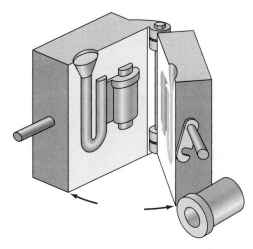

Figure 6.2
Permanent mold casting. Solidified casting with gate and metal core is shown in the left half of the mold. The completed casting is shown in front of the mold.
(After H. F. Taylor, M. C. Flemings, and J. Wulff, "Foundry Engineering," Wiley, 1959, p. 58.)

Extruded shapes such as channels and structural shapes are produced from extrusion ingots, and rod and wire are manufactured from wire bar ingots. All these products that are manufactured by hot and cold working the metal from large ingots are called *wrought alloy products.* The effects of permanent deformation on the structure and properties of metals will be treated in Secs. 6.5 and 6.6.

On a smaller scale molten metal may be cast into a mold that is in the shape of the final product, and usually only a small amount of machining or other finishing operation is required to produce the final casting. Products made in this manner are called *cast products* and the alloys used to produce them, *casting alloys.* For example, pistons used in automobile engines are usually made by casting molten metal into a permanent steel mold. A schematic diagram of a simple permanent mold containing a casting is shown in Fig. 6.2. Figure 6.3*a* shows an operator pouring an aluminum alloy into a permanent mold to produce a pair of piston castings; Fig. 6.3*b* shows the castings after they have been removed from the mold. After being trimmed, heat-treated, and machined, the finished piston (Fig. 6.3*c*) is ready for installation in an automobile engine.

6.1.2 Hot and Cold Rolling of Metals and Alloys

Hot and cold rolling are commonly used methods for fabricating metals and alloys. Long lengths of metal sheet and plate with uniform cross sections can be produced by these processes.

Hot Rolling of Sheet Ingots   **Hot rolling** of sheet ingots is carried out first since greater reductions in thickness can be taken with each rolling pass when the metal

(a)

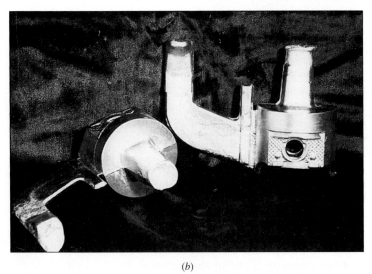

(b)

(c)

Figure 6.3
(a) Simultaneous permanent mold casting of two aluminum alloy pistons. (b) Aluminum alloy permanent mold piston castings after being removed from the mold shown in (a). (c) An automobile piston after heat treatment and machining is ready for installation in an engine.
(*Courtesy of General Motors Corporation.*)

is hot. Before hot rolling, sheet and plate ingots are preheated to a high temperature (about 1200°C). However, sometimes it is possible to hot roll the ingot-slabs directly from the caster (Fig. 6.4). After removal from the preheat furnace, the ingot sections are usually hot rolled in a reversing break-down rolling mill (Fig. 6.5).

Figure 6.4

Perfectly formed slabs are cut to required lengths by an automatic flame torch on the continuous slab caster at Bethlehem Steel Co.'s Burns Harbor, Indiana, plant. The two-strand caster converts 300 tons of molten steel into heavy slabs in 45 minutes.

(Courtesy of Bethlehem Steel Co.)

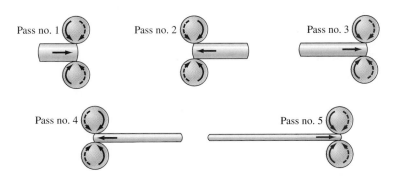

Figure 6.5

Diagrammatic representation of the sequence of hot-rolling operations involved in reducing an ingot to a slab on a reversing two-high mill.

(After H. E. McGannon [ed.], "The Making, Shaping, and Treating of Steel," 9th ed., United States Steel, 1971, p. 677.)

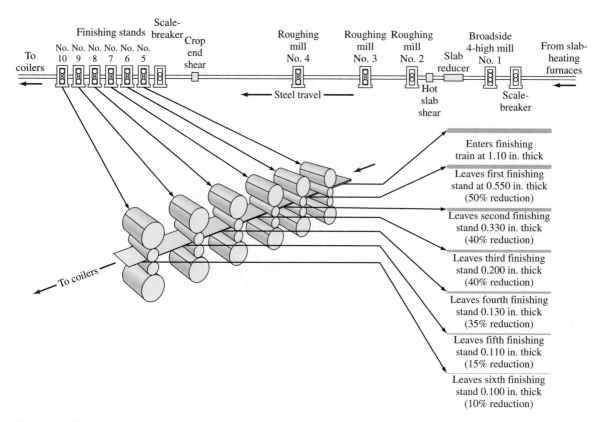

Figure 6.6
Typical reductions per pass in the finishing stands of a hot-strip mill equipped with four roughing stands and six finishing stands. Drawing is not to scale.
(*After H. E. McGannon [ed.], "The Making, Shaping, and Treating of Steel," 9th ed., United States Steel, 1971, p. 937.*)

Hot rolling is continued until the temperature of the slab drops so low that continued rolling becomes too difficult. The slab is then reheated and hot rolling is continued, usually until the hot-rolled strip is thin enough to be wound into a coil. In most large-scale operations hot rolling of the slab is carried out by using a series of four-high rolling mills alone and in series, as shown for the hot rolling of steel strip in Fig. 6.6.

Cold Rolling of Metal Sheet[3]   After hot rolling, which may also include some **cold rolling,** the coils of metal are usually given a reheating treatment called **annealing** to soften the metal to remove any cold work introduced during the hot-rolling operation. Cold rolling, which normally is done at room temperature, is again usually carried out with four-high rolling mills either alone or in series (Fig. 6.7). Figure 6.8 shows some sheet steel being cold-rolled on an industrial rolling mill.

---

[3]Cold rolling of metals is usually carried out below the recrystallization temperature of the metal and results in the strain hardening of the metal.

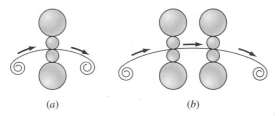

(a)   (b)

Figure 6.7
Schematic drawing illustrating the metal path during the cold rolling of metal sheet by four-high rolling mills: (a) single mill and (b) two mills in series.

Figure 6.8
Cold rolling sheet steel. Mills of this type are used for cold rolling steel strip, tin plate, and nonferrous metals.
(*Courtesy of Bethlehem Steel Co.*)

The **percent cold reduction** of a plate or sheet of metal can be calculated as follows:

$$\% \text{ cold reduction} = \frac{\text{initial metal thickness} - \text{final metal thickness}}{\text{initial metal thickness}} \times 100\%$$

(6.1)

EXAMPLE PROBLEM 6.1

Calculate the percent cold reduction in cold rolling an aluminum sheet alloy from 0.120 to 0.040 in.

■ **Solution**

$$\% \text{ cold reduction} = \frac{\text{initial thickness} - \text{final thickness}}{\text{initial thickness}} \times 100\%$$

$$= \frac{0.120 \text{ in.} - 0.040 \text{ in.}}{0.120 \text{ in.}} \times 100\% = \frac{0.080 \text{ in.}}{0.120 \text{ in.}} \times 100\%$$

$$= 66.7\%$$

EXAMPLE PROBLEM 6.2

A sheet of a 70% Cu–30% Zn alloy is cold-rolled 20 percent to a thickness of 3.00 mm. The sheet is then further cold-rolled to 2.00 mm. What is the total percent cold work?

■ **Solution**

We first determine the starting thickness of the sheet by considering the first cold reduction of 20 percent. Let x equal the starting thickness of the sheet. Then,

$$\frac{x - 3.00 \text{ mm}}{x} = 0.20$$

or

$$x - 3.00 \text{ mm} = 0.20x$$

$$x = 3.75 \text{ mm}$$

We can now determine the *total* percent cold work from the starting thickness to the finished thickness from the relationship

$$\frac{3.75 \text{ mm} - 2.00 \text{ mm}}{3.75 \text{ mm}} = \frac{1.75 \text{ mm}}{3.75 \text{ mm}} = 0.466 \text{ or } 46.6\%$$

6.1.3 Extrusion of Metals and Alloys

Extrusion is a plastic forming process in which a material under high pressure is reduced in cross section by forcing it through an opening in a die (Fig. 6.9). For most metals the extrusion process is used to produce cylindrical bars or hollow tubes. For the more readily extrudable metals, such as aluminum and copper and some of

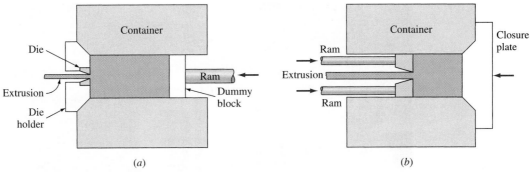

(a)   (b)

Figure 6.9
Two basic types of extrusion processes for metals: (*a*) direct and (*b*) indirect.
(*After G. Dieter, "Mechanical Metallurgy," 2nd ed., McGraw-Hill, 1976, p. 639.*)

Animation

their alloys, shapes with irregular cross sections are also commonly produced. Most metals are extruded hot since the deformation resistance of the metal is lower than if it is extruded cold. During extrusion the metal of a billet in the container of an extrusion press is forced by a ram through a die so that the metal is continuously deformed into a long length of metal with a uniform desired cross section.

The two main types of extrusion processes are *direct extrusion* and *indirect extrusion.* In direct extrusion the metal billet is placed in a container of an extrusion press and forced directly through the die by the ram (Fig. 6.9*a*). In indirect extrusion a hollow ram holds the die, with the other end of the container of the extrusion press being closed by a plate (Fig. 6.9*b*). The frictional forces and power requirements for indirect extrusion are lower than those for direct extrusion. However, the loads that can be applied by using a hollow ram in the indirect process are more limited than those that can be used for direct extrusion.

The extrusion process is used primarily for producing bar shapes, tube, and irregular shapes of the lower-melting nonferrous metals such as aluminum and copper and their alloys. However, with the development of powerful extrusion presses and improved lubricants such as glass, some carbon and stainless steels can also be hot-extruded.

6.1.4 Forging

Forging is another primary method for working metals into useful shapes. In the forging process the metal is hammered or pressed into a desired shape. Most forging operations are carried out with the metal in the hot condition, although in some cases the metal may be forged cold. There are two major types of forging methods: *hammer* and *press forging.* In hammer forging a drop hammer repeatedly exerts a striking force against the surface of the metal. In press forging the metal is subjected to a slowly moving compressive force (Fig. 6.10).

Forging processes can also be classified as *open-die forging* or *closed-die forging.* Open-die forging is carried out between two flat dies or dies with very simple shapes such as vees or semicircular cavities (Fig. 6.11) and is particularly useful for

Figure 6.10
Heavy-duty manipulator holding an ingot in position while a 10,000-ton
press squeezes the hot steel into the rough shape of the finished product.
(After H. E. McGannon [ed.], "The Making, Shaping, and Treating of Steel," 9th ed.,
United States Steel Corporation, 1971, p. 1044.)

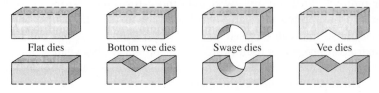

Flat dies   Bottom vee dies   Swage dies   Vee dies

Figure 6.11
Basic shapes for open-die forging.
(After H. E. McGannon [ed.], "The Making, Shaping, and Treating of Steel,"
9th ed., United States Steel, 1971, p. 1045.)

producing large parts such as steel shafts for electric steam turbines and generators.
In closed-die forging the metal to be forged is placed between two dies that have
the upper and lower impressions of the desired shape of the forging. Closed-die
forging can be carried out by using a single pair of dies or multiple-impression dies.
An example of a closed-die forging in which multiple-impression dies are used is
the automobile engine connecting rod (Fig. 6.12).

In general the forging process is used for producing irregular shapes that require
working to improve the structure of the metal by reducing porosity and refining the
internal structure. For example, a wrench that has been forged will be tougher and

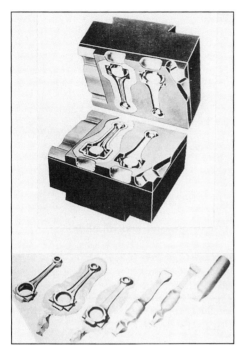

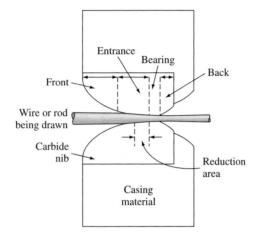

Figure 6.12
A set of closed forging dies used to produce an automobile connecting rod.
(*Courtesy of the Forging Industry Association.*)

Figure 6.13
Section through a wire-drawing die.
(*After "Wire and Rods, Alloy Steel," Steel Products Manual, American Iron and Steel Institute, 1975.*)

less likely to break than one that is simply cast into shape. Forging is also sometimes used to break down the as-cast ingot structure of some highly alloyed metals (e.g., some tool steels) so that the metal is made more homogeneous and less likely to crack during subsequent working.

6.1.5 Other Metal-Forming Processes

There are many types of secondary metal-forming processes whose descriptions are beyond the scope of this book. However, two of these processes, *wire drawing* and *deep drawing* of sheet metal, will be briefly described.

Wire drawing is an important metal-forming process. Starting rod or wire stock is drawn through one or more tapered wire-drawing dies (Fig. 6.13). For steel wire drawing, a tungsten carbide inner "nib" is inserted inside a steel casing. The hard carbide provides a wear-resistant surface for the reduction of the steel wire. Special precautions must be taken to make sure the surface of the stock to be drawn into wire is clean and properly lubricated. Intermediate softening heat treatments are sometimes necessary when the drawn wire work hardens during processing. The procedures used vary considerably, depending on the metal or alloy being drawn and the final diameter and temper desired.

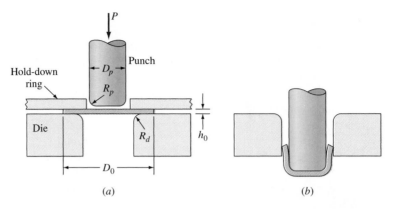

Figure 6.14

Deep drawing of a cylindrical cup (*a*) before drawing and (*b*) after drawing.

(*After G. Dieter, "Mechanical Metallurgy," 2nd ed., McGraw-Hill, 1976, p. 688.*)

EXAMPLE PROBLEM 6.3

Calculate the percent cold reduction when an annealed copper wire is cold-drawn from a diameter of 1.27 mm (0.050 in.) to a diameter of 0.813 mm (0.032 in.).

■ **Solution**

$$
\% \text{ cold reduction} = \frac{\text{change in cross-sectional area}}{\text{original area}} \times 100\%
$$

$$
= \frac{(\pi/4)(1.27 \text{ mm})^2 - (\pi/4)(0.813 \text{ mm})^2}{(\pi/4)(1.27 \text{ mm})^2} \times 100\% \qquad \textbf{(6.2)}
$$

$$
= \left[1 - \frac{(0.813)^2}{(1.27)^2} \right](100\%)
$$

$$
= (1 - 0.41)(100\%) = 59\% \blacktriangleleft
$$

Deep drawing is another metal-forming process and is used for shaping flat sheets of metal into cup-shaped articles. A metal blank is placed over a shaped die and then is pressed into the die with a punch (Fig. 6.14). Usually a hold-down device is used to allow the metal to be pressed smoothly into the die to prevent wrinkling of the metal.

6.2 STRESS AND STRAIN IN METALS

In the first section of this chapter we briefly examined most of the principal methods by which metals are processed into semifinished wrought and cast products. Let us now investigate how the mechanical properties of strength and ductility are evaluated for engineering applications.

6.2.1 Elastic and Plastic Deformation

When a piece of metal is subjected to a uniaxial tensile force, deformation of the metal occurs. If the metal returns to its original dimensions when the force is removed, the metal is said to have undergone **elastic deformation.** The amount of elastic deformation a metal can undergo is small, since during elastic deformation the metal atoms are displaced from their original positions but not to the extent that they take up new positions. Thus, when the force on a metal that has been elastically deformed is removed, the metal atoms return to their original positions and the metal takes back its original shape. If the metal is deformed to such an extent that it cannot fully recover its original dimensions, it is said to have undergone *plastic deformation.* During plastic deformation, the metal atoms are *permanently* displaced from their original positions and take up new positions. The ability of some metals to be extensively plastically deformed without fracture is one of the most useful engineering properties of metals. For example, the extensive plastic deformability of steel enables automobile parts such as fenders, hoods, and doors to be stamped out mechanically without the metal fracturing.

Virtual Lab

6.2.2 Engineering Stress and Engineering Strain

Engineering Stress   Let us consider a cylindrical rod of length l_0 and cross-sectional area A_0 subjected to a uniaxial tensile force F, as shown in Fig. 6.15. By definition,

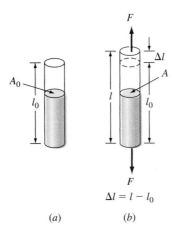

$$\Delta l = l - l_0$$

(a)   (b)

Figure 6.15
Elongation of a cylindrical metal rod subjected to a uniaxial tensile force F.
(*a*) The rod with no force on it; and (*b*) the rod subjected to a uniaxial tensile force F, which elongates the rod from length l_0 to l.

the **engineering stress** σ on the bar is equal to the average uniaxial tensile force F on the bar divided by the original cross-sectional area A_0 of the bar. Thus,

$$\text{Engineering stress } \sigma = \frac{F \text{ (average uniaxial tensile force)}}{A_0 \text{ (original cross-sectional area)}} \tag{6.3}$$

The units for engineering stress are:

U.S. customary: pounds force per square inch ($\text{lb}_f/\text{in.}^2$, or psi);
lb_f = pounds force
SI: newtons per square meter (N/m^2) or pascals (Pa), where $1 \text{ N/m}^2 = 1 \text{ Pa}$

The conversion factors for psi to pascals are

$$1 \text{ psi} = 6.89 \times 10^3 \text{ Pa}$$

$$10^6 \text{ Pa} = 1 \text{ megapascal} = 1 \text{ MPa}$$

$$1000 \text{ psi} = 1 \text{ ksi} = 6.89 \text{ MPa}$$

EXAMPLE PROBLEM 6.4

A 0.500-in.-diameter aluminum bar is subjected to a force of 2500 lb_f. Calculate the engineering stress in pounds per square inch (psi) on the bar.

■ **Solution**

$$\sigma = \frac{\text{force}}{\text{original cross-sectional area}} = \frac{F}{A_0}$$

$$= \frac{2500 \text{ lb}_f}{(\pi/4)(0.500 \text{ in})^2} = 12{,}700 \text{ lb}_f/\text{in.}^2 \blacktriangleleft$$

EXAMPLE PROBLEM 6.5

A 1.25-cm-diameter bar is subjected to a load of 2500 kg. Calculate the engineering stress on the bar in megapascals (MPa).

■ **Solution**

The load on the bar has a mass of 2500 kg. In SI units the force on the bar is equal to the mass of the load times the acceleration of gravity (9.81 m/s^2), or

$$F = ma = (2500 \text{ kg})(9.81 \text{ m/s}^2) = 24{,}500 \text{ N}$$

The diameter d of the bar = 1.25 cm = 0.0125 m. Thus, the engineering stress on the bar is

$$\sigma = \frac{F}{A_0} = \frac{F}{(\pi/4)(d^2)} = \frac{24{,}500 \text{ N}}{(\pi/4)(0.0125 \text{ m})^2}$$

$$= (2.00 \times 10^8 \text{ Pa})\left(\frac{1 \text{ MPa}}{10^6 \text{ Pa}}\right) = 200 \text{ MPa} \blacktriangleleft$$

Engineering Strain   When a uniaxial tensile force is applied to a rod, such as that shown in Fig. 6.15, it causes the rod to be elongated in the direction of the force. Such a displacement is called *strain*. By definition, **engineering strain,** which is caused by the action of a uniaxial tensile force on a metal sample, is the ratio of the change in length of the sample in the direction of the force divided by the original length of sample considered. Thus, the engineering strain for the metal bar shown in Fig. 6.15 (or for a similar-type metal sample) is

$$\text{Engineering strain } \epsilon = \frac{l - l_0}{l_0} = \frac{\Delta l \text{ (change in length of sample)}}{l_0 \text{ (original length of sample)}} \qquad \textbf{(6.4)}$$

where l_0 = original length of sample and l = new length of sample after being extended by a uniaxial tensile force. In most cases engineering strain is determined by using a small length, usually 2 in., called the *gage length,* within a much longer, for example, 8 in., sample (see Example Problem 6.6).

Virtual Lab

The *units for engineering strain ϵ* are:

U.S. customary: inches per inch (in./in.)

SI: meters per meter (m/m)

Thus, engineering strain has *dimensionless units.* In industrial practice it is common to convert engineering strain into *percent strain* or *percent elongation:*

$$\% \text{ engineering strain} = \text{engineering strain} \times 100\% = \% \text{ elongation}$$

A sample of commercially pure aluminum 0.500 in. wide, 0.040 in. thick, and 8 in. long that has gage markings 2.00 in. apart in the middle of the sample is strained so that the gage markings are 2.65 in. apart (Fig. 6.16). Calculate the engineering strain and the percent engineering strain elongation that the sample undergoes.

EXAMPLE PROBLEM 6.6

■ **Solution**

$$\text{Engineering strain } \epsilon = \frac{l - l_0}{l_0} = \frac{2.65 \text{ in.} - 2.00 \text{ in.}}{2.00 \text{ in.}} = \frac{0.65 \text{ in.}}{2.00 \text{ in.}} = 0.325 \blacktriangleleft$$

$$\% \text{ elongation} = 0.325 \times 100\% = 32.5\% \blacktriangleleft$$

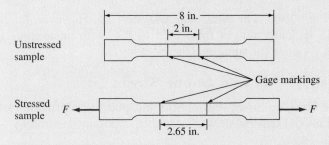

Figure 6.16
Flat tensile specimen before and after testing.

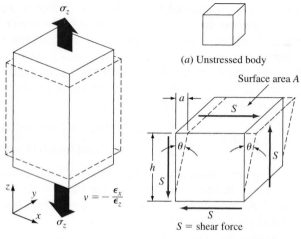

(a) Unstressed body

(b) Body under tensile stress   (c) Body under shear stress

Figure 6.17
(a) Unstressed cubic body. (b) Cubic body subjected
to tensile stress. The ratio of the elastic contraction
perpendicular to the extension is designated Poisson's
ratio v. (c) Cubic body subjected to pure shear forces S
acting over surface areas A. The shear stress τ acting
on the body is equal to S/A.

6.2.3 Poisson's Ratio

A longitudinal elastic deformation of a metal produces an accompanying lateral
dimensional change. As shown in Fig. 6.17b, a tensile stress σ_z produces an axial
strain $+\epsilon_z$ and lateral contractions of $-\epsilon_x$ and $-\epsilon_y$. For isotropic behavior,[4] ϵ_x and
ϵ_y are equal. The ratio

$$v = -\frac{\epsilon \,(\text{lateral})}{\epsilon \,(\text{longitudinal})} = -\frac{\epsilon_x}{\epsilon_z} = -\frac{\epsilon_y}{\epsilon_z} \tag{6.5}$$

is called *Poisson's ratio.* For ideal materials, $v = 0.5$. However, for real materials,
Poisson's ratio typically ranges from 0.25 to 0.4, with an average of about 0.3.
Table 6.1 lists v values for some metals and alloys.

6.2.4 Shear Stress and Shear Strain

Until now we have discussed the elastic and plastic deformation of metals and
alloys under uniaxial tension stresses. Another important method by which a metal

---

[4]*Isotropic:* exhibiting properties with the same values when measured along axes in all directions.

Table 6.1 Typical room-temperature values of elastic constants for isotropic materials

Material	Modulus of elasticity, 10^{-6} psi (GPa)	Shear modulus, 10^{-6} psi (GPa)	Poisson's ratio
Aluminum alloys	10.5 (72.4)	4.0 (27.5)	0.31
Copper	16.0 (110)	6.0 (41.4)	0.33
Steel (plain carbon and low-alloy)	29.0 (200)	11.0 (75.8)	0.33
Stainless steel (18-8)	28.0 (193)	9.5 (65.6)	0.28
Titanium	17.0 (117)	6.5 (44.8)	0.31
Tungsten	58.0 (400)	22.8 (157)	0.27

Source: G. Dieter, "Mechanical Metallurgy," 3rd ed., McGraw-Hill, 1986.

can be deformed is under the action of a **shear stress.** The action of a simple shear stress couple (shear stresses act in pairs) on a cubic body is shown in Fig. 6.17c, where a shearing force S acts over an area A. The shear stress τ is related to the shear force S by

$$\tau \text{ (shear stress)} = \frac{S \text{ (shear force)}}{A \text{ (area over which shear force acts)}} \qquad \textbf{(6.6)}$$

The units for shear stress are the same as for uniaxial tensile stress:

U.S. customary: pounds force per square inch ($lb_f/in.^2$, or psi)

SI: newtons per square meter (N/m^2) or pascals (Pa)

The **shear strain** γ is defined in terms of the amount of the shear displacement a in Fig. 5.17c divided by the distance h over which the shear acts, or

$$\gamma = \frac{a}{h} = \tan \theta \qquad \textbf{(6.7)}$$

For pure elastic shear, the proportionality between shear and stress is

$$\tau = G\gamma \qquad \textbf{(6.8)}$$

where G is the elastic modulus.

We will be concerned with shear stresses when we discuss the plastic deformation of metals in Sec. 6.5.

6.3 THE TENSILE TEST AND THE ENGINEERING STRESS-STRAIN DIAGRAM

The *tensile test* is used to evaluate the strength of metals and alloys. In this test a metal sample is pulled to failure in a relatively short time at a constant rate. Figure 6.18 is a picture of a modern tensile testing machine, and Fig. 6.19 illustrates schematically how the sample is tested in tension.

Virtual Lab

Figure 6.18
Modern Tensile Testing Machine. The force (load) on
the specimen is measured by the load cell while the
strain is measured by the clip-on extensometer. The
data is collected and analyzed by computer controlled
software.
(*Courtesy the Instron® Corporation.*)

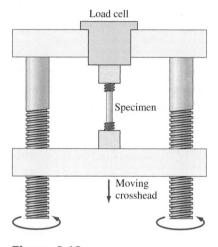

Figure 6.19
Schematic illustration showing how
the tensile machine of Fig. 6.18
operates. Note, however, that
the crosshead of the machine in
Fig. 6.18 moves up.
(*After H. W. Hayden, W. G. Moffatt, and
John Wulff, "The Structure and Properties
of Materials," vol. 3: "Mechanical
Behavior," Wiley, 1965, Fig. 1.1, p. 2.*)

The force (load) on the specimen being tested is measured by the load cell while
the strain is obtained from the extensometer attached to the specimen (Fig. 6.20) and
the data is collected in a computer control software package.

The types of samples used for the tensile test vary considerably. For metals with
a thick cross section such as plate, a 0.50-in.-diameter round specimen is commonly
used (Fig. 6.21*a*). For metal with thinner cross sections such as sheet, a flat specimen is used (Fig. 6.21*b*). A 2-in. gage length within the specimen is the most commonly used gage length for tensile tests.

The force data obtained from the chart paper for the tensile test can be converted
to engineering stress data, and a plot of engineering stress versus engineering strain

Figure 6.20
Close-up of the tensile machine extensometer that measures the strain that the sample undergoes during the tensile test. The extensometer is attached to the sample by small spring clamps.
(*Courtesy of the Instron Corporation.*)

Virtual Lab
Animation

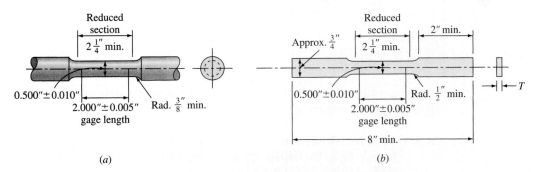

(a)   *(b)*

Figure 6.21
Examples of the geometrical shape of commonly used tension test specimens. (*a*) Standard round tension test specimen with 2-in. gage length. (*b*) Standard rectangular tension test specimen with 2-in. gage length.
(*Based on H. E. McGannon [ed.], ASTM Standards, 1968, "The Making, Shaping, and Treating of Steel," 9th ed., United States Steel, 1971, p. 1220.*)

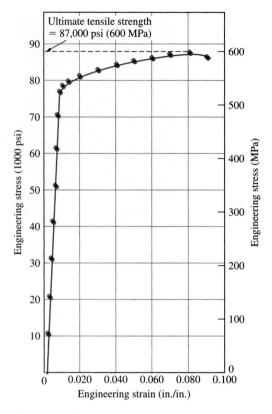

Figure 6.22
Engineering stress-strain diagram for a
high-strength aluminum alloy (7075-T6). The
specimens for the diagram were taken from
⅝-in. plate and had a 0.50-in. diameter with a
2-in. gage length.
(*Courtesy of Aluminum Company of America.*)

Virtual Lab

can be constructed. Figure 6.22 shows an **engineering stress-strain diagram** for a
high-strength aluminum alloy.

6.3.1 Mechanical Property Data Obtained from the Tensile Test and the Engineering Stress-Strain Diagram

The mechanical properties of metals and alloys that are of engineering importance
for structural design and can be obtained from the engineering tensile test are:

1. Modulus of elasticity
2. Yield strength at 0.2 percent offset

3. Ultimate tensile strength
4. Percent elongation at fracture
5. Percent reduction in area at fracture

Modulus of Elasticity In the first part of the tensile test the metal is deformed elastically. That is, if the load on the specimen is released, the specimen will return to its original length. For metals the maximum elastic deformation is usually less than 0.5 percent. In general, metals and alloys show a linear relationship between stress and strain in the elastic region of the engineering stress-strain diagram, which is described by Hooke's law:[5]

$$\sigma \text{ (stress)} = E\epsilon \text{ (strain)} \tag{6.9}$$

or

Virtual Lab

$$E = \frac{\sigma \text{ (stress)}}{\epsilon \text{ (strain)}} \quad \text{(units of psi or Pa)}$$

where E is the **modulus of elasticity,** or *Young's modulus.*[6]

The modulus of elasticity is related to the bonding strength between the atoms in a metal or alloy. Table 6.1 lists the elastic moduli for some common metals. Metals with high elastic moduli are relatively stiff and do not deflect easily. Steels, for example, have high elastic moduli values of 30×10^6 psi (207 GPa),[7] whereas aluminum alloys have lower elastic moduli of about 10 to 11×10^6 psi (69 to 76 GPa). Note that in the elastic region of the stress-strain diagram, the modulus does not change with increasing stress.

Yield Strength The **yield strength** is a very important value for use in engineering structural design since it is the strength at which a metal or alloy shows significant plastic deformation. Because there is no definite point on the stress-strain curve where elastic strain ends and plastic strain begins, the yield strength is chosen to be that strength when a definite amount of plastic strain has occurred. For American engineering structural design, the yield strength is chosen when 0.2 percent plastic strain has taken place, as indicated on the engineering stress-strain diagram of Fig. 6.23.

The 0.2 percent yield strength, also called the *0.2 percent offset yield strength,* is determined from the engineering stress-strain diagram, as shown in Fig. 6.23. First, a line is drawn parallel to the elastic (linear) part of the stress-strain plot at 0.002 in./in. (m/m) strain, as indicated on Fig. 6.23. Then at the point where this line intersects the upper part of the stress-strain curve, a horizontal line is drawn to the stress axis. The 0.2 percent offset yield strength is the stress where

---

[5]Robert Hooke (1635–1703). English physicist who studied the elastic behavior of solids.

[6]Thomas Young (1773–1829). English physicist.

[7]SI prefix G = giga = 10^9.

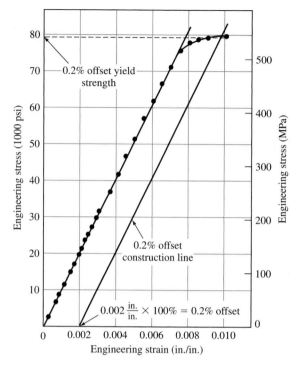

Figure 6.23

Linear part of engineering stress-strain diagram of Fig. 6.22 expanded on the strain axis to make a more accurate determination of the 0.2 percent offset yield stress.

(*Courtesy of Aluminum Company of America.*)

Virtual Lab

the horizontal line intersects the stress axis, and in the case of the stress-strain curve of Fig. 6.23, the yield strength is 78,000 psi. It should be pointed out that the 0.2 percent offset yield strength is arbitrarily chosen, and thus the yield strength could have been chosen at any other small amount of permanent deformation. For example, a 0.1 percent offset yield strength is commonly used in the United Kingdom.

Ultimate Tensile Strength The **ultimate tensile strength** (UTS) is the maximum strength reached in the engineering stress-strain curve. If the specimen develops a localized decrease in cross-sectional area (commonly called *necking*) (Fig. 6.24), the engineering stress will decrease with further strain until fracture occurs since the engineering stress is determined by using the *original* cross-sectional area of the specimen. The more ductile a metal is, the more the specimen will neck before fracture and hence the more the decrease in the stress on the stress-strain curve beyond the maximum stress. For the high-strength

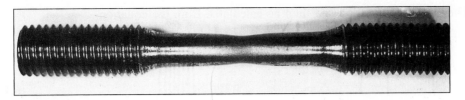

Figure 6.24
Necking in a mild-steel round specimen. The specimen was originally uniformly cylindrical. After being subjected to uniaxial tension forces up to almost fracture, the specimen decreased in cross section, or "necked" in the middle.

aluminum alloy whose stress-strain curve is shown in Fig. 6.22, there is only a small decrease in stress beyond the maximum stress because this material has relatively low ductility.

An important point to understand with respect to engineering stress-strain diagrams is that the metal or alloy continues to increase in stress up to the stress at fracture. It is only because we use the original cross-sectional area to determine engineering stress that the stress on the engineering stress-strain diagram decreases at the latter part of the test.

The ultimate tensile strength of a metal is determined by drawing a horizontal line from the maximum point on the stress-strain curve to the stress axis. The stress where this line intersects the stress axis is called the *ultimate tensile strength,* or sometimes just the *tensile strength.* For the aluminum alloy of Fig. 6.22, the ultimate tensile strength is 87,000 psi.

The ultimate tensile strength is not used much in engineering design for ductile alloys since too much plastic deformation takes place before it is reached. However, the ultimate tensile strength can give some indication of the presence of defects. If the metal contains porosity or inclusions, these defects may cause the ultimate tensile strength of the metal to be lower than normal.

Percent Elongation The amount of elongation that a tensile specimen undergoes during testing provides a value for the ductility of a metal. Ductility of metals is most commonly expressed as percent elongation, starting with a gage length usually of 2 in. (5.1 cm) (Fig. 6.21). In general the higher the ductility (the more deformable the metal is), the higher the percent elongation is. For example, a sheet of 0.062-in. (1.6-mm) commercially pure aluminum (alloy 1100-0) in the soft condition has a high percent elongation of 35 percent, whereas the same thickness of the high-strength aluminum alloy 7075-T6 in the fully hard condition has a percent elongation of only 11 percent.

As previously mentioned, during the tensile test an extensometer can be used to continuously measure the strain of the specimen being tested. However, the percent elongation of a specimen after fracture can be measured by fitting the fractured specimen together and measuring the final elongation with calipers.

The percent elongation can then be calculated from the equation

Virtual Lab

$$\% \text{ elongation} = \frac{\text{final length* } - \text{ initial length*}}{\text{initial length}} \times 100\%$$

$$= \frac{l - l_0}{l_0} \times 100\% \qquad (6.10)$$

The percent elongation at fracture is of engineering importance not only as a measure of ductility but also as an index of the quality of the metal. If porosity or inclusions are present in the metal or if damage due to overheating the metal has occurred, the percent elongation of the specimen tested may be decreased below normal.

Percent Reduction in Area   The ductility of a metal or alloy can also be expressed in terms of the percent reduction in area. This quantity is usually obtained from a tensile test using a specimen 0.50 in. (12.7 mm) in diameter. After the test, the diameter of the reduced cross section at the fracture is measured. Using the measurements of the initial and final diameters, the percent reduction in area can be determined from the equation

$$\% \text{ reduction in area} = \frac{\text{initial area } - \text{ final area}}{\text{initial area}} \times 100\%$$

$$= \frac{A_0 - A_f}{A_0} \times 100\% \qquad (6.11)$$

The percent reduction in area, like the percent elongation, is a measure of the ductility of the metal and is also an index of quality. The percent reduction in area may be decreased if defects such as inclusions and/or porosity are present in the metal specimen.

---

*The initial length is the length between the gage marks on the specimen before testing. The final length is the length between these same gage marks after testing when the fractured surface of the specimen is fitted together (see Example Problem 6.5).

EXAMPLE PROBLEM 6.7

A 0.500-in.-diameter round sample of a 1030 carbon steel is pulled to failure in a tensile testing machine. The diameter of the sample was 0.343 in. at the fracture surface. Calculate the percent reduction in area of the sample.

■ **Solution**

$$\% \text{ reduction in area} = \frac{A_0 - A_f}{A_0} \times 100\% = \left(1 - \frac{A_f}{A_0}\right)(100\%)$$

$$= \left[1 - \frac{(\pi/4)(0.343 \text{ in.})^2}{(\pi/4)(0.500 \text{ in.})^2}\right](100\%)$$

$$= (1 - 0.47)(100\%) = 53\% \blacktriangleleft$$

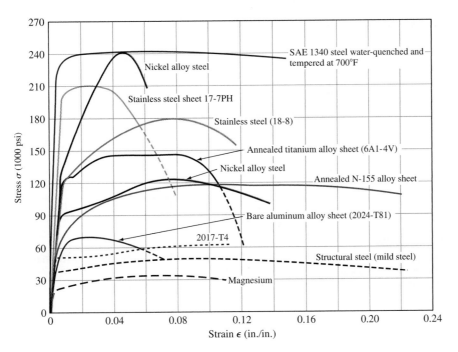

Figure 6.25
Engineering stress-strain curves for selected metals and alloys.
(After J. Marin, "Mechanical Behavior of Engineering Materials," Prentice-Hall, 1962, p. 24, and new data.)

6.3.2 Comparison of Engineering Stress-Strain Curves for Selected Alloys

Engineering stress-strain curves for selected metals and alloys are shown in Fig. 6.25. Alloying a metal with other metals or nonmetals and heat treatment can greatly affect the tensile strength and ductility of metals. The stress-strain curves of Fig. 6.25 show a great variation in ultimate tensile strength. Elemental magnesium has a UTS of 35 ksi (1 ksi = 1000 psi), whereas SAE 1340 steel water-quenched and tempered at 700°F (370°C) has a UTS of 240 ksi.

6.3.3 True Stress and True Strain

The engineering stress is calculated by dividing the applied force F on a tensile test specimen by its original cross-sectional area A_0 (Eq. 6.3). Since the cross-sectional area of the test specimen changes continuously during a tensile test, the engineering stress calculated is not precise. During the tensile test, after necking of the sample occurs (Fig. 6.24), the engineering stress decreases as the strain

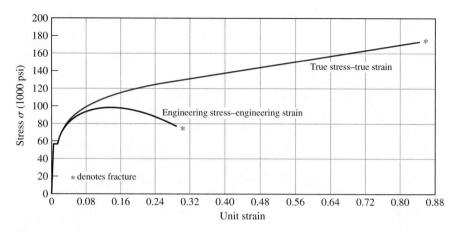

Virtual Lab

Figure 6.26

Comparison of the true stress–true strain curve with the engineering (nominal) stress-strain diagram for a low-carbon steel.

(From H. E. McGannon [ed.], "The Making, Shaping, and Treating of Steel," United States Steel, 1971.)

increases, leading to a maximum engineering stress in the engineering stress-strain curve (Fig. 6.26). Thus, once necking begins during the tensile test, the true stress is higher than the engineering stress. We define the true stress and true strain by the following:

$$\text{True stress } \sigma_t = \frac{F \text{ (average uniaxial force on the test sample)}}{A_i \text{ (instantaneous minimum cross-sectional area of sample)}}$$

(6.12)

$$\text{True strain } \epsilon_t = \int_{l_0}^{l_i} \frac{dl}{l} = \ln \frac{l_i}{l_0}$$

(6.13)

where l_0 is the original gage length of the sample and l_i is the instantaneous extended gage length during the test. If we assume constant volume of the gage-length section of the test specimen during the test, then $l_0 A_0 = l_i A_i$ or

$$\frac{l_i}{l_0} = \frac{A_0}{A_i} \qquad \text{and} \qquad \epsilon_t = \ln \frac{l_i}{l_0} = \ln \frac{A_0}{A_i}$$

Figure 6.26 compares engineering stress-strain and true stress-strain curves for a low-carbon steel.

Engineering designs are not based on true stress at fracture since as soon as the yield strength is exceeded, the material starts to deform. Engineers use instead the 0.2 percent offset engineering yield stress for structural designs with the proper safety factors. However, for research, sometimes the true stress-strain curves are needed.

Compare the engineering stress and strain with the true test and strain for the tensile test of a low-carbon steel that has the following test values.

Load applied to specimen = 17,000 lb_f Initial specimen diameter = 0.500 in.

Diameter of specimen under 17,000 lb_f load = 0.472 in.

■ **Solution**

$$\text{Area at start } A_0 = \frac{\pi}{4}d^2 = \frac{\pi}{4}(0.500 \text{ in.})^2 = 0.196 \text{ in.}^2$$

$$\text{Area under load } A_i = \frac{\pi}{4}(0.472 \text{ in.})^2 = 0.175 \text{ in.}^2$$

Assuming no volume change during extension, $A_0 l_0 = A_i l_i$ or $l_i/l_0 = A_0/A_i$.

$$\text{Engineering stress} = \frac{F}{A_0} = \frac{17,000 \text{ lb}_f}{0.196 \text{ in.}^2} = 86,700 \text{ psi} \blacktriangleleft$$

$$\text{Engineering strain} = \frac{\Delta l}{l} = \frac{l_i - l_0}{l_0} = \frac{A_0}{A_i} - 1 = \frac{0.196 \text{ in.}^2}{0.175 \text{ in.}^2} - 1 = 0.12$$

$$\text{True stress} = \frac{F}{A_i} = \frac{17,000 \text{ lb}_f}{0.175 \text{ in.}^2} = 97,100 \text{ psi} \blacktriangleleft$$

$$\text{True strain} = \ln\frac{l_i}{l_0} = \ln\frac{A_0}{A_i} = \ln\frac{0.196 \text{ in.}^2}{0.175 \text{ in.}^2} = \ln 1.12 = 0.113$$

6.4 HARDNESS AND HARDNESS TESTING

Hardness is a measure of the resistance of a metal to permanent (plastic) deformation. The hardness of a metal is measured by forcing an indenter into its surface. The indenter material, which is usually a ball, pyramid, or cone, is made of a material much harder than the material being tested. For example, hardened steel, tungsten carbide, or diamond are commonly used materials for indenters. For most standard hardness tests a known load is applied slowly by pressing the indenter at 90° into the metal surface being tested [Fig. 6.27b (2)]. After the indentation has been made, the indenter is withdrawn from the surface [Fig. 6.27b (3)]. An empirical hardness number is then calculated or read off a dial (or digital display), which is based on the cross-sectional area or depth of the impression.

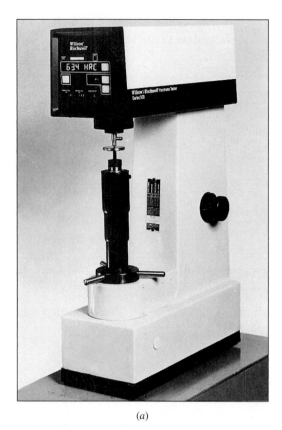

(a)

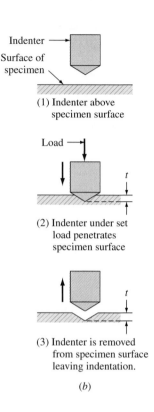

(1) Indenter above
    specimen surface

(2) Indenter under set
    load penetrates
    specimen surface

(3) Indenter is removed
    from specimen surface
    leaving indentation.

(b)

Virtual Lab

Figure 6.27
(a) A Rockwell hardness tester.
(*Courtesy of the Page-Wilson Co.*)
(b) Steps in the measurement of hardness with a diamond-cone identer. The depth t
determines the hardness of the material. The lower the value of t, the harder the
material.

Table 6.2 lists the types of indenters and types of impressions associated with
four common hardness tests: Brinell, Vickers, Knoop, and Rockwell. The hardness
number for each of these tests depends on the shape of the indentation and the
applied load. Figure 6.27 shows a modern Rockwell hardness tester, which has a
digital readout display.

The hardness of a metal depends on the ease with which it plastically deforms.
Thus a relationship between hardness and strength for a particular metal can be determined empirically. The hardness test is much simpler than the tensile test and can
be nondestructive (i.e., the small indentation of the indenter may not be detrimental
to the use of an object). For these reasons, the hardness test is used extensively in
industry for quality control.

Table 6.2 Hardness tests

Test	Indenter	Shape of indentation		Load	Formula for hardness number
		Side view	Top view		
Brinell	10 mm sphere of steel or tungsten carbide	$\leftarrow D \rightarrow$ $\leftarrow d \rightarrow$	$\leftarrow d \rightarrow$	P	$BHN = \dfrac{2P}{\pi D(D - \sqrt{D^2 - d^2})}$
Vickers	Diamond pyramid	$136°$	$d_1 \quad d_1$	P	$VHN = \dfrac{1.72P}{d_1^2}$
Knoop microhardness	Diamond pyramid	t $l/b = 7.11$ $b/t = 4.00$	b l	P	$KHN = \dfrac{14.2P}{l^2}$
Rockwell A C D	Diamond cone	$120°$ t		60 kg $R_A =$ 150 kg $R_C =$ 100 kg $R_D =$	$\left.\begin{array}{c} \\ \\ \end{array}\right\}$ 100–500f
B F G	$\frac{1}{16}$-in.-diameter steel sphere			100 kg $R_B =$ 60 kg $R_F =$ 150 kg $R_G =$ 100 kg $R_E =$	$\left.\begin{array}{c} \\ \\ \\ \end{array}\right\}$ 130–500f
E	$\frac{1}{8}$-in.-diameter steel sphere	t			

Source: After H. W. Hayden, W. G. Moffatt, and J. Wulff, "The Structure and Properties of Materials," vol. III, Wiley, 1965, p. 12.

6.5 PLASTIC DEFORMATION OF METAL SINGLE CRYSTALS

6.5.1 Slipbands and Slip Lines on the Surface of Metal Crystals

Let us first consider the permanent deformation of a rod of a zinc single crystal by stressing it beyond its elastic limit. An examination of the zinc crystal after the deformation shows that step markings appear on its surface, which are called

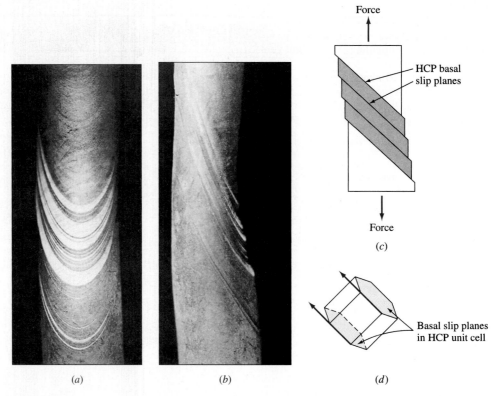

(a)   (b)   (d)

Figure 6.28
Plastically deformed zinc single crystal showing slipbands: (*a*) front view of real crystal,
(*b*) side view of real crystal, (*c*) schematic side view indicating HCP basal slip planes in
crystal, and (*d*) HCP unit cell indicating basal slip planes.
(*Zinc single-crystal photos courtesy of Prof. Earl Parker of the University of California at Berkeley.*)

slipbands (Fig. 6.28*a* and *b*). The slipbands are caused by the slip or shear deformation of metal atoms on specific crystallographic planes called *slip planes*. The
deformed zinc single-crystal surface illustrates the formation of slipbands very
clearly since **slip** in these crystals is restricted primarily to slip on the HCP basal
planes (Fig. 6.28*c* and *d*).

In single crystals of ductile FCC metals like copper and aluminum, slip occurs
on multiple slip planes, and as a result the slipband pattern on the surface of these
metals when they are deformed is more uniform (Fig. 6.29). A closer examination
of the slipped surface of metals at high magnification shows that slip has occurred
on many slip planes within the slipbands (Fig. 6.30). These fine steps are called *slip
lines* and are usually about 50 to 500 atoms apart, whereas slipbands are commonly
separated by about 10,000 atom diameters. Unfortunately, the terms slipband and
slip line are often used interchangeably.

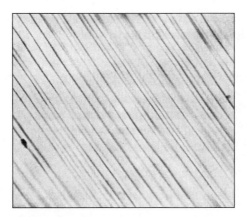

Figure 6.29
Slipband pattern on surface of copper single crystal after 0.9 percent deformation. (Magnification 100×.)
[*After F. D. Rosi. Trans. AIME,* **200:***1018 (1954).*]

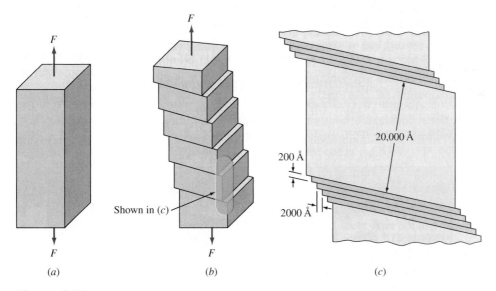

20,000 Å

200 Å

2000 Å

Shown in (*c*)

(*a*)   (*b*)   (*c*)

Figure 6.30
The formation of slipbands during plastic deformation. (*a*) A single crystal under a tensile force. (*b*) Slipbands appear when the applied stress exceeds the yield stress. Blocks of crystal slide past each other. (*c*) The shaded region of (*b*) has been magnified. Slip occurs on a large number of closely packed slip planes that are parallel. This region is called a *slipband* and appears as a line at lower magnification.
(*After M. Eisenstadt, "Introduction to Mechanical Properties of Materials," Macmillan, 1971, p. 219.*)

Animation

6.5.2 Plastic Deformation in Metal Crystals by the Slip Mechanism

Figure 6.31 shows a possible atomic model for the slippage of one block of atoms over another in a perfect metal crystal. Calculations made from this model determine that the strength of metal crystals should be about 1000 to 10,000 times greater than their observed shear strengths. Thus, this mechanism for atomic slip in large real metal crystals must be incorrect.

In order for large metal crystals to deform at their observed low shear strengths, a high density of crystalline imperfections known as *dislocations* must be present. These dislocations are created in large numbers ($\sim 10^6$ cm/cm^3) as the metal solidifies, and when the metal crystal is deformed, many more are created so that a highly deformed crystal may contain as high as 10^{12} cm/cm^3 of dislocations. Figure 6.32 shows schematically how an *edge dislocation* can produce a unit of slip under a low *shear stress*. A relatively small amount of stress is required for slip by this process since only a small group of atoms slips over each other at any instant.

An analogous situation to the movement of a dislocation in a metal crystal under a shear stress can be envisaged by the movement of a carpet with a ripple in it across a very large floor. Moving the carpet by pulling on one end may be impossible because of the friction between the floor and the carpet. However, by putting a ripple in the carpet (analogous to a dislocation in a metal crystal), the carpet may be moved by pushing the ripple in the carpet one step at a time across the floor (Fig. 6.32*d*).

Dislocations in real crystals can be observed in the transmission electron microscope in thin metal foils and appear as lines due to the atomic disarray at the dislocations that interfere with the transmission path of the electron beam of the microscope. Figure 6.33 shows a cellular wall pattern of dislocations created by lightly deforming an aluminum sample. The cells are relatively free from dislocations but are separated by walls of high dislocation density.

Figure 6.31
Large groups of atoms in large metal crystals do *not* slide over each other simultaneously during plastic shear deformation, as indicated in this figure, since the process requires too much energy. A lower-energy process involving the slippage of a small group of atoms takes place instead.

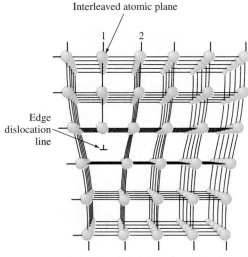

Interleaved atomic plane

Edge
dislocation
line

(a) An edge dislocation, pictured as formed
by an extra half plane of atoms.

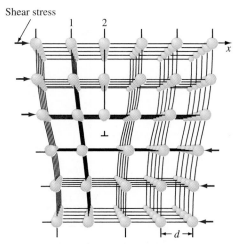

Shear stress

(b) A low stress causes a shift of atomic bonds
to free a new interleaved plane.

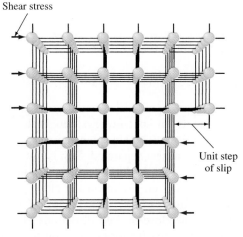

Shear stress

Unit step
of slip

(c) Repetition of this process causes the
dislocation to move across the crystal.

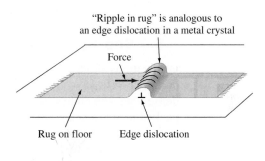

"Ripple in rug" is analogous to
an edge dislocation in a metal crystal

Force

Rug on floor   Edge dislocation

(d)

Figure 6.32
Schematic illustration of how the motion of an edge dislocation produces a unit step of slip under
a low shear stress. (a) An edge dislocation, pictured as formed by an extra half plane of atoms.
(b) A low stress causes a shift of atomic bonds to free a new interleaved plane. (c) Repetition of
this process causes the dislocation to move across the crystal. This process requires less energy
than the one depicted in Fig. 6.30.
(*After A. G. Guy, "Essentials of Materials Science," McGraw-Hill, 1976, p. 153.*)
(d) The "ripple in the rug" analogy. A dislocation moves through a metal crystal during plastic
deformation in a manner similar to a ripple that is pushed along a carpet lying on a floor. In both
cases a small amount of relative movement is caused by the passage of the dislocation or ripple,
and hence a relatively low amount of energy is expended in this process.

Animation

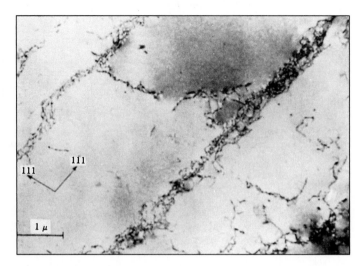

Figure 6.33
Dislocation cell structure in a lightly deformed aluminum
sample as revealed by transmission electron microscopy. The
cells are relatively free from dislocations but are separated by
walls of high dislocation density.
*(After P. R. Swann, in G. Thomas and J. Washburn, [eds.], "Electron
Microscopy and Strength of Crystals," Wiley, 1963, p. 133.)*

6.5.3 Slip Systems

Dislocations produce atomic displacements on specific crystallographic slip planes
and in specific crystallographic slip directions. The slip planes are usually the most
densely packed planes, which are also the farthest separated. Slip is favored on close-
packed planes since a lower shear stress for atomic displacement is required than for
less densely packed planes (Fig. 6.34). However, if slip on the close-packed planes
is restricted due to local high stresses, for example, then planes of lower atomic
packing can become operative. Slip in the close-packed directions is also favored
since less energy is required to move the atoms from one position to another if the
atoms are closer together.

A combination of a slip plane and a slip direction is called a **slip system.** Slip
in metallic structures occurs on a number of slip systems that are characteristic for
each crystal structure. Table 6.3 lists the predominant slip planes and slip directions
for FCC, BCC, and HCP crystal structures.

For metals with the FCC crystal structure, slip takes place on the close-packed
$\{111\}$ octahedral planes and in the $\langle 1\bar{1}0 \rangle$ close-packed directions. There are eight $\{111\}$
octahedral planes in the FCC crystal structure (Fig. 6.35). The (111) type planes at
opposite faces of the octahedron that are parallel to each other are considered the
same type of (111) slip plane. Thus, there are only four different types of $(1\underset{\;}{1}1)$ slip
planes in the FCC crystal structure. Each (111)-type plane contains three $[1\bar{1}0]$-type
slip directions. The reverse directions are not considered different slip directions.

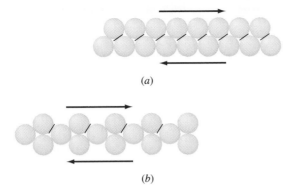

Figure 6.34
Comparison of atomic slip on (*a*) a close-packed plane and (*b*) a non–close-packed plane. Slip is favored on the close-packed plane because less force is required to move the atoms from one position to the next closest one, as indicated by the slopes of the bars on the atoms. Note that dislocations move one atomic slip step at a time.
(After A. H. Cottrell, The Nature of Metals, "Materials," Scientific American, 1967, p. 48. Copyright © by Scientific American, Inc. All rights reserved.)

Thus for the FCC lattice there are 4 slip planes $\times$ 3 slip directions = 12 slip systems (Table 6.3).

The BCC structure is *not* a close-packed structure and does not have a predominant plane of highest atomic packing like the FCC structure. The {110} planes have the highest atomic density, and slip commonly takes place on these planes. However, slip in BCC metals also occurs on {112} and {123} planes. Since the slip planes in the BCC structure are not close-packed as in the case of the FCC structure, higher shear stresses are necessary for slip in BCC than in FCC metals. The slip direction in BCC metals is always of the $\langle 111 \rangle$ type. Since there are six (110)-type slip planes of which each can slip in two [$\bar{1}11$] directions, there are $6 \times 2 = 12$ {110}$\langle \bar{1}11 \rangle$ slip systems.

In the HCP structure, the basal plane (0001) is the closest-packed plane and is the common slip plane for HCP metals such as Zn, Cd, and Mg that have high c/a ratios (Table 6.3). However, for HCP metals such as Ti, Zr, and Be that have low c/a ratios, slip also occurs commonly on prism {10$\bar{1}$0} and pyramidal {10$\bar{1}$1} planes. In all cases the slip direction remains $\langle 11\bar{2}0 \rangle$. The limited number of slip systems in HCP metals restricts their ductilities.

6.5.4 Critical Resolved Shear Stress for Metal Single Crystals

The stress required to cause slip in a pure-metal single crystal depends mainly on the crystal structure of the metal, its atomic bonding characteristics, the temperature at which it is deformed, and the orientation of the active slip planes with respect to the

Table 6.3 Slip systems observed in crystal structures

Structure	Slip plane	Slip direction	Number of slip systems	
FCC: Cu, Al, Ni, Pb, Au, Ag, γFe, ...	{111}	$\langle 1\bar{1}0 \rangle$	$4 \times 3 = 12$	
BCC: αFe, W, Mo, β brass	{110}	$\langle \bar{1}11 \rangle$	$6 \times 2 = 12$	
αFe, Mo, W, Na	{211}	$\langle \bar{1}11 \rangle$	$12 \times 1 = 12$	
αFe, K	{321}	$\langle \bar{1}11 \rangle$	$24 \times 1 = 24$	
HCP: Cd, Zn, Mg, Ti, Be, ...	{0001}	$\langle 11\bar{2}0 \rangle$	$1 \times 3 = 3$	
Ti (prism planes)	{10$\bar{1}$0}	$\langle 11\bar{2}0 \rangle$	$3 \times 1 = 3$	
Ti, Mg (pyramidal planes)	{10$\bar{1}$1}	$\langle 11\bar{2}0 \rangle$	$6 \times 1 = 6$	

Source: After H. W. Hayden, W. G. Moffatt, and J. Wulff, "The Structure and Properties of Materials," vol. III, Wiley, 1965, p. 100.

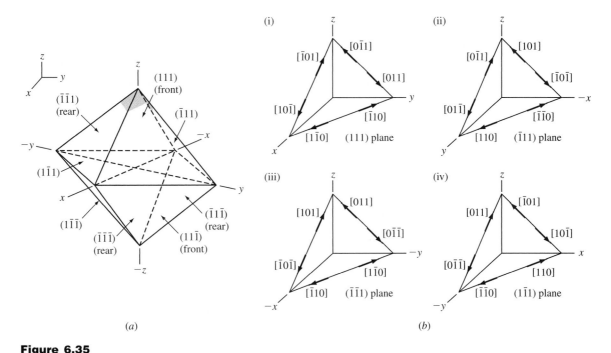

Figure 6.35
Slip planes and directions for the FCC crystal structure. (a) Only four of the eight {111} octahedral planes are considered slip planes since planes opposite each other are considered the same slip plane. (b) For each slip plane there are three ⟨1$\overline{1}$0⟩ slip directions since opposite directions are considered as only one slip direction. Note that slip directions are only shown for the upper four slip planes of the octahedral FCC planes. Thus, there are 4 slip planes × 3 slip directions, giving a total of 12 slip systems for the FCC crystal structure.

shear stresses. Slip begins within the crystal when the shear stress on the slip plane in the slip direction reaches a required level called the *critical resolved shear stress* τ_c. Essentially, this value is the yield stress of a single crystal and is equivalent to the yield stress of a polycrystalline metal or alloy determined by a stress-strain tensile test curve.

Table 6.4 lists values for the critical resolved shear stresses of some pure-metal single crystals at room temperature. The HCP metals Zn, Cd, and Mg have low critical resolved shear stresses ranging from 0.18 to 0.77 MPa. The HCP metal titanium, on the other hand, has a very high τ_c of 13.7 MPa. It is believed that some covalent bonding mixed with metallic bonding is partly responsible for this high value of τ_c. Pure FCC metals such as Ag and Cu have low τ_c values of 0.48 and 0.65 MPa, respectively, because of their multiple slip systems.

6.5.5 Schmid's Law

The relationship between a uniaxial stress acting on a cylinder of a pure metal single crystal and the resulting resolved shear stress produced on a slip system within the cylinder can be derived as follows. Consider a uniaxial tensile stress σ acting on

Table 6.4 Room-temperature slip systems and critical resolved shear stress for metal single crystals

Metal	Crystal structure	Purity (%)	Slip plane	Slip direction	Critical shear stress (MPa)
Zn	HCP	99.999	(0001)	[11$\bar{2}$0]	0.18
Mg	HCP	99.996	(0001)	[11$\bar{2}$0]	0.77
Cd	HCP	99.996	(0001)	[11$\bar{2}$0]	0.58
Ti	HCP	99.99	(10$\bar{1}$0)	[11$\bar{2}$0]	13.7
		99.9	(10$\bar{1}$0)	[11$\bar{2}$0]	90.1
Ag	FCC	99.99	(111)	[1$\bar{1}$0]	0.48
		99.97	(111)	[1$\bar{1}$0]	0.73
		99.93	(111)	[1$\bar{1}$0]	1.3
Cu	FCC	99.999	(111)	[1$\bar{1}$0]	0.65
		99.98	(111)	[1$\bar{1}$0]	0.94
Ni	FCC	99.8	(111)	[1$\bar{1}$0]	5.7
Fe	BCC	99.96	(110)	[$\bar{1}$11]	27.5
			(112)		
			(123)		
Mo	BCC	. . .	(110)	[$\bar{1}$11]	49.0

Source: After G. Dieter, "Mechanical Metallurgy," 2nd ed., McGraw-Hill, 1976, p. 129.

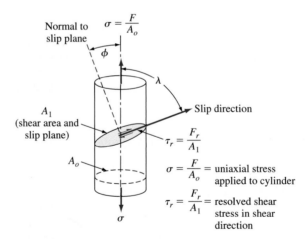

$$\sigma = \frac{F}{A_o}$$

Normal to slip plane

ϕ

λ

Slip direction

A_1 (shear area and slip plane)

A_o

$$\tau_r = \frac{F_r}{A_1}$$

$$\sigma = \frac{F}{A_o} = \text{uniaxial stress applied to cylinder}$$

$$\tau_r = \frac{F_r}{A_1} = \text{resolved shear stress in shear direction}$$

σ

Figure 6.36

Axial stress σ can produce a resolved shear stress τ_r and cause dislocation motion in slip plane A_1 in the slip direction.

a metal cylinder, as shown in Fig. 6.36. Let A_0 be the area normal to the axial force F, and A_1 the area of the slip plane or shear area on which the resolved shear force F_r is acting. We can orient the slip plane and slip direction by defining the angles ϕ and λ. ϕ is the angle between the uniaxial force F and the normal to the slip plane area A_1, and λ is the angle between the axial force and the slip direction.

In order for dislocations to move in the slip system, a sufficient resolved shear stress acting in the slip direction must be produced by the applied axial force. The resolved shear stress is

$$\tau_r = \frac{\text{shear force}}{\text{shear area (slip plane area)}} = \frac{F_r}{A_1} \tag{6.14}$$

The resolved shear force F_r is related to the axial force F by $F_r = F \cos \lambda$. The area of the slip plane (shear area) $A_1 = A_0/\cos \phi$. By dividing the shear force $F \cos \lambda$ by the shear area $A_0/\cos \phi$, we obtain

$$\tau_r = \frac{F \cos \lambda}{A_0/\cos \phi} = \frac{F}{A_0} \cos \lambda \cos \phi = \sigma \cos \lambda \cos \phi \tag{6.15}$$

which is called *Schmid's law*. Let us now consider an example problem to calculate the resolved shear stress when a slip system is acted upon by an axial stress.

EXAMPLE PROBLEM 6.9

Calculate the resolved shear stress on the (111) [0$\bar{1}$1] slip system of a unit cell in an FCC nickel single crystal if a stress of 13.7 MPa is applied in the [001] direction of a unit cell.

■ **Solution**
By geometry the angle λ between the applied stress and the slip direction is 45°, as shown in Fig. EP6.9a. In the cubic system the direction indices of the normal to a crystal plane are the same as the Miller indices of the crystal plane. Therefore, the normal to the (111)

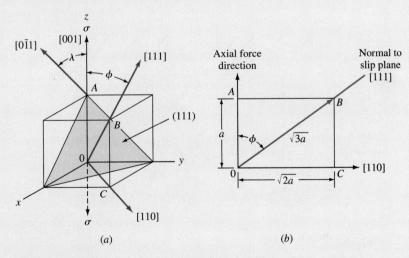

(a)   *(b)*

Figure EP6.9
An FCC unit cell is acted upon by a [001] tensile stress producing a resolved shear stress on the (111) [0$\bar{1}$1] slip system.

plane that is the slip plane is the [111] direction. From Fig. EP6.9*b*,

$$\cos \phi = \frac{a}{\sqrt{3}a} = \frac{1}{\sqrt{3}} \quad \text{or} \quad \phi = 54.74°$$

$$\tau_r = \sigma \cos \lambda \cos \phi = (13.7 \text{ MPa})(\cos 45°)(\cos 54.74°) = 5.6 \text{ MPa} \blacktriangleleft$$

6.5.6 Twinning

A second important plastic deformation mechanism that can occur in metals is **twinning.** In this process a part of the atomic lattice is deformed so that it forms a mirror image of the undeformed lattice next to it (Fig. 6.37). The crystallographic plane of symmetry between the undeformed and deformed parts of the metal lattice is called the *twinning plane.* Twinning, like slip, occurs in a specific direction called the *twinning direction.* However, in slip the atoms on one side of the slip plane all move equal distances (Fig. 6.32), whereas in twinning the atoms move distances proportional to their distance from the twinning plane (Fig. 6.37). Figure 6.38 illustrates the basic difference between slip and twinning on the surface of a metal after deformation. Slip leaves a series of steps (lines) (Fig. 6.38*a*), whereas twinning leaves small but well-defined regions of the crystal deformed (Fig. 6.38*b*). Figure 6.39 shows some deformation twins on the surface of titanium metal.

Virtual Lab

Twinning only involves a small fraction of the total volume of the metal crystal, and so the amount of overall deformation that can be produced by twinning is small. However, the important role of twinning in deformation is that the lattice

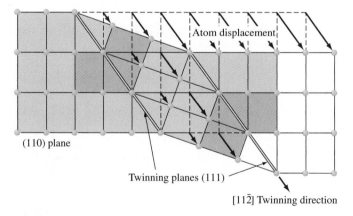

Figure 6.37
Schematic diagram of the twinning process in an FCC lattice.
(After H. W. Hayden, W. G. Moffatt, and J. Wulff, "The Structure and Properties of Materials," vol. III, Wiley, 1965, p. 111.)

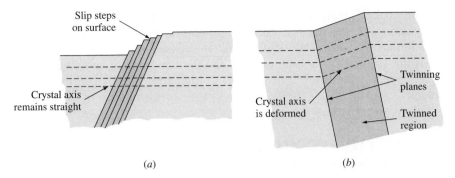

(a)   (b)

Figure 6.38
Schematic diagram of surfaces of a deformed metal after (a) slip and
(b) twinning.

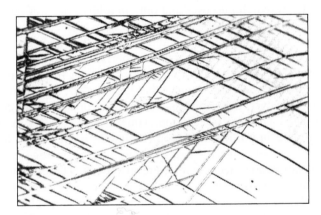

Figure 6.39
Deformation twins in unalloyed (99.77%) titanium.
(Magnification 150×.)
[*After F. D. Rosi, C. A. Dube, and B. H. Alexander, Trans. AIME,*
197:*259 (1953).*]

orientation changes that are caused by twinning may place new slip systems into
favorable orientation with respect to the shear stress and thus enable additional slip
to occur. Of the three major metallic unit-cell structures (BCC, FCC, and HCP),
twinning is most important for the HCP structure because of its small number of
slip systems. However, even with the assistance of twinning, HCP metals like zinc
and magnesium are still less ductile than the BCC and FCC metals that have more
slip systems.

Deformation twinning is observed at room temperature for the HCP metals.
Twinning is found in the BCC metals such as Fe, Mo, W, Ta, and Cr in crystals
that were deformed at very low temperatures. Twinning has also been found in

some of these BCC metal crystals at room temperature when they have been subjected to very high strain rates. The FCC metals show the least tendency to form deformation twins. However, deformation twins can be produced in some FCC metals if the stress level is high enough and the temperature sufficiently low. For example, copper crystals deformed at 4 K at high stress levels can form deformation twins.

6.6 PLASTIC DEFORMATION OF POLYCRYSTALLINE METALS

6.6.1 Effect of Grain Boundaries on the Strength of Metals

Almost all engineering alloys are polycrystalline. Single-crystal metals and alloys are used mainly for research purposes and only in a few cases for engineering applications.[8] Grain boundaries strengthen metals and alloys by acting as barriers to dislocation movement except at high temperatures, where they become regions of weakness. For most applications where strength is important, a fine grain size is desirable, and so most metals are fabricated with a fine grain size. In general, at room temperature, fine-grained metals are stronger, harder, tougher, and more susceptible to strain hardening. However, they are less resistant to corrosion and creep (deformation under constant load at elevated temperatures; see Sec. 7.4). A fine grain size also results in a more uniform and isotropic behavior of materials. In Sec. 4.5, the ASTM grain size number and a method to determine the average grain diameter of a metal using metallography techniques were discussed. These parameters allow us to make a relative comparison of grain density and therefore grain boundary density in metals. Accordingly, for two components made of the same alloy, the component that has a larger ASTM grain size number or a smaller average grain diameter is stronger. The relationship between strength and grain size is of great importance to engineers. The well known *Hall–Petch equation,* Eq. 6.16, is an empirical (based on experimental measurements and not on theory) equation that relates the yield strength of a metal σ_y to its average grain diameter d as follows:

$$\sigma_y = \sigma_0 + k/(d)^{1/2} \qquad \textbf{(6.16)}$$

where σ_0 and k are constants related to the material of interest. A similar effect exists between hardness (Vickers microhardness test) and grain size. The equation clearly shows that as grain diameter decreases, the yield strength of the material increases. Considering that the conventional grain diameters may range from a few hundred microns to a few microns, one may expect a significant change in strength

---

[8]Single-crystal turbine blades have been developed for use in gas turbine engines to avoid grain boundary cracking at high temperatures and stresses. See F. L. Ver Snyder and M. E. Shank, *Mater. Sci. Eng.,* **6**:213–247(1970).

Table 6.5 Hall–Petch relationship constants for selected materials

	σ_0 (MPa)	k (MPa $\cdot$ m$^{1/2}$)
Cu	25	0.11
Ti	80	0.40
Mild steel	70	0.74
Ni$_3$Al	300	1.70

Source: www.tf.uni-kiel.de/matwis/matv/pdf/chap_3_3.pdf

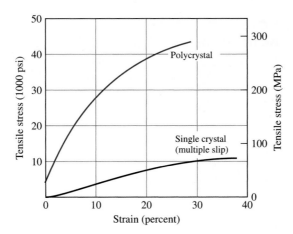

Figure 6.40
Stress-strain curves for single-crystal and polycrystalline copper. The single crystal is oriented for multiple slip. The polycrystal shows higher strength at all strains.
(*After M. Eisenstadt, "Introduction to Mechanical Properties of Materials," Macmillan, 1971, p. 258.*)

through grain refinement. The values for σ_0 and k for selected materials are given in Table 6.5. It is important to note that the Hall–Petch equation does not apply to (1) extremely coarse or extremely fine grain sizes and (2) metals used at elevated temperatures.

Figure 6.40 compares the tensile stress-strain curves for single-crystal and polycrystalline unalloyed copper at room temperature. At all strains the polycrystalline copper is stronger than the single-crystal copper. At 20 percent strain the tensile strength of the polycrystalline copper is 40 ksi (276 MPa) as compared to 8 ksi (55 MPa) for single-crystal copper.

During the plastic deformation of metals, dislocations moving along on a particular slip plane cannot go directly from one grain into another in a straight line.

Figure 6.41
Polycrystalline aluminum that has been plastically deformed. Note that
the slipbands are parallel within a grain but are discontinuous across
the grain boundaries. (Magnification 60×.)
(After G. C. Smith, S. Charter, and S. Chiderley of Cambridge University.)

As shown in Fig. 6.41, slip lines change directions at grain boundaries. Thus, each
grain has its own set of dislocations on its own preferred slip planes, which have
different orientations from those of neighboring grains. As the number of grains
increases and grain diameter becomes smaller, dislocations within each grain can
travel a smaller distance before they encounter the grain boundary, at which point
their movement is terminated (dislocation pileup). It is for this reason that fine-
grained materials possess a higher strength. Figure 6.42 shows clearly a high-angle
grain boundary that is acting as a barrier to dislocation movement and has caused
dislocations to pile up at the grain boundary.

6.6.2 Effect of Plastic Deformation on Grain Shape and Dislocation Arrangements

Grain Shape Changes with Plastic Deformation   Let us consider the plastic
deformation of annealed samples[9] of unalloyed copper that have an equiaxed grain
structure. Upon cold plastic deformation the grains are sheared relative to each

---

[9]Samples in the annealed conditions have been plastically deformed and then reheated to such an extent that
a grain structure in which the grains are approximately equal in all directions (equiaxed) is produced.

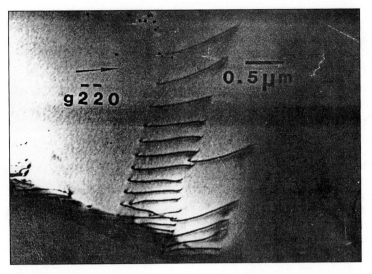

Figure 6.42
Dislocations piled up against a grain boundary as observed with a
transmission electron microscope in a thin foil of stainless steel.
(Magnification 20,000×.)
[*After Z. Shen, R. H. Wagoner, and W. A. T. Clark, Scripta Met., 20: 926 (1986).*]

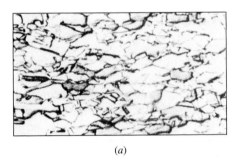

(a)

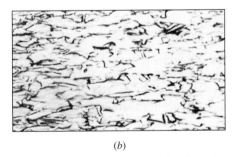

(b)

Figure 6.43
Optical micrographs of deformation structures of unalloyed copper that was cold-rolled
to reductions of (*a*) 30 percent and (*b*) 50 percent. (Etch: potassium dichromate;
magnification 300×.)
(*After J. E. Boyd in "Metals Handbook," vol. 8: "Metallography, Structures, and Phase Diagrams," 8th ed.,
American Society for Metals, 1973, p. 221. ASM International*)

other by the generation, movement, and rearrangement of dislocations. Figure 6.43
shows the microstructures of samples of unalloyed copper sheet that was cold-
rolled to reductions of 30 and 50 percent. Note that with increased cold rolling the
grains are more elongated in the rolling direction as a consequence of dislocation
movements.

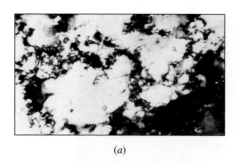
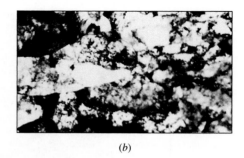

(a)   (b)

Figure 6.44
Transmission electron micrographs of deformation structures of unalloyed copper that
was cold-rolled to reductions of (a) 30 percent and (b) 50 percent. Note that these
electron micrographs correspond to the optical micrographs of Fig. 6.43. (Thin-foil
specimens, magnification 30,000×.)
(After J. E. Boyd in "Metals Handbook," vol. 8: "Metallography, Structures, and Phase Diagrams," 8th ed.,
American Society for Metals, 1973, p. 221. ASM International)

Dislocation Arrangement Changes with Plastic Deformation   The dislocations in
the unalloyed copper sample after 30 percent plastic deformation form cell-like configurations with clear areas in the centers of the cells (Fig. 6.44a). With increased
cold plastic deformation to 50 percent reduction, the cell structure becomes denser
and elongated in the direction of rolling (Fig. 6.44b).

6.6.3 Effect of Cold Plastic Deformation on Increasing the Strength of Metals

As shown by the electron micrographs of Fig. 6.44, the dislocation density increases
with increased cold deformation. The exact mechanism by which the dislocation density is increased by cold working is not completely understood. New dislocations are
created by the cold deformation and must interact with those already existing. As
the dislocation density increases with deformation, it becomes more and more difficult for the dislocations to move through the existing "forest of dislocations," and
thus the metal work or strain hardens with increased cold deformation.

When ductile metals such as copper, aluminum, and α iron that have been
annealed are cold-worked at room temperature, they strain-harden because of the dislocation interaction just described. Figure 6.45 shows how cold working at room
temperature increases the tensile strength of unalloyed copper from about 30 ksi
(200 MPa) to 45 ksi (320 MPa) with 30 percent cold work. Associated with the increase
in tensile strength, however, is a decrease in elongation (ductility), as observed in
Fig. 6.45. With 30 percent cold work, the elongation of unalloyed copper decreases
from about 52 to 10 percent elongation.

Cold working or **strain hardening** is one of the most important methods for
strengthening some metals. For example, pure copper and aluminum can be strengthened significantly only by this method. Thus, cold-drawn unalloyed copper wire can
be produced with different strengths (within certain limitations) by varying the
amount of strain hardening.

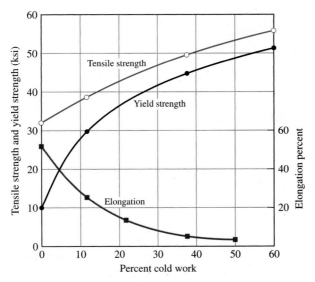

Figure 6.45
Percent cold work versus tensile strength and elongation for unalloyed oxygen-free copper. Cold work is expressed as a percent reduction in cross-sectional area of the metal being reduced.

We wish to produce a 0.040-in.-thick sheet of oxygen-free copper with a tensile strength of 45 ksi. What percent cold work must the metal be given? What must the starting thickness of the metal be before cold rolling?

**EXAMPLE
PROBLEM 6.10**

■ **Solution**
From Fig. 6.45 the percent cold work must be 25 percent. Thus, the starting thickness must be

$$\frac{x - 0.040 \text{ in.}}{x} = 0.25$$

$$x = 0.053 \text{ in.} \blacktriangleleft$$

6.7 SOLID-SOLUTION STRENGTHENING OF METALS

Another method besides cold working by which the strength of metals can be increased is by **solid-solution strengthening.** The addition of one or more elements to a metal can strengthen it by the formation of a solid solution. The structure of *substitutional* and *interstitial solid solutions* has already been discussed in Sec. 4.3 and should be referred to for review. When substitutional (solute) atoms are mixed in the solid state with those of another metal (solvent), stress fields are created around each solute atom. These stress fields interact with dislocations and make their movement more difficult, and thus the solid solution becomes stronger than the pure metal.

Two important factors in solid-solution strengthening are:

1. *Relative-size factor.* Differences in atomic size of solute and solvent atoms affect the amount of solid-solution strengthening because of the crystal lattice distortions produced. Lattice distortions make dislocation movement more difficult and hence strengthen the metallic solid solution.

2. *Short-range order.* Solid solutions are rarely random in atomic mixing, and some kind of short-range order or clustering of like atoms takes place. As a result, dislocation movement is impeded by different bonding structures.

In addition to these factors there are others that also contribute to solid-solution strengthening but will not be dealt with in this book.

As an example of solid-solution strengthening, let us consider a solid solution alloy of 70 wt % Cu and 30 wt % Zn (cartridge brass). The tensile strength of unalloyed copper with 30 percent cold work is about 48 ksi (330 MPa) (Fig. 6.45). However, the tensile strength of the 70 wt % Cu–30 wt % Zn alloy with 30 percent cold work is about 72 ksi (500 MPa) (Fig. 6.46). Thus, solid-solution strengthening in this case produced an increase in strength in the copper of about 24 ksi (165 MPa). On the other hand, the ductility of the copper by the 30 percent zinc addition after 30 percent cold work was reduced from about 65 to 10 percent (Fig. 6.46).

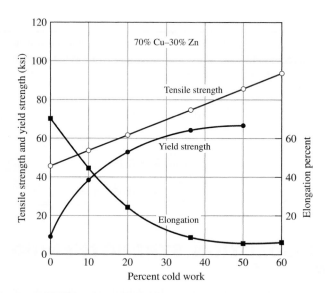

Figure 6.46
Percent cold work versus tensile strength and elongation for 70 wt % Cu–30 wt % Zn alloy. Cold work is expressed as a percent reduction in cross-sectional area of the metal being reduced (see Eq. 6.2).

6.8 RECOVERY AND RECRYSTALLIZATION OF PLASTICALLY DEFORMED METALS

In previous sections the effect of plastic deformation on the mechanical properties and microstructural features of metals was discussed. When metal-forming processes such as rolling, forging, extrusion, and others are performed cold, the work material has many dislocations and other defects, and the grains are stretched and deformed; as a result, the worked metal is significantly stronger but less ductile. Many times the reduced ductility of the cold-worked metal is undesirable, and a softer metal is required. To achieve this, the cold-worked metal is heated in a furnace. If the metal is reheated to a sufficiently high temperature for a long enough time, the cold-worked metal structure will go through a series of changes called (1) **recovery,** (2) **recrystallization,** and (3) **grain growth.** Figure 6.47 shows these structural changes schematically as the temperature of the metal is increased along with the corresponding changes in mechanical properties. This reheating treatment that softens a cold-worked metal is called **annealing,** and the terms *partial anneal* and *full anneal* are often used to refer to degrees of softening. Let us now examine these structural changes in more detail, starting with the heavily cold-worked metal structure.

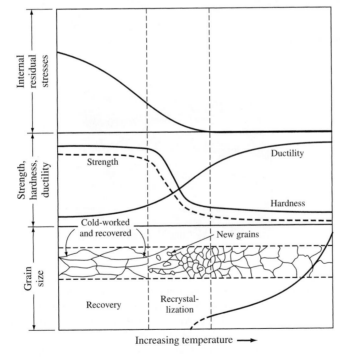

Figure 6.47

Effect of annealing on the structure and mechanical property changes of a cold-worked metal.

(Adapted from Z. D. Jastrzebski, "The Nature and Properties of Engineering Materials," 2nd ed., Wiley, 1976, p. 228.)

6.8.1 Structure of a Heavily Cold-Worked Metal before Reheating

When a metal is heavily cold-worked, much of the strain energy expended in the plastic deformation is stored in the metal in the form of dislocations and other imperfections such as point defects. Thus a strain-hardened metal has a higher internal energy than an unstrained one. Figure 6.48a shows the microstructure (100×) of an Al−0.8% Mg alloy sheet that has been cold-worked with 85 percent reduction. Note that the grains are greatly elongated in the rolling direction. At higher magnification (20,000×) a thin-foil transmission electron micrograph

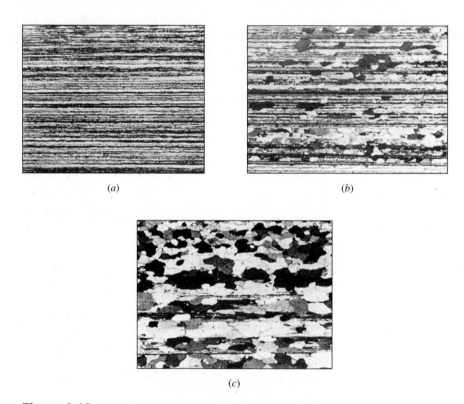

(a)

(b)

(c)

Figure 6.48
Aluminum alloy 5657 (0.8% Mg) sheet showing microstructures after cold rolling 85 percent and subsequent reheating (optical micrographs at 100× viewed under polarized light). (a) Cold-worked 85 percent; longitudinal section. Grains are greatly elongated. (b) Cold-worked 85 percent and stress-relieved at 302°C (575°F) for 1 h. Structure shows onset of recrystallization, which improves the formability of the sheet. (c) Cold-worked 85 percent and annealed at 316°C (600°F) for 1 h. Structure shows recrystallized grains and bands of unrecrystallized grains.
(After "Metals Handbook," vol. 7, 8th ed., American Society for Metals, 1972, p. 243. ASM International)

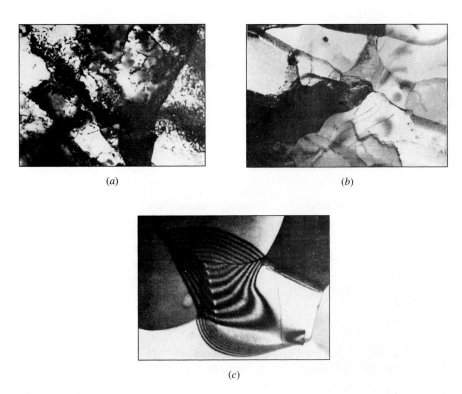

(a)　　(b)

(c)

Figure 6.49
Aluminum alloy 5657 (0.8% Mg) sheet showing microstructures after cold
rolling 85 percent and subsequent reheating. The microstructures shown in
this figure were obtained by using thin-foil transmission electron microscopy.
(Magnified 20,000×.) (a) Sheet was cold-worked 85 percent; micrograph
shows dislocation tangles and banded cells (subgrains) caused by cold
working extensively. (b) Sheet was cold-worked 85 percent and subsequently
stress-relieved at 302°C (575°F) for 1 h. Micrograph shows dislocation
networks and other low-angle boundaries produced by polygonization.
(c) Sheet was cold-worked 85 percent and annealed at 316°C (600°F) for 1 h.
Micrograph shows recrystallized structure and some subgrain growth.
(After "Metals Handbook," vol. 7, 8th ed., American Society for Metals, 1972, p. 243.
ASM International)

(Fig. 6.49) shows the structure to consist of a cellular network with cell walls of
high dislocation density. A fully cold-worked metal has a density of approximately 10^{12} dislocation lines/cm^2.

6.8.2 Recovery

When a cold-worked metal is heated in the recovery temperature range that is just
below the recrystallization temperature range, internal stresses in the metal are
relieved (Fig. 6.47). During recovery, sufficient thermal energy is supplied to allow

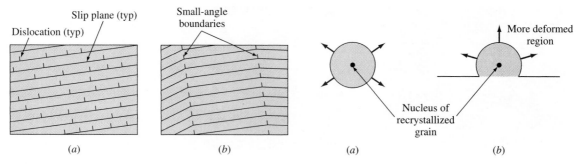

Figure 6.50
Schematic representation of polygonization in a deformed metal. (*a*) Deformed metal crystal showing dislocations piled up on slip planes. (*b*) After recovery heat treatment, dislocations move to form small-angle grain boundaries.
(After L. E. Tanner and I. S. Servi, in "Metals Handbook," vol. 8, 8th ed., American Society for Metals, 1973, p. 222.)

Figure 6.51
Schematic model of the growth of a recrystallized grain during the recrystallization of a metal. (*a*) Isolated nucleus expanded by growth within a deformed grain. (*b*) Original high-angle grain boundary migrating into a more highly deformed region of metal.

the dislocations to rearrange themselves into lower-energy configurations (Fig. 6.50). Recovery of many cold-worked metals (such as pure aluminum) produces a subgrain structure with low-angle grain boundaries, as shown in Fig. 6.49*b*. This recovery process is called *polygonization,* and often it is a structural change that precedes recrystallization. The internal energy of the recovered metal is lower than that of the cold-worked state since many dislocations are annihilated or moved into lower-energy configurations by the recovery process. During recovery the strength of a cold-worked metal is reduced only slightly but its ductility is usually significantly increased (Fig. 6.47).

6.8.3 Recrystallization

Upon heating a cold-worked metal to a sufficiently high temperature, new strain-free grains are nucleated in the recovered metal structure and begin to grow (Fig. 6.48*b*), forming a recrystallized structure. After a long enough time at a temperature at which recrystallization takes place, the cold-worked structure is completely replaced with a recrystallized grain structure, as shown in Fig. 6.48*c*.

Primary recrystallization occurs by two principal mechanisms: (1) an isolated nucleus can expand with a deformed grain (Fig. 6.51*a*) or (2) an original high-angle grain boundary can migrate into a more highly deformed region of the metal (Fig. 6.51*b*). In either case, the structure on the concave side of the moving boundary is strain-free and has a relatively low internal energy, whereas the structure on the convex side of the moving interface is highly strained with a high dislocation density and high internal energy. Grain boundary movement is therefore away from the boundary's center of curvature. Thus, the growth of an expanding new grain during primary recrystallization leads to an overall decrease in the internal energy of the metal by replacing deformed regions with strain-free regions.

The tensile strength of a cold-worked metal is greatly decreased and its ductility increased by an annealing treatment that causes the metal structure to be recrystallized. For example, the tensile strength of a 0.040-in. (1-mm) sheet of 85% Cu–15% Zn brass that had been cold-rolled to 50 percent reduction was decreased from 75 to 45 ksi (520 to 310 MPa) by annealing 1 h at 400°C (Fig. 6.52*a*). The

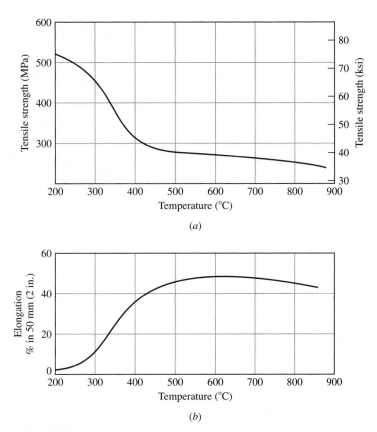

(*a*)

(*b*)

Figure 6.52
Effect of annealing temperature on (*a*) the tensile strength and
(*b*) elongation of a 50 percent cold-rolled 85% Cu–15% Zn,
0.040-in. (1 mm) thick sheet. (Annealing time was 1 h at
temperature.)
(*After "Metals Handbook," vol. 2, 9th ed., American Society for Metals,
1979, p. 320.*)

Figure 6.53a
Furnaces for annealing coils of sheet steel. The coils are placed under cylindrical covers and then a furnace top is placed over the covered coils. In this box-annealing process the coils are held at a temperature of 1200 to 1300°F (650 to 700°C) for an average of 26 h. During the cooling period, a controlled deoxidizing atmosphere is provided to protect the surface of the steel coils. (*Courtesy of United States Steel Corporation.*)

ductility of the sheet, on the other hand, was increased from 3 to 38 percent with the annealing treatment (Fig. 6.52b). Figure 6.53a shows a photo of the box annealing process for coils of steel (a discontinuous process). Figure 6.53b shows a schematic diagram of a continuous annealing process for sheet steel, and Fig. 6.53c shows a photo of a continuous annealing process line.

Important factors that affect the recrystallization process in metals and alloys are (1) amount of prior deformation of the metal, (2) temperature, (3) time, (4) initial grain size, and (5) composition of the metal or alloy. The recrystallization of a metal can take place over a range of temperatures, and the range is dependent to some extent on the variables just listed. Thus, one cannot refer to the recrystallization temperature of a metal in the same sense as the melting

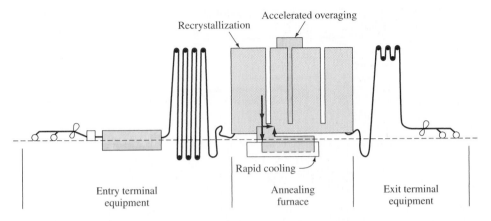

Recrystallization Accelerated overaging

Rapid cooling

Entry terminal
equipment

Annealing
furnace

Exit terminal
equipment

Figure 6.53*b*
Continuous annealing schematic diagram.
(After W. L. Roberts, "Flat Processing of Steel," Marcel Dekker, 1988.)

Figure 6.53*c*
Photo of continuous annealing line for low-carbon sheet steel.
(Courtesy of Bethlehem Steel Co.)

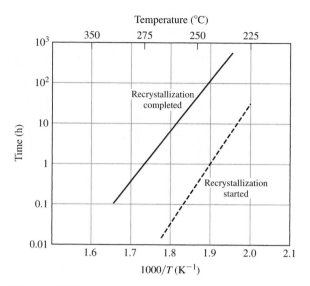

Figure 6.54

Time-temperature relations for the recrystallization of 99.0% Al cold-worked 75 percent. The solid line is for recrystallization finished and the dashed line for recrystallization started. Recrystallization in this alloy follows an Arrhenius-type relationship of ln t versus $1/T(\text{K}^{-1})$ (see Sec. 5.2).
(After "Aluminum," vol. 1, American Society for Metals, 1967, p. 98.)

temperature of a pure metal. The following generalizations can be made about the recrystallization process:

1. A minimum amount of deformation of the metal is necessary for recrystallization to be possible.

2. The smaller the degree of deformation (above the minimum), the higher the temperature needed to cause recrystallization.

3. Increasing the temperature for recrystallization decreases the time necessary to complete it (see Fig. 6.54).

4. The final grain size depends mainly on the degree of deformation. The greater the degree of deformation, the lower the annealing temperature for recrystallization and the smaller the recrystallized grain size.

5. The larger the original grain size, the greater the amount of deformation required to produce an equivalent recrystallization temperature.

6. The recrystallization temperature decreases with increasing purity of the metal. Solid-solution alloying additions always increase the recrystallization temperature.

If it takes 9.0×10^3 min to recrystallize a piece of copper at 88°C and 200 min at 135°C, what is the activation energy for the process, assuming the process obeys the Arrhenius rate equation and the time to recrystallize $= Ce^{-Q/RT}$, where $R = 8.314$ J/(mol · K) and T is in kelvins?

■ **Solution**

$$t_1 = 9.0 \times 10^3 \text{ min}; T_1 = 88°C + 273 = 361 \text{ K}$$

$$t_2 = 200 \text{ min}; T_2 = 135°C + 273 = 408 \text{ K}$$

$$t_1 = Ce^{Q/RT_1} \quad \text{or} \quad 9.0 \times 10^3 \text{ mm} = Ce^{Q/R(361 \text{ K})} \quad \text{(6.16)}$$

$$t_2 = Ce^{Q/RT_2} \quad \text{or} \quad 200 \text{ min} \qquad = Ce^{Q/R(408 \text{ K})} \quad \text{(6.17)}$$

Dividing Eq. 6.16 by 6.17 gives

$$45 = \exp\left[\frac{Q}{8.314}\left(\frac{1}{361} - \frac{1}{408}\right)\right]$$

$$\ln 45 = \frac{Q}{8.314}(0.00277 - 0.00245) = 3.80$$

$$Q = \frac{3.80 \times 8.314}{0.000319} = 99{,}038 \text{ J/mol or } 99.0 \text{ kJ/mol} \blacktriangleleft$$

6.9 SUPERPLASTICITY IN METALS

A careful examination of Fig. 6.25 shows that most metals, even those that are classified as ductile, undergo a limited amount of plastic deformation prior to fracture. For example, mild steel undergoes 22 percent elongation before fracture in uniaxial tensile tests. As discussed in Sec. 6.1, many metal-forming operations are performed at elevated temperatures in order to achieve a higher degree of plastic deformation by increasing the ductility of the metal. **Superplasticity** refers to the ability of some metal alloys, such as some aluminum and titanium alloys, to deform as much as 2000 percent at elevated temperatures and slow loading rates. These alloys do not behave superplastically when loaded at normal temperatures. For example, annealed Ti alloy (6Al-4V) elongates close to 12 percent prior to fracture in a conventional tensile test at room temperature. The same alloy when tested at elevated temperatures (840 to 870°C) and at very low loading rates ($1.3 \times 10^{-4}\text{ s}^{-1}$) can elongate as much as 750 to 1170 percent. To achieve superplasticity, the material and the loading process must meet certain conditions:

1. The material must possess very fine grain size (5–10 μm) and be highly strain-rate sensitive.
2. A high loading temperature greater that 50 percent of the melt temperature of the metal is required.
3. A low and controlled strain rate in the range of 0.01 to 0.0001 s^{-1} is required.[10]

---

[10]High strain rate ($>10^{-2}\text{ s}^{-1}$) superplasticity has been reported in some aluminum alloys.

These requirements are not easily achievable for all materials and therefore not all materials can attain superplastic behavior. In most cases Condition (1) is very difficult to achieve, i.e., ultrafine grain size.[11]

Superplastic behavior is an extremely useful property that can be used to manufacture complex structural components. The question is, "What is the deformation mechanism that accounts for this incredible level of plastic deformation?" In previous sections we discussed the role of dislocations and their movements on the plastic behavior of materials under room temperature loading. As dislocations move across the grain, plastic deformation is produced. But as grain size decreases, the movement of dislocations becomes more limited and the material becomes stronger. However, metallographical analysis of materials that undergo superplastic behavior has revealed very limited dislocation activity inside the grain. This supports the fact that superplastic materials are susceptible to other deformation mechanisms such as grain boundary sliding and grain boundary diffusion. At elevated temperatures, a large amount of strain is believed to be accumulated by the sliding and rotation of individual grains or clusters of grains relative to each other. There is also a belief that grain boundary sliding is accommodated by a gradual change in grain shape as matter is moved by diffusion across the grain boundary. Figure 6.55 shows the

---

[11]Static and dynamic recrystallization, mechanical alloying, and other techniques are used to create ultrafine grain structure.

(a)   (b)

Figure 6.55
Superplastic deformation in Pb–Sn eutectic alloy (a) before and (b) after deformation.

(a)

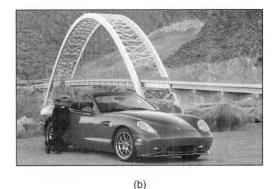

(b)

Figure 6.56
The hood for the automobile made of superplastic aluminum using the blow-molding method.
(*Courtesy of Panoz Auto.*)

microstructure of Pb–Sn eutectic alloy before (Fig. 6.55*a*) and after superplastic deformation (Fig. 6.55*b*). It is clear from the figure that grains are equiaxed before and after the deformation; sliding and rotation of the grains are noticeable.

Many manufacturing processes exist that take advantage of the superplastic behavior of materials to produce complex components. Blow forming is one such process in which a superplastic material is forced under gas pressure to deform and take the shape of a die. Figure 6.56 shows the hood of the automobile that is made from superplastically formed aluminum alloy using the blow-forming method. Also, superplastic behavior can be combined with diffusion bonding (a metal-joining method) to produce structural components with limited material waste.

6.10 NANOCRYSTALLINE METALS

In Chap. 1, the concept of nanotechnology and nanostructured materials was introduced. Any material with length scale features below 100 nm is classified as *nanostructured.* According to this definition, all metals with average grain diameters less than 100 nm are considered nanostructured or nanocrystalline. The question is, "What are the advantages of **nanocrystalline metals**?" Metallurgists have always been aware that by reducing the grain size, a harder, stronger, and tougher metal may be produced as evidenced by the Hall–Petch equation (6.16). As discussed in the previous section, it has also been known that at ultrafine grain size levels (not necessarily nanocrystalline) and under certain temperature and loading rate conditions, some materials can deform plastically to many times their conventional levels, i.e., they achieve superplasticity.

Noting the characteristics attributed to ultrafine grain sizes and extrapolating the Hall–Petch equation for nanocrystalline metals, one can foresee some extraordinary possibilities. Consider the possibility that, according to the Hall–Petch equation, if the average grain diameter of a metal decreases from 10 μm to 10 nm, its yield strength will increase by a factor of 31. Is this possible? How do nanograins affect the ductility, toughness, fatigue, and creep behavior of metals? How can we produce bulk metals with nanocrystalline structure? These questions and others like them are the driving force in research and development in the field of nanocrystalline metals. Therefore, at least in metal-manufacturing industries, the potential for improved properties with reduced grain size or dispersing secondary nanophases has been known for many decades. The difficulty has been in developing metal-forming techniques that can produce truly nanocrystalline ($d < 100$ nm) metals. In recent decades new techniques for producing such materials have been developed and older techniques have been improved. Thus, those studying these materials are enthusiastic. It should be mentioned, however, that even after decades of research many questions still remain and significant more research has to be performed to process, characterize, and evaluate such materials for practical applications.

It has been reported that the modulus of elasticity of nanocrystalline materials is comparable to bulk microcrystalline materials for grain sizes exceeding 5 nm. For d less than 5 nm, a decrease in the modulus of elasticity of metals such as nanocrystalline iron has been reported. It is not completely clear why such a drop in the modulus of elasticity occurs; one may find a reason by considering that for such small grains, the majority of the atoms are positioned on the surface of the grain (as opposed to inside the grain) and therefore along the grain boundary. This is completely opposite to what is found in microcrystalline materials.

As discussed previously, the hardness and strength of materials increase with a decrease in grain size. This increase in hardness and strength is due to dislocation pileup and the hindrance of dislocation motion for conventional grains. For nanocrystalline materials most available data are based on hardness values obtained from nanohardness tests. This is mostly due to the fact that tensile specimens with nanocrystalline structure are hard to produce. But since strength and hardness are closely correlated, nanohardness tests are acceptable at this time. It has been established that as grain size decreases to around 10 nm, the hardness increases by factors of four to six for nanocrystalline copper and six to eight for nanocrystalline nickel when compared to large grained metals ($d > 1\ \mu$m). Although this is an impressive increase, it still falls drastically short of the prediction made by the Hall–Petch equation. Additionally, there are published data that indicate a "negative Hall–Petch effect" at the finest grain size ($d < 30$ nm), indicating that a softening mechanism is at work. Some researchers believe that it is entirely possible that at such small grain levels, the concept of a moving dislocation or dislocation pileup is no longer applicable and other mechanisms, such as grain boundary sliding, diffusion, etc., are at work.

Arguments have been made that in the upper nanocrystalline range (50 nm $<$ $d < 100$ nm), dislocation-related activities similar to those seen with microcrystalline metals dominate while in the lower nanocrystalline range ($d < 50$ nm) dislocation activity (formation and movement) decreases significantly. Stresses needed to activate dislocation sources are extremely high at such small grain sizes. Some in situ HRTEM studies have been performed that support this argument. Finally, the strengthening and deformation mechanisms of nanocrystalline materials are not yet well understood, and more theoretical and experimental research is needed. In the next chapter the ductility and toughness characteristics of these materials will be discussed.

6.11 SUMMARY

Metals and alloys are processed into different shapes by various manufacturing methods. Some of the most important industrial processes are casting, rolling, extruding, wire drawing, forging, and deep drawing.

When a uniaxial stress is applied to a long metal bar, the metal deforms elastically at first and then plastically, causing permanent deformation. For many engineering designs the engineer is interested in the 0.2 percent offset yield strength, ultimate tensile strength, and elongation (ductility) of a metal or alloy. These quantities are obtained from the engineering stress-strain diagram originating from a tensile test. The hardness of a metal may also be of engineering importance. Commonly used hardness scales in industry are Rockwell B and C and Brinell (BHN).

Grain size has a direct impact on the properties of a metal. Metals with fine grain size are stronger and have more uniform properties. The strength of metal is related to its grain size through an empirical relationship called the Hall–Petch equation. Metals with grain size in the nano range (nanocrystalline metals) are expected to have ultra high strength and hardness as predicted by the Hall–Petch equation.

When a metal is plastically deformed by cold working, the metal becomes strain-hardened, resulting in an increase in its strength and a decrease in its ductility. The strain hardening can be removed by giving the metal an annealing heat treatment. When the strain-hardened metal is slowly heated to a high temperature below its melting temperature, the processes of recovery, recrystallization, and grain growth take place, and the metal is softened. By combining strain hardening and annealing, large thickness reductions of metal sections can be accomplished without fracture.

By deforming some metals at high temperature and slow loding rates, it is possible to achieve superplasticity i.e., deformation of the order of 1000–2000%. The grain size must be ultrafine to achieve superplasticity.

Plastic deformation of metals takes place most commonly by the slip process, involving the movement of dislocations. Slip usually takes place on the closest-packed planes and in the closest-packed directions. The combination of a slip plane and a slip direction constitutes a slip system. Metals with a high number of slip systems are more ductile than those with only a few slip systems. Many metals deform by twinning when slip becomes difficult.

Grain boundaries at lower temperatures usually strengthen metals by providing barriers to dislocation movement. However, under some conditions of high-temperature deformation, grain boundaries become regions of weakness due to grain boundary sliding.

6.12 DEFINITIONS

Sec. 6.1

Hot working of metals: permanent deformation of metals and alloys above the temperature at which a strain-free microstructure is produced continuously (recrystallization temperature).

Cold working of metals: permanent deformation of metals and alloys below the temperature at which a strain-free microstructure is produced continuously (recrystallization temperature). Cold working causes a metal to be strain-hardened.

Percent cold reduction:

$$\% \text{ cold reduction} = \frac{\text{change in cross-sectional area}}{\text{original cross-sectional area}} \times 100\%$$

Annealing: a heat treatment given to a metal to soften it.

Extrusion: a plastic forming process in which a material under high pressure is reduced in cross section by forcing it through an opening in a die.

Forging: a primary processing method for working metals into useful shapes in which the metal is hammered or pressed into shape.

Wire drawing: a process in which wire stock drawn through one or more tapered dies to the desired cross section.

Sec. 6.2

Elastic deformation: if a metal deformed by a force returns to its original dimensions after the force is removed, the metal is said to be elastically deformed.

Engineering stress σ: average uniaxial force divided by original cross-sectional area ($\sigma = F/A_0$).

Engineering strain ϵ: change in length of sample divided by the original length of sample ($\epsilon = \Delta l/l_0$).

Shear stress τ: shear force S divided by the area A over which the shear force acts ($\tau = S/A$).

Shear strain γ: shear displacement a divided by the distance h over which the shear acts ($\gamma = a/h$).

Sec. 6.3

Engineering stress-strain diagram: experimental plot of engineering stress versus engineering strain; σ is normally plotted as the y axis and ϵ as the x axis.

Modulus of elasticity E: stress divided by strain (σ/ϵ) in the elastic region of an engineering stress-strain diagram for a metal ($E = \sigma/\epsilon$).

Yield strength: the stress at which a specific amount of strain occurs in the engineering tensile test. In the U.S. the yield strength is determined for 0.2 percent strain.

Ultimate tensile strength (UTS): the maximum stress in the engineering stress-strain diagram.

Sec. 6.4

Hardness: a measure of the resistance of a material to permanent deformation.

Sec. 6.5

Slip: the process of atoms moving over each other during the permanent deformation of a metal.

Slipbands: line markings on the surface of a metal due to slip caused by permanent deformation.

Slip system: a combination of a slip plane and a slip direction.

Deformation twinning: a plastic deformation process that occurs in some metals and under certain conditions. In this process a large group of atoms are displaced together to form a region of a metal crystal lattice that is a mirror image of a similar region along a twinning plane.

Sec. 6.6

Hall–petch relationship: an empirical equation that relates the strength of a metal to its grain size.

Strain hardening (strengthening): the hardening of a metal or alloy by cold working. During cold working, dislocations multiply and interact, leading to an increase in the strength of the metal.

Sec. 6.7

Solid-solution hardening (strengthening): strengthening a metal by alloying additions that form solid solutions. Dislocations have more difficulty moving through a metal lattice when the atoms are different in size and electrical characteristics, as is the case with solid solutions.

Sec. 6.8

Annealing: a heat treatment process applied to a cold worked metal to soften it.

Recovery: the first stage in the annealing process that results in removal of residual stresses and formation of low energy dislocation configurations.

Recrystallization: the second stage of the annealing process in which new grains start to grow and dislocation density decreases significantly.

Grain growth: the third stage in which new grains start to grow in an equiaxed manner.

Sec. 6.9

Superplasticity: the ability of some metals to deform plastically by 1000–2000% at high temperatures and low loading rates.

Sec. 6.10

Nanocrystalline metals: metals with grain size smaller than 100 nm.

6.13 PROBLEMS

Answers to problems marked with an asterisk are given at the end of the book.

6.1   How are metal alloys made by the casting process?

6.2   Distinguish between wrought alloy products and cast alloy products.

6.3   Why are cast metal sheet ingots hot-rolled first instead of being cold-rolled?

6.4   What type of heat treatment is given to the rolled metal sheet after hot and "warm" rolling? What is its purpose?

6.5   Calculate the percent cold reduction after cold rolling 0.040-in.-thick aluminum sheet to 0.025 in.

*6.6   A 70% Cu–30% Zn brass sheet is 0.0955 cm thick and is cold-rolled with a 30 percent reduction in thickness. What must be the final thickness of the sheet?

6.7   A sheet of an aluminum alloy is cold-rolled 30 percent to a thickness of 0.080 in. If the sheet is then cold-rolled to a final thickness of 0.064 in., what is the total percent cold work done?

6.8   Describe and illustrate the following types of extrusion processes: (*a*) direct extrusion and (*b*) indirect extrusion. What is an advantage of each process?

6.9   Which process in Prob. 6.8 is used most commonly? Which metals and alloys are commonly extruded?

6.10 Describe the forging process. What is the difference between hammer and press forging?

6.11 What is the difference between open-die and closed-die forging? Illustrate. Give an example of a metal product produced by each process.

6.12 Describe the wire-drawing process. Why is it necessary to make sure the surface of the incoming wire is clean and lubricated?

*6.13 Calculate the percent cold reduction when an aluminum wire is cold-drawn from a diameter of 5.25 mm to a diameter of 2.30 mm.

*6.14 A 0.15-in.-diameter 99.5 percent copper wire is to be cold-drawn with a 30 percent cold reduction. What must be the final diameter of the wire?

6.15 A brass wire is cold-drawn 25 percent to a diameter of 1.10 mm. It is then further cold-drawn to 0.900 mm. What is the total percent cold reduction?

6.16 Distinguish between elastic and plastic deformation.

6.17 Define engineering stress. What are the U.S. customary and SI units for engineering stress?

6.18 Calculate the engineering stress in SI units on a 2.00-cm-diameter rod that is subjected to a load of 1300 kg.

*6.19 Calculate the engineering stress in SI units on a bar 15 cm long and having a cross section of 4.25 mm $\times$ 12.0 mm that is subjected to a load of 5000 kg.

*6.20 Calculate the engineering stress in SI units on a bar 25 cm long and having a cross section of 9.00 mm $\times$ 4.00 mm that is subjected to a load of 3500 kg.

6.21 Calculate the engineering stress in U.S. customary units on a 0.400-in.-diameter rod that is subjected to a force of 1500 lb.

6.22 What is the relationship between engineering strain and percent elongation?

*6.23 A tensile specimen of cartridge brass sheet has a cross section of 0.320 in. $\times$ 0.120 in. and a gage length of 2.00 in. Calculate the engineering strain that occurred during a test if the distance between gage markings is 2.35 in. after the test.

6.24 A 0.505-in.-diameter rod of an aluminum alloy is pulled to failure in a tension test. If the final diameter of the rod at the fractured surface is 0.440 in., what is the percent reduction in area of the sample due to the test?

6.25 The following engineering stress-strain data were obtained for a 0.2% C plain-carbon steel. (*a*) Plot the engineering stress-strain curve. (*b*) Determine the ultimate tensile strength of the alloy. (*c*) Determine the percent elongation at fracture.

Engineering stress (ksi)	Engineering strain (in./in.)	Engineering stress (ksi)	Engineering strain (in./in.)
0	0	76	0.08
30	0.001	75	0.10
55	0.002	73	0.12
60	0.005	69	0.14
68	0.01	65	0.16
72	0.02	56	0.18
74	0.04	51	(Fracture) 0.19
75	0.06		

6.26 Plot the data of Prob. 6.25 as engineering stress (MPa) versus engineering strain (mm/mm) and determine the ultimate strength of the steel.

6.27 The following engineering stress-strain data were obtained at the beginning of a tensile test for a 0.2% C plain-carbon steel. (*a*) Plot the engineering stress-strain curve for these data. (*b*) Determine the 0.2 percent offset yield stress for this steel. (*c*) Determine the tensile elastic modulus of this steel. (Note that these data only give the beginning part of the stress-strain curve.)

Engineering stress (ksi)	Engineering strain (in./in.)	Engineering stress (ksi)	Engineering strain (in./in.)
0	0	60	0.0035
15	0.0005	66	0.004
30	0.001	70	0.006
40	0.0015	72	0.008
50	0.0020		

6.28 Plot the data of Problem 6.27 as engineering stress (MPa) versus engineering strain (mm/mm) and determine the 0.2 percent offset yield stress of the steel.

*6.29 A 0.505-in.-diameter aluminum alloy test bar is subjected to a load of 25,000 lb. If the diameter of the bar is 0.490 in. at this load, determine (*a*) the engineering stress and strain and (*b*) the true stress and strain.

6.30 Twenty-cm-long rod with a diameter of 0.250 cm is loaded with a 5000 N weight. If the diameter decreases to 0.210 cm, determine (*a*) the engineering stress and strain at this load and (*b*) the true stress and strain at this load.

6.31 Define the hardness of a metal.

6.32 How is the hardness of a material determined by a hardness testing machine?

6.33 What types of indenters are used in (*a*) the Brinell hardness test, (*b*) Rockwell C hardness, and (*c*) Rockwell B hardness?

6.34 What are slipbands and slip lines? What causes the formation of slipbands on a metal surface?

6.35 Describe the slip mechanism that enables a metal to be plastically deformed without fracture.

6.36 Why does slip in metals usually take place on the densest-packed planes?

6.37 Why does slip in metals usually take place in the closest-packed directions?

6.38 What are the principal slip planes and slip directions for FCC metals?

*6.39 For the FCC crystal lattice, what are the four principal slip planes and three slip directions?

6.40 What are the principal slip planes and slip directions for BCC metals?

6.41 What are the principal slip planes and slip directions for HCP metals?

6.42 What other types of slip planes are important other than the basal planes for HCP metals with low *c/a* ratios?

6.43 What is the critical resolved shear stress for a pure metal single crystal?

6.44 Why do pure FCC metals like Ag and Cu have low values of τ_c?

6.45 What is believed to be responsible for the high values of τ_c for HCP titanium?

*6.46 A stress of 75 MPa is applied in the [001] direction on an FCC single crystal. Calculate (*a*) the resolved shear stress acting on the (111) [$\bar{1}$01] slip system and (*b*) the resolved shear stress acting on the (111) [$\bar{1}$10] slip system.

6.47 A stress of 55 MPa is applied in the [001] direction of a BCC single crystal. Calculate (*a*) the resolved shear stress acting on the (101) [$\bar{1}$11] system and (*b*) the resolved shear stress acting on the (110) [$\bar{1}$11] system.

*6.48 Determine the tensile stress that must be applied to the [1$\bar{1}$0] axis of a high-purity copper single crystal to cause slip on the ($\bar{1}$1 $\bar{1}$) [0$\bar{1}$1] system. The resolved shear stress for the crystal is 0.85 MPa.

*6.49 A stress of 4.75 MPa is applied in the [00$\bar{1}$] direction of a unit cell of an FCC copper single crystal. Calculate the resolved shear stress on the (11$\bar{1}$) plane in the following directions: (*a*) [$\bar{1}$0$\bar{1}$], (*b*) [0$\bar{1}$ $\bar{1}$], and (*c*) [$\bar{1}$10].

6.50 A stress of 2.78 MPa is applied in the [001] direction of a unit cell of an FCC silver single crystal. Calculate the resolved shear stress on the [1$\bar{1}$1] plane in the following directions: (*a*) [$\bar{1}$01], (*b*) [011], and (*c*) [110].

6.51 A stress of 2.34 MPa is applied in the [001] direction of a unit cell of an FCC copper single crystal. Calculate the resolved shear stress on the ($\bar{1}$11) plane in the following directions: (*a*) [101], (*b*) [110], and (*c*) [0$\bar{1}$1].

*6.52 A stress of 85 MPa is applied in the [001] direction of a unit cell of a BCC iron single crystal. Calculate the resolved shear stress for the following slip systems: (*a*) (011) [1$\bar{1}$1], (*b*) (110) [$\bar{1}$11], and (*c*) (0$\bar{1}$1) [111].

6.53 A stress of 92 MPa is applied in the [001] direction of a unit cell of a BCC iron single crystal. Calculate the resolved shear stress for the following slip systems: (*a*) (011) [$\bar{1}$ $\bar{1}$1], (*b*) (1$\bar{1}$0) [111], and (*c*) ($\bar{1}$01) [111].

6.54 Describe the deformation twinning process that occurs in some metals when they are plastically deformed.

6.55 What is the difference between the slip and twinning mechanisms of plastic deformation of metals?

6.56 What important role does twinning play in the plastic deformation of metals with regard to deformation of metals by slip?

6.57 Why is deformation by twinning especially important for HCP metals?

6.58 By what mechanism do grain boundaries strengthen metals?

6.59 What experimental evidence shows that grain boundaries arrest slip in polycrystalline metals?

6.60 Describe the grain shape changes that occur when a sheet of unalloyed copper with an original equiaxed grain structure is cold-rolled with 30 and 50 percent cold reductions.

6.61 What happens to the dislocation substructure in Prob. 6.60?

6.62 How is the ductility of a metal normally affected by cold working? Why?

6.63 An oxygen-free copper rod must have a tensile strength of 50.0 ksi and a final diameter of 0.250 in. (*a*) What amount of cold work must the rod undergo (see Fig. 6.45)? (*b*) What must the initial diameter of the rod be?

*6.64 A 70% Cu–30% Zn brass sheet is to be cold-rolled from 0.070 to 0.040 in. (*a*) Calculate the percent cold work, and (*b*) estimate the tensile strength, yield strength, and elongation from Fig. 6.46.

6.65 A 70% Cu–30% Zn brass wire is cold-drawn 20 percent to a diameter of 2.80 mm. The wire is then further cold-drawn to a diameter of 2.45 mm. (*a*) Calculate the total percent cold work that the wire undergoes. (*b*) Estimate the wire's tensile and yield strengths and elongation from Fig. 6.46.

6.66 What is solid-solution strengthening? Describe the two main types.

6.67 What are two important factors that affect solid-solution hardening?

6.68 What are the three main metallurgical stages that a sheet of a cold-worked metal such as aluminum or copper goes through as it is heated from room temperature to an elevated temperature just below its melting point?

6.69 Describe the microstructure of a heavily cold-worked metal of an Al–0.8% Mg alloy as observed with an optical microscope at 100× (see Fig. 6.48*a*). Describe the microstructure of the same material at 20,000× (see Fig. 6.49*a*).

6.70 Describe what occurs microscopically when a cold-worked sheet of metal such as aluminum undergoes a recovery heat treatment.

6.71 When a cold-worked metal is heated into the temperature range where recovery takes place, how are the following affected: (*a*) internal residual stresses, (*b*) strength, (*c*) ductility, and (*d*) hardness?

6.72 Describe what occurs microscopically when a cold-worked sheet of metal such as aluminum undergoes a recrystallization heat treatment.

6.73 When a cold-worked metal is heated into the temperature range where recrystallization takes place, how are the following affected: (*a*) internal residual stresses, (*b*) strength, (*c*) ductility, and (*d*) hardness?

6.74 Describe two principal mechanisms whereby primary recrystallization can occur.

6.75 What are five important factors that affect the recrystallization process in metals?

6.76 What generalizations can be made about the recrystallization temperature with respect to (*a*) the degree of deformation, (*b*) the temperature, (*c*) the time of heating at temperature, (*d*) the final grain size, and (*e*) the purity of the metal?

6.77 If it takes 115 h to 50 percent recrystallize an 1100-H18 aluminum alloy sheet at 250°C and 10 h at 285°C, calculate the activation energy in kilojoules per mole for this process. Assume an Arrhenius-type rate behavior.

6.78 If it takes 12.0 min to 50 percent recrystallize a piece of high-purity copper sheet at 140°C and 200 min at 88°C, how many minutes are required to recrystallize the sheet 50 percent at 100°C? Assume an Arrhenius-type rate behavior.

6.79 If it takes 80 h to completely recrystallize an aluminum sheet at 250°C and 6 h at 300°C, calculate the activation energy in kilojoules per mole for this process. Assume an Arrhenius-type rate behavior.

6.14 MATERIALS SELECTION AND DESIGN PROBLEMS

1. (*a*) How would you manufacture large propellers for ships? (*b*) Select a suitable material for this application. Explain your selection.

2. (*a*) Name five components that are manufactured by the casting process. (*b*) Why is casting so popular?

3. If you were to make a component from silver, gold, or other precious metals, what process would you use? Why?

4. If you were to make only two units of a particular design made of metal, what process would you select?

5. Consider the casting of a thick cylindrical shell made of cast iron. If the casting process is controlled such that solidification takes place from the inner walls of the tube outward, the outer layers shrink during solidification and exert compressive residual stresses on the inner walls. (*a*) What would be the advantage of developed compressive residual stresses? (*b*) Propose a process that would achieve the desired solidification path.

6. Consider casting a cube and a sphere of the same volume from the same metal. Which one would solidify faster? Why?

7. In the rolling process, the selection of the roll material is critical. (*a*) Based on your knowledge of both hot and cold rolling, what properties should the roller material have? (*b*) Select an appropriate material for this application.

8. Design a process that produces long bars with an H cross section from steel (indicate hot or cold if applicable). Draw schematics to show your procedure.

9. What process would you select to manufacture a railroad car wheel (indicate hot or cold if applicable)? Justify your choice.

10. When manufacturing complex shapes using cold forging or shape rolling operations, the mechanical properties such as yield strength, tensile strength, and ductility measure differently depending on the location and direction on the manufactured part. (*a*) How do you explain this from a micro point of view? (*b*) Will this happen during hot forging or rolling? Explain your answer.

11. What process would you select to produce ribbed rods such as those used to reinforce concrete (hot or cold if applicable)?

12. If you were to select a material for the construction of a robotic arm that would result in the smallest amount of elastic deformation (important for positional accuracy of the arm) and weight was not a critical criterion, which one of the metals given in Fig. 6.25 would you select? Why?

13. You must select the material for an industrial gear from those available in Fig. 6.25. Toughness is the main criterion at this point. Which material would you choose?

14. For a given application, a rod of copper of one inch diameter is to be used. You have copper rods of various cross sections available to you; however, all the bars are fully annealed with a yield strength of 10 ksi. The material must have a yield strength of at least 30 ksi and an elongation ability of at least 20 percent. Design a process that would achieve the expected goals. Use Fig. 6.45 for your solution.

15. The material for a rod of cross-sectional area 2.7 in.^{2} and length 75 in. must be selected such that under an axial load of 45,000 lb$_f$, it will not yield and the

elongation in the bar will remain below 0.1 in. Use Table 6.1 and Fig. 6.25 to select the proper material.

16. In the loading of a single crystal, determine the angles ϕ and λ for which the maximum resolved shear stress occurs.

17. (*a*) In the loading of a single crystal, how would you orient the crystal with respect to the loading axis to cause a resolved shear stress of zero? (*b*) What is the physical significance of this, that is, what happens to the crystal as σ increases?

18. Which of the following substitutional solid solutions would you select if the higher tensile strength was the selection criterion: Cu–30 wt % Zinc or Cu–30 wt % Ni?

19. The cupro-nickel substitutional solid solution alloys Cu–40 wt % Ni and Ni–10 wt % Cu have similar tensile strengths. For a given application for which only tensile strength is important, which one would you select?

Mechanical Properties of Metals II

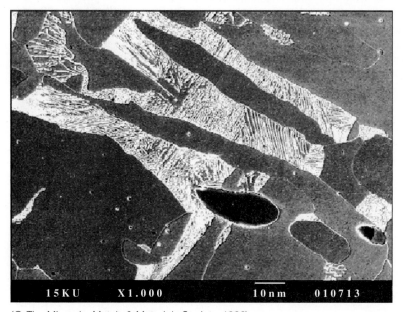

(© *The Minerals, Metals & Materials Society, 1998*)

On April 12, 1912, at 11:40 P.M., the *Titanic,* on its maiden voyage, struck a large iceberg, damaging her hull and causing six forward compartments to rupture. The seawater temperature at the time of the accident was −2°C. The ensuing flooding of these compartments resulted in complete fracture of the hull with the tragic loss of more than 1500 lives.

The *Titanic* was found on the ocean floor September 1, 1985, by Robert Ballard. She was 3.7 km below the water surface. Based on metallurgical and mechanical tests performed on the *Titanic* steel, it was determined that the ductile-brittle transition temperature of the steel used in the *Titanic* was 32°C for the longitudinal specimens made from the hull plate, and 56°C for the transverse specimens. This reveals that the steel used in the construction of the *Titanic* behaved in a highly brittle fashion when it struck the iceberg. The microstructure of the *Titanic* steel in the chapter-opening image shows ferrite grains (gray), pearlite colonies (light lamellar), and MnS particles (dark).[1] ■

[1]www.tms.org/pubs/journals/JOM/9801/Felkins-9801.html#ToC6

By the end of this chapter, students will be able to. . .

1. Describe the process of fracture of metals and differentiate between ductile and brittle fracture.

2. Describe the ductile to brittle transition of metals. What type of metals are more susceptible to DBT?

3. Define the fracture toughness of a material and explain why this property is used in engineering design instead of toughness.

4. Define fatigue loading and failure in materials. Describe the parameters that are used to characterize fluctuating stresses. Enumerate the factors that affect the fatigue strength of materials.

5. Describe creep, creep test, and the use of Larsen-Miller parameter in design for determination of time to stress rupture.

6. Describe why analysis of a failed component is important and what are the steps taken in the failure analysis process.

7. Describe the effect of nano-grain size on the strength and ductility of a metal.

This chapter continues with a study of the mechanical properties of metals. First, the fracture of metals will be discussed. Then the fatigue and fatigue crack propagation of metals, and the creep (time-dependent deformation) and stress rupture of metals are considered. A case study in fracture of a metallic component is also presented. Finally, future directions in the synthesis of nanostructured metals and their properties are discussed.

7.1 FRACTURE OF METALS

One of the important and practical aspects of materials selection in the design, development, and production of new components is the possibility of failure of the component under normal operation. Failure may be defined as the *inability* of a material or a component to (1) perform the intended function, (2) meet performance criteria although it may still be operational, or (3) perform safely and reliably even after deterioration. Yielding, wear, buckling (elastic instability), corrosion, and fracture are examples of situations in which a component has failed.

Engineers are deeply aware of the possibility of fracture in load-bearing components and its potentially detrimental effect on productivity, safety, and other economical issues. As a result all design, manufacturing, and materials engineers use safety factors in their initial analysis to reduce the possibility of fracture by essentially overdesigning the component or the machine. In many fields such as pressure vessel design and manufacturing, there exist codes and standards that are put in place by various agencies that must be followed by all designers and manufactures. Regardless of the extreme care taken in design, manufacturing, and materials selection for machines and components, failures are unavoidable, resulting in loss of property and unfortunately, sometimes, life. Every engineer should be (1) completely familiar with the concept of fracture or failure of materials and (2) able to extract information from a failed component as to the causes of failure. In most cases scientists and engineers carefully analyze the failed components to determine the cause of failure. The information gained is used to advance safe performance and minimize the possibility of failure through improvements in design, manufacturing processes, and materials synthesis and selection. From a purely mechanical performance point of view engineers are concerned with fracture failure of designed components that are made of metals, ceramics, composites, polymers, or even electronic materials. In the upcoming sections various modes of fracture and failure of metals under operation will be presented and discussed. In future chapters the fracture and failure of other classes of materials will also be discussed.

Fracture is the separation of a solid under stress into two or more parts. In general metal fractures can be classified as ductile or brittle, but a fracture can also be a mixture of the two. The **ductile fracture** of a metal occurs after extensive plastic deformation and is characterized by slow crack propagation. **Brittle fracture,** in contrast, usually proceeds along characteristic crystallographic planes called *cleavage planes* and has rapid crack propagation. Figure 7.1 shows an example of a ductile fracture in an aluminum alloy test specimen. Owing to their rapidity, brittle fractures generally lead to sudden, unexpected, catastrophic failures while the plastic deformation accompanying ductile fracture may be detectable before fracture occurs.

7.1.1 Ductile Fracture

Virtual Lab

Ductile fracture of a metal occurs after extensive plastic deformation. For simplicity let us consider the ductile fracture of a round (0.50-in.-diameter) tensile specimen. If a stress is applied to the specimen that exceeds its ultimate tensile strength and is sustained long enough, the specimen will fracture. Three distinct stages of ductile fracture can be recognized: (1) the specimen forms a neck, and cavities form within the necked region (Fig. 7.2a and b); (2) the cavities in the neck coalesce into a crack in the center of the specimen and propagate toward the surface of the specimen in a direction perpendicular to the applied stress (Fig. 7.2c); and (3) when the crack nears the surface, the direction of the crack changes to 45° to the tensile axis and a cup-and-cone fracture results (Fig. 7.2d and e). Fig. 7.3 shows a scanning electron micrograph of a ductile fracture of a spring-steel specimen, and Fig. 7.4 shows internal cracks in the necked region of a deformed specimen of high-purity copper.

In practice ductile fractures are less frequent than brittle fractures, and the main cause for their occurrence is overloading of the component. Overloading could occur as a result of (1) improper design, including the selection of materials

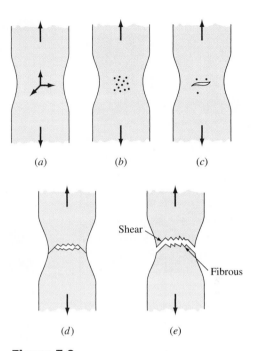

Figure 7.1
Ductile (cup-and-cone) fracture of an aluminum alloy.
(After ASM Handbook of Failure Analysis and Prevention, vol. 11. 1992. ASM International)

Animation

Figure 7.2
Stages in the formation of a cup-and-cone ductile fracture.
(After G. Dieter, "Mechanical Metallurgy," 2nd ed., McGraw-Hill, 1976, p. 278.)

(underdesigning), (2) improper fabrication, or (3) abuse (component is used at load levels above that allowed by the designer). An example of a ductile failure is given in Fig. 7.5. In this figure the rear axle shaft of a vehicle is shown that has undergone significant plastic twisting (note torsion marks on the shaft) due to applied torsion. Based on engineering analysis, the cause of this failure has been attributed to a poor choice of material. AISI type S7 tool steel was used for this component with an improperly low hardness level of 22–27 HRC. The required hardness for the metal was over 50 HRC, which is usually achieved through heat treatment processes (see Chap. 9).

7.1.2 Brittle Fracture

Many metals and alloys fracture in a brittle manner with very little plastic deformation. Figure 7.6 shows a tensile specimen that failed in a brittle manner. Comparison of this figure with Fig. 7.1 reveals the drastic differences in the deformation level prior to fracture between ductile and brittle fractures. Brittle fracture usually proceeds along specific crystallographic planes called *cleavage planes* under

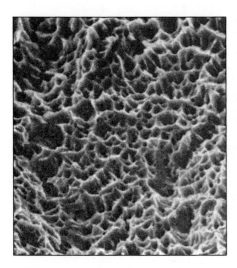

Figure 7.3
Scanning electron micrograph showing conical equiaxed dimples produced during the fracture of a spring-steel specimen. These dimples, which are formed during the microvoid coalescence of the fracture, are indicative of a ductile fracture.
(*After ASM Handbook, vol. 12—Fractography, p. 14, fig. 2a, 1987. ASM International*)

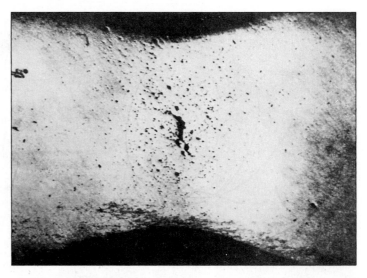

Figure 7.4
Internal cracking in the necked region of a polycrystalline specimen of high-purity copper. (Magnification 9×.)
[*After K. E. Puttick, Philas. Mag.* **4:**964(1959).]

Figure 7.5
Failed axle shaft.
(*ASM Handbook of Failure Analysis and Prevention, vol. 11. 1992. ASM International*)

a stress normal to the cleavage plane (see Fig. 7.7). Many metals with the HCP crystal structure commonly show brittle fracture because of their limited number of slip planes. A zinc single crystal, for example, under a high stress normal to the (0001) planes will fracture in a brittle manner. Many BCC metals such as α iron, molybdenum, and tungsten also fracture in a brittle manner at low temperatures and high strain rates.

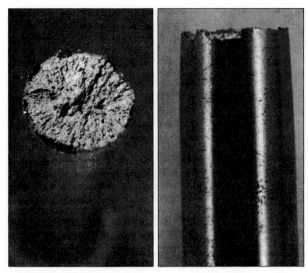
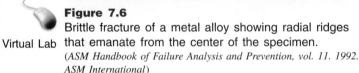

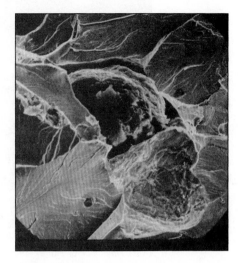

Figure 7.6

Virtual Lab

Brittle fracture of a metal alloy showing radial ridges that emanate from the center of the specimen.
(*ASM Handbook of Failure Analysis and Prevention, vol. 11. 1992. ASM International*)

Figure 7.7

Brittle cleavage fracture in ferritic ductile iron. SEM, 1000×.
(*After W. L. Bradley, Texas A&M University, From ASM Handbook, vol. 12, p. 237, fig. 97, 1987. ASM International*)

Most brittle fractures in polycrystalline metals are transgranular, i.e., the cracks propagate across the matrix of the grains. However, brittle fracture can occur in an intergranular manner if the grain boundaries contain a brittle film or if the grain boundary region has been embrittled by the segregation of detrimental elements.

Brittle fracture in metals is believed to take place in three stages:

1. Plastic deformation concentrates dislocations along slip planes at obstacles.
2. Shear stresses build up in places where dislocations are blocked, and as a result microcracks are nucleated.
3. Further stress propagates the microcracks, and stored elastic strain energy may also contribute to the propagation of the cracks.

In many cases brittle fractures occur because of the existence of defects in the metal. These defects are either formed during the manufacturing stage or develop during service. Undesirable defects such as folds, large inclusions, undesirable grain flow, poor microstructure, porosity, tears, and cracks may form during manufacturing operations such as forging, rolling, extrusion, and casting. Fatigue cracks, embrittlement due to the atomic hydrogen (see Sec. 13.5.11), and corrosion damage often result in final brittle fracture. When brittle fracture occurs, it consistently initiates at the defect location (*stress risers*) regardless of the cause for the formation of the defect. Certain defects, low operating temperatures, or high

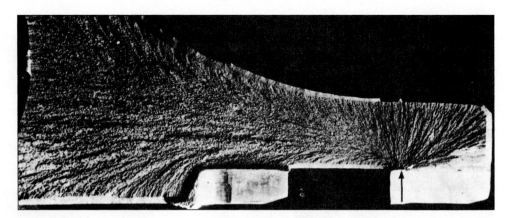

Figure 7.8
A snap ring made of 4335 steel failed in a brittle manner because of the existence of a sharp corner.
(*ASM Handbook of Failure Analysis and Prevention, vol. 11. 1992. ASM International*)

loading rates may also cause the brittle fracture of some moderately ductile materials. The transition from ductile to brittle behavior is called a **ductile to brittle transition (DBT).** Thus, ordinarily ductile materials can, under certain circumstances, fracture in brittle manner. Figure 7.8 shows the brittle fracture of a snap ring due to the existence of a sharp corner as the defect (see arrow on the figure); note the chevron pattern pointing toward the origin of the fracture (typically found in a brittle fracture surface.)

7.1.3 Toughness and Impact Testing

Toughness is a measure of the amount of energy a material can absorb before fracturing. It becomes of engineering importance when the ability of a material to withstand an impact load without fracturing is considered. One of the simplest methods of measuring toughness is to use an impact-testing apparatus. A schematic diagram of a simple impact-testing machine is shown in Fig. 7.9. One method of using this apparatus is to place a Charpy V-notch specimen (shown in the upper part of Fig. 7.9) across parallel jaws in the machine. In the impact test a heavy pendulum released from a known height strikes the sample on its downward swing, fracturing it. By knowing the mass of the pendulum and the difference between its initial and final heights, the energy absorbed by the fracture can be measured. Figure 7.10 shows the relative effect of temperature on the impact energy of some types of materials.

7.1.4 Ductile to Brittle Transition Temperature

As mentioned above, under certain conditions a marked change in the fracture resistance of some metals is observed in service, i.e., ductile to brittle transition. Low temperatures, high stress states, and fast loading rates may all cause a ductile

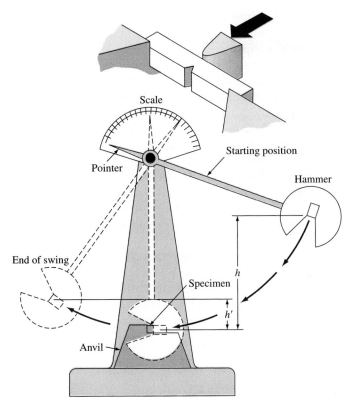

Figure 7.9
Schematic drawing of a standard impact-testing apparatus.
(After H. W. Hayden, W. G. Moffatt, and J. Wulff, "The Structure and Properties of Materials," vol. III, Wiley, 1965, p. 13.)

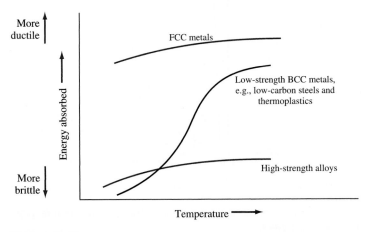

Figure 7.10
Effect of temperature on the energy absorbed upon impact by different types of materials.

material to behave in a brittle manner; however, customarily the temperature is selected as the variable that represents this transition while the load rate and stress rate are held constant. The impact-testing apparatus discussed in the previous section may be used to determine the temperature range for the transition from ductile to brittle behavior of materials. The temperature of the Charpy specimen may be set using furnace and refrigeration units. Although some metals show a distinct DBT temperature, for many this transition occurs over a range of temperatures (see Fig. 7.10). Also, Fig. 7.10 shows that FCC metals do not undergo DBT and, as a consequence, are suitable for low temperature use. Factors that influence the DBT temperature are alloy composition, heat treatment, and processing. For instance, the carbon content of annealed steels affects this transition temperature range, as shown in Fig. 7.11. Low-carbon annealed steels have a lower-temperature transition range and a narrower one than high-carbon steels. Also, as the carbon content of the annealed steels is increased, the steels become more brittle, and less energy is absorbed on impact during fracture.

Ductile to brittle transition is an important consideration in materials selection for components that operate in cold environments. For instance, ships that sail in cold waters (see the chapter opener) and offshore platforms that are located in the arctic seas are especially susceptible to DBT. For such applications, the selected materials should have a DBT temperature that is significantly lower than the operating or service temperature.

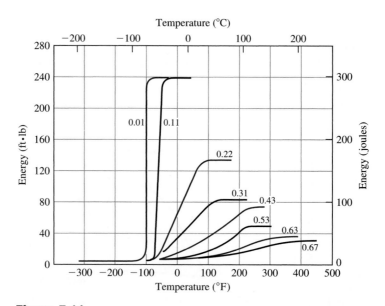

Figure 7.11
Effect of carbon content on the impact energy temperature plots for annealed steels.
[*After J. A. Rinebolt and W. H. Harris, Trans. ASM,* **43**:*1175(1951).*]

7.1.5 Fracture Toughness

Impact tests such as the one previously described give useful comparative quantitative data with relatively simple test specimens and equipment. However, these tests do not provide proper data for design purposes for material sections containing cracks or flaws. Data of this type are obtained from the discipline of fracture mechanics in which theoretical and experimental analyses are made of the fracture of structural materials containing preexisting cracks or flaws. In this book we shall focus on the fracture toughness property of fracture mechanics and show how it can be applied to some simple component designs.

The fracture of a metal (material) starts at a place where the stress concentration is the highest, which may be at the top of a sharp crack, for example. Let us consider a plate sample under uniaxial tension that contains an edge crack (Fig. 7.12*a*) or a center-through crack (Fig. 7.12*b*). The stress at the tip of a sharp crack is highest at the tip as indicated in Fig. 7.12*c*.

The stress intensity at the crack tip is found to be dependent on both the applied stress and the width of the crack. We use the stress-intensity factor K_I to express the combination of the effects of the stress at the crack tip and the crack length. The subscript I (pronounced "one") indicates mode I testing in which a tensile stress causes the crack to open. By experiment, for the case of uniaxial tension on a metal plate containing an edge or internal crack (mode I testing), we find that

$$K_I = Y\sigma\sqrt{\pi a} \tag{7.1}$$

where K_I = stress-intensity factor

σ = applied nominal stress

a = edge-crack length or half the length of an internal through crack

Y = dimensionless geometric constant of the order of 1

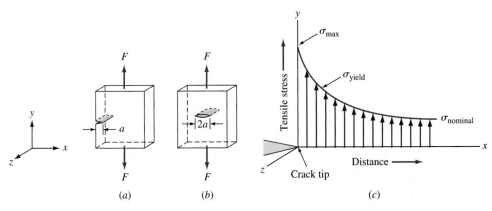

Figure 7.12
Metal alloy plate under uniaxial tension (*a*) with edge crack *a*, (*b*) with center crack 2*a*. (*c*) Stress distribution versus distance from crack tip. The stress is a maximum at the crack tip.

The critical value of the stress-intensity factor that causes failure of the plate is called the *fracture toughness* K_{IC}, (pronounced "kay-one-see") of the material. In terms of the fracture stress σ_f and the crack length a for an edge crack (or one-half of the internal crack length),

$$K_{IC} = Y\sigma_f \sqrt{\pi a} \tag{7.2}$$

Fracture-toughness (K_{IC}) values have the SI units of MPa$\sqrt{m}$ and U.S. customary units of ksi$\sqrt{in}$. Figure 7.13*a* is a schematic diagram of the compact type of fracture-toughness test specimen. To obtain constant values for K_{IC}, the base dimension B of the specimen must be relatively large compared to the notch-depth dimension a so that so-called plain-strain conditions prevail. Plain-strain conditions require that during testing there is no strain deformation in the direction of the notch (i.e., in the z direction of Fig. 7.13*a*). Plain-strain conditions generally prevail when B (specimen thickness) = 2.5 (K_{IC}/yield strength)^2. Note that the fracture-toughness specimen has a machined notch and a fatigue crack at the end of the notch of about 3 mm depth to start the fracture during the test. Figure 7.13*b* shows a real fracture-toughness test at the time of rapid fracture.

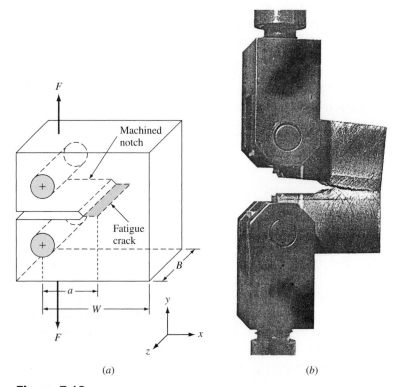

(*a*) (*b*)

Figure 7.13
Fracture-toughness test using a compact type of specimen and plain-strain conditions. (*a*) Dimensions of specimen. (*b*) Actual test at the critical stress for fracture, using a laser beam to detect this stress.
(*Courtesy of White Shell Research.*)

Table 7.1 Typical fracture-toughness values for selected engineering alloys

Material	K_{IC}		$\sigma_{\text{yield strength}}$	
	MPa$\sqrt{\text{m}}$	**ksi$\sqrt{\text{in.}}$**	**MPa**	**ksi**
Aluminum alloys:				
2024-T851	26.4	24	455	66
7075-T651	24.2	22	495	72
7178-T651	23.1	21	570	83
Titanium alloy:				
Ti-6A1-4V	55	50	1035	150
Alloy steels:				
4340 (low-alloy steel)	60.4	55	1515	220
17-7 pH (precipitation hardening)	76.9	70	1435	208
350 maraging steel	55	50	1550	225

Source: R. W. Herzberg, "Deformation and Fracture Mechanics of Engineering Materials," 3rd ed., Wiley, 1989.

Fracture-toughness values of materials are most useful in mechanical design when working with materials of limited toughness or ductility such as high-strength aluminum, steel, and titanium alloys. Table 7.1 lists some K_{IC} values for some of these alloys. Materials that show little plastic deformation before fracture have relatively low fracture toughness K_{IC} values and tend to be more brittle, whereas those with higher K_{IC} values are more ductile. Fracture-toughness values can be used in mechanical design to predict the allowable flaw size in alloys with limited ductility when acted upon by specific stresses (a factor of safety is also applied for added safety). Example Problem 7.1 illustrates this design approach.

A structural plate component of an engineering design must support 207 MPa (30 ksi) in tension. If aluminum alloy 2024-T851 is used for this application, what is the largest internal flaw size that this material can support? (Use $Y = 1$).

EXAMPLE PROBLEM 7.1

■ **Solution**

$$K_{IC} = Y\sigma_f \sqrt{\pi a} \qquad (7.2)$$

Using $Y = 1$ and $K_{IC} = 26.4$ MPa$\sqrt{\text{m}}$ from Table 7.1,

$$a = \frac{1}{\pi}\left(\frac{K_{IC}}{\sigma_f}\right)^2 = \frac{1}{\pi}\left(\frac{26.4 \text{ MPa}\sqrt{\text{m}}}{207 \text{ MPa}}\right)^2 = 0.00518 \text{ m} = 5.18 \text{ mm}$$

Thus, the largest internal crack size that this plate can support is $2a$, or $(2)(5.18 \text{ mm}) = 10.36$ mm.

7.2 FATIGUE OF METALS

In many types of service applications metal parts subjected to repetitive or cyclic stresses will fail due to **fatigue** loading at a much lower stress than that which the part can withstand under the application of a single static stress. These failures that occur under repeated or cyclic stressing are called **fatigue failures.** Examples of

Figure 7.14
Light fractograph of the fatigue-fracture surface of a keyed
shaft of 1040 steel (hardness ~ Rockwell C 30). The
fatigue crack originated at the left bottom corner of the
keyway and extended almost through the entire cross
section before final rupture occurred. (Magnified $1\frac{7}{8}\times$.)
(*After "Metals Handbook," vol. 9, 8th ed., American Society for Metals,
1974, p. 389. ASM International*)

machine parts in which fatigue failures are common are moving parts such as shafts,
connecting rods, and gears. Some estimates of failures in machines attribute about
80 percent to the direct action of fatigue failures.

A typical fatigue failure of a keyed steel shaft is shown in Fig. 7.14. A fatigue
failure usually originates at a point of stress concentration such as a sharp corner or
notch (Fig. 7.14), or at a metallurgical inclusion or flaw. Once nucleated, the crack
propagates across the part under the cyclic or repeated stresses. During this stage of
the fatigue process, clamshell or "beach" marks are created, as shown in Fig. 7.14.
Finally, the remaining section becomes so small that it can no longer support the
load, and complete fracture occurs. Thus, there are usually two distinct types of surface areas that can be recognized: (1) a smooth surface region due to the rubbing
action between the open surface region as the crack propagates across the section
and (2) a rough surface area formed by the fracture when the load becomes too high
for the remaining cross section. In Fig. 7.14 the fatigue crack had propagated almost
through the entire cross section before final rupture occurred.

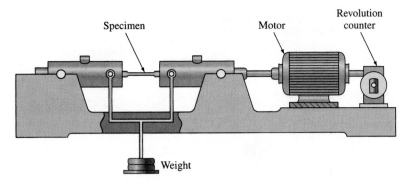

Figure 7.15
Schematic diagram of an R. R. Moore reversed-bending fatigue machine.
(After H. W. Hayden, W. G. Moffatt, and J. Wulff, "The Structure and Properties of Materials," vol. III, Wiley, 1965, p. 15.)

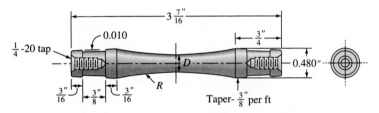

$D = 0.200$ to 0.400 in. selected on basis of ultimate strength of material
$R = 3.5$ to 10 in.

Figure 7.16
Sketch illustrating rotating-beam fatigue specimen (R. R. Moore type).
(From "Manual on Fatigue Testing," American Society for Testing and Materials, 1949).

Many types of tests are used to determine the **fatigue life** of a material. The most commonly used small-scale fatigue test is the rotating-beam test in which a specimen is subjected to alternating compression and tension stresses of equal magnitude while being rotated (Fig. 7.15). A sketch of the specimen for the R. R. Moore reversed-bending fatigue test is shown in Fig. 7.16. The surface of this specimen is carefully polished and tapered toward the center. During the testing of a fatigue sample by this apparatus, the center of the specimen is actually undergoing tension on the lower surface and compression on the upper surface by the weight attached in the center of the apparatus (Fig. 7.15), as exaggerated in Fig. 7.17. Data from this test are plotted in the form of *SN* curves in which the stress *S* to cause failure is plotted against the number of cycles *N* at which failure occurs. Figure 7.18 shows typical *SN* curves for a high-carbon steel and a high-strength aluminum alloy. For the aluminum alloy, the stress to cause failure decreases as the number of cycles is increased. For the carbon steel, there is first a decrease in fatigue strength as the number of cycles is increased

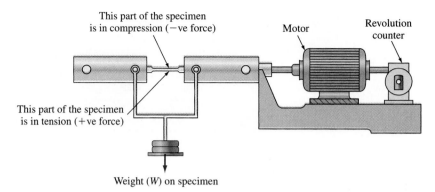

Figure 7.17

Exaggerated bending of sample to show action that produces positive tension and negative compressive forces on specimen.

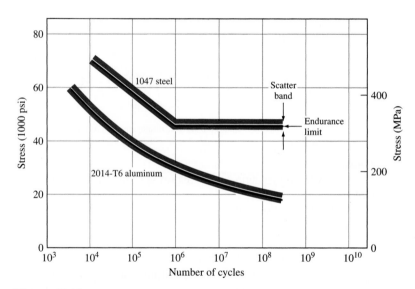

Figure 7.18

Stress versus number of cycles (*SN*) curves for fatigue failure for aluminum alloy 2014-T6 and medium-carbon steel 1047.
(*After H. W. Hayden, W. G. Moffatt, and J. Wulff, "The Structure and Properties of Materials," vol. III, Wiley, 1965, p. 15.*)

and then there is leveling off in the curve, with no decrease in fatigue strength as the number of cycles is increased. This horizontal part of the *SN* plot is called the *fatigue* or *endurance limit* and lies between 10^6 and 10^{10} cycles. Many ferrous alloys exhibit an endurance limit that is about one-half their tensile strength. Nonferrous alloys such as aluminum alloys do not have an endurance limit and may have fatigue strengths as low as one-third their tensile strength.

7.2.1 Cyclic Stresses

The applied fatigue stress may vary greatly in real cases and in fatigue tests. Many different kinds of fatigue test methods used in industry and research involve axial, torsional, and flexural stresses. Figure 7.19 shows graphs of fatigue stress versus number of fatigue cycles for three fatigue cycle tests. Figure 7.19a shows a graph of stress versus fatigue cycles for a *completely reversed stress cycle* of a sinusoidal form. This graph is typical of that produced by a rotating shaft operating at constant speed without overloads. The R. R. Moore reversed-bending fatigue machine shown in Fig. 7.15 produces similar stress versus number of fatigue cycles plots. In this fatigue cycle the maximum and minimum stresses are equal. By definition, the tensile stresses are considered positive and the compressive stresses negative, and the maximum stress has the highest numeric value and the minimum stress the lowest.

Figure 7.19b shows a *repeated stress cycle* in which the maximum stress σ_{max} and the minimum stress σ_{min} are equal. In this case both maximum and minimum stresses are tensile, but a repeated stress cycle can also have maximum and minimum stresses of opposite sign, or both in compression. Finally, a cyclic stress may

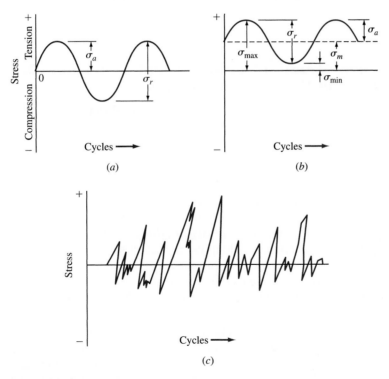

Figure 7.19
Typical fatigue stress versus cycles plots. (*a*) Completely reversed stress cycle. (*b*) Repeated stress cycle with equal σ_{max} and σ_{min}. (*c*) Random stress cycle.

vary randomly in amplitude and frequency, as shown in Fig. 7.19c. In this case there can be a spectrum of different fatigue graphs of stress versus cycles.

Fluctuating stress cycles are characterized by a number of parameters. Some of the most important ones are:

1. *Mean stress* σ_m is the algebraic mean of the maximum and minimum stresses in the fatigue cycle.

$$\sigma_m = \frac{\sigma_{max} + \sigma_{min}}{2} \tag{7.3}$$

2. *Range of stress* σ_r is the difference between σ_{max} and σ_{min}.

$$\sigma_r = \sigma_{max} - \sigma_{min} \tag{7.4}$$

3. *Stress amplitude* σ_a is one-half the stress cycle.

$$\sigma_a = \frac{\sigma_r}{2} = \frac{\sigma_{max} - \sigma_{min}}{2} \tag{7.5}$$

4. *Stress ratio* R is the ratio of minimum and maximum stresses.

$$R = \frac{\sigma_{min}}{\sigma_{max}} \tag{7.6}$$

7.2.2 Basic Structural Changes that Occur in a Ductile Metal in the Fatigue Process

When a specimen of a ductile homogeneous metal is subjected to cyclic stresses, the following basic structural changes occur during the fatigue process:

1. *Crack initiation.* The early development of fatigue damage occurs.

2. *Slipband crack growth.* Crack initiation occurs because plastic deformation is not a completely reversible process. Plastic deformation in one direction and then in the reverse direction causes surface ridges and grooves called *slipband extrusions* and *slipband intrusions* to be created on the surface of the metal specimen as well as damage within the metal along *persistent slipbands* (Figs. 7.20 and 7.21). The surface irregularities and damage along the persistent slipbands cause cracks to form at or near the surface that propagate into the specimen along the planes subjected to high shear stresses. This is called *stage I of fatigue crack growth,* and the rate of the crack growth is in general very low (for example, 10^{-10} m/cycle).

3. *Crack growth on planes of high tensile stress.* During stage I the crack may grow in a polycrystalline metal only a few grain diameters before it changes its direction to be perpendicular to the direction of the maximum tensile stress on the metal specimen. In this *stage II of crack growth,* a well-defined crack propagates at a relatively rapid rate (i.e., micrometers per cycle), and fatigue striations are created as the crack advances across the cross section of the metal specimen (Fig. 7.14). These striations are useful in fatigue failure analysis for determining the origin and direction of propagation of fatigue cracks.

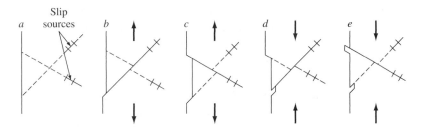

Figure 7.20
Mechanism for the formation of slipband extrusions and intrusions.
[*After A. H. Cottrell and D. Hull, Proc. R. Soc. London,* **242A:**211–213(1957).]

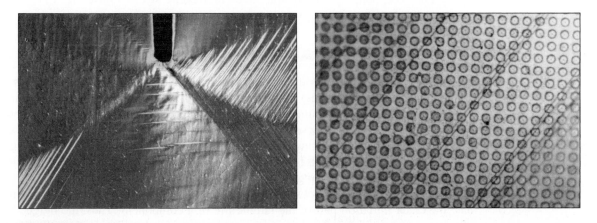

Figure 7.21
(*a*) Persistent slipbands in copper single crystal. (*b*) The polymer dots deposited on the surface are in many cases cut in half by the slipbands (dark lines on the surface) resulting in relative displacement of the two halves.
(*Courtesy of Wendy C. Crone, University of Wisconsin.*)

4. *Ultimate ductile failure.* Finally, when the crack covers a sufficient area so that the remaining metal at the cross section cannot support the applied load, the sample ruptures by ductile failure.

7.2.3 Some Major Factors that Affect the Fatigue Strength of a Metal

The fatigue strength of a metal or alloy is affected by factors other than the chemical composition of the metal itself. Some of the most important of these are:

1. *Stress concentration.* Fatigue strength is greatly reduced by the presence of stress raisers such as notches, holes, keyways, or sharp changes in cross sections. For example, the fatigue failure shown in Fig. 7.14 started at the keyway in the steel shaft. Fatigue failures can be minimized by careful design to avoid stress raisers whenever possible.

2. *Surface roughness.* In general the smoother the surface finish on the metal sample, the higher the fatigue strength. Rough surfaces create stress raisers that facilitate fatigue crack formation.

3. *Surface condition.* Since most fatigue failures originate at the metal surface, any major change in the surface condition will affect the fatigue strength of the metal. For example, surface-hardening treatments for steels, such as carburizing and nitriding, which harden the surface, increase fatigue life. Decarburizing, on the other hand, which softens a heat-treated steel surface, lowers fatigue life. The introduction of a favorable compressive residual stress pattern on the metal surface also increases fatigue life.

4. *Environment.* If a corrosive environment is present during the cyclic stress of a metal, the chemical attack greatly accelerates the rate at which fatigue cracks propagate. The combination of corrosion attack and cyclic stresses on a metal is known as *corrosion fatigue.*

7.3 FATIGUE CRACK PROPAGATION RATE

Most fatigue data for metals and alloys for high-cycle fatigue (i.e., fatigue lives of greater than 10^4 to 10^5 cycles) has been concerned with the nominal stress required to cause failure in a given number of cycles, that is, *SN* curves such as those shown in Fig. 7.18. However, for these tests smooth or notched specimens are usually used, and thus it is difficult to distinguish between fatigue crack initiation life and fatigue crack propagation life. Thus, test methods have been developed to measure fatigue life associated with preexisting flaws in a material.

Preexisting flaws or cracks within a material component reduce or may eliminate the crack initiation part of the fatigue life of a component. Thus, the fatigue life of a component with preexisting flaws may be considerably shorter than the life of one without flaws. In this section we will utilize fracture mechanics methodology to develop a relationship to predict fatigue life in a material with preexisting flaws and under stress-state conditions due to cyclic fatigue action.

A high-cycle fatigue experimental setup for measuring the crack growth rate in a compact-type metal test specimen containing a preexisting crack of known length is shown in Fig. 7.22. In this setup the cyclic fatigue action is generated in the up-and-down vertical direction and the crack length is measured by the change in electrical potential produced by the crack being further opened and extended by the fatigue action.

7.3.1 Correlation of Fatigue Crack Propagation with Stress and Crack Length

Let us now consider qualitatively how fatigue crack length varies with an increasing number of applied cyclic stresses using data obtained from an experimental setup such as that shown in Fig. 7.22. Let us use several test samples of a material each of which has a mechanical crack in its side as indicated in Fig. 7.23*a*. Now let us apply a constant-amplitude cyclic stress to the samples and measure the increase in crack length as a function of the number of applied stress cycles. Figure 7.23*b* shows qualitatively how a plot of crack length versus number of stress cycles for two levels of stress might appear for a particular material such as mild steel.

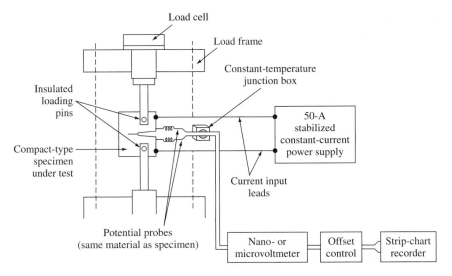

Figure 7.22
Schematic of the direct-current electrical potential crack monitoring system for the high-cycle fatigue testing of a compact test sample.
(After "Metals Handbook," vol. 8, 9th ed., American Society for Metals, 1985, p. 388.)

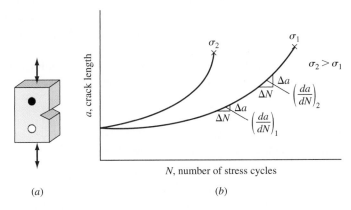

Figure 7.23
(a) Thin-plate test sample with edge crack under cyclic stress. (b) Plot of crack length versus number of stress cycles for stresses σ_1 and $\sigma_2 (\sigma_2 > \sigma_1)$.

Examination of the curves of Fig. 7.23b indicates the following:

1. When the crack length is small, the **fatigue crack growth rate *da/dN*** is also relatively small.

2. The crack growth rate *da/dN* increases with increasing crack length.

3. An increase in cyclic stress σ increases the crack growth rate.

Thus, the crack growth rate for materials under cyclic stress that behave as indicated in Fig. 7.23b shows the following relationship:

$$\frac{da}{dN} \propto f(\sigma, a) \tag{7.7}$$

which reads, "the fatigue crack growth rate da/dN varies as a function of the applied cyclic stress σ and the crack length a." After much research, it has been shown that for many materials the fatigue crack growth rate is a function of the stress-intensity factor K (mode I) of fracture mechanics, which itself is a combination of stress and crack length. For many engineering alloys, the fatigue crack growth rate expressed as the differential da/dN can be related to the stress-intensity range ΔK for a constant-amplitude fatigue stress by the equation

$$\frac{da}{dN} = A\Delta K^m \tag{7.8}$$

where da/dN = fatigue crack growth rate, mm/cycle or in./cycle

ΔK = stress-intensity factor range ($\Delta K = K_{max} - K_{min}$), MPa$\sqrt{m}$ or ksi$\sqrt{in.}$

A, m = constants that are a function of the material, environment, frequency, temperature, and stress ratio

Note in Eq. 7.8 that we use the stress-intensity factor K_I (mode I) and not the fracture toughness value K_{IC}. Thus, at the maximum cyclic stress the stress-intensity factor $K_{max} = \sigma_{max}\sqrt{\pi a}$, and at the minimum cyclic stress, $\Delta K_{min} = \sigma_{min}\sqrt{\pi a}$. For the range of stress-intensity factor, $\Delta K(\text{range}) = K_{max} - K_{min} = \Delta K = \sigma_{max}\sqrt{\pi a} - \sigma_{min}\sqrt{\pi a} = \sigma_{range}\sqrt{\pi a}$. Since the stress-intensity factor is not defined for compressive stresses, if σ_{min} is in compression, K_{min} is assigned zero value. If there is a Y geometric correction factor for the $\Delta K = \sigma_r\sqrt{\pi a}$ equation, then $\Delta K = Y\sigma_r\sqrt{\pi a}$.

7.3.2 Fatigue Crack Growth Rate versus Stress-Intensity Factor Range Plots

Usually fatigue crack length versus stress-intensity factor range data are plotted as log da/dN versus log stress-intensity factor range ΔK. These data are plotted as a log-log plot since in most cases a straight line or close to a straight-line plot is obtained. The basic reason for the straight-line plot is that the da/dN versus ΔK data closely obey the $da/dN = A\Delta K^m$ relationship, and so if the log is taken of both sides of this equation, we obtain

$$\log\frac{da}{dN} = \log(A\Delta K^m) \tag{7.9}$$

or

$$\log\frac{da}{dN} = m \log \Delta K + \log A \tag{7.10}$$

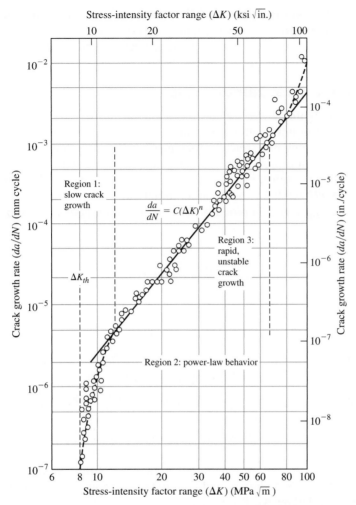

Figure 7.24
Fatigue crack growth behavior of ASTM A533 B1 steel (yield
strength 470 MPa [70 ksi]). Test conditions: $R = 0.10$:
ambient room air, 24°C.
(*After P. C. Paris et al., Stress Analysis and Growth of Cracks, STP513
ASTM, Philadelphia, 1972, pp. 141–176.*)

which is an equation of a straight line of the form $y = mx + b$. Thus, a plot of
log (da/dN) versus log ΔK produces a straight line with a slope of m.

Figure 7.24 shows a plot of log crack growth rate versus log stress-intensity
factor range for a fatigue test of an ASTM A533 B1 steel. This plot is divided into
three regions: region 1 in which the fatigue crack growth rate is very slow, region
2 in which the plot is a straight line represented by the power law $da/dn = A\Delta K^{m}$,
and region 3 in which rapid unstable crack growth takes place, approaching failure

of the sample. The limiting value of ΔK below which there is no measurable crack growth is called the *stress-intensity factor range threshold* ΔK_{th}. No crack growth should occur below this stress-intensity range level. The value of m for fatigue crack growth da/dN in region 2 usually varies from about 2.5 to 6.

7.3.3 Fatigue Life Calculations

Sometimes in designing a new engineering component using a particular material it is desirable to obtain information about the fatigue life of the part. This can be done in many cases by combining fracture toughness data with fatigue crack growth data to produce an equation that can be used to predict fatigue life.

One type of equation for calculating fatigue life can be developed by integrating Eq. 7.8, $dA/dN = A\Delta K^m$, between an initial crack (flaw) size a_o and the critical crack (flaw) size a_f, which is produced at fatigue failure after the number of cycles to failure N_f.

We start with Eq. 7.8:

$$\frac{da}{dN} = A\Delta K^m \tag{7.11}$$

Since

$$\Delta K = Y\sigma\sqrt{\pi a} = Y\sigma\pi^{1/2}a^{1/2} \tag{7.12}$$

It follows that

$$\Delta K^m = Y^m\sigma^m\pi^{m/2}a^{m/2} \tag{7.13}$$

Substituting $Y^m\sigma^m\pi^{m/2}a^{m/2}$ of Eq. 7.13 for ΔK^m of Eq. 7.11 gives

$$\frac{da}{dN} = A(Y\sigma\sqrt{\pi a})^m = A(Y^m\sigma^m\pi^{m/2}a^{m/2}) \tag{7.14}$$

After rearranging Eq. 7.14, we integrate the crack size from the initial crack size a_o to the final crack size at failure a_f and the number of fatigue cycles from zero to the number at fatigue failure N_f. Thus,

$$\int_{a_v}^{a_f} da = AY^m\sigma^m\pi^{m/2} \cdot a^{m/2}\int_0^{N_f} dN \tag{7.15}$$

and

$$\int_0^{N_f} dN = \int_{a_v}^{a_f} \frac{da}{A\sigma^m\pi^{m/2}Y^m a^{m/2}} = \frac{1}{A\sigma^m\pi^{m/2}Y^m}\int_{a_v}^{a_f}\frac{da}{a^{m/2}} \tag{7.16}$$

Using the relationship

$$\int a^n \, da = \frac{a^{n+1}}{n+1} + c \tag{7.17}$$

we integrate Eq. 7.16,

$$\int_0^{N_f} dN = N\Big|_0^{N_f} = N_f \tag{7.18a}$$

and letting $n = -m/2$,

$$\frac{1}{A\sigma^m\pi^{m/2}Y^m}\int_{a_v}^{a_f}\frac{da}{a^{m/2}} = \frac{1}{A\sigma^m\pi^{m/2}Y^m}\left(\frac{a^{-(m/2)+1}}{-m/2+1}\right)\Bigg|_{a_o}^{a_f} \qquad \textbf{(7.18b)}$$

Thus,

$$N_f = \frac{a_f^{-(m/2)+1} - a_0^{-(m/2)+1}}{A\sigma^m\pi^{m/2}Y^m[-(m/2)+1]} \qquad m \neq 2 \qquad \textbf{(7.19)}$$

Equation 7.19 assumes that $m \neq 2$ and that Y is independent of the crack length, which is not usually the case. Thus, Eq. 7.19 may or may not be the true value for the fatigue life of the component. For the more general case, $Y = f(a)$, the calculation of N_f must take into account the change in Y, and so ΔK and ΔN must be calculated for small successive amounts of length.

An alloy steel plate is subjected to constant-amplitude uniaxial fatigue cyclic tensile and compressive stresses of magnitudes of 120 and 30 MPa, respectively. The static properties of the plate are a yield strength of 1400 MPa and a fracture toughness K_{IC} of 45 MPa$\sqrt{m}$. If the plate contains a uniform through thickness *edge* crack of 1.00 mm, how many fatigue cycles are estimated to cause fracture? Use the equation da/dN (m/cycle) = $2.0 \times 10^{-12}\Delta K^3$ (MPa$\sqrt{m}$)^3. Assume $Y = 1$ in the fracture toughness equation.

EXAMPLE PROBLEM 7.2

■ **Solution**
We will assume for the plate that

$$\frac{da}{dN}\text{ (m/cycle)} = 2.0 \times 10^{-12}\,\Delta K^3\,(\text{MPa}\sqrt{m})^3$$

Thus, $A = 2.0 \times 10^{-12}$, $m = 3$, and $\sigma_r = (120 - 0)$ MPa (since compressive stresses are ignored), and $Y = 1$.

The initial crack size a_o is equal to 1.00 mm. The final crack size a_f is determined from the fracture toughness equation

$$a_f = \frac{1}{\pi}\left(\frac{K_{IC}}{\sigma_r}\right)^2 = \frac{1}{\pi}\left(\frac{45\text{ MPa}\sqrt{m}}{120\text{ MPa}}\right)^2 = 0.0449\text{ m}$$

The fatigue life in cycles N_f is determined from Eq. 7.19:

$$N_f = \frac{a_f^{-(m/2)+1} - a_o^{-(m/2)+1}}{[-(m/2)+1]A\sigma^m\pi^{m/2}Y^m} \qquad m \neq 2$$

$$= \frac{(0.0449\text{ m})^{-(3/2)+1} - (0.001\text{ m})^{-(3/2)+1}}{(-\frac{3}{2}+1)(2.0\times10^{-12})(120\text{ MPa})^3(\pi)^{3/2}(1.00)^3}$$

$$= \frac{-2}{(2\times10^{-12})(\pi^{3/2})(120)^3}\left(\frac{1}{\sqrt{0.0449}} - \frac{1}{\sqrt{0.001}}\right)$$

$$= \frac{-2\times26.88}{(2\times10^{-12})(5.56)(1.20)^3(10^6)} = 2.79\times10^6\text{ cycles} \blacktriangleleft$$

7.4 CREEP AND STRESS RUPTURE OF METALS

7.4.1 Creep of Metals

When a metal or an alloy is under a constant load or stress, it may undergo progressive plastic deformation over a period of time. This *time-dependent strain* is called **creep.** The creep of metals and alloys is very important for some types of engineering designs, particularly those operating at elevated temperatures. For example, an engineer selecting an alloy for the turbine blades of a gas turbine engine must choose an alloy with a very low **creep rate** so that the blades can remain in service for a long period of time before having to be replaced due to their reaching the maximum allowable strain. For many engineering designs operating at elevated temperatures, the creep of materials is the limiting factor with respect to how high the operating temperature can be.

Let us consider creep of a pure polycrystalline metal at a temperature above one-half its absolute melting point, $\frac{1}{2}T_M$ (high-temperature creep). Let us also consider a creep experiment in which an annealed tensile specimen is subjected to a constant load of sufficient magnitude to cause extensive creep deformation. When the change of length of the specimen over a period of time is plotted against time increments, a *creep curve,* such as the one shown in Fig. 7.25, is obtained.

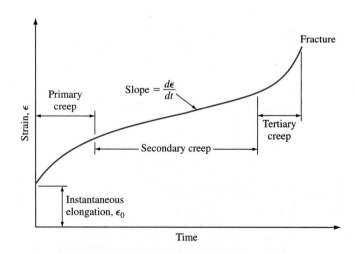

Figure 7.25

A typical creep curve for a metal. The curve represents the time versus strain behavior of a metal or alloy under a constant load at constant temperature. The second stage of creep (linear creep) is of most interest to the design engineer because extensive creep occurs under these conditions.

In the idealized creep curve of Fig. 7.25, there is first an instantaneous, rapid elongation of the specimen, ϵ_0. Following this, the specimen exhibits primary creep in which the strain rate decreases with time. The slope of the creep curve ($d\epsilon/dt$, or ϵ) is designated the *creep rate*. Thus, during primary creep the creep rate progressively decreases with time. After primary creep, a second stage of creep occurs in which the creep rate is essentially constant and is therefore also referred to as *steady-state creep*. Finally, a third or tertiary stage of creep occurs in which the creep rate rapidly increases with time up to the strain at fracture. The shape of the creep curve depends strongly on the applied load (stress) and temperature. Higher stresses and higher temperatures increase the creep rate.

During primary creep the metal strain-hardens to support the applied load and the creep rate decreases with time as further strain hardening becomes more difficult. At higher temperatures (i.e., above about $0.5T_M$ for the metal) during secondary creep, recovery processes involving highly mobile dislocations counteract the strain hardening so that the metal continues to elongate (creep) at a steady-state rate (Fig. 7.25). The slope of the creep curve ($d\epsilon/dt = \epsilon$) in the secondary stage of creep is referred to as the *minimum creep rate*. During secondary creep the creep resistance of the metal or alloy is the highest. Finally, for a constant-loaded specimen, the creep rate accelerates in the tertiary stage of creep due to necking of the specimen and also to the formation of voids, particularly along grain boundaries. Figure 7.26 shows intergranular cracking in a type 304L stainless steel that has undergone creep failure.

At low temperatures (i.e., below $0.4T_M$) and low stresses, metals show primary creep but negligible secondary creep since the temperature is too low for diffusional recovery creep. However, if the stress on the metal is above the ultimate tensile strength, the metal will elongate as in an ordinary engineering tensile test. In general, as both the stress on the metal undergoing creep and its temperature are increased, the creep rate is also increased (Fig. 7.27).

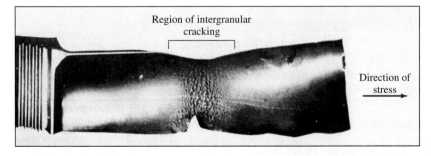

Figure 7.26
A jet engine turbine blade that has undergone creep deformation, causing local deformation and a multiplicity of intergranular cracks.
(After J. Schijve in "Metals Handbook," vol. 10, 8th ed., American Society for Metals, 1975, p. 23. ASM International)

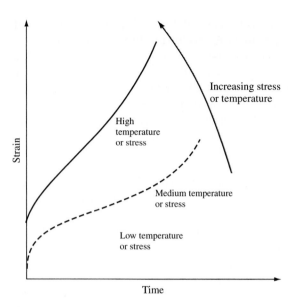

Figure 7.27
Effect of increasing stress on the shape of the
creep curve of a metal (schematic). Note that as
the stress increases, the strain rate increases.

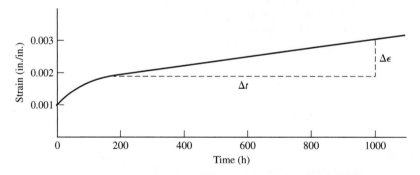

Figure 7.28
Creep curve for a copper alloy tested at 225°C and 230 MPa (33.4 ksi).
The slope of the linear part of the curve is the steady-state creep rate.

7.4.2 The Creep Test

The effects of temperatures and stress on the creep rate are determined by the creep
test. Multiple creep tests are run using different stress levels at constant temperature or different temperatures at a constant stress, and the creep curves are plotted as shown in Fig. 7.28. The minimum creep rate or slope of the second stage

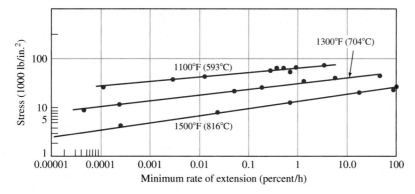

Figure 7.29
Effect of stress on the creep rate of type 316 stainless steel
(18% Cr–12% Ni–2.5% Mo) at various temperatures (1100,
1300, 1500°F [593, 704, 816°C]).
(After H. E. McGannon [ed.], "The Making, Shaping, and Treating of Steel," 9th ed.,
United States Steel, 1971, p. 1256.)

of the creep curve is measured for each curve, as indicated in Fig. 7.28. The stress to produce a minimum creep rate of 10^{-5} percent/h at a given temperature is a common standard for creep strength. In Fig. 7.29 the stress to produce a minimum creep rate of 10^{-5} percent/h for type 316 stainless steel can be determined by extrapolation.

Determine the steady-state creep rate for the copper alloy whose creep curve is shown in Fig. 7.28.

■ **Solution**
The steady-state creep rate for this alloy for the creep curve shown in Fig. 7.28 is obtained by taking the slope of the linear part of the curve as indicated in the figure. Thus,

$$\text{Creep rate} = \frac{\Delta\epsilon}{\Delta t} = \frac{0.0029 - 0.0019}{1000\ \text{h} - 200\ \text{h}} = \frac{0.001\ \text{in./in.}}{800\ \text{h}} = 1.2 \times 10^{-6}\ \text{in./in./h} \blacktriangleleft$$

7.4.3 Creep-Rupture Test

The **creep-rupture** or **stress-rupture test** is essentially the same as the creep test except that the loads are higher and the test is carried out to failure of the specimen. Creep-rupture data are plotted as log stress versus log rupture time, as shown in Fig. 7.30. In general the time for stress rupture to occur is decreased as the applied stress and temperature are increased. Slope changes as observed in Fig. 7.30 are caused by factors such as recrystallization, oxidation, corrosion, or phase changes.

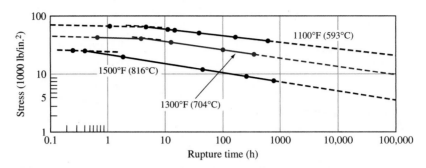

Figure 7.30
Effect of stress on the time to rupture of type 316 stainless steel
(18% Cr–12% Ni–2.5% Mo) at various temperatures (1100, 1300,
1500°F [593, 704, 816°C]).
*(After H. E. McGannon [ed.], "The Making, Shaping, and Treating of Steel," 9th ed.,
United States Steel, 1971, p. 1257.)*

7.5 GRAPHICAL REPRESENTATION OF CREEP- AND STRESS-RUPTURE TIME-TEMPERATURE DATA USING THE LARSEN-MILLER PARAMETER

Creep-stress rupture data for high-temperature creep-resistant alloys are often plotted as log stress to rupture versus a combination of log time to rupture and temperature. One of the most common time-temperature parameters used to present this kind of data is the **Larsen-Miller (L.M.) parameter,** which in a generalized form is

$$P(\text{L.M.}) = T[\log t_r + C] \tag{7.20}$$

where T = temperature, K or °R

t_r = stress-rupture time, h

C = constant, usually of the order of 20

In terms of kelvin-hours the L.M. parameter equation becomes

$$P(\text{L.M.}) = [T(°C) + 273(20 + \log t_r)] \tag{7.21}$$

In terms of Rankine-hours, the equation becomes

$$P(\text{L.M.}) = [T(°F) + 460(20 + \log t_r)] \tag{7.22}$$

According to the L.M. parameter, at a given stress level the log time to stress rupture plus a constant of the order of 20 multiplied by the temperature in kelvins or degrees Rankine remains constant for a given material.

Figure 7.31 compares L.M. parameter plots for log stress to rupture for three heat-treated high-temperature creep-resistant alloys. If two of the three variables of

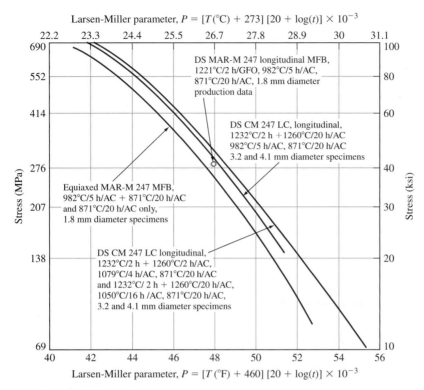

Larsen-Miller parameter, $P = [T(^\circ C) + 273][20 + \log(t)] \times 10^{-3}$

DS MAR-M 247 longitudinal MFB,
1221°C/2 h/GFO, 982°C/5 h/AC,
871°C/20 h/AC, 1.8 mm diameter
production data

DS CM 247 LC, longitudinal,
1232°C/2 h + 1260°C/20 h/AC
982°C/5 h/AC, 871°C/20 h/AC
3.2 and 4.1 mm diameter specimens

Equiaxed MAR-M 247 MFB,
982°C/5 h/AC + 871°C/20 h/AC
and 871°C/20 h/AC only,
1.8 mm diameter specimens

DS CM 247 LC longitudinal,
1232°C/2 h + 1260°C/2 h/AC,
1079°C/4 h/AC, 871°C/20 h/AC
and 1232°C/ 2 h + 1260°C/20 h/AC,
1050°C/16 h /AC, 871°C/20 h/AC,
3.2 and 4.1 mm diameter specimens

Larsen-Miller parameter, $P = [T(^\circ F) + 460][20 + \log(t)] \times 10^{-3}$

Figure 7.31
Larsen-Miller stress-rupture strength of directionally solidified (DS) CM 247
LC alloy versus DS and equiaxed MAR-M 247 alloy. MFB: machined from
blade; GFQ: gas fan-quenched; AC: air-cooled.
(After "Metals Handbook," vol. 1, 10th ed., ASM International, 1990, p. 998.)

time to rupture, temperature stressed at, and stress are known, then the third variable
that fits the L.M. parameter can be determined from log stress versus L.M. parameter
plots as indicated in Example Problem 7.4.

Using the L.M. parameter plot of Fig. 7.31 at a stress of 207 MPa (30 ksi), determine the
time to stress-rupture at 980°C for directionally solidified alloy CM 247 (uppermost graph).

**EXAMPLE
PROBLEM 7.4**

■ **Solution**
From Fig. 7.30 at a stress of 207 MPa, the value of the L.M. parameter is 27.8×10^3 K · h.
Thus,

$$P = T(K)(20 + \log t_r) \quad T = 980°C + 273 = 1253 \text{ K}$$
$$27.8 \times 10^3 = 1253(20 + \log t_r)$$
$$\log t_r = 22.19 - 20 = 2.19$$
$$t_r = 155 \text{ h} \blacktriangleleft$$

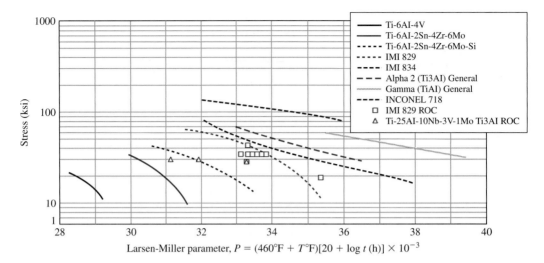

Figure 7.32
Larsen-Miller diagram for 0.2% strain, comparing ROC and IM Ti-829 and ROC Ti-25-10-3-1 to several commercially important alpha and beta alloys. ROC: rapid omnidirectional compaction.
[*After N. R. Osborne et al., SAMPE Quart, (4)22:26(1992).*]

Larsen-Miller parameter plots of a set amount of creep strain, for example, 0.2 percent, produced by variable stresses at different times and temperatures can also be used for comparing high-temperature creep properties of materials, as shown in Fig. 7.32 for various titanium alloys. Example Problem 7.5 shows how time of stress, creep time, and temperature of creep can be determined using log stress versus L.M. parameter plots.

EXAMPLE PROBLEM 7.5

Calculate the time to cause 0.2 percent creep strain in gamma titanium aluminide (TiAl) at a stress of 40 ksi and 1200°F using Fig. 7.32.

■ **Solution**
For these conditions, from Fig. 7.31, $P = 38{,}000$. Thus,

$$P = 38{,}000 = (1200 + 460)(\log t_{0.2\%} + 20)$$
$$22.89 = 20 + \log t$$
$$\log t = 2.89$$
$$t = 776 \text{ h} \blacktriangleleft$$

7.6 A CASE STUDY IN FAILURE OF METALLIC COMPONENTS

Owing to among other things, material defects, poor design, and misuse, metal components occasionally fail by fracture fatigue and creep. In some cases these failures occur during the testing of prototypes by manufacturers. In other cases

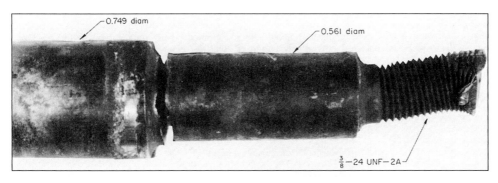

Figure 7.33
Premature failure of a fan shaft (dimensions in inches).
(*ASM Handbook of Failure Analysis and Prevention, vol. 11. 1992. ASM International*)

they occur after a product has been sold and is in use. Both cases justify a failure analysis to determine the cause of the failure. In the former case information about the cause can be used as feedback to improve design and materials selection. In the latter case failure analysis may be required as part of a product liability case. In either case engineers utilize their knowledge of the mechanical behavior of materials to perform the analysis. The analyses themselves are forensic procedures requiring documentation and preservation of evidence. An example is given in the following case study.

The first step in the failure analysis process is to determine the function of the component, the specifications required by the user, and the circumstances under which failure occurred. In this case, a fan drive support shaft was required to be made of cold-drawn 1040 or 1045 steel with a yield strength of 586 MPa. The expected life of the shaft was estimated at 6440 km. However, the shaft fractured after only 3600 km of service, Fig. 7.33.

The investigation generally starts with the visual examination of the failed component. During the preliminary visual examination care must be taken to protect the fracture surface from any additional damage. One should not try to mate the fracture surfaces since this could introduce superficial damage to the surface that could unduly influence any future analysis. In the case of the fan drive shaft, the investigation revealed that fracture had initiated at two points in the immediate vicinity of a fillet in the region with an abrupt change in the diameter of the shaft. The two initiation points were approximately 180 degrees apart, Fig. 7.34. Based on the visual analysis of the surface, the investigators determined that fractures propagated from the two initiation points toward the center of the shaft where a final catastrophic fracture had occurred. Because of the symmetry of the two initiation points and the beach mark patterns observed on the surface, the investigators concluded that the fracture was typical of reverse-bending fatigue. A combination of cyclic alternating bending and a sharp fillet radius (*stress riser*) was believed to have caused the fracture.

Figure 7.34
Fracture surface showing the symmetrically positioned fracture origins. Note the center region at which the final fracture took place.
(*ASM International*)

After the visual inspection and all nondestructive examinations were completed, other more intrusive and even destructive tests were performed. For instance, chemical analysis of the shaft material revealed that it was in fact 1040 steel as required by the user. The investigators were also able to machine a tensile test specimen out of the center of the fractured shaft. The tensile test revealed that tensile and yield strengths of the metal were 631 and 369 MPa, respectively, with an elongation at fracture of 27 percent. The yield strength of the shaft material (369 MPa) is considerably lower than that required by the user (586 MPa). A subsequent metallographic examination revealed that the grain structure was predominantly equiaxed. Recall from the discussion in the previous chapter that cold working increases the yield strength, decreases ductility, and creates a grain structure that is generally not equiaxed but elongated. The lower yield strength, relatively high ductility, and the equiaxed nature of the microstructure revealed that the 1040 steel used in the manufacturing of this shaft was not cold drawn as specified by the user. In fact the evidence suggests that the material was hot rolled.

The use of hot-rolled steel (with a lower fatigue limit) together with the stress-riser effect of the sharp fillet resulted in the failure of the component under reverse-bending loading. If the component were made of cold-drawn 1040 steel (with a 40 percent higher fatigue limit), the fracture may have been avoided or delayed. This case study shows the importance of the engineer's knowledge of material properties, processing techniques, heat treatment, and selection in the successful design and operation of components.

7.7 RECENT ADVANCES AND FUTURE DIRECTIONS IN IMPROVING THE MECHANICAL PERFORMANCE OF METALS

In previous chapters some of the structurally attractive features of nanocrystalline materials such as high strength, increased hardness, and improved resistance to wear were briefly discussed. However, all these improvements would only have anecdotal scientific value if the ductility, fracture, and damage tolerance of these materials do not meet acceptable levels for specific applications. In the following paragraphs the current knowledge and advances related to the above properties of nanocrystalline metals will be discussed. It should be mentioned that only limited information on the behavior of nanocrystalline metals exists and significantly more research is required to establish a level of understanding comparable to that of microcrystalline metals.

7.7.1 Improving Ductility and Strength Simultaneously

Pure copper in its annealed and coarse-grained state shows tensile ductility as large as 70 percent but very low yield strength. The nanocrystalline form of pure copper with grain size less than 30 nm possesses significantly higher yield strength but with tensile ductility of less than 5 percent. Early evidence indicates that this trend is typical of pure nanocrystalline metals with FCC structure such as copper and nickel. A similar trend has not been established yet for BCC and HCP nanocrystalline metals; on the contrary, nanocrystalline Co (HCP) shows a tensile elongation comparable to microcrystalline Co. Nanocrystalline FCC metals are significantly more brittle than their microcrystalline counterparts. This is schematically illustrated in Fig. 7.35. In this figure curve *A* shows

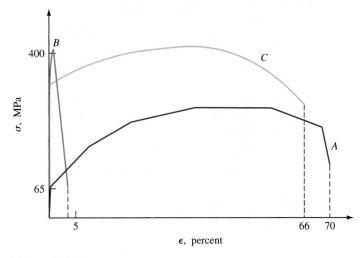

Figure 7.35
A schematic of the stress strain diagrams for microcrystalline (curve *A*), nanocrystalline (curve *B*), and mixed grain (curve *C*) pure copper.

the stress-strain behavior of annealed microcrystalline copper: A yield strength of approximately 65 MPa and a ductility of about 70 percent is observed. Curve *B* represents the nanocrystalline copper with enhanced yield strength of about 400 MPa and a ductility of less than 5 percent. The toughness of each metal may be determined by estimating the area under the corresponding stress-strain curve. It is clear from this figure that the toughness of the nanocrystalline copper is significantly lower. The reduction in ductility and therefore toughness has been attributed to the formation of localized bands of strain called *shear bands*. In the absence of dislocation activity in the nanograin (due to extremely small size), the grain does not deform in a conventional manner; instead, the deformation localizes into tiny shear bands that eventually fracture without significant deformation in the rest of the grain. Consequently, the extraordinary gains in yield strength are rendered ineffectual since low ductility levels result in reduced toughness and therefore minimal practical applications of these materials.

Fortunately, scientists have recently been able to produce nanocrystalline copper with ductility comparable to its microcrystalline counterpart. This is accomplished through a complex thermomechanical-forming process that includes a series of cold-rolling passes of the sample at liquid nitrogen temperature with additional cooling after each pass followed by a highly controlled annealing process. The resulting material contains about 25 percent micrometer-sized grains in a matrix of nanometer and ultrafine grains, (Fig. 7.36).

The cold-rolling process at liquid nitrogen temperature allows the formation of a large density of dislocations. The cold temperature does not allow the recovery of dislocations and in turn causes the density to increase beyond the levels achievable at room temperature. At this stage the severely deformed sample has a mixture of

Figure 7.36
A TEM micrograph showing the mixture of micro-, ultrafine-, and nanosized grains in pure copper.
(*Y. M. Wang, M. W. Chen, F. Zhou, E. Ma, Nature, vol. 419, Oct. 2002: Figure 3b*)

nanocrystalline and ultrafine grain structure. The sample is then annealed under highly controlled conditions. The annealing process allows recrystallization and growth of some grains to a range of 1 to 3 μm (called *abnormal grain growth* or *secondary recrystallization*). The existence of the large grains allows an elevated level of dislocation and twinning activity and therefore deformation in the overall material while the predominantly nanosized and ultrafine grains maintain high yield strength. This copper possesses both high yield strength and high ductility and therefore high toughness as schematically presented by curve *C* in Fig. 7.35. In addition to such novel thermomechanical-forming processes that allow increased toughness of nanocrystalline materials, some research efforts have shown that synthesis of two-phase nanocrystalline materials may result in improved ductility and toughness. Such technological advancements are critical in facilitating the application of nanomaterials to various fields.

7.7.2 Fatigue Behavior in Nanocrystalline Metals

Primary fatigue experiments on nanocrystalline (4 to 20 nm), ultrafine (300 nm), and microcrystalline pure nickel at a load ratio *R* of zero (zero-tension-zero) and a cycle frequency of 1 Hz has shown a significant affect on its *SN* fatigue response. Both nanocrystalline and ultrafine nickel show an increase in the tensile stress range ($\sigma_{max} - \sigma_{min}$) at which endurance limit (defined as 2-million cycles) is achieved when compared to microcrystalline nickel. The nanocrystalline nickel shows a slightly higher increase then the ultrafine nickel. However, fatigue crack growth experiments using edge notched specimens of nickel with the same grain size as above show a different picture. These experiments show that the fatigue crack growth is increased in the intermediate regime with decreasing grain size. Additionally, a lower fatigue crack growth threshold, K_{th}, is observed for the nanocrystalline metal. Overall, the results show a concurrent beneficial and detrimental effect on fatigue performance of materials in the nanoscale grain size.

It is clear from the brief synopsis give above that a great deal of research needs to be performed before we can begin to understand the behavior of these materials and to use these materials in applications required by various industries. The extraordinary promise of these materials drives the efforts to find the answers.

7.8 SUMMARY

Fracture of metallic components during service is of great importance and consequence. The proper selection of a material for a component is a critical step in avoiding unwanted failures. Fracture of metals may generally be classified as either ductile or brittle. This is easily observed by performing simple static tension tests. Ductile fracture is accompanied by severe plastic deformation prior to failure. On the contrary, brittle shows little or no deformation prior to fracture and is therefore more problematic. In some cases, under high loading rates or lowered temperatures, originally ductile materials behave in a brittle fashion called ductile to brittle transition. Therefore selection of materials for components that operate in cold temperatures should be done with care.

Since defects such as microcracks weaken a material, engineers use the concept of fracture toughness based on the assumption of pre-existing flaws (fracture mechanics) to design components that are safer. The concept of stress intensity factor, K, at a crack tip is used to represent the combined effect of stress at the crack tip and the crack length.

The failure of metallic components under cyclic or repeated loading called fatigue failure is of tremendous importance to engineers. Its importance is due to the low level of stresses at which such failures occur, the hidden nature of the damage (inside the material), and its sudden and abrupt failure. Another form of failure that occurs at high temperatures and constant loading is called creep, which is defined as progressive plastic deformation over a period of time. Engineers are very conscious of such failures and use high factors of safety to guard against them.

Engineers and scientists always search for new materials that offer higher strength, ductility, fatigue resistance, and are in general resistant to failure. Nanocrystalline materials promise to be the materials of choice for the future offering a combination of properties that will greatly enhance a material's resistance to fracture. However, more research is needed by materials scientists to achieve this goal.

7.9 DEFINITIONS

Sec. 7.1

Ductile fracture: a mode of fracture characterized by slow crack propagation. Ductile fracture surfaces of metals are usually dull with a fibrous appearance.

Brittle fracture: a mode of fracture characterized by rapid crack propagation. Brittle fracture surfaces of metals are usually shiny and have a granular appearance.

Ductile to brittle transition (DBT): observed reduced ductility and fracture resistance of a material when temperature is low.

Sec. 7.2

Fatigue: the phenomenon leading to fracture under repeated stresses having a maximum value less than the ultimate strength of the material.

Fatigue failure: failure that occurs when a specimen undergoing fatigue fractures into two parts or otherwise has been significantly reduced in stiffness.

Fatigue life: the number of cycles of stress or strain of a specific character that a sample sustains before failure.

Sec. 7.3

Fatigue crack growth rate *da/dN*: the rate of crack growth extension caused by constant-amplitude fatigue loading.

Sec. 7.4

Creep: time-dependent deformation of a material when subjected to a constant load or stress.

Creep rate: the slope of the creep-time curve at a given time.

Creep (stress)-rupture strength: the stress that will cause fracture in a creep (stress-rupture) test at a given time and in a specific environment at a particular temperature.

Sec. 7.5

Larsen Miller parameter: a time-temperature parameter used to predict stress rupture due to creep.

7.10 PROBLEMS

Answers to problems marked with an asterisk are given at the end of the book.

7.1 What are the characteristics of the surface of a ductile fracture of a metal?

7.2 Describe the three stages in the ductile fracture of a metal.

7.3 What are the characteristics of the surface of a brittle fracture of a metal?

7.4 Describe the three stages in the brittle fracture of a metal.

7.5 Describe the simple impact test that uses a Charpy V-notch sample.

7.6 How does the carbon content of a plain-carbon steel affect the ductile-brittle transition temperature range?

*7.7 Determine the critical crack length for a through crack contained within a thick plate of 7075-T751 aluminum alloy that is under uniaxial tension. For this alloy $K_{IC} = 22.0$ ksi$\sqrt{\text{in}}$. and $\sigma_f = 82.0$ ksi. Assume $Y = \sqrt{\pi}$.

7.8 Determine the critical crack length for a through crack in a thick plate of 7150-T651 aluminum alloy that is in uniaxial tension. For this alloy $K_{IC} = 25.5$ MPa$\sqrt{\text{m}}$ and $\sigma_f = 400$ MPa. Assume $Y = \sqrt{\pi}$.

*7.9 The critical stress intensity (K_{IC}) for a material for a component of a design is 23.0 ksi$\sqrt{\text{in}}$. What is the applied stress that will cause fracture if the component contains an internal crack 0.13 in. long? Assume $Y = 1$.

7.10 What is the largest size (mm) internal through crack that a thick plate of aluminum alloy 7075-T651 can support at an applied stress of (*a*) three-quarters of the yield strength and (*b*) one-half of the yield strength? Assume $Y = 1$.

7.11 A Ti-6Al-4V alloy plate contains an internal through crack of 1.90 mm. What is the highest stress (MPa) that this material can withstand without catastrophic failure? Assume $Y = \sqrt{\pi}$.

*7.12 Using the equation $K_{IC} = \sigma_f \sqrt{\pi a}$, plot the fracture stress (MPa) for aluminum alloy 7075-T651 versus surface crack size a (mm) for a values from 0.2 mm to 2.0 mm. What is the minimum size surface crack that will cause catastrophic failure?

*7.13 Determine the critical crack length (mm) for a through crack in a thick 2024-T6 alloy plate that has a fracture toughness $K_{IC} = 23.5$ MPa$\sqrt{\text{m}}$ and is under a stress of 300 MPa. Assume $Y = 1$.

7.14 Describe a metal fatigue failure.

7.15 What two distinct types of surface areas are usually recognized on a fatigue failure surface?

7.16 Where do fatigue failures usually originate on a metal section?

7.17 What is a fatigue test *SN* curve, and how are the data for the *SN* curve obtained?

7.18 How does the *SN* curve of a carbon steel differ from that of a high-strength aluminum alloy?

*7.19 A fatigue test is made with a maximum stress of 25 ksi (172 MPa) and a minimum stress of -4.00 ksi (-27.6 MPa). Calculate (*a*) the stress range, (*b*) the stress amplitude, (*c*) the mean stress, and (*d*) the stress ratio.

7.20 A fatigue test is made with a mean stress of 17,500 psi (120 MPa) and a stress amplitude of 24,000 psi (165 MPa). Calculate (*a*) the maximum and minimum stresses, (*b*) the stress ratio, and (*c*) the stress range.

7.21 Describe the four basic structural changes that take place when a homogeneous ductile metal is caused to fail by fatigue under cyclic stresses.

7.22 Describe four major factors that affect the fatigue strength of a metal.

7.23 A large flat plate is subjected to constant-amplitude uniaxial cyclic tensile and compressive stresses of 120 and 35 MPa, respectively. If before testing the largest surface crack is 1.00 mm and the plain-strain fracture toughness of the plate is $35 \, \text{MPa}\sqrt{\text{m}}$, estimate the fatigue life of the plate in cycles to failure. For the plate, $m = 3.5$ and $A = 5.0 \times 10^{-12}$ in MPa and meter units. Assume $Y = 1.3$.

*7.24 Same as Prob. 7.23 with tensile stress of 70 MPa. If the initial and critical crack lengths are 1.25 mm and 12 mm in the plate and the fatigue life is 2.0×10^6 cycles, calculate the maximum tensile stress in MPa that will produce this life. Assume $m = 3.0$ and $A = 6.0 \times 10^{-13}$ in MPa and meter units. Assume $Y = 1.20$.

7.25 Same as Prob. 7.23. Compute the final critical surface crack length if the fatigue life must be a minimum of 7.0×10^5 cycles. Assume the initial maximum edge surface crack length of 1.80 mm and a maximum tensile stress of 160 MPa. Assume $m = 1.8$ and $A = 7.5 \times 10^{-13}$ in MPa and meter units. Assume $Y = 1.25$.

*7.26 Same as Prob. 7.23. Compute critical surface edge crack if fatigue life must be 8.0×10^6 cycles and maximum tensile stress is 21,000 psi. $m = 3.5$ and $A = 4.0 \times 10^{-11}$ in ksi and inch units. Initial crack (edge) is 0.120 in. $Y = 1.15$.

7.27 What is metal creep?

7.28 For which environmental conditions is the creep of metals especially important industrially?

7.29 Draw a typical creep curve for a metal under constant load and at a relatively high temperature, and indicate on it all three stages of creep.

7.30 The following creep data were obtained for a titanium alloy at 50 ksi and 400°C. Plot the creep strain versus time (hours) and determine the steady-state creep rate for these test conditions.

Strain (in./in.)	Time (h)	Strain (in./in.)	Time (h)
0.010×10^{-2}	2	0.075×10^{-2}	80
0.030×10^{-2}	18	0.090×10^{-2}	120
0.050×10^{-2}	40	0.11×10^{-2}	160

*7.31 Equiaxed MAR-M 247 alloy is to support a stress of 276 MPa (Fig. 7.31). Determine the time to stress rupture at 850°C.

7.32 DS MAR-M 247 alloy (Fig. 7.31) is to support a stress of 207 MPa. At what temperature (°C) will the stress rupture lifetime be 210 h?

*7.33 If DS CM 247 LC alloy (middle graph of Fig. 7.31) is subjected to a temperature of 960°C for 3 years, what is the maximum stress that it can support without rupturing?

*7.34 A Ti-6Al-2Sn-4Zr-6Mo alloy is subjected to a stress of 20,000 psi. How long can the alloy be used under this stress at 500°C so that no more than 0.2% creep strain will occur? Use Fig. 6.37.

7.35 Gamma titanium aluminide is subjected to a stress of 50,000 psi. If the material is to be limited to 0.2% creep strain, how long can this material be used at 1100°F? Use Fig. 7.32.

7.11 MATERIALS SELECTION AND DESIGN PROBLEMS

1. Starting with a 2-in.-diameter rod of brass, we would like to process 0.2-in.-diameter rods that possess a minimum yield strength of 20 ksi and a minimum elongation to fracture of 10 percent (see Fig. 6.46). Design a process that achieves that. (Hint: Reduction of the diameter directly from 2 in. to 0.2 in. is not possible. Why?)

2. An ultrasonic crack detection equipment used by company A can find cracks of length 0.10 in. and longer. A lightweight component is to be designed and manufactured and then inspected for cracks using this machine. The maximum uniaxial stress applied to the component will be 60 ksi. Your choices of metals are Al 7178 T651, Ti-6Al-4V, or 4340 steel as listed in Table 7.1. (*a*) Which metal would you select for the largest margin of safety (use $Y = 1.0$)? (*b*) Which metal would you choose when considering both safety and weight?

3. In Prob. 2, if you did not consider the existence of cracks at all and only considered yielding as a failure criterion, which metal would you select that would not yield under the given conditions and would result in the lightest design (use data in Table 7.1)?

4. An alloy steel plate is subjected to a tensile stress of 120 MPa. The fracture toughness of the materials is given to be 45 MPa$\sqrt{m}$ (*a*) Determine the critical crack length to assure the plate will not fail under the static loading conditions (assume $Y = 1$). (*b*) Consider the same plate under the action of cyclic tensile/compressive stresses of 120 MPa and 30 MPa, respectively. If the component is to remain safe for three million cycles, what is the largest allowable crack length? (*c*) Compare the two critical crack lengths in (*a*) and (*b*). What is your conclusion?

5. A cylindrical rod made of directionally solidified alloy CM 247 is to carry a 10,000 N load at a temperature of 900°C and for a period of 300 h. Using the Larsen-Miller plot in Figure 7.31, design the appropriate dimensions for the cross section.

6. In the manufacturing of connecting rods, 4340 alloy steel heat treatable to 260 ksi may be used. There are two options for the manufacturing of the component: (*a*) heat-treat the component and use, and (*b*) heat-treat and grind the surface. Which option would you use and why?

7. In aircraft applications, aluminum panels are riveted together through holes drilled in the sheets. It is the industry practice to cold expand the holes to the desired diameter. (*a*) Explain why this process is done and how it benefits the structure. (*b*) Design a system that would accomplish the cold expansion process effectively and cheaply. (*c*) What are some precautions that must be taken during the cold expansion process?

8. What factors would you consider in selecting materials for coins? Suggest materials for this application.

9. The plumbing noise in drainage and waste systems is generally a result of vibration of the system and airborne noise passing through the wall. What material would you select for the piping system to reduce the vibration and noise? Why?

Phase Diagrams

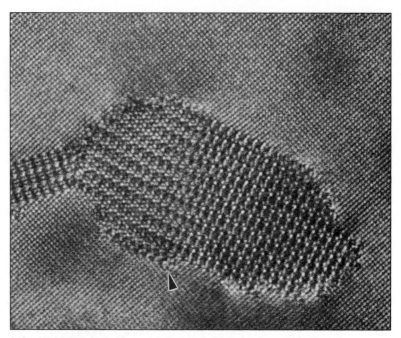

(*After W. M. Rainforth, 'Opportunities and pitfalls in characterization of nanoscale features,' Materials Science and Technology, vol. 16 (2000) 1349–1355.*)

Precipitation hardening or age hardening is a heat-treatment process used to produce a mixture of uniformly distributed hard phases in a soft matrix. The precipitated phase interferes with the movement of dislocations and, as a result, strengthens the alloy. The chapter-opening figure is a high-resolution electron microscope image of the Al_2CuMg phase in an aluminum matrix.[1]

A **phase** in a material is a region that differs in its microstructure and/or composition from another region. *Phase diagrams* are graphical representations of what phases are present in a materials system at various temperatures, pressures, and compositions. Most phase diagrams are constructed by using equilibrium conditions[2] and are used by engineers and scientists to understand and predict many aspects of the behavior of materials. ∎

---

[1]http://www.shef.ac.uk/uni/academic/D-H/em/research/centres/sorbcent.html

[2]**Equilibrium phase diagrams** are determined by using slow cooling conditions. In most cases equilibrium is approached but never fully attained.

By the end of this chapter, students will be able to. . .

1. Describe equilibrium, phase, and degrees of freedom for a materials system.

2. Describe the application of Gibbs rule in a material system.

3. Describe cooling curves and phase diagrams and the type of information that may be extracted from them.

4. Describe a binary isomorphous phase diagram and be able to draw a generic diagram showing all phase regions and relevant information.

5. Be able to apply tie line and lever rule to phase diagrams in order to determine the phase composition and phase fraction in a mixture.

6. Describe nonequilibrium solidification of metals and explain the general differences in microstructure when compared to equilibrium solidification.

7. Describe a binary eutectic phase diagram and be able to draw a generic diagram showing all phase regions and relevant information.

8. Describe the microstructure evolution during equilibrium cooling as metal solidifies at various regions of the phase diagram.

9. Define various invariant reactions.

10. Define intermediate phase compounds and intermetallics.

11. Describe ternary phase diagrams.

8.1 PHASE DIAGRAMS OF PURE SUBSTANCES

Animation

A pure substance such as water can exist in solid, liquid, or vapor phases, depending on the conditions of temperature and pressure. An example familiar to everyone of two phases of a pure substance in **equilibrium** is a glass of water containing ice cubes. In this case solid and liquid water are two separate and distinct phases that are separated by a phase boundary, the surface of the ice cubes. During the boiling of water, liquid water and water vapor are two phases in equilibrium. A graphical representation of the phases of water that exist under different conditions of temperature and pressure is shown in Fig. 8.1.

In the *pressure-temperature* (*PT*) phase diagram of water there exists a *triple point* at low pressure (4.579 torr) and low temperature (0.0098°C) where solid, liquid, and vapor phases of water coexist. Liquid and vapor phases exist along the vaporization line and liquid and solid phases along the freezing line, as shown in Fig. 8.1. These lines are two-phase equilibrium lines.

Pressure-temperature equilibrium phase diagrams also can be constructed for other pure substances. For example, the equilibrium *PT* phase diagram for pure iron is shown in Fig. 8.2. One major difference with this phase diagram is that there are three separate and distinct *solid phases*: alpha (α) Fe, gamma (γ) Fe, and delta (δ) Fe. Alpha and delta iron have BCC crystal structures, whereas gamma iron has an FCC structure. The

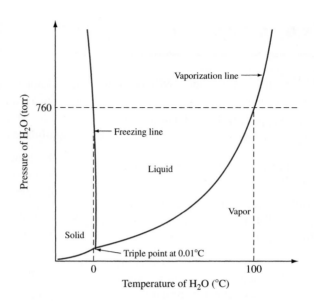

Figure 8.1
Approximate *PT* equilibrium phase diagram for pure water. (The axes of the diagram are distorted to some extent.)

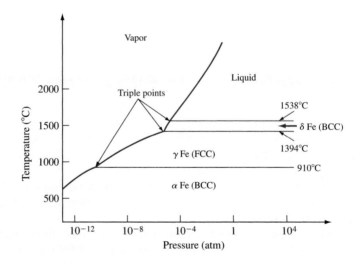

Animation

Figure 8.2
Approximate *PT* equilibrium phase diagram for pure iron.
(After W. G. Moffatt et al., "Structure and Properties of Materials," vol. 1, Wiley, 1964, p. 151.)

phase boundaries in the solid state have the same properties as the liquid and solid phase boundaries. For example, under equilibrium conditions, alpha and gamma iron can exist at a temperature of 910°C and 1 atm pressure. Above 910°C only single-phase gamma exists, and below 910°C only single-phase alpha exists (Fig. 8.2). There are also three triple points in the iron PT diagram where three different phases coexist: (1) liquid, vapor, and δFe, (2) vapor, δFe, and γFe, and (3) vapor, γFe, and αFe.

8.2 GIBBS PHASE RULE

From thermodynamic considerations, J. W. Gibbs[3] derived an equation that computes the number of phases that can coexist in equilibrium in a chosen system. This equation, called **Gibbs phase rule,** is

$$P + F = C + 2 \tag{8.1}$$

where P = number of phases that coexist in a chosen system

C = **number of components** in the system

F = degrees of freedom

Usually a component C is an element, compound, or solution in the system. F, the **degrees of freedom,** is the number of variables (pressure, temperature, and composition) that can be changed independently without changing the number of phases in equilibrium in the chosen system.

Let us consider the application of Gibbs phase rule to the PT phase diagram of pure water (Fig. 8.1). At the triple point, three phases coexist in equilibrium, and since there is one component in the system (water), the number of degrees of freedom can be calculated:

$$P + F = C + 2$$
$$3 + F = 1 + 2 \tag{8.1}$$

or

$$F = 0 \quad \text{(zero degrees of freedom)}$$

Since none of the variables (temperature or pressure) can be changed and still keep the three phases in balance, the triple point is called an *invariant point.*

Consider next a point along the liquid-solid freezing curve of Fig. 8.1. At any point along this line two phases will coexist. Thus, from the phase rule,

$$2 + F = 1 + 2$$

or

$$F = 1 \quad \text{(one degree of freedom)}$$

This result tells us that there is one degree of freedom, and thus one variable (T or P) can be changed independently and still maintain a system with two coexisting phases. Thus, if a particular pressure is specified, there is only one temperature at which both liquid and solid phases can coexist. For a third case, consider a point on the water PT phase diagram inside a single phase. Then there will be only one phase present ($P = 1$), and substituting into the phase-rule equation gives

$$1 + F = 1 + 2$$

[3]Josiah Willard Gibbs (1839–1903). American physicist. He was a professor of mathematical physics at Yale University and made great contributions to the science of thermodynamics, which included the statement of the phase rule for multiphase systems.

or

$$F = 2 \qquad \text{(two degrees of freedom)}$$

This result tells us that two variables (temperature and pressure) can be varied independently and the system will still remain a single phase.

Most binary phase diagrams used in materials science are temperature composition diagrams in which pressure is kept constant, usually at 1 atm. In this case, we have the condensed phase rule, which is given by

$$P + F = C + 1 \qquad\qquad\qquad \textbf{(8.1}a\textbf{)}$$

Equation 8.1*a* will apply to all subsequent binary phase diagrams discussed in this chapter.

8.3 COOLING CURVES

Cooling curves can be used to determine phase transition temperatures for both pure metals and alloys. A **cooling curve** is obtained by recording the temperature of a material versus time as it cools from a temperature at which it is molten through solidification and finally to room temperature. The cooling curve for a pure metal is shown in Fig. 8.3. If the metal is allowed to cool under equilibrium conditions (slow cooling), its temperature drops continuously along line *AB* of the curve. At the melting point (freezing temperature) solidification begins and the cooling curve becomes flat (horizontal segment *BC*, also called a **plateaue** or **region of thermal arrest**) and remains flat until solidification is complete. In region *BC*, the metal is in the form of a mixture of solid and liquid phases. As point *C* is approached, the weight fraction of solid in the mixture grows until solidification is complete. The temperature remains constant because there is a balance between the heat lost by the metal through the mold and the latent heat supplied by the solidifying metal. Simply stated, the latent heat keeps the mixture at the freezing temperature until complete solidification is achieved. After solidification is complete at *C*, the cooling curve will again show a drop in temperature with time (segment *CD* of the curve).

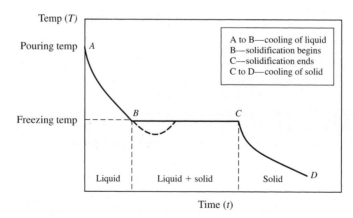

Figure 8.3
The cooling curve for a pure metal.

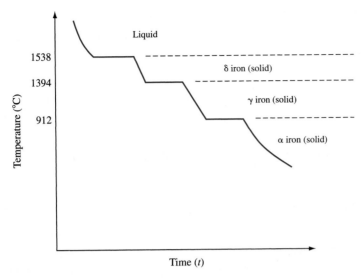

Figure 8.4
Cooling curve for pure iron at a pressure of 1 atm.

As discussed in the sections on the solidification of pure metals in Chap. 4 a degree of undercooling (cooling below the freezing temperature) is required for the formation of solid nuclei. The undercooling will appear on the cooling curve as a drop below the freezing temperature as shown in Fig. 8.3.

The cooling curve may also provide information regarding the solid state phase transformation in metals. An example of such a cooling curve would be that of pure iron. The pure iron cooling curve under atmospheric pressure conditions ($P = 1$ atm) shows a freezing temperature of 1538°C at which point a high-temperature solid of BCC structure is formed called δ *iron* (Fig. 8.4). Upon additional cooling, at a temperature of approximately 1394°C, the cooling curve shows a second plateau. At this temperature a solid-solid phase transformation of BCC δ ferrite to an FCC solid called γ *iron* (polymorphic transformation, see Sec. 3.10) takes place. With further cooling a second solid-solid phase transformation takes place at a temperature of 912°C. In this transformation the FCC γ iron reverts back to a BCC iron structure called α *iron*. This solid-solid transformation has important technological implications in steel-processing industries and will be discussed in Chap. 9.

8.4 BINARY ISOMORPHOUS ALLOY SYSTEMS

Let us now consider a mixture or alloy of two metals instead of pure substances. A mixture of two metals is called a *binary alloy* and constitutes a *two-component* system since each metallic element in an alloy is considered a separate component. Thus, pure copper is a one-component system, whereas an alloy of copper and nickel is a two-component system. Sometimes a compound in an alloy is also considered a separate component. For example, plain-carbon steels containing mainly iron and iron carbide are considered two-component systems.

In some binary metallic systems the two elements are completely soluble in each other in both the liquid and solid states. In these systems only a single type of crystal structure exists for all compositions of the components, and therefore they are called **isomorphous systems.** In order for the two elements to have complete solid solubility in each other, they usually satisfy one or more of the following conditions formulated by Hume-Rothery[4] and known as the Hume-Rothery solid solubility rules:

1. The size of the atoms of each of the two elements must not differ by more than 15 percent.
2. The elements should not form compounds with each other, i.e., there should be no appreciable difference in the electronegativities of the two elements.
3. The crystal structure of each element of the solid solution must be the same.
4. The elements should have the same valence.

The Hume-Rothery rules are not all applicable for every pair of elements that shows complete solid solubility.

An important example of an isomorphous binary alloy system is the copper-nickel system. A phase diagram of this system with temperature as the ordinate and chemical composition in weight percent as the abscissa is shown in Fig. 8.5. This diagram has been determined for slow cooling or equilibrium conditions at atmospheric pressure and does not apply to alloys that have been rapidly cooled through the solidification temperature range. The area above the upper line in the diagram, called the **liquidus,** corresponds to the region of stability of the liquid phase, and the area below the lower line, or **solidus,** represents the region of stability for the solid phase. The region between the liquidus and solidus represents a two-phase region where both the liquid and solid phases coexist.

For the binary isomorphous phase diagram of Cu and Ni, according to the Gibbs phase rule ($F = C - P + 1$), at the melting point of the pure components, the number of components C is 1 (either Cu or Ni) and the number of phases available P is 2 (liquid or solid), resulting in a degree of freedom of 0 ($F = 1 - 2 + 1 = 0$). These points are referred to as *invariant points* ($F = 0$). This means that any change in temperature will change the microstructure either into solid or liquid. Accordingly, in the single-phase regions (liquid or solid), the number of components, C, is 2 and the number of phases available, P, is 1, resulting in a degree of freedom of 2 ($F = 2 - 1 + 1 = 2$). This means that we can maintain the microstructure of the system in this region by varying either the temperature or composition independently. In the two-phase region the number of components, C, is 2 and the number of phases available, P, is 2, resulting in a degree of freedom of 1 ($F = 2 - 2 + 1 = 1$). This means that only one variable (either temperature or composition) can be changed independently while maintaining the two-phase structure of the system. If the temperature is changed, the phase composition will also change.

---

[4]William Hume-Rothery (1899–1968). English metallurgist who made major contributions to theoretical and experimental metallurgy and who spent years studying alloy behavior. His empirical rules for solid solubility in alloys were based on his alloy design work.

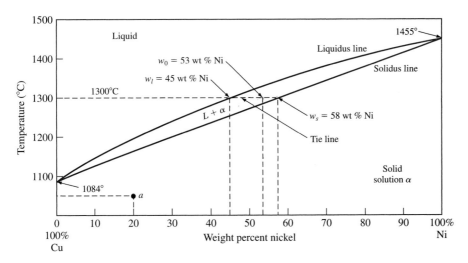

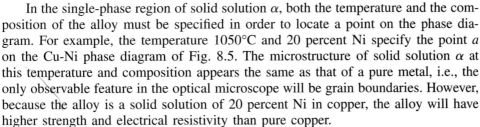

Figure 8.5
The copper-nickel phase diagram. Copper and nickel have complete liquid solubility and complete solid solubility. Copper-nickel solid solutions melt over a range of temperatures rather than at a fixed temperature, as is the case for pure metals.

IceNine

(*Adapted from "Metals Handbook," vol. 8, 8th ed., American Society for Metals, 1973, p. 294.*)

In the single-phase region of solid solution α, both the temperature and the composition of the alloy must be specified in order to locate a point on the phase diagram. For example, the temperature 1050°C and 20 percent Ni specify the point a on the Cu-Ni phase diagram of Fig. 8.5. The microstructure of solid solution α at this temperature and composition appears the same as that of a pure metal, i.e., the only observable feature in the optical microscope will be grain boundaries. However, because the alloy is a solid solution of 20 percent Ni in copper, the alloy will have higher strength and electrical resistivity than pure copper.

In the region between the liquidus and solidus lines, both liquid and solid phases exist. The amount of each phase present depends on the temperature and chemical composition of the alloy. Let us consider an alloy of 53 wt % Ni–47 wt % Cu at 1300°C in Fig. 8.5. Since this alloy contains both liquid and solid phases at 1300°C, neither of these phases can have the average composition of 53% Ni–47% Cu. The compositions of the liquid and solid phases at 1300°C can be determined by drawing a horizontal *tie line* at 1300°C from the liquidus line to the solidus line and then dropping vertical lines to the horizontal composition axis. The composition of the liquid phase (w_l) at 1300°C is 45 wt % Ni and that of the solid phase (w_s) is 58 wt % Ni, as indicated by the intersection of the dashed vertical lines with the composition axis.

Binary equilibrium phase diagrams for components that are completely soluble in each other in the solid state can be constructed from a series of liquid-solid cooling curves, as shown for the Cu-Ni system in Fig. 8.6. The cooling curves for pure metals show horizontal thermal arrests at their freezing points, as shown for pure copper and nickel in Fig. 8.6*a* at *AB* and *CD*. Binary solid solutions exhibit slope

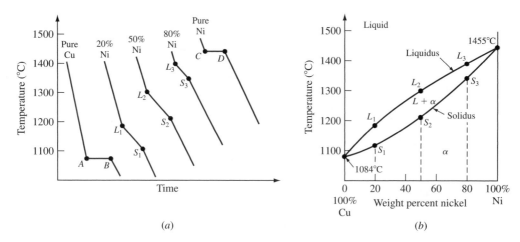

Figure 8.6
Construction of the Cu–Ni equilibrium phase diagram from liquid-solid cooling curves.
(*a*) Cooling curves and (*b*) equilibrium phase diagram.

changes in their cooling curves at the liquidus and solidus lines, as shown in Fig. 8.6*a* at compositions of 80% Cu–20% Ni, 50% Cu–50% Ni, and 20% Cu–80% Ni. The slope changes at L_1, L_2, and L_3 in Fig. 8.6*a* correspond to the liquidus points L_1, L_2, and L_3 of Fig. 8.6*b*. Similarly, the slope changes of S_1, S_2, and S_3 of Fig. 8.6*a* correspond to the points S_1, S_2, and S_3 on the solidus line of Fig. 8.6*b*. Further accuracy in the construction of the Cu-Ni phase diagram can be attained by determining more cooling curves at intermediate alloy compositions.

The cooling curve for metal alloys does not contain the thermal arrest region that one observes in the solidification of a pure metal. Instead, solidification begins at a specific temperature and ends at a lower temperature as presented by L and S symbols in Fig. 8.6. As a result, unlike pure metals, alloys solidify over a range of temperatures. Thus, when we refer to the freezing temperature of a metal alloy, we are speaking of the temperature at which the solidification process is complete.

8.5 THE LEVER RULE

The weight percentages of the phases in any two-phase region of a binary equilibrium phase diagram can be calculated by using the **lever rule.** For example, by using the lever rule the weight percent liquid and weight percent solid for any particular temperature can be calculated for any average alloy composition in the two-phase liquid-plus-solid region of the binary copper-nickel phase diagram of Fig. 8.5.

To derive the lever-rule equations let us consider the binary equilibrium phase diagram of two elements A and B that are completely soluble in each other, as shown in Fig. 8.7. Let x be the alloy composition of interest and its weight fraction of B in A be w_0. Let T be the temperature of interest and let us construct a **tie line** (LS) at temperature T from the liquidus at point L to the solidus line at point S, forming the tie line LOS. At temperature T, the alloy x consists of a mixture of liquid of w_l weight fraction of B and solid of w_s weight fraction of B.

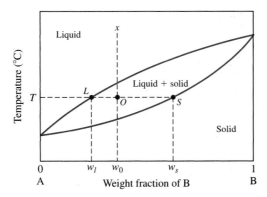

Figure 8.7
Binary phase diagram of two metals A and B
completely soluble in each other being used
to derive the lever-rule equations. At
temperature T, the composition of the liquid
phase is w_l and that of the solid is w_s.

The lever-rule equations can be derived by using weight balances. One equation for the derivation of the lever-rule equations is obtained from the fact that the sum of the weight fraction of the liquid phase, X_l, and the weight fraction of the solid phase, X_s, must equal 1. Thus,

$$X_l + X_s = 1 \tag{8.2}$$

or

$$X_l = 1 - X_s \tag{8.2a}$$

and

$$X_s = 1 - X_l \tag{8.2b}$$

A second equation for the derivation of the lever rule can be obtained by a weight balance of B in the alloy as a whole and the sum of B in the two separate phases. Let us consider 1 g of the alloy and make this weight balance:

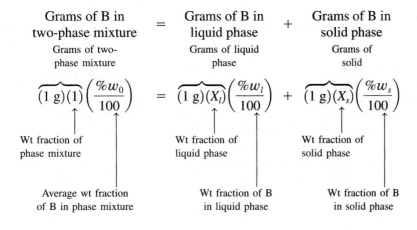

$$(8.3)$$

Thus, $$w_0 = X_l w_l + X_s w_s \tag{8.4}$$

combined with $$X_l = 1 - X_s \tag{8.2a}$$

gives $$w_0 = (1 - X_s)w_l + X_s w_s$$

or $$w_0 = w_l - X_s w_l + X_s w_s$$

Rearranging, $$X_s w_s - X_s w_l = w_0 - w_l$$

$$\boxed{\text{Wt fraction of solid phase} = X_s = \frac{w_0 - w_l}{w_s - w_l}} \tag{8.5}$$

Similarly, $$w_0 = X_l w_l + X_s w_s \tag{8.4}$$

combined with $$X_s = 1 - X_l \tag{8.2b}$$

gives $$\boxed{\text{Wt fraction of liquid phase} = X_l = \frac{w_s - w_0}{w_s - w_l}} \tag{8.6}$$

Equations 8.5 and 8.6 are the lever-rule equations. Effectively, the lever-rule equations state that to calculate the weight fraction of one phase of a two-phase mixture, one must use the segment of the tie line that is on the opposite side of the alloy of interest and is farthest away from the phase for which the weight fraction is being calculated. The ratio of this line segment of the tie line to the total tie line provides the weight fraction of the phase being determined. Thus, in Fig. 8.7 the weight fraction of the liquid phase is the ratio *OS/LS* and the weight fraction of the solid phase is the ratio *LO/LS*.

Weight fractions can be converted to weight percentages by multiplying by 100 percent. Example Problem 8.1 shows how the lever rule can be used to determine the weight percentage of a phase in a binary alloy at a particular temperature.

EXAMPLE PROBLEM 8.1

Derive the lever rule for the case shown in Fig. EP8.1.

■ Solution

To derive the lever rule, let us consider the binary equilibrium diagram of two elements A and B that are completely soluble in each other, as shown in Fig. EP8.1. Let x be the alloy composition of interest and its weight fraction of B in A be w_0. Let T be the temperature of interest and let us construct a tie line from the solidus line at point S forming the tie line *SOL*. From the solution of these equations:

The weight fraction of the liquid phase would equal

$$\frac{w_0 - w_s}{w_l - w_s} = \frac{SO}{LS}$$

The weight fraction of the solid phase would equal

$$\frac{w_l - w_0}{w_l - w_s} = \frac{OL}{LS}$$

This problem is illustrated in Example Problem 8.3 at 1200°C.

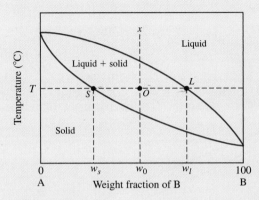

Figure EP8.1

A copper-nickel alloy contains 47 wt % Cu and 53 wt % Ni and is at 1300°C. Use Fig. 8.5 and answer the following:

EXAMPLE PROBLEM 8.2

a. What is the weight percent of copper in the liquid and solid phases at this temperature?
b. What weight percent of this alloy is liquid and what weight percent is solid?

■ **Solution**

a. From Fig. 8.5 at 1300°C, the intersection of the 1300°C tie line with the liquidus gives 55 wt % Cu in the liquid phase and the intersection of the solidus of the 1300°C tie line gives 42 wt % Cu in the solid phase.

b. From Fig. 8.5 and using the lever rule on the 1300°C tie line,

$$w_0 = 53\% \text{ Ni} \qquad w_l = 45\% \text{ Ni} \qquad w_s = 58\% \text{ Ni}$$

Wt fraction of liquid phase $= X_l = \dfrac{w_s - w_0}{w_s - w_l}$

$$= \frac{58 - 53}{58 - 45} = \frac{5}{13} = 0.38$$

Wt % of liquid phase $= (0.38)(100\%) = 38\%$ ◄

Wt fraction of solid phase $= X_s = \dfrac{w_0 - w_l}{w_s - w_l}$

$$= \frac{53 - 45}{58 - 45} = \frac{8}{13} = 0.62$$

Wt % of solid phase $= (0.62)(100\%) = 62\%$ ◄

EXAMPLE PROBLEM 8.3

Calculate the percent liquid and solid for the Ag-Pd phase diagram shown in Fig. EP8.3 at 1200°C and 70 wt % Ag. Assume $W_l = 74$ without Ag and $W_s = 64$ without Ag.

■ **Solution**

$$W(\%) \text{ liquid} = \frac{70 - 64}{74 - 64} = \frac{6}{10} = 60\%$$

$$W(\%) \text{ solid} = \frac{74 - 70}{74 - 64} = \frac{4}{10} = 40\%$$

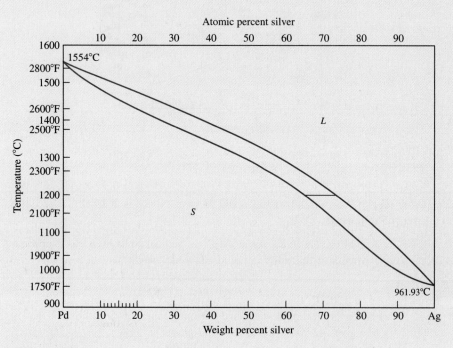

Figure EP8.3
The Ag-Pd equilibrium phase diagram.

IceNine

8.6 NONEQUILIBRIUM SOLIDIFICATION OF ALLOYS

The phase diagram for the Cu-Ni system previously referred to was constructed by using very slow cooling conditions approaching equilibrium. That is, when the Cu-Ni alloys were cooled through the two-phase liquid + solid regions, the compositions of the liquid and solid phases had to readjust continuously by solid-state diffusion as the temperature was lowered. Since atomic diffusion is very slow in the solid state, an extensive period of time is required to eliminate concentration gradients. Thus, the as-cast microstructures of slowly solidified alloys usually have a **cored structure** (Fig. 8.8) caused by regions of different chemical composition.

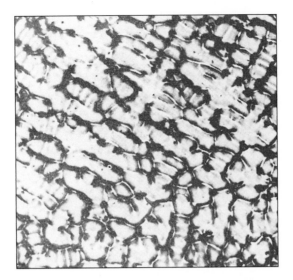

Figure 8.8
The microstructure of an as-cast 70% Cu–30%
Ni alloy showing a cored structure.
*(After W. G. Moffatt et al., "Structure and Properties of
Materials," vol. I, Wiley, 1964, p. 177.)*

The copper-nickel alloy system provides a good example to describe how such a cored structure originates. Consider an alloy of 70% Ni–30% Cu that is cooled from a temperature T_0 at a rapid rate (Fig. 8.9). The first solid forms at temperature T_1 and has the composition α_1 (Fig. 8.9). Upon further rapid cooling to T_2, additional layers of composition α_2 will form without much change in the composition of the solid primarily solidified. The overall composition at T_2 lies somewhere between α_1 and α_2 and will be designated α_2'. Since the tie line $\alpha_2'L_2$ is longer than α_2L_2, there will be more liquid and less solid in the rapidly cooled alloy than if it were cooled under equilibrium conditions to the same temperature. Thus, solidification has been delayed at that temperature by the rapid cooling.

As the temperature is lowered to T_3 and T_4, the same processes occur and the average composition of the alloy follows the *nonequilibrium solidus* $\alpha_1\alpha_2'\alpha_3' \cdots$. At T_6 the solid freezing has less copper than the original composition of the alloy, which is 30 percent Cu. At temperature T_7 the average composition of the alloy is 30 percent Cu and freezing is complete. Regions in the microstructure of the alloy will thus consist of compositions varying from α_1 to α_7' as the cored structure forms during solidification (Fig. 8.10). Figure 8.8 shows a cored microstructure of rapidly solidified 70% Cu–30% Ni alloy.

Most as-cast microstructures are cored to some extent and thus have composition gradients. In many cases this structure is undesirable, particularly if the alloy is to be subsequently worked. To eliminate the cored structure, as-cast ingots or castings are heated to elevated temperatures to accelerate solid-state diffusion. This process is called **homogenization** since it produces a homogeneous structure in the alloy.

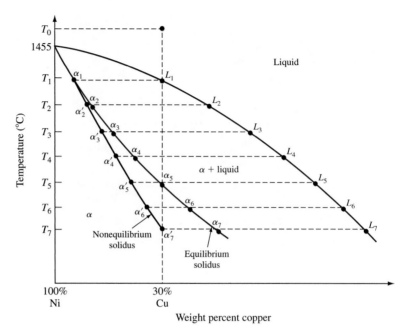

Figure 8.9
Nonequilibrium solidification of a 70% Ni–30% Cu alloy. This phase
diagram has been distorted for illustrative purposes. Note the
nonequilibrium solidus α_1 to α'_7. The alloy is not completely solidified
until the nonequilibrium solidus reaches α'_7 at T_7.

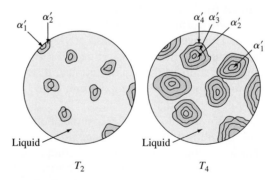

Figure 8.10
Schematic microstructures at temperature T_2
and T_4 of Fig. 8.9 for the nonequilibrium solidification of a 70% Ni–30% Cu alloy illustrating
the development of a cored structure.

Figure 8.11 shows aluminum alloy sheet ingots being loaded into a homogenizing furnace. The homogenizing heat treatment must be carried out at a temperature that is
lower than the lowest-melting solid in the as-cast alloy or else melting will occur. For
homogenizing the 70% Ni–30% Cu alloy just discussed, a temperature just below T_7

Figure 8.11
Large direct-chill cast aluminum alloy sheet ingots being loaded into a homogenizing furnace. When the as-cast ingots are reheated at a high temperature below the melting temperature of the lowest melting phase, atomic solid-state diffusion creates a more homogeneous internal structure.
(*Courtesy of Reynolds Metals Co.*)

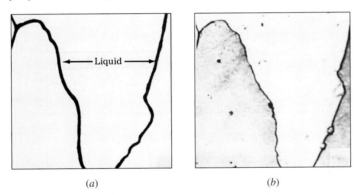

(*a*)   (*b*)

Figure 8.12
Liquation in a 70% Ni–30% Cu alloy. Heating only slightly above the solidus temperature so that melting just begins, produces a liquated structure such as shown in (*a*). In (*b*) the grain-boundary region was slightly melted, and then upon subsequent freezing, the melted zone became copper-rich and caused the grain boundaries to appear as broad dark lines.
(*Courtesy of F. Rhines.*)

indicated in Fig. 8.9 should be used. If the alloy is overheated, localized melting or *liquation* may take place. If the liquid phase forms a continuous film along the grain boundaries, the alloy will lose strength and may break up during subsequent working. Figure 8.12 shows liquation in the microstructure of a 70% Ni–30% Cu alloy.

8.7 BINARY EUTECTIC ALLOY SYSTEMS

Many binary alloy systems have components that have limited solid solubility in each other as, for example, in the lead-tin system (Fig. 8.13). The regions of restricted solid solubility at each end of the Pb-Sn diagram are designated as alpha and beta phases and are called *terminal solid solutions* since they appear at the ends of the diagram. The alpha phase is a lead-rich solid solution and can dissolve in solid solution a maximum of 19.2 wt % Sn at 183°C. The beta phase is a tin-rich solid solution and can dissolve a maximum of 2.5 wt % Pb at 183°C. As the temperature is decreased below 183°C, the maximum solid solubility of the solute elements decreases according to the **solvus** lines of the Pb-Sn phase diagram.

In simple binary eutectic systems like the Pb-Sn one, there is a specific alloy composition known as the **eutectic composition** that freezes at a lower temperature than all other compositions. This low temperature, which corresponds to the lowest

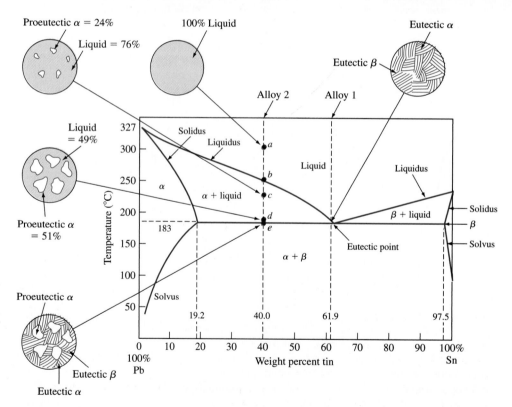

Figure 8.13

The lead-tin equilibrium phase diagram. This diagram is characterized by the limited solid solubility of each terminal phase (α and β). The eutectic invariant reaction at 61.9% Sn and 183°C is the most important feature of this system. At the eutectic point, α (19.2% Sn), β (97.5% Sn), and liquid (61.9% Sn) can coexist.

IceNine

temperature at which the liquid phase can exist when cooled slowly, is called the **eutectic temperature.** In the Pb-Sn system the eutectic composition (61.9 percent Sn and 38.1 percent Pb) and the eutectic temperature (183°C) determine a point on the phase diagram called the **eutectic point.** When liquid of eutectic composition is slowly cooled to the eutectic temperature, the single liquid phase transforms simultaneously into two solid forms (solid solutions α and β). This transformation is known as the **eutectic reaction** and is written as

$$\text{Liquid} \xrightarrow[\text{cooling}]{\text{eutectic temperature}} \alpha \text{ solid solution} + \beta \text{ solid solution} \qquad \textbf{(8.7)}$$

The eutectic reaction is called an **invariant reaction** since it occurs under equilibrium conditions at a specific temperature and alloy composition that cannot be varied (according to Gibbs rule, $F = 0$). During the progress of the eutectic reaction the liquid phase is in equilibrium with the two solid solutions α and β, and thus during a eutectic reaction, three phases coexist and are in equilibrium. Since three phases in a binary phase diagram can only be in equilibrium at one temperature, a horizontal thermal arrest appears at the eutectic temperature in the cooling curve of an alloy of eutectic composition.

Slow cooling of a Pb-Sn alloy of eutectic composition. Consider the slow cooling of a Pb-Sn alloy (alloy 1 of Fig. 8.13) of eutectic composition (61.9 percent Sn) from 200°C to room temperature. During the cooling period from 200 to 183°C, the alloy remains liquid. At 183°C, which is the eutectic temperature, all the liquid solidifies by the eutectic reaction and forms a eutectic mixture of solid solutions α (19.2 percent Sn) and β (97.5 percent Sn) according to the reaction

$$\text{Liquid (61.9\% Sn)} \xrightarrow[\text{cooling}]{183°C} \alpha\,(19.2\% \text{ Sn}) + \beta\,(97.5\% \text{ Sn}) \qquad \textbf{(8.8)}$$

After the eutectic reaction has been completed, upon cooling the alloy from 183°C to room temperature, there is a decrease in solid solubility of solute in the α and β solid solutions, as indicated by the solvus lines. However, since diffusion is slow at the lower temperatures, this process does not normally reach equilibrium, and thus solid solutions α and β can still be distinguished at room temperature, as shown in the microstructure of Fig. 8.14*a*.

Compositions to the left of the eutectic point are called **hypoeutectic,** Fig. 8.14*b*. Conversely, compositions to the right of the eutectic point are called **hypereutectic,** Fig. 8.14*d*.

Slow cooling of a 60% Pb–40% Sn alloy. Next consider the slow cooling of a 40% Sn–60% Pb alloy (alloy 2 of Fig. 8.13) from the liquid state at 300°C to room temperature. As the temperature is lowered from 300°C (point *a*), the alloy will remain liquid until the liquidus line is intersected at point *b* at about 245°C. At this temperature solid solution α containing 12 percent Sn will begin to precipitate from the liquid. The first solid to form in this type of alloy is called **primary** or **proeutectic** *alpha.* The term proeutectic alpha is

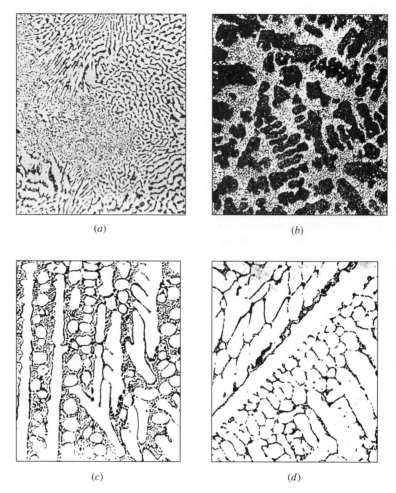

(a)   (b)

(c)   (d)

Figure 8.14
Microstructures of slowly cooled Pb–Sn alloys: (a) eutectic
composition (63% Sn–37% Pb), (b) 40% Sn–60% Pb, (c) 70%
Sn–30% Pb, (d) 90% Sn–10% Pb. (Magnification 75×.)
(*From J. Nutting and R. G. Baker, "Microstructure of Metals," Institute of Metals,
London, 1965, p. 19.*)

used to distinguish this constituent from the alpha that forms later by the
eutectic reaction.

As the liquid cools from 245°C to slightly above 183°C through the two-
phase liquid + alpha region of the phase diagram (points b to d), the composition of the solid phase (alpha) follows the solidus and varies from 12 percent
Sn at 245°C to 19.2 percent Sn at 183°C. Likewise, the composition of the
liquid phase varies from 40 percent Sn at 245°C to 61.9 percent Sn at 183°C.
These composition changes are possible since the alloy is cooling very slowly

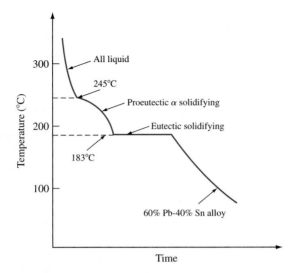

Figure 8.15
Schematic temperature-time cooling curve for a
60% Pb–40% Sn alloy.

and atomic diffusion occurs to equalize compositional gradients. At the eutectic temperature (183°C) all the remaining liquid solidifies by the eutectic reaction (Eq. 8.8). After the eutectic reaction is completed, the alloy consists of proeutectic alpha and a eutectic mixture of alpha (19.2 percent Sn) and beta (97.5 percent Sn). Further cooling below 183°C to room temperature lowers the tin content of the alpha phase and the lead content of the beta phase. However, at the lower temperatures the diffusion rate is much lower and equilibrium is not attained. Figure 8.14*b* shows the microstructure of a 40% Sn–60% Pb alloy that has been slowly cooled. Note the dark-etching dendrites of the lead-rich alpha phase surrounded by eutectic. Figure 8.15 shows a cooling curve for a 60% Pb–40% Sn alloy. Note that a slope change occurs at the liquidus at 245°C and a horizontal thermal arrest appears during the freezing of the eutectic.

Make phase analyses of the equilibrium (ideal) solidification of lead-tin alloys at the following points in the lead-tin phase diagram of Fig. 8.13:

EXAMPLE PROBLEM 8.4

a. At the eutectic composition just below 183°C (eutectic temperature).
b. The point *c* at 40% Sn and 230°C.
c. The point *d* at 40% Sn and 183°C + ΔT.
d. The point *e* at 40% Sn and 183°C − ΔT.

■ **Solution**

a. At the eutectic composition (61.9 percent Sn) just below 183°C:

Phases present:	alpha	beta
Compositions of phases:	19.2% Sn in alpha phase	97.5% Sn in beta phase
Amounts of phases:	Wt % alpha phase[5]	Wt % beta phase[5]
	$= \dfrac{97.5 - 61.9}{97.5 - 19.2}(100\%)$	$= \dfrac{61.9 - 19.2}{97.5 - 19.2}(100\%)$
	$= 45.5\%$	$= 54.5\%$

b. The point c at 40 percent Sn and 230°C:

Phases present:	liquid	alpha
Compositions of phases:	48% Sn in liquid phase	15% Sn in alpha phase
Amounts of phases:	Wt % liquid phase	Wt % alpha phase
	$= \dfrac{40 - 15}{48 - 15}(100\%)$	$= \dfrac{48 - 40}{48 - 15}(100\%)$
	$= 76\%$	$= 24\%$

c. The point d at 40 percent Sn and 183°C $+ \Delta T$:

Phases present:	liquid	alpha
Compositions of phases:	61.9% Sn in liquid phase	19.2% Sn in alpha phase
Amounts of phases:	Wt % liquid phase	Wt % alpha phase
	$= \dfrac{40 - 19.2}{61.9 - 19.2}(100\%)$	$= \dfrac{61.9 - 40}{61.9 - 19.2}(100\%)$
	$= 49\%$	$= 51\%$

d. The point e at 40 percent Sn and 183°C $- \Delta T$:

Phases present:	alpha	beta
Compositions of phases:	19.2% Sn in alpha phase	97.5% Sn in beta phase
Amounts of phases:	Wt % alpha phase	Wt % beta phase
	$= \dfrac{97.5 - 40}{97.5 - 19.2}(100\%)$	$= \dfrac{40 - 19.2}{97.5 - 19.2}(100\%)$
	$= 73\%$	$= 27\%$

---

[5]Note that in the lever-rule calculations one uses the ratio of the tie-line segment that is *farthest away* from the phase for which the weight percent is being determined to the whole tie line.

One kilogram of an alloy of 70 percent Pb and 30 percent Sn is slowly cooled from 300°C. Refer to the lead-tin phase diagram of Fig. 8.13 and calculate the following:

a. The weight percent of the liquid and proeutectic alpha at 250°C.
b. The weight percent of the liquid and proeutectic alpha just above the eutectic temperature (183°C) and the weight in kilograms of these phases.
c. The weight in kilograms of alpha and beta formed by the eutectic reaction.

■ **Solution**

a. From Fig. 8.13 at 250°C,

$$\text{Wt \% liquid}^6 = \frac{30 - 12}{40 - 12}(100\%) = 64\% \blacktriangleleft$$

$$\text{Wt \% proeutectic } \alpha^6 = \frac{40 - 30}{40 - 12}(100\%) = 36\% \blacktriangleleft$$

b. The weight percent liquid and proeutectic alpha just above the eutectic temperature, $183°C + \Delta T$, is

$$\text{Wt \% liquid} = \frac{30 - 19.2}{61.9 - 19.2}(100\%) = 25.3\% \blacktriangleleft$$

$$\text{Wt \% proeutectic } \alpha = \frac{61.9 - 30.0}{61.9 - 19.2}(100\%) = 74.7\% \blacktriangleleft$$

$$\text{Weight of liquid phase} = 1 \text{ kg} \times 0.253 = 0.253 \text{ kg} \blacktriangleleft$$

$$\text{Weight of proeutectic } \alpha = 1 \text{ kg} \times 0.747 = 0.747 \text{ kg} \blacktriangleleft$$

c. At $183°C - \Delta T$,

$$\text{Wt \% total } \alpha \text{ (proeutectic } \alpha + \text{ eutectic } \alpha) = \frac{97.5 - 30}{97.5 - 19.2}(100\%)$$

$$= 86.2\%$$

$$\text{Wt \% total } \beta \text{ (eutectic } \beta) = \frac{30 - 19.2}{97.5 - 19.2}(100\%)$$

$$= 13.8\%$$

$$\text{Wt total } \alpha = 1 \text{ kg} \times 0.862 = 0.862 \text{ kg}$$

$$\text{Wt total } \beta = 1 \text{ kg} \times 0.138 = 0.138 \text{ kg}$$

The amount of proeutectic alpha will remain the same before and after the eutectic reaction. Thus,

$$\text{Wt of } \alpha \text{ created by eutectic reaction} = \text{total } \alpha - \text{proeutectic } \alpha$$

$$= 0.862 \text{ kg} - 0.747 \text{ kg}$$

$$= 0.115 \text{ kg} \blacktriangleleft$$

$$\text{Wt of } \beta \text{ created by eutectic reaction} = \text{total } \beta$$

$$= 0.138 \text{ kg} \blacktriangleleft$$

[6]See note 5 in Example Problem 8.4.

A lead-tin (Pb-Sn) alloy contains 64 wt % proeutectic α and 36 wt % eutectic $\alpha + \beta$ at $183°C - \Delta T$. Calculate the average composition of this alloy (see Fig. 8.13).

■ **Solution**
Let x be the wt % Sn in the unknown alloy. Since this alloy contains 64 wt % proeutectic α, the alloy must be hypoeutectic, and x will therefore lie between 19.2 and 61.9 wt % Sn as indicated in Fig. EP8.6. At $183°C + \Delta T$, using Fig. EP8.6 and the lever rule gives

$$\% \text{ proeutectic } \alpha = \frac{61.9 - x}{61.9 - 19.2}(100\%) = 64\%$$

or

$$61.9 - x = 0.64(42.7) = 27.3$$
$$x = 34.6\%$$

Thus, the alloy consists of 34.6 percent Sn and 65.4 percent Pb. ◄ Note that we use the lever-rule calculation above the eutectic temperature since the percentage of the proeutectic α remains the same just above and just below the eutectic temperature.

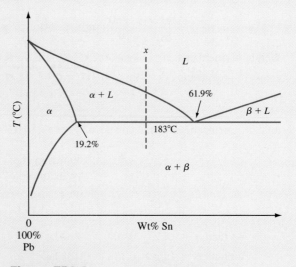

Figure EP8.6
Lead-rich end of the Pb–Sn phase diagram.

IceNine

In a binary eutectic reaction the two solid phases ($\alpha + \beta$) can have various morphologies. Figure 8.16 shows schematically some varied eutectic structures. The shape that will be created depends on many factors. Of prime importance is a minimization of free energy at the $\alpha - \beta$ interfaces. An important factor that determines the eutectic shape is the manner in which the two phases (α and β) nucleate and

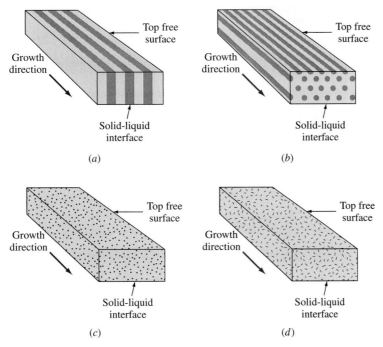

Figure 8.16
Schematic illustration of various eutectic structures: (*a*) lamellar,
(*b*) rodlike, (*c*) globular, (*d*) acicular.
(*After W. C. Winegard, "An Introduction to the Solidification of Metals," Institute of Metals, London, 1964.*)

grow. For example, rod- and plate-type eutectics form when repeated nucleation of the two phases is not required in certain directions. An example of a *lamellar eutectic structure* formed by a Pb-Sn eutectic reaction is shown in Fig. 8.17. Lamellar eutectic structures are very common. A mixed irregular eutectic structure found in the Pb-Sn system is shown in Fig. 8.14*a*.

8.8 BINARY PERITECTIC ALLOY SYSTEMS

Another type of reaction that often occurs in binary equilibrium phase diagrams is the **peritectic reaction.** This reaction is commonly present as part of more-complicated binary equilibrium diagrams, particularly if the melting points of the two components are quite different. In the peritectic reaction a liquid phase reacts with a solid phase to form a new and different solid phase. In the general form the peritectic reaction can be written as

$$\text{Liquid} + \alpha \xrightarrow[\text{cooling}]{} \beta \qquad \textbf{(8.9)}$$

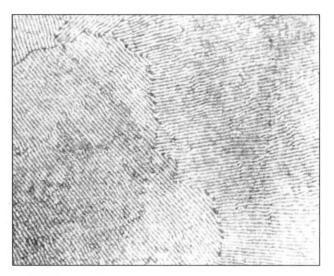

Figure 8.17
Lamellar eutectic structure formed by the Pb–Sn eutectic
reaction. (Magnification 500×.)
*(After W. G. Moffatt et al., "Structure and Properties of Materials,"
vol. I, Wiley, 1964.)*

Figure 8.18 shows the peritectic region of the iron-nickel phase diagram. In this diagram there are solid phases (δ and γ) and one liquid phase. The δ phase is a solid solution of nickel in BCC iron, whereas the γ phase is a solid solution of nickel in FCC iron. The peritectic temperature of 1517°C and the peritectic composition of 4.3 wt % Ni in iron define the peritectic point c in Fig. 8.18. This point is invariant since the three phases δ, γ, and liquid coexist in equilibrium. The peritectic reaction occurs when a slowly cooled alloy of Fe–4.3 wt % Ni passes through the peritectic temperature of 1517°C. This reaction can be written as

$$\text{Liquid (5.4 wt \% Ni)} + \delta\,(4.0 \text{ wt \% Ni}) \xrightarrow[\text{cooling}]{1517°\text{C}} \gamma\,(4.3 \text{ wt \% Ni}) \qquad \textbf{(8.10)}$$

To further understand the peritectic reaction, consider an alloy of Fe–4.3 wt % Ni (peritectic composition) that is slowly cooled from 1550°C to slightly under 1517°C (points a to c in Fig. 8.18). From 1550 to about 1525°C (points a to b in Fig. 8.18) the alloy cools as a homogeneous liquid of Fe–4.3% Ni. When the liquidus is intersected at about 1525°C (point b), solid δ begins to form. Further cooling to point c results in more and more solid δ being formed. At the peritectic temperature of 1517°C (point c) solid δ of 4.0 percent Ni and liquid of 5.4 percent Ni are in equilibrium, and at this temperature all the liquid reacts with all the δ solid phase

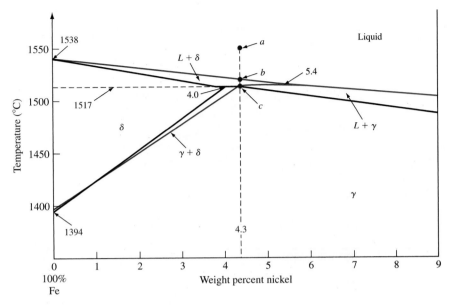

Figure 8.18
The peritectic region of the iron-nickel phase diagram. The peritectic point is
located at 4.3% Ni and 1517°C, which is point *c*.

IceNine

to produce a new and different solid phase γ of 4.3 percent Ni. The alloy remains
as single-phase γ solid solution until another phase change occurs at a lower temperature with which we are not concerned. The lever rule can be applied in the
two-phase regions of the peritectic diagram in the same way as for the eutectic
diagram.

If an alloy in the Fe-Ni system has less than 4.3 percent Ni and is slowly cooled
from the liquid state through the liquid $+\delta$ region, there will be an excess of δ phase
after the peritectic reaction is completed. Similarly, if an Fe-Ni alloy of more than
4.3 percent Ni but less than 5.4 percent Ni is slowly cooled from the liquid state
through the $\delta+$ liquid region, there will be an excess of the liquid phase after the
peritectic reaction is completed.

The silver-platinum binary equilibrium phase diagram is an excellent example of a system that has a single invariant peritectic reaction (Fig. 8.19). In this
system the peritectic reaction $L + \alpha \rightarrow \beta$ occurs at 42.4 percent Ag and 1186°C.
Figure 8.20 schematically illustrates how the peritectic reaction progresses
isothermally in the Pt-Ag system. In Example Problem 8.7, phase analyses are
made at various points on this phase diagram. However, during the natural freezing of peritectic alloys, the departure from equilibrium is usually very large
because of the relatively slow atomic diffusion rate through the solid phase created
by this reaction.

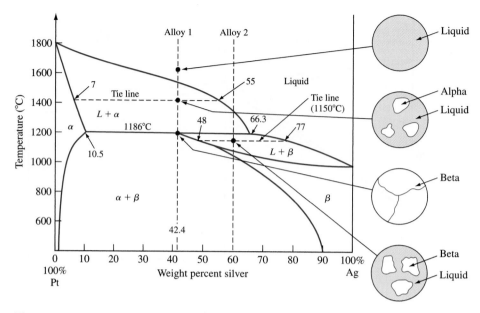

IceNine

Figure 8.19
The platinum-silver phase diagram. The most important feature of this diagram is
the peritectic invariant reaction at 42.4% Ag and 1186°C. At the peritectic point,
liquid (66.3% Ag), alpha (10.5% Ag), and beta (42.4% Ag) can coexist.

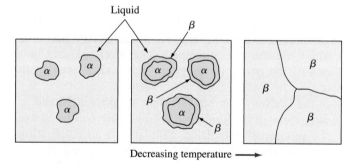

Figure 8.20
Schematic representation of the progressive development of
the peritectic reaction liquid + $\alpha \rightarrow \beta$.

**EXAMPLE
PROBLEM 8.7**

Make phase analyses at the following points in the platinum-silver equilibrium phase
diagram of Fig. 8.19.

a. The point at 42.4 percent Ag and 1400°C.
b. The point at 42.4 percent Ag and 1186°C + ΔT.
c. The point at 42.4 percent Ag and 1186°C − ΔT.
d. The point at 60 percent Ag and 1150°C.

■ **Solution**

a. At 42.4 percent Ag and 1400°C:

Phases present:	liquid	alpha
Compositions of phases:	55% Ag in liquid phase	7% Ag in alpha phase
Amounts of phases:	Wt % liquid phase	Wt % alpha phase
	$= \dfrac{42.4 - 7}{55 - 7}(100\%)$	$= \dfrac{55 - 42.4}{55 - 7}(100\%)$
	$= 74\%$	$= 26\%$

b. At 42.4 percent Ag and 1186°C $+ \Delta T$:

Phases present:	liquid	alpha
Compositions of phases:	66.3% Ag in liquid phase	10.5% Ag in alpha phase
Amounts of phases:	Wt % liquid phase	Wt % alpha phase
	$= \dfrac{42.4 - 10.5}{66.3 - 10.5}(100\%)$	$= \dfrac{66.3 - 42.4}{66.3 - 10.5}(100\%)$
	$= 57\%$	$= 43\%$

c. At 42.4 percent Ag and 1186°C $- \Delta T$:

Phase present:	beta only
Composition of phase:	42.4% Ag in beta phase
Amounts of phase:	100% beta phase

d. At 60 percent Ag and 1150°C:

Phases present:	liquid	beta
Compositions of phases:	77% Ag in liquid phase	48% Ag in beta phase
Amounts of phases:	Wt % liquid phase	Wt % beta phase
	$= \dfrac{60 - 48}{77 - 48}(100\%)$	$= \dfrac{77 - 60}{77 - 48}(100\%)$
	$= 41\%$	$= 59\%$

During the equilibrium or very slow cooling of an alloy of peritectic composition through the peritectic temperature, all the solid-phase alpha reacts with all the liquid to produce a new solid-phase beta, as indicated in Fig. 8.20. However, during the rapid solidification of a cast alloy through the peritectic temperature, a nonequilibrium phenomenon called *surrounding* or *encasement* occurs. During the peritectic reaction of $L + \alpha \rightarrow \beta$, the beta phase created by the peritectic reaction surrounds

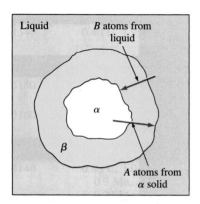

Figure 8.21
Surrounding during the peritectic reaction. The slow rate of atoms diffusing from the liquid to the alpha phase causes the beta phase to surround the alpha phase.

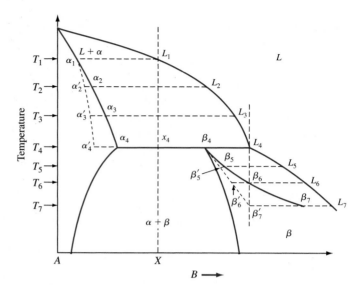

Figure 8.22
A hypothetical binary peritectic phase diagram to illustrate how coring occurs during natural freezing. Rapid cooling causes a nonequilibrium change of solidus α_1 to α_4' and β_4 to β_7' one, which lead to cored alpha phase and cored beta phase. The surrounding phenomenon also occurs during the rapid solidification of peritectic-type alloys.
(*After F. Rhines, "Phase Diagrams in Metallurgy," McGraw-Hill, 1956, p. 86.*)

or encases the primary alpha, as shown in Fig. 8.21. Since the beta phase formed is a solid phase and since solid-state diffusion is relatively slow, the beta formed around the alpha creates a diffusion barrier and the peritectic reaction proceeds at an ever-decreasing rate. Thus, when a peritectic-type alloy is rapidly cast, coring occurs during the formation of the primary alpha (Fig. 8.22 along α_1 to α_4'), and encasement of the cored α by β occurs during the peritectic reaction. Figure 8.23 schematically illustrates these combined nonequilibrium structures. The microstructure of a 60% Ag–40% Pt alloy that was rapidly cast is shown in Fig. 8.24. This structure shows cored alpha and its encasement by the beta phase.

8.9 BINARY MONOTECTIC SYSTEMS

Another three-phase invariant reaction that occurs in some binary phase diagrams is the **monotectic reaction** in which a liquid phase transforms into a solid phase and another liquid phase as

$$L_1 \xrightarrow{\text{cooling}} \alpha + L_2 \tag{8.11}$$

Over a certain range of compositions the two liquids are immiscible like oil in water and so constitute individual phases. A reaction of this type occurs in the copper-lead

Figure 8.23
Schematic representation of surrounding or encasement in a cast peritectic-type alloy. A residue of cored primary α is represented by the solid circles concentric about smaller dashed circles; surrounding the cored α is a layer of β of peritectic composition. The remaining space is filled with cored β, represented by dashed curved lines.
(After F. Rhines, "Phase Diagrams in Metallurgy," McGraw-Hill, 1956, p. 87.)

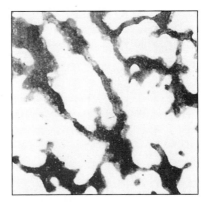

Figure 8.24
Cast 60% Ag–40% Pt hyperperitectic alloy. White and light gray areas are residual cored α; dark two-toned areas are β, the outer portions being of peritectic composition and the darkest central areas being the cored β that formed at temperatures below that of the peritectic reaction. (Magnification 1000×.)
(Courtesy of F. Rhines.)

system at 955°C and 36 percent Pb, as shown in Fig. 8.25. The copper-lead phase diagram has a eutectic point at 326°C and 99.94 percent Pb, and as a result terminal solid solutions of almost pure lead (0.007 percent Cu) and pure copper (0.005 percent Pb) are formed at room temperature. Figure 8.26 shows the microstructure of a cast monotectic alloy of Cu–36% Pb. Note the distinct separation of the lead-rich phase (dark) and the copper matrix (light).

Lead is added in small amounts up to about 0.5 percent to many alloys (e.g., the Cu-Zn brasses) to make the machining of alloys easier by reducing ductility sufficiently to cause machined metal chips to break away from the workpiece. This small addition of lead reduces the strength of the alloy only slightly. Leaded alloys are also used for bearings where small amounts of lead smear out at wear surfaces between the bearing and shaft and thus reduce friction.

8.10 INVARIANT REACTIONS

Three invariant reactions that commonly occur in binary phase diagrams have been discussed so far: the eutectic, peritectic, and monotectic types. Table 8.1 summarizes these reactions and shows their phase-diagram characteristics at their reaction

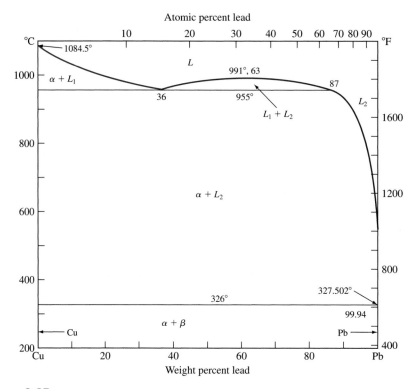

IceNine

Figure 8.25
The copper-lead phase diagram. The most important feature of this diagram is the monotectic invariant reaction at 36% Pb and 955°C. At the monotectic point α (100% Cu), L_1 (36% Pb), and L_2 (87% Pb) can coexist. Note that copper and lead are essentially insoluble in each other.
(*"Metals Handbook," vol. 8: "Metallography, Structures, and Phase Diagrams," 8th ed., American Society for Metals, 1973, p. 296.*)

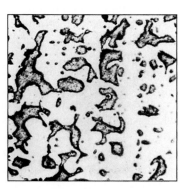

Figure 8.26
The microstructure of cast monotectic alloy Cu–36% Pb. Light areas are the Cu-rich matrix of the monotectic constituent; dark areas are the Pb-rich portion, which existed as L_2 at the monotectic temperature. (Magnification 100×.)
(*Courtesy of F. Rhines.*)

Table 8.1 Types of three-phase invariant reactions occurring in binary phase diagrams

Name of reaction	Equation	Phase-diagram characteristic
Eutectic	$L \xrightarrow{\text{cooling}} \alpha + \beta$	
Eutectoid	$\alpha \xrightarrow{\text{cooling}} \beta + \gamma$	
Peritectic	$\alpha + L \xrightarrow{\text{cooling}} \beta$	
Peritectoid	$\alpha + \beta \xrightarrow{\text{cooling}} \gamma$	
Monotectic	$L_1 \xrightarrow{\text{cooling}} \alpha + L_2$	

points. Two other important invariant reactions occurring in binary systems are the *eutectoid* and *peritectoid types*. Eutectic and eutectoid reactions are similar in that two solid phases are formed from one phase on cooling. However, in the eutectoid reaction the decomposing phase is solid, whereas in the eutectic reaction it is liquid. In the peritectoid reaction two solid phases react to form a new solid phase, whereas in the peritectic reaction a solid phase reacts with a liquid phase to produce a new solid phase. It is interesting to note that the peritectic and peritectoid reactions are the inverse of the corresponding eutectic and eutectoid reactions. The temperatures and compositions of the reacting phases are fixed for all these invariant reactions, i.e., according to the phase rule, there are zero degrees of freedom at the reaction points.

8.11 PHASE DIAGRAMS WITH INTERMEDIATE PHASES AND COMPOUNDS

The phase diagrams considered so far have been relatively simple and contained only a small number of phases and have had only one invariant reaction. Many equilibrium diagrams are complex and often show intermediate phases or compounds. In phase-diagram terminology it is convenient to distinguish between two types of solid solutions: **terminal phases** and **intermediate phases.** Terminal solid-solution phases occur at the ends of phase diagrams, bordering on pure components. The α and β solid solutions of the Pb-Sn diagram (Fig. 8.13) are examples. Intermediate solid-solution phases occur in a composition range inside the phase diagram and are separated from other phases in a binary diagram by two-phase regions. The Cu-Zn phase diagram has both terminal and intermediate phases (Fig. 8.27). In this system α and η are terminal phases and β, γ, δ, and ϵ are intermediate phases. The Cu-Zn diagram contains five invariant peritectic points and one eutectoid invariant point at the lowest point of the δ intermediate-phase region.

Intermediate phases are not restricted to binary metal phase diagrams. In the ceramic phase diagram of the $Al_2O_3-SiO_2$ system an intermediate phase called

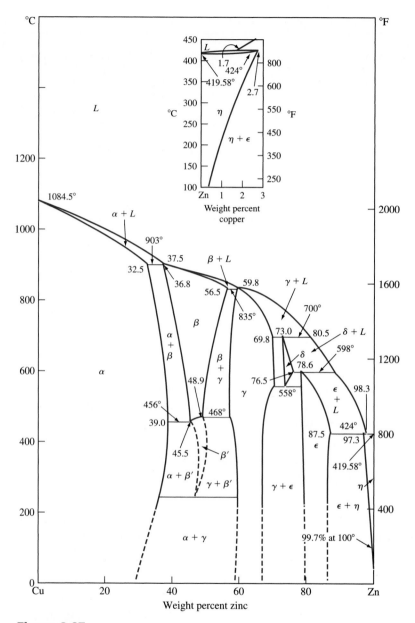

Figure 8.27

The copper-zinc phase diagram. This diagram has terminal phases α and η and intermediate phases β, γ, δ, and ϵ. There are five invariant peritectic points and one eutectoid point.

(After "Metals Handbook," vol. 8: "Metallography, Structures, and Phase Diagrams," 8th ed., American Society for Metals, 1973, p. 301.)

IceNine

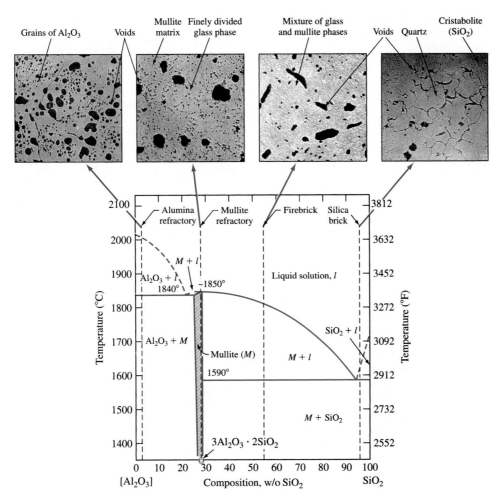

Figure 8.28
The phase diagram of the Al_2O_3–SiO_2 system, which contains mullite as an intermediate phase. Typical compositions of refractories having Al_2O_3 and SiO_2 as their main components are shown.
(*After A. G. Guy, "Essentials of Materials Science," McGraw-Hill, 1976.*)

IceNine

mullite is formed, which includes the compound $3Al_2O_3 \cdot 2SiO_2$ (Fig. 8.28). Many refractories[7] have Al_2O_3 and SiO_2 as their main components. These materials will be discussed in Chap. 11 on ceramic materials.

If the intermediate compound is formed between two metals, the resulting material is a crystalline material called an *intermetallic compound* or simply an *intermetallic*. Generally speaking, the intermetallic compounds should have a distinct chemical formula or be stoichiometric (fixed ratio of involved atoms). However, in

---

[7]A refractory is a heat-resisting ceramic material.

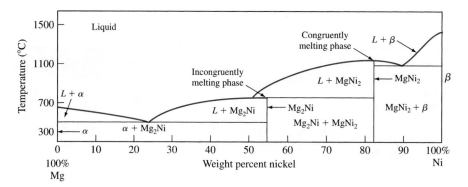

Figure 8.29

The magnesium-nickel phase diagram. In this diagram there are two intermetallic compounds, Mg_2Ni and $MgNi_2$.

(*"Metals Handbook," 8th ed., vol. 8, American Society for Metals, 1973, p. 314.*)

IceNine

many cases a certain degree of atomic substitution takes place that accommodates large deviations from stoichiometry. In a phase diagram intermetallics appear either as a single vertical line, signifying the stoichiometric nature of the compound (see $TiNi_3$ line in Fig. EP8.8), or sometimes as a range of composition, signifying a nonstoichiometric compound (for example, the substitution of Cu for Zn or Zn for Cu atoms in the β and γ phases of the Cu-Zn phase diagram shown in Fig. 8.27). The majority of the intermetallic compounds possess a mixture of metallic-ionic or metallic-covalent bonds. The percentage of ionic or covalent bonds formed in intermetallic compounds depends on the differences in the electronegativities of the elements involved (see Sec. 2.9).

The Mg-Ni phase diagram contains the intermediate compounds Mg_2Ni and $MgNi_2$, which are primarily metallically bonded and have fixed compositions and definite stoichiometries (Fig. 8.29). The intermetallic compound $MgNi_2$ is said to be a *congruently melting compound* since it maintains its composition right up to the melting point. On the other hand, Mg_2Ni is said to be an *incongruently melting compound* since, upon heating, it undergoes peritectic decomposition at 761°C into liquid and $MgNi_2$ phases. Other examples of intermediate compounds that occur in phase diagrams are Fe_3C and Mg_2Si. In Fe_3C the bonding is mainly metallic in character, but in Mg_2Si the bonding is mainly covalent.

EXAMPLE PROBLEM 8.8

Consider the titanium-nickel (Ti-Ni) phase diagram in Fig. EP8.8. This phase diagram has six points where three phases coexist. For each of these three-phase points:

a. List the coordinates of composition (weight percent) and temperature for each point.
b. Write the invariant reaction that occurs during slow cooling of the Ti-Ni alloy through each point.
c. Name the type of invariant reaction that takes place at each point.

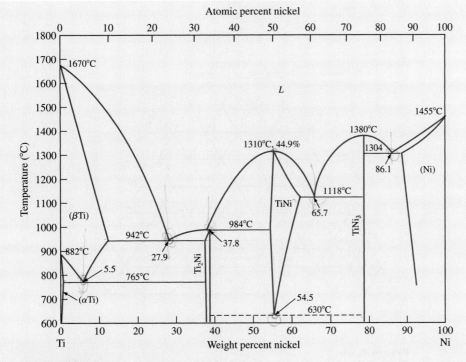

Figure EP8.8
Titanium-nickel phase diagram.
(After Binary Alloy Phase Diagrams, ASM Int., 1986, p. 1768.)

IceNine

■ **Solution**

a. (i) 5.5 wt % Ni, 765°C
    (ii) $(\beta \, \text{Ti}) \longrightarrow (\alpha \, \text{Ti}) + \text{Ti}_2\text{Ni}$
    (iii) Eutectoid reaction

b. (i) 27.9 wt % Ni, 942°C
    (ii) $L \longrightarrow (\beta \, \text{Ti}) + \text{Ti}_2\text{Ni}$
    (iii) Eutectic reaction

c. (i) 37.8 wt % Ni, 984°C
    (ii) $L + \text{TiNi} \longrightarrow \text{Ti}_2\text{Ni}$
    (iii) Peritectic reaction

d. (i) 54.5 wt % Ni, 630°C
    (ii) $\text{TiNi} \longrightarrow \text{Ti}_2\text{Ni} + \text{TiNi}_3$
    (iii) Eutectoid reaction

e. (i) 65.7 wt % Ni, 1118°C
    (ii) $L \longrightarrow \text{TiNi} + \text{TiNi}_3$
    (iii) Eutectic reaction

f. (i) 86.1 wt % Ni, 1304°C
    (ii) $L \longrightarrow \text{TiNi}_3 + (\text{Ni})$
    (iii) Eutectic reaction

8.12 TERNARY PHASE DIAGRAMS

Until now we have discussed only binary phase diagrams in which there are two components. We shall now turn our attention to ternary phase diagrams that have three components. Compositions on ternary phase diagrams are usually constructed by using an equilateral triangle as a base. Compositions of ternary systems are

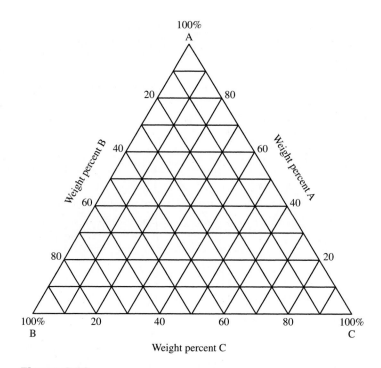

Figure 8.30
Composition base for a ternary phase diagram for a system with
pure components A, B, and C.

represented on this base with the pure component at each end of the triangle. Figure 8.30 shows the composition base of a ternary phase diagram for a ternary metal
alloy consisting of pure metals A, B, and C. The binary alloy compositions AB, BC,
and AC are represented on the three edges of the triangle.

Ternary phase diagrams with a triangular composition base are normally constructed at a constant pressure of 1 atm. Temperature can be represented as uniform
throughout the whole diagram. This type of ternary diagram is called an *isothermal
section*. To show a range of temperatures at varying compositions, a figure with temperature on a vertical axis with a triangular composition base can be constructed.
However, more commonly, temperature contour lines are drawn on a triangular composition base to indicate temperature ranges just as different elevations are shown
on a flat-page map of a terrain.

Let us now consider the determination of the composition of a ternary alloy indicated by a point on a ternary diagram of the type shown in Fig. 8.30. In Fig. 8.30 the
A corner of the triangle indicates 100 percent metal A, the *B* corner indicates 100 percent metal B, and the *C* corner indicates 100 percent metal C. The weight percent of
each pure metal in the alloy is determined in the following way: A perpendicular line
is drawn from a pure metal corner to the side of the triangle opposite that corner, and
the distance *from* the side to the corner along the perpendicular line is measured as

a fraction of 100 percent for the whole line. This percentage is the weight percent of the pure metal of that corner in the alloy. Example Problem 8.9 explains this procedure in more detail.

Determine the weight percents of metals A, B, and C for a ternary alloy ABC at point x on the ternary phase diagram grid shown in Fig. EP8.9.

■ **Solution**

The composition at a point in a ternary phase diagram grid of the type shown in Fig. EP8.9 is determined by separately determining the compositions of each of the pure metals from the diagram. To determine the percent A at point x in Fig. EP8.9, we first draw the perpendicular line AD from the corner A to point D on the side of the triangle opposite corner A. The total length of the line from D to A represents 100 percent A. At point D the percent A in the alloy is zero. The point x is on an isocomposition line at 40 percent A, and thus the percentage of A in the alloy is 40 percent. In a similar manner we draw line BE and determine that the percentage of B in the alloy is also 40 percent. A third line CF is drawn and the percentage of C in the alloy is determined to be 20 percent. Thus the composition of the ternary alloy at point x is 40 percent A, 40 percent B, and 20 percent C. Actually only two percentages need to be determined since the third can be obtained by subtracting the sum of two from 100 percent.

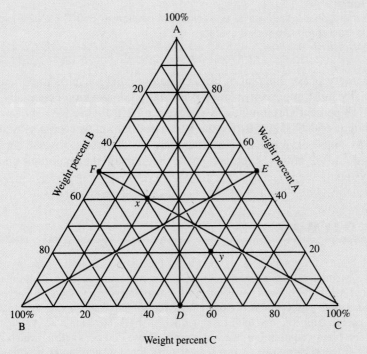

Figure EP8.9
Ternary phase diagram composition base for an ABC alloy.

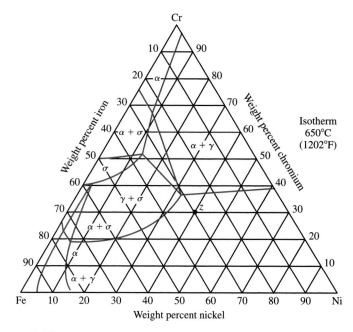

Figure 8.31

Ternary phase diagram of an isothermal section at 650°C (1202°F) for the iron-chromium-nickel system.

(After "Metals Handbook," vol. 8, 8th ed., American Society for Metals, 1973, p. 425.)

The ternary phase diagram of iron, chromium, and nickel is important since the commercially most important stainless steel has a composition essentially of 74 percent iron, 18 percent chromium, and 8 percent nickel. Figure 8.31 shows an isothermal section at 650°C (1202°F) for the iron-chromium-nickel ternary system.

Ternary phase diagrams also are important for the study of some ceramic materials. Figure 11.32 shows a ternary phase diagram of the important silica-leucite-mullite system.

8.13 SUMMARY

Phase diagrams are graphical representations of what phases are present in an alloy (or ceramic) system at various temperatures, pressures, and compositions. Phase diagrams are constructed using the information gathered from cooling curves. Cooling curves are time temperature plots generated for various alloy compositions and provide information about phase transition temperatures. In this chapter the emphasis has been placed on temperature-composition binary equilibrium phase diagrams. These diagrams tell us which phases are present at different compositions and temperatures for slow cooling or heating conditions that approach equilibrium. In two-phase regions of these diagrams the chemical compositions of each of the two phases is indicated by the intersection of the isotherm with the phase boundaries. The weight fraction of each phase in a two-phase region can be determined by using the lever rule along an isotherm (tie line at a particular temperature).

In binary equilibrium *isomorphous phase diagrams* the two components are completely soluble in each other in the solid state, and so there is only one solid phase. In binary equilibrium alloy (ceramic) phase diagrams, *invariant reactions* involving three phases in equilibrium often occur. The most common of these reactions are:

1. Eutectic reaction: $L \rightarrow \alpha + \beta$
2. Eutectoid reaction: $\alpha \rightarrow \beta + \gamma$
3. Peritectic reaction: $\alpha + L \rightarrow \beta$
4. Peritectoid reaction: $\alpha + \beta \rightarrow \gamma$
5. Monotectic reaction: $L_1 \rightarrow \alpha + L_2$

In many binary equilibrium phase diagrams, intermediate phase(s) and/or compounds are present. The intermediate phases have a range of compositions, whereas the intermediate compounds have only one composition. If the components are both metal, the intermediate compound is called an intermetallic.

During the rapid solidification of many alloys, compositional gradients are created and *cored* structures are produced. A cored structure can be eliminated by homogenizing the cast alloy for long times at high temperatures just below the melting temperature of the lowest melting phase in the alloy. If the cast alloy is overheated slightly so that melting occurs at the grain boundaries, a *liquated* structure is produced. This type of structure is undesirable since the alloy loses strength and may break up during subsequent working.

8.14 DEFINITIONS

Sec. 8.1

Equilibrium: a system is said to be in equilibrium if no macroscopic changes take place with time.

Phase: a physically homogeneous and distinct portion of a material system.

Equilibrium phase diagram: a graphical representation of the pressures, temperatures, and compositions for which various phases are stable at equilibrium. In materials science the most common phase diagrams involve temperature versus composition.

Sec. 8.2

System: a portion of the universe that has been isolated so that its properties can be studied.

Gibbs phase rule: the statement that at equilibrium the number of phases plus the degrees of freedom equals the number of components plus 2. $P + F = C + 2$. In the condensed form with pressure $\approx$ 1 atm, $P + F = C + 1$.

Degrees of freedom F: the number of variables (temperature, pressure, and composition) that can be changed *independently* without changing the phase or phases of the system.

Number of components of a phase diagram: the number of elements or compounds that make up the phase-diagram system. For example, the Fe-Fe$_3$C system is a two-component system; the Fe-Ni system is also a two-component system.

Sec. 8.3

Cooling curve: plots of temp. vs time acquired during solidification of a metal. It provides phase change information as temperature is lowered.

Thermal arrest: a region of the cooling curve for a pure metal where temperature does not change with time (plateaue) representing the freezing temperature.

Sec. 8.4

Isomorphous system: a phase diagram in which there is only one solid phase, i.e., there is only one solid-state structure.

Liquidus: the temperature at which liquid starts to solidify under equilibrium conditions.

Solidus: the temperature during the solidification of an alloy at which the last of the liquid phase solidifies.

Sec. 8.5

Lever rule: the weight percentages of the phases in any two-phase region of a binary phase diagram can be calculated using this rule if equilibrium conditions prevail.

Tie line: a horizontal working line drawn at a particular temperature between two phase boundaries (in a binary phase diagram) to be used to apply the lever rule. Vertical lines are drawn from the intersection of the tie line with the phase boundaries to the horizontal composition line. A vertical line is also drawn from the tie line to the horizontal line at the intersection point of the tie line with the alloy of interest to use with the lever rule.

Sec. 8.6

Cored structure: a type of microstructure that occurs during rapid solidification or non-equilibrium cooling of a metal.

Homogenization: a heat treatment process given to a metal to remove undesirable cored structures.

Sec. 8.7

Eutectic reaction (in a binary phase diagram): a phase transformation in which all the liquid phase transforms on cooling into two solid phases isothermally.

Eutectic temperature: the temperature at which a eutectic reaction takes place.

Eutectic composition: the composition of the liquid phase that reacts to form two new solid phases at the eutectic temperature.

Eutectic point: the point determined by the eutectic composition and temperature.

Invariant reactions: equilibrium phase transformations involving zero degrees of freedom.

Solvus: a phase boundary below the isothermal liquid + proeutectic solid phase boundary and between the terminal solid solution and two-phase regions in a binary eutectic phase diagram.

Hypoeutectic composition: one that is to the left of the eutectic point.

Hypereutectic composition: one that is to the right of the eutectic point.

Proeutectic phase: a phase that forms at a temperature above the eutectic temperature.

Primary phase: a solid phase that forms at a temperature above that of an invariant reaction and is still present after the invariant reaction is completed.

Sec. 8.8

Peritectic reaction (in a binary phase diagram): a phase transformation in which, upon cooling, a liquid phase combines with a solid phase to produce a new solid phase.

Sec. 8.9

Monotectic reaction (in a binary phase diagram): a phase transformation in which, upon cooling, a liquid phase transforms into a solid phase and a new liquid phase (of different composition than the first liquid phase).

Sec. 8.10

Invariant reactions: those reactions in which the reacting phases have fixed temperature and composition. The degree of freedom, F_1 is zero at these reaction points.

Sec. 8.11

Terminal phase: a solid solution of one component in another for which one boundary of the phase field is a pure component.

Intermediate phase: a phase whose composition range is between those of the terminal phases.

8.15 PROBLEMS

Answers to problems marked with an asterisk are given at the end of the book.

8.1 Define (*a*) a phase in a material and (*b*) a phase diagram.

8.2 In the pure water pressure-temperature equilibrium phase diagram (Fig. 8.1), what phases are in equilibrium for the following conditions?

    a. Along the freezing line

    b. Along the vaporization line

    c. At the triple point

8.3 How many triple points are there in the pure iron pressure-temperature equilibrium phase diagram of Fig. 8.2? What phases are in equilibrium at each of the triple points?

8.4 Write the equation for Gibbs phase rule and define each of the terms.

8.5 Refer to the pressure-temperature equilibrium phase diagram for pure water (Fig. 8.1) and answer the following:

    a. How many degrees of freedom are there at the triple point?

    b. How many degrees of freedom are there along the freezing line?

8.6 What is a binary isomorphous alloy system?

8.7 What are the four Hume-Rothery rules for the solid solubility of one element in another?

8.8 A number of elements along with their crystal structures and atomic radii are listed in the following table. Which pairs might be expected to have complete solid solubility in each other?

	Crystal structure	Atomic radius (nm)		Crystal structure	Atomic radius (nm)
Silver	FCC	0.144	Lead	FCC	0.175
Palladium	FCC	0.137	Tungsten	BCC	0.137
Copper	FCC	0.128	Rhodium	FCC	0.134
Gold	FCC	0.144	Platinum	FCC	0.138
Nickel	FCC	0.125	Tantalum	BCC	0.143
Aluminum	FCC	0.143	Potassium	BCC	0.231
Sodium	BCC	0.185	Molybdenum	BCC	0.136

8.9 Derive the lever rule for the amount in weight percent of each phase in two-phase regions of a binary phase diagram. Use a phase diagram in which two elements are completely soluble in each other.

8.10 Consider an alloy containing 70 wt % Ni and 30 wt % Cu (see Fig. 8.5).

a. At 1350°C make a phase analysis assuming equilibrium conditions. In the phase analysis include the following:

   (i) What phases are present?

   (ii) What is the chemical composition of each phase?

   (iii) What amount of each phase is present?

b. Make a similar phase analysis at 1500°C.

c. Sketch the microstructure of the alloy at each of these temperatures by using circular microscopic fields.

8.11 Describe how the liquidus and solidus of a binary isomorphous phase diagram can be determined experimentally.

8.12 Explain how a cored structure is produced in a 70% Cu–30% Ni alloy.

8.13 How can the cored structure in a 70% Cu–30% Ni alloy be eliminated by heat treatment?

8.14 Explain what is meant by the term *liquation.* How can a liquated structure be produced in an alloy? How can it be avoided?

8.15 Consider the binary eutectic copper-silver phase diagram in Fig. 8.32. Make phase analyses of an 88 wt % Ag–12 wt % Cu alloy at the temperatures (*a*) 1000°C, (*b*) 800°C, (*c*) 780°C + ΔT, and (*d*) 780°C − ΔT. In the phase analyses, include:

   (i) The phases present

   (ii) The chemical compositions of the phases

   (iii) The amounts of each phase

   (iv) Sketch the microstructure by using 2-cm-diameter circular fields.

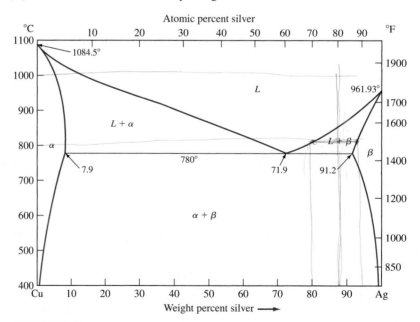

Figure 8.32
The copper-silver phase diagram.
(After "Metals Handbook," vol. 8, 8th ed., American Society for Metals, 1973, p. 253.)

IceNine

*8.16 If 500 g of a 40 wt % Ag–60 wt % Cu alloy is slowly cooled from 1000°C to just below 780°C (see Fig. 8.32):

    a. How many grams of liquid and proeutectic alpha are present at 850°C?

    b. How many grams of liquid and proeutectic alpha are present at 780°C + ΔT?

    c. How many grams of alpha are present in the eutectic structure at 780°C − ΔT?

    d. How many grams of beta are present in the eutectic structure at 780°C − ΔT?

*8.17 A lead-tin (Pb-Sn) alloy consists of 60 wt % proeutectic β and 60 wt % eutectic $\alpha + \beta$ at 183°C − ΔT. Calculate the average composition of this alloy (see Fig. 8.13).

8.18 A Pb-Sn alloy (Fig. 8.13) contains 40 wt % β and 60 wt % α at 50°C. What is the average composition of Pb and Sn in this alloy?

8.19 An alloy of 30 wt % Pb and 70 wt % Sn is slowly cooled from 250°C to 27°C (see Fig. 8.13).

    a. Is this alloy hypoeutectic or hypereutectic?

    b. What is the composition of the first solid to form?

    c. What are the amounts and compositions of each phase that is present at 183°C + ΔT?

    d. What is the amount and composition of each phase that is present at 183°C − ΔT?

    e. What are the amounts of each phase present at room temperature?

8.20 Consider the binary peritectic iridium-osmium phase diagram of Fig. 8.33. Make phase analyses of a 70 wt % Ir–30 wt % Os at the temperatures (a) 2600°C, (b) 2665°C + ΔT, and (c) 2665°C − ΔT. In the phase analyses include:

    (i) The phases present

    (ii) The chemical compositions of the phases

    (iii) The amounts of each phase

    (iv) Sketch the microstructure by using 2-cm-diameter circular fields.

IceNine

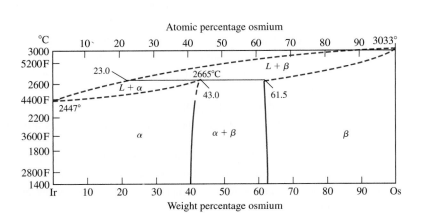

Figure 8.33

The iridium-osmium phase diagram.

(After "Metals Handbook," vol. 8, 8th ed., American Society for Metals, 1973, p. 332.)

8.21 Consider the binary peritectic iridium-osmium phase diagram of Fig. 8.33. Make phase analyses of a 40 wt % Ir–60 wt % Os at the temperatures (a) 2600°C, (b) 2665°C + ΔT, (c) 2665°C − ΔT, and (d) 2800°C. Include in the phase analyses the four items listed in Prob. 8.20.

8.22 Describe the mechanism that produces the phenomenon of *surrounding* in a peritectic alloy that is rapidly solidified through the peritectic reaction.

8.23 Can coring and surrounding occur in a peritectic-type alloy that is rapidly solidified? Explain.

*8.24 Consider an Fe–4.2 wt % Ni alloy (Fig. 8.18) that is slowly cooled from 1550 to 1450°C. What weight percent of the alloy solidifies by the peritectic reaction?

8.25 Consider an Fe–5.0 wt % Ni alloy (Fig. 8.18) that is slowly cooled from 1550 to 1450°C. What weight percent of the alloy solidifies by the peritectic reaction?

*8.26 Determine the weight percent and composition in weight percent of each phase present in an Fe–4.2 wt % Ni alloy (Fig. 8.18) at 1517°C + ΔT.

8.27 Determine the composition in weight percent of the alloy in the Fe–Ni system (Fig. 8.18) that will produce a structure of 40 wt % δ and 60 wt % γ, just below the peritectic temperature.

8.28 What is a monotectic invariant reaction? How is the monotectic reaction in the copper-lead system important industrially?

*8.29 In the copper-lead (Cu-Pb) system (Fig. 8.25) for an alloy of Cu–10 wt % Pb, determine the amounts and compositions of the phases present at (a) 100°C, (b) 955°C + ΔT, (c) 955°C − ΔT, and (d) 200°C.

8.30 For an alloy of Cu–70 wt % Pb (Fig. 8.25), determine the amounts and compositions in weight percent of the phases present at (a) 955°C + ΔT, (b) 955°C − ΔT, and (c) 200°C.

*8.31 What is the average composition (weight percent) of a Cu-Pb alloy that contains 30 wt % L_1 and 70 wt % α at 955°C + ΔT?

8.32 Write equations for the following invariant reactions: eutectic, eutectoid, peritectic, and peritectoid. How many degrees of freedom exist at invariant reaction points in binary phase diagrams?

8.33 How are eutectic and eutectoid reactions similar? What is the significance of the -*oid* suffix?

8.34 Distinguish between (a) a terminal phase and (b) an intermediate phase.

8.35 Distinguish between (a) an intermediate phase and (b) an intermediate compound.

8.36 What is the difference between a congruently melting compound and an incongruently melting one?

IceNine

8.37 Consider the Cu-Zn phase diagram of Fig. 8.27.

  a. What is the maximum solid solubility in weight percent of Zn in Cu in the terminal solid solution α?

  b. Identify the intermediate phases in the Cu-Zn phase diagram.

  c. Identify the three-phase invariant reactions in the Cu-Zn diagram.

    (i) Determine the composition and temperature coordinates of the invariant reactions.

    (ii) Write the equations for the invariant reactions.

    (iii) Name the invariant reactions.

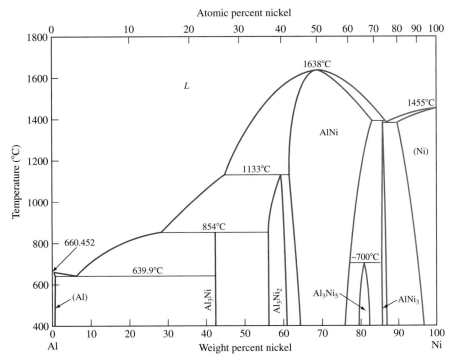

Atomic percent nickel

Figure 8.34
Aluminum-nickel phase diagram.
(After Binary Phase Diagrams, ASM Int., 1986, p. 142.)

IceNine

8.38 Consider the aluminum-nickel (Al-Ni) phase diagram of Fig. 8.34. For this phase diagram:

a. Determine the coordinates of the composition and temperature of the invariant reactions.

b. Write the equations for the three-phase invariant reactions and name them.

c. Label the two-phase regions in the phase diagram.

8.39 Consider the nickel-vanadium (Ni-V) phase diagram of Fig. 8.35. For this phase diagram repeat questions (*a*), (*b*), and (*c*) of Prob. 8.38.

8.40 Consider the titanium-aluminum (Ti-Al) phase diagram of Fig. 8.36. For this phase diagram repeat questions (*a*), (*b*), and (*c*) of Prob. 8.38.

8.41 What is the composition of point y in Fig. EP8.9?

8.16 MATERIALS SELECTION AND DESIGN PROBLEMS

1. Design a Cu-Ni alloy that will exist at a completely molten state at 1300°C and become completely solid at 1200°C (use Fig. 8.5).

2. Design a Cu-Ni alloy that will be completely solid at 1200°C (use Fig. 8.5).

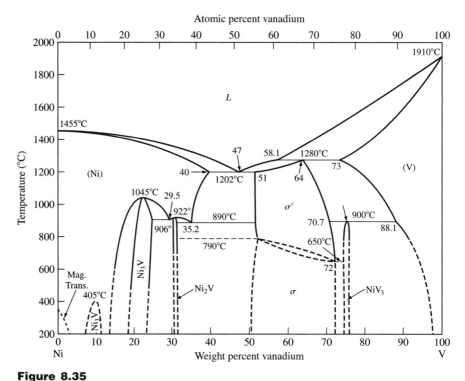

Figure 8.35
Nickel-vanadium phase diagram.
(*After Binary Phase Diagrams, ASM Int., 1986, p. 1773.*)

IceNine

3. Design a Pb-Sn alloy that exists at a completely liquid state at the lowest possible temperature (use Fig. 8.13).

4. (*a*) Design a Pb-Sn alloy that will have a 50-50 solid and liquid phase fraction at 184°C. (*b*) How many grams of each component should you use to achieve this? (use Fig. 8.13).

5. Given that Pb and Sn have similar tensile strengths, design a Pb-Sn alloy that when cast would be the strongest alloy (use Fig. 8.13). Explain your reasons for your choice.

6. Based on the Al-Ni phase diagram given in Fig. 8.34, how many grams of Ni should be alloyed with 100 grams of Al to synthesize an alloy of liquidus temperature of approximately 640°C?

7. An Al–10 wt % Ni alloy is completely liquid at 800°C. How many grams of Ni can you add to this alloy at 800°C without creating a solid phase?

8. Based on the Al_2O_3-SiO_2 phase diagram in Fig. 8.28, plot the weight percent of phases present for Al_2O_3−55 wt % SiO_2 over the 2000 to 1500°C temperature range.

9. An intermetallic compound is formed during slow cooling of Ti–63 wt % Al. (*a*) Determine the chemical formula of this compound. (*b*) Compare the melt temperature of this compound to that of Ti and Al.

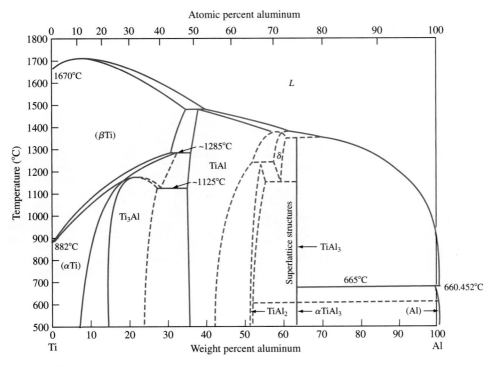

Figure 8.36
Titanium-aluminum phase diagram.
(After Binary Phase Diagrams, ASM Int., 1986, p. 1773.)

IceNine

10. Based on the Cu-Ag phase diagram in Fig. 8.32, draw the approximate cooling curve for the following alloys with appropriate temperatures and explanations:
(*a*) Pure Cu, (*b*) Cu–7.9 wt % Ag, (*c*) Cu–10 wt % Ag, (*d*) Cu–40 wt % Ag, (*e*) Cu–71.9 wt % Ag, and (*f*) Cu–95 wt % Ag.

Engineering Alloys

(© *Textron Power Transmission*)

A variety of metal alloys such as plain-carbon steels, alloy steels, stainless steels, cast iron, and copper alloys are used in manufacturing various gears. For example, chromium steels are used for automobile transmission gears, chromium-molybdenum steels are used for aircraft gas turbine gears, nickel-molybdenum steels are used for earth moving equipment, and some copper alloys are used to manufacture gears for low load levels. The choice of the gear metal and its manufacturing depends on size, stresses involved, power requirements, and the environment in which they will operate. The chapter-opening photos show gears of various sizes used in various industries.[1] ∎

---

[1]http://www.textronpt.com/cgi-bin/products.cgi?prod=highspeed&group=spcl

LEARNING OBJECTIVES

By the end of this chapter, students will be able to . . .

1. Describe steelmaking and processing of steel components. Differentiate between plain carbon steel, alloy steel, cast iron, and stainless steel.

2. Reconstruct the iron-carbon phase diagram indicating all key phases, reactions, and microstructures.

3. Describe what pearlite and marteniste are, their mechanical property differences, their microstructural differences, and how they are produced.

4. Define isothermal and continuous cooling transformations.

5. Describe annealing, normalizing, quenching, tempering, martempering, and austempering processes.

6. Describe the classification of plain carbon and alloy steels and explain the effect of various alloying elements on properties of steel.

7. Describe the classification, heat-treatability, microstructure, and general properties of aluminum alloys, copper alloys, stainless steels, and cast irons.

8. Explain the importance and applications of intermetallics, shape memory, and amorphous alloys.

9. Describe the advantages and disadvantages of alloys that are used in biomedical applications.

Metals and alloys have many useful engineering properties and so have widespread application in engineering designs. Iron and its alloys (principally steel) account for about 90 percent of the world's production of metals mainly because of their combination of good strength, toughness, and ductility at a relatively low cost. Each metal has special properties for engineering designs and is used after a comparative cost analysis with other metals and materials (see Table 9.1).

Alloys based on iron are called *ferrous alloys,* and those based on the other metals are called *nonferrous alloys*. In this chapter we shall discuss some aspects of the

Table 9.1 Approximate prices ($/lb) of some metals as of May 2001*

Steel†	0.27	Nickel	2.74
Aluminum	0.67	Tin	2.30
Copper	0.76	Titanium‡	3.85
Magnesium	3.29	Gold	3108.00
Zinc	0.45	Silver	52.00
Lead	0.22		

*Prices of metals vary with time.
†Hot-rolled plain-carbon steel sheet.
‡Titanium sponge. Prices for large quantity.

processing, structure, and properties of some of the important ferrous and nonferrous alloys. The last two sections of this chapter are devoted to advanced alloys and their application to various fields, including the biomedical field.

9.1 PRODUCTION OF IRON AND STEEL

9.1.1 Production of Pig Iron in a Blast Furnace

Most iron is extracted from iron ores in large blast furnaces (Fig. 9.1). In the blast furnace coke (carbon) acts as a reducing agent to reduce iron oxides (mainly Fe_2O_3) to produce raw pig iron, which contains about 4 percent carbon along with some other impurities according to the typical reaction

$$Fe_2O_3 + 3CO \longrightarrow 2Fe + 3CO_2$$

The pig iron from the blast furnace is usually transferred in the liquid state to a steel-making furnace.

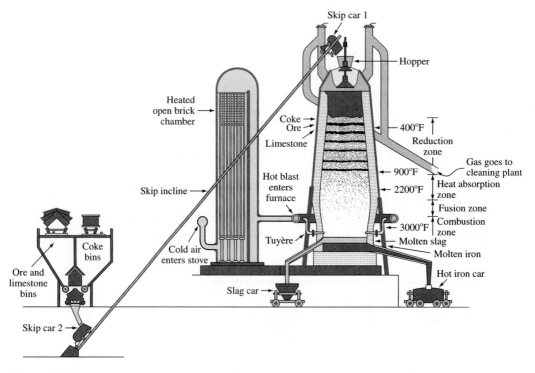

Figure 9.1
Cross section of the general operation of a modern blast furnace.
Animation   (*After A. G. Guy, "Elements of Physical Metallurgy," 2nd ed., © 1959, Addison-Wesley, Fig. 2-5, p. 21.*)

9.1.2 Steelmaking and Processing of Major Steel Product Forms

Plain-carbon steels are essentially alloys of iron and carbon with up to about 1.2 percent carbon. However, the majority of steels contain less than 0.5 percent carbon. Most steel is made by oxidizing the carbon and other impurities in the pig iron until the carbon content of the iron is reduced to the required level.

The most commonly used process for converting pig iron into steel is the basic-oxygen process. In this process pig iron and up to about 30 percent steel scrap are charged into a barrel-shaped refractory-lined converter into which an oxygen lance is inserted (Fig. 9.2). Pure oxygen from the lance reacts with the liquid bath to form iron oxide. Carbon in the steel then reacts with the iron oxide to form carbon monoxide:

$$FeO + C \longrightarrow Fe + CO$$

Immediately before the oxygen reaction starts, slag-forming fluxes (chiefly lime) are added in controlled amounts. In this process the carbon content of the steel can be drastically lowered in about 22 min along with a reduction in the concentration of impurities such as sulfur and phosphorus (Fig. 9.3).

The molten steel from the converter is either cast in stationary molds or continuously cast into long slabs from which long sections are periodically cut off. Today approximately 96 percent of the steel is cast continuously, with about 4000 ingots still being cast individually. However, about one-half of the raw steel produced is by recycling old steel, such as junk cars and old appliances.[2]

After being cast, the ingots are heated in a soaking pit (Fig. 9.4) and hot-rolled into slabs, billets, or blooms. The slabs are subsequently hot- and cold-rolled into

[2]Table 23, pp. 73–75 of the *Annual Statistical Report of the AI&SI.*

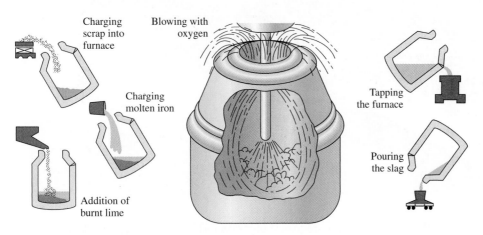

Figure 9.2
Steelmaking in a basic-oxygen furnace.
(*Courtesy of Inland Steel.*)

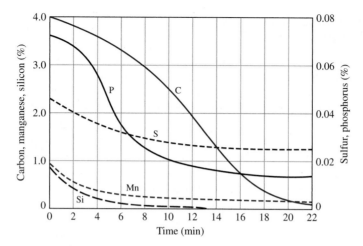

Figure 9.3

Schematic representation of progress of refining in a top-blown basic-lined vessel.

[After H. E. McGannon (ed.), "The Making, Shaping, and Treating of Steel," 9th ed., United States Steel Corp., 1971, p. 494.]

Figure 9.4

Hot rolling of steel strip. This picture shows the roughing hot-rolling mills in the background and six finishing hot-rolling mills in the foreground. A strip of steel is exiting the last finishing stand and is being water-quenched.

(Courtesy of United States Steel Corporation)

steel sheet and plate (see Figs. 9.4 and 6.4 to 6.8). The billets are hot- and cold-rolled into bars, rods, and wire, while blooms are hot- and cold-rolled into shapes such as I beams and rails. Figure 9.5 is a flow diagram that summarizes the principal process steps involved in converting raw materials into major steel product forms.

9.2 THE IRON-CARBON SYSTEM

Iron-carbon alloys containing from a very small amount (about 0.03 percent) to about 1.2 percent carbon, 0.25 to 1.00 percent manganese, and minor amounts of other elements[3] are termed *plain-carbon steels.* However, for purposes of this section of the book, plain-carbon steels will be treated as essentially iron-carbon binary alloys. The effects of other elements in steels will be dealt with in later sections.

9.2.1 The Iron–Iron-Carbide Phase Diagram

The phases present in very slowly cooled iron-carbon alloys at various temperatures and compositions of iron with up to 6.67 percent carbon are shown in the Fe-Fe_3C phase diagram of Fig. 9.6. This phase diagram is not a true equilibrium diagram since the compound iron carbide (Fe_3C) that is formed is not a true equilibrium phase. Under certain conditions, Fe_3C, which is called **cementite,** can decompose into the more stable phases of iron and carbon (graphite). However, for most practical conditions Fe_3C is very stable and will therefore be treated as an equilibrium phase.

Ice Nine

9.2.2 Solid Phases in the Fe-Fe_3C Phase Diagram

The Fe-Fe_3C diagram contains the following solid phases: α ferrite, austenite (γ), cementite (Fe_3C), and δ ferrite.

> α **ferrite.** This phase is an interstitial solid solution of carbon in the BCC iron crystal lattice. As indicated by the Fe-Fe_3C phase diagram, carbon is only slightly soluble in α ferrite, reaching a maximum solid solubility of 0.02 percent at 723°C. The solubility of carbon in α ferrite decreases to 0.005 percent at 0°C.
>
> **Austenite** (γ). The interstitial solid solution of carbon in γ iron is called *austenite.* Austenite has an FCC crystal structure and a much higher solid solubility for carbon than α ferrite. The solid solubility of carbon in austenite is a maximum of 2.08 percent at 1148°C and decreases to 0.8 percent at 723°C (Fig. 9.6).
>
> *Cementite* (Fe_3C). The intermetallic compound Fe_3C is called *cementite.* Cementite has negligible solubility limits and a composition of 6.67 percent C and 93.3 percent Fe. Cementite is a hard and brittle compound.
>
> δ *ferrite.* The interstitial solid solution of carbon in δ iron is called δ *ferrite.* It has a BCC crystal structure like α ferrite but with a greater lattice constant. The maximum solid solubility of carbon in δ ferrite is 0.09 percent at 1465°C.

---

[3]Plain-carbon steels also contain impurities of silicon, phosphorus, and sulfur as well as others.

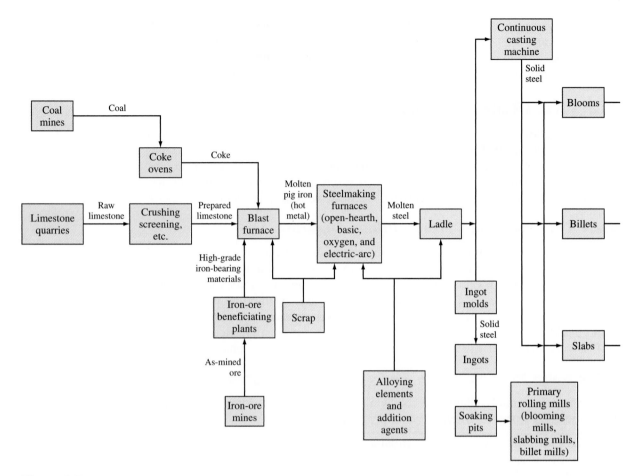

Figure 9.5
Flow diagram showing the principal process steps involved in converting raw materials into the major product forms, excluding coated products.
[*H. E. McGannon (ed.), "The Making, Shaping, and Treating of Steel," 9th ed., United States Steel Corp., 1971, p. 2.*]

9.2.3 Invariant Reactions in the Fe-Fe₃C Phase Diagram

Peritectic Reaction   At the peritectic reaction point, liquid of 0.53 percent C combines with δ ferrite of 0.09 percent C to form γ austenite of 0.17 percent C. This reaction, which occurs at 1495°C, can be written as

$$\text{Liquid } (0.53\% \text{ C}) + \delta\,(0.09\% \text{ C}) \xrightarrow{\,1495°\text{C}\,} \gamma\,(0.17\% \text{ C})$$

δ Ferrite is a high-temperature phase and so is not encountered in plain-carbon steels at lower temperatures.

Eutectic Reaction   At the eutectic reaction point, liquid of 4.3 percent forms γ austenite of 2.08 percent C and the intermetallic compound Fe₃C (cementite),

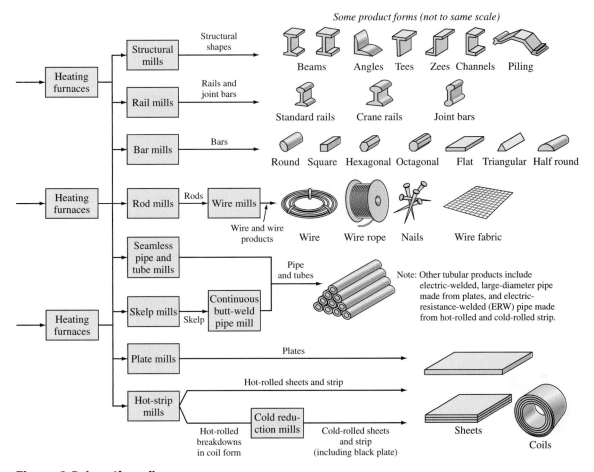

Figure 9.5 (*continued*)

which contains 6.67 percent C. This reaction, which occurs at 1148°C, can be written as

$$\text{Liquid (4.3\% C)} \xrightarrow{1148°C} \gamma \text{ austenite (2.08\% C)} + \text{Fe}_3\text{C (6.67\% C)}$$

This reaction is not encountered in plain-carbon steels because their carbon contents are too low.

Eutectoid Reaction　At the eutectoid reaction point, solid austenite of 0.8 percent C produces α ferrite with 0.02 percent C and Fe_3C (cementite) that contains 6.67 percent C. This reaction, which occurs at 723°C, can be written as

$$\gamma \text{ austenite (0.8\% C)} \xrightarrow{723°C} \alpha \text{ ferrite (0.02\% C)} + \text{Fe}_3\text{C (6.67\% C)}$$

This eutectoid reaction, which takes place completely in the solid state, is important for some of the heat treatments of plain-carbon steels.

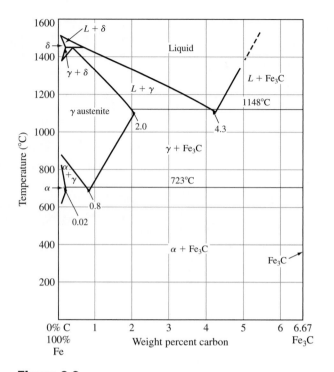

Figure 9.6
The iron–iron carbide phase diagram.

Ice Nine

A plain-carbon steel that contains 0.8 percent C is called a **eutectoid steel** since an all-eutectoid structure of α ferrite and Fe_3C is formed when austenite of this composition is slowly cooled below the eutectoid temperature. If a plain-carbon steel contains less than 0.8 percent C, it is termed a **hypoeutectoid steel,** and if the steel contains more than 0.8 percent C, it is designated a **hypereutectoid steel.**

9.2.4 Slow Cooling of Plain-Carbon Steels

Eutectoid Plain-Carbon Steels   If a sample of a 0.8 percent (eutectoid) plain-carbon steel is heated to about 750°C and held for a sufficient time, its structure will become homogeneous austenite. This process is called **austenitizing.** If this eutectoid steel is then cooled very slowly to just above the eutectoid temperature, its structure will remain austenitic, as indicated in Fig. 9.7 at point *a.* Further cooling to the eutectoid temperature or just below it will cause the entire structure to transform from austenite to a lamellar structure of alternate plates of α ferrite and cementite (Fe_3C). Just below the eutectoid temperature, at point *b* in Fig. 9.7, the lamellar structure will appear as shown in Fig. 9.8. This eutectoid structure is called **pearlite** since it resembles mother-of-pearl. Since the solubility of carbon in α ferrite and Fe_3C changes very little from 723°C to room temperature, the pearlite structure will remain essentially unchanged in this temperature interval.

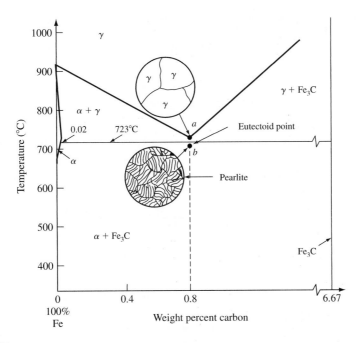

Figure 9.7
Transformation of a eutectoid steel (0.8% C) with slow cooling.
(*After W. F. Smith, "Structure and Properties of Engineering Alloys," 2nd ed., McGraw-Hill, 1981, p. 8.*)

Ice Nine

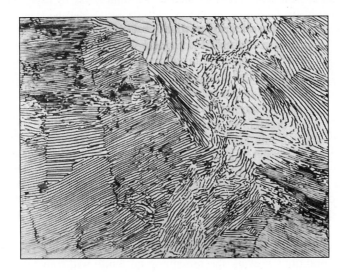

Figure 9.8
Microstructure of a slowly cooled eutectoid steel. The microstructure consists of lamellar eutectoid pearlite. The dark etched phase is cementite, and the white phase is ferrite. (Etch: picral; magnification 650×.)
(*United States Steel Corporation, as presented in "Metals Handbook," vol. 8, 8th ed., American Society for Metals, 1973, p. 188.*)

Virtual Lab

EXAMPLE PROBLEM 9.1

A 0.80 percent C eutectoid plain-carbon steel is slowly cooled from 750°C to a temperature just slightly below 723°C. Assuming that the austenite is completely transformed to α ferrite and cementite:

a. Calculate the weight percent eutectoid ferrite formed.
b. Calculate the weight percent eutectoid cementite formed.

■ **Solution**

Referring to Fig. 9.6, we first draw a tie line just below 723°C from the α ferrite phase boundary to the Fe₃C phase boundary and indicate the 0.80 percent C composition on the tie line as shown in the following figure:

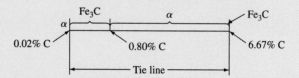

a. The weight fraction of ferrite is calculated from the ratio of the segment of the tie line to the right of 0.80 percent C over the whole length of the tie line. Multiplying by 100 percent gives the weight percent ferrite:

$$\text{Wt \% ferrite} = \frac{6.67 - 0.80}{6.67 - 0.02} \times 100\% = \frac{5.87}{6.65} \times 100\% = 88.3\% \blacktriangleleft$$

b. The weight percent cementite is calculated in a similar way by using the ratio of the segment of the tie line to the left of 0.80 percent C over the length of the whole tie line and multiplying by 100 percent:

$$\text{Wt \% cementite} = \frac{0.80 - 0.02}{6.67 - 0.02} \times 100\% = \frac{0.78}{6.65} \times 100\% = 11.7\% \blacktriangleleft$$

Hypoeutectoid Plain-Carbon Steels   If a sample of a 0.4 percent C plain-carbon steel (hypoeutectoid steel) is heated to about 900°C (point *a* in Fig. 9.9) for a sufficient time, its structure will become homogeneous austenite. Then, if this steel is slowly cooled to temperature *b* in Fig. 9.9 (about 775°C), **proeutectoid[4] ferrite** will nucleate and grow mostly at the austenitic grain boundaries. If this alloy is slowly cooled from temperature *b* to *c* in Fig. 9.9, the amount of proeutectoid ferrite formed will continue to increase until about 50 percent of the austenite is transformed. While the steel is cooling from *b* to *c*, the carbon content of the remaining austenite will be increased from 0.4 to 0.8 percent. At 723°C, if very slow cooling conditions prevail, the remaining austenite will transform isothermally

[4]The prefix *pro-* means "before," and thus the term *proeutectoid ferrite* is used to distinguish this constituent, which forms earlier, from eutectoid ferrite, which forms by the eutectoid reaction later in the cooling.

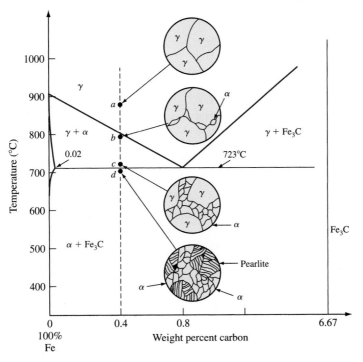

Figure 9.9

Transformation of a 0.4% C hypoeutectoid plain-carbon steel with slow cooling.

(After W. F. Smith, "Structure and Properties of Engineering Alloys," 2nd ed., McGraw-Hill, 1993, p. 10.)

into pearlite by the eutectoid reaction austenite → ferrite + cementite. The α ferrite in the pearlite is called **eutectoid ferrite** to distinguish it from the proeutectoid ferrite that forms first above 723°C. Figure 9.10 is an optical micrograph of the structure of a 0.35 percent C hypoeutectoid steel that was austenitized and slowly cooled to room temperature.

a. A 0.40 percent C hypoeutectoid plain-carbon steel is slowly cooled from 940°C to a temperature just slightly above 723°C.
  (i) Calculate the weight percent austenite present in the steel.
  (ii) Calculate the weight percent proeutectoid ferrite present in the steel.
b. A 0.40 percent C hypoeutectoid plain-carbon steel is slowly cooled from 940°C to a temperature just slightly below 723°C.
  (i) Calculate the weight percent proeutectoid ferrite present in the steel.
  (ii) Calculate the weight percent eutectoid ferrite and weight percent eutectoid cementite present in the steel.

EXAMPLE PROBLEM 9.2

■ **Solution**

Referring to Fig. 9.6 and using tie lines:

a.   (i)   Wt % austenite $= \dfrac{0.40 - 0.02}{0.80 - 0.02} \times 100\% = 50\%$ ◄

     (ii)   Wt % proeutectoid ferrite $= \dfrac{0.80 - 0.40}{0.80 - 0.02} \times 100\% = 50\%$ ◄

b.   (i)   The weight percent proeutectoid ferrite present in the steel just below 723°C will be the same as that just above 723°C, which is 50 percent.

     (ii)   The weight percent total ferrite and cementite just below 723°C are

$$\text{Wt \% total ferrite} = \frac{6.67 - 0.40}{6.67 - 0.02} \times 100\% = 94.3\%$$

$$\text{Wt \% total cementite} = \frac{0.40 - 0.02}{6.67 - 0.02} \times 100\% = 5.7\%$$

$$\text{Wt \% eutectoid ferrite} = \text{total ferrite} - \text{proeutectoid ferrite}$$

$$= 94.3 - 50 = 44.3\% \;◄$$

$$\text{Wt \% eutectoid cementite} = \text{wt \% total cementite} = 5.7\% \;◄$$

(No proeutectoid cementite was formed during cooling.)

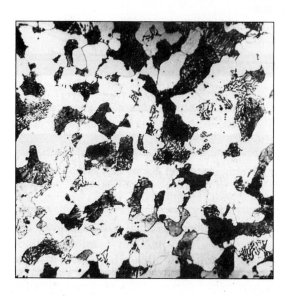

Figure 9.10

Microstructure of a 0.35% C hypoeutectoid plain-carbon steel slowly cooled from the austenite region. The white constituent is proeutectoid ferrite; the dark constituent is pearlite. (Etchant: 2% nital; magnification 500×.)

(*After W. F. Smith, "Structure and Properties of Engineering Alloys," 2nd ed., McGraw-Hill, 1993, p. 11.*)

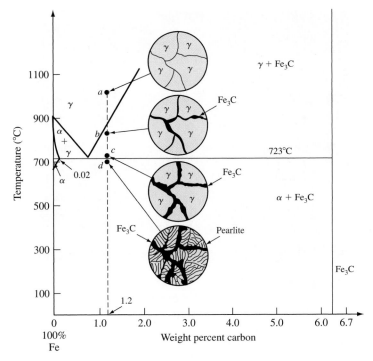

Figure 9.11
Transformation of a 1.2% C hypereutectoid plain-carbon steel
with slow cooling.
*(After W. F. Smith, "Structure and Properties of Engineering Alloys," 2nd ed.,
McGraw-Hill, 1993, p. 12.)*

Hypereutectoid Plain-Carbon Steels   If a sample of a 1.2 percent C plain-
carbon steel (hypereutectoid steel) is heated to about 950°C and held for a sufficient time, its structure will become essentially all austenite (point *a* in Fig. 9.11).
Then, if this steel is cooled very slowly to temperature *b* in Fig. 9.11, **proeutectoid
cementite** will begin to nucleate and grow primarily at the austenite grain boundaries. With further slow cooling to point *c* of Fig. 9.11, which is just above 723°C,
more proeutectoid cementite will be formed at the austenite grain boundaries. If
conditions approaching equilibrium are maintained by the slow cooling, the overall carbon content of the austenite remaining in the alloy will change from 1.2 to
0.8 percent.

With still further slow cooling to 723°C or just slightly below this temperature,
the remaining austenite will transform to pearlite by the eutectoid reaction, as indicated at point *d* of Fig. 9.11. The cementite formed by the eutectoid reaction is called
eutectoid cementite to distinguish it from the proeutectoid cementite formed at temperatures above 723°C. Similarly, the ferrite formed by the eutectoid reaction is

Figure 9.12
Microstructure of a 1.2% C hypereutectoid steel slowly
cooled from the austenite region. In this structure the
proeutectoid cementite appears as the white constituent
that has formed at the former austenite grain boundaries.
The remaining structure consists of coarse lamellar
pearlite. (Etchant: picral; magnification 1000×.)
(*Courtesy of United States Steel Corp., Research Laboratory*)

termed *eutectoid ferrite*. Figure 9.12 is an optical micrograph of the structure of a
1.2 percent C hypereutectoid steel that was austenitized and slowly cooled to room
temperature.

**EXAMPLE
PROBLEM 9.3**

A hypoeutectoid plain-carbon steel that was slow-cooled from the austenitic region to room
temperature contains 9.1 wt % eutectoid ferrite. Assuming no change in structure on cooling from just below the eutectoid temperature to room temperature, what is the carbon
content of the steel?

■ **Solution**

Let x = the weight percent carbon of the hypoeutectoid steel. Now we can use the equation
that relates the eutectoid ferrite to the total ferrite and the proeutectoid ferrite, which is

$$\text{Eutectoid ferrite} = \text{total ferrite} - \text{proeutectoid ferrite}$$

Using Fig. EP9.3 and the lever rule, we can make the equation

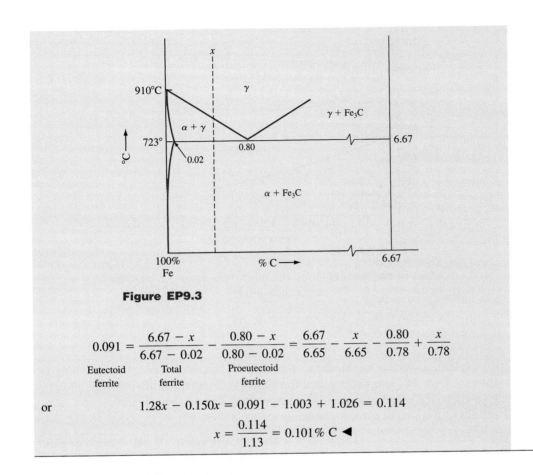

Figure EP9.3

$$0.091 = \underbrace{\frac{6.67 - x}{6.67 - 0.02}}_{\substack{\text{Total} \\ \text{ferrite}}} - \underbrace{\frac{0.80 - x}{0.80 - 0.02}}_{\substack{\text{Proeutectoid} \\ \text{ferrite}}} = \frac{6.67}{6.65} - \frac{x}{6.65} - \frac{0.80}{0.78} + \frac{x}{0.78}$$

$$\underset{\substack{\text{Eutectoid} \\ \text{ferrite}}}{}$$

or

$$1.28x - 0.150x = 0.091 - 1.003 + 1.026 = 0.114$$

$$x = \frac{0.114}{1.13} = 0.101\% \text{ C} \blacktriangleleft$$

9.3 HEAT TREATMENT OF PLAIN-CARBON STEELS

By varying the manner in which plain-carbon steels are heated and cooled, different combinations of mechanical properties for steels can be obtained. In this section we shall examine some of the structural and property changes that take place during some of the important heat treatments given to plain-carbon steels.

9.3.1 Martensite

Formation of Fe-C Martensite by Rapid Quenching If a sample of a plain-carbon steel in the austenitic condition is rapidly cooled to room temperature by quenching it in water, its structure will be changed from austenite to **martensite.** Martensite in plain-carbon steels is a metastable phase consisting of a supersaturated interstitial solid solution of carbon in body-centered cubic iron or body-centered tetragonal iron (the tetragonality is caused by a slight distortion of the BCC iron unit cell). The temperature,

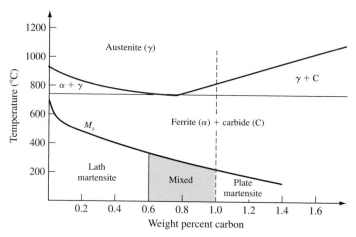

Figure 9.13

Effect of carbon content on the martensite-transformation start temperature, M_s, for iron-carbon alloys.

(After A. R. Marder and G. Krauss, as presented in "Hardenability Concepts with Applications to Steel," AIME, 1978, p. 238.)

upon cooling, at which the austenite-to-martensite transformation starts is called the *martensite start, M_s,* temperature, and the temperature at which the transformation finishes is called the *martensite finish, M_f,* temperature. The M_s temperature for Fe-C alloys decreases as the weight percent carbon increases in these alloys, as shown in Fig. 9.13.

Microstructure of Fe-C Martensites The microstructure of martensites in plain-carbon steels depends on the carbon content of the steel. If the steel contains less than about 0.6 percent C, the martensite consists of *domains* of laths of different but limited orientations through a whole domain. The structure within the laths is highly distorted, consisting of regions with high densities of dislocation tangles. Figure 9.14*a* is an optical micrograph of *lath martensite* in an Fe–0.2% C alloy at 600×, while Fig. 9.15 shows the substructure of lath martensite in this alloy in an electron micrograph at 60,000×.

As the carbon content of the Fe-C martensites is increased to above about 0.6 percent C, a different type of martensite, called *plate martensite,* begins to form. Above about 1 percent C, Fe-C alloys consist entirely of plate martensite. Figure 9.14*b* is an optical micrograph of plate martensite in an Fe–1.2% C alloy at 600×. The plates in high-carbon Fe-C martensites vary in size and have a fine structure of parallel twins, as shown in Fig. 9.16. The plates are often surrounded by large amounts of untransformed (retained) austenite. Fe-C martensites with carbon contents between about 0.6 and 1.0 percent C have microstructures consisting of both lath- and plate-type martensites.

Structure of Fe-C Martensites on an Atomic Scale The transformation of austenite to martensite in Fe-C alloys (plain-carbon steels) is considered to be *diffusionless*

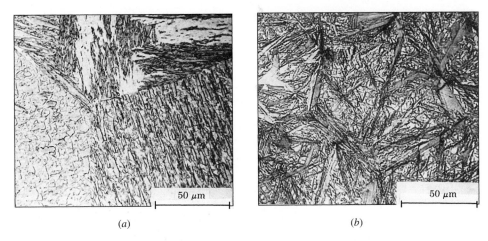

(a)   (b)

Figure 9.14
Effect of carbon content on the structure of martensite in plain-carbon steels: (a) lath type, (b) plate type. (Etchant: sodium bisulfite; optical micrographs.)
[*After A. R. Marder and G. Krauss, Trans. ASM,* **60***:651(1967).*]

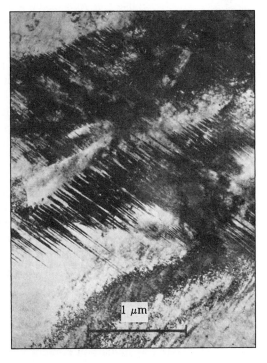

Figure 9.15
Structure of lath martensite in an Fe–0.2% C alloy. (Note the parallel alignment of the laths.)
[*After A. R. Marder and G. Krauss, Trans. ASM,* **60***:651(1967).*]

Figure 9.16
Plate martensite showing fine transformation twins.
[*After M. Oka and C. M. Wayman, Trans. ASM,* **62***:370(1969).*]

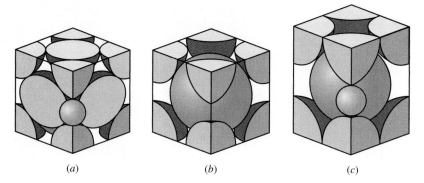

(a)   (b)   (c)

MatVis

Figure 9.17

(*a*) FCC γ iron unit cell showing a carbon atom in a large interstitial hole along the cube edge of the cell. (*b*) BCC α iron unit cell indicating a smaller interstitial hole between cube-edge atoms of the unit cell. (*c*) BCT (body-centered tetragonal) iron unit cell produced by the distortion of the BCC unit cell by the interstitial carbon atom.

(After E. R. Parker and V. F. Zackay, Strong and Ductile Steels, Sci. Am., November 1968, p. 36; Copyright © by Scientific American, Inc; all rights reserved.)

since the transformation takes place so rapidly that the atoms do not have time to intermix. There appears to be no thermal-activation energy barrier to prevent martensite from forming. Also, it is believed that no compositional change in the parent phase takes place after the reaction and that each atom tends to retain its original neighbors. The relative positions of the carbon atoms with respect to the iron atoms are the same in the martensite as they were in the austenite.

For carbon contents in Fe-C martensites of less than about 0.2 percent C, the austenite transforms to a BCC α ferrite crystal structure. As the carbon content of the Fe-C alloys is increased, the BCC structure is distorted into a BCT (*body-centered tetragonal*) crystal structure. The largest interstitial hole in the γ iron FCC crystal structure has a diameter of 0.104 nm (Fig. 9.17*a*), whereas the largest interstitial hole in the α iron BCC structure has a diameter of 0.072 nm (Fig. 9.17*b*). Since the carbon atom has a diameter of 0.154 nm, it can be accommodated in interstitial solid solution to a greater extent in the FCC γ iron lattice. When Fe-C martensites with more than about 0.2 percent C are produced by rapid cooling from austenite, the reduced interstitial spacings of the BCC lattice cause the carbon atoms to distort the BCC unit cell along its *c* axis to accommodate the carbon atoms (Fig. 9.17*c*). Figure 9.18 shows how the *c* axis of the Fe-C martensite lattice is elongated as its carbon content increases.

Hardness and Strength of Fe-C Martensites The hardness and strength of Fe-C martensites are directly related to their carbon content and increase as the carbon content is increased (Fig. 9.19). However, ductility and toughness also decrease with increasing carbon content, and so most martensitic plain-carbon steels are tempered by reheating at a temperature below the transformation temperature of 723°C.

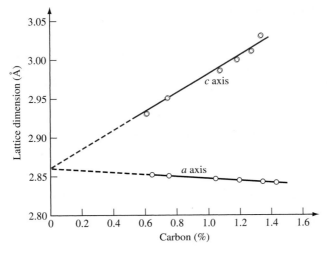

Figure 9.18

Variation of *a* and *c* axes of the Fe–C martensite lattice as a function of carbon content.
(*After E. C. Bain and H. W. Paxton, "Alloying Elements in Steel," 2nd ed., American Society for Metals, 1966, p. 36.*)

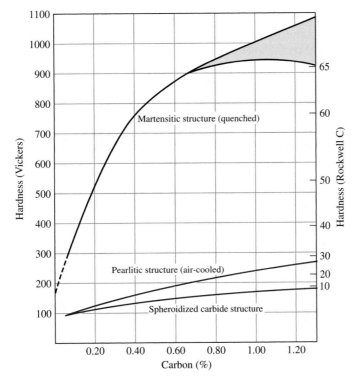

Figure 9.19

Approximate hardness of fully hardened martensitic plain-carbon steel as a function of carbon content. The shaded region indicates some possible loss of hardness due to the formation of retained austenite, which is softer than martensite.
(*After E. C. Bain and H. W. Paxton, "Alloying Elements in Steel," 2nd ed., American Society for Metals, 1966, p. 37.*)

Low-carbon Fe-C martensites are strengthened by a high concentration of dislocations being formed (lath martensite) and by interstitial solid-solution strengthening by carbon atoms. The high concentration of dislocations in networks (lath martensite) makes it difficult for other dislocations to move. As the carbon content increases above 0.2 percent, interstitial solid-solution strengthening becomes more important and the BCC iron lattice becomes distorted into tetragonality. However, in high-carbon Fe-C martensites the numerous twinned interfaces in plate martensite also contribute to the hardness.

9.3.2 Isothermal Decomposition of Austenite

Isothermal Transformation Diagram for a Eutectoid Plain-Carbon Steel In previous sections the reaction products from the decomposition of austenite of eutectoid plain-carbon steels for very slow and rapid cooling conditions have been described. Let us now consider what reaction products form when austenite of eutectoid steels is rapidly cooled to temperatures below the eutectoid temperature and then *isothermally transformed.*

Isothermal transformation experiments to investigate the microstructural changes for the decomposition of eutectoid austenite can be made by using a number of small samples, each about the size of a dime. The samples are first austenitized in a furnace at a temperature above the eutectoid temperature (Fig. 9.20*a*). The samples are then rapidly cooled (quenched) in a liquid salt bath at the desired temperature below the eutectoid temperature (Fig. 9.20*b*). After various time intervals, the samples are removed from the salt bath one at a time and quenched into water at room temperature (Fig. 9.20*c*). The microstructure after each transformation time can then be examined at room temperature.

Consider the microstructural changes that take place during the isothermal transformation of a eutectoid plain-carbon steel at 705°C, as schematically shown in

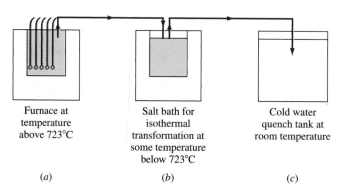

Furnace at temperature above 723°C	Salt bath for isothermal transformation at some temperature below 723°C	Cold water quench tank at room temperature
(*a*)	(*b*)	(*c*)

Figure 9.20

Experimental arrangement for determining the microscopic changes that occur during the isothermal transformation of austenite in a eutectoid plain-carbon steel.

(*After W. F. Smith, "Structure and Properties of Engineering Alloys," McGraw-Hill, 1981, p. 14.*)

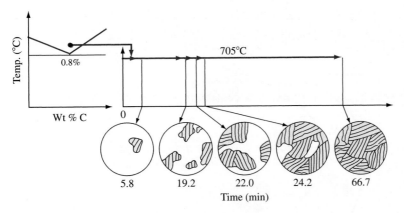

Figure 9.21

Experiment for following the microstructural changes that occur during the isothermal transformation of a eutectoid plain-carbon steel at 705°C. After austenitizing, samples are quenched in a salt bath at 705°C, held for the times indicated, and then quenched in water at room temperature.

(*After W. F. Smith, "Structure and Properties of Engineering Alloys," McGraw-Hill, 1981, p. 14.*)

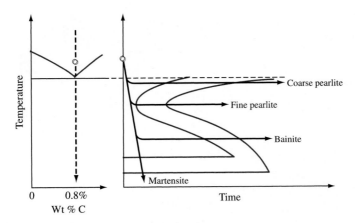

Figure 9.22

Isothermal transformation diagram for a eutectoid plain-carbon steel showing its relationship to the Fe–Fe$_3$C phase diagram.

Fig. 9.21. After being austenitized, the samples are hot-quenched into a salt bath at 705°C. After about 6 min, coarse pearlite has formed to a small extent. After about 67 min, the austenite is completely transformed to coarse pearlite.

By repeating the same procedure for the isothermal transformation of eutectoid steels at progressively lower temperatures, an **isothermal transformation (IT) diagram** can be constructed, as shown schematically in Fig. 9.22 and from experimental

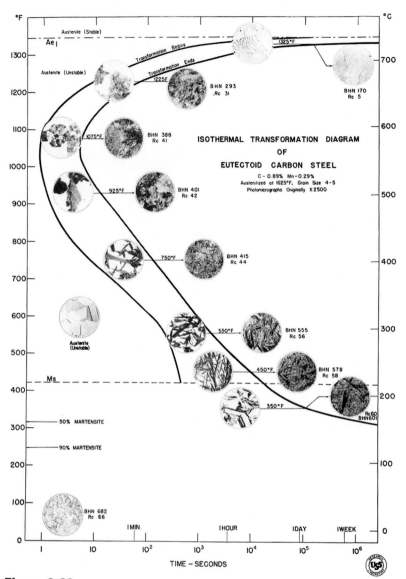

Figure 9.23

Isothermal transformation diagram of a eutectoid steel.
(*Courtesy of United States Steel Corp., Research Laboratory*)

data in Fig. 9.23. The S-shaped curve next to the temperature axis indicates the time necessary for the isothermal transformation of austenite to begin, and the second S curve indicates the time required for the transformation to be completed.

Isothermal transformations of eutectoid steels at temperatures between 723 and about 550°C produce pearlitic microstructures. As the transformation temperature is

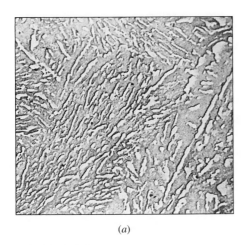
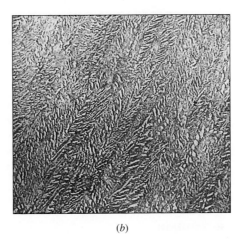

(a)   (b)

Figure 9.24
(a) Microstructure of upper bainite formed by a complete transformation of a eutectoid steel at 450°C (850°F). (b) Microstructure of lower bainite formed by a complete transformation of a eutectoid steel at 260°C (500°F). The white particles are Fe_3C, and the dark matrix is ferrite. (Electron micrographs, replica-type; magnification 15,000×.)
[*After H. E. McGannon (ed.), "The Making, Shaping, and Treating of Steel," 9th ed., United States Steel Corp., 1971.*]

decreased in this range, the pearlite changes from a coarse to a fine structure (Fig. 9.23). Rapid quenching (cooling) of a eutectoid steel from temperatures above 723°C, where it is in the austenitic condition, transforms the austenite into martensite, as has been previously discussed.

If eutectoid steels in the austenitic condition are hot-quenched to temperatures in the 550 to 250°C range and are isothermally transformed, a structure intermediate between pearlite and martensite, called **bainite,**[5] is produced. Bainite in Fe-C alloys can be defined as an austenitic decomposition product that has a *nonlamellar eutectoid structure* of α ferrite and cementite (Fe_3C). For eutectoid plain-carbon steels, a distinction is made between *upper bainite,* which is formed by isothermal transformation at temperatures between about 550 and 350°C, and *lower bainite,* which is formed between about 350 and 250°C. Figure 9.24a shows an electron micrograph (replica-type) of the microstructure of upper bainite for a eutectoid plain-carbon steel, and Fig. 9.24b shows one for lower bainite. Upper bainite has large, rodlike cementite regions, whereas lower bainite has much finer cementite particles. As the transformation temperature is decreased, the carbon atoms cannot diffuse as easily, and hence the lower bainite structure has smaller particles of cementite.

---

[5]Bainite is named after E. C. Bain, the American metallurgist who first intensively studied the isothermal transformations of steels. See E. S. Davenport and E. C. Bain, *Trans. AIME,* **90:**117 (1930).

EXAMPLE PROBLEM 9.4

Small, thin pieces of 0.25 mm thick hot-rolled strips of 1080 steel are heated for 1 h at 850°C and then given the heat treatments shown in the following list. Using the isothermal transformation diagram of Fig. 9.23, determine the microstructures of the samples after each heat treatment.

a.   Water-quench to room temperature.
b.   Hot-quench in molten salt to 690°C and hold 2 h; water-quench.
c.   Hot-quench to 610°C and hold 3 min; water-quench.
d.   Hot-quench to 580°C and hold 2 s; water-quench.
e.   Hot-quench to 450°C and hold 1 h; water-quench.
f.   Hot-quench to 300°C and hold 30 min; water-quench.
g.   Hot-quench to 300°C and hold 5 h; water quench.

■ Solution

The cooling paths are indicated on Fig. EP9.4 and the microstructures obtained are listed as follows:

a.   All martensite.
b.   All coarse pearlite.
c.   All fine pearlite.
d.   Approximately 50 percent fine pearlite and 50 percent martensite.
e.   All upper bainite.
f.   Approximately 50 percent lower bainite and 50 percent martensite.
g.   All lower bainite.

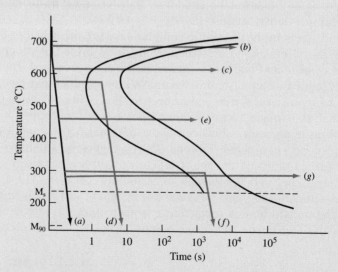

Figure EP9.4
Isothermal transformation diagram for a eutectoid plain-carbon steel indicating various cooling paths.

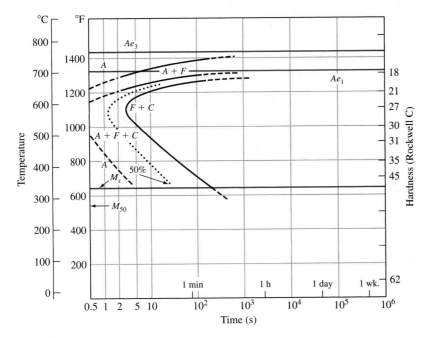

Figure 9.25
Isothermal transformation diagram for a hypoeutectoid steel containing
0.47% C and 0.57% Mn (austenitizing temperature: 843°C).
[*After R. A. Grange, V. E. Lambert, and J. J. Harrington, Trans. ASM,* **51:***377(1959).*]

Isothermal Transformation Diagrams for Noneutectoid Plain-Carbon Steels
Isothermal transformation diagrams have been determined for noneutectoid plain-carbon steels. Figure 9.25 shows an IT diagram for a 0.47 percent hypoeutectoid plain-carbon steel. Several differences are evident between the IT diagram for a noneutectoid plain-carbon steel and the diagram for a eutectoid one (Fig. 9.23). One major difference is that the S curves of the hypoeutectoid steel have been shifted to the left, so that it is not possible to quench this steel from the austenitic region to produce an entirely martensitic structure.

A second major difference is that another transformation line has been added to the upper part of the eutectoid steel IT diagram that indicates the start of the formation of proeutectoid ferrite. Thus, at temperatures between 723 and about 765°C, only proeutectoid ferrite is produced by isothermal transformation.

Similar IT diagrams have been determined for hypereutectoid plain-carbon steels. However, in this case, the uppermost line of the diagram for these steels is for the start of the formation of proeutectoid cementite.

9.3.3 Continuous-Cooling Transformation Diagram for a Eutectoid Plain-Carbon Steel

In industrial heat-treating operations, in most cases a steel is not isothermally transformed at a temperature above the martensite start temperature but is continuously

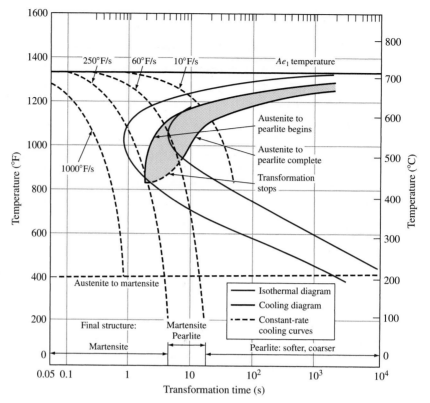

Figure 9.26

Continuous-cooling diagram for a plain-carbon eutectoid steel.
(After R. A. Grange and J. M. Kiefer as adapted in E. C. Bain and H. W. Paxton, "Alloying Elements in Steel," 2nd ed., American Society for Metals, 1966, p. 254.)

cooled from the austenitic temperature to room temperature. In continuously cooling a plain-carbon steel, the transformation from austenite to pearlite occurs over a range of temperatures rather than at a single isothermal temperature. As a result, the final microstructure after continuous cooling will be complex since the reaction kinetics change over the temperature range in which the transformation takes place. Figure 9.26 shows a **continuous-cooling transformation (CCT) diagram** for a eutectoid plain-carbon steel superimposed over an IT diagram for this steel. The continuous-cooling diagram transformation start and finish lines are shifted to longer times and slightly lower temperatures in relation to the isothermal diagram. Also, there are no transformation lines below about 450°C for the austenite-to-bainite transformation.

Figure 9.27 shows different rates of cooling for thin samples of eutectoid plain-carbon steels cooled continuously from the austenitic region to room temperature. Cooling curve *A* represents very slow cooling, such as would be obtained by shutting off the power of an electric furnace and allowing the steel to cool as the furnace cools. The microstructure in this case would be coarse pearlite. Cooling curve *B* represents

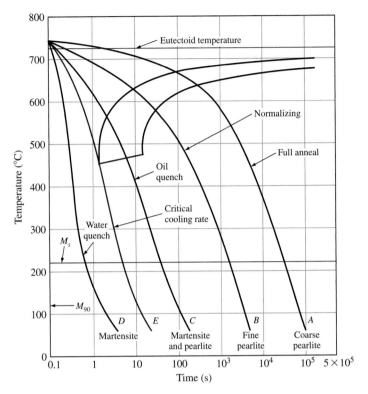

Figure 9.27
Variation in the microstructure of a eutectoid plain-carbon steel by continuously cooling at different rates.
(*From R. E. Reed-Hill, "Physical Metallurgy Principles," 2nd ed., D. Van Nostrand Co., 1973 © PWS Publishers.*)

more rapid cooling, such as would be obtained by removing an austenitized steel from a furnace and allowing the steel to cool in still air. A fine pearlite microstructure is formed in this case.

Cooling curve *C* of Fig. 9.27 starts with the formation of pearlite, but there is insufficient time to complete the austenite-to-pearlite transformation. The remaining austenite that does not transform to pearlite at the upper temperatures will transform to martensite at lower temperatures starting at about 220°C. This type of transformation, since it takes place in two steps, is called a *split transformation*. The microstructure of this steel will thus consist of a mixture of pearlite and martensite. Cooling at a rate faster than curve *E* of Fig. 9.27, which is called the *critical cooling rate*, will produce a fully hardened martensitic structure.

Continuous-cooling diagrams have been determined for many hypoeutectoid plain-carbon steels and are more complex since at low temperatures some bainitic structure is also formed during continuous cooling. The discussion of these diagrams is beyond the scope of this book.

9.3.4 Annealing and Normalizing of Plain-Carbon Steels

In Chapter 6 the cold-working and annealing processes for metals were discussed, and reference should be made to that section. The two most common types of annealing processes applied to commercial plain-carbon steels are *full annealing* and *process annealing.*

In full annealing, hypoeutectoid and eutectoid steels are heated in the austenite region about 40°C above the austenite-ferrite boundary (Fig. 9.28), held the necessary time at the elevated temperature, and then slowly cooled to room temperature, usually in the furnace in which they were heated. For hypereutectoid steels, it is customary to austenitize in the two-phase austenite plus cementite (Fe_3C) region, about 40°C above the eutectoid temperature. The microstructure of hypoeutectoid steels after full annealing consists of proeutectoid ferrite and pearlite (Fig. 9.10).

Process annealing, which is often referred to as a *stress relief,* partially softens cold-worked low-carbon steels by relieving internal stresses from cold working. This treatment, which is usually applied to hypoeutectoid steels with less than 0.3 percent C, is carried out at a temperature below the eutectoid temperature, usually between 550 and 650°C (Fig. 9.28).

Normalizing is a heat treatment in which the steel is heated in the austenitic region and then cooled in still air. The microstructure of thin sections of normalized

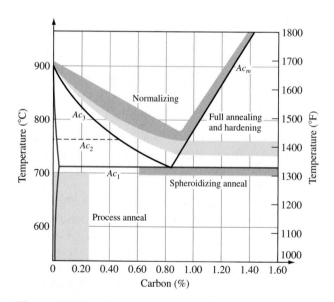

Figure 9.28

Commonly used temperature ranges for annealing plain-carbon steels.

(After T. G. Digges et al., "Heat Treatment and Properties of Iron and Steel," NBS Monograph 88, 1966, p. 10.)

hypoeutectoid plain-carbon steels consists of proeutectoid ferrite and fine pearlite. The purposes for normalizing vary. Some of these are:

1. To refine the grain structure
2. To increase the strength of the steel (compared to annealed steel)
3. To reduce compositional segregation in castings or forgings and thus provide a more uniform structure

The austenitizing temperature ranges used for normalizing plain-carbon steels are shown in Fig. 9.28. Normalizing is more economical than full annealing since no furnace is required to control the cooling rate of the steel.

9.3.5 Tempering of Plain-Carbon Steels

The Tempering Process   **Tempering** is the process of heating a martensitic steel at a temperature below the eutectoid transformation temperature to make it softer and more ductile. Figure 9.29 schematically illustrates the customary quenching and tempering process for a **plain-carbon steel.** As shown in Fig. 9.29, the steel is first austenitized and then quenched at a rapid rate to produce martensite and to avoid the transformation of austenite to ferrite and cementite. The steel is then subsequently reheated at a temperature below the eutectoid temperature to soften the martensite by transforming it to a structure of iron carbide particles in a matrix of ferrite.

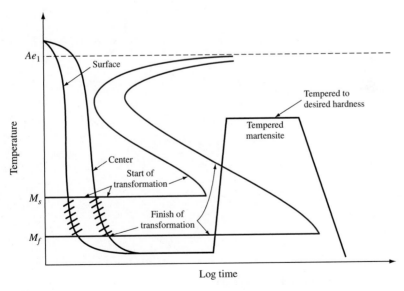

Figure 9.29
Schematic diagram illustrating the customary quenching and tempering process for a plain-carbon steel.
(*From "Suiting the Heat Treatment to the Job," United States Steel Corp., 1968, p. 34.*)

Microstructural Changes in Martensite Upon Tempering Martensite is a metastable structure and decomposes upon reheating. In lath martensites of low-carbon plain-carbon steels there is a high dislocation density, and these dislocations provide lower-energy sites for carbon atoms than their regular interstitial positions. Thus, when low-carbon martensitic steels are first tempered in the 20 to 200°C range, the carbon atoms segregate themselves to these lower-energy sites.

For martensitic plain-carbon steels with more than 0.2 percent carbon, the main mode of carbon redistribution at tempering temperatures below 200°C is by precipitation clustering. In this temperature range a very small-sized precipitate called *epsilon* (*ε*) *carbide* forms. The carbide that forms when martensitic steels are tempered from 200 to 700°C is *cementite, Fe₃C*. When the steels are tempered between 200 and 300°C, the shape of the precipitate is rodlike (Fig. 9.30). At higher tempering temperatures from 400 to 700°C, the rodlike carbides coalesce to form spherelike particles. Tempered martensite that shows the coalesced cementite in the optical microscope is called **spheroidite** (Fig. 9.31).

Figure 9.30
Precipitation of Fe_3C in Fe–0.39% C martensite tempered 1 h at 300°C. (Electron micrograph.)
[*After G. R. Speich and W. C. Leslie, Met. Trans.,* **31:***1043(1972).*]

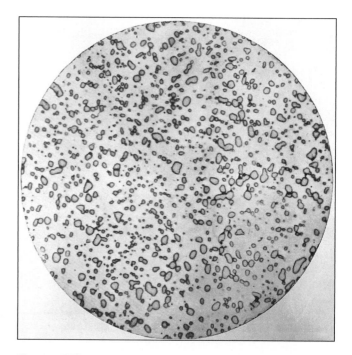

Figure 9.31
Spheroidite in a 1.1% C hypereutectoid steel. (Magnification 1000×.)
(*After J. Vilella, E. C. Bain, and H. W. Paxton, "Alloying Elements in Steel," 2nd ed., American Society for Metals, 1966, p. 101.*)

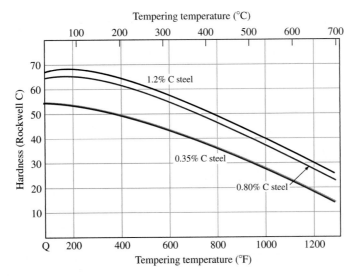

Figure 9.32
Hardness of iron-carbon martensites (0.35 to 1.2% C) tempered 1 h at indicated temperatures.
(After E. C. Bain and H. W. Paxton, "Alloying Elements in Steel," 2nd ed., American Society for Metals, 1966, p. 38.)

Effect of Tempering Temperature on the Hardness of Plain-Carbon Steels
Figure 9.32 shows the effect of increasing tempering temperature on the hardness of several martensitic plain-carbon steels. Above about 200°C, the hardness gradually decreases as the temperature is increased up to 700°C. This gradual decrease in hardness of the martensite with increasing temperature is due mainly to the diffusion of the carbon atoms from their stressed interstitial lattice sites to form second-phase iron carbide precipitates.

Martempering (Marquenching) Martempering (marquenching) is a modified quenching procedure used for steels to minimize distortion and cracking that may develop during uneven cooling of the heat-treated material. The martempering process consists of (1) austenitizing the steel, (2) quenching it in hot oil or molten salt at a temperature just slightly above (or slightly below) the M_s temperature, (3) holding the steel in the quenching medium until the temperature is uniform throughout and stopping this isothermal treatment before the austenite-to-bainite transformation begins, and (4) cooling at a moderate rate to room temperature to prevent large temperature differences. The steel is subsequently tempered by the conventional treatment. Figure 9.33 shows a cooling path for the martempering process.

The structure of the martempered steel is *martensite,* and that of the martempered (marquenched) steel that is subsequently tempered is *tempered martensite.* Table 9.2 lists some of the mechanical properties of a 0.95 percent C plain-carbon steel after martempering and tempering along with those obtained by conventional

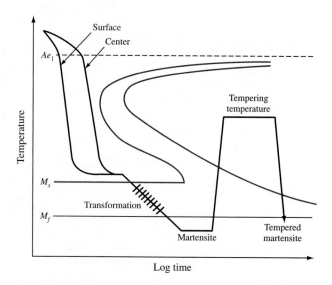

Figure 9.33

Cooling curve for martempering (marquenching) superimposed on a eutectoid plain-carbon steel IT diagram. The interrupted quench reduces the stresses developed in the metal during quenching.

(After "Metals Handbook," vol. 2, 8th ed., American Society for Metals, 1964, p. 37.)

Table 9.2 Some mechanical properties (at 20°C) of a 1095 steel developed by austempering as compared to some other heat treatments

Heat treatment	Rockwell C hardness	Impact (ft · lb)	Elongation in 1 in. (%)
Water-quench and temper	53.0	12	0
Water-quench and temper	52.5	14	0
Martemper and temper	53.0	28	0
Martemper and temper	52.8	24	0
Austemper	52.0	45	11
Austemper	52.5	40	8

Source: "Metals Handbook," vol. 2, 8th ed., American Society for Metals, 1964.

quenching and tempering. The major difference in these two sets of properties is that the martempered and tempered steel has higher impact energy values. It should be noted that the term *martempering* is misleading, and a better word for the process is *marquenching*.

Austempering   **Austempering** is an isothermal heat treatment that produces a bainite structure in some plain-carbon steels. The process provides an alternative procedure to

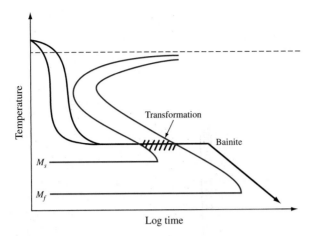

Figure 9.34
Cooling curves for austempering a eutectoid plain-carbon steel. The structure resulting from this treatment is bainite. An advantage of this heat treatment is that tempering is unnecessary. Compare with the customary quenching and tempering process shown in Fig. 9.29. M_s and M_f are the start and finish of martensitic transformation, respectively. *(From "Suiting the Heat Treatment to the Job," United States Steel Corp., 1968, p. 34.)*

quenching and tempering for increasing the toughness and ductility of some steels. In the austempering process the steel is first austenitized, then quenched in a molten salt bath at a temperature just above the M_s temperature of the steel, held isothermally to allow the austenite-to-bainite transformation to take place, and then cooled to room temperature in air (Fig. 9.34). The final structure of an austempered eutectoid plain-carbon steel is *bainite*.

The advantages of austempering are (1) improved ductility and impact resistance of certain steels over those values obtained by conventional quenching and tempering (Table 9.2) and (2) decreased distortion of the quenched material. The disadvantages of austempering over quenching and tempering are (1) the need for a special molten salt bath and (2) the fact that the process can be used for only a limited number of steels.

9.3.6 Classification of Plain-Carbon Steels and Typical Mechanical Properties

Plain-carbon steels are most commonly designated by a four-digit AISI-SAE[6] code. The first two digits are 10 and indicate that the steel is a plain-carbon steel. The last

---

[6]AISI stands for the American Iron and Steel Institute, and SAE for the Society for Automotive Engineers.

two digits indicate the nominal carbon content of the steel in hundredths of a percent. For example, the AISI-SAE code number 1030 for a steel indicates that the steel is a plain-carbon steel containing a nominal 0.30 percent carbon. All plain-carbon steels contain manganese as an alloying element to enhance strength. The manganese content of most plain-carbon steels ranges between 0.30 and 0.95 percent. Plain-carbon steels also contain impurities of sulfur, phosphorus, silicon, and some other elements.

Typical mechanical properties of some AISI-SAE type plain-carbon steels are listed in Table 9.3. The very-low-carbon plain-carbon steels have relatively low strengths but very high ductilities. These steels are used for sheet material for forming applications such as fenders and body panels for automobiles. As the carbon content of the plain-carbon steels is increased, the steels become stronger but less ductile. Medium-carbon steels (1020–1040) find application for shafts and gears. High-carbon steels (1060–1095) are used, for example, for springs, die blocks, cutters, and shear blades.

9.4 LOW-ALLOY STEELS

Plain-carbon steels can be used successfully if the strength and other engineering requirements are not too severe. These steels are relatively low in cost but have some limitations, which include the following:

1. Plain-carbon steels cannot be strengthened beyond about 100,000 psi (690 MPa) without a substantial loss in ductility and impact resistance.
2. Large-section thicknesses of plain-carbon steels cannot be produced with a martensitic structure throughout, i.e., they are not deep-hardenable.
3. Plain-carbon steels have low corrosion and oxidation resistance.
4. Medium-carbon plain-carbon steels must be quenched rapidly to obtain a fully martensitic structure. Rapid quenching leads to possible distortion and cracking of the heat-treated part.
5. Plain-carbon steels have poor impact resistance at low temperatures.

To overcome the deficiencies of plain-carbon steels, alloy steels have been developed that contain alloying elements to improve their properties. Alloy steels in general cost more than plain-carbon steels, but for many applications they are the only materials that can be used to meet engineering requirements. The principal alloying elements added to make alloy steels are manganese, nickel, chromium, molybdenum, and tungsten. Other elements that are sometimes added include vanadium, cobalt, boron, copper, aluminum, lead, titanium, and columbium (niobium).

9.4.1 Classification of Alloy Steels

Alloy steels may contain up to 50 percent of alloying elements and still be considered alloy steels. In this book low-alloy steels containing from about 1 to 4 percent

Table 9.3 Typical mechanical properties and applications of plain-carbon steels

Alloy AISI-SAE number	Chemical composition (wt %)	Condition	Tensile strength		Yield strength		Elongation (%)	Typical applications
			ksi	MPa	ksi	MPa		
1010	0.10 C, 0.40 Mn	Hot-rolled	40–60	276–414	26–45	179–310	28–47	Sheet and strip for drawing; wire, rod, and nails and screws; concrete reinforcement bar
		Cold-rolled	42–58	290–400	23–38	159–262	30–45	
1020	0.20 C, 0.45 Mn	As rolled	65	448	48	331	36	Steel plate and structural sections; shafts, gears
		Annealed	57	393	43	297	36	
1040	0.40 C, 0.45 Mn	As rolled	90	621	60	414	25	Shafts, studs, high-tensile tubing, gears
		Annealed	75	517	51	352	30	
		Tempered*	116	800	86	593	20	
1060	0.60 C, 0.65 Mn	As rolled	118	814	70	483	17	Spring wire, forging dies, railroad wheels
		Annealed	91	628	54	483	22	
		Tempered*	160	110	113	780	13	
1080	0.80 C, 0.80 Mn	As rolled	140	967	85	586	12	Music wire, helical springs, cold chisels, forging die blocks
		Annealed	89	614	54	373	25	
		Tempered*	189	1304	142	980	12	
1095	0.95 C, 0.40 Mn	As rolled	140	966	83	573	9	Dies, punches, taps, milling cutters, shear blades, high-tensile wire
		Annealed	95	655	55	379	13	
		Tempered*	183	1263	118	814	10	

*Quenched and tempered at 315°C (600°F).

Table 9.4 Principal types of standard alloy steels

13xx	Manganese 1.75
40xx	Molybdenum 0.20 or 0.25; or molybdenum 0.25 and sulfur 0.042
41xx	Chromium 0.50, 0.80, or 0.95, molybdenum 0.12, 0.20, or 0.30
43xx	Nickel 1.83, chromium 0.50 or 0.80, molybdenum 0.25
44xx	Molybdenum 0.53
46xx	Nickel 0.85 or 1.83, molybdenum 0.20 or 0.25
47xx	Nickel 1.05, chromium 0.45, molybdenum 0.20 or 0.35
48xx	Nickel 3.50, molybdenum 0.25
50xx	Chromium 0.40
51xx	Chromium 0.80, 0.88, 0.93, 0.95, or 1.00
51xxx	Chromium 1.03
52xxx	Chromium 1.45
61xx	Chromium 0.60 or 0.95, vanadium 0.13 or min 0.15
86xx	Nickel 0.55, chromium 0.50, molybdenum 0.20
87xx	Nickel 0.55, chromium 0.50, molybdenum 0.25
88xx	Nickel 0.55, chromium 0.50, molybdenum 0.35
92xx	Silicon 2.00; or silicon 1.40 and chromium 0.70
50Bxx*	Chromium 0.28 or 0.50
51Bxx*	Chromium 0.80
81Bxx*	Nickel 0.30, chromium 0.45, molybdenum 0.12
94Bxx*	Nickel 0.45, chromium 0.40, molybdenum 0.12

*B denotes boron steel.

Source: "Alloy Steel: Semifinished; Hot-Rolled and Cold-Finished Bars," American Iron and Steel Institute, 1970.

of alloying elements will be considered alloy steels. These steels are mainly automotive- and construction-type steels and are commonly referred to simply as *alloy steels.*

Alloy steels in the United States are usually designated by the four-digit AISI-SAE system. The first two digits indicate the principal alloying element or groups of elements in the steel, and the last two digits indicate the hundredths of percent of carbon in the steel. Table 9.4 lists the nominal compositions of the principal types of standard alloy steels.

9.4.2 Distribution of Alloying Elements in Alloy Steels

The way in which alloy elements distribute themselves in carbon steels depends primarily on the compound- and carbide-forming tendencies of each element. Table 9.5 summarizes the approximate distribution of most of the alloying elements present in alloy steels.

Nickel dissolves in the α ferrite of the steel since it has less tendency to form carbides than iron. Silicon combines to a limited extent with the oxygen in the steel to form nonmetallic inclusions but otherwise dissolves in the ferrite. Most of the manganese added to carbon steels dissolves in the ferrite. Some of the manganese, however, will form carbides but will usually enter the cementite as $(Fe,Mn)_3C$.

Table 9.5 Approximate distribution of alloying elements in alloy steels*

Element	Dissolved in ferrite	Combined in carbide	Combined as carbide	Compound	Elemental
Nickel	Ni			Ni_3Al	
Silicon	Si			$SiO_2 \cdot M_xO_y$	
Manganese	Mn $\longleftrightarrow$ Mn		$(Fe,Mn)_3C$	$MnS; MnO \cdot SiO_2$	
Chromium	Cr $\longleftrightarrow$ Cr		$(Fe,Cr_3)C$		
			Cr_7C_3		
			$Cr_{23}C_6$		
Molybdenum	Mo $\longleftrightarrow$ Mo		Mo_2C		
Tungsten	W $\longleftrightarrow$ W		W_2C		
Vanadium	V $\longleftrightarrow$ V		V_4C_3		
Titanium	Ti $\longleftrightarrow$ Ti		TiC		
Columbium[†]	Cb $\longleftrightarrow$ Cb		CbC		
Aluminum	Al			Al_2O_3; AlN	
Copper	Cu (small amount)				
Lead					Pb

*The arrows indicate the relative tendencies of the elements listed to dissolve in the ferrite or combine in carbides.
†Cb = Nb (niobium).
Source: E. C. Bain and H. W. Paxton, "Alloying Elements in Steel," 2nd ed., American Society for Metals, 1966.

Chromium, which has a somewhat stronger carbide-forming tendency than iron, partitions between the ferrite and carbide phases. The distribution of chromium depends on the amount of carbon present and on whether other stronger carbide-forming elements such as titanium and columbium are absent. Tungsten and molybdenum combine with carbon to form carbides if there is sufficient carbon present and if other stronger carbide-forming elements such as titanium and columbium are absent. Vanadium, titanium, and columbium are strong carbide-forming elements and are found in steels mainly as carbides. Aluminum combines with oxygen and nitrogen to form the compounds Al_2O_3 and AlN, respectively.

9.4.3 Effects of Alloying Elements on the Eutectoid Temperature of Steels

The various alloying elements cause the eutectoid temperature of the Fe-Fe_3C phase diagram to be raised or lowered (Fig. 9.35). Manganese and nickel both lower the eutectoid temperature and act as *austenite-stabilizing elements* enlarging the austenitic region of the Fe-Fe_3C phase diagram (Fig. 9.6). In some steels with sufficient amounts of nickel or manganese, the austenitic structure may be obtained at room temperature. The carbide-forming elements such as tungsten, molybdenum, and titanium raise the eutectoid temperature of the Fe-Fe_3C phase diagram to

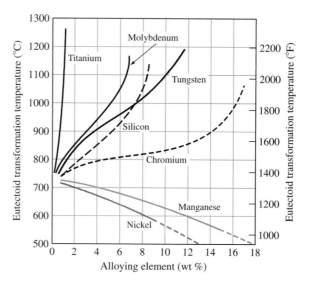

Figure 9.35
The effect of the percentage of alloying elements on the eutectoid temperature of the transformation of austenite to pearlite in the Fe-Fe$_3$C phase diagram.
(After "Metals Handbook," vol. 8, 9th ed., American Society for Metals, 1973, p. 191.)

higher values and reduce the austenitic phase field. These elements are called *ferrite-stabilizing elements.*

9.4.4 Hardenability

The **hardenability** of a steel is defined as that property which determines the depth and distribution of hardness induced by quenching from the austenitic condition. The hardenability of a steel depends primarily on (1) the composition of the steel, (2) the austenitic grain size, and (3) the structure of the steel before quenching. Hardenability should not be confused with the *hardness* of a steel, which is its resistance to plastic deformation, usually by indentation.

In industry, hardenability is most commonly measured by the **Jominy hardenability test.** In the Jominy end-quench test, the specimen consists of a cylindrical bar with a 1 in. diameter and 4 in. length and with a $\frac{1}{16}$ in. flange at one end (Fig. 9.36a). Since prior structure has a strong effect on hardenability, the specimen is usually normalized before testing. In the Jominy test, after the sample has been austenitized, it is placed in a fixture, as shown in Fig. 9.36b, and a jet of water is quickly splashed at one end of the specimen. After cooling, two parallel flat surfaces are ground on the opposite sides of the test bar, and Rockwell C

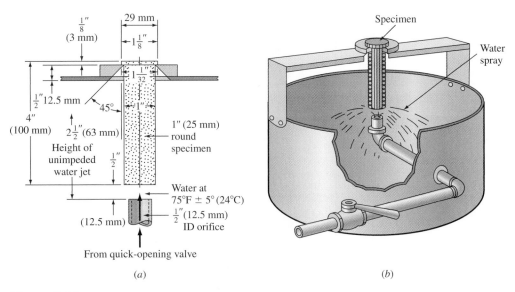

Figure 9.36

(*a*) Specimen and fixture for end-quench hardenability test.

(*After M. A. Grossmann and E. C. Bain, "Principles of Heat Treatment," 5th ed., American Society for Metals, 1964, p. 114.*)

(*b*) Schematic illustration of the end-quench test for hardenability.

[*After H. E. McGannon (ed.), "The Making, Shaping, and Treating of Steel," 9th ed., United States Steel Corp., 1971, p. 1099.*]

hardness measurements are made along these surfaces up to 2.5 in. from the quenched end.

Figure 9.37 shows a hardenability plot of Rockwell C hardness versus distance from the quenched end for a 1080 eutectoid plain-carbon steel. This steel has relatively low hardenability since its hardness decreases from a value of RC = 65 at the quenched end of the Jominy bar to RC = 50 at just $\frac{3}{16}$ in. from the quenched end. Thus, thick sections of this steel cannot be made fully martensitic by quenching. Figure 9.37 correlates the end-quench hardenability data with the continuous transformation diagram for the 1080 steel and indicates the microstructural changes that take place along the bar at four distances *A*, *B*, *C*, and *D* from the quenched end.

Hardenability curves for some 0.40 percent C alloy steels are shown in Fig. 9.38. The 4340 alloy steel has exceptionally high hardenability and can be quenched to a hardness of RC = 40 at 2 in. from the quenched end of a Jominy bar. Alloy steels thus are able to be quenched at a slower rate and still maintain relatively high hardness values.

Alloy steels such as the 4340 steel are highly hardenable because, upon cooling from the austenitic region, the decomposition of austenite to ferrite and bainite is delayed and the decomposition of austenite to martensite can be accomplished at

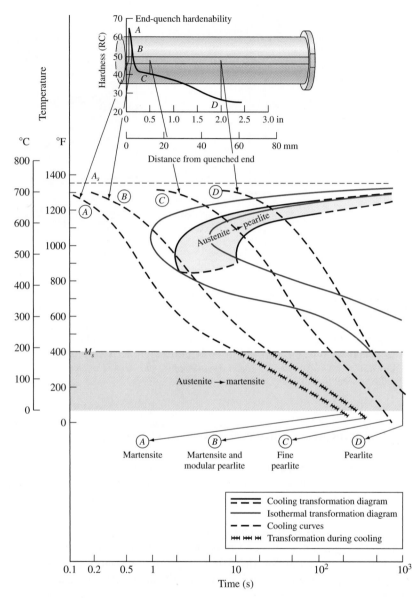

Figure 9.37
Correlation of continuous-cooling transformation diagram and end-quench hardenability test data for eutectoid carbon steel.
(After "Isothermal Transformation Diagrams," United States Steel Corp., 1963, p. 181.)

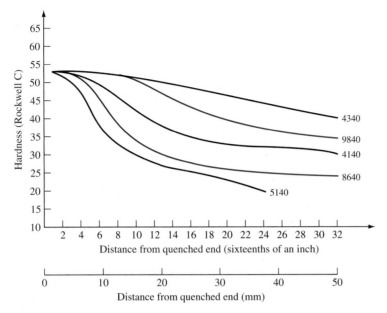

Figure 9.38
Comparative hardenability curves for 0.40% C alloy steels.
[*After H. E. McGannon (ed.), "The Making, Shaping, and Treating of Steels," United States Steel Corp., 1971, p. 1139.*]

slower rates. This delay of the austenite to ferrite plus bainite decomposition is quantitatively shown on the continuous-cooling transformation diagram of Fig. 9.39.

For most carbon and low-alloy steels a standard quench produces at the same cross-section position common cooling rates along long round steel bars of the same diameter. However, the cooling rates differ (1) for different bar diameters, (2) for different positions in the cross sections of the bars, and (3) for different quenching media. Figure 9.40 shows bar diameter versus cooling rate curves for different cross-section locations within steel bars using quenches of agitated water in agitated oil. These plots can be used to determine the cooling rate and the associated distance from the quenched end of a standard quenched Jominy bar for a selected bar diameter at a particular cross-section location in the bar using a specific quenching medium. These cooling rates and their associated distances from the end of Jominy quenched bars can be used with Jominy plots of surface hardness versus distance from the quenched end for specific steels to determine the hardness of a particular steel at a specific location in the cross section of the steel bar in question. Example Problem 9.5 shows how the plots of Fig. 9.40 can be used to predict the hardness of a steel bar of a given diameter at a specific cross-section location quenched in a given medium. It should be pointed out that the Jominy hardness versus distance from the quenched-end plots are usually plotted as bands of data rather than as lines so that hardnesses obtained from the line curves are actually values in the center of a range of values.

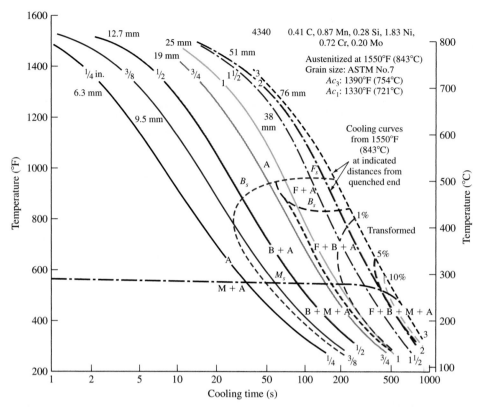

Figure 9.39

Continuous-cooling transformation diagram for AISI 4340 alloy steel. A = austenite, F = ferrite, B = bainite, M = martensite.
(After Metal Progress, September 1964, p. 106.)

EXAMPLE PROBLEM 9.5

A austenitized 40 mm diameter 5140 alloy steel bar is quenched in agitated oil. Predict what the Rockwell C(RC) hardness of this bar will be at (*a*) its surface and (*b*) its center.

■ **Solution**

a. Surface of bar. The cooling rate at the surface of the 40 mm steel bar quenched in agitated oil is found from part (ii) of Fig. 9.40 to be comparable to the cooling rate at 8 mm from the end of a standard quenched Jominy bar. Using Fig. 9.38 at 8 mm from the quenched end of the Jominy bar and the curve for the 5140 steel indicates that the hardness of the bar should be about 45 RC.

b. Center of the bar. The cooling rate at the center of the 40 mm diameter bar quenched in oil is found from part (ii) of Fig. 9.40 to be associated with 13 mm from the end of a quenched Jominy bar. The corresponding hardness for this distance from the end of a quenched Jominy bar for the 5140 alloy is found by using Fig. 9.38 to be about 32 RC.

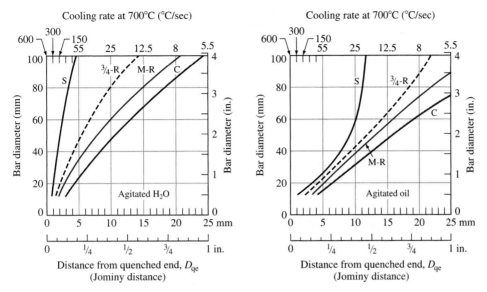

Figure 9.40
Cooling rates in long round steel bars quenched in (i) agitated water and (ii) agitated oil. Top abscissa, cooling rates at 700°C; bottom abscissa, equivalent positions on an end-quenched test bar. (C = center, M-R = midradius, S = surface, dashed line = approximate curve for $\frac{3}{4}$-radius positions on the cross section of bars.)
(*After L. H. VanVlack, "Materials for Engineering: Concepts and Applications," Addison-Wesley, 1982, p. 155.*)

9.4.5 Typical Mechanical Properties and Applications for Low-Alloy Steels

Table 9.6 lists some typical tensile mechanical properties and applications for some commonly used low-alloy steels. For some strength levels low-alloy steels have better combinations of strength, toughness, and ductility than plain-carbon steels. However, low-alloy steels cost more and so are used only when necessary. Low-alloy steels are used to a great extent in the manufacture of automobiles and trucks for parts that require superior strength and toughness properties that cannot be obtained from plain-carbon steels. Some typical applications for low-alloy steels in automobiles are shafts, axles, gears, and springs. Low-alloy steels containing about 0.2 percent C are commonly carburized or surface heat-treated to produce a hard, wear-resistant surface while maintaining a tough inner core.

9.5 ALUMINUM ALLOYS

Before discussing some of the important aspects of the structure, properties, and applications of aluminum alloys, let us examine the precipitation-strengthening (hardening) process that is used to increase the strength of many aluminum and other metal alloys.

Table 9.6 Typical mechanical properties and applications of low-alloy steels

Alloy AISI-SAE number	Chemical composition (wt %)	Condition	Tensile strength		Yield strength		Elongation (%)	Typical applications
			ksi	MPa	ksi	MPa		
Manganese steels								
1340	0.40 C, 1.75 Mn	Annealed	102	704	63	435	20	High-strength bolts
		Tempered*	230	1587	206	1421	12	
Chromium steels								
5140	0.40 C, 0.80 Cr, 0.80 Mn	Annealed	83	573	43	297	29	Automobile transmission gears
		Tempered*	229	1580	210	1449	10	
5160	0.60 C, 0.80 Cr, 0.90 Mn	Annealed	105	725	40	276	17	Automobile coil and leaf springs
		Tempered*	290	2000	257	1773	9	
Chromium-molybdenum steels								
4140	0.40 C, 1.0 Cr, 0.9 Mn, 0.20 Mo	Annealed	95	655	61	421	26	Gears for aircraft gas turbine engines, transmissions
		Tempered*	225	1550	208	1433	9	
Nickel-molybdenum steels								
4620	0.20 C, 1.83 Ni, 0.55 Mn, 0.25 Mo	Annealed	75	517	54	373	31	Transmission gears, chain pins, shafts, roller bearings
		Normalized	83	573	53	366	29	
4820	0.20 C, 3.50 Ni, 0.60 Mn, 0.25 Mo	Annealed	99	683	67	462	22	Gears for steel mill equipment, paper machinery, mining machinery, earth-moving equipment
		Normalized	100	690	70	483	60	
Nickel (1.83%)-chromium-molybdenum steels								
4340 (E)	0.40 C, 1.83 Ni, 0.90 Mn, 0.80 Cr, 0.20 Mo	Annealed	108	745	68	469	22	Heavy sections, landing gears, truck parts
		Tempered*	250	1725	230	1587	10	
Nickel (0.55%)-chromium-molybdenum steels								
8620	0.20 C, 0.55 Ni, 0.50 Cr, 0.80 Mn, 0.20 Mo	Annealed	77	531	59	407	31	Transmission gears
		Normalized	92	635	52	359	26	
8650	0.50 C, 0.55 Ni, 0.50 Cr, 0.80 Mn, 0.20 Mo	Annealed	103	710	56	386	22	Small machine axles, shafts
		Tempered*	250	1725	225	1552	10	

*Tempered at 600°F (315°C).

9.5.1 Precipitation Strengthening (Hardening)

Precipitation Strengthening of a Generalized Binary Alloy   The object of precipitation strengthening is to create in a heat-treated alloy a dense and fine dispersion of precipitated particles in a matrix of deformable metal. The precipitate particles act as obstacles to dislocation movement and thereby strengthen the heat-treated alloy.

The precipitation-strengthening process can be explained in a general way by referring to the binary phase diagram of metals A and B shown in Fig. 9.41. In order for an alloy system to be able to be precipitation-strengthened for certain alloy compositions, there must be a terminal solid solution that has a decreasing solid solubility as the temperature decreases. The phase diagram of Fig. 9.41 shows this type of decrease in solid solubility in terminal solid solution α in going from point a to point b along the indicated solvus.

Let us now consider the precipitation strengthening of an alloy of composition x_1 of the phase diagram of Fig. 9.41. We choose the alloy composition x_1 since there is a large decrease in the solid solubility of solid solution α in decreasing the temperature from T_2 to T_3. The precipitation-strengthening process involves the following three basic steps:

1.   *Solution heat treatment* is the *first step* in the precipitation-strengthening process. Sometimes this treatment is referred to as *solutionizing*. The alloy sample, which may be in the wrought or cast form, is heated to a temperature between the solvus and solidus temperatures and soaked there until a uniform solid-solution structure

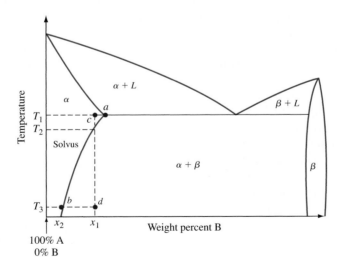

Figure 9.41
Binary phase diagram for two metals A and B having a terminal solid solution α that has a decreasing solid solubility of B in A with decreasing temperature.

is produced. Temperature T_1 at point c of Fig. 9.41 is selected for our alloy x_1 because it lies midway between the solvus and solidus phase boundaries of solid solution α.

2.   *Quenching* is the *second step* in the precipitation-strengthening process. The sample is rapidly cooled to a lower temperature, usually room temperature, and the cooling medium is usually water at room temperature. The structure of the alloy sample after water quenching consists of a supersaturated solid solution. The structure of our alloy x_1 after quenching to temperature T_3 at point d of Fig. 9.41 thus consists of a supersaturated solid solution of the α phase.

3.   *Aging* is the *third basic step* in the precipitation-strengthening process. Aging the solution heat-treated and quenched alloy sample is necessary so that a finely dispersed precipitate forms. The formation of a finely dispersed precipitate in the alloy is the objective of the precipitation-strengthening process. The fine precipitate in the alloy impedes dislocation movement during deformation by forcing the dislocations to either cut through the precipitated particles or go around them. By restricting dislocation movement during deformation, the alloy is strengthened.

Aging the alloy at room temperature is called *natural aging,* whereas aging at elevated temperatures is called *artificial aging.* Most alloys require artificial aging, and the aging temperature is usually between about 15 and 25 percent of the temperature difference between room temperature and the solution heat-treatment temperature.

Decomposition Products Created by the Aging of the Supersaturated Solid Solution   A precipitation-hardenable alloy in the supersaturated solid-solution condition is in a high-energy state, as indicated schematically by energy level 4 of Fig. 9.42. This energy state is relatively unstable, and the alloy tends to seek a lower energy state by the spontaneous decomposition of the supersaturated solid solution into metastable phases or the equilibrium phases. The driving force for the precipitation of metastable phases or the equilibrium phase is the lowering of the energy of the system when these phases form.

When the supersaturated solid solution of the precipitation-hardenable alloy is aged at a relatively low temperature where only a small amount of activation energy is available, clusters of segregated atoms called *precipitation zones,* or *GP zones,*[7] are formed. For the case of our alloy A-B of Fig. 9.41, the zones will be regions enriched with B atoms in a matrix primarily of A atoms. The formation of these zones in the supersaturated solid solution is indicated by the circular sketch at the lower energy level 3 of Fig. 9.42. Upon further aging and if sufficient activation energy is available (as a result of the aging temperature being high enough), the zones develop into or are replaced by a coarser (larger in size) intermediate metastable precipitate, indicated by the circular sketch at the still-lower energy

---

[7]Preprecipitation zones are sometimes referred to as GP zones, named after the two early scientists Guinier and Preston who first identified these structures by x-ray diffraction analyses.

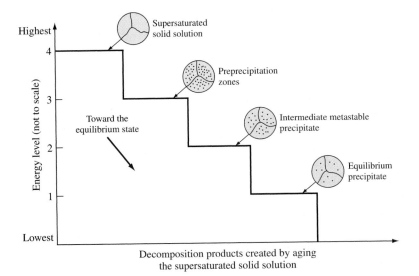

Figure 9.42
Decomposition products created by the aging of a supersaturated
solid solution of a precipitation-hardenable alloy. The highest energy
level is for the supersaturated solid solution, and the lowest energy
level is for the equilibrium precipitate. The alloy can go spontaneously
from a higher energy level to a lower one if there is sufficient
activation energy for the transformation and if the kinetic conditions
are favorable.

level 2. Finally, if aging is continued (usually a higher temperature is necessary)
and if sufficient activation energy is available, the intermediate precipitate is
replaced by the equilibrium precipitate indicated by the even-still-lower energy
level 1 of Fig. 9.42.

**The Effect of Aging Time on the Strength and Hardness of a Precipitation-
Hardenable Alloy that Has Been Solution Heat-Treated and Quenched**   The
effect of aging on strengthening a precipitation-hardenable alloy that has been solution heat-treated and quenched is usually presented as an *aging curve.* The aging
curve is a plot of strength or hardness versus aging time (usually on a logarithmic
scale) at a particular temperature. Figure 9.43 shows a schematic aging curve. At zero
time, the strength of the supersaturated solid solution is indicated on the ordinate
axis of the plot. As the aging time increases, preprecipitation zones form and their
size increases, and the alloy becomes stronger and harder and less ductile (Fig. 9.43).
A maximum strength (peak aged condition) is eventually reached if the aging temperature is sufficiently high, which is usually associated with the formation of an
intermediate metastable precipitate. If aging is continued so that the intermediate
precipitate coalesces and coarsens, the alloy overages and becomes weaker than in
the peak aged condition (Fig. 9.43).

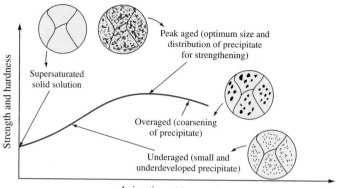

Figure 9.43
Schematic aging curve (strength or hardness versus time) at a particular temperature for a precipitation-hardenable alloy.

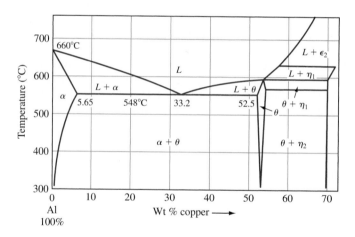

Ice Nine

Figure 9.44
Aluminum-rich end of aluminum-copper phase diagram.
[*After K. R. Van Horn (ed.), "Aluminum," vol. 1, American Society for Metals, 1967, p. 372.*]

Precipitation Strengthening (Hardening) of an Al–4% Cu Alloy Let us now examine the structure and hardness changes that occur during the precipitation heat treatment of an aluminum–4% copper alloy. The heat-treatment sequence for the precipitation strengthening of this alloy is:

1. Solution heat treatment: the Al–4% Cu alloy is solutionized at about 515°C (see the Al-Cu phase diagram of Fig. 9.44).

2. Quenching: the solution heat-treated alloy is rapidly cooled in water at room temperature.

3. Aging: the alloy after solution heat treatment and quenching is artificially aged in the 130 to 190°C range.

Structures Formed During the Aging of the Al–4% Cu Alloy   In the precipitation strengthening of Al–4% Cu alloys, five sequential structures can be identified: (1) supersaturated solid-solution α, (2) GP1 zones, (3) GP2 zones (also called θ'' phase), (4) θ' phase, and (5) θ phase, CuAl$_2$. Not all these phases can be produced at all aging temperatures. GP1 and GP2 zones are produced at lower aging temperatures, and θ' and θ phases occur at higher temperatures.

GP1 zones. These preprecipitation zones are formed at lower aging temperatures and are created by copper atoms segregating in the supersaturated solid-solution α. GP1 zones consist of segregated regions in the shape of disks a few atoms thick (0.4 to 0.6 nm) and about 8 to 10 nm in diameter and form on the {100} cubic planes of the matrix. Since the copper atoms have a diameter of about 11 percent less than the aluminum ones, the matrix lattice around the zones is strained tetragonally. GP1 zones are said to be *coherent* with the matrix lattice since the copper atoms just replace aluminum atoms in the lattice (Fig. 9.45). GP1 zones are detected under the electron microscope by the strain fields they create (Fig. 9.46*a*).

GP2 zones (θ'' phase). These zones also have a tetragonal structure and are coherent with the {100} of the matrix of the Al–4% Cu alloy. Their size ranges from about 1 to 4 nm thick to 10 to 100 nm in diameter as aging proceeds (Fig. 9.46*b*).

θ' phase. This phase nucleates heterogeneously, especially on dislocations, and is incoherent with the matrix. (An *incoherent precipitate* is one in which the precipitated particle has a distinct crystal structure different from the matrix [Fig. 9.45*a*]). θ' phase has a tetragonal structure with a thickness of 10 to 150 nm (Fig. 9.46*c*).

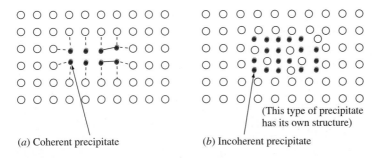

(*a*) Coherent precipitate   (*b*) Incoherent precipitate

Figure 9.45
Schematic comparison of the nature of (*a*) a coherent precipitate and (*b*) an incoherent precipitate. The coherent precipitate is associated with a high strain energy and low surface energy and the incoherent one is associated with a low strain energy and high surface energy.

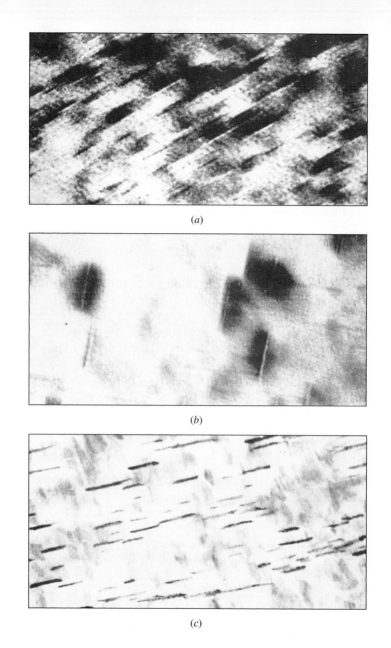

(a)

(b)

(c)

Figure 9.46

Microstructures of aged Al–4% Cu alloys. (a) Al–4% Cu, heated to 540°C, water-quenched and aged 16 h at 130°C. The GP zones have been formed as disks parallel to the {100} planes of the FCC matrix and at this stage are a few atoms thick and about 100 A in diameter. Only disks lying on one crystallographic orientation are visible. (Electron micrograph; magnification 1,000,000×.) (b) Al–4% Cu, solution-treated at 540°C, quenched in water, and aged for 1 day at 130°C. This thin-foil micrograph shows strain fields due to coherent GP2 zones. The dark regions surrounding the zones are caused by strain fields. (Electron micrograph; magnification 800,000×.) (c) Al–4% Cu alloy solution heat-treated at 540°C, quenched in water, and aged for 3 days at 200°C. This thin-foil micrograph shows the incoherent and metastable phase θ' which forms by heterogeneous nucleation and growth. (Electron micrograph; magnification 25,000×.)

(After J. Nutting and R. G. Baker, "The Microstructure of Metals," Institute of Metals, 1965, pp. 65 and 67.)

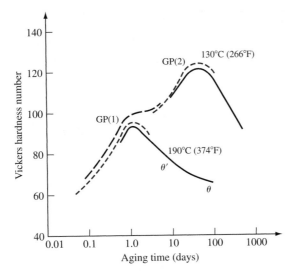

Figure 9.47
Correlation of structures and hardness of Al–4%
Cu alloy aged at 130 and 190°C.
[*After J. M. Silcock, T. J. Heal, and H. K. Hardy as
presented in K. R. Van Horn (ed.), "Aluminum," vol. 1,
American Society for Metals, 1967, p.123.*]

θ phase. The equilibrium phase θ is incoherent and has the composition $CuAl_2$. This phase has a BCT structure ($a = 0.607$ nm and $c = 0.487$ nm) and forms from θ' or directly from the matrix.

The general sequence of precipitation in binary aluminum-copper alloys can be represented by

Supersaturated solid solution $\rightarrow$ GP1 zones $\rightarrow$
$$\text{GP2 zones } (\theta'' \text{phase}) \rightarrow \theta' \rightarrow \theta \ (CuAl_2)$$

Correlation of Structures and Hardness in an Al–4% Cu Alloy   The hardness versus aging time curves for an Al–4% Cu alloy aged at 130 and 190°C are shown in Fig. 9.47. At 130°C, GP1 zones are formed and increase the hardness of the alloy by impeding dislocation movement. Further aging at 130°C creates GP2 zones that increase the hardness still more by making dislocation movement still more difficult. A maximum in hardness is reached with still more aging time at 130°C as θ' forms. Aging beyond the hardness peak dissolves the GP2 zones and coarsens the θ' phase, causing the decrease in the hardness of the alloy. GP1 zones do not form during aging at 190°C in the Al–4% Cu alloy since this temperature is above the GP1 solvus. With long aging times at 190°C the equilibrium θ phase forms.

EXAMPLE PROBLEM 9.6

Calculate the theoretical weight percent of the θ phase that could be formed at 27°C (room temperature) when a sample of Al–4.50 wt % Cu alloy is very slowly cooled from 548°C. Assume the solid solubility of Cu in Al at 27°C is 0.02 wt % and that the θ phase contains 54.0 wt % Cu.

■ **Solution**

First, we draw a tie line xy on the Al-Cu phase diagram at 27°C between the α and θ phases, as shown in Fig. EP9.5a. Next we indicate the 4.5 percent Cu composition point at z. The ratio xz divided by the whole tie line xy (Fig. EP9.5b) gives the weight fraction of the θ phase. Thus,

$$\theta \text{ wt \%} = \frac{4.50 - 0.02}{54.0 - 0.02}(100\%) = \frac{4.48}{53.98}(100\%) = 8.3\% \blacktriangleleft$$

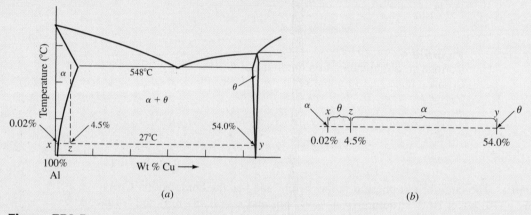

Figure EP9.5
(a) Al-Cu phase diagram with tie line xy indicated on it at 27°C and point z located at 4.5% Cu.
(b) Isolated tie line xy indicating segment xz as representing the weight fraction of the θ phase.

9.5.2 General Properties of Aluminum and Its Production

Engineering Properties of Aluminum   Aluminum possesses a combination of properties that make it an extremely useful engineering material. Aluminum has a low density (2.70 g/cm^3), making it particularly useful for transportation manufactured products. Aluminum also has good corrosion resistance in most natural environments due to the tenacious oxide film that forms on its surface. Although pure aluminum has low strength, it can be alloyed to a strength of about 100 ksi (690 MPa). Aluminum is nontoxic and is used extensively for food containers and packaging. The good electrical properties of aluminum make it suitable for many applications in the electrical industry. The relatively low price of aluminum (96¢/lb in 1989) along with its many useful properties make this metal very important industrially.

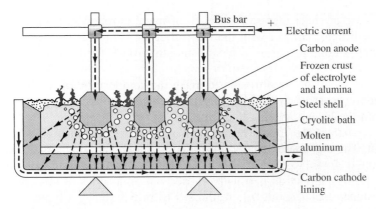

Bus bar

Electric current

Carbon anode

Frozen crust
of electrolyte
and alumina

Steel shell

Cryolite bath

Molten
aluminum

Carbon cathode
lining

Figure 9.48
Electrolytic cell used to produce aluminum.
(*Courtesy of Aluminum Company of America.*)

Animation

Production of Aluminum   Aluminum is the most abundant metallic element in the earth's crust and always occurs in the combined state with other elements such as iron, oxygen, and silicon. Bauxite, which consists mainly of hydrated aluminum oxides, is the chief commercial mineral used for the production of aluminum. In the Bayer process, bauxite is reacted with hot sodium hydroxide to convert the aluminum in the ore to sodium aluminate. After separation of the insoluble material, aluminum hydroxide is precipitated from the aluminate solution. The aluminum hydroxide is then thickened and calcined to aluminum oxide, Al_2O_3.

The aluminum oxide is dissolved in a molten bath of cryolite (Na_3AlF_6) and electrolyzed in an electrolytic cell (Fig. 9.48) by using carbon anodes and cathode. In the electrolysis process, metallic aluminum forms in the liquid state and sinks to the bottom of the cell and is periodically tapped off. The cell-tapped aluminum usually contains from 99.5 to 99.9 percent aluminum with iron and silicon being the major impurities.

Aluminum from the electrolytic cells is taken to large refractory-lined furnaces where it is refined before casting. Alloying elements and alloying-element master ingots may also be melted and mixed in with the furnace charge. In the refining operation, the liquid metal is usually purged with chlorine gas to remove dissolved hydrogen gas, which is followed by a skimming of the liquid-metal surface to remove oxidized metal. After the metal has been degassed and skimmed, it is screened and cast into ingot shapes for remelting or into primary ingot shapes such as sheet or extrusion ingots for further fabrication.

9.5.3 Wrought Aluminum Alloys

Primary Fabrication   Ingot shapes such as sheet and extrusion ingots are usually semicontinuously cast by the *direct-chill method.* Figure 4.8 shows schematically how an aluminum ingot is cast by this method, and Fig. 4.1 is a photograph of a large semicontinuously cast ingot being removed from the casting pit.

In the case of sheet ingots, about $\frac{1}{2}$ in. of metal is removed from the ingot surfaces that will make contact with the hot-rolling-mill rolls. This operation is called *scalping* and is done to ensure a clean, smooth surface for the fabricated sheet or plate. Next the ingots are *preheated* or *homogenized* at a high temperature for about 10 to 24 h to allow atomic diffusion to make the composition of the ingot uniform. The preheating must be done at a temperature below the melting point of the constituent with the lowest melting temperature.

After reheating, the ingots are *hot-rolled* by using a four-high reversing hot-rolling mill. The ingots are usually hot-rolled to about 3 in. thick and then reheated and hot-rolled down to about $\frac{3}{4}$ to 1 in. with an intermediate hot-rolling mill (Fig. 6.1). Further reduction is usually carried out on a series of tandem hot-rolling mills to produce metal about 0.1 in. thick. Figure 6.8 shows a typical cold-rolling operation. More than one intermediate anneal is usually required if thin sheet is to be produced.

Classification of Wrought Aluminum Alloys   Aluminum alloys produced in the wrought form (i.e., sheet, plate, extrusions, rod, and wire) are classified according to the major alloying elements they contain. A four-digit numerical designation is used to identify aluminum wrought alloys. The first digit indicates the alloy group that contains specific alloying elements. The last two digits identify the aluminum alloy or indicate the aluminum purity. The second digit indicates modification of the original alloy or impurity limits. Table 9.7 lists the wrought aluminum alloy groups.

Temper Designations   Temper designations for wrought aluminum alloys follow the alloy designation and are separated by a hyphen (for example, 1100-0). Subdivisions of a basic temper are indicated by one or more digits and follow the letter of the basic designation (for example, 1100-H14).

Basic Temper Designations

F—As fabricated. No control over the amount of strain hardening; no mechanical property limits.

O—Annealed and recrystallized. Temper with the lowest strength and highest ductility.

Table 9.7 Wrought aluminum alloy groups

Aluminum, 99.00% minimum and greater	1xxx
Aluminum alloys grouped by major alloying elements:	
Copper	2xxx
Manganese	3xxx
Silicon	4xxx
Magnesium	5xxx
Magnesium and silicon	6xxx
Zinc	7xxx
Other element	8xxx
Unused series	9xxx

H—Strain-hardened (see subsequent subsection for subdivisions).

T—Heat-treated to produce stable tempers other than F or O (see subsequent subsection for subdivisions).

Strain-Hardened Subdivisions

H1—Strain-hardened only. The degree of strain hardening is indicated by the second digit and varies from quarter-hard (H12) to full-hard (H18), which is produced with approximately 75 percent reduction in area.

H2—Strain-hardened and partially annealed. Tempers ranging from quarter-hard to full-hard obtained by partial annealing of cold-worked materials with strengths initially greater than desired. Tempers are H22, H24, H26, and H28.

H3—Strain-hardened and stabilized. Tempers for age-softening aluminum-magnesium alloys that are strain-hardened and then heated at a low temperature to increase ductility and stabilize mechanical properties. Tempers are H32, H34, H36, and H38.

Heat-Treated Subdivisions

T1—Naturally aged. Product is cooled from an elevated-temperature shaping process and naturally aged to a substantially stable condition.

T3—Solution heat-treated, cold-worked, and naturally aged to a substantially stable condition.

T4—Solution heat-treated and naturally aged to a substantially stable condition.

T5—Cooled from an elevated-temperature shaping process and then artificially aged.

T6—Solution heat-treated and then artificially aged.

T7—Solution heat-treated and stabilized.

T8—Solution heat-treated, cold-worked, and then artificially aged.

Non–Heat-Treatable Wrought Aluminum Alloys Wrought aluminum alloys can conveniently be divided into two groups: *non–heat-treatable* and *heat-treatable alloys*. Non–heat-treatable aluminum alloys cannot be precipitation-strengthened but can only be cold-worked to increase their strength. The three main groups of non–heat-treatable wrought aluminum alloys are the 1xxx, 3xxx, and 5xxx groups. Table 9.8 lists the chemical composition, typical mechanical properties, and applications for some selected industrially important wrought aluminum alloys.

1xxx alloys. These alloys have a minimum of 99.0 percent aluminum with iron and silicon being the major impurities (alloying elements). An addition of 0.12 percent copper is added for extra strength. The 1100 alloy has a tensile strength of about 13 ksi (90 MPa) in the annealed condition and is used mainly for sheet metal work applications.

3xxx alloys. Manganese is the principal alloying element of this group and strengthens aluminum mainly by solid-solution strengthening. The most important alloy of this group is 3003, which is essentially an 1100 alloy

Table 9.8 Typical mechanical properties and applications for aluminum alloys

Alloy number*	Chemical composition (wt %)†	Condition‡	Tensile strength		Yield strength		Elongation (%)	Typical applications
			ksi	MPa	ksi	MPa		
Wrought alloys								
1100	99.0 min Al, 0.12 Cu	Annealed (-O)	13	89 (av)	3.5	24 (av)	25	Sheet metal work, fin stock
		Half-hard (-H14)	18	124 (av)	14	97 (av)	4	
3003	1.2 Mn	Annealed (-O)	17	117 (av)	5	34 (av)	23	Pressure vessels, chemical equipment, sheet metal work
		Half-hard (-H14)	23	159 (av)	23	159 (av)	17	
5052	2.5 Mg, 0.25 Cr	Annealed (-O)	28	193 (av)	9.5	65 (av)	18	Bus, truck, and marine uses, hydraulic tubes
		Half-hard (-H34)	38	262 (av)	26	179 (av)	4	
2024	4.4 Cu, 1.5 Mg, 0.6 Mn	Annealed (-O)	32	220 (max)	14	97 (max)	12	Aircraft structures
		Heat-treated (-T6)	64	442 (min)	50	345 (min)	5	
6061	1.0 Mg, 0.6 Si, 0.27 Cu, 0.2 Cr	Annealed (-O)	22	152 (max)	12	82 (max)	16	Truck and marine structures, pipelines, railings
		Heat-treated (-T6)	42	290 (min)	35	241 (min)	10	
7075	5.6 Zn, 2.5 Mg, 1.6 Cu, 0.23 Cr	Annealed (-O)	40	276 (max)	21	145 (max)	10	Aircraft and other structures
		Heat-treated (-T6)	73	504 (min)	62	428 (min)	8	
Casting alloys								
355.0	5 Si, 1.2 Cu, 0.5 Mg	Sand cast (-T6)	32	220 (min)	20	138 (min)	2.0	Pump housings, aircraft fittings, crankcases
		Permanent mold (-T6)	37	285 (min)	...	...	1.5	
356.0	7 Si, 0.3 Mg	Sand cast (-T6)	30	207 (min)	20	138 (min)	3	Transmission cases, truck axle housings, truck wheels
		Permanent mold (-T6)	33	229 (min)	22	152 (min)	3	
332.0	9.5 Si, 3 Cu, 1.0 Mg	Permanent mold (-T5)	31	214 (min)				Automotive pistons
413.0	12 Si, 2 Fe	Die casting	43	297	21	145 (min)	2.5	Large, intricate castings

*Aluminum Association number.
†Balance aluminum.
‡O = annealed and recrystallized; H14 = strain-hardened; H34 = strain-hardened and stabilized; T5 = cooled from elevated-temperature shaping process, then artificially aged; T6 = solution heat-treated, then artificially aged.

with the addition of about 1.25 percent manganese. The 3003 alloy has a tensile strength of about 16 ksi (110 MPa) in the annealed condition and is used as a general-purpose alloy where good workability is required.

5xxx alloys. Magnesium is the principal alloying element of this group and is added for solid-solution strengthening in amounts up to about 5 percent. One of the most industrially important alloys of this group is 5052, which contains about 2.5 percent magnesium (Mg) and 0.2 percent chromium (Cr). In the annealed condition alloy 5052 has a tensile strength of about 28 ksi (193 MPa). This alloy is also used for sheet metal work, particularly for bus, truck, and marine applications.

Heat-Treatable Wrought Aluminum Alloys   Some aluminum alloys can be precipitation-strengthened by heat treatment (see pages 406 to 410). Heat-treatable wrought aluminum alloys of the 2xxx, 6xxx, and 7xxx groups are all precipitation-strengthened by similar mechanisms as described in Sec. 9.5 for aluminum-copper alloys. Table 9.8 lists the chemical compositions, typical mechanical properties, and applications of some of the industrially important wrought heat-treatable aluminum alloys. Figure 9.49 shows a rack of aluminum alloy extrusions about to be solution heat-treated in an industrial precipitation-strengthening heat treatment.

2xxx alloys. The principal alloying element of this group is *copper,* but magnesium is also added to most of these alloys. Small amounts of other elements are also added. One of the most important alloys of this group is 2024, which

Figure 9.49
Industrial precipitation strengthening (hardening of aluminum alloy extrusions).
(*Courtesy of Reynolds Metals Co.*)

contains about 4.5 percent copper (Cu), 1.5 percent Mg, and 0.6 percent Mn. This alloy is strengthened mainly by solid-solution and precipitation strengthening. An intermetallic compound of the approximate composition of Al_2CuMg is the main strengthening precipitate. Alloy 2024 in the T6 condition has a tensile strength of about 64 ksi (442 MPa) and is used, for example, for aircraft structurals.

6xxx alloys. The principal alloying elements for the 6xxx group are *magnesium* and *silicon,* which combine together to form an intermetallic compound, Mg_2Si, which in precipitate form strengthens this group of alloys. Alloy 6061 is one of the most important alloys of this group and has an approximate composition of 1.0 percent Mg, 0.6 percent Si, 0.3 percent Cu, and 0.2 percent Cr. This alloy in the T6 heat-treated condition has a tensile strength of about 42 ksi (290 MPa) and is used for general-purpose structurals.

7xxx alloys. The principal alloying elements for the 7xxx group of aluminum alloys are zinc, magnesium, and copper. Zinc and magnesium combine to form an intermetallic compound, $MgZn_2$, which is the basic precipitate that strengthens these alloys when they are heat-treated. The relatively high solubility of zinc and magnesium in aluminum makes it possible to create a high density of precipitates and hence to produce very great increases in strength. Alloy 7075 is one of the most important alloys of this group and has an approximate composition of 5.6 percent Zn, 2.5 percent Mg, 1.6 percent Cu, and 0.25 percent Cr. Alloy 7075, when heat-treated to the T6 temper, has a tensile strength of about 73 ksi (504 MPa) and is used mainly for aircraft structurals.

9.5.4 Aluminum Casting Alloys

Casting Processes   Aluminum alloys are normally cast by one of three main processes: sand casting, permanent-mold, and die casting.

Sand casting is the simplest and most versatile of the aluminum casting processes. Figure 9.50 shows how a simple sand mold for producing sand castings is constructed. The sand-casting process is usually chosen for the production of (1) small quantities of identical castings, (2) complex castings with intricate cores, (3) large castings, and (4) structural castings.

In *permanent-mold casting* the molten metal is poured into a permanent metal mold under gravity, low pressure, or centrifugal pressure only. Figure 6.2 shows an open permanent mold, while Fig. 6.3*a* shows the permanent-mold casting of two aluminum alloy automobile pistons. Castings of the same alloy and shape produced by a permanent mold have a finer grain structure and higher strength than those cast by sand molds. The faster cooling rate of permanent-mold casting produces a finer grain structure. Also, permanent-mold castings usually have less shrinkage and gas porosity than sand castings. However, permanent molds have size limitations, and complex parts may be difficult or impossible to cast with a permanent mold.

In *die casting,* identical parts are cast at maximum production rates by forcing molten metal under considerable pressure into metal molds. Two metal die halves are securely locked together to withstand high pressure. The molten aluminum is forced

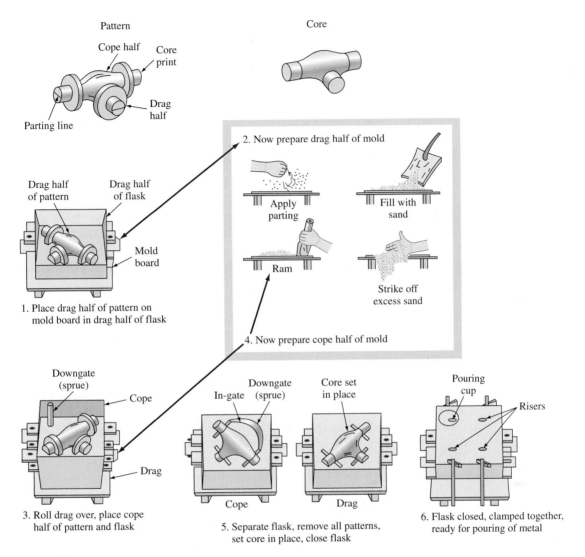

Figure 9.50

Steps in the construction of a simple sand mold for making a sand casting.

(*After H. F. Taylor, M. C. Flemings, and J. Wulff, "Foundry Engineering," Wiley, 1959, p. 20.*)

into the cavities in the dies. When the metal has solidified, the dies are unlocked and opened to eject the hot casting. The die halves are locked together again and the casting cycle is repeated. Some of the advantages of die casting are (1) parts die cast are almost completely finished and can be produced at high rates, (2) dimensional tolerances of each cast part can be more closely held than with any other major casting process, (3) smooth surfaces on the casting are obtainable, (4) rapid cooling of the casting produces a fine-grain structure, and (5) the process can be automated easily.

Table 9.9 Cast aluminum alloy groups

Aluminum, 99.00% minimum and greater	1xx.x
Aluminum alloys grouped by major alloying elements:	
Copper	2xx.x
Silicon, with added copper and/or magnesium	3xx.x
Silicon	4xx.x
Magnesium	5xx.x
Zinc	7xx.x
Tin	8xx.x
Other element	9xx.x
Unused series	6xx.x

Aluminum Casting Alloy Compositions   Aluminum casting alloys have been developed for casting qualities such as fluidity and feeding ability as well as for properties such as strength, ductility, and corrosion resistance. As a result, their chemical compositions differ greatly from those of the wrought aluminum alloys. Table 9.8 lists the chemical compositions, mechanical properties, and applications for some selected aluminum casting alloys. These alloys are classified in the United States according to the Aluminum Association system. In this system, aluminum casting alloys are grouped by the major alloying elements they contain by using a four-digit number with a period between the last two digits, as listed in Table 9.9.

Silicon in the range of about 5 to 12 percent is the most important alloying element in aluminum casting alloys since it increases the fluidity of the molten metal and its feeding ability in the mold as well as strengthens the aluminum. Magnesium in the range of about 0.3 to 1 percent is added to increase strength, mainly by precipitation strengthening through heat treatment. Copper in the range of about 1 to 4 percent is also added to some aluminum casting alloys to increase strength, particularly at elevated temperatures. Other alloying elements such as zinc, tin, titanium, and chromium are also added to some aluminum casting alloys.

In some cases, if the cooling rate of the solidified casting in the mold is sufficiently rapid, a heat-treatable alloy can be produced in the supersaturated solid condition. Thus, the solution heat-treatment and quenching steps can be omitted for precipitation strengthening the casting, and only subsequent aging of the casting after it has been removed from the mold is required. A good example of the application of this type of heat treatment is in the production of precipitation-strengthened automobile pistons. The pistons shown in Fig. 6.3a, after being removed from the mold, only require an aging treatment to be precipitation-strengthened. This heat-treatment temper is called T5.

9.6 COPPER ALLOYS

9.6.1 General Properties of Copper

Copper is an important engineering metal and is widely used in the unalloyed condition as well as combined with other metals in the alloyed form. In the unalloyed form copper has an extraordinary combination of properties for industrial applications. Some of these are high electrical and thermal conductivity, good corrosion resistance, ease of fabrication, medium tensile strength, controllable annealing properties, and general soldering and joining characteristics. Higher strengths are attained in a series of brass and bronze alloys that are indispensable for many engineering applications.

9.6.2 Production of Copper

Most copper is extracted from ores containing copper and iron sulfides. Copper sulfide concentrates obtained from lower-grade ores are smelted in a reverberatory furnace to produce a matte that is a mixture of copper and iron sulfides and is separated from a slag (waste material). The copper sulfide in the matte is then chemically converted to impure or blister copper (98% + Cu) by blowing air through the matte. The iron sulfide is oxidized first and slagged off in this operation. Subsequently most of the impurities in the blister copper are removed in a refining furnace and are removed as a slag. This fire-refined copper is called *tough-pitch copper,* and although it can be used for some applications, most tough-pitch copper is further refined electrolytically to produce 99.95 percent *electrolytic tough-pitch (ETP) copper.*

9.6.3 Classification of Copper Alloys

Copper alloys in the United States are classified according to a designation system administered by the *Copper Development Association* (CDA). In this system the numbers C10100 to C79900 designate wrought alloys and the numbers from C80000 to C99900 designate casting alloys. Table 9.10 lists the alloy groups of each major level, and Table 9.11 lists the chemical compositions, typical mechanical properties, and applications for some selected copper alloys.

Table 9.10 Classification of copper alloys (Copper Development Association System)

Wrought alloys	
C1xxxx	Coppers* and high-copper alloys†
C2xxxx	Copper-zinc alloys (brasses)
C3xxxx	Copper-zinc-lead alloys (leaded brasses)
C4xxxx	Copper-zinc-tin alloys (tin brasses)
C5xxxx	Copper-tin alloys (phosphor bronzes)
C6xxxx	Copper-aluminum alloys (aluminum bronzes), copper-silicon alloys (silicon bronzes) and miscellaneous copper-zinc alloys
C7xxxx	Copper-nickel and copper-nickel-zinc alloys (nickel silvers)
Cast alloys	
C8xxxx	Cast coppers, cast high-copper alloys, cast brasses of various types, cast manganese-bronze alloys, and cast copper-zinc-silicon alloys
C9xxxx	Cast copper-tin alloys, copper-tin-lead alloys, copper-tin-nickel alloys, copper-aluminum-iron alloys, and copper-nickel-iron and copper-nickel-zinc alloys.

*"Coppers" have a minimum copper content of 99.3 percent or higher.
†High-copper alloys have less than 99.3 percent Cu but more than 96 percent and do not fit into the other copper alloy groups.

Table 9.11 Typical mechanical properties and applications of copper alloys

Alloy number	Chemical composition (wt %)	Condition	Tensile strength		Yield strength		Elongation in 2 in. (%)	Typical applications
			ksi	MPa	ksi	MPa		
		Wrought alloys						
C10100	99.99 Cu	Annealed	32	220	10	69	45	Bus conductors, waveguides, hollow conductors, lead-in wires and anodes for vacuum tubes, vacuum seals, transistor components, glass-to-metal seals, coaxial cables and tubes, klystrons, microwave tubes, rectifiers
		Cold-worked	50	345	45	310	6	
C11000 (ETP)	99.9 Cu, 0.04 O	Annealed	32	220	10	69	45	Gutters, roofing, gaskets, auto radiators, busbars, nails, printing rolls, rivets, radio parts
		Cold-worked	50	345	45	310	6	
C26000	70 Cu, 30 Zn	Annealed	47	325	15	105	62	Radiator cores and tanks, flashlight shells, lamp fixtures, fasteners, locks, hinges, ammunition components, plumbing accessories, pins, rivets
		Cold-worked	76	525	63	435	8	
C28000	60 Cu, 40 Zn	Annealed	54	370	21	145	45	Architectural, large nuts and bolts, brazing rod, condenser plates, heat exchanger and condenser tubing, hot forgings
		Cold-worked	70	485	50	345	10	
C17000	99.5 Cu, 1.7 Be, 0.20 Co	SHT*	60	410	28	190	60	Bellows, Bourdon tubing, diaphragms, fuse clips, fasteners, lock washers, springs, switch parts, roll pins, valves, welding equipment
		SHT, CW, PH*	180	1240	155	1070	4	
C61400	95 Cu, 7 Al, 2 Fe	Annealed	80	550	40	275	40	Nuts, bolts, stringers and threaded members, corrosion-resistant vessels and tanks, structural components, machine parts, condenser tube and piping systems, marine protective sheathing and fastening
		Cold-worked	89	615	60	415	32	

Alloy	Composition	Condition						Typical applications
C71500	70 Cu, 30 Ni	Annealed	55	380	18	125	36	Communication relays, condensers, condenser plates, electrical springs, evaporator and heat exchanger tubes, ferrules, resistors
		Cold-worked	84	580	79	545	3	
Casting alloys								
C80500	99.75 Cu	As-cast	25	172	9	62	40	Electrical and thermal conductors; corrosion- and oxidation-resistant applications
C82400	96.4 Cu, 1.70 Be, 0.25 Co	As-cast	72	497	37	255	20	Safety tools, molds for plastic parts, cams, bushings, bearings, valves, pump parts, gears
		Heat-treated	150	1035	140	966	1	
C83600	85 Cu, 5 Sn, 5 Pb, 5 Zn	As-cast	37	255	17	117	30	Valves, flanges, pipe fittings, plumbing goods, pump castings, water pump impellers and housings, ornamental fixtures, small gears
C87200	89 Cu, 4 Si	As-cast	55	379	25	172	30	Bearings, belts, impellers, pump and valve components, marine fittings, corrosion-resistant castings
C90300	93 Cu, 8 Sn, 4 Zn	As-cast	45	310	21	145	30	Bearings, bushings, pump impellers, piston rings, valve components, seal rings, steam fittings, gears
C95400	85 Cu, 4 Fe, 11 Al	As-cast	85	586	35	242	18	Bearings, gears, worms, bushings, valve seats and guides, pickling hooks
		Heat-treated	105	725	54	373	8	
C96400	69 Cu, 30 Ni, 0.9 Fe	As-cast	68	469	37	255	28	Valves, pump bodies, flanges, elbows used for seawater corrosion resistance

*SHT = solution heat-treated; CW = cold-worked; PH = precipitation-hardened.

9.6.4 Wrought Copper Alloys

Unalloyed Copper Unalloyed copper is an important engineering metal, and because of its high electrical conductivity, it is used to a large extent in the electrical industry. Electrolytic tough-pitch copper is the least expensive of the industrial coppers and is used for the production of wire, rod, plate, and strip. ETP copper has a nominal oxygen content of 0.04 percent. Oxygen is almost insoluble in ETP copper and forms interdendritic Cu_2O when copper is cast. For most applications the oxygen in ETP copper is an insignificant impurity. However, if ETP copper is heated to a temperature above about 400°C in an atmosphere containing hydrogen, the hydrogen can diffuse into the solid copper and react with the internally dispersed Cu_2O to form steam according to the reaction

$$Cu_2O + H_2 \,(\text{dissolved in Cu}) \rightarrow 2Cu + H_2O \,(\text{steam})$$

The large water molecules formed by the reaction do not diffuse readily and so they form internal holes, particularly at the grain boundaries, which makes the copper brittle (Fig. 9.51).

To avoid hydrogen embrittlement caused by Cu_2O, the oxygen can be reacted with phosphorus to form phosphorus pentoxide (P_2O_5) (alloy C12200). Another way to avoid hydrogen embrittlement is to eliminate the oxygen from the copper by casting the ETP copper under a controlled reducing atmosphere. The copper

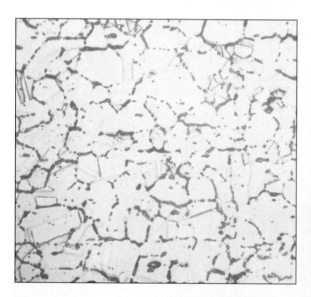

Figure 9.51
Electrolytic tough-pitch copper exposed to hydrogen
Virtual Lab at 850°C for $\frac{1}{2}$ h; structure shows internal holes developed by steam, which makes the copper brittle. (Etch: potassium dichromate; magnification 150×.)
(Courtesy of Amax Base Metals Research, Inc.)

Figure 9.52
Microstructures of cartridge brass
(70% Cu–30% Zn) in the annealed condition.
(Etchant: $NH_4OH + H_2O_2$; magnification 75×.)

produced by this method is called *oxygen-free high-conductivity (OFHC) copper* and is alloy C10200.

Copper-Zinc Alloys   The copper-zinc brasses consist of a series of alloys of copper with additions of about 5 to 40 percent zinc. Copper forms substitutional solid solutions with zinc up to about 35 percent zinc, as indicated in the all-alpha phase region of the Cu-Zn phase diagram (Fig. 8.27). When the zinc content reaches about 40 percent, alloys with two phases, alpha and beta, form.

The microstructure of the single-phase alpha brasses consists of an alpha solid solution, as shown in Fig. 9.52 for a 70% Cu–30% Zn alloy (C26000, cartridge brass). The microstructure of the 60% Cu–40% Zn brass (C28000, Muntz metal) has two phases, alpha and beta, as shown in Fig. 9.53.

Small amounts of lead (0.5 to 3 percent) are added to some Cu-Zn brasses to improve machinability. Lead is almost insoluble in solid copper and is distributed in leaded brasses in small globules (Fig. 9.54).

The tensile strengths of some selected brasses are listed in Table 9.11. These alloys are of medium strength (34 to 54 ksi; 234 to 374 MPa) in the annealed condition and can be cold-worked to increase their strength.

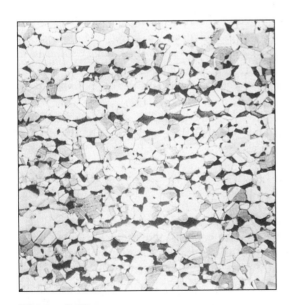

Figure 9.53
Hot-rolled Muntz metal sheet (60% Cu–40% Zn). Structure consists of beta phase (dark) and alpha phase (light). (Etchant: $NH_4OH + H_2O_2$; magnification 75×.)
(Courtesy of Anaconda American Brass Co.)

Figure 9.54
Free-cutting brass extruded rod showing elongated lead globules. Remainder of structure is α phase. (Etch: $NH_4OH + H_2O_2$; magnification 75×.)
(Courtesy of Anaconda American Brass Co.)

Copper-Tin Bronzes   Copper-tin alloys, which are properly called *tin bronzes* but often called *phosphor bronzes,* are produced by alloying about 1 to 10 percent tin with copper to form solid-solution-strengthened alloys. Wrought tin bronzes are stronger than Cu-Zn brasses, especially in the cold-worked condition, and have better corrosion resistance, but cost more. Cu-Sn casting alloys containing up to about 16 percent Sn are used for high-strength bearings and gear blanks. Large amounts of lead (5 to 10 percent) are added to these alloys to provide lubrication for bearing surfaces.

Copper-Beryllium Alloys   Copper-beryllium alloys are produced containing between 0.6 and 2 percent Be with additions of cobalt from 0.2 to 2.5 percent. These alloys are precipitation-hardenable and can be heat-treated and cold-worked to produce tensile strengths as high as 212 ksi (1463 MPa), which is the highest strength developed in commercial copper alloys. Cu-Be alloys are used for tools requiring high hardness and nonsparking characteristics for the chemical industry. The excellent corrosion resistance, fatigue properties, and strength of these alloys make them useful for springs, gears, diaphragms, and valves. However, they have the disadvantage of being relatively costly materials.

9.7 STAINLESS STEELS

Stainless steels are selected as engineering materials mainly because of their excellent corrosion resistance in many environments. The corrosion resistance of stainless steels is due to their high chromium contents. In order to make a "stainless steel" stainless, there must be at least 12 percent chromium (Cr) in the steel. According to classical theory, chromium forms a surface oxide that protects the underlying iron-chromium alloy from corroding. To produce the protective oxide, the stainless steel must be exposed to oxidizing agents.

In general there are four main types of stainless steels: ferritic, martensitic, austenitic, and precipitation-hardening. Only the first three types will be briefly discussed in this section.

9.7.1 Ferritic Stainless Steels

Ferritic stainless steels are essentially iron-chromium binary alloys containing about 12 to 30 percent Cr. They are called ferritic since their structure remains mostly ferritic (BCC, α iron type) at normal heat-treatment conditions. Chromium, since it has the same BCC crystal structure as α ferrite, extends the α phase region and suppresses the γ phase region. As a result, the "γ loop" is formed in the Fe-Cr phase diagram and divides it into FCC and BCC regions (Fig. 9.55). Ferritic stainless steels, since they contain more than 12 percent Cr, do not undergo the FCC-to-BCC transformation and cool from high temperatures as solid solutions of chromium in α iron.

Table 9.12 lists the chemical compositions, typical mechanical properties, and applications of some selected stainless steels, including ferritic type 430.

The ferritic stainless steels are relatively low in cost since they do not contain nickel. They are used mainly as general construction materials in which their special

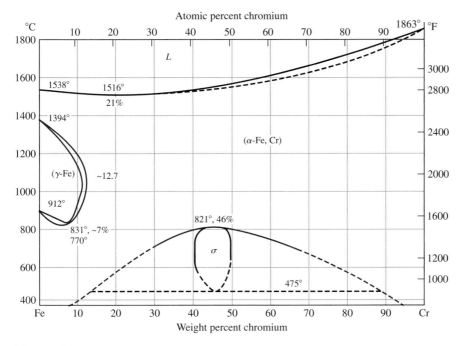

Figure 9.55
Iron-chromium phase diagram.
(*After "Metals Handbook," vol. 8, 8th ed., American Society for Metals, 1973, p. 291.*)

Ice Nine

corrosion and heat resistance is required. Figure 9.56 shows the microstructure of the ferritic stainless steel type 430 in the annealed condition. The presence of the carbides in this steel reduces its corrosion resistance to some extent. New ferritics have more recently been developed with very low carbon and nitrogen levels and so have improved corrosion resistance.

9.7.2 Martensitic Stainless Steels

Martensitic stainless steels are essentially Fe-Cr alloys containing 12 to 17 percent Cr with sufficient carbon (0.15 to 1.0 percent C) so that a martensitic structure can be produced by quenching from the austenitic phase region. These alloys are called *martensitic* because they are capable of developing a martensitic structure after an austenitizing and quenching heat treatment. Since the composition of martensitic stainless steels is adjusted to optimize strength and hardness, the corrosion resistance of these steels is relatively poor compared to the ferritic and austenitic types.

The heat treatment of martensitic stainless steels for increased strength and toughness is basically the same as that for plain-carbon and low-alloy steels. That is, the alloy is austenitized, cooled fast enough to produce a martensitic structure, and then tempered to relieve stresses and increase toughness. The high hardenability

Table 9.12 Typical mechanical properties and applications of stainless steels

Alloy number	Chemical composition (wt %)*	Condition	Tensile strength		Yield strength		Elongation in 2 in (%)	Typical applications
			ksi	MPa	ksi	MPa		
Ferritic stainless steels								
430	17 Cr, 0.012 C	Annealed	75	517	50	345	25	General-purpose, nonhardenable; uses: range hoods, restaurant equipment
446	25 Cr, 0.20 C	Annealed	80	552	50	345	20	High-temperature applications; heaters, combustion chambers
Martensitic stainless steels								
410	12.5 Cr, 0.15 C	Annealed Q & T†	75	517	40	276	30	General-purpose heat-treatable; machine parts, pump shafts, valves
440A	17 Cr, 0.70 C	Annealed Q & T†	105 265	724 1828	60 245	414 1690	20 5	Cutlery, bearings, surgical tools
440C	17 Cr, 1.1 C	Annealed Q & T†	110 285	759 1966	70 275	276 1897	13 2	Balls, bearings, races, valve parts
Austenitic stainless steels								
301	17 Cr, 7 Ni	Annealed	110	759	40	276	60	High work-hardening rate alloy; structural applications
304	19 Cr, 10 Ni	Annealed	84	580	42	290	55	Chemical and food processing equipment
304L	19 Cr, 10 Ni, 0.03 C	Annealed	81	559	39	269	55	Low carbon for welding; chemical tanks
321	18 Cr, 10 Ni, Ti = 5 × %C min	Annealed	90	621	35	241	45	Stabilized for welding; process equipment, pressure vessels
347	18 Cr, 10 Ni, Cb (Nb) = 10 × C min	Annealed	95	655	40	276	45	Stabilized for welding; tank cars for chemicals
Precipitation-hardening stainless steels								
17-4PH	16 Cr, 4 Ni, 4 Cu, 0.03 Cb (Nb)	Precipitation-hardened	190	1311	175	1207	14	Gears, cams, shafting, aircraft and turbine parts

*Balance Fe.
†Quenched and tempered.

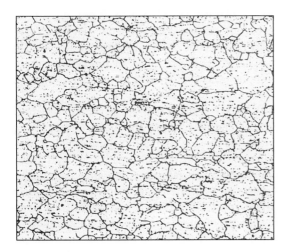

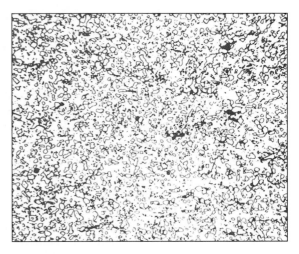

Figure 9.56

Type 430 (ferritic) stainless steel strip annealed at 788°C (1450°F). The structure consists of a ferrite matrix with equiaxed grains and dispersed carbide particles. (Etchant: picral + HCl; magnification 100×.)
(*Courtesy of United States Steel Corp., Research Laboratories.*)

Figure 9.57

Type 440C (martensitic) stainless steel hardened by austenitizing at 1010°C (1850°F) and air-cooled. Structure consists of primary carbides in martensite matrix. (Etchant: HCl + picral; magnification 500×.)
(*Courtesy of Allegheny Ludlum Steel Co.*)

of the Fe–12 to 17 percent Cr alloys avoids the need for water quenching and allows a slower cooling rate to produce a martensitic structure.

Table 9.12 includes the chemical compositions, typical mechanical properties, and applications for types 410 and 440C martensitic stainless steels. The 410 stainless steel with 12 percent Cr is a lower-strength martensitic stainless steel and is a general-purpose heat-treatable type used for applications such as machine parts, pump shafts, bolts, and bushings.

When the carbon content of Fe-Cr alloys is increased up to about 1 percent C, the α loop is enlarged. Consequently, Fe-Cr alloys with about 1 percent C can contain about 16 percent Cr and still be able to produce a martensitic structure upon austenitizing and quenching. Type 440C alloy with 16 percent Cr and 1 percent C is the martensitic stainless steel that has the highest hardness of any corrosion-resisting steel. Its high hardness is due to a hard martensitic matrix and to the presence of a large concentration of primary carbides, as shown in the 440C steel microstructure of Fig. 9.57.

9.7.3 Austenitic Stainless Steels

Austenitic stainless steels are essentially iron-chromium-nickel ternary alloys containing about 16 to 25 percent Cr and 7 to 20 percent Ni. These alloys are called *austenitic* since their structure remains austenitic (FCC, γ iron type) at all normal heat-treating temperatures. The presence of the nickel, which has an FCC crystal structure, enables the FCC structure to be retained at room temperature. The high formability of the austenitic stainless steels is due to their FCC crystal structure. Table 9.12 includes the chemical composition, typical mechanical properties, and applications for austenitic stainless steel types 301, 304, and 347.

Austenitic stainless steels normally have better corrosion resistance than ferritic and martensitic ones because the carbides can be retained in solid solution by rapid cooling from high temperatures. However, if these alloys are to be welded or slowly cooled from high temperatures through the 870 to 600°C range, they can become susceptible to intergranular corrosion, because chromium-containing carbides precipitate at the grain boundaries. This difficulty can be circumvented to some degree either by lowering the maximum carbon content in the alloy to about 0.03 percent C (type 304L alloy) or by adding an alloying element such as columbium (niobium) (type 347 alloy) to combine with the carbon in the alloy (see Sec. 12.5 on intergranular corrosion). Figure 9.58 shows the microstructure of a type 304 stainless steel that has been annealed at 1065°C and air-cooled. Note that there are no carbides visible in the microstructure, as in the case of the type 430 steel (Fig. 9.56) and type 440C steel (Fig. 9.57).

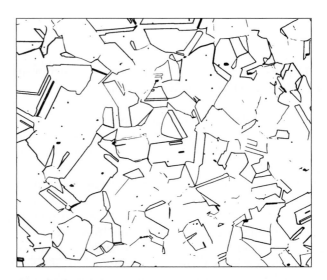

Figure 9.58

Type 304 (austenitic) stainless steel strip annealed 5 min at 1065°C (1950°F) and air-cooled. Structure consists of equiaxed austenite grains. Note annealing twins. (Etchant: HNO_3–acetic–HCl–glycerol; magnification 250×.)

(Courtesy of Allegheny Ludlum Steel Co.)

9.8 CAST IRONS

9.8.1 General Properties

Cast irons are a family of ferrous alloys with a wide range of properties, and as their name implies, they are intended to be cast into the desired shape instead of being worked in the solid state. Unlike steels, which usually contain less than about 1 percent carbon, cast irons normally contain 2 to 4 percent carbon and 1 to 3 percent silicon. Other alloying elements may also be present to control or vary certain properties.

Cast irons make excellent casting alloys since they are easily melted, are very fluid in the liquid state, and do not form undesirable surface films when poured. Cast irons solidify with slight to moderate shrinkage during casting and cooling. These alloys have a wide range of strengths and hardness and in most cases are easy to machine. They can be alloyed to produce superior wear, abrasion, and corrosion resistance. However, cast irons have relatively low impact resistance and ductility, and this limits their use for some applications. The wide industrial use of cast irons is due mainly to their comparatively low cost and versatile engineering properties.

9.8.2 Types of Cast Irons

Four different kinds of cast irons can be differentiated from each other by the distribution of carbon in their microstructures: **white, gray, malleable,** and **ductile iron.** *High-alloy cast irons* constitute a fifth type of cast iron. However, since the chemical compositions of cast irons overlap, they cannot be distinguished from each other by chemical composition analyses. Table 9.13 lists the chemical composition ranges for the four basic cast irons, and Table 9.14 presents some of their typical tensile mechanical properties and applications.

9.8.3 White Cast Iron

White cast iron is formed when much of the carbon in a molten cast iron forms iron carbide instead of graphite upon solidification. The microstructure of as-cast unalloyed white cast iron contains large amounts of iron carbides in a pearlitic matrix

Table 9.13 Chemical composition ranges for typical unalloyed cast irons

Element	Gray iron (%)	White iron (%)	Malleable iron (cast white) (%)	Ductile iron (%)
Carbon	2.5–4.0	1.8–3.6	2.00–2.60	3.0–4.0
Silicon	1.0–3.0	0.5–1.9	1.10–1.60	1.8–2.8
Manganese	0.25–1.0	0.25–0.80	0.20–1.00	0.10–1.00
Sulfur	0.02–0.25	0.06–0.20	0.04–0.18	0.03 max
Phosphorus	0.05–1.0	0.06–0.18	0.18 max	0.10 max

Source: C. F. Walton (ed.), *Iron Castings Handbook,* Iron Castings Society, 1981.

Table 9.14 Typical mechanical properties and applications of cast irons

Alloy name and number	Chemical composition (wt %)	Condition	Microstructure	Tensile strength ksi	Tensile strength MPa	Yield strength ksi	Yield strength MPa	Elongation (%)	Typical applications
Gray cast irons									
Ferritic (G2500)	3.4 C, 2.2 Si, 0.7 Mn	Annealed	Ferritic matrix	26	179	...	...	...	Small cylinder blocks, cylinder heads, clutch plates
Pearlitic (G3500)	3.2 C, 2.0 Si, 0.7 Mn	As-cast	Pearlitic matrix	36	252	...	...	...	Truck and tractor cylinder blocks, heavy gear boxes
Pearlitic (G4000)	3.3 C, 2.2 Si, 0.7 Mn	As-cast	Pearlitic matrix	42	293	...	...	...	Diesel engine castings
Malleable cast irons									
Ferritic (32510)	2.2 C, 1.2 Si, 0.04 Mn	Annealed	Temper carbon and ferrite	50	345	32	224	10	General engineering service with good machinability
Pearlitic (45008)	2.4 C, 1.4 Si, 0.75 Mn	Annealed	Temper carbon and pearlite	65	440	45	310	8	General engineering service with dimensional tolerance specified
Martensitic (M7002)	2.4 C, 1.4 Si, 0.75 Mn	Quenched and tempered	Tempered martensite	90	621	70	438	2	High-strength parts; connecting rods and universal joint yokes
Ductile cast irons									
Ferritic (60-40-18)	3.5 C, 2.2 Si	Annealed	Ferritic	60	414	40	276	18	Pressure castings, such as valve and pump bodies
Pearlitic	3.5 C, 2.2 Si	As-cast	Ferritic-pearlitic	80	552	55	379	6	Crankshafts, gears, and rollers
Martensitic (120-90-02)	3.5 C, 2.2 Si	Martensitic	Quenched and tempered	120	828	90	621	2	Pinions, gears, rollers, and slides

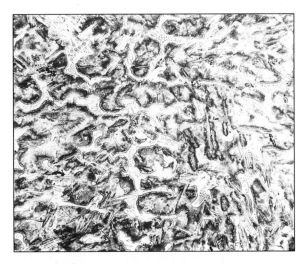

Figure 9.59
Microstructure of white cast iron. The white
constituent is iron carbide. The gray areas are
unresolved pearlite. (Etch: 2% nital; magnification
100×.)
(Courtesy of Central Foundry.)

(Fig. 9.59). White cast irons are so called because they fracture to produce a "white"
or bright crystalline fractured surface. To retain the carbon in the form of iron carbide in white cast irons, their carbon and silicon contents must be kept relatively low
(that is, 2.5–3.0 percent C and 0.5–1.5 percent Si) and the solidification rate must
be high.

White cast irons are most often used for their excellent resistance to wear and abrasion. The large amount of iron carbides in their structure is mainly responsible for their
wear resistance. White cast iron serves as the raw material for malleable cast irons.

9.8.4 Gray Cast Iron

Gray cast iron is formed when the carbon in the alloy exceeds the amount that can
dissolve in the austenite and precipitates as graphite flakes. When a piece of solidified gray iron is fractured, the fracture surface appears gray because of the exposed
graphite.

Gray cast iron is an important engineering material because of its relatively low
cost and useful engineering properties, including excellent machinability at hardness
levels that have good wear resistance, resistance to galling under restricted lubrication, and excellent vibrational damping capacity.

Composition and Microstructure   As listed in Table 9.13, unalloyed gray cast
irons usually contain 2.5 to 4 percent C and 1 to 3 percent Si. Since silicon is a
graphite stabilizing element in cast irons, a relatively high silicon content is used to

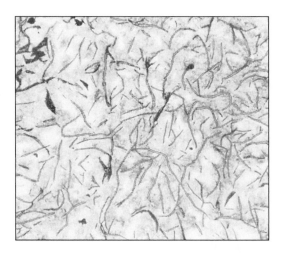

Figure 9.60
Class 30 as-cast gray iron cast in a sand mold. Structure is type A graphite flakes in a matrix of 20% free ferrite (light constituent) and 80% pearlite (dark constituent). (Etch: 3% nital; magnification 100×.)
(After "Metals Handbook," vol. 7, 8th ed., American Society for Metals, 1972, p. 82.)

Figure 9.61
Scanning electron micrograph of hypereutectic gray iron with matrix etched out to show position of type B graphite in space. (Etch: 3:1 methyl acetate–liquid bromine. Magnification 130×.)
(After "Metals Handbook," vol. 7, 8th ed., American Society for Metals, 1972, p. 82.)

promote the formation of graphite. The solidification rate is also an important factor that determines the extent to which graphite forms. Moderate and slow cooling rates favor the formation of graphite. The solidification rate also affects the type of matrix formed in gray cast irons. Moderate cooling rates favor the formation of a pearlitic matrix, whereas slow cooling rates favor a ferritic matrix. To produce a fully ferritic matrix in an unalloyed gray iron, the iron is usually annealed to allow the carbon remaining in the matrix to deposit on the graphite flakes, leaving the matrix completely ferritic.

Figure 9.60 shows the microstructure of an unalloyed as-cast gray iron with graphite flakes in a matrix of mixed ferrite and pearlite. Figure 9.61 shows a scanning electron micrograph of a hypereutectic gray iron with the matrix etched out.

9.8.5 Ductile Cast Irons

Ductile cast iron (sometimes called *nodular* or *spherulitic graphite cast iron*) combines the processing advantages of gray cast iron with the engineering advantages of steel. Ductile iron has good fluidity and castability, excellent machinability, and good wear resistance. In addition, ductile cast iron has a number of properties similar to those of steel such as high strength, toughness, ductility, hot workability, and hardenability.

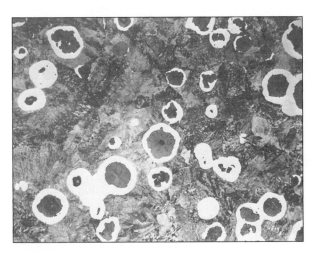

Figure 9.62
Grade 80-55-06 as-cast pearlitic ductile iron.
Graphite nodules (spherulites) in envelopes of free
ferrite (bull's-eye structure) in a matrix of pearlite.
(Etch: 3% nital; magnification 100×.)
(After "Metals Handbook," vol. 7, 8th ed., American Society for Metals, 1972, p. 88.)

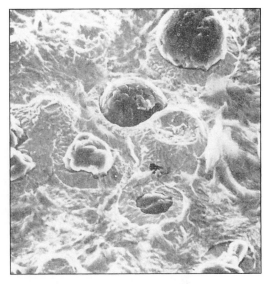

Figure 9.63
Scanning electron micrograph of as-cast
pearlitic ductile iron with matrix etched away
to show secondary graphite, and bull's-eye
ferrite around primary graphite nodules.
(Etch: 3:1 methyl acetate–liquid bromine;
magnification 130×.)
(After "Metals Handbook," vol. 7, 8th ed., American Society for Metals, 1972, p. 88.)

Composition and Microstructure   The exceptional engineering properties of ductile iron are due to the spherical nodules of graphite in its internal structure, as shown in the microstructures of Figs. 9.62 and 9.63. The relatively ductile matrix regions between the nodules allow significant amounts of deformation to take place without fracture.

The composition of unalloyed ductile iron is similar to that of gray iron with respect to carbon and silicon contents. As listed in Table 9.13, the carbon content of unalloyed ductile iron ranges from 3.0 to 4.0 percent C and the silicon content from 1.8 to 2.8 percent. The sulfur and phosphorus levels of high-quality ductile iron, however, must be kept very low at 0.03 percent S maximum and 0.1 percent P maximum, which are about 10 times lower than the maximum levels for gray cast iron. Other impurity elements also must be kept low because they interfere with the formation of graphite nodules in ductile cast iron.

The spherical nodules in ductile cast iron are formed during the solidification of the molten iron because the sulfur and oxygen levels in the iron are reduced to very low levels by adding magnesium to the metal just before it is cast. The magnesium

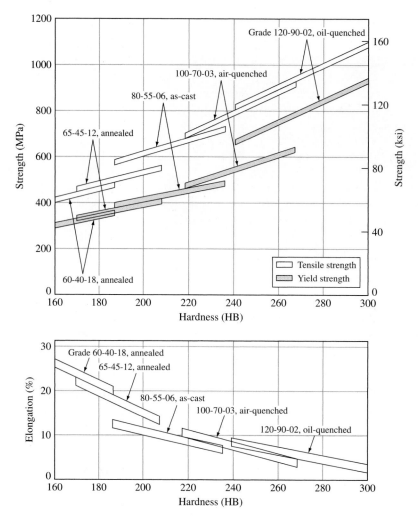

Figure 9.64
Tensile properties of ductile iron versus hardness.
(After "Metals Handbook," vol. 1, 9th ed., American Society for Metals, 1978, p. 36.)

reacts with sulfur and oxygen so that these elements cannot interfere with the formation of the sphere-like nodules.

The microstructure of unalloyed ductile cast iron is usually of the bull's-eye type shown in Fig. 9.62. This structure consists of "spherical" graphite nodules with envelopes of free ferrite around them in a matrix of pearlite. Other as-cast structures with all-ferrite and all-pearlite matrixes can be produced by alloying additions. Subsequent heat treatments can also be used to change the as-cast bull's-eye structure and hence the mechanical properties of as-cast ductile cast iron, as indicated in Fig. 9.64.

9.8.6 Malleable Cast Irons

Composition and Microstructure   Malleable cast irons are first cast as white cast irons that contain large amounts of iron carbides and no graphite. The chemical compositions of malleable cast irons are therefore restricted to compositions that form white cast irons. As listed in Table 9.13, the carbon and silicon contents of malleable irons are in the 2.0 to 2.6 percent C and 1.1 to 1.6 percent Si ranges.

To produce a malleable iron structure, cold white iron castings are heated in a malleablizing furnace to dissociate the iron carbide of the white iron to graphite and iron. The graphite in the malleable cast iron is in the form of irregular nodular aggregates called *temper carbon*. Figure 9.65 is a microstructure of a ferritic malleable cast iron that shows temper carbon in a matrix of ferrite.

Malleable cast irons are important engineering materials since they have the desirable properties of castability, machinability, moderate strength, toughness, corrosion resistance for certain applications, and uniformity since all castings are heat-treated.

Heat Treatment   The heat treatment of white irons to produce malleable irons consists of two stages:

1.   *Graphitization.* In this stage the white iron castings are heated above the eutectoid temperature, usually about 940°C (1720°F), and held for about 3 to 20 h depending on

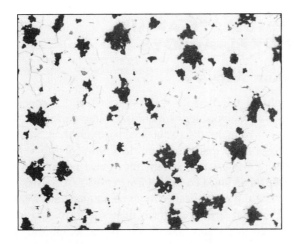

Figure 9.65
Microstructure of ferritic malleable cast iron (grade M3210), two-stage annealed by holding 4 h at 954°C (1750°F), cooling to 704°C (1300°F) in 6 h, and air cooling. Graphite (temper carbon) nodules in a matrix of granular ferrite. (Etch: 2% nital; magnification 100×.)
(After "Metals Handbook," vol. 7, 8th ed., American Society for Metals, 1972, p. 95.)

the composition, structure, and size of the casting. In this stage the iron carbide of the white iron is transformed to temper carbon (graphite) and austenite.

2. *Cooling.* In this stage the austenite of the iron can be transformed to three basic types of matrixes: ferrite, pearlite, or martensite.

Ferritic malleable iron. To produce a ferrite matrix, the casting, after the first-stage heating, is fast-cooled to 740 to 760°C (1360 to 1400°F) and then slowly cooled at a rate of about 3 to 11°C (5 to 20°F) per hour. During cooling the austenite is transformed to ferrite and graphite, with the graphite depositing on existing particles of temper carbon (Fig. 9.65).

Pearlitic malleable iron. To produce this iron, the castings are slowly cooled to about 870°C (1600°F) and are air-cooled. The rapid cooling in this case transforms the austenite to pearlite; as a result pearlitic malleable iron is formed, which consists of temper carbon nodules in a pearlite matrix.

Tempered martensitic malleable iron. This type of malleable iron is produced by cooling the castings in the furnace to a quenching temperature of 845 to 870°C (1550 to 1600°F), holding for 15 to 30 min to allow them to homogenize, and quenching in agitated oil to develop a martensitic matrix. Finally, the castings are tempered at a temperature between 590 and 725°C (1100 to 1340°F) to develop the specified mechanical properties. The final microstructure is thus temper carbon nodules in a tempered martensitic matrix.

9.9 MAGNESIUM, TITANIUM, AND NICKEL ALLOYS

9.9.1 Magnesium Alloys

Magnesium is a light metal (density = 1.74 g/cm^3) and competes with aluminum (density = 2.70 g/cm^3) for applications requiring a low-density metal. However, magnesium and its alloys have many disadvantages that limit their widespread usage. First of all, magnesium costs more than aluminum ($3.29/lb for Mg versus $0.67/lb for Al in 2001, see Table 9.1). Magnesium is difficult to cast because in the molten state it burns in air, and cover fluxes must be used during casting. Also, magnesium alloys have relatively low strength and poor resistance to creep, fatigue, and wear. In addition, magnesium has the HCP crystal structure, which makes deformation at room temperature difficult since only three major slip systems are available. On the other hand, because of their very low density, magnesium alloys are used advantageously, for example, for aerospace applications and materials-handling equipment. Table 9.15 compares some of the physical properties and costs of magnesium with some other engineering metals.

Classification of Magnesium Alloys   There are two major types of magnesium alloys: *wrought alloys,* mainly in the form of sheet, plate, extrusions, and forgings, and *casting alloys.* Both types have non–heat-treatable and heat-treatable grades.

Magnesium alloys are usually designated by two capital letters followed by two or three numbers. The letters stand for the two major alloying elements in the alloy,

Table 9.15 Some physical properties and costs of some engineering metals

Metal	Density at 20°C (g/cm^3)	Melting point (°C)	Crystal structure	Cost ($/lb)(1989)
Magnesium	1.74	651	HCP	1.63
Aluminum	2.70	660	FCC	0.96
Titanium	4.54	1675	HCP ⇌ BCC*	5.25–5.50‡
Nickel	8.90	1453	FCC	7
Iron	7.87	1535	BCC ⇌ FCC†	0.20
Copper	8.96	1083	FCC	1.45

*Transformation occurs at 883°C.
†Transformation occurs at 910°C.
‡Titanium sponge. Price is for about 50 tons.

with the first letter indicating the one in highest concentration and the second letter indicating the one in second highest concentration. The first number following the letters stands for the weight percent of the first letter element (if there are only two numbers), and the second number stands for the weight percent of the second letter element. If a letter A, B, etc., follows the numbers, it indicates that there has been an A, B, etc., modification to the alloy. The following letter symbols are used for magnesium alloying elements:

A = aluminum	K = zirconium	M = manganese
E = rare earths	Q = silver	S = silicon
H = thorium	Z = zinc	T = tin

The temper designations for magnesium alloys are the same as those for aluminum alloys and are listed in Sec. 9.5.

Explain the meaning of the magnesium alloy designations (*a*) HK31A-H24 and (*b*) ZH62A-T5.

EXAMPLE PROBLEM 9.7

■ **Solution**

a. The designation HK31A-H24 means that the magnesium alloy contains a nominal 3 wt % thorium and 1 wt % zirconium and that the alloy is the A modification. The H24 designation means that the alloy was cold-rolled and partially annealed back to the half-hard temper.

b. The designation ZH62A-T5 means that the magnesium alloy contains a nominal 6 wt % zinc and 2 wt % thorium and is the A modification. The T5 signifies that the alloy was artificially aged after casting.

Structure and Properties Magnesium has the HCP crystal structure, and thus the cold working of magnesium alloys can only be carried out to a limited extent. At elevated temperatures for magnesium, some slip planes other than the basal ones become active. Thus magnesium alloys are usually hot-worked instead of being cold-worked.

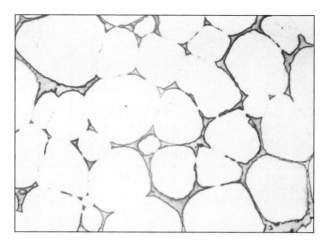

Figure 9.66
Microstructure of magnesium alloy EZ33A in the as-cast condition showing massive Mg_9R (rare-earth) compound grain-boundary network. (Etch: glycol; magnification 500×.)
(Courtesy of the Dow Chemical Co.)

Aluminum and zinc are commonly alloyed with magnesium to form wrought magnesium alloys. Aluminum and zinc both increase the strength of magnesium by solid-solution strengthening. Aluminum also combines with magnesium to form the precipitate $Mg_{17}Al_{12}$, which can be used to age-harden Mg-Al alloys. Thorium and zirconium also form precipitates in magnesium and are used to make alloys that can be used at elevated temperatures up to about 427°C (800°F).

Magnesium casting alloys are made with aluminum and zinc because these elements contribute to solid-solution strengthening. Alloying magnesium with rare-earth metals, mainly cerium, produces a rigid grain-boundary network, as shown in Fig. 9.66. Table 9.16 summarizes the mechanical properties and typical applications of some magnesium wrought and casting alloys.

9.9.2 Titanium Alloys

Titanium is a relatively light metal (density = 4.54 g/cm³) but has high strength (96 ksi for 99.0% Ti), and so titanium and its alloys can compete favorably with aluminum alloys for some aerospace applications even though titanium costs much more ($3.85/lb for Ti[8] versus $0.67/lb for Al in 2001, see Table 9.1). Titanium is also used for applications where it has superior corrosion resistance to many chemical environments such as solutions of chlorine and inorganic chloride solutions.

---

[8]Titanium sponge in quantities of about 50 tons.

Table 9.16 Typical mechanical properties and applications of some magnesium, titanium, and nickel alloys

Alloy name and number	Chemical composition (wt %)	Condition*	Tensile strength ksi	Tensile strength MPa	Yield strength ksi	Yield strength MPa	Elongation (%)	Typical applications
Wrought magnesium alloys								
AZ31B	3 Al, 1 Zn, 0.2 Mn	Annealed	32	228	...	...	11	Airborne cargo equipment; shelves and racks
		H24	36	248	23	159	7	Sheet and plate missile and aircraft uses up to 800°F (427°C)
HM21A	2 Th, 0.8 Mn	T8	32	228	20	138	6	
ZK60	6 Zn, 0.5 Zr	T5	45	310	34	235	5	Highly stressed aerospace uses; extrusions and forgings
Magnesium casting alloys								
AZ63A	6 Zn, 3 Al, 0.15 Mn	As-cast	26	179	11	76	4	Sand castings requiring good room-temperature strength
		T6	34	235	16	110	3	
EZ33A	3 RE, 3 Zn, 0.7 Zr	T5	20	138	14	97	2	Pressure-tight sand and permanent-mold castings used at 350–500°F (150–260°C)
Titanium alloys								
	99.0% Ti (α structure)	Annealed	96	662	85	586	20	Chemical and marine uses; airframe and aircraft engine parts
	Ti–5 Al–2 Sn (α structure)	Annealed	125	862	117	807	16	Weldable alloy for forging and sheet metal parts
Nickel alloys								
Nickel 200	99.5 Ni	Annealed	70	483	22	152	48	Chemical and food processing; electronic parts
Monel 400	66 Ni, 32 Cu	Annealed	80	552	38	262	45	Chemical and oil processing; marine service
Monel K500	66 Ni, 30 Cu, 2.7 Al, 0.6 Ti	Age-hardened	150	1035	110	759	25	Valves, pumps, springs, oil-well drill collars

*H24 = cold-worked and partially annealed half-hard; T5 = age-hardened; T6 = solution heat-treated and aged; T8 = solution heat-treated, cold-worked, and age-hardened.

Titanium metal is expensive because it is difficult to extract in the pure state from its compounds. At high temperatures titanium combines with oxygen, nitrogen, hydrogen, carbon, and iron, and so special techniques must be used to cast and work the metal.

Titanium has the HCP crystal structure (alpha) at room temperature, which transforms to the BCC (beta) structure at 883°C. Elements such as aluminum and oxygen stabilize the α phase and increase the temperature at which the α phase transforms to the β phase. Other elements such as vanadium and molybdenum stabilize the beta phase and lower the temperature at which the β phase is stable. Still other elements such as chromium and iron reduce the transformation temperature at which the β phase is stable by causing a eutectoid reaction that produces a two-phase structure at room temperature.

Table 9.16 lists typical mechanical properties and applications for commercially pure titanium (99.0 percent Ti) and several titanium alloys. The Ti–6 Al–4 V alloy is the most extensively used titanium alloy since it combines high strength with workability. The tensile strength of this alloy in the solution heat-treated and aged condition reaches 170 ksi (1173 MPa).

9.9.3 Nickel Alloys

Nickel is an important engineering metal mainly because of its exceptional resistance to corrosion and high-temperature oxidation. Nickel also has the FCC crystal structure, which makes it highly formable, but is relatively expensive ($7/lb in 1989) and has a high density (8.9 g/cm^3), which limit its use.

Commercial Nickel and Monel Alloys Commercially pure nickel because of its good strength and electrical conductivity is used for electrical and electronics parts and, because of its good corrosion resistance, for food-processing equipment. Nickel and copper are completely soluble in each other in the solid state at all compositions, and so many solid-solution-strengthened alloys are made with nickel and copper. Nickel is alloyed with about 32 percent copper to produce the Monel 400 alloy (Table 9.16), which has relatively high strength, weldability, and excellent corrosion resistance to many environments. The 32 percent copper strengthens the nickel to a limited extent and lowers its cost. The addition of about 3 percent aluminum and 0.6 percent titanium increases the strength of Monel (66% Ni–30% Cu) significantly by precipitation hardening. The strengthening precipitates in this case are Ni$_3$Al and Ni$_3$Ti.

Nickel-Base Superalloys A whole spectrum of nickel-base superalloys have been developed primarily for gas turbine parts that must be able to withstand high temperatures and high oxidizing conditions and be creep-resistant. Most wrought nickel-base superalloys consist of about 50 to 60 percent nickel, 15 to 20 percent chromium, and 15 to 20 percent cobalt. Small amounts of aluminum (1 to 4 percent) and titanium (2 to 4 percent) are added for precipitation strengthening. The nickel-base superalloys consist essentially of three main phases: (1) a matrix of gamma austenite,

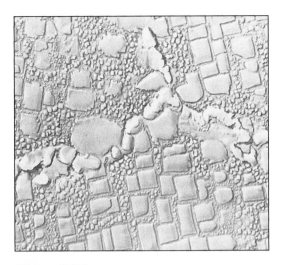

Figure 9.67
Astrology forging, solution heat-treated 4 h at 1150°C, air-cooled, aged at 1079°C for 4 h, oil-quenched, aged at 843°C for 4 h, air-cooled, aged at 760°C for 16 h, air-cooled. Intergranular gamma prime precipitated at 1079°C, fine gamma prime at 843 and 760°C. Carbide particles are also at grain boundaries. Matrix is gamma. (Electrolytic: H_2SO_4, H_3PO_4, HNO_3; magnification 10,000×.)
(After "Metals Handbook," vol. 7, 8th ed., American Society for Metals, 1972, p. 171.)

(2) a precipitate phase of Ni_3Al and Ni_3Ti called *gamma prime,* and (3) carbide particles (due to the addition of about 0.01 to 0.04 percent C). The gamma prime provides high-temperature strength and stability to these alloys, and the carbides stabilize the grain boundaries at high temperatures. Figure 9.67 shows the microstructure of one of the nickel-base superalloys after heat treatment. In this micrograph the gamma prime and carbide particles are clearly visible.

9.10 SPECIAL-PURPOSE ALLOYS AND APPLICATIONS

9.10.1 Intermetallics

Intermetallics (see Sec. 8.10) constitute a class of metallic materials that possess unique combinations of properties attractive to many industries. Examples of high-temperature structural intermetallics that have attracted a great deal attention in aircraft and jet engine applications are nickel aluminides (Ni_3Al and $NiAl$), iron aluminides (Fe_3Al and $FeAl$), and titanium aluminides (Ti_3Al and $TiAl$). These intermetallics contain aluminum: Aluminum can form a thin passive layer of alumina (Al_2O_3) in an oxidizing environment that serves to protect the alloy from corrosion damage. The densities of these intermetallics are low compared to other high-temperature alloys such as nickel superalloys and therefore are more suitable for aerospace applications. These alloys also have relatively high melting points and good high-temperature strength. The factor that has limited the application of these metals is their brittle nature at ambient temperatures. Some aluminides, for example, Fe_3Al, also exhibit environmental embrittlement at ambient temperatures. The embrittlement is due to the reaction of water vapor in the environment with elements such as aluminum to form atomic hydrogen, which diffuses into the metal and causes reduced ductility and premature fracture.

The aluminide Ni_3Al is of special interest because of its strength and corrosion resistance at high temperatures. This aluminide has also been used as a finely dispersed constituent in nickel-based superalloys to increase the alloys strength. The addition of about 0.1 wt % boron to Ni_3Al (with less than 25 at. % Al) has shown to eliminate the brittleness of the alloy; in fact the addition improves its ductility by as much as 50 percent by reducing hydrogen embrittlement. These alloys also exhibit the interesting and useful anomaly of increased yield strength with increasing temperature. In addition to boron, 6 to 9 at. % Cr is added to the alloy to reduce environmental embrittlement at elevated temperatures; Zr is added to improve the strength through solid solution hardening; and Fe is added to improve weldability. Addition of each impurity adds to the complexity of the diagram and the subsequent phase analysis. In addition to aircraft engine applications, this intermetallic is used to make furnace parts, aircraft fasteners, pistons and valves, and toolings. However, the application of intermetallics is not limited to structural uses. For instance, Fe_3Si has been developed for magnetic applications because of its superior magnetic properties and its resistance to wear; $MoSi_2$ has been used for electrical-heating elements in high-temperature furnaces because of its high electrical and thermal conductivity; and $NiTi$ (nitinol) is used as a shape-memory alloy in medical applications.

9.10.2 Shape-Memory Alloys

General Properties and Characteristics   **Shape-memory alloys** (SMA) have the ability to recover a previously defined shape when subjected to an appropriate heat-treatment procedure. In reverting back to their original shapes, they can also apply forces. There are a number of metallic alloys that exhibit such behavior, including Au-Cd, Cu-Zn-Al, Cu-Al-Ni, and Ni-Ti alloys. The most practical applications are for those SMAs that have the ability to recover a significant amount of strain (super-elasticity) or those that can apply large forces when reverting back to their original shapes.

Production of SMAs and Mechanics of Behavior   The SMA alloy may be processed using both hot- and cold-forming techniques such as forging, rolling,

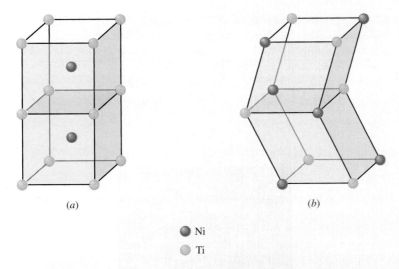

Ni

Ti

Figure 9.68
The structures in Ni-Ti alloy. (*a*) austenitic and (*b*) martensitic.

extrusion and wire drawing to produce ribbons, tubes, wires, sheets, or springs. To impart the desired shape memory, the alloy is heat treated at a temperature range of 500 to 800°C. During the heat treatment process, the SMA is restrained to the desired shape. At this temperature the material has a well-ordered cubic structure called *austenite* (*parent phase*) (Fig. 9.68*a*). Once the material is cooled, its structure changes to a highly twinned or alternatingly sheared platelets structure called *martensite* (Fig. 9.68*b*). The alternatingly sheared structure, i.e., consecutively opposite shears, maintains the overall shape of the crystal, as shown in Fig. 9.69 (actual microstructure is shown in Fig. 9.69*c*).

The shape-recovery effect in SMAs is a result of the solid-solid phase transformation between two material structures, i.e., austenite and martensite. In the martensitic state, the SMA is very easy to deform by the application of stress because of the propagation of the twin boundary (Fig. 9.70). If at this stage the load is removed, the deformation in the martensite remains, giving the appearance of a plastic deformation. However, after deformation in the martensitic state, heating will cause a martensite to austenite transformation with the component reverting back to its original shape (Fig. 9.71). The change in structure does not occur at a discrete temperature but over a range of temperatures that depends on the alloy system as schematically shown in Fig. 9.72. Upon cooling, the transformation starts at M_s (100 percent austenite) and ends at M_f (0 percent martensite) while upon heating the transformation starts at A_s (100 percent martensite) and ends at A_f (0 percent martensite). Additionally, the transformations during cooling and heating do not overlap, i.e., the system exhibits *hysteresis* (Fig. 9.72).

When the SMA is at a temperature above A_f (100 percent austenite), applied stress may deform and transform the SMA to a martensitic state. If at this point the

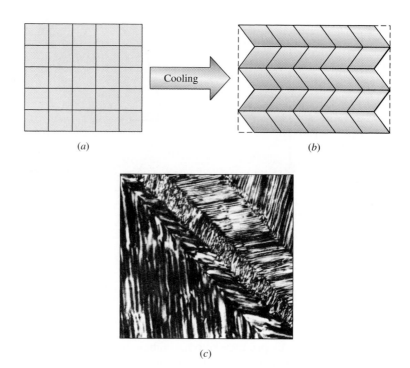

Figure 9.69
The austenite to martensite transformation upon cooling maintaining the overall shape of the crystal. (*a*) Austenite crystal, (*b*) highly twinned martensite, and (*c*) martensite showing alternatingly sheared structure.
(*After Metals Handbook, 2nd ed., ASM International, 1998.*)

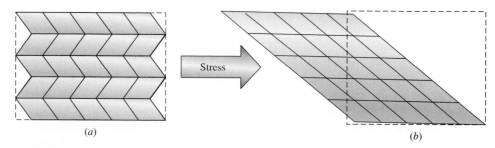

Figure 9.70
The deformation of the martensite structure due to applied stress. (*a*) Twinned martensite and (*b*) deformed martensite.

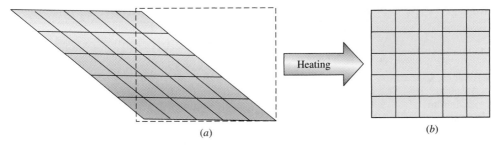

Figure 9.71
The deformed martensite to austenite transformation upon heating. (*a*) Deformed martensite and (*b*) austenite.

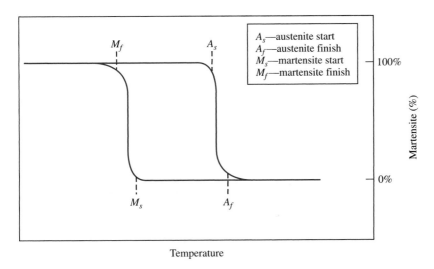

Figure 9.72
A typical transformation-temperature diagram for a stressed specimen as it is heated and cooled.

load is removed, the martensitic phase becomes thermodynamically unstable (due to the high temperature) and recovers its original structure and shape in an elastic manner. This is the basis for the superelastic behavior of SMAs. The transformation, since it was achieved at a constant temperature and under load, is called *stress-induced.*

Applications of SMAs   The intermetallic Ni-Ti (nitonol) is one of the most commonly used SMA with a composition range of Ni–49 at. % Ti to Ni–51 at. % Ti. Nitonol has a shape-memory strain of around 8.5 percent, is nonmagnetic, has excellent corrosion resistance, and has higher ductility than other SMAs (see Table 9.17). The applications for nitonol include actuation devices in which the material (1) has

Table 9.17 Some properties of nitonol

Properties	Property value
Melting temperature, °C (°F)	1300 (2370)
Density, g/cm^3 (lb/in.^{3})	6.45 (0.233)
Resistivity, $\mu\Omega \cdot$ cm	
Austenite	~100
Martensite	~70
Thermal conductivity, W/m $\cdot$ °C (Btu/ft $\cdot$ h $\cdot$ °F)	
Austenite	18 (10)
Martensite	8.5 (4.9)
Corrosion resistance	Similar to 300 series stainless steel or titanium alloys
Young's modulus, GPa (10^6 psi)	
Austenite	~83 (~12)
Martensite	~28–41 (~4–6)
Yield strength, MPa (ksi)	
Austenite	195–690 (28–100)
Martensite	70–140 (10–20)
Ultimate tensile strength, MPa (ksi)	895 (130)
Transformation temperatures, °C (°F)	−200 to 110 (−325 to 230)
Hysteresis, Δ°C (Δ°F)	~30 (~55)
Latent heat of transformation, kJ/kg $\cdot$ atom (cal/g $\cdot$ atom)	167 (40)
Shape-memory strain	8.5% maximum

Source: Metals Handbook, 2nd ed., ASM International, 1998.

the ability to freely revert back to its original shape, (2) is fully constrained so that upon shape recovery, it exerts a large force on the constraining structure, or (3) is partially constrained by the surrounding deformable material in which case the SMA performs work. Practical examples of actuation devices are vascular stents, coffeepot thermostats, and hydraulic pipe couplings. Other components such as eyeglass frames and arch wires in orthodontics are examples in which the superelasticity of the material is the desired property. Additionally, the martensitic phase has excellent energy-absorption and fatigue-resistance capabilities because of its twinned structure. Thus, the martensitic phase is used as a vibration dampener and in flexible surgical tools for open-heart surgery. Clearly, in the selection of these materials for specific applications, one has to be aware of the operating temperatures in comparison with the transformation temperatures.

9.10.3 Amorphous Metals

General Properties and Characteristics Generally speaking, the terms *amorphous* and *metal* seem to be contradictory in nature. In the previous chapters, when the concepts of crystal structure and solidification of metals were introduced, it was often stated that metals have a high affinity to form crystal structures with long-range order. However, as discussed briefly in Chap. 3, under certain conditions, even metals can form noncrystalline, highly disordered, amorphous, or glassy structures (thus, the term *metallic glass*) in which atoms are arranged in a random manner. Figure 9.73*a*

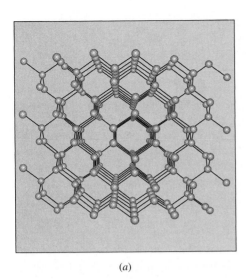
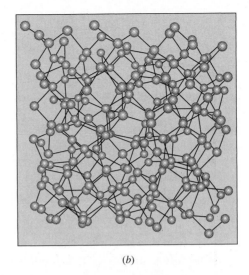

(a)   (b)

Figure 9.73
Comparison of atomic order in (a) crystalline Zr-based alloy and (b) glassy Zr-based alloy.
(*http://ceramicmaterials.rutgers.edv/firstyearintro/atomictramp.pdf*)

shows the schematic of a crystalline solid (note the ordered and parallel features) while Fig. 9.73b shows the amorphous or glassy atomic structure. One can easily note the amorphous nature of the glassy alloy when compared to the crystalline one.

Production of Amorphous Metals and Mechanics of Behavior   The concept of an amorphous metal is not new, and its study dates back to the 1960s. Amorphous metals were first developed by applying molten metal to the surface of a rapidly moving and cooled surface. This results in the rapid quenching of the metal at a rate of 10^5 K/s. The rapid quenching process allows a very short time for the molten metal to solidify. In this short period of time there is no opportunity for the diffusion of atoms and the formation of crystals; as a result, a solid with a glassy state is formed. Achieving such a rapid quenching rate is not easy. Because of the poor thermal conductivity of metallic glasses, only foils, wires, or powder forms (products with at least one very small dimension) of these materials could be developed until recently.

Because of the random arrangement of atoms in amorphous metals, dislocation activities are minimal, and as a result metals that are formed in this manner are very hard. Additionally, these metals do not strain harden but behave in an elastic, perfectly plastic manner (the plastic portion of the stress-strain curve is flat). The plastic deformation in metallic glasses is highly inhomogeneous and localizes into bands of intense shear. Thus, the structure, mechanisms of deformation, and properties of such metals are completely different than crystalline metals.

Applications of Glassy Metals   Recently, it has become possible to manufacture centimeter-scale or bulk-forms of amorphous alloys at significantly lower cooling rates, some of which are now commercially available. New discoveries have shown that if metals of considerably different atomic radii such as Ti, Cu, Zr, Be, or Ni are mixed to form an alloy, crystallization is hindered and the resulting solid has an amorphous structure. As such alloys solidify, they do not shrink significantly and high dimensional precision can be achieved. As a result, unlike conventional metals, sharp metal surfaces such as those found in knives and surgical tools could be produced without any extra sharpening processes or finishing operations. One of the main drawbacks of metallic glass is that it is metastable, i.e., if the temperature is increased to a critical level, the metal reverts to a crystalline state and regains its standard characteristics.

One commercial example of metallic glass is the cast alloy Vit-001 (Zr-based) that has a high modulus of elasticity, high yield strength (1900 MPa), and is corrosion resistant. Its density is higher than that of aluminum and titanium but lower than steel. It can undergo a recoverable strain of about 2 percent, significantly higher than conventional metals. Because of this high elastic strain limit and hardness, the early applications of glassy metals have been in sporting equipment industries such as golf. Harder and more elastic club head inserts made of glassy metal result in longer drives. With better processing techniques to produce bulk amorphous metals, the number of new applications will also grow.

9.11 METALS IN BIOMEDICAL APPLICATIONS—BIOMETALS

Metals are used extensively in numerous biomedical applications. Some applications are specific to the replacement of damaged or dysfunctional tissues in order to restore function, for instance, orthopedic applications in which all or part of a bone or a joint are replaced or reinforced by metal alloys. In dental applications, metals are used as the filler material for decayed teeth, screws for dental implant support, and dental material replacement. Metal alloys that replace damaged biological tissues, restore function, and are constantly or intermittently in contact with body fluids are collectively called *biomaterials* or in this case, because we are focusing on metals, *biometals*. Clearly, metals used in medical, dental, and surgical instruments in addition to metals used in external prosthesis are not classified as biomaterials since they are not exposed to body fluids in a continuous or intermittent manner. In this section we will discuss the most often used biometals in structurally important applications such as implants and fixation devices for various joints (such as the hip, knee, shoulder, ankle, or wrist) and bones in the body.

Biometals have specific characteristics that make them suitable for application in the human body. The internal environment of the body is highly corrosive and could degrade the implant material (orthopedic or dental), resulting in the release of harmful ions or molecules. Thus, the primary characteristic of a biometal is *biocompatibility,* which is defined as chemical stability, corrosion resistance, noncarcinogenic, and nontoxic when used in the human body. Once the biocompatibility of the metal is established, the second important characteristic of being able to endure large and variable (cyclic) stresses in the highly corrosive environment of the human body must be met. The importance of the load-bearing capacity of the metal may be appreciated when one realizes that the average person may experience 1 to 2.5 million cycles of stress on his or her hip per year (due to normal daily activities). This translates into 50 to 100 million cycles of stress over a 50-year period. Therefore, the biomaterial must be strong and fatigue- and wear-resistant in a highly corrosive environment. What metals satisfy these conditions?

Pure metals such as Co, Cu, and Ni are considered toxic in the human body. On the other hand, pure metals such as Pt, Ti, and Zr have high levels of biocompatibility. Metals such as Fe, Al, Au, and Ag have moderate biocompatibility. Some stainless steel and Co-Cr alloys also have moderate compatibility. In practice, the most often used metals for load-bearing applications in the human body are stainless steels (Sec. 9.7), cobalt-base alloys, and titanium alloys (Sec. 9.9.2). These metals have acceptable biocompatibility and load-bearing characteristics; however, none has all the required characteristics for the specific application.

9.11.1 Stainless Steels

Various classes of stainless steels were discussed in Sec. 9.7 including ferritic, martensitic, and austenitic steels. In orthopedic applications the austenitic stainless steel 316L (18 Cr, 14 Ni, 2.5 Mo-F138) is used most often. This metal is popular because it is relatively inexpensive, and can be shaped easily with existing metal-forming techniques. The suitable ASTM grain size is 5 or finer. The metal is often used in the 30 percent cold-worked state for enhanced yield, ultimate strength, and fatigue strength when compared to the annealed state. The main drawback is that this metal is not suitable for long-term use because of its limited corrosion resistance in the human body. As a result, the most effective applications are in bone screws, plates (Fig. 9.74), intramedullary bone nails, pins, and other temporary fixation devices. Figure 9.75 shows an example of a fracture in which a bone plate and numerous screws are used for stabilization. These components can be removed after sufficient healing occurs.

9.11.2 Cobalt-Based Alloys

There are four main types of cobalt-based alloys used in orthopedic implants: (1) Co-28 Cr-6 Mo cast alloy (ASTM F 75), (2) Co-20 Cr-15 W-10 Ni wrought alloy (ASTM F 90), (3) Co-28 Cr-6 Mo cast alloy, heat-treated (ASTM F 799), and (4) Co-35 Ni-20 Cr-10 Mo wrought alloy (ASTM F 562). As with stainless steels, the high percentage of Cr in these alloys promotes corrosion resistance through the formation of a passive layer. It should be noted that the long-term corrosion resistance of these alloys is far superior to that of stainless steel, making them significantly more biocompatible. The F75 alloy is a casting alloy that produces a coarse grain size and also has a tendency to create a cored microstructure (a nonequilibrium structure discussed in Chap. 5). Both of these characteristics are

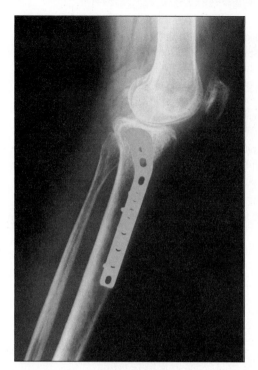

Figure 9.74
Bone plates for long bones.
(© *Science Photo Library/Photo Researchers, Inc.*)

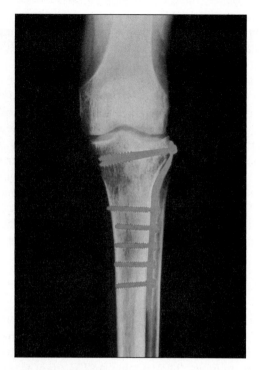

Figure 9.75
Reduction of a fibular fracture with a
compression bone plate and screws.
(© *Science Photo Library/Photo Researchers, Inc.*)

undesirable for orthopedic applications since they result in a weakened component. The F799 alloy is similar in composition to the F75 alloy, but it is forged to the final shape in a series of steps. The early forging stages are performed hot to allow for significant flow, and the final stages are performed cold to allow for strengthening. This improves the strength characteristics of the alloy when compared to F75. The F90 alloy contains a significant level of Ni and W to improve machinability and fabrication characteristics. In the annealed state its properties match those of F75, but with 44 percent cold working, its yield, ultimate strength, and fatigue strength almost double that of F75. However, care must be taken to achieve uniform properties throughout the component or it will be prone to unexpected failures. Finally, the F562 alloy has by far the most effective combination of strength, ductility, and corrosion resistance. It is both cold-worked and age-hardened to a yield strength exceeding 1795 MPa while maintaining a ductility of about 8 percent. Because of their combination of long-term corrosion resistance and strength, these alloys are often used as permanent fixation devices and joint replacement components (Fig. 9.76).

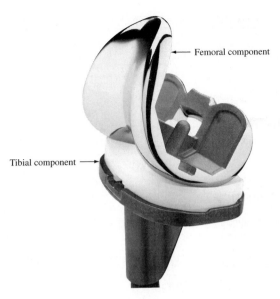

Figure 9.76
A cobalt-chromium knee replacement
prosthesis. Note the femoral component resting
on the tibial component.
(Courtesy of Zimmer, Inc.)

9.11.3 Titanium Alloys

Titanium alloys including commercially pure, alpha, beta, and alpha-beta were briefly described in Sec. 9.92. Each alloy has mechanical and forming characteristics that are attractive for different applications. What is true of these alloys is their outstanding corrosion resistance even in some aggressive environments such as the human body. The corrosion resistance of these alloys is superior to both stainless steel and cobalt-chromium alloys. Their corrosion resistance stems from their ability to form a protective oxide layer TiO_2 below 535°C (see Chap. 13). From an orthopedic point of view titanium's excellent biocompatibility, high corrosion resistance, and low modulus of elasticity are highly desirable. The commercially pure titanium (CP–F67) is a relatively low-strength metal and is used in orthopedic applications that do not require high strength such as screws and staples for spinal surgery. Alpha alloys that contain Al (alpha stabilizer), Sn, and/or Zr cannot be significantly strengthened by heat treatment and as a result do not offer significant advantages over CP alloys in orthopedic applications. Alpha-beta alloys contain both alpha (Al) and beta stabilizers (V or Mo). As a result, at room temperate a mixture of alpha and beta phases coexist. Solution treatment can increase the strength of these alloys by 30 to 50 percent when compared to the annealed state. Examples of alpha-beta alloys used in orthopedic application are Ti-6 Al-4 V (F1472), Ti-6 Al-7 Nb, and Ti-5 Al-2.5 Fe.

Table 9.18 Properties of selected metal alloys used in orthopedic applications

Material	ASTM Designation	Condition	Elastic modulus (GPa)	Yield strength (MPa)	Ultimate strength (MPa)	Endurance limit (MPa)
Stainless steels	F55, F56, F138, F139	Annealed	190	331	586	241–276
		30% cold-worked	190	792	930	310–448
		Cold forged	190	1213	1351	820
Cobalt alloys	F75	As-cast, annealed	210	448–517	655–889	207–310
		Hip	253	841	1277	725–950
	F799	Hot forged	210	896–1200	1399–1586	600–896
	F90	Annealed	210	448–648		951–1220
		44% cold-worked	210	1606	1896	586
	F562	Hot forged	232	965–1000	1206	500
		Cold-forged, aged	232	1500	1795	689–793
Titanium alloy	F67	30% cold-worked	110	485	760	300
	F136	Forged, annealed	116	896	965	620
		Forged, heat treated	116	1034	1103	620–689

Source: "Orthopedic Basic Science," American Academy of Orthopedic Science, 1999.

The alloy F1472 is the most common in orthopedic applications such as total joint replacement. The other two alloys are used in femoral hip stems, plates, screws, rods, and nails. Beta alloys (containing mostly beta stabilizers) have excellent forgeability because they do not strain harden. However, they can be solution treated and aged to strength levels that are higher than those of alpha-beta alloys. The beta alloys offer the lowest modulus of elasticity (a medical advantage—see the next section) of all titanium alloys and in fact of all alloys used in orthopedic implant manufacturing. For comparison, the mechanical properties of the popular orthopedic alloys are presented in Table 9.18. The main disadvantages of titanium alloys in orthopedic applications are their poor wear resistance and high notch sensitivity (existence of a scratch or notch reduces their fatigue life). Because of their poor wear resistance, they should not be used in articulating surfaces such as hip and knee joints unless they are surface treated through ion-implantation processes.

9.12 SOME ISSUES IN THE ORTHOPEDIC APPLICATION OF METALS

In orthopedic implants critical properties include high yield strength (to resist plastic deformation under load), fatigue strength (to resist cyclic loads), hardness (to resist wear when articulation is involved), and, interestingly, low moduli of elasticity (to achieve bone-metal load-bearing proportionality). To clearly understand this, consider that prior to a fracture, all the acting forces (muscle, tendon, and bone) are balanced. After a fracture this balance is lost, and an operation is required to attach the fractured component (including any fragments) to orthopedic implants and to stabilize the fracture. If the fracture is perfectly reconstructed, the bone will still carry a significant portion of the load and the implant acts mainly as the structure around which the fractured bone is reconstructed (carries little load). However, in many situations, due to the complexity of the fracture (i.e., some fragments may be missing) or inadequate fixation or stabilization, the implant will not only carry a disproportionate amount of the load, it could also undergo elastic (in some situations permanent) twisting and bending. All of this could give rise to fatigue-related failure of the implant. For these reasons, the yield, tensile, and fatigue strength of biometals are critical and must be most favorable.

The modulus of elasticity of the biometal, however, is a different issue. In many fracture fixation situations the high modulus of elasticity (a measure of resistance to elastic deformation) of the metal implant is a point of concern. To understand this, note that the modulus of elasticity of bone (in the load-bearing direction) is close to 17 GPa (2.5×10^6 psi). In comparison, the modulus of elasticity of titanium, stainless steel, and cobalt-based alloys are 110 (80 for beta alloys), 190, and 240 GPa, respectively. Consider a situation in which the tibial (the tibia is one of the long bones below the knee joint) shaft is broken in a simple transverse manner as shown in Fig. 9.77 (see arrow). To fix and stabilize the fracture, a nail-like metallic device is used with interlocking screws. The interlocking screws are not always necessary but help with stabilization and prevent shortening of the bone or rotation of the

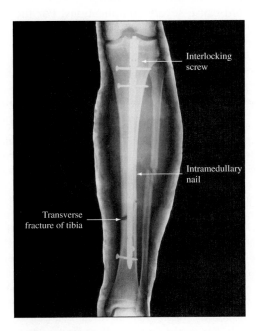

Figure 9.77
Application of intramedullary nail and interlocking screw to stabilize a fracture of the tibia.
(© *Science Photo Library/Photo Researchers, Inc.*)

fracture fragments. Because the metal has a significantly higher modulus of elasticity than the bone, the metal implant carries a disproportionate portion of the load. In other words, the metallic implant will shield the bone from carrying the load that it would support under normal conditions; a phenomenon called *stress shielding*. Although from an engineering point of view this sounds desirable and logical, from a biological point of view it is undesirable. Bone material responds to stress by remodeling (rebuilding) itself to the applied level of stress. Because of the stress shielding, the bone remodels itself to the lower load level and its quality deteriorates. It is for this reason that titanium alloys with the lowest modulus of elasticity among the three major alloys is the most desirable in such applications. This example clearly explains the challenges involved in the selection of materials for orthopedic applications.

Another important issue in orthopedic application of metals, especially in situations where articulation is involved such as the knee and hip joint, is the wear of the articulating surfaces. When a diseased or damaged hip is replaced by a metallic prosthesis, the contacting surfaces undergo wear and tear due to the large number of loading cycles involved. The wear of the surface results in formation of micron-sized metallic wear particles that house themselves in the surrounding tissue and cause loosening of the prosthesis and metal toxicity. Although wear is an eventuality and occurs regardless of the metal or material used, titanium alloys are especially prone to wear and could cause severe complications. For these applications cobalt-chromium alloys are most suitable.

9.13 S U M M A R Y

Engineering alloys can be conveniently subdivided into two types: ferrous and nonferrous. Ferrous alloys have iron as their principal alloying metal, whereas nonferrous alloys have a principal alloying metal other than iron. The steels, which are ferrous alloys, are by far the most important metal alloys, mainly because of their relatively low cost and wide range of mechanical properties. The mechanical properties of carbon steels can be varied considerably by cold working and annealing. When the carbon content of steels is increased to above about 0.3 percent, they can be heat-treated by quenching and tempering to produce high strength with reasonable ductility. Alloying elements such as nickel, chromium, and molybdenum are added to plain-carbon steels to produce low-alloy steels. Low-alloy steels have good combinations of high strength and toughness and are used extensively in the automotive industry for applications such as gears, shafts, and axles.

Aluminum alloys are the most important of the nonferrous alloys mainly because of their lightness, workability, corrosion resistance, and relatively low cost. Unalloyed copper is used extensively because of its high electrical conductivity, corrosion resistance, workability, and relatively low cost. Copper is alloyed with zinc to form a series of brass alloys that have higher strength than unalloyed copper.

Stainless steels are important ferrous alloys because of their high corrosion resistance in oxidizing environments. To make a stainless steel "stainless," it must contain at least 12 percent Cr.

Cast irons are still another industrially important family of ferrous alloys. They are low in cost and have special properties such as good castability, wear resistance, and durability. Gray cast iron has high machinability and vibration-damping capacity due to the graphite flakes in its structure.

Other nonferrous alloys briefly discussed in this chapter are magnesium, titanium, and nickel alloys. Magnesium alloys are exceptionally light and have aerospace applications, and they are also used for materials-handling equipment. Titanium alloys are expensive but have a combination of strength and lightness not available from any other metal alloy system; they are used extensively for aircraft structural parts. Nickel alloys have high corrosion and oxidation resistance and are therefore commonly used in the oil and chemical process industries. Nickel when alloyed with chromium and cobalt forms the basis for the nickel-base superalloys needed in gas turbines for jet aircraft and some electric-power generating equipment.

In this chapter we have discussed to a limited extent the structure, properties, and applications of some of the important engineering alloys. We have also introduced special purpose alloys that are growing in importance and application in various industries. Of particular importance is the use of intermetallics, amorphous metals, and advanced superalloys in the biomedical field. These materials have superior properties to conventional alloys.

9.14 DEFINITIONS

Sec. 9.2

Austenite (γ phase in Fe-Fe$_3$C phase diagram): an interstitial solid solution of carbon in FCC iron; the maximum solid solubility of carbon in austenite is 2.0 percent.

Austenitizing: heating a steel into the austenite temperature range so that its structure becomes austenite. The austenitizing temperature will vary depending on the composition of the steel.

α ferrite (α phase in the Fe-Fe$_3$C phase diagram): an interstitial solid solution of carbon in BCC iron; maximum solid solubility of carbon in BCC iron is 0.02 percent.

Cementite: the intermetallic compound Fe$_3$C; a hard and brittle substance.

Pearlite: a mixture of α ferrite and cementite (Fe$_3$C) phases in parallel plates (lamellar structure) produced by the eutectoid decomposition of austenite.

Eutectoid α ferrite: α ferrite that forms during the eutectoid decomposition of austenite; the α ferrite in pearlite.

Eutectoid cementite (Fe$_3$C): cementite that forms during the eutectoid decomposition of austenite; the cementite in pearlite.

Eutectoid (plain-carbon steel): a steel with 0.8 percent C.

Hypoeutectoid (plain-carbon steel): a steel with less than 0.8 percent C.

Hypereutectoid (plain-carbon steel): a steel with 0.8 to 2.0 percent C.

Proeutectoid α ferrite: α ferrite that forms by the decomposition of austenite at temperatures above the eutectoid temperature.

Proeutectoid cementite (Fe$_3$C): cementite that forms by the decomposition of austenite at temperatures above the eutectoid temperature.

Sec. 9.3

Martensite: a supersaturated interstitial solid solution of carbon in body-centered tetragonal iron.

Bainite: a mixture of α ferrite and very small particles of Fe$_3$C particles produced by the decomposition of austenite; a nonlamellar eutectoid decomposition product of austenite.

Spheroidite: a mixture of particles of cementite (Fe_3C) in an α ferrite matrix.

Isothermal transformation (IT) diagram: a time-temperature-transformation diagram that indicates the time for a phase to decompose into other phases isothermally at different temperatures.

Continuous-cooling transformation (CCT) diagram: a time-temperature-transformation diagram that indicates the time for a phase to decompose into other phases continuously at different rates of cooling.

Martempering (marquenching): a quenching process whereby a steel in the austenitic condition is hot-quenched in a liquid (salt) bath at above the M_s temperature, held for a time interval short enough to prevent the austenite from transforming, and then allowed to cool slowly to room temperature. After this treatment the steel will be in the martensitic condition, but the interrupted quench allows stresses in the steel to be relieved.

Austempering: a quenching process whereby a steel in the austenitic condition is quenched in a hot liquid (salt) bath at a temperature just above the M_s of the steel, held in the bath until the austenite of the steel is fully transformed, and then cooled to room temperature. With this process a plain-carbon eutectoid steel can be produced in the fully bainitic condition.

M_s: the temperature at which the austenite in a steel starts to transform to martensite.

M_f: the temperature at which the austenite in a steel finishes transforming to martensite.

Tempering (of a steel): the process of reheating a quenched steel to increase its toughness and ductility. In this process martensite is transformed into tempered martensite.

Plain-carbon steel: an iron-carbon alloy with 0.02 to 2 percent C. All commercial plain-carbon steels contain about 0.3 to 0.9 percent manganese along with sulfur, phosphorus, and silicon impurities.

Sec. 9.4

Hardenability: the ease of forming martensite in a steel upon quenching from the austenitic condition. A highly hardenable steel is one that will form martensite throughout in thick sections. Hardenability should not be confused with hardness. Hardness is the resistance of a material to penetration. The hardenability of a steel is mainly a function of its composition and grain size.

Jominy hardenability test: a test in which a 1 in. (2.54 cm) diameter bar 4 in. (10.2 cm) long is austenitized and then water-quenched at one end. Hardness is measured along the side of the bar up to about 2.5 in. (6.35 cm) from the quenched end. A plot called the *Jominy hardenability curve* is made by plotting the hardness of the bar against the distance from the quenched end.

Sec. 9.8

White cast irons: iron-carbon-silicon alloys with 1.8 to 3.6 percent C and 0.5 to 1.9 percent Si. White cast irons contain large amounts of iron carbide that make them hard and brittle.

Gray cast irons: iron-carbon-silicon alloys with 2.5 to 4.0 percent C and 1.0 to 3.0 percent Si. Gray cast irons contain large amounts of carbon in the form of graphite flakes. They are easy to machine and have good wear resistance.

Ductile cast irons: iron-carbon-silicon alloys with 3.0 to 4.0 percent C and 1.8 to 2.8 percent Si. Ductile cast irons contain large amounts of carbon in the form of graphite nodules (spheres) instead of flakes as in the case of gray cast iron. The addition of magnesium (about 0.05 percent) before the liquid cast iron is poured enables the nodules to form. Ductile irons are in general more ductile than gray cast irons.

Malleable cast irons: iron-carbon-silicon alloys with 2.0 to 2.6 percent C and 1.1 to 1.6 percent Si. Malleable cast irons are first cast as white cast irons and then are heat-treated at about 940°C (1720°F) and held about 3 to 20 h. The iron carbide in the white iron is decomposed into irregularly shaped nodules or graphite.

Sec. 9.10
Intermetallics: stoichiometric compounds of metallic elements with high hardness and high temp strength, but brittle.
Shape memory alloys: metal alloys that recover a previously defined shape when subjected to an appropriate heat treatment process.
Amorphous metal: metals with a noncrystalline structure also called glassy metal. These alloys have high elastic strain threshold.

9.15 PROBLEMS

Answers to problems marked with an asterisk are given at the end of the book.

9.1   How is raw pig iron extracted from iron oxide ores?

9.2   Write a typical chemical reaction for the reduction of iron oxide (Fe_2O_3) by carbon monoxide to produce iron.

9.3   Describe the basic oxygen process for converting pig iron into steel.

9.4   Why is the $Fe-Fe_3C$ phase diagram a metastable phase diagram instead of a true equilibrium phase diagram?

9.5   Define the following phases that exist in the $Fe-Fe_3C$ phase diagram:
(*a*) austenite, (*b*) α ferrite, (*c*) cementite, and (*d*) δ ferrite.

9.6   Write the reactions for the three invariant reactions that take place in the $Fe-Fe_3C$ phase diagram.

9.7   What is the structure of pearlite?

9.8   Distinguish between the following three types of plain-carbon steels:
(*a*) eutectoid, (*b*) hypoeutectoid, and (*c*) hypereutectoid.

9.9   Describe the structural changes that take place when a plain-carbon eutectoid steel is slowly cooled from the austenitic region just above the eutectoid temperature.

9.10   Describe the structural changes that take place when a 0.4 percent C plain-carbon steel is slowly cooled from the austenitic region just above the upper transformation temperature.

9.11   Distinguish between proeutectoid ferrite and eutectoid ferrite.

*9.12   A 0.65 percent C hypoeutectoid plain-carbon steel is slowly cooled from about 950°C to a temperature just slightly *above* 723°C. Calculate the weight percent austenite and weight percent proeutectoid ferrite in this steel.

9.13   A 0.25 percent C hypoeutectoid plain-carbon steel is slowly cooled from 950°C to a temperature just slightly *below* 723°C.

     a.   Calculate the weight percent proeutectoid ferrite in the steel.

     b.   Calculate the weight percent eutectoid ferrite and eutectoid cementite in the steel.

*9.14   A plain-carbon steel contains 93 wt % ferrite and 7 wt % Fe_3C. What is its average carbon content in weight percent?

9.15 A plain-carbon steel contains 45 wt % proeutectoid ferrite. What is its average carbon content in weight percent?

9.16 A plain-carbon steel contains 5.9 wt % hypoeutectoid ferrite. What is its average carbon content?

*9.17 A 0.90 percent C hypereutectoid plain-carbon steel is slowly cooled from 900°C to a temperature just slightly *above* 723°C. Calculate the weight percent proeutectoid cementite and austenite present in the steel.

9.18 A 1.10 percent C hypereutectoid plain-carbon steel is slowly cooled from 900°C to a temperature just slightly *below* 723°C.

  a. Calculate the weight percent proeutectoid cementite present in the steel.

  b. Calculate the weight percent eutectoid cementite and the weight percent eutectoid ferrite present in the steel.

*9.19 If a hypereutectoid plain-carbon steel contains 4.7 wt % proeutectoid cementite, what is its average carbon content?

*9.20 A hypereutectoid plain-carbon steel contains 10.7 wt % eutectoid Fe_3C. What is its average carbon content in weight percent?

9.21 A plain-carbon steel contains 20.0 wt % proeutectoid ferrite. What is its average carbon content?

9.22 A 0.55 percent C hypoeutectoid plain-carbon steel is slowly cooled from 950°C to a temperature just slightly below 723°C.

  a. Calculate the weight percent proeutectoid ferrite in the steel.

  b. Calculate the weight percent eutectoid ferrite and eutectoid cementite in the steel.

9.23 A hypoeutectoid steel contains 44.0 wt % eutectoid ferrite. What is its average carbon content?

*9.24 A hypoeutectoid steel contains 24.0 wt % eutectoid ferrite. What is its average carbon content?

9.25 A 1.10 percent C hypereutectoid plain-carbon steel is slowly cooled from 900°C to a temperature just slightly below 723°C.

  a. Calculate the weight percent proeutectoid cementite present in the steel.

  b. Calculate the weight percent eutectoid cementite and the eutectoid ferrite present in the steel.

9.26 Define an Fe-C martensite.

9.27 Describe the following types of Fe-C martensites that occur in plain-carbon steels: (*a*) lath martensite and (*b*) plate martensite.

9.28 Describe some of the characteristics of the Fe-C martensite transformation that occurs in plain-carbon steels.

9.29 What causes the tetragonality to develop in the BCC iron lattice when the carbon content of Fe-C martensites exceeds about 0.2 percent?

9.30 What causes the high hardness and strength to be developed in Fe-C martensites of plain-carbon steels when their carbon content is high?

9.31 What is an isothermal transformation in the solid state?

9.32 Draw an isothermal transformation diagram for a plain-carbon eutectoid steel and indicate the various decomposition products on it. How can such a diagram be constructed by a series of experiments?

9.33 If a thin sample of a eutectoid plain-carbon steel is hot-quenched from the austenitic region and held at 700°C until transformation is complete, what will be its microstructure?

9.34 If a thin sample of a eutectoid plain-carbon steel is water-quenched from the austenitic region to room temperature, what will be its microstructure?

9.35 What does the bainite microstructure consist of? What is the microstructural difference between upper and lower bainite?

9.36 Draw time-temperature cooling paths for a 1080 steel on an isothermal transformation diagram that will produce the following microstructures. Start with the steels in the austenitic condition at time = 0 and 850°C. (*a*) 100 percent martensite, (*b*) 50 percent martensite and 50 percent coarse pearlite, (*c*) 100 percent fine pearlite, (*d*) 50 percent martensite and 50 percent upper bainite, (*e*) 100 percent upper bainite, and (*f*) 100 percent lower bainite.

9.37 How does the isothermal transformation diagram for a hypoeutectoid plain-carbon steel differ from that of a eutectoid one?

9.38 Draw a continuous-cooling transformation diagram for a eutectoid plain-carbon steel. How does it differ from a eutectoid isothermal transformation diagram for a plain-carbon steel?

9.39 Draw time-temperature cooling paths for a 1080 steel on a continuous-cooling transformation diagram that will produce the following microstructures. Start with the steels in the austenitic condition at time = 0 and 850°C. (*a*) 100 percent martensite, (*b*) 50 percent fine pearlite and 50 percent martensite, (*c*) 100 percent coarse pearlite, and (*d*) 100 percent fine pearlite.

9.40 Describe the full-annealing heat treatment for a plain-carbon steel. What types of microstructures are produced by full annealing (*a*) a eutectoid steel and (*b*) a hypoeutectoid steel?

9.41 Describe the process-annealing heat treatment for a plain-carbon hypoeutectoid steel with less than 0.3 percent C.

9.42 What is the normalizing heat treatment for steel? What are some of its purposes?

9.43 Describe the tempering process for a plain-carbon steel.

9.44 What types of microstructures are produced by tempering a plain-carbon steel with more than 0.2 percent carbon in the temperature ranges (*a*) 20–250°C, (*b*) 250–350°C, and (*c*) 400–600°C?

9.45 What causes the decrease in hardness during the tempering of a plain-carbon steel?

9.46 Describe the martempering (marquenching) process for a plain-carbon steel. Draw a cooling curve for a martempered (marquenched) austenitized eutectoid plain-carbon steel by using an IT diagram. What type of microstructure is produced after martempering this steel?

9.47 What are the advantages of martempering? What type of microstructure is produced after tempering a martempered steel?

9.48   Why is the term *martempering* a misnomer? Suggest an improved term.

9.49   Describe the austempering process for a plain-carbon steel. Draw a cooling curve for an austempered austenitized eutectoid plain-carbon steel by using an IT diagram.

9.50   What is the microstructure produced after austempering a eutectoid plain-carbon steel? Does an austempered steel need to be tempered? Explain.

9.51   What are the advantages of the austempering process? Disadvantages?

9.52   Thin pieces of 0.3 mm thick hot-rolled strips of 1080 steel are heat-treated in the following ways. Use the IT diagram of Fig. 9.23 and other knowledge to determine the microstructure of the steel samples after each heat treatment.

   a.   Heat 1 h at 860°C; water-quench.

   b.   Heat 1 h at 860°C; water-quench; reheat 1 h at 350°C. What is the name of this heat treatment?

   c.   Heat 1 h at 860°C; quench in molten salt bath at 700°C and hold 2 h; water-quench.

   d.   Heat 1 h at 860°C; quench in molten salt bath at 260°C and hold 1 min; air-cool. What is the name of this heat treatment?

   e.   Heat 1 h at 860°C; quench in molten salt bath at 350°C; hold 1 h; air-cool. What is the name of this heat treatment?

   f.   Heat 1 h at 860°C; water-quench; reheat 1 h at 700°C.

9.53   Explain the numbering system used by the AISI and SAE for plain-carbon steels?

9.54   What are some of the limitations of plain-carbon steels for engineering designs?

9.55   What are the principal alloying elements added to plain-carbon steels to make low-alloy steels?

9.56   What is the AISI-SAE system for designating low-alloy steels?

9.57   What elements dissolve primarily in the ferrite of carbon steels?

9.58   List in order of increasing carbide-forming tendency the following elements: titanium, chromium, molybdenum, vanadium, and tungsten.

9.59   What compounds does aluminum form in steels?

9.60   Name two austenite-stabilizing elements in steels.

9.61   Name four ferrite-stabilizing elements in steels.

9.62   Which elements raise the eutectoid temperature of the Fe-Fe_3C phase diagram? Which elements lower it?

9.63   Define the hardenability of a steel. Define the hardness of a steel.

9.64   Describe the Jominy hardenability test.

9.65   Explain how the data for the plotting of the Jominy hardenability curve are obtained and how the curve is constructed.

9.66   Of what industrial use are Jominy hardenability curves?

9.67   An austenitized 55 mm diameter steel bar of a 9840 steel is quenched in agitated oil. Predict the RC hardness at $\frac{3}{4}$ R from the center of the bar and at the center of the bar.

*9.68 An austenitized 60 mm diameter 4140 long steel bar is quenched in agitated water. Predict the RC hardness at its surface and center.

*9.69 An austenitized 50 mm diameter 5140 steel bar is quenched in agitated oil. Predict what the Rockwell C hardness of the bar will be at (*a*) its surface and (*b*) midway between its surface and center (midradius).

9.70 An austenitized 80 mm diameter 4340 steel bar is quenched in agitated water. Predict what the Rockwell C hardness of the bar will be at (*a*) its surface and (*b*) its center.

9.71 An austenitized 50 mm diameter 5140 steel bar is quenched in agitated oil. Predict what the Rockwell C hardness of the bar will be at (*a*) $\frac{3}{4}$ R and (*b*) the center.

9.72 An austenitized and quenched 4140 steel bar has a Rockwell C hardness of 40 at a point on its surface. What cooling rate did the bar experience at this point?

*9.73 An austenitized and quenched 8640 steel bar has a Rockwell C hardness of 35 at a point on its surface. What cooling rate did the bar experience at this point?

9.74 An austenitized and quenched 5140 steel bar has a Rockwell C hardness of 35 at a point on its surface. What cooling rate did the bar experience at this point?

9.75 An austenitized 40 mm diameter 4340 steel bar is quenched in agitated water. Plot the Rockwell C hardness of the bar versus distance from one surface of the bar to the other across the diameter of the bar at the following points: surface, $\frac{3}{4}$ R, $\frac{1}{2}$ R (midradius), and center. This type of plot is called a *hardness profile across the diameter of the bar.* Assume the hardness profile is symmetrical about the center of the bar.

9.76 An austenitized 50 mm diameter bar of 9840 steel is quenched in agitated oil. Repeat the hardness profile of Prob. 9.75 for this steel.

9.77 An austenitized 60 mm diameter 8640 steel bar is quenched in agitated oil. Repeat the hardness profile of Prob. 9.75 for this steel.

9.78 An austenitized 60 mm diameter 8640 steel bar is quenched in agitated water. Repeat the hardness profile of Prob. 9.75 for this steel.

9.79 An austenitized 4340 standard steel bar is cooled at a rate of 5°C/s (51 mm from the quenched end of a Jominy bar). What will be the constituents in the microstructure of the bar at 200°C? See Fig. 9.39.

*9.80 An austenitized 4340 standard steel bar is cooled at a rate of 8°C/s (19.0 mm from the quenched end of a Jominy bar). What will be the constituents in the microstructure of the bar at 200°C? See Fig. 9.39.

9.81 An austenitized 4340 standard steel bar is cooled at a rate of 50°C/s (9.5 mm from the quenched end of a Jominy bar). What will be the constituents in the microstructure of the bar at 200°C? See Fig. 9.39.

9.82 Explain how a precipitation-hardenable alloy is strengthened by heat treatment.

9.83 What type of phase diagram is necessary for a binary alloy to be precipitation-hardenable?

9.84 What are the three basic heat-treatment steps to strengthen a precipitation-hardenable alloy?

9.85 In what temperature range must a binary precipitation-hardenable alloy be heated for the solution heat-treatment step?

9.86 Why is a precipitation-hardenable alloy relatively weak just after solution heat treatment and quenching?

9.87 Distinguish between natural aging and artificial aging for a precipitation-hardenable alloy.

9.88 What is the driving force for the decomposition of a supersaturated solid solution of a precipitation-hardenable alloy?

9.89 What is the first decomposition product of a precipitation-hardenable alloy in the supersaturated solid-solution condition after aging at a low temperature?

9.90 What are GP zones?

9.91 Why doesn't the equilibrium precipitate form directly from the supersaturated solid solution of a precipitation-hardenable alloy if the aging temperature is low? How can the equilibrium precipitate be formed from the supersaturated solid solution?

9.92 What is an aging curve for a precipitation-hardenable alloy?

9.93 What types of precipitates are developed in an alloy that is considerably underaged at low temperatures? What types are developed upon overaging?

9.94 What is the difference between a coherent precipitate and an incoherent one?

9.95 Describe the four decomposition structures that can be developed when a supersaturated solid solution of an Al–4% Cu alloy is aged.

*9.96 Calculate the wt % θ in an Al–5.0% Cu alloy that is slowly cooled from 548 to 27°C. Assume the solid solubility of Cu in Al at 27°C is 0.02 wt % and that the θ phase contains 54.0 wt % Cu.

Ice Nine

9.97 A binary Al–8.5 wt % Cu alloy is slowly cooled from 700°C to just below 548°C (the eutectic temperature).

a. Calculate the wt % proeutectic α present just above 548°C.

b. Calculate the wt % eutectic α present just below 548°C.

c. Calculate the wt % θ phase present just below 548°C.

9.98 What are some of the properties that make aluminum an extremely useful engineering material?

9.99 How is aluminum oxide extracted from bauxite ores? How is aluminum extracted from pure aluminum oxide?

9.100 How are aluminum wrought alloys classified?

9.101 What are the basic temper designations for aluminum alloys?

9.102 Which series of aluminum wrought alloys are non–heat-treatable? Which are heat-treatable?

9.103 What are the basic strengthening precipitates for the wrought heat-treatable aluminum alloys?

9.104 Describe the three principal casting processes used for aluminum alloys.

9.105 How are aluminum casting alloys classified? What is the most important alloying element for aluminum casting alloys? Why?

9.106 What are some of the important properties of unalloyed copper that make it an important industrial metal?

9.107 How is copper extracted from copper sulfide ore concentrates?

9.108 How are copper alloys classified by the Copper Development Association system?

9.109 Why can't electrolytic tough-pitch copper be used for applications in which it is heated above 400°C in a hydrogen-containing atmosphere?

Virtual Lab

9.110 How can the hydrogen embrittlement of ETP copper be avoided? (Give two methods.)

9.111 Describe the microstructures of the following Cu-Zn brasses at 75×:
(a) 70% Cu–30% Zn (cartridge brass) in the annealed condition and
(b) 60% Cu–40% Zn (Muntz metal) in the hot-rolled condition.

9.112 Why are small amounts of lead added to some Cu-Zn brasses? In what state is the lead distributed in brasses?

9.113 What are the highest-strength commercial copper alloys? What type of heat treatment and fabrication method makes these alloys so strong?

9.114 What alloying element and how much of it (weight percent) is necessary to make a stainless steel "stainless"?

9.115 What type of surface film protects stainless steels?

9.116 What are the four basic types of stainless steels?

9.117 What is the gamma loop in the Fe-Cr phase diagram? Is chromium an austenite- or ferrite-stabilizing element for iron? Explain the reason for your answer.

9.118 What is the basic composition of ferritic stainless steels?

9.119 Why are ferritic stainless steels considered non–heat-treatable?

9.120 What is the basic composition of martensitic stainless steels? Why are these steels heat-treatable?

9.121 What are some applications for ferritic and martensitic stainless steels?

9.122 What makes it possible for an austenitic stainless steel to have an austenitic structure at room temperature?

9.123 What makes austenitic stainless steels that are cooled slowly through the 870 to 600°C range become susceptible to intergranular corrosion?

9.124 How can the intergranular susceptibility of slow-cooled austenitic stainless steels be prevented?

9.125 What are the cast irons? What is their basic range of composition?

9.126 What are some of the properties of cast irons that make them important engineering materials? What are some of their applications?

9.127 What are the four basic types of cast irons?

9.128 Describe the as-cast microstructure of unalloyed white cast iron at 100×.

9.129 Why does the fractured surface of white cast iron appear "white"?

9.130 Describe the microstructure of a class 30 gray cast iron in the as-cast condition at 100×. Why does the fractured surface of a gray cast iron appear gray?

9.131 What are the composition ranges for the carbon and silicon in gray cast iron? Why do gray cast irons have relatively high amounts of silicon?

9.132 What are some of the applications for gray cast irons?

9.133 What casting conditions favor the formation of gray cast iron?

9.134 How can a fully ferritic matrix be produced in an as-cast gray iron after it has been cast?

9.135 What are the composition ranges for the carbon and silicon in ductile cast irons?

9.136 Describe the microstructure of an as-cast grade 80-55-06 ductile cast iron at 100×. What causes the bull's-eye structure?

9.137 Why are ductile cast irons in general more ductile than gray cast irons?

9.138 What are some applications for ductile cast irons?

9.139 Why does the graphite form spherical nodules in ductile cast irons instead of graphite flakes as in gray cast irons?

9.140 What are the composition ranges of carbon and silicon in malleable cast irons?

9.141 Describe the microstructure of a ferritic malleable cast iron (grade M3210) at 100×.

9.142 How are malleable cast irons produced?

9.143 What are some of the property advantages of malleable cast irons?

9.144 What are some applications for malleable cast irons?

9.145 What advantages do magnesium alloys have as engineering materials?

9.146 How are magnesium alloys designated?

9.147 Explain what the following magnesium alloy designations indicate: (*a*) ZE63A-T6, (*b*) ZK51A-T5, and (*c*) AZ31B-H24.

9.148 What alloying elements are added to magnesium for solid-solution strengthening?

9.149 Why is it difficult to cold-work magnesium alloys?

9.150 What alloying elements are added to magnesium to provide better high-temperature strengths?

9.151 Why are titanium and its alloys of special engineering important for aerospace applications?

9.152 Why is titanium metal so expensive?

9.153 What crystal-structure change takes place in titanium at 883°C?

9.154 What are two alpha phase stabilizing elements for titanium?

9.155 What are two beta phase stabilizing elements for titanium?

9.156 What is the most important titanium alloy?

9.157 What are some applications for titanium and its alloys?

9.158 Why is nickel an important engineering metal? What are its advantages? Disadvantages?

9.159 What are the Monel alloys? What are some of their applications?

9.160 What type of precipitates are used to strengthen the precipitation-hardenable alloy Monel K500?

9.161 In what respect are the nickel-base superalloys "super"?

9.162 What is the basic composition of most nickel-base superalloys?

9.163 What are the three main phases present in nickel-base superalloys?

9.16 MATERIALS SELECTION AND DESIGN PROBLEMS

1. In the manufacturing of forged connecting rods, an alloy steel and a plain-carbon steel are being considered. These metals are heat-treatable to 260 ksi (alloy steel) and 113 ksi (carbon steel), respectively. If the diameter of the connecting rod is to be 0.5 in., which metal would you select for this application and why?

2. Plain-carbon and alloy steels are extensively used in the manufacturing of bolts and screws. Give as many reasons as you can for this.

3. To avoid surface pitting of heavy-duty mating gears, which of these metals would you select in the manufacturing of the gear: 4140 steel (48 HRC) or cast iron (28 HRC), and why?

4. (*a*) When selecting the materials for brake or clutch friction pads, what characteristics should you consider? (*b*) Suggest materials that would satisfy your requirements.

5. (*a*) Give examples of components or products that were originally made from steel alloys and are now made from aluminum alloys. (*b*) In each case give reasons why this change took place.

6. (*a*) Give two examples of aluminum alloys that cannot be hardened by heat treating. (*b*) In these cases, how would you increase the hardness?

7. Magnesium alloys are "pyrophoric." (*a*) What does this mean? (*b*) How does that affect the selection of the manufacturing process?

8. (*a*) If you were to select the material for the manufacturing of a conventional condenser or a heat exchanger tubing based on both performance and cost, which of the following metal alloys would you select: plain-carbon steel alloys, nickel superalloys, or copper alloys, and why? (*b*) Give a specific example of the alloy that you would select.

9. (*a*) What type of an alloy is Inconel (give its composition)? (*b*) What are its important properties? (*c*) Give some applications.

10. Based on the information in Fig. 9.19, (*a*) select a plain-carbon steel alloy that is capable of reaching a hardness value of 60 HRC after quenching, (*b*) explain how this change in hardness is achieved, and (*c*) propose a process to reduce the hardness back to 20 HRC.

11. Design a heat treatment process based on the isothermal transformation diagram of eutectoid carbon steel that would result in a metal with (*a*) Rockwell hardness greater than 66 HRC, (*b*) Rockwell hardness of approximately 44 HRC, and (*c*) Rockwell hardness of 5 HRC.

12. (*a*) Design a plain-carbon steel alloy that contains 90 wt % ferrite and 10 wt % cementite at room temperature. (*b*) Is this alloy hypereutectoid or hypoeutectoid?

13. (*a*) Select a plain-carbon steel alloy that contains 80 wt % austenite and 20 wt % Fe_3C at 900°C. (*b*) Is this alloy hypereutectoid or hypoeutectoid?

14. Design a heat treatment process for the eutectoid steel that would result in a final microstructure of 50 percent martensite and 50 percent fine pearlite.

15. Design a heat treatment process for the eutectoid steel that would result in an all-bainite structure with a hardness of HRC 47.

16. Cracks often appear around sharp corners of steel components after quenching. To avoid this, sharp corners are discouraged. If both sharp corners and the heat treatment process are required, what do you suggest?

17. A machinist machines a bolt and a nut from annealed 1080 steel. He then heat-treats the components by austenitizing, quenching, and tempering to achieve a higher hardness. (*a*) After the heat treatment process, the threads no longer fit. Explain why. (*b*) How would you avoid this problem?

10

Polymeric Materials

(© *Eye of Science/Photo Researchers, Inc.*)

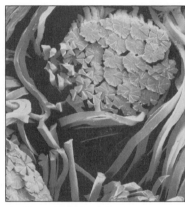

(© *Science Photo Library/Photo Researchers, Inc.*)

(© *Tim Tadder/NewSport/Corbis*)

Microfibers are man-made fibers that are significantly smaller than a human hair (finer than silk fiber) and are split many times into v-shapes, see Figure 1. Conventional fibers are significantly thicker and have a solid circular cross-section. Microfibers can be produced from a variety of polymers including polyester, nylon, and acrylic. Fabrics made of microfiber possess significantly higher surface due to smaller fiber and v-shaped crevices that can trap liquid and dirt. The water and dirt are actually trapped in v-shaped crevices of the fiber as opposed to being simply pushed away by conventional solid circular fibers. Thus, microfiber fabrics are softer and have a silky feeling (important in the clothing industry), and absorb water and dirt at significantly higher amounts (important to the cleaning industry). The above characteristics make microfiber cloths highly popular in sports garment and cleaning industries. The two main microfiber materials are polyester (scrubbing material) and polyamide (absorbing material). ■

By the end of this chapter, students will be able to . . .

1. Define and classify polymers including thermosets, thermoplastics, and elastomers.

2. Describe various polymerization reactions and steps.

3. Describe terms such as functionality, vinyl, vinylidene, homopolymer, and copolymer.

4. Describe various industrial polymerization methods.

5. Describe the structure of polymers and compare with metals.

6. Describe the glass transition temperature and the changes to the structure and properties of polymeric materials around this temperature.

7. Describe various manufacturing processes used to manufacture thermosetting and thermoplastic components.

8. Be able to name a reasonable number of general purpose thermoplastics, thermosets, elastomers, and their applications.

9. Be able to explain the deformation, strengthening, stress-relaxation, and fracture mechanisms in polymers.

10. Describe biopolymers and their use in biomedical applications.

10.1 INTRODUCTION

The word *polymer* literally means "many parts." A polymeric solid material may be considered to be one that contains many chemically bonded parts or units that themselves are bonded together to form a solid. In this chapter we shall study some aspects of the structure, properties, processing, and applications of two industrially important polymeric materials: *plastics* and *elastomers*. *Plastics*[1] are a large and varied group of synthetic materials that are processed by forming or molding into shape. Just as we have many types of metals such as aluminum and copper, we have many types of plastics such as polyethylene and nylon. Plastics can be divided into two classes, **thermoplastics** and **thermosetting** *plastics,* depending on how they are structurally chemically bonded. *Elastomers* or rubbers can be elastically deformed a large amount when a force is applied to them and can return to their original shape (or almost) when the force is released.

---

[1]The word *plastic* has many meanings. As a noun, plastic refers to a class of materials that can be molded or formed into shape. As an adjective, plastic can mean capable of being molded. Another use of plastic as an adjective is to describe the continuous permanent deformation of a metal without rupture, as in the "plastic deformation of metals."

Thermoplastics

Thermoplastics require heat to make them formable and after cooling, retain the shape they were formed into. These materials can be reheated and reformed into new shapes a number of times without significant change in their properties. Most thermoplastics consist of very long main chains of carbon atoms covalently bonded together. Sometimes nitrogen, oxygen, or sulfur atoms are also covalently bonded in the main molecular chain. Pendant atoms or groups of atoms are covalently bonded to the main-chain atoms. In thermoplastics the long molecular chains are bonded to each other by secondary bonds.

Thermosetting Plastics (Thermosets)

Thermosetting plastics formed into a permanent shape and cured or "set" by a chemical reaction cannot be remelted and reformed into another shape but degrade or decompose upon being heated to too high a temperature. Thus, thermosetting plastics cannot be recycled. The term *thermosetting* implies that heat (the Greek word for heat is *thermē*) is required to permanently set the plastic. There are, however, many so-called thermosetting plastics that set or cure at room temperature by a chemical reaction only. Most thermosetting plastics consist of a network of carbon atoms covalently bonded to form a rigid solid. Sometimes nitrogen, oxygen, sulfur, or other atoms are also covalently bonded into a thermoset network structure.

Plastics are important engineering materials for many reasons. They have a wide range of properties, some of which are unattainable from any other materials, and in most cases they are relatively low in cost. The use of plastics for mechanical engineering designs offers many advantages, which include elimination of parts through engineering design with plastics, elimination of many finishing operations, simplified assembly, weight savings, noise reduction, and in some cases elimination of the need for lubrication of some parts. Plastics are also very useful for many electrical engineering designs mainly because of their excellent insulative properties. Electrical-electronic applications for plastic materials include connectors, switches, relays, TV tuner components, coil forms, integrated circuit boards, and computer components. Figure 10.1 shows some examples of the use of plastic materials in engineering designs.

The amount of plastic materials used by industry has increased markedly. A good example of the increased industrial use of plastics is in the manufacture of automobiles. Engineers designing the 1959 Cadillac were amazed to find that they had put as much as 25 lb of plastics in that vehicle. In 1980 there was an average of about 200 lb of plastic used per car. The use of plastics in the 1990 car was about 300 lb per car. Certainly not all industries have increased their usage of plastics like the auto industry, but there has been an overall increased usage of plastics in industry in recent decades. Let us now look into the details of the structure, properties, and applications of plastics and elastomers.

(*a*) (*b*) (*c*)

Figure 10.1
Some applications for engineering plastics. (*a*) TV remote control casing using advanced styrenic resin to meet requirements for gloss, toughness, and crack resistance.
(*Photo: Dow Chemical U.S.A.*)
(*b*) Semiconductor wafer wands made of Vitrex PEEK (polyetheretherketone) thermoplastic. (*c*) Nylon thermoplastic reinforced with 30 percent glass fiber to replace aluminum in the air intake manifold of the turbodiesel engine for the Ford Transit. (DSM Engineering Plastics, The Netherlands.)
[(*a*) © *Royalty-Free/CORBIS.* (*b*) © *Charles O'Rear/CORBIS.* (*c*) © *Tom Pantages.*]

10.2 POLYMERIZATION REACTIONS

Most thermoplastics are synthesized by the process of *chain-growth polymerization.* In this process many (there may be thousands) small molecules are covalently bonded together to form very long molecular chains. The simple molecules that are covalently bonded into long chains are called **monomers** (from the Greek words *mono,* meaning "one," and *meros*, meaning "part"). The long-chain molecule formed from the monomer units is called a **polymer** (from the Greek words *polys,* meaning "many," and *meros,* meaning "part").

10.2.1 Covalent Bonding Structure of an Ethylene Molecule

The ethylene molecule, C_2H_4, is chemically bonded by a double covalent bond between the carbon atoms and by four single covalent bonds between the carbon and hydrogen atoms (Fig. 10.2). A carbon-containing molecule that has one or more

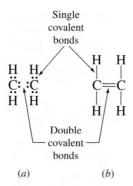

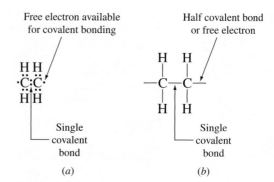

Figure 10.2
Covalent bonding in the ethylene molecule illustrated by (a) electron-dot (dots represent valence electrons) and (b) straight-line notation. There is one double carbon-carbon covalent bond and four single carbon-hydrogen covalent bonds in the ethylene molecule. The double bond is chemically more reactive than the single bonds.

Figure 10.3
Covalent bonding structure of an activated ethylene molecule. (a) Electron-dot notation (where dots represent valence electrons). Free electrons are created at each end of the molecule that can be covalently bonded with free electrons from other molecules. Note that the double covalent bond between the carbon atoms has been reduced to a single bond. (b) Straight-line notation. The free electrons created at the ends of the molecule are indicated by half bonds that are only attached to one carbon atom.

carbon-carbon double bonds is said to be an *unsaturated molecule.* Thus, ethylene is an unsaturated carbon-containing molecule since it contains one carbon-carbon double bond.

10.2.2 Covalent Bonding Structure of an Activated Ethylene Molecule

When the ethylene molecule is activated so that the double bond between the two carbon atoms is "opened up," the double covalent bond is replaced by a single covalent bond, as shown in Fig. 10.3. As a result of the activation, each carbon atom of the former ethylene molecule has a free electron for covalent bonding with another free electron from another molecule. In the following discussion we shall see how the ethylene molecule can be activated and, as a result, how many ethylene monomer units can be covalently bonded together to form a polymer. This is the process of **chain polymerization.** The polymer produced by the polymerization of ethylene is called *polyethylene.*

10.2.3 General Reaction for the Polymerization of Polyethylene and the Degree of Polymerization

The general reaction for the chain polymerization of ethylene monomer into polyethylene may be written as

The repeating subunit in the polymer chain is called a **mer.** The mer for polyethylene is $-CH_2-CH_2-$ and is indicated in the preceding equation. The n in the equation is known as the **degree of polymerization** (DP) of the polymer chain and is equal to the number of subunits or mers in the polymer molecular chain. The average DP for polyethylene ranges from about 3500 to 25,000, corresponding to average molecular masses ranging from about 100,000 to 700,000 g/mol.

If a particular type of polyethylene has a molecular mass of 150,000 g/mol, what is its degree of polymerization?

EXAMPLE PROBLEM 10.1

■ **Solution**

The repeating unit or mer for polyethylene is $-CH_2-CH_2-$. This mer has a mass of 4 atoms $\times$ 1 g = 4 g for the hydrogen atoms plus a mass of 2 atoms $\times$ 12 g = 24 g for the carbon atoms, making a total of 28 g for each polyethylene mer.

$$DP = \frac{\text{molecular mass of polymer (g/mol)}}{\text{mass of a mer (g/mer)}} \qquad (10.1)$$

$$= \frac{150,000 \text{ g/mol}}{28 \text{ g/mer}} = 5357 \text{ mers/mol} \blacktriangleleft$$

10.2.4 Chain Polymerization Steps

The reactions for the chain polymerization of monomers like ethylene into linear polymers like polyethylene can be divided into the following steps: (1) initiation, (2) propagation, and (3) termination.

Initiation   For the chain polymerization of ethylene, one of many types of catalysts can be used. In this discussion we shall consider the use of organic peroxides that act as free-radical formers. A *free radical* can be defined as an atom, often part of a larger group, that has an unpaired electron (free electron) that can covalently bond to an unpaired electron (free electron) of another atom or molecule.

Let us first consider how a molecule of hydrogen peroxide, H_2O_2, can decompose into two free radicals, as shown by the following equations. Using electron-dot notation for the covalent bonds,

$$H:\overset{..}{\underset{..}{O}}:\overset{..}{\underset{..}{O}}:H \xrightarrow{\text{heat}} H:\overset{..}{\underset{..}{O}}\cdot + \cdot\overset{..}{\underset{..}{O}}:H$$

Hydrogen peroxide   Free radicals

Using straight-line notation for the covalent bonds,

$$H{-}O{-}O{-}H \xrightarrow{\text{heat}} 2H{-}O\overset{\displaystyle\cdot}{}$$

Hydrogen peroxide   Free radicals

Free electron

In the free-radical chain polymerization of ethylene, an organic peroxide can decompose in the same way as hydrogen peroxide. If $R{-}O{-}O{-}R$ represents an organic peroxide, where R is a chemical group, then upon heating, this peroxide can decompose into two free radicals in a manner similar to that of hydrogen peroxide, as

$$R{-}O{-}O{-}R \longrightarrow 2R{-}O\overset{\displaystyle\cdot}{}$$

Organic peroxide   Free radicals

Free electron

Benzoyl peroxide is an organic peroxide used to initiate some chain polymerization reactions. It decomposes into free radicals as it follows:[2]

Benzoyl peroxide   Free radicals

Free electron

One of the free radicals created by the decomposition of the organic peroxide can react with an ethylene molecule to form a new longer-chain free radical, as shown by the reaction

Free electron

Free radical   Ethylene   Free radical

Free electron

---

[2]The hexagonal ring represents the benzene structure, as indicated below. Also see Sec. 2.6.

The organic free radical in this way acts as an initiator catalyst for the polymerization of ethylene.

Propagation   The process of extending the polymer chain by the successive addition of monomer units is called *propagation*. The double bond at the end of an ethylene monomer unit can be "opened up" by the extended free radical and be covalently bonded to it. Thus, the polymer chain is further extended by the reaction

$$R-CH_2-CH_2^{\bullet} + CH_2{=}CH_2 \longrightarrow R-CH_2-CH_2-CH_2-CH_2^{\bullet}$$

The polymer chains in chain polymerization keep growing spontaneously because the energy of the chemical system is lowered by the chain polymerization process. That is, the sum of the energies of the produced polymers is lower than the sum of the energies of the monomers that produced the polymers. The degrees of polymerization of the polymers produced by chain polymerization vary within the polymeric material. Also, the average DP varies among polymeric materials. For commercial polyethylene the DP usually averages in the range from 3500 to 25,000.

Termination   *Termination* can occur by the addition of a terminator free radical or when two chains combine. Another possibility is that trace amounts of impurities may terminate the polymer chain. Termination by the coupling of two chains can be represented by the reaction

$$R(CH_2-CH_2)_m^{\bullet} + R'(CH_2-CH_2)_n^{\bullet} \longrightarrow R(CH_2-CH_2)_m-(CH_2-CH_2)_nR'$$

10.2.5 Average Molecular Weight for Thermoplastics

Thermoplastics consist of chains of polymers of many different lengths, each of which has its own molecular weight and degree of polymerization. Thus, one must speak of an average molecular weight when referring to the molecular mass of a thermoplastic material.

The average molecular weight of a thermoplastic can be determined by using special physical-chemical techniques. One method commonly used for this analysis is to determine the weight fractions of molecular weight ranges. The average molecular weight of the thermoplastic is then the sum of the weight fractions times their mean molecular weight for each particular range divided by the sum of the weight fractions. Thus,

$$\overline{M}_m = \frac{\sum f_i M_i}{\sum f_i} \tag{10.2}$$

where $\overline{M}_m$ = average molecular weight for a thermoplastic

M_i = mean molecular weight for each molecular weight range selected

f_i = weight fraction of the material having molecular weights of a selected molecular weight range

Calculate the average molecular weight $\overline{M}_m$ for a thermoplastic material that has the mean molecular weight fractions f_i for the molecular weight ranges listed in the following table:

Molecular weight range, g/mol	M_i	f_i	$f_i M_i$
5000–10,000	7500	0.11	825
10,000–15,000	12,500	0.17	2125
15,000–20,000	17,500	0.26	4550
20,000–25,000	22,500	0.22	4950
25,000–30,000	27,500	0.14	3850
30,000–35,000	32,500	0.10	3250
		$\sum = 1.00$	$\sum = 19,550$

■ **Solution**

First determine the mean values for the molecular weight ranges and then list these values, as in the column under M_i shown in the table. Then multiply f_i by M_i to obtain the $f_i M_i$ values. The average molecular weight for this thermoplastic is

$$\overline{M}_m = \frac{\sum f_i M_i}{\sum f_i} = \frac{19,550}{1.00} = 19,550 \text{ g/mol} \blacktriangleleft$$

10.2.6 Functionality of a Monomer

In order for a monomer to polymerize, it must have at least two active chemical bonds. When a monomer has two active bonds, it can react with two other monomers, and by repetition of the bonding, other monomers of the same type can form a long-chain or linear polymer. When a monomer has more than two active bonds, polymerization can take place in more than two directions, and thus three-dimensional network molecules can be built up.

The number of active bonds a monomer has is called the **functionality** of the monomer. A monomer that uses two active bonds for the polymerization of long chains is called *bifunctional*. Thus, ethylene is an example of a bifunctional monomer. A monomer that uses three active bonds to form a network polymeric material is called *trifunctional*. Phenol, C_6H_5OH, is an example of a trifunctional monomer and is used in the polymerization of phenol and formaldehyde, which will be discussed later.

10.2.7 Structure of Noncrystalline Linear Polymers

If we microscopically examine a short length of a polyethylene chain, we find that it takes on a zigzag configuration (Fig. 10.4) because the covalent bonding angle between single carbon-carbon covalent bonds is about 109°. However, on a larger scale, the polymer chains in noncrystalline polyethylene are randomly entangled like spaghetti thrown into a bowl. This entanglement of a linear polymer is illustrated in Fig. 10.5. For some polymeric materials, of which polyethylene is one, there can be

Carbon

Hydrogen

Figure 10.4
The molecular structure of a short length of a polyethylene chain. The carbon atoms have a zigzag arrangement because all carbon-carbon covalent bonds are directed at about 109° to each other.
(*After W. G. Moffatt, G. W. Pearsall, and J. Wulff, "The Structure and Properties of Materials," vol. I: "Structure," Wiley, 1965, p. 65.*)

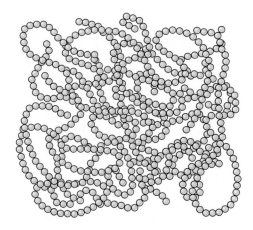

Figure 10.5
A schematic representation of a polymer. The spheres represent the repeating units of the polymer chain, not specific atoms.
(*After W. G. Moffatt, G. W. Pearsall, and J. Wulff, "The Structure and Properties of Materials," vol. I: "Structure," Wiley, 1965, p. 104.*)

both crystalline and noncrystalline regions. This subject will be discussed in more detail in Sec. 10.4.

The bonding between the long molecular chains in polyethylene consists of weak, permanent dipole secondary bonds. However, the physical entanglement of the long molecular chains also adds to the strength of this type of polymeric material. Side branches also can be formed that cause loose packing of the molecular chains and favor a noncrystalline structure. Branching of linear polymers thus weakens secondary bonds between the chains and lowers the tensile strength of the bulk polymeric material.

10.2.8 Vinyl and Vinylidene Polymers

Many useful addition (chain) polymeric materials that have carbon main-chain structures similar to polyethylene can be synthesized by replacing one or more of the hydrogen atoms of ethylene with other types of atoms or groups of atoms. If only one hydrogen atom of the ethylene monomer is replaced with another atom or group of atoms, the polymerized polymer is called a *vinyl polymer.* Examples of vinyl polymers are polyvinyl chloride, polypropylene, polystyrene, acrylonitrile, and polyvinyl acetate. The general reaction for the polymerization of the vinyl polymers is

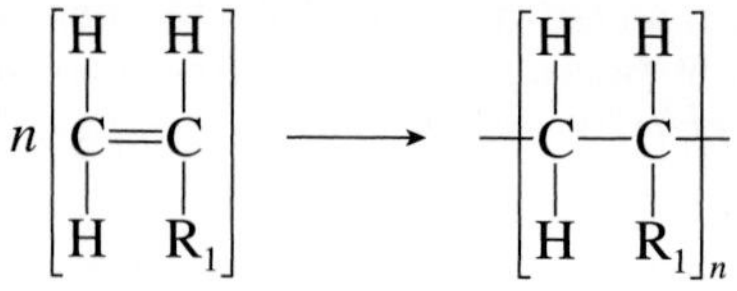

$$n \begin{bmatrix} H & H \\ | & | \\ C = C \\ | & | \\ H & R_1 \end{bmatrix} \longrightarrow \begin{bmatrix} H & H \\ | & | \\ C - C \\ | & | \\ H & R_1 \end{bmatrix}_n$$

where R_1 can be another type of atom or group of atoms. Figure 10.6 shows the structural bonding of some vinyl polymers.

If both hydrogen atoms on one of the carbon atoms of the ethylene monomer are replaced by other atoms or groups of atoms, the polymerized polymer is called a *vinylidene polymer.* The general reaction for the polymerization of vinylidene polymers is

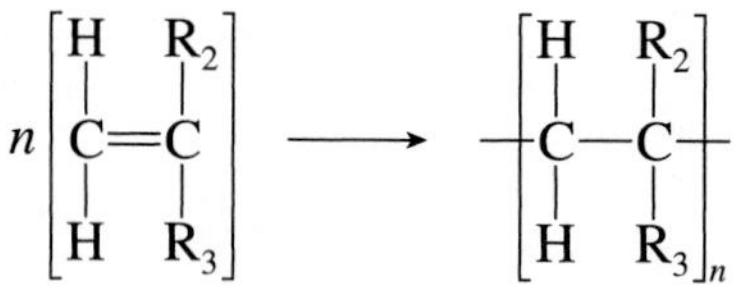

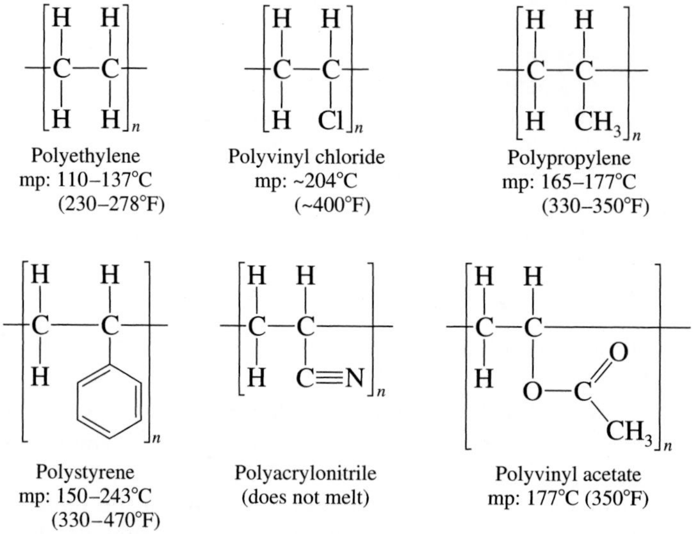

$$n \begin{bmatrix} H & R_2 \\ | & | \\ C = C \\ | & | \\ H & R_3 \end{bmatrix} \longrightarrow \begin{bmatrix} H & R_2 \\ | & | \\ C - C \\ | & | \\ H & R_3 \end{bmatrix}_n$$

Polyethylene
mp: 110–137°C
(230–278°F)

Polyvinyl chloride
mp: ~204°C
(~400°F)

Polypropylene
mp: 165–177°C
(330–350°F)

Polystyrene
mp: 150–243°C
(330–470°F)

Polyacrylonitrile
(does not melt)

Polyvinyl acetate
mp: 177°C (350°F)

Figure 10.6
Structural formulas for some vinyl polymers.

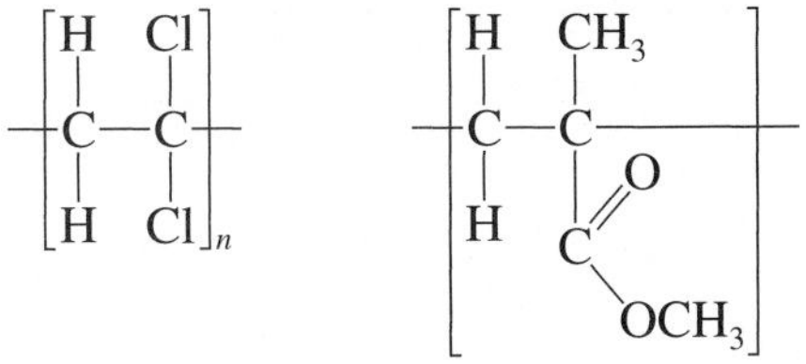

$$\begin{bmatrix} H & Cl \\ | & | \\ -C-C- \\ | & | \\ H & Cl \end{bmatrix}_n$$

Polyvinylidene chloride
mp: 177°C (350°F)

Polymethyl methacrylate
mp: 160°C (320°F)

Figure 10.7
Structural formulas for some vinylidene
polymers.

where R_2 and R_3 can be other types of atoms or atomic groups. Figure 10.7 shows the structural bonding for two vinylidene polymers.

10.2.9 Homopolymers and Copolymers

Homopolymers are polymeric materials that consist of polymer chains made up of single repeating units. That is, if A is a repeating unit, a homopolymer chain will have a sequence of AAAAAAA ··· in the polymer molecular chain. **Copolymers,** in contrast, consist of polymer chains made up of two or more chemically different repeating units that can be in different sequences.

Although the monomers in most copolymers are randomly arranged, four distinct types of copolymers have been identified: random, alternating, block, and graft (Fig. 10.8).

Random copolymers. Different monomers are randomly arranged within the polymer chains. If A and B are different monomers, then an arrangement might be (Fig. 10.8a)

AABABBBBAABABABAAB ···

Alternating copolymers. Different monomers show a definite ordered alternation as (Fig. 10.8b)

ABABABABABAB ···

Block copolymers. Different monomers in the chain are arranged in relatively long blocks of each monomer as (Fig. 10.8c)

AAAAA—BBBBB— ···

Graft copolymers. Appendages of one type of monomer are grafted to the long chain of another as (Fig. 10.8d)

AAAAAAAAAAAAAAAAAAAAA
     B   B
     B   B
     B   B

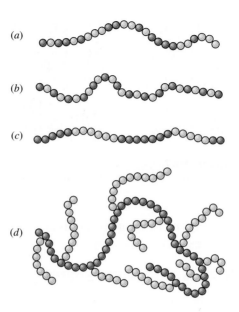

Figure 10.8
Copolymer arrangements. (*a*) A copolymer in which the different units are randomly distributed along the chain. (*b*) A copolymer in which the units alternate regularly. (*c*) A block copolymer. (*d*) A graft copolymer.
(After W. G. Moffatt, G. W. Pearsall, and J. Wulff, "The Structure and Properties of Materials," vol. I: "Structure," Wiley, 1965, p. 108.)

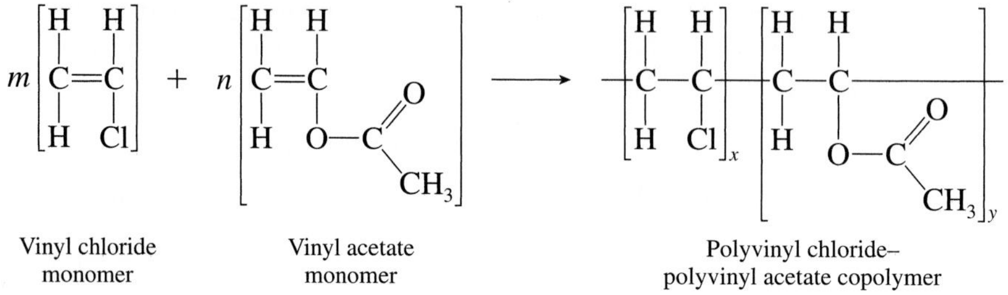

Figure 10.9
Generalized polymerization reaction of vinyl chloride and vinyl acetate monomers to produce a copolymer of polyvinyl chloride–polyvinyl acetate.

Chain-reaction polymerization can take place between two or more different monomers if they can enter the growing chains at the same relative energy level and rates. An example of an industrially important copolymer is one formed with polyvinyl chloride and polyvinyl acetate, which is used as coating material for cables, pools, and cans. A generalized polymerization reaction for the production of this copolymer is given in Fig. 10.9.

A copolymer consists of 15 wt % polyvinyl acetate (PVA) and 85 wt % polyvinyl chloride (PVC). Determine the mole fraction of each component.

■ **Solution**

Let the basis be 100 g of copolymer; therefore we have 15 g of PVA and 85 g of PVC. First we determine the number of moles of each component that we have, and then we calculate the mole fractions of each.

Moles of polyvinyl acetate. The molecular weight of the PVA mer is obtained by adding up the atomic masses of the atoms in the structural formula for the PVA mer (Fig. EP10.3a):

$$4 \text{ C atoms} \times 12 \text{ g/mol} + 6 \text{ H atoms} \times 1 \text{ g/mol} + 2 \text{ O atoms} \times 16 \text{ g/mol} = 86 \text{ g/mol}$$

$$\text{No. of moles of PVA in 100 g of copolymer} = \frac{15 \text{ g}}{86 \text{ g/mol}} = 0.174$$

Moles of polyvinyl chloride. The molecular weight of the PVC mer is obtained from Fig. EP10.3b.

$$2 \text{ C atoms} \times 12 \text{ g/mol} + 3 \text{ H atoms} \times 1 \text{ g/mol} + 1 \text{ Cl atoms} \times 35.5 \text{ g/mol} = 62.5 \text{ g/mol}$$

$$\text{No. of moles of PVC in 100 g of copolymer} = \frac{85 \text{ g}}{62.5 \text{ g/mol}} = 1.36$$

$$\text{Mole fraction of PVA} = \frac{0.174}{0.174 + 1.36} = 0.113$$

$$\text{Mole fraction of PVC} = \frac{1.36}{0.174 + 1.36} = 0.887$$

(a)   *(b)*

Figure EP10.3
Structural formulas for the mers of
(a) polyvinyl acetate and (b) polyvinyl chloride.

EXAMPLE PROBLEM 10.4

Determine the mole fractions of vinyl chloride and vinyl acetate in a copolymer having a molecular weight of 10,520 g/mol and a degree of polymerization of 160.

■ **Solution**

From Example Problem 10.3, the molecular weight of the PVC mer is 62.5 g/mol and that of the PVA mer is 86 g/mol.

Since the sum of the mole fractions of polyvinyl chloride, f_{vc}, and polyvinyl acetate, $f_{va} = 1$, $f_{va} = 1 - f_{vc}$. Thus, the average molecular weight of the copolymer mer is

$$MW_{av}(mer) = f_{vc}MW_{vc} + f_{va}MW_{va} = f_{vc}MW_{vc} + (1 - f_{vc})MW_{va}$$

The average molecular weight of the copolymer mer is also

$$MW_{av}(mer) = \frac{MW_{av}(polymer)}{DP} = \frac{10,520 \text{ g/mol}}{160 \text{ mers}} = 65.75 \text{ g/(mol} \cdot \text{mer)}$$

The value of f_{vc} can be obtained by equating the two equations of $MW_{av}(mer)$.

$$f_{vc}(62.5) + (1 - f_{vc})(86) = 65.75 \quad \text{or} \quad f_{vc} = 0.86$$

$$f_{va} = (1 - f_{vc}) = 1 - 0.86 = 0.14$$

EXAMPLE PROBLEM 10.5

If a vinyl chloride–vinyl acetate copolymer has a ratio of 10:1 vinyl chloride to vinyl acetate mers and a molecular weight of 16,000 g/mol, what is its degree of polymerization?

■ **Solution**

$$MW_{av}(mer) = \tfrac{10}{11}MV_{vc} + \tfrac{1}{11}MW_{va} = \tfrac{10}{11}(62.5) + \tfrac{1}{11}(86) = 64.6 \text{ g/(mol} \cdot \text{mer)}$$

$$DP = \frac{16,000 \text{ g/mol (polymer)}}{64.6 \text{ g/(mol} \cdot \text{mer)}} = 248 \text{ mers}$$

10.2.10 Other Methods of Polymerization

Stepwise Polymerization   In **stepwise polymerization,** monomers chemically react with each other to produce linear polymers. The reactivity of the functional groups at the ends of a monomer in stepwise polymerization is usually assumed to be about the same for a polymer of any size. Thus monomer units can react with each other or with produced polymers of any size. In many stepwise polymerization reactions a small molecule is produced as a by-product, so these types of reactions are sometimes called *condensation polymerization reactions.* An example of a stepwise polymerization reaction is the reaction of hexamethylene diamine with adipic acid to produce nylon 6,6 and water as a by-product, as shown in Fig. 10.10 for the reaction of one molecule of hexamethylene diamine with another of adipic acid.

Network Polymerization   For some polymerization reactions that involve a chemical reactant with more than two reaction sites, a three-dimensional network plastic

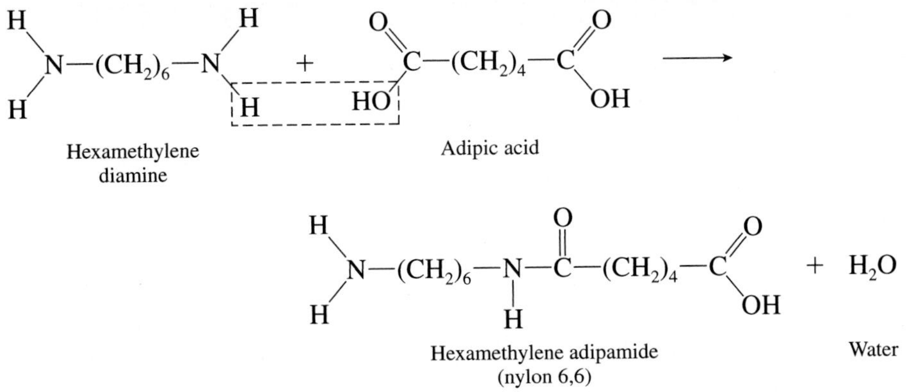

Figure 10.10
Polymerization reaction of hexamethylene diamine with adipic acid to produce a unit of nylon 6,6.

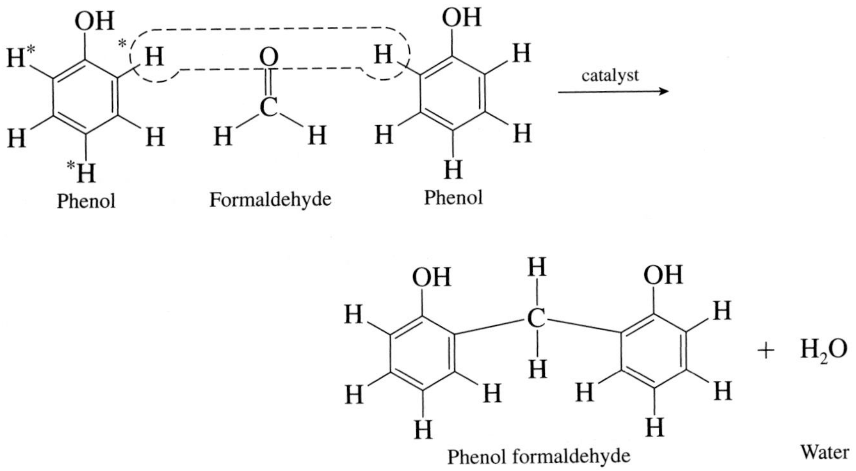

Figure 10.11
Polymerization reaction of phenol (asterisks represent reaction sites) with formaldehyde to produce a phenolic resin unit linkage.

material can be produced. This type of polymerization occurs in the curing of thermosetting plastics such as the phenolics, epoxies, and some polyesters. The polymerization reaction of two phenol molecules and one formaldehyde molecule is shown in Fig. 10.11. Note that a molecule of water is formed as a by-product of the reaction. The phenol molecule is trifunctional, and in the presence of a suitable catalyst and sufficient heat and pressure, it can be polymerized with formaldehyde into a network thermosetting phenolic plastic material that is sometimes referred to by the trade name *Bakelite.*

10.3 INDUSTRIAL POLYMERIZATION METHODS

At this stage one must certainly be wondering how plastic materials are produced industrially. The answer to this question is not simple since many different processes are used and new ones are constantly being developed. To start with, basic raw materials such as *natural gas, petroleum,* and *coal* are used to produce the basic chemicals for the polymerization processes. These chemicals are then polymerized by many different processes into plastic materials such as granules, pellets, powders, or liquids that are further processed into finished products. The chemical polymerization processes used to produce plastic materials are complex and diverse. The chemical engineer plays a major role in their development and industrial utilization. Some of the most important polymerization methods are outlined in the following paragraphs and illustrated in Figs. 10.12 and 10.13.

Bulk polymerization (Fig. 10.12*a*). The monomer and activator are mixed in a reactor that is heated and cooled as required. This process is used extensively for condensation polymerization where one monomer may be charged into the reactor and another added slowly. The bulk process can be

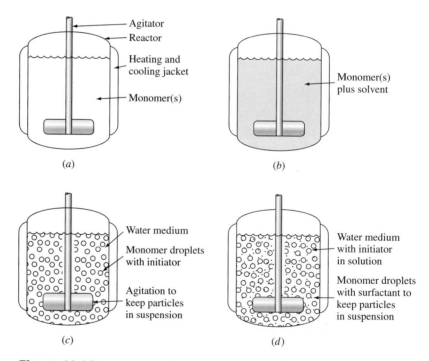

Figure 10.12
Schematic illustration of some commonly used industrial polymerization methods: (*a*) bulk, (*b*) solution, (*c*) suspension, and (*d*) emulsion.
(*After W. E. Driver, "Plastics Chemistry and Technology," Van Nostrand Reinhold, 1979, p. 19.*)

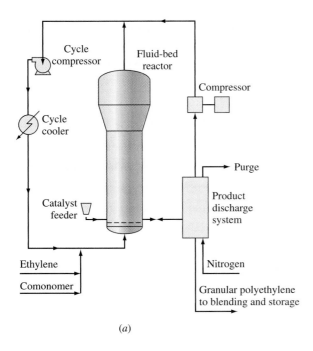

Figure 10.13

Gas-phase polymerization process for low-density polyethylene. (*a*) Flow diagram outlining the basic steps of the process. (*b*) Twin reactors used in the process.
(*After Chemical Engineering, Dec. 3, 1979, pp. 81, 83.*)

used for many condensation polymerization reactions because of their low heats of reaction.

Solution polymerization (Fig. 10.12*b*). The monomer is dissolved in a nonreactive solvent that contains a catalyst. The heat released by the reaction is absorbed by the solvent, and so the reaction rate is reduced.

Suspension polymerization (Fig. 10.12*c*). The monomer is mixed with a catalyst and then dispersed as a suspension in water. In this process the heat released by the reaction is absorbed by the water. After polymerization, the polymerized product is separated and dried. This process is commonly used to produce many of the vinyl-type polymers such as polyvinyl chloride, polystyrene, polyacrylonitrile, and polymethyl methacrylate.

Emulsion polymerization (Fig. 10.12*d*). This polymerization process is similar to the suspension process since it is carried out in water. However, an emulsifier is added to disperse the monomer into very small particles.

In addition to the batch polymerization processes just described, many types of mass continuous polymerization processes have been developed, and research and development in this area continues. One very important process[3] is Union Carbide's gas-phase Unipol process for producing low-density polyethylene. In this process gaseous ethylene monomer along with some comonomer are fed continuously into a fluidized-bed reactor into which a special catalyst is added (Fig. 10.13a). The advantages of this process are lower temperature for polymerization (100°C instead of the older process's 300°C) and lower pressure (100 psi instead of 300 psi for the older process). Many industrial plants are already using the Unipol process.

10.4 CRYSTALLINITY AND STEREOISOMERISM IN SOME THERMOPLASTICS

A thermoplastic, when solidified from the liquid state, forms either a noncrystalline or a partly crystalline solid. Let us investigate some of the solidification and structural characteristics of these materials.

10.4.1 Solidification of Noncrystalline Thermoplastics

Let us consider the solidification and slow cooling to low temperatures of a non-crystalline thermoplastic. When noncrystalline thermoplastics solidify, there is no sudden decrease in specific volume (volume per unit mass) as the temperature is lowered (Fig. 10.14). The liquid, upon solidification, changes to a supercooled liquid that is in the solid state and shows a gradual decrease in specific volume with decreasing temperature, as indicated along the line ABC in Fig. 10.14.

Upon cooling this material to lower temperatures, a change in slope of the specific volume versus temperature curve occurs, as indicated by C and D of curve $ABCD$ of Fig. 10.14. The average temperature within the narrow temperature range over which the slope in the curve changes is called the **glass transition temperature** T_g. Above T_g, noncrystalline thermoplastics show viscous (rubbery or flexible leathery) behavior, and below T_g, these materials show glass-brittle behavior. In some ways T_g might be considered a ductile-brittle transition temperature. Below T_g, the material is glass-brittle because molecular chain motion is very restricted. Figure 10.15 shows an experimental plot of specific volume versus temperature for noncrystalline polypropylene that indicates a slope change for the T_g of this material at -12°C. Table 10.1 lists T_g values for some thermoplastics.

10.4.2 Solidification of Partly Crystalline Thermoplastics

Let us now consider the solidification and cooling to low temperatures of a partly crystalline thermoplastic. When this material solidifies and cools, a sudden decrease in specific volume occurs, as indicated by the line BE in Fig. 10.14. This decrease in specific volume is caused by the more-efficient packing of the polymer chains into crystalline regions. The structure of the partly crystalline thermoplastic at E will thus be that of crystalline regions in a supercooled liquid (viscous solid) noncrystalline

---

[3]*Chemical Engineering*, Dec. 3, 1979, p. 80.

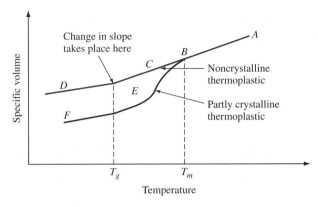

Figure 10.14

Solidification and cooling of noncrystalline and partly crystalline thermoplastics showing change in specific volume with temperature (schematic). T_g is the glass transition temperature and T_m is the melting temperature. Noncrystalline thermoplastic cools along line *ABCD*, where *A* = liquid, *B* = highly viscous liquid, *C* = supercooled liquid (rubbery), and *D* = glassy solid (hard and brittle). Partly crystalline thermoplastic cools along line *ABEF*, where *E* = solid crystalline regions in supercooled liquid matrix and *F* = solid crystalline regions in glassy matrix.

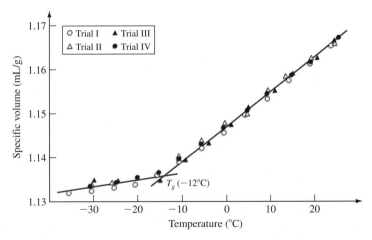

Figure 10.15

Experimental data of specific volume versus temperature for the determination of the glass transition temperature of atactic polypropylene. T_g is at $-12°C$.

[After D. L. Beck, A. A. Hiltz, and J. R. Knox, Soc. Plast. Eng. Trans, **3:**279(1963).]

Table 10.1 Glass transition temperature T_g* (°C) for some thermoplastics

Polyethylene	−110	(nominal)
Polypropylene	−18	(nominal)
Polyvinyl acetate	29	
Polyvinyl chloride	82	
Polystyrene	75–100	
Polymethyl methacrylate	72	

*Note that the T_g of a thermoplastic is not a physical constant like the melting temperature of a crystalline solid but depends to some extent on variables such as degree of **crystallinity,** average molecular weight of the polymer chains, and rate of cooling of the thermoplastic.

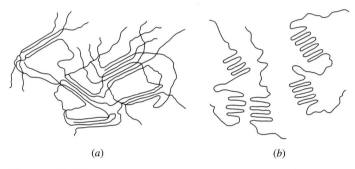

(a)   (b)

Figure 10.16
Two suggested crystallite arrangements for partly crystalline thermoplastic materials: (*a*) fringed-micelle model and (*b*) folded-chain model.
(*After F. Rodriguez, "Principles of Polymer Systems," 2nd ed., McGraw-Hill, 1982, p. 42.*)

matrix. As cooling is continued, the glass transition is encountered, as indicated by the slope change of specific volume versus temperature in Fig. 10.14 between *E* and *F.* In going through the glass transition, the supercooled liquid matrix transforms to the glassy state, and thus the structure of the thermoplastic at *F* consists of crystalline regions in a glassy noncrystalline matrix. An example of a thermoplastic that solidifies to form a partly crystalline structure is polyethylene.

10.4.3 Structure of Partly Crystalline Thermoplastic Materials

The exact way in which polymer molecules are arranged in a crystalline structure is still in doubt, and more research is needed in this area. The longest dimension of crystalline regions or crystallites in polycrystalline polymeric materials is usually about 5 to 50 nm, which is a small percentage of the length of a fully extended polymer molecule that may be about 5000 nm. An early model called the *fringed-micelle model* pictured long polymer chains of about 5000 nm wandering successively through a series of disordered and ordered regions along the length of the polymer molecule (Fig. 10.16*a*). A newer model called the *folded-chain model* pictures

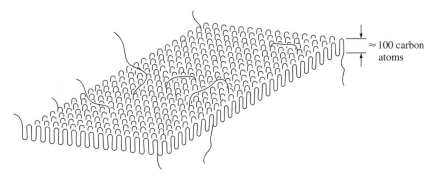

Figure 10.17
Schematic folded-chain structure of a lamella of low-density polyethylene.
[*(After R. L. Boysen, Olefin Polymers (High-Pressure Polyethylene), in "Encyclopedia of Chemical Technology," vol. 16, Wiley, 1981, p. 405.)*]

sections of the molecular chains folding on themselves so that a transition from crystalline to noncrystalline regions can be formed (Fig. 10.16*b*).

There has been an intensive study over the past years on partly crystalline thermoplastics, especially polyethylene. Polyethylene is believed to crystallize in a folded-chain structure with an orthorhombic cell, as shown in Fig. 10.17. Each length of chain between folds is about 100 carbon atoms, with each single layer of the folded-chain structure being referred to as a *lamella.* Under laboratory conditions, low-density polyethylene crystallizes in a spherulitic-type structure, which is shown in Fig. 10.18. The spherulitic regions, which consist of crystalline lamellae, are the dark areas, and the regions between the spherulitic structures are noncrystalline white areas. The spherulitic structure shown in Fig. 10.18 grows only under carefully controlled stress-free laboratory conditions.

The degree of crystallinity in partly crystalline linear polymeric materials ranges from about 5 to 95 percent of their total volume. Complete crystallization is not attainable even with polymeric materials that are highly crystallizable because of molecular entanglements and crossovers. The amount of crystalline material within a thermoplastic affects its tensile strength. In general as the degree of crystallinity increases, the strength of the material increases.

10.4.4 Stereoisomerism in Thermoplastics

Stereoisomers are molecular compounds that have the same chemical compositions but different structural arrangements. Some thermoplastics such as polypropylene can exist in three different stereoisomeric forms:

1. **Atactic stereoisomer.** The pendant methyl group of polypropylene is randomly arranged on either side of the main-carbon chain (Fig. 10.19*a*).
2. **Isotactic stereoisomer.** The pendant methyl group is always on the same side of the main-carbon chain (Fig. 10.19*b*).
3. **Syndiotactic stereoisomer.** The pendant group regularly alternates from one side of the main chain to the other side (Fig. 10.19*c*).

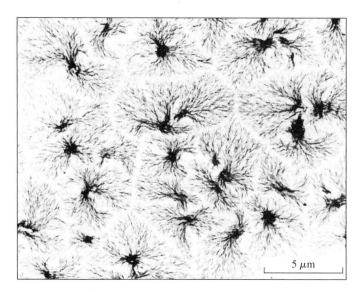

Figure 10.18
Cast-film spherulitic structure of low-density polyethylene, density 0.92 g/cm^3.
[After R. L. Boysen, Olefin Polymers (High-Pressure Polyethylene), in "Encyclopedia of Chemical Technology," vol. 16, Wiley, 1981, p. 406.]

The discovery of a catalyst that made possible the industrial polymerization of isotactic linear-type polymers was a great breakthrough for the plastics industry. With a **stereospecific catalyst,** isotactic polypropylene could be produced on a commercial scale. Isotactic polypropylene is a highly crystalline polymeric material with a melting point of 165 to 175°C. Because of its high crystallinity, isotactic propylene has higher strengths and higher heat-deflection temperatures than atactic polypropylene.

10.4.5 Ziegler and Natta Catalysts

Karl Ziegler and Giulio Natta were awarded the 1963 Nobel Prize (jointly)[4] for their work on linear polyethylenes and stereoisomers of polypropylene. Many books have been written on this subject, and the details are beyond the scope of this book, but references are given for those who wish to pursue this subject. Briefly, metallocene catalysts are used in conjunction with the product. Thus, the metallocenes are not true catalysts in the sense that they *do* take part in the reactions and are consumed to a small extent by the reactions. Therefore, metallocene catalysts have opened a new era for polyolefin polymerization.

---

[4]1963 Nobel laureates in Chemistry were Karl Ziegler and Giulio Natta for their work on the controlled polymerization of hydrocarbons through the use of novel organometallic catalysts.

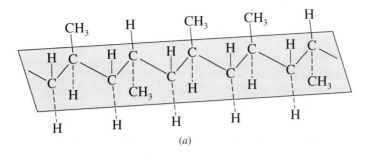

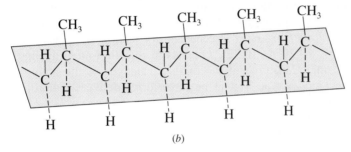

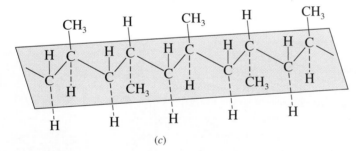

Figure 10.19

Polypropylene stereoisomers. (*a*) Atactic isomer in which the pendant CH₃ groups are randomly arranged on either side of main-chain carbons. (*b*) Isotactic isomer in which the pendant CH₃ groups are all on the same side of the main-chain carbons. (*c*) Syndiotactic isomer in which the pendant CH₃ groups are regularly alternating from one side to the other of main-chain carbons.

[*After G. Crespi and L. Luciani, Olefin Polymers (Polyethylene), in "Encyclopedia of Chemical Technology," vol. 16, Wiley, 1982, p. 454.*]

10.5 PROCESSING OF PLASTIC MATERIALS

Many different processes are used to transform plastic granules and pellets into shaped products such as sheet, rods, extruded sections, pipe, or finished molded parts. The process used depends to a certain extent on whether the plastic is a thermoplastic or thermosetting one. Thermoplastics are usually heated to a soft condition and then reshaped before cooling. On the other hand, thermosetting materials, not having been completely polymerized before processing to the finished shape, use a process by which a chemical reaction occurs to cross-link polymer chains into a network polymeric material. The final polymerization can take place by the application of heat and pressure or by catalytic action at room temperature or higher temperatures.

In this section we shall discuss some of the most important processes used for thermoplastic and thermosetting materials.

10.5.1 Processes Used for Thermoplastic Materials

Injection Molding **Injection molding** is one of the most important processing methods used for forming thermoplastic materials. The modern injection-molding machine uses a reciprocating-screw mechanism for melting the plastic and injecting it into a mold (Figs. 10.20 to 10.22). Older-type injection-molding machines use a plunger for melt injection. One of the main advantages of the reciprocating-screw method over the plunger type is that the screw drive delivers a more homogeneous melt for injection.

In the injection-molding process, plastic granules from a hopper are fed through an opening in the injection cylinder onto the surface of a rotating screw drive, which carries them forward toward the mold (Fig. 10.22a). The rotation of the screw forces

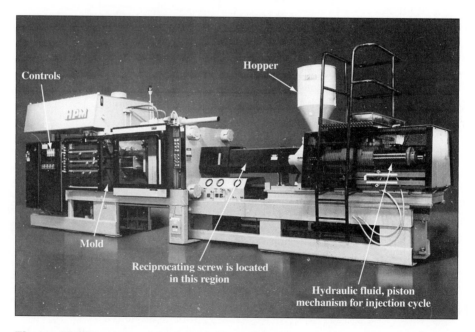

Figure 10.20
Front view of a 500-ton reciprocating-screw injection-molding machine for plastic materials.
(*Courtesy of HPM Corporation.*)

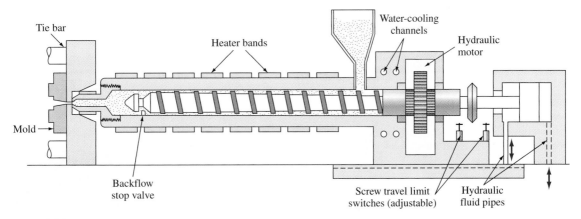

Figure 10.21
Cross section of reciprocating-screw injection-molding machine for plastic materials.
(After J. Bown, "Injection Molding of Plastic Components," McGraw-Hill, 1979, p. 28.)

the granules against the heated walls of the cylinder, causing them to melt due to the heat of compression, friction, and the hot walls of the cylinder (Fig. 10.22*b*). When sufficient plastic material is melted at the mold end of the screw, the screw stops and by plungerlike motion injects a "shot" of melted plastic through a runner-gate system and then into the closed mold cavities (Fig. 10.22*c*). The screw shaft maintains pressure on the plastic material fed into the mold for a short time to allow it to become solid and then is retracted. The mold is water-cooled to rapidly cool the plastic part. Finally, the mold is opened and the part is ejected from the mold with air or by spring-loaded ejector pins (Fig. 10.22*d*). The mold is then closed and ready for another cycle.

The main advantages of injection molding are:

1. High-quality parts can be produced at a high production rate.
2. The process has relatively low labor costs.
3. Good surface finishes can be produced on the molded part.
4. The process can be highly automated.
5. Intricate shapes can be produced.

The main disadvantages of injection molding are:

1. High cost of the machine means that a large volume of parts must be made to pay for the machine.
2. The process must be closely controlled to produce a quality product.

Extrusion   **Extrusion** is another of the important processing methods used for thermoplastics. Some of the products manufactured by the extrusion process are pipe, rod, film, sheet, and shapes of all kinds. The extrusion machine is also used for

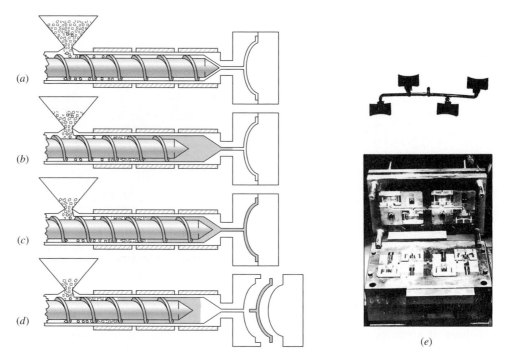

Figure 10.22

Sequence of operations for the reciprocating-screw injection-molding process for plastic materials. (*a*) Plastic granules are delivered by a revolving-screw barrel. (*b*) Plastic granules are melted as they travel along the revolving screw, and when sufficient material is melted at the end of the screw, the screw stops rotating. (*c*) The screw barrel is then driven forward with a plungerlike motion and injects the melted plastic through an opening into a runner and gate system and then into a closed-mold cavity. (*d*) The screw barrel is retracted and the finished plastic part ejected. (*e*) Open mold dies showing removed plastic part above.

(*Courtesy of Plastics Engineering Co., Sheboygan, Wisc.*)

making compounded plastic materials for the production of raw shapes such as pellets and for the reclamation of scrap thermoplastic materials.

In the extrusion process, thermoplastic resin is fed into a heated cylinder, and the melted plastic is forced by a rotating screw through an opening (or openings) in an accurately machined die to form continuous shapes (Fig. 10.23). After exiting from the die, the extruded part must be cooled below its glass transition temperature to ensure dimensional stability. The cooling is usually done with an air-blast or water-cooling system.

Blow Molding and Thermoforming   Other important processing methods for thermoplastics are blow molding and thermoforming of sheet. In **blow molding,** a cylinder or tube of heated plastic called a *parison* is placed between the jaws of a mold (Fig. 10.24*a*). The mold is closed to pinch off the ends of the cylinder (Fig. 10.24*b*),

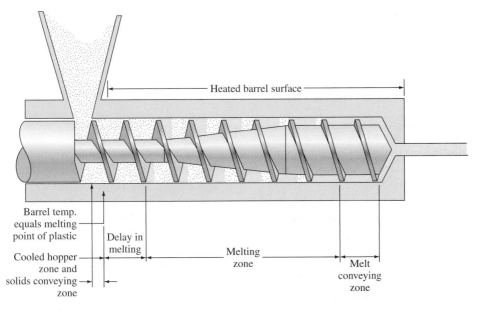

Heated barrel surface

Barrel temp. equals melting point of plastic

Delay in melting

Melting zone

Melt conveying zone

Cooled hopper zone and solids conveying zone

Figure 10.23

Schematic drawing of an extruder, showing the various functional zones: hopper, solids-conveying zone, delay in melting start, melting zone, and melt-pumping zone.

[*After H. S. Kaufman and J. J. Falcetta (eds.), "Introduction to Polymer Science and Technology," Society of Plastic Engineers, Wiley, 1977, p. 462.*]

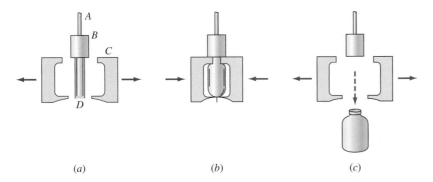

(*a*)   (*b*)   (*c*)

Figure 10.24

Sequence of steps for the blow molding of a plastic bottle. (*a*) A section of tube is introduced into the mold. (*b*) The mold is closed, and the bottom of the tube is pinched together by the mold. (*c*) Air pressure is fed through the mold into the tube, which expands to fill the mold, and the part is cooled as it is held under air pressure. *A* = air line, *B* = die, *C* = mold, *D* = tube section.

(*After P. N. Richardson, Plastics Processing, in "Encyclopedia of Chemical Technology," vol. 18, Wiley, 1982, p. 198.*)

and compressed air is blown in, forcing the plastic against the walls of the mold (Fig. 10.24c).

In *thermoforming,* a heated plastic sheet is forced into the contours of a mold by pressure. Mechanical pressure may be used with mating dies, or a vacuum may be used to pull the heated sheet into an open die. Air pressure may also be used to force a heated sheet into an open die.

10.5.2 Processes Used for Thermosetting Materials

Compression Molding   Many thermosetting resins such as the phenol-formaldehyde, urea-formaldehyde, and melamine-formaldehyde resins are formed into solid parts by the compression-molding process. In **compression molding,** the plastic resin, which may be preheated, is loaded into a hot mold containing one or more cavities (Fig. 10.25a). The upper part of the mold is forced down on the plastic resin, and the applied pressure and heat melts the resin and forces the liquefied plastic to fill the cavity or cavities (Fig. 10.25b). Continued heating (usually a minute or two) is required to complete the cross-linking of the thermosetting resin, and then the part is ejected from the mold. The excess flash is trimmed later from the part.

The advantages of compression molding are:

1. Because of the relative simplicity of the molds, initial mold costs are low.
2. The relatively short flow of material reduces wear and abrasion on molds.
3. Production of large parts is more feasible.

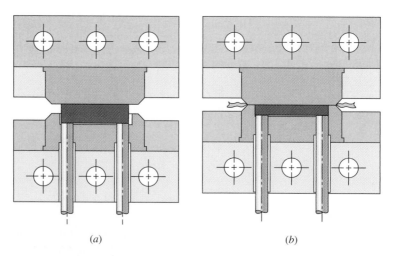

(a)   *(b)*

Figure 10.25
Compression molding. (*a*) Cross section of open mold containing preformed powdered shape in mold cavity. (*b*) Cross section of closed mold showing molded specimen and excess flash.
(After R. B. Seymour, Plastics Technology, in "Encyclopedia of Chemical Technology," vol. 15, Wiley, 1968, p. 802.)

4. More-compact molds are possible because of the simplicity of the mold.

5. Expelled gases from the curing reaction can escape during the molding process.

The disadvantages of compression molding are:

1. Complicated part configurations are difficult to make with this process.

2. Inserts may be difficult to hold to close tolerances.

3. Flash must be trimmed from the molded parts.

Transfer Molding   **Transfer molding** is also used for molding thermosetting plastics such as the phenolics, ureas, melamines, and alkyd resins. Transfer molding differs from compression molding in how the material is introduced into the mold cavities. In transfer molding the plastic resin is not fed directly into the mold cavity but into a chamber outside the mold cavities (Fig. 10.26*a*). In transfer molding,

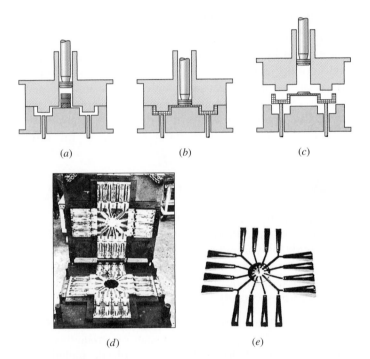

Figure 10.26
Transfer molding. (*a*) Preformed plastic shape is forced into a preclosed mold by a plunger. (*b*) Pressure on plastic shape is applied, and plastic is forced through a system of runners and gates into the mold cavities. (*c*) After the plastic has cured, the plunger is removed and the mold cavity is opened. The part is then ejected. (*d*) Open mold showing mold cavities. (*e*) Removed plastic part.
(*Courtesy of Plastics Engineering Co., Sheboygan, Wisc.*)

when the mold is closed, a plunger forces the plastic resin (which is usually preheated) from the outside chamber through a system of runners and gates into the mold cavities (Fig. 10.26*b*). After the molded material has had time to cure so that a rigid network polymeric material is formed, the molded part is ejected from the mold (Fig. 10.26*c*). Figure 10.26*d* shows the mold cavities of an open transfer mold, and Fig. 10.26*e* shows the as-molded plastic material containing multiple thermoset plastic parts.

The advantages of transfer molding are:

1. Transfer molding has the advantage over compression molding in that no flash is formed during molding, and thus the molded part requires less finishing.

2. Many parts can be made at the same time by using a runner system (Fig. 10.26*e*).

3. Transfer molding is especially useful for making small, intricate parts that would be difficult to make by compression molding.

Injection Molding   Using modern technology, some thermosetting compounds can be injection molded by reciprocating-screw injection-molding machines. Special heating and cooling jackets have been added to the standard-type injection-molding machines so that the resin can be cured in the process. Good venting of the mold cavities is required for some thermosetting resins that give off reaction products during curing. In the future, injection molding will probably become more important for producing thermosetting parts because of the efficiency of this process.

10.6 GENERAL-PURPOSE THERMOPLASTICS

In this section some important aspects of the structure, chemical processing, properties, and applications of the following thermoplastics will be discussed: polyethylene, polyvinyl chloride, polypropylene, polystyrene, ABS, polymethyl methacrylate, cellulose acetate and related materials, and polytetrafluoroethylene.

Let us first, however, examine the sales tonnages, list prices, and some of the important properties of these materials.

Global Sales Tonnages and Bulk List Prices for Some General-Purpose Thermoplastics   Based on the global sales tonnages of certain thermoplastics in 1998, along with their 2000 bulk list prices, four major plastic materials—polyethylene, polyvinyl chloride, polypropylene, and polystyrene—accounted for most of the plastic materials sold. These materials have a relatively low cost of about 50¢/lb (2000 prices), which undoubtedly accounts for part of the reason for their extensive usage in industry and for many engineering applications. However, when special properties are required that are unobtainable with the cheaper thermoplastics, more costly plastic materials are used. For example, polytetrafluoroethylene (Teflon), which has special high-temperature and lubrication properties, cost about five to nine dollars per pound in 2000.

Table 10.2 Some properties of selected general-purpose thermoplastics

Material	Density (g/cm³)	Tensile strength (×1000 psi)*	Impact strength, Izod (ft · lb/in)†	Dielectric strength (V/mil)‡	Max-use temp. (no load)	
					°F	°C
Polyethylene:						
Low-density	0.92–0.93	0.9–2.5		480	180–212	82–100
High-density	0.95–0.96	2.9–5.4	0.4–14	480	175–250	80–120
Rigid, chlorinated PVC	1.49–1.58	7.5–9	1.0–5.6		230	110
Polypropylene, general-purpose	0.90–0.91	4.8–5.5	0.4–2.2	650	225–300	107–150
Styrene-acrylonitrile (SAN)	1.08	10–12	0.4–0.5	1775	140–220	60–104
ABS, general-purpose	1.05–1.07	5.9	6	385	160–200	71–93
Acrylic, general-purpose	1.11–1.19	11.0	2.3	450–500	130–230	54–110
Cellulosics, acetate	1.2–1.3	3–8	1.1–6.8	250–600	140–220	60–104
Polytetrafluoroethylene	2.1–2.3	1–4	2.5–4.0	400–500	550	288

*1000 psi = 6.9 MPa.
†Notched Izod test: 1 ft · lb/in = 53.38 J/m.
‡1 V/mil = 39.4 V/mm.

Source: *Materials Engineering*, May 1972.

Some Basic Properties of Selected General-Purpose Thermoplastics Table 10.2 lists the densities, tensile strengths, impact strengths, dielectric strengths, and maximum-use temperatures for some selected general-purpose thermoplastics. One of the most important advantages of many plastic materials for many engineering applications is their relatively low densities. Most general-purpose plastics have densities of about 1 compared to 7.8 for iron.

The tensile strengths of plastic materials are relatively low, and as a result this property can be a disadvantage for some engineering designs. Most plastic materials have a tensile strength of less than 10,000 psi (69 MPa) (Table 10.2). The tensile test for plastic materials is carried out with the same equipment used for metals (Fig. 6.18).

The impact test that is usually used for plastic materials is the notched Izod test. In this test a $\frac{1}{8} \times \frac{1}{2} \times 2\frac{1}{2}$ in. sample (Fig. 10.27) is normally used and is clamped to the base of a pendulum testing machine. The amount of energy absorbed per unit length of the notch when the pendulum strikes the sample is measured and is called the *notched impact strength* of the material. This energy is usually reported in foot pounds per inch (ft · lb/in) or joules per meter (J/m). The notched impact strengths of general-purpose plastic materials of Table 10.2 range from 0.4 to 14 ft · lb/in.

Plastic materials are generally good electrical insulative materials. The electrical insulative strength of plastic materials is usually measured by their *dielectric strength*, which may be defined as the voltage gradient that produces electrical breakdown

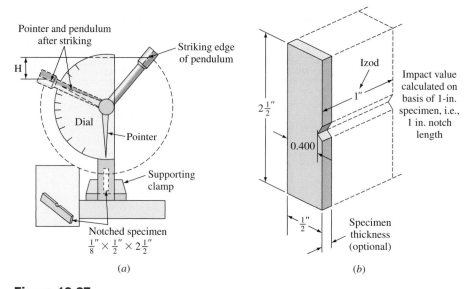

Figure 10.27

(a) Izod impact test. (b) Sample used for plastic materials for the Izod impact test.
(*After W. E. Driver, "Plastics Chemistry and Technology," Van Nostrand Reinhold, 1979, pp. 196–197.*)

through the material. Dielectric strength is usually measured in volts per mil or volts per millimeter. The dielectric strengths of the plastic materials of Table 10.2 vary from 385 to 1775 V/mil.

The maximum-use temperature for most plastic materials is relatively low and varies from 130 to 300°F (54 to 149°C) for most thermoplastic materials. However, some thermoplastics have higher maximum-use temperatures. For example, polytetrafluoroethylene can withstand temperatures up to 550°F (288°C).

10.6.1 Polyethylene

Polyethylene (PE) is a clear-to-whitish, translucent thermoplastic material and is often fabricated into clear thin films. Thick sections are translucent and have a waxy appearance. With the use of colorants a wide variety of colored products is obtained.

Repeating Chemical Structural Unit

$$
\begin{bmatrix} \ \overset{\displaystyle H}{\underset{\displaystyle H}{\overset{|}{\underset{|}{C}}}} - \overset{\displaystyle H}{\underset{\displaystyle H}{\overset{|}{\underset{|}{C}}}} \ \end{bmatrix}_n
\qquad
\begin{array}{l}\text{Polyethylene}\\ \text{mp: } 110\text{--}137°C\\ \quad (230\text{--}278°F)\end{array}
$$

Types of Polyethylene   In general there are two types of polyethylene: *low-density* (LDPE) and *high-density* (HDPE). Low-density has a branched-chain structure

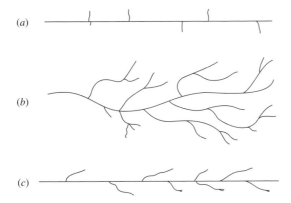

Figure 10.28
Chain structure of different types of polyethylene:
(*a*) high-density, (*b*) low-density, and (*c*) linear-low-density.

(Fig. 10.28*b*), whereas high-density polyethylene has essentially a straight-chain structure (Fig. 10.28*a*).

Low-density polyethylene was first commercially produced in the United Kingdom in 1939 by using autoclave (or tubular) reactors requiring pressures in excess of 14,500 psi (100 MPa) and a temperature of about 300°C. High-density polyethylene was first produced commercially by the Phillips and Ziegler processes by using special catalysts in 1956–1957. In these processes the pressure and temperature for the reaction to convert ethylene to polyethylene were considerably lowered. For example, the Phillips process operates at 100 to 150°C and 290 to 580 psi (2 to 4 MPa) pressure.

In about 1976 a new low-pressure simplified process for producing polyethylene was developed that uses a pressure of about 100 to 300 psi (0.7 to 2 MPa) and a temperature of about 100°C. The polyethylene produced is described as *linear-low-density polyethylene* (LLDPE) and has a linear-chain structure with short, slanting side branches (Fig. 10.28*c*). A process for producing LLDPE has been described in Sec. 10.3 (see Fig. 10.13).

Structure and Properties The chain structures of low- and high-density polyethylenes are shown in Fig. 10.28. Low-density polyethylene has a branched-chain structure that lowers its degree of crystallinity and its density (Table 10.2). The branched-chain structure also lowers the strength of low-density polyethylene because it reduces intermolecular bonding forces. High-density polyethylene, in contrast, has very little branching on the main chains, and so the chains can pack more closely together to increase crystallinity and strength (Table 10.3).

Polyethylene is by far the most extensively used plastic material. The main reason for its prime position is that it is low in cost and has many industrially important properties, which include toughness at room temperature and at low temperatures

Table 10.3 Some properties of low- and high-density polyethylenes

Property	Low-density polyethylene	Linear-low-density polyethylene	High-density polyethylene
Density (g/cm^3)	0.92–0.93	0.922–0.926	0.95–0.96
Tensile strength, ($\times$1000 psi)	0.9–2.5	1.8–2.9	2.9–5.4
Elongation (%)	550–600	600–800	20–120
Crystallinity (%)	65	$\cdots$	95

Figure 10.29
High-density polyethylene film pond liner dwarfs workers installing it. Individual sheets can be half an acre in area and weigh up to 5 tons.
(*Courtesy of Schlegel Lining Technology, Inc.*)

with sufficient strength for many product applications, good flexibility over a wide range of temperatures even down to $-73°C$, excellent corrosion resistance, excellent insulating properties, odorlessness and tastelessness, and low water-vapor transmission.

Applications   Applications for polyethylene include containers, electrical insulation, chemical tubing, housewares, and blow-molded bottles. Uses for polyethylene films include films for packaging and materials handling and water-pond liners (Fig. 10.29).

10.6.2 Polyvinyl Chloride and Copolymers

Polyvinyl chloride (PVC) is a widely used synthetic plastic that has the second largest sales tonnage in the world. The widespread use of PVC is attributed mainly to its high chemical resistance and its unique ability to be mixed with additives to produce a large number of compounds with a wide range of physical and chemical properties.

Repeating Chemical Structural Unit

$$\begin{bmatrix} \begin{array}{cc} H & H \\ | & | \\ C - C \\ | & | \\ H & Cl \end{array} \end{bmatrix}_n$$

Polyvinyl chloride
mp: ~204°C (~400°F)

Structure and Properties   The presence of the large chlorine atom on every other carbon atom of the main chain of polyvinyl chloride produces a polymeric material that is essentially amorphous and does *not* recrystallize. The strong cohesive forces between the polymer chains in PVC are due mainly to the strong dipole moments caused by the chlorine atoms. The large negative chlorine atoms, however, cause some steric hindrance and electrostatic repulsion, which reduces the flexibility of the polymer chains. This molecular immobility results in difficulty in the processing of the homopolymer, and only in a few applications can PVC be used without being compounded with a number of additives so that it can be processed and converted into finished products.

PVC homopolymer has a relatively high strength (7.5 to 9.0 ksi), along with brittleness. PVC has a medium heat-deflection temperature (57 to 82°C [135 to 180°F] at 66 psi), good electrical properties (425 to 1300 V/mil dielectric strength), and high solvent resistance. The high chlorine content of PVC produces flame and chemical resistance.

Polyvinyl Chloride Compounding   Polyvinyl chloride can only be used for a few applications without the addition of a number of compounds to the basic material so that it can be processed and converted into a finished product. Compounds added to PVC include plasticizers, heat stabilizers, lubricants, fillers, and pigments.

Plasticizers impart flexibility to polymeric materials. They are usually high-molecular-weight compounds that are selected to be completely miscible and compatible with the basic material. For PVC, phthalate esters are commonly used as plasticizers. The effect of some plasticizers on the tensile strength of PVC is shown in Fig. 10.30.

Heat stabilizers are added to PVC to prevent thermal degradation during processing and may also help to extend the life of the finished product. Typical stabilizers used may be all organic or inorganic but are usually organometallic compounds based on tin, lead, barium-cadmium, calcium, and zinc.

Lubricants aid the melt flow of PVC compounds during processing and prevent adhesion to metal surfaces. Waxes, fatty esters, and metallic soaps are commonly used lubricants.

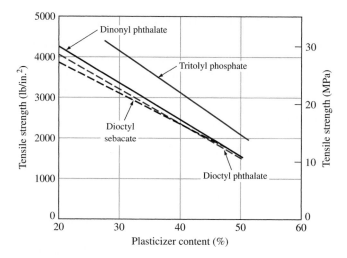

Figure 10.30

Effect of different plasticizers on the tensile strength of polyvinyl chloride.

[*After C. A. Brighton, Vinyl Chloride Polymers (Compounding), in "Encyclopedia of Polymer Science and Technology," vol. 14, Wiley, 1971, p. 398.*]

Fillers such as calcium carbonate are mainly added to lower the cost of PVC compounds.

Pigments, both inorganic and organic, are used to give color, opacity, and weatherability to PVC compounds.

Rigid Polyvinyl Chloride   Polyvinyl chloride alone can be used for some applications but is difficult to process and has low impact strength. The addition of rubbery resins can improve melt flow during processing by forming a dispersion of small, soft rubbery particles in the matrix of rigid PVC. The rubbery material serves to absorb and disperse impact energy so that the impact resistance of the material is increased. With improved properties, rigid PVC is used for many applications. In building construction rigid PVC is used for pipe, siding, window frames, gutter, and interior molding and trim. PVC is also used for electrical conduit.

Plasticized Polyvinyl Chloride   The addition of plasticizers to PVC produces softness, flexibility, and extensibility. These properties can be varied over a wide range by adjusting the plasticizer-polymer ratio. Plasticized polyvinyl chloride is used in many applications where it outperforms rubber, textiles, and paper. Plasticized PVC is used for furniture and auto upholstery, interior wall coverings, rainwear, shoes, luggage, and shower curtains. In transportation plasticized PVC is used for auto top coverings, electrical wire insulation, floor mats, and interior and exterior trim. Other applications include garden hoses, refrigerator gaskets, appliance components, and housewares.

10.6.3 Polypropylene

Polypropylene is the third most important plastic from a sales tonnage standpoint and is one of the lowest in cost since it can be synthesized from low-cost petrochemical raw materials using a Ziegler-type catalyst.

Repeating Chemical Structural Unit

$$\left[\begin{array}{c} H \quad H \\ | \quad | \\ C-C \\ | \quad | \\ H \quad CH_3 \end{array}\right]_n$$

Polypropylene
mp: 165−177°C
(330−350°F)

Structure and Properties   In going from polyethylene to polypropylene, the substitution of a methyl group on every second carbon atom of the polymer main chain restricts rotation of the chains, producing a stronger but less flexible material. The methyl groups on the chains also increase the glass transition temperature, and thus polypropylene has higher melting and heat-deflection temperatures than polyethylene. With the use of stereospecific catalysts, isotactic polypropylene can be synthesized with a melting point in the 165 to 177°C (330 to 350°F) range. This material can be subjected to temperatures of about 120°C (250°F) without deformation.

Polypropylene has a good balance of attractive properties for producing many manufactured goods. These include good chemical, moisture, and heat resistance, along with low density (0.900 to 0.910 g/cm³), good surface hardness, and dimensional stability. Polypropylene also has outstanding flex life as a hinge and can be used for products with an integral hinge. Along with the low cost of its monomer, polypropylene is a very competitive thermoplastic material.

Applications   The major applications for polypropylene are housewares, appliance parts, packaging, laboratory ware, and bottles of various types. In transportation, high-impact polypropylene copolymers have replaced hard rubber for battery housings. Similar resins are used for fender liners and splash shrouds. Filled polypropylene finds application for automobile fan shrouds and heater ducts, where high resistance to heat deflection is needed. Also, polypropylene homopolymer is used extensively for primary carpet backing and as a woven material is used for shipping sacks for many industrial products. In the film market, polypropylene is used as a bag and overwrap film for soft goods because of its luster, gloss, and good stiffness. In packaging, polypropylene is used for screw closures, cases, and containers.

10.6.4 Polystyrene

Polystyrene is the fourth-largest tonnage thermoplastic. Homopolymer polystyrene is a clear, odorless, and tasteless plastic material that is relatively brittle unless modified. Besides crystal polystyrene, other important grades are rubber-modified, impact-resistant, and expandable polystyrenes. Styrene also is used to produce many important copolymers.

Repeating Chemical Structural Unit

$$
\left[\begin{array}{cc}
H & H \\
| & | \\
C & C \\
| & | \\
H & \bigcirc
\end{array}\right]_n
\quad \text{Polystyrene} \\
\text{mp: 150–243°C} \\
\text{(330–470°F)}
$$

Structure and Properties   The presence of the phenylene ring on every other carbon atom of the main chain of polystyrene produces a rigid bulky configuration with sufficient steric hindrance to make the polymer very inflexible at room temperature. The homopolymer is characterized by its rigidity, sparkling clarity, and ease of processibility but tends to be brittle. The impact properties of polystyrene can be improved by copolymerization with the polybutadiene elastomer, which has the following chemical structure:

$$
\left[\begin{array}{cccc}
H & H & H & H \\
| & | & | & | \\
C & C = C & C \\
| & & & | \\
H & & & H
\end{array}\right]_n
\quad \text{Polybutadiene}
$$

Copolymers of impact styrene usually have rubber levels between 3 and 12 percent. The addition of the rubber to polystyrene lowers the rigidity and heat-deflection temperature of the homopolymer.

In general polystyrenes have good dimensional stability and low-mold shrinkage and are easily processed at a low cost. However, they have poor weatherability and are chemically attacked by organic solvents and oils. Polystyrenes have good electrical insulating properties and adequate mechanical properties within operating temperature limits.

Applications   Typical applications include automobile interior parts, appliance housings, dials and knobs, and housewares.

10.6.5 Polyacrylonitrile

This acrylic-type polymeric material is often used in the form of fibers, and because of its strength and chemical stability, it is also used as a comonomer for some engineering thermoplastics.

Repeating Chemical Structural Unit

$$
\left[\begin{array}{cc}
H & H \\
| & | \\
C & C \\
| & | \\
H & C \equiv N
\end{array}\right]_n
\quad \text{Polyacrylonitrile} \\
\text{(does not melt)}
$$

Structure and Properties   The high electronegativity of the nitrile group on every other carbon atom of the main chain exerts mutual electrical repulsion, causing the molecular chains to be forced into extended, stiff, rodlike structures. The regularity of the rodlike structures allows them to orient themselves to produce strong fibers by hydrogen bonding between the polymer chains. As a result, the acrylonitrile fibers have high strength and good resistance to moisture and solvents.

Applications   Acrylonitrile is used in fiber form for woollike applications such as sweaters and blankets. Acrylonitrile is also used as a comonomer for producing *styrene-acrylonitrile copolymers* (SAN resins) and *acrylonitrile-butadiene-styrene terpolymers* (ABS resins).

10.6.6 Styrene-Acrylonitrile (SAN)

Styrene-acrylonitrile thermoplastics are high-performance members of the styrene family.

Structure and Properties   SAN resins are random, amorphous copolymers of styrene and acrylonitrile. This copolymerization creates polarity and hydrogen-bond attractive forces between the polymer chains. As a result, SAN resins have better chemical resistance, higher heat-deflection temperatures, toughness, and load-bearing characteristics than polystyrene alone. SAN thermoplastics are rigid and hard, process easily, and have the gloss and clarity of polystyrene.

Applications   Major applications for SAN resins include automotive instrument lenses, dash components, and glass-filled support panels; appliance knobs and blender and mixer bowls; medical syringes and blood aspirators; construction safety glazing; and houseware tumblers and mugs.

10.6.7 ABS

ABS is the name given to a family of thermoplastics. The acronym is derived from the three monomers used to produce ABS: *a*crylonitrile, *b*utadiene, and *s*tyrene. ABS materials are noted for their engineering properties such as good impact and mechanical strength combined with ease of processing.

Chemical Structural Units   ABS contains the following three chemical structural units:

A: polyacrylonitrile   B: polybutadiene   S: polystyrene

Structure and Properties of ABS   The wide range of useful engineering properties exhibited by ABS is due to the contributing properties of each of its components. Acrylonitrile contributes heat and chemical resistance and toughness, butadiene

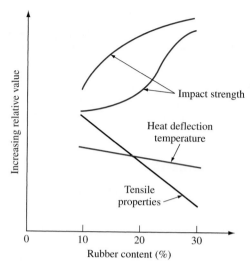

Figure 10.31
Percent rubber versus some properties of ABS.

(After G. E. Teer, ABS and Related Multipolymers, in "Modern Plastics Encyclopedia," McGraw-Hill, 1981–1982.)

Table 10.4 Some typical properties of ABS plastics (23°C)

	High-impact	**Medium-impact**	**Low-impact**
Impact strength (Izod):			
ft · lb/in	7–12	4–7	2–4
J/m	375–640	215–375	105–320
Tensile strength:			
×1000 psi	4.8–6.0	6.0–7.0	6.0–7.5
MPa	33–41	41–48	41–52
Elongation (%)	15–70	10–50	5–30

provides impact strength and low-property retention, and styrene provides surface gloss, rigidity, and ease of processing. The impact strengths of ABS plastics are increased as the percent rubber content is increased, but the tensile-strength properties and heat-deflection temperatures are decreased (Fig. 10.31). Table 10.4 lists some of the properties of high-, medium-, and low-impact ABS plastics.

The structure of ABS is *not* that of a random terpolymer. ABS can be considered a blend of a glassy copolymer (styrene-acrylonitrile) and rubbery domains (primarily a butadiene polymer or copolymer). Simply blending rubber with the glassy copolymer does not produce optimal impact properties. The best impact strength is obtained when the styrene-acrylonitrile copolymer matrix is grafted to the rubber domains to produce a two-phase structure (Fig. 10.32).

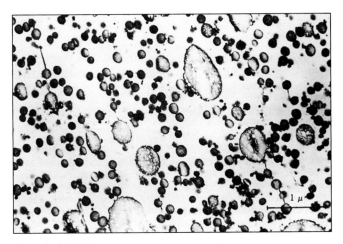

Figure 10.32
Electron micrograph of an ultrathin section of a type
G ABS resin showing the rubbery particles in a copolymer
of styrene-acrylonitrile.
[*After M. Matsuo, Polym. Eng. Sci.,* **9:**206(1969).]

Applications   The major use for ABS is for pipe and fittings, particularly drain-waste-
and-vent pipe in building. Other uses for ABS are for automotive parts, appliance parts
such as refrigerator door liners and inner liners, business machines, computer housings
and covers, telephone housings, electrical conduit, and electromagnetic interference–
radio-frequency shielding applications.

10.6.8 Polymethyl Methacrylate (PMMA)

Polymethyl methacrylate (PMMA) is a hard, rigid transparent thermoplastic that has
good outdoor weatherability and is more impact-resistant than glass. This material
is better known by the trade names Plexiglas or Lucite and is the most important
material of the group of thermoplastics known as the *acrylics.*

Repeating Chemical Structural Unit

$$\left[\begin{array}{cc} H & CH_3 \\ | & | \\ -C-C- \\ | & \diagdown \\ H & O \\ & C \\ & \diagup \\ & OCH_3 \end{array}\right]_n$$

Polymethyl methacrylate
mp: 160°C (320°F)

Structure and Properties   The substitution of the methyl and methacrylate groups
on every other carbon atom of the main carbon chain provides considerable steric hindrance and thus makes PMMA rigid and relatively strong. The random configuration
of the asymmetrical carbon atoms produces a completely amorphous structure that

has a high transparency to visible light. PMMA also has good chemical resistance to outdoor environments.

Applications   PMMA is used for glazing for aircraft and boats, skylights, exterior lighting, and advertising signs. Other uses include auto taillight lenses, safety shields, protective goggles, and knobs and handles.

10.6.9 Fluoroplastics

These materials are plastics or polymers made from monomers containing one or more atoms of fluorine. The fluoroplastics have a combination of special properties for engineering applications. As a class they have a high resistance to hostile chemical environments and outstanding electrical insulating properties. Those fluoroplastics containing a large percentage of fluorine have low coefficients of friction, which give them self-lubricating and nonstick properties.

There are many fluoroplastics produced, but the two most widely used ones, *polytetrafluoroethylene* (PTFE) (Fig. 10.33) and *polychlorotrifluoroethylene* (PCTFE), will be discussed in this subsection.

Polytetrafluoroethylene

Repeating Chemical Structural Unit

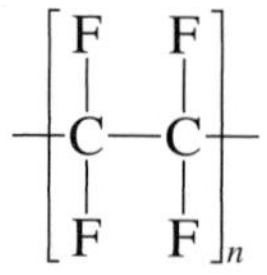

$$\left[\begin{array}{cc} F & F \\ | & | \\ C - C \\ | & | \\ F & F \end{array}\right]_n$$

Polytetrafluoroethylene
Softens at 370°C (700°F)

Figure 10.33
Structure of
polytetrafluoroethylene.

Chemical Processing   PTFE is a completely fluorinated polymer formed by the free-radical chain polymerization of tetrafluoroethylene gas to produce linear-chain polymers of —CF_2— units. The original discovery of the polymerization of tetrafluoroethylene gas into polytetrafluoroethylene (Teflon) was made by R. J. Plunkett in 1938 in a Du Pont laboratory.

Structure and Properties   PTFE is a crystalline polymer with a crystalline melting point of 327°C (620°F). The small size of the fluorine atom and the regularity of the fluorinated carbon chain polymer results in a highly dense crystalline polymeric material. The density of PTFE is high for plastic materials, 2.13 to 2.19 g/cm^3.

PTFE has exceptional resistance to chemicals and is insoluble in all organics with the exception of a few fluorinated solvents. PTFE also has useful mechanical properties from cryogenic temperatures (−200°C [−330°F]) to 260°C (550°F). Its impact strength is high, but its tensile strength, wear, and creep resistance are low when compared with other engineering plastics. Fillers such as glass fibers can be used to increase strength. PTFE is slippery and waxy to the touch and has a low coefficient of friction.

Processing   Since PTFE has such a high melt viscosity, conventional extrusion and injection-molding processes cannot be used. Parts are molded by compressing granules at room temperature at 2000 to 10,000 psi (14 to 69 MPa). After compression, the performed materials are sintered at 360 to 380°C (680 to 716°F).

Applications   PTFE is used for chemically resistant pipe and pump parts, high-temperature cable insulation, molded electrical components, tape, and nonstick coatings. Filled PTFE compounds are used for bushings, packings, gaskets, seals, O rings, and bearings.

Polychlorotrifluoroethylene

Repeating Chemical Structural Unit

$$\left[\begin{array}{cc} F & F \\ | & | \\ C - C \\ | & | \\ F & Cl \end{array} \right]_n$$ Polychlorotrifluoroethylene
mp: 218°C (420°F)

Structure and Properties   The substitution of a chlorine atom for every fourth fluorine atom produces some irregularity in the polymer chains, making the material less crystalline and more moldable. Thus, PCTFE has a lower melting point (218°C [420°F]) than PTFE and can be extruded and molded by conventional processes.

Applications   Extruded, molded, and machined products of PCTFE polymeric materials are used for chemical processing equipment and electrical applications. Other applications include gaskets, O rings, seals, and electrical components.

10.7 ENGINEERING THERMOPLASTICS

In this section some important aspects of the structure, properties, and applications of engineering thermoplastics will be discussed. The definition of an engineering plastic is arbitrary since there is virtually no plastic that cannot in some form be considered an engineering plastic. A thermoplastic in this book will be considered an engineering thermoplastic if it has a balance of properties that makes it especially useful for engineering applications. In this discussion the following families of thermoplastics have been selected as engineering thermoplastics: polyamides (nylons), polycarbonates, phenylene oxide–based resins, acetals, thermoplastic polyesters, polysulfone, polyphenylene sulfide, and polyetherimide.

Sales tonnages of engineering thermoplastics are a relatively small percentage in total compared to the general-purpose plastics. An exception might be the nylons because of their special properties. However, the figures are not available and so will not be quoted. The bulk list prices are available.

Some Basic Properties of Selected Engineering Thermoplastics   Table 10.5 lists the densities, tensile strengths, impact strengths, dielectric strengths, and maximum-use

Table 10.5 Some properties of selected engineering thermoplastics

Material	Density (g/cm³)	Tensile strength (×1000 psi)*	Impact strength, Izod (ft · lb/in)†	Dielectric strength (V/mil)‡	Max-use temp. (no load) °F	Max-use temp. (no load) °C
Nylon 6,6	1.13–1.15	9–12	2.0	385	180–300	82–150
Polyacetal, homo.	1.42	10	1.4	320	195	90
Polycarbonate	1.2	9	12–16	380	250	120
Polyester:						
PET	1.37	10.4	0.8	...	175	80
PBT	1.31	8.0–8.2	1.2–1.3	590–700	250	120
Polyphenylene oxide	1.06–1.10	7.8–9.6	5.0	400–500	175–220	80–105
Polysulfone	1.24	10.2	1.2	425	300	150
Polyphenylene sulfide	1.34	10	0.3	595	500	260

*1000 psi = 6.9 MPa.
†Notched Izod test: 1 ft · lb/in = 53.38 J/m.
‡1V/mil = 39.4 V/mm.

temperatures for some selected engineering thermoplastics. The densities of the engineering thermoplastics listed in Table 10.5 are relatively low, ranging from 1.06 to 1.42 g/cm³. The low densities of these materials are an important property advantage for many engineering designs. As for almost all plastic materials, their tensile strengths are relatively low, with those in Table 10.5 ranging from 8000 to 12,000 psi (55 to 83 MPa). These low strengths are usually an engineering design disadvantage. As for the impact strengths of the engineering thermoplastics, polycarbonate has an outstanding impact strength, having values from 12 to 16 ft · lb/in. The low values of 1.4 and 2.0 ft · lb/in. for polyacetal and nylon 6,6 are somewhat misleading since these materials are "tough" plastic materials but are notch-sensitive, as the notched Izod impact test indicates.

The electrical insulation strengths of the engineering thermoplastics listed in Table 10.5 are high, as is the case for most plastic materials, and range from 320 to 700 V/mil. The maximum-use temperatures for the engineering thermoplastics listed in Table 10.5 are from 180 to 500°F (82 to 260°C). Of the engineering thermoplastics in Table 10.5, polyphenylene sulfide has the highest use temperature of 500°F (260°C).

There are many other properties of engineering thermoplastics that make these materials industrially important. Engineering thermoplastics are relatively easy to process into a near-finished or finished shape, and their processing can be automated in most cases. Engineering thermoplastics have good corrosion resistance to many environments. In some cases engineering plastics have superior resistance to chemical attack. For example, polyphenylene sulfide has no known solvents below 400°F (204°C).

10.7.1 Polyamides (Nylons)

Polyamides or *nylons* are melt-processible thermoplastics whose main-chain structure incorporates a repeating amide group. Nylons are members of the engineering plastics family and offer superior load-bearing capability at elevated temperatures, good toughness, low-frictional properties, and good chemical resistance.

Chemical Repeating Linkage   There are many types of nylons, and the repeating unit is different for each type. They all, however, have the *amide linkage* in common:

$$\overset{O\quad\ H}{\underset{\displaystyle -C-N-}{\|\quad\ |}}\quad \text{Amide linkage}$$

Chemical Processing and Polymerization Reactions   Some types of nylons are produced by a stepwise polymerization of a dibasic organic acid with a diamine. Nylon 6,6,[5] which is the most important of the nylon family, is produced by a polymerization reaction between hexamethylene diamine and adipic acid to produce polyhexamethylene diamine (Fig. 10.10). The repeating chemical structural unit for nylon 6,6 is

$$\left[\ \overset{H}{\underset{}{\overset{|}{N}}}-(CH_2)_6-\overset{H}{\underset{\overset{|}{H}}{N}}-\overset{O}{\overset{\|}{C}}-(CH_2)_4-\overset{O}{\overset{\|}{C}}\ \right]_n \quad \begin{array}{l}\text{Nylon 6,6}\\ \text{mp: } 250-266°C\\ (482-510°F)\end{array}$$

Other important commercial nylons made by the same type of reaction are nylons 6,9, 6,10, and 6,12, which are made with hexamethylene diamine and azelaic (9 carbons), sebacic (10 carbons), or dodecanedioic (12 carbons) acids, respectively.

Nylons can also be produced by chain polymerization of ring compounds that contain both organic acid and amine groups. For example, nylon 6 can be polymerized from ϵ-caprolactam (6 carbons) as shown in the following diagram:

ϵ-Caprolactam

Nylon 6
mp: 216−225°C (420−435°F)

Structure and Properties   Nylons are highly crystalline polymeric materials because of the regular symmetrical structure of their main polymer chains. The high crystallizability of the nylons is apparent by the fact that under controlled solidification conditions spherulites can be produced. Figure 10.34 shows an excellent example of the formation of a complex spherulitic structure in nylon 9,6 grown at 210°C.

The high strength of the nylons is partly due to the hydrogen bonding between the molecular chains (Fig. 10.35). The amide linkage makes possible a NHO type of hydrogen bond between the chains. As a result the nylon polyamides have high strength, high

---

[5]The designation "6,6" of nylon 6,6 refers to the fact that there are 6 carbon atoms in the reacting diamine (hexamethylene diamine) and also 6 carbon atoms in the reacting organic acid (adipic acid).

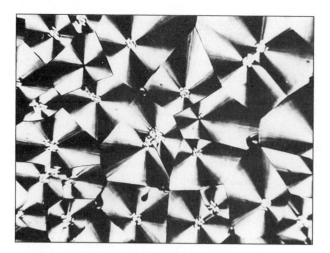

Figure 10.34
Complex spherulitic structure of nylon 9,6 grown at
210°C. The fact that spherulites can be grown in this
nylon material emphasizes the capability of nylon
materials to crystallize.
(*Courtesy of J. H. Magill, University of Pittsburgh.*)

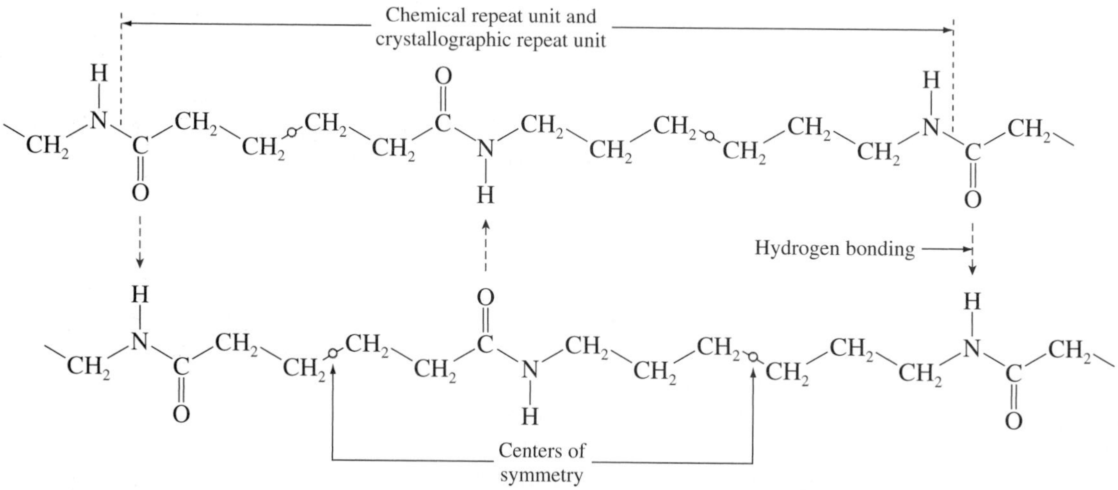

Figure 10.35
Schematic representation of hydrogen bonding between two molecular chains.
[*After M. I. Kohan (ed.), "Nylon Plastics," Wiley, 1973, p. 274.*]

heat-deflection temperatures, and good chemical resistance. The flexibility of the main carbon chains produces molecular flexibility, which leads to low melt viscosity and easy processibility. The flexibility of the carbon chains contributes to high lubricity, low surface friction, and good abrasion resistance. However, the polarity and hydrogen bonding of the amide groups causes high water absorption, which results in dimensional changes with increasing moisture content. Nylons 11 and 12 with their longer carbon chains between amide groups are less sensitive to water absorption.

Processing   Most nylons are processed by conventional injection-molding or extrusion methods.

Applications   Applications for nylons are found in almost all industries. Typical uses are for unlubricated gears, bearings, and antifriction parts, mechanical parts that must function at high temperatures and resist hydrocarbons and solvents, electrical parts subjected to high temperatures, and high-impact parts requiring strength and rigidity. Automobile applications include speedometer and windshield wiper gears and trim clips. Glass-reinforced nylon is used for engine fan blades, brake and power-steering fluid reservoirs, valve covers, and steering column housings. Electrical and/or electronic applications include connectors, plugs, hookup wire insulation, antenna mounts, and terminals. Nylon is also used in packaging and for many general-purpose applications.

Nylons 6,6 and 6 make up most of the U.S. nylon sales tonnage because they offer the most favorable combination of price, properties, and processibility. However, nylons 6,10 and 6,12 and nylons 11 and 12 as well as others are produced and sold at a premium price when their special properties are required.

10.7.2 Polycarbonate

Polycarbonates form another class of engineering thermoplastics because some of their special high-performance characteristics such as high strength, toughness, and dimensional stability are required for some engineering designs. Polycarbonate resins are manufactured in the United States by General Electric under the trade name Lexan and by Mobay under the trade name Merlon.

Basic Repeating Chemical Structural Unit

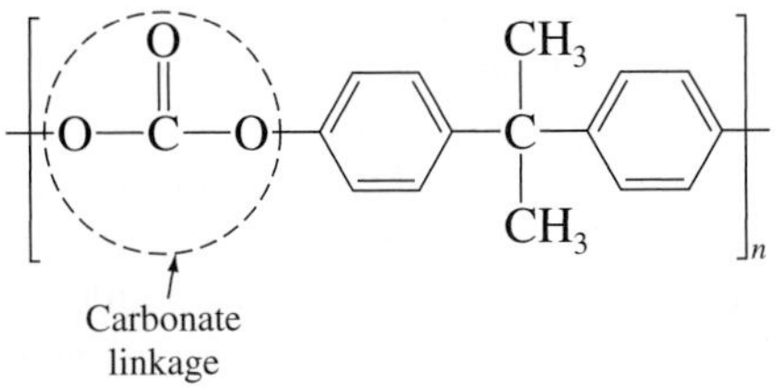

Polycarbonate
mp: 270°C (520°F)

Carbonate
linkage

Figure 10.36
Structure of polycarbonate thermoplastic.

Structure and Properties   The two phenyl and two methyl groups attached to the same carbon atom in the repeating structural unit (Fig. 10.36) produce considerable steric hindrance and make a very stiff molecular structure. However, the carbon-oxygen single bonds in the carbonate linkage provide some molecular flexibility for the overall molecular structure, which produces high-impact energy. The tensile strengths of the polycarbonates at room temperature are relatively high at about 9 ksi (62 MPa), and their impact strengths are very high at 12 to 16 ft·lb/in. (640 to 854 J/m) as measured by the Izod test. Other important properties of polycarbonates for engineering designs are their high heat-deflection temperatures, good electrical insulating properties, and transparency. The creep resistance of these materials is also good. Polycarbonates are resistant to a variety of chemicals but are attacked by solvents. Their high dimensional stability enables them to be used in precision engineering components where close tolerances are required.

Applications   Typical applications for polycarbonates include safety shields, cams and gears, helmets, electrical relay covers, aircraft components, boat propellers, traffic light housings and lenses, glazing for windows and solar collectors, and housings for handheld power tools, small appliances, and computer terminals.

10.7.3 Phenylene Oxide–Based Resins

The phenylene oxide–based resins form a class of engineering thermoplastic materials.

Basic Repeating Chemical Structural Unit

Polyphenylene oxide

Chemical Processing   A patented process for the oxidative coupling of phenolic monomers is used to produce phenylene oxide–based thermoplastic resins that have the trade name Noryl (General Electric).

Structure and Properties   The repeating phenylene rings[6] create steric hindrance to rotation of the polymer molecule and electronic attraction due to the resonating electrons in the benzene rings of adjacent molecules. These factors lead to a polymeric material with high rigidity, strength, chemical resistance to many environments, dimensional stability, and heat-deflection temperature.

There are many different grades of these materials to meet the requirements of a wide range of engineering design applications. Among the principal design advantages of the polyphenylene oxide resins are excellent mechanical properties over the temperature range from −40 to 150°C (−40 to 300°F), excellent dimensional

---

[6]A phenylene ring is a benzene ring chemically bonded to other atoms as, for example,

stability with low creep, high modulus, low water absorption, good dielectric properties, excellent impact properties, and excellent resistance to aqueous chemical environments.

Applications   Typical applications for polyphenylene oxide resins are electrical connectors, TV tuners and deflection yoke components, small appliance and business machine housings, and automobile dashboards, grills, and exterior body parts.

10.7.4 Acetals

Acetals are a class of high-performance engineering thermoplastic materials. They are among the strongest (tensile strength of 10 ksi [68.9 MPa]) and stiffest (modulus in flexure of 410 ksi [2820 MPa]) thermoplastics and have excellent fatigue life and dimensional stability. Other important characteristics include low friction coefficients, good processibility, good solvent resistance, and high heat resistance to about 90°C (195°F) with no load.

Repeating Chemical Structural Unit

$$\left[\begin{array}{c} H \\ | \\ C - O \\ | \\ H \end{array} \right]_n$$ Polyoxymethylene
mp: 175°C (347°F)

Types of Acetals   At present there are two basic types of acetals: a homopolymer (Du Pont's Delrin) and a copolymer (Celanese's Celcon).

Structure and Properties   The regularity, symmetry, and flexibility of the acetal polymer molecules produce a polymeric material with high regularity, strength, and heat-deflection temperature. Acetals have excellent long-term load-carrying properties and dimensional stability and thus can be used for precision parts such as gears, bearings, and cams. The homopolymer is harder and more rigid, and has higher tensile strength and flexural strength than the copolymer. The copolymer is more stable for long-term, high-temperature applications and has a higher elongation.

The low moisture absorption of unmodified acetal homopolymer provides it with good dimensional stability. Also, the low wear and friction characteristics of acetal make it useful for moving parts. In all moving parts, acetal's excellent fatigue resistance is an important property. However, acetals are flammable, and so their use in electrical and/or electronic applications is limited.

Applications   Acetals have replaced many metal castings of zinc, brass, and aluminum and stampings of steel because of lower cost. Where the higher strength of the metals is not required, finishing and assembly operation costs can be reduced or eliminated by using acetals for many applications.

In automobiles, acetals are used for components in fuel systems, seat belts, and window handles. Machinery applications for acetals include mechanical couplings, pump impellers, gears, cams, and housings. Acetals are also used in a wide variety of consumer products such as zippers, fishing reels, and writing pens.

10.7.5 Thermoplastic Polyesters

Polybutylene Terephthalate and Polyethylene Terephthalate Two important engineering thermoplastic polyesters are *polybutylene terephthalate* (PBT) and *polyethylene terephthalate* (PET). PET is widely used for film for food packaging and as a fiber for clothing, carpeting, and tire cord. Since 1977 PET has been used as a container resin. PBT, which has a higher-molecular-weight repeating unit in its polymer chains, was introduced in 1969 as a replacement material for some applications where thermosetting plastics and metals were used. The use of PBT is continuing to expand because of its properties and relatively low cost.

Repeating Chemical Structural Units

Polyethylene terephthalate

Polybutylene terephthalate

Structure and Properties The phenylene rings along with the carbonyl groups (C=O) in PBT form large, flat, bulky units in the polymer chains. This regular structure crystallizes quite readily in spite of its bulkiness. The phenylene ring structure provides rigidity to this material, and the butylene units provide some molecular mobility for melt processing. PBT has good strength (7.5 ksi [52 MPa] for unreinforced grades and 19 ksi [131 MPa] for 40 percent glass-reinforced grades). Thermoplastic polyester resins also have low moisture-absorption characteristics. The crystalline structure of PBT makes it resistant to most chemicals. Most organic compounds have little effect on PBT at moderate temperatures. PBT also has good electrical insulation properties that are nearly independent of temperature and humidity.

Applications Electrical-electronic applications for PBT include connectors, switches, relays, TV tuner components, high-voltage components, terminal boards, integrated circuit boards, motor brush holders, end bells, and housings. Industrial uses for PBT include pump impellers, housings and support brackets, irrigation valves and bodies, and water meter chambers and components. PBT is also used for appliance housings and handles. Automotive applications include large exterior-body components, high-energy ignition caps and rotors, ignition coil caps, coil bobbins, fuel injection controls, and speedometer frames and gears.

Polysulfone

Repeating Chemical Structural Unit

Polysulfone
mp: 315°C (600°F)

Sulfone
linkage

Structure and Properties The phenylene rings of the polysulfone repeating unit restrict rotation of the polymer chains and create strong intermolecular attraction to provide high strength and rigidity to this material. An oxygen atom in para[7] position of the phenylene ring with respect to the sulfone group provide the high-oxidation stability of the sulfone polymers. The oxygen atoms between the phenylene rings (ether linkage) provide chain flexibility and impact strength.

Properties of polysulfone of special significance to the design engineer are its high heat-deflection temperature of 174°C (345°F) at 245 psi (1.68 MPa) and its ability to be used for long times at 150 to 174°C (300 to 345°F). Polysulfone has a high tensile strength (for thermoplastics) of 10.2 ksi (70 MPa) and a relatively low tendency to creep. Polysulfones resist hydrolysis in aqueous acid and alkaline environments because the oxygen linkages between the phenylene rings are hydrolytically stable.

Applications Electrical-electronic applications include connectors, coil bobbins and cores, television components, capacitor film, and structural circuit boards. Polysulfone's resistance to autoclave sterilization makes it widely used for medical instruments and trays. In chemical processing and pollution control equipment, polysulfone is used for corrosion-resistant piping, pumps, tower packing, and filter modules and support plates.

10.7.6 Polyphenylene Sulfide

Polyphenylene sulfide (PPS) is an engineering thermoplastic that is characterized by outstanding chemical resistance along with good mechanical properties and stiffness at elevated temperatures. PPS was first produced in 1973 and is manufactured by Phillips Chemical Co. under the trade name Ryton.

Repeating Chemical Structural Unit Polyphenylene sulfide has a repeating structural unit in its main chain of para-substituted benzene rings and divalent sulfur atoms:

Polyphenylene sulfide
mp: 288°C (550°F)

Structure and Properties The compact symmetrical structure of the phenylene rings separated by sulfur atoms produces a rigid and strong polymeric material. The compact molecular structure also promotes a high degree of crystallinity. Because of the presence of the sulfur atoms, PPS is highly resistant to attack by chemicals. In fact, no chemical has been found to dissolve PPS readily below 200°C (392°F). Even at high temperatures, few materials react chemically with PPS.

Unfilled PPS has a room-temperature strength of 9.5 ksi (65 MPa), whereas when 40 percent glass-filled, its strength is raised to 17 ksi (120 MPa). Because of its crystalline structure, the loss of strength with increasing temperature is gradual, and even at 200°C (392°F) considerable strength is retained.

---

[7]Para positions are at opposite ends of the benzene ring.

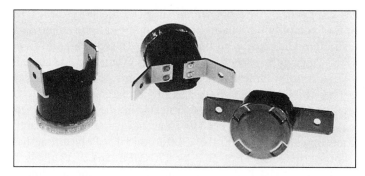

Figure 10.37
Fixed internal thermostats made partly from polyphenylene sulfide (black part of the device). The PPS provides good electrical resistance, high heat resistance, and low shrinkage during molding.
(*Courtesy of Texas Instruments.*)

Applications   Industrial-mechanical applications include chemical process equipment such as submersible, centrifugal, vane, and gear-type pumps. PPS compounds are specified for many under-the-hood automobile applications such as emission-control systems because they are impervious to the corrosive effects of engine exhaust gases as well as to gasoline and other automotive fluids. Electrical-electronic applications include computer components such as connectors, coil forms, and bobbins. Figure 10.37 shows some fixed internal thermostats that utilize the heat resistance and insulative properties of PPS. Corrosive-resistant and thermally stable coatings of PPS are used for oil field pipe, valves, fittings, couplings, and other equipment in the petroleum and chemical processing industries.

10.7.7 Polyetherimide

Polyetherimide is one of the newer amorphous high-performance engineering thermoplastics. It was introduced in 1982 and is commercially available from General Electric Co. under the trademark Ultem. Polyetherimide has the following chemical structure:

Polyetherimide   Imide linkage

The stability of the imide linkage gives this material high heat resistance, creep resistance, and high rigidity. The ether linkage between the phenyl rings provides the

Table 10.6 Some commercial polymer alloys

Polymer alloy	Trade name of material	Supplier
ABS/polycarbonate	Bayblend MC2500	Mobay
ABS/polyvinyl chloride	Cycovin K-29	Borg-Warner Chemicals
Acetal/elastomer	Celcon C-400	Celanese
Polycarbonate/polyethylene	Lexan EM	General Electric
Polycarbonate/PBT/elastomer	Xenoy 1000	General Electric
PBT/PET	Valox 815	General Electric

Source: *Modern Plastics Encyclopedia, 1984–85,* McGraw-Hill.

necessary degree of chain flexibility required for good melt processibility and flow characteristics. This material has good electrical insulation properties that are stable over a wide range of temperatures and frequencies. Uses for polyetherimide include electrical-electronic, automotive, aerospace, and specialty applications. Electrical-electronic applications include high-voltage circuit breaker housings, pin connectors, high-temperature bobbins and coils, and fuse blocks. Printed wiring boards made of reinforced polyetherimide offer dimensional stability for vapor-phase soldering conditions.

10.7.8 Polymer Alloys

Polymer alloys consist of mixtures of structurally different homopolymers or copolymers. In thermoplastic polymer alloys, different types of polymer molecular chains are bonded together by secondary intermolecular dipole forces. By contrast, in a copolymer two structurally different monomers are bonded together in a molecular chain by strong covalent bonds. The components of a polymer alloy must have some degree of compatibility or adhesion to prevent phase separation during processing. Polymer alloys are becoming more important today since plastic materials with specific properties can be created and cost and performance can be optimized.

Some of the early polymer alloys were made by adding rubbery polymers such as ABS to rigid polymers such as polyvinyl chloride. The rubbery material improves the toughness of the rigid material. Today, even the newer thermoplastics are alloyed together. For example, polybutylene terephthalate is alloyed with some polyethylene terephthalate to improve surface gloss and reduce cost. Table 10.6 lists some commercial polymer alloys.

10.8 THERMOSETTING PLASTICS (THERMOSETS)

Thermosetting plastics or thermosets are formed with a network molecular structure of primary covalent bonds. Some thermosets are cross-linked by heat or a combination of heat and pressure. Others may be cross-linked by a chemical reaction that occurs at room temperature (cold-setting thermosets). Although cured parts made from thermosets can be softened by heat, their covalent-bonding cross-links prevent them from being restored to the flowable state that existed before the plastic resin was cured. Thermosets, therefore, cannot be reheated and remelted as can thermoplastics. This is a disadvantage for thermosets since scrap produced during the processing cannot be recycled and reused.

In general, the advantages of thermosetting plastics for engineering design applications are one or more of the following:

1. High thermal stability
2. High rigidity
3. High dimensional stability
4. Resistance to creep and deformation under load
5. Light weight
6. High electrical and thermal insulating properties

Thermosetting plastics are usually processed by using compression or transfer molding. However, in some cases thermoset injection-molding techniques have been developed so that the processing cost is lowered.

Many thermosets are used in the form of molding compounds consisting of two major ingredients: (1) a resin containing curing agents, hardeners, and plasticizers and (2) fillers and/or reinforcing materials that may be organic or inorganic materials. Wood flour, mica, glass, and cellulose are commonly used filler materials.

Let us first look at the bulk list prices in the United States and at some of the important properties of some selected thermoset materials as a group for comparative purposes.

Bulk List Prices of Some Thermoset Plastics   The bulk list prices of the commonly used thermosets are in the low to medium range for plastics, ranging from $0.55 to $1.26 (2000 prices). Of all the thermosets listed, the phenolics are the lowest in price and have the largest sales tonnage. Unsaturated polyesters are also relatively low in price and have a relatively large sales tonnage. Epoxy resins, which have special properties for many industrial applications, command a premium price.

Some Basic Properties of Selected Thermoset Plastics   Table 10.7 lists the densities, tensile strengths, impact strengths, dielectric strengths, and maximum-use temperatures for some selected thermoset plastics. The densities of thermoset plastics tend to be slightly higher than most plastic materials, with those listed in Table 10.7 ranging from 1.34 to 2.3 g/cm^3. The tensile strengths of most thermosets are relatively low, with most strengths ranging from 4000 to 15,000 psi (28 to 103 MPa). However, with a high amount of glass filling, the tensile strength of some thermosets can be increased to as high as 30,000 psi (207 MPa). Glass-filled thermosets also have much higher impact strengths, as indicated in Table 10.7. The thermosets also have good dielectric strengths, ranging from 140 to 650 V/mil. Like all plastic materials, however, their maximum-use temperature is limited. The maximum-use temperature for the thermosets listed in Table 10.7 ranges from 170 to 550°F (77 to 288°C).

Let us now examine some of the important aspects of the structure, properties, and applications of the following thermosets: phenolics, epoxy resins, unsaturated polyesters, and amino resins.

Table 10.7 Some properties of selected thermoset plastics

Material	Density (g/cm³)	Tensile strength (×1000 psi)*	Impact strength, Izod (ft · lb/in)†	Dielectric strength (V/mil)‡	Max-use temp. (no load)	
					°F	°C
Phenolic:						
Wood-flour-filled	1.34–1.45	5–9	0.2–0.6	260–400	300–350	150–177
Mica-filled	1.65–1.92	5.5–7	0.3–0.4	350–400	250–300	120–150
Glass-filled	1.69–1.95	5–18	0.3–18	140–400	350–550	177–288
Polyester:						
Glass-filled SMC	1.7–2.1	8–20	8–22	320–400	300–350	150–177
Glass-filled BMC	1.7–2.3	4–10	15–16	300–420	300–350	150–177
Melamine:						
Cellulose-filled	1.45–1.52	5–9	0.2–0.4	350–400	250	120
Flock-filled	1.50–1.55	7–9	0.4–0.5	300–330	250	120
Glass-filled	1.8–2.0	5–10	0.6–18	170–300	300–400	150–200
Urea, cellulose-filled	1.47–1.52	5.5–13	0.2–0.4	300–400	170	77
Alkyd:						
Glass-filled	2.12–2.15	4–9.5	0.6–10	350–450	450	230
Mineral-filled	1.60–2.30	3–9	0.3–0.5	350–450	300–450	150–230
Epoxy (bis A):						
No filler	1.06–1.40	4–13	0.2–10	400–650	250–500	120–260
Mineral-filled	1.6–2.0	5–15	0.3–0.4	300–400	300–500	150–260
Glass-filled	1.7–2.0	10–30	· · ·	300–400	300–500	150–260

*1000 psi = 6.9 MPa.
†Notched Izod test: 1 ft · lb/in = 53.38 J/m.
‡1V/mil = 39.4 V/mm.
Source: *Materials Engineering,* May 1972.

10.8.1 Phenolics

Phenolic thermosetting materials were the first major plastic material used by industry. The original patents for the reaction of phenol with formaldehyde to make the phenolic plastic Bakelite were issued to L. H. Baekeland in 1909. Phenolic plastics are still used today because they are low in cost and have good electrical and heat insulating properties along with good mechanical properties. They are easily molded but are limited in color (usually black or brown).

Chemistry Phenolic resins are most commonly produced by the reaction of phenol and formaldehyde by condensation polymerization, with water as a by-product. However, almost any reactive phenol or aldehyde can be used. Two-stage (novolac) phenolic resins are commonly produced for convenience for molding. In the first stage a brittle thermoplastic resin is produced that can be melted but cannot cross-link to form a network solid. This material is prepared by reacting less than a mole of formaldehyde with a mole of phenol in the presence of an acid catalyst. The polymerization reaction is shown in Fig. 10.11.

The addition of *hexamethylenetetramine* (hexa), which is a basic catalyst, to the first-stage phenolic resin makes it possible to create methylene cross-linkages to form a thermosetting material. When heat and pressure are applied to the hexa-containing novolac resin, the hexa decomposes, producing ammonia, which provides methylene cross-linkages to form a network structure.

The temperature required for the cross-linking (curing) of the novolac resin ranges from 120 to 177°C (250 to 350°F). Molding compounds are made by combining the resin with various fillers, which sometimes account for up to 50 to 80 percent of the total weight of the molding compounds. The fillers reduce shrinkage during molding, lower cost, and improve strength. They also can be used to increase electrical and thermal insulating properties.

Structure and Properties The high cross-linking of the aromatic structure (Fig. 10.38) produces high hardness, rigidity, and strength combined with good heat and electrical insulating properties and chemical resistance.

Some of the various types of phenolic molding compounds manufactured are:

1. *General-purpose compounds.* These materials are usually wood flour–filled to increase impact strength and lower cost.

2. *High-impact-strength compounds.* These compounds are filled with cellulose (cotton flock and chopped fabric), mineral, and glass fibers to provide impact strengths of up to 18 ft · lb/in. (961 J/m).

3. *High electrical insulating compounds.* These materials are mineral- (e.g., mica) filled to increase electrical resistance.

4. *Heat-resistant compounds.* These are mineral- (e.g., asbestos) filled and are able to withstand long-term exposure to temperatures of 150 to 180°C (300 to 350°F).

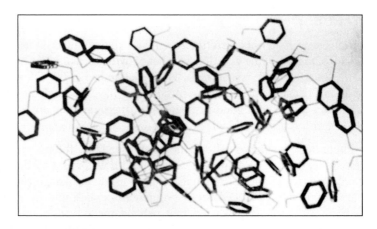

Figure 10.38
Three-dimensional model of polymerized phenolic resin.
(After E. G. K. Pritchett, in "Encyclopedia of Polymer Science and Technology," vol. 10, Wiley, 1969, p. 30.)

Figure 10.39
The three parts shown are molded of glass-reinforced phenolic material and
make up the parts for the transmission reactor for a 1985 automobile.
The phenolic reactor will replace one that was previously made of metal.
(*Courtesy of Rogers Corp.*)

Applications Phenolic compounds are widely used in wiring devices, electrical
switchgear, connectors, and telephone relay systems. Automotive engineers use
phenolic molding compounds for power-assist brake components and transmission
parts (Fig. 10.39). Phenolics are widely used for handles, knobs, and end panels for
small appliances. Because they are good high-temperature and moisture-resistant
adhesives, phenolic resins are used in laminating some types of plywood and in particleboard. Large amounts of phenolic resins are also used as a binder material for
sand in the foundry and for shell molding.

10.8.2 Epoxy Resins

Epoxy resins are a family of thermosetting polymeric materials that do not give off
reaction products when they cure (cross-link) and so have low cure shrinkage. They
also have good adhesion to other materials, good chemical and environmental resistance, good mechanical properties, and good electrical insulating properties.

Chemistry   Epoxy resins are characterized by having two or more epoxy groups per molecule. The chemical structure of an epoxide group is

$$\underset{H}{\overset{O}{CH_2-C-}} \quad \begin{array}{l} \text{Covalent half bond} \\ \text{available for bonding} \end{array}$$

Most commercial epoxy resins have the following general chemical structure:

$$CH_2-CH-CH_2 \left[O-Be-\underset{CH_3}{\overset{CH_3}{C}}-Be-O-CH_2-\underset{OH}{CH}-CH_2 \right]_n O-Be-\underset{CH_3}{\overset{CH_3}{C}}-Be-O-CH_2-CH-CH_2$$

where Be = benzene ring. For liquids the *n* in the structure is usually less than 1. For solid resins *n* is 2 or greater. There are also many other kinds of epoxy resins that have different structures than the one shown here.

To form solid thermosetting materials, epoxy resins must be cured by using cross-linking agents and/or catalysts to develop the desired properties. The epoxy and hydroxyl groups (—OH) are the reaction sites for cross-linking. Cross-linking agents include amines, anhydrides, and aldehyde condensation products.

For curing at room temperature when the heat requirements for the epoxy solid materials are low (under about 100°C), amines such as diethylene triamine and tri-ethylene tetramine are used as curing agents. Some epoxy resins are cross-linked by using a curing reagent, while others can react with their own reaction sites if an appropriate catalyst is present. In an epoxy reaction the epoxide ring is opened and a donor hydrogen from, for example, an amine or hydroxyl group bonds with the oxygen atom of the epoxide group. Figure 10.40 shows the reaction of epoxide groups at the ends of two linear epoxy molecules with ethylene diamine.

In the reaction of Fig. 10.40 the epoxy rings are opened up and hydrogen atoms from the diamine form —OH groups, which are reaction sites for further cross-linking. An important characteristic of this reaction is that no by-product is given off. Many different kinds of amines can be used for cross-linking epoxy resins.

Structure and Properties   The low molecular weight of uncured epoxide resins in the liquid state gives them exceptionally high molecular mobility during processing. This property allows the liquid epoxy resin to quickly and thoroughly wet surfaces. This wetting action is important for epoxies used for reinforced materials and adhesives. Also the ability to be poured into final form is important for electrical potting and encapsulating. The high reactivity of the epoxide groups with curing agents such as amines provides a high degree of cross-linking and produces good hardness, strength, and chemical resistance. Since no by-product is given off during the curing reaction, there is low shrinkage during hardening.

Applications   Epoxy resins are used for a wide variety of protective and decorative coatings because of their good adhesion and good mechanical and chemical resistance. Typical uses are can and drum linings, automotive and appliance primers, and

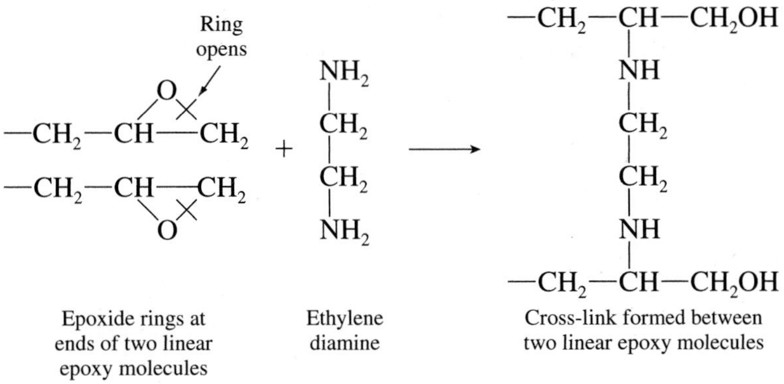

Figure 10.40
Reaction of epoxy rings at the ends of two linear epoxy molecules with ethylene diamine to form a cross-link. Note that no by-product is given off.

wire coatings. In the electrical and electronics industry epoxy resins are used because of their dielectric strength, low shrinkage on curing, good adhesion, and ability to retain properties under a variety of environments such as wet and high-humidity conditions. Typical applications include high-voltage insulators, switchgear, and encapsulation of transistors. Epoxy resins are also used for laminates and for fiber-reinforced matrix materials. Epoxy resins are the predominate matrix material for most high-performance components such as those made with high-modulus fibers (e.g., graphite).

10.8.3 Unsaturated Polyesters

Unsaturated polyesters have reactive double carbon-carbon covalent bonds that can be cross-linked to form thermosetting materials. In combination with glass fibers, unsaturated polyesters can be cross-linked to form high-strength reinforced composite materials.

Chemistry　The ester linkage can be produced by reacting an alcohol with an organic acid, as

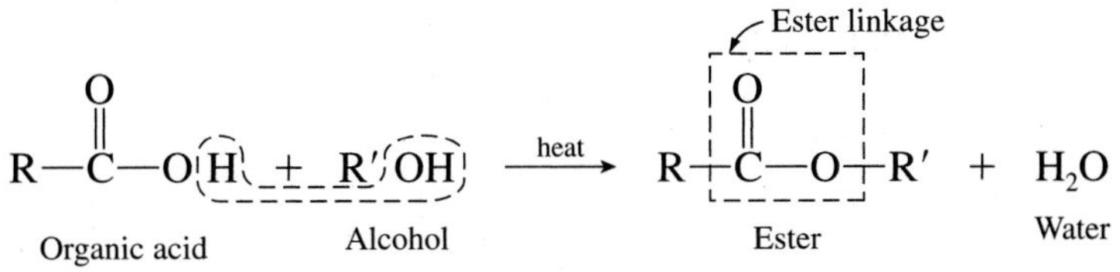

The basic unsaturated polyester resin can be formed by the reaction of a diol (an alcohol with two —OH groups) with a diacid (an acid with two —COOH groups) that contains a reactive double carbon-carbon bond. Commercial resins may

have mixtures of different diols and diacids to obtain special properties, for example, ethylene glycol can be reacted with maleic acid to form a linear polyester:

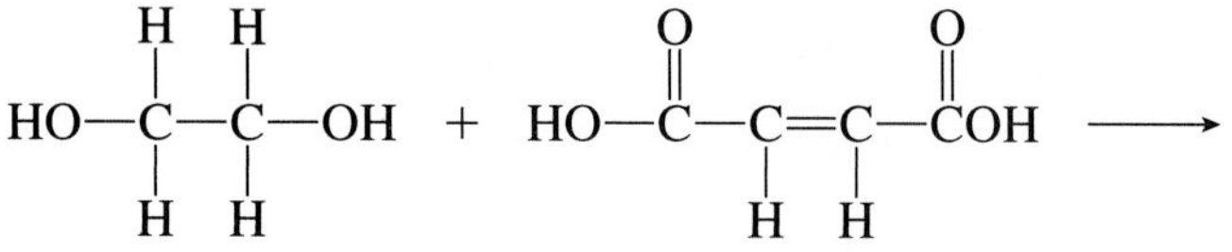

$$
\underset{\substack{\text{Ethylene glycol}\\\text{(alcohol)}}}{HO-\overset{\overset{H}{|}}{\underset{\underset{H}{|}}{C}}-\overset{\overset{H}{|}}{\underset{\underset{H}{|}}{C}}-OH} \ + \ \underset{\substack{\text{Maleic acid}\\\text{(organic acid)}}}{HO-\overset{\overset{O}{||}}{C}-\overset{\overset{H}{|}}{C}=\overset{\overset{H}{|}}{C}-\overset{\overset{O}{||}}{C}OH} \longrightarrow
$$

Ester linkage — Reactive double bond

$$
\left[-O-\overset{\overset{H}{|}}{\underset{\underset{H}{|}}{C}}-\overset{\overset{H}{|}}{\underset{\underset{H}{|}}{C}}-O-\overset{\overset{O}{||}}{C}-\overset{H}{C}=\overset{H}{C}-\overset{\overset{O}{||}}{C}- \right]_n + H_2O
$$

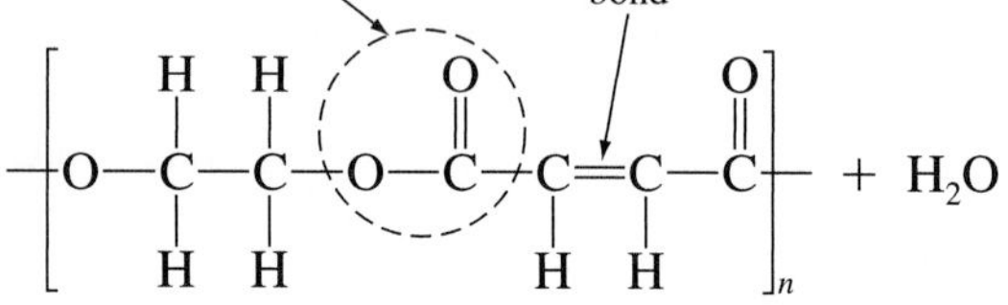

Linear polyester

The linear unsaturated polyesters are usually cross-linked with vinyl-type molecules such as styrene in the presence of a free-radical curing agent. Peroxide curing agents are most commonly used, with *methyl ethyl ketone* (MEK) peroxide usually being used for the room-temperature curing of polyesters. The reaction is commonly activated with a small amount of cobalt naphthanate.

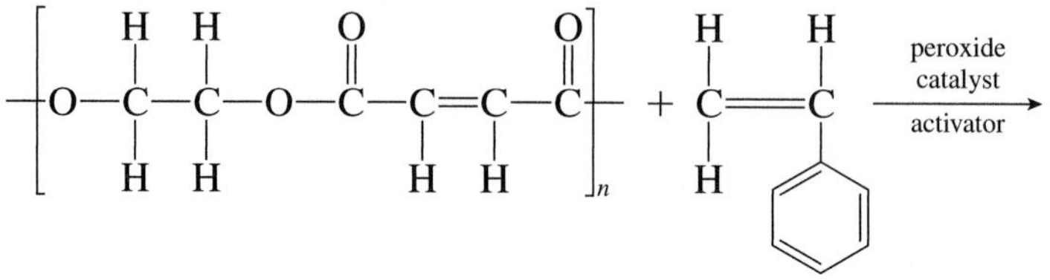

$$
\left[-O-\overset{\overset{H}{|}}{\underset{\underset{H}{|}}{C}}-\overset{\overset{H}{|}}{\underset{\underset{H}{|}}{C}}-O-\overset{\overset{O}{||}}{C}-C=C-\overset{\overset{O}{||}}{C}- \right]_n + \ \underset{\text{Styrene}}{\overset{\overset{H}{|}}{C}=\overset{\overset{H}{|}}{C}} \ \xrightarrow[\text{activator}]{\substack{\text{peroxide}\\\text{catalyst}}}
$$

Linear polyester   Styrene

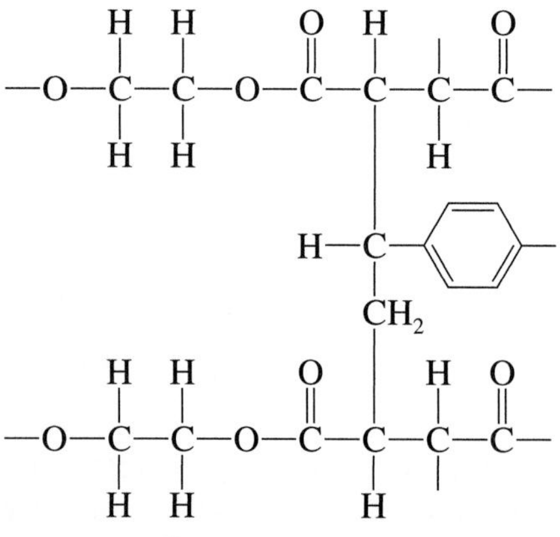

Cross-linked polyester

Structure and Properties The unsaturated polyester resins are low-viscosity materials that can be mixed with high amounts of fillers and reinforcements. For example, unsaturated polyesters may contain as high as about 80 percent by weight of glass-fiber reinforcement. Glass-fiber-reinforced unsaturated polyesters when cured have outstanding strength, 25 to 50 ksi (172 to 344 MPa), and good impact and chemical resistance.

Processing Unsaturated polyester resins can be processed by many methods, but in most cases they are molded in some way. Open-mold lay-up or spray-up techniques are used for many small-volume parts. For high-volume parts such as automobile panels, compression molding is usually used. In recent years *sheet molding compounds* (SMC) that combine resin, reinforcement, and other additives have been produced to speed up the feeding of material into molding presses made of matched metal dies.

Applications Glass-reinforced unsaturated polyesters are used for making automobile panels and body parts. This material is also used for small boat hulls and in the construction industry for building panels and bathroom components. Unsaturated reinforced polyesters are also used for pipes, tanks, and ducts where good corrosion resistance is required.

10.8.4 Amino Resins (Ureas and Melamines)

The amino resins are thermosetting polymeric materials formed by the controlled reaction of formaldehyde with various compounds that contain the amine group —NH_2. The two most important types of amino resins are urea-formaldehyde and melamine-formaldehyde.

Chemistry Both urea and melamine react with formaldehyde by condensation polymerization reactions that produce water as a by-product. The condensation reaction of urea with formaldehyde is

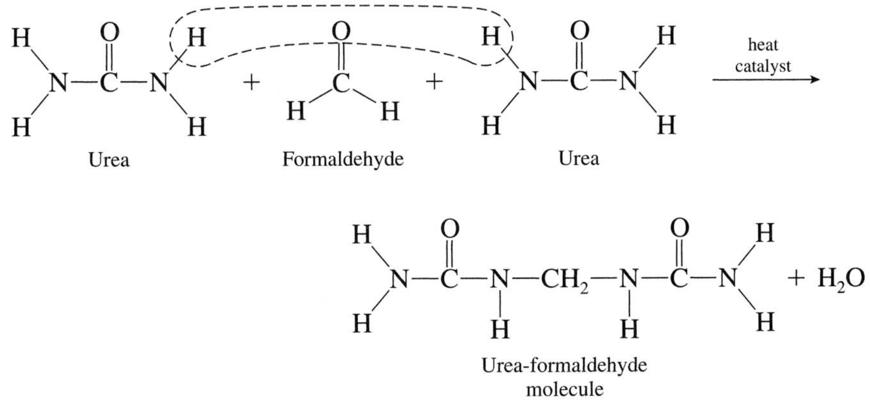

The amine groups at the ends of the molecule shown here react with more formaldehyde molecules to produce a highly rigid network polymer structure. As in the case of the phenolic resins, urea and formaldehyde are first only partially polymerized to

produce a low-molecular-weight polymer that is ground into a powder and compounded with fillers, pigments, and a catalyst. The molding compound can then be compression-molded into the final shape by applying heat (127 to 171°C [260 to 340°F]) and pressure (2 to 8 ksi [14 to 55 MPa]).

Melamine also reacts with formaldehyde by a condensation reaction, resulting in polymerized melamine-formaldehyde molecules with water being given off as a by-product:[8]

Melamine   Formaldehyde   Melamine

Melamine-formaldehyde
molecule

Structure and Properties   The high reactivity of the urea-formaldehyde and melamine-formaldehyde low-molecular-weight prepolymers enable highly cross-linked thermoset products to be made. When these resins are combined with cellulose (wood flour) fillers, low-cost products are obtained that have good rigidity, strength, and impact resistance. Urea-formaldehyde costs less than melamine-formaldehyde but does not have as high a heat resistance and surface hardness as melamine.

Applications   Cellulose-filled molding compounds of urea-formaldehyde are used for electrical wall plates and receptacles and for knobs and handles. Applications for cellulose-filled melamine compounds include molded dinnerware, buttons, control buttons, and knobs. Both urea and melamine water-soluble resins find application as adhesives and bonding resins for wood particleboard, plywood, boat hulls, flooring, and furniture assemblies. The amino resins are also used in binders for foundry cores and shell molds.

---

[8]Only one hydrogen atom is removed from each NH_2 group and one oxygen atom from a formaldehyde molecule to form the H_2O molecule.

10.9 ELASTOMERS (RUBBERS)

Elastomers, or rubbers, are polymeric materials whose dimensions can be greatly changed when stressed and which return to their original dimensions (or almost) when the deforming stress is removed. There are many types of elastomeric materials, but only the following ones will be discussed: natural rubber, synthetic polyisoprene, styrene-butadiene rubber, nitrile rubbers, polychloroprene, and the silicones.

10.9.1 Natural Rubber

Production Natural rubber is produced commercially from the latex of the *Hevea brasiliensis* tree, which is cultivated in plantations mainly in the tropical regions in southeast Asia, especially in Malaysia and Indonesia. The source of natural rubber is a milky liquid known as *latex,* which is a suspension containing very small particles of rubber. The liquid latex is collected from the trees and taken to a processing center, where the field latex is diluted to about 15 percent rubber content and coagulated with formic acid (an organic acid). The coagulated material is then compressed through rollers to remove water and to produce a sheet material. The sheets are dried either with currents of hot air or by the heat of a smoke fire (rubber-smoked sheets). The rolled sheets and other types of raw rubber are usually milled between heavy rolls in which the mechanical shearing action breaks up some of the long polymer chains and reduces their average molecular weight. Natural rubber production in 1980 accounted for about 30 percent of the total world's rubber market.

Structure Natural rubber is mainly ***cis*-1,4 polyisoprene** mixed with small amounts of proteins, lipids, inorganic salts, and numerous other components. *cis*-1,4 polyisoprene is a long-chain polymer (average molecular weight of about 5×10^5 g/mol) which has the following structural formula:

$$\begin{bmatrix} H & CH_3 & H & H \\ | & | & | & | \\ -C-C=C-C- \\ | & & & | \\ H & & & H \end{bmatrix}_n$$

cis-1,4 Polyisoprene
Repeating structural unit for natural rubber

The *cis-* prefix indicates that the methyl group and a hydrogen atom are on the same side of the carbon-carbon double bond, as shown by the dashed encirclement on the formula. The 1,4 indicates that the repeating chemical units of the polymer chain covalently bond on the first and fourth carbon atoms. The polymer chains of natural rubber are long, entangled, and coiled, and at room temperature are in a state of continued agitation. The bending and coiling of the natural rubber polymer chains is attributed to the steric hindrance of the methyl group and the hydrogen atom on the same side of the carbon-carbon double bond. The arrangement of the covalent bonds in the natural rubber polymer chain is shown schematically as follows:

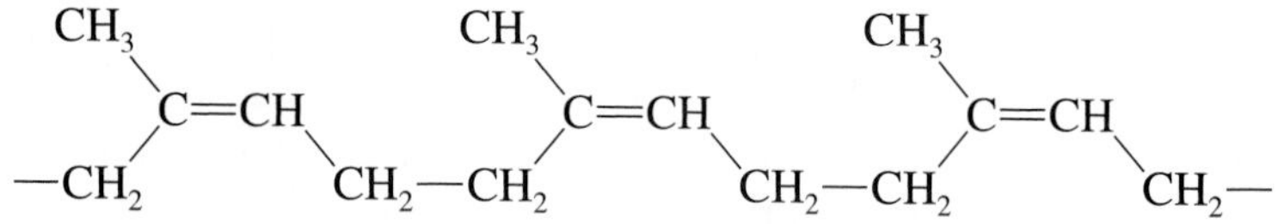

Segment of natural rubber polymer chain

There is another structural isomer[9] of polyisoprene, ***trans*-1,4 polyisoprene,**[10] called *gutta-percha,* which is not an elastomer. In this structure the methyl group and hydrogen atom covalently bonded to the carbon-carbon double bond are on opposite sides of the double bond of the polyisoprene repeating unit, as shown by the encirclement in the following diagram:

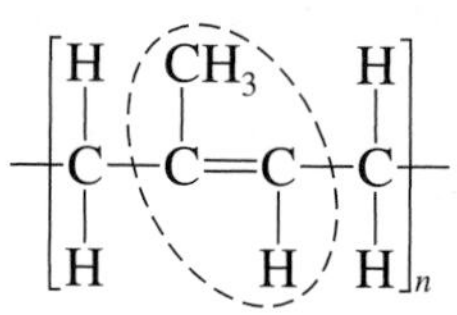

trans-1,4 Polyisoprene
Repeating structural unit for gutta-percha

In this structure the methyl group and hydrogen atom bonded to the double bond do not interfere with each other, and as a result the *trans*-1,4 polyisoprene molecule is more symmetrical and can crystallize into a rigid material.

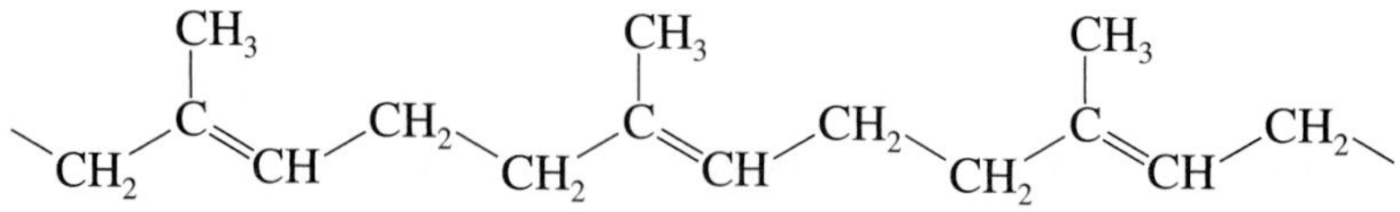

Segment of gutta-percha polymer chain

Vulcanization   **Vulcanization** is the chemical process by which polymer molecules are joined together by cross-linking into larger molecules to restrict molecular movement. In 1839 Charles Goodyear[11] discovered a vulcanization process for rubber by using sulfur and basic lead carbonate. Goodyear found that when a mixture of natural rubber, sulfur, and lead carbonate was heated, the rubber changed from a thermoplastic to an elastomeric material. Although even today the reaction of sulfur with rubber is complex and not completely understood, the final result is that some of the double bonds in the polyisoprene molecules open and form sulfur atom cross-links, as shown in Fig. 10.41.

---

[9]Structural isomers are molecules that have the same molecular formula but different structural arrangement of their atoms.

[10]The prefix *trans-* is from the Latin, meaning "across."

[11]Charles Goodyear (1800–1860). American inventor who discovered the vulcanizing process for natural rubber by using sulfur and lead carbonate as chemical agents. U.S. Patent 3633 was granted to Charles Goodyear on June 15, 1844, for an "Improvement in India-Rubber Fabrics."

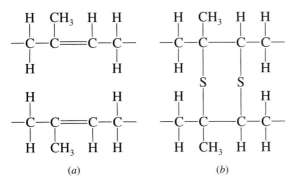

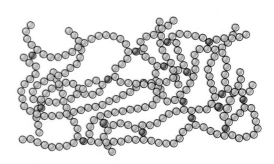

Figure 10.41

Schematic illustration of the vulcanization of rubber. In this process sulfur atoms form cross-links between chains in 1,4 polyisoprene. (*a*) *cis*-1,4 polyisoprene chain before sulfur cross-linking. (*b*) *cis*-1,4 polyisoprene chain after cross-linking with sulfur at the active double-bond sites.

Figure 10.42

Model of cross-linking of *cis*-1,4 polyisoprene chains by sulfur atoms (darker shade).

(*After W. G. Moffatt, G. W. Pearsall, and J. Wulff, "The Structure and Properties of Materials," vol. I, Wiley, 1965, p. 109.*)

Figure 10.42 shows schematically how cross-linking the sulfur atoms gives rigidity to rubber molecules, and Fig. 10.43 shows how the tensile strength of natural rubber is increased by vulcanization. Rubber and sulfur react very slowly even at elevated temperatures, so in order to shorten the cure time at elevated temperatures, accelerator chemicals are usually compounded with rubber along with other additives such as fillers, plasticizers, and antioxidants.

Usually, vulcanized soft rubbers contain about 3 wt % sulfur and are heated in the 100 to 200°C range for vulcanizing or curing. If the sulfur content is increased, the cross-linking that occurs will also increase, producing harder and less-flexible material. A fully rigid structure of hard rubber can be produced with about 45 percent sulfur.

Oxygen or ozone will also react with the carbon double bonds of the rubber molecules in a similar way to the vulcanization sulfur reaction and cause embrittlement of the rubber. This oxidation reaction can be retarded to some extent by adding antioxidants when the rubber is compounded.

The use of fillers can lower the cost of the rubber product and also strengthen the material. Carbon black is commonly used as a filler for rubber, and, in general, the finer the particle size of the carbon black, the higher the tensile strength is. Carbon black also increases the tear and abrasion resistance of the rubber. Silicas (e.g., calcium silicate) and chemically altered clay are also used for fillers for reinforcing rubber.

Properties Table 10.8 compares the tensile strength, elongation, and density properties of vulcanized natural rubber with those of some other synthetic elastomers. Note that, as expected, the tensile strengths of these materials are relatively low and their elongations extremely high.

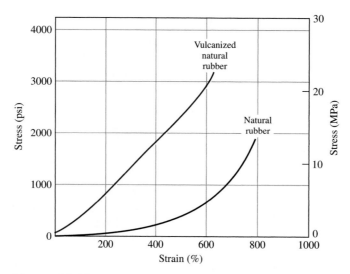

Figure 10.43
Stress-strain diagrams for vulcanized and unvulcanized
natural rubber. The cross-linking of the sulfur atoms
between the polymer chains of *cis*-1,4 polyisoprene by
vulcanization increases the strength of vulcanized rubber.
(*After M. Eisenstadt, "Introduction to Mechanical Properties of
Materials," Macmillan, 1971, p. 89.*)

Table 10.8 Some properties of selected elastomers

Elastomer	Tensile strength (ksi)†	Elongation (%)	Density (g/cm³)	Recommended operating temp.	
				°F	°C
Natural rubber* (*cis*-polyisoprene)	2.5–3.5	750–850	0.93	−60 to 180	−50 to 82
SBR or Buna S* (butadiene-styrene)	0.2–3.5	400–600	0.94	−60 to 180	−50 to 82
Nitrile or Buna N* (butadiene-acrylonitrile)	0.5–0.9	450–700	1.0	−60 to 250	−50 to 120
Neoprene* (polychloroprene)	3.0–4.0	800–900	1.25	−40 to 240	−40 to 115
Silicone (polysiloxane)	0.6–1.3	100–500	1.1–1.6	−178 to 600	−115 to 315

*Pure gum vulcanizate properties.
†1000 psi = 6.89 MPa.

10.9.2 Synthetic Rubbers

Synthetic rubbers in 1980 accounted for about 70 percent of the total world's supply
of rubber materials. Some of the important synthetic rubbers are styrene-butadiene,
nitrile rubbers, and the polychloroprenes.

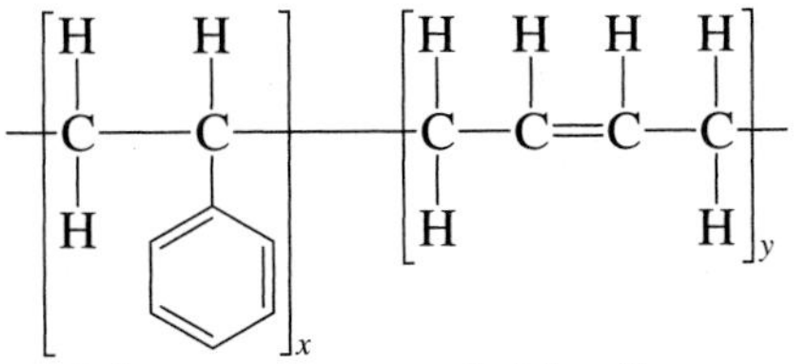

Figure 10.44
Chemical structure of styrene-butadiene synthetic rubber copolymer.

Styrene-Butadiene Rubber The most important synthetic rubber and the most widely used is *styrene-butadiene rubber* (SBR), a butadiene-styrene copolymer. After polymerization, this material contains 20 to 23 percent styrene. The basic structure of SBR is shown in Fig. 10.44.

Since the butadiene mers contain double bonds, this copolymer can be vulcanized with sulfur by cross-linking. Butadiene by itself when synthesized with a stereospecific catalyst to produce the cis isomer has even greater elasticity than natural rubber since the methyl group attached to the double bond in natural rubber is missing in the butadiene mer. The presence of styrene in the copolymer produces a tougher and stronger rubber. The phenyl side group of styrene that is scattered along the copolymer main chain reduces the tendency of the polymer to crystallize under high stresses. SBR rubber is lower in cost than natural rubber and so is used in many rubber applications. For example, for tire treads, SBR has better wear resistance but higher heat generation. A disadvantage of SBR and natural rubber is that they absorb organic solvents such as gasoline and oil and swell.

Nitrile Rubbers Nitrile rubbers are copolymers of butadiene and acrylonitrile with the proportions ranging from 55 to 82 percent butadiene and 45 to 18 percent acrylonitrile. The presence of the nitrile groups increases the degree of polarity in the main chains and the hydrogen bonding between adjacent chains. The nitrile groups provide good resistance to oils and solvents as well as improved abrasion and heat resistance. On the other hand, molecular flexibility is reduced. Nitrile rubbers are more costly than ordinary rubbers, so these copolymers are limited to special applications such as fuel hoses and gaskets where high resistance to oils and solvents is required.

Polychloroprene (Neoprene) The polychloroprene or neoprene rubbers are similar to isoprene except that the methyl group attached to the double carbon bond is replaced by a chlorine atom:

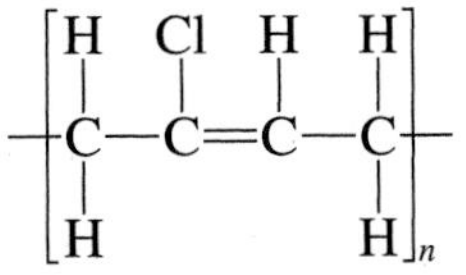

Polychloroprene (neoprene)
structural unit

The presence of the chlorine atom increases the resistance of the unsaturated double bonds to attack by oxygen, ozone, heat, light, and the weather. Neoprenes also have fair fuel and oil resistance and increased strength over that of the ordinary rubbers. However, they do have poorer low-temperature flexibility and are higher in cost. As a result, neoprenes are used in specialty applications such as wire and cable covering, industrial hoses and belts, and automotive seals and diaphragms.

10.9.3 Properties of Polychloroprene Elastomers

Neoprene is sold to its manufacturers as a raw synthetic rubber. Before it is converted to useful products, it must be compounded with selected chemicals, fillers, and processing aids. The resultant compound mixture is then shaped or molded and vulcanized. The properties of this finished product depend on the amount of raw neoprene and the compounding ingredients. Table 10.9 lists selected basic physical properties of polychloroprene as the raw rubber, gum vulcanizate, and carbon-black-filled vulcanizate.

10.9.4 Vulcanization of Polychloroprene Elastomers

The vulcanization of polychloroprene elastomers is dependent on metallic oxides rather than sulfur, which is used for many elastomeric materials. Zinc and magnesium oxides are most commonly used. The vulcanization process is believed to take place

Table 10.9 Basic physical properties of polychloroprenes

Properties	Raw polymer	Vulcanizates	
		Gum	Carbon black
Density (g/cm^3)	1.23	1.32	1.42
Coeff. of vol. exp		610	
$\beta = 1/v \cdot \delta v/\delta T$, κ^{-1}	600×10^{-6}	720×10^{-6}	
Thermal properties			
glass transition temp, K (°C)	228 (−45)	228 (−45)	230 (−43)
heat capacity, Cp [kJ/(kg · K)^b]	2.2	2.2	1.7–1.8
thermal conductivity [W/(m · K)]	0.192	0.192	0.210
Electrical			
dielectric constant (1 kHz)		6.5–8.1	
dissipation factor (1 kHz)		0.031–0.086	
conductivity (pS/m)		3 to 1400	
Mechanical			
ultimate elongation (%)		800–1000	500–600
tensile strength, MPa (ksi)		25–38 (3.6–5.5)	21–30 (3.0–4.3)
Young's modulus, MPa (psi)		1.6 (232)	3–5 (435–725)
Resilience (%)		60–65	40–50

After "Neoprene Synthetic Elastomers," *Ency. Chem. & Tech.*, 3rd ed., Vol. 8 (1979), Wiley, p. 516.

by the following reaction:

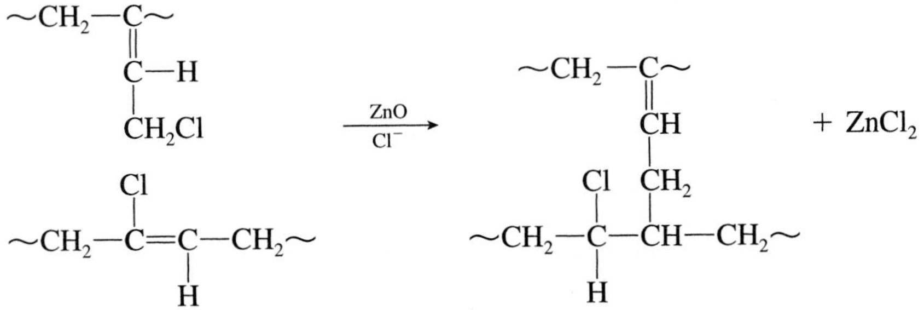

Possible chemical reaction during the vulcanization of polychloroprene

The zinc chloride formed is an active catalyst for vulcanization, and unless it is removed, it can cause problems during later processing. MgO can act as a stabilizer to remove $ZnCl_2$ as:

$$2\,ZnCl_2 + MgO \xrightarrow{H_2O} 2\,Zn\begin{smallmatrix}OH\\Cl\end{smallmatrix} + MgCl_2$$

Silicone Rubbers The silicon atom, like carbon, has a valence of 4 and is able to form polymeric molecules by covalent bonding. However, the silicone polymer has repeating units of silicon and oxygen, as shown in the following diagram:

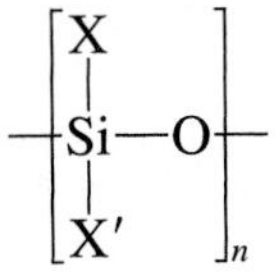

Basic repeating structural
unit for a silicone polymer

where X and X′ may be hydrogen atoms or groups such as methyl (CH_3—) or phenyl (C_6H_5—). The silicone polymers based on silicon and oxygen in the main chain are called *silicones.* Of the many silicone elastomers, the most common type is the one in which the X and X′ of the repeating unit are methyl groups:

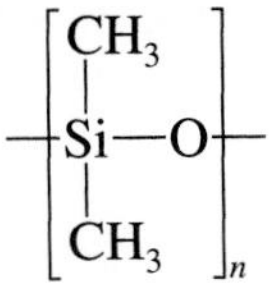

Repeating structural unit
for polydimethyl siloxane

This polymer is called *polydimethyl siloxane* and can be cross-linked at room temperature by the addition of an initiator (e.g., benzoyl peroxide), which reacts the two methyl groups together with the elimination of hydrogen gas (H_2) to form Si—CH_2—CH_2—Si bridges. Other types of silicones can be cured at higher temperatures (e.g., 50 to 150°C), depending on the product and intended use.

Silicone rubbers have the major advantage of being able to be used over a wide temperature range (that is, -100 to 250°C). Applications for the silicone rubbers include sealants, gaskets, electrical insulation, auto ignition cable, and spark plug boots.

EXAMPLE PROBLEM 10.6

How much sulfur must be added to 100 g of polyisoprene rubber to cross-link 5 percent of the mers? Assume all available sulfur is used and that only one sulfur atom is involved in each cross-linking bond.

■ **Solution**

As shown in Fig. 10.41*b*, on the average one sulfur atom will be involved with one poly-isoprene mer in the cross-linking. First we determine the molecular weight of the polyisoprene mer.

$$\text{MW (polyisoprene)} = 5 \text{ C atoms} \times 12 \text{ g/mol} + 8 \text{ H atoms} \times 1 \text{ g/mol} = 68.0 \text{ g/mol}$$

Thus, with 100 g of polyisoprene, we have 100 g/(68.0 g/mol) = 1.47 mol of polyisoprene. For 100 percent cross-linking with sulfur, we need 1.47 mol of S or

$$1.47 \text{ mol} \times 32 \text{ g/mol} = 47.0 \text{ g sulfur}$$

$$
\begin{array}{ccccc}
 & H & CH_3 & H & H \\
 & | & | & | & | \\
-&C&-C=&C&-C- \\
 & | & & & | \\
 & H & & & H
\end{array}
$$

Polyisoprene mer

For cross-linking 5 percent of the bonds, we need only

$$0.05 \times 47.0 \text{ g S} = 2.35 \text{ g S} \blacktriangleleft$$

EXAMPLE PROBLEM 10.7

A butadiene-styrene rubber is made by polymerizing one monomer of styrene with eight monomers of butadiene. If 20 percent of the cross-link sites are to be bonded with sulfur, what weight percent sulfur is required?

■ **Solution**

Basis: 100 g of copolymer

In the copolymer we have one mole of styrene combined with eight moles of polybutadiene. Thus, on a weight basis we have:

$$8 \text{ moles of polybutadiene} \times 54 \text{ g/mol} = 432 \text{ g}$$
$$1 \text{ mole of polystyrene} \times 104 \text{ g/mol} = \underline{104 \text{ g}}$$
$$\text{Total weight of copolymer} = 536 \text{ g}$$

The weight ratio of polybutadiene to copolymer = 432/536 = 0.806
Thus, in 100 g of copolymer we have 100 g × 0.806 = 80.6 g of butadiene or 80.6 g/54 g = 1.493 moles of polybutadiene.

Grams of S for 20 percent cross-linking = (1.493 moles)(32 g/mole)(0.20) = 9.55 g S.

$$\text{Weight \% S} = \left(\frac{9.55 \text{ g}}{100 \text{ g} + 9.55 \text{ g}} \right) 100\% = 8.72\% \blacktriangleleft$$

A butadiene-acrylonitrile rubber is made by polymerizing one acrylonitrile monomer with three butadiene monomers. How much sulfur is required to react with 100 kg of this rubber to cross-link 20 percent of the cross-link sites?

EXAMPLE PROBLEM 10.8

■ **Solution**

Basis: 100 g of copolymer

$$3 \text{ moles of polybutadiene} \times 54 \text{ g} = 162 \text{ g}$$
$$1 \text{ mole of polyacrylonitrile} \times 53 \text{ g} = \underline{53 \text{ g}}$$
$$\text{Total weight of copolymer} = 215 \text{ g}$$

The weight ratio of polybutadiene to copolymer = 162 g/215 g = 0.7535
    In 100 g of copolymer we have 100 g × 0.7535 = 75.35 g or 75.35 g/54 g/mol = 1.395 mol
    Weight of S for 20 percent cross-linking = (1.395 mol)(32 g/mol)(0.20) = 8.93 g S or 8.93 kg.

10.10 DEFORMATION AND STRENGTHENING OF PLASTIC MATERIALS

10.10.1 Deformation Mechanisms for Thermoplastics

The deformation of thermoplastic materials can be primarily elastic, plastic (permanent), or a combination of both types. Below their glass transition temperatures, thermoplastics deform primarily by elastic deformation, as indicated by the −40 and 68°C tensile stress-strain plots for polymethyl methacrylate of Fig. 10.45. Above their glass transition temperatures, thermoplastics deform primarily by plastic deformation, as indicated by the 122 and 140°C tensile stress-strain plots for PMMA in Fig. 10.45. Thus, thermoplastics go through a brittle-ductile transition upon being heated through their glass transition temperature. PMMA goes through a ductile-brittle transition between 86 and 104°C because the T_g of PMMA is in this temperature range.

    Figure 10.46 schematically illustrates the principal atomic and molecular mechanisms that occur during the deformation of long-chain polymers in a thermoplastic

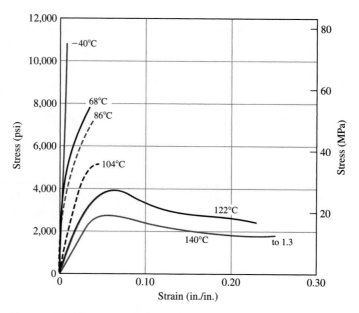

Figure 10.45
Tensile stress versus strain curves for polymethyl methacrylate at various temperatures. A brittle-ductile transition occurs between 86 and 104°C.
(*After T. Alfrey, "Mechanical Behavior of Polymers," Wiley-Interscience, 1967.*)

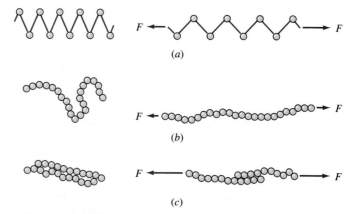

Figure 10.46
Deformation mechanisms in polymeric materials: (*a*) elastic deformation by extension of main-chain carbon covalent bonds, (*b*) elastic or plastic deformation by main-chain uncoiling, and (*c*) plastic deformation by chain slippage.
(*After M. Eisenstadt, "Introduction to Mechanical Properties of Materials," Macmillan, 1971, p. 264.*)

material. In Fig. 10.46a elastic deformation is represented as a stretching out of the covalent bonds within the molecular chains. In Fig. 10.46b elastic or plastic deformation is represented by an uncoiling of the linear polymers. Finally, in Fig. 10.46c plastic deformation is represented by the sliding of molecular chains past each other by the breaking and remaking of secondary dipole bonding forces.

10.10.2 Strengthening of Thermoplastics

Let us now look at the following factors, each of which in part determines the strength of a thermoplastic: (1) average molecular mass of the polymer chains, (2) the degree of crystallization, (3) the effect of bulky side groups on the main chains, (4) the effect of highly polar atoms on the main chains, (5) the effect of oxygen, nitrogen, and sulfur atoms in the main carbon chains, (6) the effect of phenyl rings in the main chains, and (7) the addition of glass-fiber reinforcement.

Strengthening Due to the Average Molecular Mass of the Polymer Chains   The strength of a thermoplastic material is directly dependent on its average molecular mass since polymerization up to a certain molecular mass range is necessary to produce a stable solid. However, this method is not normally used to control strength properties since in most cases after a critical molecular mass range is reached, increasing the average molecular mass of a thermoplastic material does not greatly increase its strength. Table 10.10 lists the molecular mass ranges and degrees of polymerization for some thermoplastics.

Strengthening by Increasing the Amount of Crystallinity in a Thermoplastic Material   The amount of crystallinity within a thermoplastic can greatly affect its tensile strength. In general, as the degree of crystallinity of the thermoplastic increases, the tensile strength, tensile modulus of elasticity, and density of the material all increase.

Thermoplastics that can crystallize during solidification have simple structural symmetry about their molecular chains. Polyethylenes and nylons are examples of thermoplastics that can solidify with a considerable amount of crystallization in their structure. Figure 10.47 compares the engineering stress-strain diagrams for low-density and high-density polyethylenes. The low-density polyethylene has a lower amount of crystallinity and thus lower strength and tensile modulus than the high-density polyethylene. Since the molecular chains in the low-density polyethylene are more branched and farther apart from each other, the bonding forces between the

Table 10.10 Molecular masses and degrees of polymerization for some thermoplastics

Thermoplastic	Molecular mass (g/mol)	Degree of polymerization
Polyethylene	28,000–40,000	1000–1500
Polyvinyl chloride	67,000 (average)	1080
Polystyrene	60,000–500,000	600–6000
Polyhexamethylene adipamide (nylon 6,6)	16,000–32,000	150–300

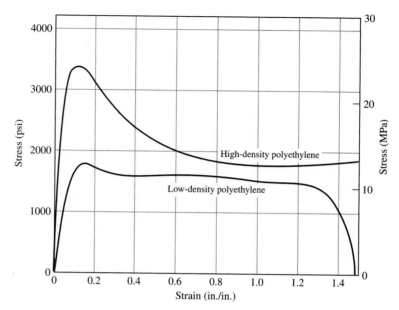

Figure 10.47

Tensile stress-strain curves for low-density and high-density polyethylene. The high-density polyethylene is stiffer and stronger due to a higher amount of crystallinity.

[After J. A. Sauer and K. D. Pae, Mechanical Properties of High Polymers, in H. S. Kaufman and J. J. Falcetta (eds.), "Introduction to Polymer Science and Technology," Wiley, 1977, p. 397.]

chains are lower, and hence the lower-density polyethylene has lower strength. The yield peaks in the stress-strain curves are due to the necking down of the cross sections of the test samples during the tensile tests.

Another example of the effect of increasing crystallinity on the tensile (yield) strength of a thermoplastic material is shown in Fig. 10.48 for nylon 6,6. The increased strength of the more highly crystallized material is due to the tighter packing of the polymer chains, which leads to stronger intermolecular bonding forces between the chains.

Strengthening Thermoplastics by Introducing Pendant Atomic Groups on the Main Carbon Chains Chain slippage during the permanent deformation of thermoplastics can be made more difficult by the introduction of bulky side groups on the main carbon chain. This method of strengthening thermoplastics is used, for example, for polypropylene and polystyrene. The tensile modulus, which is a measure of the stiffness of a material, is raised from the 0.6 to 1.5×10^5 psi range for high-density polyethylene to 1.5 to 2.2×10^5 psi for polypropylene, which has pendant methyl groups attached to the main carbon chain. The tensile elastic modulus of polyethylene is raised still further to the 4 to 5×10^5 psi range with the introduction of the more bulky pendant phenyl rings to the main carbon chain to make

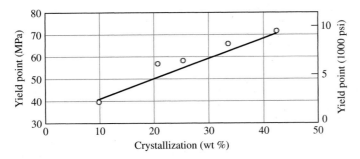

Figure 10.48
Yield point of dry polyamide (nylon 6,6) as a function of crystallinity.
[*After R. J. Welgos, Polyamides (General), in "Encyclopedia of Chemical Technology," vol. 18, Wiley, 1982, p. 331.*]

polystyrene. However, the elongation to fracture is drastically reduced from 100 to 600 percent for high-density polyethylene to 1 to 2.5 percent for polystyrene. Thus, bulky side groups on the main carbon chains of thermoplastics increase their stiffness and strength but reduce their ductility.

Strengthening Thermoplastics by Bonding Highly Polar Atoms on the Main Carbon Chain A considerable increase in the strength of polyethylene can be attained by the introduction of a chlorine atom on every other carbon atom of the main carbon chain to make polyvinyl chloride. In this case, the large, highly polar chlorine atom greatly increases the molecular bonding forces between the polymer chains. Rigid polyvinyl chloride has a tensile strength of 6 to 11 ksi, which is considerably higher than the 2.5 to 5 ksi strength of polyethylene. Figure 10.49 shows a tensile stress-strain plot for a polyvinyl chloride test sample that has a maximum yield stress of about 8 ksi. The yield peak on the curve is due to the necking down of the central part of the test sample during extension.

Strengthening Thermoplastics by the Introduction of Oxygen and Nitrogen Atoms in the Main Carbon Chain Introducing a $-\overset{\displaystyle |}{\underset{\displaystyle |}{C}}-O-\overset{\displaystyle |}{\underset{\displaystyle |}{C}}-$ ether linkage into the main carbon chain increases the rigidity of thermoplastics, as is the case for polyoxy methylene (acetal), which has the repeating chemical unit of $\left[\!\!\begin{array}{c} H \\ | \\ C-O \\ | \\ H \end{array}\!\!\right]_n$.

The tensile strength of this material is in the 9 to 10 ksi range, which is considerably higher than the 2.5 to 5.5 ksi strength of high-density polyethylene. The oxygen atoms in the main carbon chains also increase the permanent dipole bonding between the polymer chains.

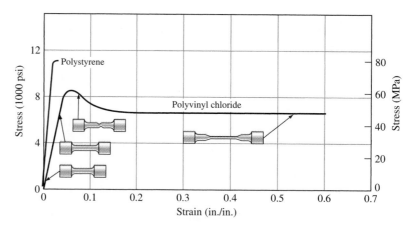

Figure 10.49

Tensile stress-strain data for the amorphous thermoplastic polyvinyl chloride (PVC) and polystyrene (PS). Sketches show modes of specimen deformation at various points on stress-strain curve.

[*After J. A. Sauer and K. D. Pae, Mechanical Properties of High Polymers, in H. S. Kaufman and J. J. Falcetta (eds.), "Introduction to Polymer Science and Technology," Wiley, 1977, p. 331.*]

By introducing nitrogen in the main chains of thermoplastics, as in the case of the amide linkage $\left(\begin{smallmatrix} & O & \\ & \| & \\ -C&-&N- \end{smallmatrix}\right)$, the permanent dipole forces between the polymer chains are greatly increased due to hydrogen bonding (Fig. 10.35). The relatively high tensile strength of 9 to 12 ksi of nylon 6,6 is a result of hydrogen bonding between the amide linkages of the polymer chains.

Strengthening Thermoplastics by Introducing Phenylene Rings into the Main Polymer Chain in Combination with Other Elements Such As O, N, and S in the Main Chain One of the most important methods for strengthening thermoplastics is the introduction of phenylene rings in the main carbon chain. This method of strengthening is commonly used for high-strength engineering plastics. The phenylene rings cause steric hindrance to rotation within the polymer chain and electronic attraction of resonating electrons between adjacent molecules. Examples of polymeric materials that contain phenylene rings are polyphenylene oxide–based materials, which have a tensile strength range of 7.8 to 9.6 ksi, thermoplastic polyesters, which have tensile strengths of about 10 ksi, and polycarbonates, which have strengths of about 9 ksi.

Strengthening Thermoplastics by the Addition of Glass Fibers Some thermoplastics are reinforced with glass fibers. The glass content of most glass-filled thermoplastics ranges from 20 to 40 wt %. The optimum glass content is a trade-off between the desired strength, overall cost, and ease of processing. Thermoplastics commonly strengthened by glass fibers include the nylons, polycarbonates, polyphenylene oxides, polyphenylene sulfide, polypropylene, ABS, and polyacetal. For example, the tensile strength of nylon 6,6 can be increased from 12 to 30 ksi with a 40 percent fiberglass content, but its elongation is reduced from about 60 to 2.5 percent by the fiberglass addition.

10.10.3 Strengthening of Thermosetting Plastics

Thermosetting plastics without reinforcements are strengthened by the creation of a network of covalent bonding throughout the structure of the material. The covalent network is produced by chemical reaction within the thermosetting material after casting or during pressing under heat and pressure. The phenolics, epoxies, and polyesters (unsaturated) are examples of materials strengthened by this method. Because of their covalently bonded network, these materials have relatively high strengths, elastic moduli, and rigidities for plastic materials. For example, the tensile strength of molded phenolic resin is about 9 ksi, that of cast polyesters about 10 ksi, and that of cast epoxy resin about 12 ksi. These materials all have low ductilities because of their covalently bonded network structure.

The strength of thermoplastics can be considerably increased by the addition of reinforcements. For example, the tensile strength of glass-filled phenolic resins can be increased to as high as 18 ksi. Glass-filled polyester-based sheet-molding compounds can have a tensile strength as high as 20 ksi. By using unidirectional laminates of carbon-fiber-reinforced epoxy-based resin, materials with tensile strengths as high as 250 ksi in one direction can be obtained. Reinforced high-strength composite materials of this type will be discussed in Chap. 11 on Composite Materials.

10.10.4 Effect of Temperature on the Strength of Plastic Materials

A characteristic of thermoplastics is that they gradually soften as the temperature is increased. Figure 10.50 shows this behavior for a group of thermoplastics. As the temperature increases, the secondary bonding forces between the molecular chains become weaker and the strength of the thermoplastic decreases. When a thermoplastic material is heated through its glass transition temperature T_g, its strength decreases greatly due to a pronounced decrease in the secondary bonding forces. Figure 10.45 shows this effect for polymethyl methacrylate, which has a T_g of about 100°C. The tensile strength of PMMA is about 7 ksi at 86°C, which is below its T_g, and decreases to about 4 ksi at 122°C, which is above its T_g. The maximum-use temperatures for some thermoplastics are listed in Tables 10.2 and 10.5.

Thermoset plastics also become weaker when heated, but since their atoms are bonded together primarily with strong covalent bonds in a network, they do not become viscous at elevated temperatures but degrade and char above their maximum-use

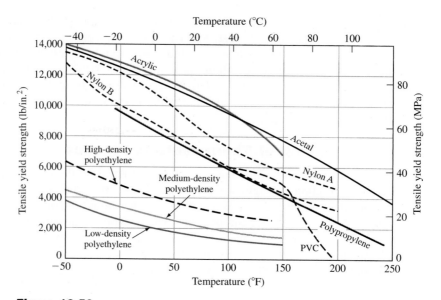

Figure 10.50
Effect of temperature on the tensile yield strength of some thermoplastics.
(*After H. E. Barker and A. E. Javitz, Plastic Molding Materials for Structural and Mechanical Applications, Electr. Manuf., May 1960.*)

temperature. In general, thermosets are more stable at higher temperatures than thermoplastics, but there are some thermoplastics that have remarkable high-temperature stability. The maximum-use temperatures of some thermosets are listed in Table 10.7.

10.11 CREEP AND FRACTURE OF POLYMERIC MATERIALS

10.11.1 Creep of Polymeric Materials

Polymeric materials subjected to a load may creep. That is, their deformation under a constant applied load at a constant temperature continues to increase with time. The magnitude of the strain increment increases with increased applied stress and temperature. Figure 10.51 shows how the creep strain of polystyrene increases under tensile stresses of 1760 to 4060 psi (12.1 to 30 MPa) at 77°F.

The temperature at which the creep of a polymeric material takes place is also an important factor determining the creep rate. At temperatures below the glass transition temperatures for thermoplastics, the creep rate is relatively low due to restricted molecular chain mobility. Above their glass transition temperatures, thermoplastics deform more easily by a combination of elastic and plastic deformation that is referred to as *viscoelastic behavior.* Above the glass transition temperature,

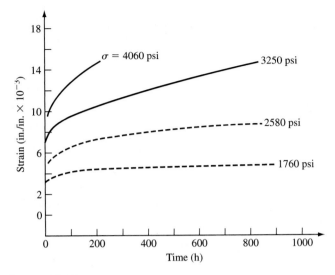

Figure 10.51
Creep curves for polystyrene at various tensile stresses at 77°F.
[*After J. A. Sauer, J. Marin, and C. C. Hsiao, J. Appl. Phys.*, **20:**507 *(1949).*]

the molecular chains slide past each other more easily, and so this type of easier deformation is sometimes referred to as *viscous flow.*

In industry the creep of polymeric materials is measured by the creep modulus, which is simply the ratio of the initial applied stress σ_0 to the creep strain $\epsilon(t)$ after a particular time and at a constant temperature of testing. A high value for the creep modulus of a material thus implies a low creep rate. Table 10.11 lists the creep moduli of various plastics at various stress levels within the 1000 to 5000 psi range. This table shows the effect of bulky side groups and strong intermolecular forces in reducing the creep rate of polymeric materials. For example, at 73°F polyethylene has a creep modulus of 62 ksi at a stress level of 1000 psi for 10 h, whereas PMMA has a much higher creep modulus of 410 ksi at the same stress level for the same time.

Reinforcing plastics with glass fibers greatly increases their creep moduli and reduces their creep rates. For example, unreinforced nylon 6,6 has a creep modulus of 123 ksi after 10 h at 1000 psi, but when reinforced with 33 percent glass fiber, its creep modulus increases to 700 ksi after 10 h at 4000 psi. The addition of glass fibers to plastic materials is an important method for increasing their creep resistance as well as for strengthening them.

10.11.2 Stress Relaxation of Polymeric Materials

The stress relaxation of a stressed polymeric material under constant strain results in a decrease in the stress with time. The cause of the stress relaxation is that viscous flow in the polymeric material's internal structure occurs by the polymer

Table 10.11 Creep modulus of polymeric materials at 73°F (23°C)

| | Test time (H) | | | |
| | 10 | 100 | 1000 | |
	Creep modulus (ksi)			Stress level (psi)
Unreinforced materials:				
Polyethylene, Amoco 31-360B1	62	36		1000
Polypropylene, Profax 6323	77	58	46	1500
Polystyrene, FyRid KS1	310	290	210	Impact-modified
Polymethyl methacrylate, Plexiglas G	410	375	342	1000
Polyvinyl chloride, Bakelite CMDA 2201	...	250	183	1500
Polycarbonate, Lexan 141-111	335	320	310	3000
Nylon 6,6, Zytel 101	123	101	83	1000, equil. at 50% RH
Acetal, Delrin 500	360	280	240	1500
ABS, Cycolac DFA-R	340	330	300	1000
Reinforced materials:				
Acetal, Thermocomp KF-1008, 30% glass fiber	1320	...	1150	5000, 75°F (24°C)
Nylon 6,6, Zytel 70G-332, 33% glass fiber	700	640	585	4000, equil. at 50% RH
Polyester, thermosetting molding compound, Cyglas 303	1310	1100	930	2000
Polystyrene, Thermocomp CF-1007	1800	1710	1660	5000, 75°F (24°C)

Source: *Modern Plastics Encyclopedia, 1984–85,* McGraw-Hill.

chains slowly sliding by each other by the breaking and reforming of secondary bonds between the chains and by mechanical untangling and recoiling of the chains. Stress relaxation allows the material to attain a lower energy state spontaneously, if there is sufficient activation energy for the process to occur. Stress relaxation of polymeric materials is thus temperature-dependent and associated with an activation energy.

The rate at which stress relaxation occurs depends on the *relaxation time* τ, which is a property of the material and which is defined as the time needed for the stress (σ) to decrease to 0.37 ($1/e$) of the initial stress σ_0. The decrease in stress with time t is given by

$$\sigma = \sigma_0 e^{-t/\tau} \tag{10.3}$$

where σ = stress after time t, σ_0 = initial stress, and τ = relaxation time.

EXAMPLE PROBLEM 10.9

A stress of 1100 psi (7.6 MPa) is applied to an elastomeric material at constant strain. After 40 days at 20°C, the stress decreases to 700 psi (4.8 MPa). (*a*) What is the relaxation-time constant for this material? (*b*) What will be the stress after 60 days at 20°C?

■ **Solution**

a.   Since $\sigma = \sigma_0 e^{-t/\tau}$ [Eq. (10.3)] or $\ln(\sigma/\sigma_0) = -t/\tau$ where $\sigma = 700$ psi,
     $\sigma_0 = 1100$ psi, and $t = 40$ days,

$$\ln\left(\frac{700 \text{ psi}}{1100 \text{ psi}}\right) = -\frac{40 \text{ days}}{\tau} \qquad \tau = \frac{-40 \text{ days}}{-0.452} = 88.5 \text{ days} \blacktriangleleft$$

b.

$$\ln\left(\frac{\sigma}{1100 \text{ psi}}\right) = -\frac{60 \text{ days}}{88.5 \text{ days}} = -0.678$$

$$\frac{\sigma}{1100 \text{ psi}} = 0.508 \qquad \text{or} \qquad \sigma = 559 \text{ psi} \blacktriangleleft$$

Since the relaxation time τ is the reciprocal of a rate, we can relate it to the temperature in kelvins by an Arrhenius-type rate equation (see Eq. [4.12]) as

$$\frac{1}{\tau} = Ce^{-Q/RT} \tag{10.4}$$

where, C = rate constant independent of temperature, Q = activation energy for the process, T = temperature in kelvins, and R = molar gas constant = 8.314 J/(mol · K). Example Problem 10.10 shows how Eq. 10.4 can be used to determine the activation energy for an elastomeric material undergoing stress relaxation.

The relaxation time for an elastomer at 25°C is 40 days, while at 35°C the relaxation time is 30 days. Calculate the activation energy for this stress-relaxation process.

EXAMPLE PROBLEM 10.10

■ **Solution**
Using Eq. 10.4, $1/\tau = Ce^{-Q/RT}$. For $\tau = 40$ days,

$$T_{25°C} = 25 + 273 = 298 \text{ K} \qquad T_{35°C} = 35 + 273 = 308 \text{ K}$$

$$\frac{1}{40} = Ce^{-Q/RT_{298}} \tag{10.5}$$

and

$$\frac{1}{30} = Ce^{-Q/RT_{308}} \tag{10.6}$$

Dividing Eq. 10.5 by Eq. 10.6 gives

$$\frac{30}{40} = \exp\left[-\frac{Q}{R}\left(\frac{1}{298} - \frac{1}{308}\right)\right] \qquad \text{or} \qquad \ln\left(\frac{30}{40}\right) = -\frac{Q}{R}(0.003356 - 0.003247)$$

$$-0.288 = -\frac{Q}{8.314}(0.000109) \qquad \text{or} \qquad Q = 22,000 \text{ J/mol} = 22.0 \text{ kJ/mol} \blacktriangleleft$$

10.11.3 Fracture of Polymeric Materials

As was the case for metals, the fracture of polymeric materials can be considered to be brittle or ductile or intermediate between the two extremes. In general, unreinforced thermosetting plastics are considered to fracture primarily in a brittle mode. Thermoplastics, on the other hand, may fracture primarily by the brittle or ductile manner. If the fracture of a thermoplastic takes place below its glass transition temperature, then its fracture mode will be primarily brittle, whereas if the fracture takes place above its glass transition temperature, its fracture mode will be ductile. Thus, temperature can greatly affect the fracture mode of thermoplastics. Thermosetting plastics heated above room temperature become weaker and fracture at a lower stress level but still fracture primarily in a brittle mode because the covalent bonding network is retained at elevated temperature. Strain rate is also an important factor in the fracture behavior of thermoplastics, with slower strain rates favoring ductile fracture because a slow strain rate allows molecular-chain realignment.

Brittle Fracture of Polymeric Materials   The surface energy required to fracture an amorphous brittle glassy polymeric material such as polystyrene or polymethyl methacrylate is about 1000 times greater than that which would be required if the fracture involved just the simple breaking of carbon-carbon bonds on a fracture plane. Thus glassy polymeric materials such as PMMA are much tougher than inorganic glasses. The extra energy required to fracture glassy thermoplastics is much higher because distorted localized regions called *crazes* form before cracking occurs. A craze in a glassy thermoplastic is formed in a highly stressed region of the material and consists of an alignment of molecular chains combined with a high density of interdispersed voids.

Figure 10.52 is a diagrammatic representation of the change in molecular structure at a craze in a glassy thermoplastic such as PMMA. If the stress is intense enough, a crack forms through the craze, as shown in Fig. 10.53 and in the photo of Fig. 10.54. As the crack propagates along the craze, the stress concentration at the tip of the crack extends along the length of the craze. The work done in aligning

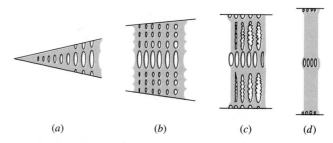

(a) (b) (c) (d)

Figure 10.52
Diagrammatic representation of the change in the microstructure of a craze in a glassy thermoplastic as it thickens.
[*After P. Beahan, M. Bevis, and D. Hull, J. Mater. Sci.,* **8***:162(1972).*]

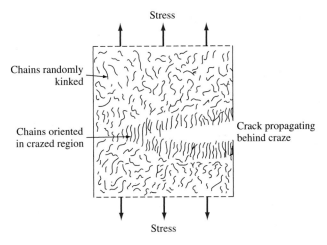

Figure 10.53
Schematic illustration of the structure of a craze near
the end of a crack in a glassy thermoplastic.

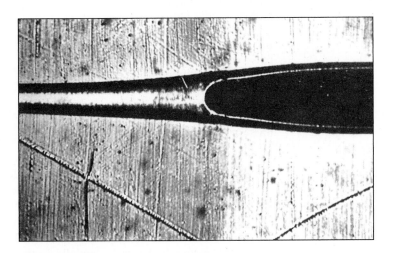

Figure 10.54
Photo of a crack through the center of a craze in a glassy
thermoplastic.
(*After D. Hull, "Polymeric Materials," American Society of Metals, 1975, p. 511.*)

the polymer molecules within the craze is the cause of the relatively high amount
of work required for the fracture of glassy polymeric materials. This explains why
the fracture energies of polystyrene and PMMA are between 300 and 1700 J/m^2
instead of about 0.1 J/m^2, which is the energy level that would be expected if only
covalent bonds were broken in the fracturing process.

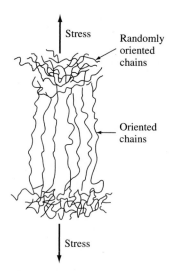

Figure 10.55
Plastic yielding of a
thermoplastic polymeric
material under stress. The
molecular chains are
uncoiled and slip past each
other so that they align
themselves in the direction
of the stress. If the stress
is too high, the molecular
chains break, causing
fracture of the material.

Ductile Fracture of Polymeric Materials   Thermoplastics above their glass transition temperatures can exhibit plastic yielding before fracture. During plastic yielding, molecular linear chains uncoil and slip past each other and gradually align closer together in the direction of the applied stress (Fig. 10.55). Eventually, when the stress on the chains becomes too high, the covalent bonds of the main chains break and fracture of the material occurs. Elastomeric materials deform essentially in the same way except they undergo much more chain uncoiling (elastic deformation), but eventually if the stress on the material is too high and the extension of their molecular chains too great, the covalent bonds of the main chains will break, causing fracture of the material.

10.12 POLYMERS IN BIOMEDICAL APPLICATIONS—BIOPOLYMERS

Polymers offer the greatest versatility as biomaterials. They have been used in various pathologies, including those of the cardiovascular, ophthalmic, and orthopedic systems, as permanent implant components. They have also been applied as provisional remedies in areas such as coronary angioplasty, hemodialysis, and wound treatment. The application of polymers in dentistry as implants, dental cements, and denture bases is also of great interest and importance. Although polymers are inferior to metals and ceramics in terms of strength properties, they possess characteristics that are very attractive in biomedical applications, including low density, ease of forming, and the possibility of modification for maximum biocompatibility. The majority of polymeric biomaterials are thermoplastics, and their mechanical properties, although inferior to metals and ceramics, are acceptable in many applications. One of the more recent developments in this field is that of **biodegradable polymers.** Biodegradable polymers are designed to perform a function and then ultimately be absorbed or integrated into the biological system. Thus, surgical removal of these components is not necessary. In the following sections the application of biopolymers to various medical areas will be discussed.[12]

10.12.1 Cardiovascular Applications of Polymers

Biopolymers are successfully being used in heart valves. Human heart valves are prone to diseases such as stenosis and incompetence. Stenosis occurs due to the stiffening of the heart valve, preventing it from opening completely. Incompetence is a condition in which the heart valve allows some back flow of blood. Both conditions are dangerous and must be treated through replacement of the damaged heart valve either with a tissue valve (animal or cadaveric) or artificial valve. A picture of the most recent design of an artificial heart valve is given in Fig. 10.56. The prosthesis consists of the flange, two semicircular leaflets, and a sewing ring. The flange and the leaflets may be made from biometals such as Ti or Co-Cr alloys. The sewing ring is made of biopolymers such as expanded PTFE (Teflon) or PET (Dacron) (see Sec. 10.7.5). It has the crucial function of connecting the

---

[12]*Sources: http://www.me.utexas.edu/~uer/heartvalves/mechanics.html;* Handbook of Materials for Medical Devices, J. R. Davis (ed.), ASM International, 2003; and Medical Device Materials, Proceedings of the Materials and Processes for Medical Devices Conference, S. Shrivastava (ed.), ASM International, 2003.

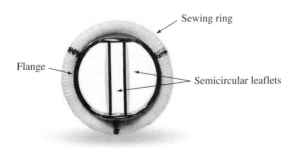

Figure 10.56
An artificial heart valve.
(*http://www.me.utexas.edu/~uer/heartvalves/mechanics.html.*
Courtesy of St. Jude Medical, Inc.)

valve to the heart tissue through the application of sutures. Polymeric materials are the only materials that make this connection possible. The leaflets allow blood flow in the position shown in the figure and block (although not completely) the back flow in the closed position. Blood clotting, which is an undesirable side effect, occurs because of the interaction of the red blood cells with the artificial valve. Patients with artificial heart valves must use anticoagulants to prevent the blood from clotting.

Vascular grafts are used in coronary artery bypass operations to bypass severely clogged arteries. These vascular grafts may be either tissue or artificial grafts. The artificial grafts must have high tensile strength and resist occlusion (clogging) of the artery caused by thrombosis (blood clotting). Either Teflon or Dacron is used for this application. However, Teflon performs better against occlusion because it minimizes shear stresses acting on the blood cells.

Blood oxygenators are designed to remove carbon dioxide from and provide oxygen to the blood. The blood from the right side of the heart (unoxygenated) is pumped through the oxygenators to produce oxygenated blood for the body during cardiopulmonary bypass surgery.[13] The oxygenators are microporous *hydrophobic* (repel water) membranes made of materials such as polypropylene (see Sec. 10.6.3). Since polypropylene is hydrophobic, the pores may be filled with gases such as O_2 instead of water. During its operation, air flows on one side of the membrane while blood flows on the other side; CO_2 is lost by diffusion through the membrane and O_2 is absorbed from the pores.

Polymers are also used in artificial hearts and heart-assist devices. These devices are critical for short-term use to maintain the patient's health until a donor heart is found. Without polymers these devices could not function in an efficient manner.

10.12.2 Ophthalmic Applications

Polymers are crucial and irreplaceable in ophthalmic (related to the eye) applications. Optical functions of the eye are corrected through eyeglasses, contact lenses (soft and hard), and intraocular implants, all of which are primarily made from polymers. Soft contact lenses are made of **hydrogel,** a soft **hydrophilic** (likes water) polymeric material that absorbs water and thus swells to a specific level. The soft hydrogel lenses are made of somewhat cross-linked polymers and copolymers. Because of their soft nature, hydrogels can take the exact shape of the cornea and allow a snug fit. However, the cornea needs oxygen that can only permeate through the lens. Hydrated hydrogels allow significant permeability of oxygen. The original material used in this application was poly-HEMA (2-hydroxyethyl methacrylate). Other newer polymers with better fabrication techniques are being developed to produce thinner soft contact lenses.

Hard lenses are loosely positioned on the cornea. The lens flexes with blinking; the lens material must have the ability to recover its shape quickly. Hard lenses were initially made of PMMA, which has excellent optical properties but suffers from a

---

[13]Surgery is performed on the artery leading from the heart to the lungs.

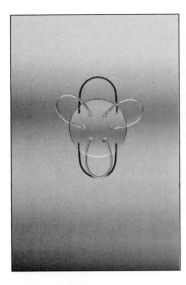 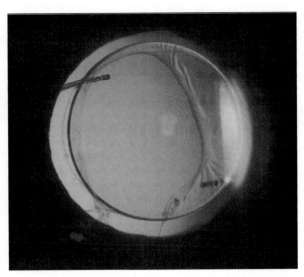

Figure 10.57
An intraocular lens.
((a) © Steve Allen/Photo Researchers, Inc. (b) Chris Barry/Phototake)

lack of permeability to oxygen. To improve the permeability to oxygen, i.e., to make *rigid gas-permeable* (RGP) lenses, copolymers of methyl methacrylate with siloxanylalkyl metacrylates were produced. However, siloxane is hydrophobic. To correct for this, hydrophilic comonomers such as metacrylic acid were added to the blend. Presently, there are a number of commercially available RGP lenses, and research is ongoing to improve these materials further.

Treatment of conditions such as cataract (clouding of the eye lens due to excess dead cells) requires surgery and removal of the opaque lens of the eye and its subsequent replacement with intraocular lens implants. The intraocular lens consists of a lens and haptics (side arms) (Fig. 10.57), which are needed to attach the lens to the suspensory ligaments to keep the lens in place. Clearly, the requirements of the intraocular lens materials are suitable optical properties and biocompatibility. As with hard lenses, both the optical portion and the side arms of most intraocular lenses are made of PMMA. Figure 10.58*a* simulates the vision of a patient with cataracts while Fig. 10.58*b* simulates the vision of the same patient after surgery. The images show the importance of materials science and engineering in improving the quality of life.

10.12.3 Drug-Delivery Systems

Biodegradable polymers such as *polylactic acid* (PLA) and *polyglycolic acid* (PGA) and their copolymers are used in implant drug-delivery systems. The polymers matrix (polymer container) contains the drug and is implanted at a location of interest inside the body. As the biodegradable polymer degrades, it releases the drug. Such delivery

(a)   (b)

Figure 10.58
Simulated vision in a patient (a) before and (b) after cataract surgery.
(© *Royalty-Free/Corbis*)

systems are especially critical if drug delivery by pills or injection is not possible because of the drug's adverse affect on other organs or tissues of the body.

10.12.4 Suture Materials

Sutures are used to close wounds and incisions. Clearly, suture materials must possess (1) sufficient tensile strength to close the wounds and (2) high knot pull strength to maintain the load in the suture after closing. Sutures may be absorbable or nonabsorbable. The nonabsorbable sutures are generally made of polypropylene, nylon, polyethylene tetraphthalate, and polyethylene. Such sutures remain intact for an indefinite period when placed in the body. Absorbable sutures are made of polyglycolic acid, which is biodegradable.

10.12.5 Orthopedic Applications

The orthopedic application of polymers is primarily as bone cement and joint prostheses. Bone cement is used as a structural material to fill the space between the implant and the bone to assure a more uniform loading condition. Thus, the term *bone cement* should not be interpreted as an adhesive function. To be most effective, the microporosity in the bone cement upon hardening must be kept to a minimum. This is achieved by using centrifuging and vacuum techniques when preparing the cement. Bone cement is also sometimes used to correct various defects in the bone. The primary polymeric material used as bone cement is PMMA. The tensile and fatigue properties of PMMA are important and can be improved by adding other agents. Application requirements call for a minimum compressive strength of 70 MPa (10 Ksi) after hardening.

As discussed in Chap. 1, an advantage of polymeric materials is that one can design and synthesize various blends to meet the objectives and requirements at hand. This is a very powerful advantage and is illuminated in the above applications. The future of polymeric materials as biomaterials is in tissue engineering. For instance, researchers are investigating the use of biodegradable polymers as scaffolding for the generation of new tissue. Biodegradable polymers such as PGA can be implanted as scaffolds together with cells and growth-promoting proteins to stimulate the generation of tissue. It is therefore plausible that, in the future, one could regenerate damaged tissues both in vivo (inside the body) for cases such as cartilage repair and in vitro (outside the body) for cases such as skin repair and replacement.

10.13 S U M M A R Y

Plastics and elastomers are important engineering materials primarily because of their wide range of properties, relative ease of forming into a desired shape, and relatively low cost. Plastic materials can be conveniently divided into two classes: *thermoplastics* and *thermosetting plastics (thermosets)*. Thermoplastics require heat to make them formable, and after cooling they retain the shape they were formed into. These materials can be reheated and reused repeatedly. Thermosetting plastics are usually formed into a permanent shape by heat and pressure during which time a chemical reaction takes place that bonds the atoms together to form a rigid solid. However, some thermosetting reactions take place at room temperature without the use of heat and pressure. Thermosetting plastics cannot be remelted after they are "set" or "cured," and upon heating to a high temperature, they degrade or decompose.

The chemicals required for producing plastics are derived mainly from petroleum, natural gas, and coal. Plastic materials are produced by the polymerizing of many small molecules called *monomers* into very large molecules called *polymers*. Thermoplastics are composed of long-molecular-chain polymers, with the bonding forces between the chains being of the secondary permanent dipole type. Thermosetting plastics are covalently bonded throughout with strong covalent bonding between all the atoms.

The most commonly used processing methods for thermoplastics are *injection molding, extrusion,* and *blow molding,* whereas the most commonly used methods for thermosetting plastics are *compression and transfer molding* and *casting.*

There are many families of thermoplastics and thermosetting plastics. Examples of some general-purpose thermoplastics are polyethylene, polyvinyl chloride, polypropylene, and polystyrene. Examples of engineering plastics are polyamides (nylons), polyacetal, polycarbonate, saturated polyesters, polyphenylene oxide, and polysulfone. (Note that the separation of thermoplastics into general-purpose and engineering plastics is arbitrary.) Examples of thermosetting plastics are phenolics, unsaturated polyesters, melamines, and epoxies.

Elastomers or *rubbers* form a large subdivision of polymeric materials and are of great engineering importance. Natural rubber is obtained from tree plantations and is still much in demand (about 30 percent of the world's rubber supply) because of its superior elastic properties. Synthetic rubbers account for about 70 percent of the world's rubber supply, with styrene-butadiene being the most commonly used type. Other synthetic rubbers such as nitrile and polychloroprene (neoprene) are used for applications where special properties such as resistance to oils and solvents are required.

Thermoplastics have a *glass transition temperature* above which these materials behave as viscous or rubbery solids and below which they behave as brittle, glasslike solids. Above the glass transition temperature, permanent deformation occurs by molecular chains sliding past each other, breaking and remaking secondary bonds. Thermoplastics used above the glass transition temperature can be strengthened by intermolecular bonding forces by using polar pendant atoms such as chlorine in polyvinyl chloride or by hydrogen bonding as in the case of the nylons. Thermosetting plastics, because they are covalently bonded throughout, allow little deformation before fracture.

The application of polymeric materials to the biomedical field has increased significantly. Polymers are used for cardiovascular, opthalmic, drug delivery, and orthopedic applications. Polymers are also the principal material used as biodegradable scaffolding in the tissue engineering field.

10.14 DEFINITIONS

Sec. 10.1

Thermoplastic (noun): a plastic material that requires heat to make it formable (plastic) and upon cooling, retains its shape. Thermoplastics are composed of chain polymers with the bonds between the chains being of the secondary permanent dipole type. Thermoplastics can be repeatedly softened when heated and harden when cooled. Typical thermoplastics are polyethylenes, vinyls, acrylics, cellulosics, and nylons.

Thermosetting plastic (thermoset): a plastic material that has undergone a chemical reaction by the action of heat, catalysis, etc., leading to a cross-linked network macromolecular structure. Thermoset plastics cannot be remelted and reprocessed since when they are heated they degrade and decompose. Typical thermoset plastics are phenolics, unsaturated polyesters, and epoxies.

Sec. 10.2

Monomer: a simple molecular compound that can be covalently bonded together to form long molecular chains (polymers). Example: ethylene.

Chain polymer: a high-molecular-mass compound whose structure consists of a large number of small repeating units called *mers*. Carbon atoms make up most of the mainchain atoms in most polymers.

Mer: a repeating unit in a chain polymer molecule.

Polymerization: the chemical reaction in which high-molecular-mass molecules are formed from monomers.

Copolymerization: the chemical reaction in which high-molecular-mass molecules are formed from two or more monomers.

Chain polymerization: the polymerization mechanism whereby each polymer molecule increases in size at a rapid rate once growth has started. This type of reaction occurs in three steps: (1) chain initiation, (2) chain propagation, and (3) chain termination. The name implies a chain reaction and is usually initiated by some external source. Example: the chain polymerization of ethylene into polyethylene.

Degree of polymerization: the molecular mass of a polymer chain divided by the molecular mass of its mer.

Functionality: the number of active bonding sites in a monomer. If the monomer has two bonding sites, it is said to be *bifunctional.*

Homopolymer: a polymer consisting of only one type of monomeric unit.

Copolymer: a polymer chain consisting of two or more types of monomeric units.

Cross-linking: the formation of primary valence bonds between polymer chain molecules. When extensive cross-linking occurs as in the case of thermosetting resins, cross-linking makes one supermolecule of all the atoms.

Stepwise polymerization: the polymerization mechanism whereby the growth of the polymer molecule proceeds by a stepwise intermolecular reaction. Only one type of reaction is involved. Monomer units can react with each other or with any size polymer molecule. The active group on the end of a monomer is assumed to have the same reactivity no matter what the polymer length is. Often a by-product such as water is condensed off in the polymerization process. Example: the polymerization of nylon 6,6 from adipic acid and hexamethylene diamine.

Sec. 10.3

Bulk polymerization: the direct polymerization of liquid monomer to polymer in a reaction system in which the polymer remains soluble in its own monomer.

Solution polymerization: in this process a solvent is used that dissolves the monomer, the polymer, and the polymerization initiator. Diluting the monomer with the solvent reduces the rate of polymerization, and the heat released by the polymerization reaction is absorbed by the solvent.

Suspension polymerization: in this process water is used as the reaction medium, and the monomer is dispersed rather than being dissolved in the medium. The polymer products are obtained in the form of small beads that are filtered, washed, and dried in the form of molding powders.

Sec. 10.4

Crystallinity (in polymers): the packing of molecular chains into a stereoregular arrangement with a high degree of compactness. Crystallinity in polymeric materials is never 100 percent and is favored in polymeric materials whose polymer chains are symmetrical. Example: high-density polyethylene can be 95 percent crystalline.

Glass transition temperature: the center of the temperature range where a heated thermoplastic upon cooling changes from a rubbery, leathery state to that of brittle glass.

Stereoisomers: molecules that have the same chemical composition but different structural arrangements.

Atactic stereoisomer: this isomer has pendant groups of atoms *randomly arranged* along a vinyl polymer chain. Example: atactic polypropylene.

Isotactic isomer: this isomer has pendant groups of atoms all on the *same side* of a vinyl polymer chain. Example: isotactic polypropylene.

Syndiotactic isomer: this isomer has pendant groups of atoms *regularly alternating* in positions on both sides of a vinyl polymer chain. Example: syndiotactic polypropylene.

Stereospecific catalyst: a catalyst that creates mostly a specific type of stereoisomer during polymerization. Example: the Ziegler catalyst used to polymerize propylene to mainly the isotactic polypropylene isomer.

Sec. 10.5

Injection molding: a molding process whereby a heat-softened plastic material is forced by a screw-drive cylinder into a relatively cool mold cavity that gives the plastic the desired shape.

Blow molding: a method of fabricating plastics in which a hollow tube (parison) is forced into the shape of a mold cavity by internal air pressure.

Extrusion: the forcing of softened plastic material through an orifice, producing a continuous product. Example: plastic pipe is extruded.

Compression molding: a thermoset molding process in which a molding compound (which is usually heated) is first placed in a molding cavity. Then the mold is closed and heat and pressure are applied until the material is cured.

Transfer molding: a thermoset molding process in which the molding compound is first softened by heat in a transfer chamber and then is forced under high pressure into one or more mold cavities for final curing.

Sec. 10.6

Plasticizers: chemical agents added to plastic compounds to improve flow and processibility and to reduce brittleness. Example: plasticized polyvinyl chloride.

Filler: a low-cost inert substance added to plastics to make them less costly. Fillers may also improve some physical properties such as tensile strength, impact strength, hardness, wear resistance, etc.

Sec. 10.9

Elastomer: a material that at room temperature stretches under a low stress to at least twice its length and then quickly returns to almost its original length upon removal of the stress.

cis-**1,4 polyisoprene:** the isomer of 1,4 polyisoprene that has the methyl group and hydrogen on the same side of the central double bond of its mer. Natural rubber consists mainly of this isomer.

trans-**1,4 polyisoprene:** the isomer of 1,4 polyisoprene that has the methyl group and hydrogen on opposite sides of the central double bond of its mer.

Vulcanization: a chemical reaction that causes cross-linking of polymer chains. Vulcanization usually refers to the cross-linking of rubber molecular chains with sulfur, but the word is also used for other cross-linking reactions of polymers such as those that occur in some silicone rubbers.

Sec. 10.12

Biodegradable polymer: a type of polymer that is absorbed by the human body after it serves its purpose; for instance, absorbable sutures.

Biopolymer: polymers that are used inside the human body for various surgical applications.

Hydrogel: a soft polymeric material that absorbs water and swells to a specific level.

Hydrophilic polymer: polymers that absorb water; like water.

10.15 PROBLEMS

Calculation Problems

Answers to problems marked with an asterisk are given at the end of the book.

*10.1 A high-molecular-weight polyethylene has an average molecular weight of 410,000 g/mol. What is its average degree of polymerization?

10.2 If a type of polyethylene has an average degree of polymerization of 10,000, what is its average molecular weight?

10.3 A nylon 6,6 has an average molecular weight of 12,000 g/mol. Calculate the average degree of polymerization (see Sec. 10.7 for its mer structure). M.W. = 226 g/mol.

*10.4 An injection-molding polycarbonate material has an average molecular weight of 25,000 g/mol. Calculate its degree of polymerization (see Sec. 10.7 for the mer structure of polycarbonate). M.W. = 254 g/mol.

10.5 Calculate the average molecular weight M_m for a thermoplastic that has the following weight fractions f_i for the molecular weight ranges listed:

Molecular weight range (g/mol)	f_i	Molecular weight range (g/mol)	f_i
0–5000	0.01	20,000–25,000	0.19
5000–10,000	0.04	25,000–30,000	0.21
10,000–15,000	0.16	30,000–35,000	0.15
15,000–20,000	0.17	35,000–40,000	0.07

*10.6 A copolymer consists of 70 wt % polystyrene and 30 wt % polyacrylonitrile. Calculate the mole fraction of each component in this material.

*10.7 An ABS copolymer consists of 25 wt % polyacrylonitrile, 30 wt % polybutadiene, and 45 wt % polystyrene. Calculate the mole fraction of each component in this material.

10.8 Determine the mole fraction of polyvinyl chloride and polyvinyl acetate in a copolymer having a molecular weight of 11,000 g/mol and a degree of polymerization of 150.

*10.9 How much sulfur must be added to 70 g of butadiene rubber to cross-link 3.0 percent of the mers? (Assume all sulfur is used to cross-link the mers and that only one sulfur atom is involved in each cross-linking bond.)

10.10 If 5 g of sulfur is added to 90 g of butadiene rubber, what is the maximum fraction of the cross-link sites that can be connected?

10.11 How much sulfur must be added to cross-link 10 percent of the cross-link sites in 90 g of polyisoprene rubber?

10.12 How many kilograms of sulfur are needed to cross-link 15 percent of the cross-link sites in 200 kg of polyisoprene rubber?

*10.13 If 3 kg of sulfur is added to 300 kg of butadiene rubber, what fraction of the cross-links are joined?

10.14 A butadiene-styrene rubber is made by polymerizing one monomer of styrene with seven monomers of butadiene. If 20 percent of the cross-link sites are to be bonded with sulfur, what weight percent sulfur is required? (See Ex. Problem 10.7.)

10.15 What weight percent sulfur must be added to polybutadiene to cross-link 20 percent of the possible cross-link sites?

10.16 A butadiene-acrylonitrile rubber is made by polymerizing one acrylonitrile monomer with five butadiene monomers. How much sulfur is required to react with 200 kg of this rubber to cross-link 22 percent of the cross-link sites? (See Ex. Problem 10.8.)

*10.17   If 15 percent of the cross-link sites in isoprene rubber are to be bonded, what weight percent sulfur must the rubber contain?

*10.18   A stress of 9.0 MPa is applied to an elastomeric material at a constant stress at 20°C. After 25 days, the stress decreases to 6.0 MPa. (*a*) What is the relaxation time τ for this material? (*b*) What will be the stress after 50 days?

10.19   A polymeric material has a relaxation time of 60 days at 27°C when a stress of 7.0 MPa is applied. How many days will be required to decrease the stress to 6.0 MPa?

*10.20   A stress of 1000 psi is applied to an elastomer at 27°C, and after 25 days the stress is reduced to 750 psi by stress relaxation. When the temperature is raised to 50°C, the stress is reduced from 1100 to 400 psi in 30 days. Calculate the activation energy for this relaxation process using an Arrhenius-type rate equation.

10.21   The stress on a sample of a rubber material at constant strain at 27°C decreases from 6.0 to 4.0 MPa in three days. (*a*) What is the relaxation time τ for this material? (*b*) What will be the stress on this material after (i) 15 days and (ii) 40 days?

*10.22   A polymeric material has a relaxation time of 100 days at 27°C when a stress of 6.0 MPa is applied. (*a*) How many days will be required to decrease the stress to 4.2 MPa? (*b*) What is the relaxation time at 40°C if the activation energy for this process is 25 kJ/mol?

Study Problems

10.23   Define the polymeric materials (*a*) plastics and (*b*) elastomers.

10.24   Define a thermoplastic plastic.

10.25   Describe the atomic structural arrangement of thermoplastics.

10.26   What types of atoms are bonded together in thermoplastic molecular chains? What are the valences of these atoms in the molecular chains?

10.27   What is a pendant atom or group of atoms?

10.28   What type of bonding exists within the molecular chains of thermoplastics?

10.29   What type of bonding exists between the molecular chains of thermoplastics?

10.30   Define thermosetting plastics.

10.31   Describe the atomic structural arrangement of thermosetting plastics.

10.32   What are some of the reasons for the great increase in the use of plastics in engineering designs over the past years?

10.33   What are some of the advantages of plastics for use in mechanical engineering designs?

10.34   What are some of the advantages of plastics for use in electrical engineering designs?

10.35   What are some of the advantages of plastics for use in chemical engineering designs?

10.36   Define the following terms: chain polymerization, monomer, and polymer.

10.37   Describe the bonding structure within an ethylene molecule by using (*a*) the electron-dot-cross notation and (*b*) straight-line notation for the bonding electrons.

10.38   What is the difference between a saturated and an unsaturated carbon-containing molecule?

10.39   Describe the bonding structure of an activated ethylene molecule that is ready for covalent bonding with another activated molecule by using (*a*) electron-dot-cross notation and (*b*) straight-line notation for the bonding electrons.

10.40   Write a general chemical reaction for the chain polymerization of ethylene monomer into the linear polymer polyethylene.

10.41   What is the repeating chemical unit of a polymer chain called? What is the chemical repeating unit for polyethylene?

10.42   Define the degree of polymerization for a polymer chain.

10.43   What are the three major reactions that occur during chain polymerization?

10.44   What is a free radical? Write a chemical equation for the formation of two free radicals from a hydrogen peroxide molecule by using (*a*) electron-dot notation and (*b*) straight-line notation for the bonding electrons.

10.45   Write an equation for the formation of two free radicals from a molecule of benzoyl peroxide by using straight-line notation for the bonding electrons.

10.46   Write an equation for the reaction of an organic free radical ($RO^{\bullet}$) with an ethylene molecule to form a new, longer-chain free radical.

10.47   What is the function of the initiator catalyst for chain polymerization?

10.48   Write a reaction for the free radical $R\!-\!CH_2\!-\!CH_2$ with an ethylene molecule to extend the free radical. What type of reaction is this?

10.49   How is it possible for a polymer chain such as a polyethylene one to keep growing spontaneously during polymerization?

10.50   What are two methods by which a linear chain polymerization reaction can be terminated?

10.51   Why must one consider the *average* degree of polymerization and the *average* molecular weight of a thermoplastic material?

10.52   Define the average molecular weight of a thermoplastic.

10.53   What is the functionality of a polymer? Distinguish between a bifunctional and trifunctional monomer.

10.54   What causes a polyethylene molecular chain to have a zigzag configuration?

10.55   What type of chemical bonding is there between the polymer chains in polyethylene?

10.56   How do side branches on polyethylene main chains affect the packing of the molecular chains in a solid polymer? How does branching of the polymer chains affect the tensile strength of solid bulk polyethylene?

10.57   Write a general reaction for the polymerization of a vinyl-type polymer.

10.58   Write structural formulas for the mers of the following vinyl polymers: (*a*) polyethylene, (*b*) polyvinyl chloride, (*c*) polypropylene, (*d*) polystyrene, (*e*) polyacrylonitrile, and (*f*) polyvinyl acetate.

10.59   Write a general reaction for the polymerization of a vinylidene polymer.

10.60   Write structural formulas for the mers of the following vinylidene polymers: (*a*) polyvinylidene chloride and (*b*) polymethyl methacrylate.

10.61   Distinguish between a homopolymer and a copolymer.

10.62   Illustrate the following types of copolymers by using filled and open circles for their mers: (*a*) random, (*b*) alternating, (*c*) block, and (*d*) graft.

10.63   Write a general polymerization reaction for the formation of a vinyl chloride and vinyl acetate copolymer.

10.64   Define stepwise polymerization of linear polymers. What by-products are commonly produced by stepwise polymerization?

10.65   Write the equation for the reaction of a molecule of hexamethylene diamine with one of adipic acid to produce a molecule of nylon 6,6. What is the by-product of this reaction?

10.66   Write the reaction for the stepwise polymerization of two phenol molecules with one of formaldehyde to produce a phenol formaldehyde molecule.

10.67   What type of by-product is produced by this reaction?

10.68   What are three basic raw materials used to produce the basic chemicals needed for the polymerization of plastic materials?

10.69   Describe and illustrate the following polymerization processes: (*a*) bulk, (*b*) solution, (*c*) suspension, and (*d*) emulsion.

10.70   In which polymerization process is the heat released by the reaction absorbed by water? In which is the heat absorbed by the solvent? Which process is used when the polymerization heat of reaction is low?

10.71   Describe the Unipol process for producing low-density polyethylene. What are the advantages of this process?

10.72   During the solidification of thermoplastics, how do the specific volume versus temperature plots differ for noncrystalline and partly crystalline thermoplastics?

10.73   Define the glass transition temperature T_g for a thermoplastic.

10.74   What are measured T_g values for (*a*) polyethylene, (*b*) polyvinyl chloride, and (*c*) polymethyl methacrylate? Are T_g values material constants?

10.75   Describe and illustrate the fringed-micelle and folded-chain models for the structure of partly crystalline thermoplastics.

10.76   Describe the spherulitic structure found in some partly crystalline thermoplastics.

10.77   Why is complete crystallinity in thermoplastics impossible?

10.78   How does the amount of crystallinity in a thermoplastic affect (*a*) its density and (*b*) its tensile strength? Explain.

10.79   What are stereoisomers with respect to chemical molecules?

10.80   Describe and draw structural models for the following stereoisomers of polypropylene: (*a*) atactic, (*b*) isotactic, and (*c*) syndiotactic.

10.81   What is a stereospecific catalyst? How did the development of a stereospecific catalyst for the polymerization of polypropylene affect the usefulness of commercial polypropylene?

10.82   In general how does the processing of thermoplastics into the desired shape differ from the processing of thermosetting plastics?

10.83   Describe the injection-molding process for thermoplastics.

10.84   Describe the operation of the reciprocal-screw injection-molding machine.

10.85   What are some advantages and disadvantages of the injection-molding process for molding thermoplastics?

10.86   What are the advantages of the reciprocating-screw injection-molding machine over the old plunger type?

10.87 Describe the extrusion process for processing thermoplastics.

10.88 Describe the blow molding and thermoforming processes for forming thermoplastics.

10.89 Describe the compression-molding process for thermosetting plastics.

10.90 What are some of the advantages and disadvantages of the compression-molding process?

10.91 Describe the transfer-molding process for thermosetting plastics.

10.92 What are some of the advantages and disadvantages of the transfer-molding process?

10.93 What are the four major thermoplastic materials that account for about 60 percent of the sales tonnage of plastic materials in the United States? What were their prices per pound in 1988? in the year 2000?

10.94 How does the molecular-chain structure differ for the following types of polyethylene: (*a*) low-density, (*b*) high-density, and (*c*) linear-low-density?

10.95 How does chain branching affect the following properties of polyethylene: (*a*) amount of crystallinity, (*b*) strength, and (*c*) elongation?

10.96 What are some of the properties that make polyethylene such an industrially important plastic material?

10.97 What are some of the industrial applications for polyethylene?

10.98 Write the general reaction for the polymerization of polyvinyl chloride.

10.99 How can the higher strength of polyvinyl chloride as compared to polyethylene be explained?

10.100 How is the flexibility of bulk polyvinyl chloride increased?

10.101 What are some of the properties of polyvinyl chloride that make it an important industrial material?

10.102 What are plasticizers? Why are they used in some polymeric materials? How do plasticizers usually affect the strength and flexibility of polymeric materials? What types of plasticizers are commonly used for PVC?

10.103 How is the processibility of PVC improved to produce rigid PVC?

10.104 What are some of the applications for plasticized PVC?

10.105 Write the general reaction for the polymerization of polypropylene from propylene gas.

10.106 Why is the use of a stereospecific catalyst in the polymerization of polypropylene so important?

10.107 How does the presence of a methyl group on every other carbon atom of the main polymer chain affect the glass transition temperature of this material when compared to polyethylene?

10.108 What are some of the properties of polypropylene that make it an industrially important material?

10.109 What are some of the applications for polypropylene?

10.110 Write the general reaction for the polymerization of polystyrene from styrene.

10.111 What effect does the presence of the phenyl group on every other carbon of the main chain have on the impact properties of polystyrene?

10.112 How can the low-impact resistance of polystyrene be improved by copolymerization?

10.113 What are some of the applications for polystyrene?

10.114 What is the repeating chemical structural unit for polyacrylonitrile?

10.115 What effect does the presence of the nitrile group on every other carbon of the main polymer chains have on the molecular structure of polyacrylonitrile?

10.116 What special properties do polyacrylonitrile fibers have? What are some applications for polyacrylonitrile?

10.117 What are SAN resins? What desirable properties do SAN thermoplastics have? What are some of the applications for SAN thermoplastics?

10.118 What do the letters A, B, and S stand for in the ABS thermoplastic?

10.119 Why is ABS sometimes referred to as a terpolymer?

10.120 What important property advantages does each of the components in ABS contribute?

10.121 Describe the structure of ABS. How can the impact properties of ABS be improved?

10.122 What are some of the applications for ABS plastics?

10.123 What is the repeating chemical structural unit for polymethyl methacrylate? By what trade names is PMMA commonly known?

10.124 What are some of the important properties of PMMA that make it an important industrial plastic?

10.125 What are the fluoroplastics? What are the repeating chemical structural units for polytetrafluoroethylene and polychlorotrifluoroethylene?

10.126 What are some of the important properties and applications of polytetrafluoroethylene?

10.127 How does the presence of the chlorine atom on every other carbon atom of the main chain of polychlorotrifluoroethylene modify the crystallinity and mold-ability of polytetrafluoroethylene?

10.128 What are some of the important properties and applications of polychlorotrifluoroethylene?

10.129 Define an engineering thermoplastic. Why is this definition arbitrary?

10.130 How do the prices of engineering thermoplastics compare with those of the commodity plastics such as polyethylene, polyvinyl chloride, and polypropylene?

10.131 How do the densities and tensile strengths of engineering thermoplastics compare with those of polyethylene and polyvinyl chloride?

10.132 What is the structural formula for the amide linkage in thermoplastics? What is the general name for polyamide thermoplastics?

10.133 Write a chemical reaction for one molecule of a dibasic acid with a diamine to form an amide linkage. What is the by-product of this reaction?

10.134 In the designation nylon 6,6, what does the 6,6 stand for?

10.135 Write a chemical reaction for one molecule of adipic acid and one molecule of hexamethylene diamine to form an amide linkage.

10.136 What is the repeating structural unit for nylon 6,6?

10.137 How can nylons 6,9, 6,10, and 6,12 be synthesized?

10.138 Write the reaction for the polymerization of nylon 6 from ϵ-caprolactam.

10.139 Illustrate the bonding between polymer chains of nylon 6,6. Why is this bonding particularly strong (see Fig. 10.35)?

10.140 What properties do nylons have that make them useful for engineering applications? What is an important undesirable property of nylon?

10.141 What are some of the engineering applications for nylons?

10.142 What is the basic repeating chemical structural unit for polycarbonates? What is the carbonate linkage? What are the common trade names for polycarbonate?

10.143 What part of the polycarbonate structure makes the molecule stiff? What part of the polycarbonate molecule provides molecular flexibility?

10.144 What are some of the properties of polycarbonates that make them useful engineering thermoplastics?

10.145 What are some engineering applications for polycarbonates?

10.146 What is the basic repeating chemical structural unit for the polyphenylene oxide–based resins? What are the trade names for these resins?

10.147 What part of the structure of polyphenylene oxide provides its relatively high strength? What part of its structure provides its molecular flexibility?

10.148 What are some of the properties that make polyphenylene oxide resins important engineering thermoplastics?

10.149 What are some of the engineering applications for polyphenylene oxide resins?

10.150 What is the repeating chemical structural unit for the acetal high-performance engineering thermoplastics? What are the two main types of acetals and what are their trade names?

10.151 What part of the structure of the acetals provides high strength?

10.152 What are some of the properties of acetals that make them important engineering thermoplastics?

10.153 What outstanding property advantage do the acetals have over nylons?

10.154 What are some of the engineering applications for acetals?

10.155 What types of materials have acetals commonly replaced?

10.156 What are the two most important engineering thermoplastic polyesters? What are their repeating chemical structural units?

10.157 What is the chemical structure of the ester linkage?

10.158 What part of the structure of the thermoplastic polyesters provides rigidity? What part provides molecular mobility?

10.159 What are some of the properties of thermoplastic polyesters that make them important engineering thermoplastics?

10.160 What are some engineering applications for PBT thermoplastics?

10.161 What is the repeating chemical structural unit for polysulfone?

10.162 What part of the polysulfone structure provides its high strength? What part provides chain flexibility and impact strength? What part provides high-oxidation stability?

10.163 What are some of the properties of polysulfone that are important for engineering designs?

10.164 What are some of the engineering applications for polysulfone?

10.165 What is the repeating chemical structural unit for polyphenylene sulfide? What engineering thermoplastic has a similar structure?

10.166 What is the trade name for polyphenylene sulfide?

10.167 What part of the structure of PPS provides its rigidity and strength? What part provides its high resistance to chemicals?

10.168 What properties make PPS a useful engineering thermoplastic?

10.169 What are some engineering applications for PPS?

10.170 What is the chemical structure of polyetherimide? What is its trade name?

10.171 What is the imide linkage?

10.172 What is the function of the ether linkage in polyetherimide?

10.173 What special properties does polyetherimide have for (*a*) electrical engineering designs and (*b*) mechanical engineering designs?

10.174 What are some applications for polyetherimide?

10.175 What are polymer alloys? How does their structure differ from copolymers?

10.176 Why are polymer alloys of great importance for engineering applications?

10.177 What type of polymer alloy is (*a*) Xenoy 1000, (*b*) Valox 815, and (*c*) Bayblend MC2500?

10.178 What are some of the advantages of thermosetting plastics for engineering design applications? What is the major disadvantage of thermosets that thermoplastics do not have?

10.179 What are the major processing methods used for thermosets?

10.180 What are the two major ingredients of thermosetting molding compounds?

10.181 What are the major advantages of phenolic plastics for industrial applications?

10.182 Using structural formulas, write the reaction for phenol with formaldehyde to form a phenol-formaldehyde molecule (use two phenol molecules and one formaldehyde molecule). What kind of molecule is condensed off in the reaction?

10.183 Why are large percentages of fillers used in phenolic molding compounds? What types of fillers are used and for what purposes?

10.184 What are some of the applications for phenolic compounds?

10.185 Write the structural formula for the epoxide group and the repeating unit of a commercial epoxy resin.

10.186 What are two types of reaction sites that are active in the cross-linking of commercial epoxy resins?

10.187 Write the reaction for the cross-linking of two epoxy molecules with ethylene diamine.

10.188 What are some of the advantages of epoxy thermoset resins? What are some of their applications?

10.189 What makes an unsaturated polyester resin "unsaturated"?

10.190 How are linear unsaturated polyesters cross-linked? Write a structural formula chemical reaction to illustrate the cross-linking of an unsaturated polyester.

10.191 How are most unsaturated polyesters reinforced?

10.192 What are some applications for reinforced polyesters?

10.193 What are elastomers? What are some elastomeric materials?

10.194 From what tree is most natural rubber obtained? What countries have large plantations of these trees?

10.195 What is natural rubber latex? Briefly describe how natural rubber is produced in the bulk form.

10.196 Write the formula for *cis*-1,4 polyisoprene. What does the prefix *cis*- stand for? What is the significance of the 1,4 in the name *cis*-1,4 polyisoprene?

10.197 What is natural rubber mainly made of? What other components are present in natural rubber?

10.198 To what structural arrangement is the coiling of the natural rubber polymer chains attributed? What is steric hindrance?

10.199 What are chemical structural isomers?

10.200 What is gutta-percha? What is the repeating chemical structural unit for gutta-percha?

10.201 What does the *trans*- prefix in the name *trans*-1,4 polyisoprene refer to?

10.202 Why does the trans isomer lead to a higher degree of crystallinity than the cis isomer for polyisoprene?

10.203 What is the vulcanization process for natural rubber? Who discovered this process and when? Illustrate the cross-linking of *cis*-1,4 polyisoprene with divalent sulfur atoms.

10.204 How does cross-linking with sulfur affect the tensile strength of natural rubber? Why is only about 3 wt % of sulfur used in the process?

10.205 What materials are used in the compounding of rubber and what is the function of each?

10.206 How can oxygen atoms cross-link the rubber molecules? How can the cross-linking of rubber molecules by oxygen atoms be retarded?

10.207 What is styrene-butadiene rubber? What weight percent of it is styrene? What are the repeating chemical structural units for SBR?

10.208 Can SBR be vulcanized? Explain.

10.209 What are some of the advantages and disadvantages of SBR? Natural rubber?

10.210 What is the composition of nitrile rubbers? What effect does the nitrile group have on the main carbon chain in nitrile rubber?

10.211 What are some applications for nitrile rubbers?

10.212 Write the repeating chemical structural unit for polychloroprene. What common name is given to polychloroprene rubber? How does the presence of the chlorine atom in polychloroprene affect some of its properties?

10.213 How are polychloroprene elastomeric materials vulcanized?

10.214 What are some engineering applications for neoprene rubbers?

10.215 What are the silicones? What is the general repeating chemical structural unit for the silicones?

10.216 What is a silicone elastomer? What is the chemical structural repeating unit of the most common type of silicone rubber? What is its technical name?

10.217 How can a silicone rubber be cross-linked at room temperature?

10.218 What are some of the engineering applications for silicone rubber?

10.219 Describe the general deformation behavior of a thermoplastic plastic above and below its glass transition temperature.

10.220 What deformation mechanisms are involved during the elastic and plastic deformation of thermoplastics?

10.221 How does the average molecular mass of a thermoplastic affect its strength?

10.222 How does the amount of crystallinity within a thermoplastic material affect (*a*) its strength, (*b*) its tensile modulus of elasticity, and (*c*) its density?

10.223 Explain why low-density polyethylene is weaker than high-density polyethylene.

10.224 Explain why bulky side groups strengthen thermoplastics.

10.225 Explain how highly polar atoms bonded to the main carbon chain strengthen thermoplastics. Give examples.

10.226 Explain how oxygen atoms covalently bonded in the main carbon chain strengthen thermoplastics. Give an example.

10.227 Explain how phenylene rings covalently bonded in the main carbon chain strengthen thermoplastics. Give an example.

10.228 Explain why thermosetting plastics have in general high strengths and low ductilities.

10.229 How does increasing the temperature of thermoplastics affect their strength? What changes in bonding structure occur as thermoplastics are heated?

10.230 Why don't cured thermoset plastics become viscous and flow at elevated temperatures?

10.231 How do increases in stress and temperature affect the creep resistance of thermoplastics?

10.232 What is viscoelastic behavior of plastic materials?

10.233 Define the creep modulus of a plastic material.

10.234 How can the creep modulus of a thermoplastic be increased?

10.235 How can the extra energy required to fracture glassy thermoplastics compared to inorganic glasses be explained?

10.236 What is a craze in a glassy thermoplastic?

10.237 Describe the structure of a craze in a thermoplastic.

10.238 Describe the molecular structure changes that occur during the ductile fracturing of a thermoplastic.

10.16 MATERIALS SELECTION AND DESIGN PROBLEMS

1. Select the material to serve as insulating cover to a conducting copper wire in an automobile engine. This material will be exposed to temperatures ranging from $-5°C$ to $120°C$. The system must have the ability to conform to the engine environment (must be flexible). Use Table 10.2.

2. (*a*) Outline the required properties of the materials to be selected in the design and manufacturing of a large travel suitcase. (*b*) Propose a number of candidate materials. (*c*) Identify your best choice and explain why.

3. What is a self-lubricating plastic? Give examples. Name some specific applications.

4. An engineer has identified epoxy, a thermoset, as an effective candidate for a specific application in humid and lightly corrosive conditions. However, the low stiffness or low modulus of elasticity of epoxy is a potential problem. Can you offer a solution to increase the stiffness of epoxy?

5. Assume that new car manufacturing regulations require that the bumpers of a car withstand a five-mile-per-hour impact without any damage to the bumper or the car. Design a system that achieves this and select the material to make the bumper.

6. In total hip replacement surgery, the femur head is replaced with a metal component, usually a Co-Cr alloy, and the acetabular cup is replaced with ultrahigh-molecular-weight polyethylene. (*a*) Give reasons for the suitability of UHMWP for this application in comparison to a metal cup. (*b*) Investigate the advantage of UHMWP over low-density polyethylene.

7. (*a*) In selecting the materials for compact discs, what factors should you consider? (*b*) What material would you select for this application?

8. (*a*) In selecting the materials for surgical gloves, what factors should you consider? (*b*) What material would you select for this application?

9. (*a*) In selecting the materials for keyboards, terminals, and other computer equipment that house the electronic components, what factors should you consider? (*b*) What material would you select for these applications?

Ceramics

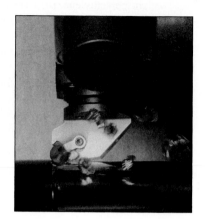

(Courtesy of Kennametal)

Due to their desirable characteristics such as high hardness, wear resistance, chemical stability, high-temperature strength, and low coefficient of thermal expansion, advanced ceramics are being selected as the preferred material for many applications. These include but are not limited to mineral processing, seals, valves, heat exchangers, metal-forming dies, adiabatic diesel engines, gas turbines, medical products, and cutting tools.

Ceramic cutting tools have many advantages over their conventional metal counterparts, including chemical stability, higher resistance to wear, higher hot hardness, and superior heat dispersal during the chip removal process. Some examples of ceramic materials that are used to manufacture cutting tools are metal oxide composites (70% Al_2O_3–30% TiC), silicon-aluminum-oxynitride (sialons), and cubic boron nitride. These tools are manufactured through powder metallurgy processes in which ceramic particles are densified into a final shape through compaction and sintering. The chapter-opening images are examples of various types of metal removal products made from advanced ceramics.[1] ∎

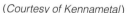

[1]*Ceramics Engineered Materials Handbook,* volume 1, ASM International.

By the end of this chapter, students will be able to . . .

1. Define and classify ceramic materials including traditional and engineering ceramics.

2. Describe various ceramic crystal structures.

3. Describe carbon and its allotropes.

4. Describe various processing methods for ceramics.

5. Describe the mechanical properties of ceramics and the corresponding mechanisms of deformation, toughening, and failure of ceramics.

6. Describe thermal properties of ceramics.

7. Describe various types of ceramic glasses, glass transition temperature, forming methods, and structure of glass.

8. Describe various ceramic coatings and applications.

9. Be able to explain the use of ceramics in biomedical applications and nanotechnology.

11.1 INTRODUCTION

Ceramic materials are inorganic, nonmetallic materials that consist of metallic and nonmetallic elements bonded together primarily by ionic and/or covalent bonds. The chemical compositions of ceramic materials vary considerably, from simple compounds to mixtures of many complex phases bonded together.

The properties of ceramic materials also vary greatly due to differences in bonding. In general, ceramic materials are typically hard and brittle with low toughness and ductility. Ceramics are usually good electrical and thermal insulators because of the absence of conduction electrons. Ceramic materials normally have relatively high melting temperatures and high chemical stability in many hostile environments because of the stability of their strong bonds. Because of these properties, ceramic materials are indispensable for many engineering designs. Two examples of the strategic importance of ceramic materials in new high technology are shown in Fig. 11.1.

In general, ceramic materials used for engineering applications can be divided into two groups: traditional ceramic materials and the engineering ceramic materials. Typically, traditional ceramics are made from three basic components: clay, silica (flint), and feldspar. Examples of traditional ceramics are glasses, bricks, and tiles used in the construction industries and electrical porcelain in the electrical industry. The engineering ceramics, in contrast, typically consist of pure or nearly pure compounds such as aluminum oxide (Al_2O_3), silicon carbide (SiC), and silicon nitride (Si_3N_4). Examples of the use of the engineering ceramics in high technology are silicon carbide in the high-temperature areas of the experimental AGT-100 automotive gas turbine engine and aluminum oxide in the support base for integrated circuit chips in a thermal-conduction module.

(*a*)

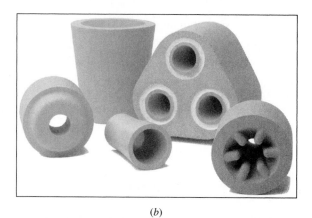

(*b*)

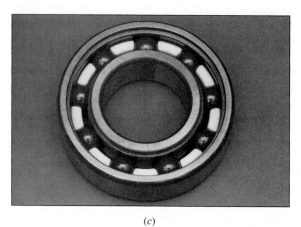

(*c*)

Figure 11.1

(*a*) Zircoa (zirconium dioxide) crucibles are used in melting superalloys. (*b*) Zircoa coarse-grained product line includes nozzles, crucibles, burner blocks, setter plates, and disks.

[*(a) and (b) After American Ceramic Bulletin, Sept., 2001. Photo Courtesy of Zircon, Inc.*]

(*c*) High-performance Ceratec ball bearing and races are made from titanium and carbon nitride feedstocks through powder metal technology.

(© *Tom Pantages photo/courtesy Bearing Works.*)

In this chapter we shall first examine some simple ceramic crystal structures and then look at some of the more complicated silicate ceramic structures. Then we shall explore some of the methods for processing ceramic materials, and following that, study some of the mechanical and thermal properties of ceramic materials. We shall examine some aspects of the structure and properties of glasses, ceramic coatings and surface engineering, and the use of ceramics in biomedical applications. Finally, we shall explore nanotechnology and ceramics.

11.2 SIMPLE CERAMIC CRYSTAL STRUCTURES

11.2.1 Ionic and Covalent Bonding in Simple Ceramic Compounds

Let us first consider some simple ceramic crystal structures. Some ceramic compounds with relatively simple crystal structures are listed in Table 11.1 with their melting points.

In the ceramic compounds listed, the atomic bonding is a mixture of ionic and covalent types. Approximate values for the percentages of ionic and covalent character for the bonds between the atoms in these compounds can be obtained by considering the electronegativity differences between the different types of atoms in the compounds and by using Pauling's equation for percent ionic character (Eq. 2.10). Table 11.2 shows that the percent ionic or covalent character varies considerably in simple ceramic compounds. The amount of ionic or covalent bonding between the

Table 11.1 Some simple ceramic compounds with their melting points

Ceramic compound	Formula	Melting point (°C)	Ceramic compound	Formula	Melting point (°C)
Hafnium carbide	HfC	4150	Boron carbide	B_4C_3	2450
Titanium carbide	TiC	3120	Aluminum oxide	Al_2O_3	2050
Tungsten carbide	WC	2850	Silicon dioxide†	SiO_2	1715
Magnesium oxide	MgO	2798	Silicon nitride	Si_3N_4	1700
Zirconium dioxide	ZrO_2*	2750	Titanium dioxide	TiO_2	1605
Silicon carbide	SiC	2500			

*Is believed to have a monoclinic fluorite (distorted) crystal structure when melted.
†Cristobalite.

Table 11.2 Percent ionic and covalent bonding in some ceramic compounds

Ceramic compound	Bonding atoms	Electronegativity difference	% ionic character	% covalent character
Zirconium dioxide, ZrO_2	Zr–O	2.3	73	27
Magnesium oxide, MgO	Mg–O	2.2	69	31
Aluminum oxide, Al_2O_3	Al–O	2.0	63	37
Silicon dioxide, SiO_2	Si–O	1.7	51	49
Silicon nitride, Si_3N_4	Si–N	1.3	34.5	65.5
Silicon carbide, SiC	Si–C	0.7	11	89

atoms of these compounds is important since it determines to some extent what type of crystal structure will form in the bulk ceramic compound.

11.2.2 Simple Ionic Arrangements Found in Ionically Bonded Solids

In ionic (ceramic) solids the packing of the ions is determined primarily by the following factors:

1. The relative size of the ions in the ionic solid (assume the ions to be hard spheres with definite radii)
2. The need to balance electrostatic charges to maintain electrical neutrality in the ionic solid

When ionic bonding between atoms takes place in the solid state, the energies of the atoms are lowered by the formation of the ions and their bonding into an ionic solid. Ionic solids tend to have their ions packed together as densely as possible to lower the overall energy of the solid as much as possible. The limitations to dense packing are the relative sizes of the ions and the necessity for maintaining charge neutrality.

Size Limitations for the Dense Packing of Ions in an Ionic Solid   Ionic solids consist of cations and anions. In ionic bonding some atoms lose their outer electrons to become *cations* and others gain outer electrons to become *anions*. Thus, cations are normally smaller than the anions they bond with. The number of anions that surround a central cation in an ionic solid is called the **coordination number (CN)** and corresponds to the number of nearest neighbors surrounding a central cation. For stability, as many anions as possible surround a central cation. However, the anions must make contact with the central cation, and charge neutrality must be maintained.

Figure 11.2 shows two stable configurations for the coordination of anions around a central cation in an ionic solid. If the anions do not touch the central cation, the structure becomes unstable because the central cation can "rattle around in its cage of anions" (third diagram, Fig. 11.2). The ratio of the radius of the central cation

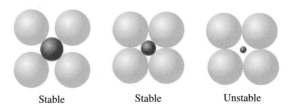

Stable   Stable   Unstable

Figure 11.2
Stable and unstable coordination configurations for ionic solids.
(After W. D. Kingery, H. K. Bowen, and D. R. Uhlmann, "Introduction to Ceramics," 2nd ed., Wiley, 1976.)

Disposition of ions about central ion	CN	Range of cation radius ratio to anion radius	
Corners of cube	8	≥0.732	
Corners of octahedron	6	≥0.414	
Corners of tetrahedron	4	≥0.225	
Corners of triangle	3	≥0.155	

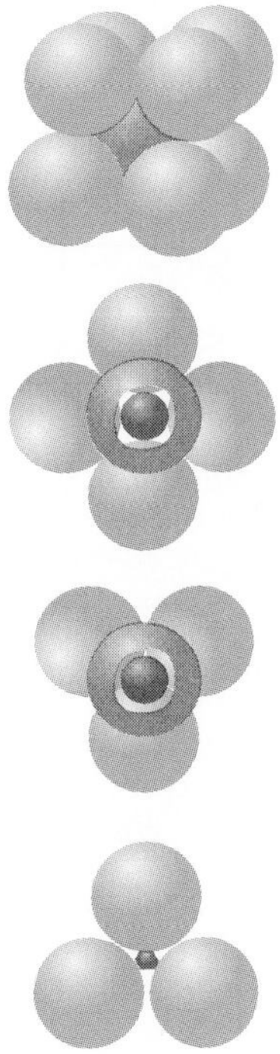

CN = Coordination number

Figure 11.3
Radius ratios for coordination numbers of 8, 6, 4, and
3 anions surrounding a central cation in ionic solids.
(*After W. D. Kingery, H. K. Bowen, and D. R. Uhlmann, "Introduction
to Ceramics," 2nd ed., Wiley, 1976.*)

to that of the surrounding anions is called the **radius ratio,** r_{cation}/r_{anion}. The radius ratio when the anions just touch each other and contact the central cation is called the **critical (minimum) radius ratio.** Allowable radius ratios for ionic solids with coordination numbers of 3, 4, 6, and 8 are listed in Fig. 11.3 along with illustrations showing the coordinations.

Calculate the critical (minimum) radius ratio r/R for the triangular coordination (CN = 3) of three anions of radii R surrounding a central cation of radius r in an ionic solid.

EXAMPLE PROBLEM 11.1

■ **Solution**
Figure 11.4*a* shows three large anions of radii R surrounding and just touching a central cation of radius r. Triangle *ABC* is an equilateral triangle (each angle = 60°), and line *AD*

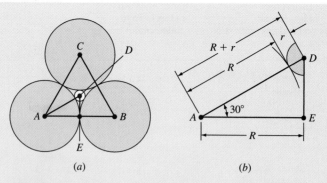

Figure 11.4
Diagram for triangular coordination.

bisects angle *CAB*. Thus, angle *DAE* = 30°. To find the relationship between R and r, triangle *ADE* is constructed as shown in Fig. 11.4*b*. Thus,

$$AD = R + r$$

$$\cos 30° = \frac{AE}{AD} = \frac{R}{R + r} = 0.866$$

$$R = 0.866(R + r) = 0.866R + 0.866r$$

$$0.866r = R - 0.866R = R(0.134)$$

$$\frac{r}{R} = 0.155 \blacktriangleleft$$

EXAMPLE PROBLEM 11.2

Predict the coordination number for the ionic solids CsCl and NaCl. Use the following ionic radii for the prediction:

$$Cs^+ = 0.170 \text{ nm} \qquad Na^+ = 0.102 \text{ nm} \qquad Cl^- = 0.181 \text{ nm}$$

■ **Solution**
The radius ratio for CsCl is

$$\frac{r(Cs^+)}{R(Cl^-)} = \frac{0.170 \text{ nm}}{0.181 \text{ nm}} = 0.94$$

Since this ratio is greater than 0.732, CsCl should show cubic coordination (CN = 8), which it does.

The radius ratio for NaCl is

$$\frac{r(Na^+)}{R(Cl^-)} = \frac{0.102 \text{ nm}}{0.181 \text{ nm}} = 0.56$$

Since this ratio is greater than 0.414 but less than 0.732, NaCl should show octahedral coordination (CN = 6), which it does.

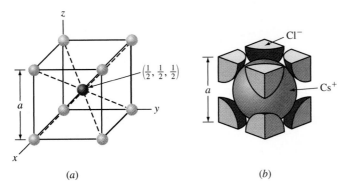

Figure 11.5
Cesium chloride (CsCl) crystal structure unit cell. (*a*) Ion-site unit cell. (*b*) Hard-sphere unit cell. In this crystal structure eight chloride ions surround a central cation in cubic coordination (CN = 8). In this unit cell there is one Cs^+ and one Cl^- ion.

MatVis

11.2.3 Cesium Chloride (CsCl) Crystal Structure

The chemical formula for solid cesium chloride is CsCl, and since this structure is principally ionically bonded, there are equal numbers of Cs^+ and Cl^- ions. Because the radius ratio for CsCl is 0.94 (see Example Problem 11.2), cesium chloride has cubic coordination (CN = 8), as shown in Fig. 11.5. Thus, eight chloride ions surround a central cesium cation at the $(\frac{1}{2}, \frac{1}{2}, \frac{1}{2})$ position in the CsCl unit cell. Ionic compounds that also have the CsCl crystal structure are CsBr, TlCl, and TlBr. The intermetallic compounds AgMg, LiMg, AlNi, and β–Cu–Zn also have this structure. The CsCl structure is not of much importance for ceramic materials but does illustrate how higher radius ratios lead to higher coordination numbers in ionic crystal structures.

Calculate the ionic packing factor for CsCl. Ionic radii are $Cs^+ = 0.170$ nm and $Cl^- = 0.181$ nm.

EXAMPLE PROBLEM 11.3

■ **Solution**
The ions touch each other across the cube diagonal of the CsCl unit cell, as shown in Fig. 11.6. Let $r = Cs^+$ ion and $R = Cl^-$ ion. Thus

$$\sqrt{3}a = 2r + 2R$$
$$= 2(0.170 \text{ nm} + 0.181 \text{ nm})$$
$$a = 0.405 \text{ nm}$$

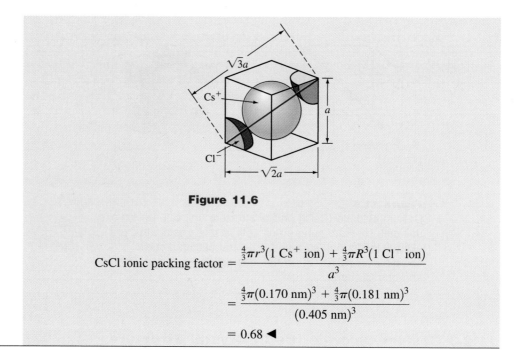

Figure 11.6

$$\text{CsCl ionic packing factor} = \frac{\frac{4}{3}\pi r^3(1 \text{ Cs}^+ \text{ ion}) + \frac{4}{3}\pi R^3(1 \text{ Cl}^- \text{ ion})}{a^3}$$

$$= \frac{\frac{4}{3}\pi(0.170 \text{ nm})^3 + \frac{4}{3}\pi(0.181 \text{ nm})^3}{(0.405 \text{ nm})^3}$$

$$= 0.68 \blacktriangleleft$$

11.2.4 Sodium Chloride (NaCl) Crystal Structure

The sodium chloride or rock salt crystal structure is highly ionically bonded and has the chemical formula NaCl. Thus, there are an equal number of Na^+ and Cl^- ions to maintain charge neutrality. Figure 11.7a shows a lattice-site NaCl unit cell and Fig. 2.13b a hard-sphere model of the NaCl unit cell. Figure 11.7a has negative Cl^- anions occupying regular FCC atom lattice sites and positive Na^+ cations occupying the interstitial sites between the FCC atom sites. The centers of the Na^+ and Cl^- ions occupy the following lattice positions, which are indicated in Fig. 11.7a:

$$Na^+: \quad (\tfrac{1}{2}, 0, 0) \quad (0, \tfrac{1}{2}, 0) \quad (0, 0, \tfrac{1}{2}) \quad (\tfrac{1}{2}, \tfrac{1}{2}, \tfrac{1}{2})$$
$$Cl^-: \quad (0, 0, 0) \quad (\tfrac{1}{2}, \tfrac{1}{2}, 0) \quad (\tfrac{1}{2}, 0, \tfrac{1}{2}) \quad (0, \tfrac{1}{2}, \tfrac{1}{2})$$

Since each central Na^+ cation is surrounded by six Cl^- anions, the structure has octahedral coordination (that is, $CN = 6$), as shown in Fig. 11.7b. This type of coordination is predicted from the radius ratio calculation of $r_{Na^+}/R_{Cl^-} = 0.102$ nm/0.181 nm $= 0.56$, which is greater than 0.414 but less than 0.732. Other examples of ceramic compounds that have the NaCl structure include MgO, CaO, NiO, and FeO.

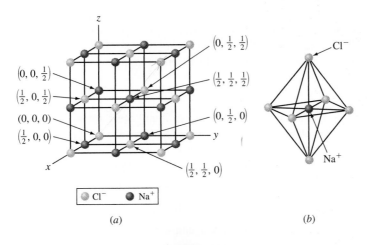

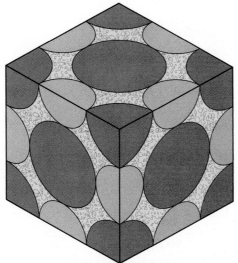

Sodium Chloride Crystal Structure
NaCl

(Light color) sodium ions—(0.102 nm radius) $\dfrac{r_{Na^+}}{R_{Cl^-}} = 0.56$
(Dark color) chloride ions—(0.181 nm radius)

(*c*)

Figure 11.7
(*a*) NaCl lattice-point unit cell indicating positions of the
Na^+ (radii = 0.102 nm) and Cl^- (radii = 0.181 nm) ions.
(*b*) Octahedron showing octahedral coordination of six Cl^-
anions around a central Na^+ cation. (*c*) NaCl unit cell
truncated.

MatVis

EXAMPLE
PROBLEM 11.4

Calculate the density of NaCl from a knowledge of its crystal structure (Fig. 11.7a), the ionic radii of Na^+ and Cl^- ions, and the atomic masses of Na and Cl. The ionic radius of $Na^+ = 0.102$ nm and that of $Cl^- = 0.181$ nm. The atomic mass of Na = 22.99 g/mol and that of Cl = 35.45 g/mol.

■ **Solution**

As shown in Fig. 11.7a, the Cl^- ions in the NaCl unit cell form an FCC-type atom lattice, and the Na^+ ions occupy the interstitial spaces between the Cl^- ions. There is the equivalent of one Cl^- ion at the corners of the NaCl unit cell since 8 corners $\times \frac{1}{8}$ ion = 1 ion, and there is the equivalent of three Cl^- ions at the faces of the NaCl unit cell since 6 faces $\times \frac{1}{2}$ ion = 3 Cl^- ions, making a total of four Cl^- ions per NaCl unit cell. To maintain charge neutrality in the NaCl unit cell, there must also be the equivalent of four Na^+ ions per unit cell. Thus, there are four Na^+Cl^- ion pairs in the NaCl unit cell.

To calculate the density of the NaCl unit cell, we shall first determine the mass of one NaCl unit cell and then its volume. Knowing these two quantities, we can calculate the density m/V.

The mass of a NaCl unit cell is

$$= \frac{(4Na^+ \times 22.99 \text{ g/mol}) + (4Cl^- \times 35.45 \text{ g/mol})}{6.02 \times 10^{23} \text{ atoms (ions)/mol}} = 3.88 \times 10^{-22} \text{ g}$$

The volume of the NaCl unit cell is equal to a^3, where a is the lattice constant of the NaCl unit cell. The Cl^- and Na^+ ions contact each other along the cube edges of the unit cell, as shown in Fig. 11.8, and thus

$$a = 2(r_{Na^+} + R_{Cl^-}) = 2(0.102 \text{ nm} + 0.181 \text{ nm}) = 0.566 \text{ nm}$$
$$= 0.566 \text{ nm} \times 10^{-7} \text{ cm/nm} = 5.66 \times 10^{-8} \text{ cm}$$
$$V = a^3 = 1.81 \times 10^{-22} \text{ cm}^3$$

The density of NaCl is

$$\rho = \frac{m}{V} = \frac{3.88 \times 10^{-22} \text{ g}}{1.81 \times 10^{-22} \text{ cm}^3} = 2.14 \frac{\text{g}}{\text{cm}^3} \blacktriangleleft$$

The handbook value for the density of NaCl is 2.16 g/cm³.

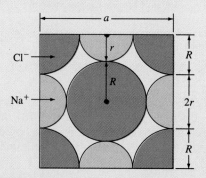

Figure 11.8
Cube face of NaCl unit cell. Ions contact along the cube edge, and thus $a = 2r + 2R = 2(r + R)$.

Calculate the linear density of Ca^{2+} and O^{2-} ions in ions per nanometer in the [110] direction of CaO, which has the NaCl structure. (Ionic radii: $Ca^{2+} = 0.106$ nm and $O^{2-} = 0.132$ nm.)

■ **Solution**

From Fig. 11.7 and Fig. EP11.5, we see that the [110] direction passes through two O^{2-} ion diameters in traversing from (0, 0, 0) to (1, 1, 0) ion positions. The length of the [110] distance across the base face of a unit cube is $\sqrt{2}a$, where a is the length of a side of the cube, or lattice constant. From Fig. 11.8 of the cube face of the NaCl unit cell, we see that $a = 2r + 2R$. Thus, for CaO,

$$a = 2(r_{Ca^{2+}} + R_{O^{2-}})$$
$$= 2(0.106 \text{ nm} + 0.132 \text{ nm}) = 0.476 \text{ nm}$$

The linear density of the O^{2-} ions in the [110] direction is

$$\rho_L = \frac{2O^{2-}}{\sqrt{2}a} = \frac{2O^{2-}}{\sqrt{2}(0.476 \text{ nm})} = 2.97 O^{2-}/\text{nm} \blacktriangleleft$$

The linear density of Ca^{2+} ions in the [110] direction is also 2.97 Ca^{2+}/nm if we shift the origin of the [110] direction from (0, 0, 0) to (0, $\frac{1}{2}$, 0). Thus, the solution to the problem is that there are 2.97(Ca^{2+} or O^{2-})/nm in the [110] direction.

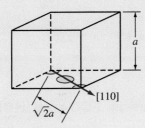

Figure EP11.5

Calculate the planar density of Ca^{2+} and O^{2-} ions in ions per square nanometer on the (111) plane of CaO, which has the NaCl structure. (Ionic radii: $Ca^{2+} = 0.106$ nm and $O^{2-} = 0.132$ nm.)

■ **Solution**

If we consider the anions (O^{2-} ions) to be located at the FCC positions of a cubic unit cell as shown for the Cl^- ions of Fig. 11.7 and Fig. EP11.6, then the (111) plane contains the equivalent of two anions. [$3 \times 60° = 180° = \frac{1}{2}$ anion + ($3 \times \frac{1}{2}$) anions at each midpoint of the sides of the (111) planar triangle of Fig. EP11.6 = a total of 2 anions within the (111) triangle.] The lattice constant for the unit cell $a = 2(r + R) = 2(0.106 \text{ nm} + 0.132 \text{ nm}) = 0.476$ nm. The planar area $A = \frac{1}{2}bh$, where $h = \frac{\sqrt{3}}{2}a^2$.

Thus,

$$A = \left(\frac{1}{2}\sqrt{2}a\right)\left(\sqrt{\frac{3}{2}}a\right) = \frac{\sqrt{3}}{2}a^2 = \frac{\sqrt{3}}{2}(0.476 \text{ nm})^2 = 0.196 \text{ nm}^2$$

The planar density for the O^{2-} anions is

$$\frac{2(O^{2-} \text{ ions})}{0.196 \text{ nm}^2} = 10.2 O^{2-} \text{ ions/nm}^2 \blacktriangleleft$$

The planar density for the Ca^{2+} cations is the same if we consider the Ca^{2+} to be located at the FCC lattice points of the unit cell, and thus

$$\rho_{\text{planar}}(\text{CaO}) = 10.2(Ca^{2+} \text{ or } O^{2-})/\text{nm}^2 \blacktriangleleft$$

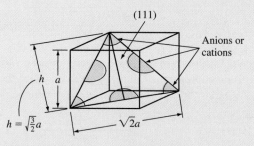

Figure EP11.6

11.2.5 Interstitial Sites in FCC and HCP Crystal Lattices

There are empty spaces or voids among the atoms or ions that are packed into a crystal-structure lattice. These voids are *interstitial sites* in which atoms or ions other than those of the parent lattice can be fitted in. In the FCC and HCP crystal structures, which are close-packed structures, there are two types of interstitial sites: **octahedral** and **tetrahedral.** In the octahedral site there are six nearest atoms or ions equidistant from the center of the void, as shown in Fig. 11.9*a*. This site is called octahedral because the atoms or ions surrounding the center of the site form an eight-sided octahedron. In the tetrahedral site there are four nearest atoms or ions equidistant from the center of the tetrahedral site, as shown in Fig. 11.9*b*. A regular tetrahedron is formed when the centers of the four atoms surrounding the void are joined.

In the FCC crystal-structure lattice the octahedral interstitial sites are located at the center of the unit cell and at the cube edges, as indicated in Fig. 11.10. There are the equivalent of four octahedral interstitial sites per FCC unit cell. Since there

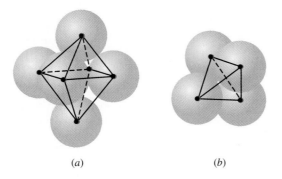

(a) (b)

Figure 11.9
Interstitial sites in FCC and HCP crystal-structure lattices. (a) Octahedral interstitial site formed at the center where six atoms contact each other. (b) Tetrahedral interstitial site formed at the center where four atoms contact each other.
(*After W. D. Kingery, H. K. Bowen, and D. R. Uhlmann, "Introduction to Ceramics," 2nd ed., Wiley, 1976.*)

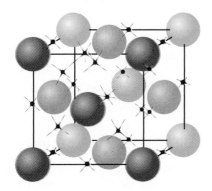

Figure 11.10
Location of octahedral and tetrahedral interstitial void sites in an FCC ionic crystal-structure unit cell. The octahedral sites are located at the center of the unit cell and at the centers of the cube edges. Since there are 12 cube edges, one-fourth of a void is located within the cube at each edge. Thus, there is the equivalent of $12 \times \frac{1}{4} = 3$ voids within the FCC unit cell at the cube edges. Therefore, there is the equivalent of four octahedral voids per FCC unit cell (one at the center and the equivalent of three at the cube edges). The tetrahedral voids are located at the $(\frac{1}{4}, \frac{1}{4}, \frac{1}{4})$-type sites, which are indicated by points with tetrahedrally directed rays. Thus, there are a total of eight tetrahedral void sites located within the FCC unit cell.
(*After W. D. Kingery, "Introduction to Ceramics," Wiley, 1960, p. 104.*)

are four atoms per FCC unit cell, there is one octahedral interstitial site per atom in the FCC lattice. Figure 11.11a indicates the lattice positions for octahedral interstitial sites in an FCC unit cell.

The tetrahedral sites in the FCC lattice are located at the $(\frac{1}{4}, \frac{1}{4}, \frac{1}{4})$-type positions, as indicated in Figs. 11.10 and 11.11b. In the FCC unit cell there are eight tetrahedral

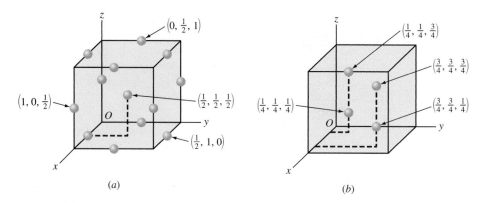

Figure 11.11
Location of interstitial sites in the FCC atom unit cell. (*a*) The octahedral sites in the FCC unit cell are located at the center of the unit cell and at the centers of the cube edges. (*b*) The tetrahedral sites in the FCC unit cell are located at the unit-cell positions indicated. Only representative positions are located in the figure.

sites per unit cell or *two* per atom of the parent FCC unit cell. In the HCP crystal structure, because of similar close packing to the FCC structure, there is also the same number of octahedral interstitial sites as atoms in the HCP unit cell and twice as many tetrahedral sites as atoms.

11.2.6 Zinc Blende (ZnS) Crystal Structure

The zinc blende structure has the chemical formula ZnS and the unit cell shown in Fig. 11.12, which has the equivalent of four zinc and four sulfur atoms. One type of atom (either S or Zn) occupies the lattice points of an FCC unit cell, and the other type (either S or Zn) occupies half the tetrahedral interstitial sites of the FCC unit cell. In the ZnS crystal-structure unit cell shown in Fig. 11.12, sulfur atoms occupy the FCC unit-cell atom positions, as indicated by the lighter circles, and Zn atoms occupy half the tetrahedral interstitial positions of the FCC unit cell, as indicated by the darker circles. The position coordinates of the S and Zn atoms in the ZnS crystal structure can thus be indicated as

$$
\begin{array}{lllll}
\text{S atoms:} & (0, 0, 0) & \left(\tfrac{1}{2}, \tfrac{1}{2}, 0\right) & \left(\tfrac{1}{2}, 0, \tfrac{1}{2}\right) & \left(0, \tfrac{1}{2}, \tfrac{1}{2}\right) \\
\text{Zn atoms:} & \left(\tfrac{3}{4}, \tfrac{1}{4}, \tfrac{1}{4}\right) & \left(\tfrac{1}{4}, \tfrac{1}{4}, \tfrac{3}{4}\right) & \left(\tfrac{1}{4}, \tfrac{3}{4}, \tfrac{1}{4}\right) & \left(\tfrac{3}{4}, \tfrac{3}{4}, \tfrac{3}{4}\right)
\end{array}
$$

According to Pauling's equation (Eq. 2.10), the Zn-S bond has 87 percent covalent character, and so the ZnS crystal structure must be essentially covalently bonded. As a result, the ZnS structure is tetrahedrally covalently bonded and the Zn and S atoms have a coordination number of 4. Many semiconducting compounds such as CdS, InAs, InSb, and ZnSe have the zinc blende crystal structure.

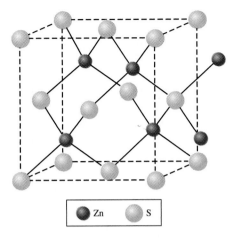

MatVis

Figure 11.12

Zinc blende (ZnS) crystal structure. In this unit cell the sulfur atoms occupy the FCC atom unit-cell sites (equivalent of four atoms). The zinc atoms occupy half the tetrahedral interstitial sites (four atoms). Each Zn or S atom has a coordination number of 4 and is tetrahedrally covalently bonded to other atoms.

(After W. D. Kingery, H. K. Bowen, and D. R. Uhlmann, "Introduction to Ceramics," 2nd ed., Wiley, 1976.)

Calculate the density of zinc blende (ZnS). Assume the structure to consist of ions and that the ionic radius of Zn^{2+} = 0.060 nm and that of S^{2-} = 0.174 nm.

EXAMPLE PROBLEM 11.7

■ **Solution**

$$\text{Density} = \frac{\text{mass of unit cell}}{\text{volume of unit cell}}$$

There are four zinc ions and four sulfur ions per unit cell. Thus

$$\text{Mass of unit cell} = \frac{(4Zn^{2+} \times 65.37 \text{ g/mol}) + (4S^{2-} \times 32.06 \text{ g/mol})}{6.02 \times 10^{23} \text{ atoms/mol}}$$

$$= 6.47 \times 10^{-22} \text{ g}$$

$$\text{Volume of unit cell} = a^3$$

From Fig. 11.13,

$$\frac{\sqrt{3}}{4}a = r_{Zn^{2+}} + R_{S^{2-}} = 0.060 \text{ nm} + 0.174 \text{ nm} = 0.234 \text{ nm}$$

$$a = 5.40 \times 10^{-8} \text{ cm}$$

$$a^3 = 1.57 \times 10^{-22} \text{ cm}^3$$

Thus,

$$Density = \frac{mass}{volume} = \frac{6.47 \times 10^{-22}\ g}{1.57 \times 10^{-22}\ cm^3} = 4.12\ g/cm^3 \blacktriangleleft$$

The handbook value for the density of ZnS (cubic) is $4.10\ g/cm^3$.

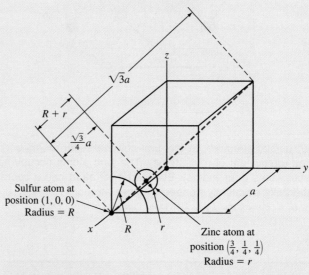

Figure 11.13
Zinc blende structure showing the relationship between the lattice constant a of the unit cell and the radii of the sulfur and zinc atoms (ions):

$$\frac{\sqrt{3}}{4}a = r_{Zn^{2+}} + R_{S^{2-}}$$

or

$$a = \frac{4}{\sqrt{3}}(r + R)$$

11.2.7 Calcium Fluoride (CaF$_2$) Crystal Structure

The calcium fluoride structure has the chemical formula CaF_2 and the unit cell shown in Fig. 11.14. In this unit cell the Ca^{2+} ions occupy the FCC lattice sites, while the F^- ions are located at the eight tetrahedral sites. The four remaining octahedral sites in the FCC lattice remain vacant. Thus, there are four Ca^{2+} ions and eight F^- ions per unit cell. Examples of compounds that have this structure are UO_2, BaF_2, $AuAl_2$, and $PbMg_2$. The compound ZrO_2 has a distorted (monoclinic) CaF_2 structure. The large number of unoccupied octahedral interstitial sites in UO_2 allows this material to be used as a nuclear fuel since fission products can be accommodated in these vacant positions.

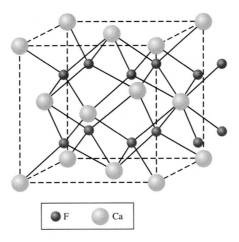

MatVis

Figure 11.14

Calcium fluoride (CaF_2) crystal structure (also called fluorite structure). In this unit cell the Ca^{2+} ions are located at the FCC unit-cell sites (four ions). Eight fluoride ions occupy all the tetrahedral interstitial sites.

(*After W. D. Kingery, H. K. Bowen, and D. R. Uhlmann, "Introduction to Ceramics," 2nd ed., Wiley, 1976.*)

Calculate the density of UO_2 (uranium oxide), which has the calcium fluoride, CaF_2, structure. (Ionic radii: $U^{4+} = 0.105$ nm and $O^{2-} = 0.132$ nm.)

EXAMPLE PROBLEM 11.8

■ **Solution**

$$\text{Density} = \frac{\text{mass/unit cell}}{\text{volume/unit cell}}$$

There are four uranium ions and eight oxide ions per unit cell (CaF_2 type). Thus,

$$\text{Mass of a unit cell} = \frac{(4U^{4+} \times 238 \text{ g/mol}) + (8O^{2-} \times 16 \text{ g/mol})}{6.02 \times 10^{23} \text{ ions/mol}}$$

$$= 1.794 \times 10^{-21} \text{ g}$$

$$\text{Volume of a unit cell} = a^3$$

From Fig. 11.13,

$$\frac{\sqrt{3}}{4}a = r_{U^{4+}} + R_{O^{2-}}$$

$$a = \frac{4}{\sqrt{3}}(0.105 \text{ nm} + 0.132 \text{ nm}) = 0.5473 \text{ nm} = 0.5473 \times 10^{-7} \text{ cm}$$

$$a^3 = (0.5473 \times 10^{-7} \text{ cm})^3 = 0.164 \times 10^{-21} \text{ cm}^3$$

$$\text{Density} = \frac{\text{mass}}{\text{volume}} = \frac{1.79 \times 10^{-21} \text{ g}}{0.164 \times 10^{-21} \text{ cm}^3} = 10.9 \text{ g/cm}^3 \blacktriangleleft$$

The handbook value for the density of UO_2 is 10.96 g/cm^3.

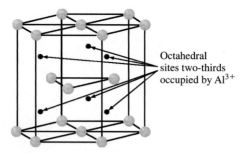

Figure 11.15
Corundum (Al$_2$O$_3$) crystal structure.
Oxygen ions (O^{2-}) occupy the HCP unit-cell sites. Aluminum ions (Al^{3+}) occupy
only two-thirds of the octahedral interstitial
sites to maintain electrical neutrality.

11.2.8 Antifluorite Crystal Structure

The antifluorite structure consists of an FCC unit cell with anions (for example, O^{2-} ions) occupying the FCC lattice points. Cations (for example, Li$^+$) occupy the eight tetrahedral sites in the FCC lattice. Examples of compounds with this structure are Li$_2$O, Na$_2$O, K$_2$O, and Mg$_2$Si.

11.2.9 Corundum (Al$_2$O$_3$) Crystal Structure

In the corundum (Al$_2$O$_3$) structure the oxygen ions are located at the lattice sites of a hexagonal close-packed unit cell, as shown in Fig. 11.15. In the HCP crystal structure as in the FCC structure there are as many octahedral interstitial sites as there are atoms in the unit cell. However, since aluminum has a valence of $+3$ and oxygen a valence of -2, there can be only *two* Al^{3+} ions for every three O^{2-} ions to maintain electrical neutrality. Thus, the aluminum ions can only occupy two-thirds of the octahedral sites of the HCP Al$_2$O$_3$ lattice, which leads to some distortion of this structure.

11.2.10 Spinel (MgAl$_2$O$_4$) Crystal Structure

A number of oxides have the MgAl$_2$O$_4$ or spinel structure, which has the general formula AB$_2$O$_4$, where A is a metal ion with a $+2$ valence and B is a metal ion with a $+3$ valence. In the spinel structure the oxygen ions form an FCC lattice and the A and B ions occupy tetrahedral and octahedral interstitial sites, depending on the particular type of spinel. Compounds with the spinel structure are widely used for nonmetallic magnetic materials for electronic applications and will be studied in more detail in Chap. 16 on Magnetic Materials.

11.2.11 Perovskite (CaTiO$_3$) Crystal Structure

In the perovskite (CaTiO$_3$) structure the Ca^{2+} and O^{2-} ions form an FCC unit cell with the Ca^{2+} ions at the corners of the unit cell and the O^{2-} ions in the centers of

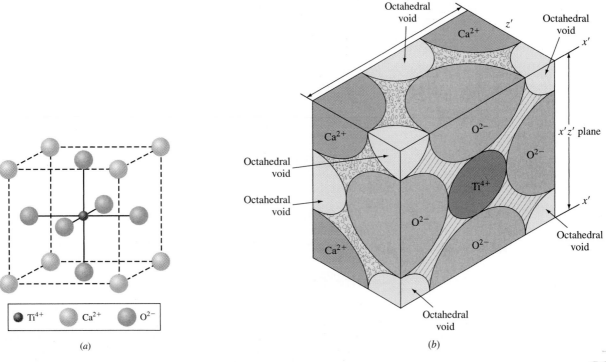

Figure 11.16

Perovskite ($CaTiO_3$) crystal structure. (*a*) Calcium ions occupy FCC unit-cell corners, and oxygen ions occupy FCC unit-cell face-centered sites. The titanium ion occupies the octahedral interstitial site at the center of the cube.

MatVis

(*After W. D. Kingery, H. K. Bowen, and D. R. Uhlmann, "Introduction to Ceramics," 2nd ed., Wiley, 1976.*)

(*b*) Mid-section of perovskite ($CaTiO_3$) crystal structure (truncated).

the faces of the unit cell (Fig. 11.16). The highly charged Ti^{4+} ion is located at the octahedral interstitial site at the center of the unit cell and is coordinated to six O^{2-} ions. $BaTiO_3$ has the perovskite structure above 120°C, but below this temperature its structure is slightly changed. Other compounds having this structure are $SrTiO_3$, $CaZrO_3$, $SrZrO_3$, $LaAlO_3$, and many others. This structure is important for piezo-electric materials (see Sec. 14.8).

11.2.12 Carbon and Its Allotropes

Carbon has many allotropes, i.e., it can exist in many crystalline forms. These allotropes have different crystal structures and have substantially different properties. Carbon and its polymorphs do not directly belong to any of the conventional classes of materials, but since graphite is sometimes considered a ceramic material, the discussion of its structure as well as some of its polymorphs are included in this section. In this section the structure and properties of graphite, diamond, buckyball, and buckytube—all of which are allotropes of carbon—will be discussed.

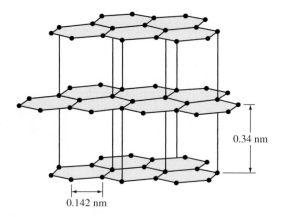

Figure 11.17
The structure of crystalline graphite. Carbon atoms form layers of strongly covalently bonded hexagonal arrays. There are weak secondary bonds between the layers.

MatVis

Graphite   The word **graphite** is based on the Greek word *graphein* (meaning to write). Graphite is formed because of trigonal sp^2 bonding of carbon atoms. Recalling the discussion of hybridized sp^3 orbitals (Chap. 2), sp^2 hybrid orbitals form only when one of the 2s electrons is promoted with two 2p electrons to form three sp^2 orbitals. The remaining electron forms an unhybridized free p orbital. The three sp^2 orbitals are in the same plane, making equal angles of 120 degrees with each other. The orbital due to the delocalized nonhybridized p electron is directed perpendicular to the plane of the three sp^2 hybrid orbitals. Accordingly, graphite has a layered structure in which the carbon atoms in the layers are strongly bonded (through sp^2 orbitals) in hexagonal arrays as shown in Fig. 11.17. The layers are bonded together by weak secondary bonds and can slide past each other easily. The free electron can easily travel from one side of the layer to the other but do not easily travel from one layer to another. Thus, graphite is *anisotropic* (properties are dependent on direction). It has a low density of 2.26 g/cm^3, is a good thermal conductor in the basal plane of graphite but not perpendicular to the plane, and is a good electrical conductor (again only on the basal plane and not perpendicular to it). Graphite can be made to form long fibers for composite materials and can also be used as a lubricant.

Diamond   The structure of diamond is explained in detail in Chap. 2. It has a cubic structure (Fig. 2.19) that is based on covalently bonded sp^3 hybrid orbitals. Its properties are significantly different from graphite. Unlike graphite, it is isotropic and has a higher density of about 3.51 g/cm^3. Diamond is the stiffest, hardest, and least compressible material made by nature. It has very high thermal conductivity (similar to graphite) but minimal electrical conductivity values (essentially an excellent insulator). Impurities such as nitrogen, however, affect its properties adversely. Natural diamond

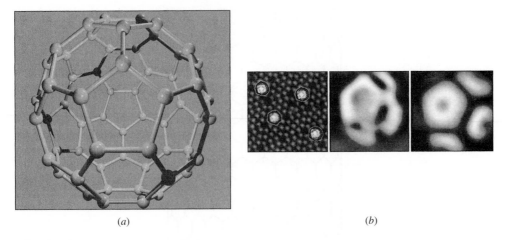

(a)   (b)

Figure 11.18
(a) A schematic of the C_{60} molecule, (b) A series of STM images showing the C_{60} molecule (the left and center images are actual STM images and the right image is a simulated image).
((a) © Tim Evans/Photo Researchers, Inc. (b) Omicron NanoTechnology GmbH)

is extremely expensive and has mostly gem value. However, synthetic (man-made) diamonds have comparable hardness, are cheaper, and are used as cutting tools, coatings, and abrasives.

Buckminster Fullerenes (Buckyball) In 1985 scientists discovered the presence of clusters of carbon atoms in a molecular range of C_{30} to C_{100}. In 1990 other scientists were able to synthesize this molecular form of carbon in the laboratory. The new structure has a form similar to the geodesic truss structures developed by the world-renowned architect, Buckminster Fuller. As a result, the new polymorph was named a *fullerene* or a **buckyball.** The buckyball looks very similar to a soccer ball that is made of 12 pentagons and 20 hexagons. At each junction point a carbon atom is covalently bonded to three other carbon atoms as shown schematically in Fig. 11.18a. The STM image of part of the C_{60} molecule is presented in Fig. 11.18b. The structure consists of a total of 60 carbon atoms; the resulting molecule is therefore C_{60}. Since 1990 other forms of this molecule such as C_{70}, C_{76}, and C_{78} have also been identified. These various forms are collectively called *fullerenes.* The diameter of the C_{60} fullerene is 0.710 nm and is therefore classified as a *nanocluster.* The aggregate form of C_{60} has an FCC structure with a C_{60} molecule at each FCC lattice point. The molecules in the FCC structure are bonded by van der Waals forces. Thus, the aggregate C_{60} and graphite have similar lubrication applications. Fullerenes are being studied for possible applications in the electronic industries and in fuel cells, lubricants, and superconductors.

Carbon Nanotubes Another recently identified carbon polymorph is the highly interesting carbon nanotube. Consider rolling a single graphite atomic layer (a

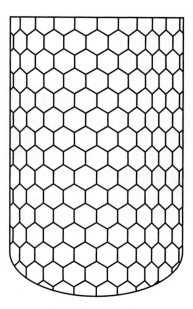

Figure 11.19
A schematic of the nanotube
showing hexagonal patterns on
the tube and pentagonal
patterns on the end cap.

graphene) with the conventional hexagonal structure into a tube making sure that the edge hexagons meet perfectly. Then by closing the ends of the tube using two hemi-fullerenes, made up of only pentagons, one obtains the structure of a carbon nanotube (Fig. 11.19). Although nanotubes of various diameters may be synthesized, the most frequently encountered diameter is 1.4 nm. The length of a nanotube can be in the micrometer or even millimeter range (a very important feature). The nanotubes can be synthesized in a *single-wall nanotube* (SWNT) or *multiwall nanotube* (MWNT) form. These nanotubes are believed to have a tensile strength 20 times that of the strongest steels. Some measurements actually show a tensile strength of 45 Gpa in the length direction of the tube. The elastic modulus of these nanotubes has been estimated at a level of 1.3 Tpa (T = Tera = 10^{12}). For comparison the strongest commercially available carbon based fiber has a strength of 7 Gpa and the highest elastic modulus available is close to 800 Gpa. In addition, carbon nanotubes have low density, high heat conductivity, and high electron conductivity. Even more importantly, one can form structures such as ropes, fibers, and thin films by aligning a large number of these tubes. The combination of these characteristics and properties has convinced many scientists that carbon nanotubes will be involved in many of the technological breakthroughs in this century. Some of the early applications are as STM tips because of their stiffness and slenderness, field emitters in flat panel display (or any device requiring an electron producing cathode), chemical sensors, and fiber material for composites.

11.3 SILICATE STRUCTURES

Many ceramic materials contain silicate structures, which consist of silicon and oxygen atoms (ions) bonded together in various arrangements. Also, a large number of naturally occurring minerals such as clays, feldspars, and micas are silicates since silicon and oxygen are the two most abundant elements in the earth's crust. Many silicates are useful for engineering materials because of their low cost, availability, and special properties. Silicate structures are particularly important for the engineering construction materials glass, portland cement, and brick. Many important electrical insulative materials also are made with silicates.

11.3.1 Basic Structural Unit of the Silicate Structures

The basic building block of the silicates is the silicate (SiO_4^{4-}) tetrahedron (Fig. 11.20). The Si—O bond in the SiO_4^{4-} structure is about 50 percent covalent and 50 percent ionic according to calculations from Pauling's equation (Eq. 2.10). The tetrahedral coordination of SiO_4^{4-} satisfies the directionality requirement of covalent bonding and the radius ratio requirement of ionic bonding. The radius ratio of the Si-O bond is 0.29, which is in the tetrahedral coordination range for stable-ion close packing. Because of the small, highly charged Si^{4+} ion, strong bonding forces are created within the SiO_4^{4-} tetrahedrons, and as a result the SiO_4^{4-} units are normally joined corner to corner and rarely edge to edge.

11.3.2 Island, Chain, and Ring Structures of Silicates

Since each oxygen of the silicate tetrahedron has one electron available for bonding, many different types of silicate structures can be produced. Island silicate structures are produced when positive ions bond with oxygens of the SiO_4^{4-} tetrahedra. For example, Fe^{2+} and Mg^{2+} ions combine with SiO_4^{4-} to form olivine, which has the basic chemical formula $(Mg,Fe)_2SiO_4$.

If two corners of each SiO_4^{4-} tetrahedron are bonded with the corners of other tetrahedra, a chain (Fig. 11.21a) or ring structure with the unit chemical formula of SiO_3^{2-} results. The mineral enstatite ($MgSiO_3$) has a chain silicate structure, and the mineral beryl [$Be_3Al_2(SiO_3)_6$] has a ring silicate structure.

11.3.3 Sheet Structures of Silicates

Silicate sheet structures form when three corners in the same plane of a silicate tetrahedron are bonded to the corners of three other silicate tetrahedra, as shown in Fig. 11.21b. This structure has the unit chemical formula of $Si_2O_5^{2-}$. These silicate sheets can bond with other types of structural sheets because there is still one unbonded oxygen on each silicate tetrahedron (Fig. 11.21b). For example, the

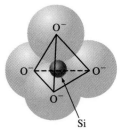

Figure 11.20
The atom (ion) bonding arrangement of the SiO_4^{4-} tetrahedron. In this structure four oxygen atoms surround a central silicon atom. Each oxygen atom has an extra electron, and thus a net negative charge, for bonding with another atom.

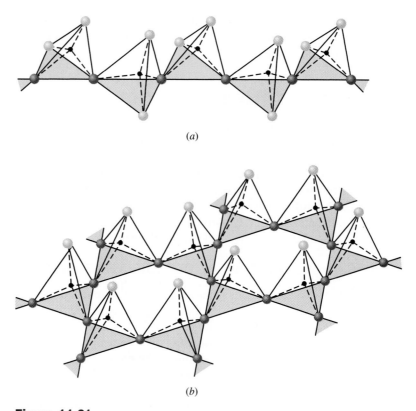

(a)

(b)

Figure 11.21
(*a*) Silicate chain structure. Two of the four oxygen atoms of the SiO_4^{4-} tetrahedra are bonded to other tetrahedra to form silicate chains.
(*b*) Silicate sheet structure. Three of the four oxygen atoms of the SiO_4^{4-} tetrahedra are bonded to other tetrahedra to form silicate sheets. The unbonded oxygen atoms are shown as lighter spheres.
(*After M. Eisenstadt, "Mechanical Properties of Materials," Macmillan, 1971, p. 82.*)

negatively charged silicate sheet can bond with a positively charged sheet of $Al_2(OH)_4^{2+}$ to form a composite sheet of kaolinite, as shown schematically in Fig. 11.22. The mineral kaolinite consists (in its pure form) of very small, flat plates roughly hexagonal in shape, with their average size being about 0.7 μm in diameter and 0.05 μm thick (Fig. 11.23). The crystal plates are made of a series (up to about 50) of parallel sheets bonded together by weak secondary bonds. Many high-grade clays consist mainly of the mineral kaolinite.

Another example of a sheet silicate is the mineral talc, in which a sheet of $Mg_3(OH)_2^{4+}$ bonds with two outer-layer $Si_2O_5^{2-}$ sheets (one on each side) to form a composite sheet with the unit chemical formula $Mg_3(OH)_2(Si_2O_5)_2$. The composite talc sheets are bonded together by weak secondary bonds, and thus this structural arrangement allows the talc sheets to slide over each other easily.

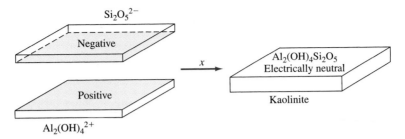

Figure 11.22
Schematic diagram of the formation of kaolinite from sheets of $Al_2(OH)_4^{2+}$ and $Si_2O_5^{2-}$. All the primary bonds of the atoms in the kaolinite sheet are satisfied.
(After M. Eisenstadt, "Mechanical Properties of Materials," Macmillan, 1971, p. 83.)

MatVis

Figure 11.23
Kaolinite crystals as observed with the electron microscope (replica technique).
(After C. E. Hall as shown in F. H. Norton, "Elements of Ceramics," 2nd ed., Addison-Wesley, 1974, p. 16.)

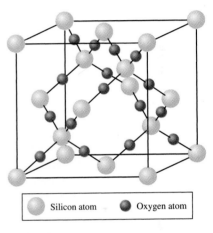

Figure 11.24
Structure of high cristobalite which is a form of silica (SiO_2). Note that each silicon atom is surrounded by four oxygen atoms and that each oxygen atom forms part of two SiO_4 tetrahedra.
(After W. D. Kingery, H. K. Bowen, and D. R. Uhlmann, "Introduction to Ceramics," 2nd ed., Wiley, 1976.)

11.3.4 Silicate Networks

Silica   When all four corners of the SiO_4^{4-} tetrahedra share oxygen atoms, an SiO_2 network called *silica* is produced (Fig. 11.24). Crystalline silica exists in several polymorphic forms that correspond to different ways in which the silicate tetrahedra are arranged with all corners shared. There are three basic silica structures: *quartz*, *tridymite*, and *cristobalite*, and each of these has two or three modifications. The

Table 11.3 Ideal silicate mineral compositions

Silica:	
Quartz	
Tridymite	Common crystalline phases of SiO_2
Cristobalite	
Alumina silicate:	
Kaolinite (china clay)	$Al_2O_3 \cdot 2SiO_2 \cdot 2H_2O$
Pyrophyllite	$Al_2O_3 \cdot 4SiO_2 \cdot H_2O$
Metakaolinite	$Al_2O_3 \cdot 2SiO_2$
Sillimanite	$Al_2O_3 \cdot SiO_2$
Mullite	$3Al_2O_3 \cdot 2SiO_2$
Alkali alumina silicate:	
Potash feldspar	$K_2O \cdot Al_2O_3 \cdot 6SiO_2$
Soda feldspar	$Na_2O \cdot Al_2O_3 \cdot 6SiO_2$
(Muscovite) mica	$K_2O \cdot 3Al_2O_3 \cdot 6SiO_2 \cdot 2H_2O$
Montmorillonite	$Na_2O \cdot 2MgO \cdot 5Al_2O_3 \cdot 24SiO_2 \cdot (6 + n)H_2O$
Leucite	$K_2O \cdot Al_2O_3 \cdot 4SiO_2$
Magnesium silicate:	
Cordierite	$2MgO \cdot 5SiO_2 \cdot 2Al_2O_3$
Steatite	$3MgO \cdot 4SiO_2$
Talc	$3MgO \cdot 4SiO_2 \cdot H_2O$
Chrysotile (asbestos)	$3MgO \cdot 2SiO_2 \cdot 2H_2O$
Forsterite	$2MgO \cdot SiO_2$

Source: O. H. Wyatt and D. Dew-Hughes, *Metals, Ceramics and Polymers,* Cambridge, 1974.

most stable forms of silica and the temperature ranges in which they exist at atmospheric pressure are low quartz below 573°C, high quartz between 573 to 867°C, high tridymite between 867 and 1470°C, and high cristobalite between 1470 and 1710°C (Fig. 11.24). Above 1710°C silica is liquid. Silica is an important component of many traditional ceramics and many different types of glasses.

Feldspars   There are many naturally occurring silicates that have infinite three-dimensional silicate networks. Among the industrially important network silicates are the feldspars, which are also among the main components of traditional ceramics. In the feldspar silicate structural network some Al^{3+} ions replace some Si^{4+} ions to form a network with a net negative charge. This negative charge is balanced with large ions of alkali and alkaline earth ions such as Na^+, K^+, Ca^{2+}, and Ba^{2+}, which fit into interstitial positions. Table 11.3 summarizes the ideal compositions of some silicate minerals.

11.4 PROCESSING OF CERAMICS

Most traditional and engineering ceramic products are manufactured by compacting powders or particles into shapes that are subsequently heated to a high enough temperature to bond the particles together. The basic steps in the processing of ceramics by the agglomeration of particles are (1) material preparation, (2) forming or casting, and (3) thermal treatment by drying (which is usually not required) and **firing** by heating the ceramic shape to a high enough temperature to bond the particles together.

11.4.1 Materials Preparation

Most ceramic products are made by the agglomeration of particles.[2] The raw materials for these products vary, depending on the required properties of the finished ceramic part. The particles and other ingredients such as binders and lubricants may be blended wet or dry. For ceramic products that do not have very "critical" properties such as common bricks, sewer pipe, and other clay products, the blending of the ingredients with water is common practice. For some other ceramic products the raw materials are ground dry along with binders and other additives. Sometimes wet and dry processing of raw materials are combined. For example, to produce one type of high-alumina (Al_2O_3) insulator, the particulate raw materials are milled with water along with a wax binder to form a slurry that is subsequently spray-dried to form small, spherical pellets (Fig. 11.25).

11.4.2 Forming

Ceramic products made by agglomerating particles may be formed by a variety of methods in the dry, plastic, or liquid conditions. Cold-forming processes are predominant in

---

[2]The production of glass products and the casting of concrete are two major exceptions.

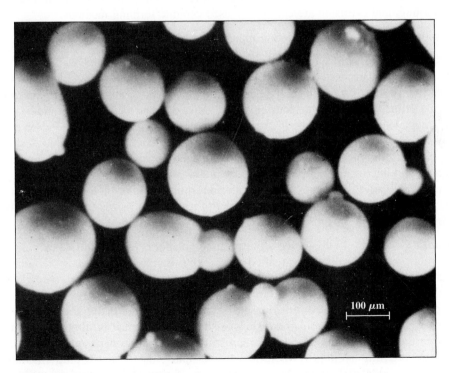

Figure 11.25
Spray-dried pellets of high-alumina ceramic body.
[*After J. S. Owens et al., American Ceramic Soc. Bull.,* **56:**437(1977).]

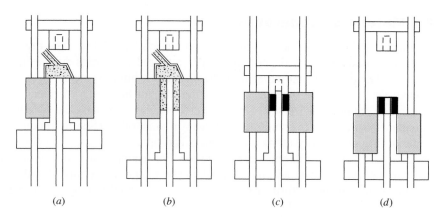

(a)   (b)   (c)   (d)

Figure 11.26
Dry pressing of ceramic particles: (*a*) and (*b*) filling, (*c*) pressing, and
(*d*) ejection.
*(After J. S. Reed and R. B. Runk, "Ceramic Fabrication Processes," vol. 9: "Treatise in
Materials Science and Technology," Academic, 1976, p. 74.)*

the ceramic industry, but hot-forming processes are also used to some extent. Pressing,
slip casting, and extrusion are commonly used ceramic forming methods.

Pressing   Ceramic particulate raw materials can be pressed in the dry, plastic, or
wet condition into a die to form shaped products.

Dry Pressing   This method is used commonly for products such as structural refractories (high-heat-resistant materials) and electronic ceramic components. **Dry pressing**
may be defined as the simultaneous uniaxial compaction and shaping of a granular powder along with small amounts of water and/or organic binder in a die. Figure 11.26
shows a series of operations for the dry pressing of ceramic powders into a simple
shape. After cold pressing, the parts are usually fired (sintered) to achieve the required
strength and microstructural properties. Dry pressing is used extensively because it can
form a wide variety of shapes rapidly with uniformity and close tolerances. For example, aluminas, titanates, and ferrites can be dry-pressed into sizes from a few mils to
several inches in linear dimensions at a rate of up to about 5000 per minute.

Isostatic Pressing   In this process the ceramic powder is loaded into a flexible (usually rubber), airtight container (called a *bag*) that is inside a chamber of hydraulic
fluid to which pressure is applied. Figure 11.27 shows a cross section of a spark plug
insulator in an isostatic pressing mold. The force of the applied pressure compacts
the powder uniformly in all directions, with the final product taking the shape of the
flexible container. After cold **isostatic pressing** the part must be fired (sintered) to
achieve the required properties and microstructure. Ceramic parts manufactured by
isostatic pressing include refractories, bricks and shapes, spark plug insulators,
radomes, carbide tools, crucibles, and bearings. Figure 11.28 shows the stages for
the manufacturing of a spark plug insulator by isostatic pressing.

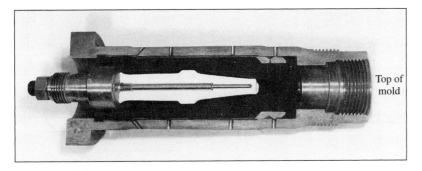

Figure 11.27
Cross section of insulator blank in isostatic pressing mold. Spray-dried
nearly spherical pellets (Fig. 11.25) are fed by gravity into top of mold
and compressed by isostatic pressure, normally in the 3000 to 6000 psi
range. The hydraulic fluid enters the side of the mold through the holes
shown in the cross section.
(*Courtesy of Champion Spark-Plug Co.*)

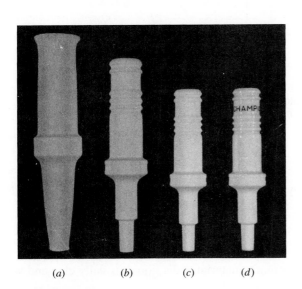
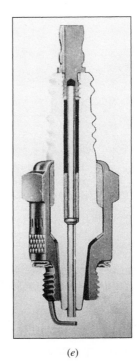

(*a*)   (*b*)   (*c*)   (*d*)   (*e*)

Figure 11.28
Stages of spark plug insulator manufacture by the isostatic processing
method. (*a*) Pressed blank. (*b*) Turned (ground) insulator. (*c*) Fired insulator.
(*d*) Glazed and decorated finished insulator. (*e*) Cross section of assembled
automotive spark plug showing position of insulator.
(*Courtesy of Champion Spark-Plug Co.*)

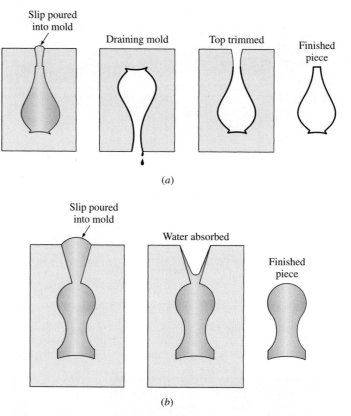

Slip poured into mold

Draining mold

Top trimmed

Finished piece

(a)

Slip poured into mold

Water absorbed

Finished piece

(b)

Figure 11.29
Slip casting of ceramic shapes. (a) Drain casting in porous plaster of paris mold. (b) Solid casting.
(*After W. D. Kingery, "Introduction to Ceramics," Wiley, 1960, p. 52.*)

Hot Pressing In this process ceramic parts of high density and improved mechanical properties are produced by combining the pressing and firing operations. Both uniaxial and isostatic methods are used.

Slip Casting Ceramic shapes can be cast by using a unique process called **slip casting,** illustrated in Fig. 11.29. The main steps in slip casting are:

1. Preparation of a powdered ceramic material and a liquid (usually clay and water) into a stable suspension called a *slip.*

2. Pouring the slip into a porous mold that is usually made of plaster of paris and allowing the liquid portion of the slip to be partially absorbed by the mold. As the liquid is removed from the slip, a layer of semihard material is formed against the mold surface.

3. When a sufficient wall thickness has been formed, the casting process is interrupted and the excess slip is poured out of the cavity (Fig. 11.29*a*). This is known as *drain casting*. Alternatively, a solid shape may be made by allowing the casting to continue until the whole mold cavity is filled, as illustrated in Fig. 11.29*b*. This type of slip casting is called *solid casting*.

4. The material in the mold is allowed to dry to provide adequate strength for handling and the subsequent removal of the part from the mold.

5. Finally the cast part is fired to attain the required microstructure and properties.

Slip casting is advantageous for forming thin-walled and complex shapes of uniform thickness. Slip casting is especially economical for development parts and short production runs. Several new variations of the slip-casting process are pressure and vacuum casting, in which the slip is shaped under pressure or vacuum.

Extrusion   Single cross sections and hollow shapes of ceramic materials can be produced by extruding these materials in the plastic state through a forming die. This method is commonly used to produce, for example, refractory brick, sewer pipe, hollow tile, technical ceramics, and electrical insulators. The means most commonly used is the vacuum-auger-type extrusion machine in which the plastic ceramic material (e.g., clay and water) is forced through a hard steel or alloy die by a motor-driven auger (Fig. 11.30). Special technical ceramics are frequently produced using a piston extrusion under high pressure so that close tolerances can be attained.

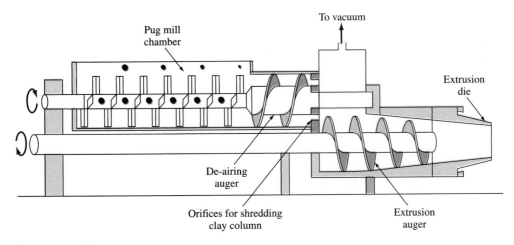

Figure 11.30
Cross section of combination mixing mill (pug mill) for ceramic materials and vacuum-auger extrusion machine.
(*After W. D. Kingery, "Introduction to Ceramics," Wiley, 1960.*)

11.4.3 Thermal Treatments

Thermal treatment is an essential step in the manufacturing of most ceramic products. In this subsection we shall consider the following thermal treatments: drying, sintering, and vitrification.

Drying and Binder Removal   The purpose of drying ceramics is to remove water from the plastic ceramic body before it is fired at higher temperatures. Generally, drying to remove water is carried out at or below 100°C and can take as long as 24 h for a large ceramic part. The bulk of organic binders can be removed from ceramic parts by heating in the range of 200 to 300°C, although some hydrocarbon residues may require heating to much higher temperatures.

Sintering   The process by which small particles of a material are bonded together by solid-state diffusion is called **sintering.** In ceramic manufacturing this thermal treatment results in the transformation of a porous compact into a dense, coherent product. Sintering is commonly used to produce ceramic shapes made of, for example, alumina, beryllia, ferrites, and titanates.

In the sintering process, particles are coalesced by solid-state diffusion at very high temperatures but below the melting point of the compound being sintered. For example, the alumina spark plug insulator shown in Fig. 11.28*a* is sintered at 1600°C (the melting point of alumina is 2050°C). In sintering, atomic diffusion takes place between the contacting surfaces of the particles so that they become chemically bonded together (Fig. 11.31). As the process proceeds, larger particles are formed at the expense of the smaller ones, as illustrated in the

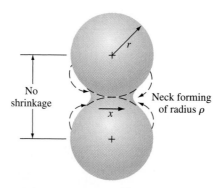

Figure 11.31
Formation of a neck during the sintering of two fine particles. Atomic diffusion takes place at the contacting surfaces and enlarges the contact area to form a neck.
(*After J. H. Brophy, R. M. Rose, and J. Wulff, "The Structure and Properties of Materials," vol. II: "Thermodynamics of Structure," Wiley, 1964, p. 139.*)

(a) (b)

(c) (d)

Figure 11.32
Scanning electron micrographs of fractured surfaces of MgO compacts (compressed powders) sintered at 1430°C in static air for (a) 30 min (fractional porosity = 0.39); (b) 303 min (f.p. = 0.14); (c) 1110 min (f.p. = 0.09); as-annealed surface of (c) is shown in (d).
[*After B. Wong and J. A. Pask, J. Am. Ceram. Soc., **62**:141(1979).*]

sintering of MgO compacts shown in Fig. 11.32a, b, and c. As the particles get larger with the time of sintering, the porosity of the compacts decreases (Fig. 11.33). Finally, at the end of the process, an "equilibrium grain size" is attained (Fig. 11.32d). The driving force for the process is the lowering of the energy of the system. The high surface energy associated with the original individual small particles is replaced by the lower overall energy of the grain-boundary surfaces of the sintered product.

Vitrification Some ceramic products such as porcelain, structural clay products, and some electronic components contain a glass phase. This glass phase serves as a reaction medium by which diffusion can take place at a lower temperature than in the rest of the ceramic solid material. During the firing of these types of ceramic materials, a process called **vitrification** takes place whereby the glass phase liquefies and fills the pore spaces in the material. This liquid glass phase may also react with some of the remaining solid refractory material. Upon cooling, the liquid phase solidifies to form a vitreous or glassy matrix that bonds the unmelted particles together.

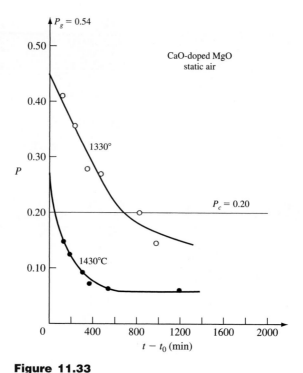

Figure 11.33
Porosity versus time for MgO compacts doped
with 0.2 wt % CaO and sintered in static air
at 1330 and 1430°C. Note that the higher
sintering temperature produces a more rapid
decrease in porosity and a lower porosity level.
[*After B. Wong and J. A. Pask, J. Am. Ceram. Soc.,*
62:*141(1979).*]

11.5 TRADITIONAL AND ENGINEERING CERAMICS

11.5.1 Traditional Ceramics

Traditional ceramics are made from three basic components: *clay, silica* (flint), and
feldspar. Clay consists mainly of hydrated aluminum silicates ($Al_2O_3 \cdot SiO_2 \cdot H_2O$)
with small amounts of other oxides such as TiO_2, Fe_2O_3, MgO, CaO, Na_2O, and K_2O.
Table 11.4 lists the chemical compositions of several industrial clays.

The clay in traditional ceramics provides workability of the material before firing hardens it and constitutes the major body material. The silica (SiO_2), also called
flint or quartz, has a high melting temperature and is the refractory component of
traditional ceramics. Potash (potassium) feldspar, which has the basic composition
$K_2O \cdot Al_2O_3 \cdot 6SiO_2$, has a low melting temperature and makes a glass when the
ceramic mix is fired. It bonds the refractory components together.

Table 11.4 Chemical compositions of some clays

Type of clay	Weight percentages of major oxides									Ignition loss
	Al_2O_3	SiO_2	Fe_2O_3	TiO_2	CaO	MgO	Na_2O	K_2O	H_2O	
Kaolin	37.4	45.5	1.68	1.30	0.004	0.03	0.011	0.005	13.9	
Tenn ball clay	30.9	54.0	0.74	1.50	0.14	0.20	0.45	0.72	⋯	11.4
Ky. ball clay	32.0	51.7	0.90	1.52	0.21	0.19	0.38	0.89	⋯	12.3

Source: P. W. Lee, *Ceramics,* Reinhold, 1961.

Table 11.5 Some triaxial whiteware chemical compositions

Type body	China clay	Ball clay	Feldspar	Flint	Other
Hard porcelain	40	10	25	25	
Electrical insulation ware	27	14	26	33	
Vitreous sanitary ware	30	20	34	18	
Electrical insulation	23	25	34	18	
Vitreous tile	26	30	32	12	
Semivitreous whiteware	23	30	25	21	
Bone china	25	⋯	15	22	38 bone ash
Hotel china	31	10	22	35	2 $CaCO_3$
Dental porcelain	5	⋯	95		

Source: W. D. Kingery, H. K. Bowen, and D. R. Uhlmann, *Introduction to Ceramics,* 2nd ed., Wiley, 1976, p. 532.

Structural clay products such as building brick, sewer pipe, drain tile, roofing tile, and floor tile are made of natural clay, which contains all three basic components. Whiteware products such as electrical porcelain, dinner china, and sanitary ware are made from components of clay, silica, and feldspar for which the composition is controlled. Table 11.5 lists the chemical compositions of some triaxial whitewares. The term *triaxial* is used since there are three major materials in their composition.

Typical composition ranges for different types of whitewares are illustrated in the silica-leucite-mullite ternary phase diagram of Fig. 11.34. The composition ranges of some whitewares are indicated by the circled areas.

The changes that occur in the structure of triaxial bodies during firing are not completely understood due to their complexity. Table 11.6 is an approximate summary of what probably occurs during the firing of a whiteware body.

Figure 11.35 is an electron micrograph of the microstructure of an electrical insulator porcelain. As observed in this micrograph, the structure is very heterogeneous. Large quartz grains are surrounded by a solution rim of high-silica glass. Mullite needles that cross feldspar relicts and fine mullite-glass mixtures are present.

Triaxial porcelains are satisfactory as insulators for 60-cycle use, but at high frequencies dielectric losses become too high. The considerable amounts of alkalies derived from the feldspar used as a flux increase the electrical conductivity and dielectric losses of triaxial porcelains.

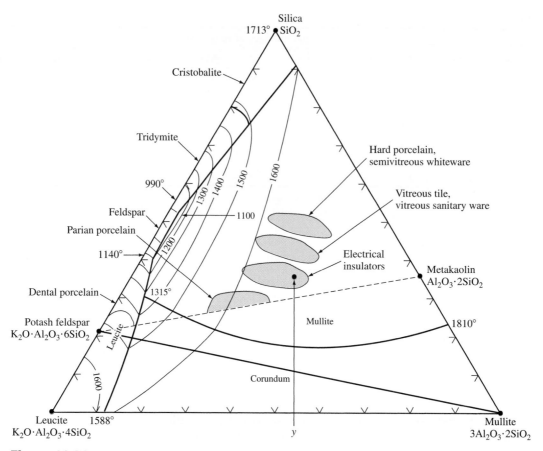

Figure 11.34

Areas of triaxial whiteware compositions shown on the silica-leucite-mullite phase-equilibrium diagram.

(*After W. D. Kingery, H. K. Bowen, and D. R. Uhlmann, "Introduction to Ceramics," 2nd ed., Wiley, 1976, p. 533.*)

Table 11.6 Life history of a triaxial body

Temperature (°C)	Reactions
Up to 100	Loss of moisture
100–200	Removal of absorbed water
450	Dehydroxylation
500	Oxidation of organic matter
573	Quartz inversion to high form. Little overall volume damage
980	Spinel forms from clay. Start of shrinkage
1000	Primary mullite forms
1050–1100	Glass forms from feldspar, mullite grows, shrinkage continues
1200	More glass, mullite grows, pores closing, some quartz solution
1250	60% glass, 21% mullite, 19% quartz, pores at minimum

Source: F. Norton, *Elements of Ceramics,* 2nd ed., Addison-Wesley, 1974, p. 140.

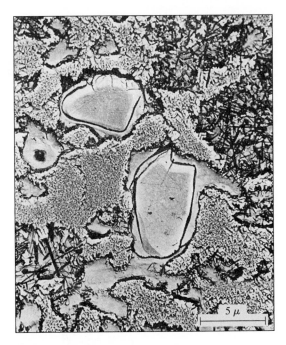

Figure 11.35
Electron micrograph of an electrical insulator porcelain (etched 10 s, 0°C, 40% HF, silica replica).
(After S. T. Lundin as shown in W. D. Kingery, H. K. Bowen, and D. R. Uhlmann, "Introduction to Ceramics," 2nd ed., Wiley, 1976, p. 539.)

11.5.2 Engineering Ceramics

In contrast to the traditional ceramics, which are mainly based on clay, engineering or technical ceramics are mainly pure compounds or nearly pure compounds of chiefly oxides, carbides, or nitrides. Some of the important engineering ceramics are alumina (Al_2O_3), silicon nitride (Si_3N_4), silicon carbide (SiC), and zirconia (ZrO_2) combined with some other refractory oxides. The melting temperatures of some of the engineering ceramics are listed in Table 11.1, and the mechanical properties of some of these materials are given in Table 11.7. A brief description of a few of the properties, processes, and applications of some of the important engineering ceramics follows.

Alumina (Al_2O_3)   Alumina was originally developed for refractory tubing and high-purity crucibles for high-temperature use and now has wide application. A classic example of the application of alumina is in spark plug insulator material (Fig. 11.28). Aluminum oxide is commonly doped with magnesium oxide, cold-pressed, and sintered, producing the type of microstructure shown in Fig. 11.36. Note the uniformity of the alumina grain structure as compared to the microstructure of the electrical porcelain of Fig. 11.35. Alumina is used commonly for high-quality electrical applications where low dielectric loss and high resistivity are needed.

Table 11.7 Mechanical properties of selected engineering ceramic materials

Material	Density (g/cm³)	Compressive strength		Tensile strength		Flexural strength		Fracture toughness	
		MPa	ksi	MPa	ksi	MPa	ksi	MPa $\sqrt{m}$	ksi $\sqrt{m}$
Al₂O₃ (99%)	3.85	2585	375	207	30	345	50	4	3.63
Si₃N₄ (hot-pressed)	3.19	3450	500	...	...	690	100	6.6	5.99
Si₃N₄ (reaction-bonded)	2.8	770	112	...	...	255	37	3.6	3.27
SiC (sintered)	3.1	3860	560	170	25	550	80	4	3.63
ZrO₂, 9% MgO (partially stabilized)	5.5	1860	270	...	...	690	100	8+	7.26+

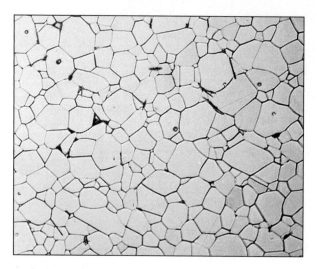

Figure 11.36
Microstructure of sintered, powdered aluminum oxide doped with magnesium oxide. The sintering temperature was 1700°C. The microstructure is nearly pore-free, with only a few pores within the grains. (Magnification 500×.)
(*Courtesy of C. Greskovich and K. W. Lay.*)

Silicon Nitride (Si_3N_4)   Of all the engineering ceramics, silicon nitride has probably the most useful combination of engineering properties. Si_3N_4 dissociates significantly at temperatures above 1800°C and so cannot be directly sintered. Si_3N_4 can be processed by reaction bonding in which a compact of silicon powder is nitrided in a flow of nitrogen gas. This process produces a microporous Si_3N_4 with moderate strength (Table 11.7). Higher-strength nonporous Si_3N_4 is made by hot-pressing with 1 to 5% MgO. Si_3N_4 is being explored for use for parts of advanced engines (Fig. 1.8a).

Silicon Carbide (SiC)   Silicon carbide is a hard, refractory carbide with outstanding resistance to oxidation at high temperatures. Although a nonoxide, SiC at high temperatures forms a skin of SiO_2 that protects the main body of material. SiC can be sintered at 2100°C with 0.5 to 1% B as a sintered aid. SiC is commonly used as a fibrous reinforcement for metal-matrix and ceramic-matrix composite materials.

Zirconia (ZrO_2)   Pure zirconia is polymorphic and transforms from the tetragonal to monoclinic structure at about 1170°C with an accompanying volume expansion and so is subject to cracking. However, by combining ZrO_2 with other refractory oxides such as CaO, MgO, and Y_2O_3, the cubic structure can be stabilized at room temperature and has found some applications. By combining ZrO_2 with 9% MgO and using special heat treatments, a partially stabilized zirconia (PSZ) can be produced with especially high fracture toughness, which has led to new ceramic applications. (See Sec. 11.6 on the fracture toughness of ceramics for more details.)

11.6 MECHANICAL PROPERTIES OF CERAMICS

11.6.1 General

As a class of materials, ceramics are relatively brittle. The observed tensile strength of ceramic materials varies greatly, ranging from very low values of less than 100 psi (0.69 MPa) to about 10^6 psi (7×10^3 MPa) for whiskers of ceramics such as Al_2O_3 prepared under carefully controlled conditions. However, as a class of materials, few ceramics have tensile strengths above 25,000 psi (172 MPa). Ceramic materials also have a large difference between their tensile and compressive strengths, with the compressive strengths usually being about five to 10 times higher than the tensile strengths, as indicated in Table 11.7 for the 99 percent Al_2O_3 ceramic material. Also, many ceramic materials are hard and have low impact resistance due to their ionic-covalent bindings. However, there are many exceptions to these generalizations. For example, plasticized clay is a ceramic material that is soft and easily deformable due to weak secondary bonding forces between layers of strongly ionic–covalently bonded atoms.

11.6.2 Mechanisms for the Deformation of Ceramic Materials

The lack of plasticity in crystalline ceramics is due to their ionic and covalent chemical bonds. In metals, plastic flow takes place mainly by the movement of line faults (dislocations) in the crystal structure over special crystal slip planes (see Sec. 7.4). In metals, dislocations move under relatively low stresses due to the nondirectional nature of the metallic bond and because all atoms involved in the bonding have an equally distributed negative charge at their surfaces. That is, there are no positive or negatively charged ions involved in the metallic bonding process.

In covalent crystals and covalently bonded ceramics, the bonding between atoms is specific and directional, involving the exchange of electron charge between pairs of electrons. Thus, when covalent crystals are stressed to a sufficient extent, they exhibit brittle fracture due to a separation of electron-pair bonds without their subsequent reformation. Covalently bonded ceramics, therefore, are brittle in both the single-crystal and polycrystalline states.

The deformation of primarily ionically bonded ceramics is different. Single crystals of ionically bonded solids such as magnesium oxide and sodium chloride show considerable plastic deformation under compressive stresses at room temperature. Polycrystalline ionically bonded ceramics, however, are brittle, with cracks forming at the grain boundaries.

Let us briefly examine some conditions under which an ionic crystal can be deformed, as illustrated in Fig. 11.37. The slip of one plane of ions over another involves ions of different charge coming into contact, and thus attractive and repulsion forces may be produced. Most ionically bonded crystals having the NaCl-type structure slip on the $\{110\}\langle1\bar{1}0\rangle$ systems because slip on the $\{110\}$ family of planes involves only ions of unlike charge, and hence the slip planes remain attracted to each other by coulombic forces during the slip process. Slip of the $\{110\}$ type is indicated by the line AA' of Fig. 11.37. On the other hand, slip on the $\{100\}$ family of planes is rarely observed because ions of the same charge come into contact, which will tend to separate the planes of ions slipping over each other. This $\{100\}$-type slip is indicated by the line BB' of Fig. 11.37. Many ceramic materials in the single-crystal form show considerable plasticity. However, in polycrystalline ceramics adjacent grains must change shape during deformation. Since there are limited slip systems in ionically bonded solids, cracking occurs at the grain boundaries and subsequent brittle fracture occurs. Since most industrially important ceramics are polycrystalline, most ceramic materials tend to be brittle.

11.6.3 Factors Affecting the Strength of Ceramic Materials

The mechanical failure of ceramic materials occurs mainly from structural defects. The principal sources of fracture in ceramic polycrystals include surface cracks produced during surface finishing, voids (porosity), inclusions, and large grains produced during processing.[3]

Pores in brittle ceramic materials are regions where stress concentrates, and when the stress at a pore reaches a critical value, a crack forms and propagates since there are no large, energy-absorbing processes in these materials such as those that operate in ductile metals during deformation. Thus, once cracks start to propagate, they continue to grow until fracture occurs. Pores are also detrimental to the strength of ceramic materials because they decrease the cross-sectional area over which a load is applied and hence lower the stress a material can support. Thus, the size and volume fraction of pores in ceramic materials are important factors affecting their strength. Figure 11.38 shows how an increasing volume fraction of pores decreases the transverse tensile strength of alumina.

Flaws in processed ceramics may also be critical in determining the fracture strength of a ceramic material. A large flaw may be the major factor affecting the strength of a ceramic. In fully dense ceramic materials in which there are no large pores, the flaw size is usually related to the grain size. For porosity-free ceramics, the strength of a pure ceramic material is a function of its grain size, with finer-grain-size ceramics

---

[3]A. G. Evans, *J. Am. Ceram. Soc.*, **65**:127(1982).

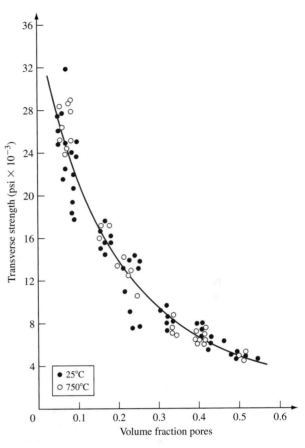

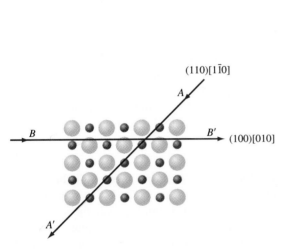

Figure 11.37
Top view of NaCl crystal structure indicating
(*a*) slip on the (110) plane and in the [110]
direction (line *AA'*) and (*b*) slip on the (100)
plane in the [010] direction (line *BB'*).

Figure 11.38
The effect of porosity on the transverse strength of
pure alumina.
[*After R. L. Coble and W. D. Kingery, J Am. Ceram. Soc.,*
39:377(1956).]

having smaller-size flaws at their grain boundaries and hence being stronger than large-grain-size ones.

The strength of a polycrystalline ceramic material is thus determined by many factors that include chemical composition, microstructure, and surface condition as major factors. Temperature and environment also are important as well as the type of stress and how it is applied. However, the failure of most ceramic materials at room temperature usually originates at the largest flaw.

11.6.4 Toughness of Ceramic Materials

Ceramic materials, because of their combination of covalent-ionic bonding, have inherently low toughness. A great amount of research has been carried out in the past years to improve the toughness of ceramic materials. By the use of processes

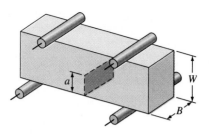

Figure 11.39
Setup for four-point beam
fracture-toughness test of a
ceramic material using a single-
edge notch.

such as hot pressing ceramics with additives and reaction bonding, engineering
ceramics with improved toughness have been produced (Table 11.7).

Fracture-toughness tests can be made on ceramic specimens to determine K_{IC}
values in a manner similar to the fracture-toughness testing of metals (see Sec. 7.3).
K_{IC} values for ceramic materials are usually obtained by using a four-point bend test
with a single-edge or chevron-notched beam specimen (Fig. 11.39). The fracture-
toughness equation,

$$K_{IC} = Y\sigma_f\sqrt{\pi a} \tag{11.1}$$

which relates fracture-toughness K_{IC} values to the fracture stress and the largest
flaw size can also be used for ceramic materials. In Eq. 11.1, K_{IC} is measured in
MPa $\sqrt{m}$ (ksi $\sqrt{in.}$), the fracture stress σ_f in MPa (ksi), and a (half the size of the
largest internal flaw) in meters (inches). Y is a dimensionless constant equal to about 1.
Example Problem 11.9 shows how this equation can be used to determine the largest-
size flaw that a particular engineering ceramic of a known fracture toughness and
strength can tolerate without fracture.

**EXAMPLE
PROBLEM 11.9**

A reaction-bonded silicon nitride ceramic has a strength of 300 MPa and a fracture toughness of 3.6 MPa $\sqrt{m}$. What is the largest-size internal crack that this material can support
without fracturing? Use $Y = 1$ in the fracture-toughness equation.

■ **Solution**

$$\sigma_f = 300 \text{ MPa} \qquad K_{IC} = 3.6 \text{ MPa-}\sqrt{m} \qquad a = ? \qquad Y = 1$$
$$K_{IC} = Y\sigma_f\sqrt{\pi a}$$

or

$$a = \frac{K_{IC}^2}{\pi\sigma_f^2} = \frac{(3.6 \text{ MPa-}\sqrt{m})^2}{\pi(300 \text{ MPa})^2}$$
$$= 4.58 \times 10^{-5} \text{ m} = 45.8 \ \mu m$$

Thus, the largest internal crack $= 2a = 2(45.8 \ \mu m) = 91.6 \ \mu m$. ◄

11.6.5 Transformation Toughening of Partially Stabilized Zirconia (PSZ)

Recently it has been discovered that phase transformations in zirconia combined with some other refractory oxides (that is, CaO, MgO, or Y_2O_3) can produce ceramic materials with exceptionally high fracture toughness. Let us now look into the mechanisms that produce transformation toughening in a ZrO_2–9 mol % MgO ceramic material. Pure zirconia, ZrO_2, exists in three different crystal structures: *monoclinic* from room temperature to 1170°C, *tetragonal* from 1170 to 2370°C, and *cubic* (the fluorite structure of Fig. 11.14) above 2370°C.

The transformation of pure ZrO_2 from the tetragonal to monoclinic structure is martensitic and cannot be suppressed by rapid cooling. Also, this transformation is accompanied by a volume increase of about 9 percent, and so it is impossible to fabricate articles from pure zirconia. However, by the addition of about 10 mol % of other refractory oxides such as CaO, MgO, or Y_2O_3, the cubic form of zirconia is stabilized so that it can exist at room temperature in the metastable state, and articles can be fabricated from this material. Cubic ZrO_2 combined with stabilizing oxides so that it retains the cubic structure at room temperature is referred to as *fully stabilized zirconia.*

Recent developments have produced zirconia-refractory oxide ceramic materials with enhanced toughness and strength by taking advantage of their phase transformations. One of the more important zirconia compound ceramics is partially stabilized zirconia (PSZ), which contains 9 mol % MgO. If a mixture of ZrO_2–9 mol % MgO is sintered at about 1800°C, as indicated in the ZrO_2-MgO phase diagram of Fig. 11.40*a*, and then rapidly cooled to room temperature, it will be in the all-metastable cubic structure. However, if this material is reheated to 1400°C and held for a sufficient time, a fine metastable submicroscopic precipitate with the tetragonal structure is precipitated, as shown in Fig. 11.40*b*. This material is known as *partially stabilized zirconia* (PSZ). Under the action of stresses that cause small cracks in the ceramic material, the tetragonal phase transforms to the monoclinic phase, causing a volume expansion of the precipitate that retards the crack propagation by a kind of crack-closing mechanism. By impeding the advances of cracks, the ceramic is "toughened" (Fig. 11.40*c*). Partially stabilized zirconia has a fracture toughness of $8+$ MPa $\sqrt{m}$, which is higher than the fracture toughness of all the other engineering ceramic materials listed in Table 11.7.

11.6.6 Fatigue Failure of Ceramics

Fatigue failure in metals occurs under repeated cyclic stresses due to the nucleation and growth of cracks within a work-hardened area of a specimen. Because of the ionic-covalent bonding of the atoms in a ceramic material, there is an absence of plasticity in ceramics during cyclic stressing. As a result, fatigue fracture in ceramics is rare. Recently, results of stable fatigue-crack growth at room temperature under compression–compression stress cycling in notched plates of polycrystalline alumina have been reported. A straight fatigue crack was produced after 79,000 compression

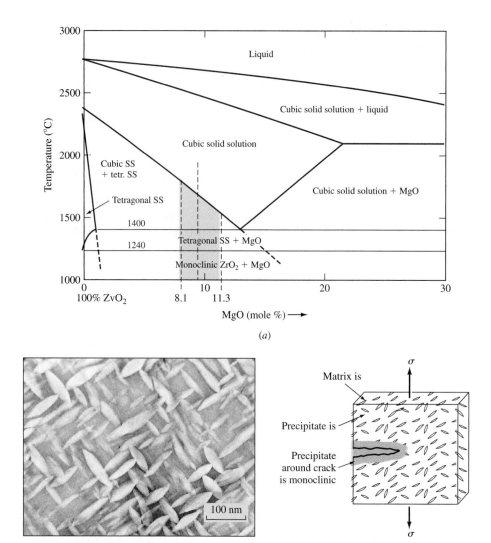

Figure 11.40

(a) Phase diagram of high-ZrO$_2$ part of the ZrO$_2$–MgO binary phase diagram. The shaded area represents the region used for combining MgO with ZrO$_2$ to produce partially stabilized zirconia.

(After A. H. Heuer, "Advances in Ceramics," vol. 3, "Science and Technology of Zirconia," American Ceramic Society, 1981.)

(b) Transmission electron micrograph of optimally aged MgO–partially stabilized ZrO$_2$ showing the tetragonal oblate spheroid precipitate. Upon the application of sufficient stress, these particles transform to the monoclinic phase with a volume expansion.

(Courtesy of A. H. Heuer.)

(c) Schematic diagram illustrating the transformation of the tetragonal precipitate to the monoclinic phase around a crack in a partially stabilized ZrO$_2$–9 mol % MgO ceramic specimen.

(a)   (b)

Figure 11.41
Fatigue cracking of polycrystalline alumina under cyclic compression. (*a*) Optical micrograph showing fatigue crack (the compression axis is vertical). (*b*) Scanning electron fractograph of the fatigue area of the same specimen where the intergranular mode of failure is evident.
[*After S. Suresh and J. R. Brockenbrough, Acta Metall.* **36:***1455 (1988).*]

cycles (Fig. 11.41*a*). Microcrack propagation along grain boundaries led to final intergranular fatigue failure (Fig. 11.41*b*). Much research is being carried out to make tougher ceramics that can support cyclic stresses for applications such as turbine rotors.

11.6.7 Ceramic Abrasive Materials

The high hardness of some ceramic materials makes them useful as abrasive materials for cutting, grinding, and polishing other materials of lower hardness. Fused alumina (aluminum oxide) and silicon carbide are two of the most commonly used manufactured ceramic abrasives. Abrasive products such as sheets and wheels are made by bonding individual ceramic particles together. Bonding materials include fired ceramics, organic resins, and rubbers. The ceramic particles must be hard with sharp cutting edges. Also, the abrasive product must have a certain amount of porosity to provide channels for air or liquid to flow through in the structure. Aluminum oxide grains are tougher than silicon carbide ones but are not as hard, and so silicon carbide is normally used for the harder materials.

By combining zirconium oxide with aluminum oxide, improved abrasives were developed[4] with higher strength, hardness, and sharpness than aluminum oxide alone. One of these ceramic alloys contains 25 percent ZrO_2 and 75 percent Al_2O_3 and another 40 percent ZrO_2 and 60 percent Al_2O_3. Another important ceramic abrasive

---

[4]ZrO_2-Al_2O_3 ceramic abrasive alloys were developed by the Norton Co. in the 1960s.

is cubic boron nitride, which has the trade name Borazon.[5] This material is almost as hard as diamond but has better heat stability than diamond.

11.7 THERMAL PROPERTIES OF CERAMICS

In general, most ceramic materials have low thermal conductivities due to their strong ionic-covalent bonding and are good thermal insulators. Figure 11.42 compares the thermal conductivities of many ceramic materials as a function of temperature. Because of

---

[5]Borazon, a product of the General Electric Co., was developed in the 1950s.

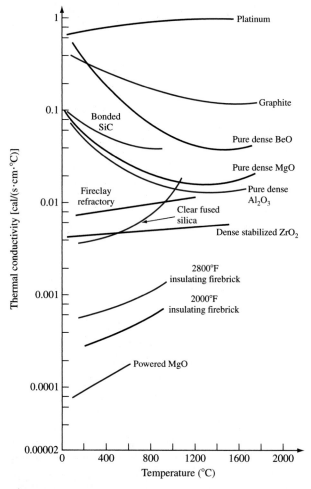

Figure 11.42

Thermal conductivity (logarithmic scale) of ceramic materials over a wide temperature range.

(After W. D. Kingery, H. K. Bowen, and D. R. Uhlmann, "Introduction to Ceramics," 2nd ed., Wiley, 1976, p. 643.)

Table 11.8 Compositions and applications for some refractory brick materials

	Composition (wt %)			
	SiO$_2$	**Al$_2$O$_3$**	**MgO**	**Other**
Acidic types:				
Silica brick	95–99			
Superduty fireclay brick	53	42		
High-duty fireclay brick	51–54	37–41		
High-alumina brick	0–50	45–99+		
Basic types:				
Magnesite	0.5–5		91–98	0.6–4 CaO
Magnesite-chrome	2–7	6–13	50–82	18–24 Cr$_2$O$_3$
Dolomite (burned)			38–50	38–58 CaO
Special types:				
Zircon	32			66 ZrO$_2$
Silicon carbide	6	2		91 SiC

Applications for some refractories:

Superduty fireclay bricks: linings for aluminum-melting furnaces, rotary kilns, blast furnaces, and hot-metal transfer ladles

High-duty fireclay brick: linings for cement and lime kilns, blast furnaces, and incinerators

High-alumina brick: boiler furnaces, spent-acid regenerating furnaces, phosphate furnaces, glass-tank refiner walls, carbon black furnaces, continuous-casting tundish linings, coal gasification reactor linings, and petroleum coke kilns

Silica brick: chemical reactor linings, glass tank parts, ceramic kilns, and coke ovens

Magnesite brick: basic-oxygen-process furnace linings for steelmaking

Zircon brick: glass-tank bottom paving and continuous-casting nozzles

Source: *Harbison-Walker Handbook of Refractory Practice,* Harbison-Walker Refractories, Pittsburgh, 1980.

their high heat resistance, ceramic materials are used as **refractories,** which are materials that resist the action of hot environments, both liquid and gaseous. Refractories are used extensively by the metals, chemical, ceramic, and glass industries.

11.7.1 Ceramic Refractory Materials

Many pure ceramic compounds with high melting points such as aluminum oxide and magnesium oxide could be used as industrial refractory materials, but they are expensive and difficult to form into shapes. Therefore, most industrial refractories are made of mixtures of ceramic compounds. Table 11.8 lists the compositions of some refractory brick compositions and gives some of their applications.

Important properties of ceramic refractory materials are low- and high-temperature strengths, bulk density, and porosity. Most ceramic refractories have bulk densities that range from 2.1 to 3.3 g/cm^3 (132 to 206 lb/ft^3). Dense refractories with low porosity have higher resistance to corrosion and erosion and to penetration by liquids and gases. However, for insulating refractories a high amount of porosity is desirable. Insulating refractories are mostly used as backing for brick or refractory material of higher density and refractoriness.

Industrial ceramic refractory materials are commonly divided into acidic and basic types. Acidic refractories are based mainly on SiO$_2$ and Al$_2$O$_3$ and basic ones on MgO, CaO, and Cr$_2$O$_3$. Table 11.8 lists the compositions of many types of industrial refractories and gives some of their applications.

11.7.2 Acidic Refractories

Silica refractories have high refractoriness, high mechanical strength, and rigidity at temperatures almost to their melting points.

Fireclays are based on a mixture of plastic fireclay, flint clay, and clay (coarse-particle) grog. In the unfired (green) condition these refractories consist of a mixture of particles varying from coarse to extremely fine. After firing, the fine particles form a ceramic bond between the larger particles.

High-alumina refractories contain from 50 to 99 percent alumina and have higher fusion temperatures than fireclay bricks. They can be used for more severe furnace conditions and at higher temperatures than fireclay bricks but are more expensive.

11.7.3 Basic Refractories

Basic refractories consist mainly of magnesia (MgO), lime (CaO), chrome ore, or mixtures of two or more of these materials. As a group, basic refractories have high bulk densities, high melting temperatures, and good resistance to chemical attack by basic slags and oxides but are more expensive. Basic refractories containing a high percentage (92 to 95 percent) of magnesia are used extensively for linings in the basic-oxygen steelmaking process.

11.7.4 Ceramic Tile Insulation for the Space Shuttle Orbiter

The development of the thermal protection system for the space shuttle orbiter is an excellent example of modern materials technology applied to engineering design. So that the space shuttle orbiter could be used for at least 100 missions, new insulative ceramic tile materials were developed.

About 70 percent of the orbiter's external surface is protected from heat by approximately 24,000 individual ceramic tiles made from a silica-fiber compound. Figure 11.43 shows the microstructure of the *high-temperature reusable-surface insulation* (HRSI) tile material, and Fig. 11.44 indicates the surface area where it is attached to the body of the orbiter. This material has a density of only 4 kg/ft^3 (9 lb/ft^3) and can withstand temperatures as high as 1260°C (2300°F). The effectiveness of this insulative material is indicated by the ability of a technician to hold a piece of the ceramic tile only about 10 s after it has been removed from a furnace at 1260°C (2300°F).

11.8 GLASSES

Glasses have special properties not found in other engineering materials. The combination of transparency and hardness at room temperature along with sufficient strength and excellent corrosion resistance to most normal environments make glasses indispensable for many engineering applications such as construction and vehicle glazing. In the electrical industry, glass is essential for various types of lamps because of its insulative properties and ability to provide a vacuumtight enclosure. In the electronics industry electron tubes also require the vacuum-tight enclosure

Figure 11.43
Microstructure of LI900 high-temperature reusable-surface insulation (ceramic tile material used for space shuttle); structure consists of 99.7% pure silica fibers. (Magnification 1200×.)
(*Courtesy of Lockheed Martin Missiles and Space Co.*)

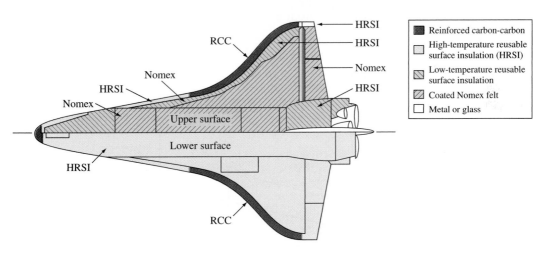

Figure 11.44
Space shuttle thermal protection systems.
(*Courtesy of NASA.*)

provided by glass along with its insulative properties for lead-in connectors. The high chemical resistance of glass makes it useful for laboratory apparatus and for corrosion-resistant liners for pipes and reaction vessels in the chemical industry.

11.8.1 Definition of a Glass

A glass is a ceramic material in that it is made from inorganic materials at high temperatures. However, it is distinguished from other ceramics in that its constituents are heated to fusion and then cooled to a rigid state without crystallization. Thus, a **glass** can be defined as *an inorganic product of fusion that has cooled to a rigid condition without crystallization.* A characteristic of a glass is that it has a noncrystalline or amorphous structure. The molecules in a glass are not arranged in a regular repetitive long-range order as exists in a crystalline solid. In a glass the molecules change their orientation in a random manner throughout the solid material.

11.8.2 Glass Transition Temperature

The solidification behavior of a glass is different from that of a crystalline solid, as illustrated in Fig. 11.45, which is a plot of specific volume (reciprocal of density) versus temperature for these two types of materials. A liquid that forms a crystalline solid upon solidifying (i.e., a pure metal) will normally crystallize at its melting point with a significant decrease in specific volume, as indicated by path *ABC* in Fig. 11.45. In contrast, a liquid that forms a glass upon cooling does not crystallize but follows a path like *AD* in Fig. 11.45. Liquid of this type becomes more viscous as its temperature is lowered and transforms from a rubbery, soft plastic state to a rigid, brittle glassy state

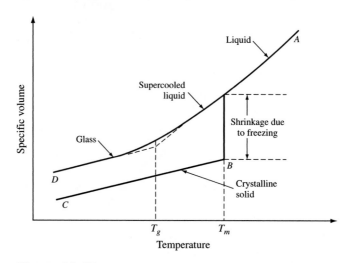

Figure 11.45
Solidification of crystalline and glassy (amorphous) materials showing changes in specific volume. T_g is the glass transition temperature of the glassy material. T_m is the melting temperature of the crystalline material.

in a narrow temperature range where the slope of the specific volume versus temperature curve is markedly decreased. The point of intersection of the two slopes of this curve define a transformation point called the **glass transition temperature** T_g. This point is structure-sensitive, with faster cooling rates producing higher values of T_g.

11.8.3 Structure of Glasses

Glass-Forming Oxides   Most inorganic glasses are based on the **glass-forming oxide** silica, SiO_2. The fundamental subunit in silica-based glasses is the SiO_4^{4-} tetrahedron in which a silicon (Si^{4+}) atom (ion) in the tetrahedron is covalently ionically bonded to four oxygen atoms (ions), as shown in Fig. 11.46a. In crystalline silica, for example, cristobalite, the Si-O tetrahedra are joined corner to corner in a regular arrangement, producing long-range order as idealized in Fig. 11.46b. In a simple silica glass the tetrahedra are joined corner to corner to form a *loose network* with no long-range order (Fig. 11.46c).

Boron oxide, B_2O_3, is also a glass-forming oxide and by itself forms subunits that are flat triangles with the boron atom slightly out of the plane of the oxygen atoms. However, in borosilicate glasses that have additions of alkali and alkaline earth oxides, BO_3^{3-} triangles can be converted to BO_4^{4-} tetrahedra, with the alkali or alkaline earth cations providing the necessary electroneutrality. Boron oxide is an important addition to many types of commercial glasses such as borosilicate and aluminoborosilicate glasses.

Glass-Modifying Oxides   Oxides that break up the glass network are known as **network modifiers.** Alkali oxides such as Na_2O and K_2O and alkaline earth oxides such as CaO and MgO are added to silica glass to lower its viscosity so that it can be worked and formed more easily. The oxygen atoms from these oxides enter the silica network at points joining the tetrahedra and break up the network, producing

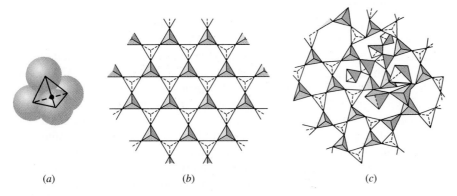

(a)   (b)   (c)

Figure 11.46
Schematic representation of (a) a silicon-oxygen tetrahedron, (b) ideal crystalline silica (cristobalite) in which the tetrahedra have long-range order, and (c) a simple silica glass in which the tetrahedra have no long-range order.
(*Courtesy of Corning Glass Works.*)

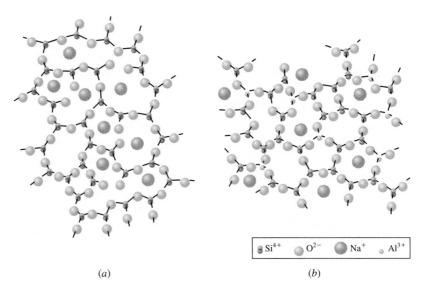

Figure 11.47
(*a*) Network-modified glass (soda-lime glass); note that the metallic
(Na^+) ions do not form part of the network. (*b*) Intermediate-oxide glass
(alumina-silica) glass; note that the small metallic (Al^{3+}) ions form part of
the network.
(*After O. H. Wyatt and D. Dew-Hughes, "Metals, Ceramics, and Polymers," Cambridge,*
1974, p. 263.)

oxygen atoms with an unshared electron (Fig. 11.47*a*). The Na^+ and K^+ ions from
the Na_2O and K_2O do not enter the network but remain as metal ions ionically
bonded in the interstices of the network. By filling some of the interstices, these ions
promote crystallization of the glass.

Intermediate Oxides in Glasses   Some oxides cannot form a glass network by
themselves but can join into an existing network. These oxides are known as **intermediate oxides.** For example, aluminum oxide, Al_2O_3, can enter the silica network
as AlO_4^{4-} tetrahedra, replacing some of the SiO_4^{4-} groups (Fig. 11.47*b*). However,
since the valence of Al is +3 instead of the necessary +4 for the tetrahedra, alkali
cations must supply the necessary other electrons to produce electrical neutrality.
Intermediate oxides are added to silica glass to obtain special properties. For example, aluminosilicate glasses can withstand higher temperatures than common glass.
Lead oxide is another intermediate oxide that is added to some silica glasses.
Depending on the composition of the glass, intermediate oxides may sometimes act
as network modifiers as well as taking part in the network of the glass.

11.8.4 Compositions of Glasses

The compositions of some important types of glasses are listed in Table 11.9 along
with some remarks about their special properties and applications. Fused silica glass,

Table 11.9 Compositions of some glasses

Glass	SiO$_2$	Na$_2$O	K$_2$O	CaO	B$_2$O$_3$	Al$_2$O$_3$	Other	Remarks
1. (Fused) silica	99.5+							Difficult to melt and fabricate but usable to 1000°C. Very low expansion and high thermal shock resistance.
2. 96% silica	96.3	<0.2	<0.2		2.9	0.4		Fabricate from relatively soft borosilicate glass; heat to separate SiO$_2$ and B$_2$O$_3$ phases; acid leach B$_2$O$_3$ phase; heat to consolidate pores.
3. Soda-lime: plate glass	71–73	12–14		10–12		0.5–1.5	MgO, 1–4	Easily fabricated. Widely used in slightly varying grades, for windows, containers, and electric bulbs.
4. Lead silicate: electrical	63	7.6	6	0.3	0.2	0.6	PbO, 21 MgO, 0.2	Readily melted and fabricated with good electrical properties.
5. High-lead	35		7.2				PbO, 58	High lead absorbs x-rays; high refractive used in achromatic lenses. Decorative crystal glass.
6. Borosilicate: low expansion	80.5	3.8	0.4		12.9	2.2		Low expansion, good thermal shock resistance, and chemical stability. Widely used in chemical industry.
7. Low electrical loss	70.0		0.5		28.0	1.1	PbO, 1.2	Low dielectric loss.
8. Aluminoborosilicate: standard apparatus	74.7	6.4	0.5	0.9	9.6	5.6	B$_2$O, 2.2	Increased alumina, lower boric oxide improves chemical durability.
9. Low alkali (E-glass)	54.5	0.5		22	8.5	14.5		Widely used for fibers in glass resin composites.
10. Aluminosilicate	57	1.0		5.5	4	20.5	MgO, 12	High-temperature strength, low expansion.
11. Glass-ceramic	40–70					10–35	MgO, 10–30 TiO$_2$, 7–15	Crystalline ceramic made by devitrifying glass. Easy fabrication (as glass), good properties. Various glasses and catalysts.

Source: O. H. Wyatt and D. Dew-Hughes, *Metals, Ceramics, and Polymers,* Cambridge, 1974, p. 261.

which is the most important single-component glass, has a high spectral transmission and is not subject to radiation damage, which causes browning of other glasses. It is therefore the ideal glass for space vehicle windows, wind tunnel windows, and optical systems in spectrophotometric devices. However, silica glass is difficult to process and expensive.

Soda-Lime Glass   The most commonly produced glass is soda-lime glass, which accounts for about 90 percent of all the glass produced. In this glass the basic composition is 71 to 73 percent SiO_2, 12 to 14 percent Na_2O, and 10 to 12 percent CaO. The Na_2O and CaO decrease the softening point of this glass from 1600 to about 730°C so that the soda-lime glass is easier to form. An addition of 1 to 4 percent MgO is used to prevent devitrification, and an addition of 0.5 to 1.5 percent Al_2O_3 is used to increase durability. Soda-lime glass is used for flat glass, containers, pressed and blown ware, and lighting products where high chemical durability and heat resistance are not needed.

Borosilicate Glasses   The replacement of alkali oxides by boric oxide in the silica glassy network produces a lower-expansion glass. When B_2O_3 enters the silica network, it weakens its structure and considerably lowers the softening point of the silica glass. The weakening effect is attributed to the presence of planar three-coordinate borons. Borosilicate glass (Pyrex glass) is used for laboratory equipment, piping, ovenware, and sealed-beam headlights.

Lead Glasses   Lead oxide is usually a modifier in the silica network but can also act as a network former. Lead glasses with high lead oxide contents are low melting and are useful for solder-sealing glasses. High-lead glasses are also used for shielding from high-energy radiation, and they find application for radiation windows, fluorescent lamp envelopes, and television bulbs. Because of their high refractive indexes, lead glasses are used for some optical glasses and for decorative-purpose glasses.

11.8.5 Viscous Deformation of Glasses

A glass behaves as a viscous (supercooled) liquid above its glass transition temperature. Under stress, groups of silicate atoms (ions) can slide past each other, allowing permanent deformation of the glass. Interatomic bonding forces resist deformation above the glass transition temperature but cannot prevent viscous flow of the glass if the applied stress is sufficiently high. As the temperature of the glass is progressively increased above its glass transition temperature, the viscosity of the glass decreases and viscous flow becomes easier. The effect of temperature on the viscosity of a glass follows an Arrhenius-type equation except that the sign of the exponential term is positive instead of negative as is usually the case (i.e., for diffusivity, the Arrhenius-type equation is $D = D_0 e^{(-Q/RT)}$). The equation relating viscosity to temperature for viscous flow in a glass is

$$\eta^* = \eta_0 e^{+Q/RT} \tag{11.2}$$

where η = viscosity of the glass, P or Pa · s;[6] η_0 = preexponential constant, P or Pa · s; Q = molar activation energy for viscous flow; R = universal molar gas

---

*η = Greek letter eta, pronounced "eight-ah."

[6]1 P (poise) = 1 dyne · s/cm^2; 1 Pa · s (pascal-second) = 1 N · s/m^2; 1 P = 0.1 Pa · s.

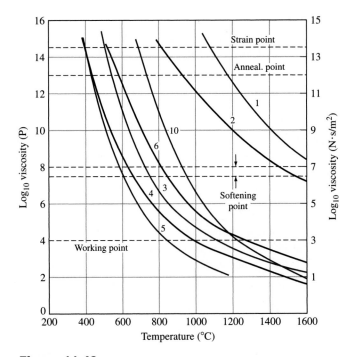

Figure 11.48
Effect of temperature on the viscosities of various types of glasses. Numbers on curves refer to compositions in Table 10.11.
(After O. H. Wyatt and D. Dew-Hughes, "Metals, Ceramics, and Polymers," Cambridge, 1974, p. 259.)

constant; and T = absolute temperature. Example Problem 11.10 shows how a value for the activation energy for viscous flow in a glass can be determined from this equation using viscosity-temperature data.

The effect of temperature on the viscosity of some commercial types of glasses is shown in Fig. 11.48. For the comparison of glasses, several viscosity reference points are used, indicated by horizontal lines in Fig. 11.48. These are working, softening, annealing, and strain points. Their definitions are:

1. Working point: viscosity = 10^4 poises (10^3 Pa · s). At this temperature glass fabrication operations can be carried out.

2. Softening point: viscosity = 10^8 poises (10^7 Pa · s). At this temperature the glass will flow at an appreciable rate under its own weight. However, this point cannot be defined by a precise viscosity because it depends on the density and surface tension of the glass.

3. Annealing point: viscosity = 10^{13} poises (10^{12} Pa · s). Internal stresses can be relieved at this temperature.

4. Strain point: viscosity = $10^{14.5}$ poises ($10^{13.5}$ Pa · s). Below this temperature the glass is rigid and stress relaxation occurs only at a slow rate. The interval between the annealing and strain points is commonly considered the annealing range of a glass.

Glasses are usually melted at a temperature that corresponds to a viscosity of about 10^2 poises (10 Pa · s). During forming, the viscosities of glasses are compared qualitatively. A *hard glass* has a high softening point, whereas a *soft glass* has a lower softening point. A *long glass* has a large temperature difference between its softening and strain points. That is, the long glass solidifies slower than a *short glass* as the temperature is decreased.

EXAMPLE PROBLEM 11.10

A 96 percent silica glass has a viscosity of 10^{13} P at its annealing point of 940°C and a viscosity of 10^8 P at its softening point of 1470°C. Calculate the activation energy in kilojoules per mole for the viscous flow of this glass in this temperature range.

■ **Solution**

$$\text{Annealing point of glass} = T_{ap} = 940°C + 273 = 1213 \text{ K} \qquad \eta_{ap} = 10^{13} \text{ P}$$

$$\text{Softening point of glass} = T_{sp} = 1470°C + 273 = 1743 \text{ K} \qquad \eta_{sp} = 10^8 \text{ P}$$

$$R = \text{gas constant} = 8.314 \text{ J/(mol · K)} \qquad Q = ? \text{ J/mol}$$

Using Eq. 10.5, $\eta = \eta_0 e^{Q/RT}$,

$$\begin{matrix} \eta_{ap} = \eta_0 e^{Q/RT_{ap}} \\ \eta_{sp} = \eta_0 e^{Q/RT_{sp}} \end{matrix} \quad \text{or} \quad \frac{\eta_{ap}}{\eta_{sp}} = \exp\left[\frac{Q}{R}\left(\frac{1}{T_{ap}} - \frac{1}{T_{sp}}\right)\right] = \frac{10^{13} \text{ P}}{10^8 \text{ P}} = 10^5$$

$$10^5 = \exp\left[\frac{Q}{8.314}\left(\frac{1}{1213 \text{ K}} - \frac{1}{1743 \text{ K}}\right)\right]$$

$$\ln 10^5 = \frac{Q}{8.314}(8.244 \times 10^{-4} - 5.737 \times 10^{-4}) = \frac{Q}{8.314}(2.507 \times 10^{-4})$$

$$11.51 = Q(3.01 \times 10^{-5})$$

$$Q = 3.82 \times 10^5 \text{ J/mol} = 382 \text{ kJ/mol} \blacktriangleleft$$

11.8.6 Forming Methods for Glasses

Glass products are made by first heating the glass to a high temperature to produce a viscous liquid and then molding, drawing, or rolling it into a desired shape.

Forming Sheet and Plate Glass   About 85 percent of the flat glass produced in the United States is made by the **float-glass** process in which a ribbon of glass moves out of the melting furnace and floats on the surface of a bath of molten tin (Fig. 11.49). The glass ribbon is cooled while moving across the molten tin and while under a chemically controlled atmosphere (Fig. 11.49). When its surfaces are

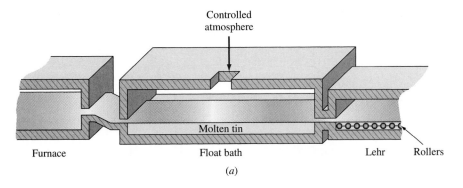

Controlled
atmosphere

Molten tin

Furnace   Float bath   Lehr   Rollers

(a)

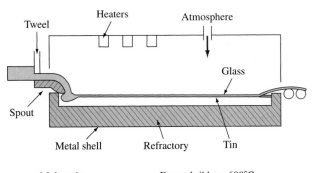

Heaters   Atmosphere

Tweel

Spout

Glass

Metal shell   Refractory   Tin

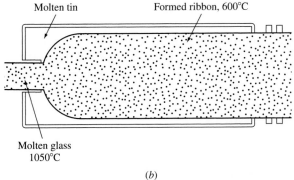

Molten tin   Formed ribbon, 600°C

Molten glass
1050°C

(b)

Figure 11.49
(a) Diagram of the float-glass process.
(After D. C. Boyd and D. A. Thompson, "Glass," vol. II: "Kirk-Othmer Encyclopedia of Chemical Technology," 3rd ed., Wiley, 1980, p. 862.)
(b) Side and top schematic views of the float-glass process.

sufficiently hard, the glass sheet is removed from the furnace without being marked by rollers and passed through a long annealing furnace called a *lehr*, where residual stresses are removed.

Blowing, Pressing, and Casting of Glass Deep items such as bottles, jars, and light bulb envelopes are usually formed by blowing air to force molten glass into

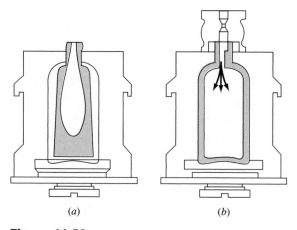

Figure 11.50

(*a*) Reheat and (*b*) final blow stages of a glass blowing machine process.

(*After W. Giegerich and W. Trier, "Glass Machines Construction and Operation of Machines for the Forming of Hot Glass," Springer-Verlag, 1969.*)

molds (Fig. 11.50). Flat items such as optical and sealed-beam lenses are made by pressing a plunger into a mold containing molten glass.

Many articles can be made by casting the glass into an open mold. A large borosilicate glass telescope mirror 6 m in diameter was made by casting the glass. Funnel-shaped items such as television tubes are formed by centrifugal casting. Gobs of molten glass from a feeder are dropped onto a spinning mold that causes the glass to flow upward to form a glass wall of approximately uniform thickness.

11.8.7 Tempered Glass

This type of glass is strengthened by rapid air cooling of the surface of the glass after it has been heated to near its softening point. The surface of the glass cools first and contracts, while the interior is warm and readjusts to the dimensional change with little stress (Fig. 11.51*a*). When the interior cools and contracts, the surfaces are rigid, and so tensile stresses are created in the interior of the glass and compressive stresses on the surfaces (Figs. 11.51*b* and 11.52). This "tempering" treatment increases the strength of the glass because applied tensile stresses must surpass the compressive stresses on the surface before fracture occurs. Tempered glass has a higher resistance to impact than annealed glass and is about four times stronger than annealed glass. Auto side windows and safety glass for doors are items that are thermally tempered.

11.8.8 Chemically Strengthened Glass

The strength of glass can be increased by special chemical treatments. For example, if a sodium aluminosilicate glass is immersed in a bath of potassium nitrate at a

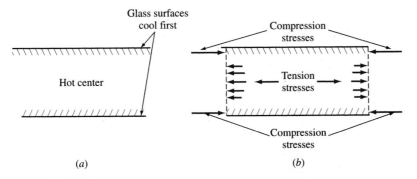

Figure 11.51
Cross section of tempered glass (*a*) after surface has cooled from high temperature near glass-softening temperature and (*b*) after center has cooled.

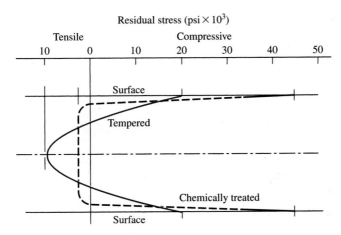

Figure 11.52
Distribution of residual stresses across the sections of glass **thermally tempered** and **chemically strengthened**.
(*After E. B. Shand, "Engineering Glass," vol. 6: "Modern Materials," Academic, 1968, p. 270.*)

temperature about 50°C below its stress point (~500°C) for 6 to 10 h, the smaller sodium ions near the surface of the glass are replaced by larger potassium ions. The introduction of the larger potassium ions into the surface of the glass produces compressive stresses at the surface and corresponding tensile stresses at its center. This chemical tempering process can be used on thinner cross sections than can thermal tempering since the compressive layer is much thinner, as shown in Fig. 11.52. Chemically strengthened glass is used for supersonic aircraft glazing and ophthalmic lenses.

11.9 CERAMIC COATINGS AND SURFACE ENGINEERING

The surface of a component is susceptible to mechanical (friction and wear), chemical (corrosion), electrical (conductivity or insulation), optical (reflectivity), and thermal (high-temperature damage) interactions. As a result, any designer in any engineering field must consider the quality and protection of the surface of any component as a major design criterion, i.e., surface engineering. One possible method of protecting the surface of a component is through the application of coatings. Coating materials may be metallic such as in the chromium plating of an automobile trim, polymeric such as paints resisting corrosion, and also ceramic. Various ceramic materials are used as coatings for applications in which high-temperature environments exist or premature wear is of concern. Ceramic coatings offer physical characteristics to the substrate material that it does not inherently possess. These coatings can transform the surface of a substrate to a chemically inert, abrasion resistant, low friction, and easy-to-clean surface over a range of temperatures. They can also offer electrical resistance and prevent hydrogen diffusion (a major source of damage in many metals). Examples of ceramic coating materials include glasses, oxides, carbides, silicides, borides, and nitrides.

11.9.1 Silicate Glasses

Silicate glass coatings have extensive industrial applications. A glass coating applied to (1) a ceramic substrate is called a **glaze,** (2) a metal surface is called a **porcelain enamel,** and (3) a glass substrate is called **glass enamel.** These coatings are used mostly for aesthetic reasons, but they also provide protections against environmental elements, largely by decreasing their permeability. Specific applications include engine exhaust ducts, space heaters, and radiators. These coatings are applied using spraying or dipping techniques. The surface of a component to be glazed or enameled must be clean (free of particles and oil) and sharp corners must be rounded to assure proper coating adhesion (i.e., avoid peeling).

11.9.2 Oxides and Carbides

Oxide coatings provide protection against oxidation and damage at elevated temperatures while carbide coatings (because of their hardness) are used in applications in which wear and seal are important. For example, zirconia (ZrO_2) is applied as a coating to an engine's moving parts. The zirconia protects the metallic substrate (Al or Fe alloys) against high temperature damage. The coating is generally applied using flame or thermal spraying technique. In this technique the ceramic particles (oxide or carbide) are heated and propelled onto the substrate surface. Figure 11.53 shows an example of such a process in which a paper-rolling mill is coated with tungsten carbide–cobalt chromium. The microstructure of the coating, which has a thickness of approximately 100 microns, is shown in Fig. 11.54. The coating protection of the roller in paper-processing industries is important because of the highly acidic or basic nature of the pulp. Ceramics are the only materials that offer resistance to abrasion

Figure 11.53
Thermal spraying of tungsten carbide–cobalt chromium coating
(WC/10Co4Cr) onto a roll for the paper manufacturing industry.
(*Gordon England www.gordonengland.co.uk/*)

Figure 11.54
The microstructure of the tungsten carbide–cobalt
chromium coating on the substrate.
[*Courtesy of TWI Ltd. (The Welding Institute, Granta Park,
Great Abington, Cambridge, UK)*]

and corrosion in such harsh environments. However, any cracks in the brittle coating layer that reaches the substrate will grow from those points, eventually resulting in the failure of the component.

11.10 CERAMICS IN BIOMEDICAL APPLICATIONS

In previous chapters the application of metallic and polymeric materials to various medical devices and instruments were discussed. Ceramics are also used extensively in the biomedical fields, including orthopedic implants, eyeglasses, laboratory ware, thermometers, and most importantly dental applications. The factors that make ceramic biomaterials an excellent candidate for biomedical applications are biocompatibility, corrosion resistance, high stiffness, resistance to wear in applications where surface articulation exists (denture materials and hip and knee implants), and low friction. In addition, the main advantage of some ceramic biomaterials is that they bond well to bone (implant-tissue attachment), which is important in many orthopedic and dental applications. Consider the situation in which a femoral stem (in a hip replacement operation) or a tibial component (in a knee replacement operation) is in direct contact with the cancellous bone (bone material just inside the surface); the problem of tissue-implant attachment is clearly an important issue in maintaining the stability of the joint (i.e., no loosening of the prostheses). However, implant loosening occurs frequently, is painful, and in many cases requires expensive secondary operations to correct the problem. This is a significant burden from both a healthcare cost viewpoint and the patient's quality of life. In the following paragraphs, the application of ceramics to various areas in the biomedical field will be discussed, including implant-tissue attachment.

11.10.1 Alumina in Orthopedic Implants

High purity alumina has excellent corrosion resistance, high wear resistance, high strength, and is biocompatible. Because of these characteristics, it is increasingly being used as the hip replacement material of choice. In total hip replacement surgery the diseased or damaged femur head and the cup on which it articulates, the *acetabular* (AC) cup, are being replaced with artificial prostheses. Figure 11.55a shows the hip of a patient that is damaged due to advanced arthritis, which manifests itself as an abnormal femur head shape and a deformed cup, both of which are replaced by artificial prostheses as shown in Fig. 11.55b. The artificial AC cup consists of a metallic base and a cup insert on which femoral articulation takes place (Fig. 11.56). The metallic base containing the cup insert is fixed to the pelvic bone using screws. As discussed in the opener of Chap. 10, the femoral head component is generally made of cobalt-chromium alloys and the AC cup is made of ultrahigh molecular weight polyethylene (metal on polymer). This combination of materials unfortunately results in wear of the polyethylene surface and ultimately in the loosening of the prostheses (see Chap. 10 opener). To avoid formation of wear particles, and therefore loosening of the prostheses, manufacturers are using alumina for both

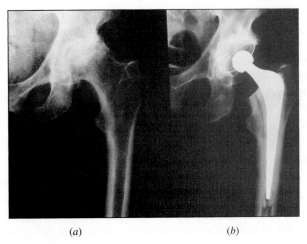

(*a*)   (*b*)

Figure 11.55
(*a*) A hip showing extensive arthritis damage. (*b*) The
same hip after *total hip replacement* (THR).
(© *Princess Margaret Rose Orthopaedic Hospital/Photo
Researchers, Inc.*)

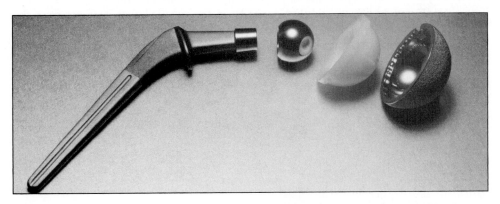

Figure 11.56
Various components of total hip prostheses including the stem with an alumina femoral
head, an alumina AC cup, and a metal base for the AC cup.
(*PhotosDisc/Getty Images*)

the femoral head and the AC cup as shown in Fig. 11.56 (ceramic on ceramic)
because of alumina's high wear resistance and surface hardness. The outstanding
properties of alumina are dependent on its grain size and purity. For orthopedic
implant applications the purity must be greater than 99.8 percent and the grain size
must be in the 3 to 6 μm range. Additionally, the articulating surfaces (ball on cup)
must have a high degree of symmetry and tight tolerances, which is accomplished

by grinding and polishing the paired head and cup components on each other. The friction coefficient of a ceramic-on-ceramic hip can approach that of a normal hip, and as a result the wear debris generation in such hips is ten times lower than in metal-on-polymer combinations. The negative side effect of the ceramic-on-ceramic hips is the stress shielding (see Sec. 10.12) due to the high modulus of elasticity of the ceramic. The **stress shielding** could result in bone mass loss and loosening in older patients. Thus, for older patients, the metal-on-polymer hip replacements may be a better choice because of their reduced stress shielding.

11.10.2 Alumina in Dental Implants

A dental implant serves as an artificial root surgically anchored to the jaw bone. The artificial root can then support a replacement tooth or a crown as shown in Fig. 11.57. Although titanium has been the material of choice for dental implants because of its biocompatibility and low modulus of elasticity, alumina is being utilized more frequently for this application. The crown is generally made from porcelain, which is also a ceramic material (see Sec. 11.5), although it could also be made of metals such as silver and gold.

11.10.3 Ceramic Implants and Tissue Connectivity

In those operations in which the implant is in direct contact with bone as shown in Figs. 11.55*b* and 11.57, its stability depends on the reaction that it elicits from the surrounding tissue. In general, four types of responses may be observed from the tissue surrounding an implant: a *toxic* response in which the tissue surrounding the implant dies; a *biologically inactive* response in which a thin fibrous tissue forms

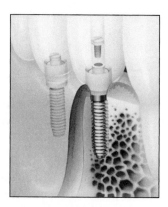

Figure 11.57
The dental implant
components.
(© *Custom Medical Stock Photo*)

around the implant; a *bioactive* response in which an interfacial bond between the bone and the prosthesis forms; and a *resorption* (dissolving) response in which the surrounding tissue replaces the implant material or portions of it. In this respect ceramic implants (for both orthopedic and dental applications) may be rated as nearly inert (type I), porous (type II), bioactive (type III), or resorbable (type IV). Alumina is rated as a type I bioceramic because of its nearly inert characteristics. Thus, alumina implants elicit the formation of a thin fibrous tissue that is acceptable in situations where the implant is fitted tightly and is under compression such as dental implants. However, in situations where the implant-tissue interface is loaded such that interfacial movement can occur as in orthopedic implants, the fibrous region grows thick and the implant becomes loose. Under these conditions type II bioceramics such as porous alumina and calcium phosphates can serve as a scaffold or a bridge for bone formation. The bone material grows into the pores available in the ceramic (osteoconductivity) and provides some load-bearing support. In these materials the pore size should be greater than 100 μm to facilitate vascular tissue growth in the pores, thus allowing blood supply to the newly formed cells. Microporous bioceramics are used specifically in situations where load bearing is not the main requirement because of their reduced strength due to porosity. Type III bioceramics or bioactive ceramics are those that promote and facilitate the formation of a bond between the implant material and the surrounding tissue. These materials develop an adherent interface that is very strong and can support a load. Glasses such as those containing SiO_2, Na_2O, CaO, and P_2O_5 were the first materials to show bioactivity. Such glasses differ from conventional soda lime glasses in their compositional ratios: less than 60 mol % silica, high Na_2O and CaO content, and high CaO to P_2O_5 ratio. These specific compositions allow high reactivity of the implant surface and therefore bonding to the bone in an aqueous medium. Finally, type IV bioceramics or resorbable ceramics are those that degrade over a period of time and are replaced by the bone material. Tricalcium phosphate, $Ca_3(PO_4)_2$, is an example of a resorbable ceramic. The challenges when using these materials are (1) to assure that the implant-bone interface remains strong and stable during the degradation-repair period and (2) to match the resorption rate with the repair rate. There is still a great deal of research and development needed to apply these materials with optimum performance; however, it is clear that ceramic materials are a strong candidate in cases where implant and human tissue are in direct contact.

11.11 NANOTECHNOLOGY AND CERAMICS

Considering the range and variety of the applications for ceramic materials, one realizes that their potential can never be fully realized because of one major disadvantage, that is, their brittle nature and therefore low toughness. Nanocrystalline ceramics may improve on this inherent weakness of these materials. The following paragraphs are intended to describe the state-of-the-art in the production of bulk nanocrystalline ceramics.

Bulk nanocrystalline ceramics are produced using the standard powder metallurgy techniques described in Sec. 11.4. The difference is that the starting powder is in the nanosize range, being smaller than 100 nm. However, the nanocrystalline ceramic powders have a tendency to chemically or physically bond together to form bigger particles called *agglomerates* or *aggregates*. Agglomerated powders, even if their size is in the nanosize or near nanosize range, do not pack as well as non-agglomerated powders. In a nonagglomerated powder after compaction, the available pore sizes are between 20 to 50 percent of the size of the nanocrystal. Because of this small pore size, the sintering stage and densification proceed quickly and at lower temperatures. For instance, in the case of nonagglomerated TiO_2 (powder size less than 40 nm), the compact densifies to almost 98 percent of theoretical density at approximately 700°C with a sintering time of 120 min. Conversely for an agglomerated powder of average size 80 nm consisting of 10 to 20 nm crystallites, the compact densifies to 98 percent of theoretical density at about 900°C with a sintering time of 30 min. The major reason for the difference in sintering temperature is the existence of larger pores in the agglomerated compact. Because higher sintering temperatures are required, the compacted nanocrystallites eventually grow to a micro-crystalline range, which is undesirable. The grain growth is drastically influenced by sintering temperature and only modestly by sintering time. Thus, the main issue in successfully producing bulk nanocrystalline ceramics is to start with nonagglomer-ated powder and to optimize the sintering process. However, this is very difficult to achieve.

To remedy the difficulty in producing nanocrystalline bulk ceramics, engineers are utilizing pressure-assisted sintering, a sintering process with externally applied pressure that is similar to *hot isostaic pressing* (HIP), hot extrusion, and sinter forg-ing (see Sec. 11.4). In these processes, the ceramic compact is deformed and densi-fied simultaneously. The primary advantage of sinter forging in producing nanocrys-talline ceramics is in its mechanism of pore shrinkage. As discussed in Sec. 11.4.3, in conventional microcrystalline ceramics the pore shrinkage process is based on atomic diffusion. Under sinter forging, pore shrinkage of the nanocrystalline compact is nondiffusional and is based on the plastic deformation of the crystals. Nanocrys-talline ceramics are more ductile at elevated temperatures (around 50 percent of melt) than their microcrystalline counterparts. It is believed that nanocrystalline ceramics are more ductile as a result of superplastic deformation. As discussed in previous chapters, superplasticity occurs because of sliding and rotation of the grains under high load and temperature. Because they are able to deform plastically, the pores are squeezed shut by plastic flow as shown in Fig. 11.58 rather than by diffusion.

Because of this ability to close large pores, even agglomerated powders can be densified to near their theoretical values. Additionally, the application of pressure pre-vents the grains from growing to above the nanoscale region. For instance, sinter forg-ing of agglomerated TiO_2 for 6h at a pressure of 60 MPa and a temperature of 610°C produces a true strain of 0.27 (extremely high for ceramics), density of 91 percent of theoretical value, and an average grain size of 87 nm. The same powder when sin-tered without pressure requires a sintering temperature of 800°C to achieve the same density while producing an average grain size of 380 nm (not nanocrystalline). It is important to note that superplastic deformation in nanocrystalline ceramics occurs in a limited range of pressures and temperatures and one must be aware of the range.

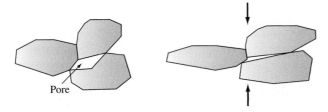

Figure 11.58
A schematic showing pore shrinkage through plastic flow (grain boundary sliding) in nanocrystalline ceramics.

If the treatment is outside this range, the diffusional mechanism of pore shrinkage may take over, which results in a microcrystalline product with low density.

In conclusion, advances in nanotechnology will potentially lead to the production of nanocrystalline ceramics with exceptional levels of strength and ductility, and therefore improved toughness. Specifically, the improvements in ductility allow for better bonding of ceramics to metals in coating technologies, and the increased toughness allows for better resistance to wear. Such advances could revolutionize the use of ceramics in a wide variety of applications.

11.12 SUMMARY

Ceramic materials are inorganic, nonmetallic materials consisting of metallic and nonmetallic elements bonded together primarily by ionic and/or covalent bonds. As a result, the chemical compositions and structures of ceramic materials vary considerably. They may consist of a single compound such as, for example, pure aluminum oxide, or they may be composed of a mixture of many complex phases such as the mixture of clay, silica, and feldspar in electrical porcelain.

The properties of ceramic materials also vary greatly due to differences in bonding. In general, most ceramic materials are typically hard and brittle with low impact resistance and ductility. Consequently, in most engineering designs, high stresses in ceramic materials are usually avoided, especially if they are tensile stresses. Ceramic materials are usually good electrical and thermal insulators due to the absence of conduction electrons, and thus many ceramics are used for electrical insulation and refractories. Some ceramic materials can be highly polarized with electric charge and are used for dielectric materials for capacitors. Permanent polarization of some ceramic materials produces piezoelectric properties that permit these materials to be used as electromechanical transducers. Other ceramic materials, for example, Fe_3O_4, are semiconductors and find application for thermistors for temperature measurement. Graphite, diamond, buckyballs, & buckytubes are all allotropes of carbon and are disscussed in this chapter because Graphite is sometimes considered a ceramic material. These allotropes have significantly different properties that are directly related to differences in atomic structure and positioning. Buckyballs & buckytubes are becoming more important in nanotechnology applications.

The processing of ceramic materials usually involves the agglomeration of small particles by a variety of methods in the dry, plastic, or liquid states. Cold-forming processes predominate in the ceramics industry, but hot-forming processes are also used. Pressing, slip casting, and extrusion are commonly used ceramic-forming processes. After forming, ceramic materials are usually given a thermal treatment such as sintering or vitrification. During sintering, the small particles of a formed article are bonded together by solid-state diffusion at high temperatures. In vitrification, a glass phase serves as a reaction medium to bond the unmelted particles together.

Glasses are inorganic ceramic products of fusion that are cooled to a rigid solid without crystallization. Most inorganic glasses are based on a network of ionically covalently bonded silica (SiO_2) tetrahedra. Additions of other oxides such as Na_2O and CaO modify the silica network to provide a more workable glass. Other additions to glasses create a spectrum of properties. Glasses have special properties such as transparency, hardness at room temperature, and excellent resistance to most environments that make them important for many engineering designs. One of the important applications of ceramics is in coating of component surface to protect it from corrosion or wear. Glasses, oxides, and carbides are all used as coating materials for various applications. Ceramics are also being applied in the biomedical field as implant materials. Their chemical stability and biocompatibility is perfectly suitable for the harsh environment of the human body in joint replacement and other orthopedic applications. Nanotechnology research is promising to improve on the major drawback of ceramic materials: their brittleness. Early research shows that nanocrystalline ceramics possess higher ductility. This may allow for cheaper production of more complex ceramic parts.

11.13 DEFINITIONS

Sec. 11.1

Ceramic materials: inorganic, nonmetallic materials that consist of metallic and nonmetallic elements bonded together primarily by ionic and/or covalent bonds.

Sec. 11.2

Coordination number (CN): the number of equidistant nearest neighbors to an atom or ion in a unit cell of a crystal structure. For example, in NaCl, CN = 6 since six equidistant Cl^- anions surround a central Na^+ cation.

Radius ratio (for an ionic solid): the ratio of the radius of the central cation to that of the surrounding anions.

Critical (minimum) radius ratio: the ratio of the central cation to that of the surrounding anions when all the surrounding anions just touch each other and the central cation.

Octahedral interstitial site: the space enclosed when the nuclei of six surrounding atoms (ions) form an octahedron.

Tetrahedral interstitial site: the space enclosed when the nuclei of four surrounding atoms (ions) form a tetrahedron.

Graphite: a layered structure of carbon atoms covalently bonded to three others inside the layer. Various layers are the bonded through secondary bonds.

Buckyball: also called Buckminister Fullerene is soccer ball shaped molecule of carbon atoms (C_{60}).

Buckytube: a tubular structure made of carbon atoms covalently bonded together.

Sec. 11.4

Dry pressing: the simultaneous uniaxial compaction and shaping of ceramic granular particles (and binder) in a die.

Isostatic pressing: the simultaneous compaction and shaping of a ceramic powder (and binder) by pressure applied uniformly in all directions.

Slip casting: a ceramic shape-forming process in which a suspension of ceramic particles and water are poured into a porous mold and then some of the water from the cast material diffuses into the mold, leaving a solid shape in the mold. Sometimes excess liquid within the cast solid is poured from the mold, leaving a cast shell.

Sintering (of a ceramic material): the process in which fine particles of a ceramic material become chemically bonded together at a temperature high enough for atomic diffusion to occur between the particles.

Firing (of a ceramic material): heating a ceramic material to a high enough temperature to cause a chemical bond to form between the particles.

Vitrification: melting or formation of a glass; the vitrification process is used to produce a viscous liquid glass in a ceramic mixture upon firing. Upon cooling, the liquid phase solidifies and forms a vitreous or glassy matrix that bonds the unmelted particles of the ceramic material together.

Sec. 11.7

Refractory (ceramic) material: a material that can withstand the action of a hot environment.

Sec. 11.8

Glass: a ceramic material that is made from inorganic materials at high temperatures and is distinguished from other ceramics in that its constituents are heated to fusion and then cooled to the rigid condition without crystallization.

Glass transition temperature: the center of the temperature range in which a noncrystalline solid changes from being glass-brittle to being viscous.

Glass-forming oxide: an oxide that forms a glass easily; also an oxide that contributes to the network of silica glass when added to it, such as B_2O_3.

Network modifiers: an oxide that breaks up the silica network when added to silica glass; modifiers lower the viscosity of silica glass and promote crystallization. Examples are Na_2O, K_2O, CaO, and MgO.

Intermediate oxides: an oxide that may act either as a glass former or as a glass modifier, depending on the composition of the glass. Example, Al_2O_3.

Glass reference points (temperatures).

 Working point: at this temperature the glass can easily be worked.

 Softening point: at this temperature the glass flows at an appreciable rate.

 Annealing point: at this temperature stresses in the glass can be relieved.

 Strain point: at this temperature the glass is rigid.

Float glass: flat glass that is produced by having a ribbon of molten glass cool to the glass-brittle state while floating on the top of a flat bath of molten tin and under a reducing atmosphere.

Thermally tempered glass: glass that has been reheated to near its softening temperature and then rapidly cooled in air to introduce compressive stresses near its surface.

Chemically tempered glass: glass that has been given a chemical treatment to introduce large ions into its surface to cause compressive stresses at its surface.

Sec. 11.9
Glaze: a glass coating applied to a ceramic substrate.
Porcelain enamel: a glass coating applied to a metal substrate.
Glass enamel: a glass coating applied to a glass substrate.

Sec. 11.10
Stress shielding: a condition in which the applied implant is carrying a significant portion of load carried by a fractured bone. It has negative side effects.

11.14 PROBLEMS

Answers to problems marked with an asterisk are given at the end of the book.

11.1 Define a ceramic material.

11.2 What are some properties common to most ceramic materials?

11.3 Distinguish between traditional and engineering ceramic materials and give examples of each.

11.4 Using Pauling's equation (Eq. 2.10), compare the percent covalent character of the following compounds: hafnium carbide, titanium carbide, tantalum carbide, boron carbide, and silicon carbide.

11.5 What two main factors affect the packing of ions in ionic solids?

11.6 Define (*a*) coordination number and (*b*) critical radius ratio for the packing of ions in ionic solids.

*11.7 Using Fig. 11.59, calculate the critical radius ratio for octahedral coordination.

11.8 Predict the coordination number for (*a*) BaO and (*b*) LiF. Ionic radii are $Ba^{2+} = 0.143$ nm, $O^{2-} = 0.132$ nm, $Li^+ = 0.078$ nm, $F^- = 0.133$ nm.

11.9 Calculate the density in grams per cubic centimeter of CsI, which has the CsCl structure. Ionic radii are $Cs^+ = 0.165$ nm and $I^- = 0.220$ nm.

*11.10 Calculate the density in grams per cubic centimeter of CsBr, which has the CsCl structure. Ionic radii are $Cs^+ = 0.165$ nm and $Br^- = 0.196$ nm.

11.11 Calculate the linear densities in ions per nanometer in the [110] and [111] directions for (*a*) NiO and (*b*) CdO. Ionic radii are $Ni^{2+} = 0.078$ nm, $Cd^{2+} = 0.103$ nm, and $O^{2-} = 0.132$ nm.

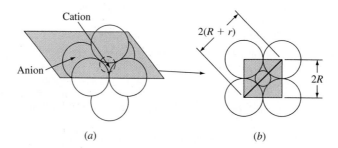

(*a*) (*b*)

MatVis

Figure 11.59
(*a*) Octahedral coordination of six anions (radii = *R*) around a central cation of radius *r*. (*b*) Horizontal section through center of (*a*).

11.12 Calculate the planar densities in ions per square nanometer on the (111) and (110) planes for (*a*) CoO and (*b*) LiCl. Ionic radii are $Co^{2+} = 0.082$ nm, $O^{2-} = 0.132$ nm, $Li^+ = 0.078$ nm, and $Cl^- = 0.181$ nm.

*11.13 Calculate the density in grams per cubic centimeter of (*a*) SrO and (*b*) VO. Ionic radii are $V^{2+} = 0.065$ nm, $Co^{2+} = 0.082$ nm, and $O^{2-} = 0.132$ nm.

*11.14 Calculate the ionic packing factor for (*a*) MnO and (*b*) SrO. Ionic radii are $Mn^{2+} = 0.091$ nm, $Sr^{2+} = 0.127$ nm, and $O^{2-} = 0.132$ nm.

11.15 ZnTe has the zinc blende crystal structure. Calculate the density of ZnTe. Ionic radii are $Zn^{2+} = 0.083$ nm and $Te^{2-} = 0.211$ nm.

*11.16 BeO has the zinc blende crystal structure. Calculate the density of BeO. Ionic radii are $Be^{2+} = 0.034$ nm and $O^{2-} = 0.132$ nm.

11.17 Draw the unit cell for BaF_2, which has the fluorite (CaF_2) crystal structure. If the Ba^{2+} ions occupy the FCC lattice sites, which sites do the F^- ions occupy?

*11.18 Calculate the density in grams per cubic centimeter of ZrO_2, which has the CaF_2 crystal structure. Ionic radii are $Zr^{4+} = 0.087$ nm and $O^{2-} = 0.132$ nm.

11.19 What fraction of the octahedral interstitial sites are occupied in the CaF_2 structure?

11.20 Calculate the linear density in ions per nanometer in the [111] and [110] directions for CeO_2, which has the fluorite structure. Ionic radii are $Ce^{4+} = 0.102$ nm and $O^{2-} = 0.132$ nm.

*11.21 Calculate the planar density in ions per square nanometer in the (111) and (110) planes of ThO_2, which has the fluorite structure. Ionic radii are $Th^{4+} = 0.110$ nm and $O^{2-} = 0.132$ nm.

MatVis

11.22 Calculate the ionic packing factor for SrF_2, which has the fluorite structure. Ionic radii are $Sr^{2+} = 0.127$ nm and $F^- = 0.133$ nm.

11.23 What is the antifluorite structure? What ionic compounds have this structure? What fraction of the tetrahedral interstitial sites are occupied by cations?

11.24 Why are only two-thirds of the octahedral interstitial sites filled by Al^{3+} ions when the oxygen ions occupy the HCP lattice sites in Al_2O_3?

11.25 Describe the perovskite structure. What fraction of the octahedral interstitial sites are occupied by the tetravalent cation?

11.26 Calculate the ionic packing factor for $CaTiO_3$, which has the perovskite structure. Ionic radii are $Ca^{2+} = 0.106$ nm, $Ti^{4+} = 0.064$ nm, and $O^{2-} = 0.132$ nm. Assume the lattice constant $a = 2(r_{Ti^{4+}} + r_{O^{2-}})$.

*11.27 Calculate the density in grams per cubic centimeter of $SrSnO_3$, which has the perovskite structure. Ionic radii are $Sr^{2+} = 0.127$ nm, $Sn^{4+} = 0.074$ nm, and $O^{2-} = 0.132$ nm. Assume $a = 2(r_{Sn^{4+}} + r_{O^{2-}})$.

11.28 What is the spinel crystal structure?

11.29 Draw a section of the graphite structure. Why are the layers of graphite able to slide past each other easily?

11.30 Describe and illustrate the following silicate structures: (*a*) island, (*b*) chain, and (*c*) sheet.

11.31 Describe the structure of a sheet of kaolinite.

11.32 Describe the bonding arrangement in the cristobalite (silica) network structure.

11.33 Describe the feldspar network structure.

11.34 What are the basic steps in the processing of ceramic products by the agglomeration of particles?

11.35 What types of ingredients are added to ceramic particles in preparing ceramic raw materials for processing?

11.36 Describe two methods for preparing ceramic raw materials for processing.

11.37 Describe the dry-pressing method for producing such ceramic products as technical ceramic compounds and structural refractories. What are the advantages of dry-pressing ceramic materials?

11.38 Describe the isostatic-pressing method for producing ceramic products.

11.39 Describe the four stages in the manufacture of a spark of plug insulator.

11.40 What are the advantages of hot-pressing ceramics materials?

11.41 Describe the steps in the slip-casting process for ceramic products.

11.42 What is the difference between (*a*) drain and (*b*) solid slip casting?

11.43 What are the advantages of slip casting?

11.44 What types of ceramic products are produced by extrusion? What are the advantages of this process? Limitations?

11.45 What are the purposes of drying ceramic products before firing?

11.46 What is the sintering process? What occurs to the ceramic particles during sintering?

11.47 What is the vitrification process? In what type of ceramic materials does vitrification take place?

11.48 What are the three basic components of traditional ceramics?

11.49 What is the approximate composition of kaolin clay?

11.50 What is the role of clay in traditional ceramics?

11.51 What is flint? What role does it have in traditional ceramics?

11.52 What is feldspar? What role does it have in traditional ceramics?

11.53 List some examples of whiteware ceramic products.

11.54 Why is the term *triaxial* used to describe some whitewares?

11.55 Determine the composition of the ternary compound at point *y* in Fig. 11.34.

11.56 Why are triaxial porcelains not satisfactory for use at high frequencies?

11.57 What kinds of ions cause an increase in the conductivity of electrical porcelain?

11.58 What is the composition of most technical ceramics?

11.59 How are pure single-compound technical ceramic particles processed to produce a solid product? Give an example.

11.60 What causes the lack of plasticity in crystalline ceramics?

11.61 Explain the plastic deformation mechanism for some single-crystal ionic solids such as NaCl and MgO. What is the preferred slip system?

11.62 What structural defects are the main cause of failure of polycrystalline ceramic materials?

11.63 How do (*a*) porosity and (*b*) grain size affect the tensile strength of ceramic materials?

11.64 A reaction-bonded silicon nitride ceramic has a strength of 250 MPa and a fracture toughness of 3.4 MPa $- \sqrt{m}$. What is the largest-sized internal flaw that this material can support without fracturing? (Use $Y = 1$ in the fracture-toughness equation.)

11.65 The maximum-sized internal flaw in a hot-pressed silicon carbide ceramic is 25 μm. If this material has a fracture toughness of 3.7 MPa $-$ $\sqrt{m}$, what is the maximum stress that this material can support? (Use $Y = \sqrt{\pi}$.)

11.66 A partially stabilized zirconia advanced ceramic has a strength of 352 MPa and a fracture toughness of 7.5 MPa $-$ $\sqrt{m}$. What is the largest-sized internal flaw (expressed in micrometers) that this material can support? (Use $Y = \sqrt{\pi}$.)

11.67 A fully stabilized, cubic polycrystalline ZrO_2 sample has a fracture toughness of $K_{IC} = 3.8$ MPa $-$ $\sqrt{m}$ when tested on a four-point bend test.

    a.   If the sample fails at a stress of 450 MPa, what is the size of the largest surface flaw? Assume $Y = \sqrt{\pi}$.

    b.   The same test is performed with a partially stabilized ZrO_2 specimen. This material is transformation-toughened and has a $K_{IC} = 12.5$ MPa $-$ $\sqrt{m}$. If this material has the same flaw distribution as the fully stabilized sample, what stress must be applied to cause failure?

11.68 What are the two most important industrial abrasives?

11.69 What are important properties for industrial abrasives?

11.70 Why do most ceramic materials have low thermal conductivities?

11.71 What are refractories? What are some of their applications?

11.72 What are the two main types of ceramic refractory materials?

11.73 Give the composition and several applications for the following refractories: (*a*) silica, (*b*) fireclay, and (*c*) high-alumina.

11.74 What do most basic refractories consist of? What are some important properties of basic refractories? What is a main application for these materials?

11.75 What is the high-temperature reusable-surface insulation that can withstand temperatures as high as 1260°C made of?

11.76 Define a glass.

11.77 What are some of the properties of glasses that make them indispensable for many engineering designs?

11.78 How is a glass distinguished from other ceramic materials?

11.79 How does the specific volume versus temperature plot for a glass differ from that for a crystalline material when these materials are cooled from the liquid state?

11.80 Define the glass transition temperature.

11.81 Name two glass-forming oxides. What are their fundamental subunits and their shape?

11.82 How does the silica network of a simple silica glass differ from crystalline (cristobalite) silica?

11.83 How is it possible for BO_3^{3-} triangles to be converted to BO_4^{4-} tetrahedra and still maintain neutrality in some borosilicate glasses?

11.84 What are glass network modifiers? How do they affect the silica-glass network? Why are they added to silica glass?

11.85 What are glass intermediate oxides? How do they affect the silica-glass network? Why are they added to silica glass?

11.86 What is fused silica glass? What are some of its advantages and disadvantages?

11.87 What is the basic composition of soda-lime glass? What are some of its advantages and disadvantages? What are some applications for soda-lime glass?

11.88   What is the purpose of (*a*) MgO and (*b*) Al_2O_3 additions to soda-lime glass?

11.89   Define the following viscosity reference points for glasses: working, softening, annealing, and strain.

11.90   Distinguish between hard and soft glasses and long and short glasses.

11.91   A soda-lime plate glass between 500°C (strain point and 700°C (softening point) has viscosities between $10^{14.2}$ and $10^{7.5}$ P, respectively. Calculate a value for the activation energy in this temperature region.

11.92   A soda-lime glass has a viscosity of $10^{14.6}$ P at 560°C. What will be its viscosity at 675°C if the activation energy for viscous flow is 430 kJ/mol?

*11.93   A soda-lime glass has a viscosity of $10^{14.3}$ P at 570°C. At what temperature will its viscosity be $10^{9.9}$ P if the activation energy for the process is 430 kJ/mol?

11.94   A borosilicate glass between 600°C (annealing point) and 800°C (softening point) has viscosities of $10^{12.5}$ P and $10^{7.4}$ P, respectively. Calculate a value for the activation energy for viscous flow in this region, assuming the equation $\eta = \eta_0 e^{Q/RT}$ is valid.

11.95   Describe the float-glass process for the production of flat-glass products. What is its major advantage?

11.96   What is tempered glass? How is it produced? Why is tempered glass considerably stronger in tension than annealed glass? What are some applications for tempered glass?

11.97   What is chemically strengthened glass? Why is chemically strengthened glass stronger in tension than annealed glass?

11.15 MATERIALS SELECTION AND DESIGN PROBLEMS

1.   (*a*) Discuss the advantages and disadvantages of using advanced ceramics in the structure of internal combustion engines. (*b*) Propose some methods of overcoming the shortcomings of ceramics for this application.

2.   Investigate the application of ceramics in the electronics industry. (*a*) What are these applications? (*b*) Why are ceramics selected?

3.   Some ceramic parts are manufactured using a process called Sol-Gel processing. (*a*) Explain what is meant by Sol. (*b*) Explain what is meant by Gel. (*c*) Explain how this process works. (*d*) What is the advantage of this process over conventional powder metallurgy techniques?

4.   For insulation purposes, you would like to cover the surface of a substrate with an extremely thin layer of Si_3N_4. (*a*) Propose a process that could achieve this. (*b*) Can the proposed process be used to form large objects with complex shapes? Explain.

5.   (*a*) What are the elements used in the structure of the ceramic carborandum? (*b*) How is this ceramic manufactured in bulk form? (*c*) What are some of its important properties?

6.   (*a*) What are Sialons? (*b*) How are they manufactured? (*c*) What are their applications?

7.   Explain, from an atomic structure point of view, why metals can be plastically deformed to form large and complex shapes while complex ceramic parts cannot be manufactured in this way.

8. Alumina (Al_2O_3) and chromium oxide (Cr_2O_3) are ceramic materials that form an isomorphous phase diagram. (*a*) What does this tell you about the solubility limit of one component in the other? (*b*) What type of a solid solution is formed? (*c*) Explain what substitution takes place.

9. (*a*) How are the ceramic tiles used in the thermal protection system of the space shuttle attached to its frame? (*b*) Why is the thermal protection system of the space shuttle made from small tiles (15 to 20 cm width) and not larger, more contoured ones?

10. The nose cap and the wing leading edges of the space shuttle may reach temperatures of up to 1650°C. (*a*) Would the silica fiber compound (HRSI) be a suitable candidate for these sections of the shuttle? (*b*) If not, select a suitable material for this location. (*c*) What are the important properties of the selected material that satisfy the design needs?

11. Carbon/carbon composites have desirable properties at very high temperatures that make suitable materials for many aerospace applications. However, the carbon in the material can react with the oxygen in the atmosphere at temperatures above 450°C, and gaseous oxides form. (*a*) Develop solutions for this problem. (*b*) Design a process to accommodate your solution.

12. Low toughness is the major problem for many structural ceramics. Many cutting tools are made of ceramics with improved toughness. For example, tungsten carbide (WC) particles are embedded in a metal matrix such as nickel or cobalt. (*a*) Explain how this improves the toughness of the tool. (*b*) How is the choice of matrix material important? Would aluminum work as the matrix material?

13. It is very difficult to machine ceramic components to a desired shape. This is because ceramics are strong and brittle. The stresses produced due to cutting forces could create surface cracks and other damages that, in turn, would weaken the component. Propose a technique that will reduce the cutting forces and the possibility of cracking during the machining of ceramics.

14. Concrete is an important construction material and is classified as a ceramic (or a ceramic composite) material. It has excellent strength characteristics in compression but is extremely weak in tension. (*a*) Propose ways of improving the tension-bearing characteristics of concrete. (*b*) What problems do you anticipate in your process?

15. Conventional taps are prone to dripping because rubber washers are susceptible to wear and the metal seat (brass) is susceptible to pitting corrosion. (*a*) Which class of materials would be a suitable replacement for the rubber/metal combination that would reduce the dripping problem? (*b*) Select a specific material for this problem. (*c*) What problems would you anticipate in using or manufacturing these components?

16. A major problem in the selection of ceramic materials for various applications is the thermal shock resistance. (*a*) What is a thermal shock? (*b*) What factors control thermal shock resistance of a material? (*c*) Which specific ceramics have the best thermal shock resistance?

17. Consider the joining of a silicon chip to a metal lead frame in electronic devices. These ceramic/metal joints do not function very well as temperature is increased. (*a*) Can you explain why? (*b*) Can you propose a solution?

18. (*a*) In selecting the material for the windshields of automobiles, what type of glass would you use? (*b*) Propose a process that would shatterproof your selection.

Composite Materials

(© AFP/CORBIS)

Carbon-carbon composites have a combination of properties that renders them uniquely superior in operating temperatures as high as 2800°C. For example, unidirectional high modulus surface-treated carbon-carbon composites (55 vol % fiber) have a tensile modulus of 180 GPa at room temperature to 175 GPa at 2000°C. The tensile strength is also remarkably constant, varying from 950 MPa at room temperature to 1100 MPa at 2000°C. In addition, properties such as high thermal conductivity and low coefficient of thermal expansion, in conjunction with high strength and modulus, indicate a material that is thermal-shock resistant. The combination of these properties make this material suitable for re-entry applications, rocket motors, and aircraft brakes. The more commercial application of this material is in the brake pads of racing cars.[1] ■

---

[1]*ASM Engineered Materials Handbook, Composites,* Volume 1, ASM International, 1991.

By the end of this chapter, students will be able to . . .

1. Define a composite material, the major constituents, and various classifications.

2. Describe the role of the particulate (fiber) and the matrix (resin). Name various forms of each.

3. Define multidirectional laminated composite and its advantages over unidirectional laminated composites.

4. Describe how one would estimate the material properties of a fiber reinforced composite material based on material properties and volume fraction of the matrix and fiber constituents.

5. Describe various processes used to produce a component made of composite materials.

6. Describe the properties, characteristics, and classifications of concrete, asphalt, and wood, composite materials that are used extensively in structural and construction applications.

7. Define a sandwich structure.

8. Define PMC, MMC, and CMC composites and name advantages and disadvantages for each.

9. Be able to describe the structure of bone, its constituents, mechanical properties, and its viscoelastic behavior.

12.1 INTRODUCTION

What is a composite material? Unfortunately there is no widely accepted definition of what a composite material is. A dictionary defines a composite as something made up of distinct parts (or constituents). At the atomic level materials such as some metal alloys and polymeric materials could be called composite materials since they consist of different and distinct atomic groupings. At the microstructural level (about 10^{-4} to 10^{-2} cm) a metal alloy such as a plain-carbon steel containing ferrite and pearlite could be called a composite material since the ferrite and pearlite are distinctly visible constituents as observed in the optical microscope. At the macrostructural level (about 10^{-2} cm or greater) a glass-fiber-reinforced plastic, in which the glass fibers can be distinctly recognized by the naked eye, could be considered a composite material. Now we see that the difficulty in defining a composite material is the size limitations we impose on the constituents that make up the material. In engineering design a composite material usually refers to a material consisting of constituents in the micro- to macroscale range, and even favors the macros. For the purposes of this book, the following is a definition of a composite material:

A **composite material** is a materials system composed of a suitably arranged mixture or combination of two or more micro- or macroconstituents with an interface separating them that differ in form and chemical composition and are essentially insoluble in each other.

The engineering importance of a composite material is that two or more distinctly different materials combine to form a composite material that possesses

(*a*)

(*b*)

Figure 12.1

(*a*) A finished duct test article, with cured flanges.

(*After page 37 step 7 of High-Performance Composites Magazine, Jan. 2002.*)

(*b*) A carbon-fiber placement won a key role on the joint-strike fighter of Lockheed-Martin's X-35. A carbon-fiber placed inlet duct is manufactured four times faster than the conventional part with significantly fewer fasteners required.

(*From cover of High-Performance Composites Magazine, Jan. 2002.*)

properties that are superior, or important in some other manner, to the properties of the individual components. A multitude of materials fit into this category, and thus a discussion of all these composite materials is far beyond the scope of this book. In this chapter only some of the most important composite materials used in engineering will be discussed. These are fiber-reinforced plastics, concrete, asphalt, wood, and several miscellaneous types of composite materials. Examples of the use of composite materials in engineering designs are shown in Fig. 12.1.

12.2 FIBERS FOR REINFORCED-PLASTIC COMPOSITE MATERIALS

Three main types of synthetic fibers are used in the United States to reinforce plastic materials: glass, aramid,[2] and carbon. Glass is by far the most widely used reinforcement fiber and is the lowest in cost. Aramid and carbon fibers have high strengths and low densities and so are used in many applications, particularly aerospace, in spite of their higher cost.

12.2.1 Glass Fibers for Reinforcing Plastic Resins

Glass fibers are used to reinforce plastic matrices to form structural composites and molding compounds. Glass-fiber plastic composite materials have the following favorable characteristics: high strength-to-weight ratio; good dimensional stability; good resistance to heat, cold, moisture, and corrosion; good electrical insulation properties; ease of fabrication; and relatively low cost.

The two most important types of glass used to produce glass fibers for composites are *E* (*electrical*) and *S* (*high-strength*) *glasses.*

E glass is the most commonly used glass for continuous fibers. Basically, E glass is a lime-aluminum-borosilicate glass with zero or low sodium and potassium levels. The basic composition of E glass ranges from 52 to 56 percent SiO_2, 12 to 16 percent Al_2O_3, 16 to 25 percent CaO, and 8 to 13 percent B_2O_3. E glass has a tensile strength of about 500 ksi (3.44 GPa) in the virgin condition and a modulus of elasticity of 10.5 Msi (72.3 GPa).

S glass has a higher strength-to-weight ratio and is more expensive than E glass and is used primarily for military and aerospace applications. The tensile strength of S glass is over 650 ksi (4.48 GPa), and its modulus of elasticity is about 12.4 Msi (85.4 GPa). A typical composition for S glass is about 65 percent SiO_2, 25 percent Al_2O_3, and 10 percent MgO.

Production of Glass Fibers and Types of Fiberglass Reinforcing Materials
Glass fibers are produced by drawing monofilaments of glass from a furnace containing molten glass and gathering a large number of these filaments to form a strand of glass fibers (Fig. 12.2). The strands then are used to make glass-fiber yarns (Fig. 12.3a) or rovings that consist of a collection of bundles of continuous filaments

---

[2]Aramid fiber is an aromatic polyamide polymer fiber with a very rigid molecular structure.

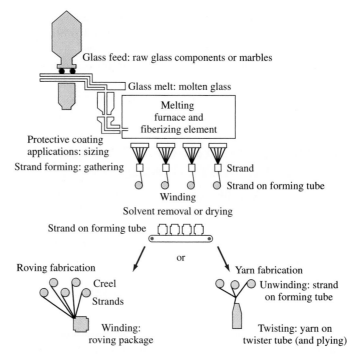

Glass feed: raw glass components or marbles

Glass melt: molten glass

Melting furnace and fiberizing element

Protective coating applications: sizing

Strand forming: gathering

Strand

Strand on forming tube

Winding

Solvent removal or drying

Strand on forming tube

or

Roving fabrication

Creel

Strands

Winding: roving package

Yarn fabrication

Unwinding: strand on forming tube

Twisting: yarn on twister tube (and plying)

Figure 12.2
Glass-fiber manufacturing process.
(After M. M. Schwartz, "Composite Materials Handbook," McGraw-Hill, 1984, pp. 2–24.)

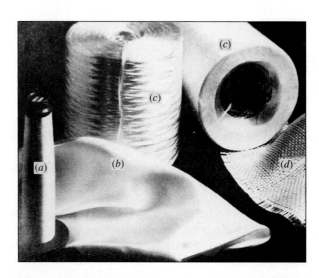

Figure 12.3
Fiberglass reinforcements for plastics: (*a*) fiberglass yarn, (*b*) woven fabric of fiberglass yarn, (*c*) continuous-strand **roving,** and (*d*) woven roving.
(Courtesy of Owens/Corning Fiberglas Co.)

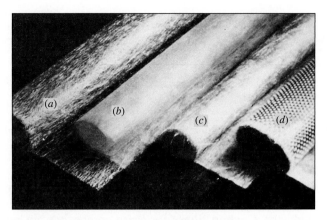

Figure 12.4
Fiberglass reinforcing mats: (*a*) continuous-strand mat,
(*b*) surfacing mat, (*c*) chopped-strand mat, and
(*d*) combination of woven roving and chopped-strand mat.
(*Courtesy of Owens/Corning Fiberglas Co.*)

Table 12.1 Comparative yarn properties for fiber reinforcements for plastics

Property	Glass (E)	Carbon (HT)	Aramid (Kevlar 49)
Tensile strength, ksi (MPa)	450 (3100)	500 (3450)	525 (3600)
Tensile modulus, Msi (GPa)	11.0 (76)	33 (228)	19 (131)
Elongation at break (%)	4.5	1.6	2.8
Density (g/cm³)	2.54	1.8	1.44

(Fig. 12.3*c*). The rovings may be in continuous strands or woven to make woven roving (Fig. 12.3*d*). Glass-fiber reinforcing mats (Fig. 12.4) are made of continuous strands (Fig. 12.4*a*) or chopped strands (Fig. 12.4*c*). The strands are usually held together with a resinous binder. Combination mats are made with woven roving chemically bonded to chopped-strand mat (Fig. 12.4*d*).

Properties of Glass Fibers   The tensile properties and density of E-glass fibers are compared to those of carbon and aramid fibers in Table 12.1. It is noted that glass fibers have a lower tensile strength and modulus than carbon and aramid fibers but a higher elongation. The density of glass fibers is also higher than that of carbon and aramid fibers. However, because of their low cost and versatility, glass fibers are by far the most commonly used reinforcing fibers for plastics (Table 12.1).

12.2.2 Carbon Fibers for Reinforced Plastics

Composite materials made by using carbon fibers for reinforcing plastic resin matrices such as epoxy are characterized by having a combination of light weight, very high strength, and high stiffness (modulus of elasticity). These properties make the use of carbon-fiber-plastic composite materials especially attractive for aerospace

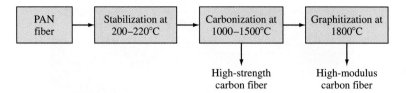

Figure 12.5
Processing steps for producing high-strength, high-modulus carbon fibers from polyacrylonitrile precursor material.

applications, such as the aircraft part shown in Fig. 12.1. Unfortunately, the relatively high cost of the carbon fibers restricts their use in many industries such as the automotive industry. **Carbon fibers** for these composites are produced mainly from two sources, *polyacrylonitrile* (PAN) (see Sec. 10.8) and pitch, which are called *precursors.*

In general, carbon fibers are produced from PAN precursor fibers by three processing stages: (1) stabilization, (2) carbonization, and (3) graphitization (Fig. 12.5). In the *stabilization stage* the PAN fibers are first stretched to align the fibrillar networks within each fiber parallel to the fiber axis, and then they are oxidized in air at about 200 to 220°C (392 to 428°F) while held in tension.

The second stage in the production of high-strength carbon fibers is *carbonization.* In this process the stabilized PAN-based fibers are pyrolyzed (heated) until they become transformed into carbon fibers by the elimination of O, H, and N from the precursor fiber. The carbonization heat treatment is usually carried out in an inert atmosphere in the 1000 to 1500°C (1832 to 2732°F) range. During the carbonization process, turbostratic graphitelike fibrils or ribbons are formed within each fiber that greatly increase the tensile strength of the material.

A third stage, or *graphitization treatment,* is used if an increase in the modulus of elasticity is desired at the expense of high tensile strength. During graphitization, which is carried out above about 1800°C (3272°F), the preferred orientation of the graphitelike crystallites within each fiber is increased.

The carbon fibers produced from PAN precursor material have a tensile strength that ranges from about 450 to 650 ksi (3.10 to 4.45 GPa) and a modulus of elasticity that ranges from about 28 to 35 Msi (193 to 241 GPa). In general the higher-moduli fibers have lower tensile strengths, and vice versa. The density of the carbonized and graphitized PAN fibers is usually about 1.7 to 2.1 g/cm^3, while their final diameter is about 7 to 10 μm. Figure 12.6 shows a photograph of a group of about 6000 carbon fibers called a **tow.**

12.2.3 Aramid Fibers for Reinforcing Plastic Resins

Aramid fiber is the generic name for aromatic polyamide fibers. Aramid fibers were introduced commercially in 1972 by Du Pont under the trade name of Kevlar, and at the present time there are two commercial types: Kevlar 29 and 49. Kevlar 29 is a low-density, high-strength aramid fiber designed for such applications as ballistic

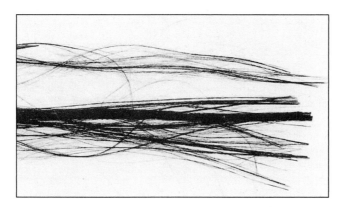

Figure 12.6
Photograph of a tow of about 6000 carbon fibers.
(*Courtesy of the Fiberite Co., Winona, Minn.*)

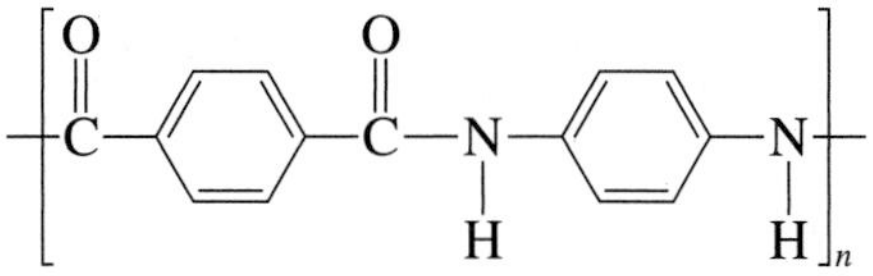

Figure 12.7
Repeating chemical structural unit for
Kevlar fibers.

protection, ropes, and cables. Kevlar 49 is characterized by a low density and high strength and modulus. The properties of Kevlar 49 make its fibers useful as reinforcement for plastics in composites for aerospace, marine, automotive, and other industrial applications.

The chemical repeating unit of the Kevlar polymer chain is that of an aromatic polyamide, shown in Fig. 12.7. Hydrogen bonding holds the polymer chains together in the transverse direction. Thus, collectively these fibers have high strength in the longitudinal direction and weak strength in the transverse direction. The aromatic ring structure gives high rigidity to the polymer chains, causing them to have a rodlike structure.

Kevlar aramid is used for high-performance composite applications where light weight, high strength and stiffness, damage resistance, and resistance to fatigue and stress rupture are important. Of special interest is that Kevlar-epoxy material is used for various parts of the space shuttle.

12.2.4 Comparison of Mechanical Properties of Carbon, Aramid, and Glass Fibers for Reinforced-Plastic Composite Materials

Figure 12.8 compares typical stress-strain diagrams for carbon, aramid, and glass fibers, and one can see that the fiber strength varies from about 250 to 500 ksi

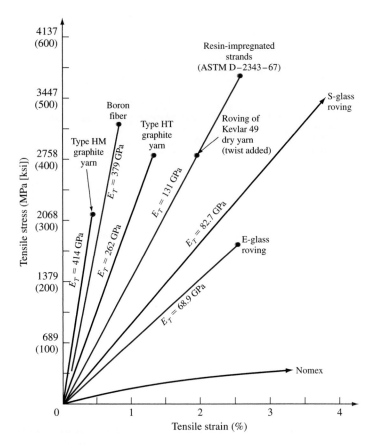

Figure 12.8
Stress-strain behavior of various types of reinforcing fibers.
(After "Kevlar 49 Data Manual," E. I. du Pont de Nemours & Co., 1974.)

(1720 to 3440 MPa), while the fracture strain ranges from 0.4 to 4.0 percent. The tensile modulus of elasticity of these fibers ranges from 10×10^6 to 60×10^6 psi (68.9 to 413 GPa). The carbon fibers provide the best combination of high strength, high stiffness (high modulus), and low density, but have lower elongations. The aramid fiber Kevlar 49 has a combination of high strength, high modulus (but not as high as the carbon fibers), low density, and high elongation (impact resistance). The glass fibers have lower strengths and moduli and higher densities (Table 12.1). Of the glass fibers, the S-glass fibers have higher strengths and elongations than the E-glass fibers. Because the glass fibers are much less costly, they are more widely used.

Figure 12.9 compares the strength to density and the stiffness (tensile modulus) to density of various reinforcing fibers. This comparison shows the outstanding strength-to-weight and stiffness-to-weight ratios of carbon and aramid (Kevlar 49)

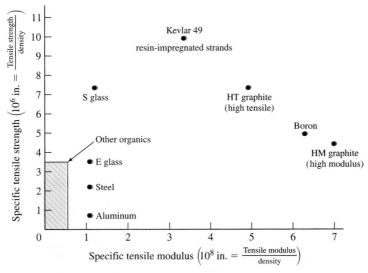

Figure 12.9

Specific tensile strength (tensile strength to density) and **specific tensile modulus** (tensile modulus to density) for various types of reinforcing fibers.

(Courtesy of E. I. du Pont de Nemours & Co., Wilmington, Del.)

fibers as contrasted to those properties of steel and aluminum. Because of these favorable properties, carbon- and aramid-reinforced-fiber composites have replaced metals for many aerospace applications.

12.3 FIBER-REINFORCED-PLASTIC COMPOSITE MATERIALS

12.3.1 Matrix Materials for Fiber-Reinforced-Plastic Composite Materials

Two of the most important matrix plastic resins for fiber-reinforced plastics are unsaturated polyester and epoxy resins. The chemical reactions responsible for the cross-linking of these thermosetting resins have already been described in Sec. 10.9.

Some of the properties of unfilled cast rigid polyester and epoxy resins are listed in Table 12.2. The polyester resins are lower in cost but are usually not as strong as the epoxy resins. Unsaturated polyesters are widely used for matrices of fiber-reinforced plastics. Applications for these materials include boat hulls, building panels, and structural panels for automobiles, aircraft, and appliances. Epoxy resins cost more but have special advantages such as good strength properties and lower shrinkage after curing than polyester resins. Epoxy resins are commonly used as matrix materials for carbon- and aramid-fiber composites.

Table 12.2 Some properties of unfilled cast polyester and epoxy resins

	Polyester	**Epoxy**
Tensile strength, ksi (MPa)	6–13 (40–90)	8–19 (55–130)
Tensile modulus of elasticity, Msi (GPa)	0.30–0.64 (2.0–4.4)	0.41–0.61 (2.8–4.2)
Flexural yield strength, ksi (MPa)	8.5–23 (60–160)	18.1 (125)
Impact strength (notched-bar Izod test), ft · lb/in. (J/m) of notch	0.2–0.4 (10.6–21.2)	0.1–1.0 (5.3–53)
Density (g/cm^3)	1.10–1.46	1.2–1.3

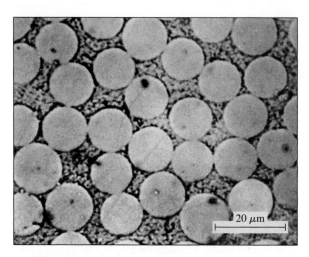

Figure 12.10
Photomicrograph of a cross section of a unidirectional fiberglass-polyester composite material.
(After D. Hull, "An Introduction to Composite Materials," Cambridge, 1981, p. 63.)

12.3.2 Fiber-Reinforced-Plastic Composite Materials

Fiberglass-Reinforced Polyester Resins   The strength of fiberglass-reinforced plastics is mainly related to the glass content of the material and the arrangement of the glass fibers. In general, the higher the weight percent glass in the composite, the stronger the reinforced plastic is. When there are parallel strands of glass, as may be the case for filament winding, the fiberglass content may be as high as 80 wt %, which leads to very high strengths for the composite material. Figure 12.10 shows a photomicrograph of a cross section of a fiberglass-polyester-resin composite material with unidirectional fibers.

Any deviation from the parallel alignment of the glass strands reduces the mechanical strength of the fiberglass composite. For example, composites made with

Table 12.3 Some mechanical properties of fiberglass-polyester composites

	Woven cloth	Chopped roving	Sheet-molding compound
Tensile strength, ksi (MPa)	30–50 (206–344)	15–30 (103–206)	8–20 (55–138)
Tensile modulus of elasticity, Msi (GPa)	1.5–4.5 (103–310)	0.80–2.0 (55–138)	
Impact strength notched bar, Izod ft · lb/in. (J/m) of notch	5.0–30 (267–1600)	2.0–20.0 (107–1070)	7.0–22.0 (374–1175)
Density (g/cm^3)	1.5–2.1	1.35–2.30	1.65–2.0

Table 12.4 Some typical mechanical properties of a commercial unidirectional composite laminate of carbon fibers (62% by volume) and epoxy resin

Properties	Longitudinal (0°)	Transverse (90°)
Tensile strength, ksi (MPa)	270 (1860)	9.4 (65)
Tensile modulus of elasticity, Msi (GPa)	21 (145)	1.36 (9.4)
Ultimate tensile strain (%)	1.2	0.70

Source: Hercules, Inc.

woven fiberglass fabrics because of their interlacing have lower strengths than if all the glass strands were parallel (Table 12.3). If the roving is chopped, producing a random arrangement of glass fibers, the strength is lower for a specific direction but equal in all directions (Table 12.3).

Carbon-Fiber-Reinforced-Epoxy Resins   In carbon-fiber composite materials, the fibers contribute the high tensile properties for rigidity and strength, while the matrix is the carrier for the alignment of the fibers and contributes some impact strength. Epoxy resins are by far the most commonly used matrixes for carbon fibers, but other resins such as polyimides, polyphenylene sulfides, or polysulfones may be used for certain applications.

The major advantage of carbon fibers is that they have very high strength and moduli of elasticity (Table 12.1) combined with low density. For this reason, carbon-fiber composites are replacing metals in some aerospace applications where weight saving is important (Fig. 12.1). Table 12.4 lists some typical mechanical properties of one type of carbon-fiber-epoxy composite material that contains 62 percent by volume of carbon fibers. Figure 12.11 shows the exceptional fatigue properties of unidirectional carbon (graphite)-epoxy composite material as compared to those of aluminum alloy 2024-T3.

In engineering-designed structures the carbon-fiber-epoxy material is **laminated** so that different tailor-made strength requirements are met (Fig. 12.12). Figure 12.13 shows a photomicrograph of a five-layer bidirectional carbon-fiber-epoxy composite material after curing.

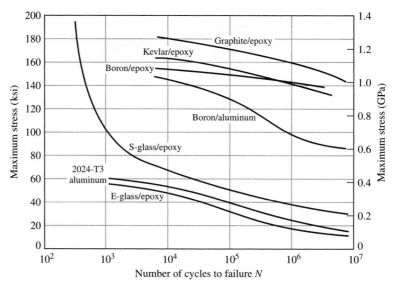

Figure 12.11

Fatigue properties (maximum stress versus number of cycles to failure) for carbon (graphite)-epoxy unidirectional composite material as compared to the fatigue properties of some other composite materials and aluminum alloy 2024-T3. *R* (the minimum stress–maximum stress for tension-tension cyclic test) = 0.1 at room temperature.
(*Courtesy of Hercules, Inc.*)

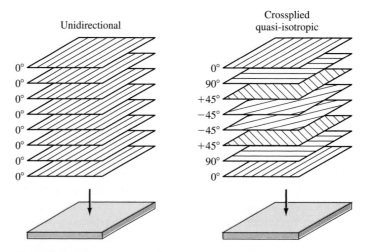

Figure 12.12

Unidirectional and **multidirectional laminate** plies for a composite laminate.
(*Courtesy of Hercules, Inc.*)

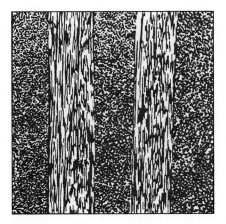

Figure 12.13
Photomicrograph of five-layer bidirectional composite of carbon-fiber-epoxy composite material.
(*After J. J. Dwyer, Composites, Am. Mach., July 13, 1979, pp. 87–96.*)

A unidirectional Kevlar 49 fiber-epoxy composite contains 60 percent by volume of Kevlar 49 fibers and 40 percent epoxy resin. The density of the Kevlar 49 fibers is 1.48 Mg/m^3 and that of the epoxy resin is 1.20 Mg/m^3. (*a*) What are the weight percentages of Kevlar 49 and epoxy resin in the composite material, and (*b*) what is the average density of the composite?

EXAMPLE PROBLEM 12.1

■ **Solution**

Basis is 1 m³ of composite material. Therefore, we have 0.60 m³ of Kevlar 49 and 0.40 m³ of epoxy resin. Density = mass/volume, or

$$\rho = \frac{m}{V} \quad \text{and} \quad m = \rho V$$

a. Mass of Kevlar 49 $= \rho V = (1.48 \text{ Mg/m}^3)(0.60 \text{ m}^3) = 0.888 \text{ Mg}$
    Mass of epoxy resin $= \rho V = (1.20 \text{ Mg/m}^3)(0.40 \text{ m}^3) = \underline{0.480 \text{ Mg}}$
    Total mass $= 1.368 \text{ Mg}$

$$\text{Wt \% Kevlar 49} = \frac{0.888 \text{ Mg}}{1.368 \text{ Mg}} \times 100\% = 64.9\%$$

$$\text{Wt \% epoxy resin} = \frac{0.480 \text{ Mg}}{1.368 \text{ Mg}} \times 100\% = 35.1\%$$

b. Average density of composite is

$$\rho_c = \frac{m}{V} = \frac{1.368 \text{ Mg}}{1 \text{ m}^3} = 1.37 \text{ Mg/m}^3 \ ◀$$

12.3.3 Equations for Elastic Modulus of a Lamellar Continuous-Fiber-Plastic Matrix Composite for Isostrain and Isostress Conditions

Isostrain Conditions   Let us consider an idealized lamellar composite test sample with alternate layers of continuous fibers and matrix materials, as shown in Fig. 12.14. In this case the stress on the material causes uniform strain on all the composite layers. We shall assume that the bonding between the layers remains intact during the stressing. This type of loading on the composite sample is called the *isostrain condition.*

Let us now derive an equation relating the elastic modulus of the composite in terms of the elastic moduli of the fiber and matrix and their volume percentages. First, the load on the composite structure is equal to the sum of the load on the fiber layers plus the load on the matrix layers, or

$$P_c = P_f + P_m \tag{12.1}$$

Since $\sigma = P/A$, or $P = \sigma A$,

$$\sigma_c A_c = \sigma_f A_f + \sigma_m A_m \tag{12.2}$$

where σ_c, σ_f, and σ_m are the stresses and A_c, A_f, and A_m are the fractional areas of the composite, fiber, and matrix, respectively. Since the lengths of the layers of matrix and fiber are equal, the areas A_c, A_f, and A_m in Eq. 12.2 can be replaced by the volume fractions V_c, V_f, and V_m, respectively:

$$\sigma_c V_c = \sigma_f V_f + \sigma_m V_m \tag{12.3}$$

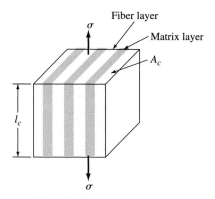

Figure 12.14

Composite structure consisting of layers of fiber and matrix under isostrain conditions of loading. (Volume of composite V_c = area A_c × length l_c.)

Since the volume fraction of the total composite is 1, then $V_c = 1$, and Eq. 12.3 becomes

$$\sigma_c = \sigma_f V_f + \sigma_m V_m \qquad (12.4)$$

For isostrain conditions and assuming a good bond between the composite layers,

$$\epsilon_c = \epsilon_f = \epsilon_m \qquad (12.5)$$

Dividing Eq. 12.4 by Eq. 12.5, since all the strains are equal, gives

$$\frac{\sigma_c}{\epsilon_c} = \frac{\sigma_f V_f}{\epsilon_f} + \frac{\sigma_m V_m}{\epsilon_m} \qquad (12.6)$$

Now we can substitute the modulus of elasticity E_c for σ_c/ϵ_c, E_f for σ_f/ϵ_f, and E_m for σ_m/ϵ_m, giving

$$\boxed{E_c = E_f V_f + E_m V_m} \qquad (12.7)$$

This equation is known as the *rule of mixtures for binary composites* and enables a value for the elastic modulus of a composite to be calculated knowing the elastic moduli of the fiber and matrix and their volume percentages.

Equations for the Loads on the Fiber and Matrix Regions of a Lamellar Composite Structure Loaded Under Isostrain Conditions The ratio of the loads on the fiber and matrix regions of a binary composite material stressed under isostrain conditions can be obtained from their $P = \sigma A$ ratios. Thus, since $\sigma = E\epsilon$ and $\epsilon_f = \epsilon_m$,

$$\frac{P_f}{P_m} = \frac{\sigma_f A_f}{\sigma_m A_m} = \frac{E_f \epsilon_f A_f}{E_m \epsilon_m A_m} = \frac{E_f A_f}{E_m A_m} = \frac{E_f V_f}{E_m V_m} \qquad (12.8)$$

If the total load on a specimen stressed under isostrain conditions is known, then the following equation applies:

$$P_c = P_f + P_m \qquad (12.9)$$

where P_c, P_f, and P_m are the loads on the total composite, fiber region, and matrix region, respectively. By combining Eq. 12.9 with Eq. 12.8, the load on each of the fiber and matrix regions can be determined if values for E_f, E_m, V_f, V_m, and P_c are known.

Calculate (*a*) the modulus of elasticity, (*b*) the tensile strength, and (*c*) the fraction of the load carried by the fiber for the following composite material stressed under isostrain conditions. The composite consists of a continuous glass-fiber-reinforced-epoxy resin produced by using 60 percent by volume of E-glass fibers having a modulus of elasticity of $E_f = 10.5 \times 10^6$ psi and a tensile strength of 350,000 psi and a hardened epoxy resin with a modulus of $E_m = 0.45 \times 10^6$ psi and a tensile strength of 9000 psi.

EXAMPLE PROBLEM 12.2

■ Solution

a. Modulus of elasticity of the composite is

$$E_c = E_f V_f + E_m V_m \tag{12.7}$$
$$= (10.5 \times 10^6 \text{ psi})(0.60) + (0.45 \times 10^6 \text{ psi})(0.40)$$
$$= 6.30 \times 10^6 \text{ psi} + 0.18 \times 10^6 \text{ psi}$$
$$= 6.48 \times 10^6 \text{ psi} \ (44.6 \text{ GPa}) \blacktriangleleft$$

b. Tensile strength of the composite is

$$\sigma_c = \sigma_f V_f + \sigma_m V_m \tag{12.4}$$
$$= (350,000 \text{ psi})(0.60) + (9000 \text{ psi})(0.40)$$
$$= 210,000 + 3600 \text{ psi}$$
$$= 214,000 \text{ psi or } 214 \text{ ksi } (1.47 \text{ GPa}) \blacktriangleleft$$

c. Fraction of load carried by fiber is

$$\frac{P_f}{P_c} = \frac{E_f V_f}{E_f V_f + E_m V_m}$$
$$= \frac{(10.5 \times 10^6 \text{ psi})(0.60)}{(10.5 \times 10^6 \text{ psi})(0.60) + (0.45 \times 10^6 \text{ psi})(0.40)}$$
$$= \frac{6.30}{6.30 + 0.18} = 0.97 \blacktriangleleft$$

Isostress Conditions Let us now consider the case of an idealized lamellar composite structure consisting of layers of fiber and matrix in which the layers are perpendicular to the applied stress, as shown in Fig. 12.15. In this case the stress on

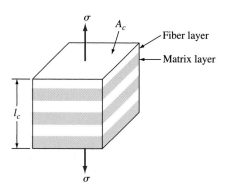

Figure 12.15
Composite structure consisting of layers of fiber and matrix under isostress conditions of loading. (Volume of composite V_c = area A_c × length l_c.)

the composite structure produces an equal stress condition on all the layers and so is called the *isostress condition.*

To derive an equation for the elastic modulus for the layered composite for this type of loading, we shall begin with an equation that states that the stress on the total composite structure is equal to the stress on the fiber layers and the stress on the matrix layers. Thus,

$$\sigma_c = \sigma_f = \sigma_m \tag{12.10}$$

The total strain for the composite in the directions of the stresses is thus equal to the sum of the strains in the fiber and matrix layers,

$$\epsilon_c = \epsilon_f + \epsilon_m \tag{12.11}$$

Assuming that the area perpendicular to the stress does not change after the stress is applied and assuming unit length for the composite after being stressed, then

$$\epsilon_c = \epsilon_f V_f + \epsilon_m V_m \tag{12.12}$$

where V_f and V_m are the volume fractions of the fiber and matrix laminates, respectively.

Assuming Hooke's law is valid under loading, then

$$\epsilon_c = \frac{\sigma}{E_c} \qquad \epsilon_f = \frac{\sigma}{E_f} \qquad \epsilon_m = \frac{\sigma}{E_m} \tag{12.13}$$

Substituting equations of Eq. 12.13 into Eq. 12.12 gives

$$\frac{\sigma}{E_c} = \frac{\sigma V_f}{E_f} = \frac{\sigma V_m}{E_m} \tag{12.14}$$

Dividing each term of Eq. 12.14 by σ gives

$$\frac{1}{E_c} = \frac{V_f}{E_f} + \frac{V_m}{E_m} \tag{12.15}$$

Obtaining a common denominator yields

$$\frac{1}{E_c} = \frac{V_f E_m}{E_f E_m} + \frac{V_m E_f}{E_m E_f} \tag{12.16}$$

Rearranging,

$$\frac{1}{E_c} = \frac{V_f E_m + V_m E_f}{E_f E_m}$$

or

$$E_c = \frac{E_f E_m}{V_f E_m + V_m E_f} \tag{12.17}$$

Calculate the modulus of elasticity for a composite material consisting of 60 percent by volume of continuous E-glass fiber and 40 percent epoxy resin for the matrix when stressed under *isostress conditions* (i.e., the material is stressed perpendicular to the continuous fiber). The modulus of elasticity of the E glass is 10.5×10^6 psi and that of the epoxy resin is 0.45×10^6 psi.

■ Solution

$$E_c = \frac{E_f E_m}{V_f E_m + V_m E_f} \tag{12.17}$$

$$= \frac{(10.5 \times 10^6\,\text{psi})(0.45 \times 10^6\,\text{psi})}{(0.60)(0.45 \times 10^6) + (0.40)(10.5 \times 10^6)}$$

$$= \frac{4.72 \times 10^{12}\,\text{psi}^2}{0.27 \times 10^6\,\text{psi} + 4.20 \times 10^6\,\text{psi}}$$

$$= 1.06 \times 10^6\,\text{psi (7.30 GPa)} \blacktriangleleft$$

Note that the isostress condition of stressing the composite material results in a modulus of elasticity for the 60 percent E-glass fiber–40 percent epoxy composite material that is about six times lower than that obtained by stressing under the isostrain condition.

A schematic representation comparing isostrain loading to isostress loading of a composite layered structure (Fig. 12.16) shows that higher-modulus values are obtained with isostrain loading for equal volumes of fibers.

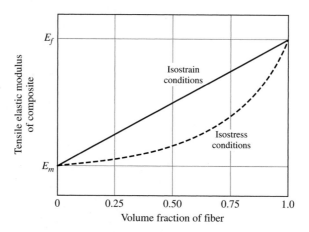

Figure 12.16
Schematic representation of the tensile elastic modulus as a function of the volume fraction of fiber in a reinforced-fiber-plastic matrix unidirectional composite laminate loaded under isostrain and isostress conditions. For a given volume fraction of fiber in the composite, the material loaded under isostrain conditions has a higher modulus.

12.4 OPEN-MOLD PROCESSES FOR FIBER-REINFORCED-PLASTIC COMPOSITE MATERIALS

There are many open-mold methods used for producing fiber-reinforced plastics. Some of the most important of these will now be discussed briefly.

12.4.1 Hand Lay-Up Process

This is the simplest method of producing a fiber-reinforced part. To produce a part with the **hand lay-up** process by using fiberglass and a polyester, a gel coat is first applied to the open mold (Fig. 12.17a). Fiberglass reinforcement that is normally in the form of a cloth or mat is manually placed in the mold. The base resin mixed with catalysts and accelerators is then applied by pouring, brushing, or spraying. Rollers (Fig. 12.17b) or squeegees are used to thoroughly wet the reinforcement with the resin and to remove entrapped air. To increase the wall thickness of the part being produced, layers of fiberglass mat or woven roving and resin are added. Applications for this method include boat hulls, tanks, housings, and building panels.

12.4.2 Spray-Up Process

The **spray-up method** of producing fiber-reinforced-plastic shells is similar to the hand lay-up method and can be used to make boat hulls, tub-shower units, and other

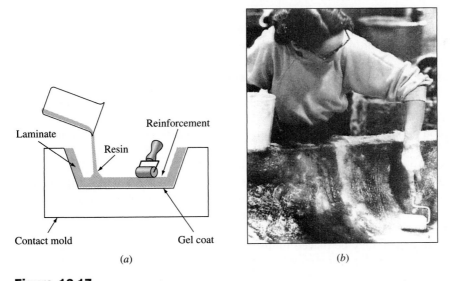

(a)   (b)

Figure 12.17
Hand lay-up method for molding fiber-reinforced-plastic composite materials.
(a) Pouring the resin over the reinforcement in the mold. (b) Use of a roller to densify the laminate to remove entrapped air.
(*Courtesy of Owens/Corning Fiberglas Co.*)

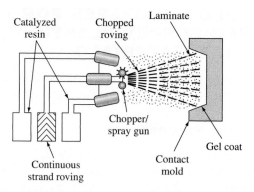

Figure 12.18
Spray-up method for molding fiber-reinforced-plastic composite materials; advantages of this method include greater shape complexity for the molded part and an ability to automate the process.
(*Courtesy of Owens/Corning Fiberglas Co.*)

medium- to large-size shapes. In this process, if fiberglass is used, continuous-strand roving is fed through a combination chopper and spray gun (Fig. 12.18) that simultaneously deposits chopped roving and catalyzed resin into the mold. The deposited laminate is then densified with a roller or squeegee to remove air and to make sure the resin impregnates the reinforcing fibers. Multiple layers may be added to produce the desired thickness. Curing is usually at room temperature, or it may be accelerated by the application of a moderate amount of heat.

12.4.3 Vacuum Bag–Autoclave Process

The **vacuum bag molding** process is used to produce high-performance laminates usually of fiber-reinforced-epoxy systems. Composite materials produced by this method are particularly important for aircraft and aerospace applications (Fig. 12.1).

We shall now look at the various steps of this process required to produce a finished part. First, a long, thin sheet, which may be about 60 in. (152 cm) wide or **prepreg** carbon-fiber-epoxy material is laid out on a large table (Fig. 12.19). The prepreg material consists of unidirectional long carbon fibers in a partially cured epoxy matrix. Next, pieces of the prepreg sheet are cut out and placed on top of each other on a shaped tool to form a laminate (Fig. 12.20). The layers, or *plies* as they are called, may be placed in different directions to produce the desired strength pattern since the highest strength of each ply is in the direction parallel to the fibers (Fig. 12.12).

After the laminate is constructed, the tooling and attached laminate are vacuum-bagged, with a vacuum being applied to remove entrapped air from the laminated part. Finally, the vacuum bag enclosing the laminate and the tooling is put into an

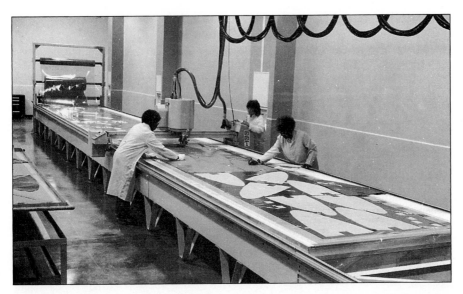

Figure 12.19
Carbon-fiber-epoxy prepreg sheet being cut with computerized cutter at
McDonnell Douglas composite facility.
(*Courtesy of McDonnell Douglas Corp.*)

Figure 12.20
Smooth-contour tooling mold for
forming a laminate from multiple layers
of prepreg fiber-reinforced-plastic
composite material.
(*Courtesy of Northrop Co.*)

autoclave for the final curing of the epoxy resin (Fig. 12.21). The conditions for curing vary depending on the material, but the carbon-fiber-epoxy composite material
is usually heated at about 190°C (375°F) at a pressure of about 100 psi. After being
removed from the autoclave, the composite part is stripped from its tooling and is
ready for further finishing operations.

Figure 12.21
Carbon-fiber-epoxy laminate of AV-8B wing section and tooling being put into autoclave for curing at McDonnell Aircraft Co. plant.
(*Courtesy of McDonnell Douglas Corp.*)

Carbon-fiber-epoxy composite materials are used mainly in the aerospace industry where the high strength, stiffness, and lightness of the material can be fully utilized. For example, this material is used for airplane wings, elevator and rudder parts, and the cargo bay doors of the space shuttle. Cost considerations have prevented the widespread use of this material in the auto industry.

12.4.4 Filament-Winding Process

Another important open-mold process to produce high-strength hollow cylinders is the **filament-winding process.** In this process the fiber reinforcement is fed through a resin bath and then wound on a suitable mandrel (Fig. 12.22). When sufficient layers have been applied, the wound mandrel is cured either at room temperature or at an elevated temperature in an oven. The molded part is then stripped from the mandrel.

The high degree of fiber orientation and high fiber loading with this method produce extremely high tensile strengths in hollow cylinders. Applications for this process include chemical and fuel storage tanks, pressure vessels, and rocket motor cases (Fig. 12.23).

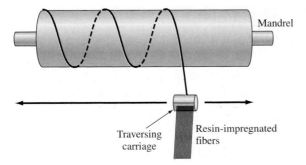

Figure 12.22

Filament-winding process for producing fiber-reinforced-plastic composite materials. The fibers are first impregnated with plastic resin and then wound around a rotating mandrel (drum). The carriage containing the resin-impregnated fibers traverses during the winding, laying down the impregnated fibers.

(After H. G. DeYoung, Plastic Composites Fight for Status, High Technol., October 1983, p. 63. © High Technology Publishing Co. Used with permission.)

Figure 12.23

Filament winding of a rocket motor case using Kevlar 49 fiber and epoxy resin.

(Courtesy of the Du Pont Co.)

12.5 CLOSED-MOLD PROCESSES FOR FIBER-REINFORCED-PLASTIC COMPOSITE MATERIALS

There are many closed-mold methods used for producing fiber-reinforced-plastic materials. Some of the most important of these will now be discussed briefly.

12.5.1 Compression and Injection Molding

These are two of the most important high-volume processes used for producing fiber-reinforced plastics with closed molds. These processes are essentially the same as those discussed in Sec. 10.6 for plastic materials except that the fiber reinforcement is mixed with the resin before processing.

12.5.2 The Sheet-Molding Compound (SMC) Process

The **sheet-molding compound** process is one of the newer closed-mold processes used to produce fiber-reinforced-plastic parts, particularly in the automotive industry. This process allows excellent resin control and good mechanical strength properties (Table 12.3) to be obtained while producing high-volume, large-size, highly uniform products.

The sheet-molding compound is usually manufactured by a highly automated continuous-flow process. Continuous-strand fiberglass roving is chopped in lengths of about 2 in. (5.0 cm) and deposited on a layer of resin-filler paste that is travelling on a polyethylene film (Fig. 12.24). Another layer of resin-filler paste is deposited later over the first layer to form a continuous sandwich of fiberglass and resin filler.

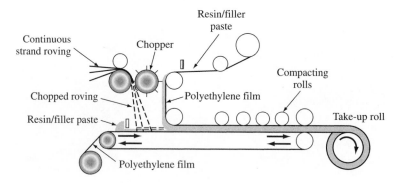

Figure 12.24
Manufacturing process for sheet-molding compound. The machine shown produces a sandwich of fiberglass and resin-filler paste between two thin-film sheets of polyethylene. The sheet-molding compound produced must be aged before being pressed into a finished product.
(Courtesy of Owens/Corning Fiberglas Co.)

The sandwich with top and bottom covers of polyethylene is compacted and rolled into package-sized rolls (Fig. 12.24).

The rolled-up SMC is next stored in a maturation room for about one to four days so that the sheet can carry the glass. The SMC rolls are then moved to near the press and cut into the proper charge pattern for the specific part and placed in a matched metal mold that is hot (300°F [149°C]). The hydraulic press then is closed, and the SMC flows uniformly under pressure (1000 psi) throughout the mold to form the final product. Sometimes an in-mold coating may be injected in the middle of the pressing operation to improve the surface quality of the SMC part.

The advantages of the SMC process over the hand lay-up or spray-up processes are more-efficient high-volume production, improved surface quality, and uniformity of product. The use of SMC is particularly advantageous in the automotive industry for the production of front-end and grille-opening panels, body panels, and hoods. For example, the front hood of the 1984 Chevrolet Corvette was made of SMC (Fig. 12.25). This clamshell hood was made by adhesively bonding a 0.080-in. (0.20-cm) inner panel to an outer mold-coated 0.10-in. (0.25-cm) panel.

Figure 12.25
The outer panel of the front hood of the 1984 Chevrolet Corvette. This panel was made of sheet-molding compound by pressing at 300°F (149°C) at 1000 psi (6.89 MPa) for 60 to 90 s.
(*Courtesy of General Tire and Rubber Co., Southfield, Mich.*)

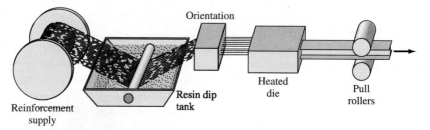

Figure 12.26
The pultrusion process for producing fiber-reinforced-plastic composite materials. Fibers impregnated with resin are fed into a heated die and then are slowly drawn out as a cured composite material with a constant cross-sectional shape.
(After H. G. De Young, Plastic Composites Fight for Status, High Technol., October 1983, p. 63. © High Technology Publishing Co. Used with permission.)

12.5.3 Continuous-Pultrusion Process

Continuous **pultrusion** is used for the manufacturing of fiber-reinforced plastics of constant cross section such as structural shapes, beams, channels, pipe, and tubing. In this process, continuous-strand fibers are impregnated in a resin bath and then are drawn through a heated steel die that determines the shape of the finished stock (Fig. 12.26). Very high strengths are possible with this material because of the high fiber concentration and orientation parallel to the length of the stock being drawn.

12.6 CONCRETE

Concrete is a major engineering material used for structural construction. Civil engineers use concrete, for example, in the design and construction of bridges, buildings, dams, retainer and barrier walls, and road pavement. In 1982 about 50×10^7 tons of concrete was produced in the United States, which is considerably greater than the 6×10^7 tons of steel that was produced in the same year. As a construction material concrete offers many advantages, including flexibility in design since it can be cast, economy, durability, fire resistance, ability to be fabricated on site, and aesthetic appearance. Disadvantages of concrete from an engineering standpoint include low tensile strength, low ductility, and some shrinkage.

Concrete is a ceramic composite material composed of coarse granular material (the **aggregate**) embedded in a hard matrix of a cement paste (the binder), which is usually made from portland cement[3] and water. Concrete varies considerably in

---

[3]Portland cement takes its name from the small peninsula on the south coast of England where the limestone is similar to some extent to portland cement.

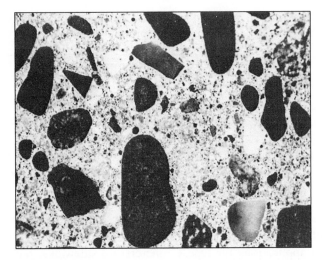

Figure 12.27
Cross section of hardened concrete. A cement and
water paste completely coats each aggregate particle
and fills the spaces between the particles to make a
ceramic composite material.
*(After "Design and Control of Concrete Mixtures," 12th ed.,
Portland Cement Association, 1979.)*

composition but usually contains (by absolute volume) 7 to 15 percent portland
cement, 14 to 21 percent water, 0.5 to 8 percent air, 24 to 30 percent fine aggregate, and 31 to 51 percent coarse aggregate. Figure 12.27 shows a polished section of a hardened concrete sample. The cement paste in concrete acts as a "glue"
to hold the aggregate particles together in this composite material. Let us now
look at some of the characteristics of the components of concrete and examine
some of its properties.

12.6.1 Portland Cement

Production of Portland Cement   The basic raw materials for **portland cement** are
lime (CaO), silica (SiO_2), alumina (Al_2O_3), and iron oxide (Fe_2O_3). These components are appropriately proportioned to produce various types of portland cement.
The selected raw materials are crushed, ground, and proportioned for the desired
composition and then blended. The mixture is then fed into a rotary kiln where it is
heated to temperatures from up to 1400 to 1650°C (2600 to 3000°F). In this process
the mixture is chemically converted to cement clinker, which is subsequently cooled
and pulverized. A small amount of gypsum ($CaSO_4 \cdot 2H_2O$) is added to the cement
to control the setting time of the concrete.

Chemical Composition of Portland Cement From a practical standpoint, portland cement can be considered to consist of four principal compounds, which are:

Compound	Chemical formula	Abbreviation
Tricalcium silicate	$3CaO \cdot SiO_2$	C_3S
Dicalcium silicate	$2CaO \cdot SiO_2$	C_2S
Tricalcium aluminate	$3CaO \cdot Al_2O_3$	C_3A
Tetracalcium aluminoferrite	$4CaO \cdot Al_2O_3 \cdot Fe_2O_3$	C_4AF

Types of Portland Cement Various types of portland cement are produced by varying its chemical composition. In general there are five main types whose basic chemical compositions are listed in Table 12.5.

Type I is the normal general-purpose portland cement. It is used where the concrete is not exposed to a high sulfate attack from soil or water, or where there is not an objectionable temperature increase due to heat generated by hydration of the cement. Typical applications for type I concrete are sidewalks, reinforced concrete buildings, bridges, culverts and tanks, and reservoirs.

Type II portland cement is used where moderate sulfate attack can occur such as in drainage structures where sulfate concentrations in groundwaters are higher than normal. Type II cement is normally used in hot weather for large structures such as large piers and heavy retaining walls since this cement has a moderate heat of hydration.

Type III portland cement is an early-strength type that develops high strength in an early period. It is used when concrete forms are to be removed early from a structure that must be put into use soon.

Type IV is a low-heat-of-hydration portland cement for use when the rate and amount of heat generated must be minimized. Type IV cement is used for massive concrete structures such as large gravity dams where the heat generated by the setting cement is a critical factor.

Type V is a sulfate-resisting cement used when the concrete is exposed to severe sulfate attack such as in concrete exposed to soils and groundwaters containing a high sulfate content.

Table 12.5 Typical compound compositions of portland cement

Cement type	ASTM C150 designation	Compositions (wt %)*			
		C_3S	C_2S	C_3A	C_4AF
Ordinary	I†	55	20	12	9
Moderate heat of hydration, moderate sulfate resistance	II	45	30	7	12
Rapid hardening	III	65	10	12	8
Low heat of hydration	IV	25	50	5	13
Sulfate-resistant	V	40	35	3	14

*Missing percentages consist of gypsum and minor components such as MgO, alkali sulfate, etc.
†This is the most common of all cement types.

Source: J. F. Young, *J. Educ. Module Mater. Sci.,* **3:**410 (1981). Used by permission, the *Journal of Materials Education,* University Park, PA.

Hardening of Portland Cement Portland cement hardens by reactions with water that are called **hydration reactions.** These reactions are complex and not completely understood. Tricalcium silicate and dicalcium silicate constitute about 75 percent of portland cement by weight, and when these compounds react with water during the hardening of the cement, the principal hydration product is *tricalcium silicate hydrate.* This material is formed in extremely small particles (less than 1 μm) and is a colloidal *gel.* Calcium hydroxide is also produced by the hydration of C_3S and C_2S and is a crystalline material. These reactions are

$$2C_3S + 6H_2O \rightarrow C_3S_2 \cdot 3H_2O + 3Ca(OH)_2$$
$$2C_2S + 4H_2O \rightarrow C_3S_2 \cdot 3H_2O + Ca(OH)_2$$

Tricalcium silicate hardens rapidly and is mostly responsible for the early strength of portland cement (Fig. 12.28). Most of the hydration of C_3S takes place in about two days, and thus early-strength portland cements contain higher amounts of C_3S.

Dicalcium silicate (C_2S) has a slow hydration reaction with water and is mainly responsible for strength increases beyond one week (Fig. 12.28). Tricalcium aluminate hydrates rapidly with a high rate of heat liberation. C_3A contributes slightly to early-strength development and is kept to a low level in sulfate-resisting (type V)

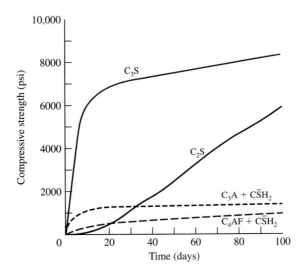

Figure 12.28
Compressive strength of pure-cement compound pastes as a function of curing time. Note that $C\bar{S}H_2$ is the abbreviated formula for $CaSO_4 \cdot 2H_2O$.
[*After J. F. Young, J. Educ. Module Mater. Sci.,* **3:**420(1981). *Used by permission, the Journal of Materials Education, University Park, Pa.*]

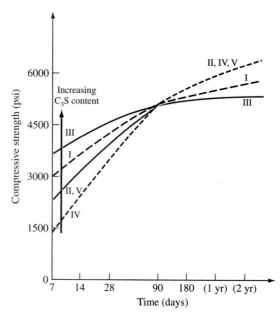

Figure 12.29
Compressive strengths of concretes made with different ASTM types of portland cements as a function of curing time.
[*After J. F. Young, J. Educ. Module Mater. Sci.,* **3**:*420(1981).*
Used by permission, the Journal of Materials Education,
University Park, Pa.]

cements (Fig. 12.28). Tetracalcium aluminoferrite is added to the cement to reduce the clinkering temperature during the manufacturing of the cement.

The extent to which the hydration reactions are completed determines the strength and durability of the concrete. Hydration is relatively rapid during the first few days after fresh concrete is put in place. It is important that water is retained by the concrete during the early curing period and that evaporation is prevented or reduced.

Figure 12.29 shows how the compressive strength of concretes made with different ASTM-type cements increases with curing time. Most of the compressive strength of the concretes is developed in about 28 days, but strengthening may continue for years.

12.6.2 Mixing Water for Concrete

Most natural water that is drinkable can be used as mixing water for making concrete. Some water unsuitable for drinking can also be used for making cement, but if the impurity content reaches certain specified levels, the water should be tested for its effect on the strength of the concrete.

12.6.3 Aggregates for Concrete

Aggregates normally make up to about 60 to 80 percent of the concrete volume and greatly affect its properties. Concrete aggregate is usually classified as either fine or coarse. Fine aggregates consist of sand particles up to $\frac{1}{4}$ in. (6 mm), while coarse aggregates are those particles retained on a no. 16 sieve (1.18 mm opening). Thus, there is some overlapping in particle-size ranges for fine and coarse aggregates. Rocks compose most of the coarse aggregate, while minerals (sand) compose most of the fine aggregates.

12.6.4 Air Entrainment

Air-entrained concretes are produced to improve the resistance of the concrete to freezing and thawing and also to improve the workability of some concretes. Air-entraining agents are added to some types of portland cements, and these are classified with the letter A following the type number as, for example, type IA and type IIA. Air-entraining agents contain surface-active agents that lower the surface tension at the air-water interface so that extremely small air bubbles form (90 percent of them are less than 100 μm) (Fig. 12.30). For satisfactory frost protection, air-entrained concretes must have between 4 and 8 percent air by volume.

12.6.5 Compressive Strength of Concrete

Concrete, which is basically a ceramic composite material, has a much higher compressive strength than tensile strength. In engineering designs, therefore, concrete is primarily loaded in compression. The tensile loading capability of concrete can be increased by reinforcing it with steel rods. This subject will be discussed later.

As shown in Fig. 12.29, the strength of concrete is time-dependent since its strength develops by hydration reactions that take time to complete. The compressive strength of concrete is also greatly dependent on the water-to-cement ratio, with high ratios of water to cement producing lower-strength concrete (Fig. 12.31). However, there is a limit to how low the water-to-cement ratio can be since less water makes it more difficult to work the concrete and to have it completely fill the concrete forms. With air entrainment, the concrete is more workable, and thus a lower water-to-cement ratio can be used.

Concrete compressive-strength test specimens are usually cylinders 6 in. (15 cm) in diameter and 12 in. (30 cm) high. However, cores or other types of specimens can be cut out of existing concrete structures.

12.6.6 Proportioning of Concrete Mixtures

The design of concrete mixtures should include consideration of the following factors:

1. Workability of the concrete. The concrete must be able to flow or be compacted into the shape of the form it is poured into.

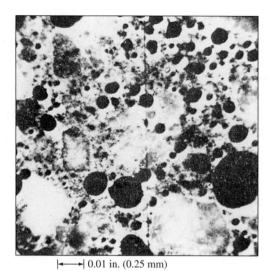

|←——→| 0.01 in. (0.25 mm)

Figure 12.30
Polished section of air-entrained concrete as observed in the optical microscope. Most of the air bubbles in this sample appear to be about 0.1 mm in diameter.
(After "Design and Control of Concrete Mixtures," 12th ed., Portland Cement Association, 1979.)

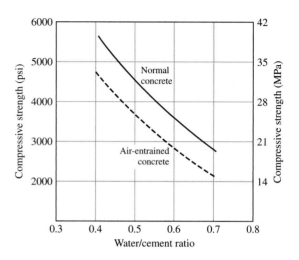

Figure 12.31
Effect of water-cement ratio by weight on the compressive strength of normal and air-entrained concrete.
(After "Design and Control of Concrete Mixtures," 12th ed., Portland Cement Association, 1979, p. 61.)

2. Strength and durability. For most applications the concrete must reach certain strength and durability specifications.

3. Economy of production. For most applications cost is an important factor and thus must be considered.

Modern concrete-mixture design methods have evolved from the early-1900s arbitrary volumetric method of 1:2:4 for the ratios of cement to fine aggregate to coarse aggregate. Today, weight and absolute-volume methods of proportioning concrete mixtures are set forth by the American Concrete Institute. Example Problem 12.4 outlines a method of determining the required amounts of cement, fine and coarse aggregate, and water for a particular volume of concrete, given the weight ratios for the components, their specific gravities, and the required volume of water per unit weight of cement.

Figure 12.32 shows the ranges for the proportions of materials used in concrete by the absolute-volume method for normal and air-entrained concrete mixtures. Normal concrete has a range by volume of 7 to 15 percent cement, 25 to 30 percent fine aggregate, 31 to 51 percent coarse aggregate, and 16 to 21 percent water. The air content of normal concrete ranges from 0.5 to 3 percent but ranges from 4 to 8 percent for air-entrained concrete. As previously stated the water-to-cement ratio is a determining factor for the compressive strength of concrete. A water-to-cement ratio above about 0.4 decreases the compressive strength of concrete significantly (Fig. 12.31).

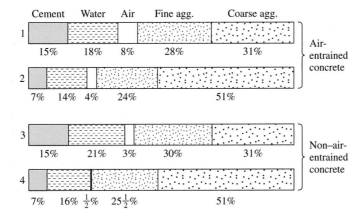

Figure 12.32

Ranges of proportions of components of concrete by absolute volume. The mixes for bars 1 and 3 have higher water and fine-aggregate contents, while the mixes for bars 2 and 4 have lower water and higher coarse-aggregate contents.

(After "Design and Control of Concrete Mixtures," 12th ed., Portland Cement Association, 1979, p. 7.)

Seventy-five cubic feet of concrete with a ratio of 1:1.8:2.8 (by weight) of cement, sand (fine aggregate), and gravel (coarse aggregate) is to be produced. What are the required amounts of the constituents if 5.5 gal of water per sack of cement is to be used? Assume the free moisture contents of the sand and gravel are 5 and 0.5 percent, respectively. Give answers in the following units: the amount of cement in sacks, the sand and gravel in pounds, and the water in gallons. Data are given in the following table:

EXAMPLE PROBLEM 12.4

Constituent	Specific gravity	Saturated surface-dry density (lb/ft^3)*
Cement	3.15	$3.15 \times 62.4 \text{ lb/ft}^3 = 197$
Sand	2.65	$2.65 \times 62.4 \text{ lb/ft}^3 = 165$
Gravel	2.65	$2.65 \times 62.4 \text{ lb/ft}^3 = 165$
Water	1.00	$1.00 \times 62.4 \text{ lb/ft}^3 = 62.4$

Saturated surface-dry density (SSDD) in lb/ft^3 = specific gravity $\times$ weight of 1 ft^3 water = specific gravity $\times$ 62.4 lb/ft^3. One sack of cement weighs 94 lb. 7.48 gal = 1 ft^3 of water.

■ Solution

First we shall calculate the absolute volumes of the constituents of the concrete per sack of cement since the problem asks for a certain number of cubic feet of concrete. Also, we shall first calculate the weight of sand and gravel required on a dry basis and then later make a correction for the moisture content of the sand and gravel.

Constituent	Ratio by weight	Weight	SSDD (lb/ft^3)	Absolute vol. per sack of cement
Cement	1	1×94 lb $= 94$ lb	197	94 lb/197 lb/ft^3 = 0.477 ft^3
Sand	1.8	1.8×94 lb $= 169$ lb	165	169 lb/165 lb/ft^3 = 1.024 ft^3
Gravel	2.8	2.8×94 lb $= 263$ lb	165	263 lb/165 lb/ft^3 = 1.594 ft^3
Water	(5.5 gal)			5.5 gal/7.48 gal/ft^3 = 0.735 ft^3
				Tot. abs. vol. per sack of cement = 3.830 ft^3

Thus, 3.830 ft^3 of concrete is produced per sack of cement on a dry basis. On this basis, the following amounts of cement, sand, and gravel are required for 75 ft^3 of concrete:

1. Required amount of cement = 75 ft^3/(3.83 ft^3/sack of cement) = 19.58 sacks ◀
2. Required amount of sand = (19.58 sacks)(94 lb/sack)(1.8) = 3313 lb
3. Required amount of gravel = (19.58 sacks)(94 lb/sack)(2.8) = 5153 lb
4. Required amount of water = (19.58 sacks)(5.5 gal/sack) = 107.7 gal

On a wet basis with correction for moisture in the sand and gravel,

$$\text{Water in sand} = (3313 \text{ lb})(1.05)$$
$$= 3479 \text{ lb}; \; 3479 - 3313 \text{ lb} = 166 \text{ lb water}$$
$$\text{Water in gravel} = (5153 \text{ lb})(1.005)$$
$$= 5179 \text{ lb}; \; 5179 - 5153 \text{ lb} = 26 \text{ lb water}$$

1. Required weight of wet sand = 3313 lb + 166 lb = 3479 lb ◀
2. Required weight of wet gravel = 5153 lb + 26 lb = 5179 lb ◀

The required amount of water equals the calculated amount on the dry basis less the amount of water in the sand and gravel. Thus,

$$\text{Total gal water in sand and gravel} = (166 \text{ lb} + 26 \text{ lb})\left(\frac{7.48 \text{ gal}}{\text{ft}^3}\right)\left(\frac{1 \text{ ft}^3}{62.4 \text{ lb}}\right)$$
$$= 23.0 \text{ gal}$$

Thus, the water required on the dry basis minus the water in the sand and gravel equals the amount of water required on the wet basis:

$$107.7 \text{ gal} - 23.0 \text{ gal} = 84.7 \text{ gal} ◀$$

12.6.7 Reinforced and Prestressed Concrete

Since the tensile strength of concrete is about 10 to 15 times lower than its compressive strength, concrete is mainly used in engineering designs so that it is primarily in compression. However, when a concrete member is subjected to tensile forces such as occur in a beam, the concrete is usually cast containing steel reinforcing bars, as shown in Fig. 12.33. In this reinforced concrete, the tension forces are transferred from the concrete to the steel reinforcement through bonding. Concrete containing steel reinforcements in the form of rods, wires, wire mesh, etc., is referred to as **reinforced concrete.**

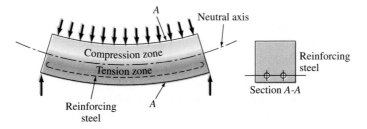

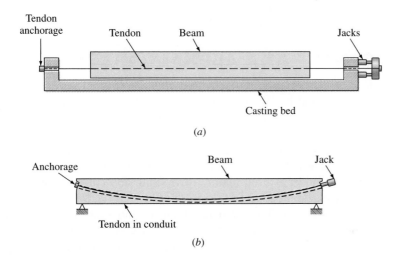

Figure 12.33
Exaggerated effect of a heavy load on a reinforced-concrete beam. Note that the reinforcing steel rods are placed in the tension zone to absorb tensile stresses.
(*After G. B. Wynne, "Reinforced Concrete Structures," Reston Publishing, 1981, p. 2.*)

Figure 12.34
Schematic drawing showing the arrangement for producing (*a*) a **pretensioned concrete** beam and (*b*) a posttensioned concrete beam.
(*After A. H. Nilson, "Design of Prestressed Concrete," Wiley, 1978, pp. 14 and 17.*)

12.6.8 Prestressed Concrete

The tensile strength of reinforced concrete can be further improved by introducing compressive stresses into the concrete by *pretensioning* or *posttensioning* using steel reinforcements called *tendons*. The tendon may be a tensioned steel rod or cable, for example. The advantage of prestressed concrete is that the compressive stresses introduced by the steel tendons have to be counteracted before the concrete is subjected to tensile stresses.

Pretensioned (Prestressed) Concrete In the United States most **prestressed concrete** is pretensioned. In this method the tendons, which are usually in the form of multiple-wire stranded cables, are stretched between an external tendon anchorage and an adjustable jack for applying the tension (Fig. 12.34*a*). The concrete is then poured over the tendons, which are in the tensioned state. When the concrete reaches the required strength, the jacking pressure is released. The steel strands would like to shorten elastically but cannot because they are bonded to the concrete. In this way compressive stresses are introduced into the concrete.

Posttensioned (Prestressed) Concrete In this process, usually hollow conduits containing steel tendons are placed in the concrete (a beam, for example) form before pouring the concrete (Fig. 12.34b). The tendons may be stranded cable, bundled parallel wires, or solid steel rods. The concrete is then poured, and when it is sufficiently strong, each tendon is anchored at one end of the cast concrete and jacking tension is applied at the other end. When the jacking pressure is sufficiently high, a fitting is used to replace the jack, which retains the tension in the tendon. The space between the tendons and the conduit is normally filled with cement grout by forcing it in as a paste at one end of the conduit under high pressure. The grouting improves the flexural stress capacity of a concrete beam.

12.7 ASPHALT AND ASPHALT MIXES

Asphalt is a *bitumen,* which is basically a hydrocarbon with some oxygen, sulfur, and other impurities, and has the mechanical characteristics of a thermoplastic polymeric material. Most asphalt is derived from petroleum refining, but some is processed directly from bitumen-bearing rock (rock asphalt) and from surface deposits (lake asphalt). The asphalt content of crude oils usually ranges from about 10 to 60 percent. In the United States about 75 percent of the asphalt consumed is used for paving roads, while the remainder is used mainly for roofing and construction.

Asphalts consist chemically of 80 to 85 percent carbon, 9 to 10 percent hydrogen, 2 to 8 percent oxygen, 0.5 to 7 percent sulfur, and small amounts of nitrogen and other trace metals. The constituents of asphalt greatly vary and are complex. They range from low-molecular-weight to high-molecular-weight polymers and condensation products consisting of chain hydrocarbons, ring structures, and condensed rings.

Asphalt is used primarily as a bituminous binder with aggregates to form an **asphalt mix,** most of which is used for road paving. The Asphalt Institute in the United States has designated eight paving mixtures based on the proportion of aggregate that passes through a no. 8 sieve.[4] For example, a type IV asphalt mix for road paving has a composition of 3.0 to 7.0 percent asphalt with 35 to 50 percent of the aggregate passing through a no. 8 sieve.

The most stable asphalt mixes are made with a dense-packed angular aggregate with just enough asphalt to coat the aggregate particles. If the asphalt content gets too high, in hot weather, asphalt may concentrate on the road surface and reduce skid resistance. Angular aggregate that does not polish easily and that interlocks produces better skid resistance than soft, easily polished aggregate. The aggregate also should bond well with the asphalt to avoid separation.

---

[4]A no. 8 sieve has nominal openings of 0.0937 in. (2.36 mm).

12.8 WOOD

Wood (timber) is the most widely used engineering construction material in the United States, with its annually produced tonnage exceeding all other engineering materials, including concrete and steel (Fig. 1.12). In addition to the use of wood for timber and lumber for the construction of houses, buildings, and bridges, wood is also used to make composite materials such as plywood, particleboard, and paper.

Wood is a naturally occurring composite material that consists mainly of a complex array of cellulose cells reinforced by a polymeric substance called **lignin** and other organic compounds. The discussion of wood in this section will first look at the macrostructure of wood, and then a brief examination will be made of the microstructure of softwoods and hardwoods. Finally, some of the properties of wood will be correlated with its structure.

12.8.1 Macrostructure of Wood

Wood is a natural product with a complex structure, and thus we cannot expect a homogeneous product for engineering designs such as with an alloy steel bar or an injection-molded thermoplastic part. As we all know, the strength of wood is highly anisotropic, with its tensile strength being much greater in the direction parallel to the tree stem.

Layers in the Cross Section of a Tree   Let us first examine the cross section of a typical tree, as shown in Fig. 12.35. Important layered regions in this figure are

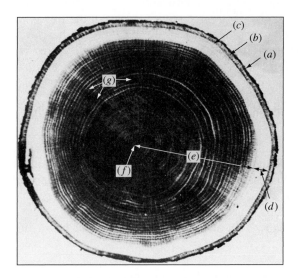

Figure 12.35
Cross section of a typical tree. (*a*) outer bark;
(*b*) inner bark; (*c*) cambium layer; (*d*) sapwood;
(*e*) heartwood; (*f*) pith; (*g*) wood rays.
*(After U.S. Department of Agriculture Handbook No. 72,
revised 1974, p. 2-2.)*

indicated by the letters A to F. The name and function of each of these layers is listed as follows:

1. *Outer bark* layer is composed of dry, dead tissue and provides external protection for the tree.
2. *Inner bark* layer is moist and soft and carries food from the leaves to all the growing parts of the tree.
3. **Cambium layer** is the tissue layer between the bark and wood that forms the wood and bark cells.
4. **Sapwood** is the light-colored wood that forms the outer part of the tree stem. The sapwood contains some living cells that function for food storage and carry sap from the roots to the leaves of the tree.
5. **Heartwood** is the older inner region of the tree stem that is no longer living. The heartwood is usually darker than the sapwood and provides strength for the tree.
6. *Pith* is the soft tissue at the center of the tree around which the first growth of the tree takes place.

Also indicated in Fig. 12.35 are the wood rays that connect the tree layers from the pith to the bark and are used for food storage and transfer of food.

Softwoods and Hardwoods Trees are classified into two major groups called **softwoods** (gymnosperms) and **hardwoods** (angiosperms). The botanical basis for their classification is that if the tree seed is exposed, the tree is a softwood type and if the seed is covered, the tree is a hardwood type. With a few exceptions, a softwood tree is one that retains its leaves and a hardwood tree is one that sheds its leaves annually. Softwood trees are often referred to as *evergreen* trees and hardwood trees as *deciduous* trees. Most softwood trees are physically soft and most hardwood trees are physically hard, but there are exceptions. Examples of softwood trees native to the United States are fir, spruce, pine, and cedar, while examples of hardwood trees are oak, elm, maple, birch, and cherry.

Annual Growth Rings During each growth season in temperate climates such as in the United States, a new layer of wood is formed annually around the tree stem. These layers are called *annual growth rings* and are particularly evident in softwood tree sections (Fig. 12.36). Each ring has two subrings: *earlywood* (spring) and *latewood* (summer). In softwoods the earlywood has a lighter color and the cell size is larger.

Axes of Symmetry for Wood It is important to be able to correlate the direction in a tree with microstructure. To do this a set of axes has been chosen, as indicated in Fig. 12.37. The axis parallel to the tree stem is called the *longitudinal axis* (*L*), while the axis perpendicular to the annual growth ring of the tree is called the *radial axis* (*R*). The third axis, the *tangential axis* (*T*), is parallel to the annual ring and is perpendicular to both the radial and longitudinal axes.

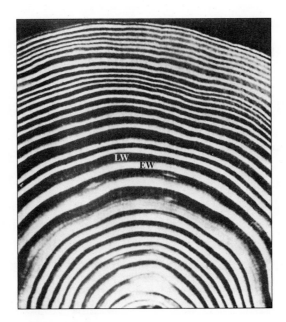

Figure 12.36
Annual growth rings in a softwood tree. The
earlywood (EW) part of the annual ring is
lighter in color than the *latewood* (LW) part.
[*After R. J. Thomas, J. Educ. Module Mater. Sci.,*
2:56(1980). *Used by permission, the Journal of Materials
Education, University Park, Pa.*]

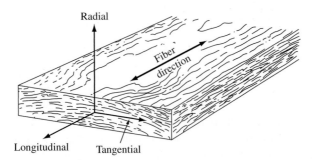

Figure 12.37
Axes in wood. The longitudinal axis is parallel to the
grain, the tangential axis is parallel to the annual
growth ring, and the radial axis is perpendicular to
the annual growth ring.
(*After U.S. Department of Agriculture Handbook No. 72, revised
1974, p. 4-2.*)

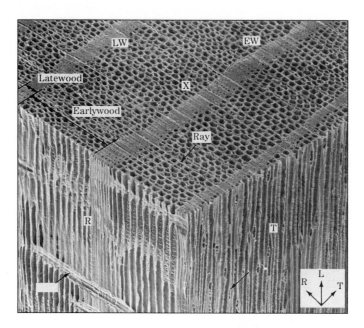

Figure 12.38
Scanning electron micrograph of softwood (longleaf pine) block showing three complete growth rings in cross-section surface. Note that the individual cells are larger in the earlywood than in the latewood. Rays that consist of food-storing cells run perpendicular to the longitudinal direction. (Magnification 75×.)
(*Courtesy of the N. C. Brown Center for Ultrastructure Studies, SUNY College of Environmental Science and Forestry.*)

12.8.2 Microstructure of Softwoods

Figure 12.38 shows the microstructure of a small block of a softwood tree at 75× in which three complete growths rings can be seen. The larger cell size of the earlywood is clearly visible in this micrograph. Softwood consists mainly of long, thin-walled tubular cells called **tracheids,** which can be observed in Fig. 12.38. The large open space in the center of the cells is called the **lumen** and is used for water conduction. The length of a longitudinal tracheid is about 3 to 5 mm and its diameter is about 20 to 80 μm. Holes or pits at the ends of the cells allow liquid to flow from one cell to another. The longitudinal tracheids constitute about 90 percent of the volume of the softwood. The earlywood cells have a relatively large diameter, thin walls, and a large-size lumen. The latewood cells have a smaller diameter and thick walls with a smaller lumen than the earlywood cells.

Wood rays that run in the transverse direction from the bark to the center of the tree consist of an aggregate of small *parenchyma cells* that are bricklike in shape. The parenchyma cells, which are used for food storage, are interconnected along the rays by pit pairs.

12.8.3 Microstructure of Hardwoods

Hardwoods, in contrast to softwoods, have large-diameter **vessels** for the conduction of fluids. The vessels are thin-walled structures consisting of individual elements called *vessel elements* and are formed in the longitudinal direction of the tree stem.

The wood of hardwood trees is classified as either *ring-porous* or *diffuse-porous,* depending on how the vessels are arranged in the growth rings. In a ring-porous hardwood the vessels formed in the earlywood are larger than those formed in the latewood (Fig. 12.39). In a diffuse-porous hardwood, the vessel diameters are essentially the same throughout all the growth ring (Fig. 12.40).

The longitudinal cells responsible for the support of the hardwood tree stem are fibers. Fibers in hardwood trees are elongated cells with close-pointed ends and are usually thick-walled. Fibers range in length from about 0.7 to 3 mm and average less than about 20 μm in diameter. The wood volume of hardwoods made up of fibers varies considerably. For example, the volume of fibers in a sweetgum hardwood is 26 percent, whereas the volume for hickory is 67 percent.

The food-storage cells of the hardwoods are the ray (transverse) and longitudinal **parenchyma,** which are brick- or box-shaped. The rays for hardwoods are usually much larger than for softwoods, having many cells across their width.

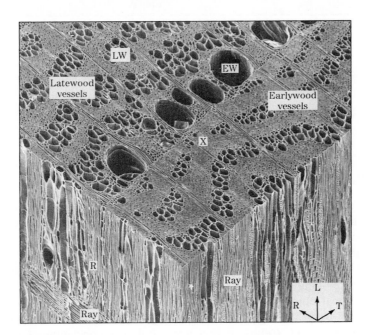

Figure 12.39

Scanning electron micrograph of ring-porous hardwood (American elm) block showing the abrupt change in the diameter of the earlywood and latewood vessels as observed in the cross-section surface. (Magnification 54×.)
(Courtesy of the N. C. Brown Center for Ultrastructure Studies, SUNY College of Environmental Science and Forestry.)

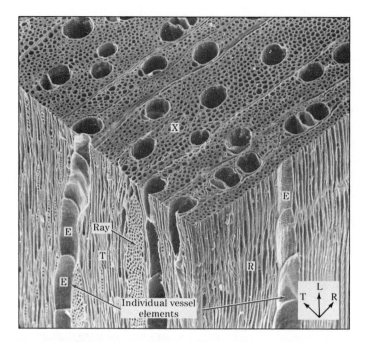

Figure 12.40
Scanning electron micrograph of diffuse-porous hardwood (sugar maple) block showing the rather uniform diameter of the vessels throughout the growth ring. The formation of the vessel from individual vessel elements is clearly visible. (Magnification 100×.)
(*Courtesy of the N. C. Brown Center for Ultrastructure Studies, SUNY College of Environmental Science and Forestry.*)

12.8.4 Cell-Wall Ultrastructure

Let us now examine the structure of a wood cell at high magnification, such as the one shown telescoped in Fig. 12.41. The initial cell wall that is formed during cell division during the growth period is called the *primary wall.* During its growth the primary wall enlarges in the transverse and longitudinal directions, and after it reaches full size, the *secondary wall* forms in concentric layers growing into the center of the cell (Fig. 12.41).

The principal constituents of the wood cell are *cellulose, hemicellulose,* and *lignin.* Cellulose crystalline molecules make up between 45 and 50 percent of the solid material of wood. Cellulose is a linear polymer consisting of glucose units (Fig. 12.42) with a degree of polymerization ranging from 5000 to 10,000. The covalent bonding within and between the glucose units creates a straight and stiff molecule with high tensile strength. Lateral bonding between the cellulose molecules is by hydrogen and permanent dipolar bonding. Hemicellulose makes up 20 to 25 percent by weight of the solid material of wood cells and is a branched amorphous

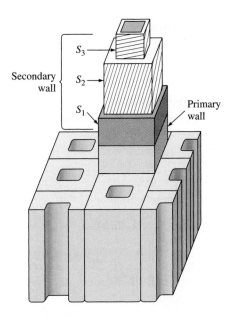

Figure 12.41

Schematic drawing of a telescoped wood cell in a multiwood cell structure showing the relative thicknesses of the primary and secondary walls of a wood cell. The lines on the primary and secondary walls indicate the microfibril orientations.

*[After R. J. Thomas, J. Educ. Module Mater. Sci., **2**:85(1980). Used by permission, the Journal of Materials Education, University Park, Pa.]*

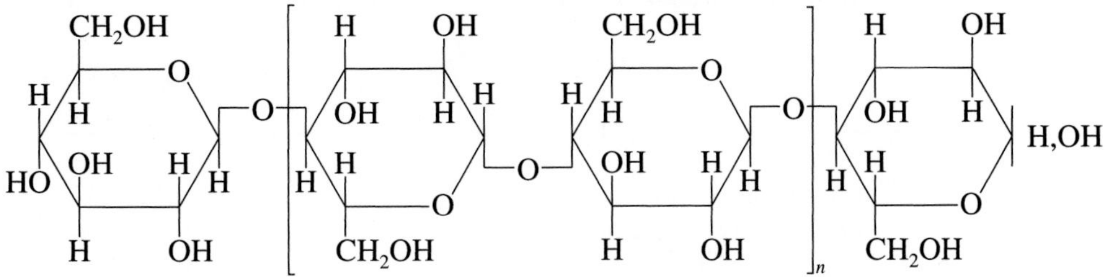

Figure 12.42

The structure of a cellulose molecule.

(After J. D. Wellons, "Adhesive Bonding of Woods and Other Structural Materials," University Park, Pa., Materials Education Council, 1983.)

molecule containing several types of sugar units. Hemicellulose molecules have a degree of polymerization between 150 and 200. The third main constituent of wood cells is lignin, which constitutes about 20 to 30 percent by weight of the solid material. Lignins are very complex, cross-linked, three-dimensional polymeric materials formed from phenolic units.

The cell wall consists mainly of **microfibrils** bonded together by a lignin cement. The microfibrils themselves are believed to consist of a crystalline core of cellulose surrounded by an amorphous region of hemicellulose and lignin. The arrangement and orientations of the microfibrils vary in different layers of the cell wall, as indicated in Fig. 12.41. The lignins provide rigidity to the cell wall and allow it to resist compressive forces. In addition to solid materials, the wood cells may absorb up to about 30 percent of their weight in water.

12.8.5 Properties of Wood

Moisture Content   Wood, unless oven-dried to a constant weight, contains some moisture. Water occurs in wood either adsorbed in the fiber walls of the cells or as unbound water in the cell-fiber lumen. By convention, the percent of water in wood is defined by the equation

$$\text{Wood moisture content (wt \%)} = \frac{\text{wt of water in sample}}{\text{wt of dry wood sample}} \times 100\% \qquad \textbf{(12.18)}$$

Because the percentage of water is on a dry basis, the moisture content of wood may exceed 200 percent.

EXAMPLE PROBLEM 12.5

A piece of wood containing moisture weighs 165.3 g and after oven drying to a constant weight, weighs 147.5 g. What is its percent moisture content?

■ **Solution**
The weight of water in the wood sample is equal to the weight of the wet wood sample minus its weight after oven drying to a constant weight. Thus,

$$\% \text{ Moisture content} = \frac{\text{weight of wet wood} - \text{weight of dry wood}}{\text{weight of dry wood}} \times 100\% \qquad \textbf{(12.18)}$$

$$= \frac{165.3 \text{ g} - 147.5 \text{ g}}{147.5 \text{ g}} \times 100\% = 12.1\% \blacktriangleleft$$

The moisture condition of wood in the living tree is referred to as the *green condition.* The average moisture content of the sapwood of softwoods in the green condition is about 150 percent, while that of the heartwood of the same species is about 60 percent. In hardwood the difference in green-condition moisture content between the sapwood and heartwood is usually much less, with both averaging about 80 percent.

Mechanical Strength   Some typical mechanical properties of some types of woods grown in the United States are listed in Table 12.6. In general, woods that are

Table 12.6 Typical mechanical properties of some commercially important woods grown in the United States

| Species | Condition | Specific gravity | Static bending | | Compression parallel to grain; maximum crushing strength (psi)* | Compression perpendicular to grain; fiber stress at prop. limit (psi)* | Shear parallel to grain; maximum shearing strength (psi)* |
			Modulus of rupture (psi)*	Modulus of elasticity (10^6 psi)*			
Hardwoods:							
Elm,	Green	0.46	7,200	1.11	2910	360	1000
American	Kiln dry†	0.50	11,800	1.34	5520	690	1510
Hickory,	Green	0.60	9,800	1.37	3990	780	1480
pecan	Kiln dry†	0.66	13,700	1.73	7850	1720	2080
Maple, red	Green	0.49	7,700	1.39	3280	400	1150
	Kiln dry†	0.54	13,400	1.64	6540	1000	1850
Oak, white	Green	0.60	8,300	1.25	3560	670	1250
	Kiln dry†	0.68	15,200	1.78	7440	1070	2000
Softwoods:							
Douglas fir,	Green	0.45	7,700	1.56	3780	380	900
coast	Kiln dry†	0.48	12,400	1.95	7240	800	1130
Cedar,	Green	0.31	5,200	0.94	2770	240	770
western red	Kiln dry†	0.32	7,500	1.11	4560	460	990
Pine,	Green	0.34	4,900	0.99	2440	220	680
eastern white	Kiln dry†	0.35	8,600	1.24	4800	440	900
Redwood,	Green	0.34	5,900	0.96	3110	270	890
young growth	Kiln dry†	0.35	7,900	1.10	5220	520	1110

*To obtain MPa, multiply psi by 6.89×10^{-3}.
†Kiln-dried to 12% moisture.

Source: *The Encyclopedia of Wood*, Sterling Publishing Co., 1980, pp. 68–75.

botanically classified as softwoods are physically soft, and those classified as hardwoods are physically hard, although there are some exceptions. For example, balsa wood, which is physically very soft, is classified botanically as a hardwood.

The compressive strength of wood parallel to the grain is considerably higher than that of wood perpendicular to the grain by a factor of about 10. For example, the compressive strength of kiln-dried (12 percent moisture content) eastern white pine is 4800 psi (33 MPa) parallel to the grain but only 440 psi (3.0 MPa) perpendicular to the grain. The reason for this difference is that in the longitudinal direction, the strength of the wood is due primarily to the strong covalent bonds of the cellulose microfibrils that are oriented mainly longitudinally (Fig. 12.42). The strength of the wood perpendicular to the grain is much lower because it depends on the strength of the weaker hydrogen bonds that bond the cellulose molecules laterally.

As observed in Table 12.6, wood in the green condition is weaker than kiln-dried wood. The reason for this difference is that the removal of water from the less-ordered regions of the cellulose of the microfibril allows the cell molecular structure to compact and form internal bridges by hydrogen bonding. Thus, upon losing moisture, the wood shrinks and becomes denser and stronger.

Shrinkage   Greenwood shrinks as moisture is eliminated from it and causes distortion of the wood, as shown in Fig. 12.43 for the radial and tangential directions of a cross section of a tree. Wood shrinks considerably more in the transverse

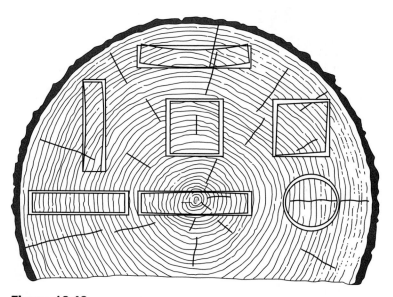

Figure 12.43
Shrinkage and distortion of tree section in the tangential, radial, and annual ring directions.
[*After R. T. Hoyle, J. of Educ. Module Mater. Sci.,* **4**:*88(1982). Used by permission, the Journal of Materials Education, University Park, Pa.*]

direction than in the longitudinal direction, with the transverse shrinkage usually ranging from 10 to 15 percent as compared to only about 0.1 percent in the longitudinal direction.

When water is eliminated from the amorphous regions on the exterior part of the microfibrils, they become closer together and the wood becomes denser. Because the long dimension of the microfibrils is mainly oriented in the longitudinal direction of the tree stem, wood shrinks transversely for the most part.

12.9 SANDWICH STRUCTURES

Composite materials made by sandwiching a core material between two thin outer layers are commonly used in engineering designs. Two types of these materials are (1) honeycomb sandwich and (2) cladded sandwich structures.

12.9.1 Honeycomb Sandwich Structure

Bonded honeycomb sandwich construction has been used as a basic construction material in the aerospace industry for over 30 years. Most aircraft flying today depend on this construction material. Most honeycomb in use today is made of aluminum alloys such as 5052 and 2024 or glass-reinforced phenolic, glass-reinforced polyester, and aramid-fiber-reinforced materials.

Aluminum honeycomb panels are fabricated by adhesively bonding aluminum alloy face sheets to aluminum alloy honeycomb core sections, as shown in Fig. 12.44. This type of construction provides a stiff, rigid, strong, and light-weight sandwich panel.

12.9.2 Cladded Metal Structures

The cladded structure is used to produce composites of a metal core with thin outer layers of another metal or metals (Fig. 12.45). Usually, thin outer metal

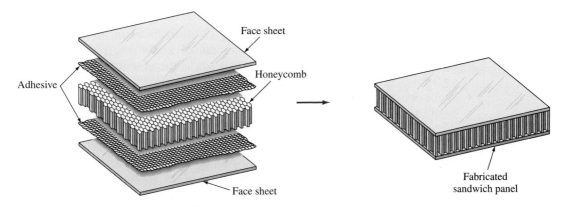

Figure 12.44
Sandwich panel fabricated by bonding aluminum faces to aluminum alloy honeycomb core.
(*Courtesy of the Hexcel Co.*)

Figure 12.45
Cross section of a cladded metal structure.

layers are hot-roll-bonded to the inner core metal to form metallurgical (atomic diffusion) bonds between the outside layers and the inner core metal. This type of composite material has many applications in industry. For example, high-strength aluminum alloys such as 2024 and 7075 have relatively poor corrosion resistance and can be protected by a thin layer of soft, highly corrosion-resistant aluminum cladding. Another application of cladded metals is the use of relatively expensive metals to protect a less-expensive metal core. For example, the U.S. 10¢ and 25¢ coins have a cladding of a Cu–25% Ni alloy over a core of less-expensive copper.

12.10 METAL-MATRIX AND CERAMIC-MATRIX COMPOSITES

12.10.1 Metal-Matrix Composites (MMCs)

Metal-matrix composite materials have been so intensely researched over the past years that many new high-strength-to-weight materials have been produced. Most of these materials have been developed for the aerospace industries, but some are being used in other applications such as automobile engines. In general, according to reinforcement, the three main types of MMCs are continuous-fiber, discontinuous-fiber, and particulate reinforced.

Continuous-Fiber Reinforced MMCs Continuous filaments provide the greatest improvement in stiffness (tensile modulus) and strength for MMCs. One of the first developed continuous-fiber MMCs was the aluminum alloy matrix–boron fiber reinforced system. The boron fiber for this composite is made by chemically vapor depositing boron on a tungsten-wire substrate (Fig. 12.46a). The Al-B composite is made by hot pressing layers of B fibers between aluminum foils so that the foils deform around the fibers and bond to each other. Figure 12.46b shows a cross section of a continuous-B-fiber–aluminum alloy matrix composite. Table 12.7 lists some mechanical properties for some B fiber reinforced–aluminum alloy composites. With the addition of 51 vol % B, the axial tensile strength of aluminum alloy 6061 was increased from 310 to 1417 MPa, while its tensile modulus was increased from 69 to 231 GPa. Applications for Al-B composites include some of the structural members in the midfuselage of the space shuttle orbiter.

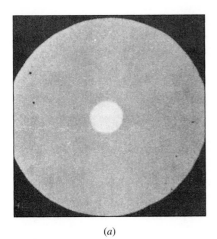

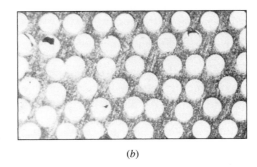

(a)   (b)

Figure 12.46
(a) 100-μm-diameter boron filament surrounding a 12.5-μm diameter tungsten-wire core. (b) Micrograph of a cross section of an aluminum alloy–boron composite (magnification 40×).
(*After "Engineered Materials Handbook," vol. 1, ASM International, 1987, p. 852.*)

Table 12.7 Mechanical properties of metal-matrix composite materials

	Tensile strength		Elastic modulus		Strain to failure (%)
	MPa	ksi	GPa	Msi	
Continuous-fiber MMCs:					
Al 2024-T6 (45% B) (axial)	1458	211	220	32	0.810
Al 6061-T6 (51% B) (axial)	1417	205	231	33.6	0.735
Al 6061-T6 (47% SiC) (axial)	1462	212	204	29.6	0.89
Discontinuous-fiber MMCs:					
Al 2124-T6 (20% SiC)	650	94	127	18.4	2.4
Al 6061-T6 (20% SiC)	480	70	115	17.7	5
Particulate MMCs:					
Al 2124 (20% SiC)	552	80	103	15	7.0
Al 6061 (20% SiC)	496	72	103	15	5.5
No reinforcement:					
Al 2124-F	455	66	71	10.3	9
Al 6061-F	310	45	68.9	10	12

Other continuous-fiber reinforcements that have been used in MMCs are silicon carbide, graphite, alumina, and tungsten fibers. A composite of Al 6061 reinforced with SiC continuous fibers is being evaluated for the vertical tail section for an advanced fighter aircraft. Of special interest is the projected use of SiC continuous-fiber reinforcements in a titanium aluminide matrix for hypersonic aircraft such as the National Aerospace plane (Fig. 1.1).

Discontinuous-Fiber and Particulate Reinforced MMCs   Many different kinds of discontinuous and particulate reinforced MMCs have been produced. These materials have the engineering advantage of higher strength, greater stiffness, and better dimensional stability than the unreinforced metal alloys. In this brief discussion of MMCs we will focus on aluminum alloy MMCs.

Particulate reinforced MMCs are low-cost aluminum alloy MMCs made by using irregular-shaped particles of alumina and silicon carbide in the range of about 3 to 200 μm in diameter. The particulate, which is sometimes given a proprietary coating, can be mixed with the molten aluminum alloy and cast into remelt ingots or extrusion billets for further fabrication. Table 12.7 indicates that the ultimate tensile strength of Al alloy 6061 can be increased from 310 to 496 MPa with a 20 percent SiC addition, while the tensile modulus can be increased from 69 to 103 GPa. Applications for this material include sporting equipment and automobile engine parts.

Discontinuous-fiber reinforced MMCs are produced mainly by powder metallurgy and melt infiltration processes. In the powder metallurgy process, needlelike silicon carbide whiskers about 1 to 3 μm in diameter and 50 to 200 μm long (Fig. 12.47) are mixed with metal powders, consolidated by hot pressing, and then extruded or forged into the desired shape. Table 12.7 shows that the ultimate tensile strength of Al alloy 6061 can be increased from 310 to 480 MPa with a 20 percent SiC whisker addition, while the tensile modulus can be raised from 69 to 115 GPa. Although greater increases in strength and stiffness can be achieved with the whisker additions than with the particulate material, the powder metallurgy and melt infiltration processes are more costly. Applications for discontinuous-fiber reinforced aluminum alloy MMCs include missile guidance parts and high-performance automobile pistons.

(*a*)

(*b*)

Figure 12.47
Micrographs of single-crystal silicon carbide whiskers used to reinforce metal-matrix composites. The whiskers are 1 to 3 μm in diameter and 50 to 200 μm long.
(*Courtesy of American Matrix Corp.*)

A metal-matrix composite is made from a boron (B) fiber–reinforced aluminum alloy (Fig. EP12.6). To form the boron fiber, a tungsten (W) wire ($r = 10\ \mu m$) is coated with boron, giving a final radius of $75\ \mu m$. The aluminum alloy is then bonded around the boron fibers, giving a volume fraction of 0.65 for the aluminum alloy. Assuming that the rule of binary mixtures (Eq. 12.7) applies also to ternary mixtures, calculate the effective tensile elastic modulus of the composite material under isostrain conditions. Data: $E_W = 410$ GPa; $E_B = 379$ GPa; $E_{Al} = 68.9$ GPa.

EXAMPLE PROBLEM 12.6

■ **Solution**

$$E_{comp} = f_W E_W + f_B E_B + f_{Al} E_{Al} \qquad f_{W+B} = 0.35$$

$$f_W = \frac{\text{area of W wire}}{\text{area of B fiber}} \times f_{W+B}$$

$$f_W = \frac{\pi (10\ \mu m)^2}{\pi (75\ \mu m)^2} \times 0.35 = 6.22 \times 10^{-3} \qquad f_{Al} = 0.65$$

$$f_B = \frac{\text{area of B fiber} - \text{area of W wire}}{\text{area of B fiber}} \times f_{W+B}$$

$$= \frac{\pi (75\ \mu m)^2 - \pi (10\ \mu m)^2}{\pi (75\ \mu m)^2} \times 0.35 = 0.344$$

$$E_{comp} = f_W E_W + f_B E_B + f_{Al} E_{Al}$$
$$= (6.22 \times 10^{-3})(410\ \text{GPa}) + (0.344)(379\ \text{GPa}) + (0.65)(68.9\ \text{GPa})$$
$$= 178\ \text{GPa} \blacktriangleleft$$

Note that the tensile modulus (stiffness) of the composite is about 2.5 times that of the unreinforced aluminum alloy.

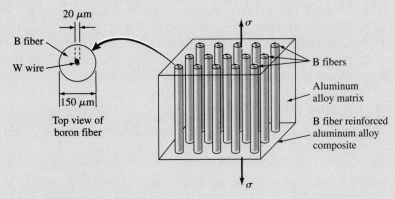

Figure EP12.6

A metal-matrix composite is made with 80 percent by volume of aluminum alloy 2124-T6 and 20 percent by volume of SiC whiskers. The density of the 2124-T6 alloy is 2.77 g/cm^3 and that of the whiskers is 3.10 g/cm^3. Calculate the average density of the composite material.

■ **Solution**

Basis is 1 m^3 of material; thus, we have 0.80 m^3 of 2124 alloy and 0.20 m^3 of SiC fiber in 1 m^3 of material.

$$\text{Mass of 2124 alloy in 1 m}^3 = (0.80 \text{ m}^3)(2.77 \text{ Mg/m}^3) = 2.22 \text{ Mg}$$

$$\text{Mass of SiC whiskers in 1 m}^3 = (0.20 \text{ m}^3)(3.10 \text{ Mg/m}^3) = \underline{0.62 \text{ Mg}}$$

$$\text{Total mass in 1 m}^3 \text{ of composite } = \underline{2.84 \text{ Mg}}$$

$$\text{Ave. density} = \frac{\text{mass}}{\text{unit volume}} = \frac{\text{total mass of material in 1 m}^3}{1 \text{ m}^3} = 2.84 \text{ Mg/m}^3 \blacktriangleleft$$

12.10.2 Ceramic-Matrix Composites (CMCs)

Ceramic-matrix composites have been developed recently with improved mechanical properties such as strength and toughness over the unreinforced ceramic matrix. Again, the three main types according to reinforcement are continuous-fiber, discontinuous-fiber, and particulate reinforced.

Continuous-Fiber Reinforced CMCs Two kinds of continuous fibers that have been used for CMCs are silicon carbide and aluminum oxide. In one process to make a ceramic-matrix composite, SiC fibers are woven into a mat and then chemical vapor deposition is used to impregnate SiC into the fibrous mat. In another process, SiC fibers are encapsulated by a glass-ceramic material (see Example Problem 12.8). Applications for these materials include heat-exchanger tubes, thermal protection systems, and materials for corrosion-erosion environments.

Discontinuous (Whisker) and Particulate Reinforced CMCs Ceramic whiskers (Fig. 12.47) can significantly increase the fracture toughness of monolithic ceramics (Table 12.8). A 20 vol % SiC whisker addition to alumina can increase the fracture toughness of the alumina ceramic from about 4.5 to 8.5 MPa $\sqrt{m}$. Short-fiber and particulate reinforced ceramic-matrix materials have the advantage of being able to be fabricated by common ceramic processes such as hot isostatic pressing (*HIP*ing).

Ceramic-matrix composites are believed to be toughened by three main mechanisms, all of which result from the reinforcing fibers interfering with crack propagation in the ceramic. These mechanisms are:

1. *Crack deflection.* Upon encountering the reinforcement, the crack is deflected, making its propagating path more meandering. Thus, higher stresses are required to propagate the crack.

2. *Crack bridging.* Fibers or whiskers can bridge the crack and help keep the material together, thus increasing the stress level needed to cause further cracking (Fig. 12.48).

3. *Fiber pullout.* The friction caused by fibers or whiskers being pulled out of the cracking matrix absorbs energy, and thus higher stresses must be applied to produce

Table 12.8 Mechanical properties of SiC whisker reinforced ceramic-matrix composites at room temperature

Matrix	SiC whisker content (vol %)	Flexural strength		Fracture toughness	
		MPa	**ksi**	**MPa$\sqrt{m}$**	**ksi$\sqrt{in.}$**
Si$_3$N$_4$	0	400–650	60–95	5–7	4.6–6.4
	10	400–500	60–75	6.5–9.5	5.9–8.6
	30	350–450	50–65	7.5–10	6.8–9.1
Al$_2$O$_3$	0	…	…	4.5	4.1
	10	400–510	57–73	7.1	6.5
	20	520–790	75–115	7.5–9.0	6.8–8.2

Source: "Engineered Materials Handbook," vol. 1, Composites, ASM International, 1987, p. 942.

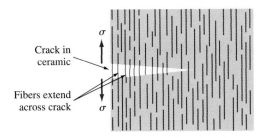

Figure 12.48
Schematic diagram showing how reinforcing fibers can inhibit crack propagation in ceramic-matrix materials by crack bridging and fiber pullout energy absorption.

further cracking. Therefore, a good interfacial bond is required between the fibers and the matrix for higher strengths. There also should be a good match of coefficient of expansion between the matrix and fibers if the material is to be used at high temperatures.

A ceramic-matrix composite is made with continuous SiC fibers embedded in a glass-ceramic matrix (Fig. EP12.8). (*a*) Calculate the tensile elastic modulus of the composite under isostrain conditions, and (*b*) calculate the stress σ at which the cracks start to grow. Data are as follows:

EXAMPLE PROBLEM 12.8

Glass-ceramic matrix:
  $E = 94$ GPa
  $K_{IC} = 2.4$ MPa $\sqrt{m}$
  Largest preexisting flaw is 10 μm
    in diameter

SiC fibers:
  $E = 350$ GPa
  $K_{IC} = 4.8$ MPa $\sqrt{m}$
  Largest surface notches are 5 μm
    deep

■ **Solution**

a. Calculation of E for the composite. Assuming Eq. 12.7 for isostrain conditions is valid,

$$E_{comp} = f_{GC}E_{GC} + f_{SiC}E_{SiC}$$

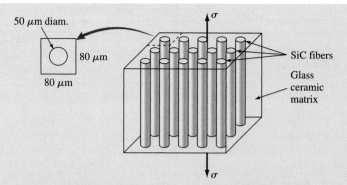

Figure EP12.8

Since the length of the fibers is the same, we can calculate the volume fraction of the SiC fibers by calculating the area fraction occupied by the fibers on the surface at the ends of the fibers. From Fig. EP12.8, each 50-μm-diameter fiber is enclosed by an 80 μm $\times$ 80 μm area. Thus,

$$f_{SiC} = \frac{\text{area of fiber}}{\text{selected tot. area}} = \frac{\pi (25 \ \mu m)^2}{(80 \ \mu m)(80 \ \mu m)} = 0.307$$

$$f_{GC} = 1 - 0.307 = 0.693$$

$$E_{comp} = (0.693)(94 \ \text{GPa}) + (0.307)(350 \ \text{GPa}) = 172 \ \text{GPa} \blacktriangleleft$$

b. Stress at which cracks first start to form in the composite. For isostrain conditions, $\epsilon_{comp} = \epsilon_{GC} = \epsilon_{SiC}$. Since $\sigma = E\epsilon$ and $\epsilon = \sigma/E$,

$$\frac{\sigma_{comp}}{E_{comp}} = \frac{\sigma_{GC}}{E_{GC}} = \frac{\sigma_{SiC}}{E_{SiC}}$$

Fracture begins in a given component when $\sigma = K_{IC}/\sqrt{\pi a}$ (Eq. 11.4), assuming $Y = 1$. We shall calculate the minimum stress to cause crack formation in both materials and then compare our results. The component that cracks with the lower stress will determine at what stress the composite will begin to crack.

(i) Glass ceramic. For this material the largest preexisting flaw is 10 μm in diameter. This value is $2a$ for Eq. 11.4, and thus $a = 10 \ \mu m/2$ or $a = 5 \ \mu m$.

$$\frac{\sigma_{comp}}{E_{comp}} = \frac{\sigma_{GC}}{E_{GC}} = \left(\frac{K_{IC,GC}}{\sqrt{\pi a}}\right)\left(\frac{1}{E_{GC}}\right)$$

$$\sigma_{comp} = \left(\frac{E_{comp}}{E_{GC}}\right)\left(\frac{K_{IC,GC}}{\sqrt{\pi a}}\right) = \left(\frac{172 \ \text{GPa}}{94 \ \text{GPa}}\right)\left[\frac{2.4 \ \text{MPa} \sqrt{m}}{\sqrt{\pi (5 \times 10^{-6} \ m)}}\right] = 1109 \ \text{MPa}$$

(ii) SiC fibers. For this material, $a = 5 \ \mu m$ for surface cracks.

$$\sigma_{comp} = \left(\frac{E_{comp}}{E_{SiC}}\right)\left(\frac{K_{IC,SiC}}{\sqrt{\pi a}}\right) = \left(\frac{172 \ \text{GPa}}{350 \ \text{GPa}}\right)\left[\frac{4.8 \ \text{MPa} \sqrt{m}}{\sqrt{\pi (5 \times 10^{-6} \ m)}}\right] = 596 \ \text{MPa}$$

Thus, the SiC fibers in the ceramic composite will begin to crack first at an applied stress of 596 MPa. $\blacktriangleleft$

12.10.3 Ceramic Composites and Nanotechnology

Recently, nanotechnology researchers have developed ceramics matrix composites with improved mechanical, chemical, and electrical properties by integrating carbon nanotubes in the microstructure of conventional alumina. This is an exciting new development that further increases the importance of nanotechnology in materials science. Researchers have been able to create a ceramic composite made of alumina, 5 to 10 percent carbon nanotubes, and 5 percent finely milled niobium. The compact was sintered and densified to produce a solid that possesses a fracture toughness up to five times greater than pure alumina. The new material is also able to conduct electricity at a rate ten trillion times higher than pure alumina. Finally, it can either conduct heat (if the nanotubes are aligned parallel to the direction of heat flow) or act as a thermal protection barrier (if the nanotubes are aligned perpendicular to the direction of heat flow). This makes the new ceramic an outstanding candidate for thermal protection coating applications.

12.11 BONE: A NATURAL COMPOSITE MATERIAL

12.11.1 Composition

Bone is a structural material found in many organisms and is another example of a complex natural composite material, consisting of a mixture of organic and inorganic materials. The inorganic component consists of **hydroxyapatite** (HA) and has a composition of $Ca_{10}(PO_4)_6(OH)_2$. Hydroxyapatite is platelike, 20 to 80 nm long and 2 to 5 nm thick. This inorganic component, which gives bone its solid and hard consistency, constitutes 60 to 70 percent of the dry weight of bone. The organic portion of the bone consists mainly of a protein called **collagen** (*Type I*) and a small amount of noncollagenous proteins. Collagen is fibrous, tough, flexible, and highly inextensible; it provides bone with its flexibility and resilience. Collagen constitutes 25 to 30 percent of the weight of dry bone. The remaining weight of dry bone is due to water at about 5 percent. This seems very similar to the properties of other fibrous composite materials that were discussed in previous sections: a mixture of two or more materials with significantly different composition and properties that produces a new material with its own unique properties.

12.11.2 Macrostructure

The microstructure of bone is complex, containing many constituents in both the micro- and the nanorange scale. A discussion of the microstructure falls outside the scope of this textbook. However, the macrostructure of bone is important to discuss because it affects the mechanical properties of the bone. Although different bones in the body have different properties and structure, the structure of all bones at the macroscopic level may be divided into two distinct types of osseous (bony) tissues: (1) **cortical or compact** and (2) **cancellous or trabecular** (Fig. 12.49). The cortical portion is dense (ivory like) and comprises the outer structure or cortex of the bone

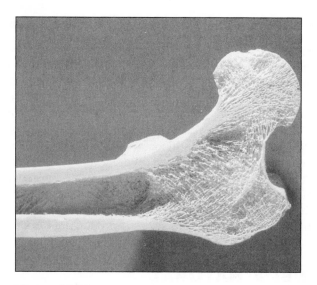

Figure 12.49
A longitudinal section through an adult femur.
(© *Lester V. Bergman/Corbis*)

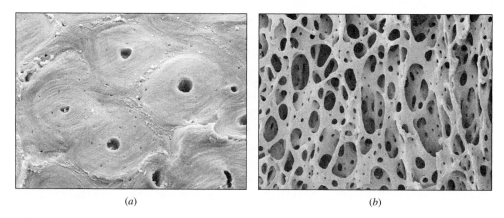

(*a*)   (*b*)

Figure 12.50
(*a*) The SEM image of the cortical bone from a human tibia. (© *Andrew Syred/Photo Researchers, Inc.*)
(*b*) A photomicrograph of cancellous bone. (© *Susumu Nishinaga/Photo Researchers, Inc.*)

as shown in Fig. 12.50*a*. The internal portion of the bone consists of the cancellous tissue, which is composed of thin plates or trabeculae loosely meshed and porous as shown in Fig. 12.50*b*. The pores in the cancellous region are filled with red marrow. Various bones have different functional requirements and therefore have different ratios of cortical to cancellous tissue, thereby affecting their properties.

12.11.3 Mechanical Properties

Bone is a two-phase composite of organic and inorganic materials. Just like any other material, its mechanical properties may be determined by performing a uniaxial tension test on it. As with other materials, an elastic range, a yield point, a plastic region, and a failure point will appear on the corresponding stress-strain curve. Cortical and cancellous bones have completely different mechanical properties. The cortical bone has a higher density and is stronger and stiffer than the cancellous bone; it is however more brittle. It yields and fractures when the strain exceeds 2.0 percent. Cancellous (trabecular) bone, however, is less dense, may sustain a strain level of 50 percent before it fractures, and, because of its porous structure, absorbs large amounts of energy before it fractures. Typical stress curves for cortical and trabecular bones of two different densities are given in Figure 12.51. One can clearly observe the differences in the modulus of elasticity, yield point, ductility, toughness, and failure strength of various bones in Fig. 12.51.

In Sec. 12.3, the anisotropic behavior of fiber-reinforced composites was discussed. For instance, Table 12.4 shows the difference in the mechanical properties in the longitudinal and transverse directions for carbon-fiber-epoxy composites. The same anisotropic behavior is observed in bone. When tensile test samples of cortical bone from the human femoral shaft were dissected in various orientations and tested uniaxially, the corresponding stress-strain curves were completely different as

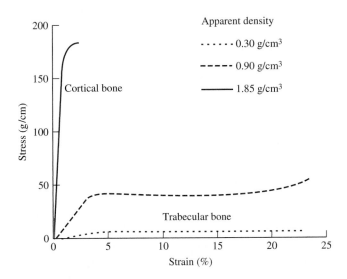

Figure 12.51
The stress-strain curves for cortical and trabecular bones. Note that the toughness (the corresponding area under the curve) of the 0.9 g/cm^3 trabecular bone is higher than that of the cortical bone.

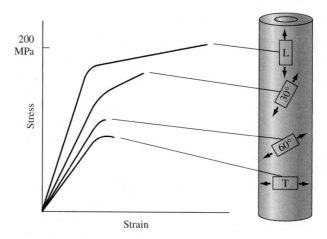

Figure 12.52
The stress-strain curves for cortical bone samples of various orientations along the bone showing the anisotropic nature of bone.

shown in Fig. 12.52. The sample that was aligned with the longitudinal axis (L) produced the highest stiffness, strength, and ductility. In contrast, the sample that was aligned transverse (T) to the longitudinal axis produced the lowest modulus of elasticity, strength, and ductility. This is a clear indication of the anisotropic behavior of bone. It can by and large be stated that various bones are strongest and most rigid (stiff) in the direction that they are generally loaded during normal daily activities. It is very important to note that bones are generally much stronger in compression than tension; for instance, cortical bone has a tensile strength of 130 Mpa and a compressive strength of 190 Mpa. Cancellous bones behave in a similar manner under tension and compression.

12.11.4 Biomechanics of Bone Fracture

In the course of normal daily activities a human bone supports various modes of loading, including tensile, compressive, bending, torsional, shear, and combined. Tensile fractures commonly occur in bones that are highly cancellous such as the bone adjacent to the achilles tendon because of the large tensile forces that the calf muscle can exert on this bone. Shear fractures also occur more commonly in highly cancellous bones. Fractures under compressive loading are found mostly in the vertebrae and are more common in older patients suffering from osteoporosis (bone porosity). Bending will cause both compressive and tensile stresses in the bone. The long bones in the body such as the femur or the tibia are most susceptible to this type of loading. Skiers are susceptible to this type of fracture ("boot top" fracture) when they fall over their boots: The proximal tibia (top part of the tibia) bends forward; the distal tibia (lower part of the tibia) undergoes the same moment due to the constrained nature of the foot and ankle in the boot; and the tibia then fractures at the top of the boot where motion is resisted by the boot contact (three-point bending). The fracture always starts on the side of the femur that is under tension since the bone is weaker in tension. Fractures due to torsion also occur in the long bones of the human body. Such fractures start almost parallel to the longitudinal axis of the bone and extend at a 30 degree angle relative to the longitudinal axis. Most fractures occur under a combined state of loading in which two or more of the loading types stated above may be present.

12.11.5 Viscoelasticity of the Bone

Another important biomechanical behavior of bone lies in its variable response to loading rate or strain rate. For example, during normal walking, the strain rate of the femur has been measured to be approximately 0.001 s^{-1} while during slow running it is measured at approximately 0.03 s^{-1}. During impact trauma the strain rate can reach as high as 1 s^{-1}. Bone reacts differently under these different loading rate conditions. As the strain rate increases, a bone becomes stiffer and stronger (fails at a higher load). Over the full range of applied strain rates, cortical bone becomes stronger by a factor of three and its modulus of elasticity increases by a factor of two. At very high strain rates (impact trauma), the bone also becomes more brittle as well as stores larger amounts of energy before fracturing. This is an important issue in trauma: Under low-energy fracture the energy is expended in the fracture of the bone, and the surrounding tissue does not experience significant damage. However, under high-energy fracture the excess available energy causes significant damage to the surrounding tissue. The dependence of the mechanical behavior of the bone on strain rate is called **viscoelasticity.** For comparison, polymeric materials also behave in a viscoelastic manner under similar loading rates.

Finally, a bone may also undergo fatigue fracture. This happens, as with other materials, when repeated cyclic loading is applied. This may be the case for an athlete undergoing weight training. After numerous loading repetitions, the muscles become tired and, as a result, the bone carries a larger portion of the load. Because of the higher stresses supported by the bone, fatigue failure may occur after many cycles.

12.11.6 Bone Remodeling

In Chap. 1, the concept of smart materials was introduced. These materials have the ability to sense environmental stimuli and respond accordingly. Bone is an example of a complex biological smart material. Bone has the ability to alter its size, shape, and structure based on the mechanical demands placed upon it. The ability of bone to gain cortical or cancellous bone mass due to the elevated level of stress is called **bone remodeling** and is named by Wolf's law. It is for this reason that older individuals with reduced physical activity and astronauts working in the weightless environment of space for long periods of time suffer bone loss. Moderate exercise with low weights is believed to reduce the bone-loss phenomenon in the aging population.

12.11.7 Nanotechnology and Bone Repair

In Chap. 11, the concept of bioactive and resorbable ceramics and their role in repairing damaged or diseased bone and in establishing strong interfaces between tissue and implant (*osseointegration*) were discussed. Specifically, the ceramic tricalcium phosphate was named as one such material. Present research efforts are concentrated on developing nanophase ceramics based on calcium phosphate and/or calcium phosphate derivatives such as hydroxyapatite, calcium carbonate, and bioactive glasses. Having already explained that a large portion of the bone is made of nanorange-size HA, the importance of nanotechnology in this field can be appreciated. Calcium phosphate nanomaterials with grain size less than 100 nm exhibit osteoinduction in various research animal models. However, the questions remain, "Does the newly generated bone have the same properties as the original bone?" and "Is there a way that we can synthesize nanoceramics that, upon resorption, produce bone-quality material?" These questions still remain unanswered, and many more years of research is required before we understand the behavior of these nanoceramics.

12.12 S U M M A R Y

A composite material with respect to materials science and engineering can be defined as a materials system composed of a mixture or combination of two or more micro- or macro-constituents that differ in form and chemical composition and are essentially insoluble in each other.

Some fiber-reinforced-plastic composite materials are made with synthetic fibers of glass, carbon, and aramid. Of these three fibers, glass fibers are the lowest in cost and have intermediate strength and highest density when compared to the others. Carbon fibers have high strength, high modulus, and low density but are expensive and so are used only for applications requiring their especially high strength-to-weight ratio. Aramid fibers have high strength and low density but are not as stiff as carbon fibers. Aramid fibers are also relatively expensive and so are used for applications where a high strength-to-weight ratio is required along with better flexibility than carbon fibers. The most commonly used matrices for glass fibers for fiber-reinforced-plastic composites are the polyesters, whereas the most commonly used matrices for carbon-fiber-reinforced plastics are the epoxies. Carbon-fiber-reinforced-epoxy composite materials are used extensively for aircraft and aerospace applications. Glass-fiber-reinforced-polyester composite materials have a much wider usage and find application in the building, transportation, marine, and aircraft industries, for example.

Concrete is a ceramic composite material consisting of aggregate particles (i.e., sand and gravel) in a hardened cement paste matrix usually made with portland cement. Concrete as a construction material has advantages that include usable compressive strength, economy, castability on the job, durability, fire resistance, and aesthetic appearance. The low tensile strength of concrete can be increased significantly by reinforcement with steel rods. A further improvement in the tensile strength of concrete is attainable by introducing residual compressive stresses in the concrete at positions of high tensile loading by prestressing with steel reinforcements.

Wood is a natural composite material consisting essentially of cellulose fibers bonded together by a matrix of polymeric material made up mainly of lignin. The macrostructure of wood consists of sapwood, which is made up primarily of living cells and which carries the nutrients, and heartwood, which is composed of dead cells. The two main types of woods are softwoods and hardwoods. Softwoods have exposed seeds and narrow (needlelike) leaves, whereas hardwoods have covered seeds and broad leaves. The microstructure of wood consists of arrays of cells mainly in the longitudinal direction of the tree stem. Softwoods have long, thin-walled tubular cells called tracheids, whereas hardwoods have a dense cell structure that contains large vessels for the conduction of fluids. Wood as a construction material has advantages that include usable strength, economy, ease of workability, and durability if properly protected.

Bone is the natural composite material in the human body. It is made up of a mixture of organic (collagen) and inorganic (hydroxyapatite) materials. The macrostructure of bone consists of two distinct types of bony tissues: cortical and cancellous. Bone behaves in an anisotropic manner consistent with composite materials. It is also important to note that bone is stronger in compression.

12.13 DEFINITIONS

Sec. 12.1

Composite material: a materials system composed of a mixture or combination of two or more micro- or macroconstituents that differ in form and chemical composition and are essentially insoluble in each other.

Sec. 12.2

Fiber-reinforced plastics: composite materials consisting of a mixture of a matrix of a plastic material such as a polyester or epoxy strengthened by fibers of high strength such as glass, carbon, or aramid. The fibers provide the high strength and stiffness, and the plastic matrix bonds the fibers together and supports them.

E-glass fibers: fibers made from E (electrical) glass, which is a borosilicate glass and which is the most commonly used glass for fibers for fiberglass-reinforced plastics.

S-glass fibers: fibers made from S glass, which is a magnesia-alumina-silicate glass and which is used for fibers for fiberglass-reinforced plastics when extra-high-strength fibers are required.

Roving: a collection of bundles of continuous fibers twisted or untwisted.

Carbon fibers (for a composite material): carbon fibers produced mainly from polyacrylonitrile (PAN) or pitch that are stretched to align the fibrillar network structure within each carbon fiber and which are heated to remove oxygen, nitrogen, and hydrogen from the starting or precursor fibers.

Tow (of fibers): a collection of numerous fibers in a straight-laid bundle, specified according to the number of fibers it contains—e.g., 6000 fibers/tow.

Aramid fibers: fibers produced by chemical synthesis and used for fiber-reinforced plastics. Aramid fibers have an aromatic (benzene ring type) polyamide linear structure and are produced commercially by the Du Pont Co. under the trade name of Kevlar.

Specific tensile strength: the tensile strength of a material divided by its density.

Specific tensile modulus: the tensile modulus of a material divided by its density.

Sec. 12.3

Laminate: a product made by bonding sheets of a material together, usually with heat and pressure.

Unidirectional laminate: a fiber-reinforced-plastic laminate produced by bonding together layers of fiber-reinforced sheets that all have continuous fibers in the same direction in the laminate.

Multidirectional laminate: a fiber-reinforced-plastic laminate produced by bonding together layers of fiber-reinforced sheets with some of the directions of the continuous fibers of the sheets being at different angles.

Laminate ply (lamina): one layer of a multilayer laminate.

Sec. 12.4

Hand lay-up: the process of placing (and working) successive layers of reinforcing material in a mold by hand to produce a fiber-reinforced composite material.

Spray lay-up: a process in which a spray gun is used to produce a fiber-reinforced product. In one type of spray-up process, chopped fibers are mixed with plastic resin and sprayed into a mold to form a composite material part.

Vacuum bag molding: a process of molding a fiber-reinforced-plastic part in which sheets of transparent flexible material are placed over a laminated part that has not been cured. The sheets and the part are sealed, and a vacuum is then applied between the cover sheets and the laminated part so that entrapped air is mechanically worked out of the laminate. Then the vacuum-bagged part is cured.

Prepreg: a ready-to-mold plastic resin-impregnated cloth or mat that may contain reinforcing fibers. The resin is partially cured to a "B" stage and is supplied to a fabricator who uses the material as the layers for a laminated product. After the layers are laid up to produce a final shape, the layers are bonded together, usually with heat and pressure, by the curing of the laminate.

Filament winding: a process for producing fiber-reinforced plastics by winding continuous reinforcement previously impregnated with a plastic resin on a rotating mandrel. When a sufficient number of layers have been applied, the wound form is cured and the mandrel removed.

Sec. 12.5

Sheet-molding compound (SMC): a compound of plastic resin, filler, and reinforcing fiber used to make fiber-reinforced-plastic composite materials. SMC is usually made with about 25 to 30 percent fibers about 1 in. (2.54 cm) long, of which fiberglass is the most commonly used fiber. SMC material is usually pre-aged to a state so that it can support itself and then cut to size and placed in a compression mold. Upon hot pressing, the SMC cures to produce a rigid part.

Pultrusion: a process for producing a fiber-reinforced-plastic part of constant cross section continuously. The pultruded part is made by drawing a collection of resin-dipped fibers through a heated die.

Sec. 12.6

Concrete (portland cement type): a mixture of portland cement, fine aggregate, coarse aggregate, and water.

Portland cement: a cement consisting predominantly of calcium silicates that react with water to form a hard mass.

Aggregate: inert material mixed with portland cement and water to produce concrete. Larger particles are called coarse aggregate (e.g., gravel), and smaller particles are called fine aggregate (e.g., sand).

Hydration reaction: reaction of water with another compound. The reaction of water with portland cement is a hydration reaction.

Air-entrained concrete: concrete in which there exists a uniform dispersion of small air bubbles. About 90 percent of the air bubbles are 100 μm or less.

Reinforced concrete: concrete containing steel wires or bars to resist tensile forces.

Prestressed concrete: reinforced concrete in which internal compressive stresses have been introduced to counteract tensile stresses resulting from severe loads.

Pretensioned (prestressed) concrete: prestressed concrete in which the concrete is poured over pretensioned steel wires or rods.

Sec. 12.7

Asphalt: a bitumen consisting mainly of hydrocarbons having a wide range of molecular weights. Most asphalt is obtained from petroleum refining.

Asphalt mixes: mixtures of asphalt and aggregate that are used mainly for road paving.

Sec. 12.8

Wood: a natural composite material consisting mainly of a complex array of cellulose fibers in a polymeric material matrix made up primarily of lignin.

Lignin: a very complex cross-linked three-dimensional polymeric material formed from phenolic units.

Sapwood: the outer part of the tree stem of a living tree that contains some living cells that store food for the tree.

Heartwood: the innermost part of the tree stem that in the living tree contains only dead cells.

Cambium: the tissue that is located between the wood and bark and is capable of repeated cell division.

Softwood trees: trees that have exposed seeds and arrow leaves (needles). Examples are pine, fir, and spruce.

Hardwood trees: trees that have covered seeds and broad leaves. Examples are oak, maple, and ash.

Parenchyma: food-storing cells of trees that are short with relatively thin walls.

Wood ray: a ribbonlike aggregate of cells extending radially in the tree stem; the tissue of the ray is primarily composed of food-storing parenchyma cells.

Tracheids (longitudinal): the predominating cell found in softwoods; tracheids have the function of conduction and support.

Wood vessel: a tubular structure formed by the union of smaller cell elements in a longitudinal row.

Microfibrils: elementary cellulose-containing structures that form the wood cell walls.

Lumen: the cavity in the center of a wood cell.

Sec. 12.11

Bone: the structural material of the human body.

Hydroxyapatite: the inorganic constituent of the bone.

Collagen: the organic component of the bone.

Cortical (compact) bone: dense bone comprising the outer structure.

Cancellous (trabecular) bone: the bone material that is porous and soft comprising the internal structure of the bone.

Viscoelasticity: the type of mechanical response in a material that depends on loading rate or strain rate.

Bone remodelling: the ability of bone to alter its size and structure based on external stress.

12.14 PROBLEMS

Answers to problems marked with an asterisk are given at the end of the book.

12.1   Define a composite material with respect to a materials system.

12.2   What are the three main types of synthetic fibers used to produce fiber-reinforced-plastic composite materials?

12.3   What are some of the advantages of glass-fiber-reinforced plastics?

12.4   What are the differences in the compositions of E and S glasses? Which is the strongest and the most costly?

12.5   How are glass fibers produced? What is a glass-fiber roving?

12.6   What properties make carbon fibers important for reinforced plastics?

12.7   What are two materials used as precursors for carbon fibers?

12.8   What are the processing steps for the production of carbon fibers from polyacrylonitrile? What reactions take place at each step?

12.9   What is a tow of carbon fibers?

12.10   What processing steps are carried out if a very-high-strength type of carbon fiber is desired? If a very-high-modulus type of carbon fiber is desired, what processing steps are carried out?

12.11   What is an aramid fiber? What are two types of commercially available aramid fibers?

12.12   What type of chemical bonding takes place within the aramid fibers? What type of chemical bonding takes place between the aramid fibers?

12.13   How does the chemical bonding within and between the aramid fibers affect their mechanical strength properties?

12.14   Compare the tensile strength, tensile modulus of elasticity, elongation, and density properties of glass, carbon, and aramid fibers (Table 12.1 and Fig. 12.8).

12.15   Define specific tensile strength and specific tensile modulus. What type of reinforcing fibers of those shown in Fig. 12.9 has the highest specific modulus and what type has the highest specific tensile strength?

12.16   What are two of the most important matrix plastics for fiber-reinforced plastics? What are some advantages of each type?

12.17   How does the amount and arrangement of the glass fibers in fiberglass-reinforced plastics affect their strength?

12.18   What are the main property contributions of the carbon fibers in carbon-fiber-reinforced plastics? What are the main property contributions of the matrix plastic?

12.19   What are some carbon-fiber-epoxy composite laminates designed with the carbon fibers of different layers oriented at different angles to each other?

*12.20   A unidirectional carbon-fiber-epoxy-resin composite contains 68 percent by volume of carbon fiber and 32 percent epoxy resin. The density of the carbon fiber is 1.79 g/cm^3 and that of the epoxy resin is 1.20 g/cm^3. (*a*) What are the weight percentages of carbon fibers and epoxy resin in the composite? (*b*) What is the average density of the composite?

12.21   The average density of a carbon-fiber-epoxy composite is 1.615 g/cm^3. The density of the epoxy resin is 1.21 g/cm^3 and that of the carbon fibers is 1.74 g/cm^3. (*a*) What is the volume percentage of carbon fibers in the composite? (*b*) What are the weight percentages of epoxy resin and carbon fibers in the composite?

12.22 Derive an equation relating the elastic modulus of a layered composite of unidirectional fibers and a plastic matrix that is loaded under isostrain conditions.

*12.23 Calculate the tensile modulus of elasticity of a unidirectional carbon-fiber-reinforced-plastic composite material that contains 64 percent by volume of carbon fibers and is stressed under isostrain conditions. The carbon fibers have a tensile modulus of elasticity of 54.0×10^6 psi and the epoxy matrix a tensile modulus of elasticity of 0.530×10^6 psi.

12.24 If the tensile strength of the carbon fibers of the 64 percent carbon-fiber-epoxy composite material of Prob. 12.23 is 0.31×10^6 psi and that of the epoxy resin is 9.20×10^3 psi, calculate the strength of the composite material in psi. What fraction of the load is carried by the carbon fibers?

12.25 Calculate the tensile modulus of elasticity of a unidirectional Kevlar 49-fiber-epoxy composite material that contains 63 percent by volume of Kevlar 49 fibers and is stressed under isostrain conditions. The Kevlar 49 fibers have a tensile modulus of elasticity of 27.5×10^6 psi and the epoxy matrix a tensile modulus of elasticity of 0.550×10^6 psi.

*12.26 If the tensile strength of the Kevlar 49 fibers is 0.550×10^6 psi and that of the epoxy resin is 11.0×10^3 psi, calculate the strength of the composite material of Prob. 12.25. What fraction of the load is carried by the Kevlar 49 fibers?

12.27 Derive an equation relating the elastic modulus of a layered composite of unidirectional fibers and a plastic matrix that is stressed under isostress conditions.

12.28 Calculate the tensile modulus of elasticity for a laminated composite consisting of 62 percent by volume of unidirectional carbon fibers and an epoxy matrix under isostress conditions. The tensile modulus of elasticity of the carbon fibers is 340 GPa and that of the epoxy is 4.50×10^3 MPa.

*12.29 Calculate the tensile modulus of elasticity of a laminate composite consisting of 62 percent by volume of unidirectional Kevlar 49 fibers and an epoxy matrix stressed under isostress conditions. The tensile modulus of elasticity of the Kevlar 49 fibers is 170 GPa and that of the epoxy is 3.70×10^3 MPa.

12.30 Describe the hand lay-up process for producing a fiberglass-reinforced part. What are some advantages and disadvantages of this method?

12.31 Describe the spray-up process for producing a fiberglass-reinforced part. What are some advantages and disadvantages of this method?

12.32 Describe the vacuum bag-autoclave process for producing a carbon-fiber-reinforced-epoxy part for an aircraft.

12.33 Describe the filament-winding process. What is a distinct advantage of this process from an engineering design standpoint?

12.34 Describe the sheet-molding compound manufacturing process. What are some of the advantages and disadvantages of this process?

12.35 Describe the pultrusion process for the manufacture of fiber-reinforced plastics. What are some advantages of this process?

12.36 What advantages and disadvantages does concrete offer as a composite material?

12.37 What are the principal components of most concretes?

12.38 What are the basic raw materials for portland cement? Why is portland cement so called?

12.39 How is portland cement made? Why is a small amount of gypsum added to portland cement?

12.40 What are the names, chemical formulas, and abbreviations for the four principal compounds of portland cement?

12.41 List the five main ASTM types of portland cement and give the general conditions for which each is used and their applications.

12.42 What type of chemical reactions occur during the hardening of portland cement?

12.43 Write the chemical reactions for C_3S and C_2S with water.

12.44 Which component of portland cement hardens rapidly and is mostly responsible for early strength?

12.45 Which component of portland cement reacts slowly and is mainly responsible for the strengthening after about one week?

12.46 Which compound is kept to a low level for sulfate-resisting portland cements?

12.47 Why is C_4AF added to portland cement?

12.48 Why is it important that during the first few days of the curing of concrete the evaporation of water from its surface be prevented or reduced?

12.49 What is air-entrained concrete? What is an advantage of air-entrained concrete?

12.50 What method is used to make air-entrained concrete? What volume percent of air is used in the concrete for frost protection?

12.51 How does the water-cement ratio (by weight) affect the compressive strength of concrete? What ratio gives a compressive strength of about 5500 psi to normal concrete? What is the disadvantage of too high a water-cement ratio? Of too low a water-cement ratio?

12.52 What major factors should be taken into account in the design of concrete mixtures?

12.53 What are the absolute volume percent ranges for the major components of normal concrete?

*12.54 We want to produce 100 ft^3 of concrete with a ratio of 1:1.9:3.8 (by weight) of cement, sand, and gravel. What are the required amounts of the components if 5.5 gal of water per sack of cement is to be used? Assume the free moisture contents of the sand and gravel are 3 and 0 percent, respectively. The specific gravities of the cement, sand, and gravel are 3.15, 2.65, and 2.65, respectively. (One sack of cement weighs 94 lb and 1 ft^3 water = 7.48 gal.) Give answers for the cement in sacks, the sand and gravel in pounds, and the water in gallons.

12.55 We want to produce 50 ft^3 of concrete with a ratio of 1:1.9:3.2 (by weight) of cement, sand, and gravel. What are the required amounts of the components if 5.5 gal of water per sack of cement is to be used? Assume the free moisture contents of the sand and gravel are 4 and 0.5 percent, respectively. The specific gravities of the cement, sand, and gravel are 3.15, 2.65, and 2.65, respectively. (One sack of cement weighs 94 lb and 1 ft^3 water = 7.48 gal.) Give answers for the cement in sacks, the sand and gravel in pounds, and the water in gallons.

12.56 Why is concrete mainly used in compression in engineering designs?

12.57 What is reinforced concrete? How is it made?

12.58 What is the main advantage of prestressed concrete?

12.59 Describe how compressive stresses are introduced in pretensioned prestressed concrete.

12.60 Describe how compressive stresses are introduced in posttensioned prestressed concrete.

12.61 What is asphalt? Where is asphalt obtained?

12.62 What are the chemical composition ranges for asphalts?

12.63 What does an asphalt mix consist of? What is the asphalt content of a type IV road-paving asphalt?

12.64 What characteristics are desirable for the aggregate for a road-paving asphalt?

12.65 Describe the different layers in the cross section of a tree stem. Also, give the functions of each layer.

12.66 What is the difference between softwoods and hardwoods? Give several examples of both. Are all hardwoods physically hard?

12.67 What are the subrings of the annual growth rings of trees?

12.68 What axis is parallel to the annual rings? What axis is perpendicular to the annual ring?

12.69 Describe the microstructure of a softwood tree.

12.70 What are the functions of the wood rays of a tree?

12.71 Describe the microstructure of a hardwood tree. What is the difference between ring-porous and diffuse-porous tree microstructures?

12.72 Describe the cell-wall ultrastructure of a wood cell.

12.73 Describe the constituents of a wood cell.

12.74 A piece of wood containing moisture weighs 210 g, and after oven drying to a constant weight, it weighs 125 g. What is its percent moisture content?

12.75 A piece of wood contains 15 percent moisture. What must its weight have been before oven drying if it has a constant weight of 125 g after drying?

12.76 A piece of wood contains 45 percent moisture. What must its final weight be after oven drying if it weighed 165 g before drying?

12.77 What is the reason for the relatively high strength of wood in the longitudinal direction of the tree stem as compared to the transverse direction?

12.78 What is the green condition for wood? Why is wood much weaker in the green condition than in the kiln-dried condition?

12.79 Why does wood shrink much more in the transverse direction than in the longitudinal direction?

12.80 A metal-matrix composite is made of a 6061-Al alloy matrix and continuous boron fibers. The boron fibers are produced with an 11.5 μm diameter tungsten-wire core that is coated with boron to make a final 107 μm diameter fiber. A unidirectional composite is made with 51 vol % of the boron fibers in the Al 2024 matrix. Assuming the law of mixtures applies to isostrain conditions, calculate the tensile modulus of the composite in the direction of the fibers. Data are $E_B = 370$ GPa, $E_W = 410$ GPa, and $E_{Al} = 70.4$ GPa.

*12.81 A newly developed metal-matrix composite is made for the National Aerospace plane with a matrix of the intermetallic compound titanium aluminide (Ti_3Al) and continuous silicon carbide fibers. A unidirectional composite is made with the SiC continuous fibers all in one direction. If the modulus of the composite is 220 GPa and assuming isostrain conditions, what must the volume percent of SiC fibers in the composite be if $E(SiC) = 390$ GPa and $E(Ti_3Al) = 145$ GPa?

*12.82 A metal-matrix composite is made with a matrix of Al 6061 alloy and 47 vol % Al_2O_3 continuous fibers all in one direction. If isostrain conditions prevail, what is the tensile modulus of the composite in the direction of the fibers? Data are $E(Al_2O_3) = 395$ GPa and $E(Al\ 6061) = 68.9$ GPa.

12.83 An MMC is made with an Al 2024 alloy with 20 vol percent SiC whiskers. If the density of the composite is 2.90 g/cm^3 and that of the SiC fibers is 3.10 g/cm^3, what must the density of the Al 2024 alloy be?

12.84 A ceramic-matrix composite is made with continuous SiC fibers embedded in a *reaction-bonded silicon nitride* (RBSN) matrix with all the SiC fibers aligned in one direction. Assuming isostrain conditions, what is the volume fraction of the SiC fibers in the composite if the composite has a tensile modulus of 250 GPa? Data are $E(SiC) = 395$ GPa and $E(RBSN) = 155$ GPa.

12.85 The largest preexisting flaw in the reaction-bonded silicon nitride matrix of Prob. 12.84 is 6.0 μm in diameter, and the largest surface notch on the SiC fibers is 3.5 μm deep. Calculate the stress at which cracks first form in the composite when stress is slowly applied under isostrain conditions and in the direction of the fibers. Data are $K_{IC}(RBSN) = 3.5$ MPa $- \sqrt{m}$ and $K_{IC}(SiC) = 4.8$ MPa $- \sqrt{m}$.

12.86 A ceramic-matrix composite is made with an aluminum oxide (Al_2O_3) matrix and continuous silicon carbide fiber reinforcement with all the SiC fibers in one direction. The composite consists of 30 vol % of SiC fibers. If isostrain conditions exist, calculate the tensile modulus of the composite in the direction of the fibers. If a load of 8 MN is applied to the composite in the direction of the fibers, what is the elastic strain in the composite if the surface area over which the load is applied is 55 cm^2? Data are $E(Al_2O_3) = 350$ GPa and $E(SiC) = 340$ GPa.

12.87 For Prob. 12.86, the Al_2O_3 matrix has flaws up to 10 μm in diameter and the largest surface notch of the SiC fibers is 4.5 μm. (*a*) Will the matrix or the fibers crack first? (*b*) What stress on the composite in the direction of the fibers will cause the first crack to form? Data are $K_{IC}(Al_2O_3) = 3.8$ MPa $- \sqrt{m}$ and $K_{IC}(SiC) = 4.6$ MPa $- \sqrt{m}$.

12.15 MATERIALS SELECTION AND DESIGN PROBLEMS

1. It is generally true that fibers are stronger (in the length direction) than the bulk material from which they are made. Can you explain why?

2. A beam of rectangular cross-section ($b = 0.3$ in., $h = 0.6$ in.) is made of 1040 annealed steel. (*a*) Calculate the product *EI*, called the *flexural rigidity* of the beam, where *E* is the modulus of elasticity and *I* the cross-sectional moment of inertia). (*b*) If you were to make this beam from aluminum 6061-T6 with the same flexural rigidity, given that *b* cannot change, what value of *h* would you choose? (*c*) If you were to make this beam from unidirectional carbon fiber composites (use Table 12.4), what would the value of *h* be? (*d*) In all three cases, calculate the weight of the beam if the length is given to be 6 ft (analyze your results).

3. (*a*) In Prob. 2, if the beam is under pure bending of 400 lb·in. and the normal stress acting on the cross section is not to exceed 30 ksi, design the cross section for the given loading. (*b*) Determine the weight of each beam. (*c*) Determine the material cost of each beam.

4. (*a*) In Prob. 2, if all three beams are to have the same cross-sectional area, compare their flexural rigidities. (*b*) Which material gives you the highest *EI*? (*c*) Which material will give you the highest resistance to normal stress?

5. A stress-bearing component is made of aluminum 2024-T4. The company would like to reduce the weight of this component by manufacturing it from unidirectional carbon fiber composite laminate (without changing its dimensions). (*a*) Design the fiber content of this laminate such that its modulus of elasticity is comparable to that of aluminum. (*b*) Will your design satisfy the requirement of saving weight? (Assume isostrain conditions and no voids will exist in your material.)

6. Repeat Prob. 5 but assume isostress conditions. Compare the results with Prob. 5 and draw a conclusion.

7. When designing a tubular fishing rod, one should consider the hoop strength (to prevent the collapsing straw effect) and the axial stiffness of the rod (to prevent excessive elastic deformation in the rod). (*a*) Design a process to make tubular fishing rods from fiber-reinforced composites. (*b*) What steps would you take to assure that (i) hoop stresses are properly supported and (ii) the axial stiffness is sufficient? (*c*) Suggest suitable materials for this application.

8. (*a*) When designing the shaft of a golf club, what mechanical loading conditions should we consider (assume a novice player is using the club)? (*b*) How can fiber-reinforced composites support these loading conditions (propose a process)? (*c*) Identify the advantages of replacing stainless steel shafts with composite shafts.

9. (*a*) Compare and contrast the properties of the following fiber-reinforced composite materials: Kevlar 49/epoxy, S glass/epoxy, SiC/tungsten, carbon/carbon. (*b*) Give one general application for each material.

10. One can increase the toughness of a ceramic material by introducing fibers to the structure of the ceramic and create a ceramic-matrix composite. However, unlike other fiber-reinforced composites, in this case the bonding between the matrix and fiber should not be very strong. Explain how this process toughens the ceramic material.

11. Give examples of metal-matrix composites, and identify their advantages over polymer-matrix composites.

13

Corrosion

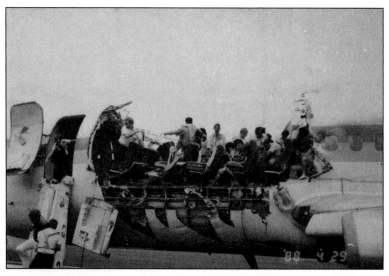

(© *AP Photo/Robert Nichols*)

On April 28, 1988, an Aloha Airline Boeing 737 lost a major portion of its upper fuselage while in flight at 24,000 feet.[1] The pilot was successful at landing the aircraft without any additional catastrophic damage to the structure of the plane. The fuselage panels that are joined together along lap joints using rivets were corroded resulting in cracking and debonding over the life of the aircraft (in this case 19 years). As a result, structural failure of the fuselage occurred in midflight due to corrosion-accelerated fatigue.[1,2]

The two aluminum alloys used most often in fuselage skin, 2024-T3 and 7075-T6, possess excellent static and fatigue strengths; however, unfortunately, they are also more prone to corrosion damage such as pitting and exfoliation. To avoid the reoccurrence of this problem, more rigorous inspection guidelines are followed to detect corrosion damage. ■

---

[1]http://www.aloha.net/~icarus/

[2]http://www.corrosion-doctors.org

By the end of this chapter, students will be able to . . .

1. Define corrosion and the corresponding electrochemical reactions often associated with it.

2. Be able to rate the reactivity (cathodic versus anodic) of some important pure metals based on standard half-cell potentials.

3. Define a galvanic cell, its important elements, the role of electrolyte in it, and various circumstances under which a galvanic cell is created in real life.

4. Explain the basic aspects of corrosion kinetics and define polarization, passivation, and the galvanic series.

5. Define various types of corrosion and the circumstances under which each occurs in everyday life.

6. Define oxidation and how it can protect metals.

7. Name various ways that one can protect against corrosion.

13.1 GENERAL

Corrosion may be defined as the deterioration of a material resulting from chemical attack by its environment. Since corrosion is caused by chemical reaction, the rate at which the corrosion takes place will depend to some extent on the temperature and the concentration of the reactants and products. Other factors such as mechanical stress and erosion may also contribute to corrosion.

When we speak of corrosion, we are usually referring to an electrochemical attack process on metals. Metals are susceptible to this attack because they possess free electrons and can set up electrochemical cells within their structure. Most metals are corroded to some extent by water and the atmosphere. Metals can also be corroded by direct chemical attack from chemical solutions and even liquid metals.

The corrosion of metals can be regarded in some ways as reverse extractive metallurgy. Most metals exist in nature in the combined state, for example, as oxides, sulfides, carbonates, or silicates. In these combined states the energies of the metals are lower. In the metallic state the energies of metals are higher, and thus there is a spontaneous tendency for metals to react chemically to form compounds. For example, iron oxides exist commonly in nature and are reduced by thermal energy to iron, which is in a higher energy state. There is, therefore, a tendency for the metallic iron to spontaneously return to iron oxide by corroding (rusting) so that it can exist in a lower energy state (Fig. 13.1).

Nonmetallic materials such as ceramics and polymers do not suffer electrochemical attack but can be deteriorated by direct chemical attack. For example,

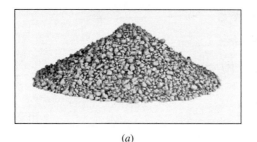
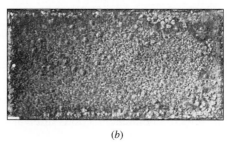

(a)   (b)

Figure 13.1
(a) Iron ore (iron oxide). (b) Corrosion products in the form of rust (iron oxide) on a steel
(iron) sample that has been exposed to the atmosphere. By rusting, metallic iron in the
form of steel has returned to its original lower energy state.
(*Courtesy of the LaQue Center for Corrosion Technology, Inc.*)

ceramic refractory materials can be chemically attacked at high temperatures by
molten salts. Organic polymers can be deteriorated by the chemical attack of organic
solvents. Water is absorbed by some organic polymers, which causes changes in
dimensions or property changes. The combined action of oxygen and ultraviolet
radiation will deteriorate some polymers even at room temperature.

Corrosion, therefore, is a destructive process as far as the engineer is concerned
and represents an enormous economic loss. Thus, it is not surprising that the engineer working in industry must be concerned about corrosion control and prevention.
The purpose of this chapter is to serve as an introduction to this important subject.

13.2 ELECTROCHEMICAL CORROSION OF METALS

Quiz

13.2.1 Oxidation-Reduction Reactions

Since most corrosion reactions are electrochemical in nature, it is important to understand the basic principles of electrochemical reactions. Consider a piece of zinc metal
placed in a beaker of dilute hydrochloric acid, as shown in Fig. 13.2. The zinc dissolves or corrodes in the acid, and zinc chloride and hydrogen gas are produced as
indicated by the chemical reaction

$$Zn + 2HCl \rightarrow ZnCl_2 + H_2 \tag{13.1}$$

This reaction can be written in a simplified ionic form, omitting the chloride ions,
as

$$Zn + 2H^+ \rightarrow Zn^{2+} + H_2 \tag{13.2}$$

This equation consists of two half-reactions: one for the oxidation of the zinc and the
other for the reduction of the hydrogen ions to form hydrogen gas. These half-cell

Figure 13.2
Reaction of hydrochloric acid with zinc to produce
hydrogen gas.
(Courtesy of the LaQue Center for Corrosion Technology, Inc.)

reactions can be written as

$$Zn \rightarrow Zn^{2+} + 2e^- \quad \text{(oxidation half-cell reaction)} \quad \textbf{(13.3a)}$$

$$2H^+ + 2e^- \rightarrow H_2 \quad \text{(reduction half-cell reaction)} \quad \textbf{(13.3b)}$$

Some important points about the oxidation-reduction half-cell reactions are:

1. *Oxidation reaction.* The oxidation reaction by which metals form ions that go into aqueous solution is called the *anodic* reaction, and the local regions on the metal surface where the oxidation reaction takes place are called *local* **anodes.** In the anodic reaction electrons are produced that remain in the metal, and the metal atoms form cations (for example, $Zn \rightarrow Zn^{2+} + 2e^-$).

2. *Reduction reaction.* The reduction reaction in which a metal or nonmetal is reduced in valence charge is called the *cathodic* reaction. The local regions on the metal surface where metal ions or nonmetal ions are reduced in valence charge are called *local* **cathodes.** In the cathode reaction there is a *consumption of electrons.*

3. Electrochemical corrosion reactions involve oxidation reactions that produce electrons and reduction reactions that consume them. Both oxidation and reduction reactions must occur at the same time and same overall rate to prevent a buildup of electric charge in the metal.

13.2.2 Standard Electrode Half-Cell Potentials for Metals

Every metal has a different tendency to corrode in a particular environment. For example, zinc is chemically attacked or corroded by dilute hydrochloric acid, whereas gold is not. One method for comparing the tendency for metals to form ions in aqueous solution is to compare their half-cell oxidation or reduction potentials (voltages) to that of a standard hydrogen–hydrogen-ion half-cell potential. Figure 13.3 shows an experimental setup for the determination of half-cell standard electrode potentials.

For this determination, two beakers of aqueous solutions are used that are separated by a salt bridge so that mechanical mixing of the solutions is prevented (Fig. 13.3). In one beaker an electrode of the metal whose standard potential is to be determined is immersed in a 1 M solution of its ions at 25°C. In Fig. 13.3 an electrode of Zn is immersed in a 1 M solution of Zn^{2+} ions. In the other beaker a platinum electrode is immersed in a 1 M solution of H^+ ions into which hydrogen gas is bubbled. A wire in series with a switch and a voltmeter connects the two electrodes.

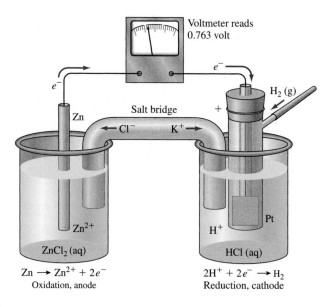

Voltmeter reads 0.763 volt

$Zn \rightarrow Zn^{2+} + 2e^-$
Oxidation, anode

$2H^+ + 2e^- \rightarrow H_2$
Reduction, cathode

Animation

Figure 13.3
Experimental setup for the determination of the standard emf of zinc. In one beaker a Zn electrode is placed in a solution of 1 M Zn^{2+} ions. In the other there is a standard hydrogen reference electrode consisting of a platinum electrode immersed in a solution of 1 M H^+ ions which contains H_2 gas at 1 atm. The overall reaction that occurs when the two electrodes are connected by an external wire is

$Zn(s) + 2H^+(ag) \rightarrow Zn^{2+}(ag) + H_2(g)$

(After R. E. Davis, K. D. Gailey, and K. W. Whitten, "Principles of Chemistry," CBS College Publishing, 1984, p. 635.)

When the switch is just closed, the voltage between the half-cells is measured. The potential due to the hydrogen half-cell reaction $H_2 \rightarrow 2H^+ + 2e^-$ is arbitrarily assigned zero voltage. Thus, the voltage of the metal (zinc) half-cell reaction $Zn \rightarrow Zn^{2+} + 2e^-$ is measured directly against the hydrogen standard half-cell electrode. As indicated in Fig. 13.3, the standard half-cell electrode potential for the $Zn \rightarrow Zn^{2+} + 2e^-$ reaction is -0.763 V.

Table 13.1 lists the standard half-cell potentials of some selected metals. Those metals that are more reactive than hydrogen are assigned negative potentials and are said to be *anodic to hydrogen*. In the standard experiment shown in Fig. 13.3, these metals are oxidized to form ions, and hydrogen ions are reduced to form hydrogen gas. The equations for the reactions involved are

$$M \rightarrow M^{n+} + ne^- \qquad \text{(metal oxidized to ions)} \qquad \textbf{(13.4a)}$$

$$2H^+ + 2e^- \rightarrow H_2 \qquad \text{(hydrogen ions reduced to hydrogen gas)} \qquad \textbf{(13.4b)}$$

Those metals that are less reactive than hydrogen are assigned positive potentials and are said to be *cathodic to hydrogen*. In the standard experiment of Fig. 13.3, the ions of such a metal are reduced to the atomic state (and may plate out on the

Table 13.1 Standard electrode potentials at 25°C*

	Oxidation (corrosion) reaction	Electrode potential ($E°$) (volts versus standard hydrogen electrode)
More cathodic (less tendency to corrode)	$Au \rightarrow Au^{3+} + 3e^-$	+1.498
	$2H_2O \rightarrow O_2 + 4H^+ + 4e^-$	+1.229
	$Pt \rightarrow Pt^{2+} + 2e^-$	+1.200
	$Ag \rightarrow Ag^+ + e^-$	+0.799
	$2Hg \rightarrow Hg_2^{2+} + 2e^-$	+0.788
	$Fe^{2+} \rightarrow Fe^{3+} + e^-$	+0.771
	$4(OH)^- \rightarrow O_2 + 2H_2O + 4e^-$	+0.401
	$Cu \rightarrow Cu^{2+} + 2e^-$	+0.337
	$Sn^{2+} \rightarrow Sn^{4+} + 2e^-$	+0.150
	$H_2 \rightarrow 2H^+ + 2e^-$	0.000
	$Pb \rightarrow Pb^{2+} + 2e^-$	−0.126
	$Sn \rightarrow Sn^{2+} + 2e^-$	−0.136
	$Ni \rightarrow Ni^{2+} + 2e^-$	−0.250
	$Co \rightarrow Co^{2+} + 2e^-$	−0.277
	$Cd \rightarrow Cd^{2+} + 2e^-$	−0.403
More anodic (greater tendency to corrode)	$Fe \rightarrow Fe^{2+} + 2e^-$	−0.440
	$Cr \rightarrow Cr^{3+} + 3e^-$	−0.744
	$Zn \rightarrow Zn^{2+} + 2e^-$	−0.763
	$Al \rightarrow Al^{3+} + 3e^-$	−1.662
	$Mg \rightarrow Mg^{2+} + 2e^-$	−2.363
	$Na \rightarrow Na^+ + e^-$	−2.714

*Reactions are written as anodic half-cells. The more negative the half-cell reaction, the more anodic is the reaction and the greater the tendency for corrosion or oxidation to occur.

metal electrode), and hydrogen gas is oxidized to hydrogen ions. The equations for the reactions involved are

$$M^{n+} + ne^- \rightarrow M \qquad \text{(metal ions reduced to atoms)} \qquad \textbf{(13.5a)}$$

$$H_2 \rightarrow 2H^+ + 2e^- \qquad \begin{array}{l}\text{(hydrogen gas oxidized to} \\ \text{hydrogen ions)}\end{array} \qquad \textbf{(13.5b)}$$

13.3 GALVANIC CELLS

Animation
Quiz

13.3.1 Macroscopic Galvanic Cells with Electrolytes That Are One Molar

Since most metallic corrosion involves electrochemical reactions, it is important to understand the principles of the operation of an electrochemical **galvanic couple (cell).** A macroscopic galvanic cell can be constructed with two dissimilar metal electrodes each immersed in a solution of their own ions. A galvanic cell of this type is shown in Fig. 13.4, which has a zinc electrode immersed in a 1 M solution of Zn^{2+} ions and another of copper immersed in a 1 M solution of Cu^{2+} ions with the solutions at 25°C. The two solutions are separated by a porous wall to prevent their mechanical mixing, and an external wire in series with a switch and a voltmeter connects the two electrodes. When the switch is just closed, electrons flow from the zinc electrode through the external wire to the copper electrode, and a voltage of -1.10 V shows on the voltmeter.

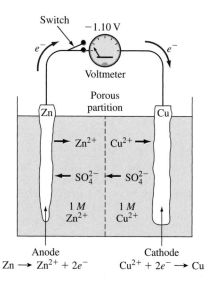

Figure 13.4
A macroscopic galvanic cell with zinc and copper electrodes. When the switch is closed and the electrons flow, the voltage difference between the zinc and copper electrodes is -1.10 V. The zinc electrode is the anode of the cell and corrodes.

In an electrochemical galvanic-couple reaction for two metals each immersed in a 1 M solution of its own ions, the electrode that has the more negative oxidation potential will be the electrode that is oxidized. A reduction reaction will take place at the electrode that has the more positive potential. Thus, for the Zn-Cu galvanic cell illustrated in Fig. 13.4, the Zn electrode will be oxidized to Zn^{2+} ions, and Cu^{2+} ions will be reduced to Cu at the Cu electrode.

Let us now calculate the electrochemical potential of the Zn-Cu galvanic cell when the switch connecting the two electrodes is just closed. First, we write the oxidation half-cell reactions for zinc and copper, using Table 13.1:

$$Zn \rightarrow Zn^{2+} + 2e^- \qquad E° = -0.763 \text{ V}$$
$$Cu \rightarrow Cu^{2+} + 2e^- \qquad E° = +0.337 \text{ V}$$

We see that the Zn half-cell reaction has the more negative potential (-0.763 V for Zn versus $+0.337$ V for Cu). Thus, the Zn electrode will be oxidized to Zn^{2+} ions, and Cu^{2+} ions will be reduced to Cu at the Cu electrode. The overall electrochemical potential of the cell, the **electromotive force** (emf), is obtained by adding the oxidation half-cell potential of the Zn to the reduction half-cell potential of the Cu. Note that the sign of oxidation half-cell potential must be changed to opposite polarity when the half-cell reaction is written as a reduction reaction.

                                                    ┌Note change
                                                        of sign

Oxidation:	$Zn \rightarrow Zn^{2+} + 2e^-$	$E° = -0.763 \text{ V}$
Reduction:	$Cu^{2+} + 2e^- \rightarrow Cu$	$E° = -0.337 \text{ V}$
Overall reaction: (by adding)	$Zn + Cu^{2+} \rightarrow Zn^{2+} + Cu$	$E°_{cell} = -1.100 \text{ V}$

For a galvanic couple, the electrode that is oxidized is called the *anode* and the electrode where the reduction takes place is called the *cathode*. At the anode *metal ions and electrons are produced,* and since the electrons remain in the metal electrode, the *anode is assigned negative polarity.* At the cathode *electrons are consumed,* and it is assigned positive polarity. In the case of the Zn-Cu cell previously described, copper atoms are plated out on the copper cathode.

A galvanic cell consists of an electrode of zinc in a 1 M $ZnSO_4$ solution and another of nickel in a 1 M $NiSO_4$ solution. The two electrodes are separated by a porous wall so that mixing of the solutions is prevented. An external wire with a switch connects the two electrodes. When the switch is just closed:

a. At which electrode does oxidation occur?
b. Which electrode is the anode of the cell?
c. Which electrode corrodes?
d. What is the emf of this galvanic cell when the switch is just closed?

EXAMPLE PROBLEM 13.1

■ **Solutions**

The half-cell reactions for this cell are

$$Zn \rightarrow Zn^{2+} + 2e^- \qquad E° = -0.763 \text{ V}$$
$$Ni \rightarrow Ni^{2+} + 2e^- \qquad E° = -0.250 \text{ V}$$

a. Oxidation occurs at the zinc electrode since the zinc half-cell reaction has a more negative $E°$ potential of -0.763 V as compared to -0.250 V for the nickel half-cell reaction.
b. The zinc electrode is the anode since oxidation occurs at the anode.
c. The zinc electrode corrodes since the anode in a galvanic cell corrodes.
d. The emf of the cell is obtained by adding the half-cell reactions together:

Anode reaction:	$Zn \rightarrow Zn^{2+} + 2e$	$E° = -0.763$ V
Cathode reaction:	$Ni^{2+} + 2e^- \rightarrow Ni$	$E° = +0.250$ V
Overall reaction:	$Zn + Ni^{2+} \rightarrow Zn^{2+} + Ni$	$E°_{cell} = -0.513$ V ◄

13.3.2 Galvanic Cells with Electrolytes That Are Not One Molar

Most electrolytes for real corrosion galvanic cells are not 1 M but are usually dilute solutions that are much lower than 1 M. If the concentration of the ions in an electrolyte surrounding an anodic electrode is less than 1 M, the driving force for the reaction to dissolve or corrode the anode is greater since there is a lower concentration of ions to cause the reverse reaction. Thus, there will be a more negative emf half-cell anodic reaction:

$$M \rightarrow M^{n+} + ne^- \tag{13.6}$$

The effect of metal ion concentration C_{ion} on the standard emf $E°$ at 25°C is given by the *Nernst*[3] *equation*. For a half-cell anodic reaction in which only one kind of ion is produced, the Nernst equation can be written in the form

$$E = E° + \frac{0.0592}{n} \log C_{ion} \tag{13.7}$$

where E = new emf of half-cell
$E°$ = standard emf of half-cell
n = number of electrons transferred (for example, $M \rightarrow M^{n+} + ne^-$)
C_{ion} = molar concentration of ions

For the cathode reaction the sign of the final emf is reversed. Example Problem 13.2 shows how the emf of a macroscopic galvanic cell in which the electrolytes are not 1 M can be calculated by using the Nernst equation.

---

[3]Walter Hermann Nernst (1864–1941). German chemist and physicist who did fundamental work on electrolyte solutions and thermodynamics.

A galvanic cell at 25°C consists of an electrode of zinc in a 0.10 M ZnSO$_4$ solution and another of nickel in a 0.05 M NiSO$_4$ solution. The two electrodes are separated by a porous wall and connected by an external wire. What is the emf of the cell when a switch between the two electrodes is just closed?

EXAMPLE PROBLEM 13.2

■ **Solution**

First, assume the dilutions from 1 M solutions will not affect the order of the potentials of Zn and Ni in the standard electrode potential series. Thus, zinc with a more negative electrode potential of −0.763 V will be the anode of the Zn-Ni electrochemical cell and nickel will be the cathode. Next, use the Nernst equation to modify the standard equilibrium potentials.

$$E_{cell} = E° + \frac{0.0592}{n} \log C_{ion} \tag{13.7}$$

Anode reaction: $E_A = -0.763 \text{ V} + \dfrac{0.0592}{2} \log 0.10$

$$= -0.763 \text{ V} - 0.0296 \text{ V} = -0.793 \text{ V}$$

Cathode reaction: $E_C = -\left(-0.250 \text{ V} + \dfrac{0.0592}{2} \log 0.05\right)$

$$= +0.250 \text{ V} + 0.0385 \text{ V} = +0.288 \text{ V}$$

Emf of cell $= E_A + E_C = -0.793 \text{ V} + 0.288 \text{ V} = -0.505 \text{ V}$ ◄

13.3.3 Galvanic Cells with Acid or Alkaline Electrolytes with No Metal Ions Present

Let us consider a galvanic cell in which iron and copper electrodes are immersed in an aqueous acidic electrolyte in which there are no metal ions initially present. The iron and copper electrodes are connected by an external wire, as shown in Fig. 13.5. The standard electrode potential for iron to oxidize is −0.440 V and that for copper is +0.337 V. Therefore, in this couple iron will be the anode and will oxidize since it has the more negative half-cell oxidation potential. The half-cell reaction at the iron anode will therefore be

$$\text{Fe} \rightarrow \text{Fe}^{2+} + 2e^- \quad \text{(anodic half-cell reaction)} \tag{13.8a}$$

Since there are no copper ions in the electrolyte to be reduced to copper atoms for a cathode reaction, hydrogen ions in the acid solution will be reduced to hydrogen atoms that will subsequently combine to form diatomic hydrogen (H$_2$) gas. The overall reaction at the cathode will thus be

$$2\text{H}^+ + 2e^- \rightarrow \text{H}_2 \quad \text{(cathodic half-cell reaction)} \tag{13.8b}$$

However, if the electrolyte also contains an oxidizing agent, the cathode reaction becomes

$$\text{O}_2 + 4\text{H}^+ + 4e^- \rightarrow 2\text{H}_2\text{O} \tag{13.8c}$$

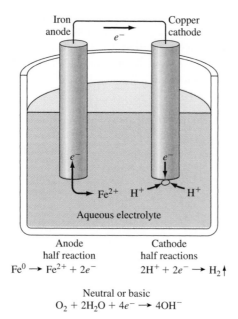

Figure 13.5
Electrode reactions in an iron-copper galvanic cell in which there are no metal ions initially present in the electrolyte.
(After J. Wulff et al., "The Structure and Properties of Materials," vol. II, Wiley, 1964, p. 164.)

Table 13.2 Some common cathode reactions for aqueous galvanic cells

Cathode reaction	Example
1. Metal deposition: $M^{n+} + ne^- \rightarrow M$	Fe-Cu galvanic couple in aqueous solution with Cu^{2+} ions; $Cu^{2+} + 2e^- \rightarrow Cu$
2. Hydrogen evolution: $2H^+ + 2e^- \rightarrow H_2$	Fe-Cu galvanic couple in acid solution with no copper ions present
3. Oxygen reduction (acid solutions): $O_2 + 4H^+ + 4e^- \rightarrow 2H_2O$	Fe-Cu galvanic couple in oxidizing acidic solution with no copper ions present
4. Oxygen reduction (neutral or basic solutions): $O_2 + 2H_2O + 4e^- \rightarrow 4OH^-$	Fe-Cu galvanic couple in neutral or alkaline solution with no copper ions present

If the electrolyte is neutral or basic and oxygen is present, oxygen and water molecules will react to form hydroxyl ions, with the cathode reaction becoming

$$O_2 + 2H_2O + 4e^- \rightarrow 4OH^- \qquad \textbf{(13.8\textit{d})}$$

Table 13.2 lists four common reactions that occur in aqueous galvanic cells.

13.3.4 Microscopic Galvanic Cell Corrosion of Single Electrodes

Quiz

If a single electrode of zinc is placed in a dilute solution of air-free hydrochloric acid, it will be corroded electrochemically since microscopic *local anodes and cathodes* will develop on its surface due to inhomogeneities in structure and composition (Fig 13.6*a*). The oxidation reaction that will occur at the local anodes is

$$Zn \rightarrow Zn^{2+} + 2e^- \quad \text{(anodic reaction)} \tag{13.9a}$$

and the reduction reaction that will occur at the local cathodes is

$$2H^+ + 2e^- \rightarrow H_2 \quad \text{(cathodic reaction)} \tag{13.9b}$$

Both reactions will occur simultaneously and at the same rate on the metal surface.

Another example of single-electrode corrosion is the *rusting of iron*. If a piece of iron is immersed in oxygenated water, ferric hydroxide [$Fe(OH)_3$] will form on its surface as indicated in Fig. 13.6*b*. The oxidation reaction that occurs at microscopic local anodes is

$$Fe \rightarrow Fe^{2+} + 2e^- \quad \text{(anodic reaction)} \tag{13.10a}$$

Since the iron is immersed in oxygenated neutral water, the reduction reaction occurring at the local cathodes is

$$O_2 + 2H_2O + 4e^- \rightarrow 4OH^- \quad \text{(cathodic reaction)} \tag{13.10b}$$

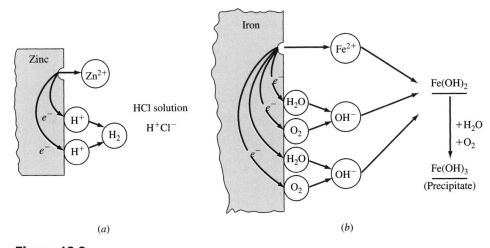

Figure 13.6
Electrochemical reactions for (*a*) zinc immersed in dilute hydrochloric acid and (*b*) iron immersed in oxygenated neutral water solution.

The overall reaction is obtained by adding the two reactions 13.10a and 13.10b to give

$$2Fe + 2H_2O + O_2 \rightarrow 2Fe^{2+} + 4OH^- \rightarrow 2Fe(OH)_2 \downarrow \quad \textbf{(13.10c)}$$
Precipitate

The ferrous hydroxide, $Fe(OH)_2$, precipitates from solution since this compound is insoluble in oxygenated aqueous solutions. It is further oxidized to ferric hydroxide, $Fe(OH)_3$, which has the red-brown rust color. The reaction for the oxidation of ferrous to ferric hydroxide is

$$2Fe(OH)_2 + H_2O + \tfrac{1}{2}O_2 \rightarrow 2Fe(OH)_3 \downarrow \quad \textbf{(13.10d)}$$
Precipitate (rust)

EXAMPLE PROBLEM 13.3

Write the anodic and cathodic half-cell reactions for the following electrode-electrolyte conditions. Use $E°$ values in Table 13.1 as basis for answers.

a. Copper and zinc electrodes immersed in a dilute cupric sulfate ($CuSO_4$) solution.
b. A copper electrode immersed in an oxygenated water solution.
c. An iron electrode immersed in an oxygenated water solution.
d. Magnesium and iron electrodes connected by an external wire immersed in an oxygenated 1% NaCl solution.

■ **Solution**

a. Anode reaction: $Zn \rightarrow Zn^{2+} + 2e^-$ $E° = -0.763$ V (oxidation)
   Cathode reaction: $Cu^{2+} + 2e^- \rightarrow Cu$ $E° = -0.337$ V (reduction)
   Comment: Zinc has a more negative potential and is thus the anode. It is therefore oxidized.

b. Little or no corrosion takes place since the potential difference between that for the oxidation of copper (0.337 V) and that for the formation of water from hydroxyl ions (0.401 V) is so small.

c. Anode reaction: $Fe \rightarrow Fe^{2+} + 2e^-$ $E° = -0.440$ V (oxidation)
   Cathode reaction: $O_2 + 2H_2O + 4e^- \rightarrow 4OH^-$ $E° = -0.401$ V
   Comment: Fe has a more negative potential and thus is the anode. Fe is therefore oxidized.

d. Anode reaction: $Mg \rightarrow Mg^{2+} + 2e^-$ $E° = -2.36$ V
   Cathode reaction: $O_2 + 2H_2O + 4e^- \rightarrow 4OH^-$ $E° = -0.401$ V
   Comment: Magnesium has a more negative oxidation potential and is thus the anode. Mg is therefore oxidized.

13.3.5 Concentration Galvanic Cells

Ion-Concentration Cells Consider an **ion-concentration cell** consisting of two iron electrodes, one immersed in a dilute Fe^{2+} electrolyte and the other in a concentrated Fe^{2+} electrolyte, as shown in Fig. 13.7. In this galvanic cell the electrode in the dilute electrolyte will be the anode since according to the Nernst equation this electrode will have a more negative potential with respect to the other.

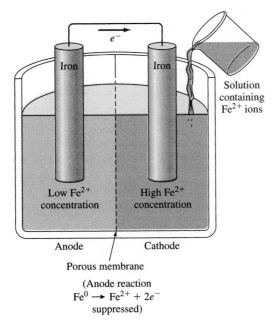

Figure 13.7

An ion-concentration galvanic cell composed of two iron electrodes. When the electrolyte is of different concentrations at each electrode, the electrode in the more dilute electrolyte becomes the anode.
(*After J. Wulff et al., "The Structure and Properties of Materials," vol. II, Wiley, 1964, p. 163.*)

Let us compare the half-cell potential for an iron electrode immersed in a 0.001 *M* dilute Fe^{2+} electrolyte with the half-cell potential of another iron electrode immersed in a more concentrated 0.01 *M* dilute Fe^{2+} electrolyte. The two electrodes are connected by an external wire, as shown in Fig. 13.7. The general Nernst equation for a half-cell oxidation reaction for $Fe \rightarrow Fe^{2+} + 2e^-$, since $n = 2$, is

$$E_{Fe^{2+}} = E° + 0.0296 \log C_{ion} \qquad \textbf{(13.11)}$$

For 0.001 *M* solution: $E_{Fe^{2+}} = -0.440 \text{ V} + 0.0296 \log 0.001 = -0.529 \text{ V}$

For 0.01 *M* solution: $E_{Fe^{2+}} = -0.440 \text{ V} + 0.0296 \log 0.01 = -0.499 \text{ V}$

Since -0.529 V is more negative than -0.499 V, the iron electrode in the more dilute solution will be the anode of the electrochemical cell and hence will be oxidized and corroded. Thus, the ion-concentration cell produces corrosion in the region of the more dilute electrolyte.

EXAMPLE PROBLEM 13.4

One end of an iron wire is immersed in an electrolyte of 0.02 M Fe^{2+} ions and the other in an electrolyte of 0.005 M Fe^{2+} ions. The two electrolytes are separated by a porous wall.

a. Which end of the wire will corrode?
b. What will be the potential difference between the two ends of the wire when it is just immersed in the electrolytes?

■ **Solution**

a. The end of the wire that will corrode will be the one immersed in the more dilute electrolyte, which is the 0.005 M one. Thus, the wire end in the 0.005 M solution will be the anode.

b. Using the Nernst equation with $n = 2$ (Eq. 13.11) gives

$$E_{Fe^{2+}} = E° + 0.0296 \log C_{ion} \qquad (13.11)$$

For 0.005 M solution: $\quad E_A = -0.440 \text{ V} + 0.0296 \log 0.005$

$$= -0.508 \text{ V}$$

For 0.02 M solution: $\quad E_C = -(-0.440 \text{ V} + 0.0296 \log 0.02)$

$$= +0.490 \text{ V}$$

$$E_{cell} = E_A + E_C = -0.508 \text{ V} + 0.490 \text{ V} = -0.018 \text{ V} \blacktriangleleft$$

Oxygen-Concentration Cells **Oxygen-concentration cells** can develop when there is a difference in oxygen concentration on a moist surface of a metal that can be oxidized. Oxygen-concentration cells are particularly important in the corrosion of easily oxidized metals such as iron that do not form protective oxide films.

Consider an oxygen-concentration cell consisting of two iron electrodes, one in a water electrolyte with a low oxygen concentration and another in an electrolyte with a high oxygen concentration, as shown in Fig. 13.8. The anode and cathode reactions for this cell are

Anode reaction: $\qquad Fe \rightarrow Fe^{2+} + 2e^-$ $\qquad\qquad$ (13.12a)

Cathode reaction: $\qquad O_2 + 2H_2O + 4e^- \rightarrow 4OH^-$ $\qquad\qquad$ (13.12b)

Which electrode is the anode in this cell? Since the cathode reaction requires oxygen and electrons, the high concentration of oxygen must be at the cathode. Also since electrons are required at the cathode, they must be produced by the anode that will have the low oxygen concentration.

In general, therefore, for an oxygen-concentration cell, the regions that are low in oxygen will be anodic to the cathode regions that are high in oxygen. Thus, corrosion will be accelerated in the regions of a metal surface where the oxygen content is relatively low such as in cracks and crevices and under accumulations of surface deposits. The effects of oxygen-concentration cells will be discussed further in Sec. 13.5 dealing with different types of corrosion.

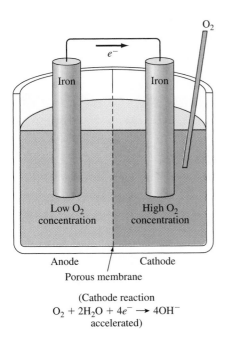

Figure 13.8

An oxygen-concentration cell. The anode in this cell is the electrode that has the low oxygen concentration surrounding it.

(After J. Wulff et al., "The Structure and Properties of Materials," vol. II, Wiley, 1964, p. 165.)

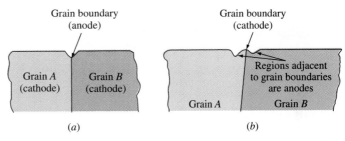

(a)　　　　　　　(b)

Figure 13.9

Corrosion at or near grain boundaries. (*a*) Grain boundary is the anode of galvanic cell and corrodes. (*b*) Grain boundary is the cathode, and regions adjacent to the grain boundary serve as anodes.

Animation

13.3.6 Galvanic Cells Created by Differences in Composition, Structure, and Stress

Microscopic galvanic cells can exist in metals or alloys because of differences in composition, structure, and stress concentrations. These metallurgical factors can seriously affect the corrosion resistance of a metal or alloy. They create anodic and cathodic regions of varying dimensions that can cause galvanic-cell corrosion. Some of the important metallurgical factors affecting corrosion resistance are:

1. Grain–grain-boundary galvanic cells
2. Multiple-phase galvanic cells
3. Impurity galvanic cells

Grain–Grain-Boundary Electrochemical Cells　In most metals and alloys grain boundaries are more chemically active (anodic) than the grain matrix. Thus, the grain boundaries are corroded or chemically attacked, as illustrated in Fig. 13.9*a*. The reason for the anodic behavior of the grain boundaries is that they have higher

Figure 13.10
Class 30 gray cast iron. Structure consists of graphite flakes in a matrix of pearlite (alternating lamellae of light-etching ferrite and darker cementite).
(After "Metals Handbook," vol. 7, 8th ed., American Society for Metals, 1972, p. 83.)

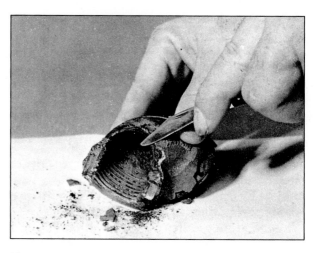

Figure 13.11
Graphite residue remaining as a result of corrosion of cast-iron elbow.
(Courtesy of the LaQue Center for Corrosion Technology, Inc.)

energies due to the atomic disarray in that area and also because solute segregation and impurities migrate to the grain boundaries. For some alloys the situation is reversed, and chemical segregation causes the grain boundaries to become more noble or cathodic than the regions adjacent to the grain boundaries. This condition causes the regions adjacent to the grain boundaries to corrode preferentially, as illustrated in Fig. 13.9*b*.

Multiple-Phase Electrochemical Cells   In most cases a single-phase alloy has higher corrosion resistance than a multiple-phase alloy since electrochemical cells are created in the multiphase alloy due to one phase being anodic relative to another that acts as the cathode. Hence, corrosion rates are higher for the multiphase alloy. A classic example of multiphase galvanic corrosion can occur in pearlitic gray cast iron. The microstructure of pearlitic gray cast iron consists of graphite flakes in a matrix of pearlite (Fig. 13.10). Since graphite is much more cathodic (more noble) than the surrounding pearlite matrix, highly active galvanic cells are created between the graphite flakes and the anodic pearlite matrix. In an extreme case of galvanic corrosion of pearlitic gray cast iron, the matrix can corrode to such an extent that the cast iron is left as a network of interconnected graphite flakes (Fig. 13.11).

Another example of the effect of second phases in reducing the corrosion resistance of an alloy is the effect of tempering on the corrosion resistance of a 0.95 percent carbon steel. When this steel is in the martensitic condition after quenching

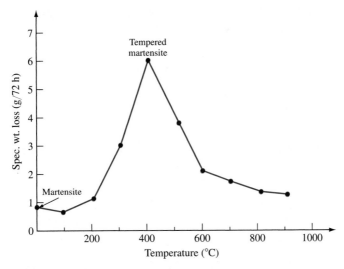

Figure 13.12
Effect of heat treatment on corrosion of 0.95% C steel in
1% H_2SO_4. Polished specimens 2.5 × 2.5 × 0.6 cm, with
tempering time probably 2 h.
(*Heyn and Bauer.*)

from the austenitic condition, its corrosion rate is relatively low (Fig. 13.12) because
the martensite is a single-phase supersaturated solid solution of carbon in interstitial
positions of a body-centered tetragonal lattice of iron. After tempering in the 200 to
500°C range, a fine precipitate of ϵ carbide and cementite (Fe_3C) is formed. This
two-phase structure sets up galvanic cells that accelerate the corrosion rate of the
steel, as observed in Fig. 13.12. At higher tempering temperatures above about 500°C,
the cementite coalesces into larger particles and the corrosion rate decreases.

Impurities   Metallic impurities in a metal or alloy can lead to the precipitation of
intermetallic phases that have different oxidation potentials than the matrix of the
metal. Thus, very small anodic or cathodic regions are created that can lead to galvanic corrosion when coupled with the matrix metal. Higher corrosion resistance is
obtained with purer metals. However, most engineering metals and alloys contain a
certain level of impurity elements since it costs too much to remove them.

13.4 CORROSION RATES (KINETICS)

Until now our study of the corrosion of metals has been centered on equilibrium conditions and the *tendency* of metals to corrode, which has been related to the standard
electrode potentials of metals. However, corroding systems are *not at equilibrium,* and
thus thermodynamic potentials do not tell us about the rates of corrosion reactions.
The kinetics of corroding systems are very complex and not completely understood.
In this section we will examine some of the basic aspects of corrosion kinetics.

13.4.1 Rate of Uniform Corrosion or Electroplating of a Metal in an Aqueous Solution

The amount of metal uniformly corroded from an anode or electroplated on a cathode in an aqueous solution in a time period can be determined by using Faraday's[4] equation of general chemistry, which states

$$w = \frac{ItM}{nF} \tag{13.13}$$

where w = weight of metal, g, corroded or electroplated in an aqueous solution in time t, s

I = current flow, A

M = atomic mass of the metal, g/mol

n = number of electrons/atom produced or consumed in the process

F = Faraday's constant = 96,500 C/mol or 96,500 A · s/mol

Sometimes the uniform aqueous corrosion of a metal is expressed in terms of a current density i, which is often expressed in amperes per square centimeter. Replacing I by iA converts Eq. 13.13 to

$$w = \frac{iAtM}{nF} \tag{13.14}$$

where i = current density, A/cm^2, and A = area, cm^2, if the centimeter is used for length. The other quantities are the same as in Eq. 13.13.

---

[4]Michael Faraday (1791–1867). English scientist who made basic experiments in electricity and magnetism. He made experiments to show how ions of a compound migrated under the influence of an applied electric current to electrodes of opposite polarity.

EXAMPLE PROBLEM 13.5

A copper electroplating process uses 15 A of current by chemically dissolving (corroding) a copper anode and electroplating a copper cathode. If it is assumed that there are no side reactions, how long will it take to corrode 8.50 g of copper from the anode?

■ **Solution**

The time to corrode the copper from the anode can be determined from Eq. 13.13:

$$w = \frac{ItM}{nF} \quad \text{or} \quad t = \frac{wnF}{IM}$$

In this case,

$$w = 8.5 \text{ g} \quad n = 2 \text{ for Cu} \rightarrow Cu^{2+} + 2e^- \quad F = 96,500 \text{ A} \cdot \text{s/mol}$$

$$M = 63.5 \text{ g/mol for Cu} \quad I = 15 \text{ A} \quad t = ? \text{ s}$$

or

$$t = \frac{(8.5 \text{ g})(2)(96,500 \text{ A} \cdot \text{s/mol})}{(15 \text{ A})(63.5 \text{ g/mol})} = 1722 \text{ s or } 28.7 \text{ min} \blacktriangleleft$$

A mild steel cylindrical tank 1 m high and 50 cm in diameter contains aerated water to the 60 cm level and shows a loss in weight due to corrosion of 304 g after six weeks. Calculate (a) the corrosion current and (b) the current density involved in the corrosion of the tank. Assume uniform corrosion on the tank's inner surface and that the steel corrodes in the same manner as pure iron.

■ **Solution**

a.   We will use Eq. 13.13 to solve for the corrosion current:

$$I = \frac{wnF}{tM}$$

$w = 304$ g $n = 2$ for Fe $\rightarrow$ Fe^{2+} + 2e^- $F = 96{,}500$ A · s/mol

$M = 55.85$ g/mol for Fe $t = 6$ wk $I = ?$ A

We must convert the time, six weeks, into seconds and then we can substitute all the values into Eq. 13.13:

$$t = 6 \text{ wk} \left(\frac{7 \text{ days}}{\text{wk}}\right)\left(\frac{24 \text{ h}}{\text{day}}\right)\left(\frac{3600 \text{ s}}{\text{h}}\right) = 3.63 \times 10^6 \text{ s}$$

$$I = \frac{(304 \text{ g})(2)(96{,}500 \text{ A} \cdot \text{s/mol})}{(3.63 \times 10^6 \text{ s})(55.85 \text{ g/mol})} = 0.289 \text{ A} \blacktriangleleft$$

b.   The current density is

$$i \text{ (A/cm}^2) = \frac{I \text{ (A)}}{\text{area (cm}^2)}$$

Area of corroding surface of tank = area of sides + area of bottom

$$= \pi Dh + \pi r^2$$
$$= \pi(50 \text{ cm})(60 \text{ cm}) + \pi(25 \text{ cm})^2$$
$$= 9420 \text{ cm}^2 + 1962 \text{ cm}^2 = 11{,}380 \text{ cm}^2$$

$$i = \frac{0.289 \text{ A}}{11{,}380 \text{ cm}^2} = 2.53 \times 10^{-5} \text{ A/cm}^2 \blacktriangleleft$$

In experimental corrosion work the uniform corrosion of a metal surface exposed to a corrosive environment is measured in a variety of ways. One common method is to measure the weight loss of a sample exposed to a particular environment and then after a period of time express the corrosion rate as a weight loss per unit area of exposed surface per unit time. For example, uniform surface corrosion is often expressed as *milligram weight loss per square decimeter per day* (mdd). Another commonly used method is to express corrosion rate in terms of loss in depth of material per unit time. Examples of corrosion rate in this system are millimeters per year (mm/yr) and mils per year (mils/yr).[5] For uniform electrochemical corrosion in

---

[5]1 mil = 0.001 in.

aqueous environments, the corrosion rate may be expressed as a current density (see Example Problem 13.8).

The wall of a steel tank containing aerated water is corroding at a rate of 54.7 mdd. How long will it take for the wall thickness to decrease by 0.50 mm?

■ **Solution**

The corrosion rate is 54.7 mdd, or 54.7 mg of metal is corroded on each square decimeter of surface per day.

$$\text{Corrosion rate in } g/(cm^2 \cdot day) = \frac{54.7 \times 10^{-3} \, g}{100 \, (cm^2 \cdot day)} = 5.47 \times 10^{-4} \, g/(cm^2 \cdot day)$$

The density of Fe = 7.87 g/cm^3. Dividing the corrosion rate in $g/(cm^2 \cdot day)$ by the density gives the depth of corrosion per day as

$$\frac{5.47 \times 10^{-4} \, g/(cm^2 \cdot day)}{7.87 \, g/cm^3} = 0.695 \times 10^{-4} \, cm/day$$

The number of days required for a decrease in 0.50 mm can be obtained by ratio as

$$\frac{x \, days}{0.50 \, mm} = \frac{1 \, day}{0.695 \times 10^{-3} \, mm}$$
$$x = 719 \, days \blacktriangleleft$$

A sample of zinc corrodes uniformly with a current density of $4.27 \times 10^{-7} \, A/cm^2$ in an aqueous solution. What is the corrosion rate of the zinc in milligrams per decimeter per day? The reaction for the oxidation of zinc is $Zn \rightarrow Zn^{2+} + 2e^-$.

■ **Solution**

To make the conversion from current density to mdd, we will use Faraday's equation (Eq. 13.14) to calculate the milligrams of zinc corroding on an area of 1 dm^2/day (mdd).

$$w = \frac{iAtM}{nF} \tag{13.14}$$

$w \, (mg)$

$$= \left[\frac{(4.27 \times 10^{-7} \, A/cm^2)(100 \, cm^2)(24 \, h \times 3600 \, s/h)(65.38 \, g/mol)}{(2)(96,500 \, A \cdot s/mol)} \right] \left(\frac{1000 \, mg}{g} \right)$$

$$= 1.25 \, mg \text{ of zinc which corrodes on an area of 1 } dm^2 \text{ in 1 day}$$

or the corrosion rate is 1.25 mdd. ◄

13.4.2 Corrosion Reactions and Polarization

Let us now consider the electrode kinetics of the corrosion reaction of zinc being dissolved by hydrochloric acid, as indicated by Fig. 13.13. The anodic half-cell reaction for this electrochemical reaction is

$$Zn \rightarrow Zn^{2+} + 2e^- \quad \text{(anodic reaction)} \quad \textbf{(13.15a)}$$

The electrode kinetics for this reaction can be represented by an electrochemical potential E (volts) versus log current density plot, as shown in Fig. 13.14. The zinc electrode in equilibrium with its ions can be represented by a point that represents its equilibrium potential $E° = -0.763$ V and a corresponding exchange current density $i_0 = 10^{-7}$ A/cm^2 (point A in Fig. 13.14). The exchange current density i_0 is the rate of oxidation and reduction reactions at an equilibrium electrode expressed in terms of current density. Exchange current densities must be determined experimentally when there is no net current. Each electrode with its specific electrolyte will have its own i_0 value.

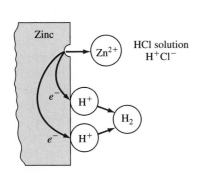

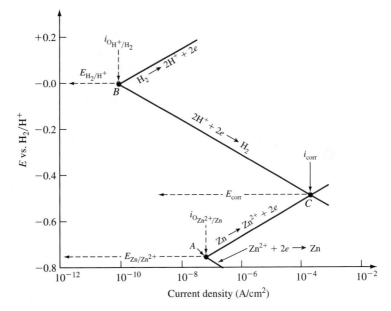

Animation

Figure 13.13
Electrochemical dissolution of zinc in hydrochloric acid.
$Zn \rightarrow Zn^{2+} + 2e^-$
  (anodic reaction)
$2H^+ + 2e^- \rightarrow H_2$
  (cathodic reaction)

Figure 13.14
Electrode kinetic behavior of pure zinc in acid solution (schematic).
(After M. G. Fontana and N. D. Greene, "Corrosion Engineering," 2nd ed., McGraw-Hill, 1978, p. 314.)

The cathodic half-cell reaction for the corrosion reaction of zinc being dissolved in hydrochloric acid is

$$2H^+ + 2e^- \rightarrow H_2 \qquad \text{(cathodic reaction)} \qquad \textbf{(13.15}b\textbf{)}$$

The hydrogen-electrode reaction occurring on the zinc surface under equilibrium conditions can also be represented by the reversible hydrogen electrode potential $E° = 0.00$ V, and the corresponding exchange current density for this reaction on a zinc surface is 10^{-10} A/cm^2 (point B in Fig. 13.14).

When the zinc begins to react with the hydrochloric acid (corrosion starts), since the zinc is a good electrical conductor, the zinc surface must be at a constant potential. This potential is E_{corr} (Fig. 13.14, point C). Thus, when the zinc starts to corrode, the potential of the cathodic areas must become more negative to reach about -0.5 V (E_{corr}) and that of the anodic areas more positive to reach -0.5 V (E_{corr}). At point C in Fig. 13.14 the rate of zinc dissolution is equal to the rate of hydrogen evolution. The current density corresponding to this rate of reaction is called i_{corr} and therefore is equal to the rate of zinc dissolution or corrosion. Example Problem 13.8 shows how a current density for a uniformly corroding surface can be expressed in terms of a certain weight loss per unit area per unit time (for example, mdd units).

Thus, when a metal corrodes by the short-circuiting of microscopic galvanic-cell action, net oxidation and reduction reactions occur on the metal surface. The potentials of the local anodic and cathodic regions are no longer at equilibrium but change their potential to reach a constant intermediate value of E_{corr}. The displacement of the electrode potentials from their equilibrium values to a constant potential of some intermediate value and the creation of a net current flow is called **polarization.** Polarization of electrochemical reactions can conveniently be divided into two types: *activation polarization* and *concentration polarization.*

Activation Polarization   Activation polarization refers to electrochemical reactions that are controlled by a slow step in a reaction sequence of steps at the metal-electrolyte interface. That is, there is a critical activation energy needed to surmount the energy barrier associated with the slowest step. This type of activation energy is illustrated by considering the cathodic hydrogen reduction on a metal surface, $2H^+ + 2e^- \rightarrow H_2$. Figure 13.15 shows schematically some of the intermediate steps possible in the hydrogen reduction at a zinc surface. In this process hydrogen ions must migrate to the zinc surface, and then electrons must combine with the hydrogen ions to produce hydrogen atoms. The hydrogen atoms must combine to form diatomic hydrogen molecules that in turn must combine to form bubbles of hydrogen gas. The slowest of these steps will control the cathodic half-cell reaction. There is also an activation-polarization barrier for the anodic half-cell reaction that is the barrier for zinc atoms to leave the metal surface to form zinc ions and go into the electrolyte.

Concentration Polarization   Concentration polarization is associated with electrochemical reactions controlled by the diffusion of ions in the electrolyte. This type of polarization is illustrated by considering the diffusion of hydrogen ions to a metal

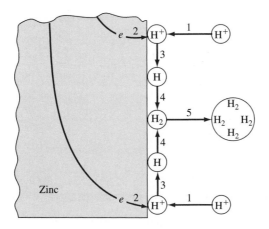

Figure 13.15
Hydrogen-reduction reaction at zinc cathode under activation polarization. The steps in the formation of hydrogen gas at the cathode are: (1) migration of hydrogen ions to zinc surface, (2) flow of electrons to hydrogen ions, (3) formation of atomic hydrogen, (4) formation of diatomic hydrogen molecules, and (5) formation of hydrogen gas bubble, which breaks away from zinc surface. The slowest of these steps will be the rate-limiting step for this activation-polarization process.
(After M. G. Fontana and N. D. Greene, "Corrosion Engineering," 2nd ed., McGraw-Hill, 1978, p. 15.)

surface to form hydrogen gas by the cathodic reaction $2H^+ + 2e^- \rightarrow H_2$, as shown in Fig. 13.16. In this case the concentration of hydrogen ions is low, and thus the reduction rate of the hydrogen ions at the metal surface is controlled by the diffusion of these ions to the metal surface.

For concentration polarization, any changes in the system that increase the diffusion rate of the ions in the electrolyte will decrease the concentration-polarization effects and increase the reaction rate. Thus, stirring the electrolyte will decrease the concentration gradient of positive ions and increase the reaction rate. Increasing the temperature will also increase the diffusion rate of ions and hence increase the reaction rate.

The total polarization at the electrode in an electrochemical reaction is equal to the sum of the effects of activation polarization and concentration polarization. Activation polarization is usually the controlling factor at low reaction rates and concentration polarization at higher reaction rates. When polarization occurs mostly at the anode, the corrosion rate is said to be *anodically controlled,* and when polarization occurs mostly at the cathode, the corrosion rate is said to be *cathodically controlled.*

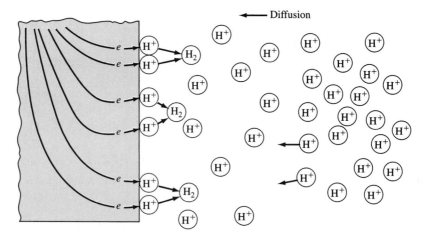

Figure 13.16
Concentration polarization during the cathodic hydrogen-ion reduction
reaction $2H^+ + 2e^- \rightarrow H_2$. The reaction at the metal surface is
controlled by the rate of diffusion of hydrogen ions to the metal surface.
(*After M. G. Fontana and N. D. Greene, "Corrosion Engineering," 2nd ed., McGraw-Hill,
1978, p. 15.*)

13.4.3 Passivation

The **passivation** of a metal as it pertains to corrosion refers to the formation of a protective surface layer of reaction product that inhibits further reaction. In other words,
the passivation of metals refers to their loss of chemical reactivity in the presence of
a particular environmental condition. Many important engineering metals and alloys
become passive and hence very corrosion-resistant in moderate to strong oxidizing
environments. Examples of metals and alloys that show passivity are stainless steels,
nickel and many nickel alloys, and titanium and aluminum and many of their alloys.

There are two main theories regarding the nature of the passive film: (1) the
oxide-film theory and (2) the adsorption theory. For the oxide-film theory it is
believed that the passive film is always a diffusion-barrier layer of reaction products
(e.g., metal oxides or other compounds) that separate the metal from its environment
and slow down the reaction rate. For the adsorption theory it is believed that passive metals are covered by chemisorbed films of oxygen. Such a layer is supposed
to displace the normally adsorbed H_2O molecules and slow down the rate of anodic
dissolution involving the hydration of metal ions. The two theories have in common
a protective film that forms on the metal surface to create the passive state, which
results in increased corrosion resistance.

The passivation of metals in terms of corrosion rate can be illustrated by a polarization curve that shows how the potential of a metal varies with current density, as
shown in Fig. 13.17. Let us consider the passivation behavior of a metal M as the
current density is increased. At point A in Fig. 13.17 the metal is at its equilibrium

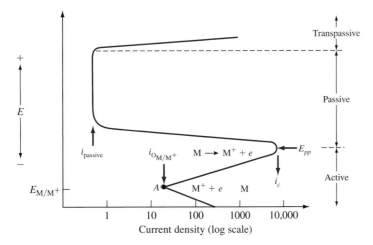

Figure 13.17
Polarization curve of a passive metal.
(After M. G. Fontana and N. D. Greene, "Corrosion Engineering," 2nd ed.,
McGraw-Hill, 1978, p. 321.)

potential E and its exchange current density i_0. As the electrode potential is made more positive, the metal behaves as an active metal, and its current density and hence its dissolution rate increase exponentially.

When the potential becomes more positive and reaches the potential E_{pp}, primary passive potential, the current density and hence the corrosion rate decrease to a low value indicated as $i_{passive}$. At the potential E_{pp}, the metal forms a protective film on its surface that is responsible for the decreased reactivity. As the potential is made still more positive, the current density remains at $i_{passive}$ over the passive region. A still further increase in potential beyond the passive region makes the metal active again, and the current density increases in the transpassive region.

13.4.4 The Galvanic Series

Since many important engineering metals form passive films, they do not behave in galvanic cells as the standard electrode potentials would indicate. Thus, for practical applications where corrosion is an important factor, a new type of series called the **galvanic series** has been developed for anodic-cathodic relationships. Thus, a galvanic series should be determined experimentally for every corrosive environment. A galvanic series for metals and alloys exposed to flowing seawater is listed in Table 13.3. The different potentials for active and passive conditions of some stainless steels are shown. In this table zinc is shown to be more active than the aluminum alloys, which is the reverse of the behavior shown in the standard electrode potentials of Table 13.1.

Table 13.3 Galvanic series in flowing seawater

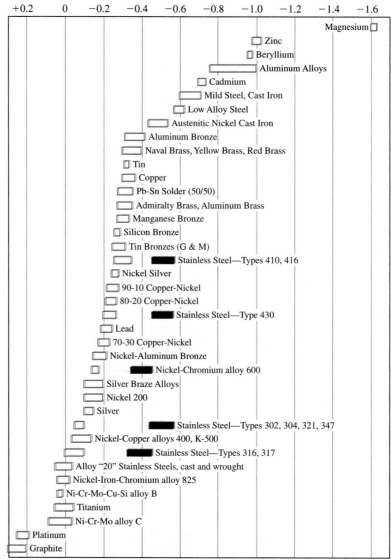

Corrosion Potentials in Flowing Seawater
(8 to 13 ft/sec) Temp. Range 50°–80°F

Volts: Saturated Calomel Half-Cell Reference Electrode

+0.2	0	−0.2	−0.4	−0.6	−0.8	−1.0	−1.2	−1.4	−1.6

Magnesium

Zinc

Beryllium

Aluminum Alloys

Cadmium

Mild Steel, Cast Iron

Low Alloy Steel

Austenitic Nickel Cast Iron

Aluminum Bronze

Naval Brass, Yellow Brass, Red Brass

Tin

Copper

Pb-Sn Solder (50/50)

Admiralty Brass, Aluminum Brass

Manganese Bronze

Silicon Bronze

Tin Bronzes (G & M)

Stainless Steel—Types 410, 416

Nickel Silver

90-10 Copper-Nickel

80-20 Copper-Nickel

Stainless Steel—Type 430

Lead

70-30 Copper-Nickel

Nickel-Aluminum Bronze

Nickel-Chromium alloy 600

Silver Braze Alloys

Nickel 200

Silver

Stainless Steel—Types 302, 304, 321, 347

Nickel-Copper alloys 400, K-500

Stainless Steel—Types 316, 317

Alloy "20" Stainless Steels, cast and wrought

Nickel-Iron-Chromium alloy 825

Ni-Cr-Mo-Cu-Si alloy B

Titanium

Ni-Cr-Mo alloy C

Platinum

Graphite

Alloys are listed in the order of the potential they exhibit in flowing seawater. Certain alloys indicated by the symbol ■■■ in low-velocity or poorly aerated water, and at shielded areas, may become active and exhibit a potential near −0.5 volts.

Source: Courtesy of the LaQue Center for Corrosion Technology, Inc.

13.5 TYPES OF CORROSION

The types of corrosion can be conveniently classified according to the appearance of the corroded metal. Many forms can be identified, but all of them are interrelated to varying extents. These include:

Uniform or general attack corrosion   Stress corrosion

Galvanic or two-metal corrosion   Erosion corrosion

Pitting corrosion   Cavitation damage

Crevice corrosion   Fretting corrosion

Intergranular corrosion   Selective leaching or dealloying

13.5.1 Uniform or General Attack Corrosion

Uniform corrosion attack is characterized by an electrochemical or chemical reaction that proceeds uniformly on the entire metal surface exposed to a corrosive environment. On a weight basis, uniform attack represents the greatest destruction of metals, particularly steels. However, it is relatively easy to control by (1) protective coatings, (2) inhibitors, and (3) cathodic protection. These methods will be discussed in Sec. 13.7 on Corrosion Control.

13.5.2 Galvanic or Two-Metal Corrosion

Galvanic corrosion between dissimilar metals has been discussed in Secs. 13.2 and 13.3. Care must be taken in attaching dissimilar metals together because the difference in their electrochemical potential can lead to corrosion.

Galvanized steel, which is steel coated with zinc, is an example where one metal (zinc) is sacrificed to protect the other (steel). The zinc that is hot-dipped or electroplated on the steel is anodic to the steel and hence corrodes and protects the steel, which is the cathode in this galvanic cell (Fig. 13.18*a*). Table 13.4 shows typical

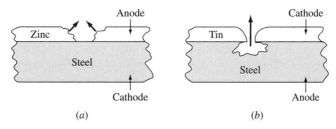

        (a)   *(b)*

Figure 13.18
Anodic-cathodic behavior of steel with zinc and tin outside layers exposed to the atmosphere. (*a*) Zinc is anodic to steel and corrodes (standard emf of Zn = −0.763 V and Fe = −0.440 V). (*b*) Steel is anodic to tin and corrodes (the tin layer was perforated before the corrosion began) (standard emf of Fe = −0.440 V and Sn = −0.136 V).
(*After M. G. Fontana and N. D. Greene, "Corrosion Engineering," 2nd ed., McGraw-Hill, 1978.*)

Table 13.4 Change in weight (in grams) of coupled and uncoupled steel and zinc

Environment	Uncoupled		Coupled	
	Zinc	**Steel**	**Zinc**	**Steel**
0.05 M MgSO$_4$	0.00	−0.04	−0.05	+0.02
0.05 M Na$_2$SO$_4$	−0.17	−0.15	−0.48	+0.01
0.05 M NaCl	−0.15	−0.15	−0.44	+0.01
0.005 M NaCl	−0.06	−0.10	−0.13	+0.02

Source: M. G. Fontana and N. D. Greene, *Corrosion Engineering,* 2nd ed., McGraw-Hill, 1978.

weight losses for uncoupled and coupled zinc and steel in aqueous environments. When the zinc and steel are uncoupled, they both corrode at about the same rate. However, when they are coupled, the zinc corrodes at the anode of a galvanic cell and hence protects the steel.

Another case of the use of two dissimilar metals in an industrial product is in the tin plate used for the "tin can." Most tin plate is produced by electrodepositing a thin layer of tin on steel sheet. The nontoxic nature of tin salts makes tin plate useful for food-container material. Tin (standard emf of −0.136 V) and iron (standard emf of −0.441 V) are close in electrochemical behavior. Slight changes in oxygen availability and ion concentrations that build up on the surface will change their relative polarity. Under conditions of atmospheric exposure, tin is normally cathodic to steel. Thus, if the outside of a piece of perforated tin plate is exposed to the atmosphere, the steel will corrode and not the tin (Fig. 13.18b). However, in the absence of atmospheric oxygen, tin is anodic to steel, which makes tin a useful container material for food and beverages. As can be seen in this example, oxygen availability is an important factor in galvanic corrosion.

Another important consideration in galvanic two-metal corrosion is the ratio of the cathodic to anodic areas. This is called the *area effect.* An unfavorable cathodic–anodic area ratio consists of a large cathodic area and a small anodic area. With a certain amount of current flow to a metal couple, such as copper and iron electrodes of different sizes, the current density is much greater for the smaller electrode than for the larger one. Thus, the smaller anodic electrode will corrode much faster. Table 13.5 shows that as the cathode-to-anode ratio for an iron-copper couple was increased from 1 to 18.5, the weight loss of iron at the anode rose from 0.23 to 1.25 g. This area effect is also illustrated in Fig. 13.19 for copper-steel couples immersed in seawater. Copper rivets (cathodes) with a small area caused only a slight increase in the corrosion of steel plates (Fig. 13.19a). On the other hand, copper plates (cathodes) caused severe corrosion of steel rivets (anodes), as shown in Fig. 13.19b. Thus, a large cathode area to a small anode area ratio should be avoided.

13.5.3 Pitting Corrosion

Pitting is a form of localized corrosive attack that produces holes or pits in a metal. This form of corrosion is very destructive for engineering structures if it

Table 13.5 Effect of area on galvanic corrosion of iron coupled to copper in 3% sodium chloride

Relative areas		Anode (iron) loss (g)*
Cathode	**Anode**	
1.01	1	0.23
2.97	1	0.57
5.16	1	0.79
8.35	1	0.94
11.6	1	1.09
18.5	1	1.25

*Tests run at 86°F in aerated and agitated solution for about 20 h. Anode area was 14 cm².

(a)

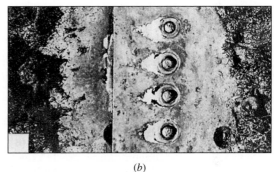

(b)

Figure 13.19
Effect of area relationships between cathode and anode for copper-steel couples immersed in seawater. (a) Small cathode (copper rivets) and large anode (steel plates) cause only slight damage to steel. (b) Small anode (steel rivets) and large cathode (copper plates) cause severe corrosion of steel rivets.
(*Courtesy of the LaQue Center for Corrosion Technology, Inc.*)

causes perforation of the metal. However, if perforation does not occur, minimum pitting is sometimes acceptable in engineering equipment. Pitting is often difficult to detect because small pits may be covered by corrosion products. Also the number and depth of pits can vary greatly, and so the extent of pitting damage may be difficult to evaluate. As a result, pitting, because of its localized nature, can often result in sudden, unexpected failures.

Figure 13.20 shows an example of pitting in a stainless steel exposed to an aggressive corrosive environment. The pitting for this example was accelerated, but in most service conditions, pitting may require months or years to perforate a metal

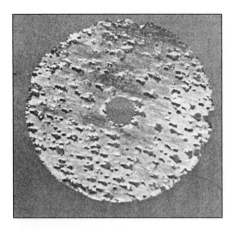

Figure 13.20
Pitting of a stainless steel in an
aggressive corrosive environment.
(*Courtesy of the LaQue Center for Corrosion*
Technology, Inc.)

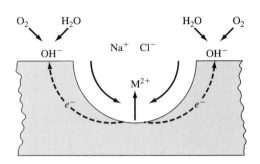

Figure 13.21
A schematic diagram of the growth of a pit
in stainless steel in an aerated salt solution.
(*After M. G. Fontana and N. D. Greene, "Corrosion*
Engineering," 2nd ed., McGraw-Hill, 1978.)

section. Pitting usually requires an initiation period, but once started, the pits grow
at an ever-increasing rate. Most pits develop and grow in the direction of gravity and
on the lower surfaces of engineering equipment.

Pits are initiated at places where local increases in corrosion rates occur. Inclusions, other structural heterogeneities, and compositional heterogeneities on the
metal surface are common places where pits initiate. Differences in ion and oxygen
concentrations create concentration cells that can also initiate pits. The propagation
of a pit is believed to involve the dissolution of the metal in the pit while maintaining a high degree of acidity at the bottom of the pit. The propagation process for
a pit in an aerated saltwater environment is illustrated in Fig. 13.21 for a ferrous
metal. The anodic reaction of the metal at the bottom of the pit is $M \rightarrow M^{n+} + e^-$.
The cathodic reaction takes place at the metal surface surrounding the pit and is the
reaction of oxygen with water and the electrons from the anodic reaction:
$O_2 + 2H_2O + 4e^- \rightarrow 4OH^-$. Thus, the metal surrounding the pit is cathodically
protected. The increased concentration of metal ions in the pit brings in chloride ions
to maintain charge neutrality. The metal chloride then reacts with water to produce
the metal hydroxide and free acid as

$$M^+Cl^- + H_2O \rightarrow MOH + H^+Cl^- \tag{13.16}$$

In this way a high acid concentration builds up at the bottom of the pit, which makes
the anodic reaction rate increase, and the whole process becomes *autocatalytic*.

To avoid pitting corrosion in the design of engineering equipment, materials that
do not have pitting-corrosion tendencies should be used. However, if this is not possible for some designs, then materials with the best corrosion resistance must be

Table 13.6 Relative pitting resistance of some corrosion-resistant alloys

Type 304 stainless steel	
Type 316 stainless steel	Increasing
Hastelloy F, Nionel, or Durimet 20	pitting resistance
Hastelloy C or Chlorimet 3	
Titanium	

Source: M. G. Fontana and N. D. Greene, *Corrosion Engineering,* 2nd ed., McGraw-Hill, 1978.

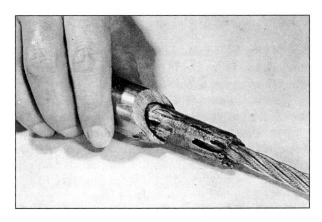

Figure 13.22
Crevice corrosion of a mooring pennant.
(*Courtesy of the LaQue Center for Corrosion Technology, Inc.*)

used. For example, if stainless steels must be used in the presence of some chloride ions, type 316 alloy with 2% molybdenum in addition to the 18% Cr and 8% Ni has better pitting resistance than type 304 alloy, which just contains the 18% Cr and 8% Ni as the main alloying elements. A qualitative guide for the pitting-resistance order of some corrosion-resistant materials is listed in Table 13.6. However, it is recommended that corrosion tests be made of various alloys before final selection of the corrosion-resistant alloy.

13.5.4 Crevice Corrosion

Crevice corrosion is a form of localized electrochemical corrosion that can occur in crevices and under shielded surfaces where stagnant solutions can exist. Crevice corrosion is of engineering importance when it occurs under gaskets, rivets, and bolts, between valve disks and seats, under porous deposits, and in many other similar situations. Crevice corrosion occurs in many alloy systems such as stainless steels and titanium, aluminum and copper alloys. Figure 13.22 shows an example of crevice-corrosion attack of a mooring pennant.

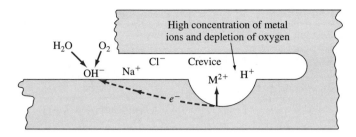

Figure 13.23
A schematic diagram of crevice-corrosion mechanism.
(After M. G. Fontana and N. D. Greene, "Corrosion Engineering," 2nd ed., McGraw-Hill, 1978.)

For crevice corrosion to occur, a crevice must be wide enough to allow liquid to enter but narrow enough to keep the liquid stagnant. Therefore, crevice corrosion usually occurs with an opening of a few micrometers (mils) or less in width. Fibrous gaskets that can act as wicks to absorb an electrolytic solution and keep it in contact with the metal surface make ideal locations for crevice corrosion.

Fontana and Greene[6] have proposed a mechanism for crevice corrosion similar to the one they proposed for pitting corrosion. Figure 13.23 illustrates this mechanism for the crevice corrosion of a stainless steel in an aerated sodium chloride solution. This mechanism assumes that initially the anodic and cathodic reactions on the surface of the crevice are

$$\text{Anodic reaction:} \qquad M \rightarrow M^+ + e^- \qquad \textbf{(13.17a)}$$

$$\text{Cathodic reaction:} \qquad O_2 + 2H_2O + 4e^- \rightarrow 4OH^- \qquad \textbf{(13.17b)}$$

Since the solution in the crevice is stagnant, the oxygen needed for the cathodic reaction is used up and not replaced. However, the anodic reaction $M \rightarrow M^+ + e^-$ continues to operate, creating a high concentration of positively charged ions. To balance the positive charge, negatively charged ions, mainly chloride ions, migrate into the crevice, forming M^+Cl^-. This chloride is hydrolyzed by water to form the metal hydroxide and free acid, as

$$M^+Cl^- + H_2O \rightarrow MOH + H^+Cl^- \qquad \textbf{(13.18)}$$

This buildup of acid breaks down the passive film and causes corrosion attack, which is autocatalytic, as in the case just discussed for pitting corrosion.

For type 304 (18% Cr–8% Ni) stainless steel, Peterson et al.[7] have concluded from their tests that the acidification within the crevice is probably mainly due to

---

[6]*Corrosion Engineering,* 2nd ed., McGraw-Hill, 1978.

[7]M. H. Peterson, T. J. Lennox, and R. E. Groover, *Mater. Prot.,* January 1970, p. 23.

the hydrolysis of chromic ions, as

$$Cr^{3+} + 3H_2O \rightarrow Cr(OH)_3 + 3H^+ \qquad\qquad (13.19)$$

since they found only traces of Fe^{3+} in the crevice.

To prevent or minimize crevice corrosion in engineering designs, the following methods and procedures can be used:

1. Use soundly welded butt joints instead of riveted or bolted ones in engineering structures.
2. Design vessels for complete drainage where stagnant solutions may accumulate.
3. Use nonabsorbent gaskets, such as Teflon, if possible.

13.5.5 Intergranular Corrosion

Intergranular corrosion is localized corrosion attack at and/or adjacent to the grain boundaries of an alloy. Under ordinary conditions if a metal corrodes uniformly, the grain boundaries will only be slightly more reactive than the matrix. However, under other conditions, the grain-boundary regions can be very reactive, resulting in intergranular corrosion that causes loss of strength of the alloy and even disintegration at the grain boundaries.

For example, many high-strength aluminum alloys and some copper alloys that have precipitated phases for strengthening are susceptible to intergranular corrosion under certain conditions. However, one of the most important examples of intergranular corrosion occurs in some austenitic (18% Cr–8% Ni) stainless steels that are heated into or slowly cooled through the 500 to 800°C (950 to 1450°F) temperature range. In this so-called *sensitizing temperature range,* chromium carbides ($Cr_{23}C_6$) can precipitate at the grain-boundary interfaces, as shown in Fig. 13.24*a*.

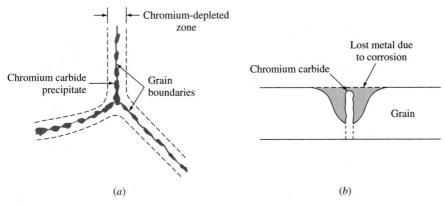

(a)   (b)

Figure 13.24
(*a*) Schematic representation of chromium carbide precipitation at grain boundaries in a sensitized type 304 stainless steel. (*b*) Cross section at grain boundary showing intergranular corrosion attack adjacent to the grain boundaries.

When chromium carbides have precipitated along the grain boundaries in austenitic stainless steels, these alloys are said to be in the *sensitized condition.*

If an 18% Cr–8% Ni austenitic stainless steel contains more than about 0.02 wt % carbon, chromium carbides ($Cr_{23}C_6$) can precipitate at the grain boundaries of the alloy if heated in the 500 to 800°C range for a long-enough time. Type 304 is an austenitic stainless steel with 18% Cr–8% Ni and between about 0.06 and 0.08 wt % carbon. Hence this alloy, if heated in the 500 to 800°C range for a sufficient time, will be put into the sensitized condition and be susceptible to intergranular corrosion. When the chromium carbides form at the grain boundaries, they deplete the regions adjacent to the boundaries of chromium so that the chromium level in these areas decreases below the 12 percent chromium level necessary for passive or "stainless" behavior. Thus, when, for example, type 304 stainless steel in the sensitized condition is exposed to a corrosive environment, the regions next to the grain boundaries will be severely attacked. These areas become anodic to the rest of the grain bodies, which are cathodic, thereby creating galvanic couples. Figure 13.24*b* shows this schematically.

Failure of welds made with type 304 stainless steel or similar alloys can occur by the same chromium carbide precipitation mechanism as previously described. This type of weld failure is called **weld decay** and is characterized by a weld decay zone somewhat removed from the centerline of the weld, as shown in Fig. 13.25. The metal in the weld decay zone was held in the temperature range for sensitizing

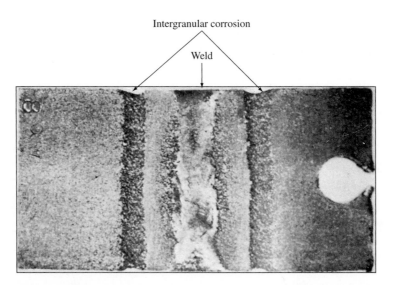

Figure 13.25
Intergranular corrosion of a stainless steel weld. The weld decay zones had been held at the critical temperature range needed for precipitation of chromium carbides during cooling.
(After H. H. Uhlig, "Corrosion and Corrosion Control," Wiley, 1963, p. 267.)

(500 to 800°C) for too long a time so that chromium carbides precipitated in the grain boundaries of the heat-affected zones of the weld. If a welded joint in the sensitized condition is not subsequently reheated to redissolve the chromium carbides, it will be subject to intergranular corrosion when exposed to a corrosive environment, and the weld could fail.

Intergranular corrosion of austenitic stainless steels can be controlled by the following methods:

1. Use a high-temperature solution heat treatment after welding. By heating the welded joint at 500 to 800°C followed by water quenching, the chromium carbides can be redissolved and returned to solid solution.

2. Add an element that will combine with the carbon in the steel so that chromium carbides cannot form. Columbium and titanium additions in alloy types 347 and 321, respectively, are used. These elements have a greater affinity for carbon than chromium. Alloys with Ti or Cb additions are said to be in the *stabilized condition*.

3. Lower the carbon content to about 0.03 wt % or less so that significant amounts of chromium carbides cannot precipitate. Type 304L stainless steel, for example, has its carbon at such a low level.

13.5.6 Stress Corrosion

Stress-corrosion cracking (SCC) of metals refers to cracking caused by the combined effects of tensile stress and a specific corrosion environment acting on the metal. During SCC the metal's surface is usually attacked very little while highly localized cracks propagate through the metal section, as shown, for example, in Fig. 13.26. The stresses that cause SCC can be residual or applied. High residual stresses that cause SCC may result, for example, from thermal stresses introduced by unequal cooling rates, poor mechanical design for stresses, phase transformations during heat treatment, cold working, and welding.

Figure 13.26
Stress-corrosion cracks in a pipe.
(*Courtesy of the LaQue Center for Corrosion Technology, Inc.*)

Table 13.7 Environments that may cause stress corrosion of metals and alloys

Material	Environment	Material	Environment

Aluminum alloys	$NaCl-H_2O_2$ solutions
	NaCl solutions
	Seawater
	Air, water vapor
		Ordinary steels	NaOH solutions
			$NaOH-Na_2SiO_3$
			  solutions
			Calcium, ammonium,
			  and sodium nitrate
			  solutions
			Mixed acids
			  ($H_2SO_4-HNO_2$)

Copper alloys	Ammonia vapors and solutions Amines Water, water vapor		
Gold alloys	$FeCl_2$ solutions Acetic acid–salt solutions		HCN solutions Acidic H_2S solutions
Inconel	Caustic soda solutions		Seawater
Lead	Lead acetate solutions		Molten Na-Pb alloys

Magnesium alloys	$NaCl-K_2CrO_4$ solutions
	Rural and coastal atmospheres
	Distilled water
		Stainless steels	Acid-chloride solutions
			  such as $MgCl_2$ and
			  $BaCl_2$

Monel	Fused caustic soda Hydrofluoric acid Hydrofluosilicic acid		$NaCl-H_2O_2$ solutions Seawater H_2S
Nickel	Fused caustic soda		$NaOH-H_2S$ solutions Condensing steam from chloride waters

		Titanium alloys	Red fuming nitric acid,
			  seawater, N_2O_4,
			  methanol-HCl

Source: M. G. Fontana and N. D. Greene, *Corrosion Engineering,* 2nd ed., McGraw-Hill, 1978, p. 100.

Only certain combinations of alloys and environments will cause SCC. Table 13.7 lists some of the alloy-environment systems in which SCC occurs. There appears to be no general pattern to the environments that produce SCC in alloys. For example, stainless steels crack in chloride environments but not in ammonia-containing ones. In contrast, brasses (Cu-Zn alloys) crack in ammonia-containing environments but not in chloride ones. New combinations of alloys and environments that cause SCC are continually being discovered.

Mechanism of Stress-Corrosion Cracking   The mechanisms involved in SCC are not completely understood since there are so many different alloy-environment systems involving many different mechanisms. Most SCC mechanisms involve crack initiation and propagation stages. In many cases the crack initiates at a pit or other discontinuity on the metal surface. After the crack has been started, the tip can advance, as shown in Fig. 13.27. A high stress builds up at the tip of the crack due to tensile stresses acting on the metal. Anodic dissolution of the metal takes place by localized electrochemical corrosion at the tip of the crack as it advances. The crack grows in a plane perpendicular to the tensile stress until the metal fractures.

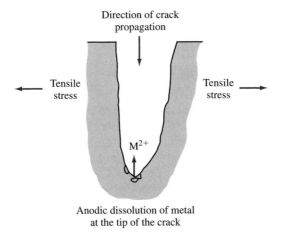

Figure 13.27
Development of a stress-corrosion crack in a
metal by anodic dissolution.
(*After R. W. Staehle.*)

If either the stress or the corrosion is stopped, the crack stops growing. A classic
experiment was made by Priest et al.,[8] who showed that an advancing crack could
be stopped by cathodic protection. When cathodic protection was removed, the crack
started to grow again.

Tensile stress is necessary for both the initiation and propagation of cracks and
is important in the rupturing of surface films. Decreasing the stress level increases
the time necessary for cracking to occur. Temperature and environment are also
important factors for stress-corrosion cracking.

Prevention of Stress-Corrosion Cracking   Since the mechanisms of SCC are not
completely understood, the methods of preventing it are general and empirical ones.
One or more of the following methods will prevent or reduce SCC in metals.

1. Lower the stress of the alloy below that which causes cracking. This may be
done by lowering the stress on the alloy or by giving the material a stress-relief
anneal. Plain-carbon steels can be stress-relieved at 600 to 650°C (1100 to 1200°F),
and austenitic stainless steels can be stress-relieved in the range 815 to 925°C (1500
to 1700°F).

2. Eliminate the detrimental environment.

3. Change the alloy if neither environment nor stress level can be changed. For
example, use titanium instead of stainless steel for heat exchangers in contact with
seawater.

---

[8]D. K. Priest, F. H. Beck, and M. G. Fontana, *Trans. ASM,* **47**:473(1955).

4. Apply cathodic protection by using consumable anodes or an external power supply (see Sec. 13.7).

5. Add inhibitors if possible.

13.5.7 Erosion Corrosion

Erosion corrosion can be defined as the acceleration in the rate of corrosion attack in a metal due to the relative motion of a corrosive fluid and a metal surface. When the relative motion of corrosive fluid is rapid, the effects of mechanical wear and abrasion can be severe. Erosion corrosion is characterized by the appearance on the metal surface of grooves, valleys, pits, rounded holes, and other metal surface damage configurations that usually occur in the direction of the flow of the corrosive fluid.

Studies of the erosion-corrosion action of silica sand slurries in mild-steel pipe have led researchers to believe that the increased corrosion rate of the slurry action is due to the removal of surface rust and salt films by the abrasive action of the silica particles of the slurry, thus permitting much easier access of dissolved oxygen to the corroding surface. Figure 13.28 shows severe wear patterns caused by the erosion corrosion of an experimental section of mild-steel pipe.

13.5.8 Cavitation Damage

This type of erosion corrosion is caused by the formation and collapse of air bubbles or vapor-filled cavities in a liquid near a metal surface. Cavitation damage occurs at metal surfaces where high-velocity liquid flow and pressure changes exist such as are encountered with pump impellers and ship propellers. Calculations indicate that rapidly collapsing vapor bubbles can produce localized pressures as high as 60,000 psi. With repeated collapsing vapor bubbles, considerable damage to a

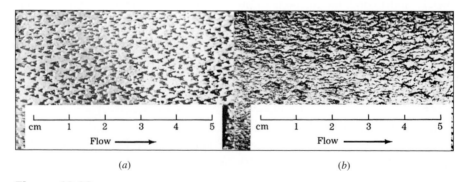

(a)   (b)

Figure 13.28
Erosion-corrosion wear pattern of silica slurry in mild-steel pipe, showing (a) pitting after 21 days and (b) irregular wavy pattern after 42 days. Slurry velocity is 3.5 m/s.
[*After J. Postlethwaite et al., Corrosion,* **34**:*245(1978).*]

metal surface can be done. By removing surface films and tearing metal particles away from the metal surface, cavitation damage can increase corrosion rates and cause surface wear.

13.5.9 Fretting Corrosion

Fretting corrosion occurs at interfaces between materials under load subjected to vibration and slip. Fretting corrosion appears as grooves or pits surrounded by corrosion products. In the case of the fretting corrosion of metals, metal fragments between the rubbing surfaces are oxidized and some oxide films are torn loose by the wearing action. As a result, there is an accumulation of oxide particles that act as an abrasive between the rubbing surfaces. Fretting corrosion commonly occurs between tight-fitting surfaces such as those found between shafts and bearings or sleeves. Figure 13.29 shows the effects of fretting corrosion on the surface of a Ti–6 Al–4 V alloy.

13.5.10 Selective Leaching

Selective leaching is the preferential removal of one element of a solid alloy by corrosion processes. The most common example of this type of corrosion is dezincification, in which the selective leaching of zinc from copper in brasses occurs. Similar processes also occur in other alloy systems such as the loss of nickel, tin, and chromium from copper alloys, iron from cast iron, nickel from alloy steels, and cobalt from stellite.

Figure 13.29
Scanning electron micrograph showing fretting corrosion on the surface of a Ti–6 Al–4 V alloy developed at 600°C using a sphere-on-flat-configuration with 40 μm amplitude of slip and after 3.5×10^6 cycles.
[*After M. M. Hamdy and R. B. Waterhouse, Wear,* **71**:*237(1981).*]

In the dezincification of a 70% Cu–30% Zn brass, for example, the zinc is preferentially removed from the brass, leaving a spongy, weak matrix of copper. The mechanism of dezincification involves the following three steps:[9]

1. Copper & zinc are dissolved by aqueous solution
2. Copper ions are replated on brass
3. Zinc ions remain in the solution

Since the copper remaining does not have the strength of the brass, the strength of the alloy is considerably lowered.

Dezincification can be minimized or prevented by changing to a brass with a lower zinc content (that is, 85% Cu–15% Zn brass) or to a cupronickel (70 to 90% Cu–10 to 30% Ni). Other possibilities are to change the corrosive environment or use cathodic protection.

13.5.11 Hydrogen Damage

Hydrogen damage refers to those situations in which the load carrying capacity of a metallic component is reduced due to interaction with atomic hydrogen (H) or Molecular hydrogen (H_2), usually in conjunction with residual or externally applied tensile stresses. Because of availability of hydrogen, as one of the most abundant elements, its interaction with metal can occur during production, processing or service and over a wide variety of environments and circumstances. Many metals such as low carbon and alloy steels, martensitic and precipitation hardened stainless steels, aluminum alloys, and titanium alloys are susceptible to hydrogen damage to various degrees. The effects of hydrogen damage may be manifested in a number of ways including cracking, blistering, hydride formation and reduced ductility of the material. Of the numerous types of hydrogen damage, three directly relate to loss of ductility and are therefore designated as **hydrogen embrittlement.** These include: (1) hydrogen environment brittlement, which occurs during plastic deformation of metals such as steels, stainless steels, and titanium alloys in the presence of hydrogen-bearing gasses (H_2) or corrosion reactions; (2) hydrogen stress cracking, which is defined as brittle fracture of an originally ductile material such as carbon and low alloy steels in the presence of hydrogen and under continuous load; (3) loss in tensile ductility, in which significant reduction in elongation capability and reduction in area is observed in metals such as steels and aluminum alloys. Examples of other types of hydrogen damage that are not directly classified as hydrogen embrittlement are: (1) hydrogen attack, a high temperature mode of attack in which hydrogen enters metals such as steels and reacts with carbon (available in the solid-solution form in carbide form) to produce methane gas resulting in formation of cracks or decarburization; (2) blistering, which occurs when atomic hydrogen (H) diffuses into available internal defects in low strength alloys of steel, copper and aluminum and precipitates as molecular hydrogen. The precipitated gas produces high internal pressure resulting in local plastic deformation and blistering that often results in rupture.

---

[9]After M. G. Fontana and N. D. Greene, "Corrosion Engineering," 2nd ed., McGraw-Hill, 1978.

The diffusion of hydrogen in metal could occur when corrosion is occurring and the cathodic partial reaction is the reduction of hydrogen ion. It could also occur when the metal is exposed to fresh or seawater, to hydrogen sulfide (often occurs during drilling of oil and gas wells), and to pickling and electroplating processes. In the pickling process, the surface oxides (scale) are removed from iron and steel through immersion in sulfuric (H_2SO_4) and hydrochloric acids (HCl).

Hydrogen damage is a serious problem and is the cause of many failures in various components. Thus a designer must be aware of this type of damage and the metals that are more susceptible to it. To reverse hydrogen contamination, a process known as bakeout (a heat treatment process) is applied to the component that promotes the diffusion of hydrogen out of the metal.

13.6 OXIDATION OF METALS

Until now we have been concerned with corrosion conditions in which a liquid electrolyte was an integral part of the corrosion mechanism. Metals and alloys, however, also react with air to form external oxides. The high-temperature oxidation of metals is particularly important in the engineering design of gas turbines, rocket engines, and high-temperature petrochemical equipment.

13.6.1 Protective Oxide Films

The degree to which an oxide film is protective to a metal depends upon many factors, of which the following are important:

1. The volume ratio of oxide to metal after oxidation should be close to 1:1.
2. The film should have good adherence.
3. The melting point of the oxide should be high.
4. The oxide film should have a low vapor pressure.
5. The oxide film should have a coefficient of expansion nearly equal to that of the metal.
6. The film should have high-temperature plasticity to prevent fracture.
7. The film should have low conductivity and low diffusion coefficients for metal ions and oxygen.

The calculation of the volume ratio of oxide to metal after oxidation is a first step that can be taken to find out if an oxide of a metal might be protective. This ratio is called the **Pilling-Bedworth**[10] **(P.B.) ratio** and can be expressed in equation form as

$$\text{P.B. ratio} = \frac{\text{volume of oxide produced by oxidation}}{\text{volume of metal consumed by oxidation}} \tag{13.20}$$

If a metal has a P.B. ratio of less than 1, as is the case for the alkali metals (for example, Na has a P.B. ratio of 0.576), the metal oxide will be porous and

---

[10]N. B. Pilling and R. E. Bedworth, *J. Inst. Met.*, **29**:529 (1923).

unprotective. If the P.B. ratio is more than 1, as is the case for Fe (Fe_2O_3 has P.B. ratio of 2.15), compressive stresses will be present and the oxide will tend to crack and spall off. If the P.B. ratio is close to 1, the oxide may be protective, but some of the other factors previously listed have to be satisfied. Thus, the P.B. ratio alone does not determine if an oxide is protective. Example Problem 13.9 shows how the P.B. ratio for aluminum can be calculated.

EXAMPLE PROBLEM 13.9	Calculate the ratio of the oxide volume to metal volume (Pilling-Bedworth ratio) for the oxidation of aluminum to aluminum oxide, Al_2O_3. The density of aluminum = 2.70 g/cm^3 and that of aluminum oxide = 3.70 g/cm^3.

■ **Solution**

$$\text{P.B. ratio} = \frac{\text{volume of oxide produced by oxidation}}{\text{volume of metal consumed by oxidation}} \qquad (13.20)$$

Assuming that 100 g of aluminum is oxidized,

$$\text{Volume of aluminum} = \frac{\text{mass}}{\text{density}} = \frac{100 \text{ g}}{2.70 \text{ g/cm}^3} = 37.0 \text{ cm}^3$$

To find the volume of Al_2O_3 associated with the oxidation of 100 g of Al, we first find the mass of Al_2O_3 produced by the oxidation of 100 g of Al, using the following equation:

$$\begin{array}{cc} 100 \text{ g} & X \text{ g} \\ 4Al + 3O_2 \rightarrow & 2Al_2O_3 \\ 4 \times \dfrac{26.98 \text{ g}}{\text{mol}} & 2 \times \dfrac{102.0 \text{ g}}{\text{mol}} \end{array}$$

or

$$\frac{100 \text{ g}}{4 \times 26.98} = \frac{X \text{ g}}{2 \times 102}$$

$$X = 189.0 \text{ g } Al_2O_3$$

Then we find the volume associated with the 189.0 g of Al_2O_3 using the relationship volume = mass/density. Therefore,

$$\text{Volume of } Al_2O_3 = \frac{\text{mass of } Al_2O_3}{\text{density of } Al_2O_3} = \frac{189.0 \text{ g}}{3.70 \text{ g/cm}^3} = 51.1 \text{ cm}^3$$

Thus,

$$\text{P.B. ratio} = \frac{\text{volume of } Al_2O_3}{\text{volume of Al}} = \frac{51.1 \text{ cm}^3}{37.0 \text{ cm}^3} = 1.38 \blacktriangleleft$$

Comment:
The ratio 1.38 is close to 1, so Al_2O_3 has a favorable P.B. ratio to be a protective oxide. Al_2O_3 is a protective oxide since it forms a tight, coherent film on aluminum. Some of the Al_2O_3 molecules at the oxide-metal interface penetrate into the aluminum metal, and vice versa.

13.6.2 Mechanisms of Oxidation

When an oxide film forms on a metal by the oxidation of a metal by gaseous oxygen, it is formed by an electrochemical process and not simply by the chemical combination of a metal and oxygen as $M + \frac{1}{2}O_2 \rightarrow MO$. The oxidation and reduction partial reactions for the formation of divalent ions are

$$\text{Oxidation partial reaction:} \quad M \rightarrow M^{2+} + 2e^- \quad \textbf{(13.21}a\textbf{)}$$

$$\text{Reduction partial reaction:} \quad \tfrac{1}{2}O_2 + 2e^- \rightarrow O^{2-} \quad \textbf{(13.21}b\textbf{)}$$

In the very early stages of oxidation, the oxide layer is discontinuous and begins by the lateral extension of discrete oxide nuclei. After the nuclei interlace, the mass transport of the ions occurs in a direction normal to the surface (Fig. 13.30). In most cases the metal diffuses as cations and electrons across the oxide film, as indicated in Fig. 13.30*a*. In this mechanism the oxygen is reduced to oxygen ions at the oxide-gas interface, and the zone of oxide formation is at this surface (cation diffusion). In some other cases, for example, for some heavy metal oxides, the oxygen may diffuse as O^{2-} ions to the metal-oxide interface and electrons diffuse to the oxide-gas interface, as shown in Fig. 13.30*b*. In this case the oxide forms at the metal-oxide interface (Fig. 13.30*b*). This is anion diffusion. The oxide movements indicated in Fig. 13.30 are determined principally by the movements of inert markers at the oxide-gas interface. In the case of cation diffusion the markers are buried in the oxide, whereas in the case of anion diffusion the markers remain at the surface of the oxide.

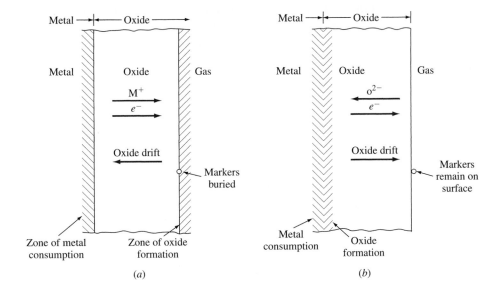

Figure 13.30
Oxidation of flat surfaces of metals. (*a*) When cations diffuse, the initially formed oxide drifts toward the metal. (*b*) When anions diffuse, the oxide drifts in the opposite direction.
[*After L. L. Shreir (ed.) "Corrosion," vol. 1, 2nd ed., Newnes-Butterworth, 1976, p. 1:242.*]

The detailed mechanisms of the oxidation of metals and alloys can be very complex, particularly when layers of different composition and defect structures are produced. Iron, for example, when oxidized at high temperatures, forms a series of iron oxides: FeO, Fe_3O_4, and Fe_2O_3. The oxidation of alloys is further complicated by the interaction of the alloying elements.

13.6.3 Oxidation Rates (Kinetics)

From an engineering standpoint, the rate at which metals and alloys oxidize is very important since the oxidation rate of many metals and alloys determines the useful life of equipment. The rate of oxidation of metals and alloys is usually measured and expressed as the weight gained per unit area. During the oxidation of different metals, various empirical rate laws have been observed; some of the common ones are shown in Fig. 13.31.

The simplest oxidation rate follows the linear law

$$w = k_L t \tag{13.22}$$

where w = weight gain per unit area

t = time

k_L = linear rate constant

Linear oxidation behavior is shown by metals that have porous or cracked oxide films, and thus the transport of reactant ions occurs at faster rates than the chemical reaction. Examples of metals that oxidize linearly are potassium, which has an oxide-to-metal volume ratio of 0.45, and tantalum, which has one of 2.50.

When ion diffusion is the controlling step in the oxidation of metals, pure metals should follow the parabolic relation

$$w^2 = k_p t + C \tag{13.23}$$

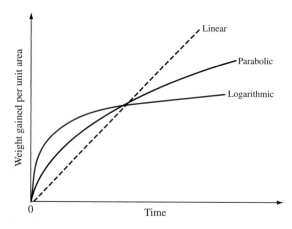

Figure 13.31
Oxidation rate laws.

where w = weight gain per unit area

   t = time

   k_p = parabolic rate constant

   C = a constant

Many metals oxidize according to the parabolic rate law, and these are usually associated with thick coherent oxides. Iron, copper, and cobalt are examples of metals that show parabolic oxidation behavior.

Some metals such as Al, Cu, and Fe oxidize at ambient or slightly elevated temperatures to form thin films that follow the logarithmic rate law

$$w = k_e \log(Ct + A) \qquad (13.24)$$

where C and A are constants and k_e is the logarithmic rate constant. These metals when exposed to oxygen at room temperature oxidize very rapidly at the start, but after a few days' exposure, the rate decreases to a very low value.

Some metals that exhibit linear rate behavior tend to oxidize catastrophically at high temperatures due to rapid exothermic reactions on their surfaces. As a result, a chain reaction occurs at their surface, causing the temperature and rate of reaction to increase. Metals such as molybdenum, tungsten, and vanadium that have volatile oxides may oxidize catastrophically. Also, alloys containing molybdenum and vanadium even in small amounts often show catastrophic oxidation that limits their use in high-temperature oxidizing atmospheres. The addition of large amounts of chromium and nickel to iron alloys improves their oxidation resistance and retards the effects of catastrophic oxidation due to some other elements.

A 1 cm^2 sample of 99.94 wt % nickel, 0.75 mm thick, is oxidized in oxygen at 1 atm pressure at 600°C. After 2 h, the sample shows a weight gain of 70 μg/cm^2. If this material shows parabolic oxidation behavior, what will the weight gain be after 10 h? (Use Eq. 13.23 with $C = 0$.)

EXAMPLE PROBLEM 13.10

■ **Solution**

First, we must determine the parabolic rate constant k_p from the parabolic oxidation rate equation $y^2 = k_p t$, where y is the thickness of the oxide produced in time t. Since the weight gain of the sample during oxidation is proportional to the growth in thickness of the oxide and can be more precisely measured, we shall replace y, the oxide thickness, with x, the weight gain per unit area of the sample during oxidation. Thus, $x^2 = k_p' t$ and

$$k_p' = \frac{x^2}{t} = \frac{(70\ \mu g/cm^2)^2}{2\ h} = 2.45 \times 10^3\ \mu g^2/(cm^4 \cdot h)$$

For time $t = 10$ h, the weight gain in micrograms per square centimeter is

$$x = \sqrt{k_p' t} = \sqrt{[2.45 \times 10^3\ (\mu g^2/(cm^4 \cdot h))](10\ h)}$$

$$= 156\ \mu g/cm^2 \blacktriangleleft$$

13.7 CORROSION CONTROL

Corrosion can be controlled or prevented by many different methods. From an indus-trial standpoint, the economics of the situation usually dictate the methods used. For example, an engineer may have to determine whether it is more economical to peri-odically replace certain equipment or to fabricate it with materials that are highly corrosion-resistant but more expensive so that it will last longer. Some of the com-mon methods of corrosion control or prevention are shown in Fig. 13.32.

13.7.1 Materials Selection

Metallic Materials   One of the most common methods of corrosion control is to use materials that are corrosion-resistant for a particular environment. When select-ing materials for an engineering design for which the corrosion resistance of the materials is important, corrosion handbooks and data should be consulted to make sure the proper material is used. Further consultation with corrosion experts of com-panies that produce the materials would also be helpful to ensure that the best choices are made.

There are, however, some general rules that are reasonably accurate and can be applied when selecting corrosion-resistant metals and alloys for engineering appli-cations. These are:[11]

1. For reducing or nonoxidizing conditions such as air-free acids and aqueous solutions, nickel and copper alloys are often used.
2. For oxidizing conditions, chromium-containing alloys are used.
3. For extremely powerful oxidizing conditions, titanium and its alloys are commonly used.

Some of the "natural" metal–corrosive environment combinations that give good cor-rosion resistance for low cost are listed in Table 13.8.

---

[11]After M. G. Fontana and N. D. Greene, "Corrosion Engineering," 2nd ed., McGraw-Hill, 1978.

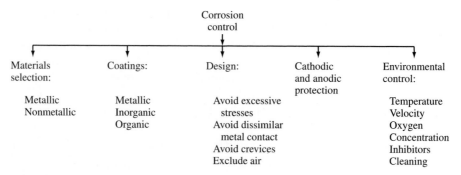

Figure 13.32
Common methods of corrosion control.

Table 13.8 Combinations of metals and environments that give good corrosion resistance for the cost

1. Stainless steels–nitric acid
2. Nickel and nickel alloys–caustic
3. Monel–hydrofluoric acid
4. Hastelloys (Chlorimets)–hot hydrochloric acid
5. Lead–dilute sulfuric acid
6. Aluminum–nonstaining atmospheric exposure
7. Tin–distilled water
8. Titanium–hot, strong oxidizing solutions
9. Tantalum–ultimate resistance
10. Steel–concentrated sulfuric acid

Source: M. G. Fontana and N. D. Greene, *Corrosion Engineering,* 2nd ed., McGraw-Hill, 1978.

One material that is often misused by fabricators not familiar with the corrosion properties of metals is stainless steel. Stainless steel is not a specific alloy but is a generic term used for a large class of steels with chromium contents above about 12 percent. Stainless steels are commonly used for corrosive environments that are moderately oxidizing, for example, nitric acid. However, stainless steels are less resistant to chloride-containing solutions and are more susceptible to stress-corrosion cracking than ordinary structural steel. Thus, great care must be taken to make sure that stainless steels are not used in applications for which they are not suited.

Nonmetallic Materials *Polymeric materials* such as plastics and rubbers are weaker, softer, and in general less resistant to strong inorganic acids than metals and alloys, and thus their use as primary materials for corrosion resistance is limited. However, as newer, higher-strength plastic materials become available, they will become more important. *Ceramic materials* have excellent corrosion and high-temperature resistance but have the disadvantage of being brittle with low tensile strengths. Nonmetallic materials are therefore mainly used in corrosion control in the form of liners, gaskets, and coatings.

13.7.2 Coatings

Metallic, inorganic, and organic coatings are applied to metals to prevent or reduce corrosion.

Metallic Coatings Metallic coatings that differ from the metal to be protected are applied as thin coatings to separate the corrosive environment from the metal. Metal coatings are sometimes applied so that they can serve as sacrificial anodes that can corrode instead of the underlying metal. For example, the zinc coating on steel to make galvanized steel is anodic to the steel and corrodes sacrificially.

Many metal parts are protected by electroplating to produce a thin protective layer of metal. In this process the part to be plated is made the cathode of an electrolytic cell. The electrolyte is a solution of a salt of the metal to be plated, and direct current is applied to the part to be plated and another electrode. The plating of a thin layer of tin on steel sheet to produce tin plate for tin cans is an example of the application of this method. The plating can also have several layers, as is the case for the chrome plate used on automobiles. This plating consists of three layers: (1) an inner flashing of copper for adhesion of the plating to the steel, (2) an intermediate layer of nickel for good corrosion resistance, and (3) a thin layer of chromium primarily for appearance.

Sometimes a thin layer of metal is roll-bonded to the surfaces of the metal to be protected. The outer thin layer of metal provides corrosion resistance to the inner core metal. For example, some steels are "clad" with a thin layer of stainless steel. This cladding process is also used to provide some high-strength aluminum alloys with a corrosion-resistant outer layer. For these *Alclad* alloys, as they are called, a thin layer of relatively pure aluminum is roll-bonded to the outer surface of the high-strength core alloy.

Inorganic Coatings (Ceramics and Glass)   For some applications it is desirable to coat steel with a ceramic coating to attain a smooth, durable finish. Steel is commonly coated with a porcelain coating that consists of a thin layer of glass fused to the steel surface so that it adheres well and has a coefficient of expansion adjusted to the base metal. Glass-lined steel vessels are used in some chemical industries because of their ease of cleaning and corrosion resistance.

Organic Coatings   Paints, varnishes, lacquers, and many other organic polymeric materials are commonly used to protect metals from corrosive environments. These materials provide thin, tough, and durable barriers to protect the substrate metal from corrosive environments. On a weight basis, the use of organic coatings protects more metal from corrosion than any other method. However, suitable coatings must be selected and applied properly on well-prepared surfaces. In many cases poor performance of paints, for example, can be attributed to poor application and preparation of surfaces. Care must also be taken *not* to apply organic coatings for applications where the substrate metal could be rapidly attacked if the coating film cracks.

13.7.3 Design

The proper engineering design of equipment can be as important for corrosion prevention as the selection of the proper materials. The engineering designer must consider the materials along with the necessary mechanical, electrical, and thermal property requirements. All these considerations must be balanced with economic limitations. In designing a system, specific corrosion problems may require the advice of corrosion experts. However, some important general design rules follow:[12]

1.   Allow for the penetration action of corrosion along with the mechanical strength requirements when considering the thickness of the metal used. This is especially important for pipes and tanks that contain liquids.

---

[12]After M. G. Fontana and N. D. Greene, *Corrosion Engineering,* 2nd ed., McGraw-Hill, 1978.

2. Weld rather than rivet containers to reduce crevice corrosion. If rivets are used, choose rivets that are cathodic to the materials being joined.

3. If possible, use galvanically similar metals for the whole structure. Avoid dissimilar metals that can cause galvanic corrosion. If galvanically dissimilar metals are bolted together, use nonmetallic gaskets and washers to prevent electrical contact between the metals.

4. Avoid excessive stress and stress concentrations in corrosive environments to prevent stress-corrosion cracking. This is especially important when using stainless steels, brasses, and other materials susceptible to stress-corrosion cracking in certain corrosive environments.

5. Avoid sharp bends in piping systems where flow occurs. Areas in which the fluid direction changes sharply promote erosion corrosion.

6. Design tanks and other containers for easy draining and cleaning. Stagnant pools of corrosive liquids set up concentration cells that promote corrosion.

7. Design systems for easy removal and replacement of parts that are expected to fail rapidly in service. For example, pumps in chemical plants should be easily removable.

8. Design heating systems so that hot spots do not occur. Heat exchangers, for example, should be designed for uniform temperature gradients.

In summary, design systems with conditions as uniform as possible and avoid heterogeneity.

13.7.4 Alteration of Environment

Environmental conditions can be very important in determining the severity of corrosion. Important methods for reducing corrosion by environmental changes are (1) lowering the temperature, (2) decreasing the velocity of liquids, (3) removing oxygen from liquids, (4) reducing ion concentrations, and (5) adding inhibitors to electrolytes.

1. Lowering the temperature of a system usually reduces corrosion because of the lower reaction rates at lower temperatures. However, there are some exceptions in which the reverse is the case. For example, boiling seawater is less corrosive than hot seawater because of the decrease in oxygen solubility with increasing temperature.

2. Decreasing the velocity of a corrosive fluid reduces erosion corrosion. However, for metals and alloys that passivate, stagnant solutions should be avoided.

3. Removing oxygen from water solutions is sometimes helpful in reducing corrosion. For example, boiler-water feed is deaerated to reduce corrosion. However, for systems that depend on oxygen for passivation, deaeration is undesirable.

4. Reducing the concentration of corrosive ions in a solution that is corroding a metal can decrease the corrosion rate of the metal. For example, reducing the chloride ion concentration in a water solution will reduce its corrosive attack on stainless steels.

5. Adding *inhibitors* to a system can decrease corrosion. Inhibitors are essentially retarding catalysts. Most inhibitors have been developed by empirical experiment,

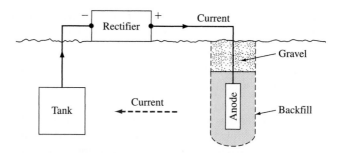

Figure 13.33

Cathodic protection of an underground tank by using impressed currents.

(After M. G. Fontana and N. D. Green, "Corrosion Engineering," 2nd ed., McGraw-Hill, 1978, p. 207.)

and many are proprietary in nature. Their actions also vary considerably. For example, the *absorption-type* inhibitors are absorbed on a surface and form a protective film. The *scavenger-type* inhibitors react to remove corrosion agents such as oxygen from solution.

13.7.5 Cathodic and Anodic Protection

Cathodic Protection   Corrosion control can be achieved by a method called **cathodic protection**[13] in which electrons are supplied to the metal structure to be protected. For example, the corrosion of a steel structure in an acid environment involves the following electrochemical equations:

$$Fe \rightarrow Fe^{2+} + 2e^-$$

$$2H^+ + 2e^- \rightarrow H_2$$

If electrons are supplied to the steel structure, metal dissolution (corrosion) will be suppressed and the rate of hydrogen evolution increased. Thus, if electrons are continually supplied to the steel structure, corrosion will be suppressed. Electrons for cathodic protection can be supplied by (1) an external DC power supply, as shown in Fig. 13.33, or by (2) galvanic coupling with a more anodic metal than the one being protected. Cathodic protection of a steel pipe by galvanic coupling to a magnesium anode is illustrated in Fig. 13.34. Magnesium anodes that corrode in place of the metal to be protected are most commonly used for cathodic protection because of their high negative potential and current density.

Anodic Protection   **Anodic protection** is relatively new and is based on the formation of protective passive films on metal and alloy surfaces by externally

---

[13]For an interesting article on the application of cathodic protection for the preservation of submerged steel in the Arabian Gulf, see R. N. Duncan and G. A. Haines, "Forty Years of Successful Cathodic Protection in the Arabian Gulf," *Mater. Perform.*, **21**:9 (1982).

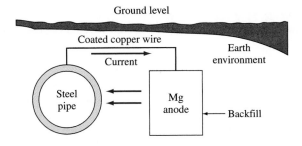

Figure 13.34
Protection of an underground pipeline with
a magnesium anode.
(*After M. G. Fontana and N. D. Greene, "Corrosion
Engineering," 2nd ed., McGraw-Hill, 1978, p. 207.*)

impressed anodic currents. Carefully controlled anodic currents by a device called a
potentiostat can be applied to protect metals that passivate, such as austenitic stainless steels, to make them passive and hence lower their corrosion rate in a corrosive
environment.[14] Advantages of anodic protection are that it can be applied in weak
to very corrosive conditions and that it uses very small applied currents. A disadvantage of anodic protection is that it requires complex instrumentation and has a
high installation cost.

---

[14]S. J. Acello and N. D. Greene, *Corrosion,* **18:**286 (1962).

A 2.2 kg sacrifical magnesium anode is attached to the steel hull of a ship. If the anode
completely corrodes in 100 days, what is the average current produced by the anode in
this period?

**EXAMPLE
PROBLEM 13.11**

■ **Solution**
Magnesium corrodes according to the reaction $Mg \rightarrow Mg^{2+} + 2e^-$. We will use Eq. 13.13
and solve for *I*, the average corrosion current in amperes:

$$w = \frac{ItM}{nF} \quad \text{or} \quad I = \frac{wnF}{tM}$$

$$w = 2.2 \text{ kg}\left(\frac{1000 \text{ g}}{\text{kg}}\right) = 2200 \text{ g} \quad n = 2 \quad F = 96{,}500 \text{ A} \cdot \text{s/mol}$$

$$t = 100 \text{ days}\left(\frac{24 \text{ h}}{\text{day}}\right)\left(\frac{3600 \text{ s}}{\text{h}}\right) = 8.64 \times 10^6 \text{ s} \quad M = 24.31 \text{ g/mol} \quad I = ? \text{ A}$$

$$I = \frac{(2200 \text{ g})(2)(96{,}500 \text{ A} \cdot \text{s/mol})}{(8.64 \times 10^6 \text{ s})(24.31 \text{ g/mol})} = 2.02 \text{ A} \blacktriangleleft$$

13.8 SUMMARY

Corrosion may be defined as the deterioration of a material resulting from chemical attack by its environment. Most corrosion of materials involves the chemical attack of metals by electrochemical cells. By studying equilibrium conditions, the tendencies of pure metal to corrode in a standard aqueous environment can be related to the standard electrode potentials of the metals. However, since corroding systems are not at equilibrium, the kinetics of corrosion reactions must also be studied. Some examples of kinetic factors affecting corrosion reaction rates are the polarization of the corrosion reactions and the formation of passive films on the metals.

There are many types of corrosion. Some of the important types discussed are uniform or general attack corrosion, galvanic or two-metal corrosion, pitting corrosion, crevice corrosion, intergranular corrosion, stress corrosion, erosion corrosion, cavitation damage, fretting corrosion, selective leaching or dealloying, and hydrogen embrittlement.

The oxidation of metals and alloys is also important for some engineering designs such as gas turbines, rocket engines, and high-temperature petrochemical installations. The study of the rates of oxidation of metals for some applications is very important. At high temperatures, care must be taken to avoid catastrophic oxidation.

Corrosion can be controlled or prevented by many different methods. To avoid corrosion, materials that are corrosion-resistant for a particular environment should be used where feasible. For many cases corrosion can be prevented by the use of metallic, inorganic, or organic coatings. The proper engineering design of equipment can also be very important for many situations. For some special cases, corrosion can be controlled by using cathodic or anodic protection systems.

13.9 DEFINITIONS

Sec. 13.1

Corrosion: the deterioration of a material resulting from chemical attack by its environment.

Sec. 13.2

Anode: the metal electrode in an electrolytic cell that dissolves as ions and supplies electrons to the external circuit.

Cathode: the metal electrode in an electrolytic cell that accepts electrons.

Sec. 13.3

Electromotive force series: an arrangement of metallic elements according to their standard electrochemical potentials.

Galvanic cell: two dissimilar metals in electrical contact with an electrolyte.

Ion-concentration cell: galvanic cell formed when two pieces of the same metal are electrically connected by an electrolyte but are in solutions of different ion concentrations.

Oxygen-concentration cell: galvanic cell formed when two pieces of the same metal are electrically connected by an electrolyte but are in solutions of different oxygen concentration.

Sec. 13.4

Cathodic polarization: the slowing down or the stopping of cathodic reactions at a cathode of an electrochemical cell due to (1) a slow step in the reaction sequence at the metal-electrolyte interface (*activation polarization*) or (2) a shortage of reactant or accumulation of reaction products at the metal-electrolyte interface (*concentration polarization*).

Passivation: the formation of a film of atoms or molecules on the surface of an anode so that corrosion is slowed down or stopped.

Galvanic (seawater) series: an arrangement of metallic elements according to their electrochemical potentials in seawater with reference to a standard electrode.

Sec. 13.5

Pitting corrosion: local corrosion attack resulting from the formation of small anodes on a metal surface.

Intergranular corrosion: preferential corrosion occurring at grain boundaries or at regions adjacent to the grain boundaries.

Weld decay: corrosion attack at or adjacent to a weld as the result of galvanic action resulting from structural differences in the weld.

Stress corrosion: preferential corrosive attack of a metal under stress in a corrosive environment.

Selective leaching: the preferential removal of one element of a solid alloy by corrosion processes.

Hydrogen embrittlement: loss of ductility in a metal due to interaction of the alloying element in the metal with atomic or molecular hydrogen.

Blistering: a type of damage due to diffusion of atomic hydrogen into internal pores in a metal creating high internal pressure resulting in rupture.

Sec. 13.6

Pilling-Bedworth (P.B.) ratio: the ratio of the volume of oxide formed to the volume of metal consumed by oxidation.

Sec. 13.7

Cathodic protection: the protection of a metal by connecting it to a sacrificial anode or by impressing a DC voltage to make it a cathode.

Anodic protection: the protection of a metal that forms a passive film by the application of an externally impressed anodic current.

13.10 PROBLEMS

Answers to problems marked with an asterisk are given at the end of the book.

13.1 Define corrosion as it pertains to materials.

13.2 What are some of the factors that affect the corrosion of metals?

13.3 Which is in a lower energy state: (*a*) elemental iron or (*b*) Fe_2O_3 (iron oxide)?

13.4 Give several examples of environmental deterioration of (*a*) ceramic materials and (*b*) polymeric materials.

13.5 What is the oxidation reaction called in which metals form ions that go into aqueous solution in an electrochemical corrosion reaction? What types of ions are produced by this reaction? Write the oxidation half-cell reaction for the oxidation of pure zinc metal in aqueous solution.

13.6 What is the reduction reaction called in which a metal or nonmetal is reduced in valence charge in an electrochemical corrosion reaction? Are electrons produced or consumed by this reaction?

13.7 What is a standard half-cell oxidation-reduction potential?

13.8 Describe a method used to determine the standard half-cell oxidation-reduction potential of a metal by using a hydrogen half-cell.

13.9 List five metals that are cathodic to hydrogen and give their standard oxidation potentials. List five metals that are anodic to hydrogen and give their standard oxidation potentials.

13.10 Consider a magnesium-iron galvanic cell consisting of a magnesium electrode in a solution of 1 M $MgSO_4$ and an iron electrode in a solution of 1 M $FeSO_4$. Each electrode and its electrolyte are separated by a porous wall, and the whole cell is at 25°C. Both electrodes are connected with a copper wire.
   a. Which electrode is the anode?
   b. Which electrode corrodes?
   c. In which direction will the electrons flow?
   d. In which direction will the anions in the solutions move?
   e. In which direction will the cations in the solutions move?
   f. Write the equation for the half-cell reaction at the anode.
   g. Write the equation for the half-cell reaction at the cathode.

13.11 A standard galvanic cell has electrodes of zinc and tin. Which electrode is the anode? Which electrode corrodes? What is the emf of the cell?

*13.12 A standard galvanic cell has electrodes of iron and lead. Which electrode is the anode? Which electrode corrodes? What is the emf of the cell?

*13.13 The emf of a standard Ni-Cd galvanic cell is −0.153 V. If the standard half-cell emf for the oxidation of Ni is −0.250 V, what is the standard half-cell emf of cadmium if cadmium is the anode?

13.14 What is the emf with respect to the standard hydrogen electrode of a cadmium electrode that is immersed in an electrolyte of 0.04 M $CdCl_2$? Assume the cadmium half-cell reaction to be $Cd \rightarrow Cd^{2+} + 2e^-$.

13.15 A galvanic cell consists of an electrode of nickel in a 0.04 M solution of $NiSO_4$ and an electrode of copper in a 0.08 M solution of $CuSO_4$ at 25°C. The two electrodes are separated by a porous wall. What is the emf of the cell?

13.16 A galvanic cell consists of an electrode of zinc in a 0.03 M solution of $ZnSO_4$ and an electrode of copper in a solution of 0.06 M $CuSO_4$ at 25°C. What is the emf of the cell?

*13.17 A nickel electrode is immersed in a solution of $NiSO_4$ at 25°C. What must be the molarity of the solution if the electrode shows a potential of −0.2842 V with respect to a standard hydrogen electrode?

13.18 A copper electrode is immersed in a solution of $CuSO_4$ at 25°C. What must be the molarity of the solution if the electrode shows a potential of +0.2985 V with respect to a standard hydrogen electrode?

*13.19 One end of a zinc wire is immersed in an electrolyte of 0.07 M Zn^{2+} ions and the other in one of 0.002 M Zn^{2+} ions, with the two electrolytes being separated by a porous wall.
   a. Which end of the wire will corrode?
   b. What will be the potential difference between the two ends of the wire when it is just immersed in the electrolytes?

13.20 Magnesium (Mg^{2+}) concentrations of 0.04 *M* and 0.007 *M* occur in an electrolyte at opposite ends of a magnesium wire at 25°C.

    a. Which end of the wire will corrode?

    b. What will be the potential difference between the ends of the wire?

13.21 Consider an oxygen-concentration cell consisting of two zinc electrodes. One is immersed in a water solution with low oxygen concentration and the other in a water solution with a high oxygen concentration. The zinc electrodes are connected by an external copper wire.

    a. Which electrode will corrode?

    b. Write half-cell reactions for the anodic reaction and the cathodic reaction.

13.22 In metals, which region is more chemically reactive (anodic), the grain matrix or the grain-boundary regions? Why?

13.23 Consider a 0.95 percent carbon steel. In which condition is the steel more corrosion-resistant: (*a*) martensitic or (*b*) tempered martensitic with ϵ carbide and Fe_3C formed in the 200 to 500°C range? Explain.

13.24 Why are pure metals in general more corrosion-resistant than impure ones?

13.25 An electroplating process uses 15 A of current by chemically corroding (dissolving) a copper anode. What is the corrosion rate of the anode in grams per hour?

*13.26 A cadmium electroplating process uses 10 A of current and chemically corrodes a cadmium anode. How long will it take to corrode 8.2 g of cadmium from the anode?

13.27 A mild steel tank 60 cm high with a 30 cm × 30 cm square bottom is filled with aerated water up to the 45 cm level and shows a corrosion loss of 350 g over a four-week period. Calculate (*a*) the corrosion current and (*b*) the corrosion density associated with the corrosion current. Assume the corrosion is uniform over all surfaces and that the mild steel corrodes in the same way as pure iron.

*13.28 A cylindrical steel tank is coated with a thick layer of zinc on the inside. The tank is 50 cm in diameter, 70 cm high, and filled to the 45 cm level with aerated water. If the corrosion current is 5.8×10^{-5} A/cm^2, how much zinc in grams per minute is being corroded?

13.29 A heated mild steel tank containing water is corroding at a rate of 90 mdd. If the corrosion is uniform, how long will it take to corrode the wall of the tank by 0.40 mm?

13.30 A mild steel tank contains a solution of ammonium nitrate and is corroding at the rate of 6000 mdd. If the corrosion on the inside surface is uniform, how long will it take to corrode the wall of the tank by 1.05 mm?

*13.31 A tin surface is corroding uniformly at a rate of 2.40 mdd. What is the associated current density for this corrosion rate?

13.32 A copper surface is corroding in seawater at a current density of 2.30×10^{-6} A/cm^2. What is the corrosion rate in mdd?

13.33 If a zinc surface is corroding at a current density of 3.45×10^{-7} A/cm^2, what thickness of metal will be corroded in 210 days?

*13.34 A galvanized (zinc-coated) steel sheet is found to uniformly corrode at the rate of 12.5×10^{-3} mm/year. What is the average current density associated with the corrosion of this material?

13.35   A galvanized (zinc-coated) steel sheet is found to uniformly corrode with an average current density of 1.32×10^{-7} A/cm^2. How many years will it take to uniformly corrode a thickness of 0.030 mm of the zinc coating?

13.36   A new aluminum container develops pits right through its walls in 350 days by pitting corrosion. If the average pit is 0.170 mm in diameter and the container wall is 1.00 mm thick, what is the average current associated with the formation of a single pit? What is the current density for this corrosion using the surface area of the pit for this calculation? Assume the pits have a cylindrical shape.

13.37   A new aluminum container develops pits right through its walls with an average current density of 1.30×10^{-4} A/cm^2. If the average pit is 0.70 mm in diameter and the aluminum wall is 0.90 mm thick, how many days will it take for a pit to corrode through the wall? Assume the pit has a cylindrical shape and that the corrosion current acts uniformly over the surface area of the pit.

13.38   What is an exchange current density? What is the corrosion current i_{corr}?

13.39   Define and give an example of (a) activation polarization and (b) concentration polarization.

13.40   Define the passivation of a metal or alloy. Give examples of some metals and alloys that show passivity.

13.41   Briefly describe the following theories of metal passivity: (a) the oxide theory and (b) the adsorption theory.

13.42   Draw a polarization curve for a passive metal and indicate on it (a) the primary passive voltage E_{pp} and (b) the passive current i_p.

13.43   Describe the corrosion behavior of a passive metal in (a) the active region, (b) the passive region, and (c) the transpassive region of a polarization curve. Explain the reasons for the different behavior in each region.

13.44   Explain the electrochemical behavior on the exterior and interior of a tin plate used as a food container.

13.45   Explain the difference in corrosion behavior of (a) a large cathode and a small anode, and (b) a large anode and a small cathode. Which of the two conditions is more favorable from a corrosion prevention standpoint and why?

13.46   What is pitting corrosion? Where are pits usually initiated? Describe an electrochemical mechanism for the growth of a pit in a stainless steel immersed in an aerated sodium chloride solution.

13.47   From an engineering standpoint, what metals would be used where pitting resistance is important?

13.48   What is crevice corrosion? Describe an electrochemical mechanism for the crevice corrosion of a stainless steel in an aerated sodium chloride solution.

13.49   From an engineering design standpoint, what should be done to prevent or minimize crevice corrosion?

13.50   What is intergranular corrosion? Describe the metallurgical condition that can lead to intergranular corrosion in an austenitic stainless steel.

13.51   For an austenitic stainless steel, distinguish between (a) the sensitized condition and (b) the stabilized condition.

13.52   Describe three methods of avoiding intergranular corrosion in austenitic stainless steels.

13.53   What is stress-corrosion cracking? Describe a mechanism of SCC.

13.54 From an engineering design standpoint, what can be done to avoid or minimize stress-corrosion cracking?

13.55 What is erosion corrosion? What is cavitation damage?

13.56 Describe fretting corrosion.

13.57 What is selective leaching of an alloy? Which types of alloys are especially susceptible to this kind of corrosion?

13.58 Describe a mechanism for the dezincification of a 70–30 brass.

13.59 What factors are important if a metal is to form a protective oxide?

13.60 Calculate the oxide-to-metal volume (Pilling-Bedworth) ratios for the oxidation of the metals listed in the following table, and comment on whether their oxides might be protective or not.

Metal	Oxide	Density of metal (g/cm^3)	Density of oxide (g/cm^3)
Tungsten, W	WO$_3$	19.35	12.11
Sodium, Na	Na$_2$O	0.967	2.27
Hafnium, Hf	HfO$_2$	13.31	9.68
Copper, Cu	CuO	8.92	6.43
Manganese, Mn	MnO	7.20	5.46
Tin, Sn	SnO	6.56	6.45

13.61 Describe the anion and cation diffusion mechanisms of oxide formation on metals.

13.62 Using equations, describe the following oxidation of metals behavior: (*a*) linear, (*b*) parabolic, (*c*) logarithmic. Give examples.

13.63 A 1 cm^2, 0.75 mm thick sample of 99.9 wt % Ni is oxidized in oxygen at 1 atm pressure at 500°C. After 7 h the sample shows a 60 μg/cm^2 weight gain. If the oxidation process follows parabolic behavior, what will the weight gain be after 20 h of oxidation?

*13.64 A sample of pure iron oxidizes according to the linear oxidation rate law. After 3 h at 720°C, a 1 cm^2 sample shows a weight gain of 7 μg/cm^2. How long an oxidation time will it take for the sample to show a weight gain of 55 μg/cm^2?

13.65 What is catastrophic oxidation? What metals are subject to this behavior? What metals when added to iron alloys retard this behavior?

13.66 What types of alloys are used for moderate oxidizing conditions for corrosion resistance?

13.67 What types of alloys are used for corrosion resistance for highly oxidizing conditions?

13.68 List six combinations of metals and environments that have good corrosion resistance.

13.69 List some of the applications of nonmetallic materials for corrosion control.

13.70 What is the function of each of the three layers of "chrome plate"?

13.71 What are Alclad alloys?

13.72 Describe eight engineering design rules that may be important to reduce or prevent corrosion.

13.73 Describe four methods of altering the environment to prevent or reduce corrosion.

13.74 Describe two methods by which cathodic protection can be used to protect a steel pipe from corroding.

*13.75 If a sacrificial zinc anode shows a 1.05 kg corrosion loss in 55 days, what is the average current produced by the corrosion process in this period?

13.76 If a sacrificial magnesium anode corrodes with an average current of 0.80 A for 100 days, what must be the loss of metal by the anode in this time period?

13.77 What is anodic protection? For what metals and alloys can it be used? What are some of its advantages and disadvantages?

13.11 MATERIALS SELECTION AND DESIGN PROBLEMS

1. A water storage tank made of steel has a small crack on its inner wall where it is submerged under water. Ultrasonic tests show that the crack is growing over time. Can you give reasons for this crack growth, considering that stresses acting on the crack are low?

2. Brass plates are to be fastened together using bolts and used in a marine environment. Would you select bolts made of steel or bolts made of nickel? Why?

3. You are to select the material for a rod to be used in an aqueous environment, and your main concern is corrosion. You have three choices for your selection: annealed Cu–20 wt % Zn, 30% cold-worked Cu–20 wt % Zn, and Cu–50 wt % Zn. Which metal would you select and why? (Consult Figure 8.25 for your answer).

4. Depending on the quality of concrete, it is possible that the reinforcing steel bars may corrode by water penetrating from the environment through small pores in the concrete. Propose ways of protecting the reinforcement from corrosion.

5. Erosion corrosion occurs in pipes that carry fluid with hard immersed particles, such as water with minerals. This is especially true at sharp elbow regions where turbulence may be created. What type of protection can you offer to reduce the erosion corrosion in pipes?

6. Corrosion is observed at the root of threads on bolts and in other areas with sharp corners. Can you explain this?

7. The fatigue life of components is significantly reduced when the component operates in a corrosive environment. This phenomenon is called corrosion fatigue. (*a*) Can you give an example of such a situation? (*b*) List important factors that influence corrosion fatigue crack propagation.

8. (*a*) Define exfoliation corrosion. (*b*) Explain how it happens. (*c*) Give examples of materials that are susceptible to exfoliation corrosion.

9. (*a*) Explain why the automobile exhaust system is especially susceptible to corrosion. (*b*) What material would you select in the manufacturing of the exhaust system?

10. How is the automobile industry protecting steel body panels from corrosion? Give specific examples.

11. Ferritic stainless steels that contain even small amounts of carbon or nitrogen are more susceptible to intergranular corrosion. Explain how carbon and nitrogen cause this.

12. It is possible to lower the intergranular corrosion in ferritic steel by adding small amounts of titanium or niobium, a process called *stabilizing*. Explain how the addition of such impurities helps reduce corrosion.

13. Rank the corrosion resistance of the following alloys in a saltwater environment (high to low): (*a*) Al alloys, (*b*) carbon steels, (*c*) nickel alloys, and (*d*) stainless steel. Give reasons for your selections.

14. Rank the corrosion resistance of the following materials in strong acid environments (high to low): (*a*) alumina Al_2O_3, (*b*) nylons, (*c*) PVC, and (*d*) cast iron. Give reasons for your selections.

15. Rank the corrosion resistance of the following classes of materials when they are exposed to UV environments (high to low): (*a*) ceramics, (*b*) polymers, and (*c*) metals. Give reasons for your selections.

16. Making simple design corrections can significantly reduce localized crevice corrosion. Based on your knowledge of crevice corrosion, give examples of such design corrections.

17. In the welding of materials, what general factors should you consider to minimize chances of corrosion?

18. In the design of parts, it is generally a good idea to avoid regions or points of stress concentration since these regions are more susceptible to corrosion. Give examples of the things that can be done to avoid stress raisers.

19. Salt (NaCl) is stored in an aluminum container. The container is used repeatedly in a humid region. (*a*) Is there a possibility of a cathodic and anodic reaction taking place? (*b*) If yes, identify the reaction.

14

Electrical Properties of Materials

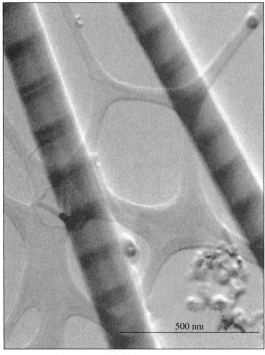

500 nm

(© *Peidong Yang/UC Berkeley*)

Researchers are constantly trying to find ways to manufacture computer chips with smaller dimensions and more devices. Today's industry focus is on developing the nanotechnology required to manufacture electronic devices on a nanowire with a diameter of approximately 100 nm.

The chapter-opening photo is the transmission electron microscope image of two heterogeneous nanowires with alternating layers of dark (silicon/germanium) and light (silicon) regions.[1]

In this chapter, electrical conduction in metals is considered first. The effects of impurities, alloy additions, and temperature on the electrical conductivity of metals are discussed. The energy-band model for electrical conduction in metals is then considered. Following this, the effects of impurities and temperature on the electrical conductivity of semiconducting materials are considered. Finally, the principles of operation of some basic semiconducting devices are examined, and some of the fabrication processes used to produce modern microelectronic circuitry are presented. An example of the complexity of recent microelectronic integrated circuitry is shown in Fig. 14.1. ■

[1]http://www.berkeley.edu/news/media/releases/2002/02/05_wires.html

By the end of this chapter, students will be able to . . .

1. Define conductivity, semiconductivity, and insulative properties of materials and be able to classify, in a general manner, how each class of materials i.e., metals, ceramics, polymers, . . . is rated according to their electrical properties.

2. Explain the concept of electrical conductivity, resistivity, drift velocity, and mean free path in metals. Describe the effect of increasing or decreasing temperature on each.

3. Describe the energy band model and be able to define electrical properties of metals, polymers, ceramics, and electronic materials based on it.

4. Define intrinsic and extrinsic semiconductors, how charge is transported in such materials.

5. Define N- and P-type semiconductors and what is the effect of temperature on their electrical behavior.

6. Name as many semiconducting devices as possible i.e., LEDs, Rectifiers, transistors, . . . and in each case explain how the device functions.

7. Define microelectronics and explain various steps in the manufacturing of ICs.

8. Explain in detail the electrical properties of ceramics as related to dielectrics, insulators, capacitors, ferroelectricity, and piezoelectric effect.

9. Project the future trends in the area of chip and computer manufacturing.

14.1 ELECTRICAL CONDUCTION IN METALS

14.1.1 The Classic Model for Electrical Conduction in Metals

In metallic solids the atoms are arranged in a crystal structure (for example, FCC, BCC, and HCP) and are bound together by their outer valence electrons by metallic bonding (see Sec. 2.7). The metallic bonds in solid metals make the free movement of the valence electrons possible since they are shared by many atoms and are not bound to any particular atom. Sometimes the valence electrons are visualized as forming an electron charge cloud, as shown in Fig. 14.2a. Other times the valence electrons are considered to be individual free electrons not associated with any particular atom, as shown in Fig. 14.2b.

In the classic model for electrical conduction in metallic solids, the outer valence electrons are assumed to be completely free to move between the positive-ion cores (atoms without valence electrons) in the metal lattice. At room temperature the positive-ion cores have kinetic energy and vibrate about their lattice positions. With increasing temperature these ions vibrate with increasing amplitudes, and there is a continuous interchange of energy between the ion cores and their valence electrons. In the absence of an electric potential, the motion of the valence electrons is random and restricted, so there is no net electron flow in any direction and thus no current

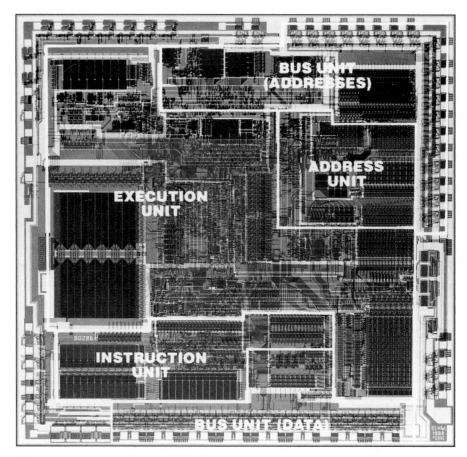

(a)

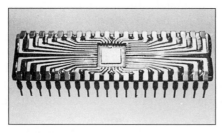

(b)

Figure 14.1

(a) The microprocessor or "computer on a chip" shown is magnified about six times on a side and incorporates about 3.1 million transistors on a single chip of silicon that is actually about 17.2 mm on each side. This microprocessor is the Intel Pentium and uses BICMOS technology with minimum design features of 0.8 μm. (b) One-half actual size microprocessor mounted in a package with connecting wire passages shown. (*Courtesy of Intel Corporation, Santa Clara, CA.*)

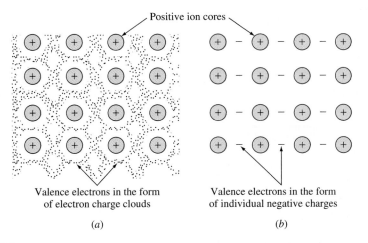

Positive ion cores

Valence electrons in the form of electron charge clouds

Valence electrons in the form of individual negative charges

(a)

(b)

Figure 14.2
Schematic arrangements of the atoms in one plane of a monovalent metal such as copper, silver, or sodium. In (a) the valence electrons are pictured as an "electron gas," and in (b) the valence electrons are visualized as free electrons of unit charge.

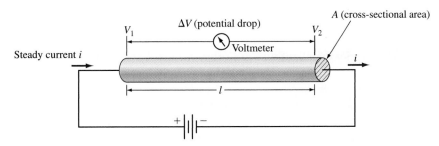

V_1 ΔV (potential drop) V_2

Voltmeter

A (cross-sectional area)

Steady current i

i

l

Figure 14.3
Potential difference ΔV applied to metal wire specimen of cross-sectional area A.

flow. In the presence of an applied electric potential the electrons attain a directed *drift velocity* that is proportional to the applied field but in the opposite direction.

14.1.2 Ohm's Law

Consider a length of copper wire whose ends are connected to a battery, as shown in Fig. 14.3. If a potential difference V is applied to the wire, a current i will flow that is proportional to the resistance R of the wire. According to *Ohm's law,* the **electric current** flow i is proportional to the applied voltage V and inversely proportional to the resistance of the wire, or

$$i = \frac{V}{R}$$

(14.1)

where i = electric current, A (amperes)

V = potential difference, V (volts)

R = resistance of wire, Ω (ohms)

The **electrical resistance** R of an electrical conductor such as the metal wire specimen of Fig. 14.3 is directly proportional to its length l and inversely proportional to its cross-sectional area A. These quantities are related by a material constant called the **electrical resistivity** ρ, as

$$R = \rho \frac{l}{A} \quad \text{or} \quad \rho = R \frac{A}{l} \tag{14.2}$$

The units for electrical resistivity, which is a constant for a material at a particular temperature, are

$$\rho = R \frac{A}{l} = \Omega \frac{m^2}{m} = \text{ohm-meter} = \Omega \cdot m$$

Often it is more convenient to think in terms of the passage of electric current instead of resistance, and so the quantity **electrical conductivity** σ^2 is defined as the reciprocal of electrical resistivity:

$$\sigma = \frac{1}{\rho} \tag{14.3}$$

The units for electrical conductivity are $(\text{ohm-meter})^{-1} = (\Omega \cdot m)^{-1}$. The SI unit for the reciprocal of the ohm is the siemens (S), but this unit is rarely used and so will not be used in this book.

Table 14.1 lists the electrical conductivities of some selected metals and nonmetals. From this table it is seen that electrical conductors such as pure metals silver, copper, and gold have the highest conductivities, about $10^7 \ (\Omega \cdot m)^{-1}$. **Electrical insulators** such as polyethylene and polystyrene, on the other hand, have very low electrical conductivities, about $10^{-14} \ (\Omega \cdot m)^{-1}$, which are about 10^{20} times less than those of the highly conductive metals. Silicon and germanium have conductivities in between those of metals and insulators and are thus classified as **semiconductors.**

$^2\sigma$ = Greek letter sigma

Table 14.1 Electrical conductivities of some metals and nonmetals at room temperature

Metals and alloys	$\sigma \ (\Omega \cdot m)^{-1}$	Nonmetals	$\sigma \ (\Omega \cdot m)^{-1}$
Silver	6.3×10^7	Graphite	10^5 (average)
Copper, commercial purity	5.8×10^7	Germanium	2.2
Gold	4.2×10^7	Silicon	4.3×10^{-4}
Aluminum, commercial purity	3.4×10^7	Polyethylene	10^{-14}
		Polystyrene	10^{-14}
		Diamond	10^{-14}

A wire whose diameter is 0.20 cm must carry a 20 A current. The maximum power dissipation along the wire is 4 W/m (watts per meter). Calculate the minimum allowable conductivity of the wire in (ohm-meters)$^{-1}$ for this application.

■ **Solution**

$$\text{Power } P = iV = i^2R \qquad \text{where } i = \text{current, A} \qquad R = \text{resistance, } \Omega$$
$$V = \text{voltage, V} \qquad P = \text{power, W (watts)}$$

$$R = \rho\frac{l}{A} \qquad \text{where } \rho = \text{resistivity, } \Omega \cdot \text{m}$$
$$l = \text{length, m}$$
$$A = \text{cross-sectional area of wire, m}^2$$

Combining these two equations gives

$$P = i^2\rho\frac{l}{A} = \frac{i^2l}{\sigma A} \qquad \text{since } \rho = \frac{1}{\sigma}$$

Rearranging gives $$\sigma = \frac{i^2l}{PA}$$

Given that $$P = 4 \text{ W (in 1 m)} \qquad i = 20 \text{ A} \qquad l = 1 \text{ m}$$

and $$A = \frac{\pi}{4}(0.0020 \text{ m})^2 = 3.14 \times 10^{-6} \text{ m}^2$$

thus

$$\sigma = \frac{i^2l}{PA} = \frac{(20 \text{ A})^2(1 \text{ m})}{(4 \text{ W})(3.14 \times 10^{-6} \text{ m}^2)} = 3.18 \times 10^7 \, (\Omega \cdot \text{m})^{-1} \blacktriangleleft$$

Therefore, for this application, the conductivity σ of the wire must be equal to or greater than $3.18 \times 10^7 \, (\Omega \cdot \text{m})^{-1}$.

If a copper wire of commercial purity is to conduct 10 A of current with a maximum voltage drop of 0.4 V/m, what must be its minimum diameter? [σ (commercially pure Cu) = $5.85 \times 10^7 \, (\Omega \cdot \text{m})^{-1}$.]

■ **Solution**

Ohm's law: $$V = iR \qquad \text{and} \qquad R = \rho\frac{l}{A}$$

Combining the two equations gives

$$V = i\rho\frac{l}{A}$$

and rearranging gives

$$A = i\rho\frac{l}{V}$$

Substituting $(\pi/4)d^2 = A$ and $\rho = 1/\sigma$ yields

$$\frac{\pi}{4}d^2 = \frac{il}{\sigma V}$$

and solving for d gives

$$d = \sqrt{\frac{4il}{\pi\sigma V}}$$

Given that $i = 10$ A, $V = 0.4$ V for 1 m of wire, $l = 1.0$ m (chosen length of wire), and conductivity of Cu wire $\sigma = 5.85 \times 10^7\ (\Omega \cdot m)^{-1}$, thus

$$d = \sqrt{\frac{4il}{\pi\sigma V}} = \sqrt{\frac{4(10\ A)(1.0\ m)}{\pi[5.85 \times 10^7\ (\Omega \cdot m)^{-1}](0.4\ V)}} = 7.37 \times 10^{-4}\ m \blacktriangleleft$$

Therefore, for this application, the Cu wire must have a diameter of 7.37×10^{-4} m or greater.

Equation 14.1 is called the *macroscopic form* of Ohm's law since the values of i, V, and R are dependent on the geometrical shape of a particular electrical conductor. Ohm's law can also be expressed in a *microscopic form,* which is independent of the shape of the electrical conductor, as

$$\mathbf{J} = \frac{\mathbf{E}}{\rho} \quad \text{or} \quad \mathbf{J} = \sigma\mathbf{E} \tag{14.4}$$

where $\mathbf{J}$ = current density, A/m^2 ρ = electrical resistivity, $\Omega \cdot m$
      $\mathbf{E}$ = electric field, V/m σ = electrical conductivity $(\Omega \cdot m)^{-1}$

The **current density J** and electric field **E** are vector quantities with magnitude and direction. Both macroscopic and microscopic forms of Ohm's law are compared in Table 14.2.

Table 14.2 Comparison of the macroscopic and microscopic forms of Ohm's law

Macroscopic form of Ohm's law	Microscopic form of Ohm's law
$i = \dfrac{V}{R}$	$\mathbf{J} = \dfrac{\mathbf{E}}{\rho}$
where i = current, A V = voltage, V R = resistance, Ω	where $\mathbf{J}$ = current density, A/m^2 $\mathbf{E}$ = electric field, V/m ρ = electrical resistivity, $\Omega \cdot m$

14.1.3 Drift Velocity of Electrons in a Conducting Metal

At room temperature the positive-ion cores in a metallic conductor crystal lattice vibrate about neutral positions and therefore possess kinetic energy. The free electrons continually exchange energy with the lattice ions by elastic and inelastic collisions. Since there is no external electric field, the electron motion is random, and since there is no net electron motion in any direction, there is no net current flow.

If a uniform electric field of intensity $\mathbf{E}$ is applied to the conductor, the electrons are accelerated with a definite velocity in the direction opposite to the applied field. The electrons periodically collide with the ion cores in the lattice and lose their kinetic energy. After a collision, the electrons are free to accelerate again in the applied field, and as a result the electron velocity varies with time in a "sawtooth manner," as shown in Fig. 14.4. The average time between collisions is 2τ, where τ is the *relaxation time*.

The electrons thus acquire an average drift velocity $\mathbf{v}_d$ that is directly proportional to the applied electric field $\mathbf{E}$. The relationship between the drift velocity and the applied field is

$$\mathbf{v}_d = \mu \mathbf{E} \tag{14.5}$$

where μ (mu), the electron mobility, $m^2/(V \cdot s)$, is the proportionality constant.

Consider the wire shown in Fig. 14.5 as having a current density $\mathbf{J}$ flowing in the direction shown. Current density by definition is equal to the rate at which charges cross any plane that is perpendicular to $\mathbf{J}$, i.e., a certain number of amperes per square meter or coulombs per second per square meter flow past the plane.

The electron flow in a metal wire subjected to a potential difference depends on the number of electrons per unit volume, the electronic charge $-e(-1.60 \times 10^{-19}\,C)$, and the drift velocity of the electrons, $\mathbf{v}_d$. The rate of charge flow per unit area

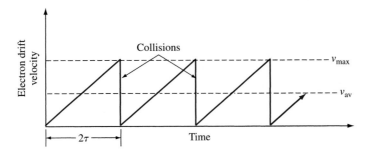

Figure 14.4

Electron drift velocity versus time for classical model for electrical conductivity of a free electron in a metal.

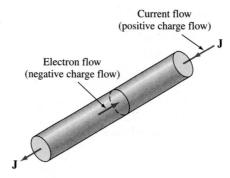

Figure 14.5
A potential difference along a copper wire causes electron flow, as indicated in the drawing. Because of the negative charge on the electron, the direction of electron flow is opposite that of conventional current flow, which assumes positive charge flow.

is $-ne\mathbf{v}_d$. However, by convention, electric current is considered to be positive charge flow, and thus the current density $\mathbf{J}$ is given a positive sign. In equation form then,

$$\mathbf{J} = ne\mathbf{v}_d \qquad (14.6)$$

14.1.4 Electrical Resistivity of Metals

The electrical resistivity of a pure metal can be approximated by the sum of two terms, a thermal component ρ_T and a residual component ρ_r:

$$\rho_{\text{total}} = \rho_T + \rho_r \qquad (14.7)$$

The thermal component arises from the vibrations of the positive-ion cores about their equilibrium positions in the metallic crystal lattice. As the temperature is increased, the ion cores vibrate more and more, and a large number of thermally excited elastic waves (called *phonons*) scatter conduction electrons and decrease the mean free paths and relaxation times between collisions. Thus as the temperature is increased, the electrical resistivities of pure metals increase, as shown in Fig. 14.6. The residual component of the electrical resistivity of pure metals is small and is caused by structural imperfections such as dislocations, grain boundaries, and impurity atoms that scatter electrons. The residual component is almost independent of temperature and becomes significant only at low temperatures (Fig. 14.7).

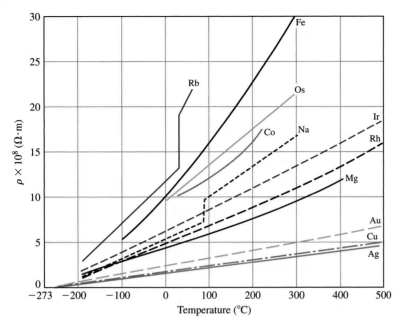

Figure 14.6
The effect of temperature on the electrical resistivity of selected metals.
Note that there is almost a linear relationship between resistivity and
temperature (°C).
*(After Zwikker, "Physical Properties of Solid Materials," Pergamon, 1954,
pp. 247, 249.)*

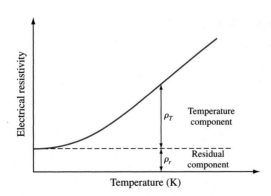

Figure 14.7
Schematic variation of electrical resistivity of
a metal with absolute temperature. Note that
at higher temperatures the electrical resistivity
is the sum of a residual component ρ_r and a
thermal component ρ_T.

Table 14.3 Temperature resistivity coefficients

Metal	Electrical resistivity at 0°C ($\mu\Omega \cdot cm$)	Temperature resistivity coefficient α_T (°C^{-1})
Aluminum	2.7	0.0039
Copper	1.6	0.0039
Gold	2.3	0.0034
Iron	9	0.0045
Silver	1.47	0.0038

For most metals at temperatures above about $-200°C$, the electrical resistivity varies almost linearly with temperature, as shown in Fig. 14.6. Thus the electrical resistivities of many metals may be approximated by the equation

$$\rho_T = \rho_{0°C}(1 + \alpha_T T) \tag{14.8}$$

where $\rho_{0°C}$ = electrical resistivity at 0°C

α_T = temperature coefficient of resistivity, °C^{-1}

T = temperature of metal, °C

Table 14.3 lists the temperature resistivity coefficients for selected metals. For these metals α_T ranges from 0.0034 to 0.0045 (°C^{-1}).

EXAMPLE PROBLEM 14.3

Calculate the electrical resistivity of pure copper at 132°C, using the temperature resistivity coefficient for copper from Table 14.3.

■ **Solution**

$$\rho_T = \rho_{0°C}(1 + \alpha_T T) \tag{14.8}$$

$$= 1.6 \times 10^{-6}\,\Omega \cdot cm \left(1 + \frac{0.0039}{°C} \times 132°C\right)$$

$$= 2.42 \times 10^{-6}\,\Omega \cdot cm$$

$$= 2.42 \times 10^{-8}\,\Omega \cdot m \blacktriangleleft$$

Alloying elements added to pure metals cause additional scattering of the conduction electrons and thus increase the electrical resistivity of pure metals. The effect of small additions of various elements on the room temperature electrical resistivity of pure copper is shown in Fig. 14.8. Note that the effect of each element varies considerably. For the elements shown, silver increases the resistivity the least and phosphorus the most for the same amount added. The addition of larger amounts of alloying elements such as 5 to 35 percent zinc to copper to make the copper-zinc brasses increases the electrical resistivity and thus decreases the electrical conductivity of pure copper greatly, as shown in Fig. 14.9.

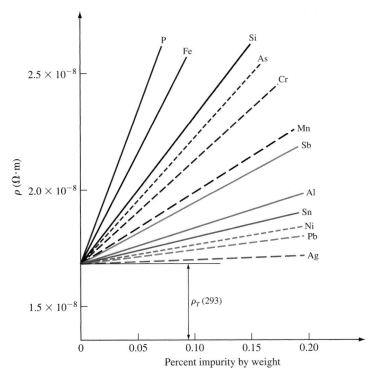

Figure 14.8
The effect of small additions of various elements on the room
temperature electrical resistivity of copper.
[*After F. Pawlek and K. Reichel, Z. Metallkd.,* **47**:*347(1956).*]

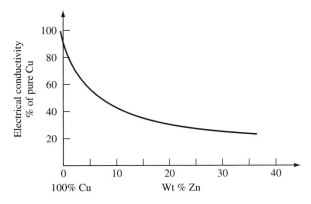

Figure 14.9
The effect of zinc additions to pure copper in
reducing the electrical conductivity of the copper.
(*After ASM data.*)

14.2 ENERGY-BAND MODEL FOR ELECTRICAL CONDUCTION

14.2.1 Energy-Band Model for Metals

Let us now consider the **energy-band model** for electrons in solid metals since it helps us to understand the mechanism of electrical conduction in metals. We use the metal sodium to explain the energy-band model since the sodium atom has a relatively simple electronic structure.

Electrons of isolated atoms are bound to their nuclei and can only have energy levels that are *sharply defined* such as the $1s^1$, $1s^2$, $2s^1$, $2s^2$, ... states as necessitated by the Pauli principle. Otherwise, it would be possible for all electrons in an atom to descend to the lowest energy state, $1s^1$! Thus, the 11 electrons in the neutral sodium atom occupy two $1s$ states, two $2s$ states, six $2p$ states, and one $3s$ state, as illustrated in Fig. 14.10a. The electrons in the lower levels ($1s^2$, $2s^2$, $2p^6$) are tightly bound and constitute the *core electrons* of the sodium atom (Fig. 14.10b). The outer $3s^1$ electron can be involved in bonding with other atoms and is called the *valence electron.*

In a solid block of metal the atoms are close together and touch each other. The valence electrons that are delocalized (Fig. 14.11a) interact and interpenetrate each other so that their original sharp atomic energy levels are broadened into wider regions called *energy bands* (Fig. 14.11b). The inner electrons, since they are shielded from the valence electrons, do not form bands.

Each valence electron in a block of sodium metal, for example, must have a slightly different energy level according to the Pauli exclusion principle. Thus, if there are N sodium atoms in a block of sodium, where N can be very large, there will be N distinct but only slightly different $3s^1$ energy levels in the $3s$ energy *band.*

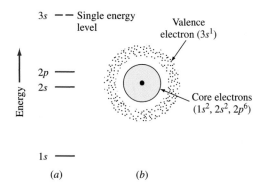

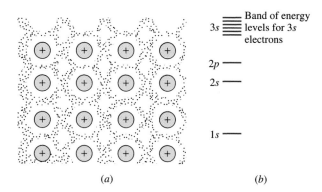

Figure 14.10
(a) Energy levels in a single sodium atom. (b) Arrangement of electrons in a sodium atom. The outer $3s^1$ valence electron is loosely bound and is free to be involved in metallic bonding.

Figure 14.11
(a) Delocalized valence electrons in a block of sodium metal. (b) Energy levels in a block of sodium metal; note the expansion of the $3s$ level into an energy band and that the $3s$ band is shown closer to the $2p$ level since bonding has caused a lowering of the $3s$ levels of the isolated sodium atoms.

Each energy level is called a *state*. In the valence energy band the energy levels are so close that they form a continuous energy band.

Figure 14.12 shows part of the energy-band diagram for metallic sodium as a function of interatomic spacing. In solid metallic sodium the 3*s* and 3*p* energy bands overlap (Fig. 14.12). However, since there is only one 3*s* electron in the sodium atom, the 3*s* band is only half-filled (Fig. 14.13*a*). As a result very little energy is required to excite electrons in sodium from the highest filled states to the lowest empty ones. Sodium is, therefore, a good conductor since very little energy is required to produce electron flow in it. Copper, silver, and gold also have half-filled outer *s* bands.

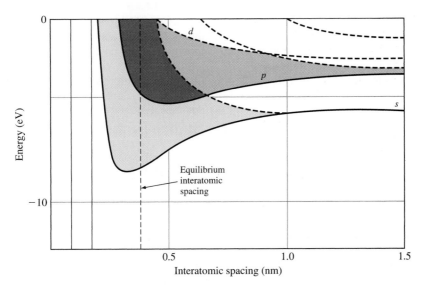

Figure 14.12
Valence energy bands in sodium metal. Note the splitting of the *s*, *p*, and *d* levels.
[*After J. C. Slater, Phys. Rev.,* **45:***794(1934).*]

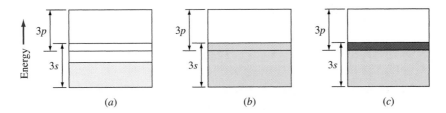

Figure 14.13
Schematic energy-band diagrams for several metallic conductors.
(*a*) Sodium, 3*s*¹: the 3*s* band is half-filled since there is only one 3*s*¹ electron. (*b*) Magnesium, 3*s*²: the 3*s* band is filled and overlaps the empty 3*p* band. (*c*) Aluminum, 3*s*²3*p*¹: the 3*s* band is filled and overlaps the partially filled 3*p* band.

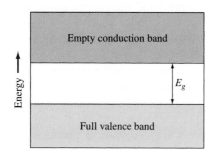

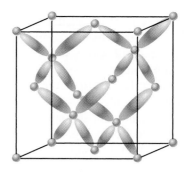

MatVis

Figure 14.14
Energy-band diagram for an insulator. The valence band is completely filled and is separated from an empty conduction band by a large energy gap E_g.

Figure 14.15
Diamond cubic crystal structure. The atoms in this structure are bonded together by sp^3 covalent bonds. Diamond (carbon), silicon, germanium, and gray tin (the tin polymorph stable below 13°C) all have this structure. There are 8 atoms per unit cell: $\frac{1}{8} \times 8$ at the corners, $\frac{1}{2} \times 6$ at the faces, and 4 inside the unit cube.

In metallic magnesium both $3s$ states are filled. However, the $3s$ band overlaps the $3p$ band and allows some electrons into it, creating a partially filled $3sp$ combined band (Fig. 14.13*b*). Thus, in spite of the filled $3s$ band, magnesium is a good conductor. Similarly, aluminum, which has both $3s$ states and one $3p$ state filled, is also a good conductor because the partially filled $3p$ band overlaps the filled $3s$ band (Fig. 14.13*c*).

14.2.2 Energy-Band Model for Insulators

In insulators, electrons are tightly bound to their bonding atoms by ionic or covalent bonding and are not "free" to conduct electricity unless highly energized. The energy-band model for insulators consists of a lower filled **valence band** and an upper empty **conduction band.** These bands are separated by a relatively large energy gap E_g (Fig. 14.14). To free an electron for conduction, the electron must be energized to "jump" the gap, which may be as much as 6 to 7 eV as, for example, in pure diamond. In diamond, the electrons are tightly held by sp^3 tetrahedral covalent bonding (Fig. 14.15).

14.3 INTRINSIC SEMICONDUCTORS

14.3.1 The Mechanism of Electrical Conduction in Intrinsic Semiconductors

Semiconductors are materials whose electrical conductivities are between those of highly conducting metals and poorly conducting insulators. **Intrinsic semiconductors** are pure semiconductors whose electrical conductivity is determined by their inherent conductive properties. Pure elemental silicon and germanium are intrinsic

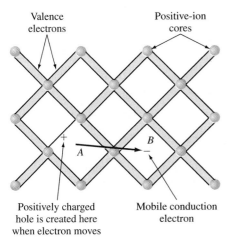

Figure 14.16
Two-dimensional representation of the diamond cubic lattice of silicon or germanium showing positive-ion cores and valence electrons. Electron has been excited from a bond at *A* and has moved to point *B*.

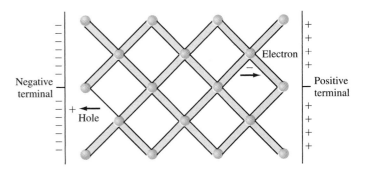

Figure 14.17
Electrical conduction in a semiconductor such as silicon showing the migration of electrons and holes in an applied electric field.

semiconducting materials. These elements, which are in group IV A in the periodic table, have the diamond cubic structure with highly directional covalent bonds (Fig. 14.15). Tetrahedral sp^3 hybrid bonding orbitals consisting of electron pairs bond the atoms together in the crystal lattice. In this structure each silicon or germanium atom contributes four valence electrons.

Electrical conductivity in pure semiconductors such as Si and Ge can be described qualitatively by considering the two-dimensional pictorial representation of the diamond cubic crystal lattice shown in Fig. 14.16. The circles in this illustration represent the *positive-ion cores* of the Si or Ge atoms, and the joining pairs of lines indicate bonding *valence electrons*. The bonding electrons are unable to move through the crystal lattice and hence to conduct electricity unless sufficient energy is provided to excite them from their bonding positions. When a critical amount of energy is supplied to a valence electron to excite it away from its bonding position, it becomes a free conduction electron and leaves behind a positively charged "hole" in the crystal lattice (Fig. 14.16).

14.3.2 Electrical Charge Transport in the Crystal Lattice of Pure Silicon

In the electrical conduction process in a semiconductor such as pure silicon or germanium, both electrons and holes are charge carriers and move in an applied electric field. Conduction **electrons** have a negative charge and are attracted to the positive terminal of an electrical circuit (Fig. 14.17). **Holes,** on the other hand,

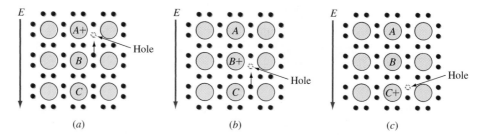

Figure 14.18
Schematic illustration of the movement of holes and electrons in a pure silicon semiconductor during electrical conduction caused by the action of an applied electric field.
(After S. N. Levine, "Principles of Solid State Microelectronics," Holt, 1963.)

behave like positive charges and are attracted to the negative terminal of an electrical circuit (Fig. 14.17). A hole has a positive charge equal in magnitude to the electron charge.

The motion of a "hole" in an electric field can be visualized by referring to Fig. 14.18. Let a hole exist at atom A where a valence electron is missing, as shown in Fig. 14.18a. When an electric field is applied in the direction shown in Fig. 14.18a, a force is exerted on the valence electrons of atom B, and one of the electrons associated with atom B will break loose from its bonding orbital and move to the vacancy in the bonding orbital of atom A. The hole will now appear at atom B and in effect will have moved from A to B in the direction of the applied field (Fig. 14.18b). By a similar mechanism the hole is transported from atom B to C by an electron moving from C to B (Fig. 14.18c). The net result of this process is that an electron is transported from C to A, which is in the direction opposite to the applied field, and a hole is transported from A to C, which is in the direction of the applied field. Thus, during electrical conduction in a pure semiconductor such as silicon, negatively charged electrons move in the direction opposite to the applied field (conventional current flow) and toward the positive terminal and positively charged holes move in the direction of the applied field toward the negative terminal.

14.3.3 Energy-Band Diagram for Intrinsic Elemental Semiconductors

Energy-band diagrams are another method of describing the excitation of valence electrons to become conduction electrons in semiconductors. For this representation only the energy required for the process is involved, and no physical picture of the electrons moving in the crystal lattice is given. In the energy-band diagram for intrinsic elemental semiconductors (for example, Si or Ge), the bound valence electrons of the covalently bonded crystal occupy the energy levels in the lower valence band, which is almost filled at 20°C (Fig. 14.19).

Above the valence band there is a forbidden energy gap in which no energy states are allowed and which is 1.1 eV for silicon at 20°C. Above the energy gap

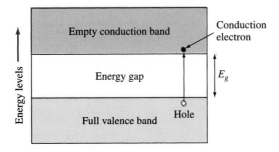

Figure 14.19
Energy-band diagram for an intrinsic elemental semiconductor such as pure silicon. When an electron is excited across the energy gap, an electron-hole pair is created. Thus, for each electron that jumps the gap, two charge carriers, an electron and a hole, are produced.

there is an almost empty (at 20°C) conduction band. At room temperature thermal energy is sufficient to excite some electrons from the valence band to the conduction band, leaving vacant sites or holes in the valence band. Thus, when an electron is excited across the energy gap into the conduction band, two charge carriers are created, a negatively charged electron and a positively charged hole. Both electrons and holes carry electric current.

14.3.4 Quantitative Relationships for Electrical Conduction in Elemental Intrinsic Semiconductors

During electrical conduction in intrinsic semiconductors, the current density $\mathbf{J}$ is equal to the sum of the conduction due to *both* electrons and holes. Using Eq. 14.6,

$$\mathbf{J} = nq\boldsymbol{v}_n^* + pq\boldsymbol{v}_p^* \tag{14.9}$$

where n = number of conduction electrons per unit volume

p = number of conduction holes per unit volume

q = absolute value of electron or hole charge, 1.60×10^{-19} C

$\boldsymbol{v}_n, \boldsymbol{v}_p$ = drift velocities of electrons and holes, respectively.

Dividing both sides of Eq. 14.9 by the electric field $\mathbf{E}$ and using Eq. 14.4, $\mathbf{J} = \sigma\mathbf{E}$,

$$\sigma = \frac{\mathbf{J}}{\mathbf{E}} = \frac{nq\boldsymbol{v}_n}{\mathbf{E}} + \frac{pq\boldsymbol{v}_p}{\mathbf{E}} \tag{14.10}$$

The quantities $\boldsymbol{v}_n/\mathbf{E}$ and $\boldsymbol{v}_p/\mathbf{E}$ are called the *electron* and *hole mobilities,* respectively, since they measure how fast the electrons and holes in semiconductors drift in an applied electric field. The symbols μ_n and μ_p are used for the mobilities of

---

*The subscript n (for negative) refers to electrons, and the subscript p (for positive) refers to holes.

Table 14.4 Some physical properties of silicon and germanium at 300 K

	Silicon	Germanium
Energy gap, eV	1.1	0.67
Electron mobility μ_n, m^2/(V · s)	0.135	0.39
Hole mobility μ_p, m^2/(V · s)	0.048	0.19
Intrinsic carrier density n_i, carriers/m^3	1.5×10^{16}	2.4×10^{19}
Intrinsic resistivity ρ, Ω · m	2300	0.46
Density, g/m^3	2.33×10^6	5.32×10^6

Source: After E. M. Conwell, "Properties of Silicon and Germanium II," *Proc. IRE,* June 1958, p. 1281.

electrons and holes, respectively. Substituting electron and hole mobilities for $\boldsymbol{v}_n/\mathbf{E}$ and $\boldsymbol{v}_p/\mathbf{E}$ in Eq. 14.10 enables the electrical conductivity of a semiconductor to be expressed as

$$\sigma = nq\mu_n + pq\mu_p \tag{14.11}$$

The units for mobility μ are

$$\frac{\boldsymbol{v}}{\mathbf{E}} = \frac{\text{m/s}}{\text{V/m}} = \frac{\text{m}^2}{\text{V} \cdot \text{s}}$$

In intrinsic elemental semiconductors, electrons and holes are created in pairs; thus, the number of conduction electrons equals the number of holes produced, so that

$$n = p = n_i \tag{14.12}$$

where n = intrinsic carrier concentration, carriers/unit volume.

Equation 14.11 now becomes

$$\sigma = n_i q(\mu_n + \mu_p) \tag{14.13}$$

Table 14.4 lists some of the important properties of intrinsic silicon and germanium at 300 K.

The mobilities of electrons are always greater than those of holes. For intrinsic silicon the electron mobility of 0.135 m^2/(V · s) is 2.81 times greater than the hole mobility of 0.048 m^2/(V · s) at 300 K (Table 14.4). The ratio of electron-to-hole mobility for intrinsic germanium is 2.05 at 300 K.

EXAMPLE PROBLEM 14.4

Calculate the number of silicon atoms per cubic meter. The density of silicon is 2.33 Mg/m^3 (2.33 g/cm^3), and its atomic mass is 28.08 g/mol.

■ **Solution**

$$\frac{\text{Si atoms}}{\text{m}^3} = \left(\frac{6.023 \times 10^{23} \text{ atoms}}{\text{mol}} \right)\left(\frac{1}{28.08 \text{ g/mol}} \right)\left(\frac{2.33 \times 10^6 \text{ g}}{\text{m}^3} \right)$$

$$= 5.00 \times 10^{28} \text{ atoms/m}^3 \blacktriangleleft$$

Calculate the electrical resistivity of intrinsic silicon at 300 K. For Si at 300 K, $n_i = 1.5 \times 10^{16}$ carriers/m³, $q = 1.60 \times 10^{-19}$ C, $\mu_n = 0.135$ m²/(V · s), and $\mu_p = 0.048$ m²/(V · s).

EXAMPLE PROBLEM 14.5

■ **Solution**

$$\rho = \frac{1}{\sigma} = \frac{1}{n_i q(\mu_n + \mu_p)} \quad \text{(reciprocal of Eq. 14.13)}$$

$$= \frac{1}{\left(\dfrac{1.5 \times 10^{16}}{m^3}\right)(1.60 \times 10^{-19} \text{ C})\left(\dfrac{0.135 \text{ m}^2}{V \cdot s} + \dfrac{0.048 \text{ m}^2}{V \cdot s}\right)}$$

$$= 2.28 \times 10^3 \ \Omega \cdot m \blacktriangleleft$$

The units for the reciprocal of Eq. 14.13 are ohm-meters as is shown by the following unit conversion:

$$\rho = \frac{1}{n_i q(\mu_n + \mu_p)} = \frac{1}{\left(\dfrac{1}{m^3}\right)(C)\left(\dfrac{1 \text{ A} \cdot s}{1 \text{ C}}\right)\left(\dfrac{m^2}{V \cdot s}\right)\left(\dfrac{1 \text{ V}}{1 \text{ A} \cdot \Omega}\right)} = \Omega \cdot m$$

14.3.5 Effect of Temperature on Intrinsic Semiconductivity

At 0 K the valence bands of intrinsic semiconductors such as silicon and germanium are completely filled and their conduction bands completely empty. At temperatures above 0 K, some of the valence electrons are thermally activated and excited across the energy gap into the conduction band, creating electron-hole pairs. Thus, in contrast to metals, whose conductivities decrease with increasing temperatures, the conductivities of semiconductors *increase* with increasing temperatures for the temperature range over which this process predominates.

Since electrons are thermally activated into the conduction band of semiconductors, the concentration of thermally activated electrons in semiconductors shows a temperature dependence similar to that of many other thermally activated processes. By analogy with Eq. 5.1, the concentration of electrons with sufficient thermal energy to enter the conduction band (and thus creating the same concentration of holes in the valence band), n_i, varies according to

$$n_i \propto e^{-(E_g - E_{\text{av}})/kT} \qquad (14.14)$$

where E_g = band energy gap

E_{av} = average energy across band gap

k = Boltzmann's constant

T = temperature, K

For intrinsic semiconductors such as pure silicon and germanium, E_{av} is half-way across the gap, or $E_g/2$. Thus Eq. 14.14 becomes

$$n_i \propto e^{-(E_g - E_g/2)/kT} \tag{14.15a}$$

or

$$n_i \propto e^{-E_g/2kT} \tag{14.15b}$$

Since the electrical conductivity σ of an intrinsic semiconductor is proportional to the concentration of electrical charge carriers, n_i, Eq. 14.15b can be expressed as

$$\sigma = \sigma_0 e^{-E_g/2kT} \tag{14.16a}$$

or in natural logarithmic form,

$$\ln \sigma = \ln \sigma_0 - \frac{E_g}{2kT} \tag{14.16b}$$

where σ_0 is an overall constant that depends mainly on the mobilities of electrons and holes. The slight temperature dependence of σ_0 on temperature will be neglected in this text.

Since Eq. 14.16b is the equation of a straight line, the value of $E_g/2k$ and hence E_g can be determined from the slope of the plot of $\ln \sigma$ versus $1/T$, K^{-1}. Figure 14.20 shows an experimental plot of $\ln \sigma$ versus $1/T$, K^{-1}, for intrinsic silicon.

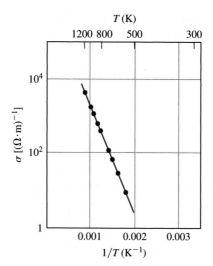

Figure 14.20
Electrical conductivity as a function of reciprocal absolute temperature for intrinsic silicon.
(*After C. A. Wert and R. M. Thomson, "Physics of Solids," 2nd ed., McGraw-Hill, 1970, p. 282.*)

The electrical resistivity of pure silicon is $2.3 \times 10^3 \ \Omega \cdot m$ at room temperature, 27°C (300 K). Calculate its electrical conductivity at 200°C (473 K). Assume that the E_g of silicon is 1.1 eV; $k = 8.62 \times 10^{-5}$ eV/K.

**EXAMPLE
PROBLEM 14.6**

■ **Solution**

For this problem we use Eq. 14.16a and set up two simultaneous equations. We then eliminate σ_0 by dividing the first equation by the second.

$$\sigma = \sigma_0 \exp \frac{-E_g}{2kT} \tag{14.16a}$$

$$\sigma_{473} = \sigma_0 \exp \frac{-E_g}{2kT_{473}}$$

$$\sigma_{300} = \sigma_0 \exp \frac{-E_g}{2kT_{300}}$$

Dividing the first equation by the second to eliminate σ_0 gives

$$\frac{\sigma_{473}}{\sigma_{300}} = \exp \left(\frac{-E_g}{2kT_{473}} + \frac{E_g}{2kT_{300}} \right)$$

$$\frac{\sigma_{473}}{\sigma_{300}} = \exp \left[\frac{-1.1 \ \text{eV}}{2(8.62 \times 10^{-5} \ \text{eV/K})} \left(\frac{1}{473 \ \text{K}} - \frac{1}{300 \ \text{K}} \right) \right]$$

$$\ln \frac{\sigma_{473}}{\sigma_{300}} = 7.777$$

$$\sigma_{473} = \sigma_{300}(2385)$$

$$= \frac{1}{2.3 \times 10^3 \ \Omega \cdot m}(2385) = 1.04 \ (\Omega \cdot m)^{-1} \ \blacktriangleleft$$

The electrical conductivity of the silicon increased by about 2400 times when the temperature was raised from 27 to 200°C.

14.4 EXTRINSIC SEMICONDUCTORS

Extrinsic semiconductors are very dilute substitutional solid solutions in which the solute impurity atoms have different valence characteristics from the solvent atomic lattice. The concentrations of the added impurity atoms in these semiconductors are usually in the range of 100 to 1000 parts per million (ppm).

14.4.1 n-Type (Negative-Type) Extrinsic Semiconductors

Consider the two-dimensional covalent bonding model for the silicon crystal lattice shown in Fig. 14.21a. If an impurity atom of a group V A element, such as phosphorus, replaces a silicon atom, which is a group IV A element, there will be one excess electron above the four needed for the tetrahedral covalent bonding

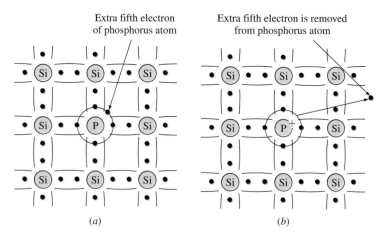

Figure 14.21

(a) The addition of a pentavalent phosphorus impurity atom to the tetravalent silicon lattice provides a fifth electron that is weakly attached to the parent phosphorus atom. Only a small amount of energy (0.044 eV) makes this electron mobile and conductive.
(b) Under an applied electric field the excess electron becomes conductive and is attracted to the positive terminal of the electrical circuit. With the loss of the extra electron, the phosphorus atom is ionized and acquires a +1 charge.

in the silicon lattice. This extra electron is only loosely bonded to the positively charged phosphorus nucleus and has a binding energy of 0.044 eV at 27°C. This energy is about 5 percent of that required for a conduction electron to jump the energy gap of 1.1 eV of pure silicon. That is, only 0.044 eV of energy is required to remove the excess electron from its parent nucleus so that it can participate in electrical conduction. When under the action of an electrical field, the extra electron becomes a free electron available for conduction and the remaining phosphorus atom becomes ionized and acquires a positive charge (Fig. 14.21b).

Group V A impurity atoms such as P, As, and Sb when added to silicon or germanium provide easily ionized electrons for electrical conduction. Since these group V A impurity atoms donate conduction electrons when present in silicon or germanium crystals, they are called *donor impurity atoms.* Silicon or germanium semiconductors containing group V impurity atoms are called **n-type (negative-type) extrinsic semiconductors** since the majority charge carriers are electrons.

In terms of the energy-band diagram for silicon, the extra electron of a group V A impurity atom occupies an energy level in the forbidden energy gap just slightly below the empty conduction band, as shown in Fig. 14.22. Such an energy level is called a **donor level** since it is provided by a donor impurity atom. A donor group V A impurity atom, upon losing its extra electron, becomes ionized and acquires a positive charge. Energy levels for the group V A impurity donor atoms Sb, P and As in silicon are shown in Fig. 14.23.

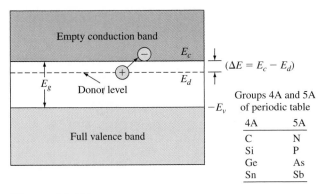

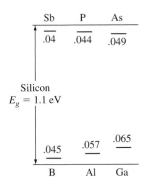

Figure 14.22
Energy-band diagram for an n-type extrinsic semiconductor showing the position of the donor level for the extra electron of a group V A element such as P, As, and Sb that is contained in the silicon crystal lattice (Fig. 14.21*a*). Electrons at the donor energy level require only a small amount of energy ($\Delta E = E_c - E_d$) to be excited into the conduction band. When the extra electron at the donor level jumps to the conduction band, a positive immobile ion is left behind.

Figure 14.23
Ionization energies (in electron volts) for various impurities in silicon.

14.4.2 p-Type (Positive-Type) Extrinsic Semiconductors

When a trivalent group III A element such as boron is substitutionally introduced in the silicon tetrahedrally bonded lattice, one of the bonding orbitals is missing, and a hole exists in the bonding structure of the silicon (Fig. 14.24*a*). If an external electric field is applied to the silicon crystal, one of the neighboring electrons from another tetrahedral bond can attain sufficient energy to break loose from its bond and move to the missing bond (hole) of the boron atom (Fig. 14.24*b*). When the hole associated with the boron atom is filled by an electron from a neighboring silicon atom, the boron atom becomes ionized and acquires a negative charge of -1. The binding energy associated with the removal of an electron from a silicon atom, thereby creating a hole, and the subsequent transfer of the electron to the boron atom is only 0.045 eV. This amount of energy is small compared to the 1.1 eV required to transfer an electron from the valence band to the conduction band. In the presence of an applied electric field, the hole created by the ionization of a boron atom behaves as a positive charge carrier and migrates in the silicon lattice toward the negative terminal, as described in Fig. 14.17.

In terms of the energy-band diagram, the boron atom provides an energy level called an **acceptor level** that is slightly higher (≈ 0.045 eV) than the uppermost level of the full valence band of silicon (Fig. 14.25). When a valence electron of a silicon atom near a boron atom fills a missing electron hole in a boron-silicon valence bond (Fig. 14.24*b*), this electron is elevated to the acceptor level and creates a negative boron ion. In this process an electron hole is created in the silicon lattice that acts as

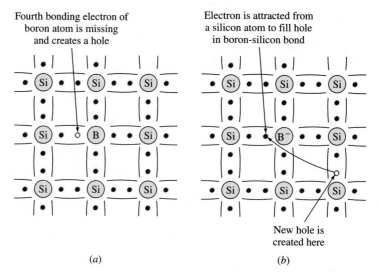

(a)   (b)

Figure 14.24

(a) The addition of a trivalent boron impurity atom into the tetravalent lattice creates a hole in one of the boron-silicon bonds since one electron is missing.

(b) Under an applied electric field only a small amount of energy (0.045 eV) attracts an electron from a nearby silicon atom to fill this hole, thereby creating an immobile boron ion with a −1 charge. The new hole created in the silicon lattice acts as a positive charge carrier and is attracted to the negative terminal of an electrical circuit.

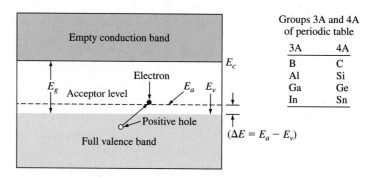

Figure 14.25

Energy-band diagram for a p-type extrinsic semiconductor showing the position of the acceptor level created by the addition of an atom of a group III A element such as Al, B, or Ga to replace a silicon atom in the silicon lattice (Fig. 14.24). Only a small amount of energy ($\Delta E = E_a - E_v$) is necessary to excite an electron from the valence band to the acceptor level, thereby creating an electron hole (charge carrier) in the valence band.

a positive charge carrier. Atoms of group III A elements such as B, Al, and Ga provide acceptor levels in silicon semiconductors and are called *acceptor atoms*. Since the majority carriers in these extrinsic semiconductors are holes in the valence bond structure, they are called **p-type (positive-carrier-type) extrinsic semiconductors.**

14.4.3 Doping of Extrinsic Silicon Semiconductor Material

The process of adding small amounts of substitutional impurity atoms to silicon to produce extrinsic silicon semiconducting material is called *doping,* while impurity atoms themselves are called *dopants.* The most commonly used method of doping silicon semiconductors is the *planar process.* In this process dopant atoms are introduced into selected areas of the silicon from one surface in order to form regions of p- or n-type material. The wafers are usually about 4 in. (10 cm) in diameter and about a few hundred micrometers[3] thick. Figure 4.13 shows some of these silicon wafers.

In the diffusion process for doping silicon wafers, the dopant atoms are typically deposited on or near the surface of the wafer by a gaseous deposition step, followed by a drive-in diffusion that moves the dopant atoms farther into the wafer. A high temperature of about 1100°C is required for this diffusion process. More details of this process will be described in Sec. 16.6 on microelectronics.

14.4.4 Effect of Doping on Carrier Concentrations in Extrinsic Semiconductors

The Mass Action Law   In semiconductors such as silicon and germanium, mobile electrons and holes are constantly being generated and recombined. At constant temperature under equilibrium conditions the product of the negative free electron and positive hole concentrations is a constant. The general relation is

$$np = n_i^2 \qquad\qquad (14.17)$$

where n_i is the intrinsic concentration of carriers in a semiconductor and is a constant at a given temperature. This relation is valid for both intrinsic and extrinsic semiconductors. In an extrinsic semiconductor the increase in one type of carrier (n or p) reduces the concentration of the other through recombination so that the product of the two (n and p) is a constant at any given temperature.

The carriers whose concentration in extrinsic semiconductors is the larger are designated the **majority carriers,** and those whose concentration is the smaller are called the **minority carriers** (see Table 14.5). The concentration of electrons in an n-type semiconductor is denoted by n_n and that of holes in n-type material by p_n. Similarly, the concentration of holes in a p-type semiconductor is given by p_p and that of electrons in p-type material by n_p.

Charge Densities in Extrinsic Semiconductors   A second fundamental relationship for extrinsic semiconductors is obtained from the fact that the total crystal must

---

[3]1 micrometer (μm) = 10^{-4} cm = 10^4 Å.

Table 14.5 Summary of the carrier concentrations in extrinsic semiconductors

Semiconductor	Majority-carrier concentrations	Minority-carrier concentrations
n-type	n_n (concentration of electrons in n-type material)	p_n (concentration of holes in n-type material)
p-type	p_p (concentration of holes in p-type material)	n_p (concentration of electrons in p-type material)

be electrically neutral. This means that the charge density in each volume element must be zero. There are two types of charged particles in extrinsic semiconductors such as Si and Ge: immobile ions and mobile charge carriers. The immobile ions originate from the ionization of donor or acceptor impurity atoms in the Si or Ge. The concentration of the positive donor ions is denoted by N_d and that of the negative acceptor ions by N_a. The mobile charge carriers originate mainly from the ionization of the impurity atoms in the Si or Ge, and their concentrations are designated by n for the negatively charged electrons and p for the positively charged holes.

Since the semiconductor must be electrically neutral, the magnitude of the total negative charge density must equal the total positive charge density. The total negative charge density is equal to the sum of the negative acceptor ions N_a and the electrons, or $N_a + n$. The total positive charge density is equal to the sum of the positive donor ions N_d and the holes, or $N_d + p$. Thus,

$$N_a + n = N_d + p \tag{14.18}$$

In an n-type semiconductor created by adding donor impurity atoms to intrinsic silicon, $N_a = 0$. Since the number of electrons is much greater than the number of holes in an n-type semiconductor (that is, $n \gg p$), then Eq. 14.18 reduces to

$$n_n \approx N_d \tag{14.19}$$

Thus, in an n-type semiconductor the free-electron concentration is approximately equal to the concentration of donor atoms. The concentration of holes in an n-type semiconductor is obtained from Eq. 14.17. Thus,

$$p_n = \frac{n_i^2}{n_n} \approx \frac{n_i^2}{N_d} \tag{14.20}$$

The corresponding equations for p-type semiconductors of silicon and germanium are

$$p_p \approx N_a \tag{14.21}$$

and

$$n_p = \frac{n_i^2}{p_p} \approx \frac{n_i^2}{N_a} \tag{14.22}$$

Typical Carrier Concentrations in Intrinsic and Extrinsic Semiconductors For silicon at 300 K the intrinsic carrier concentration n_i is equal to 1.5×10^{16} carriers/m^3. For extrinsic silicon doped with arsenic at a typical concentration of 10^{21} impurity atoms/m^3,

$$\text{Concentration of major carriers } n_n = 10^{21} \text{ electrons/m}^3$$
$$\text{Concentration of minority carriers } p_n = 2.25 \times 10^{11} \text{ holes/m}^3$$

Thus, for extrinsic semiconductors the concentration of the majority carriers is normally much larger than that of the minority carriers. Example Problem 14.7 shows how the concentrations of majority and minority carriers can be calculated for an extrinsic silicon semiconductor.

A silicon wafer is doped with 10^{21} phosphorus atoms/m^3. Calculate (*a*) the majority-carrier concentration, (*b*) the minority-carrier concentration, and (*c*) the electrical resistivity of the doped silicon at room temperature (300 K). Assume complete ionization of the dopant atoms; $n_i(\text{Si}) = 1.5 \times 10^{16}$ m^{-3}, $\mu_n = 0.135$ m^2/(V · s), and $\mu_p = 0.048$ m^2/(V · s).

EXAMPLE PROBLEM 14.7

■ **Solution**

Since the silicon is doped with phosphorus, a group V element, the doped silicon is *n-type*.

a.
$$n_n = N_d = 10^{21} \text{ electrons/m}^3 \blacktriangleleft$$

b.
$$p_n = \frac{n_i^2}{N_d} = \frac{(1.5 \times 10^{16} \text{ m}^{-3})^2}{10^{21} \text{ m}^{-3}} = 2.25 \times 10^{11} \text{ holes/m}^3 \blacktriangleleft$$

c.
$$\rho = \frac{1}{q\mu_n n_n} = \frac{1}{(1.60 \times 10^{-19} \text{ C})\left(0.135 \dfrac{\text{m}^2}{\text{V} \cdot \text{s}}\right)\left(\dfrac{10^{21}}{\text{m}^3}\right)}$$

$$= 0.0463 \ \Omega \cdot \text{m*} \blacktriangleleft$$

---

*See Example Problem 14.5 for the conversion of units.

A phosphorus-doped silicon wafer has an electrical resistivity of $8.33 \times 10^{-5} \ \Omega \cdot \text{m}$ at 27°C. Assume mobilities of charge carriers to be the constants 0.135 m^2/(V · s) for electrons and 0.048 m^2/(V · s) for holes.

EXAMPLE PROBLEM 14.8

a.   What is its majority-carrier concentration (carriers per cubic meter) if complete ionization is assumed?

b.   What is the ratio of phosphorus to silicon atoms in this material?

■ **Solution**

a.   Phosphorus produces an n-type silicon semiconductor. Therefore, the mobility of the charge carriers will be assumed to be that of electrons in silicon at 300 K, which is $0.1350 \text{ m}^2/(\text{V} \cdot \text{s})$. Thus,

$$\rho = \frac{1}{n_n q \mu_n}$$

or $n_n = \dfrac{1}{\rho q \mu_n} = \dfrac{1}{(8.33 \times 10^{-5} \ \Omega \cdot \text{m})(1.60 \times 10^{-19} \ \text{C})[0.1350 \text{ m}^2/(\text{V} \cdot \text{s})]}$

$= 5.56 \times 10^{23}$ electrons/m^3 ◄

b. Assuming each phosphorus atom provides one electron charge carrier, there will be 5.56×10^{23} phosphorus atoms/m^3 in the material. Pure silicon contains 5.00×10^{28} atoms/m^3 (Example Problem 14.4). Thus, the ratio of phosphorus to silicon atoms will be

$$\frac{5.56 \times 10^{23} \text{ P atoms/m}^3}{5.00 \times 10^{28} \text{ Si atoms/m}^3} = 1.11 \times 10^{-5} \text{ P to Si atoms} \blacktriangleleft$$

14.4.5 Effect of Total Ionized Impurity Concentration on the Mobility of Charge Carriers in Silicon at Room Temperature

Figure 14.26 shows that the mobilities of electrons and holes in silicon at room temperature are at a maximum at low impurity concentrations and then decrease with impurity concentration, reaching a minimum value at high concentrations. Example Problem 14.9 shows how neutralizing one type of charge carrier with another leads to a lower mobility for the majority carriers.

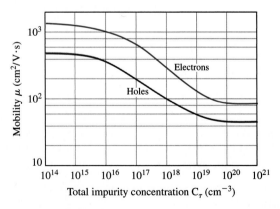

Figure 14.26
The effect of total ionized impurity concentration on the mobility of charge carriers in silicon at room temperature.
(*After A. S. Grove, "Physics and Technology of Semiconductor Devices," Wiley, 1967, p. 110.*)

A silicon semiconductor at 27°C is doped with 1.4×10^{16} boron atoms/cm³ plus 1.0×10^{16} phosphorus atoms/cm³. Calculate (a) the equilibrium electron and hole concentrations, (b) the mobilities of electrons and holes, and (c) the electrical resistivity. Assume complete ionization of the dopant atoms. n_i (Si) $= 1.50 \times 10^{10}$ cm⁻³.

■ **Solution**

a. *Majority-carrier concentration:* The net concentration of immobile ions is equal to the acceptor ion concentration minus the donor ion concentration. Thus,

$$p_p \simeq N_a - N_d = 1.4 \times 10^{16} \text{ B atoms/cm}^3 - 1.0 \times 10^{16} \text{ P atoms/cm}^3$$
$$\simeq N_a \simeq 4.0 \times 10^{15} \text{ holes/cm}^3 \blacktriangleleft$$

Minority-carrier concentration: Electrons are the minority carriers. Thus,

$$n_p = \frac{n_i^2}{N_a} = \frac{(1.50 \times 10^{10} \text{ cm}^{-3})^2}{4 \times 10^{15} \text{ cm}^{-3}} = 5.6 \times 10^4 \text{ electrons/cm}^3 \blacktriangleleft$$

b. *Mobilities of electrons and holes:* For electrons, using the total impurity concentration $C_T = 2.4 \times 10^{16}$ ions/cm³ and Fig. 14.26,

$$\mu_n = 900 \text{ cm}^2/(\text{V} \cdot \text{s}) \blacktriangleleft$$

For holes, using $C_T = 2.4 \times 10^{16}$ ions/cm³ and Fig. 14.26,

$$\mu_p = 300 \text{ cm}^2/(\text{V} \cdot \text{s}) \blacktriangleleft$$

c. *Electrical resistivity:* The doped semiconductor is p-type:

$$\rho = \frac{1}{q\mu_p p_p}$$

$$= \frac{1}{(1.60 \times 10^{-19} \text{ C})[300 \text{ cm}^2/(\text{V} \cdot \text{s})](4.0 \times 10^{15}/\text{cm}^3)}$$

$$= 5.2 \ \Omega \cdot \text{cm} \blacktriangleleft$$

14.4.6 Effect of Temperature on the Electrical Conductivity of Extrinsic Semiconductors

The electrical conductivity of an extrinsic semiconductor such as silicon that contains doped impurity atoms is affected by temperature, as shown schematically in Fig. 14.27. At lower temperatures the number of impurity atoms per unit volume activated (ionized) determines the electrical conductivity of the silicon. As the temperature is increased, more and more impurity atoms are ionized, and thus the electrical conductivity of extrinsic silicon increases with increasing temperature in the extrinsic range (Fig. 14.27).

In this extrinsic range only a relatively small amount of energy (≈ 0.04 eV) is required to ionize the impurity atoms. The amount of energy required to excite a donor electron into the conduction band in n-type silicon is $E_c - E_d$ (Fig. 14.22).

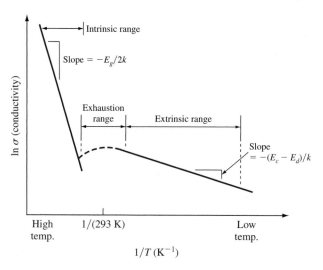

Figure 14.27
Schematic plot of ln σ (conductivity) versus $1/T$ (K^{-1}) for an n-type extrinsic semiconductor.

Thus the slope of ln σ versus $1/T$ (K^{-1}) for n-type silicon is $-(E_c - E_d)/k$. Correspondingly, the amount of energy required to excite an electron in p-type silicon into an acceptor level and thereby create a hole in the valence band is $E_a - E_v$. Thus the slope of ln σ versus $1/T$ (K^{-1}) for p-type silicon is $-(E_a - E_v)/k$ (Fig. 14.27).

For a certain temperature range above that required for complete ionization, an increase in temperature does not substantially change the electrical conductivity of an extrinsic semiconductor. For an n-type semiconductor this temperature range is referred to as the *exhaustion range* since donor atoms become completely ionized after the loss of their donor electrons (Fig. 14.27). For p-type semiconductors this range is referred to as the *saturation range* since acceptor atoms become completely ionized with acceptor electrons. To provide an exhaustion range at about room temperature (300 K), silicon doped with arsenic requires about 10^{21} carriers/m^3 (Fig. 14.28a). Donor exhaustion and acceptor saturation temperature ranges are important for semiconductor devices since they provide temperature ranges that have essentially constant electrical conductivities for operation.

As the temperature is increased beyond that of the exhaustion range, the intrinsic range is entered upon. The higher temperatures provide sufficient activation energies for electrons to jump the semiconductor gap (1.1 eV for silicon) so that intrinsic conduction becomes dominant. The slope of the ln σ versus $1/T$ (K^{-1}) plot becomes much steeper and is $-E_g/2k$. For silicon-based semiconductors with an energy gap of 1.1 eV, extrinsic conduction can be used up to about 200°C. The upper limit for the use of extrinsic conduction is determined by the temperature at which intrinsic conductivity becomes important.

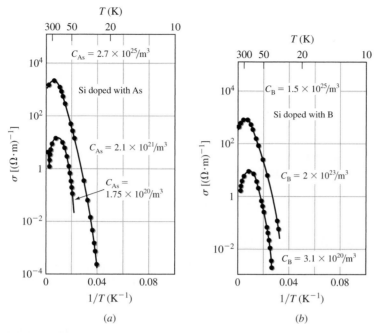

Figure 14.28

(a) Plot of ln σ versus $1/T$ (K^{-1}) for As-doped Si. At the lowest level of impurity the intrinsic contribution is slightly visible at the highest temperatures; slope of the line at 40 K gives E_i = 0.048 eV. (b) Plot of ln σ versus $1/T$ (K^{-1}) for B-doped Si. Slope of the line below 50 K gives E_i = 0.043 eV.

(*After C. A. Wert and R. M. Thomson, "Physics of Solids," 2nd ed., McGraw-Hill, 1970, p. 282.*)

14.5 SEMICONDUCTOR DEVICES

The use of semiconductors in the electronics industry has become increasingly important. The ability of semiconductor manufacturers to put extremely complex electrical circuits on a single chip of silicon of about 1 cm square or less and about 200 μm thick has revolutionized the design and manufacture of countless products. An example of the complex electronic circuitry able to be put on a silicon chip is shown in Fig. 14.1 of an advanced microprocessor or "computer on a chip." The microprocessor forms the basis for many of the latest products that use the progressive miniaturization of silicon-based semiconductor technology.

In this section we will first study the electron-hole interactions at a pn junction and then examine the operation of the pn junction diode. We shall then look at some applications of pn junction diodes. Finally, we shall briefly examine the operation of the bipolar junction transistor.

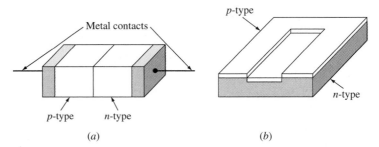

Figure 14.29
(*a*) pn junction diode grown in the form of a single crystal bar.
(*b*) Planar pn junction formed by selectively diffusing a p-type impurity into an n-type semiconductor crystal.

14.5.1 The pn Junction

Most common semiconductor devices depend on the properties of the boundary between p-type and n-type materials, and therefore we shall examine some of the characteristics of this boundary. A pn junction diode can be produced by growing a single crystal of intrinsic silicon and doping it first with n-type material and then with p-type material (Fig. 14.29*a*). More commonly, however, the pn junction is produced by solid-state diffusion of one type of impurity (for example, p-type) into existing n-type material (Fig. 14.29*b*).

The pn Junction Diode at Equilibrium   Let us consider an ideal case in which p-type and n-type silicon semiconductors are joined together to form a junction. Before joining, both types of semiconductors are electrically neutral. In the p-type material holes are the majority carriers and electrons are the minority carriers. In the n-type material electrons are the majority carriers and holes are the minority carriers.

After joining the p- and n-type materials (i.e., after a **pn junction** is formed in actual fabrication), the majority carriers near or at the junction diffuse across the junction and recombine (Fig. 14.30*a*). Since the remaining ions near or at the junction are physically larger and heavier than the electrons and holes, they remain in their positions in the silicon lattice (Fig. 14.30*b*). After a few recombinations of majority carriers at the junction, the process stops because the electrons crossing over the junction into the p-type material are repelled by the large negative ions. Similarly, holes crossing over the junction are repelled by the large positive ions in the n-type material. The immobile ions at the junction create a zone depleted of majority carriers called a *depletion region*. Under equilibrium conditions (i.e., open-circuit conditions) there exists a potential difference or barrier to majority-carrier flow. Thus, there is no net current flow under open-circuit conditions.

The pn Junction Diode Reverse-Biased   When an external voltage is applied to a pn junction, it is said to be **biased.** Let us consider the effect of applying an external voltage from a battery to the pn junction. The pn junction is said to be **reverse-biased** if the n-type material of the junction is connected to the positive terminal of the

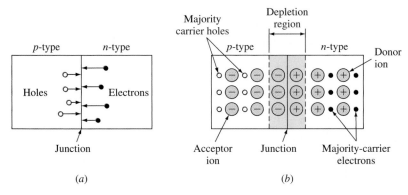

Figure 14.30
(*a*) pn junction diode showing majority carriers (holes in p-type material and electrons in n-type material) diffusing toward junction. (*b*) Formation of depletion region at and near pn junction due to loss of majority carriers in this region by recombination. Only ions remain in this region in their positions in the crystal structure.

battery and if the p-type material is connected to the negative terminal (Fig. 14.31). With this arrangement, the electrons (majority carriers) of the n-type material are attracted to the positive terminal of the battery away from the junction and the holes (majority carriers) of the p-type material are attracted to the negative terminal of the battery away from the junction (Fig. 14.31). The movement of the majority-carrier electrons and holes away from the junction increases its barrier width, and as a result current resulting from majority carriers will not flow. However, thermally generated minority carriers (holes in n-type material and electrons in p-type material) will be driven toward the junction so that they can combine and create a very small current flow under reverse-bias conditions. This minority or *leakage current* is usually of the order of microamperes (μA) (Fig. 14.32).

The pn Junction Diode Forward-Biased   The pn junction diode is said to be **forward-biased** if the n-type material of the junction is connected to the negative terminal of an external battery (or other electrical source) and if the p-type material is connected to the positive terminal (Fig. 14.33). In this arrangement the majority carriers are repelled toward the junction and can combine, that is, electrons are repelled away from the negative terminal of the battery toward the junction and holes are repelled away from the positive terminal toward the junction.

Under forward bias—that is, forward bias with respect to majority carriers—the energy barrier at the junction is reduced so that some electrons and holes can cross the junction and subsequently recombine. During the forward biasing of a pn junction, electrons from the battery enter the negative material of the diode (Fig. 14.33). For every electron that crosses the junction and recombines with a hole, another electron enters from the battery. Also, for every hole that recombines with an electron in the n-type material, a new hole is formed whenever an electron leaves the p-type material

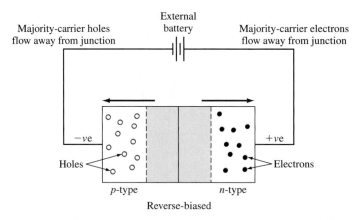

External
battery

Majority-carrier holes
flow away from junction

Majority-carrier electrons
flow away from junction

$-ve$

$+ve$

Holes

Electrons

p-type

n-type

Reverse-biased

Figure 14.31
Reverse-biased pn junction diode. Majority carriers are
attracted away from the junction, creating a wider depletion
region than when the junction is at equilibrium. Current flow
due to majority carriers is reduced to near zero. However,
minority carriers are biased forward, creating a small leakage
current, as shown in Fig. 14.32.

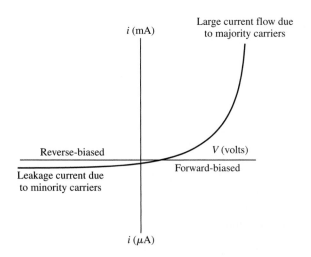

i (mA)

Large current flow due
to majority carriers

Reverse-biased

V (volts)

Forward-biased

Leakage current due
to minority carriers

i (μA)

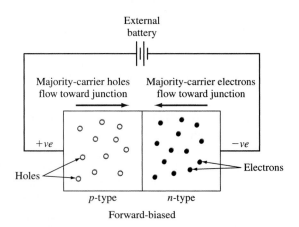

External
battery

Majority-carrier holes
flow toward junction

Majority-carrier electrons
flow toward junction

$+ve$

$-ve$

Holes

Electrons

p-type

n-type

Forward-biased

Figure 14.32
Schematic of current-voltage characteristics of a
pn junction diode. When the pn junction diode is
reverse-biased, a leakage current due to minority
carriers combining exists. When the pn junction
diode is forward-biased, a large current flows
because of recombination of majority carriers.

Figure 14.33
Forward-biased pn junction diode. Majority
carriers are repelled toward the junction and
cross over it to recombine so that a large
current flows.

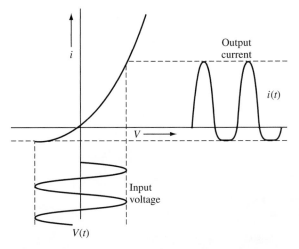

Figure 14.34
Voltage-current diagram illustrating the rectifying
action of a pn junction diode that converts
alternating current (ac) to direct current (dc). The
output current is not completely direct current but
is mostly positive. This dc signal can be smoothed
out by using other electronic devices.

and flows toward the positive terminal of the battery. Since the energy barrier to electron flow is reduced when the pn junction is forward-biased, considerable current can
flow, as indicated in Fig. 14.32. Electron flow (and hence current flow) can continue
as long as the pn junction is forward-biased and the battery provides an electron source.

14.5.2 Some Applications for pn Junction Diodes

Rectifier Diodes   One of the most important uses of pn junction diodes is to convert alternating voltage into direct voltage, a process known as *rectification.* Diodes
used for this process are called **rectifier diodes.** When an AC signal is applied to a
pn junction diode, the diode will conduct only when the p region has a positive voltage applied to it relative to the n region. As a result, half-wave rectification will be
produced, as shown in Fig. 14.34. This output signal can be smoothed out with other
electronic devices and circuits so that a steady DC signal can be produced. Solid-
state silicon rectifiers are used in a wide range of applications that require from tenths
of an ampere to several hundred amperes or more. Voltages, too, can be as high as
1000 V or more. Figure 14.35 shows some examples of silicon diode rectifiers.

Breakdown Diodes   Breakdown diodes, or *zener diodes* as they are sometimes called,
are silicon rectifiers in which the reverse current (leakage current) is small, and then
with only slightly more reverse-bias voltage, a breakdown voltage is reached upon
which the reverse current increases very rapidly (Fig. 14.36). In the so-called zener
breakdown, the electric field in the diode becomes strong enough to attract electrons

Figure 14.35
Photo of some low- to high-power silicon diode rectifiers.
(*Courtesy of International Rectifier.*)

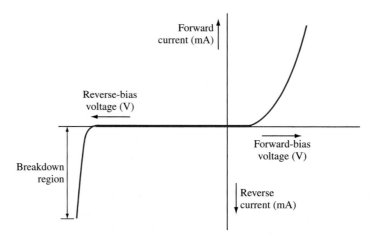

Figure 14.36
Zener (avalanche) diode characteristic curve. A large reverse
current is produced at the breakdown-voltage region.

directly out of the covalently bonded crystal lattice. The electron-hole pairs created then produce a high reverse current. At higher reverse-bias voltages than the zener break-down voltages, an avalanche effect occurs, and the reverse current is very large. One theory to explain the avalanche effect is that electrons gain sufficient energy between collisions to knock more electrons from covalent bonds, which can then reach high enough energies to conduct electricity. Breakdown diodes can be made with breakdown voltages from a few to several hundred volts and are used for voltage-limiting applica-tions and for voltage stabilizing under conditions of widely varying current.

14.5.3 The Bipolar Junction Transistor

A **bipolar junction transistor** (BJT) is an electronic device that can serve as a cur-rent amplifier. This device consists of two pn junctions that occur sequentially in a single crystal of a semiconducting material such as silicon. Figure 14.37 shows schematically an npn-type bipolar junction transistor and identifies the three main parts of the transistor: *emitter, base,* and *collector.* The emitter of the transistor emits charge carriers. Since the emitter of the npn transistor is n-type, it emits electrons. The base of the transistor controls the flow of charge carriers and is p-type for the npn transistor. The base is made very thin (about 10^{-3} cm thick) and is lightly doped so that only a small fraction of the charge carriers from the emitter will recombine with the oppositely charged majority carriers of the base. The collector of the BJT collects charge carriers mainly from the emitter. Since the collector section of the npn transistor is n-type, it collects mainly electrons from the emitter.

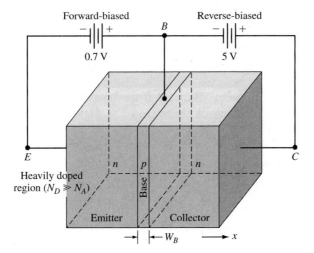

Figure 14.37
Schematic illustration of an npn bipolar junction transistor. The n region on the left is the emitter, the thin p region in the middle is the base, and the n region on the right is the collector. For normal operation the emitter-base junction is forward-biased and the collector-base junction is reverse-biased.
(After C. A. Holt, "Electronic Circuits," Wiley, 1978, p. 49.)

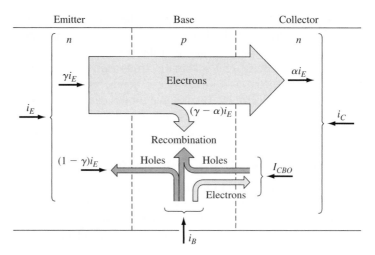

Figure 14.38

Charge carrier motion during the normal operation of an npn transistor. Most of the current consists of electrons from the emitter that go right through the base to the collector. Some of the electrons, about 1 to 5 percent, recombine with holes from the base current flow. Small reverse currents due to thermally generated carriers are also present, as indicated.

(After R. J. Smith, "Currents, Devices and Systems," 3rd ed., Wiley, 1976, p. 343.)

Under normal operation of the npn transistor, the emitter-base junction is forward-biased and the collector-base junction is reverse-biased (Fig. 14.37). The forward bias on the emitter-base junction causes an injection of electrons from the emitter into the base (Fig. 14.38). Some of the electrons injected into the base are lost by recombination with holes in the p-type base. However, most of the electrons from the emitter pass right through the thin base into the collector, where they are attracted by the positive terminal of the collector. Heavy doping of the emitter with electrons, light doping of the base with holes, and a very thin base are all factors that cause most of the emitter electrons (about 95 to 99 percent) to pass right through to the collector. Very few holes flow from the base to the emitter. Most of the current flow from the base terminal to the base region is the flow of holes to replace those lost by recombination with electrons. The current flow to the base is small and is about 1 to 5 percent of the electron current from the emitter to the collector. In some respects the current flow to the base can be thought of as a control valve since the small base current can be used to control the much larger collector current. The bipolar transistor is so named because both types of charge carriers (electrons and holes) are involved in its operation.

14.6 MICROELECTRONICS

Modern semiconductor technology has made it possible to put thousands of transistors on a "chip" of silicon about 5 mm square and 0.2 mm thick. This ability to incorporate very large numbers of electronic elements on silicon chips has greatly increased the capability of electronic device systems (Fig. 14.1).

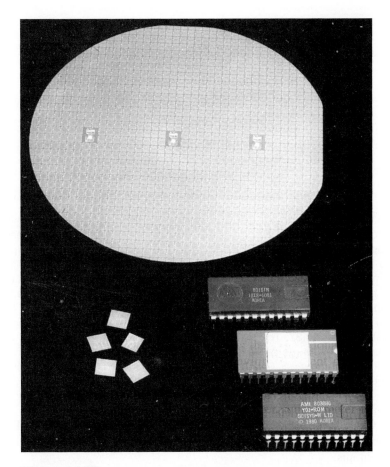

Figure 14.39
This photograph shows a wafer, individual integrated circuits, and three chip packages (the middle package is ceramic and the other two are plastic). The three larger devices along the middle of this wafer are *process control monitors* (PCMs) to monitor the technical quality of the dice on the wafer.
[*Courtesy of American Microsystems, Inc. (AMI), Santa Clara, Calif.*]

Large-scale integrated (LSI) microelectronic circuits are manufactured by starting with a silicon single-crystal wafer (n- or p-type) about 100 to 125 mm in diameter and 0.2 mm thick. The surface of the wafer must be highly polished and free from defects on one side since the semiconductor devices are fabricated into the polished surface of the wafer. Figure 14.39 shows a silicon wafer after the microelectronic circuits have been fabricated into its surface. About 100 to 1000 chips (depending on their size) can be produced from one wafer.

First, let us examine the structure of a planar-type bipolar transistor fabricated into a silicon wafer surface. Then, we will briefly look at the structure of a more compact type of transistor called the MOSFET, or *m*etal *o*xide *s*emiconductor *f*ield-*e*ffect

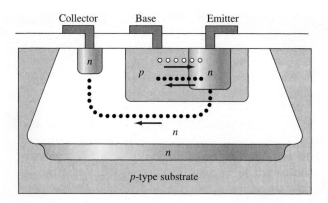

Figure 14.40
Microelectronic planar bipolar npn transistor fabricated in a single crystal of silicon by a series of operations that require access to only one surface of the silicon chip. The entire chip is doped with p-type impurities, and then islands of n-type silicon are formed. Smaller p- and n-type areas are next created within these islands in order to define the three fundamental elements of the transistor: the emitter, the base, and the collector. In this microelectronic bipolar transistor the emitter-base junction is forward-biased and the collector-base junction is reverse-biased, as in the case of the isolated npn transistor of Fig. 14.37. The device exhibits gain because a small signal applied to the base can control a large one at the collector.
(After J. D. Meindl, Microelectronic Circuit Elements, Sci. Am., September 1977, p. 75. Copyright © Scientific American Inc. All rights reserved.)

*t*ransistor, which is used in many modern semiconductor device systems. Finally, we will outline some of the basic procedures used in the manufacturing of modern microelectronic circuits.

14.6.1 Microelectronic Planar Bipolar Transistors

Microelectronic planar bipolar transistors are fabricated into the surface of a silicon single-crystal wafer by a series of operations that require access to only one surface of the silicon wafer. Figure 14.40 shows a schematic diagram of the cross section of an npn planar bipolar transistor. In its fabrication a relatively large island of n-type silicon is formed first in a p-type silicon base or substrate. Then smaller islands of p- and n-type silicon are created in the larger n-type island (Fig. 14.40). In this way the three fundamental parts of the npn bipolar transistor, the emitter, base, and collector, are formed in a planar configuration. As in the case of the individual npn bipolar transistor described in Sec. 14.5 (see Fig. 14.37), the emitter-base junction is forward-biased and the base-collector junction is reverse-biased. Thus, when electrons from the emitter are injected into the base, most of them go into the collector and only a small percentage (~1 to 5 percent) recombine with holes from the base terminal (see Fig. 14.38). The microelectronic planar bipolar transistor can therefore function as a current amplifier in the same way as the individual macroelectronic bipolar transistor.

14.6.2 Microelectronic Planar Field-Effect Transistors

In many of today's modern microelectronic systems, another type of transistor, called the *field-effect transistor,* is used because of its low cost and compactness. The most common field-effect transistor used in the United States is the n-type metal oxide semiconductor field-effect transistor. In the n-type MOSFET, or NMOS, two islands of n-type silicon are created in a substrate of p-type silicon, as shown in Fig. 14.41. In the NMOS device the contact where the electrons enter is called the *source* and the contact where they leave is called the *drain.* Between the n-type silicon of the source and the drain, there is a p-type region on whose

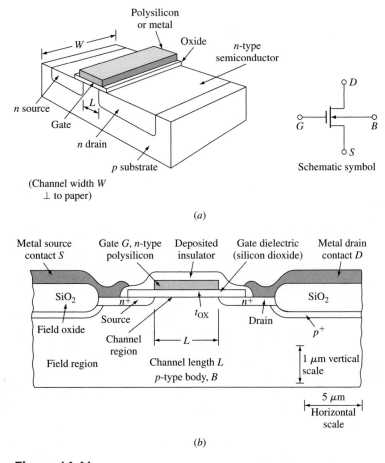

Figure 14.41

Schematic diagram of an NMOS field-effect transistor: (*a*) overall structure and (*b*) cross-sectional view.

(*After D. A. Hodges and H. G. Jackson, "Analysis and Design of Digital Integrated Circuits," McGraw-Hill, 1983, p. 40.*)

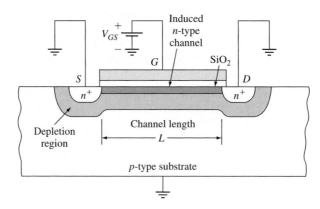

Figure 14.42
Idealized NMOS device cross section with positive
gate-source voltage (V_{GS}) applied, showing depletion
regions and the induced channel.
*(After D. A. Hodges and H. G Jackson, "Analysis and Design of
Digital Integrated Circuits," McGraw-Hill, 1983, p. 43.)*

surface a thin layer of silicon dioxide is formed that acts as an insulator. On top
of the silicon dioxide another layer of polysilicon (or metal) is deposited to form
the third contact for the transistor, called the *gate.* Since silicon dioxide is an excellent insulator, the gate connection is not in direct electrical contact with the p-type
material below the oxide.

For a simplified type of NMOS when no voltage is applied to the gate, the p-type
material under the gate contains majority carriers that are holes, and only a few
electrons are attracted to the drain. However, when a positive voltage is applied to
the gate, its electric field attracts electrons from the nearby n^+ source and drain
regions to the thin layer beneath the surface of the silicon dioxide just under the gate
so that this region becomes n-type silicon, with electrons being the majority carriers (Fig. 14.42). When electrons are present in this channel, a conducting path exists
between the source and the drain. Thus, electrons will flow between the source and
the drain if there is a positive voltage difference between them.

The MOSFET, like the bipolar transistor, is also capable of current amplification. The gain in MOSFET devices is usually measured in terms of a voltage ratio
instead of a current ratio as for the bipolar transistor. p-type MOSFETs with holes
for majority carriers can be made in a similar way, using p-type islands for the source
and drain in an n-type substrate. Since the current carriers are electrons in NMOS
devices and holes in PMOS ones, they are known as *majority-carrier devices.*

MOSFET technology is the basis for most large-scale integrated digital memory circuits, mainly because individual MOSFET take up less silicon chip area than
the bipolar transistor and hence greater densities of transistors can be obtained.

Figure 14.43
Engineer laying out an integrated circuit network.
(*Courtesy of Harris Corporation.*)

Also, the cost of fabrication of the MOSFET LSIs is less than that for the bipolar transistor types. However, there are some applications for which bipolar transistors are necessary.

14.6.3 Fabrication of Microelectronic Integrated Circuits

The design of a microelectronic integrated circuit is first laid out on a large scale, usually with computer assistance so that the most space-conserving design can be made (Fig. 14.43). In the most common fabrication process the layout is used to prepare a set of photomasks, each of which contains the pattern for a single layer of the multiple-layer finished integrated circuit (Fig. 14.44).

Photolithography The process by which a microscopic pattern is transferred from a photomask to the silicon wafer surface of the integrated circuit is called

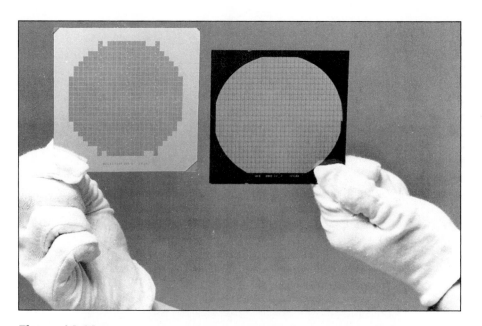

Figure 14.44
This photograph depicts two types of photolithographic masks used in the fabrication of integrated circuits. At the left is the more durable chrome mask, which is used for long production runs and can be used to produce emulsion masks like the one shown on the right. Emulsion masks are less expensive and tend to be used for shorter production runs, such as the fabrication of prototypes.
[*Courtesy of American Microsystems, Inc. (AMI), Santa Clara, Calif.*]

photolithography. Figure 14.45 shows the steps necessary to form an insulating layer of silicon dioxide on the silicon surface, which contains a pattern of regions of exposed silicon substrate. In one type of photolithographic process shown in step 2 of Fig. 14.45, an oxidized wafer is first coated with a layer of *photoresist,* a light-sensitive polymeric material. The important property of photoresist is that its solubility in certain solvents is greatly affected by its exposure to *ultraviolet* (UV) radiation. After exposure to UV radiation (step 3 of Fig. 14.45) and subsequent development, a pattern of photoresist is left wherever the mask was transparent to the UV radiation (step 4 of Fig. 14.45). The silicon wafer is then immersed in a solution of hydrofluoric acid, which attacks only the exposed silicon dioxide and not the photoresist (step 5 of Fig. 14.45). In the final step of the process the photoresist pattern is removed by another chemical treatment (step 6 of Fig. 14.45). The photolithographic process has improved so that it is now possible to reproduce surface dimensions to about 0.5 μm.

Diffusion and Ion Implantation of Dopants into the Surface of Silicon Wafers
To form active circuit elements such as bipolar and MOS transistors in integrated circuits, it is necessary to selectively introduce impurities (dopants) into the silicon

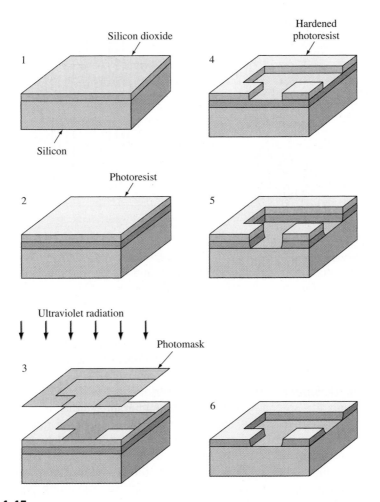

Figure 14.45

The steps for the photolithographic process. In this process a microscopic pattern can be transferred from a photomask to a material layer in an actual circuit. In this illustration a pattern is shown being etched into a silicon dioxide layer on the surface of a silicon wafer. The oxidized wafer (1) is first coated with a layer of a light-sensitive material called *photoresist* (2) and then exposed to ultraviolet light through the photomask (3). The exposure renders the photomask insoluble in a developer solution; hence a pattern of photoresist is left wherever the mask is transparent (4). The wafer is next immersed in a solution of hydrofluoric acid, which selectively attacks the silicon dioxide, leaving the photoresist pattern and the silicon substrate unaffected (5). In the final step the photoresist pattern is removed by another chemical treatment (6).

(*After W. G. Oldham, The Fabrication of Microelectronic Circuits, Sci. Am., September 1977, p. 121. Copyright © Scientific American Inc. All rights reserved.*)

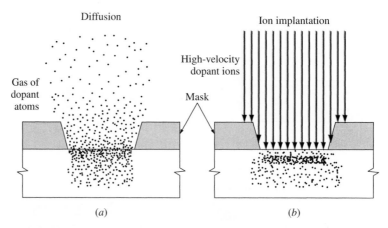

Figure 14.46
Selective doping processes for exposed silicon surfaces: (*a*) high-temperature diffusion of impurity atoms; and (*b*) ion implantation.
(*After S. Triebwasser "Today and Tomorrow in Microelectronics," from the Proceedings of an NSF workshop held at Arlie, Va., Nov. 19–22, 1978.*)

substrate to create localized n- and p-type regions. There are two main techniques for introducing dopants into the silicon wafers: (1) *diffusion* and (2) *ion implantation.*

The Diffusion Technique As previously described in Sec. 5.3, the impurity atoms are diffused into the silicon wafers at high temperatures, that is, about 1000 to 1100°C. Important dopant atoms such as boron and phosphorus move much more slowly through silicon dioxide than through the silicon crystal lattice. Thin silicon dioxide patterns can serve as masks to prevent the dopant atoms from penetrating into the silicon (Fig. 14.46*a*). Thus, a rack of silicon wafers can be placed into a diffusion furnace at 1000 to 1100°C in an atmosphere containing phosphorus (or boron), for example. The phosphorus atoms will enter the unprotected surface of the silicon and slowly diffuse into the bulk of the wafer, as shown in Fig. 14.46*a.*

The important variables that control the concentration and depth of penetration are *temperature* and *time.* To achieve the maximum control of concentration, most diffusion operations are carried out in two steps. In the first, or *predeposit,* step, a relatively high concentration of dopant atoms is deposited near the surface of the wafer. After the predeposit step, the wafers are placed in another furnace, usually at a higher temperature, for the *drive-in diffusion* step, which achieves the necessary concentration of dopant atoms at a particular depth below the surface of the silicon wafer.

The Ion Implantation Technique Another selective doping process for silicon wafers for integrated circuits is the ion implantation technique (Fig. 14.46*b*), which has the advantage that the dopant impurities can be embedded at room temperature. In this process the dopant atoms are ionized (electrons are removed from atoms to form ions) and the ions are accelerated to high energies through a high potential difference of 50 to 100 kV. When the ions strike the silicon wafer, they are embedded to varying depths, depending on their mass and energy and the type of surface protection of the silicon surface. A photoresist or silicon dioxide pattern can mask regions of the surface where ion implantation is not desired. The accelerated ions cause some damage to the crystal lattice of the silicon, but most of the damage can be healed by annealing at a moderate temperature. Ion implantation is useful wherever the doping level must be accurately controlled. Another important advantage of ion implantation is its ability to introduce dopant impurities through a thin oxide layer. This technique makes it possible to adjust the threshold voltages of MOS transistors. By means of ion implantation both NMOS and PMOS transistors can be fabricated on the same wafer.

MOS Integrated Circuit Fabrication Technology   There are many different procedures used in the fabrication of MOS integrated circuits. New innovations and discoveries for improving the equipment design and processing of ICs are constantly being made in this rapidly progressing technology. The general processing sequence for one method of producing NMOS integrated circuits is described in the following steps[4] and illustrated in Figs. 14.47 and 14.48.

1.   (See Fig. 14.47a.) A *chemical vapor deposition* (CVD) process deposits a thin layer of silicon nitride (Si_3N_4) on the entire wafer surface. The first photolithographic step defines areas where transistors are to be formed. The silicon nitride is removed outside the transistor areas by chemical etching. Boron (p-type) ions are implanted in the exposed regions to suppress unwanted conduction between the transistor sites. Next, a layer of silicon dioxide (SiO_2) about 1 μm thick is grown thermally in these inactive, or field, regions by exposing the wafer to oxygen in an electric furnace. This is known as a *selective,* or *local,* oxidation process. The Si_3N_4 is impervious to oxygen and thus inhibits growth of the thick oxide in the transistor regions.

2.   (See Fig. 14.47b.) The Si_3N_4 is next removed by an etchant that does not attack SiO_2. A clean thermal oxide about 0.1 μm thick is grown in the transistor areas, again by exposure to oxygen in a furnace. Another CVD process deposits a layer of polycrystalline silicon (poly) over the entire wafer. The second photolithographic step defines the desired patterns for the gate electrodes. The undesired poly is removed by chemical or plasma (reactive-gas) etching. An n-type dopant (phosphorus or arsenic) is introduced into the regions that will become the transistor source and drain. Either thermal diffusion or ion implantation may be used for this doping process. The thick field oxide and the poly gate are barriers to the dopant, but in the process, the poly becomes heavily doped n-type.

3.   (See Fig. 14.47c.) Another CVD process deposits an insulating layer, often SiO_2, over the entire wafer. The third masking step defines the areas in which contacts to the transistors are to be made, as shown in Fig. 14.47c. Chemical or plasma etching selectively exposes bare silicon or poly in the contact areas.

---

[4]After D. A. Hodges and H. G. Jackson, *Analysis and Design of Digital Integrated Circuits,* McGraw-Hill, 1983, pp. 16–18.

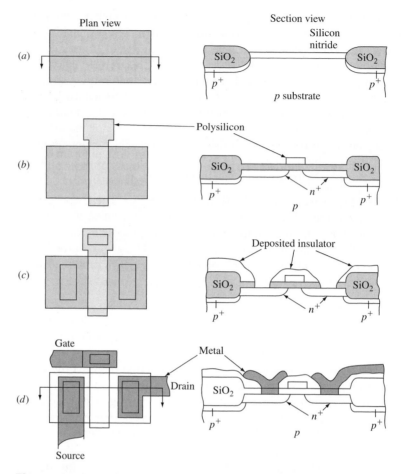

Figure 14.47
Steps in NMOS field-effect transistor fabrication: (*a*) first mask;
(*b*) second mask: polysilicon gate and source-drain diffusion;
(*c*) third mask: contact areas; and (*d*) fourth mask: metal pattern.
(After D. A. Hodges and H. G. Jackson, "Analysis and Design of Digital Integrated Circuits," McGraw-Hill, 1983, p. 17.)

4. Aluminum (Al) is deposited over the entire wafer by evaporation from a hot crucible in a vacuum evaporator. The fourth masking step patterns the Al as desired for circuit connections, as shown in Fig. 14.47*d*.

5. A protective passivating layer is deposited over the entire surface. A final masking step removes this insulating layer over the pads where contacts will be made. Circuits are tested by using needlelike probes on the contact pads. Defective units are marked, and the wafer is then sawed into individual chips. Good chips are packaged and given a final test.

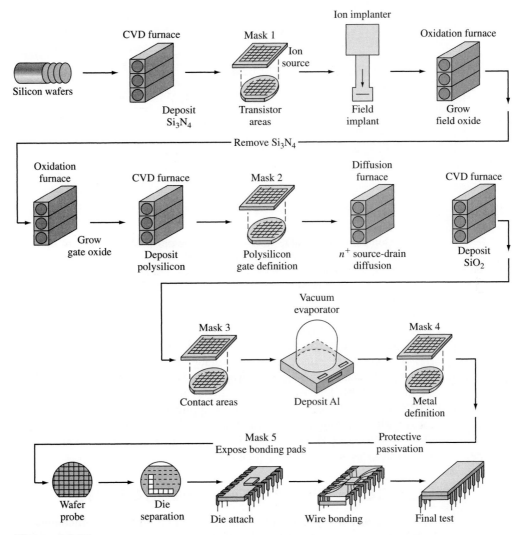

Figure 14.48

A manufacturing process for NMOS silicon-gate integrated circuits. (The processes used to make NMOS integrated circuits vary considerably from company to company. This sequence is given as an outline.)

(*Courtesy of Integrated Circuit Engineering Co.*)

This is the simplest process for forming NMOS circuits, and it is summarized schematically in Fig. 14.48. More-advanced NMOS circuit processes require more masking steps.

Complementary Metal Oxide Semiconductor (CMOS) Devices It is possible to manufacture a chip containing both types of MOSFETs (NMOS and PMOS) but only

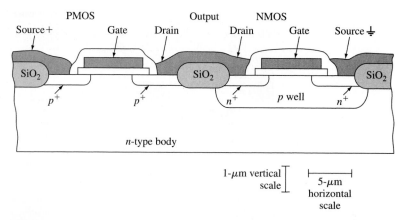

Figure 14.49
Complementary MOS field-effect transistors (CMOS). Both n- and
p-type transistors are fabricated on the same silicon substrate.
*(After D. A. Hodges and H. G. Jackson, "Analysis and Design of Digital Integrated
Circuits," McGraw-Hill, 1983, p. 42.)*

by increasing the complexity of the circuitry and lowering the density of the transistors. Circuits containing both NMOS and PMOS devices are called *complementary,* or *CMOS, circuits* and can be made, for example, by isolating all NMOS devices with islands of p-type material (Fig. 14.49). An advantage of CMOS circuits is that the MOS devices can be arranged to achieve lower power consumption. CMOS devices are used in a variety of applications. For example, large-scale integrated CMOS circuits are used in almost all the modern electronic watches and calculators. Also, CMOS technology is becoming of increasing importance for use in microprocessors and computer memories.

14.7 COMPOUND SEMICONDUCTORS

There are many compounds of different elements that are semiconductors. Among the major types of semiconducting compounds are the MX ones, where M is a more electropositive element and X a more electronegative element. Of the MX semiconductor compounds, two important groups are the III-V and II-VI compounds formed by elements adjacent to the group IV A of the periodic table (Fig. 14.50). The III-V semiconductor compounds consist of M group III elements such as Al, Ga, and In combined with X group V elements such as P, As, and Sb. The II-VI compounds consist of M group II elements such as Zn, Cd, and Hg combined with X group VI elements such as S, Se, and Te.

Table 14.6 lists some electrical properties of selected compound semiconductors. From this table the following trends can be observed.

1. When the molecular mass of a compound within a family increases by moving down in the columns of the periodic table, the energy band gap decreases, electron mobility increases (exceptions are GaAs and GaSb), and the lattice constant

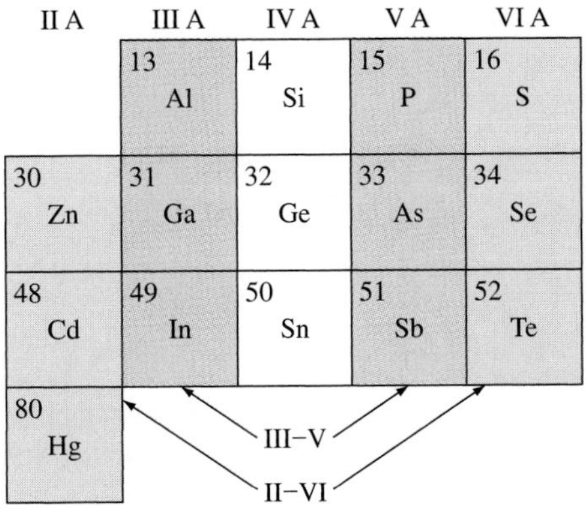

Figure 14.50
Part of the periodic table containing elements used in the formation of MX-type III–V and II–VI semiconductor compounds.

Table 14.6 Electrical properties of intrinsic semiconducting compounds at room temperature (300 K)

Group	Material	E_g eV	μ_n m²/(V · s)	μ_p m²/(V · s)	Lattice constant	n_i carriers/m³
IV A	Si	1.10	0.135	0.048	5.4307	1.50×10^{16}
	Ge	0.67	0.390	0.190	5.257	2.4×10^{19}
III A–V A	GaP	2.25	0.030	0.015	5.450	
	GaAs	1.47	0.720	0.020	5.653	1.4×10^{12}
	GaSb	0.68	0.500	0.100	6.096	
	InP	1.27	0.460	0.010	5.869	
	InAs	0.36	3.300	0.045	6.058	
	InSb	0.17	8.000	0.045	6.479	1.35×10^{22}
II A–VI A	ZnSe	2.67	0.053	0.002	5.669	
	ZnTe	2.26	0.053	0.090	6.104	
	CdSe	2.59	0.034	0.002	5.820	
	CdTe	1.50	0.070	0.007	6.481	

Source: W. R. Runyun and S. B. Watelski, in C. A. Harper (ed.), *Handbook of Materials and Processes for Electronics,* McGraw-Hill, New York, 1970.

increases. The electrons of the larger and heavier atoms have, in general, more freedom to move and are less tightly bound to their nuclei and thus tend to have smaller band gaps and higher electron mobilities.

2. By moving across the periodic table from the group IV A elements to the III-V and II-VI materials, the increased ionic bonding character causes the energy band gaps to increase and the electron mobilities to decrease. The increased ionic bonding causes a tighter binding of the electrons to their positive-ion cores, and thus II-VI compounds have larger band gaps than comparable III-V compounds.

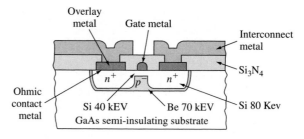

Figure 14.51
Cross-sectional view of a GaAs MESFET.
[*After A. N. Sato et al., IEEE Electron. Devices Lett.,* **9(5)**:*238 (1988).*]

Gallium arsenide is the most important of all the compound semiconductors and is used in many electronic devices. GaAs has been used for a long time for discrete components in microwave circuits. Today, many digital integrated circuits are made with GaAs. The GaAs *metal-semiconductor field-effect transistors* (MESFETs) are the most widely used GaAs transistors (Fig. 14.51).

GaAs MESFETs offer some advantages over silicon as devices for use in high-speed digital integrated circuits. Some of these are:

1. Electrons travel faster in n-GaAs as is indicated by their higher mobility in GaAs than in Si [$\mu_n = 0.720$ m^2/(V $\cdot$ s) for GaAs versus 0.135 m^2/(V $\cdot$ s) for Si].
2. Because of its larger band gap of 1.47 eV and the absence of a critical-gate oxide, GaAs devices are said to have better radiation resistance. This consideration is important for space and military applications.

Unfortunately, the major limitation of GaAs technology is that the yield for complex IC circuitry is much lower than for silicon due mainly to the fact that GaAs contains more defects in the base material than silicon. The cost of producing the base material is also higher for GaAs than silicon. However, the use of GaAs is expanding, and much research is being done in this area.

EXAMPLE PROBLEM 14.10

a. Calculate the intrinsic electrical conductivity of GaAs at (1) room temperature (27°C) and (2) 70°C.
b. What fraction of the current is carried by the electrons in the intrinsic GaAs at 27°C?

■ **Solution**
a. (1) σ at 27°C:

$$\sigma = n_i q(\mu_n + \mu_p)$$
$$= (1.4 \times 10^{12} \text{ m}^{-3})(1.60 \times 10^{-19} \text{ C})[0.720 \text{ m}^2/(\text{V} \cdot \text{s}) + 0.020 \text{ m}^2/(\text{V} \cdot \text{s})]$$
$$= 1.66 \times 10^{-7} \ (\Omega \cdot \text{m})^{-1} \blacktriangleleft$$

(2) σ at 70°C:

$$\sigma = \sigma_0 e^{-E_g/2kT} \qquad\qquad (14.16a)$$

$$\frac{\sigma_{343}}{\sigma_{300}} = \frac{\exp\{-1.47\ \text{eV}/[(2)(8.62 \times 10^{-5}\ \text{eV/K})(343\ \text{K})]\}}{\exp\{-1.47\ \text{eV}/[(2)(8.62 \times 10^{-5}\ \text{eV/K})(300\ \text{K})]\}}$$

$$\sigma_{343} = \sigma_{300} e^{3.56} = 1.66 \times 10^{-7}\ (\Omega \cdot \text{m})^{-1}(35.2)$$

$$= 5.84 \times 10^{-6}\ (\Omega \cdot \text{m})^{-1} \blacktriangleleft$$

b. $\dfrac{\sigma_n}{\sigma_n + \sigma_p} = \dfrac{n_i q \mu_n}{n_i q (\mu_n + \mu_p)} = \dfrac{0.720\ \text{m}^2/(\text{V} \cdot \text{s})}{0.720\ \text{m}^2/(\text{V} \cdot \text{s}) + 0.020\ \text{m}^2(\text{V} \cdot \text{s})} = 0.973 \blacktriangleleft$

14.8 ELECTRICAL PROPERTIES OF CERAMICS

Ceramic materials are used for many electrical and electronic applications. Many types of ceramics are used for electrical insulators for low- and high-voltage electric currents. Ceramic materials also find application in various types of capacitors, especially where miniaturization is required. Other types of ceramics called piezoelectrics can convert weak pressure signals into electrical signals, and vice versa.

Before discussing the electrical properties of the various types of ceramic materials, let us first briefly look at some of the basic properties of insulators, or **dielectrics** as they are sometimes called.

14.8.1 Basic Properties of Dielectrics

There are three important properties common to all insulators or dielectrics: (1) the *dielectric constant,* (2) the *dielectric breakdown* strength, and (3) the loss factor.

Dielectric Constant Consider a simple parallel-plate capacitor[5] with metal plates of area A separated by distance d, as shown in Fig. 14.52. Consider first the case in which the space between the plates is a vacuum. If a voltage V is applied across the

---

[5]A **capacitor** is a device that stores electric energy.

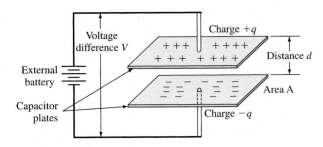

Figure 14.52
Simple parallel-plate capacitor.

plates, one plate will acquire a net charge of $+q$ and the other a net charge of $-q$. The charge q is found to be directly proportional to the applied voltage V as

$$q = CV \quad \text{or} \quad C = \frac{q}{V} \tag{14.23}$$

where C is a proportionality constant called the **capacitance** of the capacitor. The SI unit of capacitance is coulombs per volt (C/V), or the *farad* (F). Thus,

$$1 \text{ farad} = \frac{1 \text{ coulomb}}{\text{volt}}$$

Since the farad is a much larger unit of capacitance than is normally encountered in electrical circuitry, the commonly used units are the *picofarad* ($1 \text{ pF} = 10^{-12}$ F) and the *microfarad* ($1 \mu\text{F} = 10^{-6}$ F).

The capacitance of a capacitor is a measure of its ability to store electric charge. The more charge stored at the upper and lower plates of a capacitor, the higher is its capacitance.

The capacitance C for a parallel-plate capacitor whose area dimensions are much greater than the separation distance of the plates is given by

$$C = \epsilon_0 \frac{A}{d} \tag{14.24}$$

where ϵ_0 = permittivity of free space = 8.854×10^{-12} F/m.

When a dielectric (electrical insulator) fills the space between the plates (Fig. 14.53), the capacitance of the capacitor is increased by a factor κ, which is called the **dielectric constant** of the dielectric material. For a parallel-plate capacitor with a dielectric between the capacitor plates,

$$C = \frac{\kappa \epsilon_0 A}{d} \tag{14.25}$$

Table 14.7 lists the dielectric constants for some ceramic insulator materials.

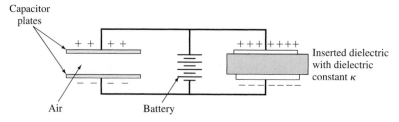

Figure 14.53
Two parallel-plate capacitors under the same applied voltage. The capacitor on the right has a dielectric (insulator inserted between the plates), and as a result the charge on the plates is increased by a factor of κ above that on the plates of the capacitor without the dielectric.

Table 14.7 Electrical properties of some ceramic insulator materials

Material	Volume resistivity ($\Omega \cdot$ m)	Dielectric strength		Dielectric constant κ		Loss factor	
		V/mil	kV/mm	60 Hz	10^6 Hz	60 Hz	10^6 Hz
Electrical porcelain insulators	10^{11}–10^{13}	55–300	2–12	6	...	0.06	
Steatite insulators	$>10^{12}$	145–280	6–11	6	6	0.008–0.090	0.007–0.025
Fosterite insulators	$>10^{12}$	250	9.8	...	6	...	0.001–0.002
Alumina insulators	$>10^{12}$	250	9.8	...	9	...	0.0008–0.009
Soda-lime glass	...	...	...	...	7.2	...	0.009
Fused silica	...	8	...	...	3.8	...	0.00004

Source: Materials Selector, *Mater. Eng.,* December 1982.

The energy stored in a capacitor of a given volume at a given voltage is increased by the factor of the dielectric constant when the dielectric material is present. By using a material with a very high dielectric constant, very small capacitors with high capacitances can be produced.

Dielectric Strength   Another property besides the dielectric constant that is important in evaluating dielectrics is the **dielectric strength.** This quantity is a measure of the ability of the material to hold energy at high voltages. Dielectric strength is defined as the voltage per unit length (electric field or voltage gradient) at which failure occurs and thus is the maximum electric field that the dielectric can maintain without electrical breakdown.

Dielectric strength is most commonly measured in volts per mil (1 mil = 0.001 in) or kilovolts per millimeter. If the dielectric is subjected to a voltage gradient that is too intense, the strain of the electrons or ions in trying to pass through the dielectric may exceed its dielectric strength. If the dielectric strength is exceeded, the dielectric material begins to break down and the passage of current (electrons) occurs. Table 14.7 lists the dielectric strengths for some ceramic insulator materials.

Dielectric Loss Factor   If the voltage used to maintain the charge on a capacitor is sinusoidal, as is generated by an alternating current, the current leads the voltage by 90 degrees when a loss-free dielectric is between the plates of a capacitor. However, when a real dielectric is used in the capacitor, the current leads the voltage by $90° - \delta$, where the angle δ is called the *dielectric loss angle.* The product of $\kappa \tan \delta$ is designated the *loss factor* and is a measure of the electric energy lost (as heat energy) by a capacitor in an ac circuit. Table 14.7 lists the loss factors for some ceramic insulator materials.

A simple parallel-plate capacitor is to be made to store 5.0×10^{-6} C at a potential of 8000 V. The separation distance between the plates is to be 0.30 mm. Calculate the area (in square meters) that the plates must have if the dielectric between the plates is (a) a vacuum ($\kappa = 1$) and (b) alumina ($\kappa = 9$). ($\epsilon_0 = 8.85 \times 10^{-12}$ F/m.)

EXAMPLE PROBLEM 14.11

■ **Solution**

$$C = \frac{q}{V} = \frac{5.0 \times 10^{-6} \, C}{8000 \, V} = 6.25 \times 10^{-10} \, F$$

$$A = \frac{Cd}{\epsilon_0 \kappa} = \frac{(6.25 \times 10^{-10} \, F)(0.30 \times 10^{-3} \, m)}{(8.85 \times 10^{-12} \, F/m)(\kappa)}$$

a. For vacuum, $\kappa = 1$: $A = 0.021 \, m^2$
b. For alumina, $\kappa = 9$: $A = 2.35 \times 10^{-3} \, m^2$

As can be seen by these calculations, the insertion of a material with a high dielectric constant can appreciably reduce the area of the plates required.

14.8.2 Ceramic Insulator Materials

Ceramic materials have electrical and mechanical properties that make them especially suitable for many insulator applications in the electrical and electronic industries. The ionic and covalent bonding in ceramic materials restricts electron and ion mobility and thus makes these materials good electrical insulators. These bondings make most ceramic materials strong but relatively brittle. The chemical compositions and microstructure of electrical- and electronic-grade ceramics must be more closely controlled than for structural ceramics such as bricks or tiles. Some aspects of the structure and properties of several insulator ceramic materials will now be discussed.

Electrical Porcelain   Typical electrical porcelain consists of approximately 50 percent clay ($Al_2O_3 \cdot 2SiO_2 \cdot 2H_2O$), 25 percent silica ($SiO_2$), and 25 percent feldspar ($K_2O \cdot Al_2O_3 \cdot 6SiO_2$). This composition makes a material that has good green-body plasticity and a wide firing temperature range at a relatively low cost. The major disadvantage of electrical insulator material is that it has a high power-loss factor compared to other electrical insulator materials (Table 14.7), which is due to highly mobile alkali ions. Figure 11.33 shows the microstructure of an electrical porcelain material.

Steatite   Steatite porcelains are good electrical insulators because they have low power-loss factors, low moisture absorption, and good impact strength and are used extensively by the electronic and electrical appliance industries. Industrial steatite compositions are based on about 90 percent talc ($3MgO \cdot 4SiO_2 \cdot H_2O$) and 10 percent clay. The microstructure of fired steatite consists of enstatite ($MgSiO_3$) crystals bonded together by a glassy matrix. Figure 14.54 shows some examples of steatite insulator parts for electrical applications.

Fosterite   Fosterite has the chemical formula Mg_2SiO_4 and thus has no alkali ions in a vitreous phase so that it has a higher resistivity and lower electrical loss with increasing temperature than the steatite insulators. Fosterite also has lower-loss dielectric properties at high frequencies (Table 14.7).

Alumina   Alumina ceramics have aluminum oxide (Al_2O_3) as the crystalline phase bonded with a glassy matrix. The glass phase, which is normally alkali-free, is

Figure 14.54
Some electronic and electrical insulator parts made of steatite.
(*Courtesy of Wisconsin Porcelain Co.*)

compounded from mixtures of clay, talc, and alkaline earth fluxes and is usually alkali-free. Alumina ceramics have relatively high dielectric strengths and low dielectric losses along with relatively high strengths. Sintered alumina (99 percent Al_2O_3) is widely used as a substrate for electronic-device applications (Fig. 11.1*b*) because of its low dielectric losses and smooth surface (Fig. 11.34). Alumina is also used for ultralow-loss applications where a large energy transfer through a ceramic window is necessary as, for example, for radomes.

14.8.3 Ceramic Materials for Capacitors

Ceramic materials are commonly used as dielectric materials for capacitors, with disk ceramic capacitors being by far the most common type of ceramic capacitor (Fig. 14.55). These very small flat-disk ceramic capacitors consist mainly of barium titanate ($BaTiO_3$) along with other additives (Table 14.8). $BaTiO_3$ is used because of its very high dielectric constant of 1200 to 1500. With additives its dielectric constant can be raised to values of many thousands. Figure 14.55*b* shows the stages in the manufacture of one type of disk ceramic capacitor. In this type of capacitor a silver layer on the top and bottom of the disk provides the metal "plates" of the capacitor. For very high capacitances with a minimum-size device, small, multilayered ceramic capacitors have been developed.

Ceramic chip capacitors are used in some ceramic-based thick-film hybrid electronic circuits. Chip capacitors can provide appreciably higher capacitance per unit area values and can be added to the thick-film circuit by a simple soldering or bonding operation.

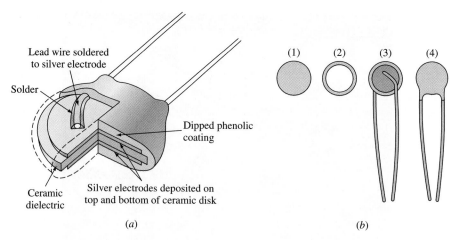

Figure 14.55
Ceramic capacitors. (*a*) Section showing construction.
(*Courtesy of Sprague Products Co.*)
(*b*) Steps in manufacture: (1) after firing ceramic disk; (2) after applying silver
electrodes; (3) after soldering leads; (4) after applying dipped phenolic coating.
(*Courtesy of Radio Materials Co.*)

Table 14.8 Representative formulations for some ceramic dielectric materials
for capacitors

Dielectric constant κ	Formulation
325	$BaTiO_3 + CaTiO_3 +$ low % $Bi_2Sn_3O_9$
2100	$BaTiO_3 +$ low % $CaZrO_3$ and Nb_2O_5
6500	$BaTiO_3 +$ low % $CaZrO_3$ or $CaTiO_3 + BaZrO_3$

Source: C. A. Harper (ed.), *Handbook of Materials and Processes for Electronics,* McGraw-Hill, 1970,
pp. 6–61.

14.8.4 Ceramic Semiconductors

Some ceramic compounds have semiconducting properties that are important for the
operation of some electrical devices. One of these devices is the **thermistor,** or thermally sensitive resistor, which is used for temperature measurement and control. For
our discussion we shall be concerned with the *negative temperature coefficient* (NTC)
type of thermistor whose resistance decreases with increasing temperature. That is, as
the temperature increases, the thermistor becomes more conductive, as in the case of
a silicon semiconductor.

The ceramic semiconducting materials most commonly used for NTC thermistors are sintered oxides of the elements Mn, Ni, Fe, Co, and Cu. Solid-solution combinations of the oxides of these elements are used to obtain the necessary range of
electrical conductivities with temperature changes.

Let us first consider the ceramic compound magnetite, Fe_3O_4, which has a rel-
atively low resistivity of about 10^{-5} $\Omega \cdot m$ as compared to a value of about 10^8 $\Omega \cdot m$

for most regular transition metal oxides. Fe_3O_4 has the inverse spinel structure with the composition $FeO \cdot Fe_2O_3$ that can be written as

$$Fe^{2+}(Fe^{3+}, Fe^{3+})O_4$$

In this structure the oxygen ions occupy FCC lattice sites, with the Fe^{2+} ions in octahedral sites and the Fe^{3+} ions half in octahedral sites and half in tetrahedral sites. The good electrical conductivity of Fe_3O_4 is attributed to the random location of the Fe^{2+} and Fe^{3+} ions in the octahedral sites so that electron "hopping" (transfer) can take place between the Fe^{2+} ions and Fe^{3+} ions while maintaining charge neutrality. The structure of Fe_3O_4 is discussed further in Sec. 16.9.

The electrical conductivities of metal oxide semiconducting compounds for thermistors can be controlled by forming solid solutions of different metal oxide compounds. By combining a low-conducting metal oxide with a high-conducting one with a similar structure, a semiconducting compound with an intermediate conductivity can be produced. This effect is illustrated in Fig. 14.56, which shows

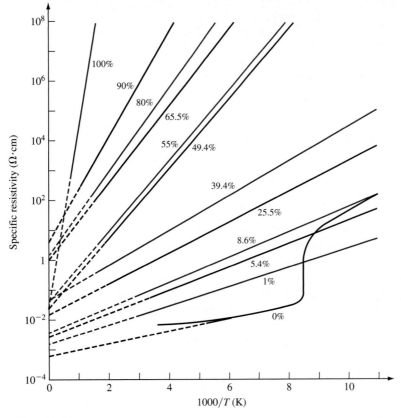

Figure 14.56
Specific resistivity of solid solution of Fe_3O_4 and $MgCr_2O_4$. Mole percent $MgCr_2O_4$ is indicated on the curves.
[*After E. J. Verwey, P. W. Haagman, and F. C. Romeijn, J. Chem. Phys.,* **15:**18(1947).]

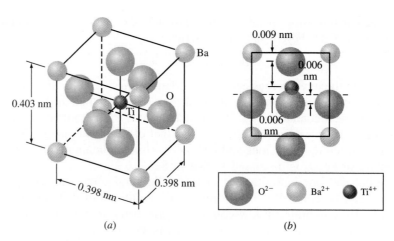

MatVis

Figure 14.57

(a) The structure of BaTiO$_3$ above 120°C is cubic. (b) The structure of BaTiO$_3$ below 120°C (its Curie temperature) is slightly tetragonal due to a slight shift of the Ti^{4+} central ion with respect to the surrounding O^{2-} ions of the unit cell. A small electric dipole moment is present in this asymmetrical unit cell.

(After K. M. Ralls, T. H. Courtney, and J. Wulff, "An Introduction to Materials Science and Engineering," Wiley, 1976, p. 610.)

how the conductivity of Fe$_3$O$_4$ is reduced gradually by adding increasing amounts in solid solution of MgCr$_2$O$_4$. Most NTC thermistors with controlled temperature coefficients of resistivity are made of solid solutions of Mn, Ni, Fe, and Co oxides.

14.8.5 Ferroelectric Ceramics

Ferroelectric Domains   Some ceramic ionic crystalline materials have unit cells that do not have a center of symmetry, and as a result their unit cells contain a small electric dipole and are called ferroelectric. An industrially important ceramic material in this class is barium titanate, BaTiO$_3$. Above 120°C, BaTiO$_3$ has the regular cubic symmetrical perovskite crystal structure (Fig. 14.57a). Below 120°C, the central Ti^{4+} ion and the surrounding O^{2-} ions of the BaTiO$_3$ unit cell shift slightly in opposite directions to create a small electric dipole moment (Fig. 14.57b). This shifting of the ion positions at the critical temperature of 120°C, called the **Curie temperature,** changes the crystal structure of BaTiO$_3$ from cubic to slightly tetragonal.

On a larger scale solid barium titanate ceramic material has a domain structure (Fig. 14.58) in which the small electric dipoles of the unit cells line up in one direction. The resultant dipole moment of a unit volume of this material is the sum of the small dipole moments of the unit cells. If polycrystalline barium titanate is slowly cooled through its Curie temperature in the presence of a strong electric field, the

Figure 14.58
Microstructure of barium titanate ceramic showing different ferroelectric domain orientations as revealed by etching. (Magnification 500×.)
[*After R. D. DeVries and J. E. Burke, J. Am. Ceram. Soc.,* **40***:200 (1957).*]

dipoles of all the domains tend to line up in the direction of the electric field to produce a strong dipole moment per unit volume of the material.

The Piezoelectric Effect[6] Barium titanate and many other ceramic materials exhibit what is called the **piezoelectric (PZT) effect,** illustrated schematically in Fig. 14.59. Let us consider a sample of a ferroelectric ceramic material that has a resultant dipole moment due to the alignment of many small unit dipoles, as indicated in Fig. 14.59*a.* In this material there will be an excess of positive charge at one end and negative charge at the other end in the direction of the polarization. Now let us consider the sample when compressive stresses are applied, as shown in Fig. 14.59*b.* The compressive stresses reduce the length of the sample between the applied stresses and thus reduce the distance between the unit dipoles, which in turn reduces the overall dipole moment per unit volume of the material. The change in dipole moment of the material changes the charge density at the ends of the sample and thus changes the voltage difference between the ends of the sample if they are insulated from each other.

---

[6]The prefix *piezo-* means "pressure" and comes from the Greek word *piezein,* which means "to press."

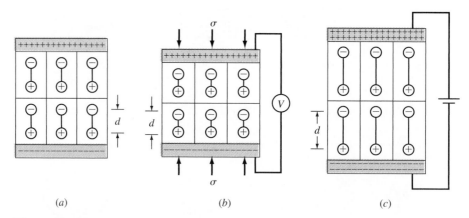

Figure 14.59

(*a*) Schematic illustration of electric dipoles within a piezoelectric material.
(*b*) Compressive stresses on material cause a voltage difference to develop due
to change in electric dipoles. (*c*) Applied voltage across ends of a sample causes
a dimensional change and changes the electric dipole moment.
(*After L. H. Van Vlack, "Elements of Materials Science and Engineering," 4th ed., Addison-Wesley,
1980, Fig. 8-6.3, p. 305.*)

On the other hand, if an electric field is applied across the ends of the sample, the charge density at each end of the sample will be changed (Fig. 14.59*c*). This change in charge density will cause the sample to change dimensions in the direction of the applied field. In the case of Fig. 14.59*c* the sample is slightly elongated due to an increased amount of positive charge attracting the negative poles of the dipoles, and the reverse at the other end of the sample. Thus, the piezoelectric effect is an electromechanical effect by which mechanical forces on a ferroelectric material can produce an electrical response, or electrical forces a mechanical response.

Piezoelectric ceramics have many industrial applications. Examples for the case of converting mechanical forces into electrical responses are the piezoelectric compression accelerometer (Fig. 14.60*a*), which can measure vibratory accelerations occurring over a wide range of frequencies, and the phonograph cartridge in which electrical responses are "picked up" from a stylus vibrating in record grooves. An example for the case of converting electrical forces into mechanical responses is the ultrasonic cleaning **transducer** that is caused to vibrate by ac power input so that it can induce violent agitation of the liquid in a tank (Fig. 14.60*b*). Another example of this type is the underwater sound transducer in which electric power input causes the transducer to vibrate to transmit sound waves.

Piezoelectric Materials   Although $BaTiO_3$ is commonly used as a piezoelectric material, it has largely been replaced by other piezoelectric ceramic materials. Of

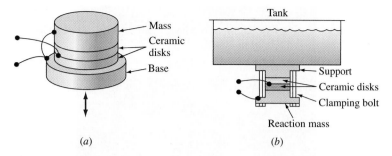

(a)   (b)

Figure 14.60

(*a*) Piezoelectric compression accelerometer. (*b*) Piezoelectric ceramic elements in an ultrasonic cleaning apparatus.
(*Courtesy of the Vernitron Piezoelectric Division, Bedford, Ohio.*)

particular importance are the ceramic materials made from solid solutions of lead zirconate ($PbZrO_3$) and lead titanate ($PbTiO_3$) to make what are called *PZT ceramics.* The PZT materials have a broader range of piezoelectric properties, including a higher Curie temperature, than $BaTiO_3$.

14.9 NANOELECTRONICS

The ability to characterize and study nanomaterials and nanodevices has improved significantly with the advent of *spanning probe microscope* (SPM) techniques (Chap. 4). Researchers have shown that by varying the imposed voltage between the STM tip and the surface, it is possible to pick up an atom (or a cluster of atoms) and manipulate its position on the surface. For instance, scientists have used the STM to create dangling (incomplete) bonds on the surface of silicon at specific positions. Then, by exposing the sample's surface to gases containing molecules of interest, these dangling bonds can become the sites of molecular adsorption. By positioning the dangling bonds and therefore the adsorbed molecules at specific points on the surface, nanoscale molecular electronics can be designed. Another example of the use of STM in nanotechnology is the formation of *quantum corrals.* STM is used to position metal atoms on the surface in a circular or elliptical form. Since electrons are confined to the path of the metal atoms, a quantum corral is formed that represents a "hot zone" of electron waves. This is similar to a hot zone of electromagnetic waves in a dish antenna. The size of the corral is in tens of nanometers. If a magnetic atom such as a cobalt atom is placed at one focus point in an elliptical corral with two focus points, some of its properties (such as a change in the surface electrons due to the cobalt's magnetism) appear at the other focus point (Fig. 14.61). On the other hand, if the single atom is placed at nonfocal positions, its properties will not

Figure 14.61
In this figure a single atom of cobalt is
placed at one of the focus points of the
elliptical corral of 36 cobalt atoms (left
peak). Some of its properties then appear
at the other focus point (right peak) where
no atoms exist.
(*IBM Research, Almaden Research Center*)

show up anywhere else in the corral. The spot at which the quantum corral is
formed is called *a quantum mirage*. A quantum mirage is envisioned as a vehicle for transfering data at the nanorange scale. Although this may be many years
in the future, the overall goal is to develop techniques that allow delivery of current in nanodevices where conventional electrical wiring is impossible because of
the small dimensions.

14.10 SUMMARY

In the classic model for electrical conduction in metals, the outer valence electrons of the
atoms of the metal are assumed to be free to move between the positive-ion cores (atoms
without their valence electrons) of the metal lattice. In the presence of an applied electric
potential the free electrons attain a directed drift velocity. The movement of electrons and
their associated electric charge in a metal constitute an electric current. By convention,

electric current is considered to be positive charge flow, which is in the opposite direction to electron flow.

In the energy-band model for electrical conduction in metals, the valence electrons of the metal atoms interact and interpenetrate each other to form energy bands. Since the energy bands of the valence electrons of metal atoms overlap, producing partially filled composite energy bands, very little energy is required to excite the highest-energy electrons so that they become free to be conductive. In insulators the valence electrons are tightly bound to their atoms by ionic and covalent bonding and are not free to conduct electricity unless highly energized. The energy-band model for an insulator consists of a lower filled valence band and a higher empty conduction band. The valence band is separated from the conduction band by a large energy gap (about 6 to 7 eV, for example). Thus, for insulators to be conductive, a large amount of energy must be applied to cause the valence electrons to "jump" the gap. Intrinsic semiconductors have a relatively small energy gap (i.e., about 0.7 to 1.1 eV) between their valence and conduction bands. By doping the intrinsic semiconductors with impurity atoms to make them extrinsic, the amount of energy required to cause semiconductors to be conductive is greatly reduced.

Extrinsic semiconductors can be n-type or p-type. The n-type (negative) semiconductors have electrons for their majority carriers. The p-type (positive) semiconductors have holes (missing electrons) for their majority charge carriers. By fabricating pn junctions in a single crystal of a semiconductor such as silicon, various types of semiconducting devices can be made. For example, pn junction diodes and npn transistors can be produced by using these junctions. Modern microelectronic technology has developed to such an extent that thousands of transistors can be placed on a "chip" of semiconducting silicon less than about 0.5 cm square and about 0.2 mm thick. Complex microelectronic technology has made possible highly sophisticated microprocessors and computer memories.

Ceramic materials are usually good electrical and thermal insulators due to the absence of conduction electrons, and thus many ceramics are used for electrical insulation and refractories. Some ceramic materials can be highly polarized with electric charge and are used for dielectric materials for capacitors. Permanent polarization of some ceramic materials produces piezoelectric properties that permit these materials to be used as electromechanical transducers. Other ceramic materials, for example, Fe_3O_4, are semiconductors and find application for thermistors for temperature measurement.

Nanotechnology research is making progress toward manufacturing electronic devices with nanometer dimensions. Quantum corrals are envisioned to deliver currents in nano devices where electrical wiring is impossible.

14.11 DEFINITIONS

Sec. 14.1

Electric current: the time rate passage of charge through material; electric current i is the number of coulombs per second that passes a point in a material. The SI unit for electric current is the ampere (1 A = 1 C/s).

Electrical resistance R: the measure of the difficulty of electric current's passage through a volume of material. Resistance increases with the length and increases with decreasing cross-sectional area of the material through which the current passes. SI unit: ohm (Ω).

Electrical resistivity ρ_e: a measure of the difficulty of electric current's passage through a *unit* volume of material. For a volume of material, $\rho_e = RA/l$, where R = resistance of material, Ω; l = its length, m; A = its cross-sectional area, m^2. In SI units, ρ_e = ohm-meters ($\Omega \cdot m$).

Electrical conductivity σ_e: a measure of the ease with which electric current passes through a unit volume of material. Units: $(\Omega \cdot m)^{-1}$. σ_e is the inverse of ρ_e.

Electrical conductor: a material with a high electrical conductivity. Silver is a good conductor and has a $\sigma_e = 6.3 \times 10^7 \, (\Omega \cdot m)^{-1}$.

Electrical insulator: a material with a low electrical conductivity. Polyethylene is a poor conductor and has a $\sigma_e = 10^{-15}$ to $10^{-17} \, (\Omega \cdot m)^{-1}$.

Semiconductor: a material whose electrical conductivity is approximately midway between the values for good conductors and insulators. For example, pure silicon is a semiconducting element and has $\sigma_e = 4.3 \times 10^{-4} \, (\Omega \cdot m)^{-1}$ at 300 K.

Electric current density J: the electric current per unit area. SI units: amperes/meter^2 (A/m^2).

Sec. 14.2

Energy-band model: in this model the energies of the bonding valence electrons of the atoms of a solid form a band of energies. For example, the 3s valence electrons in a piece of sodium form a 3s energy band. Since there is only one 3s electron (the 3s orbital can contain two electrons), the 3s energy band in sodium metal is half-filled.

Valence band: the energy band containing the valence electrons. In a conductor the valence band is also the conduction band. The valence band in a conducting metal is not full, and so some electrons can be energized to levels within the valence band and become conductive electrons.

Conduction band: the unfilled energy levels into which electrons can be excited to become conductive electrons. In semiconductors and insulators there is an energy gap between the filled lower valence band and the upper empty conduction band.

Sec. 14.3

Intrinsic semiconductor: a semiconducting material that is essentially pure and for which the energy gap is small enough (about 1 eV) to be surmounted by thermal excitation; current carriers are electrons in the conduction band and holes in the valence band.

Electron: a negative charge carrier with a charge of 1.60×10^{-19} C.

Hole: a positive charge carrier with a charge of 1.60×10^{-19} C.

Sec. 14.4

n-type extrinsic semiconductor: a semiconducting material that has been doped with an n-type element (e.g., silicon doped with phosphorus). The n-type impurities donate electrons that have energies close to the conduction band.

Donor levels: in the band theory, local energy levels near the conduction band.

p-type extrinsic semiconductor: a semiconducting material that has been doped with a p-type element (e.g., silicon doped with aluminum). The p-type impurities provide electron holes close to the upper energy level of the valence band.

Acceptor levels: in the band theory, local energy levels close to the valence band.

Majority carriers: the type of charge carrier most prevalent in a semiconductor; the majority carriers in an n-type semiconductor are conduction electrons, and in a p-type semiconductor they are conduction holes.

Minority carriers: the type of charge carrier in the lowest concentration in a semiconductor. The minority carriers in n-type semiconductors are holes, and in p-type semiconductors they are electrons.

Sec. 14.5

pn junction: an abrupt junction or boundary between p- and n-type regions within a single crystal of a semiconducting material.

Bias: voltage applied to two electrodes of an electronic device.

Forward bias: bias applied to a pn junction in the conducting direction; in a pn junction under forward bias, majority-carrier electrons and holes flow toward the junction so that a large current flows.

Reverse bias: bias applied to a pn junction so that little current flows; in a pn junction under reverse bias, majority-carrier electrons and holes flow away from the junction.

Rectifier diode: a pn junction diode that converts alternating current to direct current (AC to DC).

Bipolar junction transistor: a three-element, two-junction semiconducting device. The three basic elements of the transistor are the emitter, base, and collector. Bipolar junction transistors can be of the npn or pnp types. The emitter-base junction is forward-biased and the collector-base junction is reverse-biased so that the transistor can act as a current amplification device.

Sec. 14.8

Dielectric: an electrical insulator material.

Capacitor: an electric device consisting of conducting plates or foils separated by layers of dielectric material and capable of storing electric charge.

Capacitance: a measure of the ability of a capacitor to store electric charge. Capacitance is measured in farads; the units commonly used in electrical circuitry are the picofarad ($1 \text{ pF} = 10^{-12} \text{ F}$) and the microfarad ($1 \text{ } \mu\text{F} = 10^{-6} \text{ F}$).

Dielectric constant: the ratio of the capacitance of a capacitor using a material between the plates of a capacitor compared to that of the capacitor when there is a vacuum between the plates.

Dielectric strength: the voltage per unit length (electric field) at which a dielectric material allows conduction, that is, the maximum electric field that a dielectric can withstand without electrical breakdown.

Thermistor: a ceramic semiconductor device that changes in resistivity as the temperature changes and is used to measure and control temperature.

Ferroelectric material: a material that can be polarized by applying an electric field.

Polarization: the alignment of small electric dipoles in a dielectric material to produce a net dipole moment in the material.

Curie temperature (of a ferroelectric material): the temperature at which a ferroelectric material on cooling undergoes a crystal structure change that produces spontaneous polarization in the material. For example, the Curie temperature of $BaTiO_3$ is 120°C.

Piezoelectric effect: an electromechanical effect by which mechanical forces on a ferroelectric material can produce an electrical response and electrical forces produce a mechanical response.

Transducer: a device that is actuated by power from one source and transmits power in another form to a second system. For example, a transducer can convert input sound energy into an output electrical response.

14.12 PROBLEMS

Answers to problems marked with an asterisk are given at the end of the book.

14.1 Describe the classic model for electrical conduction in metals.

14.2 Distinguish between (*a*) positive-ion cores and (*b*) valence electrons in a metallic crystal lattice such as sodium.

14.3 Write equations for the (*a*) macroscopic and (*b*) microscopic forms of Ohm's law. Define the symbols in each of the equations and indicate their SI units.

14.4 How is electrical conductivity related numerically to electrical resistivity?

14.5 Give two kinds of SI units for electrical conductivity.

14.6 Calculate the resistance of an iron rod 0.720 cm in diameter and 0.850 m long at 20°C. [$\rho_e(20°C) = 10.0 \times 10^{-6} \, \Omega \cdot cm$.]

*14.7 A nichrome wire must have a resistance of 120 Ω. How long must it be (in meters) if it is 0.0015 in. in diameter? [$\sigma_e(\text{nichrome}) = 9.3 \times 10^5 \, (\Omega \cdot m)^{-1}$.]

14.8 A wire 0.40 cm in diameter must carry a 25 A current.

   a. If the maximum power dissipation along the wire is 0.025 W/cm, what is the minimum allowable electrical conductivity of the wire (give answer in SI units)?

   b. What is the current density in the wire?

14.9 An iron wire is to conduct a 6.5 A current with a maximum voltage drop of 0.005 V/cm. What must be the minimum diameter of the wire in meters at (20°C)?

14.10 Define the following quantities pertaining to the flow of electrons in a metal conductor: (*a*) drift velocity; (*b*) relaxation time; and (*c*) electron mobility.

14.11 What causes the electrical resistivity of a metal to increase as its temperature increases? What is a phonon?

14.12 What structural defects contribute to the residual component of the electrical resistivity of a pure metal?

14.13 What effect do elements that form solid solutions have on the electrical resistivities of pure metals?

*14.14 Calculate the electrical resistivity (in ohm-meters) of a silver wire 15 m long and 0.030 m in diameter at 160°C. [$\rho_e(\text{Fe at } 0°C) = 9.0 \times 10^{-6} \, \Omega \cdot cm$.]

*14.15 At what temperature will an iron wire have the same electrical resistivity as an aluminum one has at 35°C?

14.16 At what temperature will the electrical resistivity of an iron wire be $25.0 \times 10^{-8} \, \Omega \cdot m$?

14.17 Why are the valence-electron energy levels broadened into bands in a solid block of a good conducting metal such as sodium?

14.18 Why don't the energy levels of the inner-core electrons of a block of sodium metal also form energy bands?

14.19 Why is the 3*s* electron energy band in a block of sodium only half-filled?

14.20 What explanation is given for the good electrical conductivity of magnesium and aluminum even though these metals have filled outer 3*s* energy bands?

14.21 How does the energy-band model explain the poor electrical conductivity of an insulator such as pure diamond?

14.22 Define an intrinsic semiconductor. What are the two most important elemental semiconductors?

14.23 What type of bonding does the diamond cubic structure have? Make a two-dimensional sketch of the bonding in the silicon lattice, and show how electron-hole pairs are produced in the presence of an applied field.

14.24 Why is a hole said to be an imaginary particle? Use a sketch to show how electron holes can move in a silicon crystal lattice.

14.25 Define electron and electron hole mobility as pertains to charge movement in a silicon lattice. What do these quantities measure, and what are their SI units?

14.26 Explain, using an energy-band diagram, how electrons and electron holes are created in pairs in intrinsic silicon.

14.27 What is the ratio of the electron-to-hole mobility in silicon and germanium?

14.28 Calculate the number of germanium atoms per cubic meter.

*14.29 Calculate the electrical resistivity of germanium at 300 K.

14.30 Explain why the electrical conductivity of intrinsic silicon and germanium increases with increasing temperature.

14.31 The electrical resistivity of pure germanium is $0.46\ \Omega \cdot m$ at 300 K. Calculate its electrical conductivity at 425°C.

*14.32 The electrical resistivity of pure silicon is $2.3 \times 10^3\ \Omega \cdot m$ at 300 K. Calculate its electrical conductivity at 325°C.

14.33 Define n-type and p-type extrinsic silicon semiconductors.

14.34 Draw two-dimensional lattices of silicon of the following:

    a. n-type lattice with an arsenic impurity atom present

    b. p-type lattice with a boron impurity atom present

14.35 Draw energy-band diagrams showing donor or acceptor levels for the following:

    a. n-type silicon with phosphorus impurity atoms

    b. p-type silicon with boron impurity atoms

14.36 a. When a phosphorus atom is ionized in an n-type silicon lattice, what charge does the ionized atom acquire?

    b. When a boron atom is ionized in a p-type silicon lattice, what charge does the ionized atom acquire?

14.37 In semiconductors what are dopants? Explain the process of doping by diffusion.

14.38 What are the majority and minority carriers in an n-type silicon semiconductor? In a p-type one?

*14.39 A silicon wafer is doped with 7.0×10^{21} phosphorus atoms/m³. Calculate (*a*) the electron and hole concentrations after doping and (*b*) the resultant electrical resistivity at 300 K. [Assume $n_i = 1.5 \times 10^{16}/m^3$ and $\mu_n = 0.1350\ m^2/(V \cdot s)$.]

14.40 Phosphorus is added to make an n-type silicon semiconductor with an electrical conductivity of $250 \ \Omega \cdot m^{-1}$. Calculate the necessary number of charge carriers required.

14.41 A semiconductor is made by adding boron to silicon to give an electrical resistivity of $1.90 \ \Omega \cdot m$. Calculate the concentration of carriers per cubic meter in the material. [Assume $\mu_p = 0.048 \ m^2/(V \cdot s)$.]

14.42 A silicon wafer is doped with 2.50×10^{16} boron atoms/cm^3 plus 1.60×10^{16} phosphorus atoms/cm^3 at 27°C. Calculate (a) the electron and hole concentrations (carriers per cubic centimeter), and (b) the electron and hole mobilities (use Fig. 14.26), and (c) the electrical resistivity of the material.

14.43 A silicon wafer is doped with 2.50×10^{15} phosphorus atoms/cm^3, 3.00×10^{17} boron atoms/cm^3, and 3.00×10^{17} arsenic atoms/cm^3. Calculate (a) the electron and hole concentrations (carriers per cubic centimeter), (b) the electron and hole mobilities (use Fig. 14.26), and (c) the electrical resistivity of the material.

*14.44 An arsenic-doped silicon wafer has an electrical resistivity of $7.50 \times 10^{-4} \ \Omega \cdot cm$ at 27°C. Assume intrinsic carrier mobilities and complete ionization.

  a. What is the majority-carrier concentration (carriers per cubic centimeter)?

  b. What is the ratio of arsenic to silicon atoms in this material?

14.45 A boron-doped silicon wafer has an electrical resistivity of $5.00 \times 10^{-4} \ \Omega \cdot cm$ at 27°C. Assume intrinsic carrier mobilities and complete ionization.

  a. What is the majority-carrier concentration (carriers per cubic centimeter)?

  b. What is the ratio of boron to silicon atoms in this material?

14.46 Describe the origin of the three stages that appear in the plot of $\ln \sigma$ versus $1/T$ for an extrinsic silicon semiconductor (going from low to high temperatures). Why does the conductivity decrease just before the rapid increase due to intrinsic conductivity?

14.47 Define the term *microprocessor.*

14.48 Describe the movement of majority carriers in a pn junction diode at equilibrium. What is the depletion region of a pn junction?

14.49 Describe the movement of the majority and minority carriers in a pn junction diode under reverse bias.

14.50 Describe the movement of the majority carriers in a pn junction diode under forward bias.

14.51 Describe how a pn junction diode can function as a current rectifier.

14.52 What is a zener diode? How does this device function? Describe a mechanism to explain its operation.

14.53 What are the three basic elements of a bipolar junction transistor?

14.54 Describe the flow of electrons and holes when an npn bipolar junction transistor functions as a current amplifier.

14.55 What fabrication techniques are used to encourage electrons from the emitter of an npn bipolar transistor to go right through to the collector?

14.56 Why is a bipolar junction transistor called bipolar?

14.57 Describe the structure of a planar npn bipolar transistor.

14.58 Describe how the planar bipolar transistor can function as a current amplifier.

14.59 Describe the structure of an n-type metal oxide semiconductor field-effect transistor.

14.60 How do NMOSs function as current amplifiers?

14.61 Describe the photolithographic steps necessary to produce a pattern of an insulating layer of silicon dioxide on a silicon surface.

14.62 Describe the diffusion process for the introduction of dopants into the surface of a silicon wafer.

14.63 Describe the ion implantation process for introducing dopants into the surface of a silicon wafer.

14.64 Describe the general process for fabricating NMOS integrated circuits on a silicon wafer.

14.65 Why is silicon nitride (Si_3N_4) used in producing NMOS integrated circuits on a silicon wafer?

14.66 What are complementary metal oxide semiconductor devices? What are the advantages of CMOS devices over the NMOS or PMOS devices?

14.67 Calculate the intrinsic electrical conductivity of GaAs at 125°C. [$E_g = 1.47$ eV; $\mu_n = 0.720$ m^2/(V · s); $\mu_p = 0.020$ m^2/(V · s); $n_i = 1.4 \times 10^{12}$ m^{-3}.]

14.68 Calculate the intrinsic electrical conductivity of InSb at 60 and at 70°C. [$E_g = 0.17$ eV; $\mu_n = 8.00$ m^2/(V · s); $\mu_p = 0.045$ m^2/(V · s); $n_i = 1.35 \times 10^{22}$ m^{-3}.]

14.69 Calculate the intrinsic electrical conductivity of (a) GaAs and (b) InSb at 75°C.

*14.70 What fraction of the current is carried by (a) electrons and (b) holes in (1) InSb, (2) InB, and (3) InP at 27°C?

14.71 What fraction of the current is carried by (a) electrons and (b) holes in (1) GaSb and (2) GaP at 27°C?

14.72 What are three major applications for ceramic materials in the electrical-electronics industries?

14.73 Define the terms *dielectric*, *capacitor*, and *capacitance*. What is the SI unit for capacitance? What units are commonly used for capacitance in the electronics industry?

14.74 What is the dielectric constant of a dielectric material? What is the relationship among capacitance, dielectric constant, and the area and distance of separation between the plates of a capacitor?

14.75 What is the dielectric strength of a dielectric material? What units are used for dielectric strength? What is dielectric breakdown?

14.76 What is the dielectric loss angle and dielectric loss factor for a dielectric material? Why is a high dielectric loss factor undesirable?

14.77 A simple plate capacitor can store 7.0×10^{-5} C at a potential of 12,000 V. If a barium titanate dielectric material with $\kappa = 2100$ is used between the plates, which have an area of 5.0×10^{-5} m^2, what must be the separation distance between the plates?

*14.78 A simple plate capacitor stores 6.5×10^{-5} C at a potential of 12,000 V. If the area of the plates is 3.0×10^{-5} m^2 and the distance between the plates is 0.18 mm, what must be the dielectric constant of the material between the plates?

14.79 What is the approximate composition of electrical porcelain? What is a major disadvantage of electrical porcelain as electrical insulative material?

14.80 What is the approximate composition of steatite? What desirable electrical properties does steatite have as an insulative material?

14.81 What is the composition of fosterite? Why is fosterite an excellent insulator material?

14.82 Why is sintered alumina widely used as a substance for electronic device applications?

14.83 Why is $BaTiO_3$ used for high-value, small, flat-disk capacitors? How is the capacitance of $BaTiO_3$ capacitors varied? What are the four major stages in the manufacture of a flat-disk ceramic capacitor?

14.84 What is a thermistor? What is an NTC thermistor?

14.85 What materials are used to make NTC thermistors?

14.86 What is believed to be the mechanism for electrical conduction in Fe_3O_4?

14.87 How is the electrical conductivity of metal oxide semiconductors for thermistors changed?

14.88 What change occurs in the unit cell of $BaTiO_3$ when it is cooled below 120°C? What is this transformation temperature called?

14.89 What are ferroelectric domains? How can they be lined toward one direction?

14.90 Describe the piezoelectric effect for producing an electrical response with the application of pressure on a ferroelectric material. Do the same for producing a mechanical response by the application of an electrical force.

14.91 Describe several devices that utilize the piezoelectric effect.

14.92 What are the PZT piezoelectric materials? In what ways are they superior to $BaTiO_3$ piezoelectric materials?

14.13 MATERIALS SELECTION AND DESIGN PROBLEMS

1. Rank the following solid solutions of copper in descending order of conductivity. Give reasons for your choices. (*a*) Cu–1 wt % Zn, (*b*) Cu–1 wt % Ga, and (*c*) Cu–1 wt % Fe.

2. Consider the Cu-Ni isomorphous phase system. Draw an approximate diagram showing the conductivity versus alloy composition.

3. Investigate the electrical properties of barium titanate ($BaTiO_3$), and give an application for this material.

4. Investigate the electrical properties of lead zirconate titanate (PZT), and give an application for this material.

5. Silicon carbide (SiC) is a ceramic with semiconducting characteristics similar to those of silicon (Si). Explain the advantages of using SiC over Si as a semiconducting material.

6. Design an n-type semiconductor based on Si that allows a constant conductivity of $25\ \Omega^{-1} \cdot m^{-1}$ at room temperature.

7. Design a p-type semiconductor based on Si that allows a constant conductivity of $25\ \Omega^{-1} \cdot m^{-1}$ at room temperature.

8. Select the material for a conducting wire of diameter 20 mm that carries a current of 20 A. The maximum power dissipation is 4 W/m. (Use Table 14.1 and consider cost as a selection criterion.)

9. (*a*) What are conductive polymers? (*b*) How do they work? (*c*) Give an application.

Sources:

1. Binnig, G., H. Rohrer, et al. in *Physical Review Letters,* v 50 pp. 120–24 (1983).

2. http://ufrphy.lbhp.jussieu.fr/nano/

5. http://www.molec.com/products_consumables. html#STM

6. H. Dai, J. H. Hafner, A. G. Rinzler, D. T. Colbert, R. E. Smalley, Nature 384, 147–150 (1996).

7. http://www.omicron.de/index2.html?/results/stm_ image_of_chromium_decorated_steps_of_cu_111/ ~Omicron

8. http://www.almaden.ibm.com/almaden/media/image_ mirage.html

Optical Properties and Superconductive Materials

(Courtesy of Crystal Fibre A/S)

A photonic crystal fiber is similar to a normal crystal in structure, with the exception that the repeat pattern exists in a much larger scale (micron range) and only transverse to the length of the fiber. The fiber is manufactured by stacking a number of silica glass tubes to form a cylinder. The cylinder is then drawn at elevated temperatures to a thin fiber with diameters on the order of tens of microns. After the manufacturing process, the fiber will resemble a honeycomb. Because of their structure, light conducted through the fibers can behave in ways that are not completely understood. For example, it is possible to allow light of a certain frequency to propagate along the fiber while other frequencies are blocked. Such characteristics can be used to manufacture devices such as wavelength-tunable light sources and optical switches. The chapter-opening image shows the structure of a photonic crystal fiber. The preforms are shown in the top figure and the cross sections of selected fibers are shown at the bottom.[1] ∎

----
[1]http://www.rikei.co.jp/dbdata/products/producte249.html

By the end of this chapter, students will be able to . . .

1. Explain what phenomena can occur with light radiation as it passes from one medium into another.

2. Discuss why metallic materials are opaque to visible light.

3. Explain what determines the color of metallic materials.

4. Briefly describe the phenomenon of superconductivity.

5. Explain why amorphous materials are usually transparent.

6. Briefly describe the construction of a ruby laser.

7. Describe the mechanism of photon absorption for a semiconductor that contains electrically active defects.

8. Explain what laser means.

9. Briefly describe the advantages of high-temperature oxide superconductors.

10. Cite distinctions between opacity, translucency, and transparency.

15.1 INTRODUCTION

Optical properties of materials play an important role in much of today's high technology (Fig. 15.1). In this chapter we will first examine some of the basics of the refraction, reflection, and absorption of light with some classes of materials. Next we will investigate how some materials interact with light radiation to produce luminescence. Then we will look into the stimulated emission of radiation by lasers. In the optical fibers part of this chapter, we shall see how the development of low light-loss optical fibers has led to the new optical-fiber communications systems.

Finally, we shall look into superconducting materials that have zero electrical resistivity below their critical temperatures, magnetic fields, and current densities. Until around 1987 the highest critical temperature for a superconducting material was about 25 K. In 1987 a spectacular discovery was made that some ceramic materials could be made superconductive up to about 100 K. This discovery set off a tremendous worldwide research effort that has created high expectations for future engineering developments. In this chapter we will examine some aspects of the structure and properties of types I and II metallic superconductors as well as the new ceramic ones.

15.2 LIGHT AND THE ELECTROMAGNETIC SPECTRUM

Visible light is one form of electromagnetic radiation, with wavelengths extending from about 0.40 to 0.75 μm (Fig. 15.2). Visible light contains color bands ranging from violet through red, as shown in the enlarged scale of Fig. 15.2. The ultraviolet

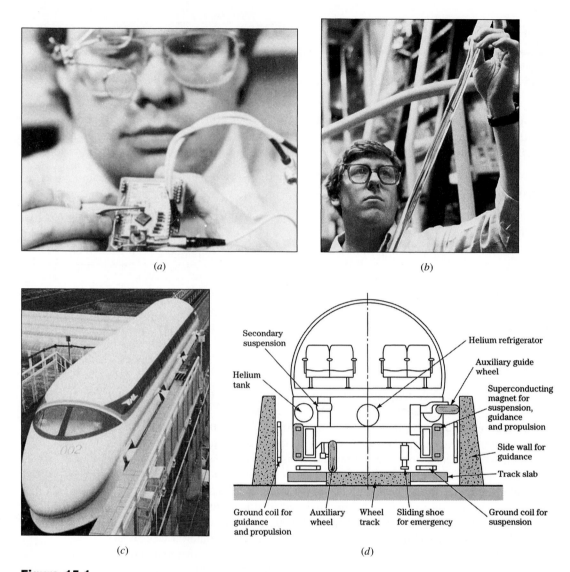

(a)

(b)

(c)

(d)

Figure 15.1

New technologies. (a) An engineer examines a laser transmitter to be used in the second transatlantic lightguide cable, which went into service in 1991. (b) A fiber-optic glass preform is examined before being drawn into hair-thin optical fiber.

[*(a) and (b) Courtesy of AT&T.*]

(c) A Japanese National Railways train that uses Nb–Ti superconducting magnets in liquid-helium cryostats. The train is designed to travel at 400 to 500 km/h on a test track.

(*After Mechanical Engineering, June 1988, p. 61.*)

(d) Cross section of an advanced levitated train design (Japanese National Railway).

(*After the "Encyclopedia of Materials Science and Engineering," MIT Press, 1986, p. 4766.*)

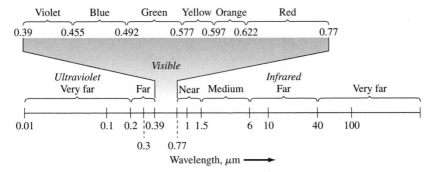

Figure 15.2
The electromagnetic spectrum from the ultraviolet to the infrared regions.

region covers the range from about 0.01 to about 0.40 μm, and the infrared region extends from about 0.75 to 1000 μm.

The true nature of light will probably never be known. However, light is considered to form waves and to consist of particles called *photons*. The energy ΔE, wavelength λ, and frequency v of the photons are related by the fundamental equation

$$\Delta E = hv = \frac{hc}{\lambda} \tag{15.1}$$

where h is Planck's constant (6.62×10^{-34} J $\cdot$ s) and c is the speed of light in vacuum (3.00×10^8 m/s). These equations allow us to consider the photon as a particle of energy E or as a wave with a specific wavelength and frequency.

A photon in a ZnS semiconductor drops from an impurity energy level at 1.38 eV below its conduction band to its valence band. What is the wavelength of the radiation given off by the photon in the transition? If visible, what is the color of the radiation? ZnS has an energy band gap of 3.54 eV.

EXAMPLE PROBLEM 15.1

■ **Solution**
The energy difference for the photon dropping from the 1.38 eV level below the conduction band to the valence band is 3.54 eV − 1.38 eV = 2.16 eV.

$$\lambda = \frac{hc}{\Delta E} \tag{15.1}$$

where $h = 6.62 \times 10^{-34}$ J $\cdot$ s
$\quad c = 3.00 \times 10^8$ m/s
$\quad 1$ eV $= 1.60 \times 10^{-19}$ J

Thus,

$$\lambda = \frac{(6.62 \times 10^{-34}\ \text{J} \cdot \text{s})(3.00 \times 10^{8}\ \text{m/s})}{(2.16\ \text{eV})(1.60 \times 10^{-19}\ \text{JeV})(10^{-9}\ \text{m/nm})} = 574.7\ \text{nm} \blacktriangleleft$$

The wavelength of this photon at 574.7 nm is the visible yellow region of the electromagnetic spectrum.

15.3 REFRACTION OF LIGHT

15.3.1 Index of Refraction

When light photons are transmitted through a transparent material, they lose some of their energy, and as a result, the speed of light is reduced and the beam of light changes direction. Figure 15.3 shows schematically how a beam of light entering from the air is slowed when entering a denser medium such as common window glass. Thus, the incident angle for the light beam is greater than the refracted angle for this case.

The relative velocity of light passing through a medium is expressed by the optical property called the **index of refraction** n. The n value of a medium is defined as the ratio of the velocity of light in vacuum, c, to the velocity of light in the medium considered, v:

$$\text{Refractive index } n = \frac{c \text{ (velocity of light in vacuum)}}{v \text{ (velocity of light in a medium)}} \qquad \textbf{(15.2)}$$

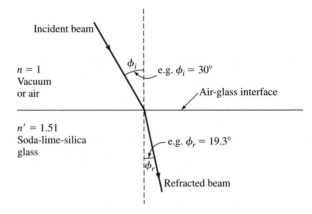

Figure 15.3
Refraction of light beam as it is transmitted from a vacuum (air) through soda-lime-silica glass.

Table 15.1 Refractive indices for selected materials

Material	Average refractive index
Glass compositions:	
Silica glass	1.458
Soda-lime-silica glass	1.51–1.52
Borosilicate (Pyrex) glass	1.47
Dense flint glass	1.6–1.7
Crystalline compositions:	
Corundum, Al_2O_3	1.76
Quartz, SiO_2	1.555
Litharge, PbO	2.61
Diamond, C	2.41
Optical plastics:	
Polyethylene	1.50–1.54
Polystyrene	1.59–1.60
Polymethyl methacrylate	1.48–1.50
Polytetrafluoroethylene	1.30–1.40

Typical average refractive indices for some glasses and crystalline solids are listed in Table 15.1. These values range from about 1.4 to 2.6, with most silicate glasses having values of about 1.5 to 1.7. The highly refractive diamond ($n = 2.41$) allows multifaceted diamond jewels to "sparkle" because of the multiple internal reflections. Lead oxide (litharge) with a value of $n = 2.61$ is added to some silicate glasses to raise their refractive indices so that they can be used for decorative purposes (see Table 11.11). It should also be noted that the refractive indices of materials are a function of wavelength and frequency. For example, the refractive index of light flint glass varies from about 1.60 at 0.40 μm to 1.57 at 1.0 μm.

15.3.2 Snell's Law of Light Refraction

The refractive indices for light passing from one medium of refractive index n through another of refractive index n' are related to the incident angle ϕ and the refractive angle ϕ' by the relation

$$\frac{n}{n'} = \frac{\sin \phi'}{\sin \phi} \qquad \text{(Snell's law)} \qquad \textbf{(15.3)}$$

When light passes from a medium with a high refractive index to one with a low refractive index, there is a critical angle of incidence ϕ_c, which if increased will result in total internal reflection of the light (Fig. 15.4). This ϕ_c angle is defined at ϕ' (refraction) = 90°.

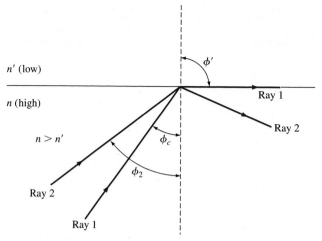

Figure 15.4
Diagram indicating the critical angle ϕ_c for total internal reflection of light passing from a high refractive index medium n to another of low refractive index n'. Note that ray 2, which has an incidence angle ϕ_2 greater than ϕ_c, is totally reflected back into the medium of high refractive index.

EXAMPLE PROBLEM 15.2

What is the critical angle ϕ_c for light to be totally reflected when leaving a flat plate of soda-lime-silica glass ($n = 1.51$) and entering the air ($n = 1$)?

■ **Solution**
Using Snell's law (Eq. 15.3),

$$\frac{n}{n'} = \frac{\sin \phi'}{\sin \phi_c}$$

$$\frac{1.51}{1} = \frac{\sin 90°}{\sin \phi_c}$$

where n = refractive index of the glass

$\quad n'$ = refractive index of air

$\quad \phi'$ = 90° for total reflection

$\quad \phi_c$ = critical angle for total reflection (unknown)

$$\sin \phi_c = \frac{1}{1.51}(\sin 90°) = 0.662$$

$$\phi_c = 41.5° \blacktriangleleft$$

Note: We shall see in Sec. 15.7 on optical fibers that by using a cladding of a low-refractive-index glass surrounding a core of high refractive index, an optical fiber can transmit light for long distances because the light is continually reflected internally.

15.4 ABSORPTION, TRANSMISSION, AND REFLECTION OF LIGHT

Every material *absorbs* light to some degree because of the interaction of light photons with the electronic and bonding structure of the atoms, ions, or molecules that make up the material **(absorptivity).** The fraction of light transmitted by a particular material thus depends on the amount of light reflected and absorbed by the material. For a particular wavelength λ, the sum of the fractions of the incoming incident light reflected, *absorbed,* and transmitted is equal to 1:

$$(\text{Reflected fraction})_\lambda + (\text{absorbed fraction})_\lambda + (\text{transmitted fraction})_\lambda = 1 \qquad \textbf{(15.4)}$$

Let us now consider how these fractions vary for some selected types of materials.

15.4.1 Metals

Except for very thin sections, metals strongly reflect and/or absorb incident radiation for long wavelengths (radio waves) to the middle of the ultraviolet range. Since the conduction band overlaps the valence band in metals, incident radiation easily elevates electrons to higher energy levels. Upon dropping to lower energy levels, the photon energies are low and their wavelengths long. This type of action results in strongly reflected beams of light from a smooth surface, as is observed for many metals such as gold and silver. The amount of energy absorbed by metals depends on the electronic structure of each metal. For example, with copper and gold there is a greater absorption of the shorter wavelengths of blue and green and a greater reflection of the yellow, orange, and red wavelengths, and thus smooth surfaces of these metals show the reflected colors. Other metals such as silver and aluminum strongly reflect all parts of the visible spectrum and show a white "silvery" color.

15.4.2 Silicate Glasses

Reflection of Light from a Single Surface of a Glass Plate   The proportion of incident light reflected by a single surface of a polished glass plate is very small. This amount depends mainly on the refractive index of the glass n and the angle of incidence of the light striking the glass. For normal light incidence (that is, $\phi_i = 90°$), the fraction of light reflected R (called the *reflectivity*) by a single surface can be determined from the relationship

$$R = \left(\frac{n - 1}{n + 1}\right)^2 \qquad \textbf{(15.5)}$$

where n is the refractive index of the reflecting optical medium. This formula may also be used with good approximation for incident light angles up to about 20 degrees.

Using Eq. 15.5, a silicate glass with $n = 1.46$ has a calculated R value of 0.035, or a percent reflectivity of 3.5 percent (see Example Problem 15.3).

EXAMPLE PROBLEM 15.3

Calculate the reflectivity of ordinary incident light from the polished flat surface of a silicate glass with a refractive index of 1.46.

■ **Solution**

Using Eq. 15.5 and $n = 1.46$ for the glass,

$$\text{Reflectivity} = \left(\frac{n-1}{n+1}\right)^2 = \left(\frac{1.46 - 1.00}{1.46 + 1.00}\right)^2 = 0.035$$

$$\% \text{ Reflectivity} = R(100\%) = 0.035 \times 100\% = 3.5\% \blacktriangleleft$$

Absorption of Light by a Glass Plate Glass absorbs energy from the light that it transmits so that the light intensity decreases as the light path increases. The relationship between the fraction of light entering, I_0, and the fraction of light exiting, I, from a glass sheet or plate of thickness t that is free of scattering centers is

$$\frac{I}{I_0} = e^{-\alpha t} \tag{15.6}$$

The constant α in this relation is called the *linear absorption coefficient* and has the units cm^{-1} if the thickness is measured in centimeters. As shown in Example Problem 15.4, there is a relatively small loss of energy by absorption through a clear silicate glass plate.

EXAMPLE PROBLEM 15.4

Ordinary incident light strikes a polished glass plate 0.50 cm thick that has a refractive index of 1.50. What fraction of light is absorbed by the glass as the light passes between the surfaces of the plate? ($\alpha = 0.03 \text{ cm}^{-1}$)

■ **Solution**

$$\frac{I}{I_0} = e^{-\alpha t} \qquad I_0 = 1.00 \qquad \alpha = 0.03 \text{ cm}^{-1}$$

$$I = ? \qquad t = 0.50 \text{ cm}$$

$$\frac{I}{1.00} = e^{-(0.03 \text{ cm}^{-1})(0.50 \text{ cm})}$$

$$I = (1.00)e^{-0.015} = 0.985$$

Thus, the fraction of light lost by absorption by the glass is: $1 - 0.985 = 0.015$, or 1.5 percent. ◄

Reflectance, Absorption, and Transmittance of Light by a Glass Plate The amount of incident light transmitted through a glass plate is determined by the

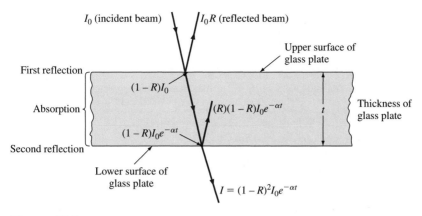

Figure 15.5
Transmittance of light through a glass plate in which reflectance takes place at upper and lower surfaces and absorption within the plate.

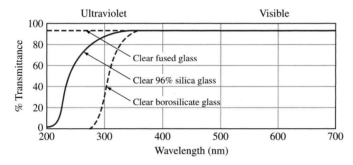

Figure 15.6
Percent transmittance versus wavelength for several types of clear glasses.

amount of light reflected from both upper and lower surfaces as well as the amount absorbed within the plate. Let us consider the transmittance of light through a glass plate, as shown in Fig. 15.5. The fraction of incident light reaching the lower surface of the glass is $(1 - R)(I_0 e^{-\alpha t})$. The fraction of incident light reflected from the lower surface will therefore be $(R)(1 - R)(I_0 e^{-\alpha t})$. Thus, the difference between the light reaching the lower surface of the glass plate and that which is reflected from the lower surface is the fraction of light transmitted I, which is:

$$I = [(1 - R)(I_0 e^{-\alpha t})] - [(R)(1 - R)(I_0 e^{-\alpha t})]$$
$$= (1 - R)(I_0 e^{-\alpha t})(1 - R) = (1 - R)^2 (I_0 e^{-\alpha t}) \qquad \textbf{(15.7)}$$

Figure 15.6 shows that about 90 percent of incident light is transmitted by silica glass if the wavelength of the incoming light is greater than about 300 nm. For shorter-wavelength ultraviolet light, much more absorption takes place and the transmittance is lowered considerably.

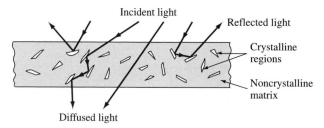

Figure 15.7
Multiple internal reflections at the crystalline-region interfaces reduce the transparency of partly crystalline thermoplastics.

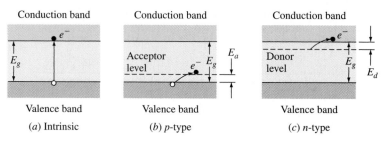

Figure 15.8
Optical absorption of photons in semiconductors. Absorption takes place in (*a*) if $h\upsilon > E_g$, (*b*) if $h\upsilon > E_a$, and (*c*) if $h\upsilon > E_d$.

15.4.3 Plastics

Many noncrystalline plastics such as polystyrene, polymethyl methacrylate, and polycarbonate have excellent transparency. However, in some plastic materials there are crystalline regions having a higher refractive index than their noncrystalline matrix. If these regions are greater in size than the wavelength of the incoming light, the light waves will be scattered by reflection and refraction, and hence the transparency of the material decreases (Fig. 15.7). For example, thin-sheet polyethylene, which has a branched-chain structure and hence a lower degree of crystallinity, has higher transparency than the higher-density, more crystalline linear-chain polyethylene. The transparencies of other partly crystalline plastics can range from cloudy to opaque, depending mainly on their degree of crystallinity, impurity content, and filler content.

15.4.4 Semiconductors

In semiconductors light photons can be absorbed in several ways (Fig. 15.8). In intrinsic (pure) semiconductors such as Si, Ge, and GaAs, photons may be absorbed to create electron-hole pairs by causing electrons to jump across the energy band gap from the valence band to the conduction band (Fig. 15.8*a*). For this to occur, the incoming light photon must have an energy value equal to or greater than the energy gap E_g. If the energy of the photon is greater than E_g, the excess energy is dissipated as heat. For semiconductors containing donor and acceptor impurities, much lower-energy (and hence much longer-wavelength) photons are absorbed in causing electrons to jump from the valence band into acceptor levels (Fig. 15.8*b*) or from donor levels into the conduction band (Fig. 15.8*c*). Semiconductors are therefore opaque to high- and intermediate-energy (short- and intermediate-wavelength) light photons and transparent to low-energy, very long wavelength photons.

Calculate the minimum wavelength for photons to be absorbed by intrinsic silicon at room temperature (E_g = 1.10 eV).

EXAMPLE PROBLEM 15.5

■ **Solution**

For absorption in this semiconductor, the minimum wavelength is given by Eq. 15.1:

$$\lambda_c = \frac{hc}{E_g} = \frac{(6.62 \times 10^{-34}\,\text{J} \cdot \text{s})(3.00 \times 10^8\,\text{m/s})}{(1.10\,\text{eV})(1.60 \times 10^{-19}\,\text{J/eV})}$$

$$= 1.13 \times 10^{-6}\,\text{m or } 1.13\,\mu\text{m} \blacktriangleleft$$

Thus, for absorption the photons must have a wavelength at least as short as 1.13 μm so that electrons can be excited across the 1.10 eV band gap.

15.5 LUMINESCENCE

Luminescence may be defined as the process by which a substance absorbs energy and then spontaneously emits visible or near-visible radiation. In this process the input energy excites electrons of a luminescent material from the valence band into the conduction band. The source of the input energy may be, for example, high-energy electrons or light photons. The excited electrons during luminescence drop to lower energy levels. In some cases the electrons may recombine with holes. If the emission takes place within 10^{-8} s after excitation, the luminescence is called **fluorescence,** and if the emission takes longer than 10^{-8} s, it is referred to as **phosphorescence.**

Luminescence is produced by materials called *phosphors* that can absorb high-energy, short-wave radiation and spontaneously emit lower-energy, longer-wavelength light radiation. The emission spectra of luminescent materials are controlled industrially by added impurities called *activators*. The activators provide discrete energy levels in the energy gap between the conduction and valence bands of the host material (Fig. 15.9). One mechanism postulated for the phosphorescent process is that excited electrons are trapped in various ways at high energy levels and must get out of the traps before they can drop to lower energy levels and emit light of a characteristic spectral band. The trapping process is used to explain the delay in light emission by excited phosphors.

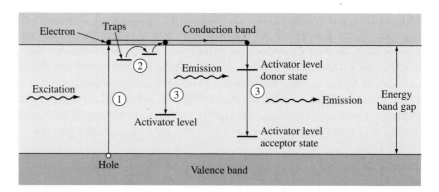

Figure 15.9
Energy changes during luminescence. (1) Electron-hole pairs are created by exciting electrons to the conduction band or to traps. (2) Electrons can be thermally excited from one trap to another or into the conduction band. (3) Electrons can drop to upper activator (donor) levels and then subsequently to lower acceptor levels, emitting visible light.

Luminescence processes are classified according to the energy source for electronic excitation. Two industrially important types are *photoluminescence* and *cathodoluminescence.*

15.5.1 Photoluminescence

In the common fluorescent lamp, photoluminescence converts ultraviolet radiation from a low-pressure mercury arc into visible light by using a halophosphate phosphor. Calcium halophosphate of the approximate composition $Ca_{10}F_2P_6O_{24}$ with about 20 percent of the F^- ions replaced with Cl^- ions is used as the host phosphor material for most lamps. Antimony (Sb^{3+}) ions provide a blue emission and manganese (Mn^{2+}) ions provide an orange-red emission band. By varying the Mn^{2+}, various shades of blue, orange, and white light may be obtained. The high-energy ultraviolet light from the excited mercury atoms causes the phosphor-coated inner wall of the fluorescent lamp tube to give off lower-energy, longer-wavelength visible light (Fig. 15.10).

15.5.2 Cathodoluminescence

This type of luminescence is produced by an energized cathode that generates a beam of high-energy bombarding electrons. Applications for this process include electron microscope, cathode-ray oscilloscope, and color television screen luminescences. The color television screen phosphorescence is especially interesting. The modern television set has very narrow (about 0.25 mm wide) vertical stripes of red-, green-, and blue-emitting phosphors deposited on the inner surface of the face plate of the television picture tube (Fig. 15.11). Through a steel shadow mask with small elongated holes (about 0.15 mm wide), the incoming television signal is scanned over

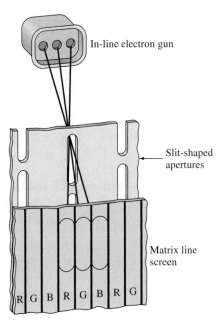

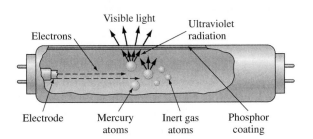

Figure 15.10
Cutaway diagram of a fluorescent lamp showing electron generation at an electrode and excitation of mercury atoms to provide the UV light to excite the phosphor coating on the inside of a lamp tube. The excited phosphor coating then provides visible light by luminescence.

Figure 15.11
Diagram showing the arrangement of the R(red), G(green), and B(blue) vertical stripes of phosphors of a color television screen. Also shown are several of the elongated apertures of the steel shadow mask.
(*Courtesy of RCA.*)

the entire screen 30 times per second. The small size and large number of phosphor areas consecutively exposed in the rapid scan of 15,750 horizontal lines per second and the persistence of an image in the human visual system make possible a clear visible picture with good resolution. Commonly used phosphors for the colors are zinc sulfide (ZnS) with an Ag^+ acceptor and Cl^- donor for the blue color, (Zn,Cd)S with a Cu^+ acceptor and Al^{3+} donor for the green color, and yttrium oxysulfide (Y_2O_2S) with 3% europium (Eu) for the red color. The phosphor materials must retain some image glow until the next scan but must not retain too much to blur the picture.

The intensity of luminescence, I, is given by

$$\ln \frac{I}{I_0} = -\frac{t}{\tau} \tag{15.8}$$

where $I_0 =$ initial intensity of luminescence and $I =$ fraction of luminescence after time t. The quantity τ is the relaxation time constant for the material.

**EXAMPLE
PROBLEM 15.6**

A color TV phosphor has a relaxation time of 3.9×10^{-3} s. How long will it take for the intensity of this phosphor material to decrease to 10 percent of its original intensity?

■ **Solution**

Using Eq. 15.8, $\ln(I/I_0) = -t/\tau$, or

$$\ln \frac{1}{10} = -\frac{t}{3.9 \times 10^{-3} \text{ s}}$$

$$t = (-2.3)(-3.9 \times 10^{-3} \text{ s}) = 9.0 \times 10^{-3} \text{ s} \blacktriangleleft$$

15.6 STIMULATED EMISSION OF RADIATION AND LASERS

Light emitted from conventional light sources such as fluorescent lamps results from the transitions of excited electrons to lower energy levels. Atoms of the same elements in these light sources give off photons of similar wavelengths independently and randomly. Consequently, the radiation emitted is in random directions and the wave trains are out of phase with each other. This type of radiation is said to be *incoherent.* In contrast, a light source called a **laser** produces a **beam** of radiation whose photon emissions are in phase, or *coherent,* and are parallel, directional, and monochromatic (or nearly so). The word "laser" is an acronym whose letters stand for "*l*ight *a*mplification by *s*timulated *e*mission of *r*adiation." In lasers some "active" emitted photons stimulate many others of the same frequency and wavelength to be emitted in phase as a coherent, intense light beam (Fig. 15.12).

To understand the mechanisms involved in laser action, let us consider the operation of a solid-state ruby laser. The ruby laser shown schematically in Fig. 15.13 is a single crystal of aluminum oxide (Al_2O_3) containing approximately 0.05 percent chromium^{3+} ions. The Cr^{3+} ions occupy substitutional lattice sites in the Al_2O_3 crystal structure and are responsible for the pink color of the laser rod. These ions act as fluorescent centers that, when excited, drop to lower energy levels, causing photon emissions at specific wavelengths. The ends of the ruby rod crystal are ground

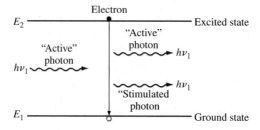

Figure 15.12
Schematic diagram illustrating the emission of a "stimulated" photon by an "active" photon of the same frequency and wavelength.

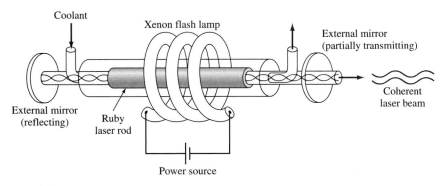

Figure 15.13
Schematic diagram of a pulsed ruby laser.

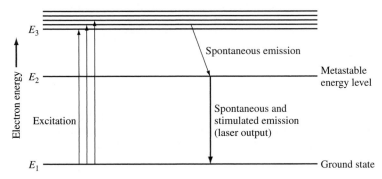

Figure 15.14
Simplified energy-level diagram for a three-level lasing system.

parallel for optical emission. A totally reflective mirror is placed parallel and near the back end of the crystal rod and another partially transmitting one at the front end of the laser, which allows the coherent laser beam to pass through.

High-intensity input from a xenon flash lamp can provide the necessary energy to excite the Cr^{3+} ion electrons from the ground state to high energy levels, as indicated by the E_3 band level of Fig. 15.14. This action in laser terminology is referred to as *pumping* the laser. The excited electrons of the Cr^{3+} ions may then drop back down to the ground state or to the metastable energy level E_2 of Fig. 15.14. However, before the stimulated emission of photons can occur in the laser, more electrons must be pumped into the high nonequilibrium metastable energy level E_2 than exist in the ground state (E_1). This condition of the laser is referred to as the **population inversion** of electron energy states as is schematically indicated in Fig. 15.15*b*; compare this condition to the equilibrium energy level condition of Fig. 15.15*a*.

The excited Cr^{3+} ions can remain in the metastable state for several milliseconds before spontaneous emission takes place by electrons dropping back to the ground state. The first few photons produced by electrons dropping from the metastable E_2

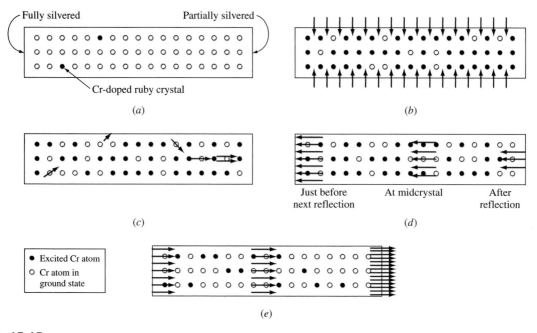

Figure 15.15
Schematic of the steps in the functioning of a pulsed ruby laser. (*a*) At equilibrium. (*b*) Excitation by xenon flash lamp. (*c*) A few spontaneously emitted photons start the stimulated emission of photons. (*d*) Reflected back, the photons continue to stimulate the emission of more photons. (*e*) The laser beam is finally emitted.
(*After R. M. Rose et al., vol. IV, "Structure and Properties of Materials," vol. IV, Wiley, 1965.*)

level of Fig. 15.14 to the ground level E_1 set off a stimulated-emission chain reaction, causing many of the electrons to make the same jump from E_2 to E_1. This action produces a large number of photons that are in phase and moving in a parallel direction (Fig. 15.15*c*). Some of the photons jumping from E_2 to E_1 are lost to the outside of the rod, but many are reflected back and forth along the ruby rod by the end mirrors. These stimulate more and more electrons to jump from E_2 and E_1, helping to build up a stronger coherent radiation beam (Fig. 15.15*d*). Finally, when a sufficiently intense coherent beam is built up inside the rod, the beam is transmitted as a high-intense-energy pulse (≈ 0.6 ms) through the partially transmitting mirror at the front end of the laser (Figs. 15.15*e* and Fig. 15.13). The laser beam produced by the Cr^{3+} doped aluminum oxide (ruby) crystal rod has a wavelength of 694.3 nm, which is a visible red line. This type of laser, which can only be operated intermittently in bursts, is said to be a *pulsed* type. In contrast, most lasers are operated with a continuous beam and are called *continuous-wave* (CW) lasers.

15.6.1 Types of Lasers

There are many types of gas, liquid, and solid lasers used in modern technology. We shall briefly describe some important aspects of several of these.

Table 15.2 Selected applications for lasers in materials processing

Applications	Type of laser	Comments
1. Welding	YAG*	High-average-power lasers for deep penetration and high-throughput welding
2. Drilling	YAG CWCO$_2$†	High peak-power densities for drilling precision holes with a minimum heat-affected zone, low taper, and maximum depths
3. Cutting	YAG CWCO$_2$	Precision cutting of complex two- and three-dimensional shapes at high rates in metals, plastics, and ceramics
4. Surface treatment	CWCO$_2$	Transformation hardening of steel surfaces by hardening them above austenitic temperatures with a scanning, defocused beam and allowing the metal to self-quench
5. Scribing	YAG CWCO$_2$	Scribing large areas of fully fired ceramics and silicon wafers to provide individual circuit substrates
6. Photolithography	Excimer	Line-narrowed and spectrally stabilized excimer photolithographic processing in the fabrication of semiconductors

*YAG = yttrium-aluminum-garnet is a crystalline host used in solid-state neodymium lasers.
†CWCO$_2$ = continuous-wave (as opposed to pulsed) carbon dioxide laser.

Ruby Laser The structure and functioning of the ruby laser has already been described. This laser is not used much today because of the difficulties in growing the crystal rods compared to the ease of making neodymium lasers.

Neodymium-YAG Lasers The *neodymium-yttrium-aluminum-garnet* (Nd:YAG) laser is made by combining one part per hundred of Nd atoms in a host of a YAG crystal. This laser emits in the near-infrared at 1.06-μm wavelength with continuous power up to about 250 W and with pulsed power as high as several megawatts. The YAG host material has the advantage of high thermal conductivity to remove excess heat. In materials processing, the Nd:YAG laser is used for welding, drilling, scribing, and cutting (Table 15.2).

Carbon Dioxide (CO$_2$) Lasers Carbon dioxide lasers are some of the most powerful lasers made. They operate mainly in the middle-infrared at 10.6 μm. They vary from a few milliwatts of continuous power to large pulses with as high as 10,000 J of energy. They operate by electron collisions exciting nitrogen molecules to metastable energy levels that subsequently transfer their energy to excite CO$_2$ molecules, which in turn give off laser radiation upon dropping to lower energy levels. Carbon dioxide lasers are used for metal processing applications such as cutting, welding, and localized heat treatment of steels (Table 15.2).

Semiconductor Lasers Semiconductor, or diode, lasers, commonly about the size of a grain of salt, are the smallest lasers produced. They consist of a pn junction made with a semiconducting compound such as GaAs that has a large enough band gap for laser action (Fig. 15.16). Originally, the GaAs diode laser was made as a homojunction laser with a single pn junction (Fig. 15.16a). The resonant cavity of the laser is created by cleaving the crystal to make two end facets. The crystal-air interfaces cause the necessary reflections for laser action due to the difference in refractive indices of air and GaAs. The diode laser achieves population inversion by a strong forward bias

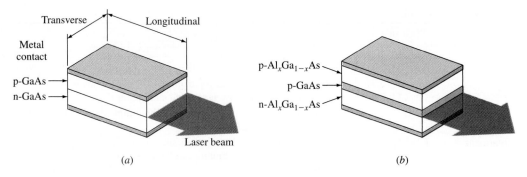

Figure 15.16
(*a*) Simple homojunction GaAs laser. (*b*) Double heterojunction GaAs laser. The p- and n-Al$_x$Ga$_{1-x}$As layers have wider band gaps and lower refractive indices and confine the electrons and holes within the active p-GaAs layer.

of a heavily doped pn junction. A great number of electron-hole pairs are generated, and many of these in turn recombine to emit photons of light.

An improvement in efficiency was achieved with the *double heterojunction* (DH) laser (Fig. 15.16*b*). In a GaAs DH laser, a thin layer of p-GaAs is sandwiched between p- and n-Al$_x$Ga$_{1-x}$As layers that confine the electrons and holes within the active p-GaAs layer. The AlGaAs layers have wider band gaps and lower refractive indices and so constrain the laser light to move in a miniature waveguide. The most widespread application of GaAs diode lasers currently is for compact disks.

15.7 OPTICAL FIBERS

Hair-thin ($\approx 1.25\ \mu$m diameter) optical fibers made primarily of silica (SiO$_2$) glass are used for modern **optical-fiber communication** systems. These systems consist essentially of a transmitter (i.e., a semiconductor laser) to encode electrical signals into light signals, optical fiber to transmit the light signals, and a photodiode to convert the light signals back into electrical signals (Fig. 15.17).

15.7.1 Light Loss in Optical Fibers

The optical fibers used for communications systems must have extremely low light loss (attenuation) so that an entering encoded light signal can be transmitted a long distance (that is, 40 km [25 mi]) and still be detected satisfactorily. For extremely low-light-loss glass for optical fibers, the impurities (particularly Fe^{2+} ions) in the SiO$_2$ glass must be very low. The light loss (**attenuation**) of an optical glass fiber is usually measured in *decibels per kilometer* (dB/km). The light loss in a light-transmitting material in dB/km for light transmission over a length l is related to the entering light intensity I_0 and the exiting light intensity I by

$$-\text{loss (dB/km)} = \frac{10}{l\,(\text{km})} \log \frac{I}{I_0} \qquad (15.9)$$

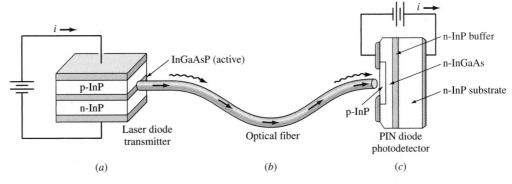

Figure 15.17
Basic elements of a fiber-optics communications system. (*a*) InGaAsP laser transmitter.
(*b*) Optical fiber for transmitting light photons. (*c*) PIN diode photodetector.

A low-loss silica glass fiber for optical transmission has a 0.20 dB/km light attenuation. (*a*) What is the fraction of light remaining after it has passed through 1 km of this glass fiber? (*b*) What is the fraction of light remaining after 40 km transmission?

EXAMPLE PROBLEM 15.7

■ **Solution**

$$\text{Attenuation (dB/km)} = \frac{10}{l\,(\text{km})} \log \frac{I}{I_0} \qquad (15.9)$$

where I_0 = light intensity at source
$\quad I$ = light intensity at detector

a. $\quad -0.20 \text{ dB/km} = \dfrac{10}{1 \text{ km}} \log \dfrac{I}{I_0} \quad$ or $\quad \log \dfrac{I}{I_0} = -0.02 \quad$ or $\quad \dfrac{I}{I_0} = 0.95 \blacktriangleleft$

b. $\quad -0.20 \text{ dB/km} = \dfrac{10}{40 \text{ km}} \log \dfrac{I}{I_0} \quad$ or $\quad \log \dfrac{I}{I_0} = -0.80 \quad$ or $\quad \dfrac{I}{I_0} = 0.16 \blacktriangleleft$

Note: Single-mode optical fibers today can transmit communication light data about 40 km without having to repeat the signal.

15.7.2 Single-Mode and Multimode Optical Fibers

Optical fibers for light transmission serve as **optical waveguides** for the light signals in optical communications. The retention of light within the optical fiber is made possible by having the light pass through the central core glass, which has a higher refractive index than the outer clad glass (Fig. 15.18). For the single-mode type, which has a core diameter of about 8 μm and an outer clad diameter of about 125 μm, there is only one acceptable guided light ray path (Fig. 15.18*a*). In the multimode-type optical glass fiber, which has a graded refractive index core, many wave modes pass through the fiber simultaneously, causing a more dispersed exiting signal than that produced by the single-mode fiber (Fig. 15.18*b*). Most new optical-fiber communication systems use

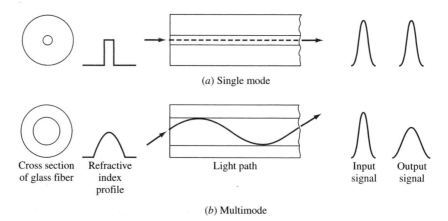

Figure 15.18

Comparison of (*a*) single-mode and (*b*) multimode optical fibers by cross section versus refractive index, light path, and signal input and output. The sharper output signal of the single-mode fiber is preferred for long-distance optical communications systems.

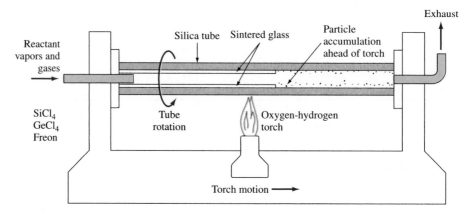

Figure 15.19

Schematic of the modified chemical vapor deposition process for the production of the glass preforms for making optical glass fibers.

[*After AT&T Tech. J.* **66**:*33(1987)*.]

single-mode fibers because they have lower light losses and are cheaper and easier to fabricate.

15.7.3 Fabrication of Optical Fibers

One of the most important methods for producing optical glass fibers for communication systems is the *modified chemical vapor deposition* (MCVD) process (Fig. 15.19). In this process, high-purity dry vapor of $SiCl_4$ with various amounts of $GeCl_4$ and fluorinated hydrocarbons vapor are passed through a rotating pure silica tube along with pure oxygen. An external oxyhydrogen torch is moved along the outer diameter of the rotating tube, allowing the contents to react to form silica glass particles doped with the desired combinations of germanium and fluorine. GeO_2 increases the refractive index of SiO_2, and fluorine lowers it. Downstream from the reaction region, the glass particles migrate to the tube wall, where they are deposited. The moving torch that caused the reaction to form the glass particles then passes over and sinters them into a thin layer of doped glass. The thickness of the doped layer depends on the number of layers that are deposited by the repeated passes of the torch. At each pass the composition of the vapors is adjusted to produce the desired composition profile so that the glass fiber subsequently produced will have the desired refractive index profile.

In the next step, the silica tube is heated to a high enough temperature so that the glass approaches its softening point. The surface tension of the glass then causes the tube with the deposited glass layers to collapse uniformly into a solid rod called a *preform* (Fig. 15.1*b*). The glass preform from the MCVD process is then inserted into a high-temperature furnace, and glass fiber about 125 μm in diameter is drawn from it (Fig. 15.20). An in-line process applies a 60-μm-thick polymer coating to

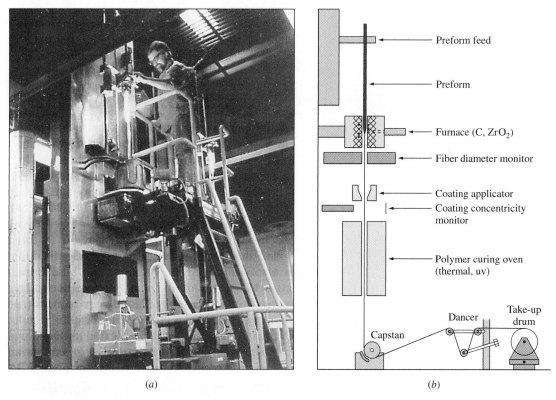

(*a*) (*b*)

Figure 15.20
Setup for drawing optical glass fiber from glass preform. (*a*) Actual fabrication.
(*Courtesy of AT&T.*)
(*b*) Schematic setup.
(*After "Encyclopedia of Materials Science and Technology," MIT Press, 1986, p. 1992.*)

Figure 15.21
Spool of optical fiber.
(*Courtesy of AT&T.*)

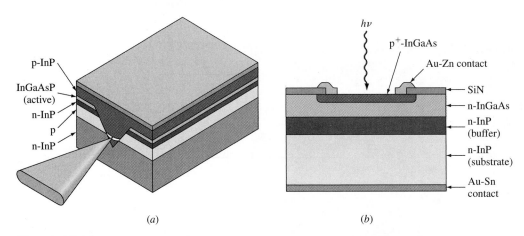

Figure 15.22
(*a*) Chemical substrate buried heterostructure InGaAsP laser diode used for long-distance fiber-optical communications systems. Note the focusing of the laser beam by the V channel.
(*b*) PIN photodetector for optical communications systems.
[*After AT&T Tech. J.* **66***(1987).*]

protect the glass fiber surface from damage. Spools of finished glass fibers are shown in Fig. 15.21. Very close tolerances for the core and outer diameter of the fiber are essential so that the fiber can be spliced (joined) without high light losses.

15.7.4 Modern Optical-Fiber Communication Systems

Most modern optical-fiber communication systems use single-mode fiber with an InGaAsP double heterojunction laser diode transmitter (Fig. 15.22*a*) operated at the infrared wavelength of 1.3 μm, where light losses are at a minimum. An

InGaAs/InP PIN photodiode is usually used for the detector (Fig. 15.22*b*). With this system optical signals can be sent about 40 km (25 mi) before the signal has to be repeated. In December 1988 the first transatlantic fiber-optic communications system began operation with a capacity of 40,000 simultaneous phone calls. By 1993 there were 289 undersea optical-fiber cable links.

Another advance in optical-fiber communication systems has been the introduction of *erbium-doped optical-fiber amplifiers* (EDFAs). An EDFA is a length (typically about 20 to 30 m [64 to 96 ft]) of optical silica fiber doped with the rare-earth element erbium to give fiber gain. When optically pumped with light from an outside semiconductor laser, the erbium-doped fiber boosts the power of all light signals passing through it with wavelengths centered on 1.55 μm. Thus, the erbium-doped optical fiber serves as both a lasing medium and a light guide. EDFAs can be used in optical transmission systems to boost the light signal power at the source (power amplifier), at the receiver (preamplifier), and along the fiber communication link (in-line repeater). The first EDFAs were used in 1993 in an AT&T network in a link between San Francisco and Point Arena, California.

15.8 SUPERCONDUCTING MATERIALS

15.8.1 The Superconducting State

The electrical resistivity of a normal metal such as copper decreases steadily as the temperature is decreased and reaches a low residual value near 0 K (Fig. 15.23). In contrast, the electrical resistivity of pure mercury as the temperature decreases drops suddenly at 4.2 K to an immeasurably small value. This phenomenon is called

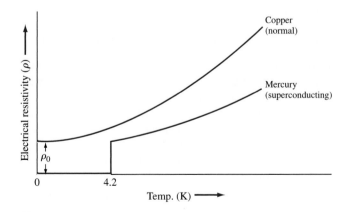

Figure 15.23
Electrical resistivity of a normal metal (Cu) compared to that of a superconductive metal (Hg) as a function of temperature near 0 K. The resistivity of the superconductive metal suddenly drops to an immeasurably small value.

Table 15.3 Critical superconducting temperatures T_c for selected metals, intermetallic and ceramic compound superconductors

Metals	T_c (K)	$H_0{}^*$ (T)	Intermetallic compounds	T_c (K)	Ceramic compounds	T_c (K)
Niobium, Nb	9.15	0.1960	Nb_3Ge	23.2	$Tl_2Ba_2Ca_2Cu_3O_x$	122
Vanadium, V	5.30	0.1020	Nb_3Sn	21	$YBa_2Cu_3O_{7-x}$	90
Tantalum, Ta	4.48	0.0830	Nb_3Al	17.5	$Ba_{1-x}K_xBiO_{3-y}$	30
Titanium, Ti	0.39	0.0100	NbTi	9.5		
Tin	3.72	0.0306				

$*H_0$ = critical field in teslas (T) at 0 K.

superconductivity, and the material that shows this behavior is called a *superconductive material.* About 26 metals are superconductive as are hundreds of alloys and compounds.

The temperature below which a material's electrical resistivity approaches zero is called the **critical temperature T_c.** Above this temperature the material is called *normal,* and below T_c it is said to be *superconducting* or *superconductive.* Besides temperature, the superconducting state also depends on many other variables, the most important of which are the magnetic field B and current density J. Thus, for a material to be superconducting, the material's critical temperature, magnetic field, and current density must not be exceeded, and for each superconducting material there exists a critical surface in T, B, J space.

The critical superconducting temperatures of some selected metals, intermetallic compounds, and new ceramic compounds are listed in Table 15.3. The extremely high T_c values (90–122 K) of the newly discovered (1987) ceramic compounds are outstanding and were a surprise to the scientific community. Some aspects of their structure and properties will be discussed later in this section.

15.8.2 Magnetic Properties of Superconductors

If a sufficiently strong magnetic field is applied to a superconductor at any temperature below its critical temperature T_c, the superconductor will return to the normal state. The applied magnetic field necessary to restore normal electrical conductivity in the superconductor is called the **critical field H_c.** Figure 15.24a shows schematically the relationship between the critical magnetic field H_c and temperature (K) at zero current. It should be pointed out that a sufficiently high electrical **(critical) current density J_c** will also destroy superconductivity in materials. The curve of H_c versus T (K) can be approximated by

$$H_c = H_0\left[1 - \left(\frac{T}{T_c}\right)^2\right] \tag{15.10}$$

where H_0 is the critical field at $T = 0$ K. Equation 15.10 represents the boundary between the superconducting and normal states of the superconductor. Figure 15.24b shows the critical field versus temperature plots for several superconducting metals.

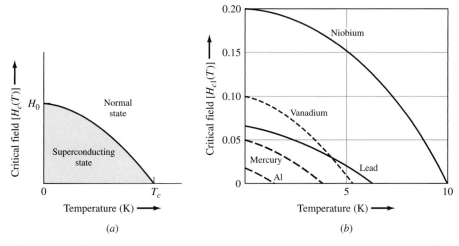

Figure 15.24
Critical field versus temperature. (a) General case. (b) Curves for several superconductors.

Calculate the approximate value of the critical field necessary to cause the superconductivity of pure niobium metal to disappear at 6 K.

■ **Solution**

From Table 15.3 at 0 K, the T_c for Nb is 9.15 K and its $H_0 = 0.1960$ T. From Eq. 15.10,

$$H_c = H_0\left[1 - \left(\frac{T}{T_c}\right)^2\right] = 0.1960\left[1 - \left(\frac{6}{9.15}\right)^2\right] = 0.112 \text{ T} \blacktriangleleft$$

According to their behavior in an applied magnetic field, metallic and intermetallic superconductors are classified into type I and type II superconductors. If a long cylinder of a **type I superconductor** such as Pb or Sn is placed in a magnetic field at room temperature, the magnetic field will penetrate normally throughout the metal (Fig. 15.25a). However, if the temperature of the type I superconductor is lowered below its T_c (7.19 K for Pb) and if the magnetic field is below H_c, the magnetic field will be expelled from the specimen except for a very thin penetration layer of about 10^{-5} cm at the surface (Fig. 15.25b). This property of a magnetic-field exclusion in the superconducting state is called the **Meissner effect.**

Type II superconductors behave differently in a magnetic field at temperatures below T_c. They are highly diamagnetic like type I superconductors up to a critical applied magnetic field, which is called the **lower critical field H_{c1}** (Fig. 15.26), and thus the magnetic flux is excluded from the material. Above H_{c1} the field starts to penetrate the type II superconductor and continues to do so until the **upper critical field H_{c2}** is reached. In between H_{c1} and H_{c2} the superconductor is in the mixed state, and above H_{c2} it returns to the normal state. In the region H_{c1} and H_{c2} the

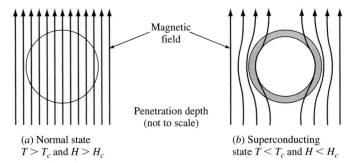

(a) Normal state
$T > T_c$ and $H > H_c$

Magnetic field

Penetration depth
(not to scale)

(b) Superconducting
state $T < T_c$ and $H < H_c$

Figure 15.25
The Meissner effect. When the temperature of a type I superconductor is lowered below T_c and the magnetic field is below H_c, the magnetic field is completely expelled from a sample except for a thin surface layer.

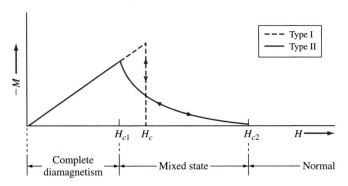

Figure 15.26
Magnetization curves for ideal type I and type II superconductors. Type II superconductors are penetrated by the magnetic field between H_{c1} and H_{c2}.

superconductor can conduct electrical current within bulk material, and thus this magnetic-field region can be used for high-current, high-field superconductors such as NiTi and Ni_3Sb, which are type II superconductors.

15.8.3 Current Flow and Magnetic Fields in Superconductors

Type I superconductors are poor carriers of electrical current since current can only flow in the outer surface layer of a conducting specimen (Fig. 15.27a). The reason for this behavior is that the magnetic field can only penetrate the surface layer, and current can only flow in this layer. In type II superconductors below H_{c1}, magnetic fields behave in the same way. However, if the magnetic field is between H_{c1} and H_{c2} (mixed state), the current can be carried inside the superconductor by filaments,

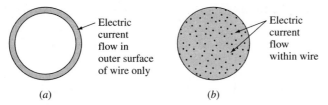

Figure 15.27
Cross section of a superconducting wire carrying an electrical current. (*a*) Type I superconductor or type II under low field ($H < H_{c1}$). (*b*) Type II superconductor under higher fields where current is carried by a filament network ($H_{c1} < H < H_{c2}$).

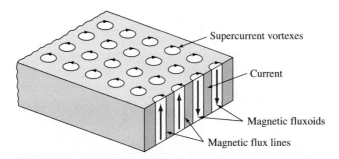

Figure 15.28
Schematic illustration showing magnetic fluxoias in a type II superconductor with the magnetic field between H_{c1} and H_{c2}.

as indicated in Fig. 15.27*b*. In type II superconductors when a magnetic field between H_{c1} and H_{c2} is applied, the field penetrates the bulk of the superconductor in the form of individual quantized flux bundles called **fluxoids** (Fig. 15.28). A cylindrical supercurrent vortex surrounds each fluxoid. With increasing magnetic-field strength, more and more fluxoids enter the superconductor and form a periodic array. At H_{c2} the supercurrent vortex structure collapses and the material returns to the normal conducting state.

15.8.4 High-Current, High-Field Superconductors

Although ideal type II superconductors can be penetrated by an applied magnetic field in the H_{c1} to H_{c2} range, they have a small current-carrying capacity below T_c since the fluxoids are weakly tied to the crystal lattice and are relatively mobile. The mobility of the fluxoids can be greatly impeded by dislocations, grain boundaries, and fine precipitates, and thus J_c can be raised by cold working and heat treatments. Heat treatment of the Nb–45 wt % Ti alloy is used to precipitate a hexagonal α phase in the BCC matrix of the alloy to help pin down the fluxoids.

The alloy Nb–45 wt % Ti and the compound Nb$_3$Sn have become the basic materials for modern high-current, high-field superconductor technology. Commercial Ni–45 wt % Ti has been produced with a $T_c \approx 9$ K and $H_{c2} \approx 6$ T and Nb$_3$Sn with a $T_c \approx 18$ K and $H_{c2} \approx 11$ T. In today's superconductor technology these superconductors are used at liquid helium temperature (4.2 K). The Nb–45 wt % Ti alloy is more ductile and easier to fabricate than the Nb$_3$Sn compound and so is preferred for many applications even though it has lower T_c's and H_{c2}'s. Commercial wires are made of many NbTi filaments, typically about 25 μm in diameter, embedded in a copper matrix (Fig. 15.29). The purpose of the copper matrix is to stabilize the superconductor wire during operation so that hot spots will not develop that could cause the superconducting material to return to the normal state.

Applications for NbTi and Nb$_3$Sn superconductors include nuclear magnetic imaging systems for medical diagnosis and magnetic levitation of vehicles such as high-speed trains (Fig. 15.1c and d). High-field superconducting magnets are used in particle accelerators in the high-energy physics field.

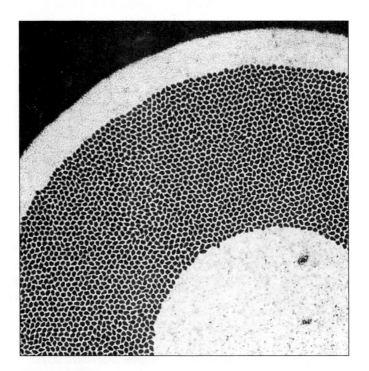

Figure 15.29
Cross section of Nb–46.5 wt % Ti–Cu composite wire made for the superconductor supercollider. The wire has a diameter of 0.0808 cm (0.0318 in), a Cu:NbTi volumetric ratio of 1.5, 7250 filaments of 6-μm diameters, and a $J_c = 2990$ A/mm^2 at 5 T and a $J_c = 1256$ A/mm^2 at 8 T (magnification 200×).
(*Courtesy of Advanced Superconductors Inc.*)

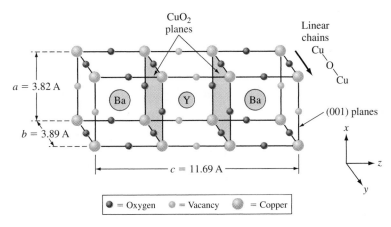

Figure 15.30
Idealized $YBa_2Cu_3O_7$ orthorhombic crystal structure. Note the location of the CuO_2 planes.

15.8.5 High Critical Temperature (T_c) Superconducting Oxides

In 1987 superconductors with critical temperatures of about 90 K were discovered, surprising the scientific community since up to that time the highest T_c for a superconductor was about 23 K. The most intensely studied high T_c material has been the $YBa_2Cu_3O_y$ compound, and so our attention will be focused on some aspects of its structure and properties. From a crystal structure standpoint, this compound can be considered to have a defective perovskite structure with three perovskite cubic unit cells stacked on top of each other (Fig. 15.30). (The perovskite structure for $CaTiO_3$ is shown in Fig. 11.16.) For an ideal stack of three perovskite cubic unit cells, the $YBa_2Cu_3O_y$ compound should have the composition $YBa_2Cu_3O_9$, in which y would equal 9. However, analyses show that y ranges from 6.65 to 6.90 for this material to be a superconductor. At $y = 6.90$, its T_c is highest ($\sim$90 K), and at $y = 6.65$, superconductivity disappears. Thus, oxygen vacancies play a role in the superconductivity behavior of $YBa_2Cu_3O_y$.

The $YBa_2Cu_3O_y$ compound, when slowly cooled from above 750°C in the presence of oxygen, undergoes a tetragonal to orthorhombic crystal structure change (Fig. 15.31a). If the oxygen content is close to $y = 7$, its T_c is about 90 K (Fig. 15.31b) and its unit cell has the constants $a = 3.82$ Å, $b = 3.88$ Å, and $c = 11.6$ Å, (Fig. 15.30). To have high T_c values, oxygen atoms on the (001) planes must be ordered so that oxygen vacancies are in the a direction. Superconductivity is believed to be confined to the CuO_2 planes (Fig. 15.30), with the oxygen vacancies providing an electron coupling between the CuO_2 planes. A transmission electron micrograph (Fig. 15.32) shows the stacking of the Ba and Y atoms of the $YBa_2Cu_3O_y$ structure.

From an engineering viewpoint, the new high T_c superconductors hold much promise for technical advances. With T_c's at 90 K, liquid nitrogen can be used as a

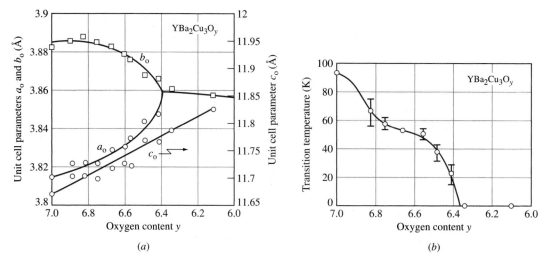

Figure 15.31

(*a*) Oxygen content versus unit cell constants for YBa$_2$Cu$_3$O$_y$. (*b*) Oxygen content versus T_c for YBa$_2$Cu$_3$O$_y$.

(*After J. M. Tarascon and B. G. Bagley, MRS Bull., January 1989, p. 55.*)

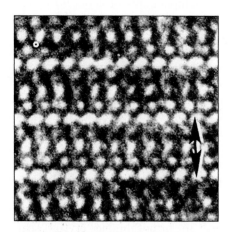

Figure 15.32

High-resolution TEM micrograph in the [100] direction down copper-oxygen chains and rows of Ba and Y atoms in the unit cell of YBa$_2$Cu$_3$O$_y$ as indicated by the arrow.

(*After J. Narayan, JOM, January 1989, p. 18.*)

refrigerant to replace the much more expensive liquid helium. Unfortunately, the high-temperature superconductors are essentially ceramics, which are brittle and in their bulk form have low current-density capability. The first applications for these materials will probably be in thin-film technology for electronic applications such as high-speed computers.

15.9 DEFINITIONS

Sec. 15.3
Index of refraction: the ratio of the velocity of light in vacuum to that through another medium of interest.

Sec. 15.4
Absorptivity: the fraction of the incident light that is absorbed by a material.

Sec. 15.5
Luminescence: absorption of light or other energy by a material and the subsequent emission of light of longer wavelength.
Fluorescence: absorption of light or other energy by a material and the subsequent emission of light within 10^{-8} s of excitation.
Phosphorescence: absorption of light by a phosphor and its subsequent emission at times longer than 10^{-8} s.

Sec. 15.6
Laser: acronym for *l*ight *a*mplification by *s*timulated *e*mission of *r*adiation.
Laser beam: a beam of monochromatic, coherent optical radiation generated by the stimulated emission of photons.
Population inversion: condition in which more atoms exist in a higher-energy state than in a lower one. This condition is necessary for laser action.

Sec. 15.7
Optical-fiber communication: a method of transmitting information by the use of light.
Light attenuation: decrease in intensity of the light.
Optical waveguide: a thin-clad fiber along which light can propagate by total internal reflection and refraction.

Sec. 15.8
Superconducting state: a solid in the superconducting state that shows no electrical resistance.
Critical temperature T_c: the temperature below which a solid shows no electrical resistance.
Critical current density J_c: the current density above which superconductivity disappears.
Critical field H_c: the magnetic field above which superconductivity disappears.
Meissner effect: the expulsion of the magnetic field by a superconductor.
Type I superconductor: one that exhibits complete magnetic-flux repulsion between the normal and superconducting states.
Type II superconductor: one in which the magnetic flux gradually penetrates between the normal and superconducting states.
Lower critical field H_{c1}: the field at which magnetic flux first penetrates a type II superconductor.

Upper critical field H_{c2}: the field at which superconductivity disappears for a type II superconductor.

Fluxoid: a microscopic region surrounded by circulating supercurrents in a type II superconductor at fields between H_{c2} and H_{c1}.

15.10 PROBLEMS

Answers to problems marked with an asterisk are given at the end of the book.

15.1   Write the equation relating the energy of radiation to its wavelength and frequency, and give the SI units for each quantity.

15.2   What are the approximate wavelength and frequency ranges for (a) visible light, (b) ultraviolet light, and (c) infrared radiation?

15.3   A photon in a ZnO semiconductor drops from an impurity level at 2.30 eV to its valence band. What is the wavelength of the radiation given off by the transition? If the radiation is visible, what is its color?

15.4   A semiconductor emits green visible radiation at a wavelength of 0.520 μm. What is the energy level from which photons drop to the valence band in order to give off this type of radiation?

15.5   If ordinary light is transmitted from air into a 1 cm thick sheet of polymethacrylate, is the light sped up or slowed down upon entering the plastic? Explain.

15.6   Explain why cut diamonds sparkle. Why is PbO sometimes added to make decorative glasses?

15.7   What is Snell's law of light refraction? Use a diagram to explain.

15.8   What is the critical angle for light to be totally reflected when leaving a flat plate of polystyrene and entering the air?

15.9   Explain why metals absorb and/or reflect incident radiation up to the middle of the ultraviolet range.

15.10   Explain why gold is yellow in color and silver is "silvery."

*15.11   Calculate the reflectivity of ordinary light from a smooth, flat upper surface of (a) borosilicate glass ($n = 1.47$) and (b) polyethylene ($n = 1.53$).

15.12   Ordinary incident light strikes the flat surface of a transparent material with a linear absorption coefficient of 0.04 cm^{-1}. If the plate of the material is 0.80 cm thick, calculate the fraction of light absorbed by the plate.

15.13   Ordinary light strikes a flat surface of a plate of a transparent material. If the plate is 0.75 thick and absorbs 5.0 percent of the entering light, what is its linear absorption coefficient?

*15.14   Calculate the transmittance for a flat glass plate 6.0 mm thick with an index of refraction of 1.51 and a linear absorption coefficient of 0.03 cm^{-1}.

15.15   Why does a sheet of 2.0 mm thick polyethylene have lower clarity than a sheet of the same thickness of polycarbonate plastic?

*15.16   Calculate the minimum wavelength of the radiation that can be absorbed by the following materials: (a) GaP, (b) GaSb, and (c) InP.

15.17   Will visible light of a wavelength of 500 nm be absorbed or transmitted by the following materials: (a) CdSb, (b) ZnSe, and (c) diamond ($E_g = 5.40$ eV)?

15.18   Explain the process of luminescence.

15.19 Distinguish between fluorescence and phosphorescence.

15.20 Explain the luminescence effect operating in a fluorescent lamp.

15.21 Explain how the color picture is produced on a color television screen.

15.22 The intensity of an Al_2O_3 phosphor activated with chromium decreases to 15 percent of its original intensity in 5.6×10^{-3} s. Determine (a) its relaxation time and (b) its retained percent intensity after 5.0×10^{-2} s.

*15.23 A Zn_2SiO_4 phosphor activated with manganese has a relaxation time of 0.015 s. Calculate the time required for the intensity of this material to decrease to 8 percent of its original value.

15.24 Distinguish between incoherent and coherent radiation.

15.25 What do the letters in the acronymn *laser* stand for?

15.26 Explain the operation of the ruby laser.

15.27 What does the term *population inversion* refer to in laser terminology?

15.28 Describe the operation and application of the following types of lasers: (a) neodymium-YAG, (b) carbon dioxide, (c) double heterojunction GaAs.

15.29 What are optical fibers?

15.30 What are the basic elements of an optical-fiber communications system?

15.31 What types of impurities are particularly detrimental to light loss in optical fibers?

15.32 If the original light intensity is reduced 6.5 percent after being transmitted 300 m through an optical fiber, what is the attenuation of the light in decibels per kilometer (dB/km) for this type of optical fiber?

15.33 Light is attenuated in an optical fiber operating at 1.55 μm wavelength at -0.25 dB/km. If 4.2 percent of the light is to be retained at a repeater station, what must the distance between the repeaters be?

*15.34 The attenuation of a 1.3 μm optical-fiber undersea transatlantic cable is -0.31 dB/km, and the distance between repeaters in the system is 40.2 km (25 mi). What is the percent of the light retained at a repeater if we assume 100 percent at the start of a repeater?

15.35 Explain how optical fibers act as waveguides.

15.36 Distinguish between single-mode and multimode types of optical fibers. Which type is used for modern long-distance communication systems and why?

15.37 How are optical fibers for communication systems fabricated? How do (a) GeO_2 and (b) F affect the refractive index of the silica glass?

15.38 What types of lasers are used in modern long-distance optical-fiber systems and why?

15.39 A single-mode optical fiber for a communications system has a core of SiO_2–GeO_2 glass with a refractive index of 1.4597 and a cladding of pure SiO_2 glass with a refractive index of 1.4580. What is the critical angle for light leaving the core to be totally reflected within the core?

15.40 What is the superconducting state for a material?

15.41 What is the significance of T_c, H_c, and J_c for a superconductor?

15.42 Describe the difference between type I and type II superconductors.

15.43 What is the Meissner effect?

15.44 Why are type I superconductors poor current-carrying conductors?

15.45 What are fluxoids? What role do they play in the superconductivity of type II superconductors in the mixed state?

15.46 How can fluxoids be pinned in type II superconductors? What is the consequence of pinning the fluxoids in a type II superconductor?

*15.47 Calculate the critical magnetic field H_{c1} in teslas for niobium at 8 K. Use Eq. 15.10 and data from Table 15.3.

15.48 If vanadium has an H_c value of 0.06 T and is superconductive, what must its temperature be?

15.49 Describe the crystal structure of $YBa_2Cu_3O_7$. Use a drawing.

15.50 Why must the $YBa_2Cu_3O_y$ compound be cooled slowly from about 750°C in order for this compound to be highly superconductive?

15.51 What are some advantages and disadvantages of the new high-temperature oxide superconductors?

15.11 MATERIALS SELECTION AND DESIGN PROBLEMS

1. Design a semiconductor that would produce photons with wavelengths corresponding to that of green light.

2. ZnS has an E_g of 2.76 eV. (*a*) What would be the color of radiation corresponding to its photon? (*b*) How would you modify the semiconductor to produce a different color of radiation?

3. Select an optical plastic that has a critical angle of 45 degrees for light to be totally reflected when leaving a flat plate and entering air (use Table 15.1).

4. (*a*) Select a material from Table 15.1 that has a reflectivity of ordinary incident light of about 5 percent. (*b*) Select the material with the highest level of reflectivity. (*c*) Select the material with the lowest level of reflectivity. (Assume all surfaces are polished.)

5. Design the thickness of a polished silicate glass that would result in (*a*) no more than 2 percent light lost due to absorption and (*b*) no more than 4 percent light lost due to absorption.

6. Explain how the relaxation time of a color TV phosphor can affect the quality of the picture.

7. Select a material for optical transmission over a length of 30 km such that at least 15 percent of light intensity at the source remains.

16 Magnetic Properties

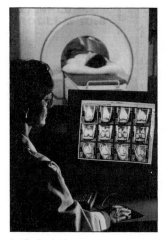
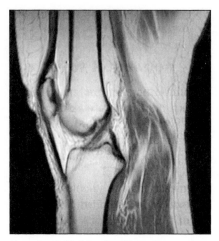
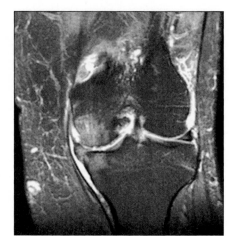

(*a* © Corbis)   (*b and c* © Neil Borden/Photo Researchers, Inc.)

Magnetic Resonance Imaging (MRI) technique is used to extract high quality images from inside the human body. It gives physicians and researchers the ability to safely investigate diseases related to the heart, brain, spine, and other organs in the human body. The images produced by an MRI are primarily due to existence of fat and water molecules which are mostly made up of hydrogen. In short, hydrogen produces a small magnetic signal which is detected by the instrument and used for mapping the tissue.

The MRI hardware is presented in the figure. It consists of a large magnet that produces the magnetic field; gradient coil that produces a gradient in the field; RF coil that detects the signal from the molecules inside the human body. The magnet is the most expensive component of the system and is usually of the superconducting type (wire of several miles length). Overall the MRI is a complex system that requires expertise from mathematics, physics, chemistry, and materials scientists. It also requires expertise from bioengineers, imaging scientists, and architects to design and implement an efficient machine with safe application.

An example of the use of the MRI technology is in orthopedics, where damage to soft tissue can be imaged accurately. The images above show the MRI images of a healthy (left) and a torn anterior cruciate ligament (right). Depending on the extent of damage, the surgeon decides to pursue arthroscopic surgery to replace the injured ACL. ∎

By the end of this chapter, students will be able to . . .

1. Briefly describe the two sources for magnetic moments in materials.

2. Describe magnetic hysteresis for a material.

3. Cite distinctive magnetic characteristics for hard magnetic and soft magnetic material.

4. Describe how increasing the temperature affects the alignment of magnetic dipoles in ferromagnetic material.

5. Describe the nature of paramagnetism.

6. Explain what alnico means.

7. Cite a few industrial applications of soft ferrites.

8. Briefly describe the source of antiferromagnetism.

9. Draw a hysteresis loop of a ferromagnetic material.

10. Describe relative magnetic permeability and magnetic permeability.

16.1 INTRODUCTION

Magnetic materials are necessary for many engineering designs, particularly in the area of electrical engineering. In general there are two main types: *soft* and *hard magnetic materials*. Soft magnetic materials are used for applications in which the material must be easily magnetized and demagnetized, such as cores for distribution power transformers (Fig. 16.1a), small electronic transformers, and stator and rotor materials for motors and generators. On the other hand, hard magnetic materials are used for applications requiring permanent magnets that do not demagnetize easily, such as the permanent magnets in loudspeakers, telephone receivers, synchronous and brushless motors, and automotive starting motors.

16.2 MAGNETIC FIELDS AND QUANTITIES

16.2.1 Magnetic Fields

Let us begin our study of magnetic materials by first reviewing some of the fundamental properties of magnetism and magnetic fields. The metals *iron, cobalt,* and *nickel* are the only three elemental metals that, when magnetized at room temperature, can produce a strong magnetic field around themselves. They are said to

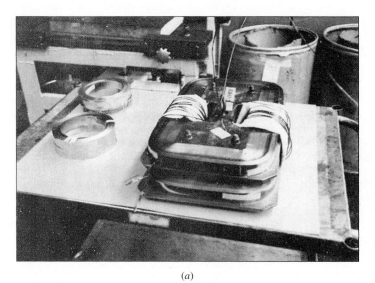
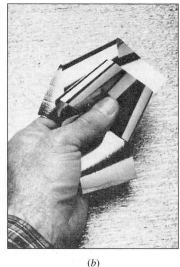

(a)   (b)

Figure 16.1

(a) A new magnetic material for engineering designs: metallic glass material is used for the magnetic cores of distribution electric power transformers. The use of highly magnetically soft amorphous metallic glass alloys for transformer cores reduces core energy losses by about 70 percent compared with those made with conventional iron-silicon alloys.
(*Courtesy of General Electric Co.*)
(b) Strip of metallic glass ribbon.
(*Courtesy of Metglas, Inc.*)

be **ferromagnetic.** The presence of a **magnetic field** surrounding a magnetized iron bar can be revealed by scattering small iron particles on a sheet of paper placed just above the bar (Fig. 16.2). As shown in Fig. 16.2, the bar magnet has two magnetic poles, and magnetic field lines appear to leave one pole and enter the other.

In general, magnetism is dipolar in nature, and no magnetic monopole has ever been discovered. There are always two magnetic poles or centers of a magnetic field separated by a definite distance, and this dipole behavior extends to the small magnetic dipoles found in some atoms.

Magnetic fields are also produced by current-carrying conductors. Figure 16.3 illustrates the formation of a magnetic field around a long coil of copper wire, called a *solenoid,* whose length is long with respect to its radius. For a solenoid of n turns and length l, the magnetic field strength H is

$$H = \frac{0.4\,\pi\,ni}{l} \tag{16.1}$$

where i is the current. The magnetic field strength H has SI units of amperes per meter (A/m) and cgs units of oersteds (Oe). The conversion equality between SI and cgs units for H is 1 A/m = $4\pi \times 10^{-3}$ Oe.

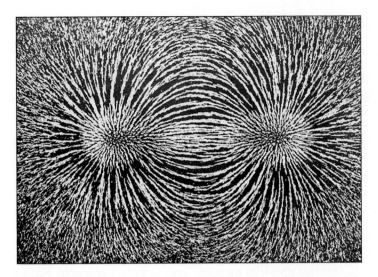

Figure 16.2
The magnetic field surrounding a bar magnet is revealed by the arrangement of iron filings lying on a sheet of paper above the magnet. Note that the bar magnet is dipolar and that magnetic lines of force appear to leave one end of the magnet and return to the other.
(Courtesy of the Physical Science Study Committee, as appearing in D. Halliday and R. Resnick, "Fundamentals of Physics," Wiley, 1974, p. 612.)

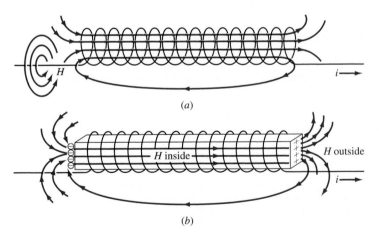

Figure 16.3
(*a*) Schematic illustration of a magnetic field created around a coil of copper wire, called a solenoid, by the passage of current through the wire. (*b*) Schematic illustration of the increase in magnetic field around the solenoid when an iron bar is placed inside the solenoid and current is passed through the wire.
(After C. S. Barrett, W. D. Nix, and A. S. Tetelman, "Principles of Engineering Materials," Prentice-Hall, 1973, p. 459.)

16.2.2 Magnetic Induction

Now let us place a demagnetized iron bar inside the solenoid and apply a magnetizing current to the solenoid, as shown in Fig. 16.3b. The magnetic field outside the solenoid is now stronger with the magnetized bar inside the solenoid. The enhanced magnetic field outside the solenoid is due to the sum of the solenoid field itself and the external magnetic field of the magnetized bar. The new additive magnetic field is called the **magnetic induction,** or *flux density,* or simply *induction,* and is given the symbol B.

The magnetic induction B is the sum of the applied field H and the external field that arises from the magnetization of the bar inside the solenoid. The induced magnetic moment per unit volume due to the bar is called the *intensity of magnetization,* or simply **magnetization,** and is given the symbol M. In the SI system of units,

$$B = \mu_0 H + \mu_0 M = \mu_0(H + M) \tag{16.2}$$

where $\mu_0 = $ *permeability of free space* $= 4\pi \times 10^{-7}$ tesla-meters per ampere $(\text{T} \cdot \text{m/A})$.[1] μ_0 has no physical meaning and is only needed in Eq. 16.2 because SI units were chosen. The SI units for B are webers[2] per square meter (Wb/m^2), or teslas (T), and the SI units for H and M are amperes per meter (A/m). The cgs unit for B is the gauss (G) and for H, the oersted (Oe). Table 16.1 summarizes these magnetic units.

An important point to note is that for ferromagnetic materials, in many cases the magnetization $\mu_0 M$ is often much greater than the applied field $\mu_0 H$, and so we can often use the relation $B \approx \mu_0 M$. Thus, for ferromagnetic materials, sometimes the quantities B (magnetic induction) and M (magnetization) are used interchangeably.

16.2.3 Magnetic Permeability

As previously pointed out, when a ferromagnetic material is placed in an applied magnetic field, the intensity of the magnetic field increases. This increase in

---

[1]Nikola Tesla (1856–1943). American inventor of Yugoslavian birth who in part developed the polyphase induction motor and invented the Tesla coil (an air transformer). 1 T = 1 Wb/m^2 = 1 V · s/m^2.

[2]1 Wb = 1 V · s.

Table 16.1 Summary of the units for the magnetic quantities

Magnetic quantity	SI units	cgs units
B (magnetic induction)	weber/meter^2 (Wb/m^2) or tesla (T)	gauss (G)
H (applied field)	ampere/meter (A/m)	oersted (Oe)
M (magnetization)	ampere/meter (A/m)	
Numerical conversion factors:		
$\quad$ 1 A/m = $4\pi \times 10^{-3}$ Oe		
$\quad$ 1 Wb/m^2 = 1.0×10^4 G		
Permeability constant		
$\quad \mu_0 = 4\pi \times 10^{-7}$ T · m/A		

magnetization is measured by a quantity called **magnetic permeability** μ, which is defined as the ratio of the magnetic induction B to the applied field H, or

$$\mu = \frac{B}{H} \qquad (16.3)$$

If there is only a vacuum in the applied magnetic field, then

$$\mu_0 = \frac{B}{H} \qquad (16.4)$$

where $\mu_0 = 4\pi \times 10^{-7}\,\text{T} \cdot \text{m/A}$ = permeability of free space, as previously stated.

An alternative method used for defining magnetic permeability is the quantity **relative permeability** μ_r, which is the ratio of μ/μ_0. Thus

$$\mu_r = \frac{\mu}{\mu_0} \qquad (16.5)$$

and

$$B = \mu_0 \mu_r H \qquad (16.6)$$

The relative permeability μ_r is a dimensionless quantity.

The relative permeability is a measure of the intensity of the induced magnetic field. In some ways the magnetic permeability of magnetic materials is analogous to the dielectric constant of dielectric materials. However, the magnetic permeability of a ferromagnetic material is not a constant but changes as the material is magnetized, as indicated in Fig. 16.4. The magnetic permeability of a magnetic material is usually

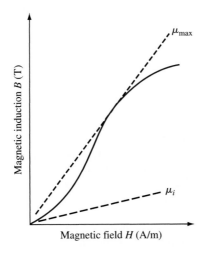

Figure 16.4
B-H initial magnetization curve for a ferromagnetic material. The slope μ_i is the initial magnetic permeability, and the slope μ_{max} is the maximum magnetic permeability.

measured by either its initial permeability μ_i or its maximum permeability $\mu_{\max}$. Figure 16.4 indicates how μ_i and $\mu_{\max}$ are measured from slopes of the *B-H* initial magnetizing curve for a magnetic material. Magnetic materials that are easily magnetized have high magnetic permeabilities.

16.2.4 Magnetic Susceptibility

Since the magnetization of a magnetic material is proportional to the applied field, a proportionality factor called the **magnetic susceptibility** χ_m is defined as

$$\chi_m = \frac{M}{H} \tag{16.7}$$

which is a dimensionless quantity. Weak magnetic responses of materials are often measured in terms of magnetic susceptibility.

16.3 TYPES OF MAGNETISM

Magnetic fields and forces originate from the movement of the basic electric charge, the electron. When electrons move in a conducting wire, a magnetic field is produced around the wire, as shown for the solenoid of Fig. 16.3. Magnetism in materials is also due to the motion of electrons, but in this case the magnetic fields and forces are caused by the intrinsic spin of electrons and their orbital motion about their nuclei (Fig. 16.5).

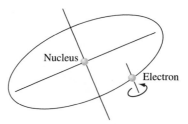

Figure 16.5
A schematic drawing of the Bohr atom indicating an electron spinning on its own axis and revolving about its nucleus. The spin of the electron on its axis and its orbital motion around its nucleus are the origins of magnetism in materials.

Table 16.2 Magnetic susceptibilities of some diamagnetic and paramagnetic elements

Diamagnetic element	Magnetic susceptibility $\chi_m \times 10^{-6}$	Paramagnetic element	Magnetic susceptibility $\chi_m \times 10^{-6}$
Cadmium	−0.18	Aluminum	+0.65
Copper	−0.086	Calcium	+1.10
Silver	−0.20	Oxygen	+106.2
Tin	−0.25	Platinum	+1.10
Zinc	−0.157	Titanium	+1.25

16.3.1 Diamagnetism

An external magnetic field acting on the atoms of a material slightly unbalances their orbiting electrons and creates small magnetic dipoles within the atoms that oppose the applied field. This action produces a negative magnetic effect known as **diamagnetism.** The diamagnetic effect produces a very small negative magnetic susceptibility of the order of $\chi_m \approx -10^{-6}$ (Table 16.2). Diamagnetism occurs in all materials, but in many its negative magnetic effect is canceled by positive magnetic effects. Diamagnetic behavior has no significant engineering importance.

16.3.2 Paramagnetism

Materials that exhibit a small positive magnetic susceptibility in the presence of a magnetic field are called *paramagnetic,* and the magnetic effect is termed **paramagnetism.** The paramagnetic effect in materials disappears when the applied magnetic field is removed. Paramagnetism produces magnetic susceptibilities in materials ranging from about 10^{-6} to 10^{-2} and is produced in many materials. Table 16.2 lists the magnetic susceptibilities of paramagnetic materials at 20°C. Paramagnetism is produced by the alignment of individual magnetic dipole moments of atoms or molecules in an applied magnetic field. Since thermal agitation randomizes the directions of the magnetic dipoles, an increase in temperature decreases the paramagnetic effect.

The atoms of some transition and rare earth elements possess incompletely filled inner shells with unpaired electrons. These unpaired inner electrons in atoms, since they are not counterbalanced by other bonding electrons in solids, cause strong paramagnetic effects and in some cases produce very much stronger ferromagnetic and ferrimagnetic effects, which will be discussed later.

16.3.3 Ferromagnetism

Diamagnetism and paramagnetism are induced by an applied magnetic field, and the magnetization remains only as long as the field is maintained. A third type of magnetism, called **ferromagnetism,** is of great engineering importance. Large magnetic fields that can be retained or eliminated as desired can be produced in ferromagnetic materials. The most important ferromagnetic elements from an industrial standpoint are iron (Fe), cobalt (Co), and nickel (Ni). Gadolinium (Gd), a rare earth element, is also ferromagnetic below 16°C but has little industrial application.

The ferromagnetic properties of the transition elements Fe, Co, and Ni are due to the way the spins of the inner unpaired electrons are aligned in their crystal lattices. The inner shells of individual atoms are filled with pairs of electrons with opposed spins, and so they do not contribute to the resultant magnetic dipole moments. In solids the outer valence electrons of atoms are combined with each other to form chemical bonds, and so there is no significant magnetic moment due to these electrons. In Fe, Co, and Ni the unpaired inner $3d$ electrons are responsible for the ferromagnetism that these elements exhibit. The iron atom has four unpaired $3d$ electrons, the cobalt atom three, and the nickel atom two (Fig. 16.6).

In a solid sample of Fe, Co, or Ni at room temperature, the spins of the $3d$ electrons of adjacent atoms align in a parallel direction by a phenomenon called *spontaneous magnetization.* This parallel alignment of atomic magnetic dipoles occurs only in microscopic regions called *magnetic domains.* If the domains are randomly oriented, then there will be no net magnetization in a bulk sample. The parallel alignment of the magnetic dipoles of atoms of Fe, Co, and Ni is due to the creation of a positive exchange energy between them. For this parallel alignment to occur the ratio of the atomic spacing to the diameter of the $3d$ orbit must be in the range from about 1.4 to 2.7 (Fig. 16.7). Thus Fe, Co, and Ni are ferromagnetic, but manganese (Mn) and chromium (Cr) are not.

Unpaired $3d$ electrons	Atom	Number of electrons	Electronic configuration $3d$ orbitals					$4s$ electrons
3	V	23	↑	↑	↑			2
5	Cr	24	↑	↑	↑	↑	↑	1
5	Mn	25	↑	↑	↑	↑	↑	2
4	Fe	26	↑↓	↑	↑	↑	↑	2
3	Co	27	↑↓	↑↓	↑	↑	↑	2
2	Ni	28	↑↓	↑↓	↑↓	↑	↑	2
0	Cu	29	↑↓	↑↓	↑↓	↑↓	↑↓	1

Figure 16.6
Magnetic moments of neutral atoms of $3d$ transition elements.

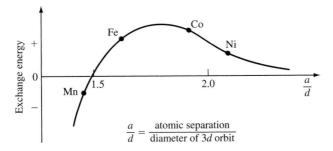

Figure 16.7
Magnetic exchange interaction energy as a function of
the ratio of atomic spacing to the diameter of the 3d
orbit for some 3d transition elements. Those elements
that have positive exchange energies are ferromagnetic;
those with negative exchange energies are
antiferromagnetic.

16.3.4 Magnetic Moment of a Single Unpaired Atomic Electron

Each electron spinning on its own axis (Fig. 16.5) behaves as a magnetic dipole and
has a dipole moment called the **Bohr magneton μ_B.** This dipole moment has a value of

$$\mu_B = \frac{eh}{4\pi m} \tag{16.8}$$

where e = electronic charge, h = Planck's constant, and m = electron mass. In
SI units $\mu_B = 9.27 \times 10^{-24}\,\text{A} \cdot \text{m}^2$. In most cases electrons in atoms are paired, and
so the positive and negative magnetic moments cancel. However, unpaired electrons
in inner electron shells can have small positive dipole moments, as is the case for
the 3d electrons of Fe, Co, and Ni.

Using the relationship $\mu_B = eh/4\pi m$, show that the numerical value for a Bohr magneton is $9.27 \times 10^{-24}\,\text{A} \cdot \text{m}^2$.

**EXAMPLE
PROBLEM 16.1**

■ **Solution**

$$\mu_B = \frac{eh}{4\pi m} = \frac{(1.60 \times 10^{-19}\,\text{C})(6.63 \times 10^{-34}\,\text{J} \cdot \text{s})}{4\pi(9.11 \times 10^{-31}\,\text{kg})}$$

$$= 9.27 \times 10^{-24}\,\text{C} \cdot \text{J} \cdot \text{s/kg}$$

$$= 9.27 \times 10^{-24}\,\text{A} \cdot \text{m}^2 \blacktriangleleft$$

The units are consistent, as follows:

$$\frac{\text{C} \cdot \text{J} \cdot \text{s}}{\text{kg}} = \frac{(\text{A} \cdot \text{s})(\text{N} \cdot \text{m})(\text{s})}{\text{kg}} = \frac{\text{A} \cdot \cancel{\text{s}}}{\cancel{\text{kg}}}\left(\frac{\cancel{\text{kg}} \cdot \text{m} \cdot \text{m}}{\cancel{\text{s}^2}}\right)(\cancel{\text{s}}) = \text{A} \cdot \text{m}^2$$

Calculate a theoretical value for the saturation magnetization M_s in amperes per meter and saturation induction B_s in teslas for pure iron, assuming all magnetic moments due to the four unpaired $3d$ Fe electrons are aligned in a magnetic field. Use the equation $B_s \approx \mu_0 M_s$ and assume that $\mu_0 H$ can be neglected. Pure iron has a BCC unit cell with a lattice constant $a = 0.287$ nm.

■ **Solution**

The magnetic moment of an iron atom is assumed to be 4 Bohr magnetons. Thus,

$$M_s = \left[\frac{\dfrac{2 \text{ atoms}}{\text{unit cell}}}{\dfrac{(2.87 \times 10^{-10} \text{ m})^3}{\text{unit cell}}} \right] \left(\frac{4 \text{ Bohr magnetons}}{\text{atom}} \right) \left(\frac{9.27 \times 10^{-24} \text{ A} \cdot \text{m}^2}{\text{Bohr magneton}} \right)$$

$$= \left(\frac{0.085 \times 10^{30}}{\text{m}^3} \right) (4)(9.27 \times 10^{-24} \text{ A} \cdot \text{m}^2) = 3.15 \times 10^6 \text{ A/m} \blacktriangleleft$$

$$B_s \approx \mu_0 M_s \approx \left(\frac{4\pi \times 10^{-7} \text{ T} \cdot \text{m}}{\text{A}} \right) \left(\frac{3.15 \times 10^6 \text{ A}}{\text{m}} \right) \approx 3.96 \text{ T} \blacktriangleleft$$

Iron has a saturation magnetization of 1.71×10^6 A/m. What is the average number of Bohr magnetons per atom that contribute to this magnetization? Iron has the BCC crystal structure with $a = 0.287$ nm.

■ **Solution**

The saturation magnetization M_s in amperes per meter can be calculated from Eq. 16.9 as

$$M_s = \left(\frac{\text{atoms}}{\text{m}^3} \right) \left(\frac{N\mu_B \text{ of Bohr magnetons}}{\text{atom}} \right) \left(\frac{9.27 \times 10^{-24} \text{ A} \cdot \text{m}^2}{\text{Bohr magneton}} \right)$$

$$= \text{ans. in A/m} \tag{16.9}$$

$$\text{Atomic density (no. of atoms/m}^3) = \frac{2 \text{ atoms/BCC unit cell}}{(2.87 \times 10^{-10} \text{ m})^3/\text{unit cell}}$$

$$= 8.46 \times 10^{28} \text{ atoms/m}^3$$

We rearrange Eq. 16.9 and solve for $N\mu_B$. After substituting values for M_s, atomic density, and μ_B, we can calculate the value of $N\mu_B$.

$$N\mu_B = \frac{M_s}{(\text{atoms/m}^3)(\mu_B)}$$

$$= \frac{1.71 \times 10^6 \text{ A/m}}{(8.46 \times 10^{28} \text{ atoms/m}^3)(9.27 \times 10^{-24} \text{ A} \cdot \text{m}^2)} = 2.18 \mu_B/\text{atom} \blacktriangleleft$$

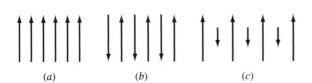

Figure 16.8
Alignment of magnetic dipoles for different types of magnetism: (*a*) ferromagnetism, (*b*) antiferromagnetism, and (*c*) ferrimagnetism.

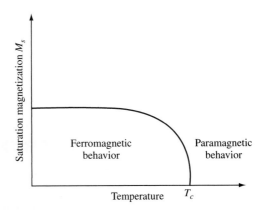

Figure 16.9
Effect of temperature on the saturation magnetization M_s of a ferromagnetic material below its Curie temperature T_c. Increasing the temperature randomizes the magnetic moments.

16.3.5 Antiferromagnetism

Another type of magnetism that occurs in some materials is **antiferromagnetism.** In the presence of a magnetic field, magnetic dipoles of atoms of antiferromagnetic materials align themselves in opposite directions (Fig. 16.8*b*). The elements manganese and chromium in the solid state at room temperature exhibit antiferromagnetism and have a negative exchange energy because the ratio of their atomic spacing to diameter of the 3*d* orbit is less than about 1.4 (Fig. 16.7).

16.3.6 Ferrimagnetism

In some ceramic materials, different ions have different magnitudes for their magnetic moments, and when these magnetic moments are aligned in an antiparallel manner, there is a net magnetic moment in one direction **(ferrimagnetism)** (Fig. 16.8*c*). As a group, ferrimagnetic materials are called *ferrites*. There are many types of ferrites. One group is based on magnetite, Fe_3O_4, which is the magnetic lodestone of the ancients. Ferrites have low conductivities that make them useful for many electronics applications.

16.4 EFFECT OF TEMPERATURE ON FERROMAGNETISM

At any finite temperature above 0 K, thermal energy causes the magnetic dipoles of a ferromagnetic material to deviate from perfect parallel alignment. Thus, the exchange energy that causes the parallel alignment of the magnetic dipoles in ferromagnetic materials is counterbalanced by the randomizing effects of thermal energy (Fig. 16.9). Finally, as the temperature increases, some temperature is reached where the ferromagnetism in a ferromagnetic material completely disappears, and the material becomes paramagnetic. This temperature is called the **Curie temperature.** When a sample of a ferromagnetic material is cooled from a temperature above its Curie temperature, ferromagnetic domains reform and the material becomes ferromagnetic again. The Curie temperatures of Fe, Co, and Ni are 770, 1123, and 358°C, respectively.

16.5 FERROMAGNETIC DOMAINS

Below the Curie temperature, the magnetic dipole moments of atoms of ferromagnetic materials tend to align themselves in a parallel direction in small-volume regions called **magnetic domains.** When a ferromagnetic material such as iron or nickel is demagnetized by slowly cooling from above its Curie temperature, the magnetic domains are aligned at random so that there is no net magnetic moment for a bulk sample (Fig. 16.10).

When an external magnetic field is applied to a demagnetized ferromagnetic material, magnetic domains whose moments are initially parallel to the applied magnetic field grow at the expense of the less favorably oriented domains

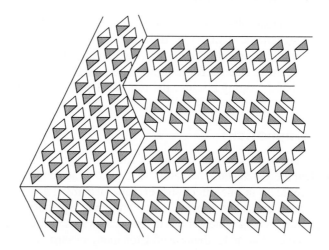

Figure 16.10
Schematic illustration of magnetic domains in a ferromagnetic metal. All the magnetic dipoles in each domain are aligned, but the domains themselves are aligned at random so that there is no net magnetization.
(*After R. M. Rose, L. A. Shepard, and J. Wulff, "Structure and Properties of Materials," vol. IV: "Electronic Properties," Wiley, 1966, p. 193.*)

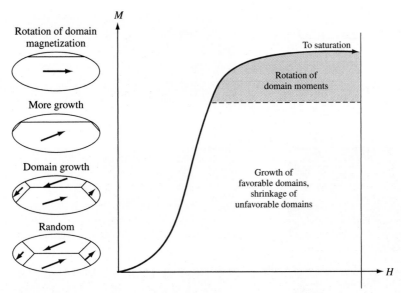

Figure 16.11

Magnetic domain growth and rotation as a demagnetized ferromagnetic material is magnetized to saturation by an applied magnetic field.

(After R. M. Rose, L. A. Shepard, and J. Wulff, "Structure and Properties of Materials," vol. IV: "Electronic Properties," Wiley, 1966, p. 193.)

(Fig. 16.11). The domain growth takes place by domain wall movement, as indicated in Fig. 16.11, and *B* or *M* increases rapidly as the *H* field increases. Domain growth by wall movement takes place first since this process requires less energy than domain rotation. When domain growth finishes, if the applied field is increased substantially, domain rotation occurs. Domain rotation requires considerably more energy than domain growth, and the slope of the *B* or *M* versus *H* curve decreases at the high fields required for domain rotation (Fig. 16.11). When the applied field is removed, the magnetized sample remains magnetized, even though some of the magnetization is lost because of the tendency of the domains to rotate back to their original alignment.

Figure 16.12 shows how the domain walls move under an applied field in iron single-crystal whiskers. The domain walls are revealed by the Bitter technique, in which a colloidal solution of iron oxide is deposited on the polished surface of the iron. The wall movement is followed by observation with an optical microscope. Using this technique, much information has been obtained about domain wall movement under applied magnetic fields.

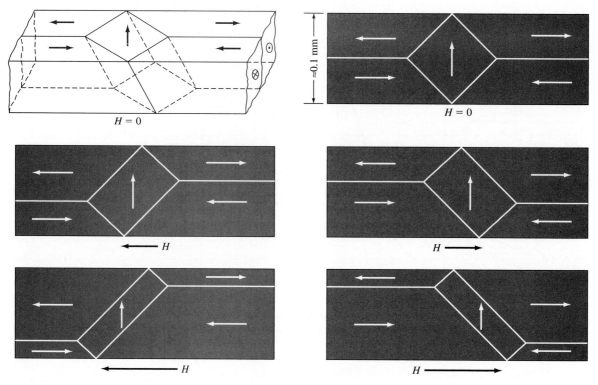

Figure 16.12

Movement of domain boundaries in an iron crystal produced by the application of an applied magnetic field. Note that as the applied field is increased, the domains with their dipoles aligned in the direction of the field enlarge and those with their dipoles opposed get smaller. (The applied fields on the left and right figures increase from top to bottom.)

(Courtesy of R. W. DeBlois, The General Electric Co., and C. D. Graham, the University of Pennsylvania.)

16.6 TYPES OF ENERGIES THAT DETERMINE THE STRUCTURE OF FERROMAGNETIC DOMAINS

The domain structure of a ferromagnetic material is determined by many types of energies, with the most stable structure being attained when the overall potential energy of the material is a minimum. The total magnetic energy of a ferromagnetic material is the sum of the contributions of the following energies: (1) exchange energy, (2) magnetostatic energy, (3) magnetocrystalline anisotropy energy, (4) domain wall energy, and (5) magnetostrictive energy. Let us now briefly discuss each of these energies.

16.6.1 Exchange Energy

The potential energy *within* a domain of a ferromagnetic solid is minimized when all its atomic dipoles are aligned in one direction (**exchange energy**). This alignment

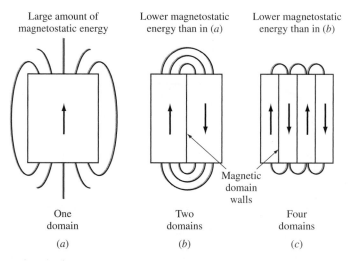

Figure 16.13
Schematic illustration showing how reducing the domain size in a magnetic material decreases the magnetostatic energy by reducing the external magnetic field. (*a*) One domain, (*b*) two domains, and (*c*) four domains.

is associated with a positive exchange energy. However, even though the potential energy within a domain is minimized, its external potential energy is increased by the formation of an external magnetic field (Fig. 16.13*a*).

16.6.2 Magnetostatic Energy

Magnetostatic energy is the potential magnetic energy of a ferromagnetic material produced by its external field (Fig. 16.13*a*). This potential energy can be minimized in a ferromagnetic material by domain formation, as illustrated in Fig. 16.13. For a unit volume of a ferromagnetic material, a single-domain structure has the highest potential energy, as indicated by Fig. 16.13*a*. By dividing the single domain of Fig. 16.13*a* into two domains (Fig. 16.13*b*), the intensity and extent of the external magnetic field is reduced. By further subdividing the single domain into four domains, the external magnetic field is reduced still more (Fig. 16.13*c*). Since the intensity of the external magnetic field of a ferromagnetic material is directly related to its magnetostatic energy, the formation of multiple domains reduces the magnetostatic energy of a unit volume of material.

16.6.3 Magnetocrystalline Anisotropy Energy

Before considering domain boundary (wall) energy, let us look at the effects of crystal orientation on the magnetization of ferromagnetic materials. Magnetization versus applied field curves for a single crystal of a ferromagnetic material can vary,

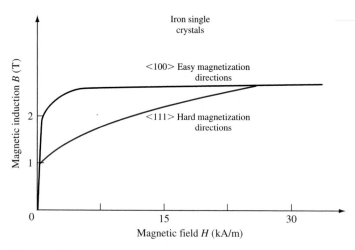

Figure 16.14
Magnetocrystalline anisotropy in BCC iron. Iron is
magnetized easier in the ⟨100⟩ directions than in the
⟨111⟩ directions.

depending on the crystal orientation relative to the applied field. Figure 16.14 shows magnetic induction B versus applied field H curves for magnetizations in the ⟨100⟩ and ⟨111⟩ directions for single crystals of BCC iron. As indicated in Fig. 16.14, saturation magnetization occurs most easily (or with the lowest applied field) for the ⟨100⟩ directions and with the highest applied field in the ⟨111⟩ directions. The ⟨111⟩ directions are said to be the hard directions for magnetization in BCC iron. For FCC nickel the easy directions of magnetization are the ⟨111⟩ directions and the ⟨100⟩ the hard directions; the hard directions for FCC nickel are just the opposite of those for BCC iron.

For polycrystalline ferromagnetic materials such as iron and nickel, grains at different orientations will reach saturation magnetization at different field strengths. Grains whose orientations are in the easy direction of magnetization will saturate at low applied fields, while grains oriented in the hard directions must rotate their resultant moment in the direction of the applied field and thus will reach saturation under much higher fields. The work done to rotate all the domains because of this anisotropy is called the **magnetocrystalline anisotropy energy.**

16.6.4 Domain Wall Energy

A *domain wall* is the boundary between two domains whose overall magnetic moments are at different orientations. It is analogous to a grain boundary where crystal orientation changes from one grain to another. In contrast to a grain boundary, at which grains change orientation abruptly and which is about three atoms wide, a domain changes orientation gradually with a domain boundary being about

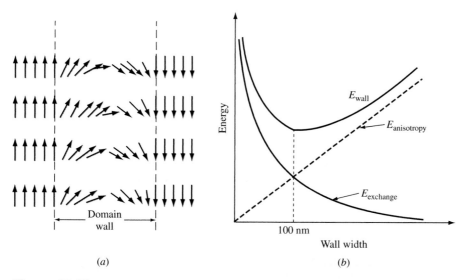

Figure 16.15

Schematic illustration of (a) magnetic dipole arrangements at domain (Bloch) wall and (b) relationship among magnetic exchange energy, magnetocrystalline anisotropy energy, and wall width. The equilibrium wall width is about 100 nm.
(Adapted from C. S. Barrett, W. D. Nix, and A. S. Tetelman, "The Principles of Engineering Materials," Prentice-Hall, 1973, p. 485.)

300 atoms wide. Figure 16.15*a* shows a schematic drawing of a domain boundary of 180 degrees change in magnetic moment direction that takes place gradually across the boundary.

The large width of a domain wall is due to a balance between two forces: exchange energy and magnetocrystalline anisotropy. When there is only a small difference in orientation between the dipoles (Fig. 16.15*a*), the exchange forces between the dipoles are minimized and the exchange energy is reduced (Fig. 16.15*b*). Thus, the exchange forces will tend to widen the domain wall. However, the wider the wall is, the greater will be the number of dipoles forced to lie in directions different from those of easy magnetization, and the magnetocrystalline anisotropy energy will be increased (Fig. 16.15*b*). Thus, the equilibrium wall width will be reached at the width where the sum of the exchange and magnetocrystalline anisotropy energies is a minimum (Fig. 16.15*b*).

16.6.5 Magnetostrictive Energy

When a ferromagnetic material is magnetized, its dimensions change slightly, and the sample being magnetized either expands or contracts in the direction of magnetization (Fig. 16.16). This magnetically induced reversible elastic strain ($\Delta l/l$) is called **magnetostriction** and is of the order of 10^{-6}. The energy due to the

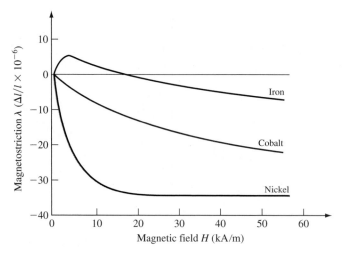

Figure 16.16

Magnetostrictive behavior of Fe, Co, and Ni ferromagnetic elements. Magnetostriction is a fractional elongation (or contraction) and in this illustration is in units of micrometers per meter.

mechanical stresses created by magnetostriction is called **magnetostrictive energy.** For iron, the magnetostriction is positive at low fields and negative at high fields (Fig. 16.16).

The cause of magnetostriction is attributed to the change in the bond length between the atoms in a ferromagnetic metal when their electron-spin dipole moments are rotated into alignment during magnetization. The fields of the dipoles may attract or repel each other, leading to the contraction or expansion of the metal during magnetization.

Let us now consider the effect of magnetostriction on the equilibrium configuration of the domain structure of cubic crystalline materials, such as those shown in Fig. 16.17*a* and *b*. Because of the cubic symmetry of the crystals, the formation of triangular-shaped domains, called *domains of closure,* at the ends of the crystal eliminates the magnetostatic energy associated with an external magnetic field and hence lowers the energy of the material. It might appear that very large domains such as those shown in Fig. 16.17*a* and *b* would be the lowest energy and most stable configuration since there is minimum wall energy. However, this is not the case since magnetostrictive stresses introduced during magnetization tend to be larger for larger domains. Smaller magnetic domains, such as those shown in Fig. 16.17*c*, reduce magnetostrictive stresses but increase domain wall area and energy. Thus, the equilibrium domain configuration is reached when the sum of the magnetostrictive and **domain wall energies** is a minimum.

In summary, the domain structure formed in ferromagnetic materials is determined by the various contributions of exchange, magnetostatic, magnetocrystalline

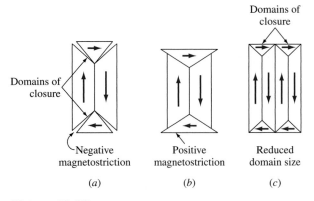

Figure 16.17
Magnetostriction in cubic magnetic materials. Schematic exaggeration of (*a*) negative and (*b*) positive magnetostriction pulling apart the domain boundaries of a magnetic material. (*c*) Lowering of magnetostrictive stresses by the creation of a smaller-domain-size structure.

anisotropic, domain wall, and magnetostrictive energies to its total magnetic energy. The equilibrium or most stable configuration occurs when the total magnetic energy is the lowest.

16.7 THE MAGNETIZATION AND DEMAGNETIZATION OF A FERROMAGNETIC METAL

Ferromagnetic metals such as Fe, Co, and Ni acquire large magnetizations when placed in a magnetizing field, and they remain in the magnetized condition to a lesser extent after the magnetizing field is removed. Let us consider the effect of an applied field H on the magnetic induction B of a ferromagnetic metal during magnetizing and demagnetizing, as shown in the B versus H graph of Fig. 16.18. First, let us demagnetize a ferromagnetic metal such as iron by slowly cooling it from above its Curie temperature. Then, let us apply a magnetizing field to the sample and follow the effect of the applied field on the magnetic induction of the sample.

As the applied field increases from zero, B increases from zero along curve OA of Fig. 16.18 until **saturation induction** is reached at point A. Upon decreasing the applied field to zero, the original magnetization curve is not retraced, and there remains a magnetic flux density called the **remanent induction B_r** (point C in Fig. 16.18). To decrease the magnetic induction to zero, a reverse (negative) applied field of the amount H_c, called the **coercive force,** must be applied (point D in Fig. 16.18). If the negative applied field is increased still more, eventually the

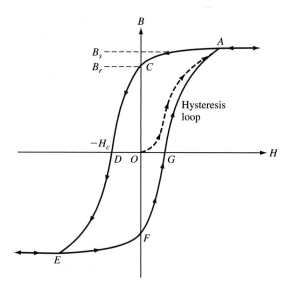

Figure 16.18
Magnetic induction *B* versus applied field *H*
hysteresis loop for a ferromagnetic material.
The curve *OA* traces out the initial *B* versus
H relationship for the magnetization of a
demagnetized sample. Cyclic magnetization and
demagnetization to saturation induction traces
out hysteresis loop *ACDEFGA.*

material will reach saturation induction in the reverse field at point *E* of Fig. 16.18.
When the reverse field is removed, the magnetic induction will return to the remanent induction at point *F* in Fig. 16.18, and upon application of a positive applied
field, the *B-H* curve will follow *FGA* to complete a loop. Further application of
reverse and forward applied fields to saturation induction will produce the repetitive
loop of *ACDEFGA*. This magnetization loop is referred to as a **hysteresis loop,** and
its internal area is a measure of energy lost or the work done by the magnetizing
and demagnetizing cycle.

16.8 SOFT MAGNETIC MATERIALS

A **soft magnetic material** is easily magnetized and demagnetized, whereas a *hard
magnetic material* is difficult to magnetize and demagnetize. In early days soft and
hard magnetic materials were physically soft and hard, respectively. Today, however,
the physical hardness of a magnetic material does not necessarily indicate that it is
magnetically soft or hard.

Soft materials such as the iron–3 to 4 percent silicon alloys used in cores for
transformers, motors, and generators have narrow hysteresis loops with low coercive

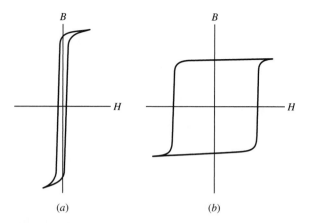

Figure 16.19
Hysteresis loops for (*a*) a soft magnetic material and (*b*) a hard magnetic material. The soft magnetic material has a narrow hysteresis loop that makes it easy to magnetize and demagnetize, whereas the hard magnetic material has a wide hysteresis loop that makes it difficult to magnetize and demagnetize.

forces (Fig. 16.19*a*). On the other hand, hard magnetic materials used for permanent magnets have wide hysteresis loops with high coercive forces (Fig. 16.19*b*).

16.8.1 Desirable Properties for Soft Magnetic Materials

For a ferromagnetic material to be soft, its hysteresis loop should have as low a coercive force as possible. That is, its hysteresis loop should be as thin as possible so that the material magnetizes easily and has a high magnetic permeability. For most applications, a high saturation induction is also an important property of soft magnetic materials. Thus, a very thin and high hysteresis loop is desirable for most soft magnetic materials (Fig. 16.19*a*).

16.8.2 Energy Losses for Soft Magnetic Materials

Hysteresis Energy Losses   **Hysteresis energy losses** are due to dissipated energy required to push the domain walls back and forth during the magnetization and demagnetization of the magnetic material. The presence of impurities, crystalline imperfections, and precipitates in soft magnetic materials all act as barriers to impede domain wall movement during the magnetization cycle and so increase hysteresis energy losses. Plastic strain, by increasing the dislocation density of a magnetic material, also increases hysteresis losses. In general, the internal area of a hysteresis loop is a measure of the energy lost due to magnetic hysteresis.

In the magnetic core of an AC electrical power transformer using 60 cycles/s, electric current goes through the entire hysteresis loop 60 times per second, and in each cycle there is some energy lost due to movement of the domain walls of the magnetic material in the transformer core. Thus, increasing the AC electrical input frequency of electromagnetic devices increases the hysteresis energy losses.

Eddy-Current Energy Losses   A fluctuating magnetic field caused by AC electrical input into a conducting magnetic core produces transient voltage gradients that create stray electric currents. These induced electric currents are called *eddy currents* and are a source of energy loss due to electrical resistance heating. **Eddy-current energy losses** in electrical transformers can be reduced by using a laminated or sheet structure in the magnetic core. An insulation layer between the conducting magnetic material prevents the eddy currents from going from one sheet to the next. Another approach to reducing eddy-current losses, particularly at higher frequencies, is to use a soft magnetic material that is an insulator. Ferrimagnetic oxides and other similar-type magnetic materials are used for some high-frequency electromagnetic applications and will be discussed in Sec. 16.9.

16.8.3 Iron-Silicon Alloys

The most extensively used soft magnetic materials are the iron–3 to 4 percent silicon alloys. Before about 1900, low-carbon plain-carbon steels were used for low-frequency (60 cycle) power application devices such as transformers, motors, and generators. However, with these magnetic materials, core losses were relatively high.

The addition of about 3 to 4 percent silicon to iron to make **iron-silicon alloys** has several beneficial effects for reducing core losses in magnetic materials:

1. Silicon increases the electrical resistivity of low-carbon steel and thus reduces the eddy-current losses.
2. Silicon decreases the magnetoanisotropy energy of iron and increases magnetic permeability and thus decreases hysteresis core losses.
3. Silicon additions (3 to 4 percent) also decrease magnetostriction and lower hysteresis energy losses and transformer noise ("hum").

However, on the detrimental side, silicon decreases the ductility of iron, so only up to about 4 percent silicon can be alloyed with iron. Silicon also decreases the saturation induction and Curie temperature of iron.

A further decrease in eddy-current energy losses in transformer cores was achieved by using a laminated (stacked-sheet) structure. For a modern power transformer core, a multitude of thin iron-silicon sheets about 0.010 to 0.014 in. (0.025 to 0.035 cm) thick are stacked on top of each other with a thin layer of insulation between them. The insulation material is coated on both sides of the iron-silicon sheets and prevents stray eddy currents from flowing perpendicular to the sheets.

Still another decrease in transformer-core energy loss was achieved in the 1940s by the production of grain-oriented iron-silicon sheet. By using a combination of

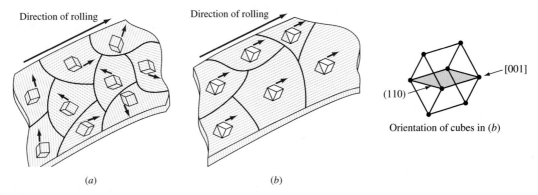

Figure 16.20
(*a*) Random and (*b*) preferred orientation (110) [001] texture in polycrystalline iron–3 to 4% silicon sheet. The small cubes indicate the orientation of each grain.
(*After R. M. Rose, L. A. Shepard, and J. Wulff, "Structure and Properties of Materials," vol. IV: "Electronic Properties," Wiley, 1966, p. 211.*)

Table 16.3 Selected magnetic properties of soft magnetic materials

Material and composition	Saturation induction B_s (T)	Coercive force H_c (A/cm)	Initial relative permeability μ_i
Magnetic iron, 0.2-cm sheet	2.15	0.88	250
M36 cold-rolled Si-Fe (random)	2.04	0.36	500
M6 (110) [001], 3.2% Si-Fe (oriented)	2.03	0.06	1,500
45 Ni-55 Fe (45 Permalloy)	1.6	0.024	2,700
75 Ni-5 Cu-2 Cr-18 Fe (Mumetal)	0.8	0.012	30,000
79 Ni-5 Mo-15 Fe-0.5 Mn (Supermalloy)	0.78	0.004	100,000
48% MnO-Fe$_2$O$_3$, 52% ZnO-Fe$_2$O$_3$ (soft ferrite)	0.36		1,000
36% NiO-Fe$_2$O$_3$, 64% ZnO-Fe$_2$O$_3$ (soft ferrite)	0.29		650

Source: G. Y. Chin and J. H. Wernick, "Magnetic Materials, Bulk," vol. 14: *Kirk-Othmer Encyclopedia of Chemical Technology*, 3rd ed., Wiley, 1981, p. 686.

cold work and recrystallization treatments, a *cube-on-edge* (COE) $\{110\}\langle 001 \rangle$ grain-oriented material was produced on an industrial scale for Fe–3% Si sheet (Fig. 16.20). Since the [001] direction is an easy axis for magnetization for Fe–3% Si alloys, the magnetic domains in COE materials are oriented for easy magnetization upon the application of an applied field in the direction parallel to the rolling direction of the sheet. Thus, the COE material has a higher permeability and lower hysteresis losses than Fe-Si sheet with a random texture (Table 16.3).

16.8.4 Metallic Glasses

Metallic glasses are a relatively new class of metallic-type materials whose dominant characteristic is a noncrystalline structure, unlike normal metal alloys, which

Table 16.4 Metallic glasses: compositions, properties, and applications

Alloy (atomic %)	Saturation induction B_s (T)	Maximum permeability	Applications
$Fe_{78}B_{13}Si_9$	1.56	600,000	Power transformers, low core losses
$Fe_{81}B_{13.5}Si_{3.5}C_2$	1.61	300,000	Pulse transformers, magnetic switches
$Fe_{67}Co_{18}B_{14}Si_1$	1.80	4,000,000	Pulse transformers, magnetic switches
$Fe_{77}Cr_2B_{16}Si_5$	1.41	35,000	Current transformers, sensing cores
$Fe_{74}Ni_4Mo_3B_{17}Si_2$	1.28	100,000	Low core losses at high frequencies
$Co_{69}Fe_4Ni_1Mo_2B_{12}Si_{12}$	0.70	600,000	Magnetic sensors, recording heads
$Co_{66}Fe_4Ni_1B_{14}Si_{15}$	0.55	1,000,000	Magnetic sensors, recording heads
$Fe_{40}Ni_{38}Mo_4B_{18}$	0.88	800,000	Magnetic sensors, recording heads

Source: Metglas Magnetic Alloys, Allied Metglas Products.

have a crystalline structure. The atoms in normal metals and alloys when cooled from the liquid state arrange themselves into an orderly crystal lattice. Table 16.4 lists the atomic compositions of eight metallic glasses of engineering importance. These materials have important soft magnetic properties and consist essentially of various combinations of ferromagnetic Fe, Co, and Ni with the metalloids B and Si. Applications for these exceptionally soft magnetic materials include low-energy core-loss power transformers, magnetic sensors, and recording heads.

Metallic glasses are produced by a rapid solidification process in which molten metallic glass is cooled very rapidly (about $10^{6\circ}C/s$) as a thin film on a rotating copper-surfaced mold (Fig. 16.21a). This process produces a continuous ribbon of metallic glass about 0.001 in. (0.0025 cm) thick and 6 in. (15 cm) wide.

Metallic glasses have some remarkable properties. They are very strong (up to 650 ksi [4500 MPa]), very hard with some flexibility, and very corrosion-resistant. The metallic glasses listed in Table 16.4 are magnetically very soft as indicated by their maximum permeabilities. Thus, they can be magnetized and demagnetized very easily. Domain walls in these materials can move with exceptional ease, mainly because there are no grain boundaries and no long-range crystal anisotropy. Figure 16.21b shows some magnetic domains in a metallic glass that were produced by bending the metallic glass ribbon. Magnetically soft metallic glasses have very narrow hysteresis loops, as is indicated in Fig. 16.21c, and thus they have very low hysteresis energy losses. This property has enabled the development of multilayered metallic glass power transformer cores that have 70 percent of the core losses of conventional iron-silicon cores (Fig. 16.1). Much research and development work in the application of metallic glasses for low-loss power transformers is in progress.

16.8.5 Nickel-Iron Alloys

The magnetic permeabilities of commercially pure iron and iron-silicon alloys are relatively low at low applied fields. Low initial permeability is not as important for power applications such as transformer cores since this equipment is operated at high magnetizations. However, for high-sensitivity communications equipment used to

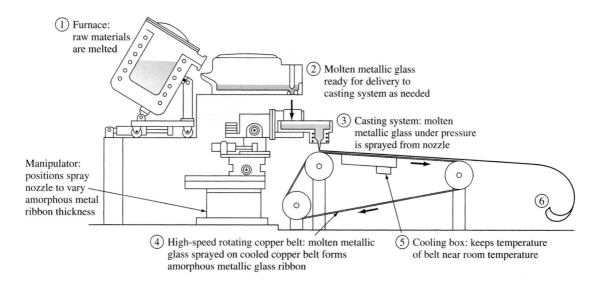

① Furnace: raw materials are melted

② Molten metallic glass ready for delivery to casting system as needed

③ Casting system: molten metallic glass under pressure is sprayed from nozzle

Manipulator: positions spray nozzle to vary amorphous metal ribbon thickness

④ High-speed rotating copper belt: molten metallic glass sprayed on cooled copper belt forms amorphous metallic glass ribbon

⑤ Cooling box: keeps temperature of belt near room temperature

⑥

(a)

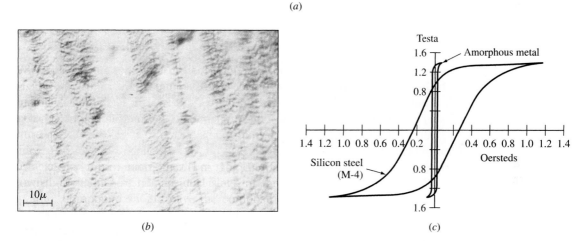

(b)

(c)

Figure 16.21
(a) Schematic drawing of the rapid solidification process for the production of metallic glass ribbon.
(*After New York Times, Jan. 11, 1989, p. D7.*)
(b) Induced magnetic domains in a metallic glass.
(*After V. Lakshmanan and J. C. M. Li, Mater. Sci. Eng., 1988, p. 483.*)
(c) Comparison of the hysteresis loops of a ferromagnetic metallic glass and M-4 iron-silicon ferromagnetic sheet.
(*Electric World, September 1985.*)

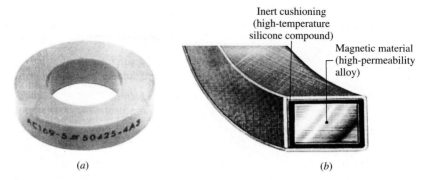

Inert cushioning
(high-temperature
silicone compound)

Magnetic material
(high-permeability
alloy)

(a)   (b)

Figure 16.22
Tape-wound magnetic cores. (*a*) Encapsulated core. (*b*) Cross section of
tape-wound core with phenolic encapsulation. Note that there is a silicone
rubber cushion between the magnetic alloy tape and the phenolic
encapsulation case. The magnetic properties of annealed high Ni-Fe tape-
wound alloys are sensitive to strain damage.
(*Courtesy of Magnetics, a diision of Spang & Company.*)

detect or transmit small signals, **nickel-iron alloys,** which have much higher permeabilities at low fields, are commonly used.

In general, two broad classes of Ni-Fe alloys are commercially produced, one
with about 50 percent Ni and another with about 79 percent Ni. The magnetic properties of some of these alloys are listed in Table 16.3. The 50 percent Ni alloy is
characterized by moderate permeability ($\mu_i = 2500$; $\mu_{max} = 25,000$) and high saturation induction [$B_s = 1.6$ T (16,000 G)]. The 79 percent Ni alloy has high permeability ($\mu_i = 100,000$; $\mu_{max} = 1,000,000$) but lower saturation induction [$B_s = 0.8$ T
(8000 G)]. These alloys are used in audio and instrument transformers, instrument
relays, and for rotor and stator laminations. Tape-wound cores, such as the cutaway
section shown in Fig. 16.22, are commonly used for electronic transformers.

The Ni-Fe alloys have such high permeabilities because their magnetoanisotropy
and magnetostrictive energies are low at the compositions used. The highest initial
permeability in the Ni-Fe system occurs at 78.5% Ni–21.5% Fe, but rapid cooling
below 600°C is necessary to suppress the formation of an ordered structure. The
equilibrium ordered structure in the Ni-Fe system has an FCC unit cell with Ni atoms
at the faces and Fe atoms at the face corners. The addition of about 5 percent Mo
to the 78.5 percent Ni (balance Fe) alloy also suppresses the ordering reaction so
that moderate cooling of the alloy from above 600°C is sufficient to prevent ordering.

The initial permeability of the Ni-Fe alloys containing about 56 to 58 percent
Ni can be increased three to four times by annealing the alloy in the presence of a
magnetic field after the usual high-temperature anneal. The magnetic anneal causes
directional ordering of the atoms of the Ni-Fe lattice and thereby increases the initial
permeability of these alloys. Figure 16.23 shows the effect of magnetic annealing on
the hysteresis loop of a 65% Ni–35% Fe alloy.

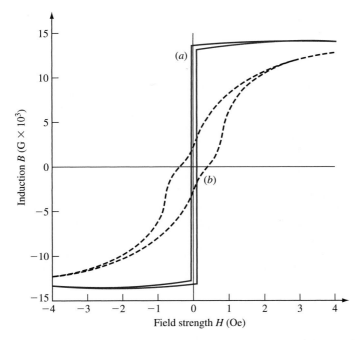

Figure 16.23
The effect of magnetic annealing on the hysteresis loop of a
65% Ni–35% Fe alloy. (*a*) 65 Permalloy annealed with field
present; (*b*) 65 Permalloy annealed with field absent.
(*After K. M. Bozorth, "Ferromagnetism," Van Nostrand, 1951, p. 121.*)

16.9 HARD MAGNETIC MATERIALS

16.9.1 Properties of Hard Magnetic Materials

Permanent or **hard magnetic materials** are characterized by a high coercive force
H_c and a high remanent magnetic induction B_r, as indicated schematically in
Fig. 16.19*b*. Thus, the hysteresis loops of hard magnetic materials are wide and high.
These materials are magnetized in a magnetic field strong enough to orient their
magnetic domains in the direction of the applied field. Some of the applied energy
of the field is converted into potential energy, which is stored in the permanent
magnet produced. A permanent magnet in the fully magnetized condition is thus in
a relatively high-energy state as compared to a demagnetized magnet.

Hard magnetic materials are difficult to demagnetize once magnetized. The demagnetizing curve for a hard magnetic material is chosen as the second quadrant of its
hysteresis loop and can be used for comparing the strengths of permanent magnets.
Figure 16.24 compares the demagnetizing curves of various hard magnetic materials.

The power or external energy of a permanent (hard) magnetic material is directly
related to the size of its hysteresis loop. The magnetic potential energy of a hard

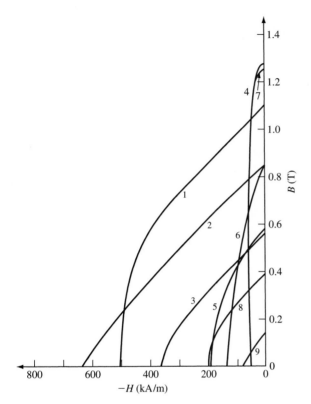

Figure 16.24

Demagnetization curves of various hard magnetic materials. 1: $Sm(Co,Cu)_{7.4}$; 2: $SmCo_5$; 3: bonded $SmCo_5$; 4: alnico 5; 5: Mn-Al-C; 6: alnico 8; 7: Cr-Co-Fe; 8: ferrite; and 9: bonded ferrite.

(After G. Y. Chin and J. H. Wernick, "Magnetic Materials, Bulk," vol. 14: "Kirk-Othmer Encyclopedia of Chemical Technology," 3rd ed., Wiley, 1981, p. 673.)

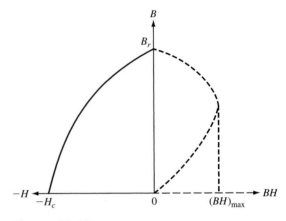

Figure 16.25

A schematic diagram of the energy product (*B* versus *BH*) curve of a hard magnetic material such as an alnico alloy is shown by the circular dashed line to the right of the *B* (induction) axis. The maximum energy product $(BH)_{max}$ is indicated at the intersection of the vertical dashed line and the *BH* axis.

magnetic material is measured by its **maximum energy product $(BH)_{max}$,** which is the maximum value of the product of *B* (magnetic induction) and *H* (the demagnetizing field) determined from the demagnetizing curve of the material. Figure 16.25 shows the external energy (*BH*) curve for a hypothetical hard magnetic material and its maximum energy product, $(BH)_{max}$. Basically, the maximum energy product of a hard magnetic material is the area occupied by the largest rectangle that can be inscribed in the second quadrant of the hysteresis loop of the material. The SI units for the energy product *BH* are kJ/m^3; the cgs system units are $G \cdot Oe$. The SI units for the energy product $(BH)_{max}$ of joules per cubic meter are equivalent units to the product of the units of *B* in teslas and *H* in amperes per meter as follows:

$$\left[B\left(T \cdot \frac{Wb}{m^2} \cdot \frac{1}{T} \cdot \frac{V \cdot s}{Wb}\right)\right]\left[H\left(\frac{A}{m} \cdot \frac{J}{V \cdot A \cdot s}\right)\right] = BH\left(\frac{J}{m^3}\right)$$

Estimate the maximum energy product $(BH)_{max}$ for the Sm $(Co,Cu)_{7.4}$ alloy of Fig. 16.24.

■ **Solution**

We need to find the area of the largest rectangle that can be located within the second-quadrant demagnetization curve of the alloy shown in Fig. 16.24. Four trial areas are listed:

Trial 1 ~ (0.8 T × 250 kA/m) = 200 kJ/m^3 (see Fig. EP16.4)

Trial 2 ~ (0.6 T × 380 kA/m) = 228 kJ/m^3

Trial 3 ~ (0.55 T × 420 kA/m) = 231 kJ/m^3

Trial 4 ~ (0.50 T × 440 kA/m) = 220 kJ/m^3

The highest value is about 231 kJ/m^3, which compares to 240 kJ/m^3 listed for the Sm (Cu,Co) alloy in Table 16.5.

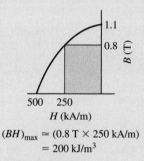

$$(BH)_{max} \simeq (0.8 \text{ T} \times 250 \text{ kA/m})$$
$$= 200 \text{ kJ/m}^3$$

Figure EP16.4
Trial 1.

Table 16.5 Selected magnetic properties of hard magnetic materials

Material and composition	Remanent induction B_r (T)	Coercive force H_c (kA/m)	Maximum energy product $(BH)_{max}$ (kJ/m^3)
Alnico 1, 12 Al, 21 Ni, 5 Co, 2 Cu, bal Fe	0.72	37	11.0
Alnico 5, 8 Al, 14 Ni, 25 Co, 3 Cu, bal Fe	1.28	51	44.0
Alnico 8, 7 Al, 15 Ni, 24 Co, 3 Cu, bal Fe	0.72	150	40.0
Rare earth–Co, 35 Sm, 65 Co	0.90	675–1200	160
Rare earth–Co, 25.5 Sm, 8 Cu, 15 Fe, 1.5 Zr, 50 Co	1.10	510–520	240
Fe-Cr-Co, 30 Cr, 10 Co, 1 Si, 59 Fe	1.17	46	34.0
MO · Fe$_2$O$_3$ (M = Ba, Sr) (hard ferrite)	0.38	235–240	28.0

Source: G. Y. Chin and J. H. Wernick, "Magnetic Materials, Bulk," vol. 14: *Kirk-Othmer Encyclopedia of Chemical Technology,* 3rd ed., Wiley, 1981, p. 686.

16.9.2 Alnico Alloys

Properties and Compositions   The **alnico (*al*uminum-*ni*ckel-*co*balt) alloys** are the most important commercial hard magnetic materials in use today. They account for about 35 percent of the hard-magnet market in the United States. These alloys are

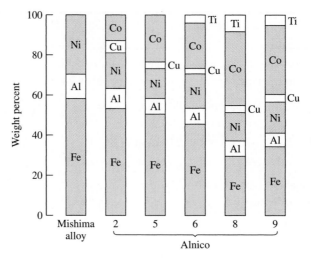

Figure 16.26
Chemical compositions of the alnico alloys. The original alloy was discovered by
Mishima in Japan in 1931.
(After B. D. Cullity, "Introduction to Magnetic Materials," Addison-Wesley, 1972, p. 566.)

characterized by a high energy product [$(BH)_{max}$ = 40 to 70 kJ/m³ (5 to 9 MG · Oe)],
a high remanent induction [B_r = 0.7 to 1.35 T (7 to 13.5 kG)], and a moderate
coercivity [H_c = 40 to 160 kA/m (500 to 2010 Oe)]. Table 16.5 lists some magnetic
properties of several alnico and other permanent magnetic alloys.

The alnico family of alloys are iron-based alloys with additions of Al, Ni, and Co
plus about 3 percent Cu. A few percent Ti is added to the high-coercivity alloys, alnicos
6 to 9. Figure 16.26 shows bar graphs of the compositions of some of the alnico alloys.
Alnicos 1 to 4 are isotropic, whereas alnicos 5 to 9 are anisotropic due to being heat-
treated in a magnetic field while the precipitates form. The alnico alloys are brittle and
so are produced by casting or by powder metallurgy processes. Alnico powders are
used primarily to produce large amounts of small articles with complex shapes.

Structure   Above their solution heat-treatment temperature of about 1250°C, the
alnico alloys are single-phase with a BCC crystal structure. During cooling to about
750 to 850°C, these alloys decompose into two other BCC phases, α and α'. The
matrix α phase is rich in Ni and Al and is weakly magnetic. The α' precipitate is
rich in Fe and Co and thus has a higher magnetization than the Ni-Al–rich α phase.
The α' phase tends to be rodlike, aligned in the $\langle 100 \rangle$ directions, and about 10 nm
in diameter and about 100 nm long.

If the 800°C heat treatment is carried out in a magnetic field, the α' precipitate
forms fine elongated particles in the direction of the magnetic field (Fig. 16.27) in
a matrix of the α phase **(magnetic anneal)**. The high coercivity of the alnicos is
attributed to the difficulty of rotating single-domain particles of the α' phase based
on shape anisotropy. The larger the aspect (length-to-width) ratio of the rods and the

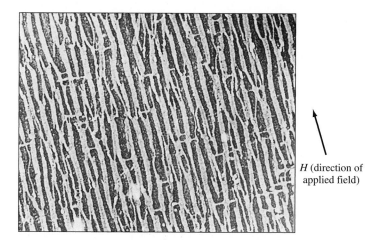

H (direction of applied field)

Figure 16.27
Replica electron micrograph showing the structure of alnico 8 (Al–Ni-Co-Fe-Ti) alloy after an 800°C heat treatment for 9 min in an applied magnetic field. The α phase (Ni-Al–rich) is light, and the α' phase (Fe-Co–rich) is dark. The α' phase is highly ferromagnetic, which is elongated in the direction of the applied field, creating anisotropy of the coercive force.
(Courtesy of K. J. deVos, 1966.)

smoother their surface, the greater the coercivity of the alloy. Thus, the forming of the precipitate in a magnetic field makes the precipitate longer and thinner and so increases the coercivity of the alnico magnetic material. It is believed that the addition of titanium to some of the highest-strength alnicos increases their coercivities by increasing the aspect ratio of the α' rods.

16.9.3 Rare Earth Alloys

Rare earth alloy magnets are produced on a large scale in the United States and have magnetic strengths superior to those of any commercial magnetic material. They have maximum energy products $(BH)_{max}$ to 240 kJ/m^3 (30 MG · Oe) and coercivities to 3200 kA/m (40 kOe). The origin of magnetism in the rare earth transition elements is due almost entirely to their unpaired $4f$ electrons in the same way that magnetism in Fe, Co, and Ni is due to their unpaired $3d$ electrons. There are two main groups of commercial rare earth magnetic materials: one based on single-phase $SmCo_5$ and the other on precipitation-hardened alloys of the approximate composition $Sm(Co,Cu)_{7.5}$.

$SmCo_5$ single-phase magnets are the most widely used type. The mechanism of coercivity in these materials is based on nucleation and/or pinning of domain walls at surfaces and grain boundaries. These materials are fabricated by powder metallurgy techniques using fine particles (1 to 10 μm). During pressing the particles are aligned in a magnetic field. The pressed particles are then carefully sintered to prevent particle growth. The magnetic strengths of these materials are high, with $(BH)_{max}$ values in the range of 130 to 160 kJ/m^3 (16 to 20 MG · Oe).

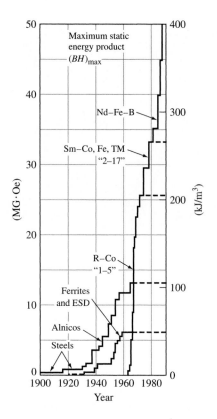

Figure 16.28
Progress in permanent-magnet quality in the twentieth century as measured by the
maximum energy product $(BH)_{max}$.
(After K. J. Strnat, Soft and Hard Magnetic Materials with Applications, ASM Inter., 1986, p. 64.)

In the precipitation-hardened $Sm(Co,Cu)_{7.5}$ alloy, part of the Co is substituted by
Cu in $SmCo_5$ so that a fine precipitate (about 10 nm) is produced at a low aging
temperature (400 to 500°C). The precipitate formed is coherent with the $SmCo_5$
structure. The coherency mechanism here is mainly based on the homogeneous pinning of domain walls at the precipitated particles. These materials are also made
commercially by powder metallurgy processes using magnetic alignment of the particles. The addition of small amounts of iron and zirconium promote the development
of higher coercivities. Typical values for an $Sm(Co_{0.68}Cu_{0.10}Fe_{0.21}Zr_{0.01})_{7.4}$ commercial alloy are $(BH)_{max} = 240$ kJ/m³ (30 MG · Oe) and $B_r = 1.1$ T (11,000 G).
Figures 16.24 and 16.28 show the outstanding improvement in magnetic strengths
achieved with rare earth magnetic alloys.

Sm-Co magnets are used in medical devices such as thin motors in implantable
pumps and valves and in aiding eyelid motion. Rare earth magnets are also used for
electronic wristwatches and traveling-wave tubes. Direct-current and synchronous
motors and generators are produced by using rare earth magnets, resulting in a size
reduction.

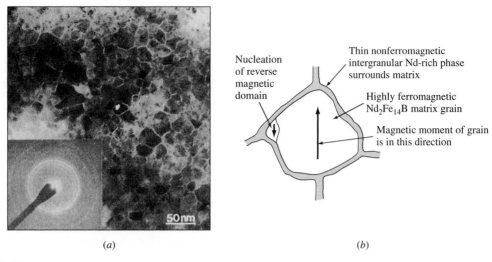

(a)   (b)

Figure 16.29

(a) Bright-field electron transmission micrograph of an optimally quenched Nd-Fe-B ribbon showing randomly oriented grains surrounded by a thin intergranular phase marked by the arrow.

(*After J. J. Croat and J. F. Herbst, MRS Bull., June 1988, p. 37.*)

(b) Single $Nd_2Fe_{14}B$ grain showing nucleation of reverse magnetic domain.

16.9.4 Neodymium-Iron-Boron Magnetic Alloys

Hard Nd-Fe-B magnetic materials with $(BH)_{max}$ products as high as 300 kJ/m^3 ($45 \text{ MG} \cdot \text{Oe}$) were discovered in about 1984, and today these materials are produced by both powder metallurgy and rapid-solidification melt-spun ribbon processes. Figure 16.29a shows the microstructure of a $Nd_2Fe_{14}B$-type rapidly solidified ribbon. In this structure highly ferromagnetic $Nd_2Fe_{14}B$ matrix grains are surrounded by a non-ferromagnetic Nd-rich thin intergranular phase. The high coercivity and associated $(BH)_{max}$ energy product of this material result from the difficulty of nucleating reverse magnetic domains that usually nucleate at the grain boundaries of the matrix grains (Fig. 16.29b). The intergranular nonferromagnetic Nd-rich phase forces the $Nd_2Fe_{14}B$ matrix grains to nucleate their reverse domains in order to reverse the magnetization of the material. This process maximizes H_c and $(BH)_{max}$ for the whole bulk aggregate of the material. Applications for Nd-Fe-B permanent magnets include all types of electric motors, especially those like automotive starting motors where reduction in weight and compactness are desirable.

16.9.5 Iron-Chromium-Cobalt Magnetic Alloys

A family of magnetic **iron-chromium-cobalt alloys** were developed in 1971 and are analogous to the alnico alloys in metallurgical structure and permanent magnetic properties, but the former are cold-formable at room temperature. A typical composition for an alloy of this type is 61% Fe–28% Cr–11% Co. Typical magnetic properties of Fe-Cr-Co alloys are $B_r = 1.0$ to 1.3 T (10 to 13 kG), $H_c = 150$ to 600 A/cm

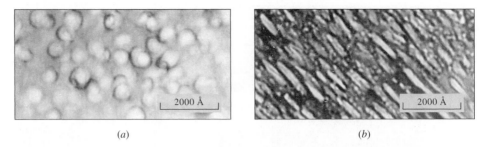

(a) (b)

Figure 16.30

Transmission electron micrographs of an Fe–34% Cr–12% Co alloy showing
(a) spherical precipitates produced before deformation and (b) elongated and aligned
particles after deformation and alignment by final heat treatment.
[*After S. Jin et al., J. Appl. Phys.,* **53**:4300(1982).]

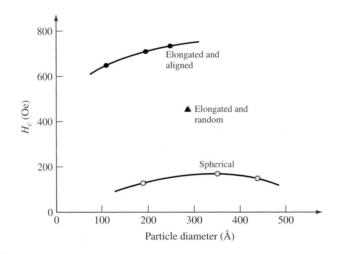

Figure 16.31

Coercivity versus particle diameter for differently shaped particles in an Fe–34%
Cr–12% Co alloy. Note the great increase in coercivity by changing from a spherical
shape to an elongated shape.
[*After S. Jin et al., J. Appl. Phys.,* **53**:4300(1982).]

(190 to 753 Oe), and $(BH)_{max} = 10$ to 45 kJ/m^3 (1.3 to 1.5 MG · Oe). Table 16.5 lists
some typical properties of an Fe-Cr-Co magnetic alloy.

The Fe-Cr-Co alloys have a BCC structure at elevated temperatures above about
1200°C. Upon slow cooling (about 15°C/h) from above 650°C, precipitates of a Cr-
rich α_2 phase (Fig. 16.30a) with particles of about 30 nm (300Å) form in a matrix
of an Fe-rich α_1 phase. The mechanism for coercivity in the Fe-Cr-Co alloys is the
pinning of domain walls by the precipitated particles since the magnetic domains
extend through both phases. Particle shape (Fig. 16.30b) is important since elongation of the particles by deformation before a final aging treatment greatly increases
the coercivity of these alloys, as clearly indicated in Fig. 16.31.

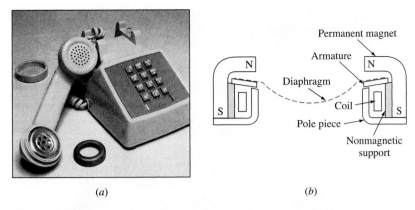

(a) (b)

Figure 16.32
Use of ductile permanent Fe-Cr-Co alloy in a telephone receiver.
(*a*) Disassembled receiver showing permanent magnet ring and
(*b*) cross-sectional view of the U-type telephone receiver indicating
position of permanent magnet.
[*After S. Jin et al., IEEE Trans. Magn.,* **17***:2935(1981).*]

Fe-Cr-Co alloys are especially important for engineering applications where their cold ductility allows high-speed room-temperature forming. The permanent magnet for many modern telephone receivers is an example of this cold-deformable permanent magnetic alloy (Fig. 16.32).

16.10 FERRITES

Ferrites are magnetic ceramic materials made by mixing iron oxide (Fe_2O_3) with other oxides and carbonates in the powdered form. The powders are then pressed together and sintered at a high temperature. Sometimes finishing machining is necessary to produce the desired part shape (Fig. 16.33). The magnetizations produced in the ferrites are large enough to be of commercial value, but their magnetic saturations are not as high as those of ferromagnetic materials. Ferrites have domain structures and hysteresis loops similar to those of ferromagnetic materials. As in the case of the ferromagnetic materials, there are *soft* and *hard ferrites.*

16.10.1 Magnetically Soft Ferrites

Soft ferrite materials exhibit ferrimagnetic behavior. In soft ferrites there is a net magnetic moment due to two sets of unpaired inner-electron spin moments in opposite directions that do not cancel each other (Fig. 16.8*c*).

Composition and Structure of Cubic Soft Ferrites Most cubic soft ferrites have the composition $MO \cdot Fe_2O_3$, where M is a divalent metal ion such as Fe^{2+}, Mn^{2+}, Ni^{2+}, or Zn^{2+}. The structure of the soft ferrites is based on the inverse spinel structure, which is a modification of the spinel structure of the mineral spinel ($MgO \cdot Al_2O_3$). Both the spinel and inverse spinel structures have cubic unit cells

Figure 16.33
Various soft ferrite parts that are used for many electrical
and electronic applications.
(*Courtesy of Magnetics, a diision of Spang & Company.*)

consisting of eight subcells, as shown in Fig. 16.34a. Each of the subcells consists
of one molecule of MO $\cdot$ Fe$_2$O$_3$. Since each subunit contains one MO $\cdot$ Fe$_2$O$_3$ molecule and since there are seven ions in this molecule, each unit cell contains a total
of 7 ions $\times$ 8 subcells = 56 ions per unit cell. Each subunit cell has an FCC crystal
structure made up of the four ions of the MO $\cdot$ Fe$_2$O$_3$ molecule (Fig. 16.34b). The
much smaller metal ions (M^{2+} and Fe^{3+}), which have ionic radii of about 0.07 to
0.08 nm, occupy interstitial spaces between the larger oxygen ions (ionic radius
$\approx$ 0.14 nm).

As previously discussed, in an FCC unit cell there are the equivalent of four octahedral and eight tetrahedral interstitial sites. In the normal spinel structure only half
the octahedral sites are occupied, and so only $\frac{1}{2}$(8 subcells $\times$ 4 sites/subcell) =
16 octahedral sites are occupied (Table 16.6). In the normal spinel structure there

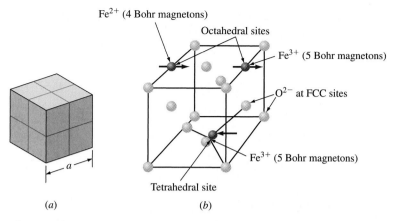

Figure 16.34
(*a*) Unit cell of soft ferrite of the type $MO \cdot Fe_2O_3$. This unit cell consists of eight subcells. (*b*) The subcell for the $FeO \cdot Fe_2O_3$ ferrite. The magnetic moments of the ions in the octahedral sites are aligned in one direction by the applied magnetic field, and those in the tetrahedral sites are aligned in the opposite direction. As a result, there is a net magnetic moment for the subcell and hence the material.

Table 16.6 Metal ion arrangements in a unit cell of a spinel ferrite of composition $MO \cdot Fe_2O_3$

Type of interstitial site	Number available	Number occupied	Normal spinel	Inverse spinel
Tetrahedral	64	8	$8\,M^{2+}$	$8\,Fe^{3+}$ $\leftarrow$
Octahedral	32	16	$16\,Fe^{3+}$	$8\,Fe^{3+},\ 8\,M^{2+}$ $\rightarrow$ $\qquad\rightarrow$

are 8×8 (tetrahedral sites per subcell) = 64 sites/unit cell. However, in the normal spinel structure only one-eighth of the 64 sites are occupied so that only *eight of the tetrahedral sites are occupied* (Table 16.6).

In the **normal spinel-structure** unit cell there are eight $MO \cdot Fe_2O_3$ molecules. In this structure the $8\,M^{2+}$ ions occupy 8 tetrahedral sites and the $16\,Fe^{3+}$ ions occupy 16 octahedral sites. However, in the **inverse spinel structure** there is a different arrangement of the ions: the $8\,M^{2+}$ ions occupy 8 octahedral sites, and the $16\,Fe^{3+}$ ions are divided so that 8 occupy octahedral sites and 8 tetrahedral sites (Table 16.6).

Net Magnetic Moments in Inverse Spinel Ferrites To determine the net magnetic moment for each $MO \cdot Fe_2O_3$ ferrite molecule, we must know the $3d$ inner-electron configuration of the ferrite ions. Figure 16.35 gives this information. When

Ion	Number of electrons	Electron configuration 3d orbitals	Ionic magnetic moment (Bohr magnetons)
Fe^{3+}	23	↑ ☐ ↑ ☐ ↑ ☐ ↑ ☐ ↑ ☐	5
Mn^{2+}	23	↑ ☐ ↑ ☐ ↑ ☐ ↑ ☐ ↑ ☐	5
Fe^{2+}	24	↑↓ ↑ ☐ ↑ ☐ ↑ ☐ ↑ ☐	4
Co^{2+}	25	↑↓ ↑↓ ↑ ☐ ↑ ☐ ↑ ☐	3
Ni^{2+}	26	↑↓ ↑↓ ↑↓ ↑ ☐ ↑ ☐	2
Cu^{2+}	27	↑↓ ↑↓ ↑↓ ↑↓ ↑ ☐	1
Zn^{2+}	28	↑↓ ↑↓ ↑↓ ↑↓ ↑↓	0

Figure 16.35
Electronic configurations and ionic magnetic moments for some 3d transition-element ions.

Table 16.7 Ion arrangements and net magnetic moments per molecule in normal and inverse spinel ferrites

Ferrite	Structure	Tetrahedral sites occupied	Octahedral sites occupied		Net magnetic moment (μ_s/molecule)
$FeO \cdot Fe_2O_3$	Inverse spinel	Fe^{3+} 5 ←	Fe^{2+} 4 →	Fe^{3+} 5 →	4
$ZnO \cdot Fe_2O_3$	Normal spinel	Zn^{2+} 0	Fe^{3+} 5 ←	Fe^{3+} 5 →	0

the Fe atom is ionized to form the Fe^{2+} ion, there are *four* unpaired 3d electrons left after the loss of two 4s electrons. When the Fe atom is ionized to form the Fe^{3+} ion, there are *five* unpaired electrons left after the loss of two 4s and one 3d electrons.

Since each unpaired 3d electron has a magnetic moment of one Bohr magneton, the Fe^{2+} ion has a moment of four Bohr magnetons and the Fe^{3+} ion a moment of five Bohr magnetons. In an applied magnetic field, the magnetic moments of the octahedral and tetrahedral ions oppose each other (Fig. 16.34b). Thus, in the case of the ferrite $FeO \cdot Fe_2O_3$, the magnetic moments of the eight Fe^{3+} ions in octahedral sites will cancel the magnetic moments of the eight Fe^{3+} ions in the tetrahedral sites. Thus, the resultant magnetic moment of this ferrite will be due to the eight Fe^{2+} ions at eight octahedral sites, which have moments of four Bohr magnetons each (Table 16.7). A theoretical value for the magnetic saturation of the $FeO \cdot Fe_2O_3$ ferrite is calculated in Example Problem 16.5 on the basis of the Bohr magneton strength of the Fe^{2+} ions.

Calculate the theoretical saturation magnetization M in amperes per meter and the saturation induction B_s in teslas for the ferrite $FeO \cdot Fe_2O_3$. Neglect the $\mu_0 H$ term for the B_s calculation. The lattice constant of the $FeO \cdot Fe_2O_3$ unit cell is 0.839 nm.

■ **Solution**

The magnetic moment for a molecule of $FeO \cdot Fe_2O_3$ is due entirely to the four Bohr magnetons of the Fe^{2+} ion since the unpaired electrons of the Fe^{3+} ions cancel each other. Since there are eight molecules of $FeO \cdot Fe_2O_3$ in the unit cell, the total magnetic moment per unit cell is

(4 Bohr magnetons/subcell)(8 subcells/unit cell) = 32 Bohr magnetons/unit cell

Thus,

$$M = \left[\frac{32 \text{ Bohr magnetons/unit cell}}{(8.39 \times 10^{-10} \text{ m})^3/\text{unit cell}} \right] \left(\frac{9.27 \times 10^{-24} \text{ A} \cdot \text{m}^2}{\text{Bohr magneton}} \right)$$

$$= 5.0 \times 10^5 \text{ A/m} \blacktriangleleft$$

B_s at saturation, assuming all magnetic moments are aligned and neglecting the H term, is given by the equation $B_s \approx \mu_0 M$. Thus,

$$B_s \approx \mu_0 M \approx \left(\frac{4\pi \times 10^{-7} \text{ T} \cdot \text{m}}{\text{A}} \right) \left(\frac{5.0 \times 10^5 \text{ A}}{\text{m}} \right)$$

$$= 0.63 \text{ T} \blacktriangleleft$$

Iron, cobalt, and nickel ferrites all have the inverse spinel structure, and all are ferrimagnetic due to a net magnetic moment of their ionic structures. Industrial soft ferrites usually consist of a mixture of ferrites since increased saturation magnetizations can be obtained from a mixture of ferrites. The two most common industrial ferrites are the nickel-zinc-ferrite ($Ni_{1-x}Zn_xFe_{2-y}O_4$) and the manganese-zinc-ferrite ($Mn_{1-x}Zn_xFe_{2+y}O_4$).

Properties and Applications of Soft Ferrites

Eddy-Current Losses in Magnetic Materials   Soft ferrites are important magnetic materials because, in addition to having useful magnetic properties, they are insulators and have high electrical resistivities. A high electrical resistivity is important in magnetic applications that require high frequencies since if the magnetic material is conductive, eddy-current energy losses would be great at the high frequencies. Eddy currents are caused by induced voltage gradients, and thus the higher the frequency, the greater the increase in eddy currents. Since soft ferrites are insulators, they can be used for magnetic applications such as transformer cores that operate at high frequencies.

Applications for Soft Ferrites   Some of the most important uses for soft ferrites are for low-signal, memory-core, audiovisual, and recording-head applications. At low signal levels, soft ferrite cores are used for transformers and low-energy inductors.

A large tonnage usage of soft ferrites is for deflection-yoke cores, flyback transformers, and convergence coils for television receivers.

Mn-Zn and Ni-Zn spinel ferrites are used in magnetic recording heads for various types of magnetic tapes. Recording heads are made from polycrystalline Ni-Zn ferrite since the operating frequencies required (100 kHz to 2.5 GHz) are too high for metallic alloy heads because of high eddy-current losses.

Magnetic-core memories based on the 0 and 1 binary logic are used for some types of computers. The magnetic core is useful where loss of power does not cause loss of information. Since magnetic-core memories have no moving parts, they are used when high shock resistance is needed, as in some military uses.

16.10.2 Magnetically Hard Ferrites

A group of **hard ferrites** that are used for permanent magnets have the general formula $MO \cdot 6Fe_2O_3$ and are hexagonal in crystal structure. The most important ferrite of this group is *barium ferrite* ($BaO \cdot 6Fe_2O_3$), which was introduced in the Netherlands by the Philips Company in 1952 under the trade name Ferroxdure. In recent years the barium ferrites have been replaced to some extent by the strontium ferrites, which have the general formula ($SrO \cdot 6Fe_2O_3$) and which have superior magnetic properties compared with the barium ferrites. These ferrites are produced by almost the same method used for the soft ferrites, with most being wet-pressed in a magnetic field to align the easy magnetizing axis of the particles with the applied field.

The hexagonal ferrites are low in cost, low in density, and have a high coercive force, as shown in Fig. 16.24. The high magnetic strengths of these materials is due mainly to their high magnetocrystalline anisotropy. The magnetization of these materials is believed to take place by domain wall nucleation and motion because their grain size is too large for single-domain behavior. Their $(BH)_{max}$ energy products vary from 14 to 28 kJ/m³.

These hard-ferrite ceramic permanent magnets find widespread use in generators, relays, and motors. Electronic applications include magnets for loudspeakers, telephone ringers, and receivers. They are also used for holding devices for door closers, seals, and latches and are used in many toy designs.

16.11 SUMMARY

Magnetic materials are important industrial materials used for many engineering designs. Most industrial magnetic materials are *ferro-* or *ferrimagnetic* and show large magnetizations. The most important ferromagnetic materials are based on alloys of Fe, Co, and Ni. More recently some ferromagnetic alloys have been made with some rare earth elements such as Sm. In ferromagnetic materials such as Fe, there exist regions called *magnetic domains* in which atomic magnetic dipole moments are aligned parallel to each other. The magnetic domain structure in a ferromagnetic material is determined by the following energies that are minimized: exchange, magnetostatic, magnetocrystalline anisotropy, domain wall, and magnetostrictive energies. When the ferromagnetic domains in a sample are at random orientations, the sample is in a demagnetized state. When a magnetic field is applied to a ferromagnetic material sample, the domains in the sample are aligned; the material becomes magnetized and remains magnetized to some extent when the field is removed. The magnetization behavior of a ferromagnetic material is recorded by a magnetic induction versus applied field graph called a *hysteresis loop*. When a demagnetized ferromagnetic material is magnetized by an applied field H, its magnetic induction B eventually reaches a saturation level called the *saturation induction B_s*. When the applied field is removed, the magnetic induction decreases to a value called the *remanent induction B_r*. The demagnetizing field required to reduce the magnetic induction of a magnetized ferromagnetic sample to zero is called the *coercive force H_c*.

A *soft magnetic material* is one that is easily magnetized and demagnetized. Important magnetic properties of a soft magnetic material are high permeability, high saturation induction, and low coercive force. When a soft ferromagnetic material is repeatedly magnetized and demagnetized, *hysteresis* and *eddy-current* energy losses occur. Examples of soft ferromagnetic materials include Fe–3 to 4% Si alloys used in motors and power transformers and generators and Ni–20 to 50% Fe alloys used primarily for high-sensitivity communications equipment.

A *hard magnetic material* is one that is difficult to magnetize and which remains magnetized to a great extent after the magnetizing field is removed. Important properties of a hard magnetic material are high coercive force and high saturation induction. The power of a hard magnetic material is measured by its maximum energy product, which is the maximum value of the product of B and H in the demagnetizing quadrant of its B-H hysteresis loop. Examples of hard magnetic materials are the alnicos, which are used as permanent magnets for many electrical applications, and some rare earth alloys that are based on $SmCo_5$ and $Sm(Co, Cu)_{7.5}$ compositions. The rare earth alloys are used for small motors and other applications requiring an extremely high energy-product magnetic material.

The *ferrites,* which are ceramic compounds, are another type of industrially important magnetic material. These materials are ferrimagnetic due to a net magnetic moment produced by their ionic structure. Most magnetically soft ferrites have the basic composition $MO \cdot Fe_2O_3$, where M is a divalent ion such as Fe^{2+}, Mn^{2+} and Ni^{2+}. These materials have the *inverse spinel structure* and are used for low-signal, memory-core, audiovisual, and recording-head applications, as examples. Since these materials are insulators, they can be used for high-frequency applications where eddy currents are a problem with alternating fields. Magnetically hard ferrites with the general formula $MO \cdot 6Fe_2O_3$, where M is usually a Ba or Sr ion, are used for applications requiring low-cost, low-density permanent magnetic materials. These materials are used for loudspeakers, telephone ringers and receivers, and for holding devices for doors, seals, and latches.

16.12 DEFINITIONS

Sec. 16.2

Ferromagnetic material: one that is capable of being highly magnetized. Elemental iron, cobalt, and nickel are ferromagnetic materials.

Magnetic field H: the magnetic field produced by an external applied magnetic field or the magnetic field produced by a current passing through a conducting wire or coil of wire (solenoid).

Magnetization M: a measure of the increase in magnetic flux due to the insertion of a given material into a magnetic field of strength H. In SI units the magnetization is equal to the permeability of a vacuum (μ_0) times the magnetization, or $\mu_0 M$. ($\mu_0 = 4\pi \times 10^{-4}$ T · m/A.)

Magnetic induction B: the sum of the applied field H and the magnetization M due to the insertion of a given material into the applied field. In SI units, $B = \mu_0(H + M)$.

Magnetic permeability μ: the ratio of the magnetic induction B to the applied magnetic field H for a material; $\mu = B/H$.

Relative permeability μ_r: the ratio of the permeability of a material to the permeability of a vacuum; $\mu_r = \mu/\mu_0$.

Magnetic susceptibility χ_m: the ratio of M (magnetization) to H (applied magnetic field); $\chi_m = M/H$.

Sec. 16.3

Diamagnetism: a weak, negative, repulsive reaction of a material to an applied magnetic field; a diamagnetic material has a small negative magnetic susceptibility.

Paramagnetism: a weak, positive, attractive reaction of a material to an applied magnetic field; a paramagnetic material has a small positive magnetic susceptibility.

Ferromagnetism: the creation of a very large magnetization in a material when subjected to an applied magnetic field. After the applied field is removed, the ferromagnetic material retains much of the magnetization.

Bohr magneton: the magnetic moment produced in a ferro- or ferrimagnetic material by one unpaired electron without interaction from any others; the Bohr magneton is a fundamental unit. 1 Bohr magneton = 9.27×10^{-24} A · m^2.

Antiferromagnetism: a type of magnetism in which magnetic dipoles of atoms are aligned in opposite directions by an applied magnetic field so that there is no net magnetization.

Ferrimagnetism: a type of magnetism in which the magnetic dipole moments of different ions of an ionically bonded solid are aligned by a magnetic field in an antiparallel manner so that there is a net magnetic moment.

Sec. 16.4

Curie temperature: the temperature at which a ferromagnetic material when heated completely loses its ferromagnetism and becomes paramagnetic.

Sec. 16.5

Magnetic domain: a region in a ferro- or ferrimagnetic material in which all magnetic dipole moments are aligned.

Sec. 16.6

Exchange energy: the energy associated with the coupling of individual magnetic dipoles into a single magnetic domain. The exchange energy can be positive or negative.

Magnetostatic energy: the magnetic potential energy due to the external magnetic field surrounding a sample of a ferromagnetic material.

Magnetocrystalline anisotropy energy: the energy required during the magnetization of a ferromagnetic material to rotate the magnetic domains because of crystalline anisotropy. For example, the difference in magnetizing energy between the hard [111] direction of magnetization and the [100] easy direction in Fe is about 1.4×10^4 J/m^3.

Domain wall energy: the potential energy associated with the disorder of dipole moments in the wall volume between magnetic domains.

Magnetostriction: the change in length of a ferromagnetic material in the direction of magnetization due to an applied magnetic field.

Magnetostrictive energy: the energy due to the mechanical stress caused by magnetostriction in a ferromagnetic material.

Sec. 16.7

Hysteresis loop: the B versus H or M versus H graph traced out by the magnetization and demagnetization of a ferro- or ferrimagnetic material.

Saturation induction B_s: the maximum value of induction B_s or magnetization M_s for a ferromagnetic material.

Remanent induction B_r: the value of B or M in a ferromagnetic material after H is decreased to zero.

Coercive force H_c: the applied magnetic field required to decrease the magnetic induction of a magnetized ferro- or ferrimagnetic material to zero.

Sec. 16.8

Soft magnetic material: a magnetic material with a high permeability and low coercive force.

Hysteresis energy loss: the work or energy lost in tracing out a B-H hysteresis loop. Most of the energy lost is expended in moving the domain boundaries during magnetization.

Eddy-current energy losses: energy losses in magnetic materials while using alternating fields; the losses are due to induced currents in the material.

Iron-silicon alloys: Fe–3 to 4% Si alloys that are soft magnetic materials with high saturation inductions. These alloys are used in motors and low-frequency power transformers and generators.

Nickel-iron alloys: high-permeability soft magnetic alloys used for electrical applications where a high sensitivity is required such as for audio and instrument transformers. Two commonly used basic compositions are 50% Ni–50% Fe and 79% Ni–21% Fe.

Sec. 16.9

Hard magnetic material: a magnetic material with a high coercive force and a high saturation induction.

Maximum energy product $(BH)_{max}$: the maximum value of B times H in the demagnetization curve of a hard magnetic material. The $(BH)_{max}$ value has SI units of J/m^3.

Alnico (aluminum-nickel-cobalt) alloys: a family of permanent magnetic alloys having the basic composition of Al, Ni, and Co, and about 25 to 50 percent Fe. A small amount of Cu and Ti is added to some of these alloys.

Magnetic anneal: the heat treatment of a magnetic material in a magnetic field that aligns part of the alloy in the direction of the applied field. For example, the α' precipitate in alnico 5 alloy is elongated and aligned by this type of heat treatment.

Rare earth alloys: a family of permanent magnetic alloys with extremely high energy products. $SmCo_5$ and $Sm(Co, Cu)_{7.4}$ are the two most important commercial compositions of these alloys.

Iron-chromium-cobalt alloys: a family of permanent magnetic alloys containing about 30% Cr–10 to 23% Co and the balance iron. These alloys have the advantage of being cold-formable at room temperature.

Sec. 16.10

Soft ferrites: ceramic compounds with the general formula $MO \cdot Fe_2O_3$, where M is a divalent ion such as Fe^{2+}, Mn^{2+}, Zn^{2+}, or Ni^{2+}. These materials are ferrimagnetic and are insulators and so can be used for high-frequency transformer cores.

Normal spinel structure: a ceramic compound having the general formula $MO \cdot M_2O_3$. The oxygen ions in this compound form an FCC lattice, with the M^{2+} ions occupying tetrahedral interstitial sites and the M^{3+} ions occupying octahedral sites.

Inverse spinal structure: a ceramic compound having the general formula $MO \cdot M_2O_3$. The oxygen ions in this compound form an FCC lattice, with the M^{2+} ions occupying octahedral sites and the M^{3+} ions occupying both octahedral and tetrahedral sites.

Hard ferrites: ceramic permanent magnetic materials. The most important family of these materials has the basic composition $MO \cdot Fe_2O_3$, where M is a barium (Ba) ion or a strontium (Sr) ion. These materials have a hexagonal structure and are low in cost and density.

16.13 PROBLEMS

Answers to problems marked with an asterisk are given at the end of the book.

16.1 What elements are strongly ferromagnetic at room temperature?

16.2 How can a magnetic field around a magnetized iron bar be revealed?

16.3 What are the SI and cgs units for the magnetic field strength H?

16.4 Define magnetic induction B and magnetization M.

16.5 What is the relationship between B and H?

16.6 What is the permeability of free-space constant μ_0?

16.7 What are the SI units for B and M?

16.8 Write an equation that relates B, H, and M, using SI units.

16.9 Why is the relation $B \approx \mu_0 M$ often used in magnetic property calculations?

16.10 Define magnetic permeability and relative magnetic permeability.

16.11 Is the magnetic permeability of a ferromagnetic material a constant? Explain.

16.12 What magnetic permeability quantities are frequently specified?

16.13 Define magnetic susceptibility. For what situation is this quantity often used?

16.14 Describe two mechanisms involving electrons by which magnetic fields are created.

16.15 Define diamagnetism. What is the order of magnitude for the magnetic susceptibility of diamagnetic materials at 20°C?

16.16 Define paramagnetism. What is the order of magnitude for the magnetic susceptibility of paramagnetic materials at 20°C?

16.17 Define ferromagnetism. Which elements are ferromagnetic?

16.18 What causes ferromagnetism in Fe, Co, and Ni?

16.19 How many unpaired $3d$ electrons are there per atom in Cr, Mn, Fe, Co, Ni, and Cu?

16.20 What are magnetic domains?

16.21 How does a positive exchange energy affect the alignment of magnetic dipoles in ferromagnetic materials?

16.22 What is the explanation given for the fact that Fe, Co, and Ni are ferromagnetic and Cr and Mn are not, even though all these elements have unpaired $3d$ electrons?

*16.23 Calculate a theoretical value for the saturation magnetization and saturation induction for nickel, assuming all unpaired $3d$ electrons contribute to the magnetization. (Ni is FCC and $a = 0.352$ nm.)

*16.24 Calculate a theoretical value for the saturation magnetization of pure cobalt metal assuming all unpaired $3d$ electrons contribute to the magnetization. (Co is HCP with $a = 0.25071$ nm and $c = 0.40686$ nm.)

16.25 Calculate a theoretical value for the saturation magnetization of pure gadolinium below 16°C assuming all seven unpaired $4f$ electrons contribute to the magnetization. (Gd is HCP with $a = 0.364$ nm and $c = 0.578$ nm.)

16.26 Cobalt has a saturation magnetization of 1.42×10^6 A/m. What is its average magnetic moment in Bohr magnetons per atom?

16.27 Nickel has an average of 0.604 Bohr magnetons/atom. What is its saturation induction?

16.28 Gadolinium at very low temperatures has an average of 7.1 Bohr magnetons per atom. What is its saturation magnetization?

16.29 Define antiferromagnetism. Which elements show this type of behavior?

16.30 Define ferrimagnetism. What are ferrites? Give an example of a ferrimagnetic compound.

16.31 What effect does increasing temperature above 0 K have on the alignment of magnetic dipoles in ferromagnetic materials?

16.32 What is the Curie temperature?

16.33 How can a ferromagnetic material be demagnetized? What is the arrangement of the magnetic domains in a demagnetized ferromagnetic material?

16.34 When a demagnetized ferromagnetic material is slowly magnetized by an applied magnetic field, what changes occur first in domain structure?

16.35 After domain growth due to magnetization of a ferromagnetic material by an applied field has finished, what change in domain structure occurs with a further substantial increase in the applied field?

16.36 What changes in domain structure occur in a ferromagnetic material when the applied field that magnetized a sample to saturation is removed?

16.37 How may the domain structure of a ferromagnetic material be revealed for observation in the optical microscope?

16.38 What are the five types of energies that determine the domain structure of a ferromagnetic material?

16.39 Define magnetic exchange energy. How can the exchange energy of a ferromagnetic material be minimized with respect to magnetic dipole alignment?

16.40 Define magnetostatic energy. How can the magnetostatic energy of a ferromagnetic material sample be minimized?

16.41 Define magnetocrystalline anisotropy energy. What are the easy directions of magnetization for (*a*) Fe and (*b*) Ni?

16.42 Define magnetic domain wall energy. What is the average width in terms of number of atoms for a ferromagnetic domain wall?

16.43 What two energies determine the domain wall width? What energy is minimized when the wall is widened? What energy is minimized when the wall is made narrower?

16.44 Define magnetostriction and magnetostrictive energy. What is the cause of magnetostriction in ferromagnetic materials?

16.45 What are domains of closure? How are magnetostrictive stresses created by domains of closure?

16.46 How does the domain size affect the magnetostrictive energy of a magnetized ferromagnetic material sample?

16.47 How does domain size affect the amount of domain wall energy in a sample?

16.48 Draw a hysteresis B-H loop for a ferromagnetic material and indicate on it (*a*) the saturation induction B_s, (*b*) the remanent induction B_r, and (*c*) the coercive force H_c.

16.49 Describe what happens to the magnetic induction when a ferromagnetic material is magnetized, demagnetized, and remagnetized by an applied magnetic field.

16.50 What happens to the magnetic domains of a ferromagnetic material sample during magnetization and demagnetization?

16.51 Define a soft magnetic material and a hard magnetic material.

16.52 What type of a hysteresis loop does a soft ferromagnetic material have?

16.53 What are desirable magnetic properties for a soft magnetic material?

16.54 What are hysteresis energy losses? What factors affect hysteresis losses?

16.55 How does the AC frequency affect the hysteresis losses of soft ferromagnetic materials? Explain.

16.56 What are eddy currents? How are they created in a ferromagnetic material?

16.57 How can eddy currents be reduced in metallic magnetic transformer cores?

16.58 Why does the addition of 3 to 4 percent silicon to iron reduce transformer-core energy losses?

16.59 What disadvantages are there to the addition of silicon to iron for transformer-core materials?

16.60 Why does a laminated structure increase the electrical efficiency of a power transformer?

16.61 Why does grain-oriented iron-silicon transformer sheet steel increase the efficiency of a transformer core?

16.62 What is the structure of a metallic glass? How are magnetic glass ribbons produced?

16.63 What are some special properties of metallic glasses?

16.64 Why are magnetic metallic glasses so easily magnetized and demagnetized?

16.65 What are the advantages of metallic glasses for power transformers? Disadvantages?

*16.66 Calculate the weight percent of the elements in the metallic glass with the atomic percent composition $Fe_{78}B_{13}Si_9$.

16.67 What are some engineering advantages of using nickel-iron alloys for electrical applications?

16.68 What compositions of Ni-Fe alloys are especially important for electrical applications?

16.69 How does ordering in an Ni-Fe alloy affect the magnetic properties of a 78.5% Ni–21.5% Fe alloy? How can ordering be prevented?

16.70 How does magnetic annealing increase the magnetic properties of a 65% Ni–35% Fe alloy?

16.71 What are important magnetic properties for a hard magnetic material?

16.72 What is the maximum energy product for a hard magnetic material? How is it calculated? What are the SI and cgs units for the energy product?

16.73 Estimate the maximum energy product for the $SmCo_5$ rare earth hard magnetic alloy (curve 2) of Fig. 16.24.

16.74 Estimate the maximum energy product for the alnico 5 alloy (curve 4) of Fig. 16.24.

16.75 Approximately how much energy in kilojoules per cubic meter would be required to demagnetize a fully magnetized 2 cm^3 block of alnico 8 alloy?

16.76 What elements are included in the alnico magnetic materials?

16.77 What two processes are used to produce alnico permanent magnets?

16.78 What is the basic structure of an alnico 8 magnetic material?

16.79 How does precipitation in a magnetic field affect the shape of the precipitates in an alnico 8 alloy? How does the shape of the precipitates affect the coercivity of this material?

16.80 What is the origin of ferromagnetism in rare earth magnetic alloys?

16.81 How do the maximum energy products of the alnicos compare with those of the rare earth magnetic alloys?

16.82 What are the two main groups of rare earth alloys?

16.83 What is believed to be the basic mechanism of coercivity for the $SmCo_5$ magnetic alloys?

16.84 What are some applications for rare earth magnetic alloys?

16.85 What fabrication advantage do the Fe-Cr-Co magnetic alloys have for making permanent magnetic alloy parts?

16.86 What is a typical chemical composition of an Fe-Cr-Co magnetic alloy?

16.87 What is the basic structure of an Fe-Cr-Co magnetic alloy? What is the mechanism for coercivity of the Fe-Cr-Co type magnetic alloys?

16.88 How does plastic deformation before the final aging treatment affect the shape of the precipitated particles and the coercivity of the Fe-Cr-Co magnetic alloys?

16.89 For what types of applications are the Fe-Cr-Co alloys particularly suited?

16.90 What are the ferrites? How are they produced?

16.91 What is the basic composition of the cubic soft ferrites?

16.92 Describe the unit cell of the spinel structure $MgO \cdot Al_2O_3$, including which ions occupy tetrahedral and octahedral interstitial sites.

16.93 Describe the unit cell of the inverse spinel structure, including which ions occupy tetrahedral and octahedral interstitial sites.

16.94 What is the net magnetic moment per molecule for each of the following ferrites: (*a*) $FeO \cdot Fe_2O_3$, (*b*) $NiO \cdot Fe_2O_3$, and (*c*) $MnO \cdot Fe_2O_3$?

16.95 Calculate the theoretical saturation magnetization in amperes per meter and saturation induction in teslas for the ferrite $NiO \cdot Fe_2O_3$ ($a = 0.834$ nm for $NiO \cdot Fe_2O_3$).

16.96 What are the compositions of the two most commonly used ferrites? Why are mixtures of ferrites used instead of a single pure ferrite?

16.97 Why is a high electrical resistivity necessary for a magnetic material that is to be used for a transformer core operating at a high frequency?

16.98 What are some industrial applications for soft ferrites?

16.99 Why are magnetic core memories particularly useful for high-shock-resistance applications?

16.100 What is the basic composition of the hexagonal hard ferrites?

16.101 What are the advantages of the hard ferrites for industrial use?

16.102 What are some applications for hard ferrite magnetic materials?

16.14 MATERIALS SELECTION AND DESIGN PROBLEMS

1. Draw the electronic structure of the $3d$ orbital of transition metals starting from Sc and ending with Cu. Based on your diagrams, compare the resulting magnetic moments.

2. Repeat Problem 1 for the $5d$ orbitals of transition metals from La to Au.

3. Amorphous metals have shown to be an attractive choice for transformer core materials. (*a*) Explain what an amorphous metal is. (*b*) Give reasons for its competitiveness.

4. (*a*) What properties should materials selected for a magnetic head recorder have? (*b*) Select a material for this application.

5. Extremely rapid magnetization can be accomplished by producing thin magnetic films over nonmagnetic substrates. (*a*) Propose a process that would achieve this structure. (*b*) Why is it important that the magnetic film be thin? (*c*) Give examples of materials used in the making of magnetic tapes and films.

6. What types of magnets would you select to make permanent magnetic poles for alternators and motors? Give a specific example.

7. Would you select soft or hard magnetic material for use in a transformer core? Why?

8. The magnetization of nickel ferrite, $NiFe_2O_4$, is due to nickel ions since ferric ions are antiferromagnetically coupled. If zinc is added to nickel ferrite, the magnetization of the crystal increases although zinc ions are not ferromagnetic themselves. Explain why the magnetization increases.

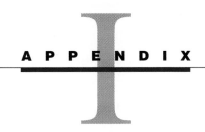

Important Properties of Selected Engineering Materials

1. ROOM TEMPERATURE DENSITY VALUES

		Density	
		g/cc	**lb/in³**
Plain Carbon and Low Alloy Steels			
Alloy 1006			
	cold drawn	7.872	0.284
	hot rolled	7.872	0.284
Alloy 1020			
	cold rolled	7.87	0.284
	annealed	7.87	0.284
Alloy 1040			
	Cold drawn	7.845	0.283
	annealed	7.845	0.283
Alloy 1090			
	annealed	7.85	0.284
	hot rolled	7.87	0.284
Alloy 4027			
	annealed	7.85	0.284
	normalized	7.85	0.284
Alloy 4140			
	annealed	7.85	0.284
	normalized	7.85	0.284
Alloy 4340			
	annealed	7.85	0.284
	normalized	7.85	0.284
Alloy 5140			
	annealed	7.85	0.284
	normalized	7.85	0.284

(continued)

Stainless Steels			
Alloy 302			
	annealed	7.86	0.284
	Cold rolled	7.86	0.284
Alloy 316			
	annealed	8	0.289
	soft tempered	8	0.289
Alloy 405			
	Cold rolled	7.8	0.282
	Annealed	7.8	0.282
Alloy 434			
	Annealed	7.8	0.282
Alloy 660			
	Cold rolled	7.92	0.286
Alloy 17-7PH			
	cold rolled	7.8	0.282
Cast Irons			
Gray Cast Irons (as cast)			
	ASTM class 20	7.15	0.258
	ASTM class 35	7.15	0.258
Ductile Cast Irons			
	grade 60-40-18	7.15	0.258
	grade 120-90-02	7.15	0.258
Aluminum Alloys			
Alloy 2024			
	Annealed	2.78	0.1
	T3 Temper	2.78	0.1
	T6 Temper	2.78	0.1
Alloy 6061			
	Annealed	2.7	0.0975
	T4 Temper	2.7	0.0975
	T6 Temper	2.7	0.0975
Alloy 7075			
	Annealed	2.81	0.102
	T6 Temper	2.81	0.102
Copper Alloys			
Electrolytic tough pitch Copper, (C11000)			
	Cold Worked	8.89	0.321
Berillium Copper, (C17000)			
	TH04 Temper	8.26	0.298
	TH01 Temper	8.26	0.298
Cartridge Brass, (C26000)			
	H00 Temper	8.53	0.308
	H04 Temper	8.53	0.308
Muntz metal, (C28000)			
	H00 Temper	8.39	0.303
	H02 Temper	8.39	0.303

(*continued*)

Free-Cutting Brass, (C36000)

	H01 Temper	8.5	0.307
	H02 Temper	8.5	0.307

Chromium Copper, (C18500)

	TH04 Temper	8.89	0.321
	TH04 Temper	8.89	0.321

90-10 Bronze, (C22000)

	H01 Temper	8.8	0.318
	H04 Temper	8.8	0.318

Magnesium Alloys		

Alloy AZ31B

	Annealed	1.77	0.0639
	Hard rolled	1.77	0.0639

Alloy AZ63

	F Temper	1.83	0.0661
	T6 Temper	1.83	0.0661

Alloy AZ91E

	F Temper	1.81	0.0654
	T4 Temper	1.81	0.0654

Nickel Alloys		

Nickel-Commercially Pure

	As Cast	8.88	0.321
	Annealed	8.88	0.321

Nickel Silver 65-18 (752)

	annealed	8.7	0.314
	hardened	8.7	0.314

Cupronickel 30(715)

	annealed	8.92	0.322
	hardened	8.92	0.322

Titanium Alloys		

Titanium-Commercially Pure

	As cast	4.5	0.163

AlloyTi-6Al-2Nb-1Ta-0.8Mo

	as-rolled	4.48	0.162
	Annealed	4.48	0.162

Miscellaneous Alloys		

	kovar Alloy	8.36	0.302

Invar

	Alloy 32-5 (Cold drawn)	8.14	0.294
	Alloy 36 Alloy (Cold Drawn)	8.05	0.291

Haynes 25

	cold rolled and annealed	9.13	0.33
	hot rolled and annealed	9.13	0.33

Biodur

	Hot Worlked	8.28	0.299
	Annealed	8.287	0.299
	Co-Cr-Ni Alloy	8.3	0.3

(*continued*)

	INCONEL (Cold Rolled)	8.05	0.291
	INCOLOY 903	8.25	0.298
	INCONEL 725 (Annealed)	8.3	0.3
	Berylium S-200	1.844	0.0666
	Tungsten	19.3	0.697
	Tantalum	16.6	0.6
	Zinc	7.1	0.257

Noble Metals			
	Gold	19.32	0.698
	Iridium	22.65	0.818
	Silver	10.491	0.379
	Platinum, Annealed	21.45	0.775

Polymers			
Elastomers			
	Styrene Butadiene Rubber (SBR)	0.93	0.0336
	Natural Rubber (Not Vulcanized)	0.93	0.0336
	Nitrile Rubber	1.25	0.0452
Epoxies			
	Epoxy (Unreinforced)	1.2	0.0434
	Nylon 66 Unreinforced	1.095	0.0396
	Phenolic (Unreinforced)	1.705	0.0695
	Polybutylene terepthalate (PBT)	1.355	0.049
Polycarbonate (PC)			
	Molded	1.31	0.0474
	Conductive	1.31	0.0474
	Polyester (Rigid)	1.65	0.0597
Polyetheretherketone (PEEK)			
	Glass Fiber Filled	1.475	0.0533
	Carbon Fiber Filled	1.436	0.0519
Polyethylene Terephthalate (PET)			
	Unreinforced	1.315	0.0475
	30% Glass Reinforced	1.595	0.0577
Polyethylene (Molded)			
	Low Density	0.925	0.0345
	Medium Density	0.938	0.0339
	High Density	1.159	0.0419
Acrylonitrile Butadiene Styrene (ABS)			
	Molded	1.115	0.0403
	40% Glass Fiber Filled	1.4275	0.0516
	40% Carbon Fiber Filled	1.305	0.0472
Polymethyl Methacrylate (PMMA)			
	Unreinforced	1.17	0.0423
	Glass Fiber Reinforced	1.445	0.0522
Polypropylene (PP)			
	Unreinforced	1.07	0.0387
	10% Glass Fiber Filled	1.18	0.0427
	10% Carbon Fiber filled	0.95	0.0343
	Polystyrene	1.06	0.0381
	Polytetrafluoroethylene (PTFE)	2.225	0.0804

(continued)

Polyvinyl Chloride (PVC)

Unreinforced	1.305	0.0472
30% Glass Fiber Reinforced	1.58	0.0571
Styrene Acrylonitrile	1.16	0.042
Cellulose Acetate	1.28	0.0463
Acrylonitrile	1.3	0.047
Polyacetal	1.39	0.0502
Polyphenylene Sulfide Molded	1.57	0.0567
Polysulfone	1.245	0.045

Ceramics, Graphite and Semiconducting Materials

Aluminum Oxide

99.9% Alumina	3.96	0.143
95% Alumina	3.68	0.1325
Concrete	1.28	0.0463

Diamond

Natural	3.515	0.127
Synthetic	3.36	0.1215
Gallium Arsenide	5.316	0.192

Glass

Borosilicate Glass	2.4	0.0867
Silica Glass	2.18	0.0788
Graphite Carbon	2.25	0.0813
Fused Silica	1.855	0.067

Silicon Carbide

sintered	3.1	0.112
sublimed	3.186	0.115
Silicon Nitride	3.2	0.116
Zirconia	5.68	0.205
Titanium Carbide	4.94	0.178
Tungsten Carbide	15.7	0.567
Zirconium Carbide	6.56	0.237
Alumina 96%	3.8	0.137
Boron Nitride	3.48	0.125
Sodium Chloride	2.17	0.0784
Calcium Flouride	3.18	0.115

Fibers

E-Glass Fiber Generic	2.57	0.0929
C-Glass Fiber Generic	2.54	0.0918
Carbon Fiber Precursor	1.81	0.0654

Composite Materials

Epoxy/Glass SMC	1.635	0.0591
Epoxy/Carbon Fiber Composite	1.53	0.0553
Polycarbonate Aramid Fiber Reinforced	1.27	0.0459
Aramid Fiber reinforced Nylon 6/6	1.22	0.0441
polycarbonate, carbon Fiber Reinforced	1.3	0.047
Plolycarbonate, carbon + Glass Fiber Reinforced	1.377	0.0498
30% Glass Fiber reinforced Clear Polycarbonate	1.43	0.0517

(continued)

Wood		
American Red Maple Wood	0.801	0.0289
American red Gum Wood	0.801	0.0289
American Alaska Cedar Wood	0.6	0.0216
American Redwood Wood	0.4	0.0145
American Red Oak Wood	0.21	0.0076
White Oak Wood	0.993	0.0359

2. ROOM TEMPERATURE HARDNESS OF MATERIALS

Materials		Hardness (Brinell unless specified)
Plain Carbon and Low Alloy Steels		
Alloy 1006		
	cold drawn	95
	hot rolled	86
Alloy 1020		
	cold rolled	121
	annealed	111
Alloy 1040		
	Cold drawn	170
	annealed	149
Alloy 1090		
	annealed	197
	hot rolled	248
Alloy 4027		
	annealed	143
	normalized	179
Alloy 4140		
	annealed	197
	normalized	302
Alloy 4340		
	annealed	217
	normalized	363
Alloy 5140		
	annealed	167
	normalized	229
Stainless Steels		
Alloy 302		
	annealed	147
Alloy 316		
	annealed	143
Alloy 405		
	Annealed	150

(*continued*)

Alloy 434

	Annealed	164
Alloy 17-7PH		
	cold rolled	378

Cast Irons		
Gray Cast Irons (as cast)		
	ASTM class 20	156
	ASTM class 35	212
Ductile Cast Irons		
	grade 60-40-18	167
	grade 120-90-02	331

Aluminum Alloys		
Alloy 2024		
	T3 Temper	120
	T6 Temper	125
Alloy 6061		
	Annealed	30
	T4 Temper	65
	T6 Temper	95
Alloy 7075		
	Annealed	60
	T6 Temper	150

Copper Alloys		
Electrolytic tough pitch Copper, (C11000)		
	Cold Worked	95
Berillium Copper, (C17000)		
	TH04 Temper	353
	TH01 Temper	320
Cartridge Brass, (C26000)		
	H00 Temper	109
	H04 Temper	154
Muntz metal, (C28000)		
	H00 Temper	102
	H02 Temper	135
Free-Cutting Brass, (C36000)		
	H01 Temper	112
	H02 Temper	143
Chromium Copper, (C18500)		
	TH04 Temper	154
	TH04 Temper	160
90-10 Bronze, (C22000)		
	H01 Temper	86
	H04 Temper	125

Magnesium Alloys		
Alloy AZ31B		
	Annealed	56
	Hard rolled	73

(continued)

Alloy AZ63

	F Temper	50
	T6 Temper	73

Alloy AZ91E

	F Temper	63
	T4 Temper	55

Nickel Alloys		

Nickel-Commercially Pure

	Annealed	79

Nickel Silver 65-18 (752)

	annealed	75
	hardened	151

Cupronickel 30(715)

	annealed	83
	hardened	147

Titanium Alloys		

Titanium-Commercially Pure

	As cast	70

AlloyTi-6Al-2Nb-1Ta-0.8Mo

	as-rolled	290
	Annealed	290

Miscellaneous Alloys		

	kovar Alloy	120

Invar

	Alloy 32-5 (Cold drawn)	135
	Alloy 36 Alloy (Cold Drawn)	184

Biodur

	Hot Worlked	400
	Annealed	290
	Co-Cr-Ni Alloy	192
	INCONEL 725 (Annealed)	170
	Tungsten	294

Noble Metals		

	Gold	25(vickers)
	Iridium	565
	Silver	25(vickers)
	Platinum, Annealed	40(Vickers)

Polymers		

Elastomers

	Styrene Butadiene Rubber (SBR)	71(Shore-A)
	Natural Rubber (Not Vulcanized)	60(Shore-A)
	Nitrile Rubber	65(Shore-A)

Epoxies

	Epoxy (Unreinforced)	95R(Rockwell-M)
	Nylon 66 Unreinforced	76R(Rockwell-M)
	Phenolic (Unreinforced)	110
	Polybutylene terepthalate (PBT)	81(Rockwell-M)

(continued)

Polycarbonate (PC)

	Molded	70(Rockwell-M)
	Conductive	110(Rockwell-R)
	Polyester (Rigid)	90(Rockwell-M)

Polyetheretherketone (PEEK)

	Glass Fiber Filled	101(Rockwell-M)
	Carbon Fiber Filled	96(Rockwell-M)

Polyethylene Terephthalate (PET)

	Unreinforced	95(Rockwell-M)
	30% Glass Reinforced	87(Rockwell-M)

Polyethylene (Molded)

	Low Density	60(Rockwell-R)
	Medium Density	59(Shore-D)
	High Density	62(Rockwell-R)

Acrylonitrile Butadiene Styrene (ABS)

	Molded	103(Rockwell-R)
	40% Glass Fiber Filled	110(Rockwell-R)
	40% Carbon Fiber Filled	110(Rockwell-R)

Polymethyl Methacrylate (PMMA)

	Unreinforced	85(Rockwell-M)
	Glass Fiber Reinforced	110(Rockwell-R)

Polypropylene (PP)

	Unreinforced	92(Rockwell-R)
	10% Glass Fiber Filled	96(Rockwell-R)
	Polystyrene	72(Rockwell-m)
	Polytetrafluoroethylene (PTFE)	58(Rockwell-R)

Polyvinyl Chloride (PVC)

	Unreinforced	99(Rockwell-R)
	30% Glass Fiber Reinforced	108(Rockwell-R)
	Styrene Acrylonitrile	84(Rockwell-M)
	Cellulose Acetate	70(Rockwell-R)
	Acrylonitrile	60(Rockwell-M)
	Polyacetal	78(Rockwell-M)
	Polyphenylene Sulfide Molded	90(Rockwell-M)
	Polysulfone	81(Rockwell-M)

Ceramics, Graphite and Semiconducting Materials

Aluminum Oxide

	99.9% Alumina	1365(Vickers)
	95% Alumina	1720(Vickers)

Diamond

	Natural	10(Mohs)
	Synthetic	10(Mohs)
	Gallium Arsenide	7500(Knoop)

Glass

	Borosilicate Glass	2(Knoop)
	Silica Glass	443
	Silicon Nitride	1600(Vickers)
	Titanium Carbide	3200(Vickers)
	Tungsten Carbide	800
	Zirconium Carbide	2600(Vickers)
	Boron Nitride	279
	Calcium Flouride	154

(continued)

Composite Materials	
Polycarbonate Aramid Fiber Reinforced	120(Rockwell-R)
polycarbonate, carbon Fiber Reinforced	119(Rockwell-R)
30% Glass Fiber reinforced Clear	
Polycarbonate	121(Rockwell-R)
Wood	
American Red Maple Wood	3(Wood Indentation)
American red Gum Wood	2.5(Wod Indentation)
American Alaska Cedar Wood	2(Wood Indentation)
American Redwood Wood	2.1(Wood Indentation)
American Red Oak Wood	4.3(Wood Indentation)
White Oak Wood	4.8(Wood Indentation)

3. ROOM TEMPERATURE ULTIMATE TENSILE STRENGTH

Materials		Tensile Strength ultimate	
		Mpa	Ksi
Plain Carbon and Low Alloy Steels			
Alloy 1006			
	cold drawn	330	47.85
	hot rolled	295	42.775
Alloy 1020			
	cold rolled	420	60.9
	annealed	395	57.275
Alloy 1040			
	Cold drawn	585	84.825
	annealed	515	74.675
Alloy 1090			
	annealed	696	100.92
	hot rolled	841	121.945
Alloy 4027			
	annealed	515	74.675
	normalized	640	92.8
Alloy 4140			
	annealed	655	94.975
	normalized	1020	147.9
Alloy 4340			
	annealed	745	108.025
	normalized	1282	185.89
Alloy 5140			
	annealed	570	82.65
	normalized	793	114.985

(*continued*)

Stainless Steels			
Alloy 302			
	annealed	620	89.9
	Cold rolled	585	84.825
Alloy 316			
	annealed	580	84.1
	soft tempered	689	99.905
Alloy 405			
	Cold rolled	469	68.005
	Annealed	448	64.96
Alloy 434			
	Annealed	517	74.965
Alloy 660			
	Cold rolled	1007	146.015
Alloy 17-7PH			
	cold rolled	1380	200.1
Cast Irons			
Gray Cast Irons (as cast)			
	ASTM class 20	152	22.04
	ASTM class 35	252	36.54
Ductile Cast Irons			
	grade 60-40-18	461	66.845
	grade 120-90-02	974	141.23
Aluminum Alloys			
Alloy 2024			
	T3 Temper	483	70.035
	T6 Temper	427	61.915
Alloy 6061			
	Annealed	124	17.98
	T4 Temper	241	34.945
	T6 Temper	310	44.95
Alloy 7075			
	Annealed	228	33.06
	T6 Temper	572	82.94
Copper Alloys			
Electrolytic tough pitch Copper, (C11000)			
	Cold Worked	345	50.025
Berillium Copper, (C17000)			
	TH04 Temper	1310	189.95
	TH01 Temper	1190	172.55
Cartridge Brass, (C26000)			
	H00 Temper	400	58
	H04 Temper	525	76.125
Muntz metal, (C28000)			
	H00 Temper	415	60.175
	H02 Temper	485	70.325

(continued)

Free-Cutting Brass, (C36000)

	H01 Temper	385	55.825
	H02 Temper	400	58

Chromium Copper, (C18500)

	TH04 Temper	530	76.85
	TH04 Temper	475	68.875

90-10 Bronze, (C22000)

	H01 Temper	310	44.95
	H04 Temper	420	60.9

Magnesium Alloys			

Alloy AZ31B

	Annealed	255	36.975
	Hard rolled	290	42.05

Alloy AZ63

	F Temper	200	29
	T6 Temper	275	39.875

Alloy AZ91E

	F Temper	165	23.925
	T4 Temper	275	39.875

Nickel Alloys			

Nickel-Commercially Pure

	As Cast	317	45.965
	Annealed	45	6.525

Nickel Silver 65-18 (752)

	annealed	386	55.97
	hardened	586	84.97

Cupronickel 30(715)

	annealed	379	54.955
	hardened	517	74.965

Titanium Alloys			

Titanium-Commercially Pure

	As cast	220	31.9

AlloyTi-6Al-2Nb-1Ta-0.8Mo

	as-rolled	830	120.35
	Annealed	830	120.35

Miscellaneous Alloys			

	kovar Alloy	517	74.965

Invar

	Alloy 32-5 (Cold drawn)	483	70.035
	Alloy 36 Alloy (Cold Drawn)	621	90.045

Haynes 25

	cold rolled and annealed	1005	145.725
	hot rolled and annealed	1015	147.175

Biodur

	Hot Worlked	1365	197.925
	Annealed	1035	150.075
	Co-Cr-Ni Alloy	860	124.7
	INCONEL (Cold Rolled)	757	109.765

(continued)

INCOLOY 903	1310	189.95
INCONEL 725 (Annealed)	855	123.975
Berylium S-200	765	110.925
Tungsten	980	142.1
Tantalum	689	99.905
Zinc	37	5.365

Noble Metals		
Gold	120	17.4
Iridium	2000	290
Silver	140	20.3
Platinum, Annealed	145	21.025

Polymers		

Elastomers

Styrene Butadiene Rubber (SBR)	30	4.35
Natural Rubber (Not Vulcanized)	28	4.06
Nitrile Rubber	17	2.465

Epoxies

Epoxy (Unreinforced)	70	10.15
Nylon 66 Unreinforced	62	8.99
Phenolic (Unreinforced)	56	8.12
Polybutylene terepthalate (PBT)	55	7.975

Polycarbonate (PC)

Molded	65	9.425
Conductive	86	12.47
Polyester (Rigid)	58	8.41

Polyetheretherketone (PEEK)

Glass Fiber Filled	140	20.3
Carbon Fiber Filled	176	25.52

Polyethylene Terephthalate (PET)

Unreinforced	55	7.975
30% Glass Reinforced	137	19.865

Polyethylene (Molded)

Low Density	16	2.32
Medium Density	25	3.625
High Density	30	4.35

Acrylonitrile Butadiene Styrene (ABS)

Molded	47	6.815
40% Glass Fiber Filled	80	11.6
40% Carbon Fiber Filled	115	16.675

Polymethyl Methacrylate (PMMA)

Unreinforced	62	8.99
Glass Fiber Reinforced	107	15.515

Polypropylene (PP)

Unreinforced	50	7.25
10% Glass Fiber Filled	36	5.22
10% Carbon Fiber filled	26	3.77
Polystyrene	45	6.525
Polytetrafluoroethylene (PTFE)	25	3.625

(*continued*)

Polyvinyl Chloride (PVC)

	Unreinforced	47	6.815
	30% Glass Fiber Reinforced	90	13.05
	Styrene Acrylonitrile	65	9.425
	Cellulose Acetate	33	4.785
	Acrylonitrile	66	9.57
	Polyacetal	60	8.7
	Polyphenylene Sulfide Molded	95	13.775
	Polysulfone	73	10.585

Ceramics, Graphite and Semiconducting Materials

Aluminum Oxide

	99.9% Alumina	300	43.5
	95% Alumina	205	29.725

Diamond

	Natural	1050	152.25
	Synthetic	1200	174
	Gallium Arsenide	66	9.57

Glass

	Borosilicate Glass	57	8.265
	Silica Glass	69	10.005
	Fused Silica	104	15.08

Silicon Carbide

	sintered	300	43.5
	Silicon Nitride	800	116
	Titanium Carbide	258	37.41
	Tungsten Carbide	344	49.88
	Zirconium Carbide	89	12.905
	Alumina 96%	200	29
	Boron Nitride	47	6.815

Fibers

	E-Glass Fiber Generic	3448	499.96
	C-Glass Fiber Generic	3310	479.95
	Carbon Fiber Precursor	5650	819.25

Composite Materials

	Epoxy/Glass SMC	243	35.235
	Epoxy/Carbon Fiber Composite	1.6	0.232
	Polycarbonate Aramid Fiber Reinforced	67	9.715
	Aramid Fiber reinforced Nylon 6/6	119	17.255
	polycarbonate, carbon Fiber Reinforced	150	21.75
	Plolycarbonate, carbon+Glass Fiber Reinforced	135	19.575

Wood

	American Red Maple Wood	4	0.58
	American red Gum Wood	4	0.58
	American Alaska Cedar Wood	2	0.29
	American Redwood Wood	2	0.29
	American Red Oak Wood	5.5	0.7975
	White Oak Wood	6	0.87

4. ROOM TEMPERATURE YIELD STRENGTH

Materials		Yield Strength	
		MPa	**Ksi**
Plain Carbon and Low Alloy Steels			
Alloy 1006			
	cold drawn	285	41.325
	hot rolled	165	23.925
Alloy 1020			
	cold rolled	350	50.75
	annealed	295	42.775
Alloy 1040			
	Cold drawn	515	74.675
	annealed	350	50.75
Alloy 1090			
	annealed	540	78.3
	hot rolled	460	66.7
Alloy 4027			
	annealed	325	47.125
	normalized	420	60.9
Alloy 4140			
	annealed	415	60.175
	normalized	655	94.975
Alloy 4340			
	annealed	470	68.15
	normalized	862	124.99
Alloy 5140			
	annealed	295	42.775
	normalized	470	68.15
Stainless Steels			
Alloy 302			
	annealed	275	39.875
	Cold rolled	255	36.975
Alloy 316			
	annealed	290	42.05
	soft tempered	515	74.675
Alloy 405			
	Cold rolled	276	40.02
	Annealed	276	40.02
Alloy 434			
	Annealed	345	50.025
Alloy 660			
	Cold rolled	703	101.935
Alloy 17-7PH			
	cold rolled	1210	175.45
Cast Irons			
Ductile Cast Irons			
	grade 60-40-18	329	47.705
	grade 120-90-02	864	125.28

(continued)

Aluminum Alloys			
Alloy 2024			
	T3 Temper	345	50.025
	T6 Temper	345	50.025
Alloy 6061			
	Annealed	55	7.975
	T4 Temper	145	21.025
	T6 Temper	276	40.02
Alloy 7075			
	Annealed	103	14.935
	T6 Temper	503	72.935
Copper Alloys			
Electrolytic tough pitch Copper, (C11000)			
	Cold Worked	310	44.95
Berillium Copper, (C17000)			
	TH04 Temper	1150	166.75
	TH01 Temper	1110	160.95
Cartridge Brass, (C26000)			
	H00 Temper	315	45.675
	H04 Temper	435	63.075
Muntz metal, (C28000)			
	H00 Temper	240	34.8
	H02 Temper	345	50.025
Free-Cutting Brass, (C36000)			
	H01 Temper	310	44.95
	H02 Temper	310	44.95
Chromium Copper, (C18500)			
	TH04 Temper	450	65.25
	TH04 Temper	435	63.075
90-10 Bronze, (C22000)			
	H01 Temper	240	34.8
	H04 Temper	370	53.65
Magnesium Alloys			
Alloy AZ31B			
	Annealed	150	21.75
	Hard rolled	220	31.9
Alloy AZ63			
	F Temper	97	14.065
	T6 Temper	130	18.85
Alloy AZ91E			
	F Temper	97	14.065
	T4 Temper	90	13.05
Nickel Alloys			
Nickel-Commercially Pure			
	As Cast	59	8.555

(*continued*)

Titanium Alloys		
Titanium-Commercially Pure		
As cast	140	20.3
AlloyTi-6Al-2Nb-1Ta-0.8Mo		
as-rolled	760	110.2
Annealed	710	102.95

Miscellaneous Alloys		
kovar Alloy	345	50.025
Invar		
Alloy 32-5 (Cold drawn)	276	40.02
Alloy 36 Alloy (Cold Drawn)	483	70.035
Haynes 25		
cold rolled and annealed	475	68.875
hot rolled and annealed	505	73.225
Biodur		
Hot Worked	930	134.85
Annealed	585	84.825
Co-Cr-Ni Alloy	520	75.4
INCONEL (Cold Rolled)	383	55.535
INCOLOY 903	1100	159.5
INCONEL 725 (Annealed)	427	61.915
Berylium S-200	414	60.03
Tungsten	750	108.75

Polymers		
Epoxies		
Epoxy (Unreinforced)	60	8.7
Nylon 66 Unreinforced	63	9.135
Phenolic (Unreinforced)	52	7.54
Polybutylene terepthalate (PBT)	67	9.715
Polycarbonate (PC)		
Molded	65	9.425
Polyester (Rigid)	70	10.15
Polyetheretherketone (PEEK)		
Glass Fiber Filled	98	14.21
Polyethylene Terephthalate (PET)		
Unreinforced	54	7.83
30% Glass Reinforced	141	20.445
Polyethylene (Molded)		
Low Density	16	2.32
Medium Density	15	2.175
High Density	16	2.32
Acrylonitrile Butadiene Styrene (ABS)		
Molded	46	6.67
40% Glass Fiber Filled	110	15.95
40% Carbon Fiber Filled	120	17.4
Polymethyl Methacrylate (PMMA)		
Unreinforced .	69	10.005
Polypropylene (PP)		
Unreinforced	28	4.06
10% Glass Fiber Filled	41	5.945
Polystyrene	46	6.67
Polytetrafluoroethylene (PTFE)	19	2.755

(continued)

Polyvinyl Chloride (PVC)			
	Unreinforced	38	5.51
	Styrene Acrylonitrile	65	9.425
	Cellulose Acetate	26	3.77
	Polyacetal	69	10.005
	Polyphenylene Sulfide Molded	6809	987.305
	Polysulfone	74	10.73
Ceramics, Graphite and Semiconducting Materials			
	polycarbonate, carbon Fiber Reinforced	110	15.95
	30% Glass Fiber reinforced Clear Polycarbonate	124	17.98

5. ROOM TEMPERATURE MODULUS OF ELASTICITY

Materials		Modulus of Elasticity	
		Gpa	**10^6 Psi**
Plain Carbon and Low Alloy Steels			
Alloy 1006			
	cold drawn	205	29.71
	hot rolled	200	28.99
Alloy 1020			
	cold rolled	205	29.71
	annealed	200	28.99
Alloy 1040			
	Cold drawn	200	28.99
	annealed	200	28.99
Alloy 1090			
	annealed	205	29.71
	hot rolled	200	28.99
Alloy 4027			
	annealed	205	29.71
	normalized	205	29.71
Alloy 4140			
	annealed	205	29.71
	normalized	205	29.71
Alloy 4340			
	annealed	205	29.71
	normalized	205	29.71
Alloy 5140			
	annealed	205	29.71
	normalized	205	29.71
Stainless Steels			
Alloy 302			
	annealed	193	27.97
	Cold rolled	193	27.97
Alloy 316			
	annealed	193	27.97
	soft tempered	193	27.97

(continued)

Alloy 405

	Cold rolled	200	28.99
	Annealed	200	28.99

Alloy 434

	Annealed	200	28.99

Alloy 660

	Cold rolled	200	28.99

Alloy 17-7PH

	cold rolled	204	29.57

Cast Irons			

Gray Cast Irons (as cast)

	ASTM class 20	83	12.03
	ASTM class 35	114	16.52

Ductile Cast Irons

	grade 60-40-18	169	24.49
	grade 120-90-02	164	23.77

Aluminum Alloys			

Alloy 2024

	Annealed	73	10.58
	T3 Temper	73	10.58
	T6 Temper	72	10.43

Alloy 6061

	Annealed	69	10.00
	T4 Temper	69	10.00
	T6 Temper	69	10.00

Alloy 7075

	Annealed	72	10.43
	T6 Temper	72	10.43

Copper Alloys			

Electrolytic tough pitch Copper, (C11000)

	Cold Worked	122	17.68

Berillium Copper, (C17000)

	TH04 Temper	115	16.67
	TH01 Temper	115	16.67

Cartridge Brass, (C26000)

	H00 Temper	110	15.94
	H04 Temper	110	15.94

Muntz metal, (C28000)

	H00 Temper	105	15.22
	H02 Temper	105	15.22

Free-Cutting Brass, (C36000)

	H01 Temper	97	14.06
	H02 Temper	97	14.06

Chromium Copper, (C18500)

	TH04 Temper	130	18.84
	TH04 Temper	130	18.84

90-10 Bronze, (C22000)

	H01 Temper	115	16.67
	H04 Temper	115	16.67

(*continued*)

Magnesium Alloys			
Alloy AZ31B			
	Annealed	45	6.52
	Hard rolled	45	6.52
Alloy AZ63			
	F Temper	45	6.52
	T6 Temper	45	6.52
Alloy AZ91E			
	F Temper	45	6.52
	T4 Temper	45	6.52
Nickel Alloys			
Nickel-Commercially Pure			
	As Cast	207	30.00
	Annealed	207	30.00
Nickel Silver 65-18 (752)			
	annealed	124	17.97
	hardened	124	17.97
Cupronickel 30(715)			
	annealed	152	22.03
	hardened	152	22.03
Titanium Alloys			
Titanium-Commercially Pure			
	As cast	116	16.81
AlloyTi-6Al-2Nb-1Ta-0.8Mo			
	as-rolled	117	16.96
	Annealed	115	16.67
Miscellaneous Alloys			
	kovar Alloy	138	20.00
Invar			
	Alloy 32-5 (Cold drawn)	145	21.01
	Alloy 36 Alloy (Cold Drawn)	148	21.45
Haynes 25			
	cold rolled and annealed	225	32.61
	hot rolled and annealed	225	32.61
	Co-Cr-Ni Alloy	190	27.54
	INCONEL (Cold Rolled)	210	30.43
	INCOLOY 903	147	21.30
	INCONEL 725 (Annealed)	204	29.57
	Berylium S-200	303	43.91
	Tungsten	400	57.97
	Tantalum	186	26.96
	Zinc	97	14.06
Noble Metals			
	Gold	77	11.16
	Iridium	524	75.94
	Silver	76	11.01
	Platinum, Annealed	1711	247.97

(*continued*)

Polymers		
Epoxies		
Epoxy (Unreinforced)	2.25	0.33
Nylon 66 Unreinforced	2.1	0.30
Phenolic (Unreinforced)	7	1.01
Polybutylene terepthalate (PBT)	12	1.74
Polycarbonate (PC)		
Molded	2.3	0.33
Conductive	5	0.72
Polyester (Rigid)	3.5	0.51
Polyetheretherketone (PEEK)		
Glass Fiber Filled	21	3.04
Carbon Fiber Filled	11	1.59
Polyethylene Terephthalate (PET)		
Unreinforced	2.7	0.39
30% Glass Reinforced	10	1.45
Polyethylene (Molded)		
Low Density	0.25	0.04
Medium Density	0.5	0.07
High Density	0.9	0.13
Acrylonitrile Butadiene Styrene (ABS)		
Molded	2.4	0.35
40% Glass Fiber Filled	9	1.30
40% Carbon Fiber Filled	22	3.19
Polymethyl Methacrylate (PMMA)		
Unreinforced	2.9	0.42
Glass Fiber Reinforced	9.2	1.33
Polypropylene (PP)		0.00
Unreinforced	2.25	0.33
10% Glass Fiber Filled	3.75	0.54
10% Carbon Fiber filled	4.75	0.69
Polystyrene	3.25	0.47
Polytetrafluoroethylene (PTFE)	0.6	0.09
Polyvinyl Chloride (PVC)		
Unreinforced	3.1	0.45
30% Glass Fiber Reinforced	5	0.72
Styrene Acrylonitrile	3.7	0.54
Cellulose Acetate	1.7	0.25
Acrylonitrile	3.6	0.52
Polyacetal	2.4	0.35
Polyphenylene Sulfide Molded	3.5	0.51
Polysulfone	2.6	0.38
Ceramics, Graphite and Semiconducting Materials		
Aluminum Oxide		
99.9% Alumina	370	53.62
95% Alumina	282	40.87
Diamond		
Natural	950	137.68
Synthetic	865	125.36
Gallium Arsenide	85.5	12.39

(continued)

Glass

	Borosilicate Glass	62	8.99
	Silica Glass	68	9.86
Silicon Carbide			
	sintered	410	59.42
	sublimed	467	67.68
	Silicon Nitride	290	42.03
	Titanium Carbide	450	65.22
	Tungsten Carbide	682	98.84
	Zirconium Carbide	395	57.25
	Alumina 96%	300	43.48
	Boron Nitride	69	10.00

Fibers		
E-Glass Fiber Generic	72	10.43
C-Glass Fiber Generic	69	10.00
Carbon Fiber Precursor	290	42.03

Composite Materials		
Epoxy/Glass SMC	55	7.97
Epoxy/Carbon Fiber Composite	225	32.61
Polycarbonate Aramid Fiber Reinforced	3.5	0.51
Aramid Fiber reinforced Nylon 6/6	9	1.30
polycarbonate, carbon Fiber Reinforced	13	1.88
Plolycarbonate, carbon+Glass Fiber Reinforced	10	1.45

Wood		
American Red Maple Wood	11.3	1.64
American red Gum Wood	11.31	1.64
American Alaska Cedar Wood	7.89	1.14
American Redwood Wood	9.24	1.34
American Red Oak Wood	12.5	1.81
White Oak Wood	12.6	1.83

6. ROOM TEMPERATURE POISSON'S RATIO

Materials	Poisson's Ratio
Plain Carbon and Low Alloy Steels	
Alloy 1006	
cold drawn	0.29
hot rolled	0.29
Alloy 1020	
cold rolled	0.29
annealed	0.29
Alloy 1040	
Cold drawn	0.29
annealed	0.29

(continued)

Alloy 1090

	annealed	0.29
	hot rolled	0.29

Alloy 4027

	annealed	0.29
	normalized	0.29

Alloy 4140

	annealed	0.29
	normalized	0.29

Alloy 4340

	annealed	0.29
	normalized	0.29

Alloy 5140

	annealed	0.29
	normalized	0.29

Stainless Steels

Alloy 302

	annealed	0.25
	Cold rolled	0.25

Alloy 316

	annealed	0.27
	soft tempered	0.27

Alloy 405

	Cold rolled	0.27
	Annealed	0.27

Alloy 434

	Annealed	0.27

Alloy 660

	Cold rolled	0.27

Alloy 17-7PH

	cold rolled	0.27

Cast Irons

Gray Cast Irons (as cast)

	ASTM class 20	0.29
	ASTM class 35	0.29

Ductile Cast Irons

	grade 60-40-18	0.29
	grade 120-90-02	0.28

Aluminum Alloys

Alloy 2024

	Annealed	0.33
	T3 Temper	0.33
	T6 Temper	0.33

Alloy 6061

	Annealed	0.33
	T4 Temper	0.33
	T6 Temper	0.33

Alloy 7075

	Annealed	0.33
	T6 Temper	0.33

(continued)

Copper Alloys		
Electrolytic tough pitch Copper, (C11000)		
	Cold Worked	0.33
Berillium Copper, (C17000)		
	TH04 Temper	0.3
	TH01 Temper	0.3
Cartridge Brass, (C26000)		
	H00 Temper	0.375
	H04 Temper	0.375
Muntz metal, (C28000)		
	H00 Temper	0.346
	H02 Temper	0.346
Free-Cutting Brass, (C36000)		
	H01 Temper	0.311
	H02 Temper	0.311
Chromium Copper, (C18500)		
	TH04 Temper	0.3
	TH04 Temper	0.3
90-10 Bronze, (C22000)		
	H01 Temper	0.307
	H04 Temper	0.307
Magnesium Alloys		
Alloy AZ31B		
	Annealed	0.35
	Hard rolled	0.35
Alloy AZ63		
	F Temper	0.35
	T6 Temper	0.35
Alloy AZ91E		
	F Temper	0.35
	T4 Temper	0.35
Nickel Alloys		
Nickel-Commercially Pure		
	As Cast	0.31
	Annealed	0.31
Nickel Silver 65-18 (752)		
	annealed	0.33
	hardened	0.33
Cupronickel 30(715)		
	annealed	0.33
	hardened	0.33
Titanium Alloys		
Titanium-Commercially Pure		
	As cast	0.34
AlloyTi-6Al-2Nb-1Ta-0.8Mo		
	as-rolled	0.31
	Annealed	0.31

(continued)

	Miscellaneous Alloys	
	kovar Alloy	0.317
Invar		
	Alloy 32-5 (Cold drawn)	0.23
	Alloy 36 Alloy (Cold Drawn)	0.23
Haynes 25		
	cold rolled and annealed	0.24
	hot rolled and annealed	0.24
	Co-Cr-Ni Alloy	0.226
	INCONEL (Cold Rolled)	0.382
	INCOLOY 903	0.239
	INCONEL 725 (Annealed)	0.31
	Berylium S-200	0.14
	Tungsten	0.28
	Tantalum	0.35
	Zinc	0.25
	Noble Metals	
	Gold	0.42
	Iridium	0.26
	Silver	0.39
	Platinum, Annealed	0.39
	Polymers	
	Nylon 66 Unreinforced	0.39
Polycarbonate (PC)		
	Molded	0.36
Polyetheretherketone (PEEK)		
	Glass Fiber Filled	0.45
	Carbon Fiber Filled	0.44
Polyethylene Terephthalate (PET)		
	30% Glass Reinforced	0.4
Polyethylene (Molded)		
	Low Density	0.38
	Polystyrene	0.33
	Polytetrafluoroethylene (PTFE)	0.46
Polyvinyl Chloride (PVC)		
	Unreinforced	0.38
	30% Glass Fiber Reinforced	0.41
	Polyacetal	0.35
	Ceramics, Graphite and Semiconducting Materials	
Aluminum Oxide		
	99.9% Alumina	0.22
Diamond		
	Natural	0.18
	Synthetic	0.2
Glass		
	Silica Glass	0.19

(continued)

Silicon Carbide

	sintered	0.14
	Silicon Nitride	0.25
	Titanium Carbide	0.18
	Zirconium Carbide	0.19
	Alumina 96%	0.22
Fibers		
	E-Glass Fiber Generic	0.2
	C-Glass Fiber Generic	0.276
Composite Materials		
	Epoxy/Glass SMC	0.36
	Epoxy/Carbon Fiber Composite	0.34
Wood		
	American Red Maple Wood	0.04
	American Redwood Wood	0.36
	American Red Oak Wood	0.33
	White Oak Wood	0.036

7. ROOM TEMPERATURE PERCENT ELONGATION AT BREAK

Materials	%Elongation at break
Plain Carbon and Low Alloy Steels	
Alloy 1006	
cold drawn	20
hot rolled	30
Alloy 1020	
cold rolled	15
annealed	36.5
Alloy 1040	
Cold drawn	12
annealed	30
Alloy 1090	
annealed	10
hot rolled	10
Alloy 4027	
annealed	30
normalized	25.8
Alloy 4140	
annealed	25.7
normalized	17.7
Alloy 4340	
annealed	22
normalized	12.2
Alloy 5140	
annealed	28.6
normalized	22.7

(continued)

Stainless Steels		
Alloy 302		
	annealed	55
	Cold rolled	57
Alloy 316		
	annealed	50
	soft tempered	50
Alloy 405		
	Cold rolled	30
	Annealed	30
Alloy 434		
	Annealed	25
Alloy 660		
	Cold rolled	25
Alloy 17-7PH		
	cold rolled	1

Cast Irons		
Ductile Cast Irons		
	grade 60-40-18	15
	grade 120-90-02	1.5

Aluminum Alloys		
Alloy 2024		
	T3 Temper	18
	T6 Temper	5
Alloy 6061		
	Annealed	25
	T4 Temper	22
	T6 Temper	12
Alloy 7075		
	Annealed	16
	T6 Temper	11

Copper Alloys		
Electrolytic tough pitch Copper, (C11000)		
	Cold Worked	12
Berillium Copper, (C17000)		
	TH04 Temper	2.5
	TH01 Temper	4.5
Cartridge Brass, (C26000)		
	H00 Temper	35
	H04 Temper	8
Muntz metal, (C28000)		
	H00 Temper	30
	H02 Temper	10
Free-Cutting Brass, (C36000)		
	H01 Temper	20
	H02 Temper	25

(continued)

Chromium Copper, (C18500)

TH04 Temper	16
TH04 Temper	26

90-10 Bronze, (C22000)

H01 Temper	25
H04 Temper	5

Magnesium Alloys	

Alloy AZ31B

Annealed	21
Hard rolled	15

Alloy AZ63

F Temper	6
T6 Temper	5

Alloy AZ91E

F Temper	2.5
T4 Temper	15

Nickel Alloys	

Nickel-Commercially Pure

As Cast	30

Nickel Silver 65-18 (752)

annealed	35
hardened	3

Cupronickel 30 (715)

annealed	45
hardened	15

Titanium Alloys	

Titanium-Commercially Pure

As cast	54

AlloyTi-6Al-2Nb-1Ta-0.8Mo

as-rolled	10
Annealed	11

Miscellaneous Alloys	

Invar

Alloy 32-5 (Cold drawn)	40
Alloy 36 Alloy (Cold Drawn)	20

Haynes 25

cold rolled and annealed	51
hot rolled and annealed	60

Biodur

Hot Worlked	22
Annealed	25
Co-Cr-Ni Alloy	38
INCONEL (Cold Rolled)	47
INCOLOY 903	14
INCONEL 725 (Annealed)	57
Berylium S-200	10

(continued)

Noble Metals		
	Gold	30
	Iridium	20
	Silver	46
	Platinum, Annealed	35

Polymers		
Elastomers		
	Styrene Butadiene Rubber (SBR)	500
	Natural Rubber (Not Vulcanized)	500
	Nitrile Rubber	320
Epoxies		
	Epoxy (Unreinforced)	5
	Nylon 66 Unreinforced	152
	Phenolic (Unreinforced)	1.3
	Polybutylene terepthalate (PBT)	148
Polycarbonate (PC)		
	Molded	62
	Conductive	6
	Polyester (Rigid)	2.4
Polyetheretherketone (PEEK)		
	Glass Fiber Filled	3
	Carbon Fiber Filled	2
Polyethylene Terephthalate (PET)		
	Unreinforced	200
	30% Glass Reinforced	3.8
Polyethylene (Molded)		
	Low Density	350
	Medium Density	550
	High Density	750
Acrylonitrile Butadiene Styrene (ABS)		
	Molded	55
	40% Glass Fiber Filled	1.5
	40% Carbon Fiber Filled	1.25
Polymethyl Methacrylate (PMMA)		
	Unreinforced	15
	Glass Fiber Reinforced	7.8
Polypropylene (PP)		
	Unreinforced	427
	10% Glass Fiber Filled	250
	10% Carbon Fiber filled	9
	Polystyrene	22.5
	Polytetrafluoroethylene (PTFE)	325
Polyvinyl Chloride (PVC)		
	Unreinforced	62
	30% Glass Fiber Reinforced	2.5
	Styrene Acrylonitrile	3.8
	Cellulose Acetate	35
	Acrylonitrile	3.5
	Polyacetal	105
	Polyphenylene Sulfide Molded	3
	Polysulfone	42.5

(*continued*)

Fibers	
E-Glass Fiber Generic	4.8
C-Glass Fiber Generic	4.8
Carbon Fiber Precursor	1.8
Composite Materials	
Epoxy/Glass SMC	0.49
Polycarbonate Aramid Fiber Reinforced	6.2
Aramid Fiber reinforced Nylon 6/6	1.6
polycarbonate, carbon Fiber Reinforced	4.5
Plolycarbonate, carbon + Glass Fiber Reinforced	2.75

8. ROOM TEMPERATURE PERCENT REDUCTION IN AREA

Materials		%Reduction of area
Plain Carbon and Low Alloy Steels		
Alloy 1006		
	cold drawn	45
	hot rolled	55
Alloy 1020		
	cold rolled	40
	annealed	66
Alloy 1040		
	Cold drawn	35
	annealed	57
Alloy 1090		
	annealed	40
	hot rolled	25
Alloy 4027		
	annealed	52.9
	normalized	60.2
Alloy 4140		
	annealed	56.9
	normalized	46.8
Alloy 4340		
	annealed	50
	normalized	36.3
Alloy 5140		
	annealed	59.2
	normalized	57.3
Stainless Steels		
Alloy 660		
	Cold rolled	36.8

(continued)

Copper Alloys		
Free-Cutting Brass, (C36000)		50
Titanium Alloys		
AlloyTi-6Al-2Nb-1Ta-0.8Mo		
	as-rolled	28
	Annealed	30
Miscellaneous Alloys		
	Alloy 36 Alloy (Cold Drawn)	60
Biodur		
	Hot Worlked	68
	Annealed	23

9. ROOM TEMPERATURE IZOD IMPACT STRENGTH

		Izod Impact	
Materials		**J**	**Ft-lb**
Plain Carbon and Low Alloy Steels			
	annealed	125	92.225
Alloy 1040			
	Cold drawn	49	36.1522
	annealed	45	33.201
Alloy 1090			
	hot rolled	4	2.9512
Alloy 4140			
	annealed	62	45.7436
	normalized	87	64.1886
Alloy 4340			
	annealed	51	37.6278
	normalized	16	11.8048
Alloy 5140			
	annealed	41	30.2498
Stainless Steels			
Alloy 302			
	annealed	136	100.3408
Alloy 316			
	annealed	129	95.1762
Alloy 405			
	Cold rolled	41	30.2498
	Annealed	41	30.2498
			(continued)

Polymers		
Epoxies		
	Epoxy (Unreinforced)	0.3	0.22134
	Nylon 66 Unreinforced	7	5.1646
	Phenolic (Unreinforced)	0.18	0.132804
	Polybutylene terepthalate (PBT)	0.27	0.199206
Polycarbonate (PC)		
	Molded	4.78	3.526684
	Conductive	0.52	0.383656
	Polyester (Rigid)	0.22	0.162316
Polyetheretherketone (PEEK)		
	Glass Fiber Filled	1.1	0.81158
	Carbon Fiber Filled	0.12	0.088536
Polyethylene Terephthalate (PET)		
	Unreinforced	1.4	1.03292
	30% Glass Reinforced	1.1	0.81158
Polyethylene (Molded)		
	Low Density	1.068	0.7879704
	Medium Density	4.8	3.54144
	High Density	4.01	2.958578
Acrylonitrile Butadiene Styrene (ABS)		
	Molded	3.2	2.36096
	40% Glass Fiber Filled	0.63	0.464814
	40% Carbon Fiber Filled	0.5	0.3689
Polymethyl Methacrylate (PMMA)		
	Unreinforced	0.16	0.118048
	Glass Fiber Reinforced	0.45	0.33201
Polypropylene (PP)		
	Unreinforced	0.16	0.118048
	10% Glass Fiber Filled	1.12	0.826336
	10% Carbon Fiber filled	0.8	0.59024
	Polystyrene	0.18	0.132804
	Polytetrafluoroethylene (PTFE)	1.6	1.18048
Polyvinyl Chloride (PVC)		
	Unreinforced	5.3	3.91034
	30% Glass Fiber Reinforced	0.8	0.59024
	Styrene Acrylonitrile	0.17	0.125426
	Cellulose Acetate	2.42	1.785476
	Acrylonitrile	2.7	1.99206
	Polyacetal	5.2	3.83656
	Polyphenylene Sulfide Molded	1.25	0.92225
	Polysulfone	0.8	0.59024
Composite Materials		
	Epoxy/Glass SMC	14.1	10.40298
	Polycarbonate Aramid Fiber Reinforced	0.6	0.44268
	Aramid Fiber reinforced Nylon 6/6	1.44	1.062432
	polycarbonate, carbon Fiber Reinforced	1.26	0.929628
	Plolycarbonate, carbon + Glass Fiber Reinforced	1.22	0.900116
	30% Glass Fiber reinforced Clear Polycarbonate	1.6	1.18048

(*continued*)

10. ROOM TEMPERATURE FRACTURE TOUGHNESS

Materials	Fracture Toughness	
	Mpa $(m)^{0.5}$	Ksi $(in)^{0.5}$
Plain Carbon and Low Alloy Steels		
Alloy 1040		
Cold drawn	54	48.6
Alloy 4140		
normalized	60	54
Alloy 4340		
normalized	55	49.5
Aluminum Alloys		
Alloy 2024		
T3 Temper	44	39.6
Alloy 6061		
T6 Temper	29	26.1
Alloy 7075		
T6 Temper	25	22.5
Polymers		
Epoxies		
Epoxy (Unreinforced)	0.6	0.54
Nylon 66 Unreinforced	2.8	2.52
Polycarbonate (PC)		
Molded	3.1	2.79
Polyester (Rigid)	0.6	0.54
Polyethylene Terephthalate (PET)		
Unreinforced	5	4.5
Polymethyl Methacrylate (PMMA)		
Unreinforced	1.2	1.08
Polypropylene (PP)		
Unreinforced	3.7	3.33
Polyvinyl Chloride (PVC)		
Unreinforced	2.3	2.07
Ceramics, Graphite and Semiconducting Materials		
Aluminum Oxide		
99.9% Alumina	4.5	4.05
95% Alumina	3.6	3.24
Concrete	0.9	0.81
Diamond		
Natural	6.5	5.85
Synthetic	8	7.2
Glass		
Borosilicate Glass	0.8	0.72
Silica Glass	0.7	0.63
Fused Silica	0.8	0.72
Silicon Carbide		
sintered	3.4	3.06
sublimed	4	3.6
Silicon Nitride	4.5	4.05
Zirconia	10	9
Tungsten Carbide	6	5.4

11. ROOM TEMPERATURE ELECTRICAL RESISTIVITY

Materials		Electrical Resistivity Ohm-cm
Plain Carbon and Low Alloy Steels		
Alloy 1006		
	cold drawn	1.74E−05
	hot rolled	1.74E−05
Alloy 1020		
	cold rolled	1.59E−05
	annealed	1.59E−05
Alloy 1040		
	Cold drawn	1.71E−05
	annealed	1.71E−05
Alloy 1090		
	annealed	1.74E−05
	hot rolled	1.74E−05
Alloy 4027		
	annealed	2.45E−05
	normalized	2.45E−05
Alloy 4140		
	annealed	2.20E−05
	normalized	2.20E−05
Alloy 4340		
	annealed	2.48E−05
	normalized	2.48E−05
Alloy 5140		
	annealed	2.28E−05
	normalized	2.28E−05
Stainless Steels		
Alloy 302		
	annealed	7.20E−05
	Cold rolled	7.20E−05
Alloy 316		
	annealed	7.40E−05
	soft tempered	7.40E−05
Alloy 405		
	Cold rolled	6.00E−05
	Annealed	6.00E−05
Alloy 434		
	Annealed	6.00E−05
Alloy 17-7PH		
	cold rolled	8.30E−05
Cast Irons		
Gray Cast Irons (as cast)		
	ASTM class 20	8.00E−05
	ASTM class 35	8.00E−05

(continued)

Ductile Cast Irons

	grade 60-40-18	6.00E−05
	grade 120-90-02	6.00E−05

Aluminum Alloys		

Alloy 2024

	Annealed	3.49E−06
	T3 Temper	5.82E−06
	T6 Temper	4.49E−06

Alloy 6061

	Annealed	3.66E−06
	T4 Temper	4.32E−06
	T6 Temper	3.99E−06

Alloy 7075

	Annealed	3.80E−06
	T6 Temper	5.50E−06

Copper Alloys		

Electrolytic tough pitch Copper, (C11000)

	Cold Worked	1.70E−06

Berillium Copper, (C17000)

	TH04 Temper	6.90E−06
	TH01 Temper	6.90E−06

Cartridge Brass, (C26000)

	H00 Temper	6.20E−06
	H04 Temper	6.20E−06

Muntz metal, (C28000)

	H00 Temper	6.16E−06
	H02 Temper	6.16E−06

Free-Cutting Brass, (C36000)

	H01 Temper	6.60E−06
	H02 Temper	6.60E−06

Chromium Copper, (C18500)

	TH04 Temper	2.16E−06
	TH04 Temper	2.16E−06

90-10 Bronze, (C22000)

	H01 Temper	2.72E−06
	H04 Temper	2.72E−06

Magnesium Alloys		

Alloy AZ31B

	Annealed	9.20E−06
	Hard rolled	9.20E−06

Alloy AZ63

	F Temper	1.15E−05
	T6 Temper	1.18E−05

Alloy AZ91E

	F Temper	1.43E−05
	T4 Temper	1.70E−05

(continued)

Nickel Alloys		
Nickel-Commercially Pure		
	As Cast	6.40E−06
	Annealed	6.40E−06
Nickel Silver 65-18 (752)		
	annealed	2.90E−05
	hardened	2.90E−05
Cupronickel 30(715)		
	annealed	3.75E−05
	hardened	3.75E−05

Titanium Alloys		
Titanium-Commercially Pure		
	As cast	5.54E−05
AlloyTi-6Al-2Nb-1Ta-0.8Mo		
	as-rolled	0.00016
	Annealed	0.00016

Miscellaneous Alloys		
	kovar Alloy	4.90E−05
Invar		
	Alloy 32-5 (Cold drawn)	8.00E−05
	Alloy 36 Alloy (Cold Drawn)	8.20E−05
Haynes 25		
	cold rolled and annealed	8.86E−05
	hot rolled and annealed	8.86E−05
Biodur		
	Hot Worlked	7.40E−05
	Annealed	7.40E−05
	Co-Cr-Ni Alloy	9.96E−05
	INCONEL (Cold Rolled)	9.85E−05
	INCOLOY 903	6.10E−05
	INCONEL 725 (Annealed)	1.30E−04
	Berylium S-200	4.30E−06
	Tungsten	5.65E−06
	Tantalum	1.24E−05
	Zinc	5.92E−06

Noble Metals		
	Gold	2.20E−06
	Iridium	4.70E−06
	Silver	1.55E−06
	Platinum, Annealed	1.06E−05

Polymers		
Elastomers		
	Styrene Butadiene Rubber (SBR)	1.00E+15
	Natural Rubber (Not Vulcanized)	1.00E+15
	Nitrile Rubber	1.00E+16

(continued)

Epoxies

	Epoxy (Unreinforced)	1.00E+15
	Nylon 66 Unreinforced	1.00E+12
	Phenolic (Unreinforced)	3.42E+11
	Polybutylene terepthalate (PBT)	1.00E+14

Polycarbonate (PC)

	Molded	1.00E+15
	Conductive	1.00E+14
	Polyester (Rigid)	1.00E+14

Polyetheretherketone (PEEK)

	Glass Fiber Filled	1.00E+15
	Carbon Fiber Filled	1.00E+09

Polyethylene Terephthalate (PET)

	Unreinforced	2.00E+15
	30% Glass Reinforced	1.00E+15

Polyethylene (Molded)

	Low Density	1.00E+15
	Medium Density	1.00E+15
	High Density	1.00E+06

Acrylonitrile Butadiene Styrene (ABS)

	Molded	1.00E+14
	40% Glass Fiber Filled	1.00E+15
	40% Carbon Fiber Filled	5.00E+04

Polymethyl Methacrylate (PMMA)

	Unreinforced	1.00E+14
	Glass Fiber Reinforced	1.00E+16

Polypropylene (PP)

	Unreinforced	1.00E+14
	10% Glass Fiber Filled	1.00E+15
	10% Carbon Fiber filled	5.00E+04
	Polystyrene	1.00E+15
	Polytetrafluoroethylene (PTFE)	1.00E+11

Polyvinyl Chloride (PVC)

	Unreinforced	6.00E+14
	Styrene Acrylonitrile	1.00E+15
	Cellulose Acetate	1.00E+12
	Polyacetal	1.00E+14
	Polyphenylene Sulfide Molded	1.00E+15
	Polysulfone	1.00E+15

Ceramics, Graphite and Semiconducting Materials		

Aluminum Oxide

	99.9% Alumina	1.00E+14
	95% Alumina	1.00E+16

Diamond

	Natural	1.00E+15
	Gallium Arsenide	1.00E+07

Glass

	Borosilicate Glass	1.00E+16
	Silica Glass	5.00E+09
	Graphite Carbon	6.00E−03
	Fused Silica	2.00E+14

(continued)

Silicon Carbide

	sintered	1.00E+00
	Silicon Nitride	1.00E+11
	Zirconia	1.00E+09
	Titanium Carbide	0.00018
	Tungsten Carbide	5.30E−04
	Zirconium Carbide	5.00E−05
	Alumina 96%	1.00E+14
	Boron Nitride	1.00E+13

Fibers	
E-Glass Fiber Generic	4.02E+12
C-Glass Fiber Generic	4.00E+10
Carbon Fiber Precursor	1.45E−03

Composite Materials	
Epoxy/Glass SMC	3.00E+14
Polycarbonate Aramid Fiber Reinforced	1.00E+16
polycarbonate, carbon Fiber Reinforced	1.00E+10
Plolycarbonate, carbon+Glass Fiber Reinforced	1.00E+06
30% Glass Fiber reinforced Clear Polycarbonate	1.00E+16

12. ROOM TEMPERATURE THERMAL CONDUCTIVITY

	Thermal conductivity	
Materials	W/m-K	Btu/ft-h-F
Plain Carbon and Low Alloy Steels		
Alloy 1006		
cold drawn	51.9	30.00
hot rolled	51.9	30.00
Alloy 1020		
cold rolled	51.9	30.00
annealed	51.9	30.00
Alloy 1040		
Cold drawn	51.9	30.00
annealed	51.9	30.00
Alloy 1090		
annealed	49.8	28.79
hot rolled	51.9	30.00
Alloy 4027		
normalized	44.5	25.72
Alloy 4140		
annealed	42.6	24.62
normalized	42.6	24.62

(continued)

Alloy 4340

	annealed	44.5	25.72
	normalized	44.5	25.72

Alloy 5140

	annealed	44.6	25.78
	normalized	44.6	25.78

Stainless Steels			

Alloy 302

	annealed	16.2	9.36
	Cold rolled	16.2	9.36

Alloy 316

	annealed	16.3	9.42
	soft tempered	16.3	9.42

Alloy 405

	Cold rolled	27	15.61
	Annealed	27	15.61

Alloy 434

	Annealed	26.1	15.09

Alloy 660

	Cold rolled	12.6	7.28

Alloy 17-7PH

	cold rolled	16.4	9.48

Cast Irons			

Gray Cast Irons (as cast)

	ASTM class 20	50	28.90
	ASTM class 35	50	28.90

Ductile Cast Irons

	grade 60-40-18	36	20.81
	grade 120-90-02	36	20.81

Aluminum Alloys			

Alloy 2024

	Annealed	193	111.56
	T3 Temper	121	69.94
	T6 Temper	151	87.28

Alloy 6061

	Annealed	180	104.05
	T4 Temper	154	89.02
	T6 Temper	167	96.53

Alloy 7075

	Annealed	173	100.00
	T6 Temper	130	75.14

Copper Alloys			

Electrolytic tough pitch Copper, (C11000)

	Cold Worked	388	224.28

Berillium Copper, (C17000)

	TH04 Temper	118	68.21
	TH01 Temper	118	68.21

(continued)

Cartridge Brass, (C26000)

	H00 Temper	120	69.36
	H04 Temper	120	69.36

Muntz metal, (C28000)

	H00 Temper	123	71.10
	H02 Temper	123	71.10

Free-Cutting Brass, (C36000)

	H01 Temper	115	66.47
	H02 Temper	115	66.47

90-10 Bronze, (C22000)

	H01 Temper	189	109.25
	H04 Temper	189	109.25

Magnesium Alloys

Alloy AZ31B

	Annealed	96	55.49
	Hard rolled	96	55.49

Alloy AZ63

	F Temper	59.2	34.22
	T6 Temper	61	35.26

Alloy AZ91E

	F Temper	72.7	42.02
	T4 Temper	72.7	42.02

Nickel Alloys

Nickel-Commercially Pure

	As Cast	60.7	35.09
	Annealed	60.7	35.09

Titanium Alloys

Titanium-Commercially Pure

	As cast	17	9.83

AlloyTi-6Al-2Nb-1Ta-0.8Mo

	as-rolled	6.4	3.70
	Annealed	6.4	3.70

Miscellaneous Alloys

	kovar Alloy	17.3	10.00

Invar

	Alloy 36 Alloy (Cold Drawn)	10.15	5.87

Haynes 25

	cold rolled and annealed	9.4	5.43
	hot rolled and annealed	9.4	5.43

Biodur

	Hot Worlked	12.766	7.38
	Annealed	12.66	7.32
	Co-Cr-Ni Alloy	12.5	7.23
	INCONEL (Cold Rolled)	12.5	7.23
	INCOLOY 903	16.7	9.65
	Berylium S-200	216	124.86
	Tungsten	163.3	94.39
	Tantalum	59.4	34.34
	Zinc	112.2	64.86

(continued)

Noble Metals		
Gold	301	173.99
Iridium	147	84.97
Silver	419	242.20
Platinum, Annealed	69.1	39.94

Polymers		

Elastomers

Natural Rubber (Not Vulcanized)	0.14	0.08

Epoxies

Epoxy (Unreinforced)	0.2	0.12
Nylon 66 Unreinforced	0.26	0.15
Phenolic (Unreinforced)	0.56	0.32
Polybutylene terepthalate (PBT)	0.2	0.12

Polycarbonate (PC)

Molded	0.2	0.12
Polyester (Rigid)	0.17	0.10

Polyetheretherketone (PEEK)

Glass Fiber Filled	0.33	0.19
Carbon Fiber Filled	0.6	0.35

Polyethylene Terephthalate (PET)

Unreinforced	0.2	0.11
30% Glass Reinforced	0.29	0.17

Polyethylene (Molded)

		0.00
Low Density	0.3	0.17
Medium Density	0.35	0.20
High Density	0.35	0.20

Acrylonitrile Butadiene Styrene (ABS)

Molded	0.158	0.09
40% Glass Fiber Filled	0.23	0.13
40% Carbon Fiber Filled	0.55	0.32

Polymethyl Methacrylate (PMMA)

Unreinforced	0.22	0.13

Polypropylene (PP)

Unreinforced	0.12	0.07
10% Glass Fiber Filled	0.24	0.14
Polystyrene	0.16	0.09
Polytetrafluoroethylene (PTFE)	0.3	0.17
Styrene Acrylonitrile	0.163	0.09
Cellulose Acetate	0.25	0.14
Acrylonitrile	0.225	0.13
Polyacetal	0.31	0.18
Polyphenylene Sulfide Molded	0.31	0.18
Polysulfone	0.18	0.10

Ceramics, Graphite and Semiconducting Materials		

Aluminum Oxide

99.9% Alumina	30	17.34
95% Alumina	20	11.56
Concrete	0.51	0.29

(continued)

Diamond

	Natural	2000	1156.07
	Synthetic	2000	1156.07
	Gallium Arsenide	50	28.90
Glass			
	Borosilicate Glass	1.1	0.64
	Silica Glass	1.38	0.80
	Graphite Carbon	24	13.87
	Fused Silica	0.8	0.46
Silicon Carbide			
	sintered	125.6	72.60
	sublimed	110	63.58
	Zirconia	1.675	0.97
	Alumina 96%	25	14.45

Fibers			
	E-Glass Fiber Generic	1.3	0.75
	C-Glass Fiber Generic	1.1	0.64
	Carbon Fiber Precursor	15	8.67

Composite Materials			
	Epoxy/Glass SMC	0.47	0.27
	Epoxy/Carbon Fiber Composite	200	115.61
	Polycarbonate Aramid Fiber Reinforced	0.22	0.13
	polycarbonate, carbon Fiber Reinforced	0.62	0.36

13. ROOM TEMPERATURE COEFFICIENT OF THERMAL EXPANSION

Materials		Coefficient of thermal expansion	
		1E−6/K	1E−6/F
Plain Carbon and Low Alloy Steels			
Alloy 1006			
	cold drawn	12.6	7.01
	hot rolled	12.6	7.01
Alloy 1020			
	cold rolled	11.7	6.51
	annealed	11.7	6.51
Alloy 1040			
	Cold drawn	11.3	6.29
	annealed	11.3	6.29
Alloy 1090			
	hot rolled	11.5	6.40
Alloy 4140			
	annealed	12.2	6.79
	normalized	12.2	6.79

(continued)

Alloy 4340

	annealed	12.3	6.84
	normalized	12.3	6.84

Alloy 5140

	annealed	12.6	7.01
	normalized	12.6	7.01

Stainless Steels			

Alloy 302

	annealed	17.2	9.57
	Cold rolled	17.2	9.57

Alloy 316

	annealed	16	8.90
	soft tempered	16	8.90

Alloy 405

	Cold rolled	10.8	6.01
	Annealed	10.8	6.01

Alloy 434

	Annealed	10.4	5.79

Alloy 660

	Cold rolled	16.9	9.40

Alloy 17-7PH

	cold rolled	11	6.12

Cast Irons			

Gray Cast Irons (as cast)

	ASTM class 20	13.5	7.51
	ASTM class 35	13.5	7.51

Ductile Cast Irons

	grade 60-40-18	13	7.23
	grade 120-90-02	13	7.23

Aluminum Alloys			

Alloy 2024

	Annealed	23.2	12.91
	T3 Temper	23.2	12.91
	T6 Temper	23.2	12.91

Alloy 6061

	Annealed	23.6	13.13
	T4 Temper	23.6	13.13
	T6 Temper	23.6	13.13

Alloy 7075

	Annealed	23.6	13.13
	T6 Temper	23.6	13.13

Copper Alloys			

Electrolytic tough pitch Copper, (C11000)

	Cold Worked	17	9.46

Berillium Copper, (C17000)

	TH04 Temper	16.7	9.29
	TH01 Temper	16.7	9.29

(continued)

Cartridge Brass, (C26000)

	H00 Temper	19.9	11.07
	H04 Temper	19.9	11.07

Muntz metal, (C28000)

	H00 Temper	20.8	11.57
	H02 Temper	20.8	11.57

Free-Cutting Brass, (C36000)

	H01 Temper	20.5	11.41
	H02 Temper	20.5	11.41

Chromium Copper, (C18500)

	TH04 Temper	17.6	9.79
	TH04 Temper	17.6	9.79

90-10 Bronze, (C22000)

	H01 Temper	18.4	10.24
	H04 Temper	18.4	10.24

Magnesium Alloys			

Alloy AZ31B

	Annealed	26	14.47
	Hard rolled	26	14.47

Alloy AZ63

	F Temper	26.1	14.52
	T6 Temper	26.1	14.52

Alloy AZ91E

	F Temper	26	14.47
	T4 Temper	26	14.47

Nickel Alloys			

Nickel-Commercially Pure

	As Cast	13.1	7.29
	Annealed	13.1	7.29

Nickel Silver 65-18 (752)

	annealed	17	9.46
	hardened	17	9.46

Cupronickel 30(715)

	annealed	15.5	8.62
	hardened	15.5	8.62

Titanium Alloys			

Titanium-Commercially Pure

	As cast	8.9	4.95

AlloyTi-6Al-2Nb-1Ta-0.8Mo

			0.00
	as-rolled	9.2	5.12
	Annealed	9.2	5.12

Miscellaneous Alloys			

	kovar Alloy	5.86	3.26

Invar

	Alloy 32-5 (Cold drawn)	0.194	0.11
	Alloy 36 Alloy (Cold Drawn)	1.3	0.72

(*continued*)

Haynes 25

	cold rolled and annealed	12.3	6.84
	hot rolled and annealed	12.3	6.84

Biodur

	Hot Worlked	12.791	7.12
	Annealed	13.18	7.33
	Co-Cr-Ni Alloy	15.17	8.44
	INCONEL (Cold Rolled)	13.46	7.49
	INCOLOY 903	7.65	4.26
	INCONEL 725 (Annealed)	13	7.23
	Berylium S-200	11.5	6.40
	Tungsten	4.4	2.45
	Tantalum	6.5	3.62
	Zinc	31.2	17.36

Noble Metals		
Gold	14.4	8.01
Iridium	6.8	3.78
Silver	19.6	10.91
Platinum, Annealed	9.1	5.06

Polymers		

Elastomers

	Natural Rubber (Not Vulcanized)	666	370.56
	Nitrile Rubber	702	390.59

Epoxies

	Epoxy (Unreinforced)	100	55.64
	Nylon 66 Unreinforced	105	58.42
	Phenolic (Unreinforced)	77	42.84
	Polybutylene terepthalate (PBT)	97	53.97

Polycarbonate (PC)

	Molded	77	42.84
	Conductive	59	32.83
	Polyester (Rigid)	135	75.11

Polyetheretherketone (PEEK)

	Glass Fiber Filled	25	13.91
	Carbon Fiber Filled	21	11.68

Polyethylene Terephthalate (PET)

	Unreinforced	83	46.18
	30% Glass Reinforced	25	13.91

Polyethylene (Molded)

	Low Density	130	72.33
	Medium Density	27	15.02
	High Density	112	62.32

Acrylonitrile Butadiene Styrene (ABS)

	Molded	107	59.53
	40% Glass Fiber Filled	23	12.80
	40% Carbon Fiber Filled	22	12.24

Polymethyl Methacrylate (PMMA)

	Unreinforced	95	52.86

(*continued*)

Polypropylene (PP)

	Unreinforced	105	58.42
	10% Glass Fiber Filled	51	28.38
	Polystyrene	87	48.41
	Polytetrafluoroethylene (PTFE)	119	66.21

Polyvinyl Chloride (PVC)

	Unreinforced	50	27.82
	30% Glass Fiber Reinforced	21	11.68
	Styrene Acrylonitrile	65	36.17
	Cellulose Acetate	159	88.47
	Acrylonitrile	66	36.72
	Polyacetal	135	75.11
	Polyphenylene Sulfide Molded	32	17.80
	Polysulfone	77	42.84

Ceramics, Graphite and Semiconducting Materials

Aluminum Oxide

	99.9% Alumina	7.4	4.12
	95% Alumina	6.2	3.45

Diamond

	Natural	1.18	0.66
	Gallium Arsenide	5.4	3.00

Glass

	Borosilicate Glass	4	2.23
	Silica Glass	0.75	0.42
	Graphite Carbon	2.4	1.34
	Fused Silica	0.6	0.33

Silicon Carbide

	sublimed	4.62	2.57
	Silicon Nitride	2.8	1.56
	Zirconia	12	6.68
	Titanium Carbide	7.7	4.28
	Tungsten Carbide	5.2	2.89
	Alumina 96%	6.4	3.56

Fibers

	E-Glass Fiber Generic	5	2.78
	C-Glass Fiber Generic	6.3	3.51
	Carbon Fiber Precursor	−0.75	−0.42

Composite Materials

	Epoxy/Glass SMC	3.6	2.00
	Epoxy/Carbon Fiber Composite	12	6.68
	Polycarbonate Aramid Fiber Reinforced	38	21.14
	polycarbonate, carbon Fiber Reinforced	22	12.24
	Plolycarbonate, carbon + Glass Fiber Reinforced	36	20.03

Wood

	American Red Oak Wood	11.1	6.18
	White Oak Wood	11.1	6.18

14. ROOM TEMPERATURE SPECIFIC HEAT CAPACITY

Materials		Specific Heat Capacity	
		J/g-K	**Btu/lb-F**
Plain Carbon and Low Alloy Steels			
Alloy 1006			
	cold drawn	0.481	0.11544
	hot rolled	0.481	0.11544
Alloy 1020			
	cold rolled	0.486	0.11664
	annealed	0.486	0.11664
Alloy 1040			
	Cold drawn	0.486	0.11664
	annealed	0.486	0.11664
Alloy 1090			
	annealed	0.472	0.11328
	hot rolled	0.472	0.11328
Alloy 4027			
	normalized	0.475	0.114
Alloy 4140			
	annealed	0.473	0.11352
	normalized	0.473	0.11352
Alloy 4340			
	annealed	0.475	0.114
	normalized	0.475	0.114
Alloy 5140			
	annealed	0.452	0.10848
	normalized	0.452	0.10848
Stainless Steels			
Alloy 302			
	annealed	0.5	0.12
	Cold rolled	0.5	0.12
Alloy 316			
	annealed	0.5	0.12
	soft tempered	0.5	0.12
Alloy 405			
	Cold rolled	0.46	0.1104
	Annealed	0.46	0.1104
Alloy 434			
	Annealed	0.46	0.1104
Alloy 660			
	Cold rolled	0.46	0.1104
Alloy 17-7PH			
	cold rolled	0.46	0.1104
Aluminum Alloys			
Alloy 2024			
	Annealed	0.875	0.21
	T3 Temper	0.875	0.21
	T6 Temper	0.875	0.21

(continued)

Alloy 6061			
	Annealed	0.896	0.21504
	T4 Temper	0.896	0.21504
	T6 Temper	0.896	0.21504
Alloy 7075			
	Annealed	0.96	0.2304
	T6 Temper	0.96	0.2304
Copper Alloys			
Electrolytic tough pitch Copper, (C11000)			
	Cold Worked	0.385	0.0924
Berillium Copper, (C17000)			
	TH04 Temper	0.42	0.1008
	TH01 Temper	0.42	0.1008
Cartridge Brass, (C26000)			
	H00 Temper	0.375	0.09
	H04 Temper	0.375	0.09
Muntz metal, (C28000)			
	H00 Temper	0.375	0.09
	H02 Temper	0.375	0.09
Free-Cutting Brass, (C36000)			
	H01 Temper	0.38	0.0912
	H02 Temper	0.38	0.0912
Chromium Copper, (C18500)			
	TH04 Temper	0.385	0.0924
	TH04 Temper	0.385	0.0924
90-10 Bronze, (C22000)			
	H01 Temper	0.376	0.09024
	H04 Temper	0.376	0.09024
Magnesium Alloys			
Alloy AZ31B			
	Annealed	1	0.24
	Hard rolled	1	0.24
Alloy AZ63			
	F Temper	1.05	0.252
	T6 Temper	1.05	0.252
Alloy AZ91E			
	F Temper	0.8	0.192
	T4 Temper	1.047	0.25128
Nickel Alloys			
Nickel-Commercially Pure			
	As Cast	0.46	0.1104
	Annealed	0.46	0.1104
Titanium Alloys			
Titanium-Commercially Pure			
	As cast	0.528	0.12672
AlloyTi-6Al-2Nb-1Ta-0.8Mo			
	as-rolled	0.552	0.13248
	Annealed	0.552	0.13248

(continued)

Miscellaneous Alloys		
kovar Alloy	0.439	0.10536
Invar		
Alloy 36 Alloy (Cold Drawn)	0.515	0.1236
Biodur		
Hot Worlked	0.452	0.10848
Annealed	0.47	0.1128
Co-Cr-Ni Alloy	0.43	0.1032
INCONEL (Cold Rolled)	0.444	0.10656
INCOLOY 903	0.435	0.1044
Berylium S-200	1.925	0.462
Tungsten	0.134	0.03216
Tantalum	0.138	0.03312
Zinc	0.3898	0.093552

Noble Metals		
Gold	0.1323	0.031752
Iridium	0.135	0.0324
Silver	0.234	0.05616
Platinum, Annealed	0.134	0.03216

Polymers		
Elastomers		
Natural Rubber (Not Vulcanized)	0.45	0.108
Epoxies		
Epoxy (Unreinforced)	1	0.24
Nylon 66 Unreinforced	2.1	0.504
Phenolic (Unreinforced)	1.6	0.384
Polybutylene terepthalate (PBT)	2	0.48
Polycarbonate (PC)		
Molded	1.1	0.264
Polyetheretherketone (PEEK)		
Glass Fiber Filled	1.9	0.456
Carbon Fiber Filled	1.9	0.456
Polyethylene Terephthalate (PET)		
30% Glass Reinforced	2.3	0.552
Polyethylene (Molded)		
Low Density	2.2	0.528
Medium Density	2.2	0.528
High Density	2.2	0.528
Acrylonitrile Butadiene Styrene (ABS)		
Molded	2	0.48
Polymethyl Methacrylate (PMMA)		
Unreinforced	1.46	0.3504
Polypropylene (PP)		
Unreinforced	2	0.48
Polystyrene	1.7	0.408
Polytetrafluoroethylene (PTFE)	1.3	0.312
Polyvinyl Chloride (PVC)		
Styrene Acrylonitrile	1.7	0.408
Cellulose Acetate	1.4	0.336
Polysulfone	1.2	0.288

(*continued*)

Ceramics, Graphite and Semiconducting Materials		
Aluminum Oxide		
99.9% Alumina	0.85	0.204
95% Alumina	0.439	0.10536
Diamond		
Natural	0.4715	0.11316
Gallium Arsenide	0.325	0.078
Glass		
Silica Glass	0.75	0.18
Graphite Carbon	0.69	0.1656
Silicon Carbide		
sintered	0.67	0.1608
Alumina 96%	0.88	0.2112
Fibers		
E-Glass Fiber Generic	0.81	0.1944
C-Glass Fiber Generic	0.787	0.18888
Composite Materials		
Epoxy/Carbon Fiber Composite	1.1	0.264

COMPOSITION OF SELECTED ALLOYS

Materials	Composition
Plain Carbon and Low Alloy Steels	
Alloy 1006	Fe-99.43 (min), C-0.08 (max), Mn-0.45 (max), P-0.04 (max), S-0.05 (max)
Alloy 1020	Fe-99.08 (min), C-0.17 (min), Mn-0.3 (min), P-0.04 (max), S-0.05 (max)
Alloy 1040	Fe-98.6 (min), C-0.37 (min), Mn-0.6 (min), P-0.04 (max), S-0.05 (max)
Alloy 1090	Fe-0.85 (min), C-0.85 (min), Mn-0.6 (min), P-0.04 (max), S-0.05 (max)
Alloy 4027	Fe-98.08 (min), C-0.25 (min), Mn-0.7 (min), Mo-0.2 (min), P-0.035 (max), S-0.04 (max), Si-0.15 (min)
Alloy 4140	Fe-96.785 (min), C-0.38 (min), Cr-0.8 (min), Mn-0.7 (min), Mo-0.15 (min), P-0.035 (max), S-0.04 (max), Si-0.15 (min)
Alloy 4340	Fe-96, C-0.37 (min), Cr-0.7 (min), Mn-0.7, Mo-0.2, Ni-1.83, P-0.035 (max), S-0.04 (max), Si-0.23
Alloy 5140	Fe-98, C-0.38 (min), Cr-0.8, Mn-0.8, P-0.035 (max), S-0.04 (max), Si-0.23

(continued)

Stainless Steels

Alloy 302

Fe-70, C-0.15 (max), Cr-18, Mn-2 (max), Ni-9, P-0.045 (max), S-0.03 (max), Si-1 (max)

Alloy 316

Fe-62, C-0.08, Cr-18 (max), Mn-2, Mo-2 (max), Ni-14 (max), P-0.045, S-0.03, Si-1

Alloy 405

Fe-85, Al-0.2, C-0.08 (max), Cr-13, Mn-1 (max), P-0.04 (max), S-0.03 (max), Si-1 (max)

Alloy 434

Fe-81, C-0.12 (max), Cr-16, Mn-1 (max), Mo-1, P-0.04 (max), S-0.03 (max), Si-1 (max)

Alloy 660

Fe-56, Al-0.035 (max), B-0.001 (min), C-0.08 (max), Cr-13.5 (min), Cu-0.25 (max), Mn-2 (max), Ni-24 (min), P-0.04 (max), S-0.03 (max), Si-1 (max), Ti-1.9 (min), V-0.1 (min)

Alloy 17-7PH

Al-0.75 (min), C-0.09 (max), Cr-16 (min), Mn-1 (max), Ni-6.5 (min), P-0.04 (max), S-0.04 (max), Si-1 (max)

Cast Irons

Gray Cast Irons (as cast)

C-3.25 (min), Cr-0.05 (min), Cu-0.15 (min), Mn-0.5 (min), Mo-0.05 (min), Ni-0.05 (min), P-0.12 (max), S-0.15 (max), Si-2.3 (max)

Ductile Cast Irons

Fe-90.68 (min), C-3 (min), Ce-0.005 (min), Cr-0.03 (min), Cu-0.15 (min), Mg-0.03 (min), Mn-0.15 (min), Mo-0.01 (min), Ni-0.05 (min), P-0.08 (max), S-0.002 (max), Si-2.8 (max)

Aluminum Alloys

Alloy 2024

Fe-0.5 (max), Al-90.7 (min), Cr-0.1 (max), Cu-3.8 (min), Mg-1.2 (max), Mn-0.3 (max), Other, each -0.05 (max), Other, total-0.15 (max), Si-0.5 (max), Ti-0.15 (max), Zn-0.25 (max)

Alloy 6061

Fe-0.7 (max), Al-95.8 (min), Cr-0.35 (max), Cu-0.15 (min), Mg-1.2 (max), Mn-0.15 (max), Other, each -0.05 (max), Other, total-0.15 (max), Si-0.8 (max), Ti-0.15 (max), Zn-0.25 (max)

Alloy 7075

Fe-0.5 (max), Al-87.1 (min), Cr-0.18 (max), Cu-1.2 (min), Mg-2.9 (max), Mn-0.3 (max), Other, each -0.05 (max), Other, total-0.15 (max), Si-0.4 (max), Ti-0.2 (max), Zn-6.1 (max)

Copper Alloys

Electrolytic tough pitch Copper, (C11000)

Cu-99.9, O-0.04

Berillium Copper, (C17000)

Be-1.6 (min), Co + Ni-0.2 (min), Co + Ni + Fe-0.6 (max), Cu-98.1

Cartridge Brass, (C26000)

Cu-68.5, Fe-0.05 (max), Other-0.15 (max), Pb-0.07 (max), Zn-28.5 (min)

Muntz metal, (C28000)

Cu-59 (min), Fe-0.07 (max), Pb-0.3 (max), Zn-40

(continued)

Free-Cutting Brass, (C36000)

Cu-60 (min), Fe-0.35 (max), Other-0.5 (max), Pb-3.7 (max), Zn-35.5 (min)

Chromium Copper, (C18500)

Ag-0.08 (min), Cr-0.4 (min), Cu-99.3, Li-0.04 (max), P-0.04 (max),
Pb-0.015 (max)

90-10 Bronze, (C22000)

Cu-89 (min), Fe-0.05 (max), Pb-0.05 (max), Zn-10

Magnesium Alloys

Alloy AZ31B

Al-2.5 (min), Ca-0.04 (max), Cu-0.05 (max), Fe-0.005 (max), Mg-97,
Mn-0.2 (min), Ni-0.005 (max), Si-0.1 (max), Zn-0.6 (min)

Alloy AZ63

Al-5.3 (min), Ca-0.25 (max), Mg-91, Mn-0.15 (min), Ni-0.01 (max),
Si-0.3 (max), Zn-2.5 (min)

Alloy AZ91E

Al-8.3 (min), Cu-0.03 (max), Fe-0.005 (max), Mg-90, Mn-0.13 (min),
Ni-0.002 (max), Si-0.1 (max), Zn-0.35 (min)

Nickel Alloys

Nickel-Commercially Pure

Ni-99.9%

Nickel Silver 65-18 (752)

Cu-53.5 (min), Fe-0.25 (max), Mn-0.5, Ni-16.5 (min), Other-0.5, Pb-0.1,
Zn-22.65 (min)

Cupronickel 30(715)

Cu-70, Ni-30

Titanium Alloys

Titanium-Commercially Pure
AlloyTi-6Al-2Nb-1Ta-0.8Mo

Al-6, Mo-0.8, Nb-2, Ta-1, Ti-90

Miscellaneous Alloys

kovar Alloy

C-0.02 (max), Co-17, Fe-53, Mn-0.3, Ni-29, Si-0.2

Invar
Alloy 32-5

C-0.02, Co-5.5, Fe-62, Mn-0.4, Ni-32, Si-0.25

Alloy 36 Alloy

C-0.02, Fe-63, Mn-0.35, Ni-36, Si-0.2

Haynes 25

C-0.1, Co-51, Cr-20, Fe-3 (max), Mn-1.5, Ni-10, Si-0.4 (max), W-15

Biodur

C-0.02 (min), Co-66, Cr-26 (min), Mo-5 (min), N-0.15 (min)

Co-Cr-Ni Alloy

Be-0.1 (max), C-0.15 (max), Co-39 (min), Cr-19 (min), Fe-16,
Mn-1.5 (min), Mo-6 (min), Ni-14 (min)

INCONEL (Cold Rolled)

Al-0.6 (max), B-0.006 (max), C-0.3 (max), Co-0.1 (max),
Cr-17.5 (max), Cu-0.35 (max), Fe-38, Mn-0.35 (max), Nb-2.5 (min),
Ni-44 (max), P-0.02, S-0.015 (max), Si-0.35 (max), Ti-0.4 (max)

(continued)

INCOLOY 903

Al-0.9, Co-15, Fe-42, Nb-3, ni-38, Ti-1.4

INCONEL 725 (Annealed)

Al-0.5 (max), C-0.03 (max), Cr-22.5 (max), Fe-9, Mn-0.35 (max), Mo-7 (min), Nb-2.75 (min), Ni-59 (max), P-0.015 (max), S-0.01 (max), Si-0.2 (max), Ti-1.7 (max)

COST AND RELATIVE COST OF SOME SELECTED MATERIALS

Materials	Cost US$/lb	Relative cost (based on 1020 steel)
Plain Carbon and Low Alloy Steels		
Alloy 1006		
cold drawn	2.91	0.43
Alloy 1020		
annealed	6.74	1.00
Alloy 1040		
Cold drawn	2.14	0.32
Alloy 1090		
annealed	7.77	1.15
Alloy 4140		
normalized	3.36	0.50
Stainless Steels		
Alloy 302		
Cold rolled	9	1.34
Alloy 316		
soft tempered	17.87	2.65
Alloy 405		
Cold rolled Annealed	394.79	58.57
Alloy 17-7PH		
cold rolled	16.39	2.43
Cast Irons		
Gray Cast Irons (as cast)		
ASTM class 35	7.42	1.10
Aluminum Alloys		
Alloy 2024		
T3 Temper	17.5	2.60
Alloy 6061		
T4 Temper	10.25	1.52
Alloy 7075		
T6 Temper	19.06	2.83

(continued)

Copper Alloys			
Electrolytic tough pitch Copper, (C11000)			
	Cold Worked	6.6	0.98
Cartridge Brass, (C26000)			
	H00 Temper	29.59	4.39
Muntz metal, (C28000)			
	H00 Temper	8.68	1.29
Free-Cutting Brass, (C36000)			
	H02 Temper	6.24	0.93
90-10 Bronze, (C22000)			
	H01 Temper	13.53	2.01
Magnesium Alloys			
Alloy AZ31B			
	Hard rolled	11.06	1.64
Nickel Alloys			
Nickel-Commercially Pure			
	As Cast	39.02	5.79
Titanium Alloys			
Titanium-Commercially Pure			
	As cast	57.77	8.57
Miscellaneous Alloys			
Invar			
	Alloy 32-5 (Cold drawn)	15	2.23
Haynes 25			
	cold rolled and annealed	40	5.93
Biodur			
	INCONEL (Cold Rolled)	38	5.64
	INCONEL 725 (Annealed)	14	2.08
	Berylium S-200	400	59.35
	Tungsten	4	0.59
	Tantalum	37	5.49
	Zinc	0.5	0.07
Noble Metals			
	Gold	7105	1054.15
	Iridium	2684	398.22
	Silver	102	15.13
	Platinum, Annealed	13531	2007.57
Polymers			
Elastomers			
	Styrene Butadiene Rubber (SBR)	6.4	0.95
	Natural Rubber (Not Vulcanized)	31	4.60
	Nitrile Rubber	5.07	0.75

(*continued*)

Epoxies

	Epoxy (Unreinforced)	5	0.74
	Nylon 66 Unreinforced	14.2	2.11
	Phenolic (Unreinforced)	10	1.48
	Polybutylene terepthalate (PBT)	5	0.74
Polycarbonate (PC)			
	Molded	7.47	1.11
	Polyester (Rigid)	7.25	1.08
Polyethylene (Molded)			
	Medium Density	11.3	1.68
Acrylonitrile Butadiene Styrene (ABS)			
	Molded	13.84	2.05
Polymethyl Methacrylate (PMMA)			
	Unreinforced	3.25	0.48
Polypropylene (PP)			
	Unreinforced	2.58	0.38
	Polystyrene	15.05	2.23
	Polytetrafluoroethylene (PTFE)	22	3.26
Polyvinyl Chloride (PVC)			
	Unreinforced	15.12	2.24
	Polysulfone	24.75	3.67

Ceramics, Graphite and Semiconducting Materials			
Aluminum Oxide			
	99.9% Alumina	129.57	19.22
Diamond			
	Natural	40000	5934.72
	Gallium Arsenide	1500	222.55
Glass			
	Borosilicate Glass	10	1.48
	Silica Glass	2.25	0.33
	Graphite Carbon	8	1.19
	Fused Silica	189	28.04
Silicon Carbide			
	sintered	20	2.97

Fibers			
	E-Glass Fiber Generic	2.2	0.33
	Carbon Fiber Precursor	30	4.45

Composite Materials			
	Epoxy/Glass SMC	18	2.67
	Epoxy/Carbon Fiber Composite	34	5.04
	Polycarbonate Aramid Fiber Reinforced	51	7.57

Notes:
1. Mechanical properties of materials are typical. The properties vary to a great extent depending upon the manufacturing process involved and testing conditions.
2. The cost of the materials are for small quantities (about 10 lbs). Most of the cost data is from a single retailer. The cost can be much less (up to 80% less) if the material is bought from a manufacturer or distributor in large quantities (thousands of pounds).

Some Properties of Selected Elements

Element	Symbol	Melting point, °C	Density,* g/cm³	Atomic radius, nm	Crystal structure† (20°C)	Lattice constants 20°C, nm	
						a	c
Aluminum	Al	660	2.70	0.143	FCC	0.40496	
Antimony	Sb	630	6.70	0.138	Rhombohedral	0.45067	
Arsenic	As	817	5.72	0.125	Rhombohedral‡	0.4131	
Barium	Ba	714	3.5	0.217	BCC‡	0.5019	
Beryllium	Be	1278	1.85	0.113	HCP‡	0.22856	0.35832
Boron	B	2030	2.34	0.097	Orthorhombic		
Bromine	Br	−7.2	3.12	0.119	Orthorhombic		
Cadmium	Cd	321	8.65	0.148	HCP‡	0.29788	0.561667
Calcium	Ca	846	1.55	0.197	FCC‡	0.5582	
Carbon (graphite)	C	3550	2.25	0.077	Hexagonal	0.24612	0.67078
Cesium	Cs	28.7	1.87	0.190	BCC		
Chlorine	Cl	−101	1.9	0.099	Tetragonal		
Chromium	Cr	1875	7.19	0.128	BCC‡	0.28846	
Cobalt	Co	1498	8.85	0.125	HCP‡	0.2506	0.4069
Copper	Cu	1083	8.96	0.128	FCC	0.36147	
Fluorine	F	−220	1.3	0.071			
Gallium	Ga	29.8	5.91	0.135	Orthorhombic		
Germanium	Ge	937	5.32	0.139	Diamond cubic	0.56576	
Gold	Au	1063	19.3	0.144	FCC	0.40788	
Helium	He	−270	. . .	. . .	HCP		
Hydrogen	H	−259	. . .	0.046	Hexagonal		
Indium	In	157	7.31	0.162	FC tetragonal	0.45979	0.49467
Iodine	I	114	4.94	0.136	Orthorhombic		
Iridium	Ir	2454	22.4	0.135	FCC	0.38389	
Iron	Fe	1536	7.87	0.124	BCC‡	0.28664	
Lead	Pb	327	11.34	0.175	FCC	0.49502	
Lithium	Li	180	0.53	0.157	BCC	0.35092	

(continued)

Element	Symbol	Melting point, °C	Density,* g/cm³	Atomic radius, nm	Crystal structure† (20°C)	Lattice constants 20°C, nm	
						a	c
Magnesium	Mg	650	1.74	0.160	HCP	0.32094	0.52105
Manganese	Mn	1245	7.43	0.118	Cubic‡	0.89139	
Mercury	Hg	−38.4	14.19	0.155	Rhombohedral		
Molybdenum	Mo	2610	10.2	0.140	BCC	0.31468	
Neon	Ne	−248.7	1.45	0.160	FCC		
Nickel	Ni	1453	8.9	0.125	FCC	0.35236	
Niobium	Nb	2415	8.6	0.143	BCC	0.33007	
Nitrogen	N	−240	1.03	0.071	Hexagonal‡		
Osmium	Os	2700	22.57	0.135	HCP	0.27353	0.43191
Oxygen	O	−218	1.43	0.060	Cubic‡		
Palladium	Pd	1552	12.0	0.137	FCC	0.38907	
Phosphorus (white)	P	44.2	1.83	0.110	Cubic‡		
Platinum	Pt	1769	21.4	0.139	FCC	0.39239	
Potassium	K	63.9	0.86	0.238	BCC	0.5344	
Rhenium	Re	3180	21.0	0.138	HCP	0.27609	0.44583
Rhodium	Rh	1966	12.4	0.134	FCC	0.38044	
Ruthenium	Ru	2500	12.2	0.125	HCP	0.27038	0.42816
Scandium	Sc	1539	2.99	0.160	FCC	0.4541	
Silicon	Si	1410	2.34	0.117	Diamond cubic	0.54282	
Silver	Ag	961	10.5	0.144	FCC	0.40856	
Sodium	Na	97.8	0.97	0.192	BCC	0.42906	
Strontium	Sr	76.8	2.60	0.215	FCC‡	0.6087	
Sulfur (yellow)	S	119	2.07	0.104	Orthorhombic		
Tantalum	Ta	2996	16.6	0.143	BCC	0.33026	
Tin	Sn	232	7.30	0.158	Tetragonal‡	0.58311	0.31817
Titanium	Ti	1668	4.51	0.147	HCP‡	0.29504	0.46833
Tungsten	W	3410	19.3	0.141	BCC	0.31648	
Uranium	U	1132	19.0	0.138	Orthorhombic‡	0.2858	0.4955
Vanadium	V	1900	6.1	0.136	BCC	0.3039	
Zinc	Zn	419.5	7.13	0.137	HCP	0.26649	0.49470
Zirconium	Zr	1852	6.49	0.160	HCP‡	0.32312	0.51477

*Density of solid at 20°C.
†b = 0.5877 nm.
‡Other crystal structures exist at other temperatures.

Ionic Radii[1] of the Elements

[1]Ionic radii can vary in different crystals due to many factors.

Atomic number	Element (symbol)	Ion	Ionic radius, nm	Atomic number	Element (symbol)	Ion	Ionic radius, nm
1	H	H^-	0.154	23	V	V^{3+}	0.065
2	He					V^{4+}	0.061
3	Li	Li^+	0.078			V^{5+}	~0.04
4	Be	Be^{2+}	0.034	24	Cr	Cr^{3+}	0.064
5	B	B^{3+}	0.02			Cr^{6+}	0.03–0.04
6	C	C^{4+}	<0.02	25	Mn	Mn^{2+}	0.091
7	N	N^{5+}	0.01–0.02			Mn^{3+}	0.070
8	O	O^{2-}	0.132			Mn^{4+}	0.052
9	F	F^-	0.133	26	Fe	Fe^{2+}	0.087
10	Ne					Fe^{3+}	0.067
11	Na	Na^+	0.098	27	Co	Co^{2+}	0.082
12	Mg	Mg^{2+}	0.078			Co^{3+}	0.065
13	Al	Al^{3+}	0.057	28	Ni	Ni^{2+}	0.078
14	Si	Si^{4-}	0.198	29	Cu	Cu^+	0.096
		Si^{4+}	0.039	30	Zn	Zn^{2+}	0.083
15	P	P^{5+}	0.03–0.04	31	Ga	Ga^{3+}	0.062
16	S	S^{2-}	0.174	32	Ge	Ge^{4+}	0.044
		S^{6+}	0.034	33	As	As^{3+}	0.069
17	Cl	Cl^-	0.181			As^{5+}	~0.04
18	Ar			34	Se	Se^{2-}	0.191
19	K	K^+	0.133			Se^{6+}	0.03–0.04
20	Ca	Ca^{2+}	0.106	35	Br	Br^-	0.196
21	Sc	Sc^{2+}	0.083	36	Kr		
22	Ti	Ti^{2+}	0.076	37	Rb	Rb^+	0.149
		Ti^{3+}	0.069	38	Sr	Sr^{2+}	0.127
		Ti^{4+}	0.064				

(continued)

Atomic number	Element (symbol)	Ion	Ionic radius, nm	Atomic number	Element (symbol)	Ion	Ionic radius, nm
39	Y	Y^{3+}	0.106	66	Dy	Dy^{3+}	0.107
40	Zr	Zr^{4+}	0.087	67	Ho	Ho^{3+}	0.105
41	Nb	Nb^{4+}	0.069	68	Er	Er^{3+}	0.104
		Nb^{5+}	0.069	69	Tm	Tm^{3+}	0.104
42	Mo	Mo^{4+}	0.068	70	Yb	Yb^{3+}	0.100
		Mo^{6+}	0.065	71	Lu	Lu^{3+}	0.099
44	Ru	Ru^{4+}	0.065	72	Hf	Hf^{4+}	0.084
45	Rh	Rh^{3+}	0.068	73	Ta	Ta^{5+}	0.068
		Rh^{4+}	0.065	74	W	W^{4+}	0.068
46	Pd	Pd^{2+}	0.050			W^{6+}	0.065
47	Ag	Ag^{+}	0.113	75	Re	Re^{4+}	0.072
48	Cd	Cd^{2+}	0.103	76	Os	Os^{4+}	0.067
49	In	In^{3+}	0.092	77	Ir	Ir^{4+}	0.066
50	Sn	Sn^{4-}	0.215	78	Pt	Pt^{4+}	0.052
		Sn^{4+}	0.074			Pt^{4+}	0.055
51	Sb	Sb^{3+}	0.090	79	Au	Au^{+}	0.137
52	Te	Te^{2-}	0.211	80	Hg	Hg^{2+}	0.112
		Te^{4+}	0.089	81	Tl	Tl^{+}	0.149
53	I	I^{-}	0.220			Tl^{3+}	0.106
		I^{5+}	0.094	82	Pb	Pb^{4-}	0.215
54	Xe					Pb^{2+}	0.132
55	Cs	Cs^{+}	0.165			Pb^{4+}	0.084
56	Ba	Ba^{2+}	0.143	83	Bi	Bi^{3+}	0.120
57	La	La^{3+}	0.122	84	Po		
58	Ce	Ce^{3+}	0.118	85	At		
		Ce^{4+}	0.102	86	Rn		
59	Pr	Pr^{3+}	0.116	87	Fr		
		Pr^{4+}	0.100	88	Ra	Ra^{+}	0.152
60	Nd	Nd^{3+}	0.115	89	Ac		
61	Pm	Pm^{3+}	0.106	90	Th	Th^{4+}	0.110
62	Sm	Sm^{3+}	0.113	91	Pa		
63	Eu	Eu^{3+}	0.113	92	U	U^{4+}	0.105
64	Gd	Gd^{3+}	0.111				
65	Tb	Tb^{3+}	0.109				
		Tb^{4+}	0.089				

*Ionic radii can vary in different crystals due to many factors.

Source: C. J. Smithells (ed.), "Metals Reference Book," 5th ed., Butterworth, 1976.

GLASS TRANSITION TEMPERATURE AND MELTING TEMPERATURE OF SELECTED POLYMERS

Polymer	Glass transition temperature (°C)	Melting temperature (°C)
Nylon 66	50	265
Nylon 12	42	179
Polybutylene terepthalate (PBT)	22	225
Polycarbonate	150	265
Polyetheretherketone (PEEK)	157	374
Polyethylene Terephthalate (PET)	69	265
Polyethylene	−78	100
Acrylonitrile Butadiene Styrene (ABS)	110	105
Polymethyl Methacrylate (PMMA)	38	160
Polypropylene (PP)	−8	176
Polystyrene	100	240
Polytetrafluoroethylene (PTFE)	−20	327
Polyvinyl Chloride (PVC)	87	227
Polyvinyl Ethyl Ether	−43	86
Polyvinyl Fluoride	40	200
Styrene Acrylonitrile	120	120
Cellulose Acetate	190	230
Acrylonitrile	100	317
Polyacetal	−30	183
Polyphenylene Sulfide Molded	118	275
Polysulfone	185	190
Polychloroprene	−50	80
Polydimethyl Siloxane	−123	−40
Polyvinyl Pyrrolidone	86	375
Polyvinylidene Chloride	−18	198

Selected Physical Quantities and Their Units

Quantity	Symbol	Unit	Abbreviation
Length	l	inch	in
		meter	m
Wavelength	λ	meter	m
Mass	m	kilogram	kg
Time	t	second	s
Temperature	T	degree Celsius	°C
		degree Fahrenheit	°F
		Kelvin	K
Frequency	v	hertz	Hz $[\text{s}^{-1}]$
Force	F	newton	N $[\text{kg} \cdot \text{m} \cdot \text{s}^{-2}]$
Stress:			
Tensile	σ	pascal	Pa $[\text{N} \cdot \text{m}^{-2}]$
Shear	τ	pounds per square inch	lb/in² or psi
Energy, work, quantity of heat		joule	J $[\text{N} \cdot \text{m}]$
Power		watt	W $[\text{J} \cdot \text{s}^{-1}]$
Current flow	i	ampere	A
Electric charge	q	coulomb	C $[\text{A} \cdot \text{s}]$
Potential difference, electromotive force	V, E	volt	V
Electric resistance	R	ohm	Ω $[\text{V} \cdot \text{A}^{-1}]$
Magnetic induction	B	tesla	T $[\text{V} \cdot \text{s} \cdot \text{m}^{-2}]$

Greek Alphabet

Name	Lowercase	Capital	Name	Lowercase	Capital
Alpha	α	A	Nu	ν	N
Beta	β	B	Xi	ξ	Ξ
Gamma	γ	Γ	Omicron	o	O
Delta	δ	Δ	Pi	π	Π
Epsilon	ϵ	E	Rho	ρ	P
Zeta	ζ	Z	Sigma	σ	Σ
Eta	η	H	Tau	τ	T
Theta	θ	Θ	Upsilon	υ	Y
Iota	ι	I	Phi	ϕ	Φ
Kappa	κ	K	Chi	χ	X
Lambda	λ	Λ	Psi	ψ	Ψ
Mu	μ	M	Omega	ω	Ω

SI Unit Prefixes

Multiple	Prefix	Symbol
10^{-12}	pico	p
10^{-9}	nano	n
10^{-6}	micro	μ
10^{-3}	milli	m
10^{-2}	centi	c
10^{-1}	deci	d
10^{1}	deca	da
10^{2}	hecto	h
10^{3}	kilo	k
10^{6}	mega	M
10^{9}	giga	G
10^{12}	tera	T

Example: 1 kilometer = 1 km = 10^3 meters.

REFERENCES FOR FURTHER STUDY BY CHAPTER

Chapter 1

Annual Review of Materials Science. Annual Reviews, Inc. Palo Alto, CA.

Bever, M. B. (ed.) *Encyclopedia of Materials Science and Engineering.* MIT Press-Pergamon, Cambridge, 1986.

Canby, T. Y. "Advanced Materials—Reshaping Our Lives." *Nat. Geog.,* 176(6), 1989, p. 746.

Engineering Materials Handbook. Vol. 1: *Composites,* ASM International, 1988.

Engineering Materials Handbook. Vol. 2: *Engineering Plastics,* ASM International, 1988.

Engineering Materials Handbook. Vol. 4: *Ceramics and Glasses,* ASM International, 1991.

Internet Sources:
www.nasa.gov
www.designinsite.dk/htmsider/inspmat.htm

Jackie Y. Ying, *Nanostructured Materials,* Academic Press, 2001.

"Materials Engineering 2000 and Beyond: Strategies for Competitiveness." *Advanced Materials and Processes* 145(1), 1994.

"Materials Issue." *Sci. Am.,* 255(4), 1986.

Metals Handbook, 2nd Edition, ASM International, 1998.

M. F. Ashby, *Materials Selection in Mechanical Design,* Butterworth-Heinemann, 1996.

M. Madou, *Fundamentals of Microfabrication,* CRC Press, 1997.

Nanomaterials: Synthesis, Properties, and Application, Editors: A. S. Edelstein and R. C. Cammarata, Institute of Physics Publishing, 2002.

National Geographic magazine, 2000–2001.

Wang, Y. et al., *High Tensile Ductility in a Nanostructured Metal, Letters to Nature,* 2002.

Chapter 2

Binnig, G., H. Rohrer, et al. in *Physical Review Letters,* v 50 pp. 120–24 (1983).

http://ufrphy.lbhp.jussieu.fr/nano/

Brown, T. L., H. E. LeMay and B. E. Bursten. *Chemistry.* 8th ed. Prentice-Hall, 2000.

Chang, R. *Chemistry.* 5th ed. McGraw-Hill, 1994.

http://www.molec.com/products_consumables. html#STM

H. Dai, J. H. Hafner, A. G. Rinzler, D. T. Colbert, R. E. Smalley, Nature 384, 147–150 (1996).

http://www.omicron.de/index2.html?/results/ stm_image_of_chromium_decorated_steps_of_ cu_111/~Omicron

http://www.almaden.ibm.com/almaden/media/ image_mirage.html

Chang, R. *General Chemistry.* 4th ed. McGraw-Hill, 1990.

Ebbing, D. D. *General Chemistry.* 5th ed. Houghton Mifflin, 1996.

McWeeny, R. *Coulson's Valence.* 3d ed. Oxford University Press, 1979.

Pauling, L. *The Nature of the Chemical Bond.* 3d ed. Cornell University Press, 1960.

Smith, W. F. *T. M. S. Fall Meeting.* October 11, 2000. Abstract only.

Chapter 3

Barrett, C. S. and T. Massalski. *Structure of Metals.* 3d ed. Pergamon Press, 1980.

Cullity, B. D. *Elements of X-Ray Diffraction.* 2d ed. Addison-Wesley, 1978.

Wilson, A. J. C. *Elements of X-Ray Cystallography.* Addison-Wesley, 1970.

Chapters 4 and 5

Flemings, M. *Solidification Processing.* McGraw-Hill, 1974.

Hirth J. P., and J. Lothe. *Theory of Dislocations.* 2d ed. Wiley, 1982.

Krauss, G. (ed.) *Carburizing: Processing and Performance.* ASM International, 1989.

Minkoff, I. *Solidification and Cast Structures.* Wiley, 1986.

Shewmon, P. G. *Diffusion in Solids.* 2d ed. Minerals, Mining and Materials Society, 1989.

Chapters 6 and 7

ASM Handbook of Failure Analysis and Prevention. Vol. 11. 1992.

ASM Handbook of Materials Selection and Design. Vol. 20. 1997.

Courtney, T. H. *Mechanical Behavior of Materials.* McGraw-Hill, 1989.

Courtney, T. H. *Mechanical Behavior of Materials.* 2d ed. 2000.

Dieter, G. E. *Mechanical Metallurgy.* 3d ed. McGraw-Hill, 1986.

Hertzberg, R. W. *Deformation and Fracture Mechanics of Engineering Materials.* 3d ed. Wiley, 1989.

http://www.wtec.org/loyola/nano/06_02.htm

Hertzberg, R. W. *Deformation and Fracture Mechanics of Materials.* 4th ed. 1972.

K. S. Kumar, H. Van Swygenhoven, S. Suresh, *Mechanical behavior of nanocrystalline metals and alloys,* Acta Materialia, 51, 5743–5774, 2003

Schaffer et al. *"The Science and Design of Engineering Materials",* McGraw-Hill, 1999.

T. Hanlon, Y. -N. Kwon, S. Suresh, *Grain size effects on the fatigue response of nanocrystalline metals,* Scripta Materialia, 49, 675–680, 2003

Wang et al., *High Tensile Ductility in a nanostructured Metal,"* Nature. Vol. 419, 2002

Wulpi, J. D., *"Understanding How Components Fail,"* ASM, 2000.

Chapter 8

Massalski, T. B. *Binary Alloy Phase Diagrams.* ASM International, 1986.

Massalski, T. B. *Binary Alloy Phase Diagrams.* 3d ed. ASM International.

Rhines, F. *Phase Diagrams in Metallurgy.* McGraw-Hill, 1956.

Chapter 9

Krauss, G. *Steels: Heat Treatment and Processing Principles.* ASM International, 1990.

The Making, Shaping and Heat Treatment of Steel. 11th ed. Vols. 1–3. The AISE Steel Foundation, 1999–2001.

Smith, W. F. *Structure and Properties of Engineering Alloys.* 2d ed. McGraw-Hill, 1993.

Steel, Annual Statistical Report. American Iron and Steel Institute, 2001.

Walker, J. L. et al. (ed.) *Alloying.* ASM International, 1988.

Chapter 10

Benedict, G. M. and B. L. Goodall. *Metallocene-Catalyzed Polymers.* Plastics Design Library, 1998.

"Engineering Plastics." Vol. 2, *Engineered Materials Handbook.* ASM International, 1988.

Kaufman, H. S., and J. J. Falcetta (eds.) *Introduction to Polymer Science and Technology.* Wiley, 1977.

Kohen, M. *Nylon Handbook.* Hanser, 1998.

Moore, E. P. *Polypropylene Handbook.* Hanser, 1996.

Moore, G. R., and D. E. Kline, *Properties and Processing of Polymers for Engineers.* Prentice-Hall, 1984.

Salamone, J. C. (ed.) *Polymeric Materials Encyclopedia.* Vols. 1 through 10. CRC Press, 1996.

Chapter 11

Barsoum, M. *Fundamentals of Ceramics.* McGraw-Hill, 1997.

Bhusan, B. (ed.) *Handbook of Nanotechnology,* Springer, 2004.

"Ceramics and Glasses," Vol. 4, *Engineered Materials Handbook.* ASM International, 1991.

Chiang, Y., D. P. Birnie, and W. D. Kingery. *Physical Ceramics.* Wiley, 1997.

Davis, J. R. (ed.) *Handbook of Materials for Medical Devices,* ASM International, 2003.

Edelstein, A. S. and Cammarata, R. C. (eds.) *Nanomaterials: Synthesis, Properties, and Application,* Institute of Physics Publishing, 2002.

Engineered Materials Handbook. Vol. 4: *Ceramics and Glasses.* ASM International, 1991.

Handbook of Materials for Medical Devices, J. R. Davis, Editor, ASM International, 2003.

Handbook of Nanotechnology, Editor: B. Bhusan, Springer, 2004.

J. A, Jacobs and T. F. Kilduf, *Engineering Materials Technology,* 5th Ed., Prentice Hall, 2004.

Jacobs, J. A. and Kilduf, T. F. *Engineering Materials Technology,* 5th ed., Prentice Hall, 2004.

Kingery, W. D., H. K. Bowen, and D. R. Uhlmann. *Introduction to Ceramics.* 2d ed. Wiley, 1976.

Medical Device Materials, Proceedings of the Materials and Processes for Medical Devices Conference, S. Shrivastava, Editor, ASM international, 2003.

Mobley, J. (ed.). *The American Ceramic Society, 100 Years.* American Ceramic Society, 1998.

Nanomaterials: Synthesis, Properties, and Application, Editors: A. S. Edelstein and R. C. Cammarata, Institute of Physics Publishing, 2002.

Nanostructured Materials, Editor: Jackie Y. Ying, Academic Press, 2001.

Shrivastava, S. (ed.) *Medical Device Materials,* Proceedings of the Materials and Processes for Medical Devices Conference, ASM International, 2003.

Wachtman, J. B. (ed.) *Ceramic Innovations in the Twentieth Century.* The American Ceramic Society, 1999.

Wachtman, J. B. (ed.) *Structural Ceramics.* Academic, 1989.

Ying, J. Y. (ed.) *Nanostructured Materials,* Academic Press, 2001.

Chapter 12

Chawla, K. K. *Composite Materials.* Springer-Verlag, 1987.

"Composites." Vol. 1, *Engineered Materials Handbook.* ASM International, 1987.

Engineered Materials Handbook. Vol. 1: *Composites.* ASM International, 1987.

Engineers' Guide to Composite Materials. ASM International, 1987.

Handbook of Materials for Medical Devices, J. R. Davis, Editor, ASM International, 2003.

Harris, B. *Engineering Composite Materials.* Institute of Metals (London), 1986.

Metals Handbook. Vol. 21: *Composites.* ASM International, 2001.

M. Nordin and V. H. Frankel, *Basic Biomechanics of the Musculoskeletal System,* 3rd Ed., Lippincot, Williams, and Wilkins, 2001.

Nanostructured Materials, Editor: Jackie Y. Ying, Academic Press, 2001.

http://silver.neep.wisc.edu/~lakes/BoneTrab.html

Chapter 13

"Corrosion." Vol. 13, *Metals Handbook.* 9th ed. ASM International, 1987.

Fontana, M. G. *Corrosion Engineering.* 3d ed. McGraw-Hill, 1986.

Jones, D. A. *Corrosion.* 2d ed. Prentice-Hall, 1996.

Uhlig, H. H. *Corrosion and Corrosion Control.* 3d ed. Wiley 1985.

Chapter 14

Binnig, G., H. Rohrer, et al. in *Physical Review Letters,* v 50 pp. 120–24 (1983).

http://ufrphy.lbhp.jussieu.fr/nano/

H. Dai, J. H. Hafner, A. G. Rinzler, D. T. Colbert, R. E. Smalley, Nature 384, 147–150 (1996).

http://www.omicron.de/index2.html?/results/stm_image_of_ chromium_decorated_steps_of_cu_111/~Omicron

http://www.almaden.ibm.com/almaden/media/image_mirage.html

Hodges, D. A. and H. G. Jackson. *Analysis and Design of Digital Integrated Circuits.* 2d ed. McGraw-Hill, 1988.

Mahajan, S. and K. S. Sree Harsha. *Principles of Growth and Processing of Semiconductors.* McGraw-Hill, 1999.

http://www.molec.com/products_consumables.html#STM

Nalwa, H. S. (ed.) *Handbook of Advanced Electronic and Photonic Materials and Devices.* Vol. 1: *Semiconductors.* Academic Press, 2001.

Sze, S. M. (ed.) *VLSI Technology.* 2d ed. McGraw-Hill, 1988.

The material property data and the glass transition temperature data were obtained from the following list of references

1. *ASM Handbooks.* Vol. 1, *Properties and selection: Irons, Steels and High performance alloys,* ASM international, Materials Park, OH.

2. *ASM Handbooks.* Vol. 2, *Properties and selection: Nonferrous alloys and special purpose metals,* ASM international, Materials Park, OH.

3. *ASM Handbooks.* Vol. 8, *Mechanical testing and evaluation,* ASM international, Materials Park, OH.

4. *ASM Handbooks.* Vol. 19, *Fatigue and Fracture,* ASM international, Materials Park, OH.

5. *ASM Handbooks.* Vol. 8, *composites,* ASM international, Materials Park, OH.

6. *ASM Metals handbook desk edition,* ASM international, Materials Park, OH.

7. *ASM Ready reference: Electrical and magnetic properties of materials.* ASM international, Materials Park, OH.

8. *ASM Engineered Materials reference Book,* ASM international, Materials Park, OH.

9. *Mechanical properties and testing of polymers: An A-Z reference (1999).* Edited by G. M. Swallowe. Kluwer Academic Publishers, Dordrecht, Netherlands.

Sze, S. M. *Semiconductor Devices.* Wiley, 1985.

Wolf, S. *Silicon Processing for the VLSI Era.* 2d ed. Lattice Press, 2000.

Chapter 15

Chafee, C. D. *The Rewiring of America.* Academic, 1988.

Hatfield, W. H. and J. H. Miller, *High Temperature Superconducting Materials.* Marcel Dekker, 1988.

Miller, S. E. and I. P. Kaminow. *Optical Fiber Communications II.* Academic, 1988.

Miller, S. E. and I. P. Kaminow. *Optical Fiber Communications II.* Academic Press, 1988.

Nalwa, H. S. (ed.) *Handbook of Advanced Electronic and Photonic Materials and Devices.* Vols. 3–8. Academic Press, 2001.

Chapter 16

Chin, G. Y. and J. H. Wernick. "Magnetic Materials, Bulk." Vol. 14, *Kirk-Othmer Encyclopedia of Chemical Technology.* 3d ed. Wiley, 1981, p. 686.

Coey, M. et al. (eds.) *Advanced Hard and Soft Magnetic Materials.* Vol. 577. Materials Research Society, 1999.

Cullity, B. D. *Introduction to Magnetic Materials.* Addison-Wesley, 1972.

Livingston, J. *Electronic Properties of Engineering Materials.* Chapter 5. Wiley, 1999.

Salsgiver, J. A. et al. (ed.) *Hard and Soft Magnetic Materials.* ASM International, 1987.

10. *Mechanical properties of polymers and composites.* Lawrence E. Nielsen and Robert F. Landel. (1994). Marcel Dekker Inc, Madison ave, New York.

11. *Engineering polymer sourcebook.* Raymond B. Seymour (1990). McGraw-Hill Inc.

12. *Mechanical properties of ceramics.* John B. Watchman (1996). John Wiley and Sons Inc.

13. *Guide to Engineered Materials* (*A Special issue of Advanced Materials and processes*). Vol. 1 (1986). ASM international, Materials Park, OH.

14. *Guide to Engineered Materials* (*A Special issue of Advanced Materials and processes*). Vol. 2 (1987). ASM international, Materials Park, OH.

15. *Guide to Engineered Materials* (*A Special issue of Advanced Materials and processes*). Vol. 3 (1988). ASM international, Materials Park, OH.

16. *ASM Ready reference: Thermal properties of materials.* ASM international, Materials Park, OH.

17. *The mechanical properties of wood.* Wangaard, Frederick Field (1950), John Wiley and Sons Inc.

18. Manufacturer data sheets.

GLOSSARY

A

absorptivity   the fraction of the incident light that is absorbed by a material.

acceptor levels   in the band theory, local energy levels close to the valence band.

activation energy   the additional energy required above the average energy for a thermally activated reaction to take place.

advanced ceramics   new generation of ceramics with improved strength, corrosion resistance, and thermal shock properties, also called engineering or structural ceramics.

aggregate   inert material mixed with portland cement and water to produce concrete. Larger particles are called coarse aggregate (e.g., gravel), and smaller particles are called fine aggregate (e.g., sand).

air-entrained concrete   concrete in which there exists a uniform dispersion of small air bubbles. About 90 percent of the air bubbles are 100 μm or less.

alloy   a mixture of two or more metals or a metal (metals) and a nonmetal (nonmetals).

alnico (aluminum-nickel-cobalt) alloys   a family of permanent magnetic alloys having the basic composition of Al, Ni, and Co, and about 25 to 50 percent Fe. A small amount of Cu and Ti is added to some of these alloys.

α ferrite (α phase in the Fe-Fe$_3$C phase diagram)   an interstitial solid solution of carbon in BCC iron; maximum solid solubility of carbon in BCC iron is 0.02 percent.

amorphous   lacking in long range atomic order

amorphous metal   metals with a noncrystalline structure also called glassy metal. These alloys have high elastic strain threshold.

anion   an ion with a negative charge.

annealing   a heat treatment given to a metal to soften it.

annealing   a heat treatment process applied to a cold worked metal to soften it.

annealing point   at this temperature stresses in the glass can be relieved.

anode   the metal electrode in an electrolytic cell that dissolves as ions and supplies electrons to the external circuit.

anodic protection   the protection of a metal that forms a passive film by the application of an externally impressed anodic current.

antiferromagnetism   a type of magnetism in which magnetic dipoles of atoms are aligned in opposite directions by an applied magnetic field so that there is no net magnetization.

aramid fibers   fibers produced by chemical synthesis and used for fiber-reinforced plastics. Aramid fibers have an aromatic (benzene ring type) polyamide linear structure and are produced commercially by the Du Pont Co. under the trade name of Kevlar.

Arrhenius rate equation   an empirical equation that describes the rate of a reaction as a function of temperature and an activation energy barrier.

asphalt   a bitumen consisting mainly of hydrocarbons having a wide range of molecular weights. Most asphalt is obtained from petroleum refining.

asphalt mixes   mixtures of asphalt and aggregate that are used mainly for road paving.

atactic stereoisomer   this isomer has pendant groups of atoms *randomly arranged* along a vinyl polymer chain. Example: atactic polypropylene.

atom   the basic unit of an element that can undergo chemical change.

atomic mass unit (u)   mass unit based on the mass of exactly 12 for $^{12}_6$C.

atomic number   the number of protons in the nucleus of an atom of an element.

atomic orbital   the region in space about the nucleus of an atom in which an electron with a given set of quantum numbers is most likely to be found. An atomic orbital is also associated with a certain energy level.

atomic packing factor (APF)   the volume of atoms in a selected unit cell divided by the volume of the unit cell.

austempering   a quenching process whereby a steel in the austenitic condition is quenched in a hot liquid (salt) bath at a temperature just above the M_s of the steel, held in the bath until the austenite of the steel is fully transformed, and then cooled to room temperature. With this process a plain-carbon eutectoid steel can be produced in the fully bainitic condition.

austenite (γ phase in Fe-Fe$_3$C phase diagram)   an interstitial solid solution of carbon in FCC iron; the maximum solid solubility of carbon in austenite is 2.0 percent.

austenitizing   heating a steel into the austenite temperature range so that its structure becomes austenite. The austenitizing temperature will vary depending on the composition of the steel.

Avogadro's number 6.023×10^{23} atoms/mol; the number of atoms in one relative gram-mole or mole of an element.

B

bainite   a mixture of α ferrite and very small particles of Fe$_3$C particles produced by the decomposition of austenite; a nonlamellar eutectoid decomposition product of austenite.

bias   voltage applied to two electrodes of an electronic device.

biodegradable polymer   a type of polymer that is absorbed by the human body after it serves its purpose; for instance, absorbable sutures.

biopolymer   polymers that are used inside the human body for various surgical applications.

bipolar junction transistor   a three-element, two-junction semiconducting device. The three basic elements of the transistor are the emitter, base, and collector. Bipolar junction transistors can be of the npn or pnp types. The emitter-base junction is forward-biased and the collector-base junction is reverse-biased so that the transistor can act as a current amplification device.

blends   mixture of two or more polymers, also called polymer alloys.

blistering   a type of damage due to diffusion of atomic hydrogen into internal pores in a metal creating high internal pressure resulting in rupture.

blow molding   a method of fabricating plastics in which a hollow tube (parison) is forced into the shape of a mold cavity by internal air pressure.

body-centered cubic (BCC) unit cell   a unit cell with an atomic packing arrangement in which one atom is in contact with eight identical atoms located at the corners of an imaginary cube.

Bohr magneton   the magnetic moment produced in a ferro- or ferri-magnetic material by one unpaired electron without interaction from any others; the Bohr magneton is a fundamental unit. 1 Bohr magneton $= 9.27 \times 10^{-24}$ A $\cdot$ m^2.

bone   the structural material of the human body.

bone remodelling   the ability of bone to alter its size and structure based on external stress.

brittle fracture   a mode of fracture characterized by rapid crack propagation. Brittle fracture surfaces of metals are usually shiny and have a granular appearance.

buckyball   also called Buckminister Fullerene is soccer ball shaped molecule of carbon atoms (C$_{60}$).

buckytube   a tubular structure made of carbon atoms covalently bonded together.

bulk polymerization   the direct polymerization of liquid monomer to polymer in a reaction system in which the polymer remains soluble in its own monomer.

C

cambium   the tissue that is located between the wood and bark and is capable of repeated cell division.

cancellous (trabecular) bone   the bone material that is porous and soft comprising the internal structure of the bone.

capacitance   a measure of the ability of a capacitor to store electric charge. Capacitance is measured in farads; the units commonly used in electrical circuitry are the picofarad (1 pF $= 10^{-12}$ F) and the micro-farad (1 μF $= 10^{-6}$ F).

capacitor   an electric device consisting of conducting plates or foils separated by layers of dielectric material and capable of storing electric charge.

carbon fibers (for a composite material)   carbon fibers produced mainly from polyacrylonitrile (PAN) or pitch that are stretched to align the fibrillar network structure within each carbon fiber and which are heated to remove oxygen, nitrogen, and hydrogen from the starting or precursor fibers.

cathode   the metal electrode in an electrolytic cell that accepts electrons.

cathodic polarization   the slowing down or the stopping of cathodic reactions at a cathode of an electrochemical cell due to (1) a slow step in the reaction sequence at the-metal-electrolyte interface (*activation polarization*) or (2) a shortage of reactant or accumulation of reaction products at the metal-electrolyte interface (*concentration polarization*).

cathodic protection   the protection of a metal by connecting it to a sacrificial anode or by impressing a DC voltage to make it a cathode.

cation   an ion with a positive charge.

cementite   the intermetallic compound Fe$_3$C; a hard and brittle substance.

ceramic materials   inorganic, nonmetallic materials that consist of metallic and nonmetallic elements bonded together primarily by ionic and/or covalent bonds.

ceramic materials   materials consisting of compounds of metals and non-metals. Ceramic materials are usually hard and brittle. Examples are clay products, glass, and pure aluminum oxide that has been compacted and densified.

chain polymer   a high-molecular-mass compound whose structure consists of a large number of small repeating units called *mers*. Carbon atoms make up most of the mainchain atoms in most polymers.

chain polymerization   the polymerization mechanism whereby each polymer molecule increases in size at a rapid rate once growth has started. This type of reaction occurs in three steps: (1) chain initiation,

(2) chain propagation, and (3) chain termination. The name implies a chain reaction and is usually initiated by some external source. Example: the chain polymerization of ethylene into polyethylene.

chemically tempered glass   glass that has been given a chemical treatment to introduce large ions into its surface to cause compressive stresses at its surface.

cis-1,4 polyisoprene   the isomer of 1,4 polyisoprene that has the methyl group and hydrogen on the same side of the central double bond of its mer. Natural rubber consists mainly of this isomer.

coercive force H_c   the applied magnetic field required to decrease the magnetic induction of a magnetized ferro- or ferrimagnetic material to zero.

cold working of metals   permanent deformation of metals and alloys below the temperature at which a strain-free microstructure is produced continuously (recrystallization temperature). Cold working causes a metal to be strain-hardened.

collagen   the organic component of the bone.

columnar grains   long, thin grains in a solidified polycrystalline structure. These grains are formed in the interior of solidified metal ingots when heat flow is slow and uniaxial during solidification.

composite material   a materials system composed of a mixture or combination of two or more micro- or macroconstituents that differ in form and chemical composition and are essentially insoluble in each other.

composite materials   materials that are mixtures of two or more materials. Examples are fiberglass-reinforcing material in a polyester or epoxy matrix.

compression molding   a thermoset molding process in which a molding compound (which is usually heated) is first placed in a molding cavity. Then the mold is closed and heat and pressure are applied until the material is cured.

concrete (portland cement type)   a mixture of portland cement, fine aggregate, coarse aggregate, and water.

conduction band   the unfilled energy levels into which electrons can be excited to become conductive electrons. In semiconductors and insulators there is an energy gap between the filled lower valence band and the upper empty conduction band.

continuous-cooling transformation (CCT) diagram   a time-temperature-transformation diagram that indicates the time for a phase to decompose into other phases continuously at different rates of cooling.

cooling curve   plots of temp. vs time acquired during solidification of a metal. It provides phase change information as temperature is lowered.

coordination number (CN)   the number of equidistant nearest neighbors to an atom or ion in a unit cell of a crystal structure. For example, in NaCl, CN $= 6$ since six equidistant Cl$^-$ anions surround a central Na$^+$ cation.

copolymer   a polymer chain consisting of two or more types of monomeric units.

copolymerization   the chemical reaction in which high-molecular-mass molecules are formed from two or more monomers.

cored structure   a type of microstructure that occurs during rapid solidification or nonequilibrium cooling of a metal.

corrosion   the deterioration of a material resulting from chemical attack by its environment.

cortical (compact) bone   dense bone comprising the outer structure.

covalent bond   a primary bond resulting from the sharing of electrons. In most cases the covalent bond involves the overlapping of half-filled orbitals of two atoms. It is a directional bond. An example of a covalently bonded material is diamond.

creep   time-dependent deformation of a material when subjected to a constant load or stress.

creep rate   the slope of the creep-time curve at a given time.

critical current density J_c   the current density above which superconductivity disappears.

critical field H_c   the magnetic field above which superconductivity disappears.

critical radius r^* of nucleus   the minimum radius that a particle of a new phase formed by nucleation must have to become a stable nucleus.

critical (minimum) radius ratio   the ratio of the central cation to that of the surrounding anions when all the surrounding anions just touch each other and the central cation.

creep (stress)-rupture strength   the stress that will cause fracture in a creep (stress-rupture) test at a given time and in a specific environment at a particular temperature.

critical temperature T_c   the temperature below which a solid shows no electrical resistance.

cross-linking   the formation of primary valence bonds between polymer chain molecules. When extensive cross-linking occurs as in the case of thermosetting resins, cross-linking makes one supermolecule of all the atoms.

crystal   a solid composed of atoms, ions, or molecules arranged in a pattern that is repeated in three dimensions.

crystal structure   a regular three-dimensional pattern of atoms or ions in space.

crystallinity (in polymers)   the packing of molecular chains into a stereoregular arrangement with a high degree of compactness. Crystallinity in polymeric materials is never 100 percent and is favored in polymeric materials whose polymer chains are symmetrical. Example: high-density polyethylene can be 95 percent crystalline.

curie temperature   the temperature at which a ferromagnetic material when heated completely loses its ferromagnetism and becomes paramagnetic.

curie temperature (of a ferroelectric material)   the temperature at which a ferroelectric material on cooling undergoes a crystal structure change that produces spontaneous polarization in the material. For example, the Curie temperature of $BaTiO_3$ is 120°C.

D

deformation twinning   a plastic deformation process that occurs in some metals and under certain conditions. In this process a large group of atoms are displaced together to form a region of a metal crystal lattice that is a mirror image of a similar region along a twinning plane.

degree of polymerization   the molecular mass of a polymer chain divided by the molecular mass of its mer.

degrees of freedom F   the number of variables (temperature, pressure, and composition) that can be changed *independently* without changing the phase or phases of the system.

diamagnetism   a weak, negative, repulsive reaction of a material to an applied magnetic field; a diamagnetic material has a small negative magnetic susceptibility.

dielectric   an electrical insulator material.

dielectric constant   the ratio of the capacitance of a capacitor using a material between the plates of a capacitor compared to that of the capacitor when there is a vacuum between the plates.

dielectric strength   the voltage per unit length (electric field) at which a dielectric material allows conduction, that is, the maximum electric field that a dielectric can withstand without electrical breakdown.

diffusivity   a measure of the rate of diffusion in solids at a constant temperature. Diffusivity D can be expressed by the equation $D = D_0 e^{-Q/RT}$, where Q is the activation energy and T is the temperature in Kelvins. D_0 and R are constants.

dislocation   a crystalline imperfection in which a lattice distortion is centered around a line. The displacement distance of the atoms around the dislocation is called the *slip* or *Burgers vector* **b.** For an *edge dislocation* the slip vector is perpendicular to the dislocation line, while for a *screw dislocation* the slip vector is parallel to the dislocation line. A *mixed dislocation* has both edge and screw components.

domain wall energy   the potential energy associated with the disorder of dipole moments in the wall volume between magnetic domains.

donor levels   in the band theory, local energy levels near the conduction band.

dry pressing   the simultaneous uniaxial compaction and shaping of ceramic granular particles (and binder) in a die.

ductile cast irons   iron-carbon-silicon alloys with 3.0 to 4.0 percent C and 1.8 to 2.8 percent Si. Ductile cast irons contain large amounts of carbon in the form of graphite nodules (spheres) instead of flakes as in the case of gray cast iron. The addition of magnesium (about 0.05 percent) before the liquid cast iron is poured enables the nodules to form. Ductile irons are in general more ductile than gray cast irons.

ductile fracture   a mode of fracture characterized by slow crack propagation. Ductile fracture surfaces of metals are usually dull with a fibrous appearance.

ductile to brittle transition (DBT)   observed reduced ductility and fracture resistance of a material when temperature is low.

E

eddy-current energy losses   energy losses in magnetic materials while using alternating fields; the losses are due to induced currents in the material.

e-glass fibers   fibers made from E (electrical) glass, which is a borosilicate glass and which is the most commonly used glass for fibers for fiberglass-reinforced plastics.

elastic deformation   if a metal deformed by a force returns to its original dimensions after the force is removed, the metal is said to be elastically deformed.

elastomer   a material that at room temperature stretches under a low stress to at least twice its length and then quickly returns to almost its original length upon removal of the stress.

electric current   the time rate passage of charge through material; electric current i is the number of coulombs per second that passes a point in a material. The SI unit for electric current is the ampere (1 A = 1 C/s).

electric current density J   the electric current per unit area. SI units: amperes/meter^2 (A/m^2).

electrical conductivity σ_e   a measure of the ease with which electric current passes through a unit volume of material. Units: $(\Omega \cdot m)^{-1}$. σ_e is the inverse of ρ_e.

electrical conductor   a material with a high electrical conductivity. Silver is a good conductor and has a $\sigma_e = 6.3 \times 10^7 (\Omega \cdot m)^{-1}$.

electrical insulator   a material with a low electrical conductivity. Polyethylene is a poor conductor and has a $\sigma_e = 10^{-15}$ to $10^{-17} (\Omega \cdot m)^{-1}$.

electrical resistance R   the measure of the difficulty of electric current's passage through a volume of material. Resistance increases with the length and increases with decreasing cross-sectional area of the material through which the current passes. SI unit: ohm (Ω).

electrical resistivity ρ_e a measure of the difficulty of electric current's passage through a *unit* volume of material. For a volume of material, $\rho_e = RA/l$, where $R =$ resistance of material, Ω; $l =$ its length, m; $A =$ its cross-sectional area, m^2. In SI units, $\rho_e =$ ohm-meters ($\Omega \cdot$ m).

electromotive force series   an arrangement of metallic elements according to their standard electrochemical potentials.

electron   a negative charge carrier with a charge of 1.60×10^{-19} C.

electron configuration   the distribution of all the electrons in an atom according to their atomic orbitals.

electron shell   a group of electrons with the same principal quantum number n.

electronic materials   materials used in electronics, especially microelectronics. Examples are silicon and gallium arsenide.

embryos   small particles of a new phase formed by a phase change (e.g., solidification) that are not of critical size and that can redissolve.

energy-band model   in this model the energies of the bonding valence electrons of the atoms of a solid form a band of energies. For example, the $3s$ valence electrons in a piece of sodium form a $3s$ energy band. Since there is only one $3s$ electron (the $3s$ orbital can contain two electrons), the $3s$ energy band in sodium metal is half-filled.

engineering strain ε change in length of sample divided by the original length of sample ($\epsilon = \Delta l/l_0$).

engineering stress σ average uniaxial force divided by original cross-sectional area ($\sigma = F/A_0$).

engineering stress-strain diagram   experimental plot of engineering stress versus engineering strain; σ is normally plotted as the y axis and ϵ as the x axis.

equiaxed grains   grains that are approximately equal in all directions and have random crystallographic orientations.

equilibrium   a system is said to be in equilibrium if no macroscopic changes take place with time.

equilibrium phase diagram   a graphical representation of the pressures, temperatures, and compositions for which various phases are stable at equilibrium. In materials science the most common phase diagrams involve temperature versus composition.

eutectic composition   the composition of the liquid phase that reacts to form two new solid phases at the eutectic temperature.

eutectic point   the point determined by the eutectic composition and temperature.

eutectic reaction (in a binary phase diagram)   a phase transformation in which all the liquid phase transforms on cooling into two solid phases isothermally.

eutectic temperature   the temperature at which a eutectic reaction takes place.

eutectoid (plain-carbon steel)   a steel with 0.8 percent C.

eutectoid α ferrite α ferrite that forms during the eutectoid decomposition of austenite; the α ferrite in pearlite.

eutectoid cementite (Fe$_3$C)   cementite that forms during the eutectoid decomposition of austenite; the cementite in pearlite.

exchange energy   the energy associated with the coupling of individual magnetic dipoles into a single magnetic domain. The exchange energy can be positive or negative.

extrusion   a plastic forming process in which a material under high pressure is reduced in cross section by forcing it through an opening in a die.

extrusion   the forcing of softened plastic material through an orifice, producing a continuous product. Example: plastic pipe is extruded.

F

face-centered cubic (FCC) unit cell   a unit cell with an atomic packing arrangement in which 12 atoms surround a central atom. The stacking sequence of layers of close-packed planes in the FCC crystal structure is *ABCABC.* . . .

fatigue   the phenomenon leading to fracture under repeated stresses having a maximum value less than the ultimate strength of the material.

fatigue crack growth rate da/dN the rate of crack growth extension caused by constant-amplitude fatigue loading.

fatigue failure   failure that occurs when a specimen undergoing fatigue fractures into two parts or otherwise has been significantly reduced in stiffness.

fatigue life   the number of cycles of stress or strain of a specific character that a sample sustains before failure.

ferrimagnetism   a type of magnetism in which the magnetic dipole moments of different ions of an ionically bonded solid are aligned by a magnetic field in an antiparallel manner so that there is a net magnetic moment.

ferroelectric material   a material that can be polarized by applying an electric field.

ferromagnetic material   one that is capable of being highly magnetized. Elemental iron, cobalt, and nickel are ferromagnetic materials.

ferromagnetism   the creation of a very large magnetization in a material when subjected to an applied magnetic field. After the applied field is removed, the ferromagnetic material retains much of the magnetization.

ferrous metals and alloys   metals and alloys that contain a large percentage of iron such as steels and cast irons.

fiber-reinforced plastics   composite materials consisting of a mixture of a matrix of a plastic material such as a polyester or epoxy strengthened by fibers of high strength such as glass, carbon, or aramid. The fibers provide the high strength and stiffness, and the plastic matrix bonds the fibers together and supports them.

Fick's first law of diffusion in solids   the flux of a diffusing species is proportional to the concentration gradient at constant temperature.

Fick's second law of diffusion in solids   the rate of change of composition is equal to the diffusivity times the rate of change of the concentration gradient at constant temperature.

filament winding   a process for producing fiber-reinforced plastics by winding continuous reinforcement previously impregnated with a plastic resin on a rotating mandrel. When a sufficient number of layers have been applied, the wound form is cured and the mandrel removed.

filler   a low-cost inert substance added to plastics to make them less costly. Fillers may also improve some physical properties such as tensile strength, impact strength, hardness, wear resistance, etc.

firing (of a ceramic material)   heating a ceramic material to a high enough temperature to cause a chemical bond to form between the particles.

float glass   flat glass that is produced by having a ribbon of molten glass cool to the glass-brittle state while floating on the top of a flat bath of molten tin and under a reducing atmosphere.

fluorescence   absorption of light or other energy by a material and the subsequent emission of light within 10^{-8} s of excitation.

fluxoid   a microscopic region surrounded by circulating supercurrents in a type II superconductor at fields between H_{c2} and H_{c1}.

forging   a primary processing method for working metals into useful shapes in which the metal is hammered or pressed into shape.

forward bias bias applied to a pn junction in the conducting direction; in a pn junction under forward bias, majority-carrier electrons and holes flow toward the junction so that a large current flows.

frenkel imperfection a point imperfection in an ionic crystal in which a cation vacancy is associated with an interstitial cation.

functionality the number of active bonding sites in a monomer. If the monomer has two bonding sites, it is said to be *bifunctional*.

G

galvanic cell two dissimilar metals in electrical contact with an electrolyte.

galvanic (seawater) series an arrangement of metallic elements according to their electrochemical potentials in seawater with reference to a standard electrode.

Gibbs phase rule the statement that at equilibrium the number of phases plus the degrees of freedom equals the number of components plus 2. $P + F = C + 2$. In the condensed form with pressure ≈ 1 atm, $P + F = C + 1$.

glass a ceramic material that is made from inorganic materials at high temperatures and is distinguished from other ceramics in that its constituents are heated to fusion and then cooled to the rigid condition without crystallization.

glass enamel a glass coating applied to a glass substrate.

glass-forming oxide an oxide that forms a glass easily; also an oxide that contributes to the network of silica glass when added to it, such as B_2O_3.

glass reference points (temperatures)

glass transition temperature the center of the temperature range where a heated thermoplastic upon cooling changes from a rubbery, leathery state to that of brittle glass.

glass transition temperature the center of the temperature range in which a noncrystalline solid changes from being glass-brittle to being viscous.

glaze a glass coating applied to a ceramic substrate.

grain a single crystal in a polycrystalline aggregate.

grain boundary a surface imperfection that separates crystals (grains) of different orientations in a polycrystalline aggregate.

grain growth the third stage in which new grains start to grow in an equiaxed manner.

grain-size number a nominal (average) number of grains per unit area at a particular magnification.

graphite a layered structure of carbon atoms covalently bonded to three others inside the layer. Various layers are the bonded through secondary bonds.

gray cast irons iron-carbon-silicon alloys with 2.5 to 4.0 percent C and 1.0 to 3.0 percent Si. Gray cast irons contain large amounts of carbon in the form of graphite flakes. They are easy to machine and have good wear resistance.

ground state the quantum state with the lowest energy.

H

Hall–petch relationship an empirical equation that relates the strength of a metal to its grain size.

hand lay-up the process of placing (and working) successive layers of reinforcing material in a mold by hand to produce a fiber-reinforced composite material.

hard ferrites ceramic permanent magnetic materials. The most important family of these materials has the basic composition $MO \cdot Fe_2O_3$, where M is a barium (Ba) ion or a strontium (Sr) ion. These materials have a hexagonal structure and are low in cost and density.

hard magnetic material a magnetic material with a high coercive force and a high saturation induction.

hardenability the ease of forming martensite in a steel upon quenching from the austenitic condition. A highly hardenable steel is one that will form martensite throughout in thick sections. Hardenability should not be confused with hardness. Hardness is the resistance of a material to penetration. The hardenability of a steel is mainly a function of its composition and grain size.

hardness a measure of the resistance of a material to permanent deformation.

hardwood trees trees that have covered seeds and broad leaves. Examples are oak, maple, and ash.

heartwood the innermost part of the tree stem that in the living tree contains only dead cells.

Heisenberg's uncertainty principle the statement that it is impossible to determine accurately at the same time the position and momentum of a small particle such as an electron.

heterogeneous nucleation (as pertains to the solidification of metals) the formation of very small regions (called *nuclei*) of a new solid phase at the interfaces of solid impurities. These impurities lower the critical size at a particular temperature of stable solid nuclei.

hexagonal close-packed (HCP) unit cell a unit cell with an atomic packing arrangement in which 12 atoms surround a central identical atom. The stacking sequence of layers of close-packed planes in the HCP crystal structure is *ABABAB. . . .*

high resolution transmission electron microscope (HRTEM) a Technique based on TEM but with significantly higher resolution by using significantly thinner samples.

hole a positive charge carrier with a charge of 1.60×10^{-19} C.

homogeneous nucleation (as pertains to the solidification of metals) the formation of very small regions of a new solid phase (called *nuclei*) in a pure metal that can grow until solidification is complete. The pure homogeneous metal itself provides the atoms that make up the nuclei.

homogenization a heat treatment process given to a metal to remove undesirable cored structures.

homopolymer a polymer consisting of only one type of monomeric unit.

hot working of metals permanent deformation of metals and alloys above the temperature at which a strain-free microstructure is produced continuously (recrystallization temperature).

hybrid orbital an atomic orbital obtained when two or more nonequivalent orbitals of an atom combine. The process of the rearrangement of the orbitals is called *hybridization*.

hydration reaction reaction of water with another compound. The reaction of water with portland cement is a hydration reaction.

hydrogel a soft polymeric material that absorbs water and swells to a specific level.

hydrogen bond a special type of intermolecular permanent dipole attraction that occurs between a hydrogen atom bonded to a highly electronegative element (F, O, N, or Cl) and another atom of a highly electronegative element.

hydrogen embrittlement loss of ductility in a metal due to interaction of the alloying element in the metal with atomic or molecular hydrogen.

hydrophilic polymer   polymers the absorb water; like water.

hydroxyapatite   the inorganic constituent of the bone.

hypereutectic composition   one that is to the right of the eutectic point.

hypereutectoid (plain-carbon steel)   a steel with 0.8 to 2.0 percent C.

hypoeutectic composition   one that is to the left of the eutectic point.

hypoeutectoid (plain-carbon steel)   a steel with less than 0.8 percent C.

hysteresis energy loss   the work or energy lost in tracing out a B-H hysteresis loop. Most of the energy lost is expended in moving the domain boundaries during magnetization.

hysteresis loop   the B versus H or M versus H graph traced out by the magnetization and demagnetization of a ferro- or ferrimagnetic material.

I

index of refraction   the ratio of the velocity of light in vacuum to that through another medium of interest.

indices for cubic crystal planes (Miller indices)   the reciprocals of the intercepts (with fractions cleared) of a crystal plane with the x, y, and z axes of a unit cube are called the Miller indices of that plane. They are designated h, k, and l for the x, y, and z axes, respectively, and are enclosed in parentheses as (hkl). Note that the selected crystal plane must *not* pass through the origin of the x, y, and z axes.

indices of direction in a cubic crystal   a direction in a cubic unit cell is indicated by a vector drawn from the origin at one point in a unit cell through the surface of the unit cell; the position coordinates (x, y, and z) of the vector where it leaves the surface of the unit cell (with fractions cleared) are the indices of direction. These indices, designated u, v, and w are enclosed in brackets as $[uvw]$. Negative indices are indicated by a bar over the index.

injection molding   a molding process whereby a heat-softened plastic material is forced by a screw-drive cylinder into a relatively cool mold cavity that gives the plastic the desired shape.

intergranular corrosion   preferential corrosion occurring at grain boundaries or at regions adjacent to the grain boundaries.

intermediate oxides   an oxide that may act either as a glass former or as a glass modifier, depending on the composition of the glass. Example, Al_2O_3.

intermediate phase   a phase whose composition range is between those of the terminal phases.

intermetallics   stochiometric compounds of metallic elements with high hardness and high temp strength, but brittle.

interstitial diffusion   the migration of interstitial atoms in a matrix lattice.

interstitial solid solution   a solid solution formed in which the solute atoms can enter the interstices or holes in the solvent-atom lattice.

interstitialcy (self-interstitial)   a point imperfection in a crystal lattice where an atom of the same kind as those of the matrix lattice is positioned in an interstitial site between the matrix atoms.

intrinsic semiconductor   a semiconducting material that is essentially pure and for which the energy gap is small enough (about 1 eV) to be surmounted by thermal excitation; current carriers are electrons in the conduction band and holes in the valence band.

invariant reactions   equilibrium phase transformations involving zero degrees of freedom.

invariant reactions   those reactions in which the reacting phases have fixed temperature and composition. The degree of freedom, F_1 is zero at these reaction points.

inverse spinal structure   a ceramic compound having the general formula MO $\cdot$ M_2O_3. The oxygen ions in this compound form an FCC lattice, with the M^{2+} ions occupying octahedral sites and the M^{3+} ions occupying both octahedral and tetrahedral sites.

ion-concentration cell   galvanic cell formed when two pieces of the same metal are electrically connected by an electrolyte but are in solutions of different ion concentrations.

ionic bond   a primary bond resulting from the electrostatic attraction of oppositely charged ions. It is a nondirectional bond. An example of an ionically bonded material is a NaCl crystal.

ionization energy   the energy required to remove an electron from its ground state in an atom to infinity.

iron-chromium-cobalt alloys   a family of permanent magnetic alloys containing about 30% Cr–10 to 23% Co and the balance iron. These alloys have the advantage of being cold-formable at room temperature.

iron-silicon alloys   Fe–3 to 4% Si alloys that are soft magnetic materials with high saturation inductions. These alloys are used in motors and low-frequency power transformers and generators.

isomorphous system   a phase diagram in which there is only one solid phase, i.e., there is only one solid-state structure.

isostatic pressing   the simultaneous compaction and shaping of a ceramic powder (and binder) by pressure applied uniformly in all directions.

isotactic isomer   this isomer has pendant groups of atoms all on the *same side* of a vinyl polymer chain. Example: isotactic polypropylene.

isothermal transformation (IT) diagram   a time-temperature-transformation diagram that indicates the time for a phase to decompose into other phases isothermally at different temperatures.

J

jominy hardenability test   a test in which a 1 in. (2.54 cm) diameter bar 4 in. (10.2 cm) long is austenitized and then water-quenched at one end. Hardness is measured along the side of the bar up to about 2.5 in. (6.35 cm) from the quenched end. A plot called the *Jominy hardenability curve* is made by plotting the hardness of the bar against the distance from the quenched end.

L

laminate   a product made by bonding sheets of a material together, usually with heat and pressure.

laminate ply (lamina)   one layer of a multilayer laminate.

Larsen Miller parameter   a time-temperature parameter used to predict stress rupture due to creep.

laser   acronym for *l*ight *a*mplification by *s*timulated *e*mission of *r*adiation.

laser beam   a beam of monochromatic, coherent optical radiation generated by the stimulated emission of photons.

lattice point   one point in an array in which all the points have identical surroundings.

lever rule   the weight percentages of the phases in any two-phase region of a binary phase diagram can be calculated using this rule if equilibrium conditions prevail.

light attenuation   decrease in intensity of the light.

lignin   a very complex cross-linked three-dimensional polymeric material formed from phenolic units.

linear density ρ_l the number of atoms whose centers lie on a specific direction on a specific length of line in a unit cube.

liquidus   the temperature at which liquid starts to solidify under equilibrium conditions.

low angle boundary (Tilt)   an array of dislocations forming angular mismatch inside a crystal.

lower critical field H_{c1} the field at which magnetic flux first penetrates a type II superconductor.

lumen   the cavity in the center of a wood cell.

luminescence   absorption of light or other energy by a material and the subsequent emission of light of longer wavelength.

M

M_f the temperature at which the austenite in a steel finishes transforming to martensite.

M_s the temperature at which the austenite in a steel starts to transform to martensite.

magnetic anneal   the heat treatment of a magnetic material in a magnetic field that aligns part of the alloy in the direction of the applied field. For example, the α' precipitate in alnico 5 alloy is elongated and aligned by this type of heat treatment.

magnetic domain   a region in a ferro- or ferrimagnetic material in which all magnetic dipole moments are aligned.

magnetic field H the magnetic field produced by an external applied magnetic field or the magnetic field produced by a current passing through a conducting wire or coil of wire (solenoid).

magnetic induction B the sum of the applied field H and the magnetization M due to the insertion of a given material into the applied field. In SI units, $B = \mu_0(H + M)$.

magnetic permeability μ the ratio of the magnetic induction B to the applied magnetic field H for a material; $\mu = B/H$.

magnetic susceptibility χ_m the ratio of M (magnetization) to H (applied magnetic field); $\chi_m = M/H$.

magnetization M a measure of the increase in magnetic flux due to the insertion of a given material into a magnetic field of strength H. In SI units the magnetization is equal to the permeability of a vacuum (μ_0) times the magnetization, or $\mu_0 M$. ($\mu_0 = 4\pi \times 10^{-4}$ T · m/A.)

magnetocrystalline anisotropy energy   the energy required during the magnetization of a ferromagnetic material to rotate the magnetic domains because of crystalline anisotropy. For example, the difference in magnetizing energy between the hard [111] direction of magnetization and the [100] easy direction in Fe is about 1.4×10^4 J/m^3.

magnetostatic energy   the magnetic potential energy due to the external magnetic field surrounding a sample of a ferromagnetic material.

magnetostriction   the change in length of a ferromagnetic material in the direction of magnetization due to an applied magnetic field.

magnetostrictive energy   the energy due to the mechanical stress caused by magnetostriction in a ferromagnetic material.

majority carriers   the type of charge carrier most prevalent in a semiconductor; the majority carriers in an n-type semiconductor are conduction electrons, and in a p-type semiconductor they are conduction holes.

malleable cast irons   iron-carbon-silicon alloys with 2.0 to 2.6 percent C and 1.1 to 1.6 percent Si. Malleable cast irons are first cast as white cast irons and then are heat-treated at about 940°C (1720°F) and held about 3 to 20 h. The iron carbide in the white iron is decomposed into irregularly shaped nodules or graphite.

martempering (marquenching)   a quenching process whereby a steel in the austenitic condition is hot-quenched in a liquid (salt) bath at above the M_s temperature, held for a time interval short enough to prevent the austenite from transforming, and then allowed to cool slowly to room temperature. After this treatment the steel will be in the martensitic condition, but the interrupted quench allows stresses in the steel to be relieved.

martensite   a supersaturated interstitial solid solution of carbon in body-centered tetragonal iron.

materials   substances of which something is composed or made. The term *engineering materials* is sometimes used to refer specifically to materials used to produce technical products. However, there is no clear demarcation line between the two terms, and they are used interchangeably.

materials engineering   an engineering discipline that is primarily concerned with the use of fundamental and applied knowledge of materials so that they can be converted into products needed or desired by society.

materials science   a scientific discipline that is primarily concerned with the search for basic knowledge about the internal structure, properties, and processing of materials.

maximum energy product $(BH)_{max}$ the maximum value of B times H in the demagnetization curve of a hard magnetic material. The $(BH)_{max}$ value has SI units of J/m^3.

Meissner effect   the expulsion of the magnetic field by a superconductor.

mer   a repeating unit in a chain polymer molecule.

metallic bond   a primary bond resulting from the sharing of delocalized outer electrons in the form of an electron charge cloud by an aggregate of metal atoms. It is a nondirectional bond. An example of a metallically bonded material is elemental sodium.

metallic Glass   metals with an amorphous atomic structure.

metallic materials (metals and metal alloys)   inorganic materials that are characterized by high thermal and electrical conductivities. Examples are iron, steel, aluminum, and copper.

microelectromechanical systems (MEMs)   any miniaturized device that performs sensing and/or actuating function.

microfibrils   elementary cellulose-containing structures that form the wood cell walls.

micromachine   MEMs that perform a specific function or task.

minority carriers   the type of charge carrier in the lowest concentration in a semiconductor. The minority carriers in n-type semiconductors are holes, and in p-type semiconductors they are electrons.

modulus of elasticity E stress divided by strain (σ/ϵ) in the elastic region of an engineering stress-strain diagram for a metal ($E = \sigma/\epsilon$).

monomer   a simple molecular compound that can be covalently bonded together to form long molecular chains (polymers). Example: ethylene.

monotectic reaction (in a binary phase diagram)   a phase transformation in which, upon cooling, a liquid phase transforms into a solid phase and a new liquid phase (of different composition than the first liquid phase).

motif   a group of atoms that or (basis) are organized relative to each other and are associated with corresponding lattice points.

multidirectional laminate   a fiber-reinforced-plastic laminate produced by bonding together layers of fiber-reinforced sheets with some of the directions of the continuous fibers of the sheets being at different angles.

N

nanocrystalline metals   metals with grain size smaller than 100 nm.

nanomaterials   materials with a characteristic length scale smaller than 100 nm.

network modifiers   an oxide that breaks up the silica network when added to silica glass; modifiers lower the viscosity of silica glass and promote crystallization. Examples are Na_2O, K_2O, CaO, and MgO.

nickel-iron alloys   high-permeability soft magnetic alloys used for electrical applications where a high sensitivity is required such as for audio and instrument transformers. Two commonly used basic compositions are 50% Ni–50% Fe and 79% Ni–21% Fe.

nonferrous metals and alloys metals and alloys that do not contain iron, or if they do contain iron, it is only in a relatively small percentage. Examples of nonferrous metals are aluminum, copper, zinc, titanium, and nickel.

non–steady-state conditions for a diffusing system the concentration of the diffusing species changes with time at different places in the system.

normal spinel structure a ceramic compound having the general formula $MO \cdot M_2O_3$. The oxygen ions in this compound form an FCC lattice, with the M^{2+} ions occupying tetrahedral interstitial sites and the M^{3+} ions occupying octahedral sites.

n-type extrinsic semiconductor a semiconducting material that has been doped with an n-type element (e.g., silicon doped with phosphorus). The n-type impurities donate electrons that have energies close to the conduction band.

nuclei small particles of a new phase formed by a phase change (e.g., solidification) that can grow until the phase change is complete.

number of components of a phase diagram the number of elements or compounds that make up the phase-diagram system. For example, the $Fe-Fe_3C$ system is a two-component system; the Fe-Ni system is also a two-component system.

O

octahedral interstitial site the space enclosed when the nuclei of six surrounding atoms (ions) form an octahedron.

optical-fiber communication a method of transmitting information by the use of light.

optical waveguide a thin-clad fiber along which light can propagate by total internal reflection and refraction.

oxygen-concentration cell galvanic cell formed when two pieces of the same metal are electrically connected by an electrolyte but are in solutions of different oxygen concentration.

P

paramagnetism a weak, positive, attractive reaction of a material to an applied magnetic field; a paramagnetic material has a small positive magnetic susceptibility.

parenchyma food-storing cells of trees that are short with relatively thin walls.

passivation the formation of a film of atoms or molecules on the surface of an anode so that corrosion is slowed down or stopped.

pauli exclusion principle the statement that no two electrons can have the same four quantum numbers.

pearlite a mixture of α ferrite and cementite (Fe_3C) phases in parallel plates (lamellar structure) produced by the eutectoid decomposition of austenite.

percent cold reduction

$$\% \text{ cold reduction} = \frac{\text{change in cross-sectional area}}{\text{original cross-sectional area}} \times 100\%$$

peritectic reaction (in a binary phase diagram) a phase transformation in which, upon cooling, a liquid phase combines with a solid phase to produce a new solid phase.

permanent dipole bond a secondary bond created by the attraction of molecules that have permanent dipoles. That is, each molecule has positive and negative charge centers separated by a distance.

phase a physically homogeneous and distinct portion of a material system.

phosphorescence absorption of light by a phosphor and its subsequent emission at times longer than 10^{-8} s.

photon a particle of radiation with an associated wavelength and frequency. Also referred to as a *quantum* of radiation.

piezoelectric ceramics materials that produce an electric field when subjected to mechanical force (and vice versa).

piezoelectric effect an electromechanical effect by which mechanical forces on a ferroelectric material can produce an electrical response and electrical forces produce a mechanical response.

pilling-Bedworth (P.B.) ratio the ratio of the volume of oxide formed to the volume of metal consumed by oxidation.

pitting corrosion local corrosion attack resulting from the formation of small anodes on a metal surface.

plain-carbon steel an iron-carbon alloy with 0.02 to 2 percent C. All commercial plain-carbon steels contain about 0.3 to 0.9 percent manganese along with sulfur, phosphorus, and silicon impurities.

planar density ρ_p the equivalent number of atoms whose centers are intersected by a selected area divided by the selected area.

plasticizers chemical agents added to plastic compounds to improve flow and processibility and to reduce brittleness. Example: plasticized polyvinyl chloride.

pn junction an abrupt junction or boundary between p- and n-type regions within a single crystal of a semiconducting material.

polarization the alignment of small electric dipoles in a dielectric material to produce a net dipole moment in the material.

polycrystalline structure a crystalline structure that contains many grains.

polymeric materials materials consisting of long molecular chains or networks of low-weight elements such as carbon, hydrogen, oxygen, and nitrogen. Most polymeric materials have low electrical conductivities. Examples are polyethylene and *polyvinyl chloride* (PVC).

polymerization the chemical reaction in which high-molecular-mass molecules are formed from monomers.

polymorphism (as pertains to metals) the ability of a metal to exist in two or more crystal structures. For example, iron can have a BCC or an FCC crystal structure, depending on the temperature.

population inversion condition in which more atoms exist in a higher-energy state than in a lower one. This condition is necessary for laser action.

porcelain enamel a glass coating applied to a metal substrate.

portland cement a cement consisting predominantly of calcium silicates that react with water to form a hard mass.

positive-ion core an atom without its valence electrons.

prepreg a ready-to-mold plastic resin-impregnated cloth or mat that may contain reinforcing fibers. The resin is partially cured to a "B" stage and is supplied to a fabricator who uses the material as the layers for a laminated product. After the layers are laid up to produce a final shape, the layers are bonded together, usually with heat and pressure, by the curing of the laminate.

prestressed concrete reinforced concrete in which internal compressive stresses have been introduced to counteract tensile stresses resulting from severe loads.

pretensioned (prestressed) concrete prestressed concrete in which the concrete is poured over pretensioned steel wires or rods.

primary phase a solid phase that forms at a temperature above that of an invariant reaction and is still present after the invariant reaction is completed.

proeutectic phase   a phase that forms at a temperature above the eutectic temperature.

proeutectoid α ferrite α ferrite that forms by the decomposition of austenite at temperatures above the eutectoid temperature.

proeutectoid cementite (Fe_3C)   cementite that forms by the decomposition of austenite at temperatures above the eutectoid temperature.

p-type extrinsic semiconductor   a semiconducting material that has been doped with a p-type element (e.g., silicon doped with aluminum). The p-type impurities provide electron holes close to the upper energy level of the valence band.

pultrusion   a process for producing a fiber-reinforced-plastic part of constant cross section continuously. The pultruded part is made by drawing a collection of resin-dipped fibers through a heated die.

Q

quantum mechanics   a branch of physics in which systems under investigation can have only discrete allowed energy values that are separated by forbidden regions.

quantum numbers   the set of four numbers necessary to characterize each electron in an atom. These are the principal quantum number n, the orbital quantum number l, the magnetic quantum number m_l, and the spin quantum number m_s.

R

radius ratio (for an ionic solid)   the ratio of the radius of the central cation to that of the surrounding anions.

rare earth alloys   a family of permanent magnetic alloys with extremely high energy products. $SmCo_5$ and $Sm(Co, Cu)_{7.4}$ are the two most important commercial compositions of these alloys.

recovery   the first stage in the annealing process that results in removal of residual stresses and formation of low energy dislocation configurations.

recrystallization   the second stage of the annealing process in which new grains start to grow and dislocation density decreases significantly.

rectifier diode   a pn junction diode that converts alternating current to direct current (AC to DC).

refractory (ceramic) material   a material that can withstand the action of a hot environment.

reinforced concrete   concrete containing steel wires or bars to resist tensile forces.

relative permeability μ_r   the ratio of the permeability of a material to the permeability of a vacuum; $\mu_r = \mu/\mu_0$.

remanent induction B_r   the value of B or M in a ferromagnetic material after H is decreased to zero.

reverse bias   bias applied to a pn junction so that little current flows; in a pn junction under reverse bias, majority-carrier electrons and holes flow away from the junction.

roving   a collection of bundles of continuous fibers twisted or untwisted.

S

sapwood   the outer part of the tree stem of a living tree that contains some living cells that store food for the tree.

saturation induction B_s   the maximum value of induction B_s or magnetization M_s for a ferromagnetic material.

scanning electron microscope (SEM)   an instrument used to examine the surface of a material at very high magnifications by impinging electrons.

scanning probe microscopy (SPM)   microscopy techniques such as STM and AFM that allow mapping of the surface of a material at the atomic level.

schottky imperfection   a point imperfection in an ionic crystal in which a cation vacancy is associated with an anion vacancy.

selective leaching   the preferential removal of one element of a solid alloy by corrosion processes.

self-diffusion   the migration of atoms in a pure material.

semiconductor   a material whose electrical conductivity is approximately midway between the values for good conductors and insulators. For example, pure silicon is a semiconducting element and has $\sigma_e = 4.3 \times 10^{-4}\,(\Omega \cdot m)^{-1}$ at 300 K.

semicrystalline   materials with regions of crystalline structure dispersed in the surrounding, amorphous region; for instance some polymers.

s-glass fibers   fibers made from S glass, which is a magnesia-alumina-silicate glass and which is used for fibers for fiberglass-reinforced plastics when extra-high-strength fibers are required.

shape memory alloys   metal alloys that recover a previously defined shape when subjected to an appropriate heat treatment process.

shape-memory alloys   materials that can be deformed but return to their original shape upon an increase in temperature.

shear strain γ   shear displacement a divided by the distance h over which the shear acts ($\gamma = a/h$).

shear stress τ   shear force S divided by the area A over which the shear force acts ($\tau = S/A$).

sheet-molding compound (SMC)   a compound of plastic resin, filler, and reinforcing fiber used to make fiber-reinforced-plastic composite materials. SMC is usually made with about 25 to 30 percent fibers about 1 in. (2.54 cm) long, of which fiberglass is the most commonly used fiber. SMC material is usually pre-aged to a state so that it can support itself and then cut to size and placed in a compression mold. Upon hot pressing, the SMC cures to produce a rigid part.

sintering (of a ceramic material)   the process in which fine particles of a ceramic material become chemically bonded together at a temperature high enough for atomic diffusion to occur between the particles.

slip   the process of atoms moving over each other during the permanent deformation of a metal.

slip casting   a ceramic shape-forming process in which a suspension of ceramic particles and water are poured into a porous mold and then some of the water from the cast material diffuses into the mold, leaving a solid shape in the mold. Sometimes excess liquid within the cast solid is poured from the mold, leaving a cast shell.

slip system   a combination of a slip plane and a slip direction.

slipbands   line markings on the surface of a metal due to slip caused by permanent deformation.

smart materials   materials with the ability to sense and respond to external stimuli.

soft ferrites   ceramic compounds with the general formula $MO \cdot Fe_2O_3$, where M is a divalent ion such as Fe^{2+}, Mn^{2+}, Zn^{2+}, or Ni^{2+}. These materials are ferrimagnetic and are insulators and so can be used for high-frequency transformer cores.

soft magnetic material   a magnetic material with a high permeability and low coercive force.

softening point   at this temperature the glass flows at an appreciable rate.

softwood trees   trees that have exposed seeds and arrow leaves (needles). Examples are pine, fir, and spruce.

solid solution   an alloy of two or more metals or a metal(s) and a nonmetal(s) that is a single-phase atomic mixture.

solid-solution hardening (strengthening)   strengthening a metal by alloying additions that form solid solutions. Dislocations have more difficulty moving through a metal lattice when the atoms are different in size and electrical characteristics, as is the case with solid solutions.

solidus   the temperature during the solidification of an alloy at which the last of the liquid phase solidifies.

solution polymerization   in this process a solvent is used that dissolves the monomer, the polymer, and the polymerization initiator. Diluting the monomer with the solvent reduces the rate of polymerization, and the heat released by the polymerization reaction is absorbed by the solvent.

solvus   a phase boundary below the isothermal liquid + proeutectic solid phase boundary and between the terminal solid solution and two-phase regions in a binary eutectic phase diagram.

space lattice   a three-dimensional array of points each of which has identical surroundings.

specific tensile modulus   the tensile modulus of a material divided by its density.

specific tensile strength   the tensile strength of a material divided by its density.

spheroidite   a mixture of particles of cementite (Fe_3C) in an α ferrite matrix.

spray lay-up   a process in which a spray gun is used to produce a fiber-reinforced product. In one type of spray-up process, chopped fibers are mixed with plastic resin and sprayed into a mold to form a composite material part.

stacking fault   a surface defect formed due to improper (out of place) stacking of atomic planes.

steady-state conditions   for a diffusing system there is no change in the concentration of the diffusing species with time at different places in the system.

stepwise polymerization   the polymerization mechanism whereby the growth of the polymer molecule proceeds by a stepwise intermolecular reaction. Only one type of reaction is involved. Monomer units can react with each other or with any size polymer molecule. The active group on the end of a monomer is assumed to have the same reactivity no matter what the polymer length is. Often a by-product such as water is condensed off in the polymerization process. Example: the polymerization of nylon 6,6 from adipic acid and hexamethylene diamine.

stereoisomers   molecules that have the same chemical composition but different structural arrangements.

stereospecific catalyst   a catalyst that creates mostly a specific type of stereoisomer during polymerization. Example: the Ziegler catalyst used to polymerize propylene to mainly the isotactic polypropylene isomer.

strain hardening (strengthening)   the hardening of a metal or alloy by cold working. During cold working, dislocations multiply and interact, leading to an increase in the strength of the metal.

strain point   at this temperature the glass is rigid.

stress corrosion   preferential corrosive attack of a metal under stress in a corrosive environment.

stress shielding   a condition in which the applied implant is carrying a significant portion of load carried by a fractured bone. It has negative side effects.

substitutional diffusion   the migration of solute atoms in a solvent lattice in which the solute and solvent atoms are approximately the same size. The presence of vacancies makes the diffusion possible.

substitutional solid solution   a solid solution in which solute atoms of one element can replace those of solvent atoms of another element. For example, in a Cu–Ni solid solution the copper atoms can replace the nickel atoms in the solid-solution crystal lattice.

superalloys   metal alloys with improved performance at elevated temperatures and high stress levels.

superconducting state   a solid in the superconducting state that shows no electrical resistance.

superplasticity   the ability of some metals to deform plastically by 1000–2000% at high temperatures and low loading rates.

suspension polymerization   in this process water is used as the reaction medium, and the monomer is dispersed rather than being dissolved in the medium. The polymer products are obtained in the form of small beads that are filtered, washed, and dried in the form of molding powders.

syndiotactic isomer   this isomer has pendant groups of atoms *regularly alternating* in positions on both sides of a vinyl polymer chain. Example: syndiotactic polypropylene.

system   a portion of the universe that has been isolated so that its properties can be studied.

T

tempering (of a steel)   the process of reheating a quenched steel to increase its toughness and ductility. In this process martensite is transformed into tempered martensite.

terminal phase   a solid solution of one component in another for which one boundary of the phase field is a pure component.

tetrahedral interstitial site   the space enclosed when the nuclei of four surrounding atoms (ions) form a tetrahedron.

thermal arrest   a region of the cooling curve for a pure metal where temperature does not change with time (plateaue) representing the freezing temperature.

thermally tempered glass   glass that has been reheated to near its softening temperature and then rapidly cooled in air to introduce compressive stresses near its surface.

thermistor   a ceramic semiconductor device that changes in resistivity as the temperature changes and is used to measure and control temperature.

thermoplastic (noun)   a plastic material that requires heat to make it formable (plastic) and upon cooling, retains its shape. Thermoplastics are composed of chain polymers with the bonds between the chains being of the secondary permanent dipole type. Thermoplastics can be repeatedly softened when heated and harden when cooled. Typical thermoplastics are polyethylenes, vinyls, acrylics, cellulosics, and nylons.

thermosetting plastic (thermoset)   a plastic material that has undergone a chemical reaction by the action of heat, catalysis, etc., leading to a cross-linked network macromolecular structure. Thermoset plastics cannot be remelted and reprocessed since when they are heated they degrade and decompose. Typical thermoset plastics are phenolics, unsaturated polyesters, and epoxies.

tie line   a horizontal working line drawn at a particular temperature between two phase boundaries (in a binary phase diagram) to be used to apply the lever rule. Vertical lines are drawn from the intersection of the tie line with the phase boundaries to the horizontal composition line. A vertical line is also drawn from the tie line to the horizontal line at the intersection point of the tie line with the alloy of interest to use with the lever rule.

tow (of fibers)   a collection of numerous fibers in a straight-laid bundle, specified according to the number of fibers it contains—e.g., 6000 fibers/tow.

tracheids (longitudinal)   the predominating cell found in softwoods; tracheids have the function of conduction and support.

trans-1,4 polyisoprene   the isomer of 1,4 polyisoprene that has the methyl group and hydrogen on opposite sides of the central double bond of its mer.

transducer   a device that is actuated by power from one source and transmits power in another form to a second system. For example, a transducer can convert input sound energy into an output electrical response.

transfer molding   a thermoset molding process in which the molding compound is first softened by heat in a transfer chamber and then is forced under high pressure into one or more mold cavities for final curing.

transmission electron microscope (TEM)   an instrument used to study, the internal defect structures based on passage of electron through a thin films of materials.

twin boundary   a mirror image misorientation of the crystal structure which is considered a surface defect.

twist boundary   an array of screw dislocations creating mismatch inside a crystal.

type I superconductor   one that exhibits complete magnetic-flux repulsion between the normal and superconducting states.

type II superconductor   one in which the magnetic flux gradually penetrates between the normal and superconducting states.

U

ultimate tensile strength (UTS)   the maximum stress in the engineering stress-strain diagram.

unidirectional laminate   a fiber-reinforced-plastic laminate produced by bonding together layers of fiber-reinforced sheets that all have continuous fibers in the same direction in the laminate.

unit cell   a convenient repeating unit of a space lattice. The axial lengths and axial angles are the lattice constants of the unit cell.

upper critical field H_{c2}   the field at which superconductivity disappears for a type II superconductor.

V

vacancy   a point imperfection in a crystal lattice where an atom is missing from an atomic site.

vacuum bag molding   a process of molding a fiber-reinforced-plastic part in which sheets of transparent flexible material are placed over a laminated part that has not been cured. The sheets and the part are sealed, and a vacuum is then applied between the cover sheets and the laminated part so that entrapped air is mechanically worked out of the laminate. Then the vacuum-bagged part is cured.

valence band   the energy band containing the valence electrons. In a conductor the valence band is also the conduction band. The valence band in a conducting metal is not full, and so some electrons can be energized to levels within the valence band and become conductive electrons.

valence electrons   electrons in the outermost shells that are most often involved in bonding.

viscoelasticity   the type of mechanical response in a material that depends on loading rate or strain rate.

vitrification   melting or formation of a glass; the vitrification process is used to produce a viscous liquid glass in a ceramic mixture upon firing. Upon cooling, the liquid phase solidifies and forms a vitreous or glassy matrix that bonds the unmelted particles of the ceramic material together.

volume density ρ_v   mass per unit volume; this quantity is usually expressed in Mg/m^3 or g/cm^3.

vulcanization   a chemical reaction that causes cross-linking of polymer chains. Vulcanization usually refers to the cross-linking of rubber molecular chains with sulfur, but the word is also used for other cross-linking reactions of polymers such as those that occur in some silicone rubbers.

W

weld decay   corrosion attack at or adjacent to a weld as the result of galvanic action resulting from structural differences in the weld.

white cast irons   iron-carbon-silicon alloys with 1.8 to 3.6 percent C and 0.5 to 1.9 percent Si. White cast irons contain large amounts of iron carbide that make them hard and brittle.

wire drawing   a process in which wire stock drawn through one or more tapered dies to the desired cross section.

wood   a natural composite material consisting mainly of a complex array of cellulose fibers in a polymeric material matrix made up primarily of lignin.

wood ray   a ribbonlike aggregate of cells extending radially in the tree stem; the tissue of the ray is primarily composed of food-storing parenchyma cells.

wood vessel   a tubular structure formed by the union of smaller cell elements in a longitudinal row.

working point   at this temperature the glass can easily be worked.

Y

yield strength   the stress at which a specific amount of strain occurs in the engineering tensile test. In the U.S. the yield strength is determined for 0.2 percent strain.

ANSWERS TO SELECTED PROBLEMS

Chapter 2

2.6. 1.82×10^{21} atoms Au

2.8. 6.27×10^{21} atoms Mo

2.9. Atomic % Sn = 65.4 at%
Atomic % Pb = 34.6 at%

2.12. Cu_3Au

2.15. $\Delta E = 6.56 \times 10^{-19}$ J = 4.1 eV

2.17. **a.** $\Delta E = 0.66$ eV $= 1.06 \times 10^{-19}$ J
b. $v = 1.6 \times 10^{14}$ Hz   **c.** $\lambda = 1876$ nm

2.19. $E = 9.30 \times 10^{-15}$ J
$v = 1.40 \times 10^{19}$ Hz

2.30. **b.** Mo   [Kr] $4d^5\, 5s^1$
            Mo^{3+} [Kr] $4d^3$
            Mo^{4+} [Kr] $4d^2$
            Mo^{6+} [Kr] $4d^3$

2.38. $F_{attractive} = 9.16 \times 10^{-9}$ N

2.39. $E_{K^+Br^-} = -6.26 \times 10^{-19}$ J

2.42. $r_{Sr^{2+}} = 0.135$ nm

2.53. % covalent bonding = 47.75%
% metallic bonding = 52.25%

2.64. While a $2 - 6$ compound typically has a higher ionic character than a $3 - 5$, the relatively high electronegativity of phosphorous causes InP to be more ionic in nature.

Chapter 3

3.11. 0.330 nm

3.13. 0.186 nm

3.16. 0.144 nm

3.18. 0.387 nm

3.25. 0.106 nm^3

3.26. 0.273 nm

3.31. **a.** Position Coordinates: (0, 0, 0), (1, 0, 0)
b. Position Coordinates: (0, 0, 0), (1, 1, 0)
c. Position Coordinates: (0, 0, 0), (1, 1, 1)

3.35. $[1\bar{4}\bar{1}]$

3.46. (100), (010), (001), ($\bar{1}$00), (0$\bar{1}$0), (00$\bar{1}$)

3.48. **a.** (1, 0, 0), (1, 0, 1), (1, 1, 0), (1, 1, 1) (1, ½, ½)
b. (1, 0, 0), (1, 0, 1), (0, 1, 0), (0, 1, 1), (½, ½, 0), (½, ½, 1)
c. (1, 0, 0), (0, 0, 1), (0, 1, 0), (½, 0, ½), (½, ½, 0), (0, ½, ½)

3.49. ($6\bar{3}4$)

3.51. ($23\bar{4}$)

3.52. ($\bar{1}\,\bar{1}2$)

3.54. ($\bar{1}22$)

3.57. **a.** 0.224 nm   **b.** 0.112 nm
c. 0.100 nm

3.58. **a.** 0.502 nm   **b.** 0.217 nm

3.63. Miller-Bravais Indices for Planes Shown in Figure P3.63(a).

The Miller indices of plane a are ($0\bar{1}10$).
The Miller indices of plane b are ($\underline{1}012$).
The Miller indices of plane c are (2200).

Miller-Bravais Indices for the Planes Shown in Figure P3.63(b).

The Miller indices of plane a are ($01\bar{1}0$).
The Miller indices of plane b are ($1\bar{1}01$).
The Miller indices of plane c are (1101).

3.65. $[\bar{1}\,\bar{1}21]$, $[\bar{2}111]$, $[\bar{1}2\bar{1}1]$,
$[11\bar{2}1]$, $[2\bar{1}\,\bar{1}1]$, $[1\bar{2}11]$

3.67. For Fig. P3.67(a), the Miller-Bravais direction indices indicated are $[\bar{2}111]$ and $[1\bar{1}\bar{2}1]$. Those associated with Fig. P3.67(b) are $[\bar{1}101]$ and $[10\bar{1}1]$.

3.71. 180.09 g/mol

3.74. **a.** 1.20×10^{13} atoms/mm^2
b. 8.50×10^{12} atoms/mm^2
c. 1.963×10^{13} atoms/mm^2

3.76. **a.** 3.29×10^6 mm   **b.** 2.33×10^6 mm
c. 3.80×10^6 mm

3.80. -4.94%

3.86. 0.3303 nm

3.89. **a.** $0.50 \Rightarrow$ BCC   **b.** 0.3296 nm
c. niobium (Nb)

3.91. **a.** $0.75 \Rightarrow$ FCC   **b.** 0.38397 nm
c. iridium (Ir)

Chapter 4

4.7. $r^* = 1.11 \times 10^{-7}$ cm

4.10. 327 atoms

4.19. **a.** low   **c.** moderate   **e.** low

4.21. 0.036 nm

4.28. 10.23

4.31. 9.64

Chapter 5

5.3. **a.** 2.77×10^{24} vacancies/m^3
     b. 2.02×10^{-5} vacancies/atom

5.13. 56.6 min.

5.16. 0.394 wt %

5.17. 340 min. = 5.67 h

5.19. 1.03 mm

5.20. 1.98×10^{-4} cm

5.23. 0.707 μm

5.25. $D = 1.71 \times 10^{-15}$ m^2/s

5.26. $D = 8.64 \times 10^{-14}$ m^2/s

5.29. 139.3 kJ/mol

Chapter 6

6.6. 0.0669 cm

6.13. 80.8%

6.14. 0.1255 in.

6.19. $\sigma = 962$ MPa

6.20. $\sigma = 954$ MPa

6.23. Engineering strain $\varepsilon = 0.175$

6.29. **a.** Engineering stress = 125,000 psi
          Engineering strain = 0.060
     b. True stress = 132,600 psi
          True strain = 0.0587

6.39. The four principal slip planes are: (111); ($\bar{1}$11);
     (1$\bar{1}$1); (11$\bar{1}$).
     The three slip directions
     are: [$\bar{1}$10]; [01$\bar{1}$]; [$\bar{1}$01].

6.46. **a.** 30.6 MPa   **b.** 0

6.48. $\sigma = 2.08$ MPa

6.49. **a.** 1.94 MPa   **b.** 1.94 MPa
     c. 0

6.52. **a.** 34.7 MPa   **b.** 0
     c. 34.7 MPa

6.64. **a.** 42.9%
     b. Ultimate Tensile Strength $\approx$ 80 ksi; Yield
          Strength $\approx$ 64 ksi; and Elongation $\approx$ 5%.

Chapter 7

7.7. 0.015 in.

7.9. 50.9 ksi

7.12. 0.76 mm

7.13. 3.90 mm

7.19. **a.** 29 ksi (199.8 MPa)   **b.** 14.5 ksi (99.9 MPa)
     c. 10.5 ksi (72.3 MPa)   **d.** -0.16

7.24. $\sigma = 149$ MPa

7.26. 0.176 in.

7.31. 1419 h

7.33. the stress is approximately 96 MPa

7.34. 186.2 h

Chapter 8

8.16. **a.** Weight of liquid phase = 364 g
          Weight of proeutectic α = 136 g
     b. Weight of liquid phase = 251 g
          Weight of proeutectic α = 249 g
     c. 307.5 g
     d. 192.5 g

8.17. 83.3% Sn and 16.7% Pb

8.24. 66.7%

8.26. Wt % liquid = 14.3%
     Wt % δ = 85.7%

8.29. **a.** Wt % α = 47.4%
          Wt % L_1 = 52.6%
     b. Wt % α = 72.2%
          Wt % L_1 = 27.8%
     c. Wt % α = 88.5%
          Wt % L_2 = 11.5%
     d. Wt % α = 90%
          Wt % β = 10%

8.31. 10.8% Pb and 89.2% Cu

Chapter 9

9.12. Wt % austenite = 80.8%
     Wt % proeutectoid ferrite = 19.2%

9.14. 0.49% C

9.17. Wt % austenite = 98.3%
     Wt % cementite = 1.7%

9.19. 1.08% C

9.20. 1.32% C

9.24. 0.232% C

9.68. about 53 RC

9.69. **a.** 40 RC   **b.** 34 RC

9.73. 12°C/sec

9.80. The constituents of the microstructure will be
     bainite, martensite and austenite.

9.96. 9.2%

Chapter 10

10.1. 14,643 mers

10.4. 98 mers

10.6. Mole fraction of polystyrene = 0.541
Mole fraction of polyacrylonitrile = 0.459

10.7. Mole fraction of polyacrylonitrile = 0.324
Mole fraction of polybutadiene = 0.382
Mole fraction of polystyrene = 0.294

10.9. 1.24 g S

10.13. 1.69%

10.17. Wt % S = 6.56%

10.18. **a.** 61.7 days   **b.** 4.0 MPa

10.20. 37.65 kJ/mol

10.22. **a.** 35.7 days   **b.** 65.95 days

Chapter 11

11.7. 0.414

11.10. 4.87 g/cm^3

11.13. **a.** 4.96 g/cm^3 **b.** 7.27 g/cm^3

11.14. **a.** 0.577   **b.** 0.524

11.16. 2.95 g/cm^3

11.18. 6.32 g/cm^3

11.21. 18.1 O^{2-} or 9.1 Th^{4+} ions/nm^2

11.27. 6.03 g/cm^3

11.93. 736.7°C

Chapter 12

12.20. **a.** Wt % carbon fibers = 76.0%
Wt % epoxy resin = 24.0%
b. 1.60 g/cm^3

12.23. 239.6 GPa

12.26. The strength of the composite material
= 30.48 GPa

The fraction of the load carried by the Kevlar
49 fibers = 0.988

12.29. 9.40 GPa

12.54. Required amount of cement = 22.43 sacks
Required amount of gravel = 8012 lb
Required weight of wet sand = 4126 lb
Required weight of water = 109 gal

12.81. 0.306

12.82. 222.2 GPa

Chapter 13

13.12. −0.314 V

13.13. −0.403 V

13.17. 0.07 M

13.19. −0.046 V

13.26. 23.5 min

13.28. $1.07 \times 10^{-2} \dfrac{\text{g}}{\text{min}}$

13.31. 4.52×10^{-7} A/cm^2

13.34. 8.34×10^{-7} A/cm^2

13.64. 23.6 h

13.75. 0.65 A

Chapter 14

14.7. 0.127 m

14.14. $2.36 \times 10^{-6}\,\Omega \times$ cm

14.15. −146.5°C

14.29. $0.45\,\Omega \cdot$ m

14.32. $17.4\,(\Omega \cdot \text{m})^{-1}$

14.39. **a.** 7.51 cm^3 **b.** $6.61 \times 10^{-3}\,\Omega \cdot$ m

14.44. **a.** 6.17×10^{18} electrons/cm^3
b. 1.24×10^{-4}

14.70. **a.** i. InSb = 0.994
ii. InP = 0.979
b. i. InSb = 0.006
ii. InP = 0.021

14.78. 3675

Chapter 15

15.11. **a.** 3.6%   **b.** 4.4%

15.14. 0.903

15.16. **a.** 0.552 μm   **b.** 1.83 μm   **c.** 0.98 μm

15.23. 3.79×10^{-2} s

15.34. 5.67% of the light is retained

15.47. 0.046 T

Chapter 16

16.23. $M_s = 1.70 \times 10^6$ A/m
the saturation induction is 2.14 T

16.24. 2.51×10^6 A/m

16.66. Wt % Fe = 91.7%
Wt % B = 2.97%
Wt % Si = 5.33%

INDEX

End of chapter definitions have bolded page numbers.

A

abnormal grain growth, 305
absorption
  of light, 859–860
  of light by a glass plate, 860–861
  of plastics, 862
absorption-type inhibitors, 768
absorptivity, 859, **883**
  of metals, 859
  of silicate glasses, 859–860
acceptor levels, 801–802, **844**
acetabular (AC) cup in orthopedic implants,
  634–636
acetals, 517
acicular structures, 333
acidic refractories, 620
acrylics, 509–510
acrylonitrile-butadiene-styrene terpolymers
  (ABS resins)
  applications of, 509
  structure and properties of, 507–508
activation energy, 173, **195**
activation polarization, 740, **771**
activators, 863
adsorption theory, 742
advanced ceramics, 12
agglomerates, 638
aggregates, 638, 674, **710**
aging
  of copper alloys, structures formed
    during, 407
  of supersaturated solid solutions, 404
aging curve, 405
aging time, 405
Ag-Pd equilibrium phase diagram, 322
air-entrained concrete, 679, **711**
AISI-SAE code numbers, 391–392
Al$_2$-SiO$_2$ system, 343
Alclad alloys, 766
allotropes, 591–595
allotropy, 102–103
alloys, 10, 446–448
  amorphous metals, 446–448
  intermetallics, 441–442
  nonequilibrium solidification of, 322–325

shape-memory alloys, 442–446
special purpose, 440–441
alloy steels
  classification of, 392–394
  comparative hardenability curves for, 399
  continuous-cooling transformation
    diagram for, 400
  distribution of alloying elements in, 394–395
  principal types of, 394
alloy systems
  binary eutectic, 326–333
  binary isomorphous, 315–318
  binary peritectic, 333–338
alnico alloys, **931**
  properties and compositions of, 917–918
  structure of, 918–919
alpha ferrite, 363, 381, **456**
alpha iron, 314
alternating copolymers, 479
alumina (Al$_2$O$_3$)
  in ceramic insulators, 834–835
  in dental implants, 636–637
  engineering ceramics, 609
  in orthopedic implants, 634–636
  porosity in, 613
aluminum. *See also* cast aluminum alloys
  engineering properties of, 410
  production of, 410–411
  properties of, 410–411
aluminum alloys, 401–418. *See also*
    wrought aluminum alloys
amide linkage, 513
amino resins (ureas and melamines), 529–530
amorphous materials, 72–73, 113–114, **115**
amorphous metals, **458**
  applications of, 447–448
  mechanics of behavior, 447
  production of, 447
  properties and characteristics, 446–448
anions, 39, **65**, 576
anisotropic properties, 592
annealing, **262, 263**
  changes during, 249
  during cold rolling, 206
  of plain-carbon steels, 386–387

temperature ranges for plain-carbon steels,
  386
annealing point, 627, **641**
annealing twin, 149
annual growth rings, 686
anodes, 725, **770**
anodically controlled corrosion rate, 741
anodic protection, 768–769, **771**
anodic reaction, 721
antiferromagnetism, 899, **930**
antifluorite crystal structure, 590
aqueous galvanic cells, 728
aramid fibers, **709**
  in composite materials, 651
  mechanical properties of, 655–657
  for reinforcing plastic resins, 654–655
area effect, 746
Arrhenius, Svante August, 175
Arrhenius plots, 193
Arrhenius rate equation, 176, **195**
artificial aging, 404
artificial heart valve, 553
asphalt and asphalt mixes, 684, **711**
ASTM grain size, 151–155
ASTM grain size method, 154
ASTM grain size number, 154, 242
atactic stereoisomer, 489, **558**
atomic bonds, 41–42
atomic conductivity, 181
atomic diffusion, 177–184
atomic mass, 26–29
atomic mass unit (u), 26, **65**
atomic numbers, 26–29, **65**
atomic orbitals, 24, 29, **65**
atomic packing factor (APF), 79, **115**
atomic size, 35
atoms, **65**
  electronic structure of, 29–41
  and ions, relative sizes of, 36
  positions in cubic unit cells, 83–84
  structure of, 25–26
attenuation, 870
austempering, 390, **457**
austenite, 363, 378–383, **456**
austenitizing, 366, **456**

autocatalytic process, 748
average molecular weight, 475–476
Avogadro's number N_A, 26, **65**

B

Bain, E. C., 381
bainite, 381, 391, **456**
Bakelite, 482
barium ferrite, 928
basal planes, 94
base, 815
basic-oxygen furnace, 361
basic refractories, 620
basis, 74
bauxite, 411
Bayer process, 411
beach marks, 282
benzene, 53–55
benzene ring, 54
bias, **845**
bifunctional monomers, 476
billets, 363
binary alloys, 315, 403–410
binary composites, 663
binary eutectic alloy systems, 326–333
binary isomorphous alloy systems, 315–318
binary monotectic systems, 338–339
binary peritectic alloy systems, 333–338
binary peritectic phase diagram, 338
binary phase diagrams
    to derive lever-rule equations, 319
    ZrO_2-MgO binary phase diagram, 616
bioactive response, 637
biocompatibility, 9, 448–449
biodegradable polymers, 553, **559**
biologically inactive response, 636
biomaterials, 448
biomechanics of bone fracture, 706–707
biomedical applications
    bioactive response, 637
    biologically inactive response, 636
    ceramics in, 634–637
    metal alloys in, 9
    orthopedic implants, 634–636
    resorption (dissolving) response, 637
    toxic response, 636
biometals
    cobalt-based alloys, 449–451
    compression bone plate, 451
    orthopedic implants, 449–450
    stainless steels, 449
    titanium alloys, 452–453
biopolymers, **559**
    artificial heart valve, 553

cardiovascular applications of, 553–554
drug-delivery systems, 555–556
hydrogel, 554
hydrophilic materials, 554
hydrophobic membranes, 554
ophthalmic applications, 554–555
orthopedic applications, 556–557
polyglycolic acid (PGA), 555, 556
polylactic acid (PLA), 555
rigid gas-permeable (RGP) lenses, 555
bipolar junction transistor (BJT), 815–816,
    845
bitumen, 684
blast furnace, 360
blends, 10
blistering, **771**
block copolymers, 479
blooms, 363
blow molding, 494–496, **559**
body-centered cubic (BCC) crystal
    structures, 75, 77–80, 98, 114, **115**
body-centered tetragonal (BCT) crystal
    structure, 376
bog, 600
Bohr, Niels Henrik, 30
Bohr equation, 30
Bohr hydrogen atom, 31
Bohr magneton, 897, **930**
Boltzmann's constant, 173
bonding energies
    of fourth-period metals, 58
    of ionic solids, 48–49
bone, **711**
    composition of, 703
    macrostructure of, 703–705
    mechanical properties of, 705–706
    viscoelasticity of, 707
bone cement, 556
bone fracture biomechanics, 706–707
bone plates, 450, 451
bone remodeling, 707, **711**
bone repair, 708
Borazon, 618
borosilicate glasses, 626
Bragg, William Henry, 107
Bragg's law, 107
Bravais, August, 74
bravais lattices and crystal systems, 74–75
breakdown diodes, 813–814
bright-field image, 158–159
Brinell hardness tests, 228–229
brittle fracture, 270, 273–276, **306,** 550
Buckyball (fullerenes), 593, **640**
buckytube, **640**

bulk free energy, 128
bulk polymerization, 484, **558**
Burgers vectors, 145, 147

C

calcium fluoride (CaF_2), 588–590
cambium layer, 686, **711**
cancellous (trabecular) bone, 703–704, **711**
capacitance, 832, **845**
capacitors, 831, 835–836, **845**
c/a ratio, 82
carbide coatings, 632–634
carbon
    and allotropes of, 591–595
    covalent bonding by, 51–53
    mechanical properties of, for reinforced
        plastic composite materials, 655–657
carbon-carbon covalent bonding, 54
carbon-containing molecules, 53
carbon dioxide lasers, 869
carbon-fiber-reinforced-epoxy resins, 659
carbon fiber reinforced plastic (CFRP), 21
carbon fibers
    for a composite material, **709**
    for reinforced plastics, 653–654
carbonization stage, 654
carbon nanotubes, 593–594
carburizing, 184–188
cardiovascular applications of biopolymers,
    553–554
carrier concentrations
    in extrinsic semiconductors, 804–805
    in intrinsic semiconductors, 804
case hardening, 184–188
castable aluminum alloys, 414
cast aluminum alloys, 416–418
cast copper alloys, 419
casting alloys, 202
casting process, 416
cast irons, 429–436. *See also* ductile iron;
    gray cast iron; malleable cast irons;
    white cast irons
cast magnesium alloys, 437–438
cast polyester, 658
cast products, 202
catalysts
    hexamethylenetetramine (hexa), 523–524
    metallocene, 490
    stereospecific, 490
    Ziegler and Natta, 490–491
cathode, 725
cathode reactions, 728
cathodically controlled corrosion rate, 741
cathodic polarization, **770**

cathodic protection, 768, **771**
cathodic reaction, 721
cathodoluminescence, 864–865
cations, **65,** 576
cavitation damage, 756–757
cellulose, 690
cell-wall ultrastructure, 690–692
cementite, **456**
  and alpha ferrite, nonlamellar eutectoid
     structure of, 381
  eutectoid, 371
  in the Fe-Fe₃C phase diagram, 363
  formation of, 389
  proeutectoid, 371
ceramic abrasive materials, 617–618
ceramic coatings and surface engineering,
    632–634
ceramic composites, 703
ceramic insulator materials, 833, 834
ceramic materials, **22, 640.** *See also*
    engineering ceramics; simple ceramic
    compounds
  for capacitors, 835–836
  for corrosion resistance, 765
  described, 11–13, 573
  factors affecting the strength of, 612–613
  fracture-toughness test of, 614
  mechanisms for the deformation of,
    611–612
  tile insulation for the space shuttle
    orbiter, 620
  toughness of, 613–615
ceramic-matrix composites (CMCs), 14,
    700–703
ceramics
  in biomedical applications, 634–637
  drain casting of, 602
  drying and binder removal of, 604
  dry pressing, 600
  electrical properties of, 830–841
  extrusion of, 602
  fatigue failure of, 615–617
  firing, 598
  forming, 599–603
  hot pressing, 602
  isostatic pressing, 600
  materials preparation, 599
  mechanical properties of, 611–613
  nanotechnology and, 637–639
  pressing, 600
  processing of, 598–603
  refractory brick materials, 619
  sintering, 604–605
  slip casting, 602

  thermal conductivity of, 618
  thermal properties of, 618–620
  thermal treatments of, 604–606
  vitrification of, 605
ceramic semiconductors, 836–837
ceramics refractory materials, 619–620
cesium chloride (CsCl), 579–580
chain-growth polymerization, 471
chain polymer, **558**
chain polymerization, 472, **558**
chain polymerization steps, 473–475
chain-reaction polymerization, 480
chain structures, 595
charge carriers, 806–807
charge neutrality restrictions, 56
chemical bonds, 41
chemical compositions
  of portland cement, 676
  of triaxial whiteware, 607
chemically tempered glass, **641**
chemical reactivity, 39–41
chemical vapor deposition (CVD)
    process, 825
cis-1,4 polyisoprene, 531, **559**
cladded metal structures, 695–696
clamshell marks, 282
classification
  of alloy steels, 392–394
  of copper alloys, 419–421
  of magnesium alloys, 436–437
  of plain-carbon steels, 391–392
  of wrought aluminum alloys, 412
clays
  chemical compositions of, 607
  traditional ceramics, 606
cleavage planes, 273, 274
closed-die forging, 209–211
closed-mold processes, 672–679
close-packed directions, 114
close-packed planes, 114
coal, 484
coatings, 765–766
cobalt-based alloys, 449–451
coercive force, 907, **931**
coherent precipitate, 407
coherent radiation, 866
cold plastic deformation, 246–247
cold-rolling
  at liquid nitrogen temperature, 304
  of metal sheet, 206
cold-worked metal structure, 250–251
cold working, 246
cold working of metals, **262**
collagen, 703, **711**

collector, 815
colloidal gel, 677
columnar grains, 133
compact tissues (bone), 703
complementary metal oxide semiconductor
    (CMOS) Devices, 827–828
completely reversed stress cycle, 285
composite laminate of carbon fibers and
    epoxy resin, 659
composite materials, **22, 709**
  bone, 703–708
  ceramic-matrix composites (CMCs),
    700–703
  cladded metal structures, 695–696
  continuous-fiber reinforced CMCs, 700
  continuous-fiber reinforced MMCs, 696
  crack propagation methods, 700–701
  defined, 649
  described, 13–15
  discontinuous (whisker) CMC's, 700
  discontinuous-fiber MMCs, 698
  fatigue properties of, 660
  fiber-reinforced-plastic, matrix materials
    for, 657–658
  fibers for reinforced-plastic, 651–657
  honeycomb sandwich structure, 695
  metal-matrix composites (MMCs),
    696–700
  multidirectional, 660
  particulate reinforced CMCs, 700
  particulate reinforced MMCs, 698
  sandwich structures, 695–696
  silicon carbide whiskers, 698
  unidirectional, 660
composition base for a ternary phase
    diagram, 346
compound semiconductors, 828–830
compression bone plate, 451
compression molding, 496–497, **559**
  for fiber-reinforced-plastic composite
    materials, 672
concentration galvanic cells, 730–732
concentration polarization, 740–741, **771**
concrete, 674–684
  aggregates for, 679
  air entrainment of, 679
  compressive strengths of, 678, 679
  mixing water for, 678–679
  portland cement type, **710**
  posttensioned, 683
  prestressed, 682–683
  pretensioned, 683
  proportioning of mixtures, 679–682
  proportions of components of, 681

reinforced, 682–683
water-cement ratio, 680
condensation polymerization reactions, 482
conduction band, **844**
conduction electrons, 793
congruently melting compound, 344
continuous annealing, 255
continuous-cooling transformation (CCT)
    diagrams, **457**
    for alloy steels, 400
    for eutectoid carbon steel, 398
    for eutectoid plain-carbon steel, 383–385
continuous-fiber-plastic matrix composite,
    662–666
continuous-fiber reinforced CMCs, 700
continuous-fiber reinforced MMCs, 696, 697
continuous-pultrusion process, 674
continuous-wave (CW) lasers, 868
cooling curves, 314–315, **349**, 389
coordination number (CN), 77, 576, **640**
copolymerization, **558**
copolymers, 479–482, **558**
copper alloys. *See also* cast copper alloys;
    wrought copper alloys
    classification of, 419–421
    copper-beryllium alloys, 424
    copper-tin bronzes, 424
    copper-zinc alloys, 423
    correlation of structures and hardness in, 409
    free-cutting brass, 423
    high-copper alloys, 419
    mechanical properties of, 420–421
    microstructure of, 408
    Muntz metal, 423
    phosphor bronzes, 424
    precipitation strengthening (hardening)
        of, 406
    production of, 419
    properties of, 418–419
    structures formed during the aging of, 407
    tin bronzes, 424
copper-beryllium alloys, 424
Copper Development Association (CDA), 419
copper-lead phase diagram, 340
copper-nickel phase diagram, 317
copper-tin bronzes, 424
copper-zinc phase diagram, 342
cored structure, 323, **350**
core electrons, 790
coring, 338
corrosion, **770**
    defined, 719
    effect of heat treatment on, 735
    galvanic, 745–746

general attack, 745
and impurities, 735
two-metal corrosion, 745–746
types of, 745–759
uniform, 745
corrosion control, 764–769
corrosion fatigue, 288
corrosion rates (kinetics), 735–744
corrosion reactions, 739
corrosion resistance
    alteration of environment for, 767–768
    design for, 766–767
    materials selection for, 765
    polymeric materials for, 765
cortical (compact) bone, 703–704, **711**
cortical tissues (bone), 705–706
corundum (Al$_2$O$_3$), 590
Coulomb, Charles Augustin, 43
coulombic forces, 43
Coulomb's law, 43
covalent bonding, 49–55, **66**
    bond energies and lengths of selected, 52
    by carbon, 51–53
    in carbon-containing molecules, 53
    in diatomic molecules, 50–51
    electron-dot notation for, 51
    in hydrogen molecule, 49–50
    straight-line notation for, 51
    structure of ethylene molecules,
        activated, 472
covalent bonds, 42
covalent-metallitivities, 57–58
crack bridging, 700–701
crack deflection, 700
crack growth, 286
crack initiation, 286
crack length, 288–293
crack monitoring system, 289
crack propagation, 272
crack propagation methods, 700–701
crazes, 550–551
creep, **306**
    of metals, 294–298
    of polymeric materials, 546–547
creep curves, 294, 296, 547
creep modulus, 548
creep rate, 294, 295, **306**
creep-rupture strength, **306**
creep-rupture test, 297–298
creep-rupture time-temperature data, 298–300
creep test, 296–297
crevice corrosion, 749–751
cristobalite, 597
critical (minimum) radius ratio, 577, **640**

critical cooling rate, 385
critical current density, **883**
critical field, 876, **883**
critical radius vs. undercooling, 129–130
critical resolved shear stress, 235–237
critical size, 128
critical superconducting temperatures, 876
critical temperature, 876, **883**
cross-linking, 532–533, **558**
crystal, **115**
crystal directions, 114
crystal lattices, 584–586
crystalline imperfections, 143–150
    line defects (dislocations), 144–147
    planar defects, 143–144
    point defects, 143–144
    volume defects, 150
crystalline material, 73
crystalline solid, 72–73
crystallinity
    in polymers, **558**
    in thermoplastics, 486–490
crystallographically equivalent directions, 85
crystallographic planes
    in hexagonal unit cells, 93–95
    miller indices for, in cubic unit cells, 88–93
crystal planes, 114
crystals, 132–133
crystal structure analysis, 103–113
    x-ray diffraction, 105–107
    x-ray sources, 104–105
crystal structures, **115**. *See also* body-
    centered cubic (BCC) crystal structures;
    face-centered cubic (FCC) crystal
    structures; hexagonal close-packed
    (HCP) crystal structures
    of antifluorite, 590
    body-centered tetragonal (BCT), 376
    of calcium fluoride (CaF$_2$), 588–590
    of cesium chloride (CsCl), 579–580
    of corundum (Al$_2$O$_3$), 590
    cubic, 615
    metallic, 75–77
    monoclinic, 615
    of perovskite (CaTiO$_3$), 590–591
    simple ceramic, 575–595
    slip systems in, 236
    of sodium chloride (NaCl), 580–584
    of spinel (MgAl$_2$O$_4$), 590
    tetragonal, 615
    x-ray diffraction analysis of, 107–112
    of zinc blende (ZnS), 586–588
crystal systems and bravais lattices, 74–75
cube-on-edge (COE) grain material, 911

cubic boron nitride, 618
cubic crystal structures, 615
cubic soft ferrites, 923–924
cubic unit cells
  atom positions in, 83–84
  diffraction conditions for, 108–110
  directions in, 84–87
  miller indices, 88–93
Cu-Ni equilibrium phase diagram, 317
cup-and-cone ductile fracture, 273
Curie temperature, **930**
  ferroelectric domains, 838
  of a ferroelectric material, **845**
  of a ferromagnetic material, 900
current density, 784, 877
cyclic stresses, 285–286
Czochralski method, 136

D

dark-field image, 158–159
debye units, 60
deciduous trees, 686
decomposition products, 404
deep drawing, 212
defects, 151–165
deformation twin, 149
deformation twinning, 241, **263**
deformed martensite, 444–445
degree of polymerization (DP), 473, **558**
degrees of freedom, 313, **349**
delocalized bonding, 55
delta ferrite, 314, 363
delta iron, 314
demagnetization, 907–914
demagnetization curves, 916
dendrites, 124
dense packing, 576
dental implants, 634–637
depletion region, 810
diamagnetism, 894–895, **930**
diamond, 52, 592–593
diatomic molecules, 50–51
die casting, 416–418
dielectric breakdown strength, 831
dielectric constant, 831, 832, **845**
dielectric loss angle, 833
dielectric loss factor, 831, 833
dielectrics, 831–832, **845**
dielectric strength, 833, **845**
diffraction conditions, 108–110
diffuse-porous wood, 689–690
diffusion
  atomic, in solids, 177–184
  of dopants, 822–824

Fick's first law of, 181
Fick's second law of, 182
of a gas into a solid, 183
impurity diffusion into silicon wafers,
    188–191
interstitial, 180
non-steady-state, 184–191
self-diffusion, 178, 179
in solids, effect of temperature on, 191–195
solute-solvent, 182
steady-state, 180–182
substitutional mechanism, 177–178
vacancy mechanism, 177–178
diffusion barrier, 338
diffusion coefficients, 181, 194
diffusionless transformation, 374
diffusion mechanisms, 177–178, 258
diffusion processes, 184–191
diffusion systems, 182
diffusivities
  data for some metallic systems, 192, 193
  for solute-solvent diffusion systems, 182
diffusivity (atomic conductivity), 181, **195**
digital video disks (DVDs), 10
dimensionless units, 215
dipole moment, 59–60
direct-chill method, 411
direct extrusion, 209
direction indices, 84, 94–95
directions
  crystallographically equivalent, 85
  in cubic unit cells, 84–87
  in hexagonal unit cells, 93–95
discontinuous (whisker) CMC's, 700
discontinuous-fiber MMCs, 698
dislocation arrangements, 244–246
dislocations, 144–147, 232
domains, 374
domains of closure, 906
domain wall, 904
domain wall energy, 904–905, 906, **930**
donor impurity atoms, 800
donor levels, 800, **844**
dopants
  diffusion of, 822–824
  in extrinsic silicon semiconductor
    material, 803
  ion implantation of, 822–824
double heterojunction (DH) lasers, 870
drain, 819
drain casting, 602
drift velocity, 779, 784
drive-in diffusion step, 824
drug-delivery systems, 555–556

drying and binder removal, 604
dry pressing, 600, **641**
dsp hybridized bonding, 57
ductile cast irons. *See* ductile iron
ductile fracture, 272–273, **306**
ductile iron, 432–434, **457**
  composition and microstructure of, 433
  mechanical properties of, 430
  properties of, versus hardness, 434
ductile metal in fatigue process, 286–287
ductile to brittle transition (DBT), 276, **306**
ductile to brittle transition temperature,
  276–278
ductility, 223, 303–305

E

earlywood, 686
eddy-current energy losses, 910, **931**
eddy currents, 910
edge dislocations, 145, 232–233
e-glass fibers, 651, **709**
elastic constants for isotropic materials, 217
elastic deformation, 213, **262**
elastomers (rubbers), 531–539, **559**
  described, 469
  natural rubber, 531–534
  properties of, 534
  synthetic rubbers, 534–536
electrical charge transport, 793–794
electrical conduction, 779–789
electrical conductivity, 782, **844**
electrical conductor, **844**
electrical insulators, 782, **844**
electrical neutrality, 48
electrical porcelain, 834–835
electrical properties
  of ceramic insulator materials, 833
  of intrinsic semiconductors, 829
electrical resistance, 782, **843–844**
electrical resistivity, 782, 786, 787, **844**
electric current, **843**
electric current density, **844**
electric current flow, 781
electric field, 784
electrochemical corrosion, 720–723
electrode potentials, 723
electrodynamic suspension (EDS), 889
electrolytic tough-pitch copper, 422
electromagnetic spectrum, 853–854
electromagnetic suspension (EMS), 889
electromotive force (emf), 725
electromotive force series, **770**
electron, **844**
electron configuration, 35–39, **65**

electron-dot notation, 51
electronegativity, 40–41
electronic materials, 8, 15, **22**
electronic structure
    and chemical reactivity, 39–41
    of multielectron atoms, 35–39
electron mobilities, 795
electronpair restrictions, 56
electrons, 25
electron shells, 35, **65**
electron spin quantum number m_s, 34
electroplating, 736
electropositive and electronegative
    elements, 39
elements
    in earth crust, 4
    electron configuration of, 35–39
    ionic radii for certain, 47
    oxidation numbers of, 40
embryo, 128
emitter, 815
emulsion polymerization, 485
encasement phenomena, 337–338, 339
end-quench hardenability test, 397, 398
endurance limit, 284
energy-band diagram, 794–795
energy-band model, **844**
    for electrical conduction, 790–792
    for insulators, 792
    for metals, 790–791
energy bands, 790
energy-level diagram, 32
energy versus separation distance, 57
engineering alloys. *See also* ferrous metals
    and alloys; nonferrous metals and alloys
    fracture-toughness values for, 281
    physical properties and costs of, 437
engineering ceramic materials, 573
engineering ceramics
    alumina (Al_2O_3), 609
    described, 12
    mechanical properties of, 610
    silicon carbide (SiC), 610
    silicon nitride (Si_3N_4), 610
    and traditional ceramics, 606–611
    Zirconia (ZrO_2), 610
engineering properties of aluminum, 410
engineering strain, 213–215, **262**
engineering stress, 213–215, **262**
engineering stress-strain curves, 225
engineering stress-strain diagrams,
    217–227, **262**
engineering thermoplastics, 511–521
    acetals, 517

phenylene oxide-based resins, 516–517
polycarbonates, 515–516
polyphenylene sulfide (PPS), 520–521
properties of, 511–512
thermoplastic polyesters, 518–519
environment, 288
epoxy resins, 525–527
    properties of, 526, 658
    structure and properties of, 526
    structure of, 526
epsilon carbide, 389
equiaxed grains, 127, 132
equilibrium, 311, **349**
equilibrium phase diagrams, 311, **349**
erbium-doped optical-fiber amplifiers
    (EDFAs), 875
erosion corrosion, 756
etching, 153
ethylene molecules
    activated, covalent bonding structure of, 472
    polymerization reactions of, 471–472
eutectic composition, 326, 327, **350**
eutectic point, 327, **350**
eutectic reaction, 364–365
eutectic reaction (in a binary phase
    diagram), **350**
eutectic structures, 333
eutectic temperature, 327, **350**
eutectoid alpha ferrite, **456**
eutectoid cementite, 371, **456**
eutectoid ferrite, 369, 371
eutectoid plain-carbon steels, **456**
    continuous-cooling transformation dia-
        gram for, 383–385, 398
    end-quench hardenability test of, 398
    isothermal transformation diagram for, 378
    slow cooling of, 366
eutectoid reactions, 341, 364–365
eutectoid steel
    microstructure of, 367
    slow cooling of, 366
    transformation of, with slow cooling, 367
eutectoid temperature of steel, 395–396
evergreen trees, 686
exchange energy, 902–903, **930**
exhaustion range, 808
experimental techniques
    for identification of microstructure and
        defects, 151–165
    optical metallography, 151–155
    scanning electron microscopy (SEM),
        156–158
    transmission electron microscopy (TEM),
        158–159

extensometer, 219, 223
extrinsic semiconductors, 799–809
    carrier concentrations in, 804–805
    charge densities in, 803–804
    doping of, 803
    effect of doping on carrier concentrations
        in, 803–804
    effect of temperature on the electrical
        conductivity of, 807–808
    n-Type (Negative-Type), 799–801
    p-Type (Positive-Type), 801–802
    p-type (positive-carrier-type), 803
extrusion, **262**
    of ceramics, 602
    defined, 133
    of metals and alloys, 208–209
    of thermoplastics, 493–494, **559**

F

face-centered cubic (FCC) crystal lattices,
    584–586
face-centered cubic (FCC) crystal structures,
    56, 75, 80–81, 96–98, 114, **115**
failure, 271, 300–302
family or form, 85, 89
farad, 832
Faraday, Michael, 736
Faraday's equation, 736
fatigue, 281–288, **306**
fatigue behavior, 305
fatigue crack growth, 286
fatigue crack growth behavior, 291
fatigue crack growth rate, 289, 290–292, **306**
fatigue crack propagation
    correlation of, with stress and crack
        length, 288–293
    rate of, 288–293
fatigue failures, 281, **306**, 615–617
fatigue life, 283, **306**
fatigue life calculations, 292–293
fatigue limit, 284
fatigue process, 286–287
fatigue properties of composite materials, 660
fatigue strength, 287–288
fatigue stress versus cycles plots, 285
Fe-C martensite
    formation of, by liquid quenching, 373–374
    hardness and strength of, 376
    microstructure of, 374
    structure of, on an atomic scale, 374–375
feldspars, 598, 606
ferrimagnetism, 899–901, **930**
ferrites, 899, 923–924
ferrite-stabilizing elements, 396

ferroelectric ceramics, 838–839
ferroelectric domains, 838
ferroelectric material, **845**
ferromagnetic domains, 900–907
ferromagnetic materials, 889–890, **929**
ferromagnetic metal, 907–909
ferromagnetism, 896–897, **930**
ferrous alloys, 359
ferrous metals and alloys, 8–10, **22**
fiberglass-polyester composites, 659
fiberglass-reinforced polyester resins, 658
fiberglass-reinforcing materials, 14, 651
fiber pullout, 700–701
fiber-reinforced-plastic composite materials,
       658–662
   closed-mold processes for, 672–679
   compression molding for, 672
   continuous-pultrusion process for, 674
   injection molding for, 672
   open-mold processes for, 667–671
   sheet-molding compound (SMC) process,
       672–674
fiber-reinforced plastics, **709**
fibrous composites, 14
Fick, Adolf Eugen, 181
Fick's first law of diffusion, 181, **195**
Fick's second law of diffusion, 182, **195**
field-effect transistor, 819
filament-winding process, 670–672, **710**
fillers, 504, **559**
fireclays, 620
firing (of a ceramic material), 598, **641**
flint, 606
float glass, **641**
float glass process, 628
fluctuating dipole bonds, 42
fluctuating dipoles, 60–61
fluorescence, 863, **883**
fluoroplastics, 510–511
flux density, 892
fluxoids, 879, **884**
folded-chain model, 488
force versus separation distance, 44
forging, 209–211, **262**
forming of ceramics, 599–603
forward bias, **845**
fosterite, 834
fractograph, 282
fracture
   bone, biomechanics of, 706–707
   defined, 272
   of metals, 271–281
   of polymeric materials, 550–552
fracture toughness, 279–281

fracture-toughness test, 280, 614
fracture-toughness values, 281
free-cutting brass, 423
free electrons, 55
free radical, 473
Frenkel, Yakov Ilyich, 144
Frenkel imperfection, 144
fretting corrosion, 757
fringed-micelle model, 488
full anneal, 249
fullerenes (Buckyball), 386, 593
fully stabilized zirconia, 615
functionality, 476, **558**
fused silica glass, 624–625

G

gage length, 215
galvanic (seawater) series, **771**
galvanic cells, 724–734, **770**
galvanic corrosion, 745–746
galvanic couple, 724
galvanic series, 743–744
gamma iron, 314
gas carburizing, 184–188
gate, 819
general attack corrosion, 745
general-purpose thermoplastics
   acrylics, 509–510
   acrylonitrile-butadiene-styrene terpoly-
       mers (ABS resins), 507–508, 509
   fluoroplastics, 510–511
   linear-low-density polyethylene
       (LLDPE), 501
   notched impact strength, 499
   polyethylene (PE), 500–502
   polymethyl methacrylate (PMMA),
       509–510
   polytetrafluoroethylene (PTFE), 510–511
   polyvinyl chloride (PVC), 503–504
   properties of, 499
   sales and prices for, 498
   styrene-acrylonitrile copolymers
       (SAN resins), 507
geometric arrangement of ions, 48
germanium, 795
Gibbs, Josiah Willard, 313–314
Gibbs phase rule, 313–314, **349**
glass enamel, **642**
glasses, 620–632, **641**
   annealing point, 627
   blowing, pressing, and casting of, 629–630
   chemically strengthened, 630–631
   compositions of, 624–625
   defined, 622

effect of temperature on the viscosities
       of, 627
   float glass process, 628
   forming methods for, 628–630
   forming sheet and plate glass, 628–630
   glass-forming oxides, 623
   glass-modifying oxides, 623–624
   glass transition temperature, 622–623
   intermediate oxides in, 624
   lehr (furnace), 629
   long-range order, 623
   loose network, 623
   network modifiers, 623
   softening point, 627
   strain point, 628
   structure of, 623–624
   tempered, 630
   thermally tempered, 631
   viscous deformation of, 626–628
   working point, 627
glass fibers
   mechanical properties of, for reinforced
       plastic composite materials, 655–657
   production of, 651
   properties of, 653
   for reinforcing plastic resins, 651–653
glass-forming oxides, 623, **641**
glass-modifying oxides, 623–624
glass reference points (temperatures), **641**
glass transition temperature (T_g), 486, 488, **641**
   glasses, 622–623
   in polymers, **558**
glaze, 632, **642**
globular structures, 333
goniometer, 108
Goodyear, Charles, 532
GP1 zones, 407
GP2 zones, 407
GP zones. *See* precipitation zones (GP zones)
graft copolymers, 479
grain boundaries, 132, 147–148, 153,
       242–244
grain boundary diffusion, 258
grain boundary sliding, 258
grain diameter determination, 151–155
grain–grain-boundary electrochemical
       cells, 733–734
grain growth, **263**
grain refiners, 132
grain shape, 244–246
grain-size number, 154
grain structure
   formation of, 132–133
   of industrial castings, 133–134

gram-mole, 26
graphene, 594
graphite, 592, **640**
graphitization, 435
graphitization treatment, 654
gray cast iron
  composition and microstructure of,
    431–432
  mechanical properties of, 430
gray cast irons, **457**
green condition, 692
ground state, 32, **65**
growth rings, 687
gutta-percha, 532

H

half-cell potentials, 722
Hall-Petch equation, 242–243, 259
Hall-Petch relationship, **263**
Hall-Petch relationship constants, 243
hammer forging, 209–211
hand lay-up process, 667, **710**
hardenability, 396–401, **457**
hardenability curves, 399
hard ferrites, **931**
hard magnetic materials, 915–923, **931**
  alnico alloys, 917–918
  demagnetization curves of, 916
  described, 889
  iron-chromium-cobalt magnetic alloys,
    921–922
  magnetic properties of, 917
  neodymium-iron-boron magnetic
    alloys, 921
  properties of, 915–916
  rare earth alloys, 919–920
hardness, 227–228, **262**, 389
hardness testing, 227–228
hardwoods, 689
hardwood trees, 686, **711**
heartwood, 686, **711**
heat stabilizers, 503
heat treatment, 435
Heisenberg, Werner Karl, 32
Heisenberg's uncertainty principle,
  32, **65**
hemicellulose, 690
heterogeneous nucleation, 131–132
Hevea brasiliensis tree, 531
hexagonal close-packed (HCP) crystal
  lattices, 584–586
hexagonal close-packed (HCP) crystal
  structures, 75, 81–83, 96–98,
  114, **115**

hexagonal close-packed (HCP) unit cells,
  94–95
hexagonal unit cells, 93–95
hexamethylenetetramine (hexa) catalyst,
  523–524
high-alumina refractories, 620
high-angle boundaries, 147
high-density polyethylenes, 500–501, 502
high-resolution transmission electron
  microscopy (HRTEM), 159–161
high-speed civil transport (HSCT), 3
high-temperature reusable-surface insulation
  (HRSI), 620
hole mobilities, 795
holes, 793–794, **844**
homogeneous nucleation, 127, 128
homogenization, 323, **350**
homogenized ingots, 412
homogenizing heat treatment, 324
homopolymers, **558**
homopolymers and copolymers, 479–482
honeycomb sandwich structure, 695
Hooke, Robert, 221
Hooke's law, 221, 665
horizontal thermal arrest, 329
hot isostaic pressing (HIP), 638
hot pressing of ceramics, 602
hot-rolled ingots, 412
hot rolling
  of sheet ingots, 203
  of steel strip, 362
hot working of metals, **262**
Hume-Rothery, William, 316
Hume-Rothery rules, 138, 316
hybridization, 51–52
hybrid orbitals, 51, **66**
hydration reactions, 677, **711**
hydrocarbons, 54
hydrogel, 554, **559**
hydrogen atom, 29–33
hydrogen bond, 61–62, **66**
hydrogen damage, 758–759
hydrogen embrittlement, 758, **771**
hydrogen molecule, 49–50
hydrophilic polymer, 554, **559**
hydrophobic membranes, 554
hydroxyapatite (HA), 703, **711**
hypereutectic compositions, 327, **350**
hypereutectoid steel, 366
hyperperitectic alloy, 339
hypoeutectic compositions, 327, **350**
hypoeutectoid plain-carbon steel,
  370–372, **456**
hypoeutectoid steel, 366, 371

hysteresis, 443
hysteresis energy losses, 909–910, **931**
hysteresis loop, 908, **931**

I

impact energy temperature, 278
impact testing and toughness, 276
impact-testing apparatus, 277
impurity diffusion, 188–191
incoherent precipitate, 407
incoherent radiation, 866
incongruently melting compound, 344
index of refraction, 856–857, **883**
indices for cubic crystal planes
  (Miller indices), **115**
indices of a family or form, 85
indices of direction in cubic crystals, **115**
indirect extrusion, 209
induction, 892
industrial castings, 133–134
industrial polymerization methods, 484–486
ingots, 412
inhibitors, 767–768
initiation of polymerization, 473–475
injection molding, 498, **558**
  for fiber-reinforced-plastic composite
    materials, 672
  of thermoplastics, 492–493
inner bark, 686
inorganic coatings (ceramics and glass), 766
inorganic glass, 113
insulators, 792
integrated circuits, 188–191
intergranular corrosion, 751–753, **771**
interionic energies, 46–47
interionic forces, 43–46
interionic separation distance, 44
intermediate oxides, 624, **641**
intermediate phases, 341, **351**
intermetallic compound, 343
intermetallics, **458**
International Space Station (ISS), 5
interplanar spacing, 91
interstices, 141
interstitialcy, 144
interstitial diffusion, 180, **195**
interstitial mechanism, 177–178
interstitial sites
  in FCC crystal lattices, 584–586
  in HCP crystal lattices, 584–586
  octahedral, 584
  tetrahedral, 584
interstitial solid solutions, 141–142, 247
interstitial solutions, 138

intramedullary nail, 450, 454
intrinsic elemental semiconductors
  energy-band diagram for, 794–795
  quantitative relationships for electrical
    conduction, 795–796
intrinsic semiconductivity, 797–799
intrinsic semiconductors, 792–799, **844**
  carrier concentrations in, 804
  electrical properties of, 829
  mechanism of electrical conduction in,
    792–793
invariant points, 313, 316
invariant reactions, 327, 339–341, **350, 351,**
    364–365
inverse spinel structure, 925–926, **931**
ion arrangements, 47–48
ion-concentration cells, 730–732, **770**
ionically bonded solids, 576–579
ionic and covalent bonding, 575–576
ionic bonds, 42–49, **65**
ionic-covalent mixed bonding, 62–63
ionic radii, 47
ionic solids
  bonding energies of, 48–49
  electrical neutrality of, 48
  geometric arrangement of ions in, 48
  ion arrangements in, 47–48
ion implantation, 822–825
ionization energy, 32, **65**
ion pair
  interionic energies for, 46–47
  interionic forces for, 43–46
ions, 576
iron
  cooling curves for, 314
  pressure-temperature (PT) phase diagram
    for, 312
  production of, 360–363
iron-carbon system, 363–372
iron-chromium-cobalt alloys, **931**
iron-chromium-cobalt magnetic alloys,
    921–923
iron-chromium phase diagram, 425
iron-iron carbide phase diagrams, 366
  invariant reactions in, 364–365
  solid phases in, 363
iron-nickel phase diagram, 335
iron-silicon alloys, 910–911, **931**
island structures, 595
isomorphous systems, 316, **350**
isostatic pressing, 600, **641**
isostrain conditions, 662, 663
isostress conditions, 664
isotactic isomer, **558**

isotactic stereoisomer, 489
isothermal decomposition of austenite,
    378–383
isothermal section, 346, 347
isothermal transformation (IT) diagrams,
    457
  of a eutectoid steel, 380
  from isothermal transformation of
    eutectoid steels, 379
  for noneutectoid plain-carbon steels, 383
isotropic materials, 217
isotropic properties, 216, 592

J

Jominy hardenability test, 396, **457**

K

Kevlar, 654
Kirkendall effect, 179
knee replacement prosthesis, 452
Knoop hardness tests, 228–229

L

lamella, 489
lamellar eutectic structure, 333, 334
lamellar structures, 333
laminated materials, 659, **709**
laminate ply (lamina), **710**
large-scale integrated (LSI) microelectronic
    circuits, 817–818
Larsen-Miller (L.M.) parameter, 298–300
laser beam, **883**
lasers, 866–870, **883**
  continuous-wave (CW), 868
  pulsed type, 868
  pumping of, 867
  types of, 868–869
latent heat of solidification, 135
latex (natural rubber), 531
lath martensite, 374, 375
lattice constants, 74
lattice energy, 49
lattice points, 74, **115**
lead glasses, 626
lead-tin equilibrium phase diagram, 326
leakage current, 811
lever rule, 318–319, **350**
lever-rule calculations, 330
Lexan (polycarbonate), 515
light
  absorption, transmission and reflection
    of, 859–863

and the electromagnetic spectrum,
    853–856
  refraction of, 856–858
light attenuation, **883**
lignin, 690, **711**
linear absorption coefficient, 860
linear atomic density, 101
linear density, **116**
linear-lowdensity polyethylene (LLDPE), 501
linear polyester, 528
line defects (dislocations), 144–147
liquation, 325
liquidus, 316, **350**
local anodes, 721, 729
local cathodes, 721, 729
local oxidation, 825
longitudinal axis (L), 686
long range order (LRO), 73, 623
loose network, 623
low-alloy steels, 392–401
  applications for, 401
  mechanical properties and applications
    of, 402
  mechanical properties for, 401
low-angle boundaries, 147
low-density polyethylenes, 500–501, 502
lower bainite, 381
lower critical field, **883**
lubricants, 503
lumen, 688, **711**
luminescence, 863–866, **883**

M

macrostructure of bone, 703–705
magnesium alloys, 436–438
  cast alloys, 436–437
  classification of, 436–437
  structures and properties of, 437–438
  wrought alloys, 436
magnesium-nickel phase diagram, 344
magnetically hard ferrites, 928
magnetically levitated (maglev) trains, 888
magnetically soft ferrites, 923–924
magnetic anneal, 918, **931**
magnetic domains, 896–897, 900, **930**
magnetic fields, 889–890, **929**
magnetic fields and quantities, 889–894
magnetic induction, 892–893, **930**
magnetic moment, 897–899
magnetic permeability, 893, **930**
magnetic quantities, 892
magnetic quantum number, 34
magnetic susceptibility, 893–894, **930**

magnetism, 894–899
magnetization, 892, 907–914, **930**
magnetocrystalline anisotropy energy, 903–904, **930**
magnetostatic energy, 903, **930**
magnetostriction, 905, **931**
magnetostrictive energy, 905–906, **931**
majority-carrier devices, 820
majority carriers, 803, **844**
malleable cast irons, **458**
    composition and microstructure of, 435
    cooling of, 436
    Ferritic malleable iron, 436
    graphitization of, 435
    heat treatment of, 435
    mechanical properties of, 430
    pearlitic malleable iron, 436
    tempered martensitic malleable iron, 436
Marcs Exploration Rover (MER), 5
marquenching. See martempering (marquenching)
martempering (marquenching), 389, 390, **457**
martensite, 373–378, **456**
    changes upon tempering, 388
    effect of carbon content on, 375
martensite finish temperature, 374, **457**
martensite lattice, 377
martensite start temperature, 383, **457**
martensite-transformation start temperature, 374
martensitic plain-carbon steel, 377
mass action law, 803
materials, **21**
    competition among, 16–17
    defined, 3
    design and selection, 20–21
    recent advances in, 18–20
    types of, 8–15
materials engineering, 6–7, **21**
materials preparation, 599
materials science, 6–7, **21**
maximum energy product (BH), 916, **931**
mean stress, 286
mechanical properties. See also properties
    of aramid for reinforced plastic composite materials, 655–657
    of austenitic stainless steels, 426
    of bone, 705–706
    of carbon for reinforced plastic composite materials, 655–657
    of castable aluminum alloys, 414
    of cast magnesium alloys, 437–438
    of ceramics, 611–613
    of continuous-fiber reinforced MMCs, 697

of copper alloys, 420–421
of ductile iron, 430
of engineering ceramics, 610
of ferritic stainless steels, 426
of glass fibers for reinforced plastic composite materials, 655–657
of gray cast iron, 430
of low-alloy steels, 401, 402
of malleable cast iron, 430
of martensitic stainless steels, 426
of nickel alloys, 437–438
obtained from engineering stress-strain diagram, 220–225
obtained from tensile tests, 220–225
of plain-carbon steels, 391–392, 393
of precipitation-hardening stainless steels, 426
of SiC whisker reinforced ceramic-matrix composites, 701
of stainless steels, 426
of titanium alloys, 437–438
of wood, 692–693
of wrought aluminum alloys, 414
of wrought magnesium alloys, 437–438
Meissner effect, 877–878, **883**
melamine-formaldehyde resins, 529–530
melamines, 529–530
mer, 473, **558**
metal alloys, 453–455
metal-forming processes, 211–212
metallic bonds, 42, 55, **66**
metallic coatings, 765–766
metallic components, failure of, 300–302
metallic-covalent mixed bonding, 63
metallic crystal structures, 75–77
metallic glasses, 113, **116**, 446, 911–912
metallic-ionic mixed bonding, 64
metallic materials, 8–10, **22**
metallic solid solutions, 138–142
metallocene catalysts, 490
metal-matrix composites (MMCs), 14, 696–700
metals
    absorptivity of, 859
    creep of, 294–296
    effect of temperature on the electrical resistivity of, 787
    electrical conduction in, 779
    electrical conductivities of, 782
    electrical resistivity of, 786
    electrochemical corrosion of, 720–721
    energy-band model for, 790–791
    fatigue of, 281–288
    half-cell potentials for, 722

improvements in mechanical performance of, 303–305
    prices ($/lb), 359
    pure, self-diffusion activation energies for, 179
    stress and strain in, 212–217
    stress rupture of, 294–298
metals and alloys
    casting of, 201–203
    extrusion of, 208–209
    hot and cold rolling of, 203–208
    processing of, 201–212
metal-semiconductor field-effect transistors (MESFETs), 830
metal single crystals
    critical resolved shear stress for, 235–237
    plastic deformation of, 229–242
    slip systems in, 238
methane molecule, 54
methyl ethyl ketone (MEK) peroxide, 528
microelectromechanical systems (MEMs), 5, 18
microelectronic integrated circuits, 821–822
microelectronic planar bipolar transistors, 818
microelectronic planar field-effect transistors, 819–821
microelectronics, 816–828
microfarad, 832
microfibrils, 692, **711**
micromachines, 18
microstructure
    experimental techniques for identification of, 151–165
    of hardwoods, 689
    of softwoods, 688
Miller-Bravais indices, 93
miller indices, 88–93
minimum creep rate, 295
minority carriers, 803, **845**
minority current, 811
mixed bonding, 62–63
mixed dislocations, 145
Mobay (polycarbonate), 515
modified chemical vapor deposition (MCVD), 873
modulus of elasticity, 217, 221, **262**
moisture content of wood, 692
mold casting, 204
mole, 26
molecular bonds, 41–42
monoclinic crystal structures, 615
monomers, 471, 476, **558**
monotectic reaction, 338, **350**

MOSFET technology, 820–821
MOS integrated circuit fabrication technology, 825–826
motif, 74, **115**
mullite, 343
multidirectional composite materials, 660
multidirectional laminate, **709**
multielectron atoms, 35–39
multiple-phase electrochemical cells, 734–735
multiwall nanotube (MWNT), 594
Muntz metal, 423

N

nanocluster, 593
nanocrystalline metals, 259–261, **263,** 305
nanoelectronics, 841–842
nanomaterials, 5, 19–20
nanotechnology
 and bone repair, 708
 ceramics and, 637–639, 703
Natta, Guilo, 490
natural aging, 404
natural gas, 484
natural rubber
 Hevea brasiliensis tree, 531
 latex, 531
 production of, 531
 properties of, 533–534
 structure of, 531–532
 vulcanization, 532
near net shape manufacturing, 200
necking, 223
negative Hall-Petch effect, 260
negative ions, 39
negative oxidation number, 39
neodymium-iron-boron magnetic alloys, 921
neodymium-yttrium-aluminum-garnet lasers, 869
neoprene rubbers, 535–536
Nernst, Walter Hermann, 726
Nernst equation, 726
network-modified glass (soda-lime glass), 624
network modifiers, 623, **641**
network polymerization, 482
neutrons, 25
nickel alloys
 commercial nickel alloys, 440
 mechanical properties and applications of, 437–438
 Monel alloys, 440
 nickel-base superalloys, 440–441
nitonol, 446
noble gases, 39, 61
nodular cast iron, 432

noncrystalline linear polymers, 476–477
noncrystalline materials, 73
noncrystalline thermoplastics, 486
nondirectional bonding, 56
nonequilibrium solidification, 322–325
nonequilibrium solidus, 323
noneutectoid plain-carbon steels, 383
nonferrous alloys, 359
nonferrous metals and alloys, 8–10, **22**
non-steady-state diffusion, 184–191, **195**
normalizing, 386–387
normal spinel structure, 925, **931**
Noryl, 516–517
npn bipolar junction transistor, 815
n-type extrinsic semiconductor, **844**
nuclei, 127–132
nylons, 512–515

O

octahedral interstitial sites, 584, **640**
offset yield strength, 221
Ohm's Law, 781–782, 784
open-die forging, 209–211
open-mold processes, 667–671
ophthalmic applications, 554–555
Opportunity, 2
optical-fiber communication, 874–875, **883**
optical fibers
 fabrication of, 873–874
 light loss in, 870–871
 single-mode and multimode, 871–872
optical metallography, 151–155
optical waveguides, 871, **883**
orbitals, 34
order of the diffraction, 107
organic coatings, 766
orthopedic applications
 bone cement, 556
 bone plates, 450
 compression bone plate and screws, 451
 intramedullary nail, 450, 454
 issues in application of, 453–455
 knee replacement prosthesis, 452
 metal alloys in, 453–455
 prostheses, 634
 stress shielding, 455
 titanium alloys, 452–453
 total hip prostheses, 635
orthopedic implants
 in acetabular (AC) cup, 634–636
 alumina (Al_2O_3) in, 634–636
 stainless steels in, 449–450
 stress shielding in, 636
osseointegration, 708

outer bark, 686
oxidation
 mechanisms of, 761–762
 of metals, 759–763
oxidation numbers of elements, 40
oxidation rates (kinetics), 762–763
oxidation reaction, 721
oxidation-reduction reactions, 720–721
oxide coatings, 632–634
oxide-film theory, 742
oxygen-concentration cells, 732–733, **770**
oxygen-free high-conductivity (OFHC) copper, 422

P

parallel spins, 37
paramagnetism, 895, **930**
parenchyma cells, 688, **711**
parison, 494
partial anneal, 249
partially stabilized zirconia (PSZ), 615
particulate composites, 14
particulate reinforced CMCs, 700
particulate reinforced MMCs, 698
partly crystalline thermoplastics
 solidification of, 486–488
 structure of, 488–489
passivation, 742, **771**
Pauli, Wolfgang, 34
Pauli exclusion principle, 34, **65**
Pauling, Linus Carl, 63
Pauling's equation, 63
Pauli principle, 790
Pb-Sn phase diagram, 332
pearlite, 366, **456**
percent cold reduction, 206, **262**
percent elongation, 215, 223
percent reduction in area, 224
percent strain, 215
periodic table of the elements, 27
peritectic reactions, 333, **350**
peritectoid reactions, 341
permanent dipole bonds, 42, **66**
permanent dipoles, 61–62
permanent magnetic materials, 915
permanent mold casting, 416
perovskite ($CaTiO_3$), 590–591
peroxide curing, 528
persistent slipbands, 286
petroleum, 484
phase, 310, **349**
phase diagram components, **349**
phase diagrams
 of Al_2-SiO_2 system, 343

defined, 310
with intermediate phases and compounds, 341–345
of pure substances, 311–313
phenolics
hexamethylenetetramine (hexa) catalyst, 523–524
molding compound types, 524
structure and properties of, 524
phenylene oxide–based resins, 516–517
phenylene ring, 516
phonons, 786
phosphor bronzes, 424
phosphorescence, 863, **883**
phosphors, 863
photolithography, 821–824
photoluminescence, 864
photons, 29, **65**, 855
photoresist, 824
picofarad, 832
piezoelectric ceramics, 18
piezoelectric effect, **845**
piezoelectric materials, 840–841
piezoelectric (PZT) ceramics, 841
piezoelectric (PZT) effect, 839
pig iron, 360
pigments, 504
piling-up fault, 150
Pilling-Bedworth (P.B.) ratio, 759–760, **771**
pith, 686
pitting corrosion, 746–749, **771**
plain-carbon steels, **457**
annealing and normalizing of, 386–387
classification of, 391–392
effect of tempering temperature on hardness, 389
eutectoid plain-carbon steels, 366
heat treatment of, 373–392
martensitic, hardness of, 377
mechanical properties of, 391–392, 393
slow cooling of, 366–372
tempering of, 387–391
planar atomic density, 99–100
planar defects, 147–150
planar density, **116**
planar process, 803
Planck, Max Ernst, 30
Planck's constant, 30, 855
planes of a family or form, 89
plastically deformed metals, 249–257
plastic deformation
described, 213
and ductile facture, 272

effect of, on dislocation arrangements, 244–246
effect of, on grain shape, 244–246
of metal single crystals, 229–242
of polycrystalline metals, 242–247
slipbands during, 230
by slip mechanism, 232–233
plasticizers, 503–504, **559**
plastic materials
deformation and strengthening of, 539–546
effect of temperature on the strength, 545–546
processing of, 491–498
plastics, 469, 653. *See also* thermoplastics; thermosetting plastics (thermosets)
plastic yielding, 552
plate, 202
plateaue, 314
plate martensite, 374, 375
platinum-silver phase diagram, 336
plies, 668
pn junction, 810–811, **845**
pn junction diode
applications for, 813–814
at equilibrium, 810
forward-biased, 811–812
reverse-biased, 810–811
point defects
Frenkel imperfection, 144
Schottky imperfection, 144
self-interstitial defects, 144
vacancies, 143–144
Poisson's ratio, 216, 217
polarization, 739, **845**
polyacrylonitrile (PAN)
carbon fibers from, 654
properties of, 506–507
structure and properties of, 506–507
structure of, 506
polyamides (nylons), 512–515
amide linkage, 513
structure and properties of, 513–514
polybutylene terephthalate (PBT), 518
polycarbonates, 515–516
polychloroprene elastomers
neoprene rubbers, 535–536
properties of, 536
silicone rubbers, 537–538
vulcanization of, 536–539
polychlorotrifluoroethylene (PCTFE), 511
polycrystalline crystals, 132, 134
polycrystalline metals, 242–247
polydimethyl siloxane, 537

polyetherimide, 520–521
polyethylene (PE)
applications of, 502
covalent bonding, 472
general reaction for the polymerization of, 473
properties of, 502
structure and properties of, 501–502
tensile stress-strain curves for low-density and high-density, 542
types of, 500–501
polyethylene terephthalate (PET), 518
polyglycolic acid (PGA), 555, 556
polygonization, 252
polylactic acid (PLA), 555
polymer, 469, 471
polymer alloys, 521
polymeric materials, **22**. *See also* elastomers (rubbers); plastics
brittle fracture of, 550–552
for corrosion resistance, 765
creep modulus of, 548
creep of, 546–547
described, 8, 10–11
effect of, on the strength of plastic materials, 546–552
fracture of, 550–552
plastic yielding of, 552
stress relaxation of, 547–549
viscoelastic behavior, 546
polymerization, **558**
degree of, 473
initiation of, 473–475
other methods of, 482–483
of polyethylene, general reaction for, 473
propagation of, 473, 475
termination of, 473, 475
polymerization methods, 484–486
polymerization reactions, 471–483
polymer matrix composite (PMC), 14
polymers, 113
polymers (biopolymers), 552–557
polymethyl methacrylate (PMMA)
bone cement, 556
structure and properties of, 509–510
tensile stress-strain plots for, 539–540
polymorphism, 102–103, 114, **116**
polypropylene, 505
polystyrene (PS), 505–506
creep curves for, 547
structure and properties of, 506
tensile stress-strain data for, 544
polysulfone, 519

polytetrafluoroethylene (PTFE)
  processing of, 511
  structure and properties of, 510–511
polyvinyl chloride (PVC)
  fillers, 504
  heat stabilizers, 503
  lubricants, 503
  plasticizers, 503–504
  polyvinyl chloride compounding, 503–504
  structure and properties of, 503
  tensile stress-strain data for, 544
polyvinyl chloride and copolymers, 503–504
population inversion, 867, **883**
porcelain enamel, 632, **642**
porosity, 613
portland cement, **710**
  chemical compositions of, 676
  compositions of, 676
  hardening of, 677
  production of, 675–678
  types of, 676
positive-ion cores, 55, **66,** 793
positive ions, 39
positive oxidation number, 39
posttensioned concrete, 683–684
potentiostat, 769
precipitation-hardenable alloy, 405
precipitation strengthening (hardening)
  of binary alloys, 403–410
  of copper alloys, 406
precipitation zones (GP zones), 404
precursors, 654
predeposit step, 824
preform, 874
prepreg carbon-fiber-epoxy material, 668, **710**
press forging, 209–211
pressing ceramics, 600
pressure-assisted sintering, 638
pressure-temperature (PT) phase diagram, 311, 312
prestressed concrete, 682–684, **711**
pretensioned (prestressed) concrete, 683, **711**
primary alpha, 327
primary atomic bonds, 42
primary phase, **350**
primary wall, 690
principal quantum number, 30, 32, 33
prism planes, 94
process annealing, 386
processing
  of ceramics, 598–603
  of metals and alloys, 201–212
  of plastic materials, 491–498
  of polytetrafluoroethylene (PTFE), 511

and processes for thermoplastic materials, 492–496
  of steel product forms and steelmaking, 361–362
  for thermosetting materials, 496–498
proeutectic alpha, 327
proeutectic phase, **350**
proeutectoid alpha ferrite, **456**
proeutectoid cementite, 371, **456**
proeutectoid ferrite, 368
propagation of polymerization, 473, 475
properties. *See also* mechanical properties
  of acetals, 517
  of acrylonitrile-butadiene-styrene terpolymers (ABS resins), 507–508
  of aluminum, 410–411
  of amorphous metals, 446–448
  of cast irons, 429
  of cast polyester, 658
  of composite laminate of carbon fibers and epoxy resin, 659
  of copper alloys, 418–419
  of ductile iron, versus hardness, 434
  of elastomers (rubbers), 534
  of engineering thermoplastics, 511–512
  of epoxy resins, 526, 658
  of fiberglass-polyester composites, 659
  of general-purpose thermoplastics, 499
  of germanium, 795
  of glass fibers, 653
  of low- and high-density polyethylenes, 502
  of magnesium alloys, 437–438
  of melamine-formaldehyde resins, 530
  of natural rubber, 533–534
  of phenolics, 524
  of phenylene oxide–based resins, 516–517
  of polyacrylonitrile, 506–507
  of polyamides (nylons), 513–514
  of polycarbonates, 515–516
  of polychloroprene elastomers, 536
  of polychlorotrifluoroethylene (PCTFE), 511
  of polyethylene (PE), 502
  of polypropylene, 505
  of polystyrene, 506
  of polysulfone, 519
  of polytetrafluoroethylene (PTFE), 510–511
  of polyvinyl chloride (PVC), 503
  of silicon, 795
  of styrene-acrylonitrile copolymers (SAN resins), 507
  of thermoplastic polyesters, 518

of thermosetting plastics (thermosets), 522–523
  of unsaturated polyesters, 529
  of wood, 692–695
prostheses, 634
protective oxide films, 759
protons, 25
p-type extrinsic semiconductor, **844**
pulsed type, lasers, 868
pultrusion, **710**

Q

quantum corrals, 841
quantum mechanics, 29, **65**
quantum mirage, **843**
quantum numbers, 33–34, **65**
quartz, 597, 606
quenching, 373–374, 404

R

radial axis (R), 686
radiation
  coherent, 866
  incoherent, 866
  stimulated emission of, 866–870
radius ratio, 577, **640**
random copolymers, 479
range of stress, 286
rare earth alloys, 919–920, **931**
rate processes, 173–176
recovery, 249, 251–252, **263**
recovery and recrystallization of plastically deformed metals, 249–257
recrystallization, 249, 252–257, **263**
rectification, 813
rectifier diodes, 813, **845**
reduction reaction, 721
reflection
  of light, 859–860
  of light by a glass plate, 860–861
reflectivity, 859
refractive indices, 857
refractories, 343, 619
refractory (ceramic) material, **641**
region of thermal arrest, 314
reinforced concrete, 682–683, **711**
reinforced plastic composite materials, 651–657
  mechanical properties of aramid for, 655–657
  mechanical properties of carbon for, 655–657
  mechanical properties of glass fibers for, 655–657

reinforced plastics, 653–654
reinforcing fibers
  specific tensile modulus of, 657
  specific tensile strength of, 656
  stress-strain behavior of, 656
reinforcing plastic resins
  aramid fibers for, 654–655
  glass fibers for, 651–653
relative permeability, 893, **930**
relaxation time, 548–549, 785
remanent induction, 907, **931**
repeated stress cycle, 285
resorption (dissolving) response, 637
reverse-bending fatigue, 301
reverse bias, **845**
reversed-bending fatigue machine, 283
rigid gas-permeable (RGP) lenses, 555
rigid polyvinyl chloride, 503–504
ring-porous wood, 689
ring structures, 595
Rockwell C hardness, 397
Rockwell hardness tests, 228–229
rodlike structures, 333
rotating-beam fatigue specimen, 283
roving, **709**
rubbers, 469
ruby lasers, 869
rule of mixtures, 663

S

sand casting, 416–417
sapwood, 686, **711**
saturation induction, 907, **931**
saturation range, 808
scalping, 412
scanning electron microscopy (SEM), 156–158
scavenger-type inhibitors, 768
Schmid's Law, 237–239
Schottky imperfection, 144
screw type dislocations, 145
secondary atomic bonds, 42
secondary bonding, 59–68
secondary molecular bonds, 42
secondary recrystallization, 305
secondary wall, 690
selective leaching, 757–758, **771**
selective oxidation, 825
self-diffusion, 178, **195**
self-diffusion activation energies, 179
self-interstitial defects, 144
semiconductor devices, 809–816
semiconductor lasers, 869–870
semiconductor materials, 15

semiconductors, 792, **844,** 862
semicrystalline materials, 113, **116**
sensitized condition, 752
sensitizing temperature range, 751
separation distance
  versus energy, 57
  versus force, 44
s-glass, 651
shape-memory alloys, 18, **458**
  alloys, 442–446
  applications of, 445
  austenitic phase, 443
  martensitic phase, 443
  nitonol, 446
  stress-induced transformation, 445
shear bands, 304
shear modulus, 217
shear strain, 216–217, **262**
shear stress, 216–217, 232, **262**
sheet, 202
sheet-molding compound (SMC), **710**
sheet-molding compound (SMC) process, 672–674
sheet structures of silicates, 595–597
short-range order (SRO), 73
shrinkage, 694
SiC whisker reinforced ceramic-matrix composites, 701
silica, 597–598
silica (flint), 606
silica-leucite-mullite, 608
silica refractories, 620
silicate glasses, 632, 859–860
silicate mineral compositions, 598
silicate networks, 597–598
silicate structures, 595–598
  basic structural unit of, 595
  island, chain, and ring structures of, 595
  sheet structures of, 595–597
silicon, 795
silicon carbide (SiC), 13, 610
silicon carbide whiskers, 698
silicone rubbers, 537–538
silicones, 537
silicon nitride (Si_3N_4), 610
silicon wafers, 188–191
simple ceramic compounds
  ionic and covalent bonding in, 575–576
  melting points of, 575
simple ceramic crystal structures, 575–595
simple ionic arrangements, 576–579
single-wall nanotube (SWNT), 594
sinter forging, 638
sintering, 600, **641**

size limitations, 576
slabs, 361
slip, 230, **262**
slipband crack growth, 286
slipband extrusions, 286
slipband intrusions, 286
slipbands, 229–231, **262**
slip casting, 602, **641**
slip lines, 229–231
slip mechanism, 232–233
slip planes, 230
slip planes and directions, 237
slip systems, 234–235, **262**
  in crystal structures, 236
  in metal single crystals, 238
slip vectors, 145, 147
slow cooling, 327
small-angle tilt boundary, 149
smart materials, 18–19
smart materials and devices, 5
Snell's law of light refraction, 857–858
soda-lime glass, 626
sodium chloride (NaCl), 580–584
softening point, 627, **641**
soft ferrite materials, 923–924, 927–928
soft ferrites, **931**
soft magnetic materials, 908–915, **931**
  described, 889
  desirable properties for, 909
  energy losses for, 909–910
  iron-silicon alloys, 910–911
  magnetic properties of, 911
  nickel-iron alloys, 912–915
softwoods, 688
softwood trees, 686, **711**
solenoid, 892
solidification
  latent heat of, 135
  of noncrystalline thermoplastics, 486–488
  of partly crystalline thermoplastics, 486
  of single crystals, 134–138
solid phases, 311
solids
  atomic diffusion in, 177–184
  diffusion in, 177
  effect of temperature on diffusion in, 191–195
  rate processes in, 173–176
solid solution, 138
solid-solution hardening (strengthening), 247–248, **263**
solid-state diffusion, 322
solidus, 316, **350**
solute-solvent diffusion systems, 182

solution heat treatment, 403
solutionizing, 403
solution polymerization, 485, **558**
solvus, **350**
solvus lines, 326
source, 819
space lattice, 73–74, 114, **115**
spanning probe microscope (SPM), 841
specific tensile modulus, 657, **709**
specific tensile strength, 657, **709**
spherical nucleus, 129
spheroidite, 389, **457**
spherulitic graphite cast iron, 432
spinel ($MgAl_2O_4$), 590
Spirit, 2
split transformation, 385
spontaneous magnetization, 896–897
spray-up process, 667–668, **710**
stabilization stage, 654
stable nuclei, 127–132
stacking faults, 147, 150
stainless steels
    austenitic stainless steels, 426, 427–428
    biometals, 449
    ferritic stainless steels, 424–425
    martensitic stainless steels, 425–427
    mechanical properties of, 426
state, 791
steady-state conditions, 180, **195**
steady-state creep, 295
steady-state diffusion, 180–182
steatite, 834
steelmaking, 364–365
steelmaking and processing, 361–363
steel product forms, 361–362
steels. *See also* alloy steels; low-alloy steels;
        plain-carbon steels; stainless steels
    effect of alloying elements on eutectoid
        temperature of, 395–396
    production of, 360–363
    refining of, 362
steel strip, 362
stepwise polymerization, 482, **558**
stereoisomerism, 486–490
stereoisomers, 489, **558**
stereospecific catalyst, 490, **558**
straight-line notation for covalent bonding, 51
strain, 294
strain hardening (strengthening), 246, **263**
strain point, 628, **641**
strength, improvements in, 303–305
strengthening
    due to the average molecular mass of the
        polymer chains, 541–545

by increasing the amount of crystallinity
        in a thermoplastic material, 541–545
    of thermoplastics by bonding highly polar
        atoms on the main carbon chain, 543
    of thermoplastics by increasing the
        amount of crystallinity in a thermo-
        plastic material, 541
    of thermoplastics by introducing pendant
        atomic groups on the main carbon
        chains, 542
    of thermoplastics by introducing pheny-
        lene rings into the main polymer
        chain, 544
    of thermoplastics by the addition of glass
        fibers, 544–545
    of thermoplastics by the introduction of
        oxygen and nitrogen atoms in the
        main carbon chain, 543
    of thermoplastics due to the average molec-
        ular mass of the polymer chains, 541
    of thermosetting plastics (thermosets), 545
stress
    fatigue crack propagation correlation with,
        288–293
    versus number of cycles (SN) for fatigue
        failure, 284
stress amplitude, 286
stress and strain, 212–217
stress concentrations, 279, 287
stress corrosion, 753–754, **771**
stress corrosion cracking
    mechanism of, 754–755
    prevention of, 755–756
stress-intensity factor, 279
stress-intensity factor range threshold, 292
stress ratio, 286
stress relaxation of polymeric materials,
        547–549
stress relief, 386
stress risers, 275, 301
stress rupture, 294–298
stress-rupture strength, **306**
stress-rupture test, 297–298
stress-rupture time-temperature data, 298–300
stress shielding, 455, 636, **642**
stress-strain curves
    for cortical tissues (bone), 705–706
    for trabecular tissues (bone), 705
structural ceramics, 12
structural isomers, 532
structure. *See also* crystal structures
    of acetals, 517
    of acrylonitrile-butadiene-styrene terpoly-
        mers (ABS resins), 507–508

of alnico alloys, 918–919
    of atoms, 25–26
    cored, 322
    of epoxy resins, 526
    of glasses, 623–624
    of natural rubber, 531–532
    of noncrystalline linear polymers,
        476–477
    of partly crystalline thermoplastic
        materials, 488–489
    of phenolics, 524
    of phenylene oxide–based resins, 516–517
    of polyacrylonitrile, 506–507
    of polyamides (nylons), 513–514
    of polycarbonates, 515–516
    of polychlorotrifluoroethylene (PCTFE),
        511
    of polypropylene, 505
    of polystyrene, 506
    of polysulfone, 519
    of polytetrafluoroethylene (PTFE),
        510–511
    of polyvinyl chloride (PVC), 503
    of styrene-acrylonitrile copolymers
        (SAN resins), 507
    of thermoplastic polyesters, 518
    of unsaturated polyesters, 529
    of urea-formaldehyde resins, 530
structures of magnesium alloys, 437–438
styrene-acrylonitrile copolymers
        (SAN resins), 507
styrene-butadiene rubber (SBR), 535
subsidiary quantum number, 33
substitutional diffusion mechanism,
        177–178, **195**
substitutional solid solutions, 138–141, 247
substitutional solutions, 138
superalloys, 8–10, 17
superconducting materials, 875–876
superconducting oxides, 881–882
superconducting state, **883**
superconductive material, 876
superconductivity, 876
superconductors
    current flow and magnetic fields in,
        878–879
    high-current, high-field, 879–880
    magnetic properties of, 876–877
    type I superconductor, 877–878
    type II superconductor, 877–878
superplasticity, 257–259, **263**
superpure aluminum, 138
supersaturated solid solutions, 404
surface condition, 288

surface energy, 127
surface engineering, 632–634
surface roughness, 288
surrounding phenomena, 337–338, 339
suspension polymerization, 485, **558**
syndiotactic isomer, 491, **558**
syndiotactic stereoisomer, 489
synthetic rubbers
    elastomers (rubbers), 534–536
    nitrile rubbers, 535
    polychloroprene (neoprene), 535–536
    styrene-butadiene rubber (SBR), 535
system, **349**

T

tangential axis (T), 686
temperature
    ductile to brittle transition, 276–278
    effect of, on energy absorbed upon impact, 277
    effect of, on the strength of plastic materials, 545–546
    effect of, on the tensile strength of thermoplastics, 546
temperature resistivity coefficients, 788
temperature-time cooling curve, 329
temper designations of wrought aluminum alloys, 412
tempering
    microstructure changes in martensite upon, 388
    of plain-carbon steels, 387–391
tempering temperature, 389
tendons, 683
tensile elastic modulus, 666
tensile specimen, 215
tensile strength, 223
tensile stress-strain data
    for polystyrene (PS), 544
    for polyvinyl chloride (PVC), 544
tensile stress-strain plots, 539–540
tensile test, 217–227
tension test specimens, 219
terminal phases, 341, **351**
terminal solid solutions, 326
termination of polymerization, 473, 476
ternary phase diagrams, 345–348
    composition base for a, 346
    of an isothermal section, 347
    of silica-leucite-mullite, 608
Tesla, Nikola, 892
tetragonal crystal structures, 615
tetrahedral covalent bonding, 52
tetrahedral interstitial sites, 584, **640**

thermal arrest, **349**
thermal conductivity of ceramics, 618
thermally tempered glass, **641**
thermal properties of ceramics, 618–620
thermal treatments of ceramics, 604–606
thermistor, **845**
thermoforming, 496
thermoplastic materials, 492–496
thermoplastic polyesters
    polybutylene terephthalate (PBT), 518
    polysulfone, 518–519
    structure and properties of, 518
thermoplastics, **558**
    average molecular weight for, 475–476
    blow molding of, 494–496
    crystallinity and stereoisomerism in, 486–490
    deformation mechanisms for, 539–541
    described, 469–470
    effect of temperature on the tensile yield strength of, 546
    engineering thermoplastics, 511–521
    extrusion of, 493–494
    general-purpose, 498–511
    glass transition temperature (T_g) for, 488
    injection molding of, 492–493
    molecular masses and degrees of polymerization for, 541
    for polystyrene (PS), 544
    polyvinyl chloride (PVC), 544
    stereoisomerism in, 489–490
    strengthening of, 541–545
    thermoforming of, 496
thermosetting materials, 496–498
thermosetting plastics (thermosets), 521–531, **558**
    amino resins (ureas and melamines), 529–530
    bulk prices of, 522
    described, 469–470
    phenolics, 523–525
    properties of, 522–523
    strengthening of, 545
    unsaturated polyesters, 527–529
theta zones, 407
three-dimensional defects, 150
tie line, 317, 318, **350**
time-dependent strain, 294
tin bronzes, 424
Titanic failure, 270
titanium alloys, 438–440
    biometals, 452–453
    gamma prime phase, 441
    knee replacement prosthesis, 452

    mechanical properties and applications of, 437–438
    orthopedic applications, 452–453
    titanium-nickel phase diagram, 344
total hip prostheses, 635
toughness
    of ceramic materials, 613–615
    and impact testing, 276
tow (of fibers), **709**
toxic response, 636
trabecular tissues (bone), 703, 705
tracheids (longitudinal), 688, **711**
traditional ceramic materials, 573
traditional ceramics, 606–608
trans-1,4 polyisoprene, 532, **559**
transducer, **845**
transfer molding, 497–498, **559**
transformation toughening, 615
transformation twins, 375
transition metals, 57
transmission
    of light, 859–860
    of light by a glass plate, 860–861
transmission electron microscopy (TEM), 158–159
triaxial whiteware, 607
tricalcium silicate hydrate, 677
tridymite, 597
trifunctional monomers, 476
triple point, 311
true stress and true strain, 225–227
true stress-true strain curve vs. engineering (nominal) stress-strain diagram, 226
twin boundaries (twins), 149
twinned martensite, 444
twinning, 240–242
twinning direction, 240
twinning plane, 240
twins, 147
twists, 147
two-component system, 315
two-metal corrosion, 745–746
type I superconductor, **883**
type II superconductor, **883**

U

Ultem (polyetherimide), 520–521
ultimate ductile failure, 287
ultimate tensile strength (UTS), 222–223, **262**
undercooling vs. critical radius, 129–130
unidirectional composite materials, 660
unidirectional laminate, **709**
uniform corrosion, 736, 745

unit cells, 73–74, 114, **115**
unit step of slip, 233
unsaturated bonds, 53
unsaturated molecule, 472
unsaturated polyesters
  linear polyester, 528
  methyl ethyl ketone (MEK) peroxide, 528
  structure and properties of, 529
upper bainite, 381
upper critical field, 877, **884**
urea-formaldehyde resins, 529–530
ureas, 529–530

V

vacancy diffusion mechanism, 177–178
vacuum bag-autoclave process, 668–670
vacuum bag molding, **710**
valence band, **844**
valence electrons, 55, **66**, 790, 793
van der Waals bonds (forces), 60
vessel elements, 689
Vickers hardness tests, 228–229
Vickers microhardness test, 242
vinylidene polymers, 478–479
vinyl polymers, 478–479
viscoelastic behavior, 546
viscoelasticity, 707, **711**
viscous deformation, 626–628
viscous flow, 547
vitrification, 605, **641**
volume (or bulk) free energy, 127
volume defects, 150
volume density, 98–99, **115**
vulcanization, 532, **559**

W

water-cement ratio, 680
water pressure-temperature (PT) phase diagram for, 312
weight fractions, 320
weld decay, 752, **771**
white cast irons, 419–431, **457**
whitewares, 607
wire drawing, 211, **262**
Wolf's law, 707

wood, **711**
  annual growth rings, 686
  axes in, 687
  cambium layer, 686
  cellulose, 690
  cell-wall ultrastructure, 690–692
  deciduous trees, 686
  diffuse-porous wood, 689–690
  earlywood, 686
  evergreen trees, 686
  green condition, 692
  growth rings, 687
  hardwoods, 686, 689
  heartwood, 686
  hemicellulose, 690
  inner bark, 686
  layers of, 685
  lignin, 690
  longitudinal axis (L), 686
  lumen, 688
  macrostructure of, 685–695
  mechanical properties, 692–693
  mechanical strength of, 692–693
  microfibrils, 692
  moisture content of, 692
  outer bark, 686
  parenchyma cells, 688
  pith, 686
  primary wall, 690
  properties of, 692–695
  radial axis (R), 686
  ring-porous wood, 689
  sapwood, 686
  secondary wall, 690
  shrinkage of, 694
  softwoods, 686, 688
  tangential axis (T), 686
  tracheids, 688
  vessel elements, 689
  wood rays, 688
wood rays, 688, **711**
wood vessel, **711**
working point, 627, **641**
wrought alloy products, 202
wrought aluminum alloy groups
  cubic unit cells, 413

  heat-treated subdivisions, 413
  non–heat-treatable, 413
  strain-hardened subdivisions, 413
  temper designations, 412
wrought aluminum alloys
  classification of, 412
  heat-treatable, 415–416
  mechanical properties and applications for, 414
  primary fabrication of, 411
  temper designations of, 412
wrought copper alloys
  classification of, 419
  electrolytic tough-pitch copper, 422
  oxygen-free high-conductivity (OFHC) copper, 422
  unalloyed copper, 422–424
wrought magnesium alloys, 437–438

X

x-ray diffraction analysis
  of crystal structures, 107–112
  data interpretation for metals with cubic crystal structures, 110–111
  powder method of, 107–108

Y

yield strength, 221, **262**
Young, Thomas, 221
Young's modulus, 221

Z

zener (avalanche) diode characteristic curve, 814
zener diodes, 813–814
0.2 percent offset yield strength, 221–222
Ziegler, Karl, 490
Ziegler and Natta catalysts, 490–491
zinc blende (ZnS), 586–588
zirconia (ZrO_2)
  described, 20
  engineering ceramics, 610
  fully stabilized, 615
  partially stabilized, 615
ZrO_2-MgO binary phase diagram, 616

Constants

Constant	Symbol	Value
Avogadro's number	N_0	6.023×10^{23} mol^{-1}
Atomic mass unit	u	1.661×10^{-24} g
Electron mass	m_e	9.110×10^{-28} g
Electronic charge (magnitude)	e	1.602×10^{-19} C
Planck's constant	h	6.626×10^{-34} J $\cdot$ s
Velocity of light	c	2.998×10^{8} m/s
Gas constant	R	1.987 cal/(mol $\cdot$ K); 8.314 J/(mol $\cdot$ K)
Boltzmann's constant	k	8.620×10^{-5} eV/K
Permittivity constant	ϵ_0	8.854×10^{-12} C^2/(N $\cdot$ m^2)
Permeability constant	μ_0	$4\pi \times 10^{-7}$ T $\cdot$ m/A
Bohr magneton	μ_B	9.274×10^{-24} A $\cdot$ m^2
Faraday	F	9.6485×10^{4} C/mol
Gravitational acceleration	g	9.806 m/s
Density of water		1 g/cm^3 = 1 Mg/m^3

Conversion Factors

Length:	1 in = 2.54 cm = 25.4 mm
	1 m = 39.37 in
	1 Å = 10^{-10} m
Mass:	1 lbm (pound-mass) = 453.6 g = 0.4536 kg
	1 kg = 2.204 lbm
Force:	1 N = 0.2248 lbf (pound-force)
	1 lbf = 4.44 N
Stress:	1 Pa = 1 N/m^2
	1 Pa = 0.145×10^{-3} lbf/in^2
	1 lbf/in^2 = 6.89×10^{3} Pa
Energy:	1 J = 1 N $\cdot$ m
	1 cal = 4.18 J
	1 eV = 1.60×10^{-19} J
Power:	1 W = 1 J/s
Temperature:	°C = K − 273
	K = °C + 273
	°C = (°F − 32)/1.8
Current:	1 A = 1 C/s
Density:	1 g/cm^3 = 62.4 lbm/ft^3

$\ln x = 2.303 \log_{10} x$